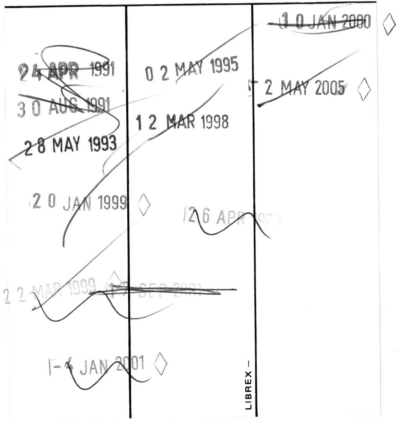

Methods of Air Sampling and Analysis

SECOND EDITION

Methods of Air Sampling and Analysis

SECOND EDITION

APHA INTERSOCIETY COMMITTEE

APHA	AIHA	APWA	AOAC
ACS	AIChE	ASCE	HPS
ACGIH	APCA	ASME	ISA
			SAE

MORRIS KATZ, PhD, *Editor*

Publication Office:

American Public Health Association
1015 Eighteenth Street NW
Washington, DC 20036

AMERICAN PUBLIC HEALTH ASSOCIATION 1872

Interdisciplinary Books & Periodicals
For the Professional and the Layman

6M5/77
Library of Congress Catalog Number: 77-6826
International Standard Book Number: 0-87553-079-6

Library of Congress Cataloging in Publication Data

Intersociety Committee.
Methods of air sampling and analysis.

Includes bibliographical references and index.
1. Air—Pollution—Measurement. 2. Air—Analysis.
I. Katz, Morris, 1901–
II. Title.
TD890.I53 1977 628.5′3 77-6826
ISBN 0-87553-079-6

Printed and bound in the United States of America
Typography: Byrd Pre-Press, Inc., Springfield VA
Set in: *Times Roman, Helvetica*
Text and Binding: R. R. Donnelley & Sons Company, Crawfordsville IN

Cover Design: Donya Melanson Assoc., Boston MA

PREFACE

This is the second edition of the manual of methods adopted by the Intersociety Committee on Methods of Air Sampling and Analysis according to its established procedures. Most of these methods were published initially in the American Public Health Association periodical "Health Laboratory Science", starting with Volume 6, No. 2 (April 1969) and continuing through subsequent issues. The first edition, published in 1972, contained 57 methods. There are 136 methods included in this second edition.

Most of these methods are designated as "tentative". Adoption as a "recommended" method, in accordance with Intersociety Committee procedures, can take place only after satisfactory completion of a cooperative test program. When accepted as "recommended", the "tentative" status of a method is elevated accordingly.

The Intersociety Committee is a joint project of the Air Pollution Control Association, American Chemical Society, American Conference of Governmental Industrial Hygienists, American Industrial Hygiene Association, American Institute of Chemical Engineers, American Public Health Association, American Society of Mechanical Engineers, American Public Works Association, American Society of Civil Engineers, Society of Automotive Engineers, Instrument Society of America, Health Physics Society and Association of Official Analytical Chemists. The main objective of the Committee has been to produce standardized methods of sampling and analysis of ambient and workplace air. It is expected that the project will provide a sound basis for improving methodology over a period of years. As new methods have been developed and found acceptable, they have replaced or have been added to existing methods in the manual.

The participating organizations each appoint a representative to the Intersociety Committee, which serves as a policy-making and editorial board. The work of the Committee is conducted by Subcommittees. Nine of these Subcommittees are responsible for preparing precise statements of procedure for methods in various substance categories. A tenth Subcommittee is responsible for sections of the manual, such as sampling techniques and laboratory precautions, which are common to all methods. The eleventh and twelfth Subcommittees are responsible for source sampling techniques and for standardization coordination. Each Subcommittee has expert representatives from the participating organizations.

During the period 1963 to 1966, Dr. Leonard Greenburg served as Chairman of the Intersociety Committee, followed by Dr. E. R. Hendrickson from 1966 to 1969 and Professor Arthur C. Stern from 1969 to 1972. The current Chairman, Dr. Bernard E. Saltzman, has served in that capacity since 1972. Dr. Moyer Thomas served as Editor from 1963 to 1968. He was succeeded by Dr. Morris Katz who has been the Editor since 1969. The Secretary of the Intersociety Committee is Dr. George Kupchik. All correspondence concerning these methods should be directed to the Secretary at the School of Health Science, Hunter College, 118 East 107th Street, New York, N.Y. 10029.

From 1963 to 1971, the Intersociety Committee was concerned only with methods for ambient air sampling and analysis. During 1971, the scope of the Committee was enlarged to include methods for air pollution source sampling and analysis. In June 1973, the Committee undertook the preparation of methods of analysis for chemicals in air of the workplace and in biological samples.

This work was supported, in part, in the past by Grant AP-00256 from the Public Health Service, by Contract Grant No. 68-02-0004 with the Environmental Protection Agency and by Contract No. HSM 99-73-89 with the National Institute for Occupational Safety and Health (NIOSH). The American Public Health Association has served as the sponsoring organization on behalf of the Intersociety Committee.

Identification of Ambient Air Methods

The SAROAD system has been used to assign numbers to methods and has the following meaning:

12345	01	74	T
Pollutant Identification (SAROAD system)*	Chronological Order of Adoption of Method for This Pollutant	Year of Adoption of Method	Tentative

Example: 42602-01-74T

1	2	3	4 5
A	B	C	D

A Denotes one of 9 major classes
eg 1—Suspended Particulate
2—Settled Particulate
3—Respirable Dust
4—Gas and Vapors
etc.

AB Denotes one of 81 subclasses
eg 42—Gas and Vapors, Inorganic
17—Suspended Particulate, Aromatic Compounds

ABC Denotes one of 729 families
eg 426—Gas and Vapors, Inorganic, Nitrogen Compounds
172—Suspended Particulate, Aromatic Compounds, Polynuclear

ABCDE Denotes one of a possible 72,171 individual pollutants
eg 42602
—Gas and Vapors, Inorganic, Nitrogen Compounds, Nitrogen Dioxide
17242—Suspended Particulate, Aromatic Compounds, Polynuclear, Benzo[a]Pyrene

---

*In the SAROAD (Storage and Retrieval of Air Quality Data) system†, in the five digit identification.

†Fair, Donald H., George B. Morgan, and C. E. Zimmer. 1968. Storage and Retrieval of Air Quality Data (SAROAD) System Description and Data Coding Manual, National Air Pollution Control Administration Publication No. APTD 68-8. Office of Technical Information and Publications, NAPCA. August, Cincinnati, Ohio, 47 p.

INTERSOCIETY COMMITTEE METHODS OF AIR SAMPLING AND ANALYSIS

INTERSOCIETY COMMITTEE

Organization*	Member	Date of Service
APCA	E. R. Hendrickson (*Chairman*, 1966–1969)	1963–
ACS	Henry Freiser	1971–
ACGIH	Robert G. Keenan	1963–1969
	John Crable	1969–
AIHA	Allen D. Brandt†	1963–1971
	Newton E. Whitman	1971–1972
	George D. Clayton	1972–
AIChE	Robert Coughlin	1972–1974
	Richard F. Toro	1974–
APHA	Leonard Greenburg (*Chairman*, 1963–1966)	1963–1972
	William J. Hausler, Jr.	1972–
APWA	James V. Fitzpatrick	1972–
ASCE	William T. Ingram	1971–
ASME	Arthur C. Stern (*Chairman*, 1969–1972)	1963–1974
	Wesley L. Hemeon	1974–
ASTM	Morris Katz	1963–1967
	Paul M. Giever	1967–1973
AOAC	Bernard E. Saltzman (*Chairman*, 1972–)	1966–
HPS	Lysle C. Schwendiman	1972–
ISA	Robert L. Chapman	1972–1974
	Robert L. Saltzman	1974–
SAE	George J. Gaudean	1972–1974
	R. T. Northrup	1974–
At-Large	Elbert C. Tabor	1970–
	Newton E. Whitman	1972–
Executive Secretary	George J. Kupchik	1963–
Editor	Moyer Thomas	1963–1969
	Morris Katz	1969–

*Participating Organizations
- APCA—Air Pollution Control Association
- ACS—American Chemical Society
- ACGIH—American Conference of Governmental Industrial Hygienists
- AIHA—American Industrial Hygiene Association
- AIChE—American Institute of Chemical Engineers
- APHA—American Public Health Association
- APWA—American Public Works Association
- ASCE—American Society of Civil Engineers
- ASME—American Society of Mechanical Engineers
- ASTM—American Society for Testing and Materials (Withdrew March 30, 1973)
- AOAC—Association of Official Analytical Chemists
- HPS—Health Physics Society
- ISA—Instrument Society of America
- SAE—Society of Automotive Engineers

†Deceased

SUBCOMMITTEE #1 (SULFUR COMPOUNDS)

	Name	Date of Service
ASTM, ISA	Donald F. Adams (*Chairman*, 1966–)	1966–1973, 1973–
APHA	Morton Corn	1966–1968
	John O. Frohliger	1968–
APCA	Charles I. Harding	1966–1969
	Dennis Falgout	1969–
ACS	Arnold M. Hartley	1971–
ACGIH	John Pate	1966–
ASME	A. L. Plumley	1966–
AOAC	Frank P. Scaringelli	1966–
AIHA	Paul Urone	1966–

SUBCOMMITTEE #2 (HALOGEN COMPOUNDS)

	Name	Date of Service
AOAC	C. Ray Thompson (*Chairman*, 1970–)	1966–
APHA	Lester V. Cralley (*Chairman*, 1966–1970)	1966–1971
	George H. Farrah	1971–
ACS	W. S. Hillman	1971–1973
	Jay S. Jacobson	1973–
ASTM	A. W. Hook	1966–1973
AIHA	Edward J. Schneider	1966–
ACGIH	John D. Strauther	1966–1974
APCA	C. R. Walbridge	1966–1967
	Lawrence V. Haff	1967–
ASME	Leonard H. Weinstein	1966–
ISA	Robert L. Saltzman	1973–1974

SUBCOMMITTEE #3 (OXIDANTS AND NITROGEN COMPOUNDS)

	Name	Date of Service
ASME	Basil Dimitriades (*Chairman*, 1974–1975)	1972–
ASTM	Paul M. Giever (*Chairman*, 1966–1967)	1966–1967
	Warren A. Cook	1967–1973
AOAC	Evaldo L. Kothny (*Chairman*, 1971–74; 1976–)	1966–
AIHA	James Long	1966–1967
	Lester Levin	1967–1972
	Gordon D. Nifong	1972–1973
APHA	Paul W. McDaniel	1966–
ACGIH	Bernard E. Saltzman (*Chairman*, 1967–1971)	1966–1972
	J. E. Cuddeback	1972–1975
APCA	Jack H. Smith	1966–1970
	Edward F. Ferrand	1969–1975
ACS	Fred T. Weiss	1971–1975
ISA	R. G. Kling	1973–1974
LIAISON	C. A. Johnson (American Refrigeration Institute)	1969–1970

SUBCOMMITTEE #4 (CARBON COMPOUNDS)

	Name	Date of Service
ACGIH	Ralph Smith (*Chairman*, 1966–)	1966–
APCA	Robert J. Bryan	1966–
AOAC	Milton Feldstein	1966–
AIHA	Joseph A. Houghton	1966–1967
	Franklin A. Miller	1967–1974
ASTM	Benjamin Levadie	1966–1973
ACS	David C. Locke	1971–
ASME	Joseph W. Mullan	1966–1968
	Edgar R. Stephans	1968–1973
APHA	Norman G. White	1966–1971
	Peter O. Warner	1971–

SUBCOMMITTEE #5 (HYDROCARBON COMPOUNDS)

	Name	Date of Service
APCA	Eugene Sawicki (*Chairman*, 1966–)	1966–
ASCE	Patrick R. Atkins	1971–1973
ASC	Theodore Belsky	1971–
ASME	Richard C. Corey	1966–1971
	R. A. Friedel	1971–
AIHA	Allan Dooley	1966–1971
	D. L. Hyde	1971–
ACGIH	J. Brennan Gisclard	1966–1969
	Lowell D. White	1969–
ASTM	J. L. Monkman	1966–1973
AOAC	Robert E. Neligan	1966–1969
	John E. Sigsby	1969–1973
	J. L. Monkman	1973–
APHA	Lyman A. Ripperton	1966–
ISA	Reinhold A. Rasmussen	1973–

SUBCOMMITTEE #6 (METALS I)

	Name	Date of Service
APHA	William G. Fredrick (*Chairman*, 1966–1970)	1966–1970
	Theo J. Kneip (*Chairman*, 1971–)	1971–
ASTM	Jacob Cholak	1966–1970
	John Driscoll	1971–1973
ACS	Henry Freiser	1971–1972
	Jarvis L. Moyers	1972–
ACGIH	Richard E. Kinser	1966–1969
	Augustine Moffitt	1969–1970
	John R. Carlberg	1971–1973
AIHA	Kathleen Kumler	1966–1972
	Fred I. Grunder	1973–
AOAC	George B. Morgan	1966–1970
	Richard J. Thompson	1971–
APCA	William M. Smith	1966–1970
	R. S. Ajemian	1971–
ASME	Lewis G. von Lossberg	1966–1970
	Lawrence Kornreich	1971–
ISA	J. West Loveland	1973–

SUBCOMMITTEE #7 (METALS II)†

	Name	Date of Service
ASTM	Elbert C. Tabor (*Chairman*, 1966–1970)	1966–1970
	John M. Bryant	1970–1972
AOAC	Richard E. Kupel, (*Chairman*, 1970–1972)	1966–1972
APCA	Moe Braverman	1966–1972
AIHA	Howard E. Bumsted	1966–1972
ASME	Arrigo Carotti	1966–1972
APHA	Harry M. Donaldson	1966–1972
ACGIH	L. Dubois	1966–1972
ACS	William W. Welbon	1971–1972

†Subcommittee terminated November 28, 1972

SUBCOMMITTEE #8 (RADIOACTIVE COMPOUNDS)

	Name	Date of Service
HPS	Andrew P. Hull (*Chairman*, 1972–)	1967–
ACS	Lloyd A. Currie	1971–
ASME	Orville G. Hansen	1966–1969
	Walter Stein	1969–
AIHA	Richard G. Heatherton	1966–
AOAC	J. B. Lockhart, Jr.	1966–1972
	L. A. Brauch	1973–
ASTM	R. S. Morse	1966–1973
APHA	Bernard Shleien (*Chairman*, 1966–1972)	1966–
APCA	Maynard E. Smith	1966–1967
	Peter F. Hildebrandt	1967–1972
ASCE	Conrad P. Straub	1971–
ACGIH	D. E. Van Farowe	1966–

SUBCOMMITTEE #9 (GENERAL TECHNIQUES AND PRECAUTIONS)

	Name	Date of Service
AIHA	A. L. Linch (*Chairman*, 1969–)	1967–
ASME	Glenn Damon	1967–1969
	Zane E. Murphy	1969–1970
	Robert A. Weidner	1970–1973
APCA	E. R. Hendrickson	1967–
APHA	Morris Katz (*Chairman*, 1967–1969)	1967–
ACS	Gary O. Nelson	1971–
AOAC	John N. Pattison	1967–
ASTM	Walter Ruch	1967–1969
	William T. Ingram	1969–1970
	Robert J. Martin	1970–1973
ACGIH	Alvin L. Vander Kolk	1967–
ISA	John N. Harman	1972–

SUBCOMMITTEE #10 (PARTICULATE MATTER)

	Name	Date of Service
AIHA	Robert A. Herrick (*Chairman*, 1971–1974)	1970–1974
	Thomas Mercer	1968–1970
APCA	Robert S. Sholtes (*Chairman*, 1968–1971)	1968–
ASME	Richard B. Engdahl	1968–1970
	Dale Lundgren	1970–
APHA	Morton Lippman	1968–1969
	Carolyn Phillips	1970–
ACS	James Lodge	1971–
AOAC	Jack Wagman	1968–1972
	Howard E. Ayer	1973–
ACGIH	J. C. Wells	1968–1969
	Edward Stein (*Acting Chairman*, 1975–)	1969–
ASTM	Paul F. Woolrich	1968–1971
	Shephard Kinsman	1971–1973

SUBCOMMITTEE #11 (STATIONARY SOURCE SAMPLING)

	Name	Date of Service
AIHA	M. Dean High (*Chairman*, 1974–)	1971–
AOAC	John S. Nader (*Chairman*, 1971–1973)	1971–1973
ASTM	Knowlton J. Caplan	1971–1973
ACGIH	Jerome Flesch	1971–
ASME	Richard Gerstle	1971–
ASCE	William T. Ingram	1971–
APCA	John B. Koogler	1971–
ASC	L. R. Perkins	1972–
APHA	Charles J. Theophil	1971–

SUBCOMMITTEE #12 (STANDARDIZATION COORDINATION)

	Name	Date of Service
ASME	Robert Spirtas (*Chairman*, 1975–)	1971–
AIHA	R. S Brief	1971–
AOAC	Luther G. Ensminger	1971–
ACS	Stever Goranson	1971–
APCA	Harrison R. Hickey	1971–1972
	Robert W. Garber (*Chairman*, 1972–1974)	1972–
ACGIH	William D. Kelley (*Chairman*, 1971–1972)	1971–
APHA	John E. Silson	1971
ASTM	Walter J. Smith	1971–1973
ISA	Oliver A. Fick	1973–

PERMISSIONS

The Intersociety Committee expresses its appreciation for permission to use data as indicated from the following:

E. I. Dupont deNemours and Company, Wilmington, Delaware for information used in Part I, Section 3 which is based in part from materials furnished by them and is reprinted from the Journal of Teflon, Reprint #41.

Metronics Associates, 3201 Porter Drive, Stanford Industrial Park, Palo Alto, California for information in Part I, Section 3.7– 3.10 which has been abstracted from Reference 10a.

The American Society of Testing Materials, 1916 Race Street, Philadelphia, Pennsylvania, 19103 for materials used in Part I, Section 5 which is based upon ASTM designation 1605–60, reapproved 1967;

for information in Part I, Section II which is based upon ASTM designation E131–68;

for information in Part I, Section 12 which is based upon ASTM designation E60–68;

and for data in Part V which is reprinted from ASTM designation 1914–68.

The American Industrial Hygiene Association Journal, Department of Industrial Health, School of Public Health, University of Michigan, Ann Arbor, Michigan 48104 for data in Part I, Section 13 which is based upon "U-V Absorptivity Curves," AIHA Analytical Abstracts, 1965 (UV);

for information in Part I, Section 14 which is based upon publication in AIHA Journal, Vol. 32, 412, June, 1971;

and for information in Part I, Section 15 which is based upon publication in AIHA Journal Vol. 31, p. 525, 1970.

The American Public Health Association, 1015 18th Street, N.W., Washington, D.C. 20036 for data in Part I, Section 18 which is based upon Part 100, Section 104A of Standard Methods for the Examination of Water and Waste Water, 14th edition, 1976.

John Wiley and Company, New York, New York for data in Part I, Section 19 which is based upon Laboratory Safety Measures described in "Treatise on Analytical Chemistry" by P. G. Elving.

The National Institute of Occupational Safety and Health (NIOSH), 1014 Broadway, Cincinnati, Ohio, 45202 for material in Part I, Section 25, which is based upon "The Industrial Environment–Evaluation and Control" by Adrian Linch.

The American Conference of Governmental Industrial Hygienists (ACGIH) P.O. Box 1937, Cincinnati, Ohio 45201 for the material in Part I, Section 26, which is based upon "Direct Reading Colorimetric Indicators" by Bernard E. Saltzman, and for the information in Part I, Section 28, based upon "Filter Media for Air Sampling" by Morton Lippmann in the ACGIH publication Air Sampling Instruments for Evaluation of Atmospheric Contaminants, 4th Edition, 1972.

TRADENAMES

Amberlite®. Rohm and Haas, distributed by Polysciences, Inc., Industrial Park, Warrington, Pa. 18976

Chromosorb GHP. Johns Manville

Deactigel®. Applied Science Laboratory, State College, Pa.

Dynacal®. Metronics Associates, Inc., registered tradename for family of gas-permeation tubes and devices.

Kel-F. Minnesota Mining & Manufacturing Co., Minneapolis, Minn.

Mylar—Dupont Co.

Poropak—Supelco Inc., Bellefonte, Pa., 16823

pyrex—refers to no specific brand, but to a general type, such as manufactured by Corning Glass Works (under the name of "Pyrex"), or Kimble Glass Co., Division of Owens Illinois ("Kimax"), or equivalent.

Saran. Dow Chemical Co.

Stractan 10®. Stein-Hall & Co. Inc., 385 Madison Ave., New York, N.Y.

Teflon. Dupont Co.

Triton X-305. Supelco Inc.

ABBREVIATIONS AND SYMBOLS

The following abbreviations and symbols have been used throughout:

Abbreviation or Symbol	Referent
Å	Angstrom(s); 10^{-7} mm
amt(s)	amount(s)
ACS	American Chemical Society
ACGIH	American Conference of Government Industrial Hygienists
AICHE	American Institute of Chemical Engineers
AIHA	American Industrial Hygiene Association
APCA	Air Pollution Control Association
APHA	American Public Health Association
APWA	American Public Works Association
ASCE	American Society Civil Engineers
ASME	American Society of Mechanical Engineers
ASTM	American Society for Testing Materials
AOAC	Association of Official Analytical Chemists
bp	boiling point
C	degree(s) Celsius; Centigrade
cc	cubic centimeter; cm^3
cf(s)	cubic foot (feet); ft^3
cfm	cubic feet per minute
cfs	cubic feet per second
cm	centimeter(s)
cpm	count per minute
conc	concentration; concentrated
diam	diameter
dpm	disintegrations per minute
F	degree(s) Farenheit
ft	foot; feet
g	gram(s)
HPS	Health Physics Society
hr	hour(s)
ID	internal diameter
ISA	Instrument Society of America
K	degree(s) Kelvin
KeV	kiloelectron volt(s)
l	liter
max	maximum
m	meter
m^3	cubic meter
M	Molar (concentration)
me	milliequivalents
MeV	megaelectron volt(s)
ml	milliliter(s)
mm	millimeter(s)
mp	melting point
mμ	millimicron(s); nanometer
mV	millivolt(s)
μ	micron(s); micrometer
μCi	microcurie(s)
μg	microgram(s)
μl	microliter(s)
μmho	microomho(s)
min	minute(s)
N	normal
nCi	nanocurie(s); 10^{-9} curie
nm	nanometer
ng	nanogram; 10^{-12} gram
OD	outside diameter
oz	ounce(s)
pH	measurement of hydrogen-ion activity
pK	(−log K)
pCi	picocurie(s); 10^{-12} curie
ppb	parts per billion (1/1000 million)
ppm	parts per million
psi	pounds per square inch
rpm	revolutions per minute
SAE	Society of Automotive Engineers
sec	second(s)
sp gr	specific gravity
sq cm	square centimeter; cm^2
sq ft	square foot (feet); ft^2
sq in	square inch(es); in^2
sq mm	square millimeter; mm^2
Torr	unit symbol pressure
Ŝ	standard taper
UV	ultraviolet
V	volt(s)
vol	volume
W	watt(s)
w/v	weight per volume
wt	weight

Notice of Adoption of Recommended Methods

The following tentative methods of analysis, published in the intersociety Committee Manual, "Methods of Air Sampling and Analysis", American Public Health Association, Washington, D.C., 1972, have been elevated to "recommended" status by the Intersociety Committee after evaluation of the results of collaborative test programs conducted for the Environmental Protection Agency (EPA) and the American Society for Testing and Materials (ASTM).

Section	Page	Adoption as Recommended Method September 1974	Collaborative Test Program
128		Method of Analysis for Carbon Monoxide Content of the Atmosphere (Nondispersive Infrared) 42101-04-69T	SWRI under EPA Contract CPA 70-40
406		Method of Analysis for Nitrogen Dioxide Content of the Atmosphere (Griess-Saltzman Reaction) 42602-01-68T	ASTM, Committee D-22 under Project Threshold
501		Method of Analysis for Suspended Particulate Matter in the Atmosphere (High Volume Method) 11101-01-70T	SWRI under EPA Contract CPA 70-40
704		Method of Analysis for Sulfur Dioxide Content of the Atmosphere (Colorimetric) 42401-01-69T	Southwest Research Institute (SWRI) under EPA Contract CPA 70-40

Classification of Industrial Hygiene Methods in Accordance with the NIOSH System

Many of the analytical methods published in **Part III** are based on procedures developed in the Physical and Chemical Analysis Branch (P & CAB) of NIOSH for industrial hygiene analyses, involving chemicals in the air of the workplace and in biological samples. Various subcommittees of the ISC have undertaken the tasks of reviewing, evaluating and revising these methods. The following system of classification has been adopted by NIOSH for industrial hygiene methods in order to give the user a guide as to the confidence one should place in the reliability of a specific method.

Class A—Recommended: A method which has been fully evaluated and successfully tested by a selected group of laboratories.

Class B—Accepted: A Method which has been subjected to a thorough evaluation procedure in the NIOSH laboratory and found to be acceptable.

Class C—Tentative: A method which is in wide use and which has been adopted as a standard method or recommended by another Government Agency or one of several professional societies such as ACGIH, AOAC, AIHA, ASTM or ISC.

Class D—Operational: A method in general use or approved by most professional industrial hygiene analysts but has not been thoroughly evaluated by NIOSH or any professional societies.

Class E—Proposed: A new, unproved or suggested method not previously used by industrial hygiene analysts but which gives promise of being suitable for the determination of a given substance.

Methods classified as "B" and lower categories are considered to be "tentative" by the ISC procedures until raised to Class A—"recommended".

TABLE OF CONTENTS

PAGE

Part III. 800—Methods for Chemicals in Air of the Workplace and in Biological Samples

---

*The number, in parenthesis, refers to the number assigned to the method by the Physical and Chemical Analysis (P&CAM) Branch of NIOSH.

PART I
GENERAL PRECAUTIONS AND TECHNIQUES

PART I

GENERAL PRECAUTIONS AND TECHNIQUES

1. Physical Precautions

1.1 HOMOGENEITY OF THE SAMPLE. No analytical result regardless of the accuracy and precision of the procedure can be any better than the quality of the sample submitted for analysis. Therefore, the primary concern of the investigator must be directed to the collection of representative samples and the homogeneity of the air mixtures employed to calibrate both the collection and analytical systems. Human sensory perceptions cannot guide the uninitiated with respect to variations of contaminant concentrations in either space or time. Almost without exception gaseous components are either colorless or are present at such low concentrations as to be effectively colorless. The human sense of smell is notoriously deceptive with respect to concentrations of even highly odoriferous components, therefore, cannot be relied upon for estimating relative quality or uniformity of the ambient atmosphere (**2, 4-Part V, 9**).

The source of the contaminant, air flow direction and velocity whether due to wind or thermal gradients, density of the contaminant, intensity of sunlight, time of day, presence of obstructions such as trees, buildings, partitions, machinery, etc., which act as baffles to produce turbulence, humidity, and half-life of the contaminant together determine the concentration at any given location. That the concentration can vary by several orders of magnitude within a relatively short radius from the point of reference has been amply confirmed by many investigators. Air may flow either in a stream line or turbulent flow and this factor alone can determine the dispersal pattern of the pollutant (**2, 3, 5, 9**).

The density of the pollutant in many cases will counteract diffusion processes to an extent which will establish stratification. This phenomenon prevails in the case of many highly toxic and irritant gases such as phosgene, mustard gas, lewisite, chlorine, etc. (**5**). Natural convective circulation and diffusion in confined spaces such as silos, coal mines, wells, coves, tanks, etc., is not sufficient to maintain a normal atmosphere containing approximately 21% oxygen.

The location of sampling sites whether on a temporary or "grab" sample basis or on a fixed station basis for continuous monitoring over long periods of time is often critical and will determine the validity of the conclusions drawn from the results (**1, 2, 3, 9**). In the vicinity of industrial atmospheres the variability may be sufficiently extreme to vitiate the results from fixed station monitors regardless of the care taken in locating the samplers. In an alkyl lead manufacturing facility, for example, no relationship could be established between exposure as indicated by urinary lead excretion and numerous fixed station samplers. In this case, correlation was not established until the monitors in the form of a small filter—microimpinger assembly—were actually worn during the entire 8 hr shift by the workmen assigned to these areas (**1**).

Several excellent treatises have been published on planning the air pollution survey and the effects of meteorological and geographical conditions on the validity of the results. This subject is outside of the field of General Techniques and Precautions; therefore, the reader is advised to consult one or more of these publications for full details (**2, 3, 6, 7, 8, 9, 37, 38**).

On a smaller scale, even greater care is required to establish uniform mixtures of pollutants in air for use as calibration mixtures whether the system be dynamic or static in principle. Passage through lengths of small diameter tubing to ensure turbulent mixing is one relatively simple expedient for dynamic mixing (**11**). Other systems employ some form of baffled chamber inserted between the mixing ports and the delivery point (**13**). Rotating paddles or even loose pieces of sheet metal or plastic (PTFE) to be activated by shaking have been inserted in containers for static mixtures for mixing by turbulence (**14**). Probably a more effective alternative is to evacuate the container then mix the contaminant with the air at a uniform rate as it is released into the void (**14**). (Inject from a hypodermic syringe through a needle inserted through a septum). Diffusion alone cannot be relied upon to produce a homogeneous mixture in most cases.

Other effects such as container wall absorption, humidity, volumetric errors, etc., discussed elsewhere in this section must be recognized and appropriate precautions taken to avoid concentration errors derived from these seemingly extraneous sources when attempts are made to produce homogeneous, known concentrations for calibration purposes.

1.2 ABSORPTION EFFECTS ON CONTAINER WALLS AND CONNECTING TUBES. Failure to recognize absorption effects can lead to serious errors in both collection of samples and in preparation of calibration mixtures especially in static systems. Losses of analysand can occur either by primary adsorption which may be reversible in response to pressure or temperature changes or by secondary adsorption which may involve chemical reaction with another material previously adsorbed (**2, 4, 10**). The affinity of glass for water, nitrogen dioxide, benzene, aniline, the isocyanates, etc., has been recognized (**11, 15, 16, 35**). The interior surface of a 5-gallon borosilicate glass carboy can adsorb as much as 50% of the benzene in the 100 to 200 ppm range from a calibration mixture in air within 30 min after preparation. These losses to some extent can be reduced by preconditioning at a concentration higher than the working level but desorption may contribute positive errors (Table 1:I). Losses by adsorption of polynuclear aromatic compounds on glass have also been reported (**17**).

This adsorption effect can be minimized in the case of collection of air samples in evacuated glass flasks (Shepherd flasks) or sampling tubes by the introduction of a suitable solvent for the vapor, such as isooctane for benzene and xylidine, directly into the container with thorough rinsing of the inner surfaces (**35**). In some cases such as the analysis for nitrogen dioxide the solvent or reagent is introduced before the sample is taken *e.g.*, 100 ml sample drawn into a syringe containing 10 ml of Saltzman's reagent (**18**).

Nitrogen oxides are so tenaciously adsorbed on glass surfaces that rinsing with Saltzman's reagent has been found necessary to obtain reliable results. Adsorption of isocyanates on glass, metal or plastic surfaces occurs to such a degree as to preclude preparation of static calibration mixtures even under strictly anhydrous conditions (**11**).

Positive errors can be included when evacuated double entry gas sampling tubes (fitted with 2 stopcocks) are used for sample collection (**10, 19**). In this case, after release of the vacuum, additional sample is drawn through the tube for 3 min before closing the stopcocks. Recovery of xylidine increased to 120% in the 50 to 70 ppm range under these conditions. This error was increased to 40% by adding glass beads to the sampling flask to increase the surface area (**35**). These conditions can be avoided by collection of the pollutant in an impinger assembly (Greenburg-Smith, midget or micro).

In dynamic systems prolonged equilibration is required before dependable calibration can be assured. Calibration of toluene diisocyanate (TDI) in the TLV range required passage of the air-TDI vapor mixture through a mixing chamber composed of a 6′ length of ¼″ copper tubing under strictly anhydrous conditions for 60 to 72 hr to reach equilibrium. This phenomenon was not entirely produced by the film of water normally adsorbed on the inner walls of the tubing as the system had been carefully heated under vacuum to remove volatile impurities before use (**11**).

Adsorption is a critical factor in making up standard mixtures for phosgene calibration at concentrations below 1 ppm even after meticulous removal of the adsorbed water film. The use of metal cylinders especially must be avoided. In the development of a continuous monitor for phosgene concentrations below 5 ppm decom-

Table 1: I. Static Calibration Mixture
Benzene Loss by Absorption on Container Walls
Borosilicate Glass Temperature 75 ± 5 F

Time Elapsed Hours	Benzene PPM Added	Benzene PPM Found	Per cent Loss	Container Conditioned Time Hours	Container Conditioned Benzene PPM	Remarks Inside Surface Approx. 800 Sq. In.
0.5	100	56	44	0	none	
0.5	100	67	33	0	none	
16	200	100	50	0	none	
0.5	100	84	16	16	100	
1	—	78	25	—	—	
0.5	100	87	13	16	200	
0.5	100	107	−7	—	—	Swept, 2nd 100 ppm, added
0.5	200	128	36	—	—	3rd 100 ppm, added
0.5	50	56	−12	—	—	Swept, 50 ppm, added
0.1	100	75	25	0	none	Idle 30 days
0.25	—	72	28	—	—	
0.5	—	55	45	—	—	
1.0	—	55	45	—	—	
4.5	—	48	52	—	—	
24	—	42	58	—	—	
192	—	0	100	—	—	

Note: Five gallon borosilicate glass carboy jacketed with a canvas cover. Calculated quantity of benzene was added and mixed by rotating a poly TFE paddle and sampled at the indicated time after addition. In two cases indicated by 16 hr (column 5), equilibrated by standing overnight, then refilled.

position in metal (steel, stainless steel, copper, aluminum, etc.) or glass tubing was found to be so extensive that their use in the sampling system could not be considered. In this case polyethylene tubing was acceptable. Water adsorbed to the contact surfaces was probably responsible for the losses as HC1 was observed to be a by-product. Catalytic decomposition on anhydrous surfaces cannot be ruled out, however (**20**).

In a study of the efficiency of midget and micro-impingers by collection of aniline from air, losses by adsorption in the glass batch-type generator (5 to 15% in the 100 μg range) was encountered unless the system was heated to 60 C in a hot water bath. Even greater were losses due to absorption by gum rubber tubing. Even 3-inch connections introduced significant errors (10% loss) but the losses were not entirely proportioned to hose length. These losses increased as the age and number of uses increased *e.g.*, recoveries decreased from 83 to 57% over a period of 8 days (**21**).

Failure to recognize the effect of adsorption on container walls delayed the determination of the half-life of SO_2 during a study of the fate of SO_2 in the atmosphere at Syracuse University. The half-life was found to be unexpectedly short evern after correction for adsorption effects (**22**).

Tubing through which high concentrations of tetraethyl lead vapor have been drawn will become a source of contamination when "clean" air is passed through the system. These effects are eliminated by use of all-glass systems connected by standard taper, ball joints, "0" ring seals or polytetrafluoroethylene (Teflon) compression connectors.

The adsorption of traces of analysands from liquid solution onto container walls is probably somewhat more widely recognized but certainly requires review to re-emphasize the importance of this source of errors. Heavy metal ions, especially lead, mercury (**25**), silver (**23, 24**), cadmium, and antimony, and readily adsorbed by glass and in some cases by plastic surfaces. Lead in urine specimens, for example, is adsorbed to an extent sufficient to introduce significant errors after storage for more than 3 days in borosilicate glass specimen bottles (**26**). This effect probably is produced by bacterial fermentation which elevates the pH by ammonia release, and co-precipitation of lead with calcium and magnesium phosphates on the glass. In the early stages the precipitate is invisible but nevertheless may contain a significant fraction of the total lead in the sample. In many cases these losses can be held within tolerable limits by reducing the pH to 2 or less with nitric acid. This adsorption loss can be a critical factor in the storage of standard reference solutions and must be taken into account when collaborative testing programs are set up. Efforts to prepare stable aqueous standardizing solutions of metallic mercury met with failure unless excess mercury maintained in contact with aqueous phase. Furthermore, centrifugation is necessary to remove dispersed microdroplets (or HgO) if shaking is employed to accelerate saturation, and a relatively large temperature coefficient of solubility must be taken into account. In spite of these limitations this saturated aqueous mercury solution can be employed as a convenient source of mercury for calibration by comparing recovery from a collecting system with direct analysis of the standard (**25**).

Filtration of samples collected in a liquid medium should in all cases be avoided if the desired contaminant is to be recovered quantitatively in the liquid phase. Not only will the contact surfaces of the filter funnel and the filtrate receiver adsorb trace materials, but the filter medium (paper, membrane, asbestos mat, glass fiber, micro metallic, etc.) may be highly adsorbant (**17**). Also, the insoluble fraction removed may retain the analysand as in the case when calcium phosphate is separated from a mixture containing traces of soluble lead compounds (**27**).

The problem of adsorption on glass surfaces previously discussed requires re-emphasis in the case of porous glass used either as a filter for particulate matter or as a diffuser for collection in liquid absorption systems. Filtration of aqueous solutions of phenanthrene, naphthalene, pyrene or anthracene through a porous glass plate reduced the hydrocarbon concentration as much as 40% (**17**). When ultraviolet (UV) absorption is employed for analysis, errors from background absorption produced form UV absorbing materials leached from the filters and light scattering from suspended particles must be taken into account (**28**). A base-line technique for correction is often required as the solute often alters the light scattering and sedimentation properties with the result the reference blank and the sample are no longer exactly comparable. If removal of sediment is necessary, filtration through polytetrafluoroethylene (PTFE) mat or decantation after centrifugation will minimize losses. In any case, potential losses at each step must be recognized and quantitated.

Rust, scale and other metallic corrosion products can introduce serious errors through adsorption of air pollution components and should be meticulously scoured out of systems used for air sampling, storage or analysis. Many of these oxidation and corrosion products are highly catalytic as illustrated by the rate at which alkyl lead compounds are oxidized by hydrated iron and bismuth oxides (**29**). Pipe dope used on threaded connections and oils used in cutting threads or lubricating moving parts also can present significant reversible absorption effect. Polytetrafluoroethylene (PTFE) ribbon thread sealant which has a very low absorption coefficient is recommended for threaded joints which require lubrication. Moving parts machined from PTFE are self-lubricating and can be sealed with modified PTFE "O" rings.

The reverse effect has been encountered also. The diffusion of HF in Conway dishes is enhanced ten-fold when silicone grease is used as a sealant for the dual concentric chambered dish cover. Fluoride can be removed quantitatively from aqueous HCl by gentle boiling in the presence of silicone fluid (**30**).

Adsorption of a secondary layer on container walls may completely alter the character of the surface presented to the collection and analytical system. Soaps and synthetic detergents are especially troublesome in this respect. The film of soap on glass is not removed by 5% trisodium phosphate, chromic acid cleaning mixture, or conc nitric acid. Only by baking out in an annealing furnace can "chemical sterility" be achieved. In the colorimetric determination of aromatic nitro and amino compounds in air, this soap film reacts with the color generating reagents. These adsorbed layers appear to retain nitrous acid or nitrogen oxides in a state which does not react with starch-iodide indicator or sulfamic acid but does react with the coupling agent (1-naphthyl 2-ethylenediamine or Chicago Acid) to produce a deep orange color. This spurious color reaction produces high results which may be sufficiently uniform to produce false calibration curves (zero intersect at some light transmittance value below 100%). To avoid this nitrite absorbing contamination and remove residual azo dyes which may be adsorbed on the coupling flasks and spectrophotometer cuvettes, soak all glassware in 5% aqueous trisodium phosphate for several hours and rinse thoroughly with distilled water. Soak Corex curvettes only in 37% nitric acid wash, however (**31**).

The nonabsorption of heavy metal ions on hydrophobic surfaces of PTFE and polyethylene (PE) may be reversed by the presence of certain organic anions. In trace quantities, cesium is adsorbed on glass surfaces by interaction with the layer of adsorbed hydroxyl ions which account for the hydrophilic character of glass surfaces. Addition of a salt of high ionic activity such as sodium nitrate reduces this adsorption effect. The ion exchange capacity of the glass surface can also become a significant factor at very low concentrations. The overall effect is orders of magnitude less (adsorption coefficient increase in the order PTFE-PE-glass) for the nonionic hydrophobic plastic surfaces. However, the absorption coefficients reverse when the tetraphenyl borate (TPB) ion is introduced. This reversal is due to adsorption of TPB ions by van der Waals forces on the hydrophobic PTFE and PE which then requester cesium ions. On the hydrophilic glass surface the adsorption is ionic in character which reduces the binding sites for the cesium ions (**32**).

Lubricants in stopcocks and ground glass joints must be avoided as these materials are a source of contamination and/or absorb those constituents of interest either directly from the air sample or from the liquid reagent system after collection. Teflon plug cocks for stopcocks and sleeves for joints provide seals which require no lubrication or liquid phase sealant for air-tight joints.

Fungus growth within an atmospheric analyzer can be a particularly vexing problem. Fungus within an SO_2 analyzer results in an indicated concentration lower than that actually present. Fungus effect can be minimized in an analyzer by: (a) Keeping the instrument running continually, (b) cleaning periodically, and (c) using a small amount of fungicide in the reagent. Permeation tubes can be inserted into the sample input to test for a fungus problem.

1.3 Solid Adsorbant Collecting Systems. Extended surface adsorbants such as activated silica gels, alumina gels, charcoals and carbons, glass beads, and spheres offer a very attractive solution to the collection and concentration of gas phase pollutants from atmospheric samples. However, the tenacity of the forces which provide quantitative extraction of the components sought may prove to be self-defeating when quantitative removal of these same components for analysis is attempted. Failure to recognize and equate these factors may, and has, produced serious errors. Mercury, either as metallic vapor or as volatile organic derivates, is not removed quantitatively either by heating or elution with liquid reagents from silica gel. The same deficiency has been encountered in attempts to use activated charcoal for collection of alkyl lead compounds.

Since the polarity of the adsorbed element or compound determines the binding strength on silica and alumina gels, components of high polarity will displace components of lesser polarity—the principle of chromatography. Therefore, in attempting to collect relatively nonpolar compounds such as benzene, the presence of coexisting polar compounds such as phenol, acetic acid, ammonium chloride must be recognized as a source of interference in the use of solid adsorbants for sample collection. Under high humidity conditions the adsorbant may be deactivated by saturation with water vapor (**28**). Activated charcoal exhibits this effect only to a minor degree. Aromatic hydrocarbons are not removed quantitatively from charcoal (**36**).

Selective displacement may be taken advantage of in eluting the sample from the adsorbant. An inert carrier gas with or without heat, for example, may be used to separate a mixture of aromatic hydrocarbons from a silica gel column. Proper choice of eluting solvent may well serve to separate components by liquid extraction also. An example would be separation of nitrobenzene from aromatic hydrocarbons for analysis by UV spectrophotometry (**28**).

Liquid elutriants must be chosen and used with care however as decrepitation of the gel may occur. The finely dispersed fragments can produce difficulties in the determination of optical density in the final step of the analysis. This problem is especially acute in UV spectroscopy. Usually this effect is caused by the heat generated when a polar solvent such as ethanol is introduced. Dilution with a nonpolar solvent such as isooctane will usually avoid this difficulty (**28**). Elution from charcoal is dependent on molecular size rather than polarity (CS_2, CCl_4, etc.)

Change in air flow resistance due to either swelling, shrinking or channeling of the adsorbant is sometimes encountered. Collection of alkyl lead vapor in crystals of iodine was plagued with channeling and shrinkage due to iodine sublimation. Addition of a head of fine inert sea sand eliminated the problem by providing enough pressure to keep the iodine crystals compacted without appreciably altering the pressure drop across the column (**33**).

Many dusts are highly adsorbant and often carry a significant gaseous component present in trace amounts such as SO_2 at concentrations below 0.5 ppm. If a prefilter is installed in the collection of trace contaminants, the particulate collected on the filter must be analyzed also if a quantitative accounting for the total contaminant present is desired (**34**). In some cases such as in the analysis for lead in air, the filter would be analyzed separately from the fraction collection in the liquid reagent to provide a differential analysis for organic and inorganic lead. This is especially important in cases where the TLV's for the two forms are different and coefficients must be used for hazard evaluation (**1**).

The use of pretreatment or guard systems such as solid dehydrating agents to remove moisture, soda lime to remove CO_2, or sulfuric acid on silica gel to remove NH_3 interference, and liquid scrubbers as an oxidizing solution to remove reducing interferences such as SO_2 and H_2S before adsorption of the desired pollutant can introduce serious errors (**28, 39**). Silica based and anhydrous $CaSO_4$ granules may partially or completely adsorb the subject element or compound as well as eliminate the interference in the air sample. Careful calibration of the collection system under use conditions is critical when interferences must be eliminated from the air sample by pretreatment before collection of the desired component occurs.

1.4 Diffusion Effects of Plastics. Diffusion outward from the air sample, into, or through plastic containers or tubing probably is one of the most frequent sources of error encountered in air analysis. These irreversible losses are in addition to the reversible surface adsorption effects noted under Section 1.2. Elastomeric materials which include natural rubber, neoprene and plasticized polyvinyl chloride (PVC) are especially troublesome. (see section 3.1 on Permeation Tubes-Dynamic Calibration).

The plasticizers employed in the manufacture of flexible tubing from PVC type polymers may act as cosolvents to remove organic components from air samples by absorption. This mechanism is often temperature sensitive and to some extent may be reversible in a fashion resembling gas chromatography column packings. In a study of the recovery of aniline vapor from air, losses as high as 1.7%/cm of exposed gum rubber tubing were encountered **(21)**. Similar experience has been obtained with alkyl lead and mercury compounds, aromatic hydrocarbons (benzene, toluene, etc.), chlorinated hydrocarbons, and aromatic nitro compounds. An all-glass system with joints and connections fitted with PTFE gaskets and the shortest possible path between air sample intake and the collecting medium to a great extent eliminate these losses. However, adsorption effects must not be neglected even in all-glass systems.

The permeability of even the inert plastics exemplified by PTFE has been employed for the constant rate generation of calibration mixtures of certain trace components such as SO_2, HF, $COCl_2$, aromatic hydrocarbons, etc. Although this property would be undesirable in the collection of air samples, the magnitude of loss by diffusion through short inert plastic sampling lines (PTFE, polyethylene, polypropylene) can be neglected. However, when samples are stored in plastic bottles or bags, diffusion may become a controlling factor. If plastic containers must be used to facilitate collection or transportation, then the diffusion rate factor must be determined for each component sought over the concentration and temperature range to be encountered and a correction factor applied to correct for the time interval between collection and analysis. Aluminum foil-lined polyester bags (Mylar) which materially reduce diffusion losses are available for CO_2, CO, water vapor and some other gases but are not recommended for the highly polar gases (SO_2, NO_2 and O_3) **(42)**. Polyester film in 2-mil thickness is considered the best all-around material for sample-bag fabrication. However, PVC (Saran) film (1-mil thickness) which has been pre-aged to reduce outgasing of compounds from the plastic into the contents of the bag has been widely accepted on the basis of inertness to many common gases and vapors, high tensile strength and heat sealability. Data relative to air transmission is summarized in the following table **(40)**.

Gas Transmission: ccs/M^2/μ/24 hr/atm at 23 C

O_2	32–43
CO_2	150–236
N_2	4.8–6.3
Air	8.3–17.3

Water Vapor Transmission: gms/M^3/24 hr.

at 24 C—0.31
at 38 C—3.05

For ultimate chemical inertness fluoroplastic film bags are available **(41)**. Laminated plastics—polyethylene on polyester, polyethylene on cellulose acetate and polypropylene on either base—used in packaging foods offer interesting alternatives for gas impervious containers and a choice of surface for contact with the air sample. Semi-flexible 10-l polyethylene bottles with greater wall thickness also offer a solution to the diffusion problem and are convenient for preparation of calibration mixtures. (See Static Calibration section 2.1).

Inward diffusion also may be encountered and may create problems in shipment or in prolonged storage. Samples taken in a low humidity environment may be altered by inward diffusion of water vapor when the plastic container is stored in a high humidity climate. Solvent vapors, gasoline vapors, gasoline and jet engine exhaust components may contaminate the sample in shipment and show up as unexplained peaks in gas chromatograph analysis or introduce interferences in wet chemical procedures. Shipment in air-tight containers would be a reasonable precaution to avoid contamination from this source.

Although obvious, the testing of all plastic containers, especially thin film bags for gross leaks before use must be emphasized. The absence of pin holes, gaps in the seams, and leakage around the sampling port and in the retaining valve or septum assembly must be verified before each use. This examination may consist of full inflation under slight positive pressure (2 to 3″ water seal) under water or in rigid sealed container with the air line introduced through a bulkhead fitting and connected to a water manometer. Soap solution stabilized with glycerine may be painted or sprayed on for leak detection to pinpoint "micro" leaks.

1.5 Mechanical Defects of Sampling Equipment. As well recognized as this source of error has become, leaks and faulty seals remain one of the major contributors to unreliable results. Any system used to collect air samples, to remove contamination components by absorption, or to calibrate such systems, must be checked for leakage, either inward or outward depending upon whether the air stream is pushed or pulled. Perhaps the most expedient procedure involves either an increase or decrease of the pressure in the system within limits of the maximum operational range of the system, by closing off tightly and measuring the rate of loss of either the pressure of vacuum on a water manometer. Pinholes in glass seals—especially ring seals—plastic tubing, and metal tubing, joints which do not seal perfectly or are insufficiently gasketed—"O" rings which may be cracked, loose or hardened—or lack a sealing lubricant, loose glass to rubber or plastic tubing connections, dirt in glass to glass joints, glass plugcocks which do not perfectly mate with the stopcock barrels, poor valve stem packing or insufficient sealing pressure, and poorly sealed threaded fittings provide the most frequent leak sources. Pinholes in stainless steel valves, reducing valve diaphragms and two stage regulators also have contributed their share of difficulties **(13)**.

Fouling of rate meters and total volume meters from particulate matter or corrosion products often is not so obvious and requires eternal vigilance to minimize. Orifice type and positive displacement meters and rotameters should be protected by an efficient filter to prevent reduction in orifice opening, erosion of moving parts, or complete pluggage. Rotameter calibration can be shifted significantly by dust accumulation and plastic tube models require protection from solvents which soften or render the bore tacky. Metal floats are attacked by corrosive gases and some reagents such as iodine may in time render the tube body opaque. Corrosive gases and aerosols must be rigidly excluded from total volume and positive displacement meters to maintain calibration validity and integrity of moving vanes or pistons. Liquid aerosols may be especially troublesome by forming a liquid film which by surface tension and capillary effects reduce air flow in small orifice throats, interfere with the movement of rotameter floats, and increase the resistance to movement of critical components in positive displacement meters. Galvanic corrosion from airborne electrolytes under high humidity conditions can be a problem with metal components.

Flow meters based on thermal conductivity (hot filament type) also are affected by dust and reactive gases, but in a different fashion. Corrosion will alter the electrical conductivity and leads ultimately to failure of the element, and coating by inert material will reduce heat transfer and ultimately reduce the response of the instrument. Chlorides and lead are especially destructive. The volumetric meter and the pump or aspirator sub-assembly of any air sampling system should be adequately protected by appropriate traps which are renewed frequently. Activated charcoal granules (5 to 10 mesh) probably offer the most efficient protection for all but a few destructive gases and vapors. Cellulose acetate membrane filters (0.8 μ pore size) will effectively remove over 99% of the dust particles which might affect instrument performance without introducing prohibitive pressure drop **(43)**.

The orifice in impinger-type collecting devices, and the interstices in porous glass diffusers are also vulnerable to partial or complete closure by particulate matter and should be protected by a filter when absorbing gases if the solid fraction is not an integral part of the sample. Porous glass diffusers may accumulate particles that are not removed by conventional cleaning procedures after repeated use **(13)**. Also irreversible retention of analysand sometimes occurs. In the collection of alkyl lead vapors from 5 to 10% of known quantities added to calibration mixtures were retained in a pencil-type diffuser substituted for the impinger tube in the standard midget impinger sampler **(33)**. The increase in pressure drop produced by the decrease in free cross-sectional area will contribute to volumetric sampling errors, and may significantly reduce the collection efficiency. In dust collection by impingement, this factor becomes critical.

Pressure drop must be taken into account when a sampling system is calibrated volumetrically. Rotameters and orifice type meters should be calibrated with the pressure reducing or elevating components in place or volume correction made to compensate for the deviations from ambient conditions **(44)**. In systems based on filter collection the pressure drop across the system may steadily increase as the filter medium loads up with retained particulates. Under these conditions an averaged correction may be within tolerable limits of volumetric error. Changes in barometric pressure usually introduce only marginal error if ignored. (Deviation of 15 from 760 Torr will introduce a 2% volumetric error).

In the use of cold traps to collect contaminated air components (freeze-out technique) consideration must be given to the possible accumulation of ice crystals in the delivery tube. This possibility can be minimized by introducing the sample through a side arm into the chamber and exhausting through the inner tube, i.e., reverse flow **(14)**.

1.6 Partial Vapor Pressure Effects. In collection systems which depend exclusively upon solubility of the component sought in a nonreactive solvent rather than retention by chemical reaction to a nonvolatile derivative, the amount absorbable is determined by the partial pressure of absorbate. The ratio of the air sample volume to the volume of absorbant is a critical factor in collection efficiency and there is a limiting value of the ratio for any system which cannot be exceeded if a given degree of efficiency is to be attained. In the case of acetone in water the limit would be approximately 5 l of air through 20 ml of water aliquoted between two bubblers **(45)**. In the case of mesityl oxide which has poor solubility in water to begin with, absorption efficiency would be very poor although relative volatility is low. This is especially critical where a low TLV is involved as large volumes of air would be required. The formation of lower boiling binary and ternary mixtures also can limit absorption. Failure to collect benzene vapor in ethanol even when chilled undoubtedly can be attributed to revaporization at the higher vapor pressure ternary alcohol-benzene-water mixture.

In general, the vapor pressure of the volatile analysand should be reduced to the lowest practical limit by refrigeration to promote efficient collection. In many cases a freeze-out trap without solvent in solid carbon dioxide, liquid air, or a mixture of liquid air and ethanol proves to be a more effective procedure **(14)**.

Sampling in the region at or close to the vapor saturation limit can produce serious errors by condensation in sampling lines. This condition is most likely to be encountered when collecting samples of air at temperatures above ambient and may involve condensation of water vapor which in turn will entrain or dissolve a significant portion of the components present in the sample under high humidity conditions.

1.7 Solubility Effects. Limitation may occur in cases where chemical reaction should, but actually does not occur readily. The classical example is the failure for SO_3 to be absorbed in pure water. The collection of aniline vapor in dilute aqueous HCl is not quantitative; therefore, chilled alcohol is the medium of choice **(16)**. Until the reaction is initiated by a catalyst, fluorine gas will bubble through distilled water without reacting (to produce FO_2 and HF) although once initiated the reaction may be violent. Other examples which can be cited are failure to obtain absorption of alkyl mercury vapor in acid permanganate and the less than quantitative absorption of alkyl mercury and lead compounds in aqueous iodine reagent **(25)**. The assumption that a given contaminant will be collected based on its reactivity with the reagents can be disastrous if calibration of the system under use conditions is omitted.

If the product from the reaction of the analysand with the collecting reagent has a very low solubility product, losses are to be expected unless care is exercised to avoid the saturation region. The precipitate may adsorb to the container walls, orifice tube, diffuser surfaces or fail to react in subsequent color development. If the solid phase is finely dispersed and has a refractive index approximately the same as the liquid system, the losses may not be observed *e.g.*, silica in chloroform.

Conversely, if the collecting train does not include a prefilter, particulate fractions may go into solution and produce false positive results or interferences, *e.g.*, lead dust or fume in the collection of tetra-alkyl lead vapor **(34)**.

1.8 TEMPERATURE EFFECTS. The physical effects of heat on the collection system under normal operating conditions is usually ignored without introducing appreciable difficulties. However, at the design and development stages, attention must be given to the upper and lower temperature limits to be anticipated. At low temperatures components of the liquid reagent may crystallize out or the solvent may freeze. Frequently, a change of solvent will solve this problem (aqueous KI_3 versus methanolic iodine reagent, **(25, 33)**. Close attention must be given to possible formation of ice or reagent crystals in orifices and porous glass diffusion beds when operating near the freezing range.

On the other end of the temperature scale, excessive evaporation may limit the choice of reagents or sampling time may be reduced below practical limits (again alcoholic versus aqueous iodine reagents). Refrigeration usually can be applied in fixed station or mobile systems to alleviate this problem. A portable assembly which derives its sampling suction from the expansion of difluorodichloromethane vapor through an aspirator also provides useful refrigeration by forming the vaporizer coil into a cavity which will receive a midget impinger **(14)**. In some cases a colorimetric system can be adapted to a nonvolatile solvent as illustrated by the use of diethyl phthalate as a solvent for the detection of phosgene in trace ranges over extended periods of time—4 to 6 hr **(46)**.

As noted in section 1.6 (Partial Vapor Pressure Effects) the temperature coefficient may be quite large and preclude collection by solution in a solvent above a critical temperature, above which no collection will be obtained, and which may be only slightly above room temperature **(45)**. In general, the adsorption capacity of solid adsorbents is decreased exponentially with rising temperature which may limit the thermal operating range. Temperature coefficients should be determined before sampling is attempted **(28)**.

Excessive evaporation at elevated temperature aggravates the scavenging problem for the metering devices and pump downstream from the collecting unit. Larger activated charcoal cartridges and more frequent replacement must be provided. Generation of volatile components of the reagent will create an even greater problem, *e.g.*, HCl from aqueous iodine monochloride solution **(1)**.

A word of caution with respect to liquid air-cooled freeze-out traps—liquid oxygen from the air sample may condense and create an explosion hazard which may not arise until the trap is removed from the cold bath. In other subnormal systems, moisture may be a mixed blessing. Condensation of water vapor often assists with the retention of contaminants, but in the end interferes with the final analysis. Analytical systems which require anhydrous conditions cannot be employed for air analysis unless dehydration can be employed before the sample enters the collecting system without loss of the analysand (*e.g.*, CO by infrared).

Although temperature extremes, sufficient to alter the mechanical performance of collecting systems, are not often encountered, this factor cannot be ignored when arctic or tropical desert conditions are encountered. Batteries for power supplies must be chosen with these conditions in mind. Piston-type pumps may bind or blow-by check valves may become inoperable, volumetric factors will require recalibration, metering valve and plug cock tolerances may shift significantly, elastomeric components lose elasticity beyond tolerable limits (tubing, "O" ring seals, check valves, gaskets, etc.), and diffusion through plastic containers, especially sample collection and storage bags may become excessive.

Instruments which require electrical circuitry or electronic components for generating analytical data will be affected by temperature extremes and must either be provided with thermal insulation or compensated for altered performance. Extrapolation of response obtained in the range of normal ambient temperature is not a reliable substitute for actual testing under use conditions.

1.9 Volumetric Errors. Rate meters of all types require frequent calibration to detect and correct drift or shifts in reference points, (see Section 1.5). Portable instruments should be checked every 40 to 50 hr of operation with at least a secondary standard (wet or dry test meter or glass rotameter) which in turn is compared frequently with a primary standard of the positive displacement total volume type (spirometer).

In volumetric calibration procedures, the pressure drop across the instrument being calibrated and across the system in which it will operate must be determined and appropriate corrections applied (see Section 1.5 for effects of fouling). When filter systems are used, the increasing pressure drop and decreasing sampling rate which develop as the filter loads must also receive consideration. Unnecessary restrictions to free sample flow also produce undesirable pressure drop increments which may limit the range of the sampling system.

Temperature effects do alter volumetric calibration significantly if the system is operated under ambient conditions more than ±20 C from the original calibration temperature conditions. A high degree of volumetric accuracy requires determination of the instrumental temperature coefficient as well as correction of the sample volume to standard temperature and pressure (25 C and 760 Torr).

Variability in certain types of pumping systems, both between units of the same model or within any given unit, may prevent constant rate sampling within tolerable limits over extended time intervals. Direct current motors are especially vulnerable to rate fluctuations unless effective control circuits are incorporated in the design. Incorporation of a pulse counter which performs as a miniature total volume meter, mechanically or electronically linked to the pump, furnishes a solution to this problem if the rate variations are within limits which do not alter collection efficiency significantly **(47)**. Manually operated squeeze-bulb devices seldom deliver reproducible volumes and should be avoided if better than semi-quantitative accuracy is required. Positive displacement syringe-type hand pumps are capable of a high degree of volumetric accuracy and are to be preferred **(48)**.

1.10 Observational Errors. The optical judgment of color shade and optical density is one, if not the most, variable factor to be considered in air analysis. Reasonably accurate results can be attained if the individual using a color comparator is willing to calibrate the instrument for his own personal use, but he must take into account the quality and intensity of the light available for illumination of the comparator, day to day variations in his own optical acuity, shade changes introduced by interferences, and stability of the color standards. Field kits based on optical color standards provide a definite, valuable service not available by any other means, but the unavoidable sacrifice of accuracy and precision must be carefully weighed when planning and evaluating investigations based on such equipment **(20, 33, 49, 50)**.

Optical acuity and judgment also enters as a factor in the detection of air contaminants with the length of stain type of detector tubes. Diffusion, migration and trailing may so obscure the stain front as to render accurate reading very questionable. Again individual calibration with knowns can be quite helpful in providing a basis for judgment.

Errors in the application of volumetric glassware are not unique to air sampling and are recognized by those conversant with analytical laboratory techniques.

Such items are "to deliver" versus "to contain", delivery time for pipets, whether or not to "touch off" or "blow out" the drop retained in the pipet tip, capillary effects in miniscus reading, error introduced by a water repellant film on the glass, chipped buret and pipet tips should be too well recognized to require further elaboration. The relative accuracy of the original glassware calibration may be overlooked. Most laboratory-ware manufacturers adhere to two standards: precision-grade (within NBS tolerances) and laboratory grade (twice precision grade tolerances). The choice of buret, volumetric or Mohr pipet or graduated cylinder will be determined by the relative accuracy required in the measurement. That is, volumetric pipets or a buret would be used to measure out standardizing solutions whereas a graduated cylinder is suitable for making up most reagents.

Reading instruments which indicate by mechanical movement such as a rotameter float, voltmeter needles, etc., involve critical judgment also. Whether or not to read the top, bottom, or center of a rotameter float, parallax effects in observing meter needles (some of the better precision meters provide a mirror background to eliminate this problem), capillary effects on the meniscus of liquid filled manometers, and scale units, illustrate several of the more common observational errors. The increasing relative error inherent in any instrument as either the minimum or maximum range limits are approached may introduce serious observational error. The upper and lower 10 per cent of any instrument's range should be avoided and if possible the capacity should be so chosen as to permit operating only in the center 50% of the overall range.

1.11 Optimum Sampling Rate. Calibration of any sample collection device should include a sufficient range of rates to delineate the optimum range consistent with the efficiency demanded of the system and to establish the penalty incurred when the rate exceeds or falls below these limits. Failure to include this factor has contributed serious errors (dust collection by impinger) which are inexcusable. Collection efficiency cannot be assumed; it must be determined by calibration **(51).**

The design of the collection in liquid reagent collection systems for gases and vapors is, in most cases, not critical if in addition to acceptable collection efficiency the design provides retention of entrainment (baffles), sufficient free board to retain bubbles and foam, interchangeability of component parts, and is easily assembled, dismantled and cleaned **(33).** Only in those cases which involve surface reactions and adsorption will a "fritted" or porous glass diffusing element perform more efficiently than a standard single orifice impinger **(13, 15).** Frequently these diffusers create problems such as retention of analysand, excessive frothing and foaming, variable pressure drop, nonphysical uniformity from unit to unit in the same batch which precludes interchangeability without recalibration, fragility, and particulate retention which are not encountered with the use of impinger-type absorbers.

References

1. Linch, A.L., E.G. Wiest, and M.D. Carter. 1970. Evaluation of Tetralkyl Lead Exposure by Personnel Monitor Surveys, A.I.H.A.J. 31:170.
2. American Society for Testing and Materials. 1967. *Industrial Water;* Atmospheric Analysis—Part 23, p. 770, ASTM Designation: D 1357–57. 1967. ASTM, Philadelphia.
3. Stern, A.C. 1968. Air Pollution and Its Effects. Vol. I: Part II (2nd Edition) Academic Press, N.Y.
4. Stern, A.C. 1968. *Air Pollution—Analysis, Monitoring and Surveying.* Vol. II: Parts IV, V and VI (2nd Ed.) Academic Press, N.Y.
5. Cralley, L.V., L.J. Cralley, and G.D. Clayton. 1968. *Industrial Hygiene Highlights* 1:297. Industrial Hygiene Foundation of America, Inc., Pittsburgh, Pa.
6. Manufacturing Chemists Association, Inc. 1952. *Air Pollution Abatement Manual.* Chap. 6, 8, 11.
7. Holmes, R.G. (ed). 1963. *Air Pollution Control District—County of Los Angeles.* Source Testing Manual.
8. Weisburd, M.I. (ed.). 1962. U.S. Department of Health, Education and Welfare, Public Health Service: *Air Pollution Control Field Operations Manual.*
9. American Industrial Hygiene Association. 1960. *Air Pollution Manual,* Part I—Evaluation American Industrial Hygiene Association.
10. American Conference of Governmental Industrial Hygienists. 1966. *Air Sampling In-*

struments for Evaluation of Atmospheric Contaminants. (3rd ed.).

11. Marcali, Kallman. 1957. Microdetermination of Toluenediisocyanates in Atmosphere. Anal. Chem. 29:552.
12. Cotabish, H.N., P.W. McConnaughey, and H.C. Messer. 1961. Making Known Concentrations for Instrument Calibration. Amer. Ind. Hyg. Assoc. J. 22:392.
13. Saltzman, B.E. 1961. Preparation and Analysis of Calibrated Low Concentrations of Sixteen Toxic Gases. Anal. Chem. 33:1100.
14. Linch, A.L., R.C. Charsha. 1960. Development of a Freeze-Out Technique and Constant Sampling Rate for the Portable Uni-Jet Air Sampler. Amer. Ind. Hyg. Assoc. J. 21:325.
15. Saltzman, B.E., and A.F. Wartburg, Jr. 1965. A Precision Flow Dilution System for Standard Low Concentrations of Nitrogen Dioxide. Anal. Chem. 37:1261.
16. Linch, A.L., and M. Corn. 1965. The Standard Midget Impinger—Design Improvement and Miniaturization. Amer. Ind. Hyg. Assoc. J. 26:601.
17. Inscoe, M.N. 1966. Losses Due to Adsorption During Filtration of Aqueous Solutions of Polycyclic Aromatic Hydrocarbons. Nature 211:1083.
18. Saltzman, B.E. 1954. Anal. Chem. 26:1949.
19. Yafee, C.D., D.H. Byers, and A.D. Mosey. 1956. *Encyclopedia of Instrumentation for Industrial Hygiene*. Ann Arbor, Michigan.
20. Linch, A.L., S.S. Lord, K.A. Kubitz, and M.R. De Brunner. 1965. Phosgene in Air—Development of Improved Detection Procedures. A.I.H.A.J. 26:465.
21. Linch, A.L., and M. Corn. 1965. The Standard Midget Impinger—Design Improvement and Miniaturization. A.I.H.A.J. 26:601.
22. Katz, M. 1970. Photochemical Reactions of Atmospheric Pollutants. Canad. J. of Chem. Eng. 48:3–11.
23. West, F.K., P.W. West, and F.A. Iddings. 1966. Adsorption of Traces of Silver on Container Surfaces. Anal. Chem. 38:1566.
24. Dyck, W. 1968. Adsorption of Silver on Borosilicate Glass. Anal. Chem. 40:454.
25. Linch, A.L., R.F. Stalzer, and D.T. Lafferts. 1968. Methyl and Ethyl Mercury Compounds—Recovery from Air and Analysis. A.I.H.A.J. 29:79.
26. Keenan, R.G. 1963. *Determination of Lead in Air and in Biological Materials. Manual of Analytical Methods*. American Conference of Governmental Industrial Hygienists.
27. Woessner, W.W., and J. Cholak. 1953. Improvements in the Rapid Screening Method for Lead in Urine. A.M.A.: Arch. Ind. Hyg. & Occup. Med. 7:249.
28. American Industrial Hygiene Analytical Guides. 1965. *Collection of Gases and Vapors—Adsorption on Silica Gel*.
29. Downing, F.B., and A.L. Linch. 1946. Process for Stabilizing or Deactivating Sludges, Precipitates and Residues Occurring or Used in the Manufacture of Tetraalkyl Leads. U.S. Patent 2,407,261.
30. Tares, D.R. 1968. Effect of Silicone Grease on Diffusion of Fluoride. Anal. Chem. 40:204.
31. Koniecki, W.B., and A.L. Linch. 1958. Determination of Aromatic Nitro-Compounds. Anal. Chem. 30:1134.
32. Skulskii, I.A., and V.V. Glasmov. 1966. Absorption of Caesium Tetraphenylborate from Aqueous Solutions on Glass, Polyethylene and Polytetrafluoroethylene. Nature 211:631.
33. Linch, A.L., R.B. Davis, R.F. Stalzer, and W.F. Anzilotti. 1964. Studies of Analytical Methods for Lead-in-Air Determination and Use with an Improved Self-Powered Portable Sampler. A.I.H.A.J. 25:81.
34. Pilat, M.J. 1968. Application of Gas-Aerosol Adsorption Data to the Solution of Air Quality Standards. J.A.P.C.A. 18:751.
35. Andrews, M.L., and D.C. Peterson. 1947. A Study of the Efficiency of Methods for Obtaining Vapor Samples in Air. J. Ind. Hyg. Toxic. 29:403.
36. Campbell, E.E., and H.M. Ide. 1966. Air Sampling and Analysis with Microcolumns of Silica Gel. A.I.H.A.J. 27:323.
37. Saltzman, B.E. 1970. *Factors in Air Monitoring Network Design*. Proceedings of the Eleventh Conference on Methods in Air Pollution and Industrial Hygiene Studies. California Air and Industrial Hygiene Laboratory, Berkeley, California. March 30.
38. Saltzman, B.E. 1970. Significance of Sampling Time in Air Monitoring. J.A.P.C.A. 10:660.
39. Nichols, R., and A. Topping. 1966. Absorption of Ethylene by Self-Indicating Soda-Lime. Nature 211:217.
40. Analytical Specialties, Inc. 4126 Packard Road, Ann Arbor, Michigan, Saran Plastic Bags for Collecting and Handling Air and Gas Samples in the Laboratory and Field.
41. Fluorodynamics, Inc. 1970. Diamond State Industrial Park, Newark, Delaware, Chemton Gas/Liquid Sampling Bags.
42. Calibrated Instruments, Inc. 1969. 17 West 60th Street, New York, N.Y. Plastic Sample Bags Preserve Gas Samples for Analysis.
43. Millipore Filter Corp., Bedford, Mass., Type Aeresol (MAW GO37AO).
44. Veillon, C., and J.Y. Park. 1970. Correct Procedures for Calibration and Use of Rotameter-Type Gas Flow Measuring Devices. Anal. Chem. 42:684.
45. Elkins, H.B. 1959. *The Chemistry of Industrial Toxicology*. (2nd Ed.) pp 284–288, John Wiley & Sons, N.Y.
46. Noweir, M.H., and E.A. Pfitzer. 1971. An Improved Method for Determination of Phosgene in Air. A.I.H.A.J., 32:163.
47. Cassella Mark I Personal Air Sampler. 1966. Wilson Air Analysis Instruments (Trade Brochure) Wilson Products Division. The Electric Storage Co. Reading, Pa.
48. Kitagawa, T. 1960. *The Rapid Measurement of Toxic Gases and Vapors*. Proceedings of the 13th International Congress on Occupational Health, New York.
49. Grimm, K.E., and A.L. Linch. 1964. Recent Isocyanate-In-Air Analysis Studies. A.I.H.A.J. 25:285.

50. Marcali, K., and A.L. Linch. 1966. Perfluoroisobutylene and Hexafluoropropylene in Air. A.I.H.A.J. 27:360.
51. Roberts, L.R., and H.C. McKee. 1959. Evaluation of Absorption Sampling Devices. J.A.P.C.A. 9:51.

2. Calibration Procedures

2.1 Preparation of Static Calibration Mixtures.

2.1.1 *Introduction.* Static calibration mixtures are prepared by introducing a known weight or volume of contaminant into a given volume of air **(1)**. Generally the mixture is held in a container of fixed dimensions, but flexible chambers may be used.

2.1.2 *Measuring the Contaminants.* Liquid contaminants are commonly introduced into the calibration system with a microsyringe or micropipet. Gaseous materials are handled with a gas-tight syringe.

A pure gas sample may be obtained from a lecture bottle by either of the methods shown in Figure 2:1. The first technique involves attaching a lecture bottle septum directly to a regulated cylinder **(2)**. Just enough gas to purge the system is bled through the pressure relief valve. The sampling chamber can then be maintained at about two atmospheres with no additional loss of gas to the atmosphere. Contaminant gas is then obtained through the septum using a gas-tight syringe.

In the second method, the pure gas is passed slowly through a T with a septum over one arm. The gas then flows from a dry midget impinger into one containing 10 ml of water. The pure gas is then withdrawn through the septum after the gas lines have been purged.

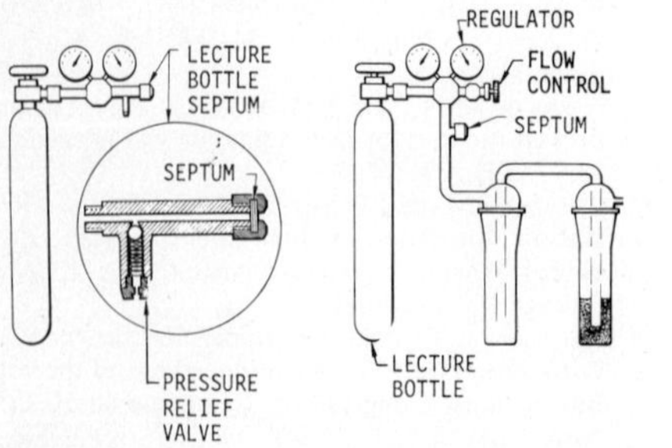

Figure 2:1—Two methods for removing small amounts of contaminant gas from a lecture bottle.

2.1.3 *Rigid Chambers.* Generally glass bottles and flasks make the best containers. However, plastic and metal chambers may be used if wall interactions with the contaminant gases are negligible.

The volume of the container is obtained by direct measurement of the chamber boundaries or volumetrically by filling with water. The usual size is on the order of 20 to 40 l which allows the removal of enough useful gas without causing excessive dilution by the replacement gas.

A typical static calibration system is shown in Figure 2:2. The vessel is first purged with fresh air to remove any residual contamination, then the contaminant is injected directly into the vessel through the rubber septum. The gases and vapors are evaporated and mixed by an externally driven stirrer, magnetic stirrer or by agitating with internally placed aluminum or Teflon strips. The gas mixture is then withdrawn after opening both valves. Placement of two or three of these vessels in series as shown in Figure 2:3 greatly reduces the dilution effect of the air as the sample gas mixture is removed **(1)**.

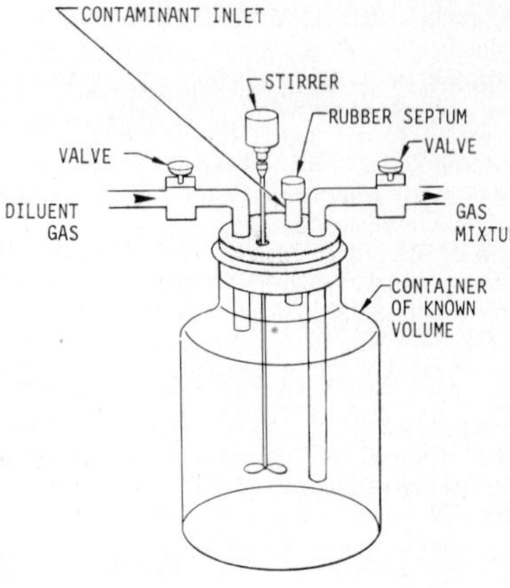

Figure 2:2—Diagram of a typical static calibration system.

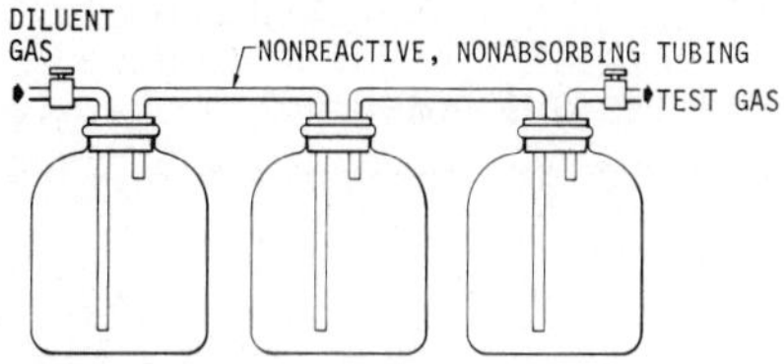

Figure 2:3—Rigid chambers connected in series.

Sometimes it is desirable to sample from a rigid container without suffering make-up gas dilution or a significant internal pressure decrease. This can be done by attaching a deflated plastic bag to the dilution-gas inlet as shown in Figure 2:4. As the mixture is sampled, dilution air fills the plastic bag rather than diluting the mixture. A bag which does not absorb appreciable quantities of the contaminant gas or vapor must be selected for this purpose.

2.1.4 *Piston Type Chambers.* Figure 2:5 illustrates a piston type container **(3).** A calibration mixture is prepared by introducing the contaminant through the inlet as the piston is raised. The inlet is then closed and 15 min are allowed for mixing.

2.1.5 *Nonrigid Chambers.* Mixtures may be prepared in plastic bags generally made from Teflon, Mylar and aluminized Mylar. The bag is alternately filled and deflated with air to remove residual contamination and final purge is accomplished by applying a vacuum to the system. The container is connected to a flow meter or wet or dry gas meter. As the bag fills, the contaminant is introduced by one of the methods shown in Figure 2:6.

The test mixture should be used as soon as possible because the initial concentration often decays with time. Such decay can be lessened if the container is first preconditioned to the test substance however.

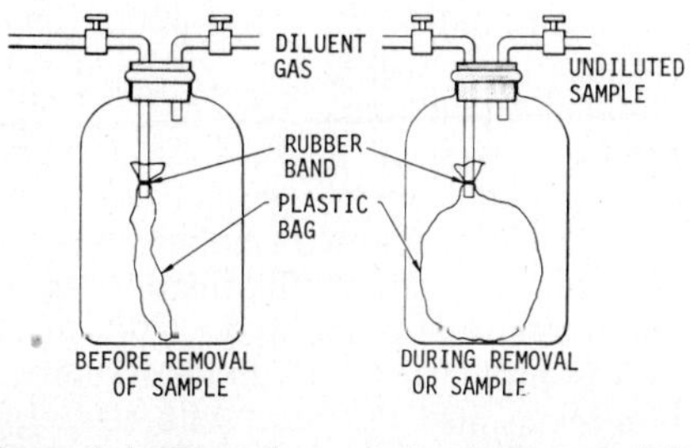

Figure 2:4—Removing a test gas from a rigid chamber without causing a sample dilution.

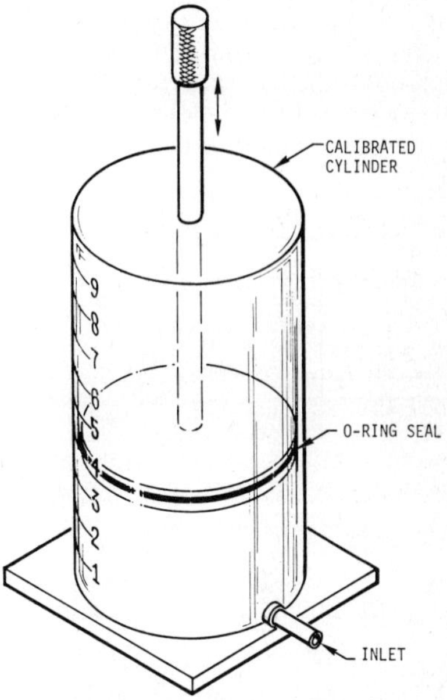

Figure 2:5—A piston-type container.

2.2 Calculations. The concentration by volume of a contaminant gas or vapor can be calculated either in per cent or in parts per million from the following equations.

For gas mixtures:

$$C\% = \frac{100V_c}{V_c + V} \text{ and} \quad (1)$$

$$C_{ppm} = \frac{10^6 V_c}{V} \quad (2)$$

where V_c and V are the contaminant and system volumes respectively. If a liquid is added to a closed system, the resulting concentration is

$$C_{ppm} = 22.4 \times 10^6 \left(\frac{pV_L}{MV}\right)\left(\frac{T}{273}\right)\left(\frac{760}{P}\right) \quad (3)$$

Where p is liquid density (g/ml), V_L is the volume of liquid (ml), T is the temperature (°K), P is the pressure (torr), M is the molecular weight (g/mole) and V is the system volume (liters).

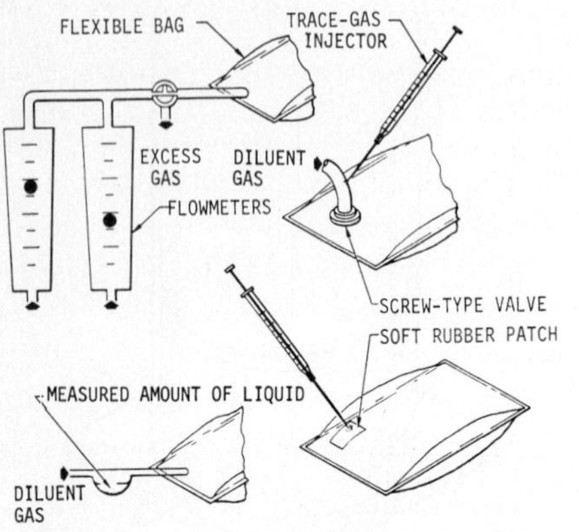

Figure 2:6—Four methods of introducing a sample into a nonrigid chamber.

2.3 Sources of Error. Lubricant greases and sealants should be avoided since they tend to absorb the trace contaminants. The use of rubber and plastic tubing, other than Teflon, should be avoided if possible. The compressed air should be of the highest quality available and a purification train should be used to remove unwanted contaminants.

For additional sources of error consult principles stated in the chapter on "Physical Precautions" which precedes this section.

References

1. Nelson, G.O. 1971. *Controlled Test Atmospheres—Principles and Techniques.* Ann Arbor Science Publishers. Inc., Ann Arbor, Mich.
2. Hamilton Co., Catalog No. H-72, Whittier, Calif.
3. Houston Atlas, Inc., Catalog No. HAI 722-K, Houston, Texas.

Subcommittee 9
A.L. Linch, **Chairman**
E.R. Hendrickson
M. Katz
R.J. Martin
G.O. Nelson
J.N. Pattison
A.L. Vander Kolk
R.B. Weidner
J.N. Harman

3. Dynamic Calibration of Air Analysis Systems

3.1 Permeation Tubes. The use of permeation tubes as primary standards for trace gas analysis was first documented by O'Keefe and Ortman in 1966 (**1**). The principle of this device is based on diffusion of gas or vapor through a plastic membrane at very slow rates. A liquefied gas or volatile liquid sealed in a section of Teflon FEP tubing when placed in a metered air stream can be used as a dynamic calibration standard. The diffusion rate is a nonlinear function of temperature; therefore, constant temperature conditions must be maintained for the permeation tube during gravimetric standardization and use as a calibration source. Also, the diffusion rate is influenced by the molecular weight of the carrier gas employed for dilution. The difference between air versus nitrogen is negligible; but if a lighter gas such as helium or a heavier gas such as argon is used, corrections or recalibration would be required.

Certain compounds react with oxygen or water vapor at the surface of the plastic tubing, namely, sulfur dioxide (**2**) and hydrogen sulfide. However, such compounds can be used if air is excluded from contact with the loaded tube by storage and standardization in a dry nitrogen atmosphere. A relatively high concentration in dry nitrogen carrier gas then is diluted stepwise with air to provide calibration mixtures.

In general, Teflon fluorocarbon resins are chemically inert. However, certain qualifications which govern their application to ambient air analysis problems must be considered. A clear distinction between chemical compatibility and purely physical properties will best serve to delineate the application limits. Very few chemicals produce chemical degradation of fluorocarbon resins under normal conditions. Reaction will occur with molten alkali metals, fluorine, strong fluorinating agents, and sodium hydroxide above 300 C. If ignition is provided finely divided fluorocarbon resins also will react with 100% oxygen, aluminum or magnesium dust. Otherwise Teflon fluorocarbons may be considered nonreactive. Examples of chemicals which Teflon resists are listed in Table 3:I (**3**). In most cases the tests included the boiling range.

Nearly all plastics absorb small quantities of certain materials on contact. Since no chemical reaction or solubility occurs in Teflon, the absorption detectable by weight increase is a result of the material filling submicroscopic voids between the polymer molecules (Table 3:II). The amount is proportional to contact time, pressure and temperature and the effect is reversible. If temperature or pressure cycling includes the boiling range, liquid inclusions within the submicroscopic pores will vaporize and thereby create internal pressure which may enlarge the pores and eventually mechanically damage the plastic. Surface blisters containing liquid provide visible evidence of this phenomena. The absorption of aqueous solutions is minimal due in part to the low degree of wettability exhibited by fluorocarbon resins (**3**). Teflon is wetted by most organic solvents.

Gases and vapors permeate through fluorocarbon resins at a considerably lower rate than for most other plastics. In general, permeation increases with temperature, pressure and surface contact area, and decreases with increased film thickness and polymer density (Figure 3:1, 2, 3). The permeability coefficient P is expressed as:

$$P = \frac{Wt \times Th}{A \times Tc \times \Delta P}$$

Where:
- Wt = Weight of permeant.
- Th = Film thickness.
- A = Area of film.
- Tc = Contact time.
- ΔP = Pressure difference.

Table 3:I Typical Chemicals with Which Teflon Resins are Compatible

Abietic acid	Cyclohexane	Hydrazine	Phosphoric acid
Acetic acid	Cyclohexanone	Hydrochloric acid	Phosphorus pentachloride
Acetic anhydride	Dibutyl phthalate	Hydrogen peroxide	Phthalic acid
Acetone	Dibutyl sebacate	Lead	Pinene
Acetophenone	Diethyl carbonate	Magnesium chloride	Piperidene
Acrylic anhydride	Diethyl ether	Mercury	Polyacrylonitrite
Allyl acetate	Dimethyl formamide	Methyl ethyl ketone	Potassium acetate
Allyl methacrylate	Di-isobutyl adipate	Methacrylic acid	Potassium hydroxide
Aluminum chloride	Dimethylformamide	Methanol	Potassium permangenate
Ammonia, liquid	Dimethyl hydrazine	Methyl methacrylate	Pyridine
Ammonium chloride	unsymmetrical	Naphthalene	Soap and detergents
Aniline	Dioxane	Naphthols	Sodium hydroxide
Benzonitrile	Ethyl acetate	Nitric acid	Sodium hypochlorite
Benzoyl chloride	Ethyl alcohol	Nitrobenzene	Sodium peroxide
Benzyl alcohol	Ethyl ether	2-Nitro-butanol	Solvents, aliphatic
Borax	Ethyl hexoate	Nitromethane	and aromatic*
Boric acid	Ethylene bromide	Nitrogen tetroxide	Stannous chloride
Bromine	Ethylene glycol	2-Nitro-2-methyl	Sulfur
n-Butyl amine	Ferric chloride	propanol	Sulfuric acid
Butyl acetate	Ferric phosphate	n-Octadecyl alcohol	Tetrabromoethane
Butyl methacrylate	Fluoronaphthalene	Oils animal and	Tetrachloroethylene
Calcium chloride	Fluoronitrobenzene	vegetable	Trichloroacetic acid
Carbon disulfide	Formaldehyde	Ozone	Trichlorethylene
Cetane	Formic acid	Perchlorethylene	Tricresyl phosphate
Chlorine	Furane	Pentachloro-	Triethanolanime
Chloroform	Gasoline	benzamide	Vinyl methacrylate
Chlorosulfonic acid	Hexachloroethane	Perfluoroxylene	Water
Chromic acid	Hexane	Phenol	Xylene
			Zinc chloride

Based on experiments conducted up to the boiling points of the liquids listed Teflon resins have normal service temperatures up to 500 F, (260 C) for TFE, 400 F, (205 C) for FEP, resins.

*Some halogenated solvents may cause moderate swelling.

Film thickness and $\triangle P$ can be combined into pressure gradient, $G = Th/\triangle P$ = atmospheres per mil (0.001 in.)

$$\text{Then: } P = \frac{Wt}{A \times Tc} \times G$$

Gases and liquids permeate plastics by molecular vibrations and motion between the plastic molecules. Therefore, an increase in plastic density for a well-molded resin will reduce the diffusion process due to reduced intermolecular spacing (Figure 3:1). Since Teflon FEP resins are processed by melt extrusion, voids larger than intermolecular are not normally present and permeation is basically molecular diffusion only **(3)**. See Figure 3:2 and Table 3:III.

Surface attractive forces between permeant and plastic barrier influence both absorption quantity and permeability constant **P**. Gases or vapors chemically related to the plastic normally show a higher permeation rate than dissimilar materials. High solubility of a material in a plastic is indicative of high permeation rate (molecular size and vapor pressure are also influential). Swelling produced by liquid absorption which increases molecular spacing of the plastic, and permeation is not a problem with fluorocarbon resins **(3)**.

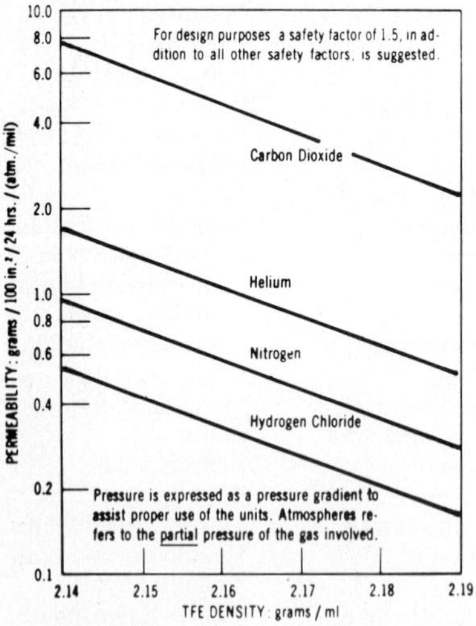

Notes: Values are averages only and not for specification purposes.
To convert the permeation values for 100 sq. in. to those for 1 sq. cm., multiply by 0.00155.

Figure 3:1—Effect of Density of "Teflon" TFE Resins on Their Permeability to Gases at 30 C. (86 F.)

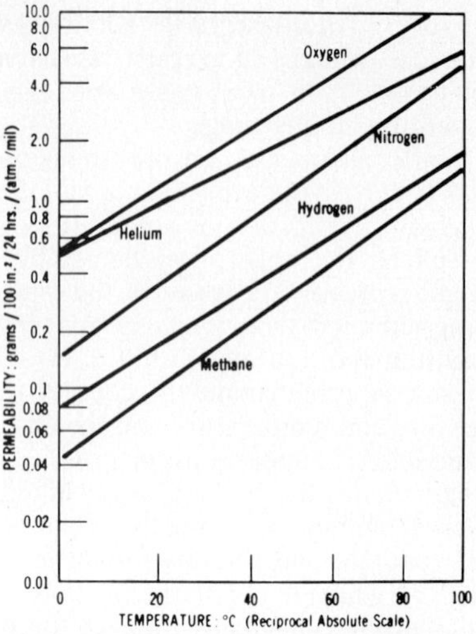

Notes. Values are averages only and not for specification purposes.
To convert the permeation values for 100 sq. in. to those for 1 sq. cm., multiply by 0.00155.

Figure 3:2—Permeability of "Teflon" FEP Resins to Gases, at Various Temperatures.

Increased temperature produces higher molecular activity of the diffusing material and hence permeation (Figure 3:2). This increase is not only due to increased solvent molecule activity but also due to the increased vapor pressure. Since the permeation rate is temperature dependent (as much as 5% per degree centigrade) precise temperature control is critical.

Perhaps the most important parameter in the standardization of permeation tubes is the time factor required for diffusion equilibrium to be reached. Two to 5 days should be allowed in a constant temperature oven (30 C ± 0.02 C) under a flowing dry air stream for accurate standardization. The time is dependent upon diffusion rate and accuracy of weighing. In use constant temperature should be maintained to ± 0.1 C for precision within ± 1% **(4)**. Gas chromatography ovens are ideal for this purpose and internal standards for calibration of field gas chromatog-

Table 3:II. Absorption of Common Solvents in Teflon Resins

Solvent	Exposure Temp* C	(F)	Exposure Time	Weight Increase %
Acetone	25	(77)	12 mo	0.30
	50	(122)	12 mo	0.4
	70	(158)	2 wk	0
Benzene	78	(172)	96 hr	0.5
	100	(212)	8 hr	0.6
	200	(392)	8 hr	1.0
Carbon tetrachloride	25	(77)	12 mo	0.6
	50	(122)	12 mo	1.6
	70	(158)	2 wk	1.9
	100	(212)	8 hr	2.5
	200	(392)	8 hr	3.7
Ethyl alcohol (95%)	25	(77)	12 mo	0
	50	(122)	12 mo	0
	70	(158)	2 wk	0
	100	(212)	8 hr	0.1
	200	(392)	8 hr	0.3
Ethyl acetate	25	(77)	12 mo	0.5
	50	(122)	12 mo	0.70
	70	(158)	2 wk	0.7
Toluene	25	(77)	12 mo	0.3
	50	(122)	12 mo	0.6
	70	(158)	2 wk	0.6

*Exposure at over the boiling point of the solvent area at its vapor pressure.
Note: Values are averages only and not for specification purposes.

Table 3:III. Permeability of Teflon Fluorocarbon Resins to Vapors (g/100 in² 24 hr)*

	TFE		FEP		
	23 C. (73 F.)	30 C. (86 F.)	23 C. (73 F.)	35 C. (95 F.)	50 C. (122 F.)
Acetic acid	—	—	—	0.42	—
Acetone	—	—	0.13	0.95	3.29
Acetophenone	0.56	—	0.47	—	—
Benzene	0.36	0.80	0.15	0.64	—
n-Butyl ether	—	—	0.08	—	0.65
Carbon tetrachloride	0.06	—	0.11	0.31	—
Decane	—	—	0.72	—	1.03
Dipentene	—	—	0.17	—	0.35
ethyl acetate	—	—	0.06	0.77	2.9
Ethyl alcohol	0.13	—	0.11	0.69	—
Hexane	—	—	—	0.57	—
HCl, 20%	<0.01	—	<0.01	—	—
Methanol	—	—	—	—	5.61
Piperidine	0.07	—	0.04	—	—
"Skydrol" hydraulic fluid	0.06	—	0.05	—	—
NaOH, 50%	5×10^{-5}	—	4×10^{-5}	—	—
H_2SO_4, 98%	1.8×10^{-5}	—	8×10^{-6}	—	—
Toluene	—	—	0.37	—	2.93
Water	—	0.35	0.09	0.45	0.89

*Test Method: ASTM E-96-53T (at vapor pressure; for 0.001" film thickness).
Notes: Values are averages only and not for specification purposes. To convert the permeation values for 100 sq in to those for 1 sq cm, multiply by 0.00155.

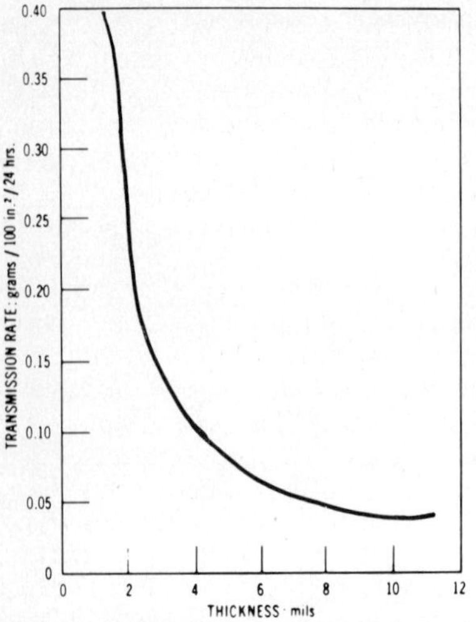

Notes: Values are averages only and not for specification purposes. To convert the permeation values for 100 sq. in. to those for 1 sq. cm., multiply by 0.00155

Figure 3:3—Water Vapor Transmission Rate of "Teflon" FEP Resins at 40 C (104 F.)

raphy apparatus can be conveniently installed inside the calibration assembly (**4**).

Care must be taken to assure tight seals at the tube ends; this is most important with low-boiling compounds. It is doubtful if tubes can be fabricated using materials that exert pressures of greater than 50 psig at normal temperatures. The method used is that described by O'Keefe and Ortman (**1**). Others are:

a. Swagelok—type plugs at each end (may be cumbersome).

b. Tight fitting caps sealed with Apiezon W—for low pressure applications.

Table 3:IV Permeation Rates.

Chemical	Permeation Rate (ng/min/cm) at 30 C*
SO_2	290
NO_2	1200
H_2S	250
Propane	80
n-Butane	2
Chlorine	1500
Ammonia	170
Methyl Mercaptan	30

*Minimum Tube Length is 2.5 cm.

3.2 Calculation of Length Required. A permeation rate of 4500 ng/min of SO_2 is required to calibrate an instrument.

1) $$L = \frac{Pr}{R}$$

where L = length (cm) of permeation tube required.

Pr = permeation rate required for calibrating instruments (ng/min).

$$L = \frac{4500}{290} = 15.5 \text{ cm.}$$

R = permeation rate in ng/min/cm.

3.3 Conversion of ng/min to ppm (vol) at 30 C. Because many instruments are calibrated in ppm (vol) and permeation rates are given in ng/min, a conversion formula is required. To convert from ng/min to ppm the required formula is:

2) $$C = \frac{Pr \cdot K}{F}$$

where C = concentration in ppm (vol).

PR = permeation rate in ng/min.

F = flow rate in cc/min.

K = constant [a]

Table 3:V K Values.

	*K Values at 30 C
SO_2	0.382
NO_2	0.532
H_2S	0.719
Propane	0.556
N-Butane	0.422
Chlorine	0.344
Ammonia	1.439
Methyl Mercaptan	0.509

*In converting ng/min to ppm (v/v) the units for $\frac{Pr}{F}$ are $\frac{ng}{cc}$ or $\frac{\mu g}{L}$. K is then 1/1000th the conversion factor usually given for converting mg/L to ppm (v/v) $\left[\frac{1}{1000} \times \frac{24.450}{\text{Mol Wt}}\right]$ (5)

Formula for calculation K

$$\mu \text{ moles} \times 22.4 \times \frac{T}{273}$$

$$\frac{\mu \text{ moles}}{L} = \frac{\frac{ng}{cc}}{\text{Mol Wt}};$$

$$T = {}^\circ\text{Kelvin} = {}^\circ\text{C} + 273$$

Example: A permeation tube containing propane and having a permeation rate of 950 ng/min is to be used at 30 C and a flow rate of 500 cc/min. What is the concentration of propane (in ppm) in the gas mixture?

$$C = \frac{950 \times 0.556}{500} = 1.06$$

$$= 1.1 \text{ ppm (vol)}$$

3.4 Permeation tubes that contain liquefied gases or volatile liquids are calibrated gravimetrically and used to prepare standard concentrations of pollutants in air as follows **(4, 5, 6, 7)**. Analyses of these known concentrations give calibration curves which simulate all of the operational conditions performed during the sampling and chemical procedure. The calibration curves include the important corrections for collection efficiency at various concentrations of pollutants.

Prepare or obtain **(4, 10)** a Teflon FEP permeation tube that emits gas or vapor at a rate consistent with the desired concentration level at standard conditions of 25 C and 1 atmosphere. A permeation tube with an effective length of 1 to 2 cm and a wall thickness of 0.76 mm, (0.030″), will yield the desired permeation rate if held at a constant temperature of 20 C for SO_2.

Permeation tubes are calibrated under a stream of dry nitrogen when reactions occur with atmospheric oxygen or moisture on or within the walls of the permeation tube (H_2S, SO_2, etc.).

3.5 To prepare standard concentrations of pollutant, select either the system designed for laboratory or field use, Figure 3:4 and 3:5, respectively. Assemble the apparatus as shown in one of these systems, consisting of a water-cooled condenser; constant-temperature water bath maintained at 20 C; cylinder containing pure, dry air with appropriate pressure regulators; needle valves and flow meters for the nitrogen, if necessary, and dry air, diluent gas streams. The diluent gases are brought to temperature by passage through a 2-m-long copper coil immersed in the water bath. Insert a calibrated permeation tube **(7)** into the central tube of the condenser maintained at 20 C by circulating water from the constant-temperature bath and pass a stream of nitrogen or air over the tube at a fixed rate of approximately 50 ml/min. Dilute this gas stream to the desired concentration by varying the flow rate of the clean, dry air. Clean, dry air may also be prepared by passing ambient air from a relatively uncontaminated outside source through absorption tubes packed with activated carbon and soda-lime, then an efficient fiber glass filter in series. This flow rate can normally be varied from 0.2 to 15 l/minute. The flow rate of the sampling system determines the lower limit for the flow rate of the diluent gases. The flow rates of the nitrogen and the diluent air must be measured to an accuracy of 1 to 2%. With a tube permeating at a rate of 0.1 μl/min (0.26 μg/min), the range of concentration will be between 0.007 to 0.4 ppm (18 to 1047 $\mu g/M^3$), a generally satisfactory range for ambient air conditions. When higher concentrations are desired, calibrate and use longer permeation tubes.

3.6 PROCEDURE FOR PREPARING SIMULATED CALIBRATION CURVES. Obviously one can prepare a multitude of curves by selecting different combinations of sampling rate and sampling time.. The following description for SO_2 represents a typical procedure for ambient air sampling of short duration with a brief mention of a modification for 24-hr sampling. The system is designed to provide an accurate measure in the 0.01 to 0.5 ppm range. It can be easily modified to meet special needs.

The dynamic range of the colorimetric procedure fixes the total volume of the sample at 30 l; then, to obtain linearity between the absorbance of the solution and the concentration in ppm, select a con-

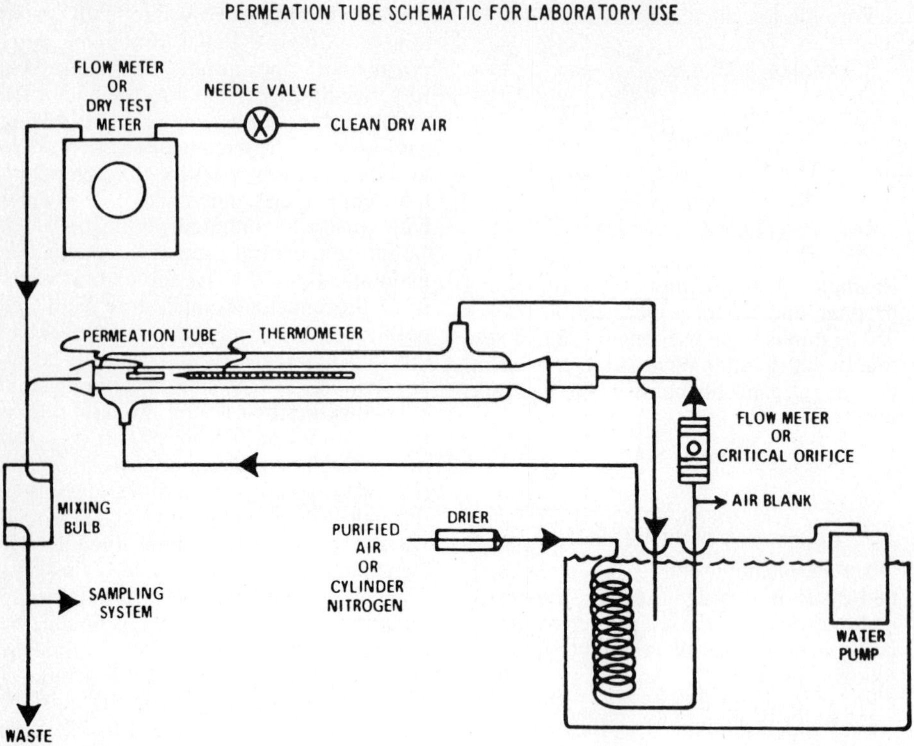

Figure 3:4—Gas dilution system for preparation of standard concentrations of sulfur dioxide for laboratory use by the permeation tube method.

stant sampling time. This fixing of sampling time is desirable also from a practical standpoint. In this case, select a sampling time of 30 min. Then to obtain a 30-l sample of air requires a flow rate of 1 l/minute. A 22-gauge hypodermic needle operating as a critical orifice will control air flow at this approximate desired rate (**11**). The concentration in air is computed as follows:

$$C = \frac{Pr \times M}{r_1 + r_2}$$

Where C = Concentration in ppm
Pr = Permeation rate in μg/min.
M = Reciprocal of vapor density, 0.382 μl/ug
r_1 = Flow rate of diluent air, liter/min.
r_2 = Flow rate of diluent nitrogen, liter/min.

Data for a typical calibration curve are listed in Table 3:VI.

Table 3:VI. Typical Calibration Data

Concentrations of SO_2 ppm	Amount of SO_2 in μl for 30 liters	Absorbance of sample
0.005	0.15	0.01
0.01	0.30	0.02
0.05	1.50	0.117
0.10	3.00	0.234
0.20	6.00	0.468
0.30	9.00	0.703
0.40	12.00	0.937

A plot of the concentration of sulfur dioxide in ppm (*x*—axis) against absorbance of the final solution (*y*—axis) will yield a straight line, the slope of which is the factor for conversion of absorbance to ppm. This factor includes the correction

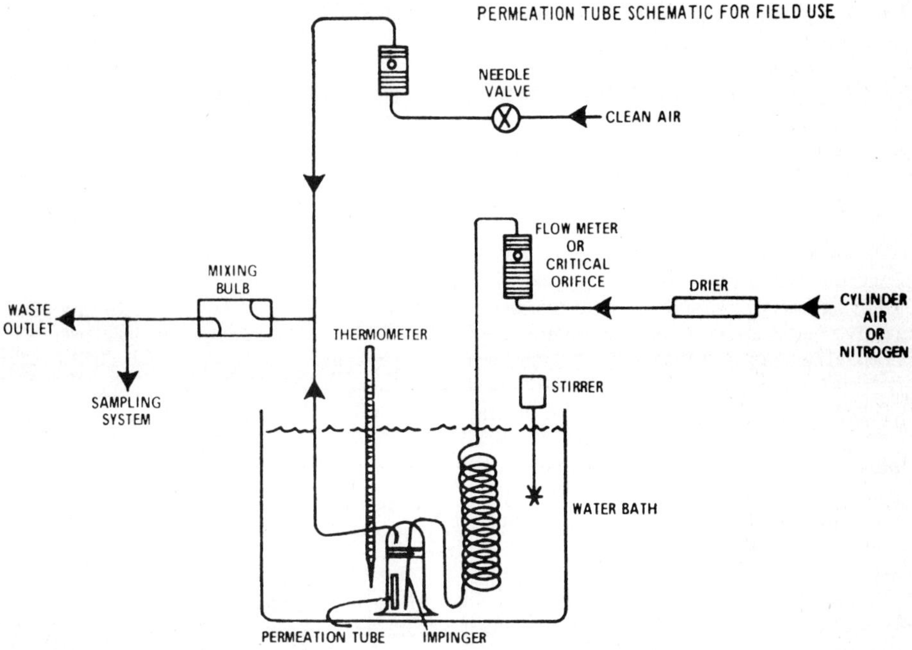

Figure 3:5—Gas dilution system for preparation of standard concentrations of sulfur dioxide for field use by the permeation tube method.

for collection efficiency. Any deviation from linearity at the lower concentration range indicates a change in collection efficiency of the sampling system. Actually, the standard concentration of 0.01 ppm and below of sulfur dioxide is slightly below the dynamic range of the method. If this is the range of interest, the total volume of air collected should be increased to obtain sufficient color within the dynamic range of the colorimetric procedure. Also, once the calibration factor has been established under simulated conditions the conditions can be modified so that the concentration of SO_2 is a simple multiple of the absorbance of the colored solution.

For long-term sampling of 24-hr duration the conditions can be fixed to collect 300 l of sample in a larger volume of tetrachloromercurate. For example, for 24 hr at 0.2 l/min, approximately 288 l of air are collected. An aliquot representing 0.1 of the entire amount of sample is taken for the analysis.

3.7 EQUILIBRIUM. Each tube's permeation rate changes with *its temperature*. Hence, time must always be allowed for the tube to reach an equilibrium condition. This is particularly true for tubes stored in a cool place and then inserted in a warm chamber for gas generation control. In such case the tube will accurately provide a gas source based on its actual temperature, rather than the temperature of the warm air flowing through the chamber. Output concentration will be lower than that predicted by a permeation rate vs. air temperature calculation. **(10a)**

3.8 VENTILATION. Each tube permeates continuously. Thus if a tube is enclosed in a chamber without ventilation, concentration buildup will occur. This is particularly true for tubes stored in a warm, closed chamber environment. In such case the actual concentration coming out of the chamber upon the first flush of ventilation air will be higher than that predicted by a permeation rate *vs* air temperature calculation. Often, considerable time must be allowed to "de-gas" a saturated system (**10a**).

3.9 CONDITIONING. When working with low concentrations of gas (*i.e.*, concentrations near nominal ambient levels) it is important to condition the system by allowing the gas to pass through the system for a period of day(s) before conducting any experiments. Conditioning is necessary even if nominally inactive materials such as glass or stainless steel are used. Active materials such as rubber should, of course, be avoided (**10a**).

3.10 TESTING *vs* CALIBRATION. Dynacal permeation tubes can be used for testing atmospheric analyzers without temperature control. However, the existing temperature needs to be measured and taken into account in calculating the approximate permeation rate. Testing is considered to be an operation to establish whether an analyzer is operating "about right". It should be performed frequently. Testing, on the other hand, is not a substitute for less frequent, periodic calibrations in which known concentrations of gas are delivered to the analyzer with precision (**10a**).

Acknowledgement. Mr. Arthur Johnston assisted with this section as Chairman of the Analytical Chemistry Committee Standardization Subcommittee.

References

1. O'KEEFE, A.E., and G.C. ORTMAN. 1966. Primary Standards for Trace Gas Analysis. Anal. Chem. 38:760.
2. American Public Health Association. 1977. Methods of Air Sampling and Analysis. 2nd. ed. p. 696. Washington, D.C. 20036.
3. DIGEL, W.A., (Ed.) 1970. The Journal of "Teflon", Reprint No. 41 DuPont, E.I. de Nemours Company.
4. ANALYTICAL INSTRUMENT DEVELOPMENT, INC., 250 S. Franklin Street, West Chester, Pa., 19380.
5. PATTY, F.A. 1963. *Industrial Hygiene and Toxicology*. 2nd Edition, Vol. II. New York. Interscience Publishers.
6. O'KEEFE, A.E., and G.C. ORTMAN. 1966. Primary Standards for Trace Gas Analysis. Anal. Chem. 38:760.
7. SCARINGELLI, F.P., A.A. FREY, and B.E. SALTZMAN. 1967. Evaluation of Teflon Permeation Tubes for Use with Sulfur Dioxide. Amer. Ind. Hyg. Assoc. J. 28:260.
8. THOMAS, M.D., and R.E. AMTOWER. 1966. Gas Dilution Apparatus for Preparing Reproducible Dynamic Gas Mixtures in any Desired Concentration and Complexity. J. Air Pollut. Contr. Assoc. 16:618.
9. SCARINGELLI, F.P., A.E. O'KEEFE, E. ROSENBERG, and J.P. BELL. 1970. Preparation of Known Concentrations of Gases and Vapors with Permeating Devices Calibrated Gravimetrically. Anal. Chem. 42:871.
10. POLYSCIENCE CORP., 909 Pitner Ave., Evanston, Ill. 60202, and Metronics, Inc., 3201 Porter Drive, Palo Alto, Calif. 94304.
10a. METRONICS ASSOCIATES, 3201 Porter Drive, Stanford Industrial Park, Palo Alto, California.
11. LODGE, J.P., JR., J.B. PATE, B.E. AMMONS, and B.A. SWANSON. 1966. The Use of Hypodermic Needles as Critical Orifices in Air Sampling. J. Air Pollut. Constr. Assoc. 16:197.

4. Sampling and Storage of Aerosols

4.1 INTRODUCTION. The collection of an aerosol sample that is not too different from the population of interest is often difficult and requires attention to details. This section deals with general information concerning the collection of aerosol samples. Specific methods of analysis are covered in other sections.

Aerosols are sampled for many reasons such as:

a. To determine whether there are hazardous concentrations of pollutants in the atmosphere or if ambient air standards have been exceeded;

b. to determine the effectiveness of control programs in reducing ambient concentrations of pollutants;

c. to determine the emission levels from a source;

d. to determine the effectiveness of control equipment;

e. to determine the sources contributing to pollution at a receptor, or

f. to identify pollutants in the atmosphere.

Aerosols are defined here to mean any material in the solid or liquid phase that is dispersed in a gas stream or the atmosphere. Dusts, smoke, soot, particles, mist, fumes, and fog are other terms used on occasion to describe certain types of aerosols. Some of the more useful terms may be defined as follows:

a. *Settleable Particulates*—larger particles that settle out of the air fairly fast such as those caught in an open jar. They are usually larger than 30 μ in diam, but in still air some particles 10 μ or smaller may settle.

b. *Suspended Particulates*—aerosol particles that tend to remain suspended in the atmosphere for long periods of time. These are generally smaller than 30 μ in diam. Commonly, the material collected by a Public Health Service High Volume Sampler (Hi-Vol) is limited by the shelter design and filter to particles less than 100 μ and consists mainly of particles below 10 μ.

c. *Condensation Nuclei*—sometimes called *Aitken nuclei,* are small particles that act as condensation sites for supersaturated vapors in the atmosphere. They are usually 0.01 to 0.1 μ diam.

d. *Agglomerates*—particles that are composed of several smaller particles that are attracted to a large particle or to each other and travel together in the atmosphere as a single particle.

It should be kept in mind that there are no sharp cutoffs between classifications and the particle sizes in each class overlaps its neighboring class. This may be due to several factors that tend to make the proper collection and classification of particles difficult and somewhat inexact. These include the shape of the particles (which affects their aerodynamic properties), their density and velocity (which affects their inertia), and electrical charge. These will be discussed in more detail later in this section.

There is a great variety of methods available for the collection of particulates and the method selected will often depend on the purpose for which the sample is being taken. For instance, if information on individual particles is desired, a few hundred particles may be collected for microscopic examination. This can determine particle shape and size distribution but does not tell all about aerodynamic properties, weight, or chemical composition. A larger amount may be collected in an inertial collector in which the particle size distribution can be determined. This does not provide separate particles that can be examined nor does it usually provide enough samples for chemical analysis. For chemical analysis, a Hi-Vol is usually employed. The average analysis of all of the particles can be determined but it does not provide information on the composition of individual particles or their sizes or shapes. Sometimes optical properties, such as visibility reduction, are of interest and none of the above methods is entirely suitable.

4.2 PRECAUTIONS IN COLLECTING AEROSOL SAMPLES.

4.2.1 The most bothersome aspect of aerosol sampling is a result of the momentum of the particles, which is a product of its mass times its velocity (MV). Thus, this problem becomes more severe with larger particles or fast moving streams. Since aerosols are much larger than gas molecules, each time the flow direction of the gas stream changes, such as a bend in a pipe or flow around an object, the larger particles tend to continue on their original line and are displaced somewhat from the original part of the gas stream they were with, as illustrated in Figure 4:1. This often leads to deposition on a nearby surface or unevenly dispersed streams where more particles will be found in one side of a pipe than another. Different sized particles will be displaced by different amounts. This makes it necessary for all parts of the stream to be sampled and properly weighted to be representative of the whole stream.

4.2.2 *Isokinetic sampling* refers to taking a sample under such conditions that there is no change in momentum so that the sample will be representative of the gases and aerosols in that portion of the stream being sampled. This is accomplished by using a thin walled tube aligned with the stream flow and drawing sample into it at the same linear velocity as the stream flow at that point. Even when these precautions are taken there can be several

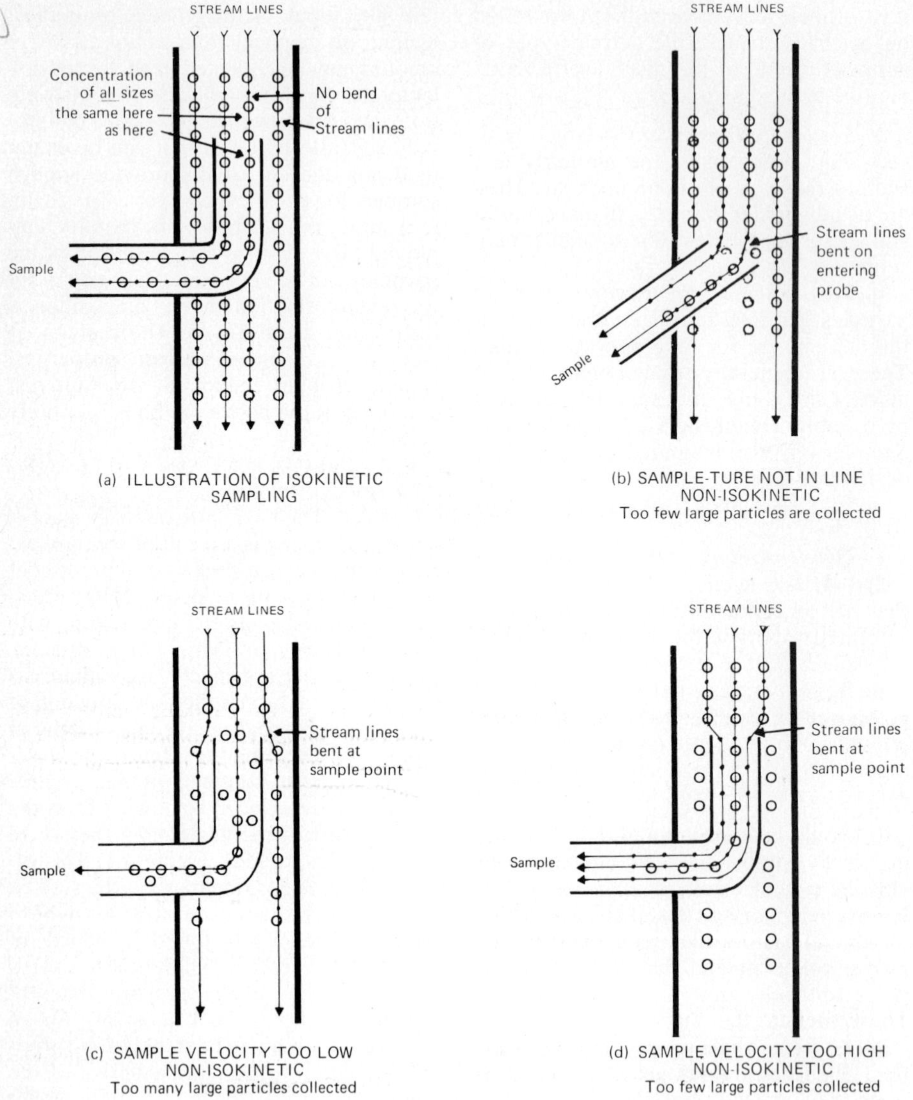

Figure 4:1—Schematic illustration of isokinetic and non-isokinetic sampling.

reasons why the sample is not truly representative:

a. The probe always has a finite wall thickness that disturbs the flow;

b. the point sampled may not be representative of the whole stream;

c. if the flow is turbulent at the point of sampling there is no way to get a truly isokinetic sample, the sample will contain smaller portions of the larger particles, or

d. sample may be lost by deposition or changed by agglomeration or deagglomeration in the sample line after it enters the

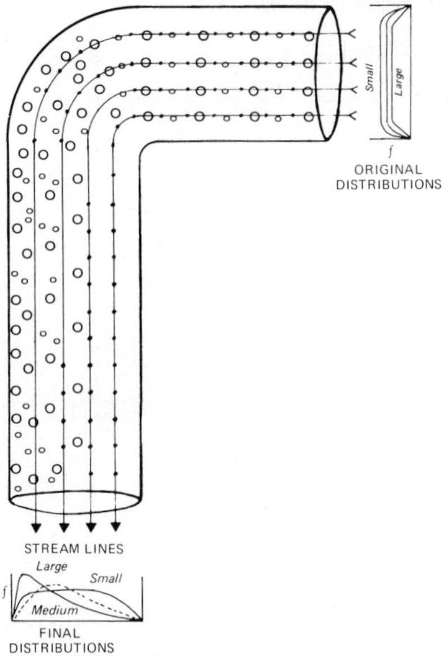

Figure 4:1(e)—Illustration of particle segregation at a pipe bend.

probe and before it enters the main sampling device.

Fortunately, it is not always necessary to use isokinetic sampling to obtain a reasonably good sample. Figure 4:2 shows the collection efficiency for particles of various sizes as a function of stream velocity. Figure 4:3 shows the effect of not aligning the sample tube in the same direction as the stream flow.

These data show that:

a. Sample velocity and alignment are both important conditions of isokinetic sampling;

b. under isokinetic conditions all particle sizes are collected efficiently;

c. small particles (under about 3 μ diameter) do not require isokinetic sampling for efficient collection—their small mass minimizes inertial effects;

d. stagnant gases such as calm ambient air do not require isokinetic conditions for efficient sampling of all particle sizes because slow moving particles have little or no momentum.

4.2.3 *Gravitational Effect*. In the case of quiescent air masses, as discussed immediately above, there will be a small effect because of gravitational settling of the particles. If the probe points up, the concentration of particles collected will be increased by a factor of $(1 + V_s/U)$ where V_s is the settling velocity of a particle and U is the sampling velocity. If the probe aims down (such as the shed opening of a Hi-Vol), the sample will contain less particles by the factor $(1 - V_s/U)$. Here again, small particles are not much of a problem. The correction factor would predict a reduction of about 1% for 11μ particles entering a Hi-Vol (assuming 100 square inches of opening and 50 CFM) but for 36μ particles there would be 10% less, for 85μ particles about 50% less, and no particles larger than 135μ would enter. Of course, turbulent winds could help a few larger particles to reach the filter.

4.2.4 *Sampling Rate*. The linear sampling rate can have an effect on several other types of aerosol collectors. For instance, the efficiency of impingement collectors such as cascade impactors, Anderson samplers, Rotorods, and jars with sticky paper all are dependent on the Stokes number:

$$StK = \frac{I_i}{R} = \frac{2V_o r^2 \surd}{9\,n\,R}$$

Where I_i is the stopping distance, R is the radius of the impingement orifice, r the radius of the particle, V_o the velocity of the stream as it leaves the orifice, $\surd$ the density of the particle, and n the viscosity of the medium.

The mass of the particle, its linear velocity and the size of the jet or object are important factors.

Other types of collectors such as electrostatic precipitators and filters also depend on velocity but to a much smaller extent.

4.2.5 *Miscellaneous Factors*. Previous mention was made of particles being lost from a sample by impingement on surfaces or bends in the sampling line which can cause errors even after a good sample

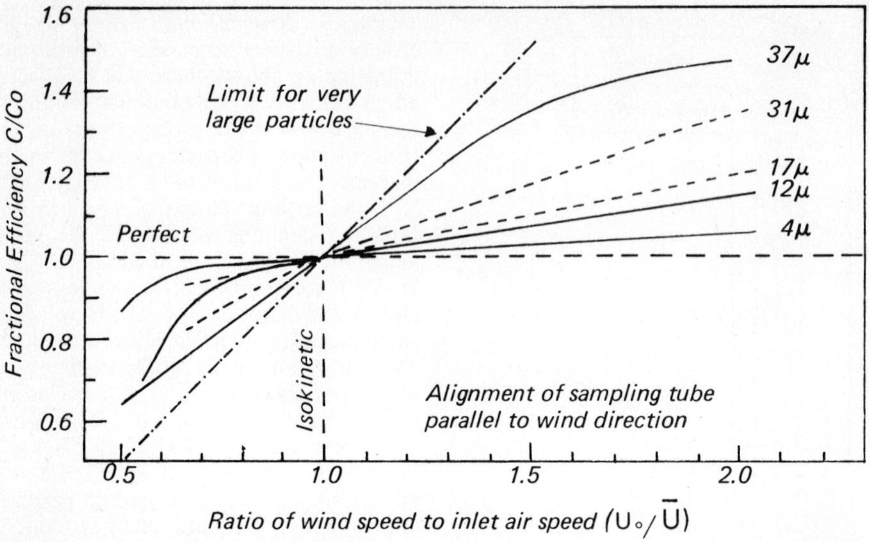

Figure 4:2—Diagram showing change in ratio of observed concentration to true concentration with departure from isokinetic conditions.
(Watson, H. 1954 Amer. Ind. Hyg. Assoc. Quart. 15:21.)

has been taken at the probe. In addition to this, particles can settle out in tubes or other components of the sampler or can be lost by electrostatic attraction to various surfaces in the sampler—especially nonconducting surfaces such as glass or plastics. Particles can also be lost due to adhesion on wet, oily, or sticky surfaces. As shown in Figure 4:4 and Table 4:I this can vary with relative humidity and surface roughness.

Several factors can result in changes in size of particles which may change the collection efficiency as discussed above. During periods of high relative humidity some particles such as sulfuric acid can absorb water vapor and get larger as illustrated in Figure 4:5. Most particles will absorb water or other vapors; therefore, collected samples are generally equilibrated to a standard relative humidity, such as 40%, before weighed. Water adsorption is generally minimal up to about 50% R.H. but can cause considerable variation and error above this value.

There is generally a tendency for small particles to agglomerate to larger particles. This can occur in the sampling system but is not fast enough to be of concern for particles larger than about 0.2μ. Even for this size, particles only agglomerate at a rate of about one %/hour. In impingement collection methods, particles that are normally agglomerated in the atmosphere may be fractured to smaller particles when they strike the collector. Some of these smaller particles may then be redispersed

Table 4:1. Effect of Surface Irregularities on the Adhesion of Quartz Particles to Pyrex Glass

Surface	Range of profilometer reading, (Å)	Root-mean-square surface irregularity (Å)	Adhesion (%)*
Pyrex flat 1	2032–2286	2159 ± 127	100
Pyrex flat 2	2032–3810	2921 ± 127	67
Pyrex microscope slide	3048–3810	3429 ± 127	59
Pyrex flat 3	4572–5080	4826 ± 127	45

*As percentage of adhesion to pyrex flat 1.

(Corn, M. 1961. J. Air Poll. Control Assoc. 11:566)

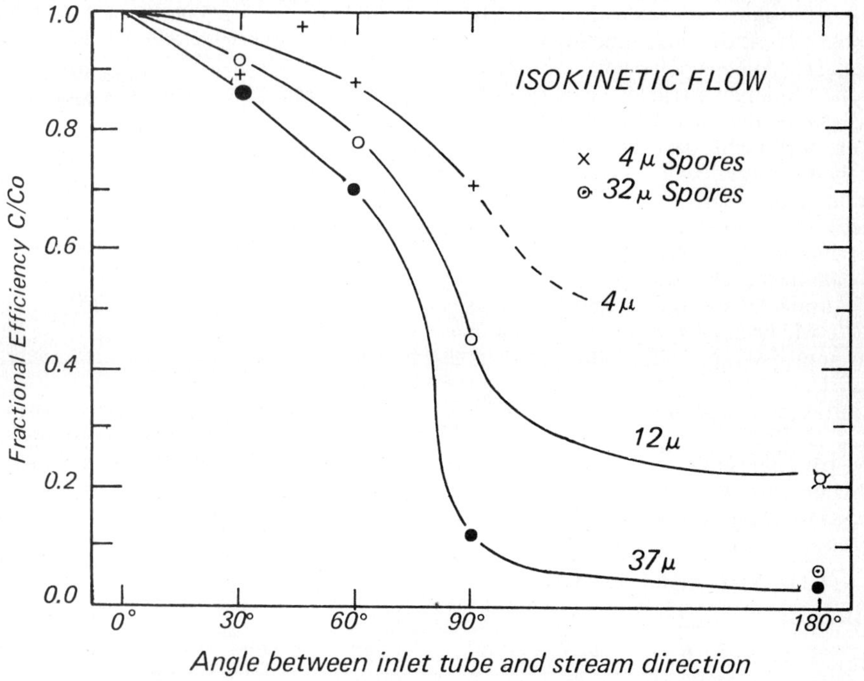

Figure 4:3—Variation of the ratio of observed concentration to true concentration with angle of yaw of inlet tube for 4, 12 and 37 micron mass median diameter diethyl phthalate droplets and for 4 and 32 micron spores.
(Watson, H. 1954. Amer. Ind. Hyg. Assoc. Quart. 15:21)

and either collected as smaller particles or lost from the collector.

The following precautions should be mentioned as possibilities but seldom create a serious problem. Collected particles may react after collection so that the material analyzed may not represent the composition of the original aerosol. For instance, sulfuric acid particles may react with cement kiln dust (C_aO) to form calcium sulfate. Particles may also react with the collection media such as Millipore filters, or in wet collectors they may react with the solvent or partially dissolve in it. Gases may adsorb on solid or liquid aerosols. Collected aerosols may also sublime or evaporate after collection. It is not uncommon for the concentration of pollutants to vary throughout the day or for the temperature to vary so that a pollutant collected at night or in the morning may be lost from the sampler after a warm afternoon. Although it is common for aerosol sampling to extend for 24 hr or more, solids generally have low vapor pressures and evaporative loss is small. For many purposes, particulate samples tend to keep well for long periods of time.

4.3 Factors Affecting Selection of Sites for Aerosol Sampling of the Atmosphere. Site selection is important in any air sampling but especially so for aerosols because they are much less uniformly dispersed in the ambient air as well as in process equipment. (Factors involved in selecting source sampling locations are discussed in the next section, 4.4). Special consideration must be given to sources, obstacles, meteorology, and time. Particles of all sizes are being continually emitted into the atmosphere. The larger particles fall out rapidly and moderate sized ones more slowly as listed in Table 4:II. At the same time the very small particles (Aitken nuclei) tend to become attached to larger particles. This process is fastest for the very small particles (80%/min. for 0.02μ particles) becoming attached to larger particles, and much slower for larger particles (14%/day for 0.2μ size). These two processes then tend to produce an equilibrium particle size distribution in the atmosphere, but because of the variable speeds of these processes there will be significant gradients near major sources, especially if they emit large particles. The height of the source, the wind velocity and turbulence, and particle size distribution will determine how fast the particles settle out. Figure 4:6 shows how the particle loading can vary with distance from the source for various particle sizes.

Table 4:II. Settling Velocity of Unit Density Spheres in Still Air

Particle Diameter, μ	Settling Velocity, V_s, cm/sec.
1000	385
500	200
220	76
100	25.1
50	7.2
40	4.8
30	2.7
10	0.30
1	0.0035

Point sources do not generally distribute their particles evenly in all directions. The particles travel with the wind which is quite variable. A sampler located directly in the path of a plume may show 10 or 100 times as much aerosol as one located a few

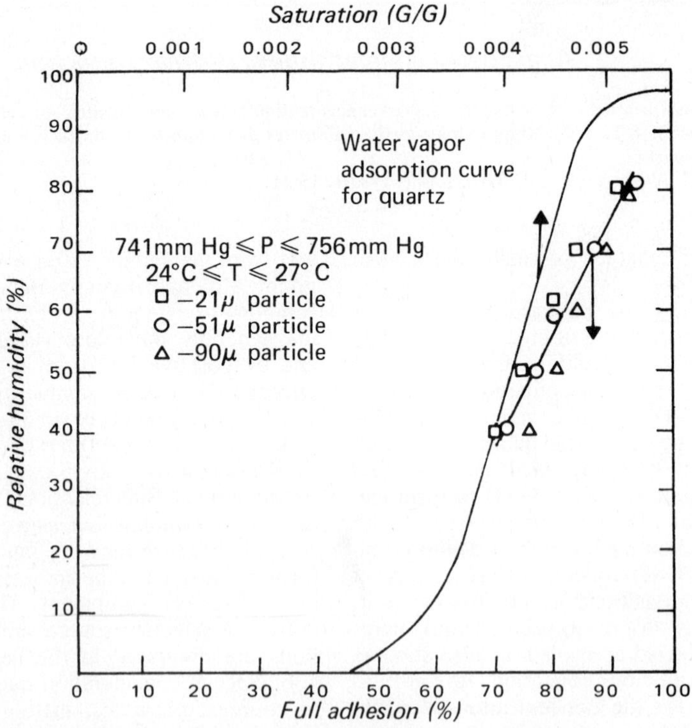

Figure 4:4—Adhesion variation with relative humidity (quartz particles to pyrex flat). (Corn, M. 1961. J. Air Poll. Control Assoc. 11:566.)

feet away. In the lee of buildings, trees, hills or other obstacles, the air is often quiescent and will contain fewer particles. Strong gradients usually exist in the vertical direction also. Because most sources are near the ground and vertical mixing is frequently limited, particulate loading usually decreases with height.

Time is also an important variable. In most cities the mixing height varies greatly throughout the day. Often there is an overnight inversion which keeps all of the pollution close to the ground and produces relatively high concentrations. When the sun comes up the inversion gradually lifts and generally breaks in the morning or early afternoon at which time there is good dispersion. Particle concentrations at a given location can easily vary by a factor of 10 or more during a 24-hr. period. Thus, a sample taken in the afternoon may give very little information on the air pollution problem. On the other hand, a 24-hr sample might average a few hours of critically severe pollution over a 24-hr period and cause little alarm although a serious problem may exist. In spite of this, 24-hr samples have become a standard practice

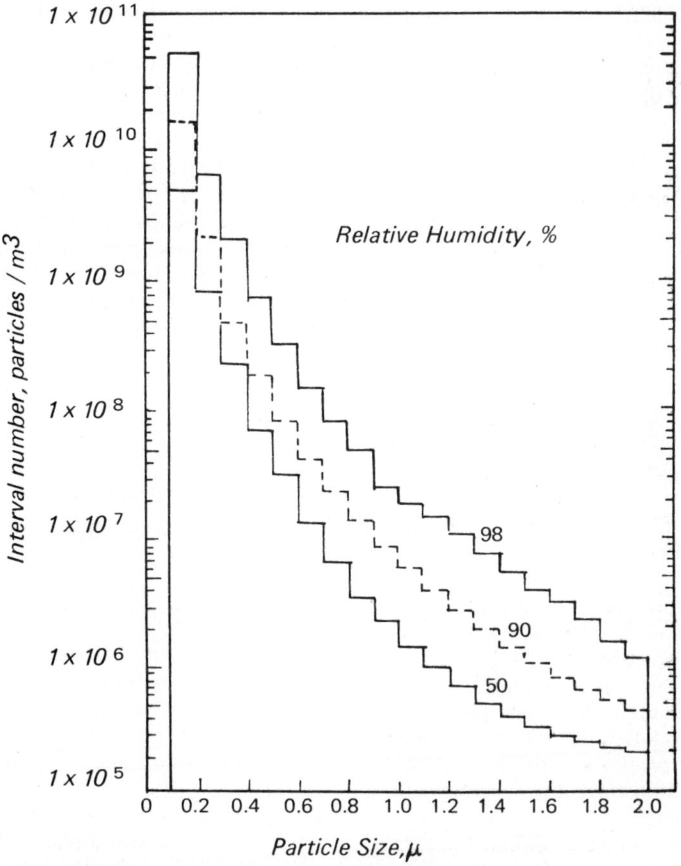

Figure 4.5—Calculated number of droplets per m³ in various size intervals at different relative humidities in a sulfuric acid mist sample having a concentration of 39μ g/m³.

(National Air Pollution Control Administration, 1969. U.S. Dept. Health, Education and Welfare. *Air Quality Criteria for Sulfur Oxides*. AP-50, pp. 1–22.)

$$V = 953 \sqrt{\frac{P_v \times T}{SP}}$$

where V = flow velocity in feet per minute at measured conditions.

P_v = velocity head in inches of water.

S = specific gravity of the gas stream compared to air.

P = absolute pressure of gas in duct, inches of Hg.

T = absolute temperature of gas in duct, ° R.

A number of measurements must be made at various points in the stream. The cross-section should be divided into equal areas of about one square foot. In a rectangular duct, a measurement should be made in the center of such rectangle with a minimum of nine measurements.

X	X	X
X	X	X
X	X	X

In the case of a circular duct, annular rings of equal area (about 1 ft.^{2}) should be visualized and measurements should be made at the center of each such ring on each of two diameters which are 90° from each other (with a minimum of 8) as shown below in Figure 4:8.

After calculating velocities, each velocity is multiplied by the area it represents and these flow rates may be added to obtain the total flow rate. Samples should be taken at each of the flow measuring points and the amounts collected and weighted together in proportion to the individual flow rates found at the corresponding points.

Particulate sampling should be carried out with probes inserted in the duct at each of the points of flow measurement. The probes must be pointed directly upstream and the sampling rate adjusted to provide isokinetic conditions at each point.

It is often more convenient to make flow measurements at the same time as the sample is collected by using a combined probe. This is especially useful in cases where the flow rate is varying and the sampling rate can be correspondingly varied to maintain isokinetic conditions.

Lack of equipment and personnel generally dictates that the points are sampled successively rather than simultaneously. However, in determining the efficiency of control equipment, it is highly desirable to take simultaneous samples before and after the control. Greater variability ($< \pm 50\%$) for flow rate or sample concentration from point to point or time to time make it advisable to increase the number of samples recommended above and place more emphasis on simultaneous sampling, whereas a somewhat smaller number can be taken if the stream is more consistent ($< \pm 20\%$). Sample concentrations determined at various cross-sectional points at a given location should be weighted together in proportion to the flow rates found at each point and the areas represented as follows:

$$M = 60(C_1V_1A_1 + C_2V_2A_2 + {}^{\circ\circ\circ}C_nV_nA_n)$$

Where M = the mass of pollutant in pounds per hour.

$C_1, C_2, {}^{\circ\circ\circ}C_n$ = the concentration of the pollutant found at sample points 1,2, n, respectively, in pounds per cubic foot at measured conditions.

$V_1, V_2, {}^{\circ\circ\circ}V_n$ = the flow velocity at sample points 1,2, ${}^{\circ\circ\circ}{}_n$ respectively in feet per minute at the same conditions used for values of C.

$A_1, A_2, {}^{\circ\circ\circ}A_n$ = the square feet of cross-sectional areas represented by sample points 1,2, ${}^{\circ\circ\circ}{}_n$, respectively.

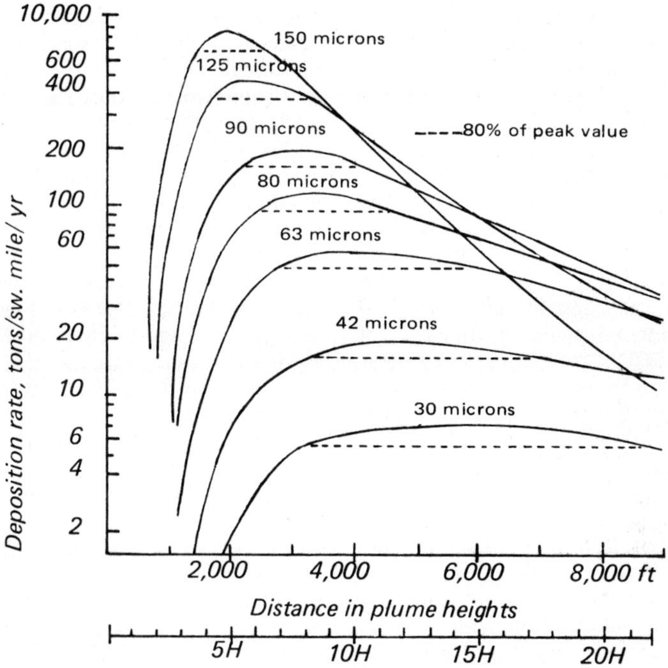

Figure 4:6—Deposition rates of particles of different diameters.[107] Effective height = 400 ft. Rate of emission of each size of particle = 0.1 ton/h. Wind speed = 20 ft./sec. Wind directional frequency, b = 0.2.

(Derived with permission from the deposition function in Stern, A.C. 1968 *Air Pollution*. 4th ed. p. 265, Academic Press, N.Y.)

and samplers are generally started at midnight by an automatic timer and run until the following midnight. From the point of record keeping this keeps the entire sample on the same date but from an air pollution point of view it produces a sample that has one day plus parts of two different nights. It would be preferable to include a single night by starting 24-hr samples at 6.00 P.M. or 6.00 A.M. or to sample for a shorter period or the critical period only.

To summarize:

a. A sample should be taken at the point of major interest, if possible, but care should be exercised in extrapolating its relevance to the ambient air of a whole city or even a nearby part of it;

b. if it is desired to get a general measure of atmospheric aerosol as it affects man, plants and animals, one should:

(1) not be directly downwind from a major point source;

(2) locate the sampler about 5′ above ground level;

(3) locate downwind from major obstacles a distance of about 10 times their height;

(4) take several such samples at different locations in the area of interest;

(5) sample the time of day of greatest interest but do not consider this representative of all day or any small part of the sampling period.

4.4 Factors Involved in Source Sampling.

4.4.1 *General Considerations.* Source sampling refers to the collection of pollutants at their source as they are being released into, and before they are diluted by the ambient atmosphere, or for design purposes at a point where controls may be installed. The objective is to collect a sample that is representative of the material that is emitted or to be controlled. The material may not always be confined in a stack or duct and the composition and amount may vary with times such as batch processes. It may be coming out of monitors, windows, etc. Collection of such samples is usually difficult and costly and requires advanced planning. Because of this, source testing is seldom done unless there is reason to suspect pollution or to determine the efficiency of control equipment or to get data for design purposes.

4.4.2 *Planning and Preparation for the Test.* The specific points of sampling are generally determined by examination of drawings or flow diagrams, and discussions with plant engineers or others who understand the process or source of emissions and its variation with time. A site visit is generally required to make the final selection. Criteria for selection of the sampling location are:

a. At the point of greatest interest (such as stack outlet);
b. in a straight section of pipe 5 or more diam downstream and 3 or more diam upstream from any bends or flow disturbances;
c. accessible to sampling personnel and equipment;
d. utilities such as electricity, water, and air available if needed; and
e. safe for test personnel.

Often an ideal site cannot be found and a compromise must be reached.

Generally the selected site must then be prepared for testing by making access holes (confined source) or a collection system (unconfined source). A hole to which a 3″ pipe coupling can be welded is generally about the right size but the particular sampling probes which are to be inserted must be considered before this decision is reached. A scaffold accessible to these holes may have to be constructed and utilities supplied. Some estimate of flow rate, temperature, pressure, dew point, particle size and gas composition must be made in advance to plan the sample collection. After the detailed procedure for collection has been worked out, the sampling and analysis equipment must be assembled and prepared for transportation to the site. The process variables must be considered in determining the proper time or times for sampling. The management or persons responsible for the source must be contacted and arrangements made to have access at the desired times. The sampling crew must be organized and briefed and the analysts alerted.

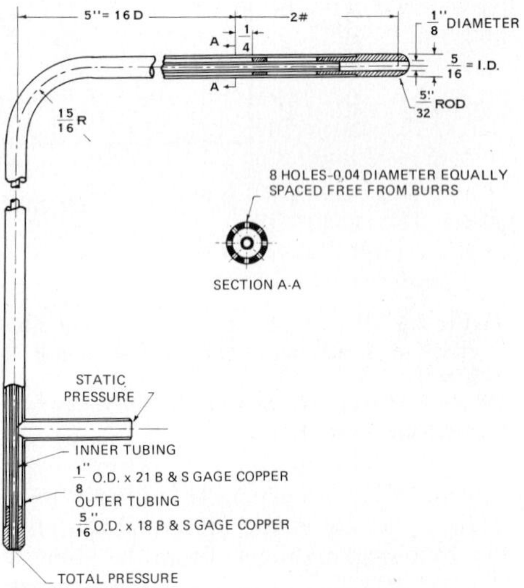

Figure 7. Standard Pitot Tube

Figure 4:7. Standard Pitot Tube.

4.4.3 *Sampling.* The equipment and crew are transported to the site. Source flow conditions must first be measured. Although a number of methods are available, pitot tubes are generally preferred. In essence a pitot tube consists of two concentric tubes (Figure 4:7) one of which has an opening facing upstream and the other has openings perpendicular to the flow. The pressure difference between these is called the velocity head and may be used to determine the flow rate from handbook tables or the following equation:

Separate calculations will need to be made for each pollutant analyzed.

Diffuse emissions such as wafts out doors, or windows, or dust from an open quarry, or construction site, generally require individual ingenuity in sampling. A useful technique applicable to some of these is referred to as variable dilution sampling. In this method a blower is operated at the site and ducted so that all of the emissions from the location of interest enter this duct. Although the emission rate may vary considerably, the fan mass-flow-rate is kept constant at a value exceeding the emission rate. A tight seal is not made so that when the effluent rate is below the blower rate, ambient air can enter freely to make up the difference so the total flow remains constant. This approach prevents creation of any appreciable unnatural suction at the source that might cause excessive emissions. At the output of the blower or some other point where the emissions and makeup air are thoroughly mixed a sample can be taken for analysis. The sample concentration (C) times the total fan rate (V) will give the emission rate. If a constant fan mass-flow-rate is used, a sample can be collected at any smaller constant mass-flow-rate and combined over any desired period of time and the combined sample concentration will be a true measure of the weighted pollution concentration over that period. This can be multiplied by the total blower flow over that period to obtain the mass of emissions for the period.

Care must be exercised to make sure that particles do not settle out in the duct prior to the sample point and that the sample point is operated isokinetically, if necessary. Another precaution of importance is to separate the particulate sample at the temperature of interest. If the temperature is too hot, liquid aerosols of interest may be vaporized and not be collected, whereas if the temperature is too low, water or other condensable vapors may form mists or aerosols which will collect with the solids and plug up the filter or give unrealistic results. The variable dilution technique tends to minimize condensation of vapors which may be beneficial or undesirable depending on the particular case. While the sample is being taken, the following data should generally be taken simultaneously:

Time and Date
Sample Location
Sample Flow Rate
Sample Pressure
Sample Temperature
Dew Point
Plant Operating Conditions
  Flow Rates and Variability
  Production Rates and Variability
  Pressures
  Temperatures
Persons Involved in Sampling

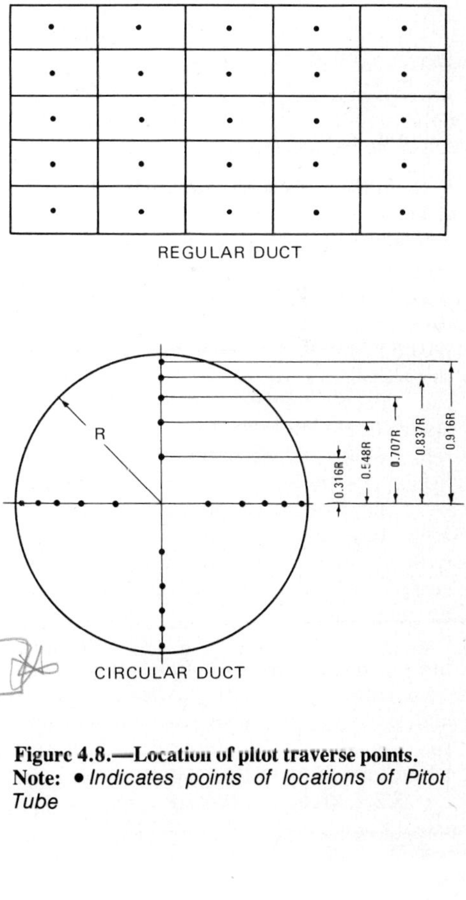

Figure 4.8.—Location of pitot traverse points.
Note: • *Indicates points of locations of Pitot Tube*

5. Sampling and Storage of Gases and Vapors

Gases and vapors follow the normal laws of diffusion and mix freely with the surrounding atmosphere. They are not affected by inertial and electrostatic forces which may disturb particulates. These characteristics make gases and vapors easier to sample than dusts or fumes.

5.1 INTERVAL BETWEEN COLLECTION AND ANALYSIS. The interval between collection and analysis is governed by the following factors:

a. Chemical changes such as chemical interactions among the components of the collected sample or photochemical decomposition;
b. adsorption of gases from the sample onto the walls of the containers; and
c. leaks.

In general, make the time interval between sample collection and analysis as short as possible, protect the sample against light and heat, and exercise precautions against leaks.

5.2 SAMPLING APPARATUS WITHOUT CONCENTRATION OF GASES AND VAPORS.

5.2.1 *Plastic Bags.* The typical materials used in plastic bags are Aluminized Scotch-pak, Scotch-pak, Mylar and Saran. Entry into the bag can be accomplished with Tygon, Teflon or glass tubing or an ordinary tire valve. It is desirable to run "decay" curves on as many substances sampled in the bag as possible so a reasonable storage time can be established. A sample is introduced into the bag by a hand operated pump or a squeeze bulb. Bags can be re-used after purging with clean air and checking for any residue components. Certain contaminants cannot be sampled or stored in any type plastic bag due to their reactivity with surrounding substances or with themselves, styrene being an example. Whether a substance can be sampled and stored in a plastic bag should be determined in the laboratory prior to field use.

5.2.2 *Rigid Containers.* An evacuated container is shown in Figure 5.1.

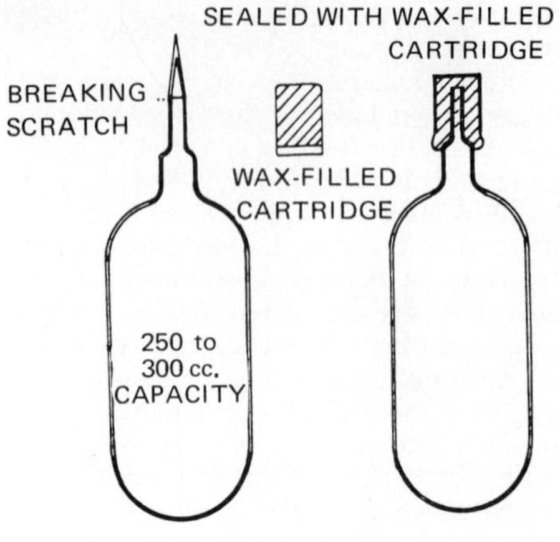

Figure 5:1.—Evacuated Sample Container.

It consists of a glass bulb from which the air has been removed by a vacuum pump and the neck of which has been sealed by heating and drawing to a tip during the final stage of evacuation. The vacuum-type sample container is suitable for sampling atmospheres for analysis of gases such as carbon dioxide, oxygen, methane, carbon monoxide, hydrogen, and nitrogen. It is not suitable for sampling for the determination of very reactive gases such as hydrogen sulfide, oxides of nitrogen, or sulfur dioxide, because such gases may react with dust particles, moisture, the wax sealing compound, and to some extent even with the glass of the container, so that by the time the sample is analyzed the proportions of these gases will have been altered.

A special evacuated container (Figure 5:2.) may be used for collecting samples directly in a chemical absorbent in which they will later be analyzed. The container and sampling method are similar to the conventional vacuum-type sample container previously described. The bulb differs from the conventional type in that it contains a liquid absorbent for the particular gas or vapor to be analyzed that dissolves such gas and preserves it in a form that may be determined by chemical analysis.

Containers filled by gas or liquid displacement include the gas sample tube shown in Figure 5:3, and ordinary heavy-wall glass bottles (Figure 5:4.).

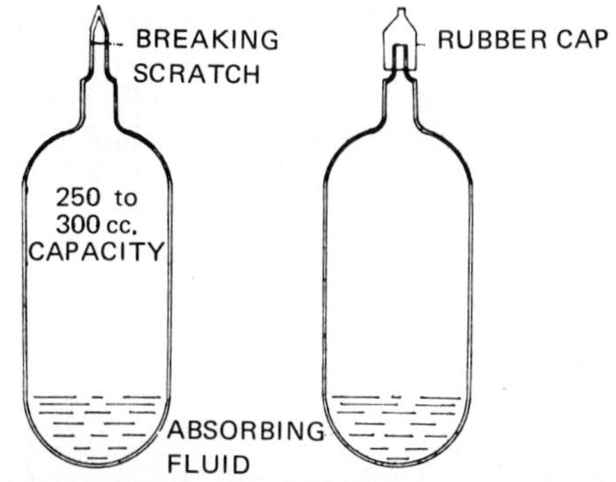

Figure 5:2.—Evacuated Sample Container for Collection of Sample in an Absorbent.

The gas sample tube (Figure 5:3.) may be metal or glass and may be closed with stopcocks or screw clamps and rubber tubing if the sample does not react with rubber. Pinch clamps are not satisfactory, as they may not have enough tension to squeeze the rubber tubing completely closed. Metal containers should be constructed of inert metal. Iron or tinned-iron sample containers are not recommended because their rusting consumes too much oxygen. Metal containers are not suitable for sampling of many substances, of which the following are typical: hydrogen sulfide, sulfur dioxide, or oxides of nitrogen. These gases react rapidly with the metal and are lost from the sample. Copper is reactive with acetylene and acts as a catalyst for a number of gas-phase reactions. The glass bottles illustrated in Figure 5:4. are closed by a rubber or plastic washer under a cap held in place by a strong spring. A supply of extra washers should be kept on hand for renewal.

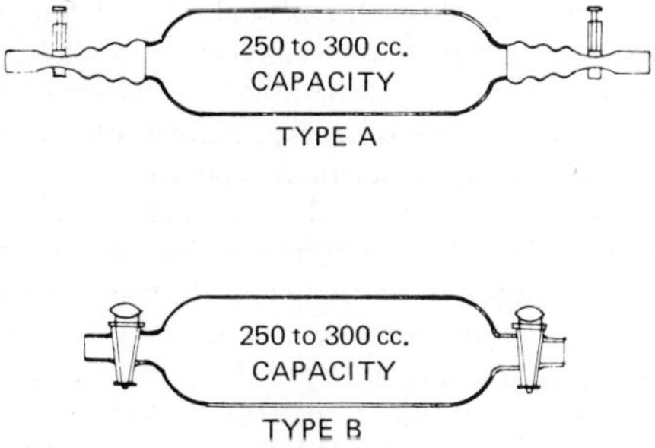

Figure 5:3.—Gas Sample Tubes for Collection of Sample by Displacement.

To fill any of the containers described here, it is necessary for its original air or gas content to be completely swept out and replaced by the air to be sampled.

For this purpose an aspirating device is usually necessary. The most convenient devices are a double-acting rubber-bulb aspirator (Figure 5:5) or a double-acting foot pump. Both may be obtained from laboratory supply houses and may be used for producing either suction or pressure. An aspirator may also be made of a bottle or can filled with water, as shown in Figure 5:6.

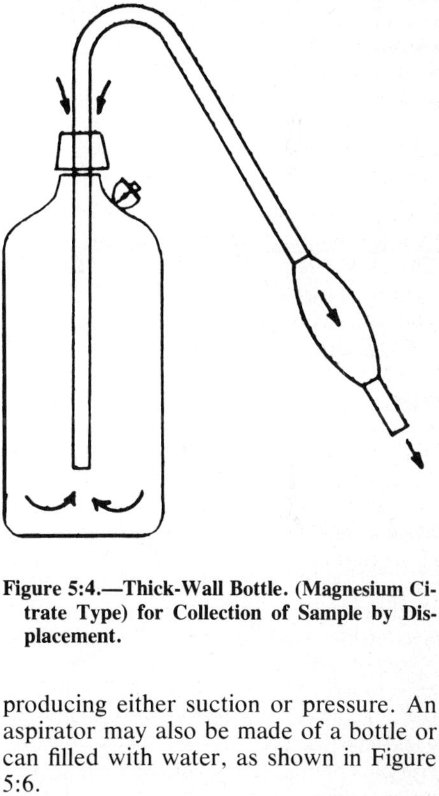

Figure 5:4.—Thick-Wall Bottle. (Magnesium Citrate Type) for Collection of Sample by Displacement.

5.2.3 *Collection of Sample.* For collection of the sample in a bulb such as shown in Figure 5:1, the bulb should be evacuated to a pressure of 1 mm of mercury or less. If it is not possible to reach such a low pressure, the bulb should be prepared in an uncontaminated atmosphere and the evacuated pressure ob-

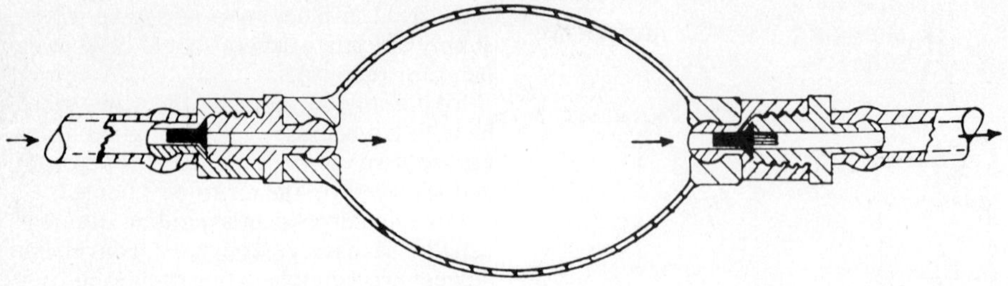

Figure 5:5.—Double-Acting Rubber-Bulb Aspirator.

tained should be measured with a mercury manometer. Then the volume of sample at barometric pressure may be calculated as follows:

$$V_s = V_b \times \frac{P_1 - P_2}{P_1}$$

where:

V_s = volume of sample at barometric pressure,
V_b = volume of bulb,
P_1 = barometric pressure, and
P_2 = pressure in evacuated bulb.

In collecting a sample of air with this type of container, all that is necessary is to break the tip at the scratch mark provided on the neck of the container. The surrounding atmosphere immediately enters and fills it to atmospheric pressure. The container shall be sealed by forcing a wax-filled cartridge over the broken neck, as shown in Figure 5:1.

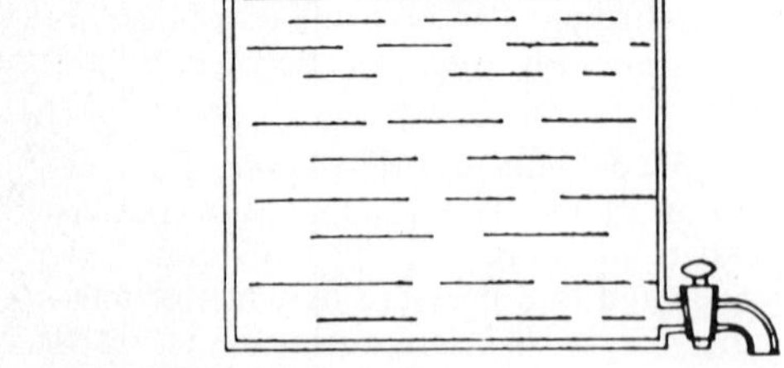

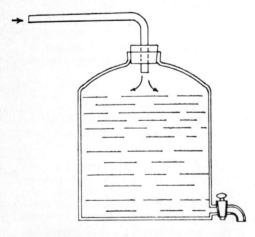

Figure 5:6.—Water-Filled Container for Collection of Sample by Displacement.

For sampling by gas displacement, in accessible locations, an aspirator may be connected so that pressure and relief on the rubber bulb will draw the original content of air out of the sample container and admit atmosphere, as shown in Figure 5:7. Thirty to forty compressions of an ordinary 75-ml capacity aspirator bulb will replace the air in a 250-ml sample container. The valves of the rubber aspirator bulb are, of necessity, small and light devices, and sometimes do not operate positively. Therefore, before using the aspirator, check its performance by compressing the bulb, closing the intake end, and noting whether there is any leakage, and then, by repeating the process, endeavoring to compress the bulb with the outflow end closed. Dust sometimes lodges on the valve seats, causing them to leak. A seemingly defective bulb often may be remedied by removing such dust and by wetting the valve seats with water. An additional precaution when using the bulb as a suction pump is to connect a hose to the outflow end and place the exit end of the hose under a water seal to 2.5 to 5 cm (1″ or 2″), thus preventing air contamination through leaky valves. In *lieu* of this, the outflow end of the bulb may be closed by the thumb or finger while the bulb is intaking.

After a sampling tube is purged, draw approximately 10 to 15 times the container volume of gas through the system. When

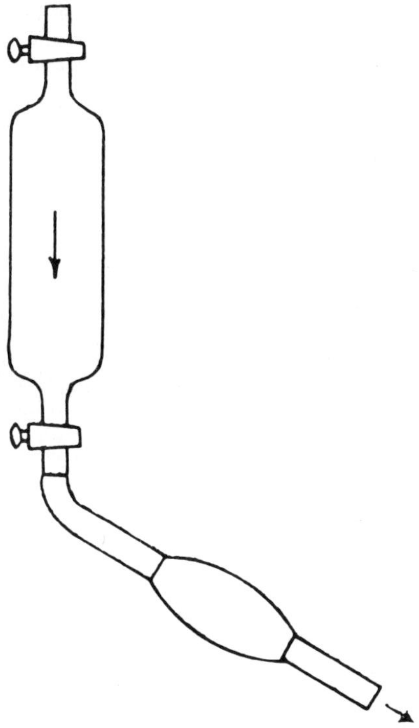

Figure 5:7—Gas Sample Tube Plus Aspirator.

using aspirating devices, sample containers of the types shown in Figure 5:3 should be between the sampling tube and aspirator to avoid contaminating the sample in case air leaks into the aspirator. Precautions described above for selecting an inert material for the container should also be applied to the sampling tube. Making a water seal at the outflow end of the aspirator bulb or pinching off the exit tube with the fingers during the intake stroke of the aspirator is an additional precaution.

When the gas-displacement method is used in filling thick-wall bottles such as magnesium citrate bottles or bottles of other types with gas from inaccessible places, place the aspirator between the container and the sampling tube (Figure 5:8).

If means for a water seal are available, insert the neck of the bottle about 2.5 cm (1″) under water and lead the gas under the water and into the bottle as shown in Figure 5:9. If a seal is not available, pinch off the hose leading into the bottle with the fingers during the intake stroke of the bulb to avoid contamination from leaky valves. Continue the aspiration while the hose is slowly withdrawn from the bottle, and play a stream of gas on the top of the bottle while closing it.

Containers of the types shown in Figure 5:3 and ordinary bottles can be filled by water displacement, but in so doing, the carbon dioxide content of the sample will be reduced somewhat because of the solubility of this gas. The containers shown in Figure 5:3 have the advantage that aspirating devices can be eliminated by filling the sample container and the sampling tube with water, if a sampling is to be taken from an inaccessible space. In sampling, simply drain the water completely from the container. If the sample is taken rather slowly, say in 2 or 3 min, the walls of the container will drain satisfactorily during the sampling period. The method is limited to applications with gases that are insoluble in and nonreactive with the displacing liquid.

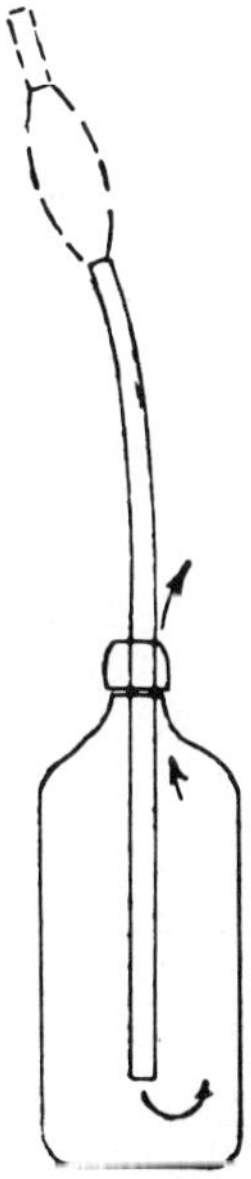

Figure 5:8—Sampling Assembly for Filling Thick-Wall (Magnesium Citrate Type) Bottles.

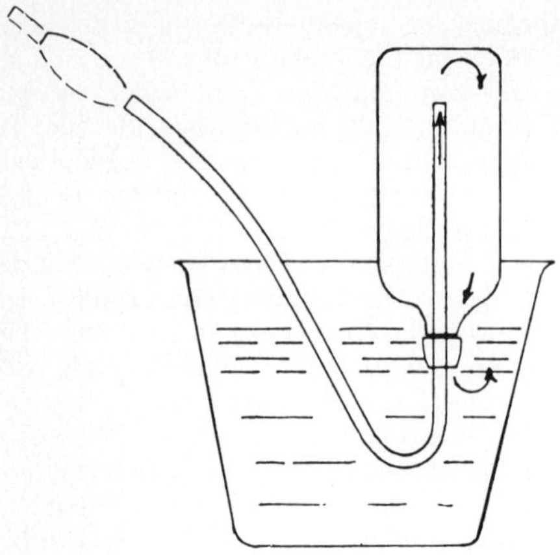

Figure 5:9—Sampling Assembly for Filling Bottles Under Water.

5.3 SAMPLING APPARATUS WITH CONCENTRATION OF GASES AND VAPORS.

5.3.1 *Gas Pumps.* Appropriate gas pumps are of several types and many sizes. They should have adequate capacity and perform uniformly. Mechanical pumps, nearly always electric, are necessary for prolonged periods of operation. Induction motors are preferred, since this type operates uniformly in spite of variations in "line load".

Hand pumps are usually large glass syringes or piston pumps. These are suitable for sampling small volumes and when gas flow need not be constant. However, one hand-operated multiple-piston pump is designed to give a uniform flow and may be used for relatively prolonged periods. All of the hand-operated pumps are customarily used without a volume meter. Accordingly, the resistance to gas flow in the sampling assembly must be constant and reproducible.

Aspirators, either water or air operated, are usable where there is an adequate supply of the latter under constant pressure.

A siphon may be used when small sample volumes are adequate and may be preferable when very low sampling rates are needed. It is desirable that the gas inlet tube of the siphon bottle extend nearly to the bottom to maintain a uniform flow as the liquid level falls.

5.3.2 *Gas Metering Device.* A gas metering device is required when the sample is pulled by means of an aspirator or a pump that acts to maintain a reduced pressure downstream of the sampler. These meters are of two types: (*a*) volume meters and (*b*) rate meters.

Volume meters may be of the wet type (wet test meters) or the dry (diaphragm) type. The wet type gives the most precise result but requires water and is heavy and cumbersome. The dry type is less precise but is somewhat more convenient. Rate meters consist of rotameters and orifice or capillary "flow-meters".

Volume meters have the advantage that variation in rate of sampling will not introduce large errors. The rate meters have the advantage of small size and convenience. These, however, require close observation and accurate timing of the sampling period and the pump must operate uniformly. With both types of meters, accuracy may suffer at very low rates of flow such as 10 ml/min and less.

5.3.3 *Pressure Gauge or Manometer.* A pressure gauge or manometer, to follow the gas metering device in the sampling assembly, should be provided.

Absorbers are classified as four basic types: (*a*) simple bubblers, (*b*) bubblers with diffusers, (*c*) spiral absorbers, and (*d*) packed towers. The performance of none has been systematically and fully investigated; with all, however, acceptable performance (90% or better efficiency) has been shown in specific applications. Their individual characteristics are summarized in Table 5:I.

5.3.4 *Sampling Lines and Probes.* It may be necessary to extend tubing and probes beyond the sampler in order to reach the desired point of sampling. In such a case, the tubing and probe must be large enough not to offer high resistance to gas flow. They must be appropriate for special conditions such as high temperature and they must not absorb or adsorb appreciable amounts of the contaminant. No single material can be specified for the line or probe; some metals, glass, and some plastics are frequently used. Every material, however, is suspect until its appropriateness has been demonstrated.

Whenever air to be sampled is conducted through equipment that continuously extracts a gas or vapor from it for analysis, it is essential to avoid any recirculation of such sampled air through the extraction equipment. Therefore, the air that has passed through the equipment must be ejected into a space from which resampling is impossible. In outdoor air sampling this is accomplished simply by exhausting downwind. In still air, conduct the exhaust by a tube or duct or blow it by a fan effectively away from the intake location.

The sampler is always used in assembly with a gas (air)-moving and a gas-metering device. A probe or sampling line may be extended beyond the sampler to a desired point and a pressure gauge or manometer may be employed in the system just after the gas meter. When sampling gas containing particulate matter that can interfere with analysis, an appropriate filter shall be inserted ahead of the sampler.

Place the sampler at the upstream end of the assembly, as close as possible to the sampling point, followed by the gas meter and the pump. While it is possible to have either or both the gas meter and pump ahead of the sampler, such an arrangement is not satisfactory since it increases the likelihood of loss of the gas constituent to be sampled and offers too much opportunity for contamination of the samples or for dilution of sample by extraneous air.

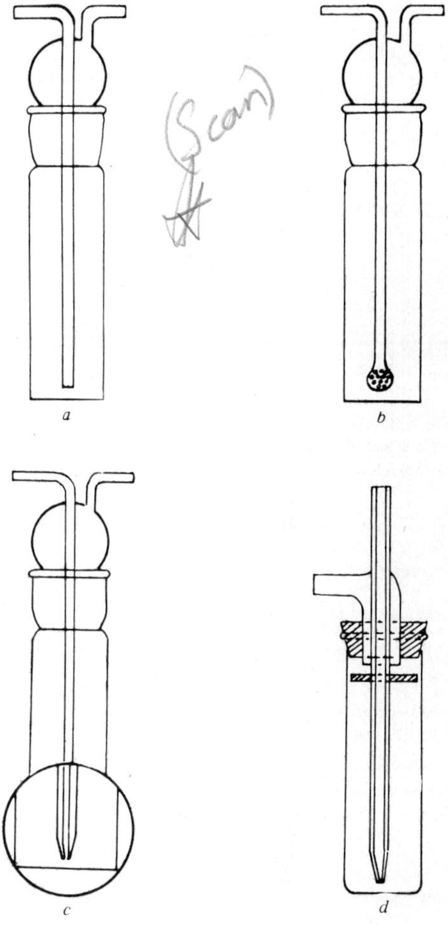

Figure 5:10—Simple Bubble Absorbers.

Table 5:I. Characteristics of absorbers—approximate ranges of use

Type of Absorber	Absorbent Capacity, ml	Sample Rate, ml per min.	Remarks
Simple bubbler (Figure 5:10)	5 to 100	5 to 3,000	Simple, non-plugging, short gas-liquid contact.
Bubbler with diffuser (Figure 5:11)	1 to 100	500 to 100,000	Easy to use, good gas-liquid contact, subject to plugging.
Spiral (Figure 5:12)	10 to 100	40 to 500	Effective only at low flow rates.
Bead-packed tower (Figure 5:13)	5 to 50	500 to 2,000	Efficient only at low flow rates. Resistance variable.

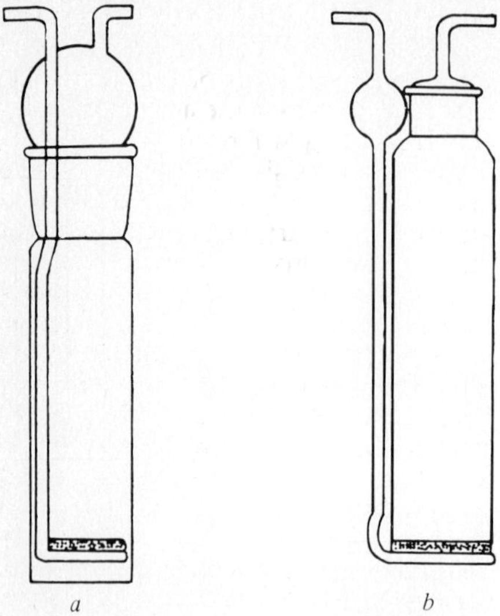

a *b*

Figure 5:11—Bubbler Absorbers with Diffusers.

A pressure gauge or manometer inserted after the meter serves several useful purposes. It enables the investigator to determine whether or not the gas volume must be corrected because of metering at too low pressure for acceptable accuracy and it furnishes the data for such correction. The pressure measurement is mandatory when there is a likelihood of increased resistance of gas flow in the absorber, as by the formation of ice or precipitate. The gauge then enables the investigator to terminate sampling when resistance becomes excessive.

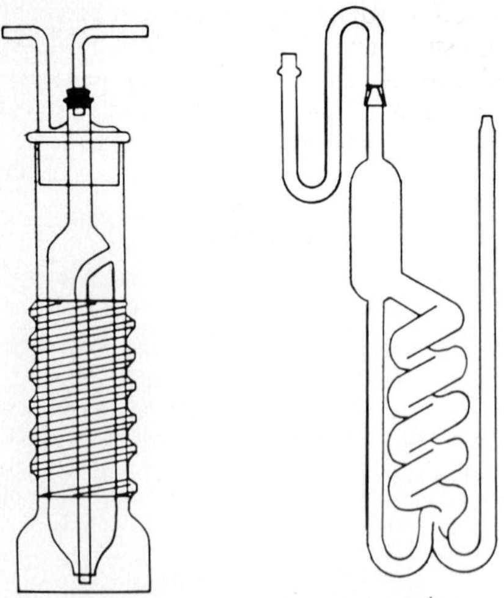

a *b*

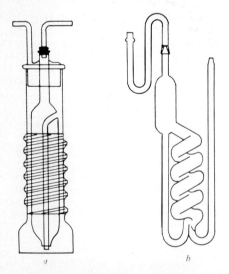

Figure 5:12—Spiral Type Absorbers.

Rate of gas flow may be measured by the pressure just ahead of the pump when the sampling assembly has a resistance to gas flow that is constant and reproducible from run to run. In such a case, rate of gas flow through the assembly as used may be calibrated against the pressure.

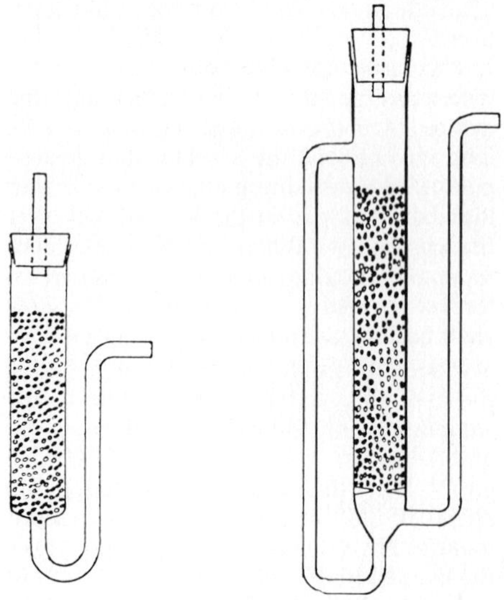

a *b*

Figure 5:13—Bead-Packed Tower Absorbers.

A constant rate of gas flow through the sampling assembly also may be obtained by means of an orifice and controlled absolute pressure. When the downstream pressure is equal to or less than 53% of the upstream pressure, the rate of flow is constant and is proportional to the upstream pressure. Accordingly, the sampling assembly may consist simply of the sampler followed by an orifice and pump or aspirator capable of maintaining a downstream pressure well below one half that of the atmosphere. The resistance of the assembly to gas flow must be reproducible for dependable results.

All meters and sampling assemblies must be calibrated regularly for assurance of accuracy. The basic apparatus for this calibration is the gasometer. Special meters, both volume and rate types, may be used for rapid checks when these are carefully maintained for the purpose.

When the air-moving device consists of a siphon arrangement, or a reciprocating hand pump, or a syringe, the air volume is determined, respectively, by the volume of fluid drained and by the cylinder or stroke volume together with the pressure downstream of the sampler.

5.3.5 *Prefiltering for Particulate Matter*. In collecting concentrated samples of gases and vapors, their separation from particulate matter must be considered. The collection of particulate matter with the bases and vapors may be undesirable because: (*a*) the particulate matter may interfere with the chemical analysis that follows the sampling, or (*b*) the particulate matter may interfere with the collection or metering by clogging orifices, etc. When particulate matter of such nature or in such quantity is anticipated it should first be removed by filtration. The filter should precede any metering or gas collecting equipment and should be nonreactive and nonabsorbing to the gases being sampled. Dry fibrous glass, porous plastic films, cellulosic paper, and asbestos are most often satisfactory. Siliceous material and cellulosic paper, if placed before the collection medium, are not permissible when hydrogen fluoride is being sampled. Nonreactive plastics (*e.g.*, polymers of vinylidene chloride, ethylene or tetrafluoroethylene) may be used to obviate this difficulty.

5.3.6 *Collection of Sample by Adsorption*. Any gas or vapor will, to some degree, adhere to any solid surface at ordinary or low temperatures. This phenomenon is called adsorption. Porous solids expose not only their exterior surface, but their interior surfaces as well, and some such solids indeed possess a vast network of extremely minute channels and submicroscopic pores within their body. Such materials have practical value as adsorbents. These solids include activated carbon, silica gel, activated alumina, and various active earths. All these differ widely in the number and kinds of substances they adsorb, as well as in the amount of sorbed substances they will retain.

In general the siliceous, metallic oxide, and active earth types of adsorbents are electrically polar; *i.e.*, their molecular structure contains an unsymmetrical electron distribution. Since polar substances have strong attraction for one another and since water is highly polar, polar adsorbents retain atmospheric moisture. Siliceous adsorbents, usually silica gel or activated alumina, may advantageously be used for short duration sampling from atmospheres that contain sufficiently high concentrations of gases or vapors to be collected or that are sufficiently dry so that the adsorbent does not become saturated with moisture before the sampling is complete. The light back-ground afforded by silica gel or activated alumina can also be used to advantage in chemisorption by permitting the direct observation of chemical changes on the adsorbent surface. Thus, for example, silver cyanide on activated alumina is used to detect hydrogen sulfide.

Activated carbon is electrically non-polar and is consequently capable of adsorbing organic gases and vapors in preference to atmospheric moisture. In fact, even activated carbon with previously adsorbed moisture will lose such moisture by displacement with organic gases and vapors from contaminated atmospheres. Such displacement of moisture is not exhibited to any significant extent by siliceous adsorbents.

The following comparison between activated carbon and silica gel will further illustrate the differences between the two types of adsorbents for purposes of sampling atmospheric gases and vapors:

a. Activated carbon has a greater overall capacity for adsorbing and retaining atmospheric gaseous and vaporous contanminants, but silica gel exhibits a greater selectivity among them;

b. activated carbon is more suitable for collecting a conglomerate mixture of gases and vapors, especially when these are present in very small concentrations and must, therefore, be collected over relatively long periods of time;

c. desorption of gases and vapors from silica gel is easier than from activated carbon, and it is in fact feasible to obtain some separation of gases and vapors by selective desorption from silica gel;

d. each adsorbent can be suitably impregnated to render it more suitable for chemisorption of specific gases or vapors.

To determine which gases and vapors may be sampled by physical adsorption using activated carbon, the following approximate criteria, based on available information, may be used: True gases, usually having critical temperatures below −50 C and boiling points below −150 C are virtually nonadsorbable at ordinary temperatures by physical means. These include, for example, hydrogen, nitrogen, oxygen, carbon monoxide, and methane. Low boiling vapors having critical temperatures between approximately 0 and 150 C and boiling points between −100 and 0 C are moderately adsorbable. The adsorption efficiency at ordinary temperatures and concentrations, however, is not quantitative. Such vapors can be concentrated from the atmosphere by using a "thick carbon bed" preferably refrigerated. These vapors include ammonia, ethylene, formaldehyde, hydrogen chloride, and hydrogen sulfide. Chemisorption methods, may also be used for such low boiling vapors. The heavier vapors (boiling above 0 C) are readily adsorbed and retained by activated carbon at ordinary temperatures; these include most of the odorous organic and inorganic substances.

Adsorbed gases and vapors may be desorbed from activated carbon by: (*a*) displacement of the adsorbed material by superheated steam and (*b*) heating the carbon under vacuum and distilling the desorbed material into cold traps.

For steam displacement, the saturated carbon is flushed with superheated steam at 300 C or higher. The effluent steam and displaced vapors and condensate are held for analysis. Flushing is continued until the condensate is substantially odorless. Upon standing, the condensate may separate into an oily and an aqueous layer. Both oily and aqueous fractions may then be subjected to qualitative chemical or spectrometric analysis for detection of inorganic ions and organic functional groups. The steam displacement method, although very effective as a means of desorption, suffers from the disadvantage that steam will hydrolyze some organic compounds, such as esters, thereby causing qualitative changes. This effect can be circumvented by using the vacuum method described in the following paragraph.

For vacuum desorption, the saturated carbon is connected to a train of three traps in series, which are immersed in ice-salt, Dry Ice—Cellosolve and liquid nitrogen, respectively. The entire system is evacuated and pumping is continued at a pressure of about 1 μ until distillation practically ceases. During the pumping, the carbon sample is held at a temperature of 200 to 250 C (**Caution:** In following this procedure, care must be exercised to avoid condensing atmospheric oxygen in the liquid nitrogen trap. If necessary, the trap should be warmed up to Dry Ice temperature before bleeding in air.) At the end of the pumping period, the system is slowly returned to atmospheric pressure by bleeding in air or nitrogen. The traps are then removed and their contents examined. Generally, most of the water and some nonaqueous matter will be found in the ice-salt trap, most other atmospheric vapors and some water in the Dry Ice trap, and some of the lighter gases in the liquid nitrogen trap. These materials, with or without further fractionation, may be subjected to chemical or spectrometric qualitative analysis.

For desorbing odorous gases or vapors from silica gel, less drastic methods are adequate. The use of superheated steam or high vacuum is unnecessary. Blowing with contaminant-free air at temperatures up to 350 C has been found to be effective in removing gases and vapors sorbed on silica gel. Other effective methods include extraction with polar solvents such as water or alcohols, and distillation with saturated steam at ambient pressures.

5.3.7 *Collection of Samples by Condensation.* The concentration of atmospheric gases and vapors by condensation at low temperatures has some advantages over other concentration methods. First, the collected material is immediately available for further separation or analysis, without requiring the removal of solvents or desorption from an adsorbent. Condensation is generally the most reliable method for preserving sampled gases and vapors without chemical reaction with any part of the collecting device or among themselves. The main disadvantages of collection by condensation are that the equipment is relatively cumbersome and requires rather frequent attention and that large quantities of water are usually condensed together with the collected vapors. However, the admixture of atmospheric gaseous and vaporous contaminants with water is not generally a serious problem because their extraction from it is experimentally simple.

A special problem in low temperature collection technique is caused by the formation of condensation mists when the sample air is cooled. Such mists are composed of solid or liquid particles of widely varying sizes, and often pass through the cold traps in sufficient proportion to reduce significantly the collection efficiency of the equipment. It is, therefore, necessary to provide a simple filter, such as a glass wool plug, within the cold trap to minimize such losses of particulate matter produced by condensation. As collection proceeds, the solid condensate which forms in the trap acts as a filter and collection efficiency becomes higher and higher until the resistance of the equipment to air flow becomes prohibitive. It should be clear that such filter is additional to the atmospheric particulates.

In selecting refrigerants for concentration of gases and vapors by condensation, it is most common to use materials which maintain a constant temperature by change of phase. A representative group of such refrigerants available for low temperatures is given in Table 5:II. Liquid oxygen generally is the coldest permissible refrigerant for the collection of atmospheric gases and vapors because any lower trap temperatures (*e.g.*, with liquid nitrogen refrigerant) may condense atmospheric oxygen in the trap, especially at low sampling rates. The important fire hazard of liquid oxygen, however, must be recognized.

For condensing a given gas or vapor, it is necessary to select a refrigerant sufficiently cold that the vapor pressure of any trapped material will be low enough to prevent significant evaporation during the sampling. Generally, the vapor pressure of condensed gases or vapors should be about 1 mm or lower at trap temperatures. The method most frequently used employs a series of traps at progressively lower temperatures, thus affording some degree of fractional condensation and, if the last trap is empty, some assurance of efficient condensation. A typical condensation trap, as reported by Shepherd (7), is the sampling tube S in Figure 5:14.

Table 5:II. Refrigerant Temperatures

Refrigerant System	Temperature, ° C
O_2 (liquid) $\rightleftarrows$ O_2 (gas)	−183.0
CS_2 (solid) $\rightleftarrows$ CS_2 (liquid)	−118.5
CO_2 (solid) $\rightleftarrows$ CO_2 (gas)	−78.5
NH_3 (liquid) $\rightleftarrows$ NH_3 (gas)	−33.4
H_2O (solid) $\rightleftarrows$ H_2O = H_2O (liquid)	0

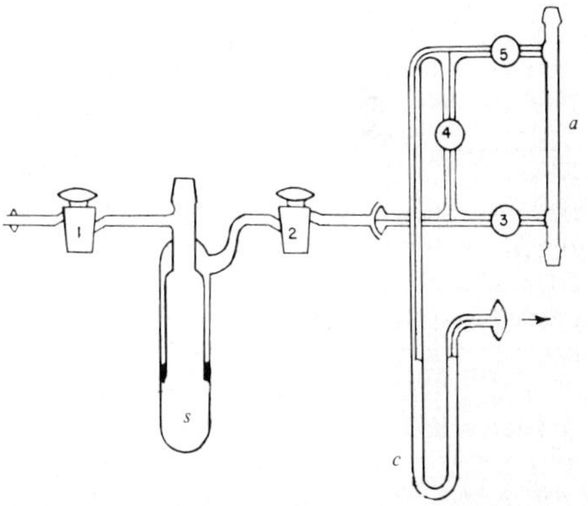

FIG. 5:14—Typical Assembly for Collection of Samples by Condensation.

For further separation of trapped components prior to analysis, the sampler S (Figure 5:14) may be used as a flask for isothermal distillation at progressively higher temperature levels. The material distilled at a given temperature is passed through drying Tube A, which contains sufficient Ascarite to remove 300 ml of atmospheric carbon dioxide, and thence into condenser C, which is immersed in liquid nitrogen (stopcocks 2, 3, and 5 open). After distillation at a given temperature is complete, the fraction in condenser C is evaporated directly into the inlet system of a mass spectrometer, the calibrated volume of whose gasometric unit includes the space in C and the connections back to closed stopcocks 4 and 5. Some experimental difficulties inherent in this procedure include the adsorption of acidic or unsaturated organic vapors by the Ascarite and the collection of some water in condenser C.

References

1. H.B. ELKINS. 1959. *The Chemistry of Industrial Toxicology*. J. Wiley & Sons, Inc., New York, N.Y.
2. H.B. ELKINS, A.K. HOBBY, and J.E. FULLER. The Determination of Atmospheric Contaminants. 1937. Part I, Organic Halogen Compounds, J. Ind. Hyg. 19:474–485.
3. J. CRABTREE, and A.R. KEMP. 1946. Accelerated Ozone Weathering Test for Rubber. Ind. and Eng. Chem. Anal. Ed. 18:769–774.
4. METHODS FOR MEASUREMENT OF GASEOUS FUEL SAMPLES (ASTM DESIGNATION: D 1071). 1966. *Book of ASTM Standards*, Part 19.
5. R.T. SANDERSON. 1948. *Vacuum Manipulation of Volatile Compounds*. Appendix, Table XVII, J. Wiley & Sons, Inc., New York, N.Y.
6. A.N. SETTERLIND. 1935. Preparation of Known Concentrations of Gases and Vapors in Air. Am. Ind. Hyg. Assn. Quarterly. 14:112–120.
7. M. SHEPHERD, *et al.* 1951. Anal. Chem. 23:1431.
8. F.M. STEAD, and G.J. TAYLOR. 1947. Calibration of Field Equipment from Air Vapor Mixtures in a Five Gallon Bottle. J. Ind. Hyg. 29:408–412.

6. Use and Care of Volumetric Glassware

6.1 GENERAL CONSIDERATIONS. The National Bureau of Standards has specified certain minimum requirements which must be adhered to in the manufacture of volumetric glassware.

6.1.1 The instrument must bear, in legible characters, the capacity, the temperature at which it is to be used, and whether it is to contain or to deliver the specified volume.

6.1.2 The instrument must bear the name or trade-mark of the manufacturer and an identification number. If the instrument has detachable parts, such as stoppers or stopcocks, such parts must also bear the same numbers.

6.1.3 The inscription and capacity mark must be etched or engraved but not scratched on the instrument.

6.2 VOLUMETRIC FLASKS. Volumetric flasks most generally employed are those which are calibrated to "contain" a definite volume and have only one mark etched on the neck of the flask. There are also flasks available which have two etched marks. The lower mark indicates the volume of solution the instrument "contains", while the upper mark indicates the volume of solution the instrument will "deliver". The flask may be provided with a ground glass or plastic stopper to prevent any spillage when inverted for mixing solutions.

The sizes of volumetric flasks generally employed in the laboratory to "contain" (TC) are 1000-ml, 500-ml, 250-ml, 100-ml, 50-ml, 25-ml, and 10-ml. Flasks calibrated to "deliver" (TD) are not sufficiently accurate since the amount of liquid adhering to the sides of the flask after emptying can only be approximated.

6.3 THE USE OF VOLUMETRIC FLASKS.

6.3.1 *Preparation of a Solution of Definite Volume*. Transfer the solution, which has been previously prepared in a beaker, by means of a funnel, to the volumetric flask. Pour the solution with the aid of a glass rod held across the lip of the beaker. Wash the beaker, when empty, thoroughly into the flask by means of a wash bottle, holding the beaker with its lip in the funnel and then directing a stream of distilled water from the wash bottle around the inside walls of the beaker. Do not use an excessive amount of wash water, so that the capacity of the flask is not exceeded. Rinse the funnel into the flask and then remove. Stopper the flask and mix the solution by shaking and inverting the flask several times. Return the flask to the upright position and add water until the level of the liquid is slightly below the graduation mark. Repeat the mixing procedure. Return again the flask to the upright position. With the eye on the level of the graduation mark, add water slowly by means of a pipet until the level of the liquid reaches exactly the graduation mark. Stopper and shake well by inverting many times, so that the contents are completely mixed.

It should be kept in mind that the solution and the diluting water must both be at room temperature.

At times, it is necessary to use a portion or an "aliquot" of a prepared solution of definite volume. The "aliquot" portion, taken from the volumetric flask, is measured out by means of a pipet, buret, or a smaller volumetric flask calibrated to "deliver".

Volumetric flasks calibrated to "deliver", should be emptied by inclining them gradually until they are almost in a vertical position. Allow to drain in this position for about 30 sec and then bring the mouth of the flask in contact with the receiving vessel so that the last adhering drop is removed from the flask.

6.3.2 *Calibration of Volumetric Flasks*. Calibrated volumetric flasks purchased from reliable manufacturers are sufficiently accurate. However, volumetric flasks should be calibrated when the accuracy of the capacity of the flask is in doubt.

Circular No. 602 of the National Bureau of Standards gives capacity tolerances for flasks. Determine from the weight of the water contained the capacity of the flask at 20 C using Table 6:I. For example, if the weight of water contained in a 50-ml volumetric flask at 23 C is 49.8310 g the capacity of the flask calculated from the table is 49.8310 x 1.0034 or 50.000 ml.

6.4 BURETS. The most common size buret employed in the laboratory is the 50-ml capacity buret, graduated to 0.1 ml in its smallest subdivision. The buret has a capillary tip of definite size so that the free outflow time of a liquid does not exceed 3 min nor is it less than 90 sec. The optimum outflow time for a 50-ml buret is about 100 sec. Other sizes of burets are of 10-, 25-, and 100-ml capacity.

The Friedman and La Mer weighing buret of somewhat different construction allows increased accuracy in titration. The amount of liquid necessary for a titration is weighed rather than measured by volume. The buret is short, of light weight, and equipped with lugs so that it can be suspended from hooks on the stirrups of a balance. It is also provided with a glass stopper and a cap which fits over the delivery tip to prevent evaporation.

There are also microburets available which have a capacity of 5 ml or less, with the error of delivery not exceeding 0.1%. The 5-ml microburet is graduated to 0.01 ml in its smallest subdivision.

The microburet is provided with a delivery tip drawn out to a fine point. Thus, the liquid is delivered in minute drops and the amount adhering to the tip is negligible. These burets are available with a variety of detachable tips that are attached to the

Table 6:I. Apparent Weights and Volumes of Water Weighed in Air for Calibration of Volumetric Glass Apparatus, Coefficient of Cubical Expansion, 0.000025 per ° C

Temperature °C	Weight of 1 ml	Volume of 1 g	Temperature °C	Weight of 1 ml	Volume of 1 g
15	0.9979	1.0021	23	0.9966	1.0034
16	0.9978	1.0022	24	0.9964	1.0036
17	0.9977	1.0023	25	0.9961	1.0039
18	0.9975	1.0025	26	0.9959	1.0041
19	0.9973	1.0027	27	0.9956	1.0044
20	0.9972	1.0028	28	0.9954	1.0046
21	0.9970	1.0030	29	0.9951	1.0049
22	0.9968	1.0032	30	0.9948	1.0052

buret by means of an adapter and can be easily removed for cleaning.

There are available microburets which discharge very small volumes of liquids with high precision and reproducibility. The Rehberg microburet does not have a stopcock but uses a mechanical device which discharges very small volumes of liquid by means of a micrometer screw pushing against a column of mercury which, in turn, is in contact with the solution to be measured. The Rehberg microburet is of 0.125- or 0.225-ml capacity and, when used, the tip is dipped below the surface of the solution to be titrated.

6.4.1 *Use of the Buret.* The buret is thoroughly cleaned before use. If the buret has a glass stopcock, it should be greased with stopcock grease, not petrolatum, so that it is well lubricated and the capillary tip allows the liquid to flow freely. The temperature of the solution withdrawn from the buret should be near 20 C.

When measuring liquids by means of a buret, the readings are taken at the lowest part of the meniscus since this area is most clearly defined. Exceptions are made for dark liquids, such as permanganate and iodine solutions, where the bottom of the meniscus cannot be seen. In this case, the top of the meniscus is taken as the reading. It is essential that the point chosen for reading the meniscus should be the same for all readings.

In observing the bottom of the meniscus, the eye should be in the same plane as the meniscus as otherwise there will be an error due to parallax. After recording the buret reading, allow about 30 sec for drainage before taking the final reading.

6.4.2 *Calibration of Burets.* Burets are generally purchased from reliable firms which guarantee their burets to meet the tolerances specified by the National Bureau of Standards. Burets are also available certified for accuracy by the National Bureau of Standards, with a Certificate of Accuracy.

Circular No. 602, of the National Bureau of Standards gives capacity tolerances for burets. The capacities of burets (even the certified ones) may change after a period of use, especially if caustic solutions are used. Therefore, burets should be calibrated when the need arises.

6.4.3 *Titrating with a Buret.* Titration is a method by which the quantity of a sample dissolved in a liquid is determined by adding a volume or weight of a reagent solution which just neutralizes or completely reacts with this sample.

A standard solution is one of definite concentration usually normal, half normal, tenth normal, etc.

The standard solution is prepared by dissolving a definite weight of a pure substance (primary standard) in a specific volume of liquid. The standard solution may also be prepared by dissolving a definite weight of a known substance (secondary standard) in a specific volume of liquid and then standardizing it, *i.e.*, determining its concentration by titrating against a primary standard.

The stoichiometric point is defined as the point at which the addition of a definite amount of substance is exactly equivalent to the dissolved substance being titrated. Actually, this is the theoretical end point which is determined from the equation of the reacting substances. However, the stoichiometric point will not always coincide with the end point of the titration.

The end point is the stage in a titration where the reaction is complete. This point is determined by means of indicators or other devices and should coincide with or be as close as possible to the stoichiometric point.

6.5 VARIATIONS OF VOLUMES WITH TEMPERATURES. The National Bureau of Standards has designated 20 C as the standard temperature at which volumetric apparatus will "contain" or "deliver" the stated volume.

Errors, although very slight, will occur if the volumetric apparatus is employed at other temperatures. Glass will expand slightly above 20 C and contract slightly at lower temperatures. The average coefficient of cubical expansion of soft glass is 0.000025 and of borosilicate glass, 0.00001/C.

The variation in volume, due to the expansion of glass, over a small temperature range from the standard 20 C is very slight and may be disregarded for ordinary work. For example, for a 50-ml buret, a variation in volume due to a decrease or increase in temperature of 5 C from the standard 20 C will be only plus or minus 0.007 ml. For accurate work, however, the volume of the vessel may be corrected to 20 C by the following formula:

$$V_{20} = V_t[1 + a\,(20 - t)]$$

where

V_{20} = volume at 20 C
V_t = recorded volume at observed temperature
t = observed temperature
a = coefficient of cubical expansion of the material of the apparatus (for soft glass 0.000025/C)

The volume of a liquid is also affected by temperature. Liquids generally expand at higher temperatures. The more concentrated the solution, the greater will be the variation. Fortunately, most standard solutions are not too concentrated and usually vary from 0.5 normal to less than 0.1 normal. Although the error involved in the variation in the volume of a 0.5 normal solution due to a 5 C change in temperature is small (about 0.05 ml for a 50-ml softglass buret), nevertheless, it is important, when employing volumetric apparatus or titrating, that the temperature of the solutions used is as close as possible to the standard temperature of 20 C.

6.6 PIPETS—Pipets of different types and capacities are available. The transfer pipets, calibrated "to deliver" at 20 C, are generally used to measure a single fixed volume.

The orifice of the tip of the pipet is of definite size so that the free outflow time should not be more than 1 min nor less than 15 sec for a 5-ml pipet; 20 sec for a 10-ml pipet; 30 sec for a 50-ml pipet; 40 sec for a 100-ml pipet and 50 sec for a 200-ml pipet.

Pipets of a capacity of 0.1 ml or less are calibrated to contain a definite volume. When a liquid has been delivered from this pipet into a receiving vessel, the pipet must be rinsed with water at least three times. These rinsings are added to the receiving vessel.

Measuring or Mohr pipets are narrow straight tubes which are graduated to deliver variable amounts of liquids. They are not ordinarily used for precise work.

6.6.1 *Calibration of Transfer Pipets.* The accuracy of a transfer pipet may be checked when necessary. The maximum tolerances allowed for pipets by the National Bureau of Standards are listed in Table 6:II.

Table 6:II. Capacity Tolerances for Transfer Pipets

Capacity (ml) Less Than and Including	Limit of Error (ml)
2	0.006
5	0.01
10	0.02
30	0.03
50	0.05
100	0.08
200	0.10

The procedure used in calibrating the pipet should be the same as that employed in the subsequent use of the pipet. A glass-stoppered weighing bottle of a capacity of about five times the volume of water to be received, is employed as the receiving vessel. The bottle should be twice as tall as wide so that the evaporation of the water during delivery will be negligible.

The bottle is weighed to the nearest milligram by substitution if a two-pan balance is used.

6.7 GRADUATED CYLINDERS. Graduated cylinders are calibrated "to deliver" or "to contain" their stated volumes at 20 C. These cylinders are not as accurate as burets, pipets, or volumetric flasks since they are graduated to a tolerance of 1%.

Graduated cylinders are generally used to deliver approximate volumes and where exact volumes are not essential. Therefore, for accurate measurements, use the other volumetric apparatus described.

When using a graduated cylinder, select a size which will be nearly filled by the volume of liquid to be measured.

6.8 CLEANING GLASSWARE. Keeping laboratory glassware clean is a necessary but tedious chore. Much of the glassware may be cleaned by rinsing with distilled water if the rinsed glassware shows satisfactory drainage of water without the appearance of droplets adhering to its walls.

Glassware can be cleaned satisfactorily by soaking directly after use in a hot soapy solution and then scrubbing with a brush. The glassware is then rinsed thoroughly in hot tap water and, finally, in distilled water.

If the glassware is greasy and dirty and resists ordinary cleaning procedures, the following general chemical methods may be used.

6.8.1 *Sulfuric Acid and Sodium Dichromate Mixture.* A solution of sodium dichromate in concentrated sulfuric acid (cleaning solution) is widely used in laboratories to clean volumetric glassware thoroughly. The oxidizing potential of this mixture, when hot, will cut most greases adhering to glass walls and remove them completely. The cleaning solution is very corrosive and must be used with great care.

To prepare the solution, moisten about 20 g of powdered sodium or potassium dichromate with water to form a thick paste in a glass container. Add about 500 ml of conc sulfuric acid to this paste and stir. Store the mixture in a glass-stoppered bottle. Some of the salt may be left undissolved in the solution. Do not remove this excess salt but use the supernatant liquid. The solution can be repeatedly used by pouring back the unused portions into the stock bottle. Continue using this red solution until it turns green. The color change is caused by the chromium being reduced to the chromous state.

Use this solution with the greatest care. Never allow it to come into contact with clothing or skin since sulfuric acid is very corrosive and bad burns will result. If some of the solution does spill, wash off immediately with running water. Never store or use this solution in metal or enameled containers. Finally, do not allow the solution to spill on floors or desks.

6.8.2 *The Use of the Cleaning Solution.* Rinse the glass vessel with the cleaning solution or immerse it in this solution. The immersion time is dependent on the stubbornness of the dirt, and may vary from a few minutes to 24 hr. A hot cleaning solution is more effective than a cold one. After cleaning, rinse the glass vessel thoroughly with tap water until all traces of the solution are removed. Finally, rinse with distilled water and allow to drain and dry. A good test for cleanliness of a glass vessel is to fill it with water and then empty. An unbroken film of water indicates that the vessel is clean.

6.8.3 *Alcoholic Sodium Hydroxide.* An alcoholic solution of sodium hydroxide is very effective for removing greasy films from glassware.

Prepare a solution by dissolving 120 g of sodium hydroxide in 120 ml of water. Allow to cool and add sufficient isopropyl alcohol to make 1 l.

Immerse the glassware in this solution for not more than 30 min since the caustic may attack the glass. Remove from the solution and scrub with a stiff brush. Rinse thoroughly with tap water and finally with distilled water. Allow to drain and dry.

6.8.4 *Synthetic Detergents.* Anionic or nonionic synthetic detergents are available under proprietary names. Their cleaning properties are due to their ability to reduce surface tension and, at the same time, to wet objects thoroughly. Synthetic detergents are compounded with alkaline salts to increase their detergency. They are efficient in that they do not form scum on glassware when used in hard water. Dilute solutions of about 0.5% concentration are sufficient to clean glassware.

Immerse the glassware in a hot or cold solution for about 30 min, remove from the solution, and scrub with a stiff brush. Rinse thoroughly with tap water and finally with distilled water. Allow to drain and dry.

6.8.5 *Cleaning Burets.* Fill the buret with the "cleaning solution" and allow to remain overnight. Drain the solution into a beaker and return to the stock bottles. Rinse the buret thoroughly with tap water and finally with distilled water. Grease the stopcock with stopcock grease, if necessary. Invert the buret and allow to drain. A test for the cleanliness of a buret is to fill it with water and then allow to drain. An unbroken film of water, without droplets, should remain on the walls of the buret.

6.8.6 *Cleaning Pipets.*

a. Routine Procedure—Secure the pipet in a clamp in a vertical position with its tip inserted into a beaker containing water or a detergent. Attach, by means of rubber tubing, the mouth of the pipet to a safety bottle which, in turn, is connected to a vacuum source. Draw up the solution by suction until it almost fills the pipet. Release the suction and allow the solution to drain. Repeat this procedure several times. Change the beaker resting under the tip of the pipet and repeat the procedure with tap water and finally with distilled water. Remove the pipet from the clamp and allow to drain and dry.

b. Cleaning Greasy Pipets—Secure the pipet in a clamp in a vertical position with its tip inserted in a beaker containing warm sodium dichromate-sulfuric acid cleaning solution. Draw up the warm solution until it fills the pipet within an inch from the top, using the procedure described earlier.

Pinch the rubber tubing with a clamp and allow the solution to remain in the pipet for about 10 min. Then remove the pinchclamp and discharge the cleaning solution. Rinse the pipet thoroughly with tap water to remove any trace of the solution and finally rinse with distilled water. Allow the pipet to drain and dry. To dry the pipet quickly, rinse it several times with ethyl alcohol, after draining the distilled water, and then rinse once with ether. Draw dry air through the pipet until dry.

7. Reagent Water

Wet chemical analysis implies that water is used in the procedure. The water used in the reagents and for effecting solubilization, dilution, and transfer is called reagent water. Assumption that the reagent water contributes insignificantly to the bias of a determination is warranted only when the method of test is not sensitive to the impurity level in the water.

7.1 PREPARATION. Preferred practice is to minimize the probability of error by preparing and storing high purity water. The preparation frequently begins with single distillation from conventional apparatus. Condensate from a conventional steam boiler is suitable if the conductivity of this condensed steam is no greater than 20 μmho/cm. A condensate of adequate quality can almost always be obtained by partial condensation. The condensing train for this purpose is arranged in steps, to (*a*) effect partial condensation in the steam pipe, (*b*) dry the steam in a separator (reject the separated condensate which contains the bulk of the more soluble gases and the entrained boiler water), (*c*) effect additional partial condensation of the steam, and (*d*) separate and store this condensate (the bulk of the less soluble gases like carbon dioxide remains in the uncondensed steam).

7.2 DEMINERALIZATION. Additional processing of the distillate or condensate is now usually accomplished by demineralization, the water being passed through a commercially available mixture of anion and cation exchange materials. Assumption of adequate upgrading by this process is unwarranted, especially as ion exchange materials vary from lot to lot, and the effluent in each instance should be discarded until the desired quality is obtained.

The successive steps of distillation and demineralization are both important. The former removes nonvolatile electrolytes and colloidal material, and the latter removes ionizable volatiles such as carbon dioxide and ammonia (which together impart the greater part of the 20 μmho/cm conductivity of the initial condensate).

7.3 FILTRATION. High purity water prepared in this way is usually satisfactory for the preparation of reagents, for the laboratory preparation and dilution of samples, and for the final rinsing of glassware and the other apparatus. It may not be adequate, however, for the determination of trace concentrations of some of the materials extracted from air samples. Organic contaminants, in a range from 1 to 5 ppb, are usually imparted to the water by its contact with the ion-exchange materials. For some determinations, such as tests for airborne micro-organisms, the "pure" water is further purified by special low porosity filtration. Examples of special purpose "purifiers" are the sized Gelman membrane filters and the Barnstead Organic Removal Cartridge.

7.4 PROTECTION AGAINST CONTAMINATION. Reagent water should at all times be protected from atmospheric contamination. Replacement air should be passed through a vent guard (for example, a drying tube filled with equal parts of 8 to 20-mesh soda lime, oxalic acid, and 4 to 8-mesh calcium chloride, each product being separated from the other by a glass wool plug). Sample containers and tubing should be made of material that has been proven to be resistant to even minor solvation by, or reaction with, the water. Such materials usually are TFE-fluorocarbon, block tin, quartz, or polyethylene—in that order of preference.

The stored water is usually considered suitable until it fails to pass the maximum electrical conductivity test (0.1 μmho/cm at 25 C). It may also be checked occasionally for its consumption of potassium permanganate, the requirement being that the permanganate color must persist for at least one hour after 0.20 ml of $KMnO_4$ solution (0.316 g/l) is added to a mixture of 500 ml of the reagent water and 1 ml of reagent grade sulfuric acid (conc) in a stoppered bottle of chemically resistant glass.

7.5 BLANK DETERMINATIONS. Despite all of these preparations and precautions, minute concentrations of impurities in reagent water may still affect the precision and accuracy of some determinations. Pretesting is a reasonable precaution. The usual practice is to run a "blank" to disclose whether or not the water contains a detectable quantity of the material for which the analysis is to be made. This "blank" is often designed to include the reactive impurities in the reagents as well as those in the water.

Occasionally, the analyst will purposely contaminate "pure" reagent water before it is used. For example, water for the preparation of the reagents used to calibrate conductivity cells is deliberately equilibrated with the atmosphere in which the calibration will be made. This tends to stabilize the content of dissolved gases, such as carbon dioxide, that affect conductivity.

When the analyst is seeking especially small concentrations of material in samples of air, he must necessarily be especially cautious in determining and correcting for any bias introduced because the water is not perfectly pure. He will be concerned, for example, with the magnitude of the "blank" in relation to the lower limit of detection of the method. He will determine the range of deviation in the "blank" to be expected from random error, and design his analysis so that the "blank" is a proper correction on the quantity he finds in the air sample. Above all, he will start with reagent water that is of a purity consistent with the projected testing and with its practical utility.

8. Common Acid, Alkali, and Other Standard Solutions

8.1 INTRODUCTION. Reagent grade chemicals shall be used in preparing and standardizing all solutions. Reagents shall conform to the current specifications of the Committee on Analytical Reagents of the American Chemical Society, where such specifications are available.* Other grades may be used, provided it is first ascertained that the reagent is of sufficiently high purity to permit its use without lessening the accuracy of the determination.

The National Bureau of Standards offers for sale certified standard samples of arsenic trioxide, benzoic acid, potassium hydrogen phthalate, potassium dichromate, and sodium oxalate. These samples

*"Reagent Chemical, American Chemical Society Specifications", Amer. Chemical Soc., Washington, D.C.

of commercially available primary standards are to be used on standardizing the volumetric solutions.

Directions are given for the preparation of the most commonly used concentrations of the standard volumetric solutions. Stronger or weaker solutions are prepared and standardized in the same general manner as described, using proportionate amounts of the reagents. Similarly, if quantities larger than 1 liter are to be prepared, proportionate amounts of the reagents should be used.

When quantities of solution larger than 1 or 2 liters are prepared, special problems are encountered in being sure that they are well mixed before being standardized. While blade stirrers with glass or metal shafts are suitable for many solutions, they are not suitable in every case. In those cases where contact of a glass or metal stirrer with the solution would be undesirable it may be possible to use a sealed polyolefin coated stirrer. In those cases where only contact of the solution with metal must be avoided, the solution can be mixed by inserting a fritted glass gas dispersion tube to the bottom of the container and bubbling nitrogen through the solution for 1 or 2 hr.

In order to make a solution of exact normality from a chemical which cannot be measured as a primary standard, a relatively concentrated stock solution may first be prepared, and then an exact dilution of this may be made to the desired strength. Another method is to make a solution of slightly stronger concentration than desired, standardize, and then make suitable adjustments in the concentration. Alternatively, the solution may be used as first standardized, with appropriate modification of the factor used in the calculation. This alternative procedure is especially useful in the case of a solution which slowly changes strength—for example, thiosulfate solution, which must be restandardized at frequent intervals. Often, however, adjustment to the exact normality specified is desirable when a laboratory runs a large number of determinations with one standard solution.

As long as the normality of a standard solution does not result in a titration volume so small as to preclude accurate measurement or so large as to cause abnormal dilution of the reaction mixture, and as long as the solution is properly standardized and the calculations are properly made, the determinations can be considered to be in accord with the instructions in this manual.

If a solution of exact normality is to be prepared by dissolving a weighed amount of a primary standard or by dilution of a stronger solution, it is necessary that the solution be brought up to exact volume in a volumetric flask.

The stock and standard solutions prescribed for the colorimetric determinations in the chemical sections of this manual should also be accurately prepared in volumetric flasks. Where the concentration does not need to be exact, it is often easier to mix the concentrated solution or the solid with measured amounts of water, using graduated cylinders for these measurements. There is usually a significant change of volume when strong solutions are mixed, so that the total volume is less than the sum of the volumes used. For approximate dilutions, the volume changes are negligible when concentrations of 6N or less are diluted.

Very thorough and complete mixing is essential when making dilutions. One of the commonest sources of error in analyses using standard solutions diluted in volumetric flasks is failure to attain complete mixing.

Glass containers are suitable for the storage of most of the standard solutions, although the use of polyolefin containers is recommended for alkali solutions.

When large quantities of solutions are prepared and standardized, it is necessary to provide protection against changes in normality due to absorption of gases or water vapor from the laboratory air. As volumes of solution are withdrawn from the container, the replacement air should be passed through a drying tube filled with equal parts of 8 to 20-mesh soda lime, oxalic acid, and 4 to 8-mesh anhydrous calcium chloride, each product being separated from the other by a glass wool plug.

8.2 SODIUM HYDROXIDE, 0.02 TO 1.0 N.

8.2.1 *Preparation of 50% NaOH Solution and of other Standard NaOH Solutions*. Dissolve 162 g of sodium hydroxide (NaOH) in 150 ml of carbon dioxide-free water. Cool the solution to 25 C and filter through a Gooch crucible, hardened filter paper, or other suitable medium. Alternatively, commercial 50% NaOH solution may be used.

To prepare a 0.1 N solution, dilute 5.45 ml of the clear solution to 1 liter with carbon dioxide-free water, mix well, and store in a tight polyolefin container.

For other normalities of NaOH solution, use the requirements given in Table 8:I.

Table 8:I Sodium Hydroxide Dilution Requirements

Desired Normality	Grams of NaOH required per 1 liter of Solution	Volume of 50 % NaOH Solution (25 C) Required per 1 liter of Solution, ml
0.02	0.8	1.09
0.04	1.6	2.18
0.05	2.0	2.73
0.1	4.0	5.45
0.2	8.0	10.90
0.25	10.0	13.63
0.5	20.0	27.25
1.0	40.0	54.54

8.2.2 *Standardization*. Crush 10 to 20 g of primary standard potassium hydrogen phthalate† ($KHC_8H_4O_4$) to 100-mesh fineness, and dry in a glass container at 120 C for 2 hrs. Stopper the container and cool in a desiccator.

To standardize a 0.1 N solution, weigh accurately 0.95 ± 0.05 g of the dried $KHC_8H_4O_4$, and transfer to a 500-ml conical flask. Add 100 ml of carbon dioxide-free water, stir gently to dissolve the sample, add 3 drops of a 1.0% solution of phenophthalein in alcohol, and titrate with NaOH solution to a color that matches that of an end point color standard.

†A primary standard grade of potassium hydrogen phthalate ($KHC_8H_4O_4$) is available from the Office of Standard Reference Materials, National Bureau of Standards, Washington, D.C. 20234.

The weights of dried $KHC_8H_4O_4$ suitable for other normalities of NaOH solution are given in Table 8:II.

Table 8:II. Weights of Dried Potassium Hydrogen Phthalate

Normality of Solution	Weight of Dried $KHC_8H_4O_4$ to be used, g*
0.02	0.19 ± 0.005
0.04	0.38 ± 0.005
0.05	0.47 ± 0.005
0.1	0.90 ± 0.005
0.2	1.90 ± 0.05
0.25	2.35 ± 0.05
0.5	4.75 ± 0.05
1.0	9.00 ± 0.05

*The listed weights are for use when a 50-ml buret is to be used. If a 100-ml buret is to be used, the weights should be doubled.

8.2.3 *pH 8.6 End Point Color Standard*. Mix 25 ml of a solution 0.2 M in boric acid (H_3BO_3) and 0.2 M in potassium chloride (KCl), with 6 ml of 0.2 M NaOH solution, add 3 drops of a 1.0% solution of phenolphthalein in alcohol, and dilute to 100 ml with carbon dioxide-free water.

8.2.4. *Calculation*. Calculate the normality of the NaOH solution, as follows:

$$A = \frac{B}{0.20423 \times C}$$

where:

A = normality of the NaOH solution,
B = grams of $KHC_8H_4O_4$ used, and
C = milliliters of NaOH solution consumed.

8.2.5 *Stability*. The use of polyolefin containers eliminates some of the difficulties attendant upon the use of glass containers, and their use is recommended. Should glass containers be used, the solution must be standardized frequently if there is evidence of action on the glass container, or if insoluble matter appears in the solution.

8.3 HYDROCHLORIC ACID, 0.02 TO 1.0 N.

8.3.1 *Preparation.* To prepare a 0.1 N solution, measure 8.3 ml of conc hydrochloric acid (HCl, sp gr 1.19) into a graduated cylinder and transfer it to a 1-l volumetric flask. Dilute to the mark with water, mix well, and store in a tightly closed glass container.

For other normalities of HCl solution, use the requirements given in Table 8:III.

Table 8:III. Hydrochloric Acid Dilution Requirements

Desired Normality	Volume of HCl to be diluted to 1 liter, ml
0.02	1.66
0.04	3.32
0.1	8.3
0.2	16.6
0.5	41.5
1.0	83.0

8.3.2 *Standardization.* Transfer 2 to 4 g of anhydrous sodium carbonate (Na_2CO_3) to a platinum dish or crucible, and dry at 250 C for 4 hr. Cool in a desiccator.

To standardize a 0.1 N solution, weigh accurately 0.22 ± 0.01 g of the dried Na_2CO_3, and transfer to a 500-ml conical flask. Add 50 ml of water, swirl to dissolve the carbonate, and add 2 drops of a 0.1% solution of methyl red in alcohol. Titrate with the HCl solution to the first appearance of a red color, and boil the solution carefully, to avoid loss, until the color is discharged. Cool to room temperature, and continue the titration, alternating the addition of HCl solution and the boiling and cooling to the first appearance of a faint red color that is not discharged on further heating.

The weights of dried Na_2CO_3 suitable for other normalities of HCl solution are given in Table 8:IV.

Table 8:IV. Weights of Dried Sodium Carbonate

Normality of Solution	Weight of dried Na_2CO_3 to be Used, g
0.02	0.088 ± 0.001*
0.04	0.176 ± 0.001*
0.1	0.22 ± 0.01†
0.2	0.44 ± 0.01†
0.5	1.10 ± 0.01†
1.0	2.20 ± 0.01†

*A 100-ml buret should be used for this standardization.

†The listed weights are for use when a 50 ml buret is used. If a 100-ml buret is to be used, the weights should be doubled.

8.3.3 *Calculation.* Calculate the normality of the HCl solution, as follows:

$$A = \frac{B}{0.053 \times C}$$

Where:

A = normality of the HCl solution,
B = grams of Na_2CO_3 used, and
C = milliliters of HCl solution consumed.

8.3.4 *Stability.* Restandardize monthly.

8.4 SULFURIC ACID, 0.02 TO 1.0 N.

8.4.1 *Preparation.* To prepare a 0.1 N solution, measure 3.0 ml of conc sulfuric acid (H_2SO_4, sp gr 1.84) into a graduated cylinder and slowly add it to one half the desired volume of water in a 600-ml beaker. Rinse the cylinder into the beaker with water. Mix the acid-water mixture, allow it to cool, and transfer to a 1-l volumetric flask. Dilute to the mark with water, mix well, and store in a tightly closed glass container.

For other normalities of the H_2SO_4 solution, use the requirements given in Table 8: V.

Table 8:V. Sulfuric Acid Dilution Requirements

Desired Normality	Volume of H_2SO_4 to be Diluted to 1 liter, ml
0.02	0.60
0.1	3.0
0.2	6.0
0.5	15.0
1.0	30.0

8.4.2 *Standardization.* Transfer 2 to 4 g of anhydrous sodium carbonate (Na_2CO_3) to a platinum dish or crucible, and dry at 25 C for 4 hr. Cool in a desiccator.

For standardization of a 0.1 N solution, weigh accurately 0.22 ± 0.01 g of the dried Na_2CO_3 and transfer to a 500-ml conical flask. Add 50 ml of water, swirl to dissolve the Na_2CO_3, and add 2 drops of a 0.1% solution of methyl red in alcohol. Titrate with the H_2SO_4 solution to the first appearance of a red color, and boil the solution carefully, to avoid loss, until the color is discharged. Cool to room temperature and continue the titration alternating the addition of H_2SO_4 solution and the boiling and cooling, to the first appearance of a faint red color that is not discharged on further heating.

The weights of dried Na_2CO_3 suitable for other normalities of H_2SO_4 solution are given in Table 8:IV.

8.4.3 *Calculation.* Calculate the normality of the H_2SO_4 solution, as follows:

$$A = \frac{B}{0.053 \times C}$$

Where:

A= normality of the H_2SO_4 solution,
B=grams of Na_2CO_3, and
C=milliliters of H_2SO_4 solution consumed.

8.4.4 *Stability.* Restandardize monthly.

8.5 IODINE (0.1 N).

8.5.1 *Preparation.* Transfer 12.7 g of iodine and 60 g of potassium iodide (KI) to an 800-ml beaker, add 30 ml of water, and stir until solution is complete. Dilute with water to 500 ml, and filter through a sintered-glass filter. Wash the filter with about 15 ml of water, transfer the combined filtrate and washing to a 1-l volumetric flask, dilute to the mark with water, and mix. Store the solution in a glass-stoppered, amber-glass bottle in a cool place.

8.5.2 *Standardization.* Transfer 1 g of arsenic trioxide (As_2O_3) to a platinum dish, and dry at 105 C for 1 hr. Cool in a desiccator. Weigh accurately 0.20 ± 0.01 g of the dried As_2O_3 and transfer to a 500-ml conical flask. Add 10 ml of sodium hydroxide solution (NaOH, 40 g/l), and swirl to dissolve. When solution is complete, add 100 ml of water and 10 ml of sulfuric acid (H_2SO_4, 1:35), and mix. Slowly add sodium bicarbonate ($NaHCO_3$) until effervescence ceases, add 2 g of $NAHCO_3$ in excess, and stir until dissolved. Add 2 ml of starch solution (10 g/l) and titrate with the iodine solution to the first permanent blue color.

8.5.3 *Calculation.* Calculate the normality of the iodine solution, as follows:

$$A = \frac{B}{0.049455 \times C}$$

Where:

A= normality of the iodine solution,
B= grams of As_2O_3 used, and
C= milliliters of iodine solution required for titration of the solution.

8.5.4 *Stability.* Restandardize sealed bottles monthly. Restandardize open bottles weekly.

8.6 SODIUM THIOSULFATE (0.1 N).

8.6.1 *Preparation.* Dissolve 25 g of sodium thiosulfate ($Na_2S_2O_3 \cdot 5H_2O$) in 500 ml of freshly boiled and cooled water, and add 0.11 g of sodium carbonate (Na_2CO_3). Dilute to 1 l with freshly boiled and cooled water, and let stand for 24 hr. Store the solution in a tightly-closed glass bottle.

8.6.2 *Standardization.* Pulverize 2 g of potassium dichromate ($K_2Cr_2O_7$) transfer to a platinum dish, and dry at 120 C for

4 hr. Cool in a desiccator. Weigh accurately 0.21 ± 0.01 g of the dried $K_2Cr_2O_7$, and transfer to a 500-ml glass-stoppered conical flask. Add 100 ml of water, swirl to dissolve, remove the stopper, and quickly add 3 g of potassium iodide (KI), 2 g of sodium bicarbonate ($NaHCO_3$), and 5 ml of hydrochloric acid (HCl). Stopper the flask quickly, swirl to ensure mixing, and let stand in the dark for 10 min. Rinse the stopper and inner walls of the flask with water and titrate with the $Na_2S_2O_3$ solution until the solution has only a faint yellow color. Add 2 ml of starch solution (10g/l), and continue the titration to the diappearance of the blue color.

8.6.3 *Calculation.* Calculate the normality of the $Na_2S_2O_3$ solution, as follows:

$$A = \frac{B}{0.04904 \times C}$$

Where:

A = normality of the $Na_2S_2O_3$ solution,
B = grams of $K_2Cr_2O_7$ used, and
C = milliliters of $Na_2S_2O_3$ solution required for titration of the solution.

8.6.4 *Stability.* Restandardize weekly.

8.7 AMMONIUM THIOCYANATE (0.1 N).

8.7.1 *Preparation.* Transfer 7.8 g of ammonium thiocyanate (NH_4SCN) to a flask, add 100 ml of water, and swirl to dissolve the NH_4SCN. When solution is complete, filter through a Gooch crucible, hardened filter paper, or other suitable medium. Dilute the clear filtrate to 1 l with water and mix. Store the solution in a tightly-stoppered glass bottle.

8.7.2 *Standardization.* Measure accurately about 40 ml of freshly standardized 0.1 N silver nitrate ($AgNO_3$), and transfer to a 250-ml conical flask. Add 50 ml of water, swirl to mix the solution, and add 2 ml of nitric acid (HNO_3) and 1 ml of ferric ammonium sulfate solution ($FeNH_4(SO_4)_2$, 80 g/liter). Titrate the $AgNO_3$ solution with the NH_4SCN solution until the first permanent reddish-brown color appears and persists after vigorous shaking for 1 min.

8.7.3 *Calculation.* Calculate the normality of the NH_4SCN solution, as follows:

$$A = \frac{B \times C}{D}$$

Where:

A = normality of the NH_4SCN solution.
B = milliliters of $AgNO_3$ solution used,
C = normality of the $AgNO_3$ solution, and
D = milliliters of NH_4SCN solution required for titration of the solution.

8.7.4 *Stability.* Restandardize monthly.

8.8 POTASSIUM DICHROMATE (0.1 N).

8.8.1 *Preparation.* Transfer 6 g of potassium dichromate ($K_2Cr_2O_7$) to a platinum dish and dry at 120 C for 4 hr. Cool in a desiccator. Place 4.9 g of the dried $K_2Cr_2O_7$ in a 1-l volumetric flask, and add 100 ml of water. Swirl to dissolve and when solution is complete, dilute to the mark with water and mix. Store the solution in a glass-stoppered bottle.

8.8.2 *Standardization.* Place 40 ml of water in a 250-ml glass-stoppered conical flask, and add 40 ml, accurately measured, of the $K_2Cr_2O_7$ solution. Stopper the flask, swirl to mix, remove the stopper, and add 3 g of potassium iodide (KI), 2 g of sodium bicarbonate ($NaHCO_3$) and 5 ml of hydrochloric acid (HCl). Stopper the flask quickly, swirl to ensure mixing, and let stand in the dark for 10 min. Rinse the stopper and inner walls of the flask with water and titrate with freshly standardized sodium thiosulfate solution ($Na_2S_2O_3$) until the solution has only a faint yellow color. Add 2 ml of starch solution (10 g/l), and continue the titration to the disappearance of the blue color.

8.8.3 *Calculation.* Calculate the normality of the $K_2Cr_2O_7$ solution, as follows:

$$A = \frac{B \times C}{D}$$

Where:
A = normality of the $K_2Cr_2O_7$ solution,
B = Milliliters of $Na_2S_2O_3$ solution required for titration of the solution,
C = normality of the $Na_2S_2O_3$ solution, and
D = milliliters of $K_2Cr_2O_7$ solution used.

8.8.4 *Stability*. Restandardize monthly.

8.9 SILVER NITRATE (0.1 N).

8.9.1 *Preparation*. Dry 17.5 g of silver nitrate ($AgNO_3$) at 105 C for 1 hr. Cool in a desiccator. Transfer 16.99 g of the dried $AgNO_3$ to a 1-l volumetric flask. Add 500 ml of water, swirl to dissolve the $AgNO_3$, dilute to the mark with water, and mix. Store the solution in a tightly-stoppered amber-glass bottle.

8.9.2 *Standardization*. Transfer 0.3 g of sodium chloride (NaCl) to a platinum dish and dry at 105 C for 2 hr. Cool in a desiccator. Weigh accurately 0.28 ± 0.01 g of the dried NaCl and transfer to a 250-ml glass stoppered conical flask. Add 25 ml of water, swirl to dissolve the NaCl, and add 2 ml of nitric acid (HNO_3). Add, from a volumetric pipet, 50 ml of the $AgNO_3$ solution, mix, and add 1 ml of ferric ammonium sulfate solution ($FeNH_4(SO_4)_2$, 80 g/l) and 5 ml of nitrobenzene (CAUTION‡). Stopper the flask and shake vigorously to coagulate the precipitate. Rinse the stopper into the flask with a few milliliters of water and titrate the excess of $AgNO_3$ with ammonium thiocyanate solution (NH_4SCN) until the first permanent reddish-brown color appears and persists after vigorous shaking for 1 min. Designate the volume of NH_4SCN solution required for the titration as Volume I.

Transfer 50 ml of the $AgNO_3$ solution to a clean, dry, 250-ml glass-stoppered conical flask. Add 25 ml of water, 2 ml of HNO_3, 1 ml of $FeNH_4(SO_4)_2$ solution (80g/l), stopper the flask, and shake vigorously. Rinse the stopper into the flask with a few milliliters of water and titrate the $AgNO_3$ solution with NH_4SCN solution until the first permanent reddish-brown color appears and persists after vigorous shaking for 1 min. Designate the volume of NH_4SCN solution consumed as Volume II.

Measure accurately, from either a buret or a volumetric pipet, 2.0 ml of the $AgNO_3$ solution, designate the exact volume as Volume III, and transfer to a 100-ml, glass-stoppered conical flask. Add 25 ml of water, 2 ml of HNO_3, 1 ml of $FeNH_4(SO_4)_2$ solution (80 g/l), and 5 ml of nitrobenzene, stopper the flask, and shake vigorously. Rinse the stopper into the flask with a few milliliters of water and titrate the $AgNO_3$ solution with NH_4SCN solution until the first permanent reddish-brown color appears and persists after vigorous shaking for 1 min. Designate the volume of NH_4SCN solution consumed as Volume IV.§

8.9.3 *Calculation*. Calculate the normality of the $AgNO_3$ solution, as follows:

$$A = \frac{B}{0.05845 \times (C-D)}$$

Where:
A = normality of the $AgNO_3$ solution,
B = grams of NaCl used,
C = volume of $AgNO_3$ solution consumed by the total chloride =

$$50 - (\text{Volume I} \times \frac{50}{\text{Volume II}})$$

and
D = volume of $AgNO_3$ solution consumed by any chloride ion in the nitrobenzene = Volume III −

$$(\text{Volume IV} \times \frac{50}{\text{Volume II}})$$

‡**Caution**—nitrobenzene, used in this section, is extremely hazardous when absorbed through the skin or when its vapor is inhaled. Such exposure may cause cyanosis; prolonged exposure may cause anemia. Do not get in eyes, on skin, or on clothing. Avoid breathing vapor. Use only with adequate ventilation.

§The ammonium thiocyanate titrate used in the three titrations must be from the same, well-mixed solution. The nitrobenzene used in each titration must also be from the same, well-mixed container.

8.9.4 *Stability*. Restandardize monthly.

8.10 POTASSIUM PERMANGANATE (0.1 N).

8.10.1 *Preparation*. Dissolve 3.2 g of potassium permanganate ($KMnO_4$) in 100 ml of water, and dilute the solution with water to 1 l. Allow the solution to stand in the dark for two weeks and then filter through a fine-porosity sintered-glass crucible. **Do not wash the filter.** Store the solution in glass-stoppered, amber-colored glass bottles.

NOTE 1. Do not permit the filtered solution to come into contact with paper, rubber, or other organic material.

8.10.2 *Standardization*. Transfer 2 g of sodium oxalate ($Na_2C_2O_4$) to a platinum dish and dry at 105 C for 1 hr. Cool in a desiccator. Weigh accurately 0.30 ± 0.01 g of the dried $Na_2C_2O_4$ and transfer to a 500-ml glass container. Add 250 ml of sulfuric acid (H_2SO_4, 1:19) that was previously boiled for 10 to 15 min and then cooled to 27 ± 3 C, and stir until the sample is dissolved. Add 39 ml of the $KMnO_4$ solution at a rate of 30 ± 5 ml per min, while stirring slowly, and let stand for about 45 sec until the pink color disappears. Heat the solution to 60 C, and complete the titration by adding $KMnO_4$ solution until a faint pink color persists for 30 sec. Add the final 0.5 to 1.0 ml dropwise, and give the solution time to decolorize before adding the next drop.

Carry out a blank determination on a second 250-ml portion of the H_2SO_4 (1:19), and make sure that the pink color at the end point matches that of the standardization solution. Correct the sample titration volume as shown to be necessary.

NOTE 2. The specified 0.30 g sample of $Na_2C_2O_4$ should consume about 44.8 ml of 0.1 N $KMnO_4$.

NOTE 3. If the pink color of the solution persists more than 45 sec after the addition of the first 39 ml of $KMnO_4$ solution is complete, discard the solution and start over with a fresh solution of the $Na_2C_2O_4$, but add less of the $KMnO_4$ solution.

NOTE 4. The blank correction usually amounts to 0.03 to 0.05 ml.

8.10.3 *Calculation*. Calculate the normality of the $KMnO_4$ solution as follows:

$$A = \frac{B}{0.06701 \times C}$$

Where:

A = normality of the $KMnO_4$ solution,
B = grams of $Na_2C_2O_4$ used and
C = milliliters of $KMnO_4$ solution required for titration of the solution.

8.10.4 *Stability*. Restandardize weekly.

9. Recovery and Internal Standard Procedures

9.1 RECOVERY PROCEDURE. A recovery procedure does not enable the analyst to apply any correction factor to the results of an analysis, however, it does give him some basis for judging the applicability of a particular method of analysis to a particular sample.

A recovery may be performed at the same time as the determination itself. Of course, recoveries would not be run on a routine basis with samples whose general composition is known or when using a method whose applicability to the sample is well established. Recovery methods are to be regarded as tools to remove doubt about the applicability of a method to a sample.

In brief, the recovery procedure involves applying the analytical method to a reagent blank, to a series of known standards covering the expected range of concentration of the sample, to the sample itself, in at least duplicate run, and to the recovery samples, prepared by adding known quantities of the substance sought to separate portions of the sample itself, each portion equal to the size of sample taken for the run. The substance sought should be added in sufficient quantity to overcome the limits of error of the analytical method, but without causing the total in the sample to exceed the range of the known standards used.

The results are first corrected by subtracting the reagent blank from each of the other determined values. The resulting known standards are then graphically represented. From this graph the amount of sought substance in the sample alone is determined. This value is then subtracted from each of the determinations consisting of sample plus known added substance. The resulting amount of substance divided by the known amount originally added and multiplied by 100 gives the percentage recovery.

The procedure outlined above may be applied to a colorimetric, flame photometric, or fluorimetric analysis. It may also be applied in a little more simple form to titrimetric, gravimetric, and other types of analyses.

Rigid rules concerning the percentage recoveries required for acceptance of results of analyses for a given sample and method cannot be stipulated. Recoveries of substances in the range of the sensitivity of the method may, of course, be very high or very low and approach a value nearer to 100 per cent recovery as the error of the method becomes small with respect to the magnitude of the amount of substance added. In general, intricate and exacting procedures for trace substances which have inherent errors due to their complexity may give recoveries that would be considered very poor and yet, from the practical viewpoint of usefulness of the result, may be quite acceptable. Poor results may reflect either interferences present in the sample or real inadequacy of the method of analysis in the range in which it is being used.

It must be stressed, however, that the judicious use of recovery methods for the evaluation of analytical procedures and their applicability to particular samples is an invaluable aid to the analyst in both routine and research investigations.

9.2 INTERNAL STANDARD PROCEDURE. The internal standard technique is used primarily for emission spectrograph and polarographic procedures. This procedure enables the analyst to compensate for electronic and mechanical fluctuations within the instrument.

In brief, the internal standard method involves the addition to the sample of known amounts of a substance which will respond to the instrument in a similar manner as the contaminant being analyzed. The ratio of the measurement of the internal standard to the measurement of the contaminant is the value used to determine concentration of contaminant present in the sample. Any changes in conditions during analyses will affect the internal standard and the contaminant the same and so will compensate for such changes. The internal standard should be of similar chemical reactivity to the contaminant, approximately the same concentration as anticipated for the contaminant, and as pure a substance as possible.

10. Interferences

10.1 INTRODUCTION. The term interference covers various effects of dissolved or suspended substances upon analytical procedures. The known interferences and related information have been included in the specific methods of analyses. Many of the methods in this manual have been developed specifically to minimize common interferences. The purpose of this section is to offer guidelines for detecting the presence of interfering substances and overcoming their effects when encountered.

In analytical chemistry, one must be prepared to use an alternate procedure if one method would be less affected than another due to the presence of the interfering substance. The most expedient method, however, for overcoming a suspected interference is attained by a smaller initial aliquot. Thus, the effect of the interfering element is diminished or eliminated through dilution of the original sample. The level of the parameter being measured is likewise reduced so care must be taken to insure an aliquot containing more than the minimum detectable concentration. If the data display a consistently increasing or decreasing pattern by dilution, interference is indicated. Dilution of final reaction volume must never be used in overcoming interferences. Certain reagents are prepared to react with a limited amount of the test element. If this limit is exceeded, a portion of the test element goes undetected. Then the analyst must repeat the entire test procedure with a smaller initial aliquot.

10.2 INTERFERENCE REACTIONS. Essentially, three reactions may occur to produce an interference:

a. An interfering substance may react with the reagents in the same manner as the element being sought. (Positive interference);
b. an interfering substance may react with the element being sought to prevent complete isolation. (Negative interference); or
c. an interfering substance may combine with the reagents to prevent further reaction with the element being sought. (Negative interference).

These reactions will produce either high or low results. An estimate of the magnitude of an interference may be obtained by the internal standard procedure. After establishing a calibration graph for the appropriate range and performing an initial analysis of the sample, this procedure may be employed.

10.3 INTERNAL STANDARD PROCEDURE.
10.3.1 Select a standard, which when added to the unknown sample, will produce a final result approximately midrange of the standard calibration graph. Example: Calibration graph: 0 to 100 ppb.

Sample analysis: 10 ppb.
Select a standard concentration, which when added to 10 ppb, will produce a total concentration of 50 ppb; *i.e.*, 40 ppb.

10.3.2 Analyze the prepared solution (sample + standard) and read the result from the graph.
10.3.3 Subtract from this result the original sample analysis.
10.3.4 The difference divided by the amount originally added and multiplied by 100 gives the percentage recovery. Example:

Sample: 10 ppb
Sample + Standard:
10 ppb + 40 ppb = 50 ppb
Sample + Standard as determined:
56 ppb
Subtract original determination:
56 − 10 = 46 ppb
Recovery equals

$$\frac{46}{40} \times 100 = 115 \text{ per cent}$$

Thus, a positive interference is present. If the result is below 100%, a negative interference is indicated. By establishing the direction of interference, the nature of the reaction taking place may be determined and appropriate steps taken for eliminating the effect. A positive error indicates the interference is reacting like the test element, and the possibility of isolation by a pH adjustment should be considered. This procedure should be applied in triplicate to both the sample and the sample plus standard and must include a reagent blank.

10.4 REMOVAL OF INTERFERENCE. If further treatment of the sample for removal of the interfering substance is necessary, several approaches are available to the analyst. These may be physical or chemical or a combination of both, such as:

a. Distillation of the sample leaving the interference behind;
b. removal of the interference by ion exchange resins;
c. addition of complexing agents;
d. extraction into organic solvents;
e. ashing;
f. pH adjustment;
g. different reaction rate;
h. change temperature.

These treatment methods serve as examples only, since a thorough review of the literature must be undertaken before proceeding to undertake a change in an established method.

10.5 SOURCES OF INTERFERENCE. In the preceding discussion, the analyst was given instructions in detecting an interference in quantitative analysis. The question of the source of the interference must be identified and corrected, if possible. The origin of an interference may be found in the following situations:

a. Present at the sampling site;
b. imparted during sample collection;
c. developed in sample storage; or
d. imparted or developed in laboratory analysis.

These four source situations may normally be avoided in applying proper sampling and laboratory techniques. These techniques are discussed in this section.

The sources may be a result of physical, chemical, or biological phenomena. The possibility of interaction (physical upon chemical, etc.) should not be overlooked. These three classifications of sources of errors are further subdivided to assist the analyst in search of the action producing the effect.

10.6 PHYSICAL SOURCES.

10.6.1 *Heat.*

a. Effect on chemical equilibria.
b. Promote side reactions at elevated temperatures—decomposition.
c. Affect rate of reaction—temperature coefficient correction. Serious errors may be encountered at low temperatures.

10.6.2 *Light.*

a. Effect of visible spectrum—accelerate photooxidation.
b. Effect of UV—Yellowing of PFIB reagent—positive error.
c. Fading of detector reagent colors (Lead and Mercury dithizonates)—negative error.

10.6.3 *Humidity.*

a. Reaction of H_2O with contaminant—TDI, $COCl_2$.
b. Reaction of H_2O with the collecting substrate.
c. Dilution of the collecting substrate.

10.6.4 *Time.*

a. Deterioration rate after sample collection, again half-life must be determined.
b. Fading of the developed color after maximum density attainment.
c. Reagents often deteriorate with time.

10.6.5 *Chemical Contamination.*

a. Contamination of sampling equipment during use.
(1) Collection in high concentration locations—T.E.L.
(2) Extraneous dust and dirt.
(3) Failure to keep collection equipment and reagents stoppered.
(4) Cross contamination in recharging collector or changing filters, especially in the field.

10.7 CHEMICAL SOURCES.

10.7.1 *pH Control in Aqueous Systems.*

a. pH sensitive collection.
b. pH sensitive color responses.
c. Buffered versus unbuffered solutions.

10.7.2 *Chemical Contamination.*

a. Impure reagents—Sulfates in gum arabic used to stabilize $BaSO_4$ colloid.
b. Failure to run blanks and controls.

10.7.3 *Interferences.*

a. *Negative*—NH_3 in Hg detector tube—redox cancellations.
b. Positive
(1) Different shade or color produced—Dithizone + oxidizing agents.
(2) Reaction the same—increased by interference.
a. Ozone—Nitrogen Oxides System.
b. Color intensified by interference.
(3) Electrolytes in conductivity and pH meters.

10.7.4 *Sensitivity.*

a. Concentration effects.
(1) Adjust to obtain maximum effect—maximum color density. Reactant ratios often nonstoichiometric.
(2) Beers law may not be followed at low concentrations.
(3) Background "noise" in electrolytic instruments.
(4) Effect of reagent concentrations on reaction rate.

(5) Effect of reagent concentrations on reaction equilibrium—CH_2O + Fuchsin reagent.

(6) Attempting to operate outside the optimum range for which the method was developed.

b. Color former structure versus extinction coefficient—Phenanthroline series for iron detection.

c. Length of stain tubes.

(1) Attempts to increase sensitivity by repeated sample aliquots—Benzene tube requires recalibration.

(2) Change in sampling rate without recalibration may produce serious errors.

10.7.5 *Catalytic Effects.*

a. Decomposition on contact with tubes and containers—ozone on metals—H_2O_2 + Heavy Metal.

b. Promotion of side reactions.

c. Failure to react in absence of catalyst.

d. Surface reactions—porous glass bubblers.

e. Inhibition of reactions.

10.7.6 *Half-life in the Ambient Atmosphere.*

a. Effect of light, H_2O vapor, and coexisting contaminants.

(1) T.E.L.
(2) TDI
(3) $COCl_2$
(4) Nitrogen Oxides
(5) Carcinogenic Hydrocarbons
(6) Polymerization
(7) Carbon Monoxide

10.8 BIOLOGICAL SOURCES.

10.8.1 *Algal Growth.*

10.8.2 *Insects.*

10.8.3 *Waste Products (Animal and Insects).*

10.9 Procedures used herein are usually well established and, when applied to the selected sample, give predictable results. If a result is suspected to be in error, samples of known composition should be analyzed by the method in use and an alternate method. If a method fails on the unknown sample while giving correct results on a known, an interference is indicated. The best way to solve a problem of this kind is by reviewing the literature. Many hours may be saved through a few hours of diligent library searching. Solving such problems depends on the skill and ingenuity of the investigator.

11. Terms and Symbols Relating to Molecular Spectroscopy*

Absorbance, A.—The logarithm to the base 10 of the reciprocal of the transmittance, (T).

$$A = \log_{10}\left(\frac{1}{T}\right) = -\log_{10}T$$

In practice the observed transmittance must be substituted for T. Absorbance expresses the excess absorption over that of a specified reference or standard. It is implied that compensation has been effected for reflectance losses, solvent absorption losses, and refractive effects, if present, and that attenuation by scattering is small compared with attenuation by absorption. Apparent deviations from the absorption laws (see *Absorptivity*) are due to inability to measure exactly the true transmittance or to know the exact concentration or to know the exact concentration of an absorbing substance.

Absorption Band—A region of the absorption spectrum in which the absorbance passes through a maximum.

Absorption Spectrum—A plot of absorbance, or any function of absorbance, against wavelength or any function of wavelength.

Absorptivity, a.—The absorbance divided by the produce of the concentration of the substance and the sample path length, $a = A/bc$. The units of b and c, or of a, shall be specified. The recommended unit for a is liters per gram-centimeter.

Absorptivity, Molar—The product of the absorptivity, a, and the molecular weight of the substance.

*This material is based on definitions adopted by ASTM in Designation E131-68 and is reprinted with the permission of the ASTM.

Analytical Curve—The graphical representation of a relation between some function of radiant power and the concentration or mass of the substance emitting or absorbing it.

Analytical Wavelength—Any wavelength at which an absorbance measurement is made for the purpose of the determination of a constituent of a sample.

Attenuated Total Reflection (ATR)—Reflection which occurs when an absorbing coupling mechanism acts in the process of total internal reflection to make the reflectance less than unity. In this process, if an absorbing sample is placed in contact with the reflecting surface, the reflectance for total internal reflection will be attenuated to some value between zero and unity ($0 < R < 1$) in regions of the spectrum where absorption of the radiant power can take place.

Background—Apparent absorption caused by anything other than the substance for which the analysis is being made.

Baseline—Any line drawn on an absorption spectrum to establish a reference point representing a function of the radiant power incident on a sample at a given wavelength.

Beer's Law—The absorbance of a homogeneous sample containing an absorbing substance is directly proportional to the concentration of the absorbing substance. See also *Absorptivity*.

Bouguer's Law—The absorbance of a homogeneous sample is directly proportional to the thickness of the sample in the optical path. Bouguer's Law is sometimes also known as Lambert's Law.

Concentration, c.—The quantity of the substance contained in a unit quantity of sample. For solution work, the recommended unit of concentration is grams of solute per liter of solution.

Derivative Absorption Spectrum—A plot of rate of change of absorbance or of any function of absorbance with respect to wavelength or any function of wavelength, against wavelength or any function of wavelength.

Difference Absorption Spectrum—A plot of the difference between two absorbances or between any function of two absorbances, against wavelength or any function of wavelength.

Dilution Factor—The ratio of the volume of a diluted solution to the volume of original solution containing the same quantity of solute as the diluted solution.

Filter—A substance that attenuates the radiant power reaching the detector in a definite manner with respect to spectral distribution.

Filter, Neutral—A filter that attenuates the radiant power reaching the detector by the same factor at all wavelengths within a prescribed wavelength region.

Fixed-Angle Internal Reflection Element—An internal reflection element that is designed to be operated at a fixed angle of incidence.

Frequency, ν.—The number of cycles per unit of time. The recommended unit is the hertz (Hz) (one cycle/sec).

Infrared—Pertaining to the region of the electromagnetic spectrum from approximately 0.78 to 300 nm.

Instrument Response Time—The time required for an indicating or detecting device to undergo a defined displacement following an abrupt change in the quantity being measured.

Internal Reflection Element (IRE)—The transparent optical element used in Internal Reflection Spectroscopy for establishing the conditions necessary to obtain the internal reflection spectra of materials. Radiant power is propagated through it by means of internal reflection. The sample material is placed in contact with the reflecting surface or it may be the reflecting surface itself. If only a single reflection takes place from the internal reflection element, the element is said to be a single reflection element; if more than one reflection takes place, the element is said to be a multiple reflection element. When the element has a recognized shape it is identified according to each shape, for example, internal reflection prism, internal reflection hemicylinder, internal reflection plate, internal reflection rod, internal reflection fiber, etc.

Internal Reflection Spectroscopy (IRS)—The technique of recording optical spectra by placing a sample material in contact with a transparent medium of greater refractive index and measuring the reflectance (single or multiple) from the interface, generally at angles of incidence greater than the critical angle.

Isoabsorptive Point—A wavelength at which the absorptivities of two or more substances are equal.

Isosbestic Point—The wavelength at which the absorptivities of two substances, one of which can be converted into the other, are equal.

Linear Dispersion—The derivative, dx/λ where x is the distance along the spectrum, in the plane of the exit slit, and λ is the wavelength.

Molar Absorptivity, E—See *absorptivity, Molar*.

Monochromator—A device or instrument that, with an appropriate energy source, may be used to provide a continuous calibrated series of electromagnetic energy bands of determinable wavelength or frequency range.

Photometer—A device so designed that it furnishes the ratio, or a function of the ratio, of the radiant power of two electromagnetic beams. These two beams may be separated in time, space, or both.

Photometric Linearity—The ability of a photometric system to yield a linear relationship between the radiant power incident on its detector and some measurable quantity provided by the system. In the case of a simple detector-amplifier combination, the relationship is a direct proportionality between incident radiant power and the deflection of a meter needle or recorder pen.

Radiant Energy—Energy transmitted as electromagnetic waves.

Radiant Power, P.—The rate at which energy is transported in a beam of radiant energy.

Reflectance, R.—The ratio of the radiant power reflected by the sample to the radiant power incident on the sample.

Resolution, Spectral—The ratio $\lambda/\Delta\lambda$, where λ is the wavelength of the region being examined, and $\Delta\lambda$ is the separation of two absorption bands that can just be distinguished. Resolution can also be defined as $\nu/\Delta\nu$, where ν and $\Delta\nu$ refer to the wave-number scale.

Sample Path Length, b.—The symbol for sample path length is b. The recommended unit for path length is centimeters.

Spectral Band Width—The wavelength or frequency interval of the radiation leaving the exit slit of a monochromator between limits set at a radiant power level half way between the continuous background and the peak of an emission line or an absorption band of negligible intrinsic width.

Spectral Position—The effective wavelength or wavenumber of an essentially monochromatic beam of radiant energy.

Spectral Slit Width—The mechanical width of the exit slit, divided by the linear dispersion in the exit slit plane.

Spectrograph—An instrument with one slit that uses photography to obtain a record of a spectral range simultaneously. The radiant power passing through the optical system is integrated over time, and the quantity recorded is a function of radiant energy.

Spectrometer—An instrument with an entrance slit and one or more exit slits, with which measurements are made either by scanning the spectral range, point by point, or by simultaneous measurements at several spectral positions. The quantity measured is a function of radiant power.

Spectrophotometer—A spectrometer with associated equipment, so designed that it furnishes the ratio, or a function of the ratio, of the radiant power of two beams as a function of spectral position. The two beams may be separated in time, space, or both.

Spectrum, Internal Reflection—The spectrum obtained by the technique of *Internal Reflection Spectroscopy*. Depending on the angle of incidence the spectrum recorded may qualitatively resemble that obtained by conventional transmission measurements, may resemble the mirror image of the dispersion in the index of refraction, or may resemble some composite of the two.

Standard Sample—A material of definite composition that closely resembles in chemical and physical nature the materials with which the analyst expects to deal, and which is employed for calibration.

Stray Radiant Energy—All radiant energy that reaches the detector at wavelengths that do not correspond to the spectral position under consideration.

Transmittance—The ratio of radiant power transmitted by the sample to the radiant power incident on the sample. In practice, the sample is often a liquid or a gas contained in an absorption cell. In this case, the observed transmittance is the ratio of the radiant power transmitted by the sample in its cell to the radiant power transmitted by some clearly specified reference material in its cell, when both are measured under the same instrument conditions such as spectral position and slit width. In the case of solids not contained in a cell, the radiant power transmitted by the sample is also measured relative to that transmitted by a clearly specified reference material. The observed transmittance is seldom equal to the true transmittance.

Ultraviolet—Pertaining to the region of the electromagnetic spectrum from approximately 10 to 380 nm. The term ultraviolet without further qualification usually refers to the region from 200 to 380 nm.

Variable-Angle Internal Reflection Element—An internal reflection element which can be operated over a range of angles of incidence.

Visible—Pertaining to radiant energy in the electromagnetic spectral range visible to the normal human eye (approximately 380 to 780 nm).

Wavelength, λ—The distance, measured along the line of propagation, between two points that are in phase on adjacent waves. The recommended unit of wavelength in the infrared region of the electromagnetic spectrum is the micrometer. The recommended unit in the ultraviolet and visible region of the electromagnetic spectrum is the nanometer or the Angstrom.

Wavenumber—The number of waves per unit length. The usual unit of wavenumber is the reciprocal centimeter, cm^{-1}. In terms of this unit, the wavenumber is the reciprocal of the wavelength, λ, when λ is expressed in centimeters.

12. Photometric Methods for Chemical Analysis*

12.1 INTRODUCTION.

12.1.1 This section covers general recommendations for photoelectric photometers and for photometric practice prescribed in ISC photometric methods for chemical analysis. A summary of the fundamental theory and practice of photoelectric photometry is given. No attempt has been made, however, to include in this section a description of every apparatus or to represent recommendations on every detail of practice in photometric methods of chemical analysis.

12.1.2 The inclusion of the following paragraph, or a suitable equivalent, in any ISC photometric method shall constitute due notification that the photometers and photometric practice prescribed in that method are subject to the recommendations set forth in this section.

"Photometers and Photometric Practice—Photometers and photometric practice prescribed in this method shall conform to the recommendations presented in **Section 12,** of Part I of the ISC Manual".

12.2 DEFINITIONS AND SYMBOLS.

12.2.1 For definitions of terms relating to absorption spectroscopy, refer to **Section 11, Terms and Symbols Relating to Molecular Spectroscopy.**

12.2.2 *Concentration Range.* The recommended concentration range shall be designated on the basis of the optical path of the cell, in centimeters, and the final volume of solution as recommended in a photometric procedure. Cells should have optical paths of 1, 2 or 5-cm. In general, the concentration range in any photometric method shall be specified as that which will produce transmittance readings between 90 and 15% under the conditions of the method.

*The material in this section is based on ASTM Designation E60-80 and is reprinted with permission of the ASTM.

12.2.3 *Photometric Reading.* The term "photometric reading" refers to the scale reading of the photometric instrument being used. Available instruments have scales calibrated in transmittance, T, **(5)** or absorbance, A, **(6)**, or even arbitrary units proportional to transmittance or absorbance.

12.2.4 *Reference Solution.* Photometric readings consist of a comparison of the intensities of the radiant energy transmitted by the absorbing solution and the radiant energy transmitted by the solvent. Any solution to which the transmittance of the absorbing solution of the substance being measured is compared shall be known as the reference solution.

12.2.5 *Initial Setting.* The initial setting is the photometric reading (usually 100 on the percentage scale or zero on the logarithmic scale) to which the photometer is adjusted with the reference solution in the absorption cell. The scale will then read directly in percentage transmittance or in absorbance.

12.2.6 *Background Absorption.* Any absorption in the solution due to the presence of absorbing ions, molecules, or complexes of elements other than that being determined is called background absorption.

12.2.7 *Reagent Blank.* The reagent blank determination is a determination of the amount of the element sought which is present as an impurity in the reagents used.

12.3 Theory.

12.3.1 Photoelectric photometry is based on Bouguer's and Beer's (or the Lambert-Beer) laws which are combined in the following expression:

$$P = P_0 10^{-abc}$$

Where

P = transmitted radiant power,

P_0 = incident radiant power, or a quantity proportional to it, as measured with pure solvent in the beam,

a = absorptivity, a constant characteristic of the solution and the frequency of the incident radiant energy,

b = internal cell length (usually in centimeters) of the column of absorbing material, and

c = concentration of the absorbing substance in grams per liter.

12.3.2 Transmittance, T, and absorbance, A, have the following values:

$$T = \frac{P}{P_0}$$

$$A = \log_{10}\frac{1}{T} = \log_{10}\frac{P_0}{P}$$

where P and P_0 have the values given in Paragraph 12.3.1.

12.3.3 From the transposed form of the Bouguer-Beer equation $A = abc$, it is evident that at constant b, a plot of A versus c gives a straight line if Beer's law is followed. This line will pass through the origin if the usual practice of cancelling out solvent reflections and absorption and other blanks is employed.

12.3.4 In photoelectric photometry it is customary to make indirect comparison with solutions of known concentration by means of calibration curves or charts. When Beer's law is obeyed and when a satisfactory photometer is employed, it is possible to dispense with the curve or chart. Thus from the transposed form of the Bouguer-Beer law, $c = A/ab$, it is evident that once a has been determined for any system, c can be obtained, since b is known and A can be measured.

12.3.5 The value for a can be obtained from the equation $a = A/cb$ by substituting the measured value of A for a given b and c. Theoretically, in the determination of a for an absorbing system, a single measurement at a given wavelength on a solution of known concentration will suffice. Actually, however, it is safer to use the average value obtained with three or more concentrations, covering the range over which the determinations are likely to be made and making several readings at each concentration. The validity of the Bouguer-Beer law for a particular system can be tested by showing that a remains constant when b and c are changed.

12.4 Apparatus.

12.4.1 *General Requirements for Photometers.*

A photoelectric photometer consists essentially of the following†:

a. An illuminant (radiant energy source);

b. a device for selecting relatively monochromatic radiant energy (consisting of a diffraction grating or a prism with selection slit, or a filter);

c. one or more absorption cells to hold the sample, standards, reagent blank or reference solution;

d. an arrangement for photoelectric measurement of the intensity of the transmitted radiant energy, consisting of one or more photocells and suitable devices for measuring current or potential.

Precision photometers that employ monochromators capable of supplying radiant energy of high purity at any chosen wavelength within their range are usually referred to as spectrophotometers. Photometers employing filters are known as filter photometers or abridged spectrophotometers, and usually isolate relatively broad bands of radiant energy. In most cases the absorption peak of the compound being measured is relatively broad, and sufficient accuracy can be obtained using a fairly broad band (10 to 75 nm) of radiant energy for th measurement. One nanometer (nm) equals one millimicron (mμ). In other cases the absorption peaks are narrow, and radiant energy of high purity (1 to 10 nm) is required. This applies particularly if accurate values are to be obtained in those systems of measurement based on the additive nature of absorbance values.

†The choice of an instrument may naturally be based on price considerations, since there is no point in using a more elaborate (and, incidentally, more expensive) photometer than is necessary. In addition to satisfactory performance from the purely physical standpoint, the instrument should be compact, rugged enough to stand routine use, and not require too much manipulation. The scales should be easily read, and the absorption cells should be easily removed and replaced, as the cleaning, refilling, and placing of the cells in the instrument consume a major portion of the time required. It is advantageous to have a photometer that permits the use of cells of different depth.

12.4.2 *Types of Photometers.*

a. Single-Photocell Photometers. In most single-photocell photometers, the radiant energy passes from the monochromator or filter through the reference solution to a photocell. The photocurrent is measured by a galvanometer or a microammeter and its magnitude is a measure of the incident radiant power, P_0. An identical absorption cell containing the solution of the absorbing component is now substituted for the cell containing the reference solution and the power of the transmitted radiant energy, P, is measured. The ratio of the current corresponding to P to that of P_0 gives the transmittance, T, of the absorbing solution, provided the illuminant and photocell are constant during the interval in which the two photocurrents are measured. It is customary to adjust the photocell output so that the galvanometer or microammeter reads 100 on the percentage scale or zero on the logarithmic scale when the incident radiant power is P_0, in order that the scale will read directly in percentage transmittance or absorbance. This adjustment is usually made in one of three ways. In the first method, the position of the cross-hair or pointer is adjusted electrically by means of a resistance in the photocell-galvanometer circuit. In the second method, adjustment is made with the aid of a rheostat in the source circuit. Kortum **(4)** has pointed out that on theoretical grounds this method of control is faulty, since the change in voltage applied to the lamp not only changes the radiant energy emitted but also alters its chromaticity. Actually, however, instruments employing this principle are giving good service in industry, so the errors involved evidently are not too great. The third method of adjustment is to control the quantity of radiant energy striking the photocell with the aid of a diaphragm somewhere in the path of radiant energy.

b. Two-Photocell Photometers. In order to eliminate the effect of fluctuation of the source, a great many types of two-photocell photometers have been proposed. Most of these are good, but some have poorly designed circuits and do not accomplish the purpose for which they are designed. Following is a brief description of two types of two-photocell photometers that are in general use and that have been found satisfactory:

(1) In the first type of two-photocell photometer, beams of radiant energy from the same source are passed through the reference solution and the sample solution and are focused on their respective photocells. Prior to insertion of the sample, the reference solution is placed in both absorption cells, and the photocells are balanced with the aid of a potentiometric bridge circuit. The reference solution and sample are then inserted and the balance reestablished by manipulation of the potentiometer until the galvanometer again reads zero. By choosing suitable resistances and by using a graduated slide wire, the scale of the latter can be made to read directly in transmittance. It is important that both photocells show linear response, and that they have identical radiation sensitivity if the light is not monochromatic.

(2) The second type of two-photocell photometer is similar to the first, except that part of the radiant energy from the source is passed through an absorption cell to the first photocell; the remainder is impinged on the second photocell without, however, passing through an absorption cell. Adjustment of the calibrated slide wire to read 100 on the percentage scale, with the reference solution in the cell, is accomplished by rotating the second photocell. The reference solution is then replaced by the sample and the slide wire is turned until the galvanometer again reads zero.

12.4.3 *Radiation Source.*

a. In most of the commercially available photoelectric photometers the illuminant is an incandescent lamp with a tungsten filament. This type of illuminant is not ideal for all work. For example, when an analysis calls for the use of radiant energy of wavelengths below 400 nm, it is necessary to maintain the filament at as high a temperature as possible in order to obtain sufficient radiant energy to ensure the necessary sensitivity for the measurements. This is especially true when operating with a photovoltaic cell, for the response of the latter falls off quickly in the near ultraviolet. The use of high-temperature filament sources may lead to serious errors in photometric work if adequate ventilation is not provided in the photometer in order to dissipate the heat. Another important source of error results from the change of the shape of the energy distribution curve with age. As a lamp is used, tungsten will be vaporized and deposited on the walls. As this condensation proceeds, there is a decrease in the radiation power emitted and, in some instances, a change in the composition of the radiant energy. This change is especially noticeable when working in the near ultraviolet range and will lead to error (unless frequent standardization is resorted to) in all except those cases where essentially monochromatic radiant energy is used. These errors have been successfully overcome in commercially available photometers. One instrument has been so designed that a very low-current lamp (or the order of 200 ma) is employed as the source. This provides for long lamp life, freedom from line fluctuation (since a storage battery is employed), stability of energy distribution, reproducibility, and low-cost operation. In addition, the stable illuminant permits operation for long periods of time without need for restandardization against known solutions.

b. In most of the commercially available photometers where relatively high-wattage lamps are used, the power is derived from the ordinary electric mains with the aid of a constant-voltage transformer. Where the line voltages vary markedly, it is necessary to resort to the use of batteries that are under continuous charge.

12.4.4 *Filters and Monochromators.*

a. Filters. Spectrophotometric methods call for the isolation of more or less narrow wavebands of radiant energy. Relatively inexpensive instruments employing filters are adequate for a large proportion of the methods used, since most of the absorbing systems show broad absorption bands. In general, filters are designed to isolate as narrow a band of the spectrum as possible. Actually, it is usually necessary, especially when the filters are to be used in conjunction with an instrument employing photovoltaic cells, to sacrifice spectral purity in order to obtain sufficient sensitivity for measurement with a rugged galvanometer or a microammeter. Glass filters are most often used because of their stability to light and heat, but gelatin filters and even aqueous solutions are sometimes used.

b. Monochromators. Two types of monochromators are in common use: the prism and the diffraction grating. Prisms have the disadvantage of exhibiting a dependence of dispersion upon wavelength. On the other hand, the elimination of stray radiation energy is less difficult when a prism is used. In a well-designed monochromator, stray radiant energy resulting from reflections from optical and mechanical members is reduced to a minimum, but some radiant energy, caused by nonspecular scatterings by the optical elements, will remain. This unwanted radiant energy can be reduced through the use of a second monochromator or a filter in combination with a monochromator. Unfortunately, any process of monochromatization is accompanied by a reduction of the radiant power, and the more complex the monochromator the greater the burden upon the measuring system.

12.4.5 *Absorption Cells.* Some photometers provide for the use of several sizes and shapes of absorption cells. Others are designed for a single type of cell. It is advantageous to have a photometer that permits the use of cells of different depths. In some single-photocell instruments there is only one receptacle for the cell; in others (and this is especially desirable in those instruments where the illuminant is unstable) a sliding carriage is provided so that two cells can be interchangeably inserted in the beam of radiant energy coming from the monochromator.

12.4.6 *Photocells.*

In photoelectric photometry, the measurement of radiant energy is usually accomplished with the aid of either photoemission or photovoltaic cells.

a. The spectral response of a photoemission cell will depend upon the alkali metal employed and upon its treatment during manufacture. The spectral response of a photovoltaic (or barrier-layer) cell is crudely similar to tht of the human eye, except that it extends from about 300 to 700 nm. In general, neither the voltage nor the current response of a photovoltaic cell is a linear function of the flux incident on the cell, but the current response is more nearly linear than the voltage response. Thus, current-measuring devices should be used with photovoltaic-cell photometers. The degree to which the response of these cells departs from linearity depends on the individual cell, its temperature, its level of illumination, the geometric distribution of this illumination on its face, and the resistance of the current-measuring circuit.

b. In order for a photocell to be useful, it must exhibit a constancy of current with time of exposure. Most commercial alkali cells in use at the present time produce a constant current after an exposure of a few minutes. The photovoltaic cells, on the other hand, frequently exhibit enough reversible fatigue to interfere with their use. The measures which improve linearity of response also tend to reduce fatigue. With most commercial photoelectric photometers, the errors due to reversible fatigue are usually less than 1%.

12.4.7 *Current Measuring Devices.*

a. The usual types of photoelectric photometers employ photovoltaic cells in conjunction with a microammeter or a moderately high-sensitivity galvanometer, as may be appropriate for the illumination level employed. The scales for the galvanometers are sometimes designed to permit reading of absorbance values but more often yield only the more conveniently read T or percentage T values. Some photometers are designed so that the current is measured potentiometrically, using the galvanometer as a null instrument. It is stated that the error due to nonlinearity of the galvanometer under load is eliminated. Actually, however, this error is usually small and, moreover, many photometers provide individual calibration of the galvanometer.

b. Where photoemission cells are used, current amplification is usually resorted to before the galvanometer or meter is used.

12.5 PHOTOMETRIC PRACTICE.

12.5.1 *Principle of Method.* Photometric methods are generally based on the measurement of the transmittance or absorbance of a solution of an absorbing salt, compound, or reaction product of the substance to be determined. It is usually desirable to perform a rather complete spectrophotometric investigation of the reaction before attempting to employ it in quantitative analysis. The investigation should include a study of the following:

a. The specificity of any reagent employed to produce absorption;

b. the validity of Beer's law;

c. the effect of salts, solvent pH, temperature, concentration of reagents, and the order of adding the reagents;

d. the time required for absorption development and the stability of the absorption;

e. the absorption curve of the reagent and the absorbing substances; and

f. the optimum concentration range for quantitative analysis.

In photometry it is necessary to decide upon the spectral region to be used in the determination. In general it is desirable to use a filter or monochromator setting such that the isolated spectral portion is in the region of the absorption maximum. In the ideal case (and, fortunately, this is true of most of the absorbing systems encountered in quantitative inorganic analysis) the absorption maximum is quite broad and flat so that deviations from Beer's law resulting from the use of relatively heterogeneous radiant energy will be negligible. Sometimes it will not be possible or desirable to work at the point of maximum absorption. In those cases where there is interference from other absorbing substances in the solution or where the absorption maximum is sharp, it is sometimes possible to find another flat portion of the curve where the measurements will be free from interference. When no flat portion free from interference can be found, it may be necessary to work on a steep portion of the curve. In this case Beer's law will not hold unless the isolated spectral band is quite narrow. There is no real objection to operation on a steep part of an absorption curve, provided the usual standard calibration curve is obtained, except that with most instruments the reproducibility of the absorbance readings will be poor unless a fixed wavelength setting of the monochromator is maintained or unless filters are used. A small change in any of a large number of conditions will decrease the accuracy by a larger amount than when observations are made where the change in absorption is more gradual.

In most photometric work it is best to prepare a calibration curve or chart rather than to rely on the assumption of linearity, since it is not at all uncommon to obtain curved lines in the calibration of solutions that are known to obey Beer's law. The two most common causes of this are the presence of stray radiant energy, and the use of filters or monochromators that isolate too broad a spectral region for the analysis. Nonlinearity will generally be more pronounced the greater the heterogeneity of the radiant energy employed. Thus, one is more likely to obtain linearity with a spectrophotometer having a prism or grating with a high resolving power than with a photometer employing rather broad-banded filters. On the other hand, high resolving power or a narrow slit width is no guarantee of linearity unless stray radiant energy is rigorously excluded. When nonlinearity is encountered at one wavelength setting, it is sometimes possible to eliminate it by changing to another wavelength (where stray radiant energy is negligible) even though the latter is less favorable from the standpoint of flatness and sensitivity. A filter photometer employing a good filter will sometimes yield a more nearly straight calibration curve than can be obtained with certain spectrophotometers. This is especially true in the violet and near ultraviolet regions where stray radiant energy is likely to be encountered in grating monochromators.

12.5.2 *Concentration Range.* The concentration of the element being determined should preferably be so regulated that the transmittance readings fall within the range that yields the minimum error for the percentage of constituent being determined. The optimum concentration range for transmittance measurement for any absorbing system with a particular instrument can be determined by preparing a Ringbom plot of % transmittance versus logarithm of concentration. The optimum concentration range corresponds to that concentration range over which occurs a nearly constant and large rate of change in transmittance as indicated by a linear portion of the curve. This is discussed by Ayers (**1**). When practicable, it is desirable to adjust the concentration so that the transmittance is between 20 and 60 per cent. This adjustment can usually be made in one of four ways: (*a*) by varying the sample path length, (*b*) by varying the wavelength of radiant energy employed, (*c*) by varying the sample weight, or (*d*) by aliquoting.

12.5.3 *Stability of Absorption.*

The absorbing compounds on which photometric methods are based vary greatly in stability. In some instances the absorption is stable indefinitely, but in the majority of methods the absorption either increases or decreases on standing. In some cases a completely (or relatively) stable absorption is obtained on standing; in other cases it is stable for a time and then changes; in still other cases it never reaches a stable intensity. In all photometric work it is desirable to measure both standards and samples during the time interval of maximum stability of the absorption, provided that this occurs reasonably soon after development of the absorption. In those cases where the absorption changes continuously, it is necessary to control rigidly the time of standing.

12.5.4 *Interfering Elements.*

In photometry there are two basic types of methods to be considered: one type in which the photometric measurement is made without previous separations, and a second type in which the element to be determined is partially or completely isolated from the other elements in the sample.

a. In the first type of method it usually happens that one or more of the elements or reagents present may cause interference with an absorbing reaction. Such interference may be due to the presence of a colored substance, to a suppressive or enhancing effect on the absorption of the substance being measured, or to the destruction or formation of a complex with the reagents thus preventing formation of the absorbing substance. The most important methods, not involving separations, used to eliminate such interferences are as follows:

(1) The use of standards whose composition matches the sample being analyzed as closely as possible;

(2) performing the measurement at a wavelength where interference is at a minimum;

(3) the use of reagents that form complexes with the interfering elements.

The question as to how much interference can be tolerated in a given method will depend upon many factors, including the degree of accuracy required in the determination. In general, it is desirable to avoid using a method where the error to be "blanked out" is appreciable. The methods involving no separations suffer from the distinct disadvantage that the analyst must often know the matrix of the sample to be analyzed and, what is more important, must be able to prepare a standard to duplicate it. This is not always easy to do, for it often happens, especially in the determination of traces, that the so-called pure substances used for preparing the synthetic standards contain more of the element to be measured than does the sample being analyzed.

b. In the second type of method, the separations may involve removal of one or more interfering elements or may provide for complete isolation of the element in question before its photometric estimation. In this type of method there is usually no attempt made to adjust the matrix of the standard solution to fit that of the sample being analyzed, since presumably all extraneous interference has been removed. The standard in this case is a standard solution of the element in question. In any photometric determination it is desirable to keep the manipulation and separations as simple as possible, for the greater the number of reagents and the more manipulation involved the greater the blank and hence the more chance of error. Very useful tabulations have been compiled of methods used to eliminate interference in photometric analysis (**2, 6**).

12.5.5 *Concentrations of Standard Solutions.*

The concentrations of standard solutions shall be expressed in milligrams or micrograms of the element per milliliter of solution.

12.5.6 *Cell Corrections.*

To correct for differences in cell paths in photometric measurements using instruments provided with multiple absorption cells, cell corrections should be determined according to the following procedure: Transfer suitable portions of the reference solution prepared in a specific method to two absorption cells (reference and "test") of approximately identical light paths. Using the reference cell, adjust the photometer to the initial setting using a light band centered at the appropriate wavelength. While maintaining this adjustment, take the absorbance reading of the "test" cell and record as the cell correction. Make certain that a positive absorbance reading is obtained. If it is negative, reverse the positions of the cells. ("Matched" cells frequently show no reading). Subtract this cell correction (as absorbance) from each absorbance value obtained in the specific method. Keep the cells in the same relative positions for all photometric measurements to which the cell correction applies.

12.5.7 *Calibration Curve or Chart.*

a. Linear relation between transmittance or absorbance and concentration is not always obtained with commerically available photometers, even though the absorbing system is known to obey Beer's law. Because of this, it is evident that the use of calibration curves or charts will be necessary with such photometers. Moreover, it is not safe, with most photometers on the market today, to use calibration curves or charts interchangeably, even though the photometers may be of the same make and model. A separate calibration curve or chart must be prepared for each photometer.

b. The use of a calibration curve or chart in photometric analysis ensures correct measurement of concentration only when the composition of the radiant energy used in the work does not change. In most cases it is necessary to restandardize from time to time to guard against change in the photocell, filters, measuring circuit, and illuminant. In addition to this, in certain photometric methods, the temperature coefficient of the reactions employed to produce absorption is large enough to require winter and summer standard curves or charts.

c. When a calibration curve is used, the usual procedure is to plot the values of A, obtained from a series of standard solutions whose concentrations adequately cover the range of the subsequent determinations, against the respective concentrations, on ordinary graph paper. When the scale of the photometer being used does not read directly in absorbance, it is then convenient to plot concentration, c, against percentage transmittance on semilogarithmic paper, using the semilogarithmic scale for the percentage T values. In some cases it is more convenient to prepare a chart of c versus A or percentage T values. In plotting, a straight line should be obtained if a good photometer is employed and if the solution obeys Beer's law. If all blanks and interference have been eliminated, the lines should pass through the origin (the point of zero concentration and zero absorbance or 100% transmittance). The use of A in the plotting is advantageous because it is directly proportional to the concentration. On the other hand, while percentage transmittance has the disadvantage of decreasing in magnitude as the concentration increases, it is more convenient to use when the photometer employed does not have a scale calibrated in absorbance.

12.5.8 *Procedure.*

a. Detailed instructions for the procedure to be followed will be found in each of the ISC photometric methods.

12.5.9 *Blanks.* In taking the photometric reading of the absorption in any solution, all the components present that absorb radiant energy in the region of interest must be taken into account. These sources of absorption are:

a. Absorption of the element sought;

b. absorption of the element sought

present as an impurity in the reagents used;

c. background;

d. absorption of all reagents used;

e. absorption produced by reaction of reagents with other elements present; and

f. turbidities.

These absorptions are additive and all or some will be included in the photometric reading, depending upon the method of preparing the calibration curve and the reference solution. Items **(e)** and **(f)** are interferences and are assumed to be eliminated by preliminary conditioning operations. Items **(b)** to **(d)** have been loosely designated as "blanks". It is less confusing to restrict the usage of the word "blank" to reagent blank, Item **(b)** in the above list. Item **(c)** is defined in Paragraph 12.2.6 and Item **(d)** is usually taken care of by the "reference solution" (Paragraph 12.2.4).

The general case is stated in the preceding paragraphs, and it is desirable that all these factors be considered in the development of a photometric method. However, it is often possible to combine some or all of these factors into the reference solution. Thus, the reference solution may, in some cases, include the reagent blank, the background, and any absorption due to the reagents used. In other cases it may be desirable to measure the reagent blank alone in order that a check may be had on the purity of the reagents. It should be noted, however, that in the case of absorbing systems that do not obey Beer's law, it may be dangerous to use the reagent blank for the reference solution, particularly if the magnitude of the absorption due to the reagent blank becomes appreciable. In such instances it is necessary to refer both reagent blank and sample solution to some arbitrary reference solution, usually water, and make suitable corrections for the absorption of the reagent blank.

The requirements for the preparation and measurement or application of these various corrections, both in the preparation of the calibration curve and in the procedure, will be found in each of the ISC photometric methods.

12.6. PRECISION AND ACCURACY.

12.6.1 The primary advantages of photometric methods are those of speed, convenience, and relatively high precision and accuracy in the determination of micro-and semi-micro-quantities of constituents. For the determination of macro-quantities, differential spectrophotometric techniques **(3)** or other analytical techniques are often preferable, since they are generally more accurate when larger quantities are involved. It should be remembered that even under the most favorable circumstances it is difficult to obtain an accuracy better than about 1% of the amount present in most photometric determinations. This does not mean that it is not practicable to analyze macro-samples photometrically. With the continued improvement in optical instruments, it has been possible to perform an increasing number of different types of determinations, especially in the cases where high accuracy is not required. In evaluating the precision and accuracy of any photometric method, the quality of the apparatus and the chemical procedure involved must be considered.

References

1. G. H. AYRES. 1949. Evaluation of Accuracy in Photometric Analysis. Anal. Chem. 21:652.
2. D. F. BOLZ. 1958. *Colorimetric Determination of Nonmetals*. Interscience Publishers, Inc., New York.
3. A. G. JONES. 1959. *Analytical Chemistry, Some New Techniques*. Academic Press, Inc., New York.
4. G. KORTUM. 1937. Photoelectric Spectrophotometry. Angewandte Chemie, 50:193.
5. M. G. MELLON. 1950. *Analytical Absorption Spectroscopy*. John Wiley and Sons, Inc., New York.
6. E. B. SANDELL. 1959. *Colorimetric Determination of Traces of Metals*. (3rd ed.) Interscience Publishers, Inc., New York.

13. Ultraviolet Absorption Spectroscopy

13.1 INTRODUCTION.

13.1.1 The property of ultraviolet absorption can be utilized for quantitative determination of a number of materials, **(1, 2, 3)**. It is particularly well suited for the analysis of aromatic compounds but any molecule containing a chromophoric group would be expected to absorb light in the ultraviolet region of the spectrum. The amount of absorbance is directly proportional to the concentration of the material present in solution.

13.2 SAMPLING.

13.2.1 Ultraviolet absorption is a convenient method for air samples which can be collected directly in an appropriate solvent (**4, 5**) or adsorbed on silica gel for subsequent leaching with the solvent. Certain precautions must be observed, however, in sampling and analysis.

13.2.2 All-glass sampling equipment must be used for samples collected in organic liquids. If two samplers are used in series, the connection should also be glass. Samples are readily contaminated by rubber or plastic connections, lubricants, and by rubber or cork stoppers; in essence, contact with any extraneous organic material should be avoided. Although samples collected in aqueous solutions are not so easily contaminated, it is advisable to prevent contact with organic substances. If a dusty atmosphere is being sampled, a prefilter should be used; suspended dust cannot be removed satisfactorily by centrifuging. Dust causes an increase in absorbance, particularly in the region below 300 nm.

13.2.3 For transport of liquid samples to the laboratory, the samples should be retained in the sampling vessel. If transfer is required, it should be either to all-glass bottles or to glass bottles with aluminum liners in the caps (**1**).

13.2.4 Adsorption of vapors on silica gel is a convenient method of collection and such samples are easy to transfer for transport. However, interferences introduced by leaching with solvents for analysis must be anticipated.

13.3 ANALYSIS.

13.3.1 Absorption maxima of specific materials will vary from one instrument to another. Thus each laboratory must determine the peaks on its own instrument. If the maximum is broad, as in the case of acetone, a precise setting is not so important. A precise setting is needed, however, for materials such as benzene and toluene that have sharp maxima; a difference of a nanometer will result in considerable error in the absorbance reading obtained.

13.3.2 Determinations are usually made at the peak wave lengths to utilize the greatest sensitivity, but the minima of the curves can also be use. When calibration curves are drawn, it is advisable to prepare curves for the various maxima and minima. With samples where the concentration is too high to use the peak wave length, it is frequently possible to use a curve based on another maximum or a minimum without need for dilution of the sample.

13.3.3 In addition to proper wave length settings, an important factor in analysis is the condition of the silica cuvettes. Not only must they be thoroughly clean but any pair used for analysis should, when filled with solvent blanks, have absorbance readings within ± 0.005 at the wave length at which they are to be used. If there is a difference greater than ± 0.005, a correction factor will have to be applied to the readings obtained with such cuvettes.

13.3.4 The reference blank should always be prepared from the same bottle of solvent that is used for sampling, as reagent-grade solvents vary widely from one bottle to another.

13.4 PLOTTING OF CURVES.

13.4.1 The curves are plotted as the logarithm of molar absorptivity versus the wave length in nanometers. The absorptivity and molar absorptivity values are given with each curve.

13.4.2 As indicated previously, the quantitative determination of materials by ultraviolet absorption is possible because the amount of absorbance is directly proportional to the concentration of the substance in solution. According to the Beer-Lambert law:

$$A = abc = (E/M)\,bc$$

where

A = absorbance (spectrophotometer reading)
a = absorptivity
b = cell length path in centimeters
c = concentration in grams per liter
E = molar absorptivity = aM
M = molecular weight

The terms and symbols used are those proposed by ASTM Committee E-13, (**6**) and are also the preferred usages of Analytical Chemistry (7).

13.4.3 By means of the foregoing equation calibration curves for a particular substance in a designated solvent can be drawn. The values "*a*" and "*b*" are known; values may be assumed for "*A*" and the "*c*" values calculated. When the absorbance values are plotted against the calculated concentrations, a straight line calibration curve will be obtained. With a fluorescent solution the true calibration curve would have a slight bend caused by the effect of the emitted light on the phototube. In the use of these calculated calibration curves, a blank must be employed to balance the instrument so that there will be zero absorbance with zero concentration.

13.4.4 As an illustration of what results might be expected from the above procedure, the case of acetone collected in water may be considered. Acetone in water is not fluorescent—hence the curve of absorbance versus concentration will be a straight line. With the value of "*a*" = 0.30 (given on the absorptivity curve) and a 1-cm cell, values for "*A*" at the maximum absorptivity, 265 nm, are selected and each "*c*" is calculated with the following results:

A	c grams per liter
0.100	0.334
0.200	0.667
0.400	1.330
0.600	2.000
0.800	2.660

13.4.5 In the data from which the molar absorptivity curve was drawn, the value of "*a*" was actually 0.304. At an absorbance of 0.600, use of this figure in the equation $A = abc$ yields a concentration of 1.975 g per l instead of 2.000 g found by using the absorptivity value given on the absorptivity curve.

13.4.6 As an example of what might happen in the case of a fluorescent solution, hydroquinone in isopropanol is representative. Although the true curve would be bent, a straight line calibration curve, at the maximum of 294 nm can be developed from calculations based on the value given for "*a*", namely, 32.4. From it, an absorbance reading of 0.800 with a 1-cm cell would indicate a concentration of 24.7 μg/ml. The actual concentration, however, determined from standard solutions, would be about 27.4 μg/ml.

13.4.7 Because of variations in spectrophotometers the greatest accuracy requires that standards be run in individual laboratories. This is especially true for fluorescent solutions.

13.5 DETERMINATION OF MIXTURES.

13.5.1 In many cases it is possible to determine two ultraviolet absorbant materials in the same solution (**8**). Each component contributes to the absorbance at a specific wave length in this manner:

$$A_0 = A_x + A_y = a_x b c_x + a_y b c_y \qquad (1)$$

where

A_0 = absorbance observed
x = one material
y = second material
b = cell length, assumed constant

The absorbance of the mixture is observed at two different wave lengths and the absorptivity of each material is determined at the same wave lengths. These data are used to form two simultaneous equations that are solved to give c_x and c_y.

13.5.2 A mixture of benzene and toluene in cyclohexane will serve to illustrate this procedure. Benzene has a maximum of 249 nm with an absorptivity of 2.3; toluene has a maximum at 270 nm with an absorptivity of 3.1. At 270 nm the absorptivity of benzene is 0.11, at 249 nm, that of toluene is 2.6. A sample of a mixture containing 0.100 g/l both of benzene and toluene in cyclohexane was found to have an absorbance of 0.319 at 270 nm and of 0.493 at 249 nm. A 1-cm cell was used.

13.5.3 The absorbance readings and the respective absorptivities are substituted in equation (1) to obtain two simultaneous equations:

$$\text{(at 270nm) } 0.319 = 3.1c_t + 0.11c_b \quad (2)$$

$$\text{(at 249nm) } 0.493 = 2.6c_t + 2.3c_b \quad (3)$$

In order to eliminate one unknown, equation (2) is multiplied by 0.839 to give:

$$0.268 = 2.6c_t + 0.092c_b \quad (4)$$

or

$$2.6c_t = 0.268 - 0.092c_b \quad (5)$$

If (5) is substituted in (3), it yields

$$0.225 = 2.2c_b \quad (6)$$

$$c_b = 0.102 \text{ g per liter}$$

This value for c_b is substituted in either (2) or (3) to give $c_t = 0.099$ g/l.

Both concentrations are very close to the theoretical values of 0.100 g/l.

13.6 ADVANTAGES OF ULTRAVIOLET ABSORPTION.

13.6.1 Although there are many difficulties and interferences in carrying out quantitative determinations by means of ultraviolet absorption, it is an easy and convenient technique. As such, it is useful as a standard procedure for many routine samples. It is especially valuable in emergency situations when it is necessary to sample for a material for which there has not been time to establish a suitable chemical procedure (**1**).

References

1. HOUGHTON, J.A., G. LEE, M.A. SHOBAKEN, and A. FOX. 1964. Practical Applications of Analysis by the Ultraviolet Absorbance Method. Amer. Ind. Hyg. Assoc. J. 25:381.
2. MAFFETT, P.A., T.F. DOHERTY, and J.L. MONKMAN. 1956. Collection and Determination of Micro Amounts of Benzene or Toluene in Air. Amer. Ind. Hyg. Assoc. Quart. 17:186.
3. HIRT, R.C., and J. GISCLARD. 1951. Determination of Parathion in Air Samples by Ultraviolet Absorption Spectroscopy. Anal. Chem. 23:185.
4. HOUGHTON, J.A., and G. LEE. 1960. Data on Ultraviolet Absorption and Fluorescence Emission. Amer. Ind. Hyg. Assoc. J. 21:219.
5. HOUGHTON, J.A., and G. LEE. 1961. Ultraviolet Spectrophotometric and Fluorescence Data. Amer. Ind. Hyg. Assoc. J. 22:296.
6. TENTATIVE DEFINITIONS OF TERMS AND SYMBOLS RELATING TO ABSORPTION SPECTROSCOPY. 1963. E131–63T, ASTM Committee E-13. American Society for Testing and Materials, Philadelphia, Pa.
7. AMERICAN CHEMICAL SOCIETY. 1963. Spectrophotometry Nomenclature. Anal. Chem. 35:2262.
8. HIRT, R.C., F.T. KING, and R.G. SCHMITT. 1954. Graphical Absorbance—Ratio Method for Rapid Two-Component Spectrophotometric Analysis. Anal. Chem. 26:1270.

14. Infrared Absorption Spectroscopy

14.1 INTRODUCTION.

14.1.1 Infrared is a powerful analytical tool because of its specificity, sensitivity, versatility, speed and simplicity. Solids, liquids and gases may all be analyzed using a few general sampling techniques; the spectra provide the necessary information for identifying and quantitating the components present.

14.2 PRINCIPLE.

14.2.1 Infrared radiation is passed through a chamber containing the sample, after which the radiation is dispersed and detected. Because each compound absorbs the radiation in a characteristic pattern, the graph of absorption versus wavelength (or spectrum) produced by the spectrometer may be used to identify each component. Furthermore, the amount of light absorbed is proportional to the concentration of the component; therefore, a quantitative analysis may also be obtained. A file of standard spectra of pure materials at known concentrations is required for the analysis.

14.3 SAMPLING.

14.3.1 The analysis of solids and liquids by infrared techniques is a well known field; the reader is referred to Potts (**1**) for further detail. Less common, and of more direct concern to the industrial hygienist, is the use of infrared to analyze gases and vapors, specifically the analysis of air for trace contaminants (100 ppm or less). Because oxygen and nitrogen are transparent to infrared radiation, multiple-reflection sampling cells with path lengths from 10 to 40 m may be utilized to achieve high sensitivity to many atmospheric contaminants.

14.3.2 The sampling cell most commonly employed is the 10-m path length cell shown in Figure 14:1.

14.3.3 Air may be sampled in a plastic bag, preferably of Saran or Mylar,

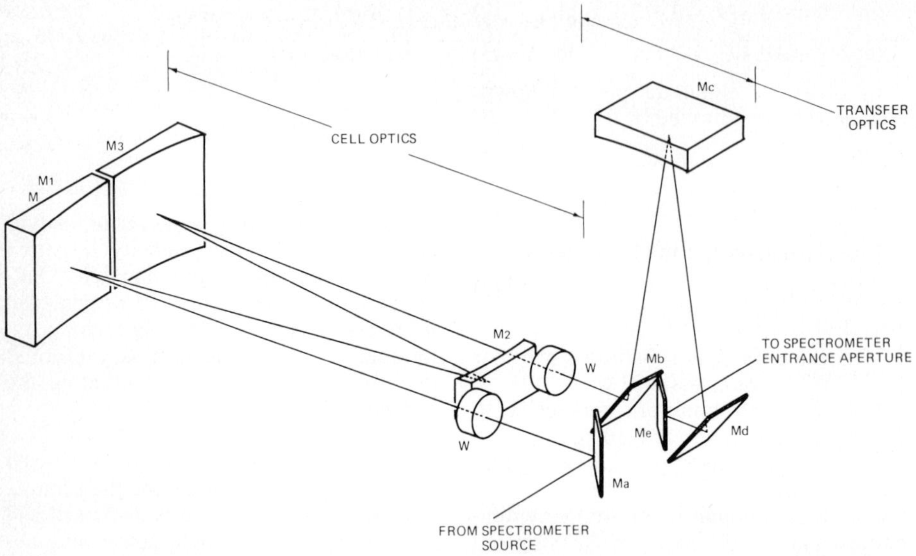

Figure 14:1—Optical diagram of multiple reflection gas cell set for 1-meter path length. The number of beam traversals can be increased by resetting M_1 and M_3 to give path lengths up to 10 meters. (Courtesy of Perkin-Elmer Corporation, Norwalk, Connecticut.)

the cell is then evacuated, the sample drawn in and the spectrum scanned. About 5 l of air are required. Some vapors may be adsorbed on silica gel and quantitatively desorbed in the laboratory **(2)**.

14.3.4 It is also possible to monitor air flowing continuously through the cell if the spectrometer is set to detect a wavelength where the contaminant being monitored has a unique absorption band.

14.3.5 Many compounds can similarly be determined in the expired breath of workers following exposure to them. In most cases the magnitude of a chemical exposure can be established as well as the identity of the toxins **(3)**.

14.4 ANALYSIS.

14.4.1 The analysis is performed by comparing the unknown spectrum with standard spectra of the compounds suspected to be present. Quantitative analysis requires the measurement of the absorbances of bands unique to each component. The following equation is used:

$$C_x = \frac{A_x}{A_s} \times \frac{b_s}{b_x} \cdot C_s$$

Where

A = absorbance
b = cell length
C = concentration
x = unknown
s = standard

14.4.2 A rough estimate of concentrations may be obtained from comparisons to published spectra; however, for accurate analysis the same instrument must be used to obtain both the sample and standard spectra. Several methods for preparing gaseous standards have been described **(3)**. Catalogs of vapor spectra are available privately **(4, 5)** or from commercial publishers **(6)**.

14.4.3 Table 14:I. lists the sensitivity of infrared analysis to a variety of gases and vapors using a 10 meter path length cell at atmospheric pressure.

Table 14:I. Infrared Analysis of Gases and Vapors in Expired Air

Compound	1969 Threshold limit value (ppm)	Infrared sensitivity (ppm)	Analytical wavelength (μ)
Acetaldehyde	200	30	8.90
Acetic acid	10	5	8.50
Acetone	1000	5	8.20
Acetonitrile	40	100	9.58
Acetylene tetrabromide	1	—[†]	—
Acrolein	0.1	10	10.43
Acrylonitrile	20	5	10.49
Allyl chloride	1	5	13.22
Ammonia*	50	20	10.77
n-Amyl acetate	100	1	8.05
Amyl alcohol	100	10	9.47
Benzene	25	20	9.62
Benzyl chloride	1	15	7.88
Bromine	0.1	—[‡]	—
Bromobenzene	—	10	9.30
1,3-Butadiene	1000	5	11.02
2-Butanone	200	10	8.52
sec-Butyl acetate	200	1	8.05
n-Butyl alcohol	100	10	9.35
tert-Butyl alcohol	100	5	10.88
Butylamine	5	10	12.85
n-Butyl glycidyl ether	50	10	8.80
Butyl mercaptan	10	5[§]	8.52
p-tert-Butyl toluene	10	20	12.25
Carbon dioxide*	5000	5	4.27
Carbon disulfide	20	20	4.57
Carbon monoxide*	50	20	4.58
Carbon tetrachloride*	10	0.5	12.60
Carbon sulfide	—	1	4.82
Chlorine	1	—[‡]	—
Chlorobenzene	75	10	9.16
Chlorobromomethane	200	10	13.35
Chloroform*	50	1	12.95
Chloropicrin	0.1	2	11.50
Cyclohexane	300	40	11.60
Cyclohexanol	50	10	9.32
Cyclohexanone	50	25	8.88
Cyclohexene	300	25	10.90
Diacetone alcohol	50	5	8.50
1,2-Dibromoethane	10	5	8.38
o-Dichlorobenzene	50	5	13.37
m-Dichlorobenzene	—	5	12.75
p-Dichlorobenzene	75	2	9.10
Dichlorodifluoromethane*	1000	1	10.85
1,1-Dichloroethane	100	5	9.42
1,2-Dichloroethane	50	10	8.18
cis-1,2-Dichloroethylene	200	5	11.58
trans,1,2-Dichloroethylene	200	2	12.05
Dichloroethyl ether	15	5	8.78
Dichlorofluoromethane	1000	1[§]	12.50
1,1-Dichloro-1-nitroethane	10	2	9.07
1,2-Dichloroethane	(See Propylene dichloride)		

Table 14:I—(continued)

Compound	1969 Threshold limit value (ppm)	Infrared sensitivity (ppm)	Analytical wavelength (μ)
Dichlorotetrafluoroethane	1000	0.5[§]	8.40
Diethylamine	25	10	8.70
Difluorodibromomethane	100	0.5[§]	12.10
Dimethyl ether	—	2[§]	8.51
Dimethyl formamide	10	5	9.22
Dioxane	100	2	8.80
Epibromohydrin	—	20	11.80
Epichlorohydrin	5	20	13.36
Ethanolamine	3	50	12.73
2-Ethoxyethylacetate	100	1	8.05
Ethyl acetate	400	1	8.02
Ethyl acrylate	25	1	8.35
Ethyl alcohol*	1000	5	9.37
Ethylamine	10	5[§]	12.95
Ethylbenzene	100	50	9.70
Ethyl bromide	200	10	7.98
Ethyl chloride	1000	10	10.18
Ethyl ether*	400	5	8.75
Ethyl formate	100	1	8.43
Ethyl mercaptan	10	40[§]	10.20
Ethyl nitrate	—	5[§]	11.75
Ethylene	—	5	10.55
Ethylene chlorohydrin	5	10	9.33
Ethylenediamine	10	15	12.85
Ethylenedibromide		(See 1,2-Dibromoethane)	
Ethylenedichloride		(See 1,2-Dichloroethane)	
Ethylene glycol ethyl ether	—	5	8.78
Ethylene imine	0.5	15	11.75
Ethylene oxide	50	5	11.48
Fluorotrichloromethane*	1000	1	11.82
FREON 11® fluorocarbon		(See Fluorotrichloromethane)	
FREON 12® fluorocarbon		(See Dichlorodifluoromethane)	
FREON 21® fluorocarbon		(See Dichlorofluoromethane)	
FREON 112® fluorocarbon		(See Tetrachlorodifluoroethane)	
Furfural	5	5	13.27
Furfuryl alcohol	50	10	9.80
Gasoline	—	5[‖]	3.40
Heptane	500	5[‖]	3.40
Hexane	500	5[‖]	3.40
2-Hexanone	100	10	8.57
Hexone (methyl isobutyl ketone)	100	10	8.52
Hydrogen cyanide	10	50[§]	3.00
4-Hydroxy-4-methylpentanone		(See Diacetone alcohol)	
Isophorone	25	—[†]	—
Isopropyl ether	500	5	8.92
Mesityl oxide	25	5	8.57
Methyl acetylene	1000	20[§]	8.05
Methylal	1000	5	8.72
Methyl alcohol*	200	5	9.45
Methylamine	20	10[§]	12.80
Methyl bromide	20	50	3.36

Table 14:I—(continued)

Compound	1969 Threshold limit value (ppm)	Infrared sensitivity (ppm)	Analytical wavelength (μ)
Methyl chloride*	100	30	3.35
Methyl chloroform*	350	2	9.20
Methyl cyclohexane	500	5[‖]	3.40
o-Methyl cyclohexanol	100	15	9.50
m-Methyl cyclohexanol	100	20	9.50
o-Methyl cyclohexanone	100	25	8.42
Methyl ethyl ketone		(See 2-Butanone)	
Methyl formate	100	5	8.53
Methyl isobutyl ketone		(See Hexone)	
Methyl mercaptan	10	100[§]	9.48
Methyl methacrylate	100	1	8.55
alpha-Methyl styrene	100	10	11.18
Methyltrimethoxysilane	—	1	9.06
Methylene bromide	—	5	8.38
Methylene chloride*	500	2	13.10
Methylene chlorobromide		(See chlorobromomethane)	
Naphtha (coal tar)	100	5[§, ‖]	3.40
Naphtha (petroleum)	—	5[‖]	3.40
Nitric oxide	25	100[§]	5.25
Nitroethane	100	30	11.43
Nitrogen dioxide	5	5[§]	7.90
Nitromethane	100	40	10.90
1-Nitropropane	25	25	12.45
2-Nitropropane	25	10	11.75
Nitrosyl chloride	—	50[§]	10.75
Nitrous oxide	—	25	7.68
Octane	500	5[‖]	3.40
Pentane	1000	5[‖]	3.40
Pentanone-2	200	10	8.52
Pentene-2	—	10	3.37
Perchloroethylene*	100	2	10.92
Phosgene	0.1	1	11.68
Propargyl bromide	—	5	8.18
Propyl acetate	200	1	8.05
i-Propyl alcohol*	200	15	10.47
n-Propyl ether	500	5	8.81
Propylene dichloride	75	15	9.80
1,2-Propylene oxide	100	10	11.96
Pyridine	5	30	9.60
Stoddard solvent	500	5[‖]	3.40
Styrene monomer	100	10	12.90
Sulfur dioxide	5	15[§]	8.55
Sulfuryl fluoride	5	1	11.32
1,1,2,2-Tetrachloro-1,2-difluoroethane	500	2	11.90
Tetrachloroethylene		(See perchloroethylene)	
Tetrahydrofuran	200	10	9.22
Toluene	200	5	13.75
1,1,2-Trichloroethane	10	10	10.60
Trichloroethylene*	100	2	11.78
1,2,3-Trichloropropane	50	15	12.44
Trimethylamine	—	5[§]	12.10

Table 14:I—(continued)

Compound	1969 Threshold limit value (ppm)	Infrared sensitivity (ppm)	Analytical wavelength (μ)
Turpentine	100	10§,‖	3.40
VIKANE fumigant®		(See Sulfuryl fluoride)	
Vinyl chloride	500	10	10.63
Vinylidine chloride	—	5	12.60
o-Xylene	100	10	13.51
m-Xylene	100	10	13.02
p-Xylene	100	10	12.58

*Materials marked * have been detected in postexposure expired air.
†no spectrum obtained from saturated vapor at 23 C, 1 atm pressure.
‡material produces no infrared spectrum.
§sensitivity estimated by extrapolating absorbance of vapor observed in 10-cm cell.
‖Aliphatic hydrocarbons can be detected at 3.40μ, but cannot usually be identified specifically at low concentration.

Note: *The above is a condensation of material found in "Detection of Volatile Organic Compounds and Toxic Gases in Humans by Rapid Infrared Techniques", by R.D. Stewart and D.S. Erley, pp. 183–200. The permission of the publisher to reprint this material is gratefully acknowledged.*

References

1. Potts, W.J., Jr. 1963. *Chemical Infrared Spectroscopy.* Vol. I: Techniques. John Wiley and Sons, New York.
2. Robertson, D.N., and D.S. Erley. 1961. Anal. Biochem 2:45.
3. Stolman, *et al.* 1965. *Progress in Chemical Toxicology.* Vol. II, Academic Press, New York.
4. Erley, D.C., and B.H. Blake. *Infrared Spectra of Gases and Vapors.* The Dow Chemical Company, Midland Michigan.
5. Pierson, R.H., A.N. Fletcher and E. Gantz. 1956. St. Cl. *Anal. Chem.* 28:1218.
6. Sadtler Research Laboratories, 1517 Vine Street, Philadelphia, Pennsylvania.

15. Atomic Absorption Spectrophotometry

15.1 Introduction. Within the past five years, there has been a dramatic acceleration in the use of atomic absorption spectrophotometry in quantitative analysis. The steadily growing list of atomic absorption analytical applications now covers some 65 elements, compared with just 20 elements recommended for analysis three short years ago.

15.2 Principle of the Method.

15.2.1 In atomic absorption, the element of interest in the sample is not excited. Rather, it is dissociated from its chemical bonds and placed into an unexcited, un-ionized "ground" state. This dissociation is most commonly achieved by burning the sample in a flame. The element is then capable of absorbing radiation at discrete lines of narrow bandwidth.

15.2.2 A hollow cathode lamp usually provides the narrow emission lines which are to be absorbed by the sample element. The lamp contains a cathode made of the element being determined and is filled with an inert atmosphere at low pressure. Such a lamp emits the spectrum of the desired element.

15.2.3 The use of a flame limits the working range for atomic absorption to the region where the atmosphere does not absorb. The resonance line for arsenic (1937 Å) is presently the lowest wavelength at which atomic absorption can be carried out. The element with its resonance line at the longest wavelength is cesium (8512 Å). (lnm = 10 Å or Angstrom units).

15.2.4 The wavelength range for atomic absorption, therefore, includes all metals and semi-metals, but excludes sulfur, phosphorus, carbon, the halogens, and gases with resonance wavelengths in the atmospheric absorption range.

15.3 Basic Atomic Absorption Unit—Single Beam.

15.3.1 In operation the light beam is passed through the dissociated element in

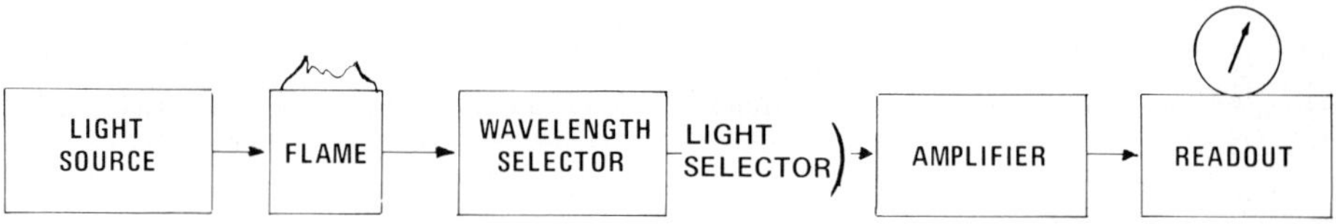

Figure 15:1—Single beam atomic absorption unit.

the flame. This radiation is then passed through a monochromator to filter out undesired emission and a photodetector finally receives the diminished radiation of specific wavelength. From the amount of radiation lost (absorbed by the element), a determination can be made as to quantity of the element in the sample (Figure 15:1).

15.3.3 *Double-Beam Instrument.* Some samples require the utmost in detection limits and/or precision. For these a double-beam instrument is preferred. In this system, the ratio is taken between two beams which share the same lamp, detector, and electronics. The sample beam is compared to a reference beam which by-

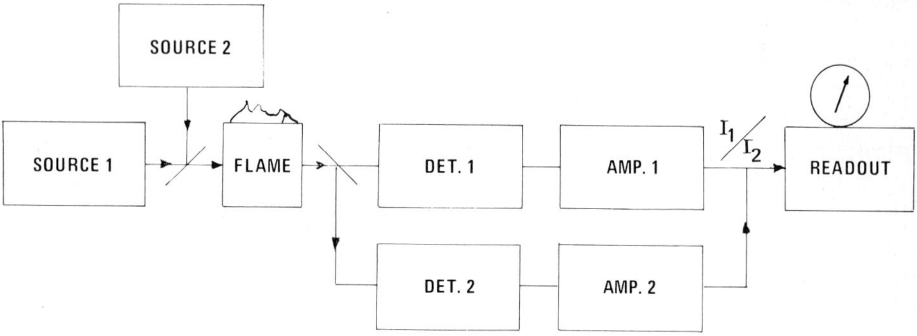

Figure 15:2—Internal standard system.

15.3.2 In the basic atomic absorption unit, variation in aspiration rate and changes in the flame characteristics cause a variation in the output. This difficulty may be overcome by comparing the sample solution with a standard. Adding a second channel for the standard and measuring the ratio of the two outputs, eliminates the effects of flame and aspiration (Figure 15:2).

passes the flame. Lamp drift is cancelled out, leading to great improvement in precision and detection limits. A double beam instrument can operate immediately with no wait for lamp warm-up.

15.4 THEORY.

15.4.1 When conditions in the flame are held constant, this (atomic) absorption obeys Beer's Law as in solution spectrophotometry, *i.e.*,

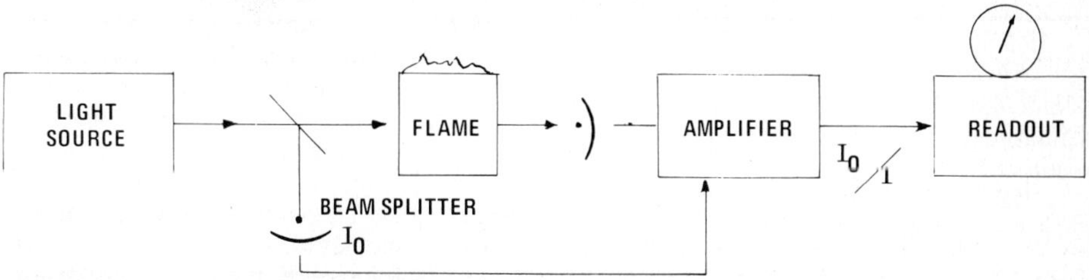

Figure 15:3—Double beam atomic absorption unit.

Table 15:I. Detection Capabilities of AA Instruments.
Detection Limit (μg/ml of solution or PPM)

Metal	Double Beam	Single Beam	Analytical Wavelength	Metal	Double Beam	Single Beam	Analytical Wavelength
			Å				Å
Ag Silver	0.0005	0.05	3281	Na Sodium	0.002	0.004	5890
Al Aluminum	0.1	0.2	3093	Nb Niobium*	3.0	—	4059
As Arsenic	0.1†	0.5†	1937	Nd Neodymium*	2.0	13.0	4634
Au Gold	0.02	0.05	2428	Ni Nickel	0.005	0.06	2320
B Boron*	6.0	—	2497	Pb Lead	0.01	0.07	2835
Ba Barium*	0.005	0.3	5536	Pd Palladium	0.03	0.2	2476
Be Beryllium*	0.002	0.01	2349	Pr Praseody-			
Bi Bismuth	0.05	0.2	2231	mium*	10.0	20.0	4951
Ca Calcium	0.002	0.01	4227	Pt Platinum	0.1	0.7	2659
Cd Cadmium	0.001†	0.003†	2288	Rb Rubidium	0.005	0.1	7800
Co Cobalt	0.005	0.03	2407	Re Rhenium*	1.0	8.0	3460
Cr Chromium	0.005	0.03	3579	Rh Rhodium	0.03	—	3435
Cs Cesium	0.05	2.0	8521	Ru Ruthenium	0.3	—	3499
Cu Copper	0.005	0.015	3247	Sb Antimony	0.1	0.2	2175
Dy Dysprosium*	0.2	0.4	4212	Sc Scandium*	0.1	—	3912
Er Erbium*	0.1	0.5	4008	Se Selenium	0.1*	0.5†	1961
Eu Europium*	0.04	0.1	4594	Si Silicon*	0.1	0.25	2516
Fe Iron	0.005	0.03	2483	Sm Samarium*	2.0		4297
Ga Gallium	0.07	—	2874	Sn Tin	0.03†	0.1†	2246
Gd Gadolinium*	4.0	—	3684	Sr Strontium	0.01	—	4607
Ge Germanium*	1.0	—	2651	Ta Tantalum*	5.0	—	2715
Hf Hafnium*	8.0	10.0	2866	Tb Terbium	2.0	3.0	4326
Hg Mercury	0.5	4.0	2537	Te Tellurium	0.1	2.0	2143
Ho Holmium*	0.1	0.3	4103	Ti Titanium*	0.1	0.4	3643
In Indium	0.05	—	3040	Tl Thallium	0.25	0.07	2768
Ir Iridium	2.0	3.0	2640	Tm Thulium*	0.15	0.25	3718
K Potassium	0.005	0.05	7665	U Uranium*	30.0	—	3514
La Lanthanum*	2.0	20.0	5501	V Vanadium*	0.02	0.2	3184
Li Lithium	0.005	0.005	6708	W Tungsten*	3.0	—	4008
Lu Lutetium*	3.0	6.0	3312	Y Yttrium*	0.3	0.5	4077
Mg Magnesium	0.0003	0.001	2852	Yb Ytterbium*	0.04	0.1	3988
Mn Manganese	0.0002	0.02	2795	Zn Zinc	0.002	0.015	2138
Mo Molybdenum	0.03	0.12	3133	Zr Zirconium*	5.0	—	3601

*Nitrous oxide flame required.
†Argon-hydrogen flame used.

$$\text{Optical Density} = \log\frac{Io}{I}$$

where,

Io = incident light intensity
I = transmitted light intensity

15.5 Detection Limits.

15.5.1 The preceding Table 15:I lists detection capabilities for double versus single beam instruments from one manufacturer.*

15.5.2 A comparison of atomic absorption detection limits with other methods is demonstrated in the following Figure 15:4. Absolute detection limits in nanograms for various elements are shown in Table 15:II.

15.6 Advantages.

15.6.1 Sensitivity and specificity are the two advantages to atomic absorption spectrophotometry. The sensitivity is high since atoms at the ground state only are detected. In the operating flame temperatures essentially all of the atoms of the analysis elements except those which form refractories in the flame, are atom-

*Perkin-Elmer Corporation, Norwalk, Connecticut.

Table 15:II. Absolute Detection Limits in Nanograms*

Element	Photometric	Atomic Absorption	Element	Photometric	Atomic Absorption
Ag	5.0	0.5	Na	8000.0	0.5
Al	0.5	10.0	Nb	5.0	2000.0
As	10.0	50.0	Ni	4.0	1.0
Au	5.0	1.0	Pb	6.0	1.0
B	50.0	600.0	Pd	10.0	50.0
Ba	100.0	10.0	Pt	10.0	50.0
Be	8.0	0.2	Rh	50.0	3.0
Bi	600.0	2.0	Ru	100.0	30.0
Ca	100.0	0.2	Sb	4.0	20.0
Cd	3.0	0.1	Sc	8.0	20.0
Co	3.0	0.7	Se	200.0	50.0
Cr	7.0	0.5	Si	100.0	10.0
Cu	2.0	0.5	Sn	60.0	10.0
Fe	200.0	1.0	Sr	10.0	1.0
Ga	40.0	7.0	Ta	40.0	600.0
Ge	20.0	100.0	Te	400.0	30.0
Hg	5.0	20.0	Ti	10.0	10.0
In	50.0	5.0	Tl	20.0	20.0
Ir	6.0	400.0	U	300.0	1200.0
K	200.0	0.5	V	10.0	2.0
La	4000.0	8.0	W	30.0	300.0
Li	3.0	0.5	Y	1000.0	30.0
Mg	2.0	0.03	Zn	100.0	0.2
Mn	5.0	0.5	Zr	50.0	500.0
Mo	10.0	10.0			

*Data for the photometric methods are on the basis of a 1-cm absorption path length. These impressive detection limits are those obtainable under *ideal* conditions, which include the absence of a large concentration of a major constituent and the satisfactory control of impurities in reagents.

ized in the ground state. The very high specificity is due to the very narrow resonance lines used. This essentially eliminates the optical interferences.

15.7 DISADVANTAGES.

15.7.1 The need to have one lamp for each analysis element is a disadvantage both in cost and required storage space. Some multi-element lamps are currently available, but the number is small compared to the number of elements of interest. Loss in sensitivity and accuracy caused by chemical interference in the atomizing and flame systems is another disadvantage. Solutions of high salt content are to be avoided to obtain the sensitivity listed by the instrument companies' literature. Due to the chemical interferences, the detection limits vary from one element to another as shown in Table 15:I.

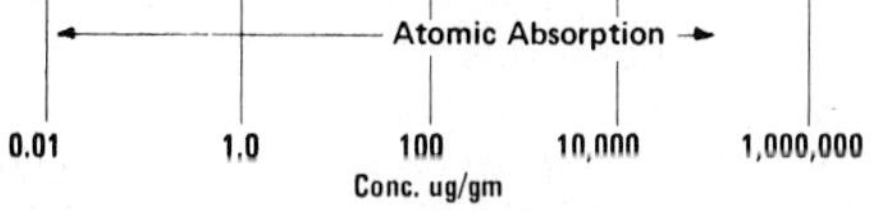

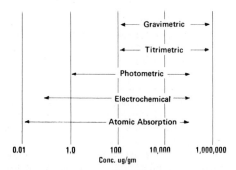

Figure 15:4—Useful ranges of chemical methods in quantitative chemical analysis.

15.8 APPLICATIONS.

15.8.1 The analytical application of atomic absorption like other analytical methods, has a lower detection limit which is specific for each individual instrument. It is therefore necessary to construct calibration curves using not only standard solutions, but also standard conditions for each individual instrument. The standard conditions include instrumental parameters, burner gas flames and aspiration rates. In routine sample analysis it is imperative that several standards be run with each set of samples so that the operating parameters are exactly the same for sample and standard. Standard procedures for analysis are supplied with most commercially available atomic absorption instruments. Standard curves must be constructed for *each* element and standards must be analyzed each time a set of samples are run. The standard curves should list all parameters of the instrument, as well as sample preparation methods.

15.8.2 Since the concentration of metals in biological specimens may be only a few micrograms, a concentration technique is frequently needed before a sample can be analyzed by atomic absorption. Concentration may be accomplished by using any of the common complexing agents such as ammonium pyrrolidine-dithiocarbamate, oxine, dithizone, cupferron, acetylacetone, and solvents such as chloroform, carbon tetrachloride, ethyl acetate, toluene and methyl-isobutylketone. The solvent solution of the metal complex can either be aspirated directly into the flame or a measured volume may be placed in a boat for analysis.† The technique of using the boat has advantages in that the detection limit is absolute, not relative as in aspirating.

15.8.3 The relative detection limit is the minimum concentration of an element which can be determined in a solution sample; the absolute detection limit is the minimum *weight* of an element which can be detected.

15.8.4 Aspirating a sample solution at a constant rate for an unlimited length of time yields the relative concentration. If a measured amount of sample is used, an absolute concentration is obtained. The sample boat technique improves the detection limits for easily atomized elements. A comparison is shown in Table 15:III.

†**Note:** *In the boat technique, the sample aliquot is dried in a thin tantalum metal container (boat) near the flame. When dry, the boat is placed quickly into the flame and the sample is rapidly volatilized producing a signal which is recorded.*

Table 15.III. Detection Limit (μg/ml)

Element	Boat	Nebulizer
Arsenic	0.02	0.1
Bismuth	0.003	0.05
Cadmium	0.0001	0.002
Lead	0.001	0.03
Mercury	0.02	0.5
Selenium	0.01	0.1
Silver	0.0001	0.005
Tellurium	0.01	0.3
Thallium	0.001	0.025
Zinc	0.00003	0.002

15.8.5 Atomic absorption methods now fill the analytical chemistry literature; many claims are made, but much work must be done to adapt most of the published methods for use in the industrial hygiene laboratory.

References

1. ROBINSON, JAMES W. 1966. *Atomic Absorption Spectroscopy*. Marcel Dekker, Inc., New York.
2. MUNOZ, JUAN RAMIREZ. 1968. *Atomic Absorption Spectroscopy and Analysis by Atomic Absorption Flame Photometry*. Elsevier Publishing Co., New York.
3. ASTM STP-443. May 1969. *Atomic Absorption Spectroscopy*. American Society for Testing and Materials, 1916 Race Street, Philadelphia, Penn.
4. KAHN, H.L., G.E. PETERSON, and J.E. SCHALLIS. March–April 1968. Atomic Absorption Microsampling with the Sampling Boat Technique. *Atomic Absorption Newsletter 7*, No. 2.
5. SANDELL, E.B. 1959. *Colorimetric Metal Analysis*. 3rd Ed., Interscience Publishers, Inc., New York, N.Y.
6. KOIRTYOHANN, S.R. 1967. Recent Developments in Atomic Absorption and Flame Emission Spectroscopy. *Atomic Absorption Newsletter*, Vol. 6, 77. Perkin-Elmer Corp.

16. Gas Chromatography

16.1 INTRODUCTION.

16.1.1 Chromatography is a technique for the separation of closely related compounds. The technique has been known for at least a century, but only since 1952 have its applications been investigated. In approximately 1956, U.S. scientists began using chromatographic methods for the separation and analysis of mixtures of gases and volatile materials. Today the field of chromatography has grown to amazing proportions with almost limitless applications (**1–4**).

In gas chromatography, the sample is vaporized (if not already in this form) and the mixture is passed by a stream of inert gas (carrier) through a rigid container (column) containing a packing material. The packing has different affinities for each particular component in the mixture and lets each component pass through at a different rate. As each component emerges from the column it is observed by a sensitive detecting device.

16.2 PRINCIPLE OF THE METHOD.

16.2.1 Basically, chromatography consists of two phases. One phase is a fixed or stationary phase. This phase may be either a solid, as in adsorption chromatography, or a liquid held by a solid, as in partition chromatography. The other phase is mobile and is generally referred to as the moving phase. This phase may be a gas, liquid or volatile solid. When phase equilibria occurs between the sample components, the moving phase and the fixed phase, the sample components are distributed or partitioned between the phases.

16.2.2 The gas chromatograph can best be compared to a fractional distillation apparatus. Chromatography is of course far more efficient than fractional distillation. A good distillation column may have 100 to 200 theoretical plates, whereas a chromatographic column operates with 1000 up to as many as 500,000 plates. Each of the above separate a mixture (liquid) into its component parts, which can then be analyzed separately.

16.2.3 There are two basic types of chromatography, gas adsorption and gas-liquid partition. To better understand the field of chromatography, a few definitions are in order,

a. gas-liquid chromatography—the moving phase is a gas and the fixed phase is a liquid distributed on an inert solid support.

b. gas-solid chromatography—the moving phase is a gas and the fixed phase is an active solid such as alumina, charcoal, silica gel, Molecular sieve, or the newer plastic granules (*i.e.*, Poropak).

c. gas chromatography—all methods where the moving phase is a gas (or vapor).

16.3 EQUIPMENT.

16.3.1 The three main components of a gas chromatograph are a sample introduction system, a column, and a detector. The sample introduction system must be capable of introducing the sample *unchanged* instantly and with reproducibility at the beginning of the column. The column is a rigid container made of metal, glass or other inert material which contains the fixed phase. This column will have various shapes depending upon the space available for housing the column (the oven). The detector is an apparatus which measures the changes in the moving phase. (Figures 16:1 and 16:2).

16.4 REAGENTS.

16.4.1 *Carrier gas.* The mobile phase used to move the sample through the column. Gases generally used are: helium, argon, hydrogen and nitrogen. Other gases may be used. For example, SF_6 is used when detecting permanent gases using a density balance detector.

16.4.2 *Liquid phase.* An essentially non-volatile (at the operating temperature used) liquid which is capable of dissolving the sample components and releasing them, preferentially by the difference in their volatility, from the solution.

16.4.3 *Solid support.* Usually an inert porous solid, either inorganic or organic, which is of a known and standard size. Active solids may be used if they are chemically inactivated. The walls of a capillary tubing may also be used as a solid support, as in Golay methods.

16.4.4 *Active solid.* Solid capable of adsorption of gaseous sample components and their preferential release, *i.e.*, the Molecular sieves.

16.5 PARTS OF THE CHROMATOGRAPH—Component parts.

16.5.1 *Sample Injection System.* Most samples are introduced to the column with a small calibrated syringe (micro-

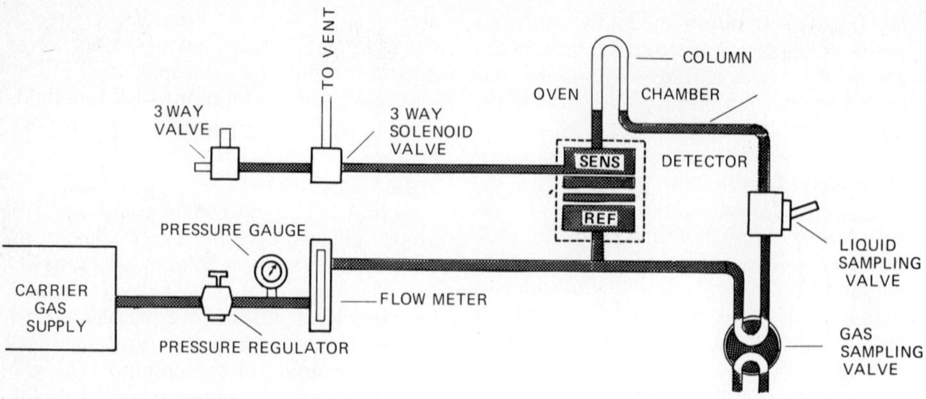

Figure 16:1—Basic diagram of a gas chromatograph—single column.

syringe), such as the Hamilton syringe shown in Figure 16:4. These syringes are made for delivering either liquid or gases. Liquid syringes are available in sizes of from 1 to 500 μl. Gas tight syringes are available from 50 to 2500 μl. For calibration with larger volumes of gases, plastic syringes are available in sizes from 0.5 to 1.5 l.

Samples may also be introduced to the column using a gas sampling valve as shown in Figure 16:5. Sample tubes are in various sizes (usually 1 to 10 ml volume).

The gas samples may be introduced to the gas sampling loop by either pressure or by drawing it into the loop using a small vacuum pump or a two-way squeeze bulb. The gas sample can be delivered to the loop from a pressure container, plastic or rubber bag or by large syringe. The only requirement is that sufficient sample is available to thoroughly purge the sample loop.

16.5.2 The columns used are of many sizes and shapes; examples are seen in Figure 16:6.

16.5.3 *Detectors.* There are essentially eight types of detectors:

a. Thermal conductivity (katharometer)—measures change in heat capacity;

b. gas density—measures change in density;

c. flame ionization—measures difference in flame ionization due to combustion of the sample;

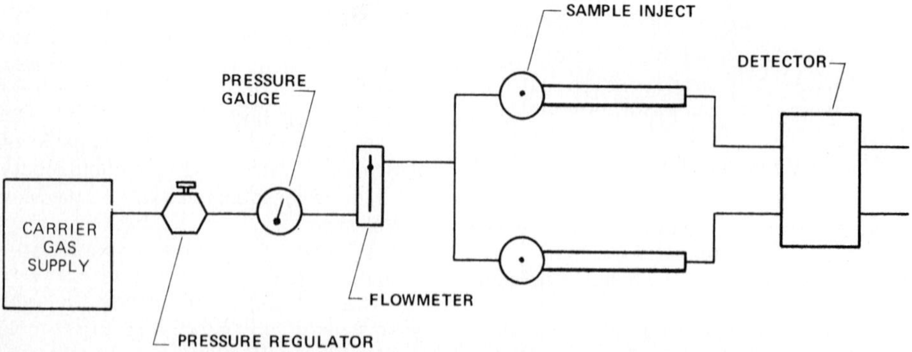

Figure 16:2—Basic diagram of a gas chromatograph—dual column.

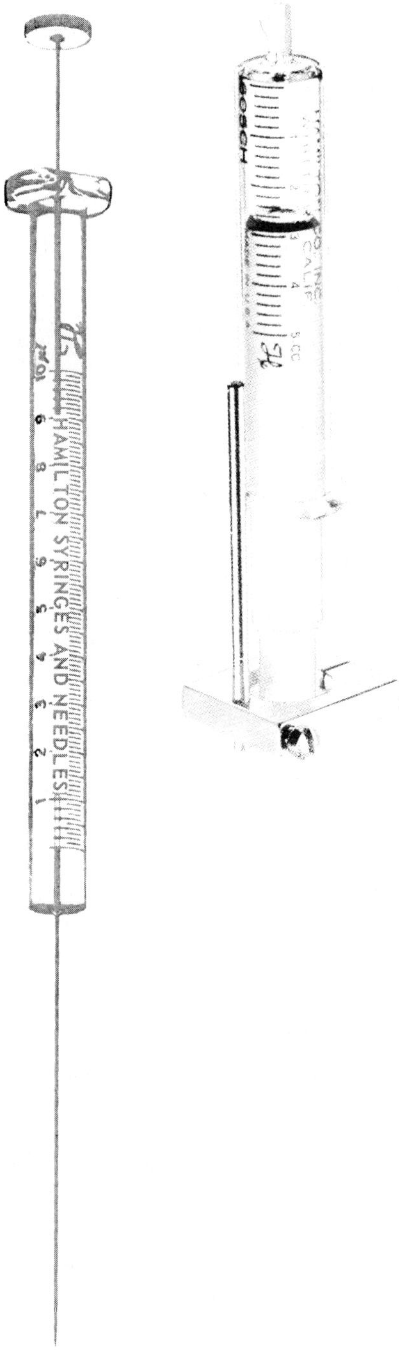

Figure 16:3—Syringe for liquids.

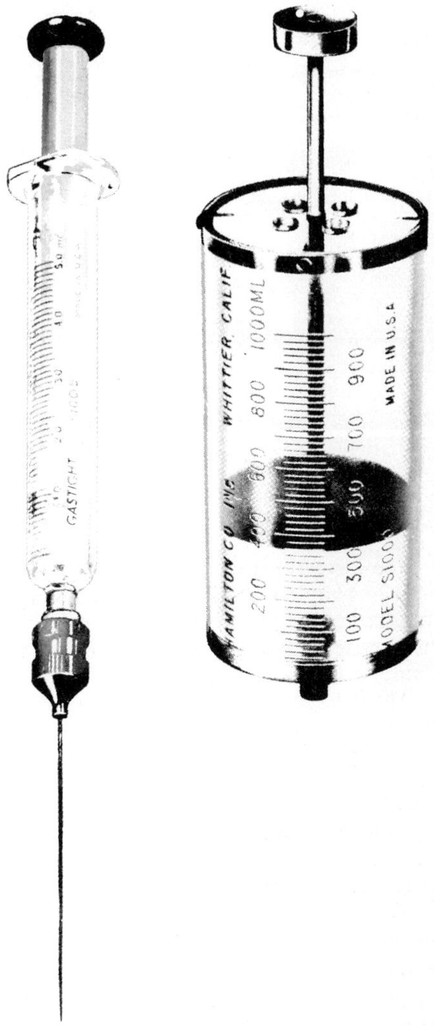

Figure 16:4—Syringe for gases.

d. beta-ray ionization—measures current flow between two electrodes caused by ionization of the gas from radioactivity;

e. photo-ionization—measures current flow between two electrodes caused by ionization of the gas from ultraviolet radiation;

f. glow-discharge—measures the voltage change between two electrodes caused by the change in discharge by different gas compositions;

g. flame temperature—measures the change in temperature caused by differ-

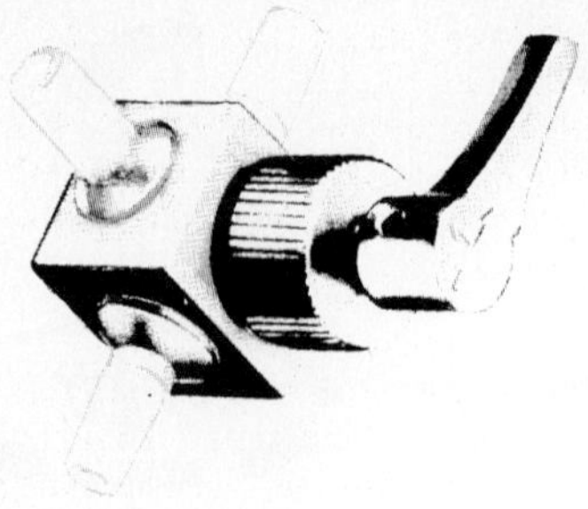

Figure 16:5—Gas sampling valve (Beckman).

Table 16:I. Sensitivity Obtained with Various Detector Types

Detector Type	Sensitivity in g/sec
Thermal conductivity	10^{-7}
Ionization detector	
ion cross section	10^{-3}
Argon diode	10^{-13}
Electron affinity	10^{-14}
Flame ionization	10^{-12}
Thermionic emission	10^{-10}

ence in gas composition in the flame; and

h. dielectric constant—measures the change in the dielectric constant caused by difference in composition of gas between plates of a capacitor.

The detector used depends upon the type of sample to be analyzed and the sensitivity needed for the analysis. Table 16:I lists the sensitivity obtainable with various detector types.

The detector output is generally presented on a recorder as a chromatogram. This chromatogram is a plot of detector response versus time. The area of the plot is used to quantitate the sample component and the time for the peak to appear is used to estimate the nature of the component. A recorder plot is shown in Figure 16:7.

16.6 GENERAL CONSIDERATIONS IN SELECTING THE PROPER COLUMN PACKING MATERIAL.

16.6.1 *Support mesh size and liquid phase concentration.* The column diameter generally determines the proper mesh size as follows:

¼″ diam column—60/80 mesh
⅛″ diam column—80/100 mesh

The liquid phase concentration is adjusted accordingly:

60/80 mesh approximately 15% by weight
80/100 mesh approximately 10% by weight

Columns using silicone gum as a liquid phase use 3 to 4% by weight.

16.6.2 *Selection of proper support material.* The most generally used support is a diatomaceous-earth. Teflon is generally used only if water, ammonia and alcohols are to be determined using a thermal conductivity detector. Glass beads can also be used in special applications.

16.6.3 *Selection of the proper column length.* The column length, of course,

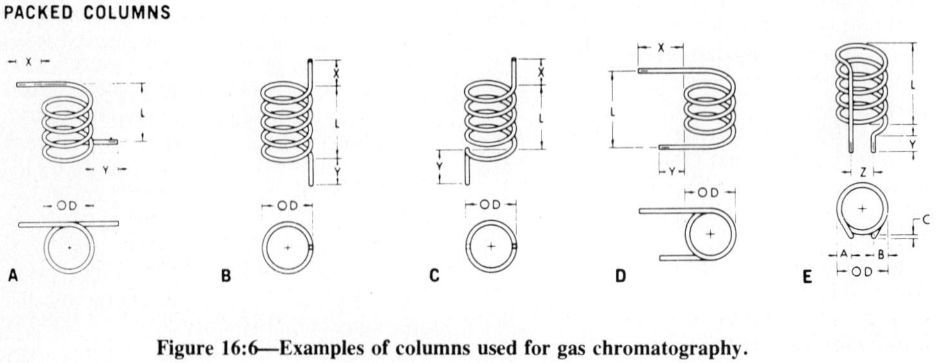

Figure 16:6—Examples of columns used for gas chromatography.

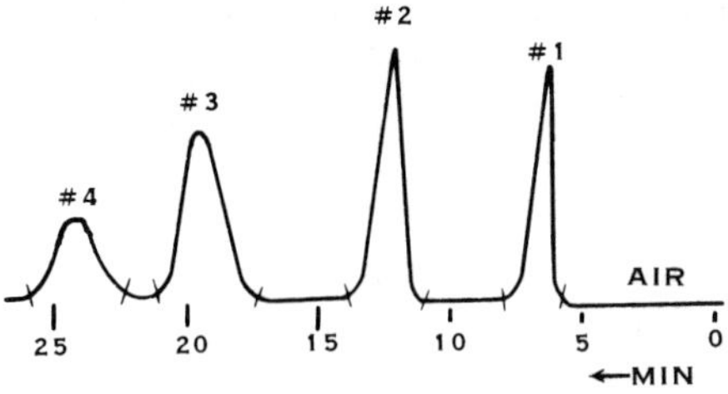

Figure 16:7—Detector response versus time.

must be determined by the analytical requirement. However, generally 6 to 12 ft columns are used. The column efficiency is increased by increasing the column length; however, the analysis time is also increased with column length. The proper column must, therefore, take both efficiency and time of analysis into account.

16.6.4 *Adsorbents.* Selection for adsorption and desorption of gases. The most widely used adsorbents and applications are:

a. Silica gel—used in the analysis of inorganic gases and light hydrocarbons;

b. molecular sieves—used for the separation of permanent gases such as H2$_1$ oxygen, nitrogen, carbon monoxide, methane and ethane. Carbon dioxide and higher hydrocarbons are irreversibly adsorbed on molecular sieves at low temperatures. Sieves are available in types: 4A, 5A and 13X;

c. activated charcoal—for separation of air, CO, CH_4, CO_2, C_2H_2, C_2H_4 C_2H_6 and nitrous oxide;

d. activated alumina—for the separation of air, CO, CH_4, CO_2, C_2H_2, C_2H_4, C_2H_6, C_3H_6, C_3H_8;

e. Chromosorb—for the separation of nitrogen, hydrocarbons, acid gases and basic gases;

f. Poropak Q—can be used to separate such widely different materials in the gas phase such as air, CO_2, SO_2, NO, N_2O, NO_2, H_2S, HGN, COS, HCI, CI_2, NH_3; and

g. Poropak N useful to separate C_2II_2 from C_2H_4 and C_2H_6.

Regeneration time and temperature for adsorbants are as follows:

a. alumina, silica gel and activated charcoal—30 min at 100 C;

b. molecular sieves—30 min at 300 C;

c. Poropak N and T—30 min at 180 C; Q,R.S,Q-S—30 min at 230 C; and

d. Chromosorb—30 min at 140 C.

16.6.5 *Solid Supports.*

The purpose of the solid support is to provide a large surface area for holding a fairly thin film of liquid phase. The main requirements for adequate support material are: chemical inertness and stability, large surface area, relatively low pressure drop and mechanical strength. The most generally used materials consist of diatomaceous earth processed or modified in various ways. Since diatomaceous earth supports are not completely inert, they are often treated chemically to inactivate them. Besides the various diatomaceous earth supports, porous polymer beads, Teflon and glass beads are used as solid supports. An example of diatomaceous earth supports are the Chromosorbs (Johns-Manville Corp.):

Chromosorb P—calcinated diatomaceous earth processed from firebrick (C-22);

Chromosorb W—flux calcinated diatomite prepared from Celite Filter Aids; and

Chromosorb G—developed especially for gas chromatographic analysis.

The various Chromosorbs are available in different qualities such as non-acid washed, acid-washed, silanized with hexamethyldisilazane, acid washed and silanized with dimethyldichlorosilane.

Another example of solid supports is the Poropak resins. The Poropak's are porous polymer beads which have the partition properties of a highly extended liquid surface without the problems of support polarity or liquid phase volatility which hamper gas-liquid chromatography.

A list of the general properties and applications of each type of Poropak follows:

Poropak N—intermediate polarity packing useful in separating formaldehyde from aqueous solution. Stable to 250 C;

Poropak P—lowest polarity of all types and has the ability to separate systems of intermediate polarity. Useable up to 300 C;

Poropak P-S—similar to type P; however, the labile sites have been deactivated by silanization which improves peak shape and the efficiency with aldehydes and glycol separations;

Poropak Q—separates hydrocarbons by vapor pressure and is useable up to 300 C;

Poropak Q-S—similar to type Q; however, labile sites are deactivated by silanization, and as a result highly polar materials such as organic acids may be analyzed in aqueous solution and show no tailing;

Poropak R—suitable for the separation of water from highly reactive inorganics such as Cl_2 and HCl;

Poropak S—suitable for the separation of normal from branched alcohols and is stable up to 250 C; and

Poropak T—the highest polarity of all the poropak resins and stable to 250 C.

Poropak may be used for the separation of most gases and compounds in the moderate boiling range (up to 200 C). High boiling aromatic and cyclic materials are strongly retained by Poropak and are very difficult to elute. When strongly polar materials such as acids or aldehydes are to be analyzed, silane treated supports should be used. All Poropak resin columns require a pretreatment before use. The column should be purged with carrier gas while heating to rid the resin of residual preparation chemicals.

16.6.6 *Liquid Phase.*

Some of the more common types of liquids used in gas chromatography are the following:

Squalane—a non-polar liquid used for the separation of aliphatic and naphthenic hydrocarbons. Stable up to 150 C; and

Apiezon L grease—a non-polar phase for general purpose columns suitable for the separation of n-alkanes, alkenes and halogenated compounds at temperatures of from 50 to 250 C.

To some degree the detector used must also be considered when selecting a column. For example, if a thermal conductivity detector is used, and the sample contains water (which the detector sees), Teflon would be a better support; otherwise, the tailing of the water peak could obscure other peaks. On the other hand, the flame ionization detector does not respond to water and therefore, the problem of tailing does not occur. When the electron capture detector is used, a liquid phase with low bleeding rate is very important. In such systems, DC-200 silicone oil with high viscosities are recommended.

The following compilations are useful in the determination of the proper column and operation conditions for analysis:

Gas Chromatographic Data, ASTM Special Technical Publication No. 343.

Gas Chromatographic Data Compilation, ASTM Special Technical Publication No. DS-25A, American Society for Testing and Materials, 1916 Race Street, Philadelphia, Pa.

16.7 Qualitative Analysis.

16.7.1 From the foregoing information, it is obvious that practically any mixture can be separated by gas chromatographic techniques. The only problem is in the qualitative determination of the mixture components. The separation achieved depends upon the column, the temperature, the detector and flow rate. Therefore, it is imperative that all these parameters be kept the same for both the sample and the standard used for the determination of component peak location. It is important to know that the unknown component is eluted in the same time as a known compound. It is also important to know that no other compound can appear at this location with the parameters used. The following Figure 16:8 illustrates the use of a known sample to determine the unknown components of the sample.

16.7.2 A more specific method for qualitative analysis is by the use of auxiliary instrumentation such as Infra-red, Ultraviolet or Mass Spectrometry. The sample components are trapped at the outlet from the gas chromatograph and are then transferred to the appropriate instrument and a qualitative analysis is performed.

16.8 Quantitative Analysis.

16.8.1 The prime application of gas chromatography is of course quantitative analysis. It is well known that the area un-

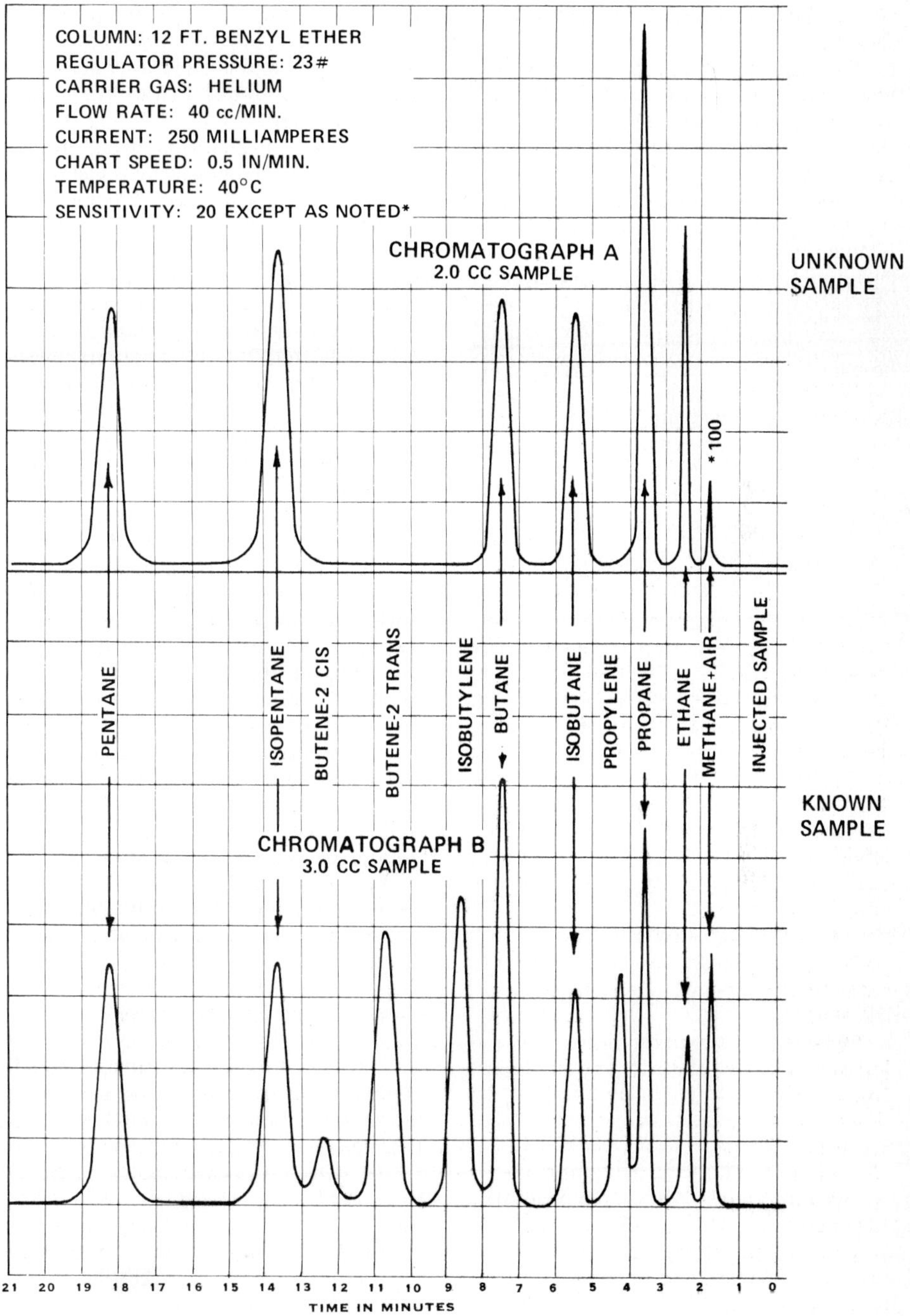

Figure 16:8—Use of known sample to determine unknown components of sample.

der a chromatographic peak is proportional to the amount of sample component in the carrier gas stream. This means that in order to be able to use gas chromatography for quantitative analysis, one has to know, first, the area of the peak and second, the proportionality factor to convert this area to concentration.

Areas can be determined by any of the following methods:

a. Automatic integrator;
b. polar planimetry;
c. by cutting out the peak and weighing the paper on an analytical balance;
d. multiplying the peak height by the peak width at half height;
e. Multiplying peak height by the retention time; or
f. calculating the area of the triangle formed by the two tangents drawn through the inflexion point of the peak, using the baseline as the base of the triangle.

16.8.2 The most commonly used of these methods are **a**, **b**, and **d**. The areas are expressed in any convenient form, the most generally used is area in cm^2. In the ideal case, where detector response is the same for all components in a mixture, a simple relationship is used to calculate percentage. As an example, assume that we have a four-component mixture of methane, ethane, propane and butane. The areas are 2.50, 1.25, 5.00, and 0.625 cm^2 respectively. The total area = 9.375 cm^2. The percentages are then; Methane = 26.67, Ethane = 13.33, Propane = 53.33 and Butane = 6.67, making the total 100.10%.

The ideal case however does not always apply. The response factors are different for individual compounds and this must be taken into consideration before calculation of percentage.

16.8.3 Standard calibration methods are numerous and only one will be discussed here. In this method, standards are run and the detector response is plotted versus concentration as shown in Figure 16:9.

16.9 CALIBRATION.

16.9.1 For accurate quantitative analysis the gas chromatographic system must be calibrated against known concentrations of the components of interest. Several methods are available for preparing known concentrations of gases and vapors for use in calibration.

Known concentrations of gases are prepared generally by dynamic methods where known amounts of the gas are mixed with diluting gas to yield the required concentration. A simple system is shown in the following sketch, Figure 16:10.

16.9.2 A second method which may be used either for the preparation of known concentrations of gases or vapor in air is the static method where a known volume of gas or volatile liquid is introduced (with an accurate volume measuring device) through a septum into a rigid container (which has been previously evacuated) of known volume. The gas or vapor (from the liquid) is then mixed with the diluting gas and stirred either by mechanical or thermal methods and samples are withdrawn from the vessel for use in calibration. This static system is shown in the sketch, Figure 16:11.

16.9.3 Samples from either calibration system can be introduced directly to the gas sampling valve of the chromatograph (if available) or may be contained in a plastic or rubber bag and samples taken from the bag with gas tight syringes and introduced to the chromatograph.

16.9.4 For accurate calibration, the concentration of the known mixture must be determined by standard chemical methods. A convenient method for determination of the concentration of many components is by infrared analysis.

16.9.5 There are several companies which market pressure containers of gases in known concentration for use in the calibration of gas chromatographic systems. These gas samples are very convenient but the concentration should be verified before the sample is used for calibration purposes.

16.10 METHOD FOR THE PRESENTATION OF CHROMATOGRAPHIC DATA. To be able to reproduce your work, or that of other investigators, all of the pertinent information concerning the analyses must be made available. The following informa-

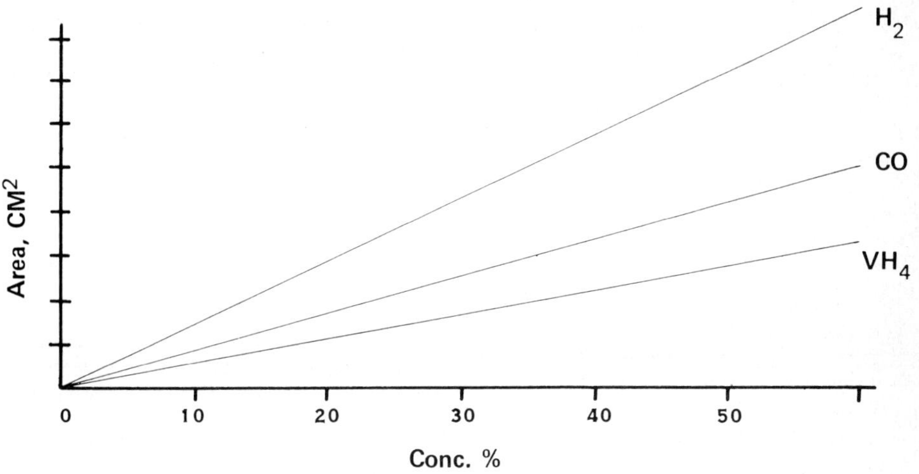

Figure 16:9—Detector response versus concentration.

tion is recommended in chromatographic reports (**5**).

Apparatus

Instrument.

Detector Type. Thermal conductivity, flame ionization, electron capture, etc.

Recorder Range.

Detector Voltage.

Bridge Current.

Column

Length and Diameter. Inside diameter is preferred.

Material. Glass, copper, stainless steel.

Packing. Weight % of liquid phase on support (give mesh size and pretreatment of support—*e.g.*, silanization).

Capillary. No packing, only liquid phase; support-coated liquid phase; or other type column.

Temperatures, C

Injection Port.

Detector.

Column (oven). Isothermal or Temperature Programmed. Give initial and final temperature and rate of temperature rise.

Flow Rate of Gases. In ml per min (ml per min) at exit port.

Carrier Gas.

Other Gases.

Sample. Volume injected in microliters (μl), milliliters (ml) if gas, concentration of solution.

Solvent Used.

Retention Time of Compounds. Minutes.

Relative Retention Time.

Quantitation

Methods. Peak area, peak height, integrator, planimeter, etc.

Precision. Repeatability.

Recovery of Added Amounts.

Accuracy. Correction, if made.

Weight. Expressed as weight, volume, or mole per cent.

Minimum Detectable Limits. Basis, 2 × noise levels or 2 × interference.

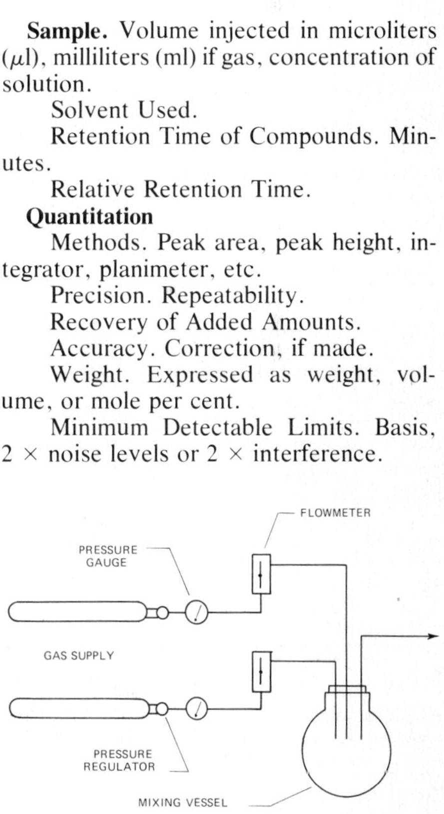

Figure 16:10—A simple mixing system.

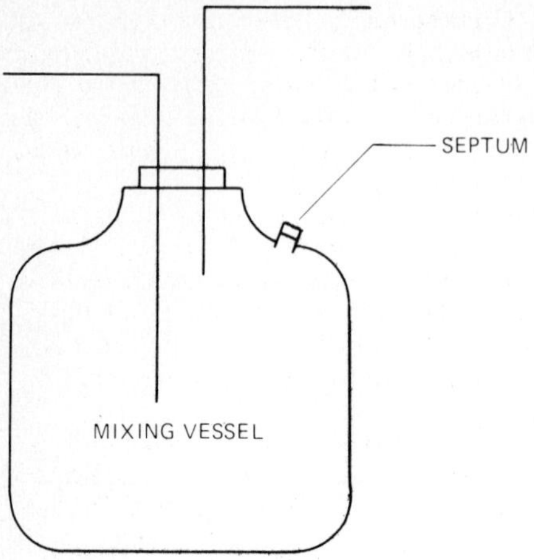

Figure 16:11—Static system.

Standards. Internal standard used for calibration.

Interference of Substrate. Baseline correction.

Chromatogram

Typical Analysis.

Retention Time (Horizontal) vs. Recorder Response (Vertical).

Labels, Peaks, attenuation, temperature program, title.

Gas holdup. Air peak, solvent front.

Special

Column conditioning procedure.

Adequate extraction (exhaustive) of compound(s) from sample.

Specificity of analysis.

Speed of analysis.

Interfering compounds related to those analyzed.

Dual-column operation, background subtraction.

Reaction gas chromatography—*e.g.*, hydrogenation, pyrolysis, etc.

Backflush technique.

Modifications. Valves, injection devices, stream splitters, trapping, flow controllers, equipment for transfer to other instruments (*e.g.*, spectrometers). Full description of material or apparatus not available commercially.

Technique for selecting representative samples.

Collaborative results.

Errors to be avoided.

How to keep apparatus clean and functional.

References

1. Ettre, Leslie S. and Albert Zlatkis. 1967. *The Practice of Gas Chromatography*. Interscience Publishers, New York, N.Y.
2. Littlewood, A. B. 1966. *Gas Chromatography—Principles, Techniques and Applications*. Academic Press, New York, N.Y.
3. Kaiser, Rudolf. 1963. *Gas Phase Chromatography* Volumes I, II, III Butterworth and Co. Ltd., Great Britain.
4. Lodding, William. 1967. *Gas Effluent Analysis* Marcel Dekker, Inc., New York, N.Y.
5. Reporting of Gas Chromatographic Methods. 1970. J. Agr. Food Chem., 18:552.

17. Radioactivity Analysis*

17.1 Introduction.

17.1.1 Radioactivity in samples of airborne particles or gases is measured by counting either *alpha*, *beta* or *gamma* emissions. The methods for detecting radiations vary, but the basic concept is to capture part or all of the energy released through decay, and to convert it into a current flow or pulse which can be recorded. A variety of sensitive instruments of the pulse recording type are available.

17.1.2 The gross count can be taken by totaling the pulses recorded on scalers. For spectral analysis, the pulses are sorted out by energy, and the number of pulses counted in each energy band is stored separately. A pulse height analyzer is used for this spectral analysis.

17.1.3 Most airborne radioactivity samples have low activity. Therefore, background radiation seen by the detector, such as natural activity in the earth and cosmic rays, must be minimized. Massive shielding and special electronic circuitry are used for this purpose.

*The material in this section is based on "Radioassay Procedures for Environmental Samples," January 1967. U.S. Department of Health, Education, and Welfare, Public Health Service Publication No. 999-RH-27.

17.1.4 Some of the important considerations in the selection of the proper instrument are:

a.The type of radiation emitted—*alpha, beta, gamma;*

b. the energy of the emission;

c. the condition of the sample when prepared for counting;

d. the activity level of the sample;

e. the radionuclide composition of the sample; and

f. the accuracy needed or desired.

17.1.5 The preparation of samples and standards for calibration of the selected instrument also depends upon the points mentioned above as well as on the type of data needed. Qualitative data identify the radionuclides present, whereas quantitative data determine how much of each is present. Spectrometry can provide both types of data depending on the extent to which the spectrum is analyzed. Without spectrometry, radiochemistry quantitatively isolates the element (a quantitative separation of the nuclide) so that the quantity of radionuclide present can be counted.

17.2 EQUIPMENT.

17.2.1 Instruments for measuring radioactivity make use of the ionizing properties of nuclear radiation. Basically they provide a medium with which the radiation can interact and a means of detecting and identifying this interaction. All instruments can be classified by their interaction medium under three general categories: gas ionization detectors, scintillation detectors or phosphors, and solid state detectors.

17.2.2 When a charged particle such as an *alpha* or *beta* particle moves through a gas at high velocity it may act on a gas molecule or atom to remove an electron and produce an ion pair. Gas ionization detectors contain an encloses volume of gas in which two collecting electrodes are placed. When a direct voltage is applied across these electrodes in the presence of ionizing radiation, five types of response will be observed as the voltage is increased.

a. In the *ion recombination region* of response many of the ion pairs recombine with each other before reaching the collecting electrode. Gas ionization instruments are not operated in this region of response.

b. In the *ionization chamber region* of response the applied voltage is sufficient to prevent significant recombination of ion pairs and the ions are collected in a one-to-one ratio with those formed. Ionization chamber instruments operate in this region of response.

c. At voltages in the *proportional region* of response the ions are accelerated to achieve enough energy to produce secondary ionization in the gas medium, thus amplifying the primary ion current. Although the ion current or pulse size is amplified, to an extent described by the "gas amplification factor," it is still proportional to the energy of the incident particle *beta* (or *gamma*) radiation. *Alpha* and *beta* proportional counters with or without windows to separate samples from the gas ionization medium operate at voltages in the proportional region of response.

d. Geiger counters, or GM counters, operate in the *Geiger-Mueller region* of response where the gas amplification factor is so large that an avalanche of electrons spreads along the entire length of the anode. Here the pulse size is independent of the number of primary ions or energy of the incident radiation. This accounts for the high sensitivity of GM detectors and their inability to distinguish between the various types of radiation.

e. In the *continuous discharge region* of response the gas amplification factor is so large that the gas begins to arc generating a series of self-perpetuating discharges. Operations of any gas ionization instrument at voltages in this region even for a few seconds will seriously damage the electrodes.

17.2.3 Scintillation detectors make use of the long-known ability of ionizing radiation to produce short-lived flashes of light (scintillations) in phosphors. The light flashes can produce photoelectrons from a photosensitive cathode in a photomultiplier tube and these photoelectrons are further amplified into a pulse which can be counted. When the scintillator is large enough to absorb all the energy of the radiation in the excitation and ionization interaction, the output pulse will be proportional to the energy of the incident radiation. This proportional response is necessary for gamma scintillation spectrometry. Scintillators are of many materials including organic crystals such as anthracene, liquid solutions of an organic scintillator such as p-terphenyl in an organic solvent, solid solutions of an organic scintillator in a plastic solvent (plastic scintillation detectors), inorganic crystals such as sodium iodide or zinc sulfide, and noble gases.

17.2.4 Solid state detectors make use of the property of some insulators such as diamond or silver chloride crystals, or semiconductors such as germanium or silicon, to show an instantaneous conductivity when high-energy particles or radiation interact with the material of which they are composed.

17.2.5 A useful description of the more common type of laboratory instruments used in radioactivity determinations may be found in Common Laboratory Instruments for Measurement of Radioactivity, June 1968, U.S. Department of Health, Education, and Welfare, Public Health Service Publication No. 999-RH-32.

17.3 CALCULATION OF RESULTS AND COUNTING ERROR.

17.3.1 *Integral Counting*.

Any of the integral counters provide a gross count rate which consists of counts due to both sample and background activity. To determine the activity from the sample alone, the background count rate of the instrument with no sample present, but otherwise identical, must be obtained. A clean sample container, similar to that used for the sample, should be in place for the background determination. The background count rate is subtracted from the sample count rate to obtain a net counting rate. This net count rate is then divided by the counting efficiency, previously determined for the particular sample type, to obtain the disintegration rate of the sample. The following equation expresses this relationship.

$$\text{Disintegration Rate (dpm)} = \frac{\text{Gross Count Rate (cpm)} - \text{Background Count Rate (cpm)}}{\text{Efficiency (cpm/dpm)}}$$

To express the result in activity units of pCi/l (or/g, etc.), the equation becomes:

$$\text{Activity (pCi/liter)} = \frac{\text{Gross Count Rate (cpm)} - \text{Background Count Rate (cpm)}}{\text{Efficiency (cpm/dpm)} \times 2.22\ \text{dpm/pCi} \times \text{Sample Volume (liters)}}$$

A large enough sample must be taken to insure that the disintegration rate will be statistically significant and not primarily the result of counting error. This counting error can be determined as follows:

$$E = Z\left[\frac{S}{t_s} + \frac{B}{t_b}\right]^{\frac{1}{2}}$$

where

E = permissible counting error (cpm)
Z = the constant associated with a given confidence level
S = gross count rate of the sample (cpm)
B = background count rate (cpm)
t_s = duration of sample count, or sample counting time (minutes)
t_b = duration of background count, or background counting time (minutes)

As an example, assume the gross count rate of the sample to be 120 counts/min and the background count rate to be 40 counts/min. If the counting time of the sample is 30 min, and the counting time of the background is 100 min, the counting error at the 95% confidence level will be

$$E = \pm 1.96\left[\frac{120}{30} + \frac{40}{100}\right]^{\frac{1}{2}}$$

$$= \pm 1.96[4 + 0.4]^{\frac{1}{2}}$$
$$= \pm 4.2 \text{ cpm}$$

The sample then has a net count rate of 80 ± 4.2 counts/min, or slightly more than a 5% counting error at the 95% confidence level. This means that we are 95% certain that the true count rate for the sample lies between 75.8 and 84.2 counts/min.

When the conversion is made from counts to disintegrations, care must be taken to obtain the proper efficiency value. For *alpha* or *beta* counting, self-absorption in the sample can be very significant, and appropriate efficiency values must be determined and applied. For particle counting, either *alpha* or *beta*, backscatter from the sample container contributes to the count rate, and the efficiency values must be obtained for the proper container, which is the same container in which the samples will be counted.

17.3.2 *Spectral analysis.*

It should be noted that the counting error for each radionuclide in a complex spectrum is difficult to determine mathematically, and that the method used for integral counting does not apply. There are errors associated with each radionuclide in its own and other photopeak areas. In general, the more radionuclides present in the spectrum, the larger the error becomes for any single nuclide, but the percentage of interference is of primary importance. It should also be noted that the low-energy gamma photons do not add to the counting error of high-energy gamma photons.

Descriptive material on the calculation of results of gamma analyses can be found in Radioassay Procedures for Environmental Samples, January 1967. U.S. Department of Health, Education, and Welfare, Public Health Service Publication No. 999-RH-27. Computer calculation of results is described in a Computer Program for the Analyses of Gamma Ray Spectra by the Method of Least Squares, August 1966. U.S. Department of Health, Education, and Welfare, Public Health Service Publication No. 999-RH-21.

17.4 Radioactive standards.

17.4.1 The proper selection and proper use of standards for calibrating any detection system are both essential to obtain reliable quantitative results.

17.4.2 The decay scheme used in standardization should be identified by the supplier, since there are discrepancies in reported decay schemes for most nuclides, and for some methods of standardization a knowledge of the decay scheme is essential. The amount and the chemical form of carrier added to the standard can also be of importance in minimizing adsorption on the container walls or volatilization of the solution. Such information is needed to make accurate dilutions for calibrating, since most standards have too high a disintegration rate to be counted on sensitive laboratory instruments without prior dilution.

17.4.3 It is essential that any standard prepared for instrument calibration be counted in the same geometrical configuration as the samples.

17.5 Quality Control.

17.5.1 Application of quality control principles to the analysis of airborne radioactivity measurements can minimize the amount of disparity within and among laboratories to a considerable extent. This is done by isolating discrepancies in analytical results and taking appropriate action to remedy the causative factor. While quality control procedures can greatly assist in obtaining consistent data, they obviously cannot guarantee accuracy. For maintaining production of precise, unbiased data, quality control is a continuous task.

17.5.2 Accurate data are data that are both precise and unbiased. Complete control over analytical measurements requires day-to-day control over instrumentation, chemical steps, and associated factors at each individual laboratory. The application of three allied but independent procedures has been found most useful for insuring accurate data. The procedures are:

a. "Blind" duplicate analysis of actual samples within one laboratory. This procedure allows a laboratory. This procedure allows a laboratory to evaluate its internal precision or reproducibility;

b. cross-check analysis of samples among several laboratories. This intercomparison procedure allows a laboratory to determine its agreement with other laboratories doing similar work;

c. special standard sample analysis. This procedure allows a laboratory to study its accuracy, *i.e.*, the agreement of its results with a known value.

17.5.3 Each of the three quality control procedures supplies the observer with different but complementary information. Depending on the number or scope of the analysis, one or more can be employed.

17.6 UNITS.

Units of Radioactivity

millicurie (mCi)	10^{-3} curies (Ci)
microcurie (μCi)	10^{-6} curies
nanocurie (nCi)	10^{-9} curies
picocurie (pCi)	10^{-12} curies
femtocurie (fCi)	10^{-15} curies

17.7 SAFETY.

17.7.1 Analyses of airborne radioactivity samples require no unusual safety measures. Although the procedures are for radioassay, the samples themselves constitute no hazard greater than that of the environment from which they were taken. The most likely source of potential radiation hazard is the undiluted radioactive solution obtained for purposes of preparing spiked samples for tracer studies or standards for calibration, performance references, or quality control.

17.7.2 People who are to work with radioactive materials should be thoroughly trained in their proper use. Basic radiological safety regulations appropriate to the license held by the laboratory for use of such materials and machines are set forth by the Atomic Energy Commission in the Code of Federal Regulations, Title 10—Atomic Energy. (United States Government. "Standards for protection against radiation," Code of Federal Regulations, Title 10—Atomic Energy, Chapter II, Section II, Part 20. U.S. Atomic Energy Commission, Revised 1966, Washington, Government Printing Office (1963).

18. Precision and Accuracy*

18.1 INTRODUCTION.

A clear distinction should be made between the terms "precision" and "accuracy" when they are applied to methods of analysis. *Precision* refers to the reproducibility of a method when it is repeated on a homogeneous sample under controlled conditions, regardless of whether or not the observed values are widely displaced from the true value as a result of systematic or constant errors present throughout the measurements. Precision can be expressed by the standard deviation. *Accuracy* refers to the agreement between the amount of a component measured by the test method and the amount actually present. *Relative error* expresses the difference between the measured and the actual amounts, as a percentage of the actual amount. A method may have very high precision but recover only a part of the element being determined; or an analysis, although precise, may be in error because of poorly standardized solutions, inaccurate dilution techniques, inaccurate balance weights, or improperly calibrated equipment. On the other hand, a method may be accurate but lack precision because of low instrument sensitivity, variable rate of biological activity, or other factors beyond the control of the analyst.

It is possible to determine both the precision and the accuracy of a test method by analyzing samples to which known quantities of standard substances have been added. It is possible to determine the precision, but not the accuracy, of such methods as those for suspended particulate matter, polycyclic aromatic hydrocarbons and many other contaminants because of the unavailability of standard substances that can be added in known quantities on which percentage recovery can be based.

*Modified and adapted from Standard Methods for the Examination of Water and Wastewater. (14th Ed.), 1976. Part 104A. pp. 20–26.

18.2 Statistical Approach.

18.2.1 *Standard deviation (σ).* Experience has shown that if a determination is repeated a large number of times under essentially the same conditions, the observed values, x, will be distributed at random about an average as a result of uncontrollable or experimental errors. If there is an infinite number of observations from a common universe of causes, a plot of the relative frequency against magnitude will produce a symmetrical bell-shaped curve known as the Gaussian or normal curve (Figure 18:1). The shape of this curve is completely defined by two statistical parameters: (1) the mean or average $\bar{x}$, of n observations; and (2) the standard deviation, σ, which fixes the width or spread of the curve on each side of the mean. The formula is:

$$\sigma = \sqrt{\frac{\Sigma\,(x - \bar{x})^2}{n - 1}}$$

The proportion of the total observations lying within any given range about the mean is related to the standard deviation. For example, 68.27% of the observations lie between $\bar{x} \pm 1\ \sigma$, 95.45% between $\bar{x} \pm 2\ \sigma$, and 99.70% between $\bar{x} \pm 3\ \sigma$.

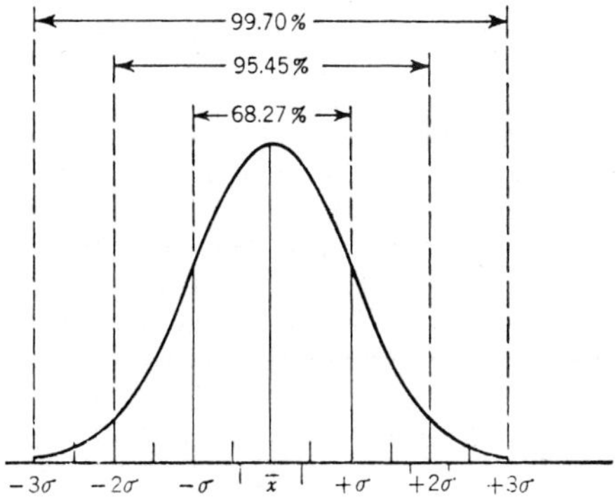

Figure 18:1—Gaussian or normal curve of frequencies.

These limits do not apply exactly for any finite sample from a normal population; the agreement with them may be expected to be better as the number of observations, n, increases.

18.2.2 *Application of standard deviation.* If the standard deviation, σ, for a particular analytical procedure has been determined from a large number of samples, and a set of n replicates on a sample gives a mean result $\bar{x}$, there is a 95% chance that the true value of the mean for this sample lies within the values $\bar{x} \pm 1.96\sigma\sqrt{n}$. This range is known as the 95% confidence interval. It provides an estimate of the reliability of the mean, and may be used to forecast the number of replicates needed to secure suitable precision.

If the standard deviation is not known and is estimated from a single small sample,† or a few small samples, the 95% confidence interval of the mean of n observations is given by the equation $\bar{x} \pm t\sigma\sqrt{n}$, where t has the following values:

n	t
2	12.71
3	4.30
4	3.18
5	2.78
10	2.26
∞	1.96

The use of t compensates for the tendency of small samples to underestimate the variability.

18.2.3 *Range (R).* The difference between the smallest and largest of n observations is also closely related to the standard deviation. When the distribution of errors is normal in form, the range, R, of n observations exceeds the standard deviation times a factor d_n only in 5% of the cases. Values for the factor d_n are:

†A "small sample" in statistical discussions means a small number of replicate determinations, n, and does not refer to the quantity used for a determination.

n	d_n
2	2.77
3	3.32
4	3.63
5	3.86
6	4.03

As it is rather general practice to run replicate analyses, use of these limits is very convenient for detecting faulty technique, large sampling errors, or other assignable causes of variation.

18.2.4 *Rejection of experimental data.* Quite often in a series of observations, one or more of the results deviate greatly from the mean whereas the other values are in close agreement with the mean value. At this point, one must decide whether to reject disagreeing values. Theoretically, no results should be rejected, since the presence of disagreeing results shows faulty techniques and therefore casts doubt on all the results. Of course the result of any test in which a known error has occurred is rejected immediately. For methods for the rejection of other experimental data, standard texts on analytical chemistry or statistical measurement should be consulted.

18.3 GRAPHICAL REPRESENTATION OF DATA.

Graphical representation of data is one of the simplest methods for showing the influence of one variable on another. Graphs frequently are desirable and advantageous in colorimetric analysis because they show any variation of one variable with respect to the other within specified limits.

18.3.1 *General.* Ordinary rectangular-coordinate paper is satisfactory for most purposes. Twenty lines per inch is often convenient. For some graphs, semilogarithmic paper is preferable.

The five rules listed by Worthing and Geffner for choosing the coordinate scales are useful. Although these rules are not inflexible, they are satisfactory. When doubt arises, common sense should prevail. The rules are:

a. The independent and dependent variables should be plotted on abscissa and ordinate in a manner that can be comprehended easily.

b. the scales should be chosen so that the value of either coordinate can be found quickly and easily;

c. the curve should cover as much of the graph paper as possible;

d. the scales should be chosen so that the slope of the curve approaches unity as nearly as possible;

e. other things being equal, the variables should be chosen to give a plot that will be as nearly a straight line as possible.

The title of a graph should describe adequately what the plot is intended to show. Legends should be presented on the graph to clarify possible ambiguities. Complete information on the conditions under which the data were obtained should be included in the legend.

18.3.2 *Method of least squares.* If sufficient points are available and the functional relationship between the two variables is well defined, a smooth curve can be drawn through the points. If the function is not well defined, as is frequently the case when experimental data are used, the method of least squares is used to fit a straight line to the pattern.

Any straight line can be represented by the equation $x = my + b$. The slope of the line is represented by the constant m and the slope intercept (on the x axis) is represented by the constant b. The method of least squares has the advantage of giving a set of values for these constants not dependent upon the judgment of the investigator. Two equations in addition to the one for a straight line are involved in these calculations:

$$m = \frac{n\,\Sigma xy - \Sigma x\,\Sigma y}{n\,\Sigma y^2 - (\Sigma y)^2}$$

$$b = \frac{\Sigma y^2\,\Sigma x - \Sigma y\,\Sigma xy}{n\,\Sigma y^2 - (\Sigma y)^2}$$

n being the number of observations (sets of x and y values) to be summed. In order to compute the constants by this method, it is necessary first to calculate Σx, Σy, Σy^2, and Σxy. These operations are carried out to more places than the number of significant figures in the experimental data because the experimental values are assumed to be exact for the purposes of the calculations.

Example: Given the following data to be graphed, find the best line to fit the points:

Absorbance	Solute Concentration mg/l
0.10	29.8
0.20	32.6
0.30	38.1
0.40	39.2
0.50	41.3
0.60	44.1
0.70	48.7

Let y equal the absorbance values, which are subject to error, and x the accurately known concentration of solute. The first step is to find the summations (Σ) of x, y, y^2, and xy:

x	y	y^2	xy
29.8	0.10	0.01	2.98
32.6	0.20	0.04	6.52
38.1	0.30	0.09	11.43
39.2	0.40	0.16	15.68
41.3	0.50	0.25	20.65
44.1	0.60	0.36	26.46
48.7	0.70	0.49	34.09
Σ = 273.8	2.80	1.40	117.81

Next substitute the summations in the equations for m and b; $n = 7$ because there are seven sets of x and y values:

$$m = \frac{7\,(117.81) - 2.80\,(273.8)}{7\,(1.40) - (2.80)^2} = 29.6$$

$$b = \frac{1.4\,(273.8) - 2.80\,(117.81)}{7\,(1.40) - (2.80)^2} = 27.27$$

To plot the line, select three convenient values of y—say, 0, 0.20, 0.60—and calculate the corresponding values of x:

$$x_0 = 29.6(0) + 27.27 = 27.27$$
$$x_1 = 29.6(0.20) + 27.27 = 33.19$$
$$x_2 = 29.6(0.60) + 27.27 = 45.04$$

When the points representing these values are plotted on the graph, they will lie in a straight line (unless an error in calculation has been made) that is the line of best fit for the given data. The points representing the latter are also plotted on the graph, as in Figure 18:2.

18.4 SELF-EVALUATION (DESIRABLE PHILOSOPHY FOR THE ANALYST)

A good analyst continually tempers his confidence with doubt. Such doubt stimulates a search for new and different methods of confirmation for his reassurance. Frequent self-appraisals should embrace every step—from collecting samples to reporting results.

The analyst's first critical scrutiny should be directed at the entire sample collection process in order to guarantee a representative sample for the purpose of the analysis and to avoid any possible losses or contamination during the act of collection. Attention should also be given to the type of container and to the manner of transport and storage, as discussed elsewhere in this volume.

A periodic reassessment should be made of the available analytical methods, with an eye to applicability for the purpose and the situation. In addition, each method selected must be evaluated by the analyst himself for sensitivity, precision, and accuracy, because only in this way can he determine whether his technique is satisfactory and whether he has interpreted the directions properly. Self-evaluation on these

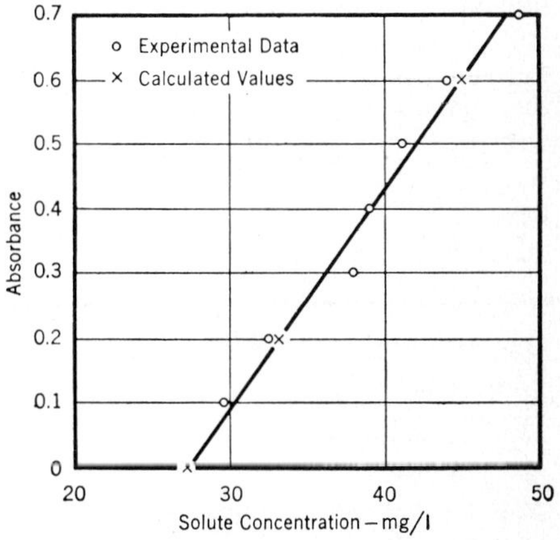

Figure 18:2—Example of least-squares method.

points can give the analyst confidence in the value and significance of his reported results.

The benefits of less rigid intralaboratory as well as interlaboratory evaluations deserve serious consideration. The analyst can regularly check standard or unknown concentrations with and without interfering elements and compare results on the same sample with results obtained by other workers in the laboratory. Such programs can uncover weaknesses in the analytical chain and permit improvements to be instituted without delay. The results can disclose whether the trouble stems from faulty sample treatment, improper elimination of interference, poor calibration practices, sloppy experimental technique, impure or incorrectly standardized reagents, defective instrumentation, or even inadvertent mistakes in arithmetic.

19. General Safety Practices

19.1 Introduction. The safety practices described below are the result of many decades of experience in academic and industrial laboratories. Some of them appear too obvious to be mentioned, but at least one unwary person has suffered at the hands of each of them.

19.2 Personal Protection. The corrosive nature of most chemicals on the eyes and the consequences of blindness are so severe that some form of eye protection is essential. This can be spectacles with specially armored lenses, which are available from most optical suppliers, either plain or with prescription lenses and with or without side shields. These glasses would offer protection against the unexpected. Where a splash or flying particles are expected, a face shield should be worn.

There are many types of gloves available: rubber for hazardous liquids, leather for physical hazards such as glassware and dry ice, asbestos for very hot objects, and even polyethylene hand guards to keep loose or dusty chemicals off the hands. Experience has shown that the majority of laboratory injuries is to the hands; proper use of gloves could have prevented most of them. All the special equipment mentioned above can be obtained from most laboratory supply houses.

It is essential that contact of the body with chemicals be kept to a minimum. The use of laboratory coats or aprons, the use of gloves, frequent showers and changes of clothing will help. If a chemical does touch the body, it should be removed promptly by thorough washing with water. If the chemical is known to be toxic, or if damage to the skin or eye results, a qualified physician should be consulted, preferably one who is familiar with the action of chemicals on the body.

For the same reason, the consumption of food or beverage in a chemical laboratory should be avoided wherever possible. Hands and face should be washed before eating. Chemicals should not be stored in refrigerators containing food.

Smoking in a laboratory not only creates a fire hazard, but offers one more possibility for ingesting chemicals. It must not be done near any flammable solvents, and the part that contacts the mouth must be protected from contamination by chemicals.

No one should work alone in a laboratory without arranging in advance for frequent contacts with some other person who can summon aid in case he is overcome.

19.3 Electrical Equipment. Electrical equipment (even 110 volts) has been known to kill people. Whenever possible, the chassis should be grounded so that any short will blow a fuse and not subject someone to electrocution. Electrical repairs should not be attempted by those unfamiliar with the equipment, and should be done with the equipment disconnected. Working on "hot" circuits is for professionals only. Some electrical equipment can cause a spark, and should not be used near flammable solvents.

19.4 Laboratory Materials. Many accidents have been caused by use of unlabeled or poorly labeled chemicals. Keep all chemicals clearly labeled and discard any that are not so.

Because of the fire hazard, the amount of flammable solvents should be kept at a minimum. Quantities in excess of five gallons should be stored outside the main building. Metal safety cans having flame arresters in the pouring spouts and a spring-loaded lid on the spout should be used for quantities from 1 liter to 5 gallons. Smaller amounts may be stored in glass, but only one bottle of each solvent should be in the laboratory.

Acid containers should be stored in an acid-proof tray that will hold the entire contents should the bottle break. Do not store chemicals that may react with each other in the same tray, *e.g.*, acids and ammonium hydroxide. To dispense small amounts more easily, 500-ml reagent bottles may be filled from the larger bottles.

Flammable solvents constitute a severe fire hazard and must be handled with care. Heat only on a steam or water bath, or spark-proof hot plates. Ground all metal containers when pouring from one to another.

Strong oxidizing agents (such as chlorates, perchlorates, permanganates, hypochlorites, peroxides, etc.) can cause a fire or explosion in contact with organic material. Clean up any spilled material promptly. Do not throw waste material in the trash can, but flush it down the sink with plenty of water. Do not use perchlorate drying agents (Dehydrite, Desichloride or Anhydrite) for drying organic materials.

The free acids of such materials as potassium permanganate and potassium chlorate are very unstable. Keep strong acids away from these materials. Strong hydrogen peroxide solutions can cause burns. Handle them with rubber gloves.

It is frequently desirable to use flammable solvents in wash bottles. Plastic or plastic-coated bottles are much safer than glass for this purpose. Flexible plastic bottles have the additional advantage that the liquid may be dispensed by squeezing the bottle.

The use of open flames for heating in the analytical laboratory is decreasing due to the hazards involved. Hot plates, particularly spark-proof units, are more convenient and much safer.

Heating of solvents in open vessels is dangerous and should be done only in a hood. Large quantities (250 ml or more) should be heated with a metal pan under the glass vessel. Certain solvents, such as ethyl ether, dioxane and Cellosolve can form peroxides. These solvents should never be evaporated to dryness. Empty glass vessels should not be heated. Severe strains can be set up, causing the vessel to fail unexpectedly.

19.5 Glassware. Glassware represents one of the chief hazards in the laboratory due to the ease with which it is broken and the razor-sharp edges which broken glass presents. Broken or chipped glassware should be repaired, including reannealing, or discarded immediately. A special container should be provided for broken glassware to prevent injury to janitorial employees.

Beakers should not be carried by the top edge. Small beakers should be carried by gripping around the side. Large beakers should be carried with one hand supporting the bottom.

Pipets obviously were designed for filling by mouth suction, but this is a very dangerous practice. Apply vacuum only by a rubber bulb or a vacuum hose.

The neck of a volumetric flask is its weakest point because of the calibration ring. Flasks over 100-ml capacity should be handled with both hands. Avoid vigorous motion which might snap the neck from the base.

Graduated cylinders tip very easily. Place them at the back of the work bench.

19.6 Laboratory Operations. When using Buchner funnels, apply vacuum *before* pouring material on the funnel. Accidents have occurred through connecting the hose to an air or gas outlet by mistake and blowing the material from the funnel.

When heating material in test tubes, do not point the mouth of the test tube at anyone. The material may spurt out. Apply heat only to test tubes of heat-resistant glass such as pyrex, and use a test tube holder while heating. Pour out the contents before leaving the test tubes for final cleaning.

Protect the hands with leather gloves when cutting glass tubing or inserting tubing or thermometers in rubber stoppers or tubing. Fire polish the sharp edges of glass tubing before using or storing. Remove frozen rubber stoppers or tubing from glass tubing by cutting with a knife. Wear leather gloves. An alternate method is to work a cork borer of the proper size between the glass and the stopper until the glass is free.

Wherever possible, substitute plastic for glass tubing and fittings. Plastic or metal "Y's" and "T's" are now available.

Patience is essential for the safe freeing of frozen stopcocks. Use a stopcock puller if available. Wear leather gloves. Tap the plug lightly with the wooden handle of a spatula and/or apply heat gently.

When glassware assemblies are erected, all glassware should be adequately supported by metal rings, tripods, or clamps. If heat is to be applied, arrange for rapid removal of the heat source if needed. When adjusting clamps, be careful not to put any lateral tension on the glass when the clamp is tightened. When arranging clamps in a vertical assembly, keep all the clamps on the same side of the vertical supporting rod. This makes backward and forward movement of the whole assembly possible. If the assembly is to contain dangerous solvents or mercury, a catch pan should be placed under it to contain the liquids in case of breakage.

Rubber tubing attached to glassware or plastic should be secured with wire or several wraps of a rubber band to keep it from coming off.

19.7 Dewar Flasks. If a Dewar flask should break, the resulting implosion can scatter sharp glass fragments over a wide area. To guard against this hazard, Dewar flasks should be encased in metal jackets or covered with tape.

The tape to be used should have a cloth body (medical type) to give it sufficient strength. Masking tape or plastic electrical tape is not recommended. A 25% minimum overlap between rows should be used. The flask should be completely taped over the outside wall whenever possible. If it is necessary to view through the wall, the viewing area should be minimized and be shielded by taping a piece of clear plastic in place under the tape. The tape should be cut carefully to expose only as much plastic as necessary.

19.8 Chemical Effects. Chemicals can affect the human body in a variety of ways. Some of these are:

Dermatitis. This manifests itself as a reddening of the skin or even eruptions. This has been known to result in death in extreme cases. Immediate treatment by a physician familiar with chemical dermatitis is essential.

Cyanosis. This causes a bluish cast to the lips and fingernails, dizziness and headache, and may not appear until several hours after exposure. The consumption of alcoholic beverages will often aggravate the condition. The patient should be kept quiet and a physician summoned. Oxygen should be administered to help breathing. Cyanosis can be fatal, prompt treatment is essential.

19.9 Hoods. Laboratory fume hoods provide not only efficient ventilation, but also shielding for potentially dangerous reactions. However, their safe use requires proper attention to several factors.

Hoods should not be used for disposal of large amounts of noxious gases. Use a scrubbing or absorption system. Most analytical operations are conducted at benchtop level. Therefore, the hood should be equipped with a baffle in the rear to allow most of the draft to be at bench top. If manipulations require the hood window to be open during use, place all equipment at least 10″ in from the edge of the bench to make sure that no fumes are carried out of the hood by stray air currents.

Hoods should be tested at least once a year to insure their effectiveness. This is best done by producing a smoke (titanium tetrachloride ampoules are very convenient) and observing the hood's action visually. A satisfactory hood will not release any smoke into the room when the smoke is released 10″ in from the hood face. Even cross currents from windows or people walking should not cause smoke to escape from the hood.

19.10 Pressure Precautions. Any reactions that may endanger personnel through splashes or explosion should be shielded. The shield should completely surround the reaction if it is set up on an open bench. All shields should be well supported to contain flying debris in case of an explosion.

Pressure vessels should be equipped with suitable relief devices, and should be tested hydrostatically at least once a year to 2 ½ times their working pressure.

Large cylinders (over 5 l) should be stored outside the laboratory and their contents piped in. Cylinders should be anchored at all times to prevent tipping. They should not be stored near radiators or other sources of heat. If it is necessary to heat a cylinder, first connect it to a suitable pressure relief device. Use water or steam baths or infrared lamps for heating. If gas from a cylinder is being released below the surface of a liquid, be sure to have a trap in the line to prevent suck-back into the cylinder. Systems to carry oxygen under pressure should be carefully cleaned of any oxidizable substance such as oil.

19.11 New Equipment. New equipment received in the laboratory can constitute a hazard because of the unfamiliarity of personnel with it. Also, the manufacturers do not engineer their products for safety as well as they might. Therefore, new equipment merits special attention.

Electrical equipment should have its chassis grounded by a three-wire cord. Hot areas, which might cause burns, or sharp edges which might cause cuts should be adequately guarded. Electrical connections should not be exposed. Any pressure equipment should have a suitable pressure relief device in working order.

After the preliminary examination is complete, a "dummy" run should be made, with special attention to any possible hazards. This, also, familiarizes personnel with the equipment. After all safety precautions have been taken, a run with a sample may be made.

19.12 Safety in Unit Operations of Analytical Chemistry.

19.12.1 *Introduction.* Fortunately, the hazards associated with analytical chemistry are minimized by the use of small samples and known reactions. Experience indicates, however, that analysts are subject to many hazards which can cause severe consequences if precautions are not taken. Problems arise from two major sources—the chemical nature of the sample and hazards associated with the procedure (reagents, reactions and equipment).

Effective communication of safety information is very important for safe operation. In many cases, an analyst is called upon to analyze a sample prepared by someone else using a procedure developed by another analytical chemist. Accordingly, the analyst may have no first hand knowledge of the hazards associated with either the sample or the procedure. This problem is most acute in an analytical laboratory providing non-routine analyses on samples from research groups. In order to ensure communication of safety information, personal contacts between the submitter and analyst are encouraged. It is also advisable to include safety information in formal written procedures or publications and on request-for-analysis forms.

Analytical procedures which are written in a standard format should include a definite safety section which reminds the writer to consider these problems and enables the reader to find the information conveniently. This section should include, when appropriate, the type and degree of hazard, routes for exposure, need for special medical department clearance, safety equipment to be used, instructions for special disposal of sample, decontamination of equipment and procedure to be followed in case of an accident. Brief safety instructions should also be included where pertinent in the procedure section of the analytical method. If an analytical procedure is used to analyze many different kinds of samples, it is cumbersome to include sufficient information on sample hazards in the procedure and this information is then indicated on request-for-analysis forms. These forms should include a check list which reminds the submitter to make recommendations regarding precautions to be taken in handling the sample.

When an analytical research chemist is studying a new technique or an unknown sample, he may have no source of firsthand pertinent safety information. In this case, he should give considerable thought to the hazards that might be foreseen from general safety and chemical knowledge.

19.12.2 *Preparation of Sample.* Hazards associated with preparing liquid or solid samples for analysis are usually derived from the chemical nature of the sample. Before drying or grinding a sample, it is well to make sure that it is not sensitive to impact, friction, or heat. Many samples are corrosive or can cause dermatitis; contact with these chemicals must be avoided.

Materials which are gases at room temperature provide additional handling problems. Small samples at or below atmospheric pressure are usually taken in glass sampling bulbs. Hazards associated with the use of these containers are (*a*) the possibility of implosion during evacuation, (*b*) explosion from too great an internal pressure, and (*c*) contamination of laboratory air by leakage of toxic contents. Bulbs should be covered with a metal screen and evacuated behind a shield. Filling should be monitored with a manometer. Stopcock holders should be used and bulbs containing toxic materials should be stored in a hood.

Large gas samples at pressures exceeding 760 Torr are best handled in metal cylinders. The common method of transferring gaseous materials to cylinders involves their condensation at atmospheric pressure with a carbon ice-solvent or liquid nitrogen bath for cooling followed by their transferral into suitable cylinders. In addition to the possible toxicity of the gas, the following dangers are inherent in these procedures:

a. If the vapor pressure of the compound at ambient temperatures exceeds the service pressure of the cylinder, rupture of the cylinder may result. Interstate Commerce Commission (ICC) regulations require that a product's vapor pressure at 21 C (70 F) not exceed the service pressure of the cylinder and that the vapor pressure P at 55 C (130 F) not exceed 1.25 times the cylinder's service pressure (**3**). These regulations also require periodic hydrostatic testing of the cylinders;

b. if too much sample is loaded into a cylinder at a low temperature, expansion of the liquid during warming to room temperature may cause the cylinder to become liquid-full and rupture explosively. This hazard is avoided by determining the maximum weight of sample which the cylinder will hold safely and weighing the cylinder before and immediately after filling to make sure that this weight has not been exceeded;

c. oxygen (boiling point, −183 C) from the air tends to condense in cylinders) cooled with liquid nitrogen (bp −196 C) and creates a possible explosion hazard due to excess pressure build-up or explosive reaction with organic compounds. For this reason, a carbon ice-solvent bath is used where possible (*i.e.*, for compounds boiling above −50 C). When liquid nitrogen is used, precautions must be taken to avoid introduction of air into the system. Liquid air should not be used as a coolant since the less volatile oxygen is concentrated during evaporation and has been known to react violently with organic material which leaked from a broken glass trap;

d. common high-pressure cylinders are constructed of steel and may be weakened by crystallization at liquid nitrogen temperatures. Suitable stainless steel cylinders are available commercially;

e. during the transfer of a liquefied gas from a glass cold trap to a sampling cylinder, pressure buildup may occur due to a restriction in the lines or insufficient cooling. An open-end manometer should be included in the system to indicate such a problem so corrective measures can be taken.

19.12.3 *Decomposition and Disolution of Samples.*

a. Inorganic Samples—Dissolution of inorganic compounds by heating with acids or fusion fluxes requires careful handling of very corrosive reagents. Use of a hood is required if toxic gases are evolved. Perchloric acid must be used cautiously. Use of this acid with metallic bismuth and antimony (**1**), strong reducing agents (**12**) and finely divided metals requires special precautions. In general, dilution of the perchloric acid with water or nitric acid tempers the reaction (**1**), but specific information should be obtained for each system. Shields are recommended for perchloric acid dissolutions and fusions with strong oxidants. Leather gloves and face shields are sometimes appropriate.

Hydrofluoric acid is sometimes used for dissolution of siliceous materials and volatilization of silicon tetrafluoride. Both concentrated and dilute hydrofluoric acid will cause painful skin burns which heal slowly. Burns caused by dilute acid may not appear until many hours after exposure. Vapors of hydrofluoric acid will cause extreme irritation of eyes, skin, mucous membranes and lungs. Moderate concentrations of the vapor will cause first degree skin burns; large concentrations will cause pulmonary edema and death. Containers of the acid should be stored in a lead tray in a cool and well-protected location. Analysts should wear rubber gloves and a rubber apron when handling the reagent. Work should be carried out in a fume hood and the hood window or a safety shield should be used for protection of the analyst when the reagent is being used.

b. Organic Samples—The initial step in analysis of organic compounds is often destruction of the organic matter. It is usually at this step that the most concentrated reagents and most vigorous reactions are employed. As a result, this is often the most hazardous operation in the procedure. Prior knowledge of the physical and chemical nature of the sample and information on hazards of the reagents and reactions are very important for safe operation.

c. Dry Ashing—Dry ashing is often the least hazardous way of destroying organic materials. Heat-sensitive explosive compounds, *e.g.*, nitro compounds, must not be ashed until it is established that this can be done safely. Experiments with questionable materials should be carried out on a small scale and behind a barricade. Preliminary ashing of nitrogenous materials should be carried out in a hood since hydrocyanic acid is sometimes evolved.

Use of an oxygen bomb for combustion of organic materials with oxygen under pressure involves the potential hazard of explosive rupture of the bomb. However, if the equipment is kept in good condition and operating instructions are followed carefully, the technique can be utilized quite safely. Bombs should be hydrostatically tested periodically and the sample size should never exceed the amount specified by the manufacturers of the bomb. A barricade and a normally open ignition switch should be used.

The oxygen flask technique for combusting organic samples has been reported to be perfectly safe by some investigators; however, a few isolated explosions have occurred with the hazard of flying glass from the flask. Two systems have been proposed for igniting the sample while the flask is behind a shield. One involves ignition with an electrically heated platinum wire (**7**), while the other utilizes the heat from a projector bulb for the ignition (**8**). Both procedures are simple and convenient, and certainly reduce the hazard when burning thermally unstable samples. If flame ignition is used, the flask should be held behind a ¼″ plastic shield and a heavy leather glove should be worn. Not more than 100 mg of sample should be burned in a 500-ml flask. The flask should be made of heavy-walled heat-resistant glass.

d. Wet Ashing—Wet ashing of organic samples may involve the possibility of explosions due to too rapid oxidation, formation of explosive intermediates (*e.g.*, nitro compounds or organic perchlorates), or rapid depolymerization of polymers. In general, digestions should be carried out in fume hoods in order to remove the acid fumes and confine any vigorous reactions or explosions that may occur. Safety spectacles, rubber gloves and a rubber apron should be worn by the analyst. A shield should be used for experiments with untested reactions. Use of micro methods reduces the hazards considerably.

The choice of reagents for wet combustion has an important influence on the hazards involved. Nitric acid is hazardous when the sample may be nitrated. Carius digestions with nitric acid also involve the difficulty of obtaining a good strain-free seal in the tube. Several procedures for sealing have been recommended (**2, 9**). Again, the amounts of sample and reagent should not exceed amounts known to be safe. Tubes should be enclosed in proper steel sheaths and should not be removed until the tubes have been cooled and opened.

Most organic materials can be mineralized conveniently by charring with sulfuric acid followed by heating with small incremental additions of nitric acid and finally heating with 30% hydrogen peroxide. The order of addition of reagents must be strictly followed and each reaction should be complete before the following reagent is added.

Perchloric acid is a very useful reagent for digestions, but its hazards must be understood and precautions followed exactly if explosions are to be avoided. Most explosions are probably attributable to formation of anhydrous perchloric acid or organic perchlorates. Ethyl alcohol, glycerol, cellulose, sugars, carbohydrates and pyridine form explosive compounds (**10, 12, 15**). Precautions for the use of perchloric acid are beyond the scope of this chapter; special storage, hoods and handling procedures are required. In spite of the hazards, perchloric acid is being used without incident in many laboratories and has been accepted for use in recommended standards procedures (**1, 4**). Safe use of this material demands an understanding of the hazards and exact compliance with operating instructions.

e. Special Reagents—Fusion of organic materials with sodium peroxide in a closed bomb presents an explosion hazard if too much organic material is charged into the bomb, the bomb has been weakened by previous use, the gasket is worn, or the heating is not done properly. Use of a shield (at least 3/16″ thick steel) is imperative. Heat should be applied locally; it should be kept away from the gasket and should be removed as soon as the reaction occurs. Liquid samples which attack gelatin capsules should be handled in glass weighing bulbs. Double gelatin capsules may be satisfactory if the charging operation is carried out promptly. Hazards are reduced by operation on a micro scale. However, the macro procedure is quite safe if precautions are observed.

Fusions of organic material with metallic sodium have been performed in glass tubes, but nickel bombs are preferable—particularly for high-temperature fusions. Potassium should never be melted in glass tubes since it reacts with glass with explosive violence. Properly designed bombs and shields eliminate the danger from explosions. The remaining hazards are those associated with the use of the alkali metals:

(1) Fire and explosion may result if alkali metals contact moisture;

(2) eye injuries and flesh burns may be caused by small pieces of metals or their oxides;

(3) irritation of eyes and nasal tissues can be caused by exposure to products from burning alkali metals. Fumes from burning potassium are toxic;

(4) sodium detonates in carbon tetrachloride. Carbon dioxide-propelled sodium chloride or soda ash extinguishers or dry sand may be used on sodium fires. Potassium reacts explosively with sand and forms explosive carbonyl with carbon dioxide. The only recommended extinguisher for potassium fires is sodium chloride powder or a mixture of sodium chloride and sodium carbonate. Potassium is also more hazardous than sodium since it forms an explosive superoxide KO_2 which can cause explosions even when the metal is stored under "inert" solvents. Scrap sodium should be disposed of by covering it with N-butyl alcohol. The safest means of disposal for potassium is removal to a safe open area where the potassium is allowed to react with atmospheric oxygen and moisture, then washed with a large excess of water.

19.12.4 *Separations.*

a. Distillation—The hazards associated with laboratory distillation at atmospheric pressure or under vacuum are as follows:

(1) The flask may implode due to a weakness or a blow. If the contents of the flask are flammable a fire may result if a source of sparks or a hot surface is nearby;

(2) toxic substances may be released into the atmosphere if the condenser is inadequate or the coolant flow is interrupted;

(3) a pressure buildup and explosion may result if the column or head is plugged with polymeric material or high-melting solids; and

(4) an explosion may result if thermally unstable compounds are distilled.

Distillation flasks containing flammable materials should never be heated with open flames or electric heaters with exposed elements. A heating mantle controlled with a variable transformer is the preferred equipment. The flask and heater should be placed in a metal container large enough to confine any spills from a broken flask.

Any large (over 250-ml) glass distilling flask being used under vacuum should either be completely enclosed or placed behind a shield. This rule also applies to distillations at atmospheric pressure if thermally sensitive compounds are being distilled. In this case, the addition of a high-boiling chaser is advisable.

Solvents which may peroxidize (especially hydrocarbons, aldehydes and ethers) should always be analyzed for active oxygen prior to distillation **(6)**. If the active oxygen content exceeds 0.02%, the peroxide should be removed **(14, 11)** before distillation. In distillation of materials which have a tendency to polymerize, it is well to add an inhibitor. In some cases, it is advisable to add a solution of the inhibitor dissolved in part of the distillate dropwise through the reflux condenser during the entire distillation.

b. Centrifugation—Use of a centrifuge involves the hazards of injury from whirling tubes, breakage of tubes due to imbalance and fire or explosion if a flammable solvent is spilled in a centrifuge which is not of explosion-proof construction. Centrifuges are built with a shield over the moving parts, but care must be taken to stop the equipment completely before opening the cover. Tubes with their contents should be carefully balanced. The safe speed recommended by the centrifuge manufacturer for a given type of tube should not be exceeded.

c. Extractions—If volatile solvents or high temperatures are used in extractions, care should be used to relieve the pressure which is generated by shaking; otherwise the analyst may be exposed to toxic solvents and sample may be lost by spurts which accompany the sudden release of pressure by lifting of the glass stopper. The preferred technique is to invert the extraction funnel, then cautiously open the stopcock to relieve the pressure. This should be done frequently, especially at the start of the extraction.

d. Ion Exchange—The volume of ion exchange resin in a column will change, depending on the ion associated with it and/or the solvent being used. If glass columns are used, plenty of room should be allowed for expansion or broken tubes may result.

e. Liquid-Solid Chromatography—The use of fairly large quantities of flammable solvents requires that this procedure be performed in an area which is free from flames, hot surfaces or sparking electrical equipment. Glass joints, tubing connections and stopcocks should be clamped. The column and reservoir should be placed over a tray which will confine any spill which does occur. Explosion-proof fraction collectors should be used and the apparatus should be operated in a hood with sufficient air flow to prevent the concentration of solvent vapor from reaching the explosive range.

f. Electrophoresis and Electrochromatography—The use of relatively high voltages requires shielding of electrodes and care to avoid physical contact.

19.12.5 *Miscellaneous.*

a. Hydrogenation—Analytical hydrogenations are usually carried out in glass equipment at pressures not exceeding 30 psia. The possibility of an unsuspected leak or a break in the equipment makes it advisable to ban any smoking in the vicinity of the experiment. Some hydrogenation catalysts are pyrophoric and must be kept wet or fires may result. If mercury is used as a confining liquid, the whole apparatus should be over a large tray which will confine any mercury spills.

b. Pressure Bottle Reactions—Saponifications, acetylations, hydrolyses and similar reactions are often carried out in citrate bottles or other pressure bottles. The use of volumetric flasks as pressure bottles is not recommended since they occasionally have areas of thin glass in their bulbs and they may lose their calibration during heating. Heavy-walled pyrex pressure bottles are recommended. They should be well annealed and inspected for strains with polarized light. If facilities are available, it is advisable to perform a hydrostatic bursting test on several bottles and then use the remaining bottles at pressures not exceeding 40% of the minimum bursting pressure. The bottles should be as small as possible, but not filled more than half full. They should be wrapped with strong adhesive tape or enclosed in metal screening. The heating bath should have a safety high-temperature cut off which will not allow the temperature to rise sufficiently to produce dangerous pressures if the primary temperature control sticks in the "on" position. An enclosed constant temperature bath or a shield should be used. The bottle closure should be well designed, sturdy and positive in action.

19.13 INSTRUMENTAL METHODS.

19.13.1 *Gas Chromatography*. The principle hazards associated with gas chromatography are connected with injection of liquid samples. At least one serious injury has resulted from accidental injection of a chemical when a technician's finger was inadvertently jabbed with a hypodermic syringe needle. A severe facial chemical burn resulted in another instance when a corrosive chemical was being injected into a poorly designed injection port and the pressure in the gas chromatograph forced the plunger and the sample out of the back of the syringe onto the face of the chemist.

Injection ports should be designed and located so injections can be made without having the analyst's face close to the syringe. Unnecessarily high operating pressure in the instrument should be avoided where possible. Permanently attached needles eliminate the hazard of the needle becoming detached from the syringe with possible exposure to the sample. Short needles reduce the hazard of breaking needles. Unguarded needles should not be left on a bench top or in a drawer, but should be kept in a tray or box. The tips should be covered with tubing or inserted into cork or rubber stoppers. Needles should not be discarded in waste baskets. Syringes must be handled carefully; they should never be pointed at the operator or other workers. When making an injection into a gas chromatograph, the analyst should keep his thumb or the palm of his hand firmly against the back of the plunger so that it will not be dislodged from the barrel by the pressure in the instrument. If the sample is corrosive, rubber gloves should be worn by the analyst. Long needles should be guided with the fingers to prevent bending or breakage. Bent needles should not be used. They are more difficult to control and are prone to break. Sharp needles decrease the danger of breakage.

In case a chemical is injected into the body accidentally, the following steps should be taken as quickly as possible:

a. Squeeze or suck out (slight vacuum) the injected material;

b. wash the area with a large amount of water;

c. where possible, apply a light tourniquet above the puncture wound, and

d. go to a physician promptly (with an escort) and report the nature and amount of material injected and the depth of needle penetration.

19.13.2 *X-Ray Diffraction and Fluorescence Spectrography*. The use of X-ray analytical equipment involves possible exposure to hazardous X-rays and high voltages. Care must be taken to avoid exposure of any part of the body to a direct X-ray beam or to the secondary (emitted or scattered) radiation which occurs when the primary beam strikes or passes through any material. Precautions should be taken to make it impossible for unqualified personnel to operate the instrument. The available range of potential and current provided by the instrument and specified in the manufacturer's instructions should not be exceeded, or the provided shielding may not be sufficient for prevention of penetration or emergence of unsafe amounts of radiation.

Operating personnel should stay as far away from the instrument as consistent with proper manipulation. Radiation badges or dosimeters should be worn and checked at monthly or more frequent intervals. Persons not involved in the operation of the equipment should not be permitted to remain in the area.

Primary and secondary radiation around the equipment should be checked periodically with a monitoring instrument. The cause of any unusual amount of radiation should be ascertained and removed. If there is a radiation protection officer at the location of the laboratory, he should also make periodic inspections of the X-ray facilities.

When a diffraction apparatus is used, care should be taken to see that all ports to the X-ray tube are covered by a camera, a shutter or a cap. If the instructions call for radiation baffles on the cameras, they must be put in place to minimize side leakage. Shields designed to prevent accidental exposure to X-rays should not be removed. Instruments should be provided with automatic shutters which will prevent accidental exposure of the analyst's hands to the X-ray beam while changing samples.

During alignment of the goniometer in a diffraction instrument or adjustment of the counter arms and crystal holders in a spectrograph, it may be necessary to operate without all the shields and guards in position. These are hazardous operations. Great care should be taken to avoid placing hands or other parts of the body in the direct path of primary or secondary radiation. It is often feasible to reduce power to the X-ray tube during these adjustments to minimize the effects of any accidental exposures. In some spectrographs, removal of the specimen chamber causes an extremely dangerous concentration of X-rays to stream out of the opening. The power to the X-ray tube must be shut off when the specimen chamber or turret head is removed (**5**).

Before changing tubes or making internal adjustments in the instrument, the power must be shut off. Access door interlocks designed to prevent exposure to high voltages should not be by-passed or wired out. Certain parts of the equipment will store high voltages even after the power is shut off. The charge must be removed with an insulated wire which has previously been connected to ground before service work is attempted.

19.13.3 *Electron Microscopy.* High voltages (50 to 100 kilovolts) and secondary emission of X-rays are the hazards associated with electron diffraction and electron microscopy. The precautions in the use of high-voltage equipment specified in the section on X-ray diffraction and spectrometry are equally applicable in electron microscopy. Circuit checking and trouble shooting should be performed by no fewer than two technicians working together. A grounding rod or wire must be used to remove high voltages which are stored in capacitors even after the power is turned off. In addition to grounding, it is well to check the voltage in each circuit with a voltmeter and to leave the grounding rod touching a conducting surface in the circuit under investigation before any physical contact is made. Interlocks and automatic grounding devices should not be tampered with.

Under certain conditions, significant amounts of X-rays may be emitted while operating continuously at 100 kilovolts. Each instrument should be checked for X-ray emission after installation and at periodic intervals thereafter. If significant emission of X-rays is observed when high voltages are use, it may be necessary to add additional shielding and additional lead glass in front of the viewing windows.

When metal evaporation under vacuum is used for shadow casting, the glass bell jar should be covered with a plastic shield and welder's goggles should be used to observe the metal filament.

19.13.4 *Flame Spectrophotometry and Atomic Absorption.* In the use of hydrogen or acetylene as fuel, it is important that the system be free from leaks and that the fuel be ignited promptly after the valve is opened. The oxygen should be turned on before the fuel when igniting the flame and it should be turned off after the fuel when extinguishing the flame. Oxygen lines, gauges and fittings must be free of oil, grease and pipe dope. When many samples containing large amounts of mineral acids or significant amounts of toxic metals are to be analyzed, a fume exhaust line should be provided over the burner. If samples are dissolved in volatile flammable solvents, they should be prepared and stored at a location away from the flame. Atomic absorption burners which have sample-air mixing chambers should always be shielded and the mixing chamber drain should have a water seal.

19.13.5 *Polarography.* The major hazard associated with polarography is the use of fairly large quantities of mercury. The entire dropping mercury electrode assembly should be in a tray which will confine any mercury spills.

19.13.6 *Infrared Spectrography.* The use of a high-pressure press for forming potassium bromide pellets has potential hazards for finger injuries, but no incidents should occur if reasonable care is observed.

Carbon disulfide, which is a popular solvent for infrared work, is extremely flammable. Preparation of solutions in carbon disulfide and other flammable solvents should be carried out in a fume hood and not in the vicinity of the spectrometer. Some spectrometers have no panel enclosing the underside of the instrument and contact with 110 volts can result if an operator's fingers extend under the chassis while lifting or tilting the instrument with the power on. The power should be turned off before carrying out this operation, or leather gloves should be worn. The chart drive gear shield should not be removed during operation of the spectrometer.

The use of prisms or other optical parts made of thallium salts involves a hazard, since thallium is a cumulative poison. Any broken pieces should be collected and carefully discarded; they should not be allowed to lie on the laboratory bench or floor where they may become pulverized to a dust and, thus, present an inhalation hazard.

19.13.7 *Emission Spectrography.* The use of arc and spark sources provides the major hazards associated with emission spectrography. The electrodes should be protected by shields as much as possible in order to avoid inadvertent contact with high voltages. If many samples of toxic metals are to be analyzed, a fume vent should be provided over the source. Proper eye protection in the form of a shield should be provided in front of the arc source.

References

1. A.S.T.M. COMMITTEE. 1956. *Recommended Practices for Apparatus and Reagents for Chemical Analysis of Metals.* A.S.T.M. Designation E50-53, American Society for Testing Materials. Philadelphia, Pennsylvania.
2. GORDON, C.L. 1943. *J. of Research, Natl. Bur. of Standards*, 30:107–111.
3. *Interstate Commerce Commission Regulations.* 1963. Sections 73.301e, 73.304e, and 73.308, T.C. George, New York.
4. JOLLY, S.C. 1963. *Official, Standardized and Recommended Methods of Analysis*, pp. 3–19 and 42–44, The Society for Analytical Chemistry, Heffer, Cambridge, England.
5. MANUFACTURING CHEMISTS' ASSOCIATION, SAFETY AND FIRE PROTECTION COMMITTEE. 1962. *Case Histories of Accidents in the Chemical Industry* Vol. 1, Case Histories No. 547, 590, 592, Manufacturing Chemists' Association, Inc.
6. MARTIN, A.J., 1960. IN J. MITCHELL, JR., ET AL. *Organic Analysis*, p. 1–64, Interscience, New York.
7. MARTIN, A.J. and H. DEVERAUX. 1959. *Anal. Chem.* 31:1932.
8. OGG, C.L., R.B. KELLY, and J.A. CONNELLY. 1961. Design of Apparatus for Safe Oxygen Filled Combustion. *Abstracts of Papers from International Symposium on Microchemical Techniques*, Paper No. 51, p. 34. The Pennsylvania State University, University Park, Pa.
9. PARR OXYGEN COMBUSTION BOMBS. Specifications No 1100, Parr Instrument Company, Moline, Illinois.
10. NATIONAL SAFETY COUNCIL, Perchloric Acid. Data Sheet D-311 (D-Chem. 44). Chicago, Illinois.

11. Ramsey, J.B., and F.T. Aldrich. 1955. *J. Am. Chem. Soc.* 77:2561.
12. Smith, G.F. 1953. *Anal. Chim. Acta* 8:397–421.
13. Smith, G.F. 1942. *Mixed Perchloric, Sulfuric and Phsophoris Acids and their Applications in Analysis* (pamphlet). 2nd ed., The G. Fredrick Smith Chemical Company, Columbus, Ohio.
14. Williams, F.E. 1951. Distillation in Weissbergers' *Technique of Organic Chemistry*, Vol. IV, pp. 300–302. Interscience, New York.
15. Zackerl, M.K. 1948. *Mikrochemie ver. Mikrochim, Acta* 33:387–388, (Through Chem. Abstracts. 42, 6538b 1948).

20. Air Purification*

20.1 Introduction. Laboratory compressed air is the most common source of diluent gas for low-concentration, high-volume standard gas mixtures. It is continuously supplied as needed, usually by diesel or electric compressors at pressures of 80 to 125 psi, and it is stored in holding tanks. Several undesirable contaminants can be introduced during compression and storage. Oil mists are a common by-product, as are substantial amounts of carbon dioxide, nitrogen dioxide, aldehydes, carbon monoxide, and unburned hydrocarbons. Acid gases as well as dust particles and pipe scales of all sizes are also a problem. Even if the air is known to be 99.9% pure it can still contain up to 1000 ppm of undesirable materials. Before any quality low-concentration work can be done, the air-supply system must be scrupulously cleaned to prevent contamination and possible chemical reaction. The composition of clean, dry air is given in Table 20:I.

20.1.1. This section describes the basic methods of removing contaminants from flowing air streams. General multipurpose filtering devices are discussed, as are methods for removing excess water vapor, oil mists, extraneous gases, and particulate matter. The air-purification procedures that are described can also be applied to such relatively stable gases as nitrogen and oxygen as well as to inert gases.

*Abstracted from G. O. Nelson, "Controlled Test Atmospheres," Chap. 2, Air Purification. Ann Arbor Science Publishers, Inc., P.O. Box 1425, Ann Arbor, Michigan 48106.

20.2 Removal of water vapor. Moisture can be removed from gases by a variety of methods. Chief among these are adsorption, absorption, cooling, compression and combined compression and cooling. Usually, only the first three methods are used in the laboratory.

20.3 Solid Desiccants.

20.3.1 Solid desiccants constitute the most conventional method of removing water vapor in the laboratory. They remove moisture either by chemical reaction (absorption) or by capillary condensation (adsorption) (**1**). Solid absorbing agents include calcium chloride, calcium sulfate, and magnesium perchlorate; solid adsorbing agents include activated alumina and silica gel. Solid desiccants are one of the most practical tools for drying gases because they are commercially available, they are easy to store, they can be regenerated by heating, and they often indicate their condition by their color.

Table 20:I. Composition of clean, dry air

Nitrogen	78.08%
Oxygen	20.95%
Argon	0.934%
Carbon dioxide	0.033%
Neon	18.2 ppm
Helium	5.24 ppm
Methane	2.0 ppm
Krypton	1.14 ppm
Hydrogen	0.5 ppm
Nitrous oxide	0.5 ppm
Xenon	0.087 ppm

20.3.2 Solid desiccants are generally evaluated by comparing their drying efficiencies and capacities. The efficiencies of drying agents (*i.e.*, the degree of dryness achieved) can be compared by measuring the water vapor remaining in a gas after it passes through the desiccant at the equilibrium velocity (**2**). The drying efficiencies of several solid desiccants are compared in Table 20:II. Barium oxide and magnesium perchlorate are the most efficient desiccants of those compared, whereas copper sulfate and granular calcium chloride are the least efficient. Recently, an extensive investigation by Trusell and Diehl evaluated the efficiencies of 21 desiccants in drying a stream of nitrogen (**3**). Their re-

Table 20:II. Comparative efficiencies of various solid desiccants used in drying air*

Desiccant	Granular Form	Residual Water† (mg/liter)
BaO	—	0.00065
$Mg(ClO_4)_2$	—	0.002
CaO	—	0.003
$CaSO_4$	Anhydrous	0.005
Al_2O_3	—	0.005
KOH	Sticks	0.014
Silica gel	—	0.030
$Mg(ClO_4)_2{\cdot}3H_2O$	—	0.031
$CaCl_2$	Dehydrated	0.36
NaOH	Sticks	0.80
$Ba(ClO_4)_2$	—	0.82
$ZnCl_2$	Sticks	0.98
$CaCl_2$	Technical anhydrous	1.25
$CaCl_2$	Granular	1.5
$CuSO_4$	Anhydrous	2.8

*Data taken from Reference 2.
†After drying to equilibrium.

sults, which are listed in Table 20:III, show that the most efficient desiccant is anhydrous magnesium perchlorate.

20.3.3 The capacity of a desiccant is the amount of water it is able to remove per unit of the desiccant's dry weight. Often, a drying agent is efficient but is unsuitable for drying large quantities of gas because of its limited capacity. Anhydrous calcium chloride is an example of such a desiccant. The capacity of a desiccant depends not only on the kind of material of which it is composed, but also on the size of the grains, the amount of surface area exposed to the gas, and the thickness through which the gas flows. Additional factors include the type of gas being dried as well as its velocity, temperature, pressure, and moisture content (**7**). The capacities of several desiccants as a function of relative humidity are shown in Figure 20.1. The relative capacities of the most common drying agents can be determined from Table 20:III by comparing the volumes of gas each desiccant is able to dry. The materials with the highest capacities are anhydrous magnesium perchlorate, calcium sulfate, and phosphorus pentoxide.

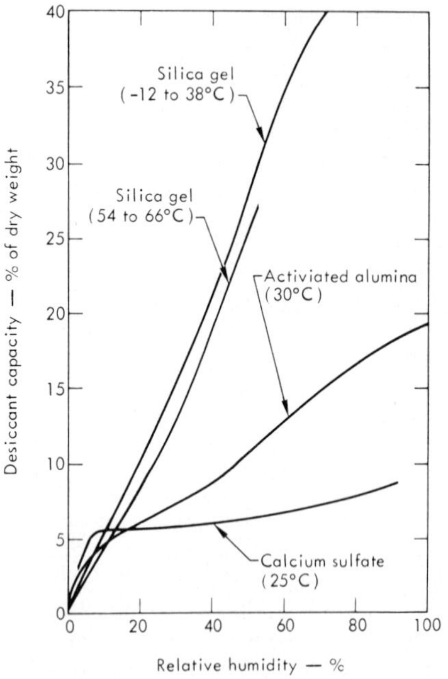

Figure 20:1—Desiccant capacity versus relative humidity for silica gel, activated alumina, and calcium sulfate (Drierite). The calcium sulfate temperature is estimated. (Taken from Reference 5.)

20.3.4 One should not indiscriminately choose a desiccant simply because it has the proper efficiency and capacity. The geometry and size of the drying train must be considered to allow enough residence time to achieve equilibrium. In addition, many drying agents heat up violently when they are exposed to too much moisture over too short a time, and the pressure drop through the desiccant can be a problem at high flow rates. The most acceptable desiccants are anhydrous magnesium perchlorate, calcium sulfate, silica gel, and activated alumina. These are described below.

a. Anhydrous magnesium perchlorate (Anhydrone or Dehydrite) has the highest efficiency as well as the greatest capacity **(3)**. It is hygroscopic but not deliquescent, and it can absorb up to 35% of its own weight without evolving corrossive fumes as phosphorus pentoxide does. Since the monohydrate does not dissociate to liberate water until 134 C, it can be used to dry gases at high temperatures. Hydration continues until the hexahydrate, which has a theoretical capacity of 48.4% is formed. Magnesium perchlorate is available in either the regular or the indicating form, the latter containing about 1% potassium permanganate.

The chief disadvantages of anhydrous magnesium perchlorate are its relatively high cost (roughly four times the cost of the other three desiccants), and the difficulty of regeneration. The temperature must be raised slowly while the perchlorate is dried in a vacuum in order to prevent the crystals from fusing. A final temperature of about 245 C is recommended to return the perchlorate to its anhydrous state **(3)**. A further disadvantage of this and other perchlorate desiccants is the *tendency to form explosive compounds in the presence of organic materials,* especially when they are heated **(6)**. Oil mists and other organic vapors must therefore be removed before such desiccants are used.

b. Calcium sulfate (Drierite or Anhydrocel) has an average efficiency of about 0.1 mg/liter and a capacity of 7 to 14% at 25 C. It is stable, inert and does not deliquesce even at peak capacity. It is easily regenerated (1 to 2 hr at 200 C), and it operates at an almost constant efficiency over a wide range of temperatures. However, continued regeneration is difficult because the constant formation and destruction of the hemihydrate breaks down the grains. Calcium sulfate is available in sizes from 4 to 20 mesh and in either the regular or the indicating form.

c. Silica gel has a moderately high efficiency and capacity because of its large number of capillary pores, which occupy about 50% of the gel's specific volume **(1)**. The capacity of the gel varies from batch to batch because of differences in the size and shape of the pores. The gel maintains its efficiency until it has absorbed 20% of its weight, and it can be regenerated indefinitely at 120 C. At this relatively low regeneration temperature, however, the gel cannot be used for high-temperature drying. If it is regenerated above 260 C, it loses some of its capacity. Silica gel is available in sizes from 2 to 300 mesh. The addition of cobalt chloride to the surface of the gel provides an indicating ability.

d. Activated alumina has a higher efficiency than silica gel but offers less capacity (12 to 14%), especially at high humidities. It can be regenerated between 180 and 400 C without losing much of its capacity. It is available in sizes from 14 mesh to 1 inch and in either the regular or the indicating form.

20.4 Liquid Desiccants.

20.4.1 If solid desiccants are not practical, then liquid desiccants can be used. Liquid desiccants have a much higher capacity than their solid counterparts (see Figure 20:2), and they can be continuously regenerated via spraying, pumping, or recirculating. On the other hand, their efficiencies are normally very low unless the anhydrous forms are used, and they usually cannot produce relative humidities below about 20% **(5)**. Some of the more common liquid desiccants are described in Table 20:IV. Note that although the strong acids and bases achieve the best efficiencies, they also emit corrosive vapors.

Table 20:III. Comparative efficiencies and capacities of various solid desiccants in drying a stream of nitrogen*

Desiccant	Initial Composition	Regeneration Requirements Drying Time (hr)	Regeneration Requirements Temperature (C)	Average Efficiency† (mg/liter)	Relative Capacity‡ (liters)
Anhydrous magnesium perchlorate§	$Mg(ClO_4)_2 \cdot 0.12H_2O$	48‖	245‖	0.0002	1168
Anhydrone§#	$Mg(ClO_4)_2 \cdot 1.48H_2O$	—	240**	0.0015	1157
Barium oxide	96.2% BaO	—	1000#	0.0028	244
Activated alumina	Al_2O_3	6–8‡‡	175, 400#	0.0029	263
Phosphorus pentoxide§§	P_2O_5	—	—	0.0035	566
Molecular sieve 5A#	Calcium aluminum silicate	—	—	0.0039	215
Indicating anhydrous magnesium perchlorate§	88% $Mg(ClO_4)_2$ and 0.86% $KMnO_4$	48‖	240‖**	0.0044	435
Anhydrous lithium perchlorate§§	$LiClO_4$	12‖, 12	70‖, 110	0.013	267
Anhydrous calcium chloride§§	$CaCl_2 \cdot 0.18H_2O$	16‖	127‖	0.067	33
Drierite#	$CaSO_4 \cdot 0.02H_2O$	1–2	200–225‖‖	0.067	232
Silica gel	. . . .	12	118–127††	0.070	317
Ascarite#	91.0% NaOH	—	—	0.093	44
Calcium chloride§§	$CaCl_2 \cdot 0.28H_2O$	—	200‖	0.099	57
Anhydrous calcium chloride§§	$CaCl_2$	16‖	245‖	0.137	31
Anhydrocel#	$CaSO_4 \cdot 0.21H_2O$	1–2	200–225‖‖	0.207	683
Sodium hydroxide§§	$NaOH \cdot 0.03H_2O$	—	—	0.513	178
Anhydrous barium perchlorate	$Ba(ClO_4)_2$	16	127	0.599	28
Calcium oxide	CaO	6	500, 900**	0.656	51
Magnesium oxide	MgO	6	800	0.753	22
Potassium hydroxide§§	$KOH \cdot 0.52H_2O$	—	—	0.939	18.4
Mekohbite#§§	68.7% NaOH	—	—	1.378	68

*Nitrogen at an average flow rate of 225 ml/min was passed through a drying train consisting of three Swartz drying tubes (14 mm i. d. × 150 mm deep) maintained at 25C. Except as noted in columns 3 and 4, the data in this table are taken from Reference **3**.

†The average amount of water remaining in the nitrogen after it was dried to equilibrium.

‡The average maximum volume of nitrogen dried at the specified efficiency for a given volume of desiccant.

§Hygroscopic.

‖Dried in a vacuum.

#Trade name.

Taken from Reference **4.

††Taken from Reference **5**.

‡‡Taken from Reference **6**.

§§Deliquescent.

‖‖Taken from Reference **2**.

20.5 Cooling.

20.5.1 Cooling is the most efficient laboratory method of removing water from a stream of gas. The gas is directed through a vessel in a low-temperature bath, and the excess water condenses on the cold walls of the vessel. As an example, a bath of dry ice and acetone at −70 C removes all but about 0.01 mg/liter of water in air at equilibrium. A liquid-nitrogen bath at −194 C removes all but about 1×10^{-23} mg/liter of water in air at equilibrium. This is about 19 orders of magnitude more efficient than anhydrous magnesium perchlorate, the best solid desiccant.

20.6 Removal of Particulates.

20.6.1 There are a number of inline filters that remove micron-sized particles. One example is a sintered-bronze mesh 1″ in diam and 2.5″ long that can filter out 2-μm particles at a flow rate of 1 ft^3/min and an operating pressure of 110 psi **(8)**. Filters made of metal fibers (pore size = 5 to 750 μm) **(9)**, sintered stainless steel (Pore size = 2 × 150 μm) **(10)**, and foamed metals (pore size = 5 μm to 0.1″) **(11)** are available. Porous Teflon and Kel-F filters that can remove particles as small as 2 μm are also available **(12)**.

20.6.2 If removal down to certain precise particle sizes is required, membrane filters are often useful. These are available in sizes from 13 to 293 mm with pore sizes from 75 Å (7.5 nm) to 8 μm **(13–15)**. These filters are constructed from a wide selection of microporous materials (*e.g.*, regenerated cellulose, polyvinyl chloride, glass fibers, and polypropylene) whose capacities are accurately known **(13)**. The flow rate per unit of filter area at a given temperature is a function of the pore size and the upstream pressure. The relationship between pressure, pore size, and flow rate for a typical membrane filter is shown in Figure 20:3.

20.7 Removal of Organic Vapors.

20.7.1 Organic vapors such as unburned hydrocarbons are sometimes present in a compressor-generated gas even after it has been passed through an in-line filter. The concentrations of such organic vapors can be further reduced either by passing the gas through an activiated-charcoal filter or by continuously burning the vapors in a combustion furnace.

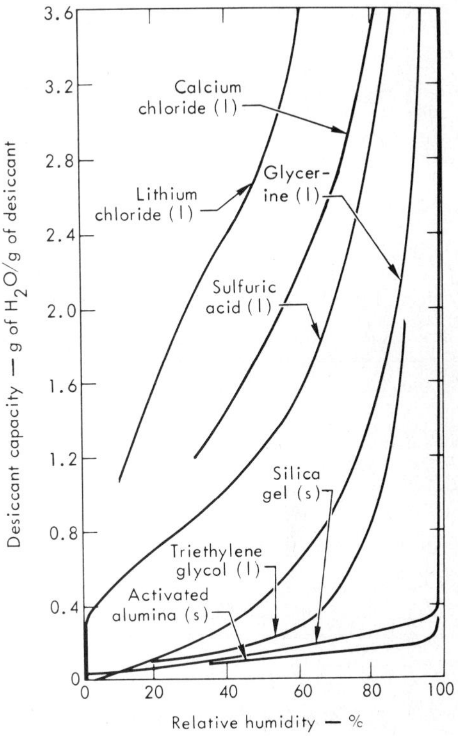

Figure 20:2—Desiccant capacity versus relative humidity for two solid desiccants and five liquid desiccants. (Taken from Reference 5.)

20.7.2 Activated coconut charcoal has long been the most popular material for removing organic vapors. Instead of water vapor being adsorbed on the filter, the organic vapors tend to displace any water that may be present. A filter containing fine grains of coconut charcoal uniformly dispersed throughout a web matrix is also available **(16)**. Not all organic vapors are completely adsorbed by charcoal filters, even when large filter areas and low flow rates are used. For example, such low molecular weight compounds as acetylene, ethane, ethylene, methane, hydrogen, carbon monoxide, and carbon dioxide have almost no affinity for activated charcoal **(17)**.

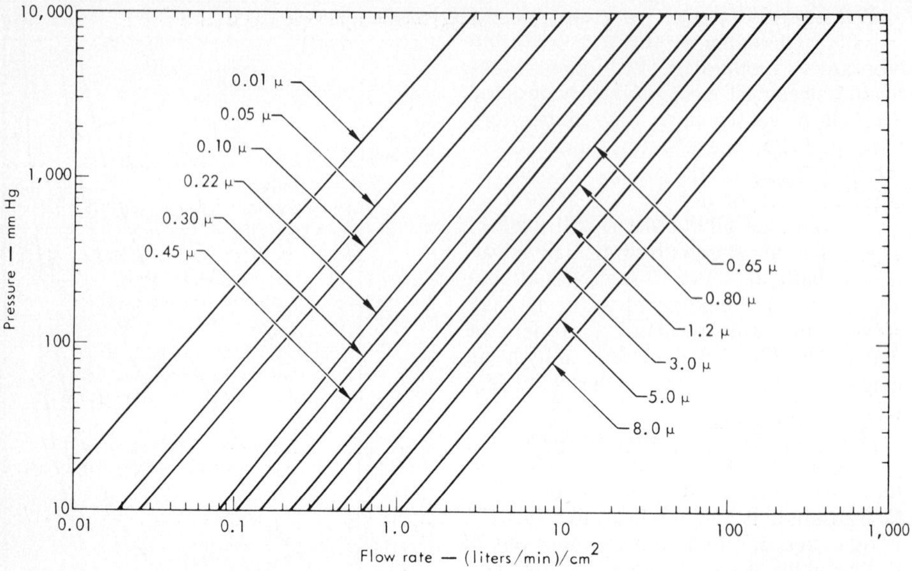

Figure 20:3—Pressure versus flow rate for 12 mean pore sizes at 25 C. (Taken from Reference 15.)

20.7.3 As for the second method of removing organic vapors, Kusnetz *et al*. recommend passing the gas through a 2″-diam Mullite tube that is filled with copper shavings and surrounded by a combustion furnace maintained at 1250 F **(18)**. The exiting gas is cooled to 800 F by a finned brass tube and a water-cooled condenser. To remove the combustion products, the gas is bubbled through a sulfuric acid-dichromate solution and passed through filters of Ascarite, activated charcoal, and glass wool, in that order.

20.8 Removal of Miscellaneous Contaminants.

20.8.1 A host of in-line filters are available for treating compressed gases in the laboratory. These filters can remove both water and oil mists as well as particles. Drain plugs are provided for periodically removing the collected liquid, and the filters themselves can be exchanged when they become clogged or excessively loaded. Automatic drain traps are also available **(19)**. Most compressed-gas filters operate via a two-stage separation that involves centrifugal or inertial separation followed by diffusion through a filter **(20–22)** or a special absorption material **(23)**.

20.8.2 Compressed gases may also contain acid gases, carbon monoxide, and carbon dioxide. The acid gases are those that produce hydrogen ions either by direct dissociation or by hydrolysis and include hydrogen cyanide, hydrogen chloride, and sulfur dioxide. Both acid gases and carbon dioxide can be filtered out by soda lime, a mixture of calcium and sodium hydroxides. Soda lime with various moisture contents is available in sizes from 4 to 14 mesh, in either the regular or the indicating form, and can absorb up to 25% of its weight of carbon dioxide. The rate at which acid gases are absorbed by soda lime depends on the condition of the lime. As the lime becomes spent, a thin film of calcium carbonate covers the surface of the soda lime particles and cannot be removed by regeneration.

Table 20:IV. Comparative properties of eight liquid desiccants*

Desiccant	Relative Humidity Achieved at 21 C (%)	Solution Concentration (%)	Operating Temperature Range (C)	Remarks
Calcium chloride	20–25	40–50	32–49	None.
Diethylene glycol	5–10	70–95	16–43	Can be regenerated with heating to 150 C.
Glycerol	30–40	70–80	21–38	Oxidizes and decomposes at high temperatures; can be regenerated with vacuum evaporation.
Lithium chloride	10–20	30–45	21–38	None.
Phosphoric acid	5–20	80–95	16–38	Corrosive; fumes carried over during the drying process; does not fume during regeneration.
Sodium and potassium hydroxides	10–20	Saturated	29–49	Corrosive; frequently used to remove CO_2 and water simultaneously.
Sulfuric acid	5–20	60–70	21–49	Corrosive; most efficient liquid desiccant.
Triethylene glycol	5–10	70–96	16–43	None.

* Data taken from Reference **1**.
† 5% with anhydrous diethylene glycol.

20.8.3 Both carbon dioxide and water vapor can be absorbed by Ascarite, which consists of sodium hydroxide in a woven asbestos matrix. Ascarite is available in sizes from 8 to 30 mesh.

20.8.4 Carbon monoxide is relatively unaffected by its passage through soda lime, activated carbon, or the previously described desiccants, so special provisions for its removal must be made. Hopcalite, usually a mixture of copper and manganese oxides, has traditionally been the most practical agent for removing carbon monoxide. It operates as a catalytic oxidizing agent, converting the carbon monoxide to carbon dioxide which can ultimately be absorbed by Ascarite or soda lime. The main requirement for using Hopcalite is that it be kept scrupulously dry, for it loses its catalytic ability in the presence of water. Other carbon monoxide filters are also available commercially **(23)**.

20.8.5 For the removal of specific substances, the appropriate sorbents are listed in Reference 24.

References

1. Perry, J.H. 1950. *Chemical Engineers Handbook.* McGraw-Hill Book Company, Inc., New York.
2. Hammond, W.A. 1958. Drierite, the Versatile Desiccant, and its Applications in the Drying of Solids, Liquids, and Gases. The Stoneman Press, Columbus, Ohio.
3. Trusell, F., and Diehl, H. 1963. Anal. Chem. 35:674.
4. Skoog, D.A.,and D.M. West. 1963. *Fundamentals of Analytical Chemistry.* Holt, Rinehart & Winston, Inc., New York.
5. Hougen, O.A., and F.W. Dodge. 1947. *The Drying of Gases.* Edwards Brothers, Inc., Ann Arbor, Michigan.
6. Morton, A.A. 1938. *Laboratory Techniques in Organic Chemistry.* McGraw-Hill Book Company, Inc., New York.
7. Nonhebel, G. 1964. *Gas Purification Processes.* George Newnes, Ltd., London.
8. Catalog No. 200, Permanent Filter Corporation, Compton, California.
9. Bulletin Nos. FM-1100, -1200, and -1300, Huyck Metals Company, Milford, Connecticut.
10. Bulletin No. 1, Sintered Specialties, Janesville, Wisconsin.
11. Brochure, General Electric Company, Detroit, Michigan.

12. BROCHURE, PALL CORPORATION, Glen Cove, New York.
13. 1969 catalog, Pure Aire Corporation of America, Van Noys, California.
14. BROCHURE, ARTHUR H. THOMAS COMPANY, Philadelphia, Pennsylvania.
15. CATALOG NO. MF-64, Millipore Corporation, Bedford, Massachusetts.
16. BROCHURE, THE DEXTER CORPORATION, Windsor Locks, Connecticut.
17. BARNEBEY, H.L. 1958. Heating, Piping, Air Conditioning 30, 153.
18. KUSNETZ, H.L, B.E. SALTZMAN, and M.E. LANIER. 1960. Am. Ind. Hyg. Assoc. J. 21:361.
19. CATALOG NO. 600, King Engineering Corporation, Ann Arbor, Michigan.
20. BULLETIN NO. 118, R.P. Adams Company, Inc., Buffalo, New York.
21. CIRCULAR NO. 1066, Wilkerson Corporation, Englewood, Colorado.
22. BULLETIN NO. 200, Dollinger Corporation, Rochester, New York.
23. FORM NO. 101-E, Deltech Engineering, Inc., New Castle, Delaware.
24. RESPIRATORY PROTECTIVE DEVICES MANUAL. 1963. Chapter 5. Braun and Brumfield, Inc., Ann Arbor, Michigan.

Subcommittee 9
A. L. LINCH, *Chairman*
E. R. HENDRICKSON
M. KATZ
J. R. MARTIN
G. O. NELSON
J. N. PATTISON
A. L. VANDER KOLK
R. B. WEIDNER

21. Liquid Chromatography

21.1 INTRODUCTION. In comparison to gas chromatography, LC has several advantages. Foremost is the fact that roughly 85% of all known chemical compounds are not sufficiently volatile or stable to be separated by GC. Thus, liquid chromatography is applicable to a much wider range of materials than GC—particularly to those materials with no or low volatility and to labile or unstable compounds. Liquid chromatography is especially useful to separate mixture compounds or identification by supplementary techniques, such as infrared and mass spectroscopy.

21.2 In spite of its advantages, LC always has been associated with one significant drawback—lack of speed. Separations involve a mobile liquid traveling through a packed column containing some immiscible stationary phase. The sample mixture, introduced into the mobile liquid, undergoes a series of interactions with the stationary phase or sorbent as it moves through the column. Separated components emerge in the order of increasing interaction with the stationary material—the least retarded component elutes first, the most strongly retained material elutes last.

21.3 At the low flow rates traditionally used, the rate of distribution of these components between the two largely is diffusion-controlled. Since diffusion in liquids is extremely slow, compared to that in gases, liquid chromatography always was slow and was used by many researchers only as a last resort for chemical analysis.

21.4 With the aid of chromatographic theory gained from research in gas chromatography and with innovations in hardware and instrumentation, it is now possible to refine liquid chromatographic procedures and practices to a point where the time and efficiency of LC separations rival that of GC.

21.5 One of the most important considerations in obtaining faster, more efficient separations involves the particles used in the column packing. For high-speed liquid chromatography, the packing should be finely divided and have regularity of shape, preferably spherical, to allow for optimum homogeneity and packing density.

21.6 Because high pressure is necessary to obtain reasonably rapid flow rates in densely-packed columns, these particles should have good mechanical stability under high pressure. To minimize the time required for the movement of sample components to and from the interaction sites in the column, it is necessary to have the stationary sorbent in the form of a thin uniform film with no stagnant pools.

21.2 SUPPORT COLUMNS.

21.2.1 Several column supports or packings tailored to these general requirements have been introduced commercially. One of these is "Zipax," a controlled surface porosity support developed by E. I. du Pont de Nemours & Co., Wilmington, Del. Zipax consists of a solid core with a thin porous shell.

21.2.2 For liquid chromatography, this material has an overall average diameter of about 30 μ, with the thickness of the porous shell being about 1 μ. The porous surface shell is relatively open, with a high surface area on which thin films of sorbent are uniformly dispersed. The solid center eliminates the possibility of deep pools of sorbent and/or stagnant carrier and prevents column efficiency losses due to slow distribution of sample components between the sorbent phase and the carrier liquid.

21.2.3 The easy handling of hard spherical particles allows highly reproducible, efficient columns of 2 to 4 mm ID to be conveniently prepared with a minimum of operator skill. This column diameter range is most frequently used for high-speed LC studies at present.

21.2.4 The importance of a porous surface-layer support becomes evident when compared to conventional chromatographic supports, such as the diatomaceous earths. The latter have irregular particle shapes with deep pools of stationary phase. In addition, stagnant pockets of carrier phase are formed in these deep pores.

21.2.5 Sample components, traveling through such a LC column, cross back and forth many times between the mobile and stationary phase. When deep pools of immobile liquid (stationary or carrier) exist in the support, molecules spend more time diffusing randomly in these pools. This results in a greater probability that all the molecules will not spend an equal time in the immobile liquid.

21.2.6 At faster flow rates, such as those used in high-speed LC, molecules "loitering" in the pools lag even further behind those that cross back and forth promptly. This effect creates wide component peaks and the "efficiency" of such separations is poor from a standpoint of time and effort. Porous surface layer supports minimize these deep pools of immobile liquid and give rise to a higher column efficiency at greater carrier velocities.

21.2.7 Columns of densely-packed small particles create high back-pressures. It becomes necessary to have a pump that can deliver reproducible flow rates under such high pressure conditions. At present, most high-speed liquid chromatography can be accomplished with pressures of 3,000 psi or less. Pressures of 500 to 1,000 psi most commonly are used, providing flow rates of 1 to 2 ml/min in columns of about 2 to 3 mm ID.

21.2.8 The thin uniform sorbent films, which promote efficient interaction between sample components and stationary phase, impose a limit on the sample capacity or size that can be used. This necessitates detectors with very high sensitivity to detect these small amounts of sample, which are usually in the microgram range.

21.2.9 If full advantage is to be taken of the increase in separation efficiency, there must be an absolute minimum of "dead volume" between the injection port and the detection point. Sample inlet and detector cell design, therefore, is critical.

21.2.10 Fortunately, these high-speed LC hardware requirements generally can be met by current technology. Components that satisfy the demands are available now and liquid chromatographs suitable for high-speed liquid chromatography recently have been introduced by a number of firms, including E. I. du Pont de Nemours & Co., Nester-Faust, Varian Aerograph, and Waters Associates. These instruments permit liquid chromatography to be carried out with the time and convenience normally associated with gas chromatography. Hupe-Busch chromatographs, made in West Germany and distributed in the United States by Tracor, also contain high-speed components.

21.3 MONITORING OF THE COLUMN EFFLUENT.

21.3.1 One of the most critical features of a high-speed liquid chromatographic apparatus is the component that continuously monitors the chromatographic column effluent. Unfortunately, there is no universal detector. Ultraviolet and refractive index detectors are most commonly used. Both are simple in concept but have limitations.

21.3.2 The UV detector records absorbance at a given wavelength. While materials, which have no ultraviolet absorbance cannot be detected, UV absorbing compounds do not have to exhibit an absorption maximum at the detecting wavelength. It is necessary only that the component have some UV absorbance at the analytical wavelength to be monitored.

21.3.3 The differential refractive index detector sometimes is considered a universal detector. However, this is only true when the refractive index of the sample components and the carrier are sufficiently different. This detector also is subject to several important limitations, the first being its severe sensitivity to temperature.

21.3.4 To avoid baseline drifting, a particularly important consideration if quantitative results are required, very precise detecting cell temperature control is mandatory. Detector baseline changes usually also occur with flow or temperature programing and gradient elution (changing the carrier composition during the chromatographic separation).

21.3.5 While the sensitivity limit of the refractive index detector is insufficient for many trace or minor component analyses, it can be quite useful when employed in tandem with a UV detector. Such a combination covers a very wide range of materials and allows the choice of the more applicable method of column effluent monitoring.

21.3.6 Other detectors for high-speed liquid chromatography include those based on flame ionization, differential heat of absorption, and polarography. The first two are classified as general detectors, the latter as selective. These detectors, and others, have particular areas of application.

21.3.7 Because many present detectors are deficient in terms of reproducibility, sensitivity, linearity, and so forth, it is probable that the modern liquid chromatographer will need both general and selective detectors, to enable him to carry out the desired chemical analyses. It is expected that intense current activity in liquid chromatography will improve LC detection capabilities.

21.4 Types of Absorbents.

21.4.1 When mixture components are resolved by liquid chromatography, there can be one or more mechanisms responsible. By using the best mechanism, difficult separations can be achieved with the greatest efficiency. Early liquid chromatographic separations were achieved with columns of a common adsorbent, such as silica gel or alumina. The differing degrees of sample component adsorption on the column dictated the elution order. Today, adsorption, or liquid-solid chromatography, still is widely used. However, the separations often are carried out with adsorbents especially designed for high-speed work.

21.4.2 When attempting column chromatography with adsorbents, one often can use thin-layer chromatography to quickly pinpoint possible solvent systems for a given mixture of components. The separation then can be refined and optimized by the column technique, which permits much better resolution and precise quantitative interpretation.

21.4.3 A minor drawback when using adsorbents is the necessity of maintaining constant "activity" or adsorptive power of the column packing to ensure reproducible chromatographic performance. This often is accomplished by maintaining a highly polar molecule at a constant concentration in the mobile phase so that it is preferentially adsorbed at the most active sites on the packing. Water is an excellent polar modifier for adsorption chromatography.

21.4.4 Many important separations, particularly those in biochemical systems, are carried out by ion-exchange chromatography. Ion exchangers may be considered a molecular network carrying a positive or negative surplus electrical charge that is compensated by mobile counter ions of opposite sign.

21.4.5 Counter ions may be exchanged for others of the same sign and the exchanger is usually selective, taking up certain counter ions in preference to others. In addition to the reversible exchange process, ion exchangers also can sorb solvent and solutes. Ion-exchange separation has been extremely useful for both organic and inorganic systems.

21.4.6 The combination of high-efficiency column supports, such as Zipax and special ion-exchange coatings, has resulted in some very interesting and useful separations. For example, the nucleic acid bases have been resolved in a small fraction of the time necessary by classical ion-exchange methods.

21.5 LIQUID PARTITIONING.

21.5.1 A third mechanism that has been used in many LC separations is liquid-liquid partitioning. In this case, two essentially immiscible phases are involved, the mobile liquid and the stationary liquid, the latter situated as a thin film on the column support.

21.5.2 As the solute molecules travel through the column in the mobile liquid they can "cross over" or spend part of their time dissolved in the stationary liquid. Differing solubilities of components in the stationary liquid result in differing retention times. This mechanism is similar to that in gas-liquid chromatography, except that in the case of liquid-liquid partitioning, changes in the type and composition of the mobile liquid offer additional versatility for separations.

21.5.3 Liquid-liquid chromatography does have a significant drawback, however. Because there is usually some solubility of the stationary phase in the mobile liquid, the latter must be presaturated with stationary liquid to avoid gradual stripping of this phase from the column. In addition, with the rather high flow rates used in high-speed separations, shear forces sometimes are generated in narrow-bore columns that tend to mechanically remove the stationary liquid from the support.

21.5.4 To overcome the disadvantages of this useful liquid-partitioning mechanism of separation, column packings with chemically-bonded stationary phases now have been developed. Zipax support has been reacted with organic modifiers to produce non-extractable, stable polymeric coatings having a variety of functional groups with which interactions can take place.

21.5.5 These "Permaphase" chromatographic packings provide excellent column efficiency and stability, and eliminate the problems associated with the loss of partitioning phase from the support during the operation of conventional liquid-liquid chromatographic columns. In addition, many highly polar carriers, such as chloroform and alcohol-water mixtures, can be used to rapidly elute strongly retained sample components without effecting a change in the properties of the chromatographic column.

21.6 APPLICATIONS.

21.6.1 The applications of high-speed LC are varied in scope and include both qualitative and quantitative analyses. The retention times of peaks frequently are used as one means of qualitative identification. Components, even in trace quantities, can be trapped and submitted for further characterization by mass spectrometry or other techniques.

21.6.2 Quantitative analyses by LC can be accomplished with good precision and accuracy. When an electronic integrator is used to measure peak areas of separated components, LC analyses with a standard deviation of less than 0.5% are not uncommon. The recent development of high pressure sampling valves, which ensure reproducible introduction of very small sample aliquots, may eliminate the need for internal standards, which frequently are used in quantitative work. Need for standards depends also on purity of solvents and sensitivity of detector.

21.6.3 Techniques such as temperature, flow, and carrier programing, whereby one of these parameters is varied with time, are useful in high-speed LC. For separations of particularly complex mixtures, gradient elution appears to be the most valuable approach.

21.6.4 In this process, the composition of the mobile phase is stepwise or constantly changed throughout the chromatogram by increasing polarity, ion strength, pH, and so forth. The overall effect is to resolve the early-eluting compounds as well as the more strongly retained ones and to complete the elution of all components in a reasonably short time. Programing techniques are particularly attractive in ion-exchange and adsorption chromatography, and in liquid-partition chromatography with packings having chemically-bonded or ''permanent'' stationary phases.

21.6.5 Although relatively little has been done to date, high-speed preparative liquid chromatography can be carried out by increasing the internal diameter of the column. Additional research and knowledge of large column operation, however, is needed to expand this approach.

21.6.6 Because high-speed liquid chromatographic equipment frequently involves flammable organic solvents at high pressures in closed spaces, some manufacturers wisely have chosen to provide safety devices in their instruments. Examples include nitrogen flushing of the column ovens, a nitrogen blanket over the solvent reservoir, and suitable interlocks to disable the instrument in case of a solvent vapor buildup or a leak in the system.

References

1. Leitch, R.E., and J.J. Kirkland, 1970. Ind. Res. 12:36–39.

Subcommittee 9
A. L. Linch, *Chairman*
E. R. Hendrickson
M. Katz
J. R. Martin
G. O. Nelson
J. N. Pattison
A. L. Vander Kolk
R. B. Weidner

22. Thin-Layer Chromatography*

22.1 Introduction.

22.1.1 One of the basic problems of analytical chemistry has been to find good methods for the separation of the individual components of complex mixtures or natural products. In virtually any analytical method, the problem of interference must be considered. Many analytical methods suitable for relatively pure compounds or designed for a given matrix fail when applied to other systems, unless adequate separation schemes can be developed. Most analytical procedures involve a dual approach of prior separation of the desired material, preferably in a reproducible and high yield, followed by analysis of the pure material. Often the separation step is the most challenging to the scientist.

21.1.2 Of the many separation procedures available, those involving chromatography remain among the most versatile and practical. Each chromatographic technique itself has advantages and limitations. Gas chromatography is useful for separating volatile, relatively nonpolar materials; the distribution of members of a homologous series can often be evaluated more quickly with this technique than with others. Ion-exchange chromatography excels in the separation of charged inorganic ions; gel-permeation chromatography provides useful information about the molecular-weight distribution of various polymer systems. Although every separation method has merits, few have the versatility and adaptability of thin-layer chromatography.

22.1.3 Thin-layer chromatography (TLC) has been an accepted separation technique for over ten years, and has now reached the point of relative maturity. As with all new techniques, when TLC was first introduced it was viewed by many as the ultimate problem solver. The growth of the literature was explosive in the early '60's; it abounded with TLC separations involving materials, matrices, and quantities not previously possible. The profusion of these articles and the optimistic tone of their conclusions might have led a person to believe that TLC would eventually solve all the remaining separation problems. Of course, the early workers were overly optimistic. The current TLC literature continues to be substantial, but is now more practical. It seeks to recognize TLC as a useful separation tool, unique in many respects. Workers today are attempting to explain and understand

*This material has been prepared by T. N. Tischer and A. D. Baitsholts, Eastman Kodak Company and is based on an article published in American Laboratory, 2, 69–78, May, 1970.

the principles of TLC and how to adjust its variables to obtain better separations. There are relatively few references to the theory; TLC has always been and, perhaps, will remain a practical technique. Adapting existing procedures and broadening separation experience are the best ways to adjust the experimental conditions to obtain optimum separations in new systems. Reliance on practice rather than on theory has been borne out by the continued application of the workshop approach in laboratory classrooms to introduce new workers to this technique.

22.2 ADVANTAGES OF TLC.

22.2.1 Within the last ten years, the approach to the selection of equipment used for TLC has also changed. Thin-layer chromatography is one of the few new, noninstrumental analytical techniques developed in recent years. Compared to equipment costing as much as hundreds of thousands of dollars, such as is required for mass spectrometry and neutron activation, that required for TLC is astonishingly inexpensive. A few hundred dollars or less will provide the practitioner with suitable equipment for carrying out separations. (This makes TLC a practical tool rather than merely a demonstration experiment in schools or colleges.) A wide variety of commercial or hand-built equipment is available to make the manipulations easier. However, often times the experimenter does better to evaluate his own needs as they arise. Convenience has also extended the TLC field. Most early work was done with hand-coated plates. The development of precoated products offers not only convenience, but also a level of reproducibility and purity not previously found.

22.2.2 Thin-layer chromatography offers many other advantages. It is both sensitive and selective. Separations of microgram quantities of materials have been made; yet, when necessary, the same separation can usually be scaled up to 100 mg or more. When radioactive materials or special monitoring techniques are used even nanogram separations are practical. Many factors of TLC separations can be varied to produce a high degree of selectivity; these include type of sorbent, binder, layer thickness, activation of the layer, development technique, nature of the solvent, chamber saturation, visualization, and recovery procedure. One might think that such an extensive list of variables could make TLC a very inconsistent or "touchy" technique. On the contrary, this diversity is the prime reason for its versatility and selectivity.

22.2.3 As a practical laboratory technique, TLC is relatively rapid. An adequate separation can often be obtained in a few hours even when a completely new problem is being studied. Each separation requires 10 to 45 min to complete, and once the experimenter has become familiar with the basic principles and the nature of the system, adjustments in the experimental variables can usually be made in a relatively short time. Again, experience coupled with trial and error may sometimes be more rewarding than a detailed literature survey.

22.2.4 TLC is a versatile technique that can be used to separate inorganic or organic materials, and low-molecular-weight species up to high polymers. There are few volatility restrictions, as opposed to gas chromatography; however, highly volatile materials are difficult to migrate and to recover. There are virtually no polarity restrictions on materials to be separated. If a solvent is available for the material, generally a TLC separation system can be found to resolve it. The extensive use of relatively stable sorbents makes TLC separations feasible for many reactive species that are difficult to separate by other means. The further adaptability of TLC is shown, for example, by use of techniques involving ion exchange, ion exclusion, and radiometry.

22.2.5 A variety of reference books is available that describe the variables of thin-layer chromatography. The book by Bobbitt **(1)** and Randerath **(2)** offer concise information for those who have little experience with the technique. The compilations of Kirchner **(3)** and Mikes **(4)**, plus the comprehensive manual by Stahl **(5)**, not only describe the general techniques, but contain many literature references on detailed separations of a variety of materials. The books by Truter (**6**) and Gordon and Eastoe (**7**) also furnish much useful information.

22.2.6 Having considered some of the features of thin-layer chromatography and "why" it should be considered for separation, we now turn to some of the "how," or practical considerations important in its use.

22.3 Media and solvent selection.

22.3.1 The two most important variables in thin-layer chromatography are media and solvents. Many workers tend to approach this point from the semi-theoretical view and try to define rigidly the specific separation process. Although this may be advantageous sometimes, the practical view, *i.e.*, asking just what is to be separated is often better.

22.3.2 In devising a new separation procedure for an unknown system, the physical or chemical differences between the materials to be separated should be understood. For separation problems where even the functional groups are unknown, an evaluation of solubility is highly desirable. The old rule "like dissolves like" is helpful. Learning what will dissolve the sample is a good starting point for selecting the basic thin-layer separation process and type of media. As an example, one would take a completely different approach if the compounds to be separated had maximum solubility in benzene or ether as opposed to solubility in water.

22.3.3 If we neglect the special separation techniques such as ion exchange, ion exclusion, and electrophoresis, most TLC separations involve either adsorption or partition processes. A continuous spectrum of conditions can apply between the two extremes; some systems cannot be defined as applying completely to either partition or adsorption.

22.4 Adsorbents and Solvents.

22.4.1 Adsorption processes are generally associated with the separation of relatively nonpolar, hydrophobic materials, in which case a strong adsorbent, such as alumina or silica gel with a moderately nonpolar (low dielectric constant) organic solvent or mixture is used. The activity of the layer and the polarity of the solvent are the prime variables of these systems. The adsorbent serves as a stationary phase; the organic solvent is the mobile phase of this two-phase system. As materials in the liquid phase are brought into contact with the solid phase, a concentration gradient is established at the interface. Components in the mixture separate because of differences in adsorption coefficients. The "benzene-soluble" material mentioned previously is relatively hydrophobic; thus separation involving adsorption chromatography should be considered.

22.4.2 Silica gel is probably the most widely used adsorbent. It has high-capacity and is useful for separating a variety of acidic or neutral materials. For basic materials, such as amines, alumina is often used, but it has the disadvantage that it may catalyze the decomposition of many organic compounds.

22.4.3 A third adsorbent, and one of the strongest known, is charcoal; however, it is not too suitable for TLC because of visualization difficulties.

22.4.4 Other specialized TLC adsorbents include kieselguhr, magnesium silicate, talc, calcium sulfate (also used as a binder for stronger adsorbents), magnesium oxide, and calcium carbonate. However, there are few adsorption separations that cannot be adapted to alumina or silica; if in doubt, silica is the first choice.

22.4.5 The activity of the adsorbent layer is an important variable. Adsorbent layers should generally be activated in an oven before use. After treatment, precautions must be taken to prevent the uncontrolled adsorption of water vapor from the air. This occurs rapidly with both silica gel and alumina. Separations on a highly active layer (silica used directly after treatment in an oven at 120 C) will be markedly different from those where the layer has been allowed to equilibrate with the water vapor in normal room atmospheres. Depending upon the specific separation, a highly active adsorbent may not be required, or desired, but to obtain reproducible results the activity must be controlled. Complete saturation of silica gel with water produces a relatively inert adsorbent (suitable for partition chromatography).

22.4.6 In practice, it is best to pick a specific adsorbent and then vary the solvent. Solvent polarity is the dominant variable in adsorption chromatography. An eluotropic series is a good aid in selecting a solvent (see Table 22:I).

22.4.7 Generally, the more polar the solvent, the lower is the adsorption affinity of a given material and the greater is its migration distance. A rapid evaluation of solvents can often be made by spotting the unknown on a series of small strips and effecting the migration in a series of six or so solvents contained in small jars. Adjustments involving solvent mixtures to obtain more subtle separations can often be made in the same way. If a separation can be accomplished with a single solvent, the use of mixed solvent systems (particularly those consisting of components of widely varying volatility) should be avoided because of "second-order" factors.

22.4.8 To evaluate solvents, or for diagnostic work, plates in the form of 1″ × 3″ microscope slides are handy. For preparative work, 20 × 20-cm (also 5 × 20 or 10 × 20 cm) plates are considered standard. Early workers had to hand-coat plates by pouring, dipping, or using a knife-edge coater. Not only was this very time consuming, but it was difficult for the novice to produce defect-free smooth-surfaced coated plates. Today, hand-coating of plates is no longer necessary because a variety of high-purity, uniform commercial coatings are available on glass, aluminum, and polyester supports.

The normal layer thickness ranges from 100 to 250 μm, although layers 1 to 2 mm thick may be used for preparative work.

22.5 Sample Application.

22.5.1 Many devices are useful for applying the sample solution to the TLC sorbent layer. These include capillary tubes, syringes, and micropipets of all types. Simple applicators are the most efficient, even for precise quantitative analysis. The applied zone should be very compact. It is helpful to use as nonpolar a solvent as possible to minimize sample diffusion during the spotting operation. Optimum chromatographic resolution is obtained when 1 to 5 μl of material is applied per spot. A 1.0% solution of the sample mixture is a convenient working concentration for spotting. The use of a volatile solvent is advisable to ensure its complete removal before development begins. (Figure 22:1)

Table 22:I. Eluotropic solvent series

Solvent	Dielectric Constant
n-Hexane	1.9 (20 C)
Cyclohexane	2.0 (20 C)
Carbon tetrachloride	2.2 (20 C)
Benzene	2.3 (20 C)
Toluene	2.4 (25 C)
Trichloroethylene	3.4 (16 C)
Diethyl ether	4.3 (20 C)
Chloroform	4.8 (20 C)
Ethyl acetate	6.0 (25 C)
l-Butanol	17. (25 C)
l-Propanol	20. (25 C)
Acetone	21. (25 C)
Ethanol	24. (25 C)
Methanol	33. (25 C)
Water	80. (20 C)

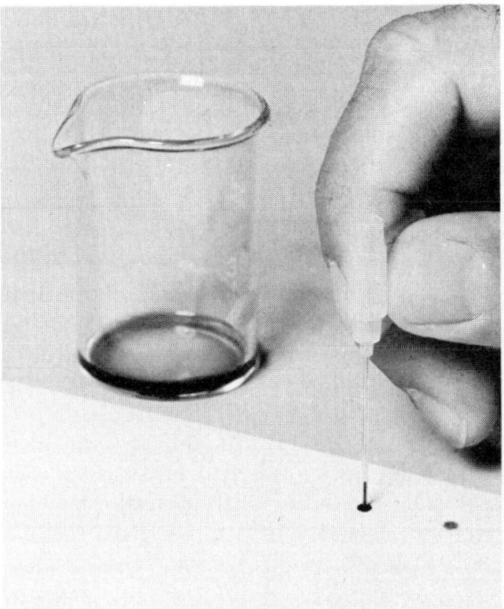

Figure 22:1—Sheets can be spotted most easily and conveniently using capillary tube applicators.

22.5.2 If a compound migrates with the solvent front, even with a relatively nonpolar solvent such as hexane, a stronger adsorbent should be considered. Conversely, compounds that fail to migrate from the point of application with very polar solvents such as methanol should be separated on a weaker adsorbent or perhaps by partition chromatography. Nonpolar sample mixtures are separated best with nonpolar solvents on strong adsorbents. More polar materials may require more polar solvents on weaker or less highly activated adsorbents or may be considered for separation by partition chromatography.

22.5.3 Partition chromatography is used primarily to separate relatively polar organic compounds or inorganic salts. The layer, as such, plays no active part in the separation. Two immiscible liquid phases are present. The mobile phase is organic and hydrophobic. The stationary phase often consists merely of water held on cellulose fibers. The components of the mixture to be separated distribute in different ratios between the two liquid phases. Although partition chromatography has been practiced for many years in the form of paper chromatography, the experimental variables are harder to define because the adsorption activity rules, the solubility rules, and the eluotropic series of solvents are inapplicable. Most existing separations originally designed for paper chromatography are easily adaptable to thin cellulose layers without extensive modifications. The separation on TLC plates produces sharper zones, owing to a reduction in diffusion, and migration time is markedly reduced (30 to 90 min instead of up to 16 hr) compared with classical paper chromatography. When it is necessary to apply large amounts of the sample, narrow horizontal bands are preferred over more concentrated spots. Under ideal conditions, as much as about 1 mg of material can be separated on an 8″ chromatographic plate. When thick layers are used, even larger quantities can be separated for semipreparative work.

22.5.4 For comparison, standards should always be run alongside the sample on the same sheet or plate. It is important that sample and standard be applied in the same solvent in the same manner.

22.6 DEVELOPMENT.

22.6.1 Nonreproducible results and difficulties in duplicating literature separations can usually be traced to differences in developing conditions. The type of chamber, the initial degree of saturation and changes in conditions·during the course of development, significantly influence the results.

22.6.2 The ascending method of development is by far the most common in TLC. The two most common types of chamber in use are the jar (usually lined with filter paper to promote saturation) and the sandwich chamber (a minimum-volume chamber which saturates rapidly, see Figure 22:2). It is difficult to predict in advance which chamber will yield the best separations, since both the sample mixture and the solvent system behave differently in each type of chamber. Trial and error are still necessary, and time spent experimenting with various chamber environments is almost always well invested.

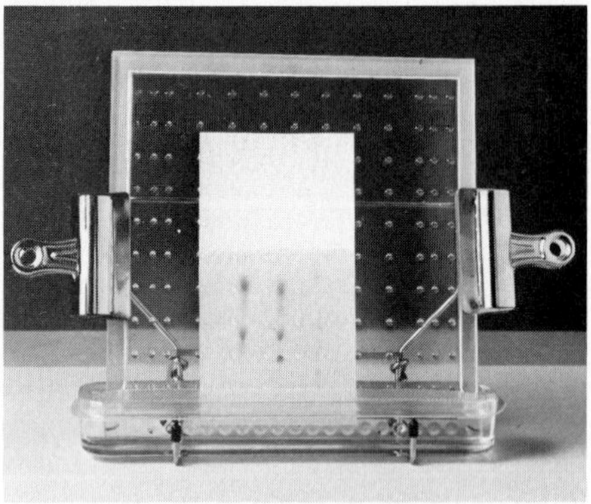

Figure 22:2—The sandwich chamber is a simple and efficient piece of laboratory apparatus in which to develop a chromatogram. After the sheet has been spotted, it is clipped into the glass plate sandwich and immersed in a trough in which the solvent level is adjusted to immerse just the bottom of the sheet. As the solvent rises through the sheet by capillary action, components of the mixture to be separated are carried in a differential manner to produce discrete zones.

22.6.3 The time required for a development distance of 10 cm above spot origin (the most commonly reported and the most practical development distance for 20 × 20 cm plates) varies between 10 min and 2 hr, depending on the type of sorbent, solvent system, and developing conditions.

22.6.4 Special techniques of stepwise, continuous, multiple, and two-dimensional development, as well as a combination of TLC and thin-layer electrophoresis (TLE), have also been used to advantage in the separation of the more complex mixtures.

22.6.5 Stepwise development consists of repeated migrations on a single chromatographic plate using a series of solvents (usually of increasing polarity). It is a useful technique for complex mixtures that do not separate with a single migration. Not only are recovery and respotting steps eliminated, but it may be possible to avoid a time-consuming evaluation of solvents to find the best one.

22.6.6 In continuous development the solvent is allowed to evaporate from the solvent front to produce enhanced resolution of materials that migrate poorly.

22.6.7 Multiple development involves repeated migration with a single solvent. It can be used to remove a small amount of material from a relatively insoluble zone.

22.6.8 Complex mixtures such as amino acids have been separated by two-dimensional development. A sample is spotted in one corner and caused to migrate with one solvent. The chromatogram is then rotated 90° and a second solvent is used for the second migration.

22.6.9 A variety of other specialized development techniques have also been devised but are of limited utility.

22.7 Visualization Techniques.

22.7.1 Separations involving dyes and pigments can be viewed directly. However, when compounds are not detectable by their visible color, one must search for a method of locating these materials on the chromatogram. The immediate tendency is to spray the plate with a nonselective color-forming reagent (see Figure 22:3). This practice should be resisted

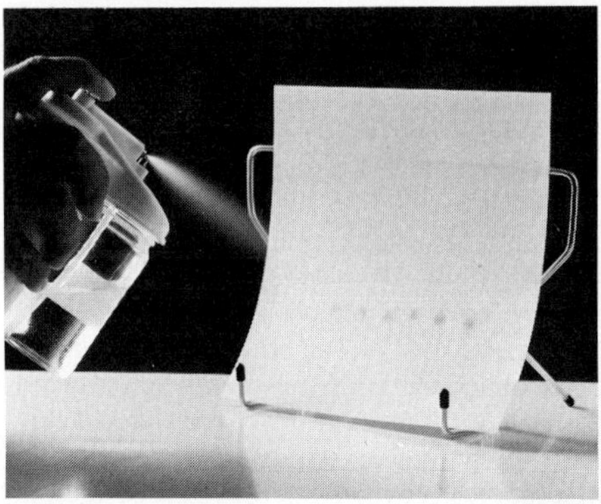

Figure 22:3—Materials that are not colored can be visualized by spraying with a reagent that reacts to form a colored product.

at least until some thought is given to the many nondestructive visualization methods available. Often, compounds can be located by their fluorescence under either long- or short-wavelength ultraviolet light (366 nm and 254 nm, respectively) or by their ability to quench the fluorescence of an inorganic phosphor that is added to the sorbent layer. A light spray of water is often repelled by hydrophobic compounds on the chromatogram to a point at which white spots appear on a translucent background. Iodine vapor is reversibly absorbed by many organic compounds and thus may be used as a nondestructive visualization technique. Autoradiography is a highly sensitive method, useful for the monitoring of separations involving radioactive compounds.

22.7.2 If all the previous techniques fail, then it may be necessary to turn to one of the many destructive methods. These range in sophistication from highly sensitive and specific sprays for forming fluorescent derivatives with nanogram quantities of material to sprays of sulfuric acid or other reagents which, because of their strong oxidizing capabilities, non-selectively turn all organic matter present to black carbon. Chemical spot tests are known for scores of functional groups and hundreds of different compounds **(8, 9)**. Most of these are directly applicable to use on thin layers. These color reactions are adequately described in the literature (**3, 5**).

22.8 Chemistry on the layer.

22.8.1 Carrying out chemistry on the sorbent layer can be extended beyond visualization reactions. Derivatives can be formed after spot application and before development. Other chemical conversions can be carried out to enhance multiple, stepwise, or two-dimensional development. Layers can be chemically altered either during preparation or by modifying the solvent system to control ionization or to complex certain species under study. The analyst's imagination and inventiveness set the upper limit of what can be accomplished.

22.9 Elution of separated zones.

22.9.1 After the TLC separation is complete, it may be desired to remove and isolate the pure material contained in a single zone. One may then wish to quantitate this material, confirm its identity by some additional instrumental or chemical technique, use it in a subsequent experiment as a pure reference compound, determine one of its physical constants, or even isolate milligram quantities for special studies.

22.9.2 TLC zones are conveniently removed from the sorbent by elution with the proper solvent. Some procedural guidelines follow:

a. Prewash the plate or precoated sheet with the eluting solvent prior to activation and sample application. This removes impurities that would also be eluted with the separated zones after chromatography;

b. use precoated flexible TLC media that allow the separated zones to be cut from the sheet for elution. This technique is particularly useful if ultraviolet spectrophotometry or liquid scintillation counting is to be used to quantitate the separated material;

c. when scraping separated zones for the removal of sorbent by filtration, apply a cake of calcium sulfate to the filter from a slurry of the finely divided powder before carrying out the filtration. This procedure aids in removing the last traces of finely divided sorbent, which often interfere, particularly if an infrared spectrum is being used for characterization of unknown zones;

d. when attempting to elute very polar compounds from a strong adsorbent, add a material to the eluting solvent that will be more strongly adsorbed to the solid phase than is the compound under study. Pyridine or quaternary ammonium compound is often selected for this type of displacement;

e. obtain useful concentrations of the materials being separated in the horizontal bands described earlier by carrying out chromatography a second time at a 90° angle to the first development. A very polar solvent is used that will carry the compound of interest with the moving front. Scraping channels in the sorbent layer or cutting the flexible TLC sheet can be advantageous. At the end of the second development the ratio of separated sample to sorbent should be very favorable.

22.10 Recording results.

22.10.1 Keeping records in chromatography may present some special problems, although at first it appears to be quite simple. Often, sufficient attention is not paid to this subject until serious problems arise. At that point, the data needed may be irretrievable.

22.10.2 The most important and perhaps the most often neglected aspect of TLC record-keeping is recording the *exact* chromatographic conditions. If work is to be reproduced at a later date, it is essential to know all system variables: the type of adsorbent, activation conditions, sample application, developing solvent and technique, type of developing chamber, degree and method of chamber saturation, type and distance of development, temperature and humidity conditions (for critical work or in laboratories where seasonal variations are common), and visualization details. Other special conditions, particularly those unique to the sample being analyzed, should also be recorded. The best way to proceed for routine work is to prepare a form which the analyst fills out and files. Better still, a photograph of the chromatogram can be made using either visible or ultraviolet light.

22.10.3 Abrasion-resistant plates and precoated sheets lend themselves well to file storage. Although at first this may seem like the ultimate in record retention, it must be remembered that most zones fade or at best change with time. If the chromatograms themselves are to be kept for permanent records, outlining the separated zones with a pencil to avoid later confusion is recommended (see Figure 22:4).

Figure 22:4—Specific areas of a flexible, precoated sheet can be cut out for inclusion in laboratory notebooks or for further tests which require the spots to be removed from the sheet.

22.11 R_f VALUES.

22.11.1 The results of a separation of a mixture of compounds are often described by calculating their R_f values. The ratio of the distance traveled by the sample component divided by the distance traveled by the solvent is referred to as the R_f value. Materials that do not move from the origin have an R_f value of 0; migration with the solvent front means an R_f value of 1.0. Most useful separations are those that produce widely spaced R_f values between 0.3 and 0.7.

22.11.2 Although R_f values are useful in describing results, they are quite inadequate as absolute "standards" and their importance is often overemphasized. When they are used to communicate observed effects, they should be heavily qualified, because they are affected by so many experimental variables.

22.12 QUANTITATION.

22.12.1 Although great strides have been made in instrumentation and although quantitation of many TLC separations is possible within the accuracy and precision required, TLC is usually not considered a good quantitative tool. Quantitation without removal of the material from the layer is generally difficult, and the accuracy that can be achieved by direct measurement techniques leaves much to be desired. Therefore, most chemists prefer to remove the separated material and apply a conventional analytical technique for quantitation. The various quantitative procedures are described by Shellard **(10)**.

22.13 LIMITATIONS. TLC is subject to limitations as are other separation techniques. However, in addition to certain quantitation problems, the following difficulties remain: (a) it is difficult to separate high polymers and closely related materials, such as isomers and homologs; (b) volatile materials are not easily handled; (c) quantities greater than several grams are unwieldy to handle; and (d) reference standards or other characterization techniques are required to verify the identity of the separated materials.

22.14 APPLICATIONS.

22.14.1 Thin-layer chromatography is used routinely in industrial, biomedical, and pollution-control laboratories. The papers now appearing in technical journals, although perhaps slightly fewer in number than a few years earlier, seem to be very practical in their scope. The technique has found use in air-pollution monitoring, especially with regard to the detection and determination of polynuclear hydrocarbons. Some representative compounds that have been detected by this technique include:

7-H-Benz[*de*]anthracen-7-one	**(11, 12)**
Phenalen-1-one	**(11, 12)**
Benzo[*a*]pyrene	**(12)**
Benzo[*c*]acridine	**(13)**
Benzo[*h*]quinoline	**(13)**

Fatiadi **(14)** has studied the oxidation products of pyrene using TLC. The direct reaction of azulene on a TLC chromatogram has been applied to identify 5-hydroxymethylfurfural in the effluent from a coffee-roasting plant **(15)**. Components in automobile exhaust, including benzaldehyde, salicylaldehyde, and acetophenone, have been separated and identified by paper and thin-layer chromatography **(16)**.

22.14.2 In the study of airborne particulates, other techniques have been combined with thin-layer chromatography to enhance the detection limits or improve the quantitative aspects. Complex mixtures of aromatic hydrocarbons have been separated by TLC and subsequently identified by mass spectrometry **(17)**. Minute quantities of benzo[*a*]pyrene are detected by combining TLC and spectrofluorimetry **(18)**. A method for the direct spectrofluorimetric determination of polycyclic-hydrocarbons and azaheterocyclics on chromatograms has been devised **(19)**. Two-dimensional TLC and fluorimetry have been combined to determine 9-acridanone in urban air particulates **(20)**. There are numerous other applications of the use of TLC in the field of air pollution.

22.14.3 The analysis of pesticide residues is another area where thin layer chromatography can be used to practical advantage. TLC separation data of 60 different pesticides using 19 different solvents have been summarized by Walker and Beroza **(21)**. TLC is an ideal technique for separating the diversity of moderately nonpolar materials encountered in this field. Kirchner **(3)** presents a summary of references for the TLC separation of various fungicides, herbicides, and insecticides such as pyrethrins and thiocarbamates, in addition to those mentioned above.

22.14.4 The application of TLC are endless. A pharmaceutical research worker detects which component in a natural product mixture is responsible for certain biological activity. An analyst in a nearby crime laboratory is learning to detect the presence of illicit drugs by modifying a technique he has mastered for the comparison of writing inks. In biomedical research the ability to subdivide metabolic products present in complex blood and body fluids offers great promise not only in understanding what changes have occurred in these fluids and associating them with certain diseases, but in offering a possible technique for the early detection of these diseases.

22.14.5 The growing consumption of TLC supplies and particularly precoated media further indicates the fact that use of this technique is being established in numerous routine situations. Simplification of the test methods, design of new equipment, and more direct ways to interpret results promise to extend the applications of TLC testing, particularly in hospital clinical laboratories and industrial quality control. Commercial suppliers of materials and equipment have been quick to recognize these needs and have been introducing a wide variety of prepared systems ranging from modifications in coated layers to complete kits containing all the materials required to perform specific routine tests. New designs in spotting aids, developing chambers, and measurement devices also promise to extend the use of TLC to analysis of larger numbers of samples of a more routine nature. Like so many other techniques, TLC is being used in our daily lives, and is no longer a tool limited to research applications.

References

1. Bobbitt, J.M. 1963. *Thin-Layer Chromatography*, Reinhold Publishing Co., New York.
2. Randerath, K. 1966. *Thin-Layer Chromatography*, 2nd Ed. Academic Press, Inc., New York.
3. Kirchner, J.G. 1967. *Thin-Layer Chromatography*, Vol. XII of Techniques of Organic Chemistry, Interscience Publishers, New York.
4. Mikes, O. 1966. *Laboratory Handbook of Chromatographic Methods*, D. Van Nostrand Co., Inc., New York.
5. Stahl, E. 1969. *Thin-Layer Chromatography—A Laboratory Handbook*, 2nd Ed. Academic Press, Inc., New York.

6. TRUTER, E.V. 1963. *Thin-Film Chromatography*, Cleaver-Hume Press, Ltd., London.
7. GORDON, A.H., AND J.E. EASTOE, 1964. *Practical Chromatographic Techniques*, D. Van Nostrand Co., Inc., New York.
8. FEIGL, F. 1966. *Spot Tests in Inorganic Analysis*, Seventh English Ed. American Elsevier Publishing Co., Inc., New York.
9. FEIGL, F. 1958. *Spot Tests in Inorganic Analysis*, Fifth English Ed. D. Van Nostrand Co., Inc., Princeton, New Jersey.
10. SHELLARD, E.J. 1968. *Quantitative Paper and Thin-Layer Chromatography*, Academic Press, Inc., New York.
11. ENGEL, C.R., AND E.J. SAWICKI, 1968. Chromatog. 37:508.
12. STANLEY, T.W., M.J. MORGAN, AND E.M. GRISBY, 1967. Environ. Sci. Technol. 1:927.
13. SAWICKI, E., T.W. STANLEY, AND W.C. ELBERT, 1967. J. Chromatog. 26:72.
14. FATIADI, A.J. 1968. Environ. Sci. Technol. 2:464.
15. ENGEL, C.R., AND E. SAWICKI, 1968. Microchem. J. 13:202.
16. BARBER, E.D., AND E. SAWICKI, 1968. Anal. Chem. 40:984.
17. WALLCAVE, L. 1969. Environ. Sci. Technol. 3:948.
18. BENDER, D.F. 1968. Environ. Sci. Technol. 2:204.
19. SAWICKI, E., T.W. STANLEY, AND H. JOHNSON. 1964. Microchem. J. 8:257.
20. SAWICKI, E., T.W. STANLEY, AND W.C. ELBERT, 1967. J. Chromatog. 14:431.
21. WALKER, K.C., AND M. BEROZA, 1963. J. Assoc. Official Analyt. Chemists. 46:250.

Subcommittee 9
A. L. LINCH, *Chairman*
E. R. HENDRICKSON
M. KATZ
J. R. MARTIN
G. O. NELSON
J. N. PATTISON
A. L. VANDER KOLK
R. B. WEIDNER

23. Nondestructive Neutron Activation Analysis of Air Pollution Particulates

23.1 INTRODUCTION

23.1.1 Elemental analysis of pollution aerosols requires a precise yet sensitive method if the results are to be used for study of source identification or atmospheric transport processes. For studies involving large numbers of samples speed and ease of analysis are necessary, and selectivity in the detection of many elements at the nanogram to microgram level is desirable in order to permit sampling of only a few cubic meters of air. Zoller and Gordon (**2**) have presented a discussion of the principles and merits of nondestructive neutron activation analysis in this application, together with an outline of a procedure and results of several analyses of Cambridge, Massachusetts, air pollution particulate samples for more than 20 elements.

23.1.2 The National Air Sampling Network (**3**) has used an emission spectrographic technique to analyze for 16 elements—Be, Bi, Cd, Co, Cr, Cu, Fe, Mn, Mo, Ni, Pb, Sb, Sn, Ti, V, Zn—using a procedure involving ashing and extracting in nitric acid. The method requires skilled operators, is not highly sensitive, and is often limited by high blank values for several elements. Of the reported values, 45% are given only as upper limits. When the high neutron fluxes of nuclear reactors are used, neutron activation analysis is an extremely sensitive method and no blanks due to chemical reagents are introduced. Several authors (**4, 5, 6**) have applied NaI γ spectrometry coupled with radiochemistry for resolution of the γ spectra to the determination of several elements in aerosols. High resolution Ge(Li) detectors greatly extend the scope of nondestructive activation analysis so that a very large number of isotopes can be counted simultaneously. If computer assisted, this technique can become almost completely automated. Several authors (**2, 7, 9**) have used Ge(Li) detectors for destructive or nondestructive analysis of aerosols.

23.2 SAMPLING PROCEDURE

23.2.1 Aerosols may be sampled by passing air through a filter which allows a high flow rate and at the same time has good particle retentivity down to 0.1 μm size. Glass fiber filters have often been used for the analysis of organics, sulfate, nitrate, and total particulate, but this filter must be ruled out for nondestructive activation analysis because of its high concentrations of trace metals. Although it is not an ideal filter, polystyrene (**10**) can be used, combining a good filtering performance with fairly low blank values. However, the sensitivity of a number of elements, such as Cl, Na, Al, Ca, Mn, Zn, and Sb, is still limited by the magnitude of the blank (**10**), and values for Cl are not reported in the test results described below for this reason.

23.2.2 Insofar as pumps are concerned, those in common use for air pollution monitoring are low vacuum, high volume types used with 20 × 25 cm (8″ × 10″) filters. High vacuum, low free air capacity pumps equipped with 25 mm (or 47 mm) diam holders can also be used. Where the high volume pumps generate a flow rate of 4.5 l/min-cm^2 the high vacuum pumps and 25 mm holders reach 12 l/min-cm^2. In spite of the fact that the latter holder has an unexposed waste zone at the edge of the filter (25% of the total area) which decreases the signal to blank ratio somewhat, this figure is still twice as high as for the high volume sampler.

23.3 NONDESTRUCTIVE NEUTRON ACTIVATION ANALYSIS

23.3.1 A procedure for the nondestructive analysis of air pollution particulate matter for up to 33 elements in solid form has been developed (**1**) though it can also be used for the analysis of other types of environmental samples as well.

23.3.2 For the analysis of elements giving rise to short-lived isotopes, each sample is packaged in a polyethylene vial, then placed in a rabbit which carries it through a pneumatic tube to a position near the core of the Nuclear Reactor, where it is irradiated for five minutes at a flux of 2 × 10^{12} neutrons/cm^2 sec. At the end of this period it rapidly returns to the laboratory where it is manually transferred to a counting vial. At 3 min after irradiation a count of 400 sec live-time duration is begun, and this is followed by a count of 1000 sec live-time starting 15 min after irradiation. Both these and subsequent counts are performed on a 30 cm^3 Ge(Li) detector coupled to a 4096 channel analyzer. The detector is housed in an iron shield and operated in an air conditioned room at a gain setting of 1 keV/channel. The observed resolution is 2.5 keV FWHM for the ^{60}Co 1332 keV photopeak and a peak to Compton ratio of 18/1. Table 23:I shows the isotopes determined by the first two counts.

23.3.3 All spectra are recorded on 7-track magnetic tape for future data analy-

Table 23:I. Nuclear Properties and Measurement of Short-Lived Isotopes

Element	Isotope	Half-life	Irradiate	Cool	Count	Gamma-Rays Used, keV
Al	^{28}Al	2.31 min	5 min	3 min	400 sec	1778.9
S	^{37}S	5.05 min	"	"	"	3102.4
Ca	^{49}Ca	8.8 min	"	"	"	3083.0
Ti	^{51}Ti	5.79 min	"	"	"	320.0
V	^{52}V	3.76 min	"	"	"	1434.4
Cu	^{66}Cu	5.1 min	"	"	"	1039.0
Na	^{24}Na	15 min	"	15 min	1000 sec	1368.4; 2753.6
Mg	^{27}Mg	9.45 min	"	"	"	1014.1
Cl	^{38}Cl	37.3 min	"	"	"	1642.0; 2166.8
Mn	^{56}Mn	2.58 hr	"	"	"	846.9; 1810.7
Br	^{80}Br	17.6 min	"	"	"	617.0
In	^{116m}In	54 min	"	"	"	417.0; 1097.1
I	^{128}I	25 min	"	"	"	442.7

sis and can also be printed on paper tape. Conversion of counting rates under the various peaks to concentrations is accomplished by subjecting a few standard solutions containing well-known mixtures of these same elements to the same irradiation and counting sequences. To avoid possible errors due to coincidence summing or to broadening of peaks at high counting rates, sample sizes are generally adjusted to make counting rates of sample and standard of comparable magnitude.

23.3.4 Though all the short irradiations are performed at the same site in the reactor and under conditions of nearly constant power, small corrections for variations of both the neutron flux and rabbit placement from irradiation to irradiation are accomplished by coirradiation of a titanium metal foil flux monitor with each sample. It is counted for 20 sec at 13 min after the end of irradiation, between the two sample counts. If the analysis rate does not exceed one sample in 40 min, the same flux monitor may be used repeatedly, with less than 1% of the original 5.8-min ^{51}Ti remaining in the next count. Net counting rates of the sample spectrum are normalized to an arbitrary Ti activity, equivalent to a reference neutron flux. During one 20-day reactor operating cycle the standard deviation determined by this flux monitor was 3%. However, variations of up to 8% were experienced between the neutron flux at the irradiation site during different reactor cycles.

23.3.5 The same example, or another portion of the same air filter, is then irradiated at a higher flux (1.5×10^{13} neutrons/cm^2 sec) 2 to 5 hr in the reactor core. Each is individually heat sealed in a polyethylene tube and irradiated together with eight others plus a standard mixture of elements in a polyethylene bottle, 4 cm in diam, lowered into the reactor pool. Cooling of the samples during irradiation is achieved by allowing the pool water to circulate through several holes cut in the container bottle. Standards are prepared by depositing 100μl each of two well-balanced mixtures of the appropriate elements onto a highly pure substrate (ashless filter paper) and allowing to dry, then sealing immediately inside polyethylene tubes. After irradiation, the samples and standards are transferred to clean containers and counted once for 2000 sec live-time after 20 to 30 hr of cooling and then for 4000 sec live-time after 20 to 30 days of cooling. Table 23:II shows elements determined from these counts. Horizontal and vertical flux gradients at the irradiation site in the core have been measured, and errors due to thermal neutron flux gradients over the bottle dimensions appear to be less than 5% provided the samples are confined to a single horizontal layer of vertically oriented tubes at the bottom of the bottle and the bottle is rotated 180° at half of the irradiation time. Fast neutron flux gradients at this site are about twice as large as thermal gradients, but the only fast neutron reaction used in our procedure is in the determination of nickel, ^{58}Ni (n,p) ^{58}Co.

23.3.6 The entire irradiation and counting procedure is illustrated in Figure 23:1, and Tables 23:I and 23:II list the γ transitions used. Sometimes the most prominent photopeak of an isotope cannot be used because of interferences by neighboring peaks of other isotopes. Examples of unusable peaks include 844 KeV of ^{27}Mg (846 keV ^{56}Mn) and 559 keV of ^{76}As (555 keV ^{82}Br and 564 keV ^{122}Sb). The monoenergetic ^{203}Hg (279.1 keV) is interfered with by ^{75}Se (279.6 keV), but a correction based on the spectrum of pure ^{75}Se can be applied because the interference is usually less than 20% of the ^{203}Hg activity in the air pollution samples we have analyzed so far. The measurement of ^{64}Cu (511.0 keV) is interfered with by external pair production of high energy γ=rays. In typical samples of 15 hr ^{24}Na is the most important source of high energy γ-rays after a decay period of 20 hr, and a correction, usually less than 10% can be applied to the apparent ^{64}Cu activity.

23.3.7 The ratio of thermal to fast neutron flux was determined at both irradiation sites using the reactions ^{31}P (n,γ) ^{32}P and ^{32}S (n,p) ^{32}P. The ratios obtained were 7.5 for pneumatic tube and 4.5 for pool irradiation sites. Interferences by threshold reactions were calculated and checked experimentally. In typical aerosol samples the only reaction affecting a calcu-

Table 23:II. Nuclear Properties and Measurement of Long-Lived Isotopes

Element	Isotope	Half-life	Irradiate	Cool	Count	Gamma-Rays Used, keV
K	^{42}K	12.52 hr	2–5 hr	20–30 Hr	2000 sec	1524.7
Cu	^{64}Cu	12.5 hr	"	"	"	511.0
Zn	^{69m}Zn	13.8 hr	"	"	"	438.7
Br	^{82}Br	35.9 hr	"	"	"	776.6; 619.0; 1043.9
As	^{76}As	26.3 hr	"	"	"	657.0; 1215.8
Ga	^{72}Ga	14.3 hr	"	"	"	630.1; 834.1; 1860.4
Sb	^{122}Sb	2.75 day	"	"	"	564.0; 692.5
La	^{140}La	40.3 hr	"	"	"	486.8; 1595.4
Sm	^{153}SM	47.1 hr	"	"	"	103.2
Eu	^{152m}Eu	9.35 hr	"	"	"	121.8; 963.5
W	^{187}W	24.0 hr	"	"	"	479.3; 685.7
Au	^{198}Au	2.70 day	"	"	"	411.8
Sc	^{46}Sc	83.9 day	"	20–30 day	4,000 sec	889.4; 1120.3
Cr	^{51}Cr	27.8 day	"	"	"	320.0
Fe	^{59}Fe	45.1 day	"	"	"	1098.6; 1291.5
Co	^{60}Co	5.2 yr	"	"	"	1173.1; 1332.4
Ni	^{58}Co	71.3 day	"	"	"	810.3
Zn	^{65}Zn	245 day	"	"	"	1115.4
Se	^{75}Se	121 day	"	"	"	136.0; 264.6
Ag	^{110m}Ag	253 day	"	"	"	937.2; 1384.0
Sb	^{124}Sb	60.9 day	"	"	"	602.6; 1690.7
Ce	^{141}Ce	32.5 day	"	"	"	145.4
Hg	^{203}Hg	46.9 day	"	"	"	279.1
Th	^{233}Pa	27.0 day	"	"	"	311.8

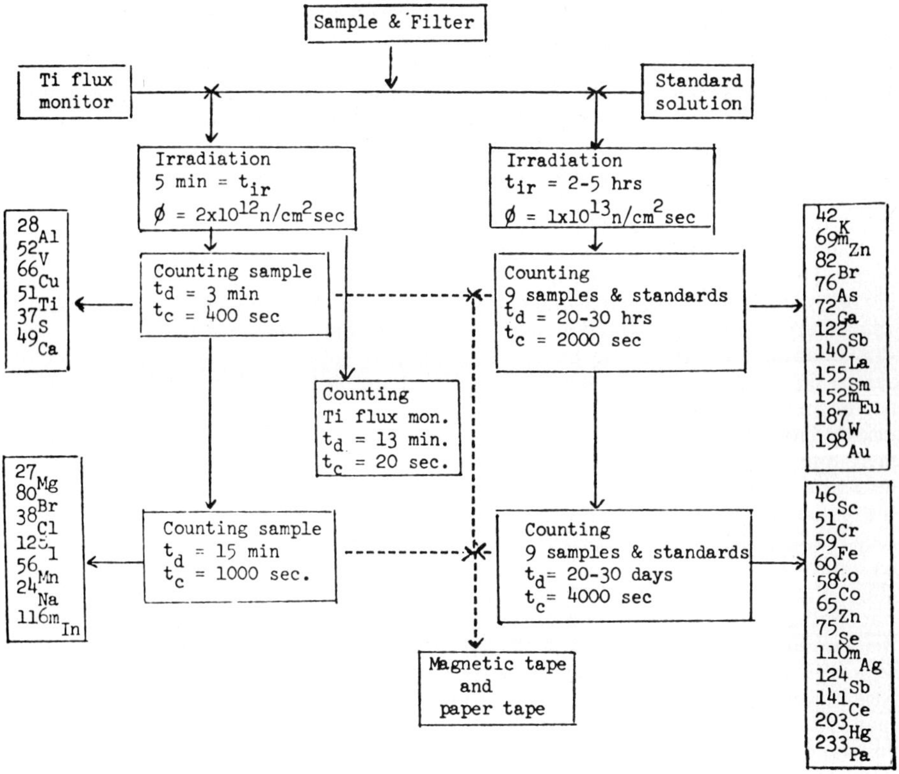

Figure 23:1—Irradiation-counting scheme.

lated concentration by more than 2 per cent is ^{27}Al (n.p) ^{27}Mg. Once the aluminum concentration is known, the appropriate correction can be applied to the magnesium concentration.

23.4 Automated Data Reduction

23.4.1 In order that non destructive neutron activation analysis procedure should be applicable to large numbers of samples, such as routine monitoring, automatic data reduction is necessary. Accuracy as well as speed is increased by elimination of human errors, but human judgment should be retained in the examination of the data and in devising procedures for checking data quality.

23.4.2 In the present investigation a computer program was developed and used for the following to: (a) qualitatively determine the presence of isotopes, (b) calculate net peak areas, (c) convert areas to weights of trace elements, (d) subtract analytical blanks due to filter materials, (e) calculate concentrations of trace elements in the original air sample.

23.5 The Program

23.5.1 The magnetic tape on which up to 100 γ-ray spectra, each of 4096 channels, are stored is submitted to the IBM 360/67 computer together with a Fortran IV program typed on cards. The program makes an inventory of the tape and creates a new compact error-free version, providing a list with position and tagword of each spectrum on the final tape.

23.5.2 Initially an approximate energy calibration of the spectrometer is required. On this basis the program can refine and update the calibration for each spectrum by comparing the observed positions of some prominent peaks with the true γ-ray energies, present as a data library in the program. This calibration fit can in principle be a polynomial of any order, but usually a linear or at most quadratic fit suffices.

23.5.3 The main program for obtaining trace element concentrations from spectra requires a data set specifying γ transition energies of expected peaks. From this and the above energy calibration an approximate peak location is computed. The channel with the greatest number of counts is sought in a 7-channel interval centered on this calculated location. If this channel occurs at the end of the search interval, such as when the expected peak is masked by the edge of a nearby large peak, it fails to qualify as a peak and the program moves on to the next peak search; otherwise it is taken as the actual peak location. The net peak area is evaluated by summing counts over an interval of seven channels centered about the actual peak position and subtracting base-line counts. The estimate of the base line under peaks is accomplished by a regression technique as described by Ralston and Wilcox (**11**). Not all maxima found within the scan interval are associated with real peaks. Included in the output is a measure of the statistical significance of each net peak area which enables the user to assess data quality at a glance.

23.5.4 Further data sets are provided in the program. For the short counts (Table 23:I) these include conversion factors from peak areas to concentrations (calculated from standard short-lived spectra), and for the longer count data (Table 23:II) they include concentrations of the elements in the standards, filter blank values, and isotope half lives. Apart from the above data sets, the program requires only one card per spectrum, containing information about the irradiation and count type, filter type and fraction of sample irradiated, whether the sample is a standard, flux monitor, or unknown, and finally factors used in converting weights of elements to concentrations in air, expressed in any desired units.

23.5.5 If the spectrum is a flux monitor, only one net peak area is determined (320 keV ^{51}Ti). This value is used to normalize all peak areas of the unknown (400 and 1000 second counts) to a reference neutron flux. In an unknown sample the peak areas are converted to weights. If the spectrum is from a long count (2000 or 4000 seconds), peak areas are compared to those of the standard and after decay cor-

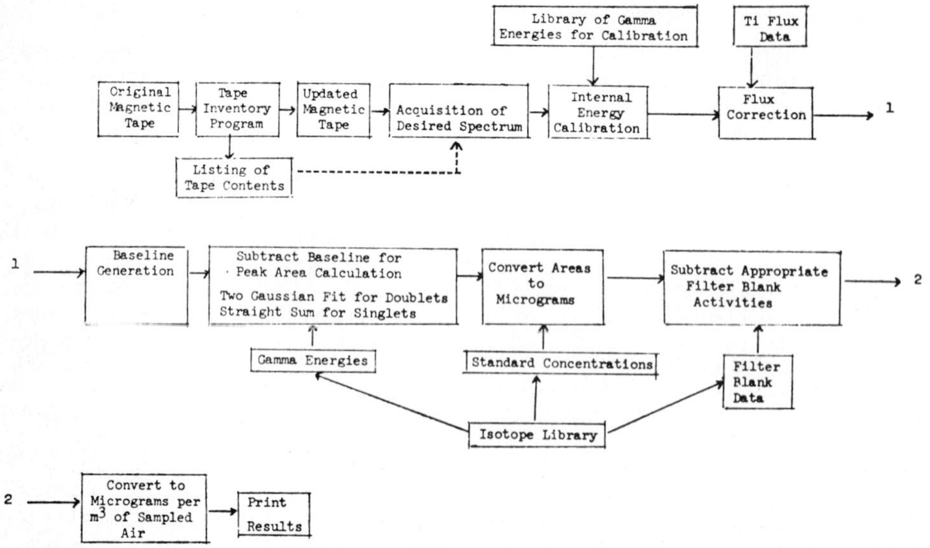

Figure 23:2—A simplified flow diagram of automated spectrum analysis.

Table 23:III. Sensitivities for Determination of Trace Elements in Aerosols

	Neutron Activation			Emission Spectography
Element	Decay Time*	Detection Limit (μg)	Minimum Concentration in Urban Air (μg/m³) 24-hour Sample	Minimum Concentration in Urban Air (μg/m³) 24-hour Sample
Al	3 min	0.04	0.008	—
S	"	25.0	5.0	—
Ca	"	1.0	0.2	—
Ti	"	0.2	0.04	0.0024
V	"	0.001	0.002	0.0032
Cu	"	0.1	0.02	0.01
Na	15 min	0.2	0.04	—
Mg	"	3.0	0.6	—
Cl	"	0.5	0.1	—
Mn	"	0.003	0.0006	0.011
Br	"	0.02	0.004	—
In	"	0.0002	0.00004	—
I	"	0.1	0.02	—
K	20–30 hr	0.075	0.0075	—
Cu	"	0.05	0.005	0.01
Zn	"	0.2	0.02	0.24
Br	"	0.025	0.0025	—
As	"	0.04	0.004	—
Ga	"	0.01	0.001	—
Sb	"	0.03	0.003	0.040
La	"	0.002	0.0002	—
Sm	"	0.00005	0.000005	—
Eu	"	0.0001	0.00001	—
W	"	0.005	0.0005	—
Au	"	0.001	0.0001	—
Sc	20–30 day	0.003	0.000004	—
Cr	"	0.02	0.00025	0.0064
Fe	"	1.5	0.02	0.084
Co	"	0.002	0.000025	0.0064
Ni	"	1.5	0.02	0.0064
Zn	"	0.1	0.001	0.24
Se	"	0.01	0.0001	—
Ag	"	0.1	0.001	—
Sb	"	0.08	0.001	0.040
Ce	"	0.02	0.00025	—
Hg	"	0.01	0.0001	—
Th	"	0.003	0.00004	—

*Decay time before counting. See Tables 23:I and 23:II.

rections (arising because sample and standard are not counted at the same time after irradiation) weights of the trace elements are calculated. Appropriate blanks are subtracted, followed by division by volume of air sampled to give concentrations. A detailed error analysis gives standard deviations of concentrations based on counting statistics and uncertainties in blank and standard values. Figure 23:2 shows schematically the outline of the computer program.

23.5.6 The program does not resolve doublets. A Guassian fit treating these cases has been tried successfully in some cases but the increased running time and consequent expense does not justify inclusion as a permanaent feature. For similar reasons the computer technique is supplemented by manual calculations for ^{76}As,

Table 23:IV. Elements Detected in Suspended Particulate from East Chicago, Indiana, ng/m^3

Element	1st Anal.*	2nd Anal.*	3rd Anal.*	4th Anal.*	Mean†
Ca	7700 (1200)	6600 (1000)	6650 (1000)	—	7000 (700)
Ti	225 (60)	165 (55)	170 (60)	—	190 (40)
V	20 (1.4)	16.1 (1.2)	18.2 (1.3)	—	18.1 (1.5)
Cu	1020 (100)	1050 (100)	1140 (100)	—	1070 (80)
Al	2370 (150)	1980 (130)	2200 (150)	—	2175 (170)
S	11,000 (9,000)	—	15,000 (9,000)	—	13,000 (8,000)
Na	485 (36)	405 (35)	470 (40)	—	455 (40)
Mg	2600 (750)	1650 (700)	2850 (800)	—	2400 (600)
Mn	245 (14)	222 (12)	305 (17)	—	255 (40)
In	0.13 (0.07)	0.06 (0.06)	0.11 (0.07)	—	0.10 (0.05)
Br	81 (9)	74 (9)	94 (10)	—	83 (9)
K	1510 (100)	1380 (70)	1600 (160)	1210 (120)	1415 (150)
La	5.8 (0.7)	6.0 (0.5)	6.5 (0.6)	5.6 (0.7)	5.9 (0.4)
Sm	0.53 (0.06)	0.35 (0.04)	0.47 (0.04)	0.39 (0.05)	0.41 (0.05)
Eu	0.165 (0.03)	0.12 (0.03)	0.17 (0.05)	0.10 (0.03)	0.135 (0.02)
Cu	1150 (150)	1125 (150)	1200 (150)	1100 (150)	1150 (100)
Zn	1440 (150)	1370 (140)	1420 (140)	1365 (140)	1400 (100)
W	7.6 (2)	5.0 (1)	5.6 (1)	5.2 (2)	5.8 (1)
Ga	1.3 (0.7)	1.4 (0.6)	1.2 (0.8)	1.3 (0.7)	1.3 (0.4)
As	14.5 (0.7)	8 (2.5)	14 (2)	12.5 (3)	12 (3)
Sb	25 (3)	24 (3)	30 (3)	23 (4)	25 (2)
Br	63 (7)	68 (6)	77 (8)	60 (7)	67 (4)
Sc	3.8 (0.3)	3.2 (0.5)	2.1 (0.2)	—	3.1 (0.5)
Ce	15.2 (2.0)	14.2 (2.0)	8.6 (1.5)	—	13 (3)
Th	1.4 (0.3)	1.2 (0.5)	—	—	1.3 (0.4)
Cr	137 (10)	112 (15)	88 (10)	—	113 (20)
Fe	17,000 (1200)	13,500 (1000)	10,500 (1000)	—	13,800 (3000)
Co	3.35 (0.3)	2.6 (0.2)	1.9 (0.2)	—	2.6 (0.6)
Ni	—	<60	—	—	<60
Ag	3.1 (2.5)	1.8 (2.0)	2.2 (2.4)	—	2.4 (1.5)
Zn	2210 (240)	1450 (150)	1430 (140)	—	1690 (300)
Hg	5.3 (1.4)	4.4 (1.3)	—	—	4.8 (1.0)
Sb	43 (4)	28 (3)	23 (3)	—	32 (9)
Se	3.0 (0.8)	4.6 (1.4)	—	—	3.8 (1.0)

*Standard deviations are based on counting statistics of sample, blank, and standard.
†Standard deviations are based on dispersion of replicate analyses.

^{72}Ga, ^{122}Sb, and ^{187}W where very small photopeaks are located on tails of neighboring large peaks. Processing of one 4096 channel spectrum takes about 15 sec of computer time.

23.6 SENSITIVITY

23.6.1 The sensitivity for the different elements is often determined by the degree of interference from other substances. Column 3 of Table 23:III shows the sensitivity obtained by the present procedure for the analysis of typical inland aerosol samples, expressed in weights of the elements. The sensitivity obtained by the National Air Sampling Network (**3**) applying emission spectrography and using 90 cm^2 (14″^{2}) of a 24 hr high volume sample is expressed in concentrations of the elements in urban air (column 5). For convenience, the sensitivities obtained by the present neutron activation technique were also converted to concentrations in urban air (column 4) when counting irradiated samples after the decay times indicated, where 0.8, 0.8, 1.6 and 13 cm^2 of the filter were used for the four counts, respectively, corresponding approximately to 5, 5, 10 and 80 m^3 of air collected during 24 hr by a high volume sampler. The sensitivities obtainable in non-urban areas are better for both methods. It should be borne in mind that the sensitivities given are not fixed values for all urban aerosols

Table 23:V. Elements Detected in Suspended Particulate from Niles, Michigan, ng/m^3

Element	1st Anal.*	2nd Anal.*	3rd Anal.*	4th Anal.*	Mean†
Ca	600 (200)	100 (300)	1150 (280)	—	1000 (200)
Ti	105 (35)	105 (35)	150 (40)	—	120 (25)
V	5.0 (0.5)	5.2 (0.5)	4.9 (0.5)	—	5.0 (0.3)
Cu	250 (30)	290 (35)	325 (37)	—	290 (30)
Al	1100 (100)	1240 (8)	1280 (85)	—	1200 (70)
S	9,000 (6,000)	9,000 (6,000)	16,000 (8,000)	—	11,000 (5,000)
Na	140 (18)	180 (20)	175 (20)	—	170 (20)
Mg	350 (400)	680 (400)	—	—	500 (300)
Mn	58 (3)	62 (4)	67 (5)	—	62 (4)
In	0.04 (0.03)	0.04 (0.03)	0.03 (0.03)	—	0.04 (0.03)
Br	34 (4)	28 (4)	34 (5)	—	32 (3)
K	740 (40)	930 (50)	600 (30)	700 (30)	750 (100)
La	1.4 (0.2)	1.7 (0.2)	1.1 (0.2)	1.0 (0.2)	1.3 (0.3)
Sm	0.26 (0.01)	0.30 (0.01)	0.21 (0.02)	0.20 (0.02)	0.24 (0.03)
Eu	0.05 (0.017)	0.065 (0.021)	0.045 (0.009)	0.06 (0.010)	0.055 (0.008)
Cu	260 (35)	260 (35)	245 (35)	305 (35)	270 (25)
Zn	124 (33)	186 (40)	132 (20)	113 (20)	140 (20)
W	0.4 (0.2)	0.2 (0.2)	0.55 (0.2)	0.4 (0.2)	0.4 (0.1)
Ga	1.1 (0.6)	1.0 (0.6)	0.7 (0.4)	0.9 (0.4)	0.9 (0.4)
As	4.4 (2)	5.2 (2)	4.8 (2)	4.0 (2)	4.6 (2)
Sb	5.4 (1)	7.0 (1)	5.7 (1)	4.7 (0.8)	5.8 (0.6)
Br	46 (2)	55 (3)	35 (2)	38 (2)	43 (5)
Sc	1.2 (0.1)	1.2 (0.1)	—	—	1.2 (0.1)
Ce	0.77 (0.1)	0.86 (0.1)	—	—	0.82 (0.10)
Th	0.28 (0.04)	0.26 (0.04)	—	—	0.27 (0.03)
Cr	9.1 (0.8)	10.0 (0.8)	—	—	9.5 (0.8)
Fe	1900 (100)	1840 (100)	—	—	1900 (100)
Co	1.0 (0.1)	0.90 (0.08)	—	—	0.95 (0.10)
Ni	—	—	—	—	—
Ag	<1	<1	—	—	<1
Zn	185 (10)	172 (10)	—	—	180 (10)
Hg	2.0 (0.6)	1.7 (0.6)	—	—	1.9 (0.3)
Sb	6.6 (0.7)	6.1 (0.05)	—	—	6.3 (0.5)
Sc	2.5 (0.6)	2.5 (0.6)	—	—	2.5 (0.5)

*Standard deviations are based on counting statistics of sample, blank, and standard.
†Standard deviations are based on dispersion of replicate analyses.

because they also depend to a certain extent on the composition of the sample.

23.7 APPLICATION

23.7.1 The data given in Tables 23:IV and 23:V are drawn from results of a one-day area-wide survey of the Northwest Indiana region, June 11, 1969 (full account will be published separately). Each sample was 24 hr in length, taken on a polystyrene 20 × 25 cm (8″ × 10″) filter. The East Chicago, Indiana, location was illustrative of a heavily polluted industrial area, whereas the Niles, Michigan, station was in a rural location some 100 km to the ENE. During the sampling period the sky was generally overcast (0.6 to 0.9 fractional cloud cover), and scattered traces of rain fell in the metropolitan area but not at the rural location. Winds were nearly constant from the south at 5 to 10 m/sec during the 24 hr.

23.7.2 Tables 23.IV and 23:V show that the concentrations of 30 elements in total were investigated, and 4 of these (Cu, Zn, Sb, Br) were duplicated in different counts. Each analysis was replicated from two to four times. Standard deviations are given both for the single measurements as returned by the computer (based on counting statistics and uncertainties in blanks and standards) and for the mean of the values reported (based on the dispersion of the replicate analyses).

23.8 DISCUSSION

23.8.1 Tables 23:IV and 23:V show the 29 elements determined. Some of these are present only in very low concen-

Table 23:VI. Ratios of Concentrations Found, East Chicago/Niles

3 & 15 Min Decay		20–30 Hr Decay		20–30 Day Decay	
Ca	7.0 ± 1.5	K	1.9 ± 0.3	Sc	2.6 ± 0.4
Ti	1.6 ± 0.6	La	4.5 ± 1.2	Ce	16.0 ± 4.0
V	3.6 ± 0.4	Sm	1.7 ± 0.3	Th	4.8 ± 1.4
Cu	3.7 ± 0.4	Eu	2.5 ± 0.4	Cr	12.0 ± 2.0
Al	1.8 ± 0.2	Cu	4.3 ± 0.6	Fe	7.3 ± 1.6
S	1.1 ± 0.8	Zn	10.0 ± 2.0	Co	2.7 ± 0.6
Na	2.7 ± 0.3	W	14.5 ± 4.5	Ag	> 2.4 ± 1.5
Mg	4.8 ± 3.0	Ga	1.4 ± 0.7	Zn	9.4 ± 1.7
Mn	4.1 ± 0.7	As	2.6 ± 1.3	Hg	2.5 ± 0.7
In	2.5 ± 2.0	Sb	4.3 ± 0.5	Sb	5.1 ± 1.5
Br	2.6 ± 0.4	Br	1.5 ± 0.2	Se	1.5 ± 0.2

trations. Although a much smaller amount of the filter was used, the sensitivities compare favorably with those obtained by emission spectrography as performed by the National Air Sampling Network (**3**). For most elements at least ten times better sensitivity is obtained; exceptions are Ti and Ni. In another application (published elsewhere), after a sampling time of only 90 min in a rural location, 15 elements could routinely be determined. In an industrial or urban area the same number of elements may be detected after a much shorter sampling time.

23.8.2 For some elements the sensitivity was limited by the purity of the filter paper, *e.g.*, Cl, Br, Na, and Zn. For at least 15 elements the sensitivity is affected by the composition of the sample. The abundant elements Al, Na, and Br give rise to a large amount of radioactivity, and other elements may be difficult to detect in their presence. It appears that the large amount of V found by Zoller and Gordon (**2**) in their samples caused a limitation on sensitivity in their short runs. A destructive technique involving chemical separations may improve the sensitivity for some elements, *e.g.*, Cu, Zn, Ga, As, Se, Ce, Sm, Eu, W, Au, Hg, but the nondestructive sensitivities for these elements appear to be adequate for the 24 hr samples from urban and rural areas. Some increase in sensitivity may be achieved by increasing the counting time or the neutron dose, although some chemical separations may still be required to detect elements such as Sr, Mo, Cd, I, and additional rare earths.

23.8.3 Tables 23:IV and 23:V reveal only trace quantities were determined, from 10^{-3} to 10^{-10} g. The reproducibility of the determinations was quite adequate, for the differences are generally within the calculated standard deviations (67% confidence level) of the single values. These standard deviations are high (>40%) only when the concentrations determined are near the limit of detection. Where the determination of an element is duplicated in different counts or in irradiations of different portions of the air filter sample, good agreement was usually obtained. This included the reproducibility of the chemical analysis, the accuracy in measurement of the filter area taken for analysis and the possible nonuniform air-flow through the filter. For the determination of most elements an accuracy of 25% can be obtained from one analysis. This is sufficient for many monitoring purposes.

23.8.4 A test of the adequacy of the analytical precision obtained is given in Table 23:VI which compares the ratios of concentrations found in East Chicago to those in Niles, together with the standard deviations of the ratios. In nearly every case the ratio was significantly greater than unity, and in general the ratio was not statistically the same for every element. A study involving a number of sampling locations over a wide area may lead to the identification of local sources characteristic for each element or group of elements.

References

1. Dams, R., J.A. Robbins, K.A. Rahn, and J.W. Winchester. 1970. Anal. Chem. 42:861–867.

2. ZOLLER, W.H., and G.E. GORDON. 1970. Anal. Chem. **42**:256.
3. U.S. PUBLIC HEALTH SERVICE. 1968. Air Quality Data from the National Air Sampling Networks and Contributing State and Local Network, 1966 Edition, Durham, N.C.
4. BRAR, S.S., D.M. NELSON, E.L. KANABROCKI, C.E. MOORE, C.D. BURNHAM, and D.M. HATTORI. 1969. Proceedings 1968 International Conference "Modern Trends in Activation Analysis," Gaithersburg, Maryland, p. 43.
5. LOUCKS, R.M., J.W. WINCHESTER, W.R. MATSON, and M.A. TIFFANY. 1969. Proceedings 1968 International Conference "Modern Trends in Activation Analysis," Gaithersburg, Maryland. p. 36.
6. KEANE, J.R., and E.M.R. FISHER. 1968. Atmospheric Environ., **2**:603.
7. DUDLEY, N.D., L.E. ROSS, and V.E. NOSHKIN. 1969. Proceedings 1968 International Conference "Modern Trends in Activation Analysis," Gaithersburg, Maryland, p. 55.
8. HOFFMAN, G.I., R.A. DUCE, and W.H. ZOLLER. 1969. Environ. Sci. Technol. **3**:1207.
9. PILLAY, K.K.S., and C.C. THOMAS, JR. 1969. Report WNY-046, Western New York Nuclear Research Center, Buffalo.
10. DAMS, R., K. RAHN, and J.W. WINCHESTER. 1970. Unpublished data.
11. RALSTON, H.R., and G.E. WILCOX. 1969. Proceedings 1968 International Conference "Modern Trends in Activation Analysis," Gaithersburg, Maryland, p. 1238.

Subcommitte 9
A. L. LINCH, *Chairman*
E. R. HENDRICKSON
M. KATZ
R. J. MARTIN
G. O. NELSON
J. N. PATTISON
A. L. VANDER KOLK
R. B. WEIDNER
J. N. HARMAN

24. Use of Selective Ion Electrodes to Determine Air Pollutant Species (Ambient and Source Level)

24.1 INTRODUCTION

4.1.1 *Definition*. Selective ion electrodes measure the activity of an ionic species in solution by developing an absolute potential which is related to the ionic activity; it may be determined experimentally by measuring the difference in potential of an electrode pair consisting of the selective ion electrode and a suitable reference electrode (such as a calomel or silver-silver chloride reference electrode) with a high input impedance voltmeter. The potential response of a selective ion electrode to an ionic species A is defined by the Nernst equation

$$E = E^0 + \frac{RT}{nF} \operatorname{Log} A$$

Where:

E = equilibrium electrode potential

$\frac{RT}{nF}$ = 59.15 mv @ 25 C for n = 1

A = ionic activity of species of interest

n = charge on ion of interest

E^0 = constant dependent on choice of working electrode—reference electrode system

Selective ion electrode systems are usually calibrated against known standards to account for behavioral anomalies and to insure that the system is functional for a given determination.

The first selective ion electrode to find widespread use was the glass pH electrode; it responds to changes in hydrogen ion activity according to the Nernst equation, in that for each tenfold change in hydrogen ion activity at 25 C the change of the electrodes absolute potential will be 59.15 mV (the value of RT/nF at 25 C). This is the logical point to distinguish the selective ion electrode from the only theoretically achievable "specific" ion electrode which does not suffer from interference by other ionic species; the pH electrode is predominately sensitive to hydrogen ion, but in highly alkaline solution, where the activity of $(H_3O)^+$ is significantly less than such other monovalent cations as sodium and potassium, it will respond to these ions. This phenomenon is known as the "sodium ion error" and when capitalized upon lead to the development of the technique to determine NH_4^+, K^+ and Na^+ in alkaline aqueous media by the use of the pH electrode.

24.1.2 *Construction—Mechanism*. Construction is based on sealing a membrane of a substance such as glass, an insoluble inorganic salt or an immobilized organic ion exchanger or ion exchange membrane across the end of a tube which serves to contain a filling solution and an internal half cell assembly. The membrane serves as a filter, allowing only ions of the test species (in the ideal case) to diffuse across the membrane separating the test solution and the inner solution. Such diffusion derives from the chemical potential difference of the test ion existing between the two solutions. Diffusional transport of the ions continues until a back EMF sufficient to retard further diffusion is produced—this is the variable which is normally measured when the equilibrium potential of a test electrode is determined and is related to the activity of the ionic species being determined.

24.1.3 *Problems.* Selective ion electrodes exhibit the advantage or disadvantage, depending on one's point of view, of determining ionic activities, not ionic concentrations. Activities are generally most meaningful in kinetic determinations, but are of limited value to the determination ionic concentrations—for assessment of concentrations, a conventional titration technique should be employed (selective ion electrodes may serve as indicator electrodes). Along with the inherent problems of activity measurement are the difficulties caused because the ionic activity coefficient of a particular species of interest is a function of the total ionic strength of the solution and because the selective ion electrode may exhibit anomalous values due to the effects of ionic interferents or its inability to determine species present in a complexed form. The electrodes are more suited for use in the laboratory environment than the continuous process analytical utilization where, limited life of liquid ion exchanger based systems, the problems of interferences, and the lack of long term stability contribute to lessened utility.

24.1.4 *References and Current State of the Art.* Thorough descriptive treatments of selective ion electrode principles are given in the articles by Weber, **(1)** Frant **(2)** and the publications of Orion Research Inc. **(3, 4, 5)**. The reader is referred to these publications for a more fundamental and general description of selective ion electrodes and for the emphasis which these articles place on the more conventional aqueous sample analysis role of this device. The use of selective ion electrodes has gained widespread acceptance for analytical determinations in chemical, industrial, pharmaceutical and general analytical chemistry areas. The bulk of such usage has been for the determination of ionic species already present in the aqueous sample.

The success of selective ion electrodes in providing accurate and sensitive determinations over a wide range of analyte concentrations in a wide variety of sample matrices has lead to the extension of this basic detection and quantification device to the determination of gaseous species which are soluble in a collection medium.

A listing of commercially available selective ion electrodes by detector type along with detection limits, where available, is given in Table 24:I .

24.2 Detection of Gaseous Species with Selective Ion Electrodes Background

Any gaseous species which is readily soluble in an aqueous collection medium may be determined by a selective ion electrode if a selective ion electrode exists for the ionic hydrolysis product of that species. The determination may be made on either a batch or continuous basis and the detection limit of the gaseous species will be an inverse function of the collection time in the case of batch analyses or a function of the gaseous sample flow rate/collection medium volume for a continous analysis. Orion Research **(6)** has detailed the demonstrated and possible systems suitable for analysis by selective ion electrode; these are noted in Table 24:II.

24.2.1 *Applications of Solid State and Glass Electrodes to the Measurement of Gaseous Species.*

a. Fluoride. The fluoride selective ion electrode was one of the first reliable selective ion electrodes which sensed a species of particular importance to the analysis of source level and ambient level air pollution. Several writers **(7, 8, 9)** have com-

Table 24:I. Classification of Commercially Available Selective Ion Electrodes by Species Determined and Associated Detection Limit

Glass	Solid State	Liquid Membrane	Gas Sensing
H^+ (10^{-14}M)	Br^- (5 × 10^{-6}M)	Cu^{++} 10^{-5}M)	NH_3 (10^{-6}M)
Na^+ (10^{-7}M)	Cd^{++} (10^{-7}M)	Cl^- (10^{-5}M)	CO_2 (10^{-7}M)
K^+ (5 × 10^{-5}M)	Cl^- (10^{-5}M)	NO_3^- (10^{-5}M)	NO_2 (2 × 10^{-5}M)
NH_4^+ (10^{-5}M)	Cu^{++} (10^{-7}M)	ClO_4^- (10^{-5}M)	SO_2 (10^{-7}M)
Ag^+ (10^{-2}M)	Cr^- (10^{-6}M)	K^+ (10^{-5}M)	
	F^- (10^{-7}M)	Mg^{++} + Ca^{++} (10^{-10}M)	
	I^- (1.5 × 10^{-10}M)		
	Pb^{++} (10^{-7}M)		
	Ag^+ (10^{-14}M)		
	CNS^- (10^{-5}M)		

Principal Sources:
Beckman Instruments, Leeds & Northrup, London Co., Orion Phillips, Radiometer Sensorex.

pared the accuracy of fluoride ion determination by selective ion electrode to that of the SPADNS distillation technique, where ambient or source levels of HF were determined. Purcell **(10)** has determined HF and SiF_4 being emitted from a combustion source with a fluoride ion electrode based continuous monitoring system. Thompson and Johnston **(11)** have developed a continuous fluoride monitor employing a counter current absorber and selective ion electrode detection system and used it to determine ambient levels of fluorides.

b. NO_XNO + NO_2). NO_2 is a product of air oxidation of NO liberated by most combustion processes. It has been determined by a selective ion electrode in several fashions. Driscoll **(12)** found that by collecting the NO_X NO + NO_2) sample in a ~10^{-3}M H_2O_4 + 3 × 10^{-2}M H_2O_2 solution, it was possible to obtain a correlation coefficient of 0.987 over a wide range of input concentrations between the selective ion electrode and the conventional Saltzman analyses. L. A. Dee **(13, 14)** employed the NO_3^- (solid state) electrode to determine NO_2 concentrations in stack gases by allowing this pollutant to react with PbO_2 in the collection medium to form Pb $(NO_3)_2$, which is fully ionized and readily detectable. Di Martini **(15)** has used gas phase oxidation of NO → NO_2 with O_3 and subsequent bubbling through a collection medium to determine NO_x with a NO_3^- selective ion electrode.

24.2.2 *Application of Gas Sensing Selective Ion Electrodes to the Measurement of Gaseous Species.* Gas sensing electrode consists of specific ion electrode and companion reference electrode immersed in an appropriate internal electrolyte isolated from the gaseous test sample by a gas permeable membrane. The species to be measured diffuses across the membrane

Table 24:II. Systems Suitable for Analysis by Selective Ion Electrode

Measured Species	Gaseous Species	Equilibrium in Electrolyte	Sensing Electrode
NH_3 or NH_4^+	NH_3	$NH_3 + H_2O \rightleftarrows NH_4 + OH^-$	H^{-+}
SO_2, H_2SO_4or SO_4	SO_2	$SO_2 + H_2O \rightleftarrows H^+ + HSO_4^-$	H^+
NO_2, NO_2^-	NO_2	$2 NO_2 + H_2O \rightleftarrows NO_3^- + NO_2^- + H^+$	NO_3^-, H^+
S^+, HS^-, H_2S	H_2S	$H_2S + H_2O \rightleftarrows HS^- + H^+$	$S^=$
CN^-, HCN	HCN	$Ag(CN_2)^- \rightleftarrows Ag^+ + 2 CN^-$	Ag^+
CO_2, H_2CO_3, HCO_3^-, $CO_3^=$	CO_2	$CO_2 + H_2O \rightleftarrows H^- + HCO_3^-$	H^-
HOAC, OAC (possible)	HOAC	$HOAC \rightleftarrows H^+ + OAC^-$	H^+
F^-, HF	HF	$HF \rightleftarrows H^+ + F^-$	F^-
Cl_2, OCl^-, Cl^-	Cl_2	$Cl_2 + H_2O \rightleftarrows 2H^+ + ClO^- + Cl^-$	$(H^+)(Cl^-)$
X_2, OX^-, X^-	X_2	$X_2 + H_2O \rightleftarrows 2H^+ + XO^- + X^-$	X^-( )

across a thin layer of electrolyte to the selective ion electrode; as the species of interest dissolves in this thin film of electrolyte, the concentration of the hydrolyzed product species changes in a fashion directly related to the gas phase concentration. The potential change of the electrode with changes in gas phase concentration approximates Nernstian behavior very closely.

Advantages of the membrane isolated electrode system over a selective ion electrode—conventional reference electrode system include virtual elimination of liquid junction potentials, isolation of the electrode from interferences which do not exhibit high vapor pressures in aqueous solution and elimination of interferences from ionic species which enter into redox couples at the electrode surface (since there is no ionic transport).

a. The Nitrogen Oxide Gas Sensing Electrode. This electrode responds to dissolved nitrous acid or nitrite and is subject to few interferences except formic and acetic acid, when present in great excess. SO_2 and CO_2 do not interfere when the procedures specified by the manufacturer are followed. NO_2 in air samples can be determined by collecting the NO_2 in an appropriate collection medium where the NO_2 hydrolyses to nitrite which is measured by the electrode. The dissolved NO_2 and NO in equilibrium with HNO_2 in any acidified sample diffuse through the gas-permeable membrane of the electrode until equilibrium is reached between the level of HNO_2 in the test solution and in the inner filing solution of the electrode. The hydrogen ion response is measured and exhibits a 59 mV/decade span over the concentration range of 5×10^{-7} M NO_2 to 10^{-2}M NO_2 (2×10^{-2} to 460 ppm NO_2). Response time is faster at the higher concentrations and an upper temperature limitation of 50 C is to be observed.

This electrode would serve as the basis of assessing NO_X concentrations with the use of a suitable prior oxidation medium, such as acidified $KMnO_4$, $K_2Cr_2O_7/H_2SO_4$ or O_3 to oxidize NO to NO_2 which could be collected in an appropriate collection medium with subsequent determination by the NO_2 gas sensing electrode.

b. Sulfur Dioxide Gas Sensing Electrode. The sulfur dioxide gas sensing electrode responds to dissolved SO_2 and is subject to interference only from large excess concentration (>30 fold) of weak acids such as acetic acid and HF. Cl_2, NO_2 and CO_2 do not interfere, if the recommendations about collection media given by the manufacturer are followed. SO_2 at ambient and source levels can be determined by collection in an appropriate medium such as sodium hydroxide, which is subsequently acidified to release the gaseous SO_2 to be measured by the selective ion electrode. Sensitivities as high as 0.02 ppm SO_2 are claimed (based on the sampling of 30 liters of air through the basic collection medium and a final test solution volume of 25 ml). The SO_2 gas measured by the electrode creates hydrogen ions in the inner filling solution of the electrode; these H^+ ions are measured and a 59 mV/decade span for each tenfold SO_2 concentration change is observed. The accurate concentration range for correct electrode behavior is $\sim 10^{-6}$M to 10^{-1}M (0.1 to 1000 ppm SO_2). Response time is faster at more concentrated solutions and an upper temperature limitation of 50 C is to be observed.

c. Ammonia Gas Sensing Electrode. The ammonia gas sensing electrode responds to dissolved ammonia and is subject to few other interferences other than volatile amines. Dissolved ammonia in a collection medium diffuses through the gas permeable membrane of the electrode until the concentration in the inner filling solution of the electrode is equal to that in the external solution. The (OH^-) level in this filling solution is directly proportional to the external ammonia concentration and a response of 59 mV/decade ammonia change is observed. The accurate concentration range for correct electrode behavior is 10^{-6}M to 10^{0}M ammonia (0.017 to 17,000 ppm NH_3). Response time is faster in more concentrated solutions and an upper temperature limitation of 50 C is to be observed.

24.3 PRECAUTIONS

24.3.1 *Calibration Standards and Techniques*. Calibration of selective ion electrodes used to monitor the concentration of gaseous species should be conducted according to the recommendations of the manufacturer or based upon the theoretical concepts and good laboratory practice as detailed in the articles by both Frant and Weber (**1, 2**) and Eynon (**16**). Attempts should be made to match the composition and ionic strength of the analytical standards to that of the collection medium employed. The best approach to this is to make incremental additions of the test ion of known value to the collection medium in order to develop an electrode calibration curve. This technique is detailed in most manufacturer's literature. One should observe the sequence of immersion in low concentration standards working toward high concentration standards to minimize gross contamination and absorption/desorption effects. Calibration of gas sensing electrodes may be conducted by the above mentioned techniques or may be accomplished by exposure to a gas stream containing a known concentration of the gas being determined. Such a calibration gas standard may be prepared by dynamic dilution, the use of permeation tubes or by the use of analyzed standards or mixtures. A good account of these techniques is given in Methods of Air Sampling Analysis pp. 18–25.

24.3.2 MEASUREMENT TECHNIQUES

a. Calibration.

b. Electrical Considerations. Potential measurements of the test electrode must be made with the ultimate accuracy possible; at 25 C a 0.5 mV error will account for a 1.9% error for monovalent ions, 3.8% for divalent ions and 4.7% for trivalent ions. Care should be taken to provide adequate shielding of the sensing electrode so that spurious potentials are not measured.

c. These potential measurements should be taken with a high input impedance device (typically Ω 10^{13}r) in order to minimize loading of the selective ion electrode—reference electrode pair and the observation of erroneous open circuit potentials due to current flow in the selective ion electrode—reference electrode—measuring instrument—solution loop. Care should be exercised in grounding the electrode measurement system to avoid ground loop conditions and to observe proper shielding techniques—such precautions are explained in detail in pertinent references (**17, 18, 19, 20**). Generally, reference electrodes which exhibit impedances on the order of 1 to 20K, without significant bleed rates of internal solution into the test solution (which may be variable and contribute to higher and variable ionic strength of the solution being measured) are most suitable for use in such systems.

d. Temperature Effects and Temperature Compensation. This subject has been treated in a significant article by Negus *et al.* (**21**). They classify temperature effects on selective ion electrodes into three categories:

1) The change in the potential of a selective ion electrode system due to nonuniform response of the selective ion electrode and reference electrode to a temperature excursion.
2) the temperature coefficient slope term of the Nernst equation, RT/nF. This is the parameter usually compensated for by the use of a thermocompensator in a process selective ion electrode monitoring system; the thermocompensator corrects for the slope change in the potential difference versus ion activity relationship due to the change in the slope factor RT/nF with T. This is the major thermalcompensation characteristic to be corrected.
3) the solution temperature coefficient term which arises from temperature dependent equilibria and activities existing in the test solution. The magnitude of these effects is usually so small it is neglected.

In order to minimize the effects of temperature or temperature variations on the accuracy of results obtained from a given selective ion electrode based system, calibration of the system should be conducted with standards maintained at the same temperature as the test sample which is measured. Sufficient time should be allowed when changing from calibration standard to calibration standard so that thermal equilibrium is achieved. Where practicable, it is recommended that the sample temperature be held at a constant value by thermostatic control and the electrode pair and sensing cell be maintained in a temperature controlled environment held at the same temperature as the input aqueous sample. If this is not possible and reduced readout accuracy is allowable, it is sufficient to either automatically compensate for temperature excursions with a thermocompensator or to use manual thermocompensation or some mean sample temperature when temperature excursions are small and of long diurnal nature.

24.4 Summary-Conclusions

Selective ion electrodes show great promise in the determination of certain air pollutant species. This technique has not yet been implemented into commercially available continuous air pollutant monitoring instrumentation, but may be in the near future. The use of these electrodes for the determination of air pollutant species is currently primarily oriented toward batch analysis of collected samples in a laboratory setting, but reliable continuous monitoring instrumentation based on such electrodes is under development. If the limitations exhibited by the electrodes are understood and techniques for compensating for these limitations are employed, these electrodes offer a powerful, inexpensive and reliable technique for the determination of ambient and source level air pollutant species over a wide dynamic range. For specific applications, when the user is knowledgeable in the art of application of the electrode to the system chosen, selective ion electrodes offer a viable alternative technique for the determination of ambient and source level air pollutant species.

References

1. H. Weber, S.J. 1970. Selective Ion Electrodes in Pollution Control, American Laboratory, p. 15, July.
2. Frant, M.S. 1974. Detecting Pollutants with Chemical Sensing Electrodes. Environmental Science and Technology, 8:224.
3. Anon, Orion Research Analytical Methods Guide. 1973. Orion Research Inc., Cambridge, Massachusetts.
4. Anon, Orion Newsletter, Orion Research Inc., Cambridge, Massachusetts (1970–present).
5. Anon, Applications Bulletins, Orion Research Inc., Cambridge, Massachusetts.
6. Anon, Orion Newsletter. 1973. Vol. V, No. 2 Orion Research, Cambridge, Massachusetts.
7. Elfers, L.A. *et al.* 1968. Determination of Fluoride in Air and Stack Gas Samples by use of an Ion Specific Electrode. Analytical Chemistry, 40:1658.
8. MacLeod, K.E. et al. 1973. Comparison of the SPADNS-Zirconium Lake and Specific Ion Electrode Methods of Fluoride Determination in Stack Emission Samples. Analytical Chemistry, 45:1272.
9. Thompson, R.J. 1971. Fluoride Concentrations in the Ambient Air. Journal of the Air Pollution Control Association, 21:484.
10. Purcell, R.A. 1973. A Method for the Continuous On-Site Measurement of Fluorides in Stack Gases and Emissions for Periods of up to Five Hours. Atmospheric Environment, 7:169.
11. Thomson, C.R. 1969. University of California, Riverside, California and Johnston, G.W., Jr. Beckman Instrumemts Inc.—Private Communication.
12. Driscoll, J.N. 1971. "Determination of Oxides of Nitrogen in Combustion Effluents with a Nitrate Ion Selective Electrode". Walden Research Corporation, #71-149.
13. Dee, L.A. 1972. An Improved Manual Method for NO_x Emission Measurement. EPA-R2-72-067.
14. Dee, L.A. et al. 1973. New Manual Method for Nitrogen Oxides Emission Measurement. Analytical Chemistry, 45:1477.
15. Di Martini, R. 1970. Determination of Nitrogen Dioxide and Nitric Oxide in the Parts per Million Range in Flowing Gaseous Mixtures by Means of the Nitrate-Specific-Ion Electrode. Analytical Chemistry, 42:1102.
16. Eynon, J.V. 1970. Known Increment and Known Decrement Methods of Measurement With Ion Selective Electrodes, American Laboratory, 59.
17. Mattoch, G. 1961. pH Measurement and Titration. New York, The MacMillan Co.
18. Bleak, T.M. *et al.* 1969. A New Process pH Analyzer Design. Paper presented at National ISA Meeting, Dallas, Texas.
19. Morrison, R. 1970. Grounding and Shielding Techniques, John Wiley & Sons, New York.
20. Moore, F. 1960. Taking the Errors out of pH Measurement by Grounding and Shielding". ISA Journal.
21. Negus, L.E. *et al.* 1972. Temperature Coefficients and Their Compensation in Ion-Selective Systems. Instrumentation Technology 23.

Subcommittee 9

A. L. Linch, Chairman
J. N. Harman
E. R. Hendrickson
M. Katz
R. J. Martin
G. O. Nelson
J. N. Pattison
A. L. Vander Kolk

25. Quality Control For Sampling and Analysis

25.1 INTRODUCTION.

The measurement of physical entities such as length, volume, weight electromagnetic radiation and time involves uncertainties which cannot be eliminated entirely, but when recognized can be reduced to tolerable limits by meticulous attention to detail and close control of the significant variables. In addition, errors often unrecognized, are introduced by undersirable physical or chemical effects and by interferences in chemical reaction systems. In many cases, absolute values are not directly attainable; and, therefore, standards from which the desired result can be derived by comparison must be established. Errors are inherent in the measurement system. Although the uncertainties cannot be reduced to zero, methods are available by which reliable estimates of the probable true value and the range of measurement error can be made.

In this section the fundamental procedures for the administration of an effective quality control program are presented with sufficient explanation to enable the investigator to both understand the principles and to apply the techniques. First, the detection and control of determinate and indeterminate error will be considered. Based on this foundation, the types of errors and meanings of the common terms used to define an accurate method are discussed as a basis for application of quality control to sampling and analysis. The theory, construction, applications and limitations of control charts are developed in sufficient depth to provide practical solutions to actual quality control problems. Additional statistical approaches are included to support those systems which may require further refinement of precision and accuracy to evaluate and control sampling and analysis reliability.

Finally, a discussion of collaborative testing projects and intra-laboratory quality control programs designed to improve and test the integrity of the laboratory's performance completes the survey of quality control principles and practices.

25.2 QUALITY CONTROL PRINCIPLES

25.2.1 *Total Quality Control.* A quality control program concerned with sampling and laboratory analysis is a systematic attempt to assure the precision and accuracy of future analyses by detecting determinate errors in analysis and preventing their recurrence. Confidence in the accuracy of analytical results and improvements in analysis precision are established by identification of the determinate sources of error. The precision will be governed by the indeterminate error inherent in the procedure, and can be estimated by statistical techniques. For a result to be acceptable, the procedure must not only be precise, but must also be without bias. Techniques have been developed for the elimination of bias. The quality control program should cover instrumental control as well as total analysis control. The use of replicates submitted in support of the quality control program provides assurance that the procedure will remain in statistical control.

Quality must be defined in terms of the characteristic being measured. Control must be related to the source of variation which may be either systematic or random. Usually the basic variable is continuous (any value within some limit is possible). A numerical value of an analysis for which the range of uncertainty inherent in the method has not been established cannot be reliably considered a reasonable estimate of the true or actual value. The basic quality control program incorporates the concepts of:

a. Calibration to attain accuracy;

b. replication to establish precision limits, and

c. correlation of quantitatively related tests to confirm accuracy, where appropriate.

Evaluation of the overall effectives of the quality control program encompasses a number of parameters:

a. Equipment and instruments;

b. the current state of the art;

c. expected ranges of analytical results;

d. precision of the analytical method itself;

e. control charts to determine trends as well as gross errors;

f. data sheets and procedures adopted for control of sample integrity in the laboratory, and

g. quality control results on a short term basis (daily if appropriate) as well as on an accumulated basis.

The manipulative operations which are directly influenced by quality control include:

a. Sampling techniques;

b. preservation of sample integrity (identification, shipping and storage conditions, contaminations, desired component losses, etc.);

c. aliquoting procedures;

d. dilution procedures;

e. chemical or physical concentration, separation and purification, and

f. instrument operation.

25.2.2 *Statistical Quality Control.* Statistical quality control involves application of the laws of probability to systems where chance causes operate. The technique is employed to detect and separate assignable (determinate) from random (indeterminate) causes of variation. "Statistics" is the science of uncertainty; therefore, any conclusions based on statistical inference contain varying degrees of uncertainty, which are expressed in terms of probability statements. Uncertainty can be quantified in terms of well defined statistical probability distributions, which can be applied directly to quality control. The application of statistical quality control can most efficiently indicate when a given procedure is in statistical control, and a continuing program that covers sampling, instrumentation and overall analysis quality will assure the validity of the analytical program.

25.2.3 *Quality Control Charts* **(1)**. The Shewhart Control Chart **(2)** is one of the most generally applicable and easily adapted statistical quality control techniques which can be applied to almost any phase of production, research or analysis. Control charts originally were developed for control of production lines where large numbers of manufactured articles were inspected on a continuous basis. Since analyses frequently are produced on an intermittent basis, or on a greatly reduced scale, less data are available to work with. Therefore, certain concessions must be made in order to respond quickly to objectionable changes in the analytical procedure.

This control chart may serve several functions:

a. To determine empirically and to define acceptable levels of quality;

b. to achieve the acceptable level established, and

c. to maintain performance at the established quality level.

Certain assumptions reside in this technique. The first and major assumption is that there will be variation. No process or procedure has been so well perfected, or so unaffected by its environment that exactly the same result will always be produced. Either the device used for measurement is not sufficiently sensitive or the operator performing the measurement is not sufficiently skilled. The sources of variation present in analytical work include:

a. Differences among analysts;

b. instrumental differences;

c. variations in reagents and related supplies;

d. effect of time on the differences found in Items **a, b,** and **c;** and

e. variations in the interrelationship of Items **a, b** and **c** with each other and with time.

A "system of chance causes" is inherent in the nature of processes and procedures and will produce a pattern of variation. When this pattern is stable, the process or procedure is considered to be "in statistical control" or just "in control." Any result which falls outside of this pattern will have an assignable cause which can be determined and corrected.

The control chart technique provides a means for separating the assignable cause variant from the stable pattern. The chart is a graphical presentation of the process or procedure test data which compares the variability of all results with the average or expected variability from small arbitrarily defined groups of the data. The technique in effect is a graphical analysis of variance.

The data from such a system can be plotted with vertical scale in test result units and the horizontal scale in units of time or sequence of results. The average value or mean, and the limits of the dispersion (spread, or range of results) can be calculated.

25.3 ERRORS

25.3.1 *Introduction*. Numbers are employed to either enumerate objects or to delineate quantities. If sixteen air samples are taken simultaneously at different locations in a warehouse where gasoline-powered fork lift trucks are in motion, the number, *i.e.*, the count, would be the same regardless of who counted them, when the count was made or how the count was made. However, if each individual sample is analyzed for carbon monoxide, sixteen different numbers, *i.e.*, the concentration, undoubtedly would be obtained. Furthermore, when replicate determinations are made on each sample, a range of carbon monoxide concentrations would be found **(3)**.

Experimental errors are classified as determinate or indeterminate. A fifteen count of the warehouse samples would be a determinate error quickly disclosed by recount. An indeterminate error would be encountered due to the inherent variability in repetitive determinations of carbon monoxide by gas chromatography, infrared or a colorimetric technique.

If the estimation of carbon monoxide concentration is made with a length of stain detector tube and a 6.5-mm stain length equivalent to 57 ppm is recorded by the observer, whereas the true stain length is 6.0 mm and equivalent to 50 ppm, the observational error would be $(57 - 50) \times 100 \div 50 = 14\%$.

All analytical methods are subject to errors. The determinate ones contribute constant error or bias while the indeterminate ones produce random fluctuations in the data. The concepts of accuracy and precision as applied to the detection and control of error have been clearly defined and should be used exactly.

Accuracy. Accuracy relates the amount of an element or compound recovered by the analytical procedure to the amount actually present. For results to be accurate, the analysis must yield values close to the true value.

Precision. Precision is a measure of the method's variability when repeatedly applied to a homogeneous sample under controlled conditions, without regard to the magnitude of displacement from the true value as the result of systematic or constant determinate errors which are present during the entire series of measurements. Stated conversely, precision is the degree of agreement among results obtained by repeated measurements or "checks" on a single sample under a given set of conditions **(4, 5)**.

25.3.2 *Detection and Elimination of Determinate Error*. The terms "determinate" error, "assignable" error, and "systematic" error are synonymous. A determinate error contributes constant error or bias to results which may agree precisely among themselves.

Sources of Determinate Error. A method may be capable of reproducing results to a high degree of precision, but only a fraction of the component sought is recovered. A precise analysis may be in error due to inadequate standardization of solutions, inaccurate volumetric measurements, inaccurate balance weights, improperly calibrated instruments or personal bias (color estimation). Method errors that are inherent in the procedure are the most serious and most difficult to detect and correct. The contribution from interferences is discussed later.

Personal errors other than inherent physical visual acuity deficiencies (color judgment) include consistent carelessness, lack of knowledge and personal bias which are exemplified by calculation errors, use of contaminated or improper reagents, nonrepresentative sampling or poorly calibrated standards and instruments.

Types of Determinate Error. *Additive*: An additive error occurs when the mean error has a constant value regardless of the amount of the constituent sought in the sample. A plot of the analytical value versus the theoretical value will disclose an intercept somewhere other than zero. *Proportional*: A proportional error is a determinate error in which magnitude is changed according to the amount of constituent present in the sample. A plot of the analytical value versus the theoretical value not only fails to pass through zero, but discloses a curvilinear rather than a linear function.

Recovery of "Spiked" Sample Procedures. A recovery procedure in which spiked samples are used provides a technique for the detection of determinate errors. The technique provides a basis for evaluating the applicability of a particular method to any given sample. It allows derivation of analytical quality control from the results, thus providing the basis for an excellent quality control program.

The recovery technique applies the analytical method to a reagent blank, to the sample itself (in at least duplicate), and to "spiked" samples prepared by adding known quantities of the substance sought, to separate aliquots of the sample which are equal in size to the unspiked sample taken for analysis. The substance sought should be added in sufficient quantity to exceed in magnitude the limits of analytical error, but the total should not exceed the range of the standards selected.

The results are first corrected for reagent influence by subtracting the reagent blank from each standard, sample, and "spiked" sample result. The average unspiked sample result is then subtracted from each of the "spiked" determinations, the remainder divided by the known amount originally added, and expressed as percentage covery. Table 25:I illustrates an application of this technique to the analysis of blood for lead content.

Specifications for acceptance of analytical results usually are determined by the state of the art and the final disposition of the results. Recoveries of substances within the range of the method may be very high or very low and approach 100% as the errors diminish and as the upper limit of the calibration range is approached. Trace analysis procedures which inherently have relatively large errors when operated near the limits of sensitivity deliver poor recoveries based on classical analytical criteria and yet, from a practical viewpoint of usefulness may be quite acceptable. Poor recovery may reflect interferences present in the sample, excessive manipulative loss-

TABLE 25:I. Lead in Blood Analysis

Basis: 10.0 g blood from blood bank pool, ashed and lead determined by double extraction, mixed color, dithizone procedure.

		Analyst: DJM		
μg Pb added	Optical Density	μg Pb found Total	Recovered	Recovery, %
None-blank	0.0969	—	—	—
5-Calibration Point	0.2596	—	—	—
None	0.1427	1.6	—	—
None	0.1337	1.3	—	—
None	0.1397	1.4	—	—
None	0.1397	1.4	—	—
Average	0.1389	1.4	—	—
2.0	0.1805	2.9	1.5	75
4.0	0.2636	5.4	4.0	100
6.0	0.3372	7.8	6.4	107
8.0	0.3925	9.4	8.0	100
10.0	0.4437	11.4	10.0	100
30.0 Total	—	36.9	29.9	96

Calculation of mean error (**5**)

Mean error =− 36.9 − (30.0 + 5 × 1.4) = 0.1 μg for entire set
= 2.9 − (2.0 + 1.4) = 0.5 μg for 2 μg spike

Calculation of relative error

Relative Error = (0.1 × 100)/ 37.0 = 0.27% for entire set
= (0.5 × 100)/ 3.4 = 14.7% for 2 μg spike

es, or the method's technical inadequacy in the range of application. The limit of sensitivity may be considered the point beyond which indeterminate error is a greater quantity than the desired result.

Control Charts. Trends and shifts in control chart responses also may indicate determinate error. The standard deviation is calculated from spiked samples, and control limits (usually ± 3 standard deviations) for the analysis are established. In some cases, such as BOD and pesticide samples, spiking to resemble actual conditions is not possible. However, techniques for detecting bias under these conditions have been developed (**6**).

Control charts may be prepared even for samples which cannot be spiked or for which the recovery technique is impractical. A reference value is obtained from the average of a series of replicate determinations performed on a composite or pooled sample which has been stabilized to maintain a constant concentration during the control period.

Change in Methodology. Analysis of a sample for a particular constituent by two or more methods that are entirely unrelated in principle may aid in the resolution of determinate error.

In Table 25:II, an interlaboratory evaluation of three different methods for the determination of lead concentration in ashed urine specimens (mixed color dithizone, atomic absorption and polarography) is summarized. If the highly specific polarographic method was selected as the primary standard, then the dithizone procedure is subject to a + 7.4 μg/l bias as compared with a + 3.6 μg/l bias in the atomic absorption method for lead.

Quantity of Sample. If the determinate error is additive, the magnitude may be estimated by plotting the analytical results versus a range of sample volumes or weights. If the error has a constant value regardless of the amount of the component sought, then a straight line fitted to the plotted points will not pass through the origin.

Elimination. a. *Physical.* In many cases error can be reduced to tolerable levels by quantitating the magnitude over the operating range and developing either a corrective manipulation directly in the procedure or a mathematical correction in the final calculation. Temperature coefficients (parameter change per degree) are widely applied to both physical and chemical measurements. For example, the stain length produced by carbon monoxide in a detector tube is dependent on the temperature as well as the air sampling rate and CO concentration. Therefore, when these tubes are used outside the median temperature range, a correction must be applied to the observed stain length (Table 25:III) (**7**).

As a general rule, most instruments exhibit maximum reliability over the center 70% of their range (midpoint ± 35%). As the extreme to either side is approached the response and reading errors become increasingly greater. Optical density meas-

Table 25:II. An Interlaboratory Study of the Determinate Error in the Dithizone Procedure for the Determination of Lead in Nine Urine Specimens

Polarographic Method	Mixed Color Found	Dithizone Difference	Atomic Found	Absorption Difference
10	25	15	10	0
14	28	14	22	8
12	12	0	16	4
15	20	5	16	1
21	20	−1	22	1
22	30	8	24	2
27	40	13	36	9
19	22	3	22	3
12	22	10	16	4
Mean	—	+7.4	—	+3.6

Table 25:III. Kitagawa Carbon Monoxide Detector Tube No. 100 (7)
Temperature Correction Table

Chart Readings	Correction Concentration (ppm)				
(ppm)	0 C	10 C	20 C	30 C	40 C
1,000	800	900	1,000	1,060	1,140
900	720	810	900	950	1,030
800	640	720	800	840	910
700	570	640	700	740	790
600	490	550	600	630	680
500	410	470	500	520	560
400	340	380	400	420	440
300	260	290	300	310	320
200	180	200	200	200	210
100	100	100	100	100	100

urements, for example, should be confined to the range 0.045 to 0.80 by concentration adjustment or cell path choice. Extrapolation to limits outside the range of response established for the analytical method or instrumental measurement may introduce large errors as many chemical and physical responses are linear only over a relatively narrow band in their total response capability. In absorption spectrophotometric measurements, Beer's law relating optical density to concentration may not be linear outside of rather narrow limits in some instances (colorimetric determination of formaldehyde at high dilution by the chromatropic acid method).

b. *Internal Standard.* The internal standard technique is used primarily for emission spectrograph, polarographic, and chromatographic (liquid or vapor phase) procedures. This technique enables the analyst to compensate for electronic and mechanical fluctuations within the instrument.

In brief, the internal standard method involves the addition to the sample of known amounts of substance to which the instrument will respond in a manner similar to the contaminant in the system. The ratio of the internal standard response to the contaminant response determines the concentration of contaminant in the sample. Conditions during analyses will affect the internal standard and the contaminant identically, and thereby compensate for any changes. The internal standard should be of similar chemical composition to the contaminant, of approximately the same concentration anticipated for the contaminant, and of the purest attainable quality. A detailed discussion of the sources of physical error, magnitude of their effects, and suggestions for minimizing their contribution to determine bias and error will be found in the literature (**8**).

c. *Chemical Interference.* The term "interference" relates to the effects of dissolved or suspended materials on analytical procedures. A reliable analytical procedure must anticipate and minimize interferences.

The investigator must be aware of possible interferences and be prepared to use an alternate or modified procedure to avoid errors. Analyzing a smaller initial aliquot may suppress or eliminate the effect of the interfering element through dilution. The concentration of the substance sought is likewise reduced; therefore, the aliquot must contain more than the minimum detectable amount. When the results display a consistently increasing or decreasing pattern by dilution, then interference is indicated.

An interfering substance may produce one of three effects:

(a) React with the reagents in the same manner as the component being sought (positive interference); (b) react with the component being sought to prevent complete isolation (negative interference); (c) combine with the reagents to prevent further reaction with the component being sought (negative interference).

The sampling and analytical technique employed for the surveillance of airborne toluene diisocyanate (TDI) in the manufacturing environment furnishes a good example in which all three factors can be encountered. The TDI vapor is absorbed and quantitatively hydrolyzed in an aqueous acetic acid-hydrogen chloride mixture to toluene diamine (MTD) which then is diazotized by the addition of sodium nitrite. The excess nitrous acid is destroyed with sulfamic acid and the diazotized MTD coupled with N-l-Naphthylethylenediamine to produce a bluish-red azo dye

(**9**). In the phosgenation section of the operations, the starting material (MTD) may coexist with TDI in the atmosphere sampled. If so, then a positive interference will occur as the method cannot distinguish between free MTD and MTD from the hydrolyzed TDI.

This problem can be resolved by collecting simultaneously a second sample in ethanol. The TDI reacts with ethanol to produce urethane derivatives which do not produce color in the coupling stage of the analytical procedure. The MTD is determined by the same diazotization and coupling procedure after boiling off the ethanol from the acidified scrubber solution. Then the difference represents the TDI fraction in the air sampled.

On the other hand, if the relative humidity is high or alcohol vapors are present, negative interference will reduce the TDI recovered by formation of the carbanilide (Dimer) or the urethane derivative which will not produce color in the final coupling stage. Alternative methods have not been developed for these conditions. If high concentrations of phenol are absorbed, then a negative interference will arise from side reactions with the nitrous acid required to diazotize the MTD. This loss can be avoided by testing for excess nitrous acid in the diazotization stage and adding additional sodium nitrite reagent if a deficiency is indicated.

An estimate of the magnitude of an interference may be obtained by the recovery procedure.

If recoveries of known quantities exceed 100%, a positive interference is present (Condition 1). If the results are below 100%, a negative interference is indicated (Condition 2, or 3; see reference **8** for details).

25.3.3 *Indeterminate Error and Its Control.* **Nature.** Even though all determinate errors are removed from a sampling or analytical procedure, replicate analyses will not produce identical results. This erratic variation arises from random error. Examples of this type of variation would be variation in reagent addition, instrument response, line voltage transients and physical measurement of volume and mass. In environmental analysis the sample itself is subject to a great variety of variability. Although indeterminate errors appear to be random in nature, they do conform to the laws of chance, therefore statistical measures of precision can be employed to quantitate their effects.

A measure of the degree of agreement (precision) among results can be ascertained by analyzing a given sample repeatedly under conditions controlled as closely as conditions permit. The range of these replicate results (difference between highest and lowest value) provides a measure of the indeterminate variations.

Quantification. a. *Distribution of Results.* Indeterminate error can be estimated by calculation of the standard deviation (σ) after determinate errors have been removed. When indeterminate or experimental errors occur in a random fashion, the observed results (x) will be distributed at ramdom around the average or arithmetic mean ($\bar{x}$) for a normal distribution.

Given an infinite number of observations, a graph of the relative frequency of occurrence plotted against magnitude will describe a bell-shaped curve known as the Gaussian or normal curve. However, if the results are not from a normal distribution, the curve may be flattened (no peak), skewed (unsymmetrical), narrowed, or exhibit more than one peak (multi-modal). In these cases the arithmetic mean will be misleading, and unreliable conclusions with respect to deviation ranges (σ) will be drawn from the data.

In any event the investigator should confirm the normalcy of the data at hand. Various procedures are available to test this assumption. One method is to construct a histogram, if the sample is large enough, and then to plot a normal curve having the same mean and standard deviation with the histogram to see how well the normal curve fits. This is an imprecise method at best and, unless there is an extremely good fit of a normal curve laid over the resulting histogram or polygon, the cumulative distribution should be plotted on normal probability paper before proceeding. One way of testing for normality is to use the X^2 test for normality of data. Standard

X^2 computer programs are commonly available, but judgment must be used to weigh the cost of getting an accurate determination against the value of the information.

The distribution of results within any given range about the mean is a function of σ. The proportion of the total observations which reside within $\bar{x} \pm 1\,\sigma$, $\bar{x} \pm 2\,\sigma$ and $\bar{x} \pm 3\,\sigma$ have been thoroughly established. Although these limits do not define exactly any finite sample collected from a normal group, the agreement with the normal limits improves as n increases. As an example, suppose an analyst were to analyze a composite urine specimen 1000 times for lead content. He could reasonably expect 50 results would exceed $\bar{x} \pm 2\,\sigma$ and only 3 results would exceed $\bar{x} \pm 3\,\sigma$. However, the corollary condition presents a more useful application. In the preceding example, the analyst has found $\bar{x}$ to be 0.045 mg per liter with $\sigma = \pm 0.005$ mg/l. Any result which fell outside the range $0.035 - 0.055$ mg/l ($0.045 \pm 2\,\sigma$) would be questionable as the normal distribution curve indicates this should occur only 5 times in 100 determinations. This concept provides the basis for tests of significance, a concept which is discussed in detail in any good statistical reference such as those cited in this Section.

b. *Range of Results.* The difference between the maximum and minimum of n results (range) also is related closely to σ. The range (R) for n results will exceed σ multiplied by a factor d_n only 5% of the time when a normal distribution of errors prevails.

Values for d_n:

n	d_n
2	2.77
3	3.32
4	3.63
5	3.86
6	4.03

Since the custom of analyzing replicate (usually duplicate) samples is a general practice, application of these estimated limits can provide detection of faulty technique, large sampling errors, inaccurate standardization and calibration, personal judgment and other determinate errors. However, resolution of the question whether the error occurred in sampling or in analysis can be answered more confidently when single determinations on each of three samples rather than duplicate determinations on each of two samples are made. This approach also reduces the amount of analytical work required **(6)**. Additional information relative to the evaluation of the precision of analytical methods will be found in ASTM Standards **(10)**.

c. *Collaborative Studies or "Round Robins."* After an analytical method has been evaluated fully for precision and accuracy, collaborative testing should be initiated. The values for precision and accuracy as determined by the results from a number of laboratories can be expected to be inferior when compared with the performance of the originating laboratory. Because technicians in different laboratories apply to their procedure their own characteristic determinate and indeterminate errors which may differ significantly from the original technique, the values for precision and accuracy will disclose the true reliability (ruggedness, or immunity to minor changes) or the method. Participation in collaborative programs will aid the investigator in evaluating his laboratory's performance in relation to other similar facilities and in locating sources of error.

Duplicate analyses are employed for the determination and control of precision within the laboratory and between laboratories. Initially, approximately 20% of the routine samples, with a minimum of 20 samples, should be analyzed in duplicate to establish internal reproducibility. A standard or a repeatedly analyzed control, if available, should be included periodically for long-term accuracy control. The control chart technique is directly applicable, and appropriate control limits can be established by arbitrarily subgrouping the accumulated results or by using appropriate estimates of precision from an evaluation of the procedure.

25.4 CONTROL CHARTS

25.4.1 *Description and Theory.* The control chart provides a tool for distinguishing the pattern of indeterminate (stable) variation from the determinate (assignable cause) variation. This technique displays the test data from a process or method in a form which graphically compares the variability of all test results with the average or expected variability of small groups of data—in effect, a graphical analysis of variance, and a comparison of the "within groups" variability versus the "between group" variability.

The data from a series of analytical trials can be plotted with the vertical scale in units of the test result and the horizontal scale in units of time or sequence of analyses. The average or mean value can be calculated and the spread (dispersion or range) can be established.

The determination of appropriate control limits can be based on the capability of the procedure itself or can be arbitrarily established at any desirable level. Common practice sets the limits at $\pm 3\ \sigma$ on each side of the mean. If the distribution of the basic data exhibits a normal form, the probability of results falling outside of the control limits can be readily calculated.

The control chart is actually a graphical presentation of quality control efficiency. If the procedure is "in control", the results will fall within the established control limits. Further, the chart will disclose trends and cycles from assignable causes which can be corrected promptly. Chances of detecting small changes in the process average are improved when several values for a single control point (an x chart) are used. As the sample statistical size increases, the chance that small changes in the average will not be detected is decreased.

The basic procedure of the control chart is to compare "within group" variability to "between group" variability. For a single analyst running a procedure, the "within group" may well represent one day's output and the "between group" represents between days or day-to-day variability. When several analysts or several instruments or laboratories are involved, the selection of the subgroup unit is critical. Assignable causes of variation should show up as "between group" and not "within group" variability. Thus, if the differences between analysts should provide assignable causes of variation, their results may not be lumped together in a "within group" subgrouping.

24.4.2 *Application and Limitations.* In order for quality control to provide a means for separating the determinate from indeterminate sources of variation, the analytical method must clearly emphasize those details which should be controlled to minimize variability. A check list would include:

a. Sampling procedures;
b. preservation of the sample;
c. aliquoting methods;
d. dilution techniques;
e. chemical or physical separations and purifications;
f. instrumental procedures; and
g. calculation and reporting results.

The next step to be considered is the application of control charts for evaluations and control of these unit operations. Decisions relative to the basis for construction of a chart are required:

a. Choose method of measurement;
b. select the objective
(1) precision or accuracy evaluation
(2) observe test results, or the range of results
(3) measurable quality characteristics;
c. select the variable to be measured (from the check list above);
d. basis of subgroup, if used
(1) Size—A minimum subgroup size of n = 4 is frequently recommended. The chance that small changes in the process average remain undetected decreases as the statistical sample size increases.
(2) Frequency of subgroup sampling—Changes are detected more quickly as the sampling frequency is increased;
e. Control Limits—Control limits (CL) can be calculated, but judgment must be exercised in determining whether or not the values obtained satisfy criteria established for the method, *i.e.*, does the deviation range fall within limits consistent with the solution or control of the problem. Af-

ter the mean ($\bar{X}$) or the individual results (X), and the mean of the range ($\bar{R}$) of the replicate result differences (R) have been calculated, then CL can be calculated from data established for this purpose (Table 25:IV) (**5**).

Mean of the Means ($\bar{\bar{X}}$) = $\Sigma\bar{X}/k$

CL's on Mean = $\bar{\bar{X}} \pm A_2$

Range ($\bar{R}$) = $\Sigma R/k$, or d_2 σ

Upper Control Limit (UCL) on Range = $D_4\bar{R}$

Lower Control Limit (LCL) on Range = $D_3\bar{R}$

Where: k = number of subgroups A_2, D_4 and D_3 are obtained from Table 25:IV. R may be calculated directly from the data, or from the standard deviation (σ) using factor d_2. The lower control limit for R is zero when $n \leqq 6$.

The calculated CL's include approximately the entire data under "in control" conditions and, therefore, are equivalent to ± 3 σ limits which are commonly used in place of the more laborious calculation. Warning limits (WL) set at ± 2 σ limits (95%) of the normal distribution serve a very useful function in quality control. The upper warning limit (UWL) can be calculated by:

$$UWL = \bar{R} + 2\sigma_R$$

$$UWL = \bar{R} \pm 2/3\ (D_4\bar{R})$$

Where the subgrouping is n = 2, UWL reduces to

$$UWL = 2.51\ \bar{R}.$$

25.5 CONSTRUCTION OF CONTROL CHARTS

25.5.1 *Precision Control Charts.* The use of range (R) in place of standard deviation (σ) is justified for limited sets of data $n \leqq 10$ since R is approximately as efficient and is easier to calculate. The average range ($\bar{R}$) can be calculated from accumulated results, or from a known or selected σ (d_2 σ). $LCL_R = 0$ when $n \leqq 6$. (LCL = lower control limit).

Table 25:IV. Factors for Computing Control Chart Lines*

Observations in Subgroup (n)	Factor A_2	Factor d_2	Factor D_4	Factor D_3
2	1.88	1.13	3.27	0
3	1.02	1.69	2.58	0
4	0.73	2.06	2.28	0
5	0.58	2.33	2.12	0
6	0.48	2.53	2.00	0
7	0.42	2.70	1.92	0.08
8	0.37	2.85	1.86	0.14

*ASTM Manual on Quality Control of Materials (**5**)

The steps employed in the construction of a precision control chart for an automatic analyzer illustrate the technique (Table 25:V):

Table 25:V. Precision (Duplicates) Data

Date	Data	Range (R)
9/69	# 8 25.1 24.9	0.2
	#16 25.0 24.5	0.5
	#24 10.9 10.6	0.3
10/69	# 7 12.6 12.4	0.2
	#16 26.9 26.2	0.7
	#24 4.7 5.1	0.4
2/70	# 6 9.2 8.9	0.3
	#12 13.2 13.1	0.1
	#16 16.2 16.3	0.1
	#22 8.8 8.8	0.0
4/70	# 6 14.9 14.9	0.0
	#12 17.2 18.1	0.9
	#18 21.9 22.2	0.3
5/70	# 6 34.8 32.6	2.2
	#12 37.8 37.4	0.4
6/70	# 6 40.8 39.8	1.0
	#10 46.0 43.5	2.5
	#17 40.8 41.2	0.4
	#24 38.1 36.1	2.0
7/70	# 6 12.2 12.5	0.3
	#12 25.4 26.9	1.5
	#18 20.4 19.8	0.6
		$\bar{R}$ = 14.9/22 = 0.68
	UCL = 3.27 × 0.68 = 2.2	
	UWL = 2.51 × 0.68 = 1.7	

a. Calculate R for each set of side-by-side duplicate analyses of identical aliquots;

b. calculate $\bar{R}$ from the sum of R values divided by the number (n) of sets of duplicates;

c. calculate the upper control limit (UCL_R) for the range:

$$UCL_R = D_4\bar{R}$$

Since the analyses are in duplicates, $D_4 = 3.27$ (from Table 25:IV);

d. calculate the upper warning limit (UWL): UWL_R =

$$\bar{R} + 2\sigma_R = \bar{R} \pm 2/3\ (D_4\bar{R}) = 2.51\ \bar{R}$$

25.6 ACCURACY OF CONTROL CHARTS—MEAN OR NOMINAL VALUE BASIS

$\bar{X}$ charts simplify and render more exact the calculation of CL since the distribution of data which conforms to the normal curve can be completely specified by $\bar{X}$ and σ. Stepwise construction of an accuracy control chart for the automatic analyzer based on duplicate sets of results obtained from consecutive analysis of knowns serves as an example (Table 25:VI):

a. Calculate $\bar{X}$ for each duplicate set;

b. group the $\bar{X}$ values into a consistent reference scale (in groups by orders of magnitude for the full range of known concentrations);

c. calculate the UCL and lower control limit (LCL by the equation: $CL = \pm A_2\bar{R}$ (A_2 from Table 25:IV);

d. calculate the Warning Limit (WL) by the equation: $WL = \pm 2/3\ A_2\bar{R}$;

e. chart CL's and WL's on each side of the standard which is set at zero, and

f. plot the difference between the nominal value and $\bar{X}$ (see Table 25:VI) and take action on points which fall outside of the control limits.

25.7 CONTROL CHARTS FOR INDIVIDUAL RESULTS

In many instances a rational basis for sub-grouping may not be available, or the analysis may be so infrequent as to require action on the basis of individual results. In such cases X charts are employed. However, the CLs must come from some subgrouping to obtain a measure of "within group" variability. This alternative has the advantage of displaying each result with respect to tolerance, or specification limits. The disadvantages must be recognized when considering this approach:

a. The chart does not respond to changes in the average;

b. changes in dispersion are not detected unless an R chart is included;

c. the distribution of results must approximate normal if the control limits remain valid.

Additional refinements, variations and control charts for other variables will be found in standard texts **(1, 11, 12)**.

25.8 MOVING AVERAGES AND RANGES

The $\bar{X}$ control chart is more efficient for disclosing moderate changes in the average as the subgrouping size increases. A logical compromise between the X and $\bar{X}$ approach would be application of the moving average. For a given series of analyses, the moving average is plotted. (See data shown in Table 25:VII). The moving range serves well as a measure of acceptable variation when no rational basis for subgrouping is available or when results are infrequent or expensive to gather.

TABLE 25:VI. Accuracy Data

Date	Calibration Range	Nominal (N)	Values	$\bar{X}$	N − X
9/69	10–400 ppm	100 ppm	22.9, 21.5/	22.2	−0.7
			22.7, 22.3	22.5	−0.4
	1.7–69.7 scale	22.9			
10/69	10–400	100	21.6, 21.3/	21.5	0.0
	1.5–67.6	21.5			
2/70	10–400	100	23.6, 24.1/	23.9	−0.6
	1.4–62.5	24.5			
4/70	10–400	100	25.8, 26.5/	26.2	+0.2
	1.6–59.4	26.0	26.0, 26.7	26.4	+0.4
5/70	10–150	100	72.2, 70.2/	71.2	+1.2
	6.3–83.0	70.0			
6/70	10–150	100	71.0, 70.8/	71.1	+0.1
			71.0, 71.3	71.2	+0.2
	6.6–85.0	71.0			
7/70	10–150	60	14.9, 14.7/	14.8	−0.2
			15.1, 14.4	14.8	−0.2
	1.8–33.5	15.0			

Table 25:VII. Moving Average and Range Table (N = 2)

Sample No.	Assay Value	Sample Nos. Included	Moving Average	Moving Range
1	17.09	—	—	—
2	17.35	1–2	17.22	+0.26
3	17.40	2–3	17.38	+0.05
4	17.23	3–4	17.32	−0.17
5	17.00	4–5	17.12	−0.23
6	16.94	5–6	16.97	−0.16
7	16.68	6–7	16.81	−0.26
8	17.11	7–8	16.90	+0.43
9	18.47	8–9	17.79	+1.36
10	17.08	9–10	17.78	−1.39
11	17.08	10–11	17.08	0.00
12	16.92	11–12	17.00	−0.16
13	18.03	12–13	17.48	+1.11
14	16.81	13–14	17.42	−1.22
15	17.15	14–15	16.98	+0.34
16	17.34	15–16	17.25	+0.19
17	16.71	16–17	17.03	−0.63
18	17.28	17–18	17.00	+0.57
19	16.54	18–19	16.91	−0.74
20	17.30	19–20	16.92	+0.76

25.9 Other Control Charts for Variables

Although the standard $\bar{X}$ and R control chart for variables is the most common, it does not always do the best job. Several examples follow where other charts are more applicable.

25.9.1 *Variable Subgroup Size.* The standard $\bar{X}$ and R chart is applicable for a constant size subgroup of $n = 2,3,4,5$. In some cases such a situation does not exist. Control limit values must be calculated for each sample size. Plotting is done in the usual manner with the control size limits drawn in for each subgroup depending on its size.

25.9.2 *R or σ Charts.* In some situations the dispersion is equal over a range of assay values. In this case, a control chart for either range or standard deviation is appropriate.

When the dispersion is a function of concentration, control limits can be expressed in terms of a percentage of the mean. In practice such control limits would be given as in the example below:

± 5 units/liter for 0–100 units/liter concentration

± 5% for > 100 units/liter concentration

An alternative procedure involves transformation of the data (**13**). For example, logarithms would be the appropriate transformation.

25.9.3 *$\bar{X}$ and σ Charts.* If the subgroup size exceeds 10, the Range Chart becomes inefficient. The use of a σ chart would then be appropriate. Where the cost of obtaining the test data is high, the increase in efficiency using σ rather than R may be worthwhile.

25.10 Other Statistical Tools

25.10.1 *Rejection of Questionable Results.* The question whether or not to reject results which deviate greatly from $\bar{X}$ in a series of otherwise normal (closely agreeing) results frequently arises. On a theoretical basis, no result should be rejected, as the one or more errors which render the entire series doubtful may be determinate errors that can be resolved. Tests which are known to involve mistakes, however, should not be reported exactly as analyzed. Mathematical basis for rejection of "outliers" from experimental data may be found in statistics text books (**14**).

25.10.2 *Correlated Variables–Regression Analysis.* A major objective in scientific investigations is the determination of the effect that one variable exerts on another. For example a quantity of sample (x) is reacted with a reagent to produce a result (y). The quantity x represents the independent variable over which the investigator can exert control.

The dependent variable (y) is the direct response to changes made in x, and varies in a random fashion about the true value. If the relationship is linear, the equation for a straight line will describe the effect of changes in x on the response y: $y = a + b\ x$, in which a is the intercept with the y axis and b is the slope of the line (the change in y per unit change in x). In chemical analysis a is a measure of constant error arising from a colorimetric determination, trace impurity, blank, or other determinate source. The slope b may be controlled by reaction rate, equilibrium shift or the resolution of the method. The term "regression analysis" is applied to this statistical tool. Additional useful information can be obtained by certain transformations and shortcuts (**14, 15, 16**).

25.11 GRAPHIC ANALYSIS FOR CORRELATIONS

Useful shortcuts may be elected to determine whether a significant relationship exists between x and y factors in the equation for a straight line ($y = a + b X$). The data are plotted on linear cross section paper and a straight line drawn by inspection through the points with an equal number on each side or fitted by the least squares method. If the intercept a must be zero (a blank correction may produce such a situation), the fitting is greatly simplified. Then on each side equidistant from this line draw parallel lines corresponding to the established deviation (σ) of the analytical procedure, tally up the points falling inside of the band formed by the $\pm \sigma$ lines and calculate percent correlation (conformance = No. within band x 100/total points plotted). (See **17** for illustration of this technique).

Curvilinear functions can be accommodated, especially if a log normal function is involved and a plot of the data on semi-log paper yields a straight line (**18**). Log-Log paper also is available for plotting complex functions.

A combination of curvilinear and bar charts in some cases will reveal correlations not readily detected by mathematical processes. The data derived from an industrial cyanosis control program (**19, 20**) illustrate an application which revealed a rather significant relationship between abnormal blood specimens and the frequency of cyanosis cases on a long-term basis.

Grouping data on a graph and approximating relationships by the quadrant sum test (rapid corner test for association) can provide useful results with a minimum expenditure of time (**21**).

In those cases where application of mathematical tools are tedious or completely impractical, a system of ranking is sometimes applicable to the restoration of order out of chaos. Again with reference to the cyanosis control program, a relationship between causative agent structure and biochemical potential for producing cyanosis and anemia was needed. Ten factors (categories) common to some degree for each of the 13 compounds under study had been recognized. The 13 compounds were ranked in each category in reverse order of activity (No. 1 most, No. 13 least active) and the sum of the rankings obtained for each compound. These sums then were divided by the number of categories used in the total ranking to obtain the 'score." The scores were then arranged in increasing numerical order in columnar form. The most potent cyanogenic and anemiagenic compounds then appeared at the top of the table and the least at the bottom (**20**).

25.12 CHI SQUARE TEST

Control charts are a convenient tool for daily checking with reference standards, but the answers are not always as nearly quantitative as needed. Periodic checking of the accumulated daily reference results to determine more rigorously whether all of the data belong to the same normal distribution may become necessary. One approach to this question and to assign a probability to the answer is provided by the Chi Square (X^2) test.

The Chi Square distribution describes the probability distribution of the sums of the squares of independent variables that are normally or approximately normally distributed. The general form of the expression provides a comparison of observed versus expected frequencies **(22)**. The Chi Square test is applied to variables which fall within the Poisson distribution **(23)**.

25.13 THE ANALYSIS OF VARIANCE (ANOVA)

The analysis of variance is one of the most useful statistical tools. Variation in a set of results may be analyzed in such a way as to disclose and evaluate the important sources of the variation. For a detailed description of this techniques consult standard statistics textbooks.

25.14 YOUDEN'S GRAPHICAL TECHNIQUE **(6, 24)**

Dr. W.J. Youden has devised an approach to test for determinate errors with a minimum of effort on the part of the analyst.

Two different test samples (X and Y) are prepared and distributed for analysis to as many individuals or laboratories as possible. Each participant is asked to perform only one determination on each sample (**NOTE:** It is important that the samples are relatively similar in concentration of the constituent being measured.)

Each part of laboratory results can then be plotted as a point on a graph.

A vertical line is drawn through the average of all the results obtained on sample X; a horizontal line is drawn through the average of all the results obtained on sample Y. If the ratio of the bias to standard deviation is close to zero for the determinations submitted by the participants, then one would expect the distribution of the paired values (or points) to be close to equal among the four quadrants. The fact that the majority of the points fall in the (+, +) and (−, −) quadrants indicates that the results have been influenced by some source of bias.

Furthermore, one can even learn something about a participant's precision. If all participants had perfect precision (no indeterminate error), then all the paired points would fall on a 45° line passing through the origin. Consequently the distance from such a 45° line to each participant's point provides an indication of that participant's precision.

25.15 INTRALABORATORY QUALITY CONTROL PROGRAM

25.15.1 *Responsibilities.* The attainment and maintenance of a quality control program in the laboratory is the direct responsibility of the laboratory manager or supervisor. The fundamental quality control techniques are based on:

a. Calibration to ensure accuracy;

b. duplication to ensure precision; and

c. correlation of quantitatively related tests to confirm accuracy and continual scrutiny to maintain the integrity of the results reported.

The individual technician can contribute significant assistance in this effort by his desire to deliver the best possible answers within the inherent limits of the equipment and procedure. Part of supervision's responsibility is adequate instruction to provide the "man on the bench" with sufficient "know how" to apply the principles on a routine basis.

The guidelines established by the American Industrial Hygiene Association for Accreditation of Industrial Hygiene Analytical Laboratories **(12)** delineate the minimum requirements which must be satisfied in order to qualify for proficiency recognition.

25.15.2 *Precision Quality Control.* In addition to the use of internal standards, recovery procedures and statistical evaluation of routine results, the laboratory should subscribe to a reference sample service to confirm precision and accuracy within acceptable limits. Apparatus should be calibrated directly or by comparison with National Bureau of Standards (NBS) certified equipment or its equivalent, reagents should meet or exceed ACS standards, calibration standards should be prepared from AR (analytical reagent) grade chemicals **(25)**, and standardized with NBS standards if available. To illustrate, in a laboratory engaged in an exposure control program based on biological monitoring by trace analysis of blood and urine for lead content, at least two calibration points, blanks and a recovery should be included in each batch analyzed by the dithizone procedure. In addition, the wavelength integrity and optical density response of the spectrophotometer should be checked and adjusted—if necessary by calibration with NBS cobalt acetate standard solution. Until the standard deviation for the analytical procedure has been established within acceptable limits, replicate determinations should be made on at least two samples in each batch (either aliquot each sample or take duplicate samples), and thereafter with a frequency sufficient to ensure continued operation within these limits.

Control charts are probably the most widely recognized application of statistics. They provide "instant" quality control status when plotted daily, or at other intervals sufficiently short to disclose trends without undue oscillations from over-refinement of the data.

25.15.3 *Accuracy Quality Control.* A standard or well defined control sample should be analyzed periodically to confirm accuracy of a procedure. The control chart technique is directly applicable to long-term evaluation of the reliability of the analyst as well as the accuracy of the procedure. To attain and maintain the high level of analytical integrity presented earlier in this section, the major sources of "assignable cause" errors must be reduced to a minimum level which is consistent with cost penalties and the objective of the study for which the analytical service is rendered.

25.15.4 *Interlaboratory Reference Systems.* Participation in interlaboratory studies whether by subscription from a certified laboratory supplying such a service or from a voluntary program initiated by a group of laboratories in an attempt to improve analytical integrity **(11)** is highly recommended. Evaluation of the analytical method as well as evaluation of the individual laboratory's performance can be derived by specialized statistical methods applied to the data collected from such a study. However, inasmuch as most investigators will not be called upon to conduct or evaluate interlaboratory surveys, the reader is referred to the literature in the event such specialized information is needed **(11, 26, 27, 28).** In the absence of such programs, the investigator or laboratory supervisor, should make every effort to locate colleagues engaged in similar sampling and analytical activity and arrange exchange of standards, techniques and samples to establish integrity and advance the art.

25.16 SUMMARY

Identification of the determinate sources of error of a procedure provides the information required to reduce assignable error to a minimum level. The remaining (residual) indeterminate errors then determine the precision of analyses produced by the procedure. Statistical techniques have been developed to estimate efficiently the precision. For a procedure to be accurate, the results must be not only precise, but bias must be absent. Several approaches are available to eliminate bias both within the laboratory and between laboratories by collaborative testing. Quality control programs based on appropriate control charts must be employed on a routine basis to assure adherence to established performance standards. The total analysis control program must include instrumental control, procedural control and elimination of personal errors. The use of replicate determinations, "spiked" sample techniques, reference samples, standard samples and quality control charts will provide assurance that the procedure remains in control.

The guidelines established by the American Industrial Hygiene Association for Accreditation of Industrial Hygiene Analytical Laboratories **(12)** summarizes in a succinct fashion the requirements for proficiency.

References

1. KELLEY, W.D. 1968. Statistical Method—Evaluation and Quality Control for the Laboratory. Training Course Manual in Computational Analysis. U.S. Dept. of Health, Education and Welfare Public Health Service.
2. SHEWHART, W.A. 1931. Economic Control of Quality of Manufactured Products. Bell Telephone Laboratories.
3. LINCH, A.L., H.V. PFAFF. 1971. Carbon Monox-

ide—Evaluation of Exposure by Personnel Monitor Surveys. Am. Ind. Hyg. Assoc. J. 32.
4. AMERICAN CHEMICAL SOCIETY. 1963. Guide for Measures of Precision and Accuracy. Anal. Chem. 35:2262.
5. AMERICAN SOCIETY FOR TESTING MATERIALS. 1951. ASTM Manual on Quality Control of Materials—Special Technical Publication 15-C. Philadelphia, Pa.
6. YOUDEN, W.J. 1951. Statistical Methods for Chemists. John Wiley and Sons, New York, N.Y.
7. KITAGAWA, T. 1971. Carbon Monoxide Detector Tube No. 100. National Environmental Instruments, Inc., P.O. Box 590, Fall River, Mass.
8. AMERICAN PUBLIC HEALTH ASSOCIATION. 1977. Methods for Air Sampling and Analysis. Part I. Washington, D.C. 20036.
9. GRIM, K., A.L. LINCH. 1964. Recent Isocyanate-In-Air Analysis Studies. Am Ind. Hyg. Assoc. J. 25:285.
10. AMERICAN SOCIETY FOR TESTING AND MATERIALS. 1966. Proposed Procedure for Determination of Precision of Committee D-19 Methods—Manual on Industrial Water and Industrial Wastewater. 2nd edition. 1016 Race Street, Philadelphia, Pa. 19103.
11. KEPPLER, J.F., M.E. MAXFIELD, W.D. MOSS, G. TIETJEN and A.L. LINCH. 1970. Interlaboratory Evaluation of the Reliability of Blood Lead Analysis. Am. Ind. Hyg. Assoc. J. 31:412.
12. CRALLEY, L.J., C.M. BERRY, E.D. PALMES, C.F. REINHARDT, and T.L. SHIPMAN. 1970. Guidelines for Accreditation of Industrial Hygiene Analytical Laboratories. AIHA J. 31:335.
13. COWDEN, D.J. 1957. Statistical Methods in Quality Control. Prentice Hall Inc., Englewood Cliffs, New Jersey.
14. BAUER, E.J. 1971. A Statistical Manual for Chemists. 2nd Edition. Academic Press, New York, N.Y.
15. AMERICAN PUBLIC HEALTH ASSOCIATION. 1976. Standard Methods for the Examination of Water and Wastewater. 14th edition, Washington, D.C. 20036.
16. HINCHEN, J.D. 1969. Practical Statistics for Chemical Research. Methuen and Co. Ltd., London.
17. LINCH, A.L., E.G. WIEST and M.D. CARTER. 1970. Evaluation of Tetraalkyl Lead Exposure by Personnel Monitor Surveys. Am. Ind. Hyg. Assoc. J. 31:170.
18. LINCH, A.L. and M. CORN. 1965. The Standard Midget Impinger-Design Improvement and Miniaturization. Am. Ind. Hyg. Assoc. J. 26:601.
19. WETHERHOLD, J.M., A.L. LINCH and R. C. CHARSHA. 1959. Hemoglobin Analysis for Aromatic Nitro and Amino Compound Exposure Control. Am. Ind. Hyg. Assoc. J. 20:396.
20. STEERE, N.V. (Editor). 1971. Handbook of Laboratory Safety. 2nd edition. The Chemical Rubber Co., Cleveland, Ohio.
21. WILCOXON, F. 1949. Some Rapid Approximate Statistical Procedures. Insecticide and Fungicide Section—American Cyanamid Co., Agricultural Chemicals Division, New York, N.Y.
22. MAXWELL, A.E. 1946. Analyzing Qualitative Data. Chap. 1, pp. 11–37, John Wiley & Sons, Inc., New York.
23. DUNCAN, A.J. 1965. Quality Control and Industrial Statistics, 3rd edition. R.D. Irwin, Inc., Homewood, Illinois.
24. YOUDEN, W.V. 1960. The Sample, the Procedure and the Laboratory. Anal. Chem. 32:23A–37A.
25. AMERICAN CHEMICAL SOCIETY. 1968. Reagent Chemicals, 4th edition, American Chemical Society Publications, Washington, D.C.
26. AMERICAN SOCIETY FOR TESTING AND MATERIALS. 1963. ASTM Manual for Conducting an Interlaboratory Study of a Test Method. Technical Publication No. 335.
27. WEIL, C.S. 1971. Critique of Laboratory Evaluation of the Reliability of Blood-Lead Analyses. Am. Ind. Hyg. Assoc. J. 32:304.
28. SNEE, R.D. and P.E. SMITH. 1971. Statistical Analysis of Interlaboratory Studies. Paper prepared for presentation to the Am. Ind. Hyg. Conference in San Francisco, Calif.

26. DIRECT READING COLORIMETRIC INDICATORS

26.1 INTRODUCTION

Three types of direct reading colorimetric indicators have been in use for the determination of contaminant concentrations in air: liquid reagents, chemically treated papers, and glass indicating tubes containing solid chemicals. A comprehensive bibliography in this area was prepared by Campbell and Miller (**1**).

Convenient laboratory procedures using liquid reagents have been simplified and packaged for field use. Reagents are supplied in sealed ampoules or tubes, frequently in concentrated or even solid form which is diluted or dissolved for use. Unstable mixtures may be freshly prepared when needed by breaking an ampoule containing one ingredient inside a plastic tube or bottle containing the other. Commercial apparatus of this type is available for tetraethyl lead and tetramethyl lead. Certain liquid reagents, such as the nitrogen dioxide sampling reagent, produce a direct color upon exposure without requiring additional chemicals or manipulations. These permit simplified sampling equipment. Thus, relatively high concentrations of nitrogen dioxide may be directly determined by drawing an air sample into a 50 or 100 ml glass syringe containing a measured quantity of absorbing liquid reagent, capping, and shaking. Liquids containing indicators have been used for determining acid or alkaline gases by measuring the volume of air required to produce a color change. These liquid methods are somewhat inconvenient and bulky to transport and require a degree of skill to use. However, they are capable of good accuracy, as measurement of color in liquids is inherently more reproducible and accurate than measurement of color on solids.

Chemically treated papers have been employed to detect and determine gases because of their convenience and compactness. An early example of this is the Gutzeit method in which arsine blackens a paper strip previously impregnated with mercuric bromide. Such papers may be freshly prepared and used wet, or stored and used in the dry state. Special chemical chalks or crayons have been used **(2)** to sensitize ordinary paper for phosgene, hydrogen cyanide, and other war gases. Semi-quantitative determinations may be made by hanging the paper in contaminated air. Inexpensive detector tabs are commercially available which darken upon exposure to carbon monoxide **(3)**. The accuracy of such procedures is limited by the fact that the volume of the air sample is rather indefinite and the degree of color change in the paper is influenced by air currents and temperature. More quantitative results may be obtained by employing a sampling device capable of passing a measured volume of air over or through a definite area of paper at a controlled rate, as is done in a commercial device for hydrogen fluoride. Particulate contaminants such as chromic acid and lead may be similarly determined, usually by addition of liquid reagents to the sample on a filter paper. Evaluation of the stains on the paper may be made visually by comparison with color charts or by photoelectric instruments. Recording photoelectric instruments utilizing sensitized paper tapes operate in this manner and are described in another section. Accuracy of these methods requires uniform sensitivity of the paper, stability of all chemicals employed, and careful calibration. In the case of particulate analysis it may be necessary to calibrate with the specific dust being sampled if the degree of chemical solubility is an important factor.

Glass indicating tubes containing solid chemicals are another type of convenient and compact direct reading device. The early detection tubes were made for carbon monoxide, **(4, 5)** hydrogen sulfide, **(6, 7)** and benzene. **(8, 9)** During the past decade, there has been a great expansion in the development and use of these tubes **(10–30)**, and more than a hundred different types are now commercially available. Because of the great popularity and wide use of glass indicator tubes, the bulk of this introduction will deal with them, although much of the information will be applicable to the liquid and paper indicators as well.

There are many uses for indicator tubes. They are convenient for evaluation of toxic hazards in industrial atmospheres. They are also useful for air pollution studies, although in most cases currently available tubes do not have the required sensitivity. Indicator tubes may be used for detection of explosive hazards, as well as for process control of gas composition. Confirmation of carbon monoxide poisoning may be made by determining carbon monoxide in exhaled breath or in gas released from a sample of blood *after an appropriate procedure*. Indicator tubes may be used for law enforcement purposes, such as determining alcohol in the breath, or gasoline in soil in cases of suspected incendiarism. Minute quantities of ions in aqueous solutions also may be determined, such as sulfide in waste water from pulp manufacturing, chromic acid in electrolytic plating waste water, and nickel ion in waste water of refineries.

Indicator tubes have been widely advertised as being capable of use by unskilled personnel. While it is true that the operating procedures are simple, rapid, and convenient, many limitations and potential errors are inherent in this method. The results may be dangerously misleading unless the sampling procedure is supervised and the findings interpreted by an adequately trained industrial hygienist.

26.2 Operating Procedures

The use of detecting tubes is extremely simple. After its two sealed ends are broken open, the glass tube is placed in the manufacturer's holder which is fitted with a calibrated squeeze bulb or piston pump. The recommended air volume is then drawn through the tube by the operator. Adequate time must be allowed for each stroke. Even if a squeeze bulb is fully expanded, it may still be under a partial vacuum and may not have drawn its full volume of air. The manufacturer's sampling instructions must be closely followed.

The observer then evaluates the concentration in the air by examining the exposed tube. Some of the earlier types of tube are provided with charts of color tints to be matched by the solid chemical in the indicating portion of the tube. This visual judgment depends, of course, upon the color vision of the observer and the lighting conditions. In an attempt to reduce the errors due to variations between observers, most recent types of tubes are based upon producing a variable length of stain on the indicator gel. Although in a few tubes a variable volume of sample is collected until a standard length of stain is obtained, in most cases a fixed volume of sample is passed through the tube and the length of stain is measured against a calibration scale. The scale may be printed either directly upon the tube or on a chart which is provided. In a few tubes, such as those for arsine and stibine, a variable volume of sample is drawn through the tube until the first visible discoloration is noted. This is a very difficult judgment which must be made retrospectively. There is a large range in the interpretation of tubes by different observers, since in many cases the stain fronts are not sharp. Experience in sampling known concentrations is of great value in training the operator to know whether to measure the length up to the beginning or end of the stain front, or what portion of an irregularly shaped stain to use as the limit. In some cases the stains change with time, and thus the reading should not be unduly delayed.

Care must be taken to see that pump valves and connections are maintained leak-proof. A leakage test may be made by inserting an unopened detector tube into the holder, and squeezing the bulb; at the end of two minutes any appreciable bulb expansion is evidence of a leak. If the apparatus is fitted with a calibrated piston pump, the handle is pulled back and locked. Two minutes later, it is released cautiously and the piston allowed to pull back in; it should remain out no more than 5% of its original distance. Leakage indicates the necessity of replacing check valves, tube connections, or the squeeze bulb, or of greasing the piston.

At periodic intervals the flow rate of the apparatus should be checked and maintained within specifications for the tube calibrations. This may be done simply by timing the period of squeeze bulb expansion. A more accurate method is to place a used detector tube in the holder, and to draw an air sample through a calibrated rotameter. Alternatively, the air may be drawn from a buret in an inverted vertical position, which is sealed with a soap film, and the motion of the film past the graduations timed with a stop watch (**31**). The latter method also provides a check on the total volume of the sample which is drawn. In some devices the major resistance to the air flow is in the chemical packing of the tube; thus, each batch might require checking. An incorrect flow rate indicates a partially clogged strainer or orifice which should be cleaned or replaced.

With most types of squeeze bulbs and hand pumps a variable sample air flow rate is produced, being high initially and low towards the end when the bulb or pump is almost filled. This has been claimed to be an advantage inasmuch as the initially high rate gives a long stain and the final low rate sharpens the stain front. Flow patterns for 6 commonly used pumps were found to be different (**32**). When 5 popular brands of carbon monoxide tubes were used with pumps other than their own, grossly erroneous results were obtained, even if the sample volumes were identical. The stains may depend more on flow rate than on concentrations. It should be noted that accuracy requires a close reproduction of the flow rate pattern for the calibrations to be correct.

A number of special techniques may be used in appropriate cases. When sampling in inaccessible places the indicator tube may be placed directly at the sampling point, and the pump operated at some distance away. An extension rubber tube of the same inside diameter as the indicator tube may be inserted between the pump and indicator tube. Lengths as great as 60 ft have been successfully used without appreciable error, provided that more time is allowed between strokes of the pump to compensate for the reservoir effect and obtain the full volume of sample. This method has the disadvantage that the detector tube cannot be observed during the sampling.

A second arrangement may be employed when sampling hot gases such as from a furnace stack or engine exhaust. Cooling the sample is essential, otherwise the calibration would be inaccurate and the volume of the gas sample uncertain. A probe of glass or metal may be attached to the inlet end of the detector tube with a short piece of flexible tubing. If this tube is cold initially, a length of as little as 4″ outside of the furnace is sufficient to cool the gas sample from 250 C to about 30 C. Such a probe has to be employed with caution since in some cases serious adsorption errors occur either on the tube or in condensed moisture. The dead volume of the probe should be negligible in comparison to volume of sample taken. Solvent vapors should not be sampled with this method. Other special techniques may also be employed. Some symmetrical tubes can be reversed in the holder and used for a second test. In certain special cases tubes may be re-used if a negative test was previously obtained, or after the color has faded. Two tubes also may be connected in series in special cases, such as passing crude gas through first a Kitagawa hydrogen sulfide tube and then a phosgene tube to get two simultaneous determinations and remove interferences. These techniques may be used only after testing to demonstrate that they do not impair the validity of the results.

Tubes also have been used in pressures as high as several atmospheres, for example, in underwater stations. If both the tube and pump are in the chamber, the calibrations and sample volumes are altered. It has been reported **(33)** that only the latter effect occurs for the following Draeger tubes: ammonia 5/a, arsine 0.05/a, CO_2 0.1%/a, CO 5/c, 10/b, H_2S 1/c, 5/b. For these tubes, the corrected concentration is equal to the scale reading (ppm or vol %) divided by the ambient pressure (in atmospheres) at the pump. When tube tips are broken in a pressure chamber, the tube filling should be checked for possible displacement.

The specificity of the tubes is a major consideration for determining applicability and interpreting results of the tubes. Most tubes are not specific. Reduction of chromate is a common reaction used in tubes for organic compounds. In the presence of mixtures, the uncritical acceptance of such readings can be grossly misleading. A comprehensive listing of reactions was given by Linch **(34)**, as well as a discussion of other major aspects. Six common reactions are listed in Table 26:I, with the tube types in which they are utilized. It can be seen that the name of the compound listed on the tube often refers to its calibration scale rather than to its contents.

The lack of specificity of some tubes may be used to advantage for determining substances other than those indicated by the manufacturer. Tubes for hydrocarbons and chlorinated hydrocarbons are widely applicable in this respect. Thus, the Draeger trichloroethylene tube is also applicable to chloroform, o-dichlorobenzene, dichloroethylene, ethylene chloride, methylene chloride, and perchloroethylene. The methyl bromide tube may be used for chlorobromomethane and methyl chloro-

Table 26:I. Common Colorimetric Reactions in Gas Detector Tubes

1. Reduction of chromate or dichromate to chromous ion:

Draeger: Alcohol 100/a, cyclohexane 100/a, diethyl ether 100/a, ethyl acetate 200/a, n-pentane 100/a.

Gastec: sulfur dioxide 5H, LP gas 100A, gasoline 101, n-hexane 102H, 102L, cyclohexane 103, butane 104, methanol 111*, ethanol 112*, isopropanol 113*, vinyl chloride 131, ethyl acetate 141, butyl acetate 142, acetone 151, methyl ethyl ketone 152, methyl isobutyl ketone 153, ethyl ether 161.

Kitagawa: acetone 102A, sulfur dioxide 103A, ethyl alcohol 104A, ether 107, ethyl acetate 111, n-hexane 113, cyclohexane 115, methyl alcohol 119, ethylene oxide 122, dimethyl ether 123, acrylonitrile 128A, 128B, vinyl chloride 132, butyl acetate 138, methyl ethyl ketone 139B, methyl acetate 148, isopropyl acetate 149, isopropanol 150, propyl acetate 151, isobutyl acetate 153, dioxane 154, methyl isobutyl ketone 155, furan 161, tetrahydrofuran 162, propylene oxide 163, butadiene 168A.

MSA: *Part 95097* for n-amyl alcohol, iso-amyl alcohol, sec-amyl alcohol, tert-amyl alcohol, 2-butoxyethanol (butyl cellosolve), n-butyl alcohol, iso-butyl alcohol, sec-butyl alcohol, tert-butyl alcohol, cyclohexanol, 2-ethoxyethanol (cellosolve), ethyl alcohol (ethanol), ethylene glycol monomethyl ether, furfuryl alcohol, 2-methoxyethanol, methyl alcohol (methanol), 2-methylcyclohexanol, methyl iso-butyl carbinol (methyl amyl alcohol), n-propyl alcohol, iso-propyl alcohol. *Part 460423* for acetone, methyl methacrylate.

2. Reduction of iodine pentoxide + fuming sulfuric acid to iodine:

Draeger: carbon monoxide 5/c, 8/a, 10/a, 10/b, 0.1%/a, 0.3%/a, 0.5%/a, hydrocarbon 0.1%, polytest, toluene 5/a, 25/a.

Gastec: benzene 121, toluene 122, xylene 123, ethylbenzene 125, acetylene 171.

MSA: *Part 93074* for benzene (benzol), chlorobenzene, monobromobenzene, toluene (toluol), xylene (xylol).

3. Reduction of ammonium molybdate + palladium sulfate to molybdenum blue:

Gastec: ethylene 172.

Kitagawa: acetylene 101, carbon monoxide 106A, 106B, 106C*, ethylene 108B, hydrogen sulfide and sulfur dioxide 120C, butadiene 168B.

MSA: *Part 47134* for carbon monoxide (NBS color change). *Part 85802* for acetylene, ethylene, propylene.

4. Reaction with potassium palladosulfite:

Gastec: carbon monoxide 1L, 1La, 1LL, hydrogen cyanide 12H.

MSA: *Part 91229* for carbon monoxide (length of stain).

5. Color change of pH indicators (*e.g.*, bromphenol blue, phenol red, thymol blue, methyl orange):

Draeger: ammonia 5/a, 0.5%/a, hydrazine 0.25/a, hydrochloric acid 1/a, hydrogen cyanide 2/a ($HgCl_2$ + methyl red), sulfur dioxide 0.1/a (Na_2HgCl_4 + methyl red), triethylamine 5/a.

Gastec: carbon dioxide 2H, 2L, ammonia 3M, 3L, sulfur dioxide 5M, 5L, hydrogen cyanide 12L* ($HgCl_2$ + methyl orange), carbon disulfide 13*, hydrogen chloride 14L, nitric acid 15L, acetic acid 81, trichloroethylene 132H*, 132L*, perchloroethylene 133*, acrylonitrile 191*.

Table 26:I (continued)

Kitagawa: ammonia 105B, hydrogen cyanide 112B, carbon dioxide 126A, 126B, acetaldehyde 133.

MSA: *Part 85976* for carbon dioxide. *Part 91636* for hydrogen chloride. *Part 92030*† for 1-chloro-1,1-difluoroethane (Genetron 142B), chlorotrifluoromethane (Freon 13), 1,2-dichloroethane (ethylene dichloride), dichloroethylene (trans-1,2), ethyl chloride, fluorotrichloromethane (Freon 11), methyl chloride, methylene chloride (dichloromethane), propylene dichloride (1,2-dichloropropane), 1,1,2-trichloro-1,2-trifluoroethane (Freon 113), vinyl chloride (chloroethylene). *Part 92115* for ammonia, n-butylamine, cyclohexylamine, diisopropylamine, di-n-propylamine, ethylamine, ethylene imine, m-ethylmorpholine, isopropylamine, methylamine, propylene imine, triethylamine, trimethylamine. *Part 92623* for sulfur dioxide. *Part 93865* for ozone. *Part 95739*† for dimethyl sulfoxide. *Part 460021* for acetic acid. *Parts 460103 and 460158* for ammonia. *Part 460425* for hydrazine, monomethyl hydrazine, unsymmetrical dimethyl hydrazine.

6a. Reaction with o-tolidine:

Draeger: chlorine 0.2/a, 50/a, perchloroethylene 10/a*, trichloroethane*, trichloroethylene 10/a*.

Gastec: chlorine 8L, nitrogen dioxide 9L, nitrogen oxides 10*, 11L*, methyl chloroform 135*, methyl bromide 136*.

Kitagawa: chlorine 109, bromine 114, chlorine dioxide 116, nitrogen dioxide 117.

MSA: *Part 93262* for hydrogen cyanide.

6b. Reaction with tetraphenylbenzidine:

MSA: *Part 82399* for bromine, chlorine, chlorine dioxide. *Part 83099* for nitrogen dioxide. *Part 85833** for chlorobromomethane, 1,1-dichloroethane, dichloroethylene (cis-1, 2 and trans-1,2), ethyl bromide, ethyl chloride, perchloroethylene (tetrachloroethylene) trichloroethylene, 1,2,3-trichloropropane, vinyl chloride (chloroethylene). *Part 85834** for chlorobenzene (mono), 1,2-dibromoethane (ethylene dibromide), dichlorobenzene (ortho), 1,2-dichloroethane (ethylene dichloride), dichloroethyl ether, 1,1-dichloroethylene (vinylidine chloride), methyl bromide, methylene chloride (dichloromethane), propylene dichloride (1,2-dichloropropane), 1,1,2,2-tetrabromoethane, 1,1,2,2-tetrachloroethane, 1,1,3,3-tetrachloropropane, trichloroethane (beta 1,1,2), vinyl chloride (chloroethylene). *Part 87042* for bromine, chlorine. *Part 88536*† for carbon tetrachloride, chlorobromomethane, 1-chloro-1,1-difluoroethane (Genetron 142B), chlorodifluoromethane (Freon 22), chloroform (trichloromethane), chloropentafluoroethane (Freon 115), chlorotrifluoromethane (Freon 13), 1,2-dibromoethane (ethylene dibromide), dichlorodifluoromethane (Freon 12), 1,1-dichloroethylene (vinylidine chloride), dichloroethylene (cis-1,2), dichlorotetrafluoroethane (Freon 114), fluorotrichloromethane (Freon 11), Freon 113, Freon 502, methyl bromide, methyl chloroform (1,1,1-trichloroethane), methylene chloride (dichloromethane), perchloroethylene (tetrachloroethylene), trichloroethane (beta 1,1,2), trichloroethylene, 1,1,2-trichloro-1,2,2-trifluoroethane (Freon 113), trifluoromonobromoethane (Freon 13B1). *Part 91624*† for acetonitrile, acrylonitrile, 1-chloro-1-nitropropane, cyanogen, 1,1-dichloro-1-nitroethane, dimethylacetamide, dimethylformamide, fumigants (Acritet, Insect-O-Fume, Fumi-I-Gate, Termi-Gas, Termi-Nate), methacrylonitrile, nitroethane, nitromethane, 1-nitropropane, 2-nitropropane, n-propyl nitrate, pyridine, vinyl chloride. *Part 460225* for chlorine. *Part 460424** for nitric oxide.

---

*Multiple layer tube for improved specificity or preliminary reaction.
†Pyrolyzer required.

form. The chlorine tube may be used for bromine and chlorine dioxide. The toluene tube may be used for xylene. Such use requires specific knowledge of the identity of the reagent and of the proper corrections to the calibration scales.

In some brands of indicator tubes the units of the calibration scales are in mg/cc. Although it has been said that this method of expression eliminates the necessity of making temperature and pressure corrections, such a claim is debatable since the scale calibrations themselves may be highly dependent upon these variables. Units of ppm or %/vol are most common for industrial hygiene purposes and are used on most of the newer tubes. Conversions may be made from mg/liter to ppm by the formula given below:

$$\text{P.P.M.} = \frac{(\text{mg/L}) \times 24{,}470}{\text{Molecular Wt.}} \qquad (1)$$

The constant in the numerator is the gram molecular volume in milliliters at 25 C and 760 Torr. Should the sampling conditions deviate widely from these standard conditions, the appropriate constant derived from the perfect gas law for the actual temperature and pressure should be substituted.

Although detector tubes are generally designed for relatively high gas concentrations found in industrial workplaces, recently some have been applied to the much lower outdoor air pollutant concentration. Kitagawa (**35**) determined 0.01 to 2 ppm of NO_2 by means of two glass tubes in series, thermostated at 40 C. The first contained diatomaceous earth impregnated with sulfuric acid of a definite concentration to regulate the humidity to the air sample. The second tube, 120 mm long × 2.4 mm inside diam, contained white silica gel impregnated with ortho-tolidine. (It is not clear whether or not this is identical with the commercial No. 117). Air was drawn through the tubes for 30 min at 180 ml/min by an electric pump with a stainless steel orifice plate at its inlet. Accuracy was ± 10%; no comments on the specificity were given. Grosskopf (**36, 37**) determined 0.007 to 0.5 ppm NO_2 by drawing air through a Draeger 0.5/a nitrous gas tube for 10 to 40 min at the rate of 0.5 lpm with a diaphragm pump. Readings were not affected by flow rates if they exceeded 0.5 lpm. No comments were given on the specificity, except that humidity from 30 l of air at 70% relative humidity did not impair the sensitivity. This tube responds to nitric oxide and to oxidants, both of which commonly may be present. Leichnitz (**38**) reported a new tube (Draeger 0.1/a) capable of measuring 0.1 to 3 ppm of sulfur dioxide. This tube requires 100 strokes of a hand bellows pump (each taking 7 to 14 sec), or use of the Draeger Model D41 electric pump, in which a motor-driven crank operates a bellows, and is controlled by a timer or counter. This pump is described in the text of the manufacturers listing.

Less success has been attained when detector tubes were used for sampling periods of 4 hr or longer with continuous pumps. It was found that at low concentrations, after an initial period, the stain lengths cease to increase (**23**). However, at higher concentrations a new calibration could be made (**39**) (for 3 or 5 hr samples at 8 ml/min through a Kitagawa 100 tube in the range 30 to 100 ppm of carbon monoxide). The latter investigator hypothesized that the oxygen in air bleached the black palladium stain and caused the front produced by low concentrations to remain stationary after the first 20 to 30 min. Effects of water vapor and of other contaminants also must be considered in this application. A new calibration is essential under the flow conditions to be used.

Greater accuracy can be obtained when several detector tubes are used for replicate sampling. A simplified statistical approach based on an assumed normal distribution of values was recommended (**40**) for 3 to 10 samples. However, subsequent work indicated that most of the variations were due to the environmental fluctuations rather than to the relatively small analytical errors, and that a lognormal distribution was more appropriate. A step by step procedure was presented (**64**), which categorized the results into non-compliance (less than 5% chance of erroneously citing when actually compliance exists), no decision, and compliance (less than 5% chance of failing to cite when actually non-compliance exits).

26.3 PROBLEMS IN THE MANUFACTURE OF INDICATOR TUBES

The accuracy, limitations, and applications of indicator tubes are highly dependent upon the skill with which they were manufactured. Generally, the supporting material is silica gel, alumina, ground glass, pumice, or resin. This is impregnated with an indicating chemical which should be stable, specific, sensitive, and produce a color which strongly contrasts with the unexposed color and is non-fading for at least an hour. If the reaction with the test gas is relatively slow, a color is produced throughout the length of the tube, since the gas is incompletely absorbed and the concentration at the exiting end is an appreciable fraction of that at the entrance. Such a color must be matched against a chart of standard tints, with the attendant disadvantages previously cited. A rapidly reacting indicating chemical is much more desirable, and yields a length of stain type of tube in which the test gas is completely absorbed in the stained portion.

There is a very wide and unpredictable variation in the properties of different batches of indicating gel. Since the major portion of the chemical reaction probably occurs upon the surface, the number of active centers, which is highly sensitive to trace impurities, affects the reaction speed. These problems are well known in the preparation of various catalysts. Very close controls must be kept on the purity and quality of the materials, the method of preparation, the cleanliness of the air in the factory or glove box in which the tubes are assembled, the inside diameter of the glass tubes, and even upon the size analysis of the impregnated gel, which in some cases is important in controlling the flow rate. The manufacturer also must accurately calibrate each batch of indicating gel.

Some tube types are constructed with multiple layers of different impregnated gels with inert separators. Generally, the first layer is a precleansing chemical to remove interfering gases and improve the specificity of the indication. Thus, in the case of some carbon monoxide tubes, chemicals are provided to remove interfering hydrocarbons and nitrogen oxides. In carbon disulfide tubes, hydrogen sulfide is first removed. In hydrogen cyanide tubes, hydrogen chloride or sulfur dioxide are first removed. In other cases, the entrance layer provides a preliminary reaction essential to the indicating reaction. Thus, in some trichloroethylene tubes, the first oxidation layer liberates halogen which is indicated in the subsequent layer. In some tubes for nitrogen gases, a mixture of chromium trioxide and concentrated sulfuric acid is used to oxidize nitric oxide to nitrogen dioxide, which is the form to which the sensitive indicating layer responds. While such multiple layer tubes are advantageous when properly constructed, they frequently have a shorter shelf life because of diffusion of chemicals between layers and consequent deterioration.

A shelf-life of at least two years is highly desirable for practical purposes. A great deal of disappointment with various tube performances is no doubt due to inadequate shelf life. Since many tubes have only been on the market for a short time, the manufacturer himself may have inadequate experience as to the shelf life of his product. Small variations in impurities, such as the moisture content, may have a large effect upon the shelf lives of different batches. The storage temperature, of course, greatly affects the shelf life, and it is highly desirable to store these tubes in a refrigerator. In some cases, shelf life has been estimated by accelerated tests at higher temperatures. Such a variation of shelf life (length of time within which the calibration accuracy is maintained at plus and minus 30%) is illustrated by the data below received in a personal communication from Dr. Karl Grosskopf of the Draeger Company as follows:

Shelf Life of Draeger Carbon Monoxide Tubes

Temp C	25	50	80	100	125	150
Shelf Life	> 2 yr	≥ 1/2 yr	weeks	1 week	3 days	1 day

These data plot as an approximately straight line when the logarithm of the shelf life time is plotted against a linear scale of the reciprocal of absolute temperature. Such a plot is usual for the reaction rate of a simple chemical reaction. Of course, in other cases, relationships may be more complex.

The shipping properties of tubes must also be carefully controlled. Loosely packed indicating gels may shift, causing an error in the zero point of scales printed directly upon the tube as well as an error in total stain length. When the size analysis includes an appreciable range, the fines may segregate to one side of the bore causing different flow resistances and different flow rates on the different sides of the tube. This may cause oval stain fronts which are not perpendicular to the tube bore. If the indicating gel is friable, the size analysis may change during shipping.

Obviously, satisfactory results can be obtained only if the manufacturers take great pains in the design, production, and calibration of tubes.

26.4 THEORY OF CALIBRATION SCALES

Calibration scales have been entirely empirical up to the present time. The variables which can affect the length of stain are concentration of test gas, volume of air sample, sampling flow rate, temperature, and pressure, as well as a number of factors related to tube construction. There is a striking similarity in the fact that most of the calibration scales are logarithmic with respect to concentration in spite of the widely differing chemicals employed in different tube types. Although very little data are available for these relationships, a basic mathematical analysis was made by Saltzman (**41**). These theoretical formulae discussed below will, of course, have to be modified as more data become available. The relationships were also studied by Grosskopf, (**37**) and Leichnitz (**42**).

In the usual case, although the test gas is completely sorbed, equilibrium is not reached between the gas and the absorbing indicator gel, because the sampling period is relatively short and the flow rate relatively high. The length of stain is determined by the kinetic rate at which the gas either reacts with the indicating chemical or is adsorbed on the silica gel. The theoretical analysis shows that the stain length is proportional to the logarithm of the product of gas concentration and sample volume as shown in the following simplified equation.

$$L/H = \ln (CV) + \ln (K/H) \qquad (6)$$

Where

- L is the stain length in centimeters
- C is the gas concentration in parts per million
- V is the air sample volume in cubic centimeters
- K is a constant for a given type of indicator tube and test gas
- H is a mass transfer proportionality factor having the dimension of centimeters, and known as the height of a mass transfer unit.

The factor H varies with the sampling flow rate raised to an exponent of between 0.5 and 1, depending upon the nature of the process which limits the kinetic rate of sorption. This process may be diffusion of the test gas through a stagnant gas film surrounding the gel particles, the rate of surface chemical reaction, or diffusion in the solid gel particles. If the indicator tube follows this mathematical model, a plot of stain length, L, on a linear scale, versus the logarithm of product CV (for a fixed constant flow rate) will be a straight line of slope H. The equation indicates the importance of controlling flow rate, as it may affect stain lengths more than gas concentration.

If larger samples are taken at low concentrations and the value of L/H exceeds 4, the gel approaches equilibrium saturation at the inlet end, and calibration relationships are modified. The solution to the equations for this case has been presented by Saltzman (**41**) graphically in a generalized chart. However, there is little advantage to be gained in greatly increasing the sample size, since the stain front is greatly broadened and various errors are greatly increased.

For some types of tubes such as hydrogen sulfide and ammonia the reaction rate is fast enough so that equilibrium can be attained between the indicating gel and the test gas. Under these conditions there is a stoichiometric relationship between the volume of discolored indicating gel and the quantity of test gas absorbed. In the simplest case the stain length is proportional to the product of concentration and volume sampled:

$$L = K' C V \qquad (7)$$

If adsorption is important, the exponent of concentration may differ from unity:

$$L = K'' C^{(1-n)} V \qquad (8)$$

The value of n is the same as that in the Freundlich isotherm equation for equilibrium adsorption, which states that the mass of gas adsorbed per unit mass of gel is proportional to the gas concentration raised to the power n. If the value of n is unity, which is not unusual, equation 8 indicates that stain length is proportional to sample volume, but is independent of concentration. The physical meaning of this is that all concentrations of gas are completely adsorbed by a fixed depth of gel. Such a tube is obviously of no practical value.

Equilibrium conditions may be assumed for a given type of indicator tube if stain lengths are directly proportional to the volume of air sampled (at a fixed concentration), and are not affected by air sampling flow rate. A log-log plot then may be made of stain length versus concentration for a fixed sample volume. A straight line with a slope of unity indicates equation 7 applies; if another value of slope is obtained, equation 8 applies.

In some of the narrower indicator tubes, manufacturing variation in tube diameters produces an appreciable percentage variation in tube cross-sectional areas. This results in an error in the calibration as high as 50%, if for no other reason than because the volume of sample per unit cross-sectional area is different from that under standard test conditions. An additional complicating factor is the variation produced in flow rate per unit cross-sectional area. If an exactly equal quantity of indicating gel is put into each tube, variations in cross-sectional areas will be indicated by corresponding variations in the filled tube lengths. Correction charts are provided by one manufacturer on which the tube is positioned according to the filled length and a scale is given for reading stain lengths. Although the theoretical corrections are rather complex, practically linear corrections are very close approximations which can reduce the errors to 10%. In some tubes, the tube diameters are controlled closely enough so that no correction is regarded as necessary.

Temperature is another very important variable for tube calibrations. The effect is different for different types of tubes. Since the color tint type of tube depends upon the degree of reaction, it is most sensitive to temperature. For example, some types of carbon monoxide tubes require correction by a factor of two for each deviation of 10 C. from the standard calibration conditions.

Errors in judging stain lengths produce equal percentage errors in concentration derived from the calibration scale. Errors in measuring sample volume and in flow rate may also result in errors in the final value, although the exact relationships might vary according to the tubes.

Many other complications can be expected in calibration relationships. Thus, for nitrogen dioxide the proportion of side reactions is changed at different flow rates. Changing sample volumes freely from calibration conditions is not recommended unless the tube is known to be thoroughly free from the effects of interfering gases and humidity in the air.

A crucial factor in the accuracy of the calibration is the apparatus used for preparing known low concentrations of the test gas. This subject is more fully discussed in the introductory articles of this manual. Some manufacturers have used static methods, but in our experience losses of 50% or more by adsorption are not uncommon. Low concentrations or reactive gases and vapors are best prepared in a dynamic system. This has further advantages of compactness and ability to rapidly change concentrations as required. With either type of apparatus it is highly desirable to check the concentrations using chemical methods of known adequacy. Some successful systems have been described **(43, 46)**.

A simple and company dynamic apparatus for accurately diluting tank gas (which may be either pure or a mixture) is illustrated in Figure 26:1, taken from Saltzman **(44)**. The asbestos plug flowmeter both meters and controls gas flows in the range of a few hundredths of a milliliter to a few milliliters per min. The tank valve setting is not critical, so long as at least a few bubbles per minute of excess gas are maintained at the waste outlet. Flows are calibrated by attaching a graduated 1- or 0.1-ml. pipet to the meter outlet and timing with a stopwatch the movement of a drop of water (or oil) past the graduations. To obtain uniform motion, the pipet must be carefully cleaned and the delivery tip kept dry (or preferably cut off). A linear calibration plot is obtained of flow rate vs water depth (or ml) over the outlet of the waste tube. This is reasonably constant for most gases so long as the asbestos plug is kept dry. It may be checked occasionally with little trouble. A 1-mm bore in the stopcock plug is satisfactory for most of the lower rates of flow. The asbestos plug (acid washed fiber of the type used for Gooch crucibles) is tamped by trial and error, running test calibrations, until the desired flow range is obtained. For accurate work the back pressue on the downstream face of the asbestos plug may be measured with a manometer and deducted from the water

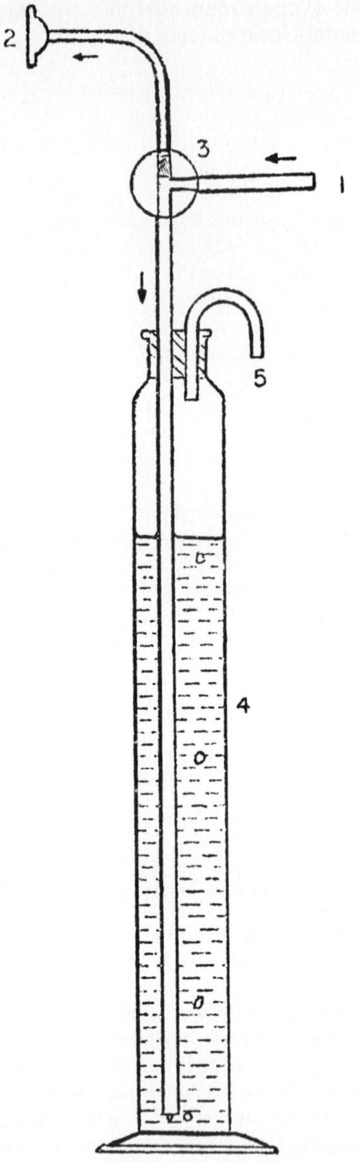

Figure 26:1—Asbestos plug flowmeter: 1 Test gas inlet (may be connected to a tank). 2 Metered gas outlet, 12/2 ball joint. 3 Capillary three-way T-shaped stopcock, with asbestos packing in upper leg of stopcock plug bore. (Caution! Never turn stopcock with tank valve open, as excessive pressure may develop in glass.) 4 250 ml graduated cylinder containing water or oil. 5 Waste gas outlet.

level in the cylinder to give the true flow from the calibration chart.

The flow dilution device, Figure 26:2, is designed to minimize back pressure and also the metered gas dead space. If the test gas is being metered at 0.1 ml per min, a dead space of even 1 ml. from the asbestos plug to the point of dilution is excessive, as 50 min will be required to flush out the residual air. The nozzle inlet should be pulled down to about 0.5 mm inside diameter. Even a small leakage at the ball joint is very serious, and can be detected by partially evacuating the system with exterior openings closed and observing whether a manometer connected to the apparatus

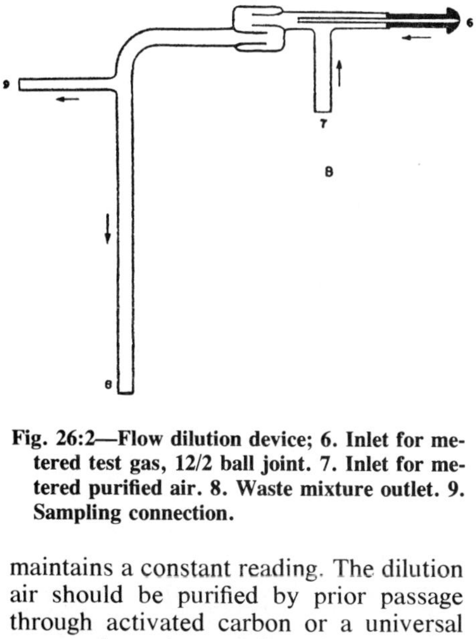

Fig. 26:2—Flow dilution device; 6. Inlet for metered test gas, 12/2 ball joint. 7. Inlet for metered purified air. 8. Waste mixture outlet. 9. Sampling connection.

maintains a constant reading. The dilution air should be purified by prior passage through activated carbon or a universal gas mask canister, followed by a filter. The excess mixture is passed from the waste outlet to the hood through a section of tubing, providing a slight positive pressure at the sampling tap. There is no danger of air diffusing or being drawn into the system if the mixture flows at a sufficiently high velocity and only a small fraction is drawn from the side arm. Partially clamping the waste tube can raise the pressure if needed for a two-stage dilution system.

Air-vapor mixtures of volatile organic liquids may be prepared in a similar apparatus using a motor-driven hypodermic syringe. High quality gears, bearings, and screws are needed in the motor drive to provide the uniform slow motion. Some commercial devices have been found unsatisfactory in this regard. Many types of permeation tubes now available also have proven useful.

It is highly desirable for the user as well as the manufacturer to have facilities available for checking calibrations. Only in this manner may the user be confident that the tubes and his technique are adequate for his purposes. Tubes may also be applied to gases other than those for which they have been calibrated by the manufacturer, in certain special cases, if the user can prepare his own calibration.

26.5 PERFORMANCE EVALUATION AND CERTIFICATION

Evaluations by users of some types of tubes have been reported (**29, 47–53**). Accuracy has been found highly variable. In some cases, the tubes were completely satisfactory; in others, completely unsatisfactory. Manufacturers, in their efforts to improve the range and sensitivity of their products, are rapidly changing the contents of their tubes, and these reports are frequently obsolete before they appear in print. Improved quality control, and perhaps greater self-policing of the industry, would greatly increase the value of the tubes, especially for the small consumer who is not in a position to check calibrations.

After reviewing to this need, a joint ACGIH-AIHA Committee made the following recommendations (**54**):

a. Manufacturers should supply a calibration chart (ppm) for each batch of tubes;

b. length of stain tubes are preferable to those exhibiting change in hue or intensity of color;

c. tests of calibrations should be made at 1/2, 1, 2, and 5 times the Threshold Limit Value (ACGIH);

d. the manufacturer should specify the methods of tests. Values should be checked by two independent methods;

Table 26:II. Performance Evaluations of Detector Tubes by NIOSH

SUBSTANCE	REF.	TLV, ppm	TEST CONCENTRATIONS (ppm)	CRITERIA*	TUBE TYPES PASSING TEST
Benzene	56	25	20,35,60,160 @ 50% r.h.	±25%	None
				±50%	Draeger CH 248, Kitagawa 118-A, MSA 93074
Carbon Monoxide	55	50	25,50,100 @ 20% 50%, 80% r.h.	±25%	None
				±50%	Bacharach 190195, 190211, Draeger CH256, CH289, Kitagawa 100, 106a, MSA 47134, 91229
Carbon Tetrachloride	58	10	10,30,50 @ 50% r.h.	±50%	None
Chlorine	60	1	0.5, 1,2,5 (dry air)	±25% @ 1,2, 5 TLV ±35% @ 0.5 TLV	MSA 82399
				±25% @ 1,2, 5 TLV	Bacharach 19-0239
				±50%	Draeger CH243 Kitagawa 109
Hydrogen Sulfide	61	10	5,10,20,50 (dry air)	±25% @ 1,2, 5 TLV ±35% @ 0.5 TLV	Draeger CH298, 5/b
				±50%	Bacharach 19-0198, Kitagawa 120b
Perchloroethylene	59	100	50,100,200,400 @ 50% r.h.	±25%	Draeger CH307, Kitagawa 134, 135
Sulfur Dioxide	57	5	2.5,5,10,25 @ 5% r.h. (to avoid conversion to H_2SO_4)	±25% at the higher concentrations	Draeger CH317, Kitagawa 103C, 103D, MSA 92623

*At 95% confidence level

e. calibration at each test point should be accurate within ±25% (95% confidence limit);

f. allowable ranges and corrections should be listed for temperature, pressure, and relative humidity;

g. each batch of tubes should be labelled with a number and an expiration date. Instructions for proper storage should be given;

h. tolerable concentrations of interferents should be listed;

i. pumping volumes should be accurate within ±5%, and flow rates should be indicated, and

j. special calibrations should be provided for extended sampling for low concentrations, and flow rates should be specified.

A performance evaluation program was initiated by the National Institute for Occupational Safety and Health (NIOSH). It soon was discovered that few tubes achieved an accuracy of ±25%, and an alternative limit of ±35% at 0.5 TLV levels was accepted. Subsequently ±50% in the range 0.5 to 5 times the TLV was allowed. Known concentrations of test substances were generated in flow systems, from sources such as cylinder mixtures, vapor

Table 26:III. Certifications of Detector Tubes by NIOSH

SUBSTANCE	DATE	TUBE TYPE	PUMP MODEL
Ammonia	4/3/74	Gastec 3M	400
	5/10/74	Drager CH 20501 (5/a)	31
	5/14/84	Kitagawa 105C, 105Sc	400
	5/28/74	MSA 460103	83499
Benzene	12/6/74	Drager 67 25541 (5/b)	31
	12/6/74	Gastec 121	400
Carbon Dioxide	2/25/74	Gastec 2L	400
	4/18/74	MSA 85976	83499
	4/19/74	Kitagawa 126A, 126Sa	400
	5/6/74	Drager CH 23501 (0.1%/a)	31
	4/4/75	Bacharach 19-0359	19-7026
Carbon Monoxide	8/31/73	Drager CH 25601 (5/c), CH 20601 (10/b)	31
	9/6/73	Gastec 1La	400
	1/30/74	Kitagawa 106S	400
	10/11/73	MSA 91229	83499
	1/15/75	Kitagawa 100	400
Carbon Tetrachloride	7/17/74	Gastec 134	400
Chlorine	11/12/74	Gastec 8La	400
	11/12/74	MSA 460225	83499
Hydrogen Cyanide	4/21/75	Drager CH 25701 (2/a)	31
	4/22/75	Kitagawa 112Sb	400
Hydrogen Sulfide	2/13/74	Gastec 4LL	400
	3/13/74	MSA 460058	83499
	4/17/74	Drager 67-19001 (1/c)	31
	7/23/74	Kitagawa 120b	400
Nitric Oxide	4/7/75	Gastec 10	400
Nitrogen Dioxide	10/31/73	Drager CH 30001 (0.5/c)	31
	1/15/74	Gastec 9L	400
	10/29/74	MSA 83099	83499
Nitrogen Oxides	4/4/75	Drager CH 31001 (2/a)	31
Sulfur Dioxide	2/11/74	Gastec 5LA	400
	4/19/74	Kitagawa 103Sd	400
	5/10/74	Drager CH 31701 (1/a)	31
	6/28/74	Kitagawa 103d	400
	1/15/75	MSA 92623	83499
Toluene	4/21/75	Drager CH 23001 (5/a)	31
	7/28/75	Gastec 122	400
Trichloroethylene	8/9/74	Gastec 132H	400
	10/1/74	Drager CH 24401 (10/a)	31

pressure equilibration at known temperatures, or permeation tubes. Results, given in Table 26:II, showed that not many tubes achieved ±25%. All tubes for carbon tetrachloride were rejected since none achieved even ±50%. Many types showed accuracies in the range ±25 to 35%.

A formal certification program (**62**) was the next step. In addition to passing performance evaluation tests at the Morgantown, W. Va. laboratory of NIOSH, manufacturers were required to provide information on the contents of the tubes, and to conduct a specified quality control program. Because of the dependence of the calibrations on the pumps used with the tubes, certifications were issued (**63**) for specified combinations of tubes and pumps, as listed in Table 26:III.

The requirements for certification generally follow the recommendations of the joint committee, listed above. However, the accuracy requirement was modified to ±35% at 0.5 TLV, and ±25% at 1, 2, and 5 times TLV, to be maintained until the expiration date if the tubes are stored according to the manufacturer's instructions. At the TLV concentration either the stain length must be 15 mm or greater, or the relative standard deviation of the readings of the same tube by three or more independent tube readers must be less than 10%. If the stain front is not exactly perpendicular to the tube axis (because of channeling of the air flow), the difference between the longest and shortest stain length measurements to the front must be less than 20% of the mean length. Color intensity tubes must have sufficient charts and sampling volume combinations to provide scale values including at least the following multiples of the TLV: 0.5, 0.75, 1.0, 1.5, 2.0, 2.5, 3.0, 4.0, and 5.0; the relative standard deviation for readings of a tube by independent readers must be <10%. Tests are to be conducted generally at 65 to 85 F (18.3 to 29.5 C), and at relative humidities of 50%, unless the humidity has to be reduced to avoid disturbing the test system. The manufacturer must file a quality control plan and keep records of his inspections of raw materials, finished tubes, and calibration and test equipment. Acceptable statistical quality levels for defects in finished tubes are as follows: critical 0% where tests are nondestructive, otherwise 1.0%, major 2.5%, minor 4.0%, accuracy 6.5%. Certification seals may be affixed to approved devices. NIOSH reserves the right to withdraw certification for cause.

Since important legal and economic consequences depend upon the accuracy of measurements of contaminant concentrations, enforcement agencies will most likely prefer certified equipment. This program should provide a stimulus for further improvements in detector tube technology.

26.6 CONCLUSIONS

Use of indicating tubes for analysis of toxic gas and vapor concentrations in air is a very rapid, convenient and inexpensive technique which can be performed by semi-skilled operators. These tubes are in various stages of development, and highly variable results have been obtained. Accuracy is dependent upon a high degree of skill in the manufacture of the tubes. At present, results may be regarded as only range-finding and approximate in nature. The best accuracy which can be expected from indicator tubes of the best types is of the order of plus and minus 25 to 50%. Since many of the tubes are far from specific, an accurate knowledge of the possible interfering gases present is very important. The quantitative effect of these interferences depends upon the volume sampled in an irregular way. In order to avoid dangerously misleading results, the operation and interpretation should be under the supervision of a skilled industrial hygienist.

References

1. CAMPBELL, E.E., and H.E. MILLER. 1961; 1964. Chemical Detectors. A Bibliography for the Industrial Hygienist with Abstracts and Annotations. Los Alamos Scientific Laboratory, LAMS-2378 (Vol. I, II).
2. DEPT. OF THE ARMY TECHNICAL MANUAL, 1953. Individual Protective and Detection Equipment. TM 3-290; Dept. of the Air Force Technical Order, TO 39C-10C-1, pp. 56–80, Sept.
3. MCFEE, D.R., R.E. LAVINE, R.J. SULLIVAN, 1970. Carbon Monoxide, a Prevalent Hazard Indicated by Detector Tabs. Am. Ind. Hyg. Assoc. J. 31:749–53.
4. LAMB, A.B., W.C. BRAY, and J.C. FRAZER 1921, Ind. Eng. Chem. 12:213 (1920); C.W. HOOVER: Ibid 13, 770.
5. SHEPHERD, M. 1947. Rapid Determination of Small Amounts of Carbon Monoxide, Preliminary Report on the NBS Colorimetric Indicating Gel. Anal. Chem. 19:77–81.
6. LITTLEFIELD, J.B., W.P. YANT, and L.B. BERGER. 1935. U.S. Bur. Min. Rep. Inv., No. 3276.
7. KITAGAWA, T. 1951. Rapid Analysis of Phosphine and Hydrogen Sulfide in Acetylene, J. Japan Chem. Ind. Soc. No. 33, (Feb.).
8. HUBBARD, B.R., L. SILVERMAN 1950. Arch. Ind. Hyg. and Occ. Med. 2:49.
9. GROSSKOPF, K. 1951. Agnew. Chem. 63:306.
10. KITAGAWA, T. 1952. Rapid Method of Quantitative Gas-Analysis by Means of Detector Tubes, Kagaku no Ruoiki 6:386.

11. SACKS, V. 1956. Carbon Monoxide Detection by Means of the Colorimetric Gas Analyzer. (German). Deutsche Zeitschrift Fur gerichtliche Medizin 45:68–71.
12. KINOSIAN, J.R., B.R. HUBBARD, 1958. Nitrogen Dioxide Indicator. Amer. Ind. Hyg. Assoc. J. 19:453.
13. GROSSKOPF, K. 1958. Detector Tubes as Detectors in Gas Chromatography. (German). Erdohl und Kohle. 11:304–6.
14. GROSSKOPF, K. 1959. Vaporous Reagents in the Detector Tube Technique for Measurement of Vapors and Gases. (German). Zeitschrift fur analytische Chemie. 170:271–7.
15. GROSSKOPF, K. 1959. Systox Detection. (German). Chemiker-Zeitung-Chemische Apparatus 83:115–7.
16. HETZEL, K.W. 1959. Poisonous Action and Detection of Injurious Gases and Vapors in Mining Operations. (German). Brennstoff-Chemie 41:115–22.
17. KITAGAWA, T. 1959. Detection of Underground Spontaneous Combustion in Its Early Stage With Detector. Tenth International Conference of Directors of Safety in Mines Research, Pittsburgh, Pennsylvania.
18. KITAGAWA, T. 1960. The Rapid Measurement of Toxic Gases and Vapors. The 13th International Congress on Occupational Health, New York, New York, July 25–9.
19. BRETZKE, W. 1960. The Determination of Carbon Dioxide Content in the Atmosphere of Silos and Fermenters. (German). Die Berufsgenossenschaft. (May).
20. KETCHAM, N.H. 1962. Practical Experiences with Routine Use of Field Indicators, Am. Ind. Hyg. Assoc. J. 23:127–31.
21. SILVERMAN, L. 1962. Panel Discussion of Field Indicators in Industrial Hygiene, Am. Ind. Hyg. Assoc. J. 23:108–11.
22. SILVERMAN, L. 1963. Techniques to Improve the Accuracy of Detector Tubes, American Industrial Hygiene Association, Cincinnati, Ohio, (May).
23. INGRAM, W.T. 1965. Personal Air-Pollution Monitoring Devices, Am. Ind. Hyg. Assoc. J. 25:298–303.
24. SILVERMAN, L. and G.R. GARDNER. 1965. Potassium Pallado Sulfite Method for Carbon Monoxide Detection, Am. Ind. Hyg. Assoc. J. 26:97–105.
25. LINCH, A.L., S.S. LORD, JR., K.A. KUBITZ, and DEBRUNNER, M.R. 1965. Phosgene in Air—Development of Improved Detection Procedures, Am. Ind. Hyg. Assoc. J. 26:465–73.
26. LEICHNITZ, K. 1971. Determination of Arsine in Air in the Work Place (German). Die Berufsgenossenschaft, (Sept.).
27. LEICHNITZ, K. 1968. Cross-Sensitivity of Detector Tube Procedures for the Investigation of Air in the Work Place. (German). Zentralblatt fur arbeitsmedizin und Arbeitsschutz. 18:97–101.
28. LINCH, A.L., R.F. STALZER, D.T. LEFFERTS. 1968. Methyl and Ethyl Mercury Compounds Recovery from Air and Analysis, Am. Ind. Hyg. Assoc. J. 29:79–86.
29. LINCH, A.L. 1965. Oxygen in Air Analyses—Evaluation of a Length of Stain Detector, Am. Ind. Hyg. Assoc. J. 26:645.
30. GRUBNER, O., J.J. LYNCH, J.W. CARES, W.A. BURGESS, 1972. Collection of Nitrogen Dioxide by Porous Polymer Beads. Am. Ind. Hyg. Assoc. J. 33:201–6.
31. KUSNETZ, J.L. 1960. Air Flow Calibration of Direct Reading Colorimetric Gas Detecting Devices. Amer. Ind. Hyg. Assoc. J. 21:340–1.
32. COLEN, F.H. 1973. A Study of the Interchangeability of Gas Detector Tubes and Pumps, Report TR-71, National Institute for Occupational Safety and Health, Morgantown, W.V., June 15.
33. LEICHNITZ, K. 1973. Effect of Pressure and Temperature on the Indication of Dräger Tubes. Dräger Review 31:1–7 (Sept.), Drägerwerk AG, D-24 Lübeck 1, P.O. 1339 (W. Germany).
34. LINCH, A.L. 1974. Evaluation of Ambient Air Quality by Personnel Monitoring, CRC Press, Inc., Cleveland, Ohio.
35. KITAGAWA, T. 1965, Detector Tube Method for Rapid Determination of Minute Amounts of Nitrogen Dioxide in the Atmosphere. Yokohama National Univ. (July).
36. INFORMATION SHEET No. 44. 1960. 0.5a Nitrous Gas/Detector Tube, Drägerwerk, Lübeck, Germany (November).
37. GROSSKOPF, K.1963. A Tentative Systematic Description of Detector Tube Reactions. (German). Chemiker Zeitung-Chemische Apparatus 87:270–5.
38. LEICHNITZ, K. 1973. Determination of Low SO_2 Concentrations by Means of Detector Tubes. Dräeger Review 30:1–4. Drägerwerk AG, D-24 Lübeck 1, P.O. Box 1339 (W. Germany).
39. LINCH, A.L., H.V. PFAFF, 1971. Carbon Monoxide—Evaluation of Exposure Potential by Personnel Monitor Surveys. Am. Ind. Hyg. Assoc. J. 32:745–52.
40. NATIONAL INSTITUTE OCCUPATIONAL SAFETY. Criteria for a Recommended Standard Occupational Exposure to Carbon Monoxide. 1972. HSM 73-11000, Dept. Health, Educ. and Welfare, Rockville, Md.
41. SALTZMAN, B.E. 1962. Basic Theory of Gas Indicator Tube Calibrations. Am. Ind. Hyg. Assoc. J. 23:112–26.
42. LEICHNITZ, K. 1967. Attempt at Explanation of Calibration Curves of Detector Tubes. (German). Chemiker-Ztg./Chem. Apparatur 91:141–8.
43. SCHERBERGER, R.F., G.P. HAPP, F.A. MILLER and FASSETT, D.W. 1958. A Dynamic Apparatus for Preparing Air-Vapor Mixtures of Known Concentrations. Am. Ind. Hyg. Assoc. J. 19:494–8.
44. SALTZMAN, B.E. 1961. Preparation and Analysis of Calibrated Low Concentrations of Sixteen Toxic Gases. Anal. Chem. 33:1100–12.
45. COTABISH, H.N., P.W. MCCONNAUGHEY, and H.C. MESSER, 1961. Making Known Concentrations for Instrument Calibration. Am. Ind. Hyg. Assoc. J. 22:392–402.
46. HERSCH, P.A. 1969. Controlled Addition of Experimental Pollutants to Air. J. Air Poll. Control Assoc. 19:164–72.
47. DITTMAR, P., and G. STRESE. 1959. The Suit-

ability of Detector Tubes for the Detection of Toxic Substances in the Air. I. Hydrogen Sulfide Detector Tubes. (German). Arbeitsschutz 8:173–7.

48. HESELTINE, J.K. 1959. The Detection and Estimation of Low Concentrations of Methyl Bromide in Air. Pest Technology (England), July/August.
49. KUSNETZ, J.L., B.E. SALTZMAN, and M.E. LANIER,, 1960. Calibration and Evaluation of Gas Detecting Tubes. Am. Ind. Hyg. Assoc. J. 21:361–73.
50. BANKS, O.M., and D.R. NELSON. 1961. Evaluation of Commercial Detector Tubes. Presented at 22nd Annual Meeting, Am. Ind. Hyg. Assoc., Detroit, Michigan, April 13.
51. LANIER, M.B., H.L. KUSNETZ. 1963. Practices in the Field Use of Detector Tubes. Arch. Environmental Health 6:418–21.
52. HAY III, E.B. 1964. Exposure to Aromatic Hydrocarbons in a Coke Oven By-Product Plant, Am. Ind. Hyg. Assoc. J. 25:386–91.
53. LARSEN, L.B., R.H. HENDRICKS. 1969. An Evaluation of Certain Direct Reading Devices for the Determination of Ozone. Am. Ind. Hyg. Assoc. J. 30:620–3.
54. JOINT COMM. ON DIRECT READING GAS DETECTING SYSTEMS. ACGIH-AIHA. 1971. Direct Reading Gas Detecting Tube Systems. Am. Ind. Hyg. Assoc. J. 32:488–9.
55. MORGANSTERN, A.S., R.A. ASH, J.R. LYNCH. 1970. The Evaluation of Gas Detector Tube Systems: I. Carbon Monoxide, Am. Ind. Hyg. Assoc. J. 31:630–2.
56. ASH, R.M., J.R. LYNCH. 1971. The Evaluation of Gas Detector Tube Systems: Benzene. Am. Ind. Hyg. Assoc. J. 32:410–11.
57. ASH, R.M., and J.R. LYNCH. 1971; 1972. The Evaluation of Detector Tube Systems: Sulfur Dioxide. Am. Ind. Hyg. Assoc. J. 32:490–. Ibid 33:11.
58. ASH, R.M., and J.R. LYNCH. 1971. The Evaluation of Detector Tube Systems: Carbon Tetrachloride. Am. Ind. Hyg. Asso. J. 32:552–3.
59. ROPER, C.P. 1971. An Evaluation of Perchloroethylene Detector Tube. Am. Ind. Hyg. Assoc. J. 32:847–9.
60. JOHNSON, B.A., C.P. ROPER. 1972. The Evaluation of Gas Detector Tube Systems: Chlorine. Am. Ind. Hyg. Assoc. J. 33:533–4.
61. JOHNSON, B.A. 1972. The Evaluation of Gas Detector Tube Systems: Hydrogen Sulfide. Am. Ind. Hyg. Assoc. J. 33:811–12.
62. NATIONAL INSTITUTE FOR OCCUPATIONAL SAFETY AND HEALTH. 1973. Certification of Gas Detector Tube Units, Federal Register 38:11458–63 (May 8,); also 43CFR pt. 84.
63. NATIONAL INSTITUTE FOR OCCUPATIONAL SAFETY AND HEALTH. NIOSH Certified Personal Protective Equipment, Morgantown, W.Va., July, 1974. HEW Publication No. (NIOSH) 75-119; also Cumulative Supplement, January, 1975.
64. LEIDEL, N.A., K.A. BUSCH. 1975. Statistical Methods for Determination of Noncompliance with Occupational Health Standards, National Institute for Occupational Safety and Health, Cincinnati, Ohio. (April), HEW Publication No. (NIOSH) 75-159.

27. Fluorescence Spectrophotometry

27.1 BASIC PRINCIPLES

Fluorescence spectrophotometry is the measurement of "fluorescent" light emitted by certain molecules when excited by a radiation source of appropriate energy or wavelength. Since energy is lost in the transition, the fluorescent or secondary light is of lower energy, and consequently longer wavelength, than the exciting or primary light. For this reason, and because intense light sources are available in this region, the ultraviolet and lower visible wavelengths are most useful as a source of excitation. The exciting and fluorescing wavelengths for a given compound are characteristic and permit identification in many cases, particularly when the fluorimeter is of the scanning type.

Organic molecules containing conjugated double bonds (alternating single and double bonds) are the most commonly encountered fluorescing materials. Substitutions in the molecule may alter the excitation or emission spectrum or substantially enhance or reduce fluorescence. Although solids may be fluorescent, quantitative measurements are usually made in solution. Emission intensity is dependent on the total number of excited molecules, and is thus theoretically directly proportional to concentration. This holds true at very low concentrations; however, as concentration increases, absorption by the sample of both primary and secondary light becomes significant, finally resulting in a phenomenon called "concentration quenching." The fluorescence may be visually intense in the first portion irradiated by the primary light but it is absorbed by the sample to such an extent that the light reaching the detector is reduced. The linear plot of concentration *vs*. instrumental emission intensity may thus reach a plateau and return towards the base line. For this reason, it is good practice to observe samples visually, and dilute those which appear to be self-quenching. Alternatively each sample may be routinely analyzed at two different dilutions to check for self-quenching.

27.2 Instrumentation

The basic fluorimeter (Figure 27:1) consists of a primary light source which is directed through an optical filter or monochromator to isolate a specific wavelength for excitation of the sample. The secondary filter or monochromator allows only the wavelength of light due to the fluorescence from the sample to reach the phototube, while blocking out any stray primary light or fluorescence of a different wavelength due to other substances in the sample. Detectors are usually photomultiplier tubes placed at 90° to the primary light source to reduce the possibility of interference from primary light. Variable slits may be employed to control resolving power or sensitivity.

Instruments using only filters for light transmission are referred to as filter fluorimeters while those using monochromators are called spectrofluorimeters. The spectrofluorimeters, which are much higher priced than the filter fluorimeters, offer the advantage of more precise control of wavelength of both primary and secondary light. This permits scanning of samples both for identification and for more precise determination of exciting and fluorescing wavelengths. The value of the more reasonably priced filter fluorimeters should not be overlooked, as they offer excellent sensitivity and reproducibility for routine quantitative work.

Monochromators for spectrofluorimeters are usually diffraction gratings. Quartz prisms may be used for greater dispersion of primary light to obtain better resolution, but at the expense of some loss of intensity.

Light sources for filter type instruments are commonly of the mercury vapor type, which emit line spectra. The most prominent useful lines are 254, 312, 334, 365, 405, and 436 nm. The shorter wavelengths are obtainable only when the mercury lamp is encased in clear quartz or synthetic silica, since glass absorbs over 40% of radiation at 320 nm, and virtually 100% at 254 nm. For the longer wavelengths, glass envelopes may be used, frequently coated or colored to filter out undesirable wavelengths. Mercury lamps require a relatively long warm-up time for stabilization. Their intensity decreases with age, so that continued reference to standards is of great importance. The xenon arc lamp is usually used in spectrofluorimeters, because it is more stable and gives an intense continuous spectrum over the ultraviolet and visible regions. The intensity varies with wavelength, and tends to diminish at short wavelengths, however, this may be corrected for instrumentally.

Where the excitation wavelength is above 320 nm ordinary borosilicate glass tubes may be used as sample containers. At lower excitation wavelengths it is necessary to use synthetic silica cells (usually square) in order to transmit the primary light.

Under optimum conditions fluorimetric methods are among the most sensitive and specific available. Detection limits of less than 1 ng/ml are common.

27.3 Application

Fluorimetry finds most use in detection or determination of organic compounds, although inorganic compounds may be determined by formation of a fluorescent organic complex.

In the field of environmental analysis fluorimetric methods have been used for the determination of coproporphyrin in urine as an index of lead exposure; the determination of uranium in air, soil, or biological materials by fluorescence of a bead obtained by fluxing the sample with sodium-lithium fluoride; and for the identification and determination of polynuclear aromatic hydrocarbons. Due to the extreme sensitivity of detection of the fluorescent dyes rhodamine B and fluorescein, they have been used as tracers in water systems. Uranine has been used as a tracer in air specifically to determine the efficiency of filters for aerosols. Beryllium and selenium have been determined with great sensitivity by fluorimetric methods.

27.4 Factors Affecting Results

- Intensity of primary light
- Band width of primary and secondary light
- Primary and secondary wavelengths used
- Cell depth

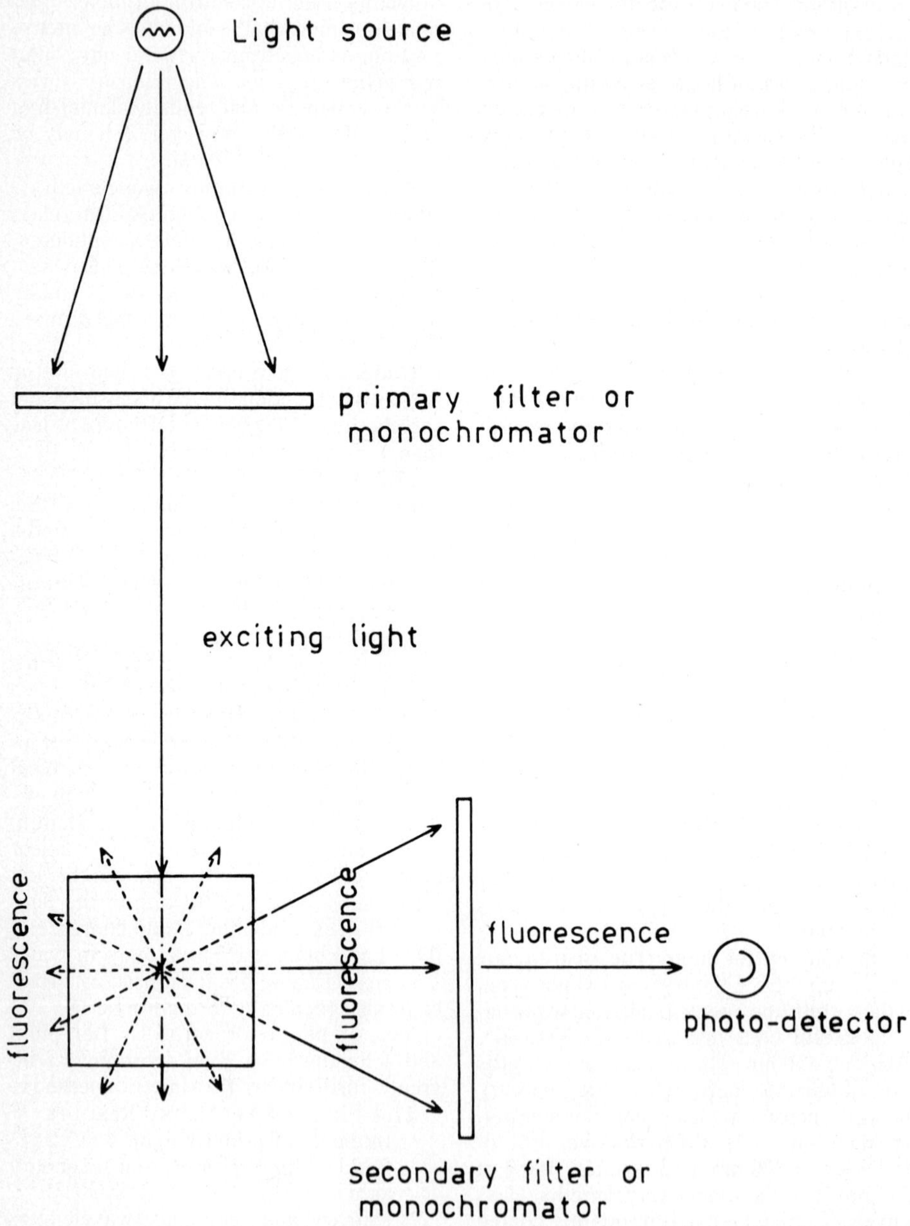

Figure 27:1—Basic Fluoremeter.

pH of sample
Spectral sensitivity of phototube
Sample solvent
Self-absorption
Temperature of sample
Quenching

27.5 ANALYTICAL APPLICATIONS

Fluorimetry is used frequently in biochemistry, as many vitamins, hormones, etc. are fluorescent. It is extremely sensitive for determination of beryllium, selenium, boron, etc. It is widely used for determination of uranium after fluxing with solid NaF-LiF and is useful in filter efficiency studies with uranine aerosol. It is applied to tracer studies: rhodamine-B, fluorescein, in water; uranine in air. It is used for determination of polycyclic aromatic hydrocarbons in air pollution.

27.6 ADVANTAGES AND LIMITATIONS

Fluorimetry is one of the most sensitive methods of analysis. It is, however, relatively non-specific, and fluorescence of many compounds fades rapidly.

References

A. *General*

1. RADLEY, J.A. and J. GRANT. 1959. Fluorescence Analysis in Ultraviolet Light, Chapman and Hall, London, 4th Ed.
2. UDENFRIEND, S. 1962, 1969. Fluorescence Spectra in Biology and Medicine. Academic Press, New York, Vol. I, Vol. II.
3. PARKER, C.A., and W.T. REES. 1962. Fluorescence Spectrometry, Analyst, 87:83.
4. HERCULES, D.M. (Ed.) 1960. Fluorescence and Phosphorescence Analysis, Interscience, N.Y.
5. HOUGHTON, J.A., and G. LEE. 1960 Data on Ultraviolet Absorption and Fluorescence Emission, AIHAJ, 21:219.
6. BERLMAN, I.B. 1971. Handbook of Fluorescence Spectra of Aromatic Molecules, Academic Press, New York, 2nd Ed.
7. KALLMAN, H. and G.M. SPRUCK. 1962. Luminescence of Organic and Inorganic Materials, Wiley, New York.

B. *Specific Analyses*

Beryllium:

8. WELFORD, G.A., and J.H. HARLEY. 1952. AIHAQ, 13:232.
9. WALKLEY, J. 1959. AIHAJ, 20:241.
10. SILL, C.W., and C.P. WILLIS. 1959. Anal. Chem., 31:598.

Boron:

11. WHITE, C.E., ET AL. 1947. Anal. Chem., 19:802.

Selenium:

12. WATKINSON, J.H. 1960. Anal. Chem., 32: 981.

Uranium:

13. CENTANNI, F.A., A.M. ROSS, and M.A. DE SESA. 1956. Anal. Chem., 28:1651.

Polycyclic Hydrocarbons:

14. DUBOIS, L. and J.L. MONKMAN. 1965. Int. J. Air and Water Pollution, 9:131.
15. SAWICKI, E., T.R. HAUSER, and T.W. STANLEY. 1960. Int. J. Air Poll., 2:253.
16. CHAUDET, J.J., and W.I. KAYE. 1961. Anal. Chem., 33:113.
17. AMERICAN PUBLIC HEALTH ASSOCIATION, INC. 1976. Methods of Air Sampling and Analysis. 2nd ed. pp. Washington, D.C.
18. IBID. pp.

28. FILTER MEDIA FOR AIR SAMPLING

28.1 INTRODUCTION

Filtration has become the most widely used technique for aerosol sampling in recent years, primarily because of its low cost and simplicity. The samples obtained usually occupy a relatively small volume, and may often be stored for subsequent analysis without deterioration. By appropriate choice of air mover, filter medium and filter size, almost any sample quantity desired can be collected in a given sampling interval.

Figure 28:1 is a schematic representation of the elements of a filter sampling system. It shows the arrangement of the component parts. These may include either all or some of the following: a sampling nozzle, filter holder, filter, flowmeter, air mover, and a means of regulating the flow. A nozzle is needed only when sampling from a moving stream, *e.g.*, a duct or stack. For these applications, careful attention must be given to its shape, size, and orientation in order to obtain representative samples. The factors affecting the entry of particles into a sampling tube, *i.e.*, particle inertia, gravity, flow convergence, and the inequality of ambient wind and suction velocity, have been critically evaluated by Davies (**1**). Errors can also arise from particle deposition between the probe inlet and the filter due to impaction at the bends and turbulent diffusion (**2**).

The filter should be upstream of everything in the system but the nozzle, so that

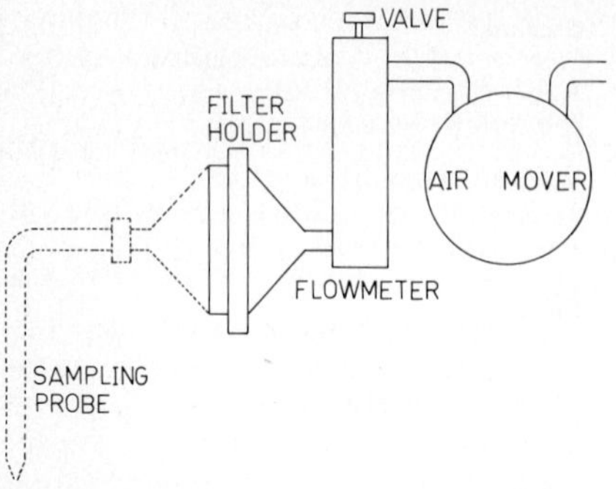

Figure 28:1—Elements of a filter sampling system.

any dirt in the system, manometer, liquid, or pump oil will not be carried accidentally onto it. The filter should be as close as possible to the sampling point, and all sampling lines must be free of contamination and obstructions.

The filter holder, designed for the specific filter size used, must provide a positive seal at the edge. A screen or other mechanical support may be required to prevent rupture or displacement of the filter in service. With a properly designed holder, the air velocity will be uniform across the cross section of the filter holder. Uniform flow distribution is especially desirable when analyses are to be performed directly on the filter.

The accurate measurement of flowrate and sampling time or sample volume is as important as the measurement of sample quantity, since aerosol concentration is determined by the ratio of sample quantity to sampled volume. Unfortunately, air volume measurements are often inaccurate. When the volumetric capacity of the air mover is highly pressure dependent, as it is for turbine blowers, ejectors and some other types of air movers, the flow cannot be metered by techniques which introduce a significant pressure drop themselves. This precludes the use of most meters which require the passage of the full volume through them, and limits the choice to low resistance flowmeters. These include by-pass meters, which measure the flowrate of a small volume fraction of the sampled air, and meters utilizing very sensitive measurements of vane displacement or pressure drop. These can provide sufficiently accurate measurements, but often require more careful maintenance and more frequent calibration and adjustment than they are likely to receive in field use.

Most flowmeters are calibrated at atmospheric pressure, and many require pressure corrections when used at other pressures. Such corrections must be based on the static pressure measured at the inlet of the flowmeter. The flowmeter should be downstream of the filter, to preclude the possibility of sample losses in the flowmeter. It will therefore be metering air at a pressure below atmospheric, due to the pressure drop across the filter. Furthermore, if the filter resistance increases due to loading, as is often the case, the pressure correction will not be a constant factor.

If the sampling flowrate is to be controlled with a throttling valve, this valve should be downstream of the flowmeter to avoid adding to the pressure correction for the flowmeter. Flowrate adjustments can be made either with a throttling valve or by speed control of the air mover motor, and can be either manual or automatic. Automatic control requires pressure or flow transducers and appropriate feed-back and control circuitry.

The discussion which follows was designed to provide the background necessary for the proper selection of filters for particular applications. Filtration theory is outlined, the various kinds of commercial filter media used for air sampling are described, and the criteria which limit the selection in various sampling situations are discussed.

28.2. FILTRATION THEORY

28.2.1 *Collection Mechanisms.* Filters remove particles from a gas stream by a number of mechanisms. These include direct interception, inertial deposition, diffusional deposition, electrical attraction, and gravitational attraction. The one or more of these mechanisms which predominate in a given case will depend on the flowrate, the nature of the filter, and the nature of the aerosol.

a. Direct interception occurs when the radius of a particle moving with a gas stream-line is greater than the distance from the stream-line to the surface. This mechanism is important only when the ratio of the particle size to the void or pore size of the filter is relatively large.

b. Inertial collection results from a change in direction of the gas flow. The particles, due to their relatively greater inertia, tend to remain on their original course and strike a surface. Capture is favored by high gas velocities and dense fiber packing.

A theoretical basis for the effect of fiber packing density has been developed and confirmed experimentally by Stenhouse, *et al.* (**3**) using a model filter with uniform fiber spacing. In commercial fibrous filter media, the fibers are not uniformly spaced. According to Benarie (**4**), the pore size in a given filter has a log-normal distribution, with a σ_g characteristics to the particular filter. The operation of the inertial mechanism in a variety of commercially available fibrous filters has been demonstrated experimentally by Ramskill and Anderson (**5**).

c. Diffusion is most effective for small particles at low flowrates. It depends on the existence of a concentration gradient. Particles diffuse from the gas stream to the surfaces of the fibers where the concentration is zero. Diffusion is favored by low gas velocities and high concentration gradients. The root mean square particle displacement of the particles, and hence the collection efficiency, increases with decreasing particle size. A theoretical basis for predicting diffusional deposition of aerosols in fibrous filters has been developed by Fuchs and coworkers (**6, 7**).

d. Electrical forces may contribute greatly to particle collection efficiency if the filter or the aerosol has a static charge. The flow of air used may induce charges on the filter.

Lundgren and Whitby (**8**) have shown that image forces, *i.e.*, the force between a charged particle and its electrical image in a neutral fiber, can strongly influence particle collection. The factors controlling particle deposition on a filter suspended in a uniform electric field, and the influence of such a field on the deposition of both charged and uncharged particles have been described by Zebel (**9**). Unfortunately, the data needed to predict the effect of electrostatic charges on the collection efficiency of sampling filters are seldom available.

e. Gravitational forces may usually be neglected when considering filter sampling. The settling velocities of airborne particles of hygienic significance are too low, and the horizontal components of the surface areas in the filters too small, for gravitational attraction to have any significant effect on particle collection efficiency, unless the face velocity through the filter is very low, *e.g.*, < 5 cm/sec.

Since a variety of collection mechanisms are involved in filtration, it is not surprising that, for a given aerosol and a given filter, the, the collection efficiency should vary with face velocity and particle size. The efficiency of a given filter for a given particle size could be high at low flows, due primarily to diffusion effects. With increasing velocity, it could first fall off and then, with still higher velocities begin to rise due to increased inertial deposition. This pattern has been observed in several experimental penetration tests (**10, 11**). At very high velocities, the retention could decrease because of re-entrainment. Additional data showing these effects are presented in Table 28:I.

Filter retention by the impaction and diffusion mechanisms are also strongly influenced by particle size as shown by experimental data from Spurny, *et al.*, (**12**) for Nuclepore filters with 5 μ diameter pores at a face velocity of 5 cm/sec. Nuclepore filters have a very different structure than other types of filters, as will be discussed, and exhibit a more extremen size dependence. Rimberg (**10**) has demonstrated experimentally that there are sizes for maximum penetrations in IPC 1478 and H-V 5G fibrous filters, and that these sizes increase with decreasing face velocity.

28.2.2 *Forces of Adhesion and Re-Entrainment.* The collection mechanisms discussed above act to arrest the motion of the particles in a gas stream as the gas

Table 28:I. Flow Rate and Collection Efficiency Characteristics of Selected Air Filter Media[a]

Filter Type	Filter	Characteristics at Indicated Face Velocities (cm/sec)							Flow Reduction Due to Loading %/M³/cm²
		mm Hg Pressure Drop			Percent Penetration of 0.3 μ DOP				
		53	106	211	26.7	53	106	211	
Cellulose	Whatman 1	86	175	350	7	0.95	0.061	0.001	17.9
	41	36	72	146	28	16	2	0.30	5.0
	541	30	61	123	56	40	22	9	10.4
	IPC 1478	1.5	3	5.5	90	90	90	85	<<0.1
Cellulose-Asbestos	H-V H-70 (9 mil)	64	127	254	1.8	0.8	0.20	0.05	1.7
Cellulose-Glass	H-V 5-G	5	10	21	32	32	26	16	0.20
Glass	MSA 1106BH	30	61	120	0.068	0.048	0.022	0.005	0.43
	Gelman A	33	65	129	0.019	0.018	0.011	0.001	0.50
	E	28	57	114	0.036	0.030	0.014	0.004	0.53
	Hurlbut 934AH	37	74	150	0.010	0.006	0.003	0.001	0.47
	Whatman GF/A	29	60	118	0.018	0.015	0.008	0.001	0.37
Polystyrene	Delbag Microsorban	44	89	176	0.45	0.40	0.20	0.05	0.29
Membrane	Millipore AA (0.8 μ)	142	285	570	0.015	0.020	—	—	1.6
	Polypore AM-1 (5 μ)	23	46	95	12	8	5	2	2.4
	AM-3 (2 μ)	84	190	380	0.36	0.22	0.090	0.015	3.1

[a]Data extracted from NRL Report No. 6054[11].

flows through the voids of a filter. The particles removed from the gas stream are then subject to forces of adhesion. If the forces of adhesion on a particle are greater than the forces which tend to push the particle free, then that particle is "collected," and will be available for analysis. However, the forces exerted on the particle by the flowing gas stream may be greater than the forces of adhesion, resulting in re-entrainment of the particle. At present, it is at least as difficult to predict forces of adhesion from theoretical considerations as it is to predict the effectiveness of the collection mechanisms. One reason is that it is usually not possible to determine whether particles which penetrate a filter were blown off after collection due to inadequate adhesion, or whether they underwent elastic rebound upon initial contact with the filter fibers. This question has been theoretically and experimentally investigated by Löffler (**13**). He concluded that the measured forces of adhesion were in good agreement with the Van der Waals forces calculated theoretically and that the flow velocity required for blowing collected particles off fibers is much higher than those normally used in air filtration. An increase in particle penetration with increasing velocity will usually be due to increased rebound or to the resuspension of particle flocs rather than individual particles.

28.3 Commercial Filter Media

Filter media of many different types and with many different properties have been designed for or adapted to air sampling requirements. For purposes of discussion, they have been divided into groups determined by their composition. Air flow resistance and collection efficiency characteristics of some commonly used filters are tabulated in Table 28:I. Table 28:II summarizes the physical characteristics of commercially available filter media based on vendor supplied or approved data. These media have been subdivided on the basis of their composition and/or structure.

28.3.1 *Cellulose Filter Papers.* Most filters of this type are used primarily by analytical chemists for liquid-solid separations. They are made of purified cellulose pulp, are low in ash content, and are usually less than 0.25 mm thick. These papers are relatively inexpensive, are obtainable in an almost unlimited range of sizes, have excellent tensile strength, show little tendency to fray during handling, and are universally obtainable. Their disadvantages include non-uniformity, resulting in variable flow resistance and collection efficiency, and hygroscopicity, which makes accurate gravimetric determinations very difficult.

For air sampling, Whatman No. 41 is the most widely used filter paper of this type. It has the advantages of low cost, high mechanical strength, and high purity typical of these papers and in addition, has a moderate flow resistance. This group also includes hardened papers, such as Whatman No. 50, from which collected particles can be removed by washing. A pleated cellulose filter (MSA Type "S"), described by Silverman and Viles (**14**), is often used for high volume sampling. Concentric fluting permits exposure of about 60 square inches of surface within a total diameter of 4 inches. The pressure drop across such filters is low (16.5 cm H_2O at Vf = 50 cm/sec.) as a result of the large filtering area and low face velocity. Counterbalancing these advantages is the large bulk of the paper, and its high and variable ash content (about 1.3%). These filters are fabricated with an organic binder which makes them unsuitable for analytical work involving organic solvents, and the paper's hygroscopicity prevents accurate gravimetric analysis.

28.3.2 *Glass Fiber Filters.* Glass fiber filters are in most cases more expensive and have poorer mechanical properties than cellulose papers. They also have many advantages, *i.e.*, reduced hygroscopicity, ability to withstand higher temperatures, and higher collection efficiencies at a comparable pressure drop. These properties, combined with the ability to make benzene, water and nitric acid extracts from particulates collected on them led to the selection of a high efficiency glass fiber filter as the standard collection medium

Table 28:II. Summary of Air Sampling Filter Characteristics

Filter	Void Size μ	Fiber Diam. μ	Thick-ness μ	Weight / Area mg/cm²	Ash Content %	Max. Oper. Temp. °C	Tensile Strength gm/cm	ΔP‡ 100 in H_2O	Source
A. CELLULOSE FIBER FILTER CHARACTERISTICS									
Whatman 1	2+	NA	180	8.7	0.06	150	4700	40.5	RA
4	4+	NA	200	9.2	0.06	150	NA	11.5	RA
32	1−	NA	200	10.0	0.025	150	NA	NA	RA
40	2	NA	200	9.5	0.01	150	4600	54	RA
41	4+	NA	190	9.1	0.01	150	4600	8.1	RA
42	>1	NA	200	10.0	0.01	150	NA	NA	RA
44	>1	NA	180	8.0	0.01	150	NA	NA	RA
50	1	NA	120	10.0	0.025	150	NA	NA	RA
541	4+	NA	160	7.6	0.008	150	7600	NA	RA
S and S 589 Green Ribbon	NA	NA	330	NA	0.008	NA	NA	NA	S & S
MSA									
Type S	NA	NA	1000	NA	NA	120	NA	NA	MSA
Cellulose	NA	NA	1000	NA	NA	120	NA	NA	MSA
Corrugated Cellulose	NA	NA	1000	NA	NA	120	NA	NA	MSA
IPC 1478	NA	av. 17	760	14.6	0.04	120	Woven Backing	0.31	IPC
B. GLASS FIBER FILTER CHARACTERISTICS									
MSA 1106B*	NA	NA	180–270	6.1	~95	540	625	19.8	MSA
1106BH†	NA	NA	180–460	5.8	~100	540	270	19.8	MSA
Gelman A†	NA	NA	380	9.3	NA	480	NA	18.9	G
E*	NA	NA	380	10.0	NA	480	NA	18.9	G
Whatman GF/A†	<1	0.5–0.75	250	5.2	100	540	500	NA	RA
GF/B†	<1	0.5–0.75	640	15.0	100	540	1000	NA	RA
GF/C†	<1	0.2–0.5	260	5.5	100	540	500	NA	NA
Reeve Angel 934AH†	NA	NA	180–320	5.7	100	540	180	24.4	RA
H & V H-93	NA	0.6	460–560	9.3	96–99	540	450	14.0	H & V
H-94	NA	0.5–3	380	8.2	96–99	480	450	NA	H & V
S & S 23*	NA	NA	320	NA	NA	540	470	NA	S & S
25†	NA	NA	320	NA	NA	540	280	NA	S & S
Dräger G100	NA	0.3–0.8	500	10	97	500	300	14	DW

C. MIXED FIBER FILTER CHARACTERISTICS

Filter	Composition	Void Size μ	Fiber Diam. μ	Thick-ness μ	Weight / Area mg/cm^2	Ash Content %	Max. Oper. Temp. ° C	Tensile Strength gm/cm	$\Delta P\ddagger_{100}$ in H_2O	Source
H & V										
H-70, 9 mil	Cellulose-asbestos	NA	0.1–35	230	8.2	20–25	150	450	21.3	H & V
H-70, 18 mil	Cellulose-asbestos	NA	0.1–35	460	15.4	20–25	715	715	26	H & V
H-64	Cellulose-asbestos	NA	0.1–35	830–1090	22.7	15–20	360	360	15	H & V
H-90	Cellulose-glass	NA	9–35	685	13.4	70	150	575	0.50	H & V
H-91	Cellulose-glass	NA	1.5–35	710	13.5	80	150	625	1.10	H & V
N-15	Synthetic fiber and glass	NA	0.5–15	1270	24.9	15	150	180	9.9	H & V
5-G	Synthetic fiber, glass and cotton	NA	0.5–15	685	14.5	4–6	150	1310 Gauze	2.0	H & V
MSA										
Glass and cellulose	Glass and cellulose	NA	NA	1000	NA	NA	120	NA	NA	MSA
Whatman ACG/B	Glass, cellulose and activated carbon	<1	NA	900	19.5	NA	150	500	NA	RA

D. PLASTIC FIBER FILTER CHARACTERISTICS

Microsorben	Polystyrene	NA	0.6–0.8	1500	24.9	<0.10	96	150	13	DL

Table 28:II. (continued)

E. MEMBRANE FILTER CHARACTERISTICS

Filter	Composition	Pore Size μ	Thick-ness μ	Weight / Area mg/cm²	Ash Content %	Max. Oper. Temp. °C	Tensile Strength psi	Refractive Index	ΔP‡₁₀₀ in H_2O	Source
Millipore										
SC	Mixed	8.0	130	5.2	<0.0001	125	175	1.515	20	M
SM	Cellulose	5.0	130	2.8	<0.0001	125	160	1.495	32	M
SS	Esters	3.0	150	3.0	<0.0001	125	150	1.495	56	M
RA§		1.2	150	4.2	<0.0001	125	300	1.512	75	M
AA§, **		0.80	150	4.7	<0.0001	125	350	1.510	102	M
DA§		0.65	150	4.8	<0.0001	125	400	1.510	112	M
HA§		0.45	150	4.9	<0.0001	125	450	1.510	250	M
PH		0.30	150	5.3	<0.0001	125	500	1.510	300	M
GS		0.22	135	5.5	<0.0001	125	700	1.510	450	M
VC		0.10	130	5.6	<0.0001	125	800	1.500	2290	M
VM		0.05	130	5.7	<0.0001	125	1000	1.500	3610	M
VS		0.025	130	5.8	<0.0001	125	1500	1.500	5100	M
WS	Cellulose ester-nylon reinforced	3.0	130	4.9	<0.0001	125	150	NA	158	M
WH		0.45	130	5.7	<0.0001	125	450	NA	250	M
BS	Polyvinyl chloride	2.0	135	4.7	NA	65	430	1.528	112	M
BD		0.6	135	5.7	NA	65	740	1.528	280	M
EA	Cellulose acetate	1.0	130	5.2	NA	75	700	1.470	102	M
EH		0.5	130	5.4	NA	75	800	1.470	224	M
EG		0.2	130	5.7	NA	75	900	1.470	448	M
NC	Nylon	14.0	150	5.4	NA	75	315	1.515‡‡	8.6	M
NS		7.0	150	5.3	NA	75	350	1.515‡‡	22	M
NR		1.0	150	5.8	NA	75	680	1.515‡‡	86	M
LC	Teflon	10.0	125	8.0	NA	260	250	NA	125	M
LS		5.0	125	8.0	NA	260	150	NA	187	M

Filter	Composition	Pore Size μ	Thick-ness μ	Weight / Area mg/cm^2	Ash Content %	Max. Oper. Temp. °C	Tensile Strength psi	Refractive Index	$\Delta P\ddagger_{100}$ in H_2O	Source
Metricel										
GA-1	Cellulose triacetate	5.0	140	NA	NA	150	NA	NA	NA	G
3		1.2	140	NA	NA	150	NA	NA	NA	G
4		0.8	140	NA	NA	150	NA	NA	NA	G
6§		0.45	140	NA	NA	150	NA	NA	NA	G
7§		0.3	140	NA	NA	150	NA	NA	NA	G
8		0.2	140	NA	NA	150	NA	NA	NA	G
9		0.1	140	NA	NA	150	NA	NA	NA	G
P.E.M.		0.0075	100	NA	NA	150	NA	NA	NA	G
VM 1	Polyvinylchloride	5.0	140	NA	NA	68	NA	NA	NA	G
4§		0.8	140	NA	NA	68	NA	NA	NA	G
6§		0.45	140	NA	NA	68	NA	NA	NA	G
Green 4N‖	Mixed Cellulose esters	0.8	140	NA	NA	125	NA	NA	NA	G
6N‖		0.45	140	NA	NA	125	NA	NA	NA	G
Black 4N§		0.8	140	NA	NA	125	NA	NA	NA	G
6N§		0.45	140	NA	NA	125	NA	NA	NA	G
Bac-T-Flex										
B1	Nitro-cellulose	8	170	6.3–8.7	0.01	100	NA	NA	NA	S & S
B2		3	170	5.6–7.3	0.008	100	NA	NA	NA	S & S
B2-1		1.2	160	5.6–7.3	0.008	100	NA	NA	NA	S & S
B3		0.8	160	5.6–7.3	0.008	100	NA	NA	NA	S & S
B9*		0.45	140	NA	NA	100	NA	NA	NA	S & S
Selas										
FM 5.0	Silver	5.0	50	32	NA	200	NA	NA	NA	SF
FM 3.0		3.0	50	32	NA	200	NA	NA	NA	SF
FM 1.2		1.2	45	32	NA	200	NA	NA	NA	SF
FM 0.8		0.8	45	32	NA	200	NA	NA	NA	SF
FM 0.45		0.45	45	32	NA	200	NA	NA	NA	SF
FM 0.2		0.2	45	32	NA	200	NA	NA	NA	SF

Table 28:II. (continued)

F. NUCLEPORE FILTERS

Filter	Composition	Pore Size μ	Thick-ness μ	Weight / Area mg/cm²	Ash Content %	Max. Oper. Temp. °C	Tensile Strength psi	Refractive Index	$\Delta P\ddagger_{100}$ in H_2O	Source
Nuclepore										
8 μ	Polycarbonate	8.0	10	1	0.7	140	20,000	1.584 & 1.614	6	GE
5 μ		5.0	10	1	0.7	140	15,000	1.584 & 1.614	7.5	GE
2 μ		2.0	10	1	0.7	140	10,000	1.584 & 1.614	11	GE
1 μ		1.0	10	1	0.7	140	8,000	1.584 & 1.614	30	GE
8/10 μ		0.8	10	1	0.7	140	5,000	1.584 & 1.614	50	GE
5/10 μ		0.5	10	1	0.7	140	3,000	1.584 & 1.614	75	GE

G. FILTER THIMBLE CHARACTERISTICS

Designation	Composition	Size	Void Size μ	Max. Oper. Temp. °C	Source	Remarks
D 1013	Cellulose	43 × 123 mm	NA	120	WP	Use with WP's D1012 Paper Thimble Holder
D 1016	Glass Cloth	2-3/16 × 14″	NA	400	WP	Use with WP's D1015 Glass Cloth Thimble Holder
RA-98	Alundum	NA	Standard	High	WP	Use with WP's D1021 Alundum Thimble Holder
RA-360	Alundum	NA	Fine	High	WP	Use with WP's D1021 Alundum Thimble Holder
RA-84	Alundum	NA	Extra Fine	High	WP	Use with WP's D1021 Alundum Thimble Holder
S & S 603	Cellulose	from 6 × 60 mm to 100 × 180 mm	NA	120	S & S	

Designation	Composition	Size	Void Size μ	Max. Oper. Temp. °C	Source	Remarks
S & S 703	Sintered Glass Fiber	from 15 × 80 mm to 55 × 180 mm	NA	300	S & S	
Whatman	Cellulose	from 10 × 50 mm to 90 × 200 mm	NA	120	RA	

*with organic binder
†without organic binder
‡pressure drop at face velocity of 100 ft/min (~ 50 cm/sec)
§available with or without imprinted grid lines
||available with or without imprinted grid lines
**available with black color
‡‡filters become translucent but not transparent at this index
NA Information Not Available or Not Applicable

H. SOURCES

Symbol	Address
DL	Delbag-Luftfilter, 11-16 Schweidnitzer Strasse, 1 Berlin 31 (Halensee), West Berlin
DW	Drägerwerk, Moislinger Allee 53/55, Lübeck, West Germany
G	Gelman Instrument Company, P.O. Box 1448, Ann Arbor, Michigan 48106
GE	General Electric Company, P.O. Box 846, Pleasanton, California 94566
H & V	Hollingsworth and Vose Company, East Walpole, Massachusetts 02032
IPC	The Institute of Paper Chemistry, Appleton, Wisconsin
M	Millipore Corporation, Bedford, Massachusetts 01730
MSA	Mine Safety Appliances Company, 201 N. Braddock Avenue, Pittsburgh, Pennsylvania 15208
RA	H. Reeve Angel and Company, Inc., 9 Bridewell Place, Clifton, New Jersey
SF	Selas Flotronics, Spring House, Pennsylvania 19477
S & S	Schleicher and Schuell Inc., 543 Washington Street, Keene, New Hampshire 03431
WP	Western Precipitation Division of Joy Manufacturing Co., 1000 W. Ninth Street, Los Angeles, California 90054

for high-volume samplers in the U.S. Public Health Service's National Air Sampling Network (NASN).

As described by Pate and Tabor (**15**), a large number of tests are routinely performed on NASN filters. Nondestructive tests, e.g., weighing, gross β-activity, and reflectance are performed prior to the chemical extractions. Also, portions of the filters are stored untreated for possible use at later times to obtain background data on air concentrations whose need was not anticipated at the time of sample collection.

The types of chemical analyses that can be performed on extracts from the NASN filters are determined by the sensitivity of the analyses and by the magnitude and variability of the extractable filter blank for the particular ion or molecule involved. One of the major continuing problems in the operation of the Network has been the variability in the composition and properties of the glass fiber filters used. The quality control requirements of NASN have proved to be beyond the capability of the filter vendor. The filter characteristics are determined by the process variables in at least four production stages, i.e., the production of the glass, the production of glass fiber from the bulk glass, the production of the fiber mat from the glass fiber, and the packaging of the individual filters. In the case of the MSA 1106BH filter, the most widely used for NASN and related applications, Pate and Tabor (**15**) describe four different types produced sequentially between 1956 and 1962, which differed in softening temperature, and chemical composition and extractability, and which resulted from manufacturing changes beyond MSA's control.

Pate and Tabor (**15**) present data on the physical and chemical properties of MSA, Gelman, Reeve-Angel, and S&S papers more extensive in some respects than those presented in Table II. They concluded that all of these media possess analytical characteristics suitable for most sampling programs, and that the MSA 1106BH and the Gelman Type A were preferred for the analytical program of NASN. They also conclude that the user should maintain his own quality control program.

One of the determinations made by NASN is gross mass of particulate by gravimetric analyses. Many other investigators use the same types of high-volume sampler and filters for routine monitoring and analyze only for gross mass concentration. The validity of these determinations is suspect. The potential errors arising from inaccurate sample volume determinations, from inadequate temperature and humidity conditioning prior to weighing, and from the precision of the weighing procedure are well known and have been discussed by Kramer and Mitchel (**16**). An additional serious source of error is the loss of filter fibers drawn through the support screen into the air mover during sample collection. The flash-fired binder-free filters used by NASN are soft and friable and the loose filter content is variable. Some NASN filters returned from the field have lower than tare weights despite the presence of visible deposits on the filter face. If gross mass concentration analyses are to be performed, other non-hygroscopic filter media, which are both mechanically strong and efficient should be used, or the filter should include a backing layer which will prevent the loss of filter fibers.

All of the preceding discussion applied to glass fiber filters which are virtually 100% efficient for all particle sizes. For some applications, e.g. a filter-pack sampler designed to provide data on particle size distribution, less efficient glass fiber filters may be desirable. Shleien, Cochran and Friend (**17**) have described the physical and collection efficiency characteristics of four less efficient glass fiber filters produced for gas cleaning and air conditioning applications, which they selected for their filter pack.

28.3.3 *Mixed Fiber Papers.* This group includes cellulose-asbestos, cellulose-glass, and glass-asbestos mixtures. Filters of this type find extensive application in air cleaning, where their characteristics of high collection efficiency and low pressure drop are especially important. However, since it is extremely difficult to remove collected dust from these media, they have limited value for air sampling. The high ash content resulting from the mineral components of these filters often interferes with chemical analysis of the deposited material. Mixed fiber filters are used for sampling when simple gravimetric analyses are to be performed and also when sampling radioactive particles where the activity can be counted without removing the sample from the filter.

28.3.4 *Plastic Fiber Filters.* Plastic fiber filters have also been used for air sampling applications. The most widely used of these has been the Microsorban **(18)** filter, made of mats of polystyrene fibers of submicron diameter. Its flow resistance is relatively low, being comparable to Whatman No. 41, while its efficiency of collection is relatively high, i.e., comparable to that of glass and cellulose-asbestos filters. Polystyrene filters are soluble in aromatic hydrocarbon solvents. Their mechanical strength is poor, and they must be well supported by a firm backup in the filter holder.

Air sample filters composed fo PVC fibers of micron size have been described by Berka **(19)**.

28.3.5 *Membrane Filters.* Filters consisting of porous membranes can be used for many applications where fibrous filters cannot. Organic membranes are produced by the formation of a gel from an organic colloid, with the gel in the form of a thin ($\sim$ 150 μ) sheet with uniform pores. Membrane filters made from cellulose nitrate achieved widespread use for air sampling in the early 1950's as discussed by Paulus, et al. **(20)**. In recent years, membrane filters made of cellulose triacetate, regenerated cellulose, polyvinyl chloride, nylon, polypropylene, teflon, and silver have become available. Silver membranes are produced by a different technique and will be discussed separately at the end of this section.

Cellulose nitrate and cellulose triacetate membranes are the most widely used and, as indicated in Table 28: II, are available in the widest range of pore sizes.

The mass of these filters is very low and their ash content is negligible. Some are completely soluble in organic solvents. Cellulose nitrate filters dissolve in methanol, acetone, and many other organic solvents. Cellulose triacetate, nylon, and PVC filters dissolve in fewer solvents, while filters composed of teflon and regenerated cellulose do not dissolve in common solvents. The ability to completely dissolve a filter in a solvent permits the concentration of the collected material within a small volume for subsequent chemical and/or physical analyses.

Collection efficiency increases with decreasing pore size, but even the large pore size filters have relatively high collection efficiencies for particles much smaller than their pores. Membrane filters do not behave at all like sieves. As in fibrous filters, particles are removed primarily by impaction and diffusion. Early investigators believed that electrostatic forces played a major role in particle deposition in membrane filters, but experimental studies by Spurny and Pich **(21)** and Megaw and Wiffen **(22)** demonstrate that diffusional and impaction deposition account for most of the observed collection and that the contribution of direct interception and electrostatic deposition, if present, are less important.

Membrane filters differ from fibrous filters in that a much greater proportion of the deposit is concentrated at or close to the front surface. Lindeken, et al., **(23)** and Lössner **(24)** have measured the penetration depth using test aerosols tagged with alpha emitters. Lindekin, et al., were interested primarily in the use of the filters for measuring the concentration of α-emitters in air. If the deposit were truly at the surface, there would be no need for correcting for differences in distance from the detector face or for absorption of α-energy in the filter. They found that, on a microscopic scale, the filter surfaces were not smooth, and that the surface roughness varied among different brands and for Millipore Co. filters, from the front surface to the back. They concluded that the smooth face of an SM Millipore was suitable for their application. Lössner **(24)** demonstrated the effect of pore size and face velocity on penetration depth for 0.55 μ SiO_2 particles.

The fact that particle collection takes place at or near the surface of the filter accounts for most of the advantages of membrane filters and also some of their disadvantages. The advantages arising from this property are:

a. It is possible to examine solid particles microscopically without going through a transfer step which might change the state or form of the particles. Examination can be by optical microscopy using immersion oil having the same index of refraction as the filter. The oil renders the filter transparent to light rays. Electron microscopy can be performed on a replica of the filter surface produced by vacuum evaporation techniques.

b. Direct measurements of the deposit can be made on the surface without interference caused by absorption in the filter itself. This is advantageous in radiometric counting of air dust, and in soiling index measurements made by reflectance.

c. Autoradiographs of radioactive particles can be produced by a technique whereby photographic emulsion is placed in contact with the membrane filter sample **(25)**.

The disadvantage arising from surface collection is that the amount of sample which can be collected is limited. When more than a single layer of dust particles is collected on a membrane filter, the resistance rapidly increases and there is a tendency for the deposit to slough off the paper.

Silver membranes have become available within recent years and air sampling applications are beginning to appear in the literature. They are made by sintering uniform metallic silver particles, and possess a structure basically similar to that of the organic membranes previously described. They have a uniform pore size, and for a given pore size, about the same flow characteristics. For filters up to 47 mm in diameter, they are 50 μ thick. According to the manufacturer, the membrane is an integral structure of permanently interconnected particles of pure silver, contains no binding agent or fibers, and is resistant to chemical attack by all fluids which do not attack pure silver. Thermal stability extends from −200 to +700 F. (−129 to 407 C).

Richards, Donovan and Hall **(26)** describe the use of silver membranes for sampling coal tar pitch volatiles. Other filter media evaluated were not suitable because of the high weight losses of blank filters in the benzene extraction step in the analysis, including 1106BH glass, cellulose acetate membrane, and Whatman 41 cellulose. The weight loss for the silver membrane was negligible. Another application of silver membranes is for sampling airborne quartz for x-ray diffraction analysis, as described by Knauber and VonderHeiden **(27)**. Instruments satisfying the new ACGIH criteria for respirable dust samplers operate at low flowrates, and the sample masses on the backup filters are too small for conventional analyses. Using silver membranes, the X-ray diffraction background is very consistent, and quartz determinations have a lower limit of sensitivity of 0.06 mg.

28.3.6 *Nuclepore Filters.* Nuclepore filters are similar to membrane filters in that both contain uniform sized pores in a solid matrix. However, they differ in structure and method of manufacture. Nuclepore filters are made by placing 10 μ thick polycarbonate sheets in contact with sheets of uranium into a nuclear reactor. The neutron flux causes U-235 fission, and the fission fragments bore holes in the plastic. Subsequent treatment in an etch solution enlarges the holes to a size determined by the temperature and strength of the bath, and the time within it. Commercial Nuclepore (R) filters are available with pore diameters between 0.2 and 8 μ.

Nuclepore filters possess many of the attributes erroneously attributed to membrane filters in earlier days. They have a smooth filtering surface, the pores are cylindrical, almost all uniform in diameter, and perpendicular to the filter surface. Also, the filters are transparent, even without immersion oil.

The structure and air paths are so simple that, as demonstrated by Spurny and Lodge (**12**), it is possible to predict their particle collection efficiency on the basis of measured dimensions and basic particle collection theory.

Although their pore volume is much lower, Nuclepore filters have about the same flowrate-pressure drop relations as membrane filters of comparable pore diameter. However, the filter penetrations at 5 cm/sec, reported by Spurny and Lodge, are much greater than those of membrane filters with the same pore sizes. Nuclepore filters have a lower and more uniform weight, and since they are non-hygroscopic, they can be used for sensitive gravimetric analyses.

The polycarbonate base is very strong and Nuclepore filter tapes do not require extra mechanical backing. They can be analyzed by light transmittance, or filter segments can be cut from the tape for microscopy.

The very smooth surface makes Nuclepore a good collector for particles to be analyzed by electron microscopy. Spurny et al. (**28**) show high resolution electron micrographs made from silicon monoxide replicas of the filter surface.

28.3.7 *Granular Beds.* Ground crystals of salicylic acid, sugar, and naphthalene have been used as aerosol filters. The particles are recovered either by volatilization or solution of the crystals. One difficulty with this type of filter is the high impurity level of most available crystals. The efficiency of these filters depends on the size of the crystals, the depth of the bed, etc. Usually, low flowrates are necessary in order to obtain high efficiencies through diffusional separation.

28.3.8 *Filters Occasionally Used For Air Sampling.*

a. Respirator filters. Respirator filters of felt and/or cellulose fiber can be, and have been, used for air sampling. In many of them, the filter is manufactured in a pleated form, which increases the surface area without increasing the overall diameter. Filters of this type have the same advantages and disadvantages as the mixed fiber filters previously discussed.

b. Filter thimbles. These thimbles are available in glass fiber, paper and cloth. They are sometimes filled with loose cotton packing to reduce clogging. The advantage is that large samples can be collected.

c. Alundum thimbles and sintered glass filters are manufactured with a variety of porosities. They have considerably higher resistance to air flow than comparable paper filters, but can be used for very high temperature sampling.

28.4 Filter Selection Criteria

28.4.1 *General Considerations.* The selection of a particular filter type for a specific application is invariably the result of a compromise among many factors. These factors include cost, availability, collection efficiency, the requirements of the analytical procedures, and the ability of the filter to retain its filtering properties and physical integrity under the ambient sampling conditions. The increasing variety of commercially available filter media sometimes makes the choice seem somewhat more difficult, but more importantly, increases the possibility of a selection which satisfies all important criteria.

28.4.2 *Efficiency of Collector.* Before discussing experimental efficiency data, it is important that a distinction be made between particle collection efficiency and mass collection efficiency. The former refers to fractions of the total number of particles, while the latter refers to fractions of the total mass of the particles. These efficiencies will be numerically equivalent on ly when all the particles are the same size, as in some laboratory investigations of filter efficiency. In almost all other cases, the mass collection efficiency will be larger than the corresponding particle collection efficiency. When sampling for total mass concentration of particulate matter, or for the mass concentration of a component of an aerosol, the efficiency of interest is mass efficiency. Submicron particles often contribute only a small fraction of the total mass of an industrial dust, even when they represent the majority of the particles. Therefore, it is not always essential that an air sampling filter have a high efficiency for the smallest particles. Insistence on high efficiency for all size particles may restrict the selection to media with other limitations, such as high flow resistance, high cost, and fragility.

Collection efficiency data for a variety of filter media are given in Table I for 0.3 μ diameter DOP droplets at various face velocities (**11**). This is a commonly used particle size for a test aerosol, since it is close to the size for maximum filter penetration for many commonly used sampling media operating at representative flowrates. On this basis, it is reasonable to assume that penetration of both smaller and larger particles would be lower, i.e., the collection efficiency would be higher.

On the basis of the data shown in Table I, it can be seen that the same paper can be inefficient at some face velocities and highly efficient at others. For example, Whatman-41 penetration below 26 cm/sec exceeds 28%, while at 100 cm/sec it is only about 2%, and at higher flowrates it is much less than that. This filter is often used in industrial hygiene surveys with both low and high volume samplers. When sampling with a 25 mm filter head at 25 lpm, the face velocity (based on an effective filtration area of 3.68 cm^2) is 113 cm/sec. When sampling with a 102 mm (4 inch) filter head at 500 lpm (17.7 cfm), the face velocity (based on effective filtration area of 60 cm^2) is 139 cm/sec. On the other hand, when sampling at lower flowrates, as in personal air samplers, Whatman-41 would not be a good choice. With a 25 mm filter head and a flowrate of 2.5 lpm, the face velocity would only be 11.3 cm/sec. For such an application other papers, more efficient at this flowrate, would be preferred.

There is need for caution in interpreting filter efficiency data in the literature. The most reliable data appear to be those of Rimberg (**10**) and Lockhart et al. (**11**), which are in reasonably good agreement with one another, for both Whatman-41 and TFA-41, which is Whatman-41 packaged and sold by the Staplex Company. The differences between the two sets of data are presumably the differences to be expected from randomly selected batches. Smith and Surprenant (**29**) obtained data based on the same technique as the data of Lockhart, i.e., lightscattering measurements of 0.3 μ DOP droplets, but their results show a large discrepancy.

Rimberg (**10**) measured the penetration of charge neutralized polystyrene latex spheres using a lightscattering photometer. The 0.3 μ points are actually interpolated from the corresponding data for 0.365 and 0.264 μ particles. Lindekin et al. (**30**) used a similar technique except that they did not neutralize the electrical charge on their polystyrene test aerosols. Their results are similar to those of Posner (**31**), who used a fluorescein dye test aerosol with a mass median diameter of 0.27 μ, and determined penetration by comparing the dye collections on a test filter and efficient back-up filter. The aerosol was produced with an atomizer-impactor generator and the electrical charge on the dry residue particles was only partly neutralized. Thus, the data of Lindekin, et al. and Posner appear to reflect the influence of particle charge on filter penetration.

In interpreting filter efficiency data, it is also important to consider that the test data are usually based on the efficiency of a "clean" filter. For most filters, collection efficiency increases with the accumulation of solid particles on the filter surfaces. The resistance to flow also increases with increasing loading, but usually at a much slower rate. A theoretical basis for these phenomena has been developed by Davies (**32**). A practical implication is that even with reliable published filter efficiency and aerosol size distribution data it is not possible to know precisely what the collection efficiency of a filter will be for a given sampling interval. The filter efficiency data can only provide an estimate of the minimum collection efficiency. The actual collection efficiency will usually be higher.

28.4.3 *Requirements of Analytical Procedures.*

a. Sample Quantity. In many instances the limited sensitivity of an analytical method, when combined with a low aerosol concentration, makes it necessary for large volumes of air to be sampled in order to collect sufficient material for an accurate analysis. In addition to the material being studied, background dust and co-contaminants must, unavoidably, also be collected. Therefore, it is highly desirable that the filter medium selected have the capacity to collect and retain large sample masses. Furthermore, it is usually desirable to have the sampling rate nearly uniform over the length of the sampling period. The flow resistance of all filters increases with increased loading, but some do so at much lower rates than others. Table I shows the rates of resistance increase for a variety of filters when sampling the ambient air outside the Naval Research Laboratory. The loading rate would certainly differ for other aerosols and these data would not be generally applicable. However, they do indicate the relative loading characteristics of these filters. Those with low values load much more slowly than those with high values. Those filters with the lowest resistance build-up rate are most useful for collecting high-volume samples, especially when using the pressure-sensitive turbine-type blowers as air movers. In general, deep-bed fibrous filters have the lowest rates of resistance pressure increase.

b. Sample Configuration. Some analyses require that the sample be collected or mounted in a particular form. For example, microscopic particle size analysis can be performed only when the particles are on a flat surface. This is due to the limited depth of focus of the objective lens. In order to use fibrous filters for collecting samples for size analysis, it must be possible to remove the sample quantitatively and transfer it to a microscope stage without altering it. For such applications, the membrane and Nuclepore filters offer significant advantages over other filters. First, the samples can be analyzed directly on the filter surface. Second, since the sample does not have to be transferred, there is a greater likelihood that the sample observed is in the same form as when it was airborne.

Another situation in which the sample configuration may be important to the analysis is the determination of airborne radio activity. Many radiation detectors such as Geiger-Mueller tubes and scintillation detectors are designed to view a limited surface area, usually a 1-inch diameter circle. Thus, to make efficient use of the detector, the effective filtering area should be limited to a similar size. An additional consideration in radiometric analysis is the depth of penetration of the particles into the filter, especially for alpha and beta emitters. The activity observed by the detector will be affected by the distance of the particles from the detector, and by absorption of radiation by intervening filter fibers.

c. Sample Recovery from Filter. High collection efficiency is valueless if all of the sample is not available for analysis. For most chemical analyses, it is necessary to either remove the sample from the filter, or to destroy the filter. Inorganic particles are usually recovered from cellulose paper filters by wet ashing (digesting in concentrated acid) or muffling (incinerating) the filter. Samples collected on glass fiber paper and cellulose-asbestos paper can only be recovered by leaching or dissolving the sample from the paper. Samples can be recovered from membrane filters, polystyrene filters and soluble granular beds by dissolving the filter in a suitable solvent.

d. Interferences Introduced by Filters. Before selecting a filter for a particular application, the filter's blank count or background level of the material to be analyzed must be determined. All filters contain various elements as major, minor, and trace constituents, and the filter medium of choice for analyzing particular elements must be one with little or no background level for the elements being analyzed.

Another type of interference is inaccessibility of the sample to a measurement or sensing device. For instance, in determining reflectance of filtered particulate matter, the more the particles penetrate the surface the less they will be visible. In such an application, the sensitivity of measurement on a membrane filter surface would be greater than a fibrous filter.

e. Size or Mass of Filter. The mass of the filter itself may be important in gravimetric determinations. In determining the mass of collected aerosol, the mass of the filter should be as small as possible, relative to the mass of the sample. Also, other things being equal, the less the filter weighs and/or the smaller it is, the simpler the sample handling and processing. Collecting the sample on a smaller filter may save a concentrating step in the analysis, and make it possible to use smaller analytical equipment and/or glassware.

28.4.4 *Limitations Introduced by Ambient Conditions.*

a. Temperature. The temperature stability of a filter must be considered when sampling hot gases such as stack effluents. For such applications combustible materials cannot be used, and a selection must be made from the several types of mineral, glass or other refractory media. In order to select the appropriate medium, the peak temperature and duration of sampling must be known. Glass fiber papers are widely used for temperatures up to about 500 C.

b. Moisture Content. For sampling under the conditions of high humidity, filter media which are relatively non-hygroscopic must be chosen. Some papers pick up moisture and this may affect their filtering properties. If their efficiency is partially dependent on electrostatic effects, moisture may reduce it. Also when paper picks up moisture it may become mechanically weaker and rupture more easily. For gravimetric analyses, it is important that a non-hygroscopic paper which will maintain a constant weight be used.

28.4.5 *Limitations Introduced by Filter Holder.*

a. Filter Size. In order to use any filter, it must be held securely and without leakage in an appropriate filter holder. This limits the diameter of a filter disc to a particular size, unless the filter holder is fabricated especially for the filter. Most filter media can be obtained in any desired size, but some, such as respirator filters, are pre-formed on molds and are available in only one size.

b. Mechanical Properties of Filters. Some filter holders can only be used with papers of high mechanical strength. A strong paper such as Whatman #41 can be used in a simple head without a back-up screen while soft papers such as glass fiber or brittle papers such as the membrane filter require a more elaborate holder with a firm back-up screen or mesh support to prevent rupture.

28.4.6 *Availability and Cost.* There are great variations in the unit cost of filter media. For example, cellulose-asbestos and glass cost about twice as much as cellulose filter paper, while membrane filters may cost ten times as much. For large scale sampling programs, such price differentials can add up to significant annual cost increments. The less expensive paper should be chosen when the differences in performance are marginal. Ready availability is another factor to be considered. The cellulose and glass papers can be obtained from any chemical supply house, while other types may be available from a limited number of suppliers.

28.5 SUMMARY AND CONCLUSIONS

The advantages of sampling by filtration have been discussed, filtration theory has been outlined, commercial filter media have been described, and the criteria for selecting appropriate filters for particular applications have been reviewed.

Of all the particle collection techniques, filter sampling is the most versatile. With appropriate filter media, samples can be collected in almost any form, quantity and state. Sample handling problems are usually minimal, and many analyses can be performed directly on the filter. No single filter medium is appropriate to all problems, but a filter appropriate to any immediate problem can usually be found.

References

1. Davies, C.N. 1968. The Entry of Aerosols into Sampling Tubes and Heads. Brit J Appl Physics, Ser 2, 1:921–932.
2. Davies, C.N. 1966. Deposition from Moving Aerosols. pp 393–446 in "AEROSOL SCIENCE" edited by C.N. Davies. Academic Press, London.
3. Stenhouse, J.I.T., J.A. Harrop, and D.C. Freshwater. The Mechanisms of Particle Capture in Gas Filters. Aerosol Science, 1:41–52.
4. Benarie, M. 1969. Influence of Pore Structure upon Separation Efficiency in Fiber Filters. Staub (English Trans.), 29:37–42.
5. Ramskill, E.A., and W.L. Anderson. 1951. The Inertial Mechanism in the Mechanism Filtration of Aerosols. J Coll Sci, 6:416–428.
6. Kirsch, A.A., and N.A. Fuchs. 1968. Studies on

Fibrous Aerosol Filters, III. Diffusional Deposition of Aerosols in Fibrous Filters. Ann Occup Hyg, 11:299–304.

7. STECHKINA, I.B., A.A. KIRSCH, and N.A. FUCHS. 1969. Studies on Fibrous Aerosol Filters, IV. Calculation of Aerosol Deposition in Model Filters in the Range of Maximum Penetration. Ann Occup Hyg, 12:1–8.
8. LUNDGREN, D.A., and K.T. WHITBY. 1965. Effect of Particle Electrostatic Charge on Filtration by Fibrous Filters. I & EC Process Des and Develop, 4:345–350.
9. ZEBEL, G. 1965. Deposition of Aerosol Flowing Past a Cylindrical Fiber in a Uniform Electric Field. J. Colloid Sci, 20:522–543.
10. RIMBERG, D. 1969. Penetration of IPC 1478, Whatman 41, and Type 5G Filter Paper as a Function of Particle Size and Velocity. Amer Industr Hyg Assoc J, 30:394–401.
11. LOCKHART, L.B., JR., R.L. PATTERSON, JR., and W.L. ANDERSON. Characteristics of Air Filter Used for Monitoring Airborne Radioactivity. NRL Report No. 6054, U.S. Naval Research Laboratory, Washington, D.C.
12. SPURNY, K.R., J.P. LODGE, JR., E.R. FRANK, and D.C. SHEESLEY. 1969. Aerosol Filtration by Means of Nuclepore Filters: Structural and Filtration Properties. Env Sci and Tech, 3:453–464.
13. LÖFFLER, F. 1968. The Adhesion of Dust Particles to Fibrous and Particulate Surfaces. Staub (English Trans.) 28:39–37.
14. SILVERMAN, L., and F.J. VILES. 1948. A High Volume Air Sampling and Filter Weighing Method for Certain Aerosols. J of Industr Hyg and Toxicol, 30:124.
15. PATE, T.B., and E.C. TABOR. 1962. Analytical Aspects of the Use of Glass Fiber Filters for the Collection and Analysis of Atmospheric Particulate Matter. Amer. Industr Hyg Assoc J, 23:145–150.
16. KRAMER, D.N., and P.W. MITCHEL. 1967. Evalu ation of Filters for High-Volume Sampling of Atmospheric Particulates. Amer Industr Hyg Assoc J, 28:224–228.
17. SHLEIEN, B., J.A. COCHRAN, and A.G. FRIEND. 1966. Calibration of Glass Fiber Filters for Particle Size Studies. Amer Industr Hyg Assoc J, 27:353–359.
18. WINKEL, A. 1959. Über neue Methode zur Straub messung. Staub, 19:253.
19. BERKA I. 1968. Organic Microfiber Filters for Sampling of Industrial Dusts. Staub (Engl. Trans.), 28:27–28.
20. PAULUS, H.J., N.A. TALVITIE, D.A. FRASER, and R.G. KEENAN. 1957. Use of Membrane Filters in Air Sampling. AIHA Quart, 18:267.
21. SPURNY, K., and J. PICH. 1964. The Separation of Aerosol Particles by Means of Membrane Filters by Diffusion and Inertial Impaction. Int J Air Wat Poll, 8:193–196.
22. MEGAW, W.J., and R.D. WIFFEN. 1964. The Efficiency of Membrane Filters. Int J Air Wat Poll, 1:501–509.
23. LINDEKEN, C.L., F.K. PETROCK, W.A. PHILLIPS, and R.D. TAYLOR. 1964. Surface Collection Efficiency of Large-Pore Membrane Filters. Health Physics, 10:495–499.
24. LÖSSNER, V. 1964. Die Bestimmung der Eindringtiefe von Aerosolen in Filtern. Staub, 24:217–221.
25. GEORGE, L.A. 1961. II. Electron Microscopy and Autoradiography. Science, 133:1423.
26. RICHARDS, R.T., D.T. DONOVAN, and J.R. HALL. 1967. A Preliminary Report on the Use of Silver Metal Membrane Filters in Sampling Coal Tar Pitch Volatiles. Amer Industr Hyg Assoc J, 28:590–594.
27. KNAUBER, J.W., and F.H. VONDERHEIDEN. 1969. A Silver Membrane X-Ray Diffraction Technique for Quartz Samples. Presented at 1969 Amer. Industr. Hyg. Assoc. Meeting, Denver, Colo.
28. SPURNY, K.R., J.P. LODGE, JR., E.R. FRANK, and D.C. SHEESLEY. 1969. Aerosol Sampling by Means of Nuclepore Filters: Aerosols Sampling and Measurement. Env Sci and Tech, 3:464–468.
29. SMITH, W.J., and N.F. SURPRENANT. 1963. Properties of Various Filtering Media for Atmospheric Dust Sampling. Presented to the American Society for Testing Materials at Philadelphia, Pennsylvania.
30. LINDEKEN, C.L., R.L. MORGIN, and K.F. PETROCK. 1963. Collection Efficiency of Whatman 41 Filter Paper for Submicron Aerosols. Health Physics, 9:305–308.
31. POSNER, S. 1961. Air Sampling Filter Retention Studies Using Solid Particles. Proc. 7th AEC Air Cleaning Conf. TID-7627.
32. DAVIES, C.N. 1970. The Clogging of Fibrous Aerosol Filters. Aerosol Science, 1:35–39.

PART II
TENTATIVE AND RECOMMENDED METHODS FOR AMBIENT AIR SAMPLING AND ANALYSIS

101.

Tentative Method of Analysis for C_1 Through C_5 Atmospheric Hydrocarbons*

43101-01-69T

1. Principle of the Method

1.1 Atmospheric hydrocarbons are identified and quantitated in this method by gas chromatography. The air sample is introduced into a chromatograph column containing activated alumina coated with B,B′ oxydipropionitrile (maintained at 0 C) using helium as a carrier gas. This gas stream is then introduced into a hydrogen-oxygen flame ionization detector. The hydrocarbon molecules are ionized in the intense heat of this flame. The ions are collected at the electrodes and the resulting current (which is proportional to the number of carbon atoms in each hydrocarbon molecule) is measured by an electrometer.

1.2 The electrometer is connected to a strip chart recorder which records the response (as a function of time) of the flame ionization detector as each hydrocarbon is eluted from the chromatographic column. The heights of the narrow peaks on the strip chart recorder are used to quantitate the C_1 through C_3 hydrocarbons and the areas of the broader peaks calculated as height times the width at one half the peak height, are used to quantitate the C_4 through C_5 hydrocarbons.

1.3 This method is rapid and is especially applicable to routine sampling and analysis for both grab and integrated samples. The elution of the following 17 hydrocarbons can be accomplished within 16 min: methane, ethane, ethylene, acetylene, propane, propylene, n-butane, isobutane, butene-1, 2, 2 dimethyl propane, isobutylene, trans-2-butene, cis-2-butene, isopentane, n-pentane, 3 methyl butene-1 and 1,3 butadiene.

2. Range and Sensitivity

2.1 The lower limit of measurement for the various hydrocarbons is 0.01 ppm by volume (*i.e.*, 0.01 μl/l). The lower limit of measurement of the C_3 through C_5 hydrocarbons can be extended to 0.1 parts/billion (ppb) by concentrating 100 ml of the gas sample in a freeze trap. (Liquid oxygen (BP-181 C) is used in the freeze trap rather than liquid nitrogen (BP-195 C to avoid condensation of oxygen in freezing the air sample.) Methane (BP-164 C) is not collected in the freeze trap and the resolution of ethane and ethylene chromatographic peaks is reduced in the concentrated sample. The upper limit of measurement can be considerably greater than the anticipated concentrations of the various hydrocarbons in the ambient air. **(1)**

3. Interferences

3. At the present time there are no known common pollutants in the ambient atmosphere in sufficient concentrations to interfere with the listed hydrocarbons. However, in order to obtain representative atmospheric samples, the selection of the sampling site is most important. The selected site should be reasonably distant from local sources of hydrocarbon emissions, and in a locale where the surrounding topography is conducive to adequate mixing of the ambient atmosphere. Under such conditions, local source emissions will not bias the hydrocarbon data toward a specific individual hydrocarbon or hydrocarbon group.

*Based on studies by P.W. Mueller and Associates at the Air and Industrial Hygiene Laboratory, California State Department of Public Health, Berkeley, California. August 1966.

4. Precision and Accuracy

4.1 Replicate analysis of aliquots of uniform air samples and known standard hydrocarbon mixtures should not deviate by more than 10% relative standard deviation.

5. Apparatus

5.1 A diagram of the gas chromatography assembly is given in Figure 101:1. The principal components are:

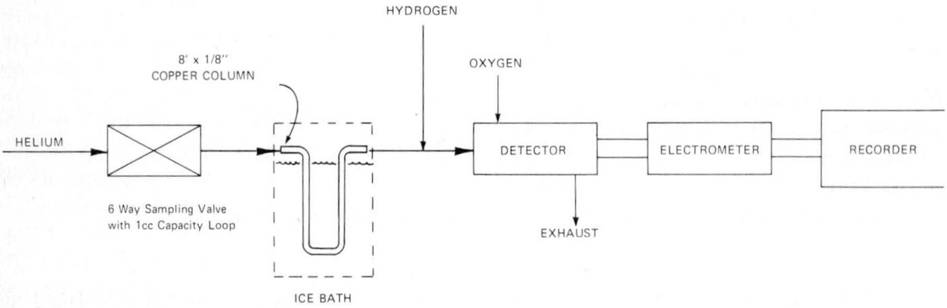

Figure 101:1—Gas chromatograph assembly for measuring C_1—C_5 atmospheric hydrocarbons.

5.1.1 *Chromatograph column.* A copper or stainless steel tubing 2.4 meters long x 3 mm OD and 1.6 mm ID, packed with activated alumina coated with 17% by weight B,B′ oxydiproprionitrile. The column is prepared by placing about 25 to 30 g of alumina in a 15-cm porcelain evaporating dish and washing several times with deionized water. After the excess water has been poured off, the alumina is heated (and activated) at 400 C for 9 hr. It is allowed to cool in a desicator for about 1 hr. 16.6 g of the activated alumina are weighed out and immediately poured into another 6″ porcelain evaporating dish containing 3.4 g B,B′ oxydipropionitrile dissolved in 40 ml dichloromethane. The solvent is evaporated under a reflector infrared heat lamp with frequent stirring; residual solvent is removed at 70 C at about 250 Torr for 5 hr. The column is filled by gravity flow and continual tapping with the column packing. The column is coiled and conditioned by passing the carrier gas through it overnight at room temperature. The column temperature should be maintained at 0 C and the helium carrier gas maintained at a flow rate of 30 ml/min with an inlet pressure of 80 psig.

5.1.2 *Sample Injector.* A 6-way gas sampling valve with a 1-ml stainless steel sampling loop.

5.1.3 *Detector.* Hydrogen flame ionization detector; hydrogen flow to the detector is maintained at 27.6 ml/min by means of a restricted stainless steel capillary; oxygen flow to the detector is 300 ml/min.

5.1.4 *Electrometer.* Such as Varian Aerograph Model 500 B, or other suitable instrument.

5.1.5 *Recorder.* A strip chart recorder with a range of zero to one millivolt and one second response; chart speed of 1.7 cm/min.

5.1.6 A bath for thermostating the column at 0 C.

5.1.7 The freeze trap consists of a 20-cm U-tube of 1-mm LD stainless steel tubing packed with Chromosorb P. Each end of the U-tube is fitted with a toggle valve.

5.2 Air samples are collected and calibration standards are prepared in aluminized plastic bags, fabricated from the following materials:

5.2.1 Scotchpak brand, heat sealable, aluminized polyester film manufactured by Minnesota Mining and Manufacturing Company, Catalog number 20A20, or similar material from another source.

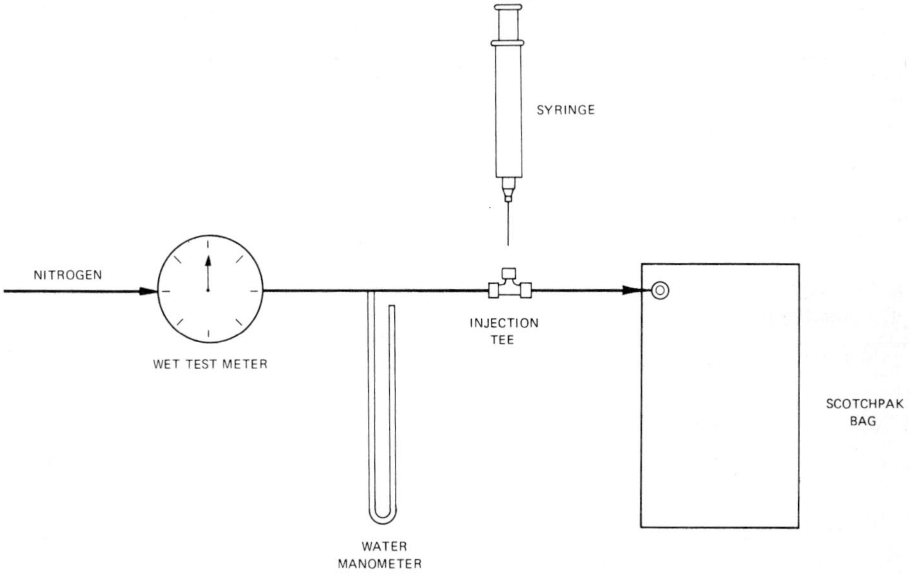

Figure 101:2—Assembly for preparation of calibration standards.

5.2.2 Super sealer heat sealing iron with Teflon cover.

5.2.3 Clamp-on valves and neoprene gaskets.

5.2.4 Osborne Arch Punch, 9.5 mm, or similar punch, available in hardware stores.

5.3 The assembly of equipment used for preparation of calibration standards is shown in Figure 101:2. The principal components are:

5.3.1 *Precision wet test gas meter.*

5.3.2 *Stainless steel injection tee.* A Swagelok all-tube tee, 6 mm, with the center arm fitted with a silicone rubber injection gasket.

5.3.3 *Gas-tight syringes,* 50-ml and 10-ml.

5.3.4 *Two sizes of aluminized Scotchpak bags fitted with Roberts valves,* 61 × 91 cm and 30 × 46 cm.

5.4 The following equipment is used for collecting grab samples of air:

5.4.1 *Aluminized Scotchpak bags fitted with Roberts valves,* Bag size 23 × 23 cm.

5.4.2 *One atomizer rubber bulb set or automatic buret bulb.* Available from all laboratory supply houses.

5.5 The following equipment is used for collecting integrated samples of air:

5.5.1 *An air sampling pump.* A suitable pump for this purpose is a diaphragm actuated pump with constant flow.

5.5.2 *Aluminized Scotchpak bags fitted with Roberts valves.* Bag size 46 × 46 cm for 15-l samples; 46 × 92 cm for 30-l samples.

5.5.3 *Micrometering valve and check valve.* This is needed for control of gas sample flow.

Limiting orifice needles. 30-gauge hypodermic needles, 1.9 cm long, of hyperchrome stainless steel are suitable for this purpose, instead of a micrometering valve as mentioned in **5.5.3**.

5.5.4 *A timer.* A useful timer for this purpose is an electric timer which automatically shuts off the sampling pump at the end of the sampling interval. A single pole, double throw time switch may be employed for this purpose.

5.5.5 *A rubber septum.* This is needed when limiting orifice needles are used for flow control. A sleeve type serum bottle stopper may be used for this purpose.

5.5.6 *Flowmeter or rotometer,* equipped with stainless steel ball.

5.5.7 A filter system upstream of the pump is recommended for the entrapment of particulate matter. An aerosol field monitor fitted with an all glass or a 0.8 μ pore membrane filter is useful for this purpose.

6. Reagents

6.1 PURITY. All chemicals should be analytical reagent grade.

6.2 CYLINDER GASES. All cylinder gases should be prepared from water pumped sources and *not oil pumped*.

6.3 REQUIRED CHEMICALS AND GASES. B,B′ oxydipropionitrile; alumina, 100 to 150 mesh; dichloromethane, hydrogen, 99.8% purity; helium, highest purity; hydrocarbons with a guaranteed minimum purity of 99 mole %; pure grade nitrogen; commercial grades of nitrogen and oxygen; chromosorb P, 45 to 60 mesh.

7. Procedure

7.1 COLLECTION OF FIELD SAMPLES.

7.1.1 As indicated in Section 3 the location of the field sampling site can have an important effect on the sample composition unless precautions are taken to avoid local sources of hydrocarbons.

a. Generally, for a homogeneous, urban air mass, hydrocarbon contaminants would be expected to fall within the ranges given as a guide in Table 101:I.

b. If there are large variances from the extremes of the range, a critical review of the data should be made. The review should consider such factors as: (a) Are there errors in the calculation of instrument response factors? (b) Are there errors in calculation of peak areas? (c) Was there possible contamination of the sample collection bags or sample valves and lines by atmosphere containing high hydrocarbon concentrations?

c. During collection of the sample, was there an intermittent source of hydrocarbon emissions such as an automobile starting, stopping or passing near the sampling site; the filling of a gasoline fuel tank at a nearby service station? If the sample site is near a refinery or petroleum processing plant, was there a possible operational upset that resulted in venting unusual amounts of hydrocarbon compounds?

d. Depending upon the sampling site, other emission sources may be a possibility.

e. As a general "rule of thumb", if methane, ethane or propane concentrations are high relative to other identified hydrocarbons (See Table 101: I), look for local source emissions of natural gas. If acetylene and the butenes are high, look for a nearby automobile exhaust source. If n-butane, n-pentane or isopentane are high, look for a source of gasoline evaporation.

7.1.2 Grab samples are collected as follows:

a. Flush the bag out 3 times with the ambient air to be sampled. This is done with a rubber buret bulb connected to the valve of the bag.

b. Fill the bag approximately three-fourths full, close the air valve securely and remove the atomizer bulb. Do not fill the bag to capacity. Some space must be allowed for expansion due to temperature and pressure variations.

Table 101:I. Ranges of Hydrocarbon Values Expected in Urban Air Masses.

	Range, ppm in air	
Component	Minimum	Maximum
Methane	1.2	15
Ethane	0.005	0.5
Propane	0.003	0.3
Isobutane	0.001	0.1
n-Butane	0.004	0.4
Isopentane	0.002	0.2
n-Pentane	0.002	0.2
Ethylene	0.004	0.3
Propene	0.001	0.1
Butene-1	0.000	0.02
Isobutylene	0.000	0.02
Trans-2-butene	0.000	0.01
Cis-2-butene	0.000	0.01
1,3 Butadiene	0.000	0.01
Acetylene	0.000	0.2

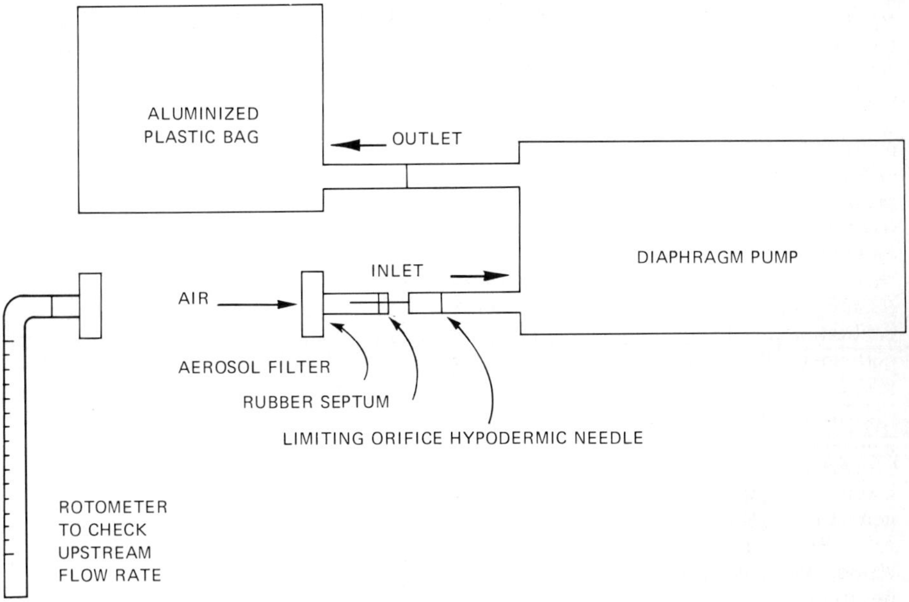

Figure 101:3—Assembly for integrated hydrocarbon sampling.

c. Give the bag an identifying number and record appropriate field information, *e.g.*, date and time of collection, location of sample collection, weather and air pollution conditions, etc.

d. The samples should be sent to the laboratory and analyzed as soon as possible.

7.1.3 A diagram of an assembly of equipment for collecting integrated samples is given in Figure 101:3.

a. With the equipment as assembled in Figure 101:3, the air flow rate into the plastic bag is measured by attaching a flowmeter to the upstream side of the diaphragm pump. If a micrometering valve is used instead of a limiting orifice, the valve is adjusted to the desired flow rate.

b. When the desired flow rate is obtained, disconnect the flowmeter. Set the electric timer for the time period over which the sample is to be collected.

c. Near the end of the sampling period, the flow rate is measured again.

d. The bag sample is identified and appropriate field data recorded.

e. It is not necessary to know the exact volume of the sample collected. A 1-ml aliquot of the sample is taken for gas chromatographic analysis and the concentration of hydrocarbons in the air is based on the 1-ml aliquot. A constant flow rate should be maintained during the sampling period or the change in flow rate during the period should be measured in order to insure a valid integrated sample.

7.2 ANALYSIS OF SAMPLES.

7.2.1 Air samples collected in the field and returned to the laboratory are analyzed according to the following procedure.

a. Turn on recorder.

b. Set electrometer attenuation at 10x.

c. Connect the sample bag to the inlet of the sampling valve on the gas chromatograph and flush 20 ml of sample through the loop.

d. Inject sample.

e. After the elution of methane (approximately 30 sec), reset attenuation to 1x. The sample should be eluted from the column in 15 min. A chromatogram of a known mixture is given in Figure 101:4, indicating the relative positions of the elution peaks of the 17 hydrocarbons.

8. Calibration and Standards

8.1 A calibration standard should be prepared for each of the 17 hydrocarbons to be measured. Retention time is used for identification of the hydrocarbon and either peak height or peak area is used for quantitation of the hydrocarbon. Standard calibration mixtures using the assembly of equipment illustrated in Figure 101:2 are prepared as follows:

8.1.1 Purge wet test meter with nitrogen for one-half hr.

8.1.2 After purge, connect the plastic bag to the system and start metering nitrogen into the bag. Bags fabricated to 61 cm × 91 cm are used for the 40-liter dilutions; bags fabricated to 30 cm × 46 cm are used for the 10-l dilutions.

8.1.3 Inject the predetermined amount of each hydrocarbon into the gas stream through the stainless steel tee. A 50-ml gas-tight glass syringe is used for quantities of gas greater than 10 ml; for quantities less than 10 ml, a 10-ml gas-tight syringe is used.

8.1.4 To make standards containing less than 10 ppm of a hydrocarbon, a double dilution is required. First a 1000 ppm standard is prepared. Aliquots of this standard are diluted to give standards with less than 10 ppm hydrocarbons (See Table 101:II).

Table 101:II. Dilutions for Preparation of Calibration Standards.

Hydrocarbon Concentration in ppm	Aliquot of Hydrocarbon	Final Dilution Volume in Liters
1000	40 ml pure hydrocarbon	40
10	0.4 ml pure hydrocarbon	40
1	10 ml of 1000 ppm standard mixture	10
0.1	1 ml of 1000 ppm standard mixture	10

8.2 Calibration of the Chromatograph.

8.2.1 Approximately 20 ml of standard gas are flushed through the 1-ml stainless steel sampling loop.

8.2.2 One ml of standard gas is then injected into the gas chromatograph.

8.2.3 The response of the hydrogen flame ionization detector is linear from 0.01 to 10 ppm.

9. Calculation

9.1 Air samples collected in the field and returned to the laboratory are calculated according to the following procedure.

9.1.1 The response of the flame ionization detector is recorded on the strip chart recorder as the hydrocarbon is eluted from the column.

9.1.2 The time of elution is recorded for each hydrocarbon.

9.1.3 Peak height is used for quantitating the C_1 through C_3 hydrocarbons.

9.1.4 Peak area is used for quantitating the C_4 through C_5 hydrocarbons. Peak area is defined as the product of the peak height, H, multiplied by the peak width, W, measured at one half the peak height. For example:

$$W \times H = \text{Area}$$

9.1.5 Determine the concentration of each hydrocarbon present in the sample, using the calibration response factor:

ppm hydrocarbon = f × peak response × attenuation

where *f* is the calibration response factor for each hydrocarbon in units of ppm per mm peak height or ppm per mm^2 peak area;

peak response is the peak height in mm for C_1 through C_3 hydrocarbons or the peak area in mm^2 for C_4 through C_5 hydrocarbons; and

attenuation is the electrometer attenuation setting which is generally 10x for methane and 1x for the other hydrocarbons.

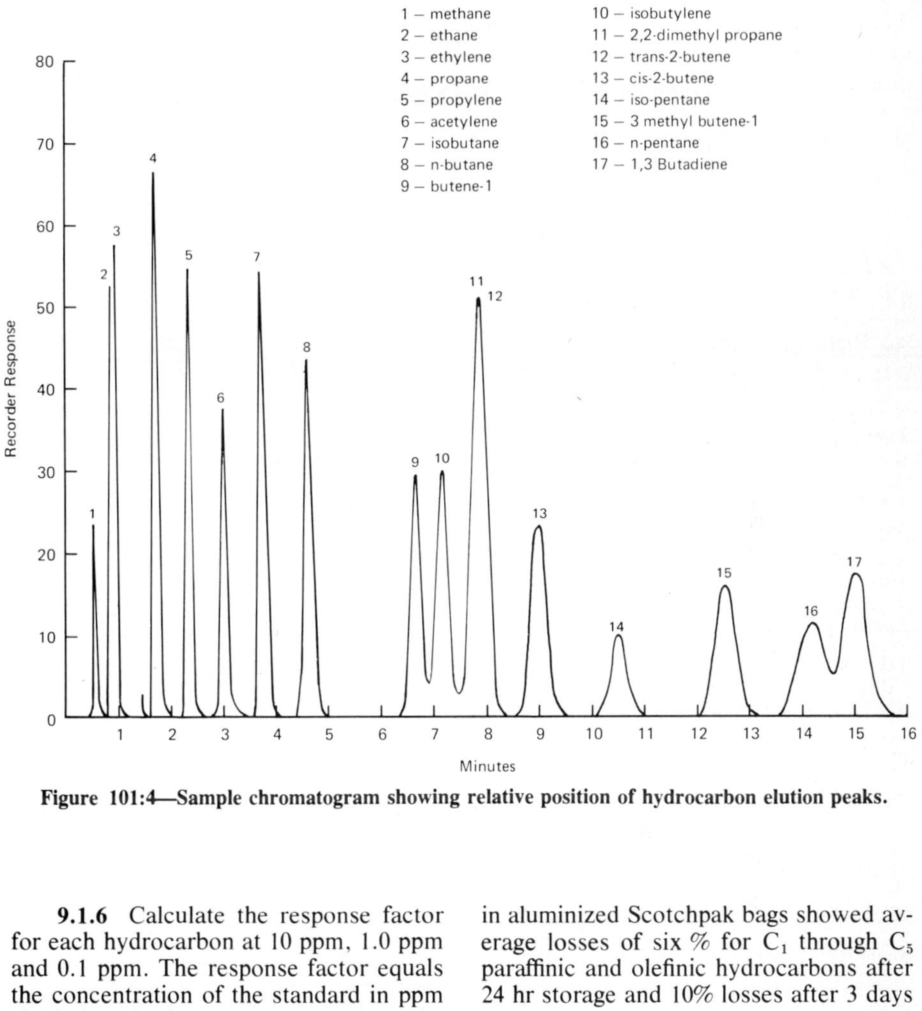

Figure 101:4—Sample chromatogram showing relative position of hydrocarbon elution peaks.

9.1.6 Calculate the response factor for each hydrocarbon at 10 ppm, 1.0 ppm and 0.1 ppm. The response factor equals the concentration of the standard in ppm divided by the detector response times the attenuation. The detector response for C_1 through C_3 hydrocarbons is peak height measured in millimeters. Peak height is the most convenient measure for quantitation of gases with short retention times when the peaks are narrow and high. Peak area must be used when the peaks are broad and unsymmetrical. The detector response for C_4 through C_5 hydrocarbons is peak area in square millimeters.

10. Effects of Storage

10.1 Storage experiments on samples in aluminized Scotchpak bags showed average losses of six % for C_1 through C_5 paraffinic and olefinic hydrocarbons after 24 hr storage and 10% losses after 3 days storage.

11. References

1. HALASZ, I., and E.E. WEGNER. 1961. Gas Chromatographic Separation of Low Boiling Hydrocarbons Using Activated Alumina as Support for Liquid Phase. *Nature*, 189:570.

Subcommittee 5
E. SAWICKI, *Chairman*
R. C. COREY
A. E. DOOLEY
J. B. GISCLARD
J. L. MONKMAN
R. E. NELIGAN
L. A. RIPPERTON

102.

Tentative Method of Microanalysis for Benzo[a]pyrene in Airborne Particulates and Source Effluents

17242-01-69T

1. Principle of the Method

1.1 The particulate collected from the urban atmosphere is extracted with an organic solvent and then separated alongside pure benzo[a]pyrene (BaP) with alumina thin-layer chromatography. The unknown and standard spots are eluted, their solutions are evaporated, and the residues are dissolved in concentrated sulfuric acid. An intensely yellow fluorescent cationic salt is obtained. Readings of standard and test spot solutions are taken at an excitation wavelength of 470 nm and an emission of 540 nm, with the spectrophotofluorimeter or with a filter fluorimeter containing a primary filter peaking at 460 nm (**2**).

2. Range and Sensitivity

2.1 The lower limit of determination of the spectrophotofluorimetric method is about 3 ng of BaP with an instrument capable of its full potential of sensitivity. The range of analysis is then 3 to 200 ng of BaP. This means that for moderately contaminated urban air samples ($<$ 500 to 1000 μg BaP/g benzene-soluble fraction) more than 6 μg of benzene-soluble fraction or more than 0.6 m^3 of air is necessary for an analysis. For samples of low contamination ($<$ 50 μg BaP/g benzene-soluble fraction) at least 60 μg of benzene-soluble fraction or 6 m^3 of air are necessary for analysis.

2.2 The lower limit of determination for the filter fluorimetric method is 10 ng of BaP. The range of analysis is 10 to 300 ng. The lower limit of material necessary for analysis is about 3 to 4 times higher than in the SPF method (**2**).

3. Interferences

3.1 Because of the sensitivity of the method the laboratory air must be clean. Cigarette smoking cannot be allowed since it contributes to the background and worsens the sensitivity of the method.

3.2 The method is highly selective for BaP. Hydrocarbons found with or near BaP in alumina chromatographic fractions do not interfere *e.g.*, benzo[k]-fluoranthene, benzo[g,h,i]perylene, benzo[e]pyrene, perylene and anthanthrene (**3**). In fact, benzo[a]pyrene can be determined by the spectrophotofluorimetric method in the presence of 50 hydrocarbons (over 40 of which are polynuclear compounds ranging in size from 2 to 7 rings) (**1**).

4. Precision and Accuracy

4.1 Eleven micromethods have been compared for the estimation of BaP in a composite benzene soluble fraction of airborne particulates from over 100 communities (**2**). The methods gave values ranging from 720 to 1000 μg BaP/g samples. An average value of 870 was obtained. In com-

---

Warning: Some polynuclear aromatic hydrocarbons, benzo[a]pyrene in particular, are carcinogenic. Consequently, care must be taken in handling laboratory equipment to avoid spilling solutions or PAH material on hands or other parts of the skin. Manipulations involving the PAH compounds should be carried out in a fume hood and must not be allowed to contaminate the laboratory air.

parison the spectrophotofluorimetric method gave a range of 800 ± 50 in 8 determinations. Pure BaP was added to an airborne particulate and recoveries of 90, 93, 94 and 100 were obtained averaging 95 ± 5%.

4.2 The composite sample analysed by the filter fluorimetric method gave a value of 950 ± 100 μg BaP/g sample in eight determinations. Recoveries of 80, 84, 84, 88, 88, 90, 100 and 100 (averaging 90 ± 10%), were obtained from the airborne particulate sample.

4.3 The fluorescence intensity is stable for at least 20 min **(1, 3)**. However, this stability is worsened if the solution is in a bright light or is repeatedly illuminated in the instrument.

4.4 The time necessary for separation and analysis has been compared for 11 methods **(2)**. For the two instrumental methods in this paper about 1 ½ hr is necessary for the separation and assay of an organic fraction.

5. Apparatus

5.1 All laboratory ware must be cleaned and tested to insure absence of contamination by organic material. This procedure is described in Intersociety Committee Method **(5)**, Section 5.1.

5.2 Thin-layer plates are coated with alumina, 250 μ in thickness, activated by heating at 100 C for 30 min. The activated plates are stored in a vacuum desiccator adjusted to 45% relative humidity with aqueous sulfuric acid. The plates should not be stored for more than 3 weeks and should be checked for fluorescent impurities before use.

5.3 THIN-LAYER CHROMATOGRAPHIC CHAMBER.

5.4 SPECTROPHOTOFLUORIMETER, such as the Aminco Bowman SPF, or any instrument of similar performance; or a

5.5 FILTER FLUORIMETER, equipped with a primary filter peaking at λ 460 nm (*e.g.* Aminco Filter No. 4-7114) and a secondary filter peaking at λ 565 nm (*e.g.* Aminco Filter No. 4-7119).

5.6 CELLS. Cylindrical cells of 10 mm path length, calibrated to hold 1.0 ml volume, with 100% transmission in wavelength range of 400 to 600nm.

5.7 CHROMATO-VUE CABINET, or some other long wavelength ultraviolet source.

5.8 MICROSOXHLET EXTRACTOR, with 30 ml-boiling flask capacity (Fisher Scientific No. 20650 or similar apparatus from any other source).

5.9 MICROPIPETS, A calibrated set to cover the range of 10 μl to 1.0 ml.

6. Reagents

6.1 PURITY. All chemicals should be of analytical reagent grade.

6.2 METHYLENE CHLORIDE, redistilled before use in an all glass, clean apparatus.

6.3 PENTANE, distilled before use in clean all glass apparatus. The fraction boiling at 35 to 36 C is used for analysis.

6.4 ETHER, ANHYDROUS, treated with sodium-lead alloy and redistilled as above before use.

6.5 CONC SULFURIC ACID.

6.6 BENZO[A]PYRENE. A standard of highest purity is required. Zone refined BaP of 99.9% purity is available from James Hinton, Newport News, Va. Other sources of lesser purity are Eastman, Aldrich, K & K. Removal of impurities from such samples is described in Section 8.2.

7. Procedure

7.1 The particulate collected from several m^3 of urban atmosphere is extracted in a microsoxhlet extractor (boiling flask capacity 30 ml) with 10 ml of methylene chloride. The combined extracts are evaporated to dryness in a current of nitrogen at room temperature or by vacuum and the residue is dissolved in 0.2 ml of methylene chloride. An aliquot of this solution, anywhere from 10 μl to the entire sample, is placed on an alumina thin-layer plate 1.50 cm from the bottom. (The benzene-soluble fraction (0.02 to 1 mg) can also be used.)

7.2 Standards (0.02 μg BaP to a spot) are also placed on the plate at the origin. The plate is transferred to a thin-layer chromatographic chamber which contains 200 ml of pentane-ether (19:1). After development to 15 cm in dim light, the plate is quickly examined and marked under UV light. The absorbent in each spot is transferred quantitatively to a throwaway pipet containing a small wad of glass wool or to a small fine-porosity fritted glass funnel. The absorbent is eluted with 70 to 100 ml of ether. The eluent is collected in a 25 ml test tube fitted with a sidearm connected to a vacuum system. The ether is evaporated by vacuum and the residue is dissolved in 1 ml of concentrated sulfuric acid.

7.3 Readings of standard and test spot solutions are taken at an excitation wavelength of 470 nm and an emission wavelength of 540 nm with the spectrophotofluorimeter or with a filter fluorimeter containing a primary filter peaking at 460 nm and a secondary filter peaking at 565 nm (**2**).

8. Comparison with Standards

8.1 A straight line relation through the origin between the concentration and the fluorescence intensity is obtained for 3 to 200 ng of BaP in the spectrophotofluorimetric method. A similar relation is found for 10 to 300 ng of BaP in the filter fluorimetric method. Comparisons are made of the standard spots and samples in order to derive the concentration from fluorescence intensity measurements.

8.2 A purer sample of BaP is obtained by thick-layer chromatography on cellulose acetate with ethanol-toluene-water (17:4:4) as developer. BaP is thus separated from some of its impurities, *e.g.* benzo[e]pyrene and perylene (**1, 4**).

9. Calculations

9.1 Even though the relationship between the concentration and the product of the meter multiplier and transmittance readings is linear, it is advisable to run standards at the same time. On this basis the amount of benzo[a]pyrene in the test mixture is readily calculated for both pro-

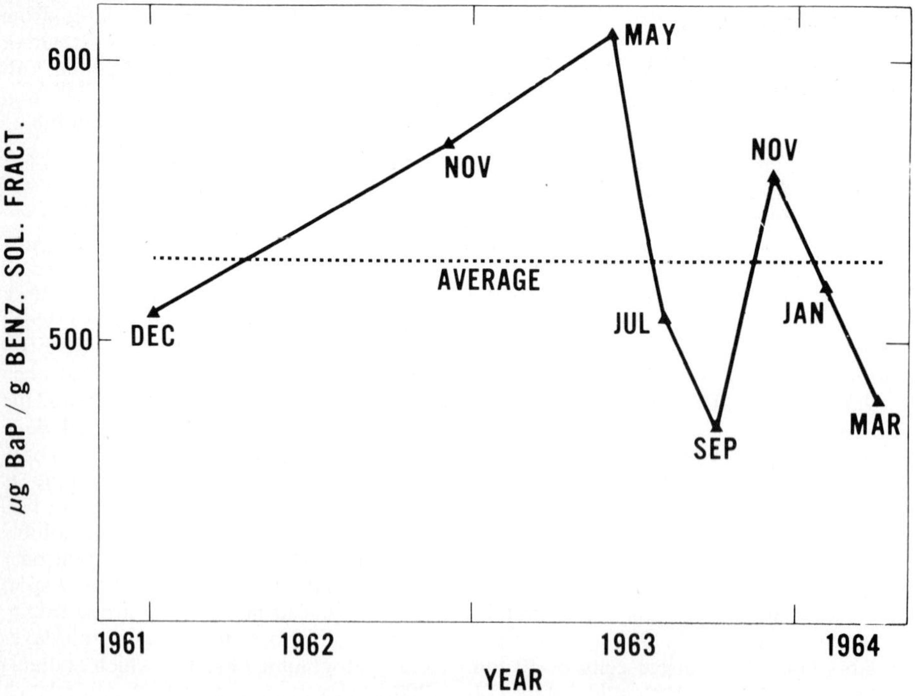

Figure 102:1—Stability of BaP in benzene-soluble fraction.

cedures by the following equations (**2**), where:

Cs = nanograms of Standard BaP analyzed;
Rs = product of the meter multiplier and transmittance readings of eluted BaP spot dissolved in 1 ml of sulfuric acid;
Rx = product of the meter multiplier and transmittance readings of the eluted unknown spot dissolved in 1 ml of sulfuric acid;
Wt = weight in mg of analyzed organic airborne particulate sample;
V = volume in m^3 of analyzed air.

$$\text{ng BP/g organic mixture} = \frac{Cs.Rx}{1000\ Wt.Rs}$$

and

$$\text{ng BaP/m}^3\text{ air} = \frac{Cs.Rx}{V.Rs}$$

10. Effect of Storage

10.1 Sample solutions should be analyzed the same day. If they cannot be analyzed until the next day, the solutions should be protected from light and stored in a cold box. The organic residues can be kept in a refrigerator in the dark for longer periods of time. Thus, the benzene-soluble fraction, stored in this way, is stable for years as far as the BaP is concerned, Figure 102:1. The variations in value are what would be expected of the column chromatographic absorption spectral method.

11. References

1. SAWICKI, E., HAUSER, T.R., and STANLEY, T.W. 1960. Int. J. Air Poll. 2:253.
2. SAWICKI, E., T.W. STANLEY, W.C. ELBERT, J. MEEKER, and S. MCPHERSON. 1967. Atmospheric Environment. 1:131.
3. SAWICKI, E., T.W. STANLEY, W.C. ELBERT, and J.D. PFAFF. 1967. Anal. Chem. 36:497.
4. SAWICKI, E., T.W. STANLEY, J.D. PFAFF, and W.C. ELBERT. 1964. Chemist-Analyst. 53.6.
5. AMERICAN PUBLIC HEALTH ASSOCIATION, INC. 1977. Methods of Air Sampling and Analysis 2nd. ed. p. 235. Washington, D.C.

Subcommittee 5

E. SAWICKI, *Chairman*
R. C. COREY
A. E. DOOLEY
J. B. GISCLARD
J. L. MONKMAN
R. E. NELIGAN
L. A. RIPPERTON

103.

Tentative Method of Spectrophotometric Analysis for Benzo[a]pyrene in Atmospheric Particulate Matter

17242-03-69T

1. Principle of the Method

1.1 Airborne particulates collected from polluted atmospheres on glass-fiber filters are extracted exhaustively with an organic solvent. These extracts are carefully reduced to tarry residues through evaporation in vacuo and then separated on thin-layers of activated alumina. The fluorescent area of pure benzo[a]pyrene and a corresponding area of the sample are removed from the TLC plate and eluted. Solutions of the pure compound and the unknown are spectrophotometrically measured between 400 and 240 nm. Wavelengths 390, 382, and 374 nm are used to determine the amount of benzo[a]pyrene present in the sample (**3, 5**).

2. Range and Sensitivity

2.1 The determination limit for this spectrophotometric method is about 0.2 μg for pure benzo[a]pyrene. For a polluted atmopshere containing 500 μg BaP/g organic extract, more than 0.4 mg of extract from approximately 40 m^3 of air would be necessary for analysis. As much as 5 mg of extract can be separated and concentrations as low as 50 μg BaP/g extract can be determined. Also ½ to 1-hr particulate samples can be analyzed if a High Volume sampler is employed to filter air at a rate of about 1 to 1.5 m^3/min.

3. Interferences

3.1 Any compound absorbing wavelengths between 390 to 375 nm would be an interference in this procedure. Benzo[g,h,i]-perylene would be a serious interference if not separated completely from the benzo[a]pyrene fraction. However, in this procedure the benzo [g,h,i]perylene (Gee) is well separated from the BaP fraction, Figure 103:1. Benzo[k]fluoranthene is generally separated in an area between the BaP and Gee fractions (**3, 5**).

4. Precision and Accuracy

4.1 A large composite air sample was analyzed by the column chromatographic procedure and found to contain 560 μg BaP/g extract. When the same sample was analyzed by this procedure, the benzo[a]pyrene concentration was found to average 546 μg/g extract and the relative standard deviation was calculated to be ± 7% from 11 determinations.

4.2 The elution of pure benzo[a]pyrene from activated alumina was 95% (± 5%) efficient based on more than 20 determinations (**5**).

5. Apparatus

5.1 THIN-LAYER PLATES. (Alumina, 250 μ thick), activated and stored as described in Sections 8.3 and 10.3.

5.2 THIN-LAYER CHROMATOGRAPHIC CHAMBER.

Warning: Some polynuclear aromatic hydrocarbons, benzo[a]pyrene in particular, are carcinogenic. Consequently, care must be taken in handling laboratory equipment to avoid spilling solutions or PAH material on hands or other parts of the skin. Manipulations involving the PAH compounds should be carried out in a fume hood and must not be allowed to contaminate the laboratory air.

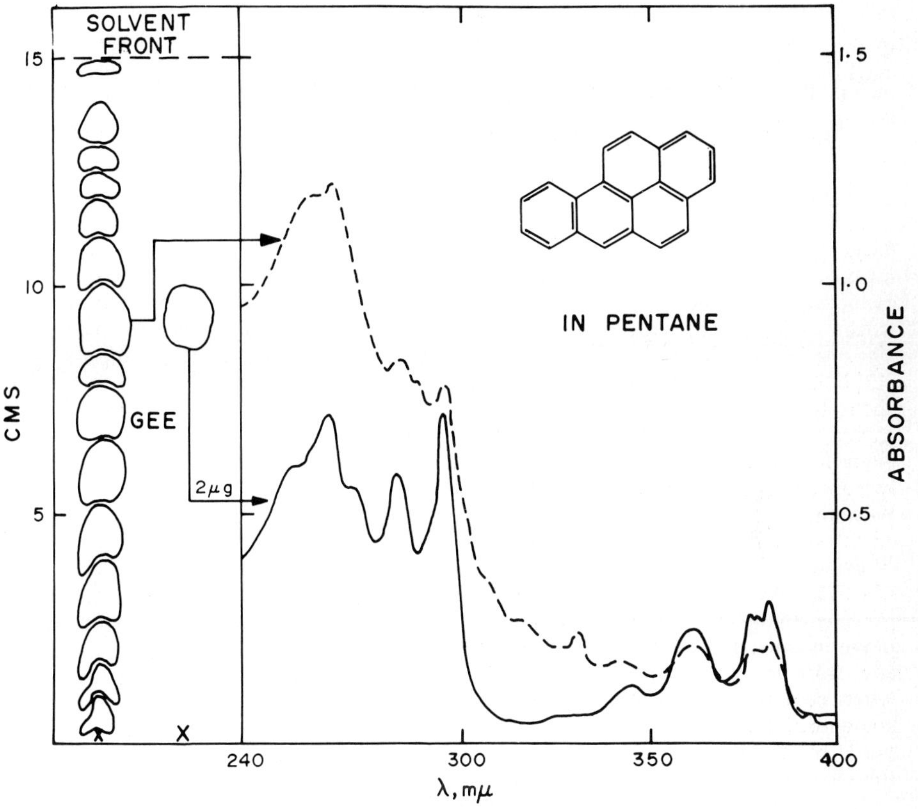

Figure 103:1—Separation of fractions.

5.3 Mechanical Mixer.

5.4 Water bath (±2 C).

5.5 Chromato-Vue. (Or other source of ultraviolet light).

5.6 Spectrophotometer. Ratio recording.

5.7 Matched 1 ml-one cm pathlength cells for use in the ultraviolet.

5.8 Microsoxhlet extractor. With 30 ml-boiling flask capacity.

5.9 All laboratory ware must be cleaned and tested to insure absence of contamination by organic material. This procedure is described in Intersociety Committee Method (**6**) Section **5.1**.

6. Reagents

6.1 Purity. All chemicals should be analytical reagent grade.

6.2 Methylene chloride. Grade distilled before use in clean all glass apparatus.

6.3 Pentane. Distilled before use in clean all glass apparatus. The fraction boiling at 35 to 36 C is used for analysis.

6.4 Ether, anhydrous diethyl ether. Treated with sodium-lead alloy and distilled as above before use.

6.5 Benzo[a]pyrene, recrystallized. A standard of highest purity is required. Zone refined BaP of 99.9% purity is available from James Hinton, Newport News, Va. Other sources of lesser purity are Eastman, Aldrich and K & K. Removal of impurities from such samples is discussed in Section **8.2**.

7. Procedure

7.1 Samples and Standard

7.1.1 Benzo[a]pyrene can be quantitatively extracted from airborne particulates with several organic solvents **(4)**. The sample of particulate matter (collected by means of a High Volume sampler or similar air filtering equipment) is extracted with 10 ml of redistilled methylene chloride in a microsoxhlet extractor until all soluble matter has been removed. This extract is transferred to a weighed 25 ml-flask by washing the Soxhlet flask with 1 ml portions of methylene chloride and is evaporated to dryness by a current of nitrogen or *in vacuo* at room temperature. The organic residue from the extraction of particulates is carefully weighed using an accurate analytical balance and redissolved in a minimum volume of methylene chloride (1.0 to 5.0 ml measured accurately from a microburet) just before analysis. More than 50 mg of organic residue can be dissolved in as little as 1 ml of methylene chloride.

7.1.2 Weigh accurately 10.0 mg of pure benzo[a]pyrene. Dissolve in 10.0 ml of methylene chloride in a volumetric flask and store in a refrigerator in the dark, when not in use, as a stock solution. A standard solution containing 2 μg/ml is prepared by measuring accurately 2.0 ml of this stock solution and diluting to 1000 ml.

7.2 SEPARATION. 100 ml of distilled pentane are poured into a development chamber (30.7 × 9.8 × 27.5 cm) and allowed to equilibrate for a minimum of 1 hr. A 0.4-ng aliquot of the methylene chloride solution of the organic extract is spotted (proportioned among 5 to 10 spots) 1.5 cm from the bottom of the thin-layer plate. A 2-μg aliquot of a methylene chloride solution of pure benzo[a]pyrene (proportioned between two spots) is carefully added to the same plate. (**Note**. Edges of all plates must be free of adsorbent and all spots be kept 1.5 cm from outer edge of plate). Precaution must be observed to ensure integrity of each spot, keep spot areas small, and avoid exposure to direct light. The spotted plates are then placed in the development chamber and the solvent allowed to travel 15 cm from the origin. The developed chromatogram is observed under a 360 nm wavelength light source and the fluorescent area of the pure standard and a corresponding area of the sample are quickly scored with a sharp stylus. Exposure to ultraviolet radiation must be kept to a minimum. The absorbent areas containing the benzo[a]pyrene fraction and the pure standard are removed from the plate and placed in test tubes for subsequent elution and analysis.

7.3 ELUTION AND ANALYSIS. 10 ml of anhydrous ether are added to the test tubes containing sample and standard and the slurries mixed mechanically for three minutes. The mixtures are then quantitatively transferred to fine-porosity sintered-glass funnels positioned in 50-ml filtering flasks attached to a vacuum line and maintained at 50 ± 2 C in a water bath. Washing of the adsorbents is continued at a rate approximately equal to the rate of evaporation, until 100 ml of ether has been used. Immediately when evaporation is complete, the flasks are removed from the water bath and the residues quantitatively transferred to optical cells with distilled pentane. Volumes are carefully adjusted to one ml and the absorbances are determined between wavelengths of 400 and 240 nm. The micrograms of benzo[a]pyrene in the sample are determined by comparison with the pure compound using wavelengths 390, 382, and 375 nm by the base line technique **(1, 2)** and Intersociety Committee Method **(6)**, Section **9.1**.

8. Concentration and Absorbance of Standards

8.1 14 μg of pure benzo[a]pyrene in 1 ml of solution yield an absorbance of 1.0 when measured using the base line technique. Also a concentration vs absorbance curve is linear from 14 μg through the origin; however, values obtained using such a curve have to be corrected for a predetermined elution efficiency. The method has better precision when a pure standard is used for each analysis.

8.2 Benzo[a]pyrene after several recrystallizations may contain about 0.06% perylene and a trace of anthanthrene.

8.3 Proper activation of the aluminum oxide thin-layer plates is achieved by heating at 100 C for 30 min. Activated plates are stored in a vacuum desiccator adjusted to 45% relative humidity with aqueous sulfuric acid. (This procedure may change with changes in climate or laboratory conditions).

9. Calculations

9.1 All calculations are based on values obtained with pure BaP run at the same time and on the same plate as the sample. The amount of benzo[a]pyrene present in the sample is calculated using the following equations:

μg BaP found/gram organic extract =

$$\frac{2000.A_x}{W_a.A_s}$$

and

μg BaP found/1000 m^3 of air =

$$\frac{W_f.W_t}{V}$$

Where:

A_s = Absorbance of 2 μg BaP (eluted from plate).

A_x = Absorbance of BaP area (eluted from plate).

W_a = Weight of organic extract analyzed in mg.

W_t = Total weight of organic extract in mg.

W_f = μg BaP found/gram organic extract.

V = Volume of air sampled in m^3

10. Effect of Storage

10.1 After evaporation of the solvent used to extract the organic fraction of air particulates, the residue can be stored in a cold dark area for several years without further decomposition in terms of benzo[a]pyrene. Methylene chloride solutions of the extracts are stable for several months when stored in a dark cold area.

10.2 Methylene chloride solutions of the pure benzo[a]pyrene change from colorless to pale yellow in 8 hr under normal laboratory conditions. However, standard solutions, which are kept cold, in the dark, and refrigerated immediately after each use, are stable for one week. After 1 week, absorbance values are about 10% below normal.

10.3 Aluminum oxide thin-layer plates when stored for long periods of time become contaminated and usually give poorer separations. Generally plates prepared in the laboratory should not be stored for more than three weeks. Also plates should be checked for fluorescent impurities before application of sample and standard.

11. References

1. COMMINS, B.T. 1958. Analyst 83:386.
2. COOPER, R.L. 1954. Analyst 79:573.
3. SAWICKI, E., T.W. STANLEY, W.C. ELBERT, and J.D. PFAFF. 1967. Anal. Chem. 36:497.
4. STANLEY, T.W., J.E. MEEKER, and M.J. MORGAN. 1967. Environ. Sci. Tech.
5. STANLEY, T.W., M.J. MORGAN, and J.E. MEEKER. 1967. Anal. Chem. 39:1327.
6. AMERICAN PUBLIC HEALTH ASSOCIATION. 1977. Methods of Air Sampling and Analysis. 2nd ed. p. 235. Washington, D.C.

Subcommittee 5
E. SAWICKI, *Chairman*
R. C. COREY
A. E. DOOLEY
J. B. GISCLARD
J. L. MONKMAN
R. E. NELIGAN
L. A. RIPPERTON

104.

Tentative Method of Chromatographic Analysis For Benzo[a]pyrene and Benzo[k]fluoranthene in Atmospheric Particulate Matter

17242-02-69T

1. Principle of the Method

1.1 This rapid method, for the measurement of benzo[a]pyrene (BaP) and benzo[k]fluoranthene (BkF) in air samples, is based on the chromatographic separation of air sample extracts on an activated alumina column using a polar solvent, toluene, as the eluting agent. The concentration of hydrocarbon in the eluates is determined from fluorescence emission measurements made on the eluates (**1**).

2. Range and Sensitivity

2.1 The method can measure concentrations in a prepared air sample extract or fraction over the range of 0 to 0.25 micrograms of BaP or BkF/ml of solution.

3. Interferences

3.1 In the chromatographic separation used, BaP and BkF will be found in about 7 to 8 eluate fractions. These fractions are likely to contain both BaP and BkF, but unlikely to contain other strongly fluorescing compounds. It then becomes a question of ability to measure BaP (or BkF) accurately in the presence of the other hydrocarbon.

Warning: Some polynuclear aromatic hydrocarbons, benzo[a]pyrene in particular, are carcinogenic. Consequently, care must be taken in handling laboratory equipment to avoid spilling solutions or PAH material on hands or other parts of the skin. Manipulations involving the PAH compounds should be carried out in a fume hood and must not be allowed to contaminate the laboratory air.

4. Precision and Accuracy

4.1 0.25 μg of BaP/1 ml can be measured with an accuracy of better than 0.002 μg. 0.25 μg of BkF can be measured with an accuracy of better than 0.001 μg. If the concentration of BaP found is more than twice the BkF concentration, the BkF results will be in error by a factor of 10%. In such a case, suitable mathematical corrections can be made.

5. Apparatus

5.1 LABORATORY WARE. All glassware is cleaned before use by rinsing with dilute hydrochloric acid, then with distilled water, alcohol and finally with redistilled or spectrograde cyclohexane to remove all traces of organic matter such as oil, grease, and detergent residues, etc. No grease shall be used on stopcocks or joints. Examine the glassware under ultraviolet light to detect any fluorescent contamination. Some polynuclear hydrocarbons are very susceptible to photooxidation, consequently all operations should be carried out in subdued light or the apparatus should be covered with aluminum foil.

5.2 SAMPLING EQUIPMENT. The sample of particulate matter is usually collected on flash fired glass fiber filter using a high volume air sampler. Other methods of collecting particulate matter may also be employed. Before use, the filter should be washed thoroughly with spectrograde cyclohexane to make the blank as low as possible. (Refer to Intersociety Committee Method (**2**), Section **7.1** for details of collection of particulate matter with a high volume sampler).

5.3 SPECTROPHOTOFLUORIMETER. A fluorimeter equipped with motor-driven excitation and emission monochromators is required in the wavelength range of about 250 to 550 nm. The Aminco Bowman spectrophotofluorimeter or any instrument with similar performance may be used.

5.4 EXTRACTORS. Microsoxhlet extractors with boiling flask capacity of 30 ml are used.

5.5 COLUMNS. Borosilicate glass chromatographic columns are required with the dimensions of 30-cm length and 10.0 mm ID. The columns should be provided with an integral reservoir at the top and a polytetrafluoroethylene plug stopcock of 2.0-mm bore.

6. Reagents

6.1 CYCLOHEXANE. Spectrograde, may be purchased, but in the quantities used this is too expensive. Cyclohexane suitable for either fluorescence or ultraviolet measurements is readily and cheaply prepared by percolating a good technical grade of cyclohexane through a column of activated carbon. When prepared, this is stored in glass stoppered borosilicate bottles of 4-l capacity. Technical cyclohexane produced from benzene is unsuitable starting material as it may contain up to 10% benzene.

6.2 Activated carbon suitable for the preparation of spectrograde cyclohexane, iso-octane or n-pentane is available from several sources, including Pittsburgh Chemical Co. A column of 12 × 30 material 2.5″ in diam and 18″ in depth allows the preparation of spectrograde cyclohexane at the rate of several l/hr.

6.3 ALUMINA, 100 to 200 mesh size, is heated for 24 hr in an oven at 140 C before use. It is used in the activated condition. Peter Spence type H alumina, or equivalent, is satisfactory.

6.4 TOLUENE. The so-called "nanograde" material as supplied by Mallinckrodt is satisfactory. Other grades may be used if free from fluorescence.

6.5 BENZO[A]PYRENE (BAP). This material is available from Eastman Kodak, Rochester, Aldrich Chemical Co., Milwaukee or James Hinton, Newport News, Va. Material from the latter source is zone refined and 99.9% purity.

6.6 BENZO[K]FLUORANTHENE (BKF). This material is apparently not available commercially at the present time. Limited quantities for analytical purposes may be obtained from the Occupational Health Division, Environmental Health Centre, Ottawa, Canada.

6.7 STANDARD SOLUTIONS OF BAP. Solutions of BaP are prepared containing 0.005, 0.010, 0.015, 0.020 and 0.025 μg/1 ml in fluorescence-free toluene. Weigh accurately 1.25 mg of BaP on a micro balance and dissolve in 250 ml of spectrograde toluene. Measure accurately 1.0 ml of this stock solution and dilute to 1000 ml with spectrograde toluene. Repeat this with 2.0, 3.0, 4.0 and 5.0 ml portions of stock solution, diluted in each case to 1000 ml.

6.8 STANDARD SOLUTIONS OF BKF. These solutions are prepared in the same manner as that described in Section **6.7**.

7. Procedure

7.1 The chromatographic columns are set up in a fume hood and the complete preparation of column and elution procedure is carried out there. Depending upon the experience of the operator, up to 10 columns (10 samples) can be processed simultaneously. Fluorescence readings are made while the elutions are in progress.

7.2 Using a clean metal punch or cork borer, 4 circles of 35.5 mm diameter are cut from the high volume glass fiber sample. These represent 1/10.5 of the total effective area of the filter. These are placed in a microsoxhlet extractor on top of a wad of glass wool which prevents the carbon particles from being washed over into the extract and avoids subsequent filtration. The area chosen, in this case, amounts to *ca* 10% of the filter. After 6 to 8 hr extraction with fluorescence-free cyclohexane, the solvent extract is evaporated carefully in a current of nitrogen to 2 ml. **(Caution:—do not allow the mixture to become dry)**

7.3 A chromatographic column is then prepared by slurrying the activated alumina with toluene and filling the tube to a depth of 12.0 cm. The concentrated extract (no more than 2 ml) is placed on the alumina and after rinsing with a further 1 ml of toluene, elution is carried out using toluene from beginning to end. The first 25 ml of eluate are discarded and 3 ml fractions are collected thereafter up to a total of about 30 fractions (90 ml). Each fraction is scanned separately in the fluorimeter and those fractions containing BaP and BkF are combined for a further measurement. BaP and BkF are usually eluted between fractions 15–21 but this can vary depending upon the activity of the particular batch of alumina in use. See Figures 104:1–5.

7.4 Having combined all the fractions containing BaP and BkF, these are carefully evaporated and made up to a final volume of 5.0 ml in toluene. Fluorescence emission measurements of peak heights are now made using excitation wavelengths 307 and 384 nm. The method used to measure peak heights is shown in Figure 104:2. These wavelengths had been established by running the excitation spectrum of both hydrocarbon standards at the 0.015 μg/1 ml concentration.

7.5 A blank determination is carried out on the glass fiber filter, glassware and reagents.

8. Calibration and Standards

8.1 Standard curves of fluorescence emission, in arbitrary units, are prepared for the various concentrations of both BaP and BkF at the two exciting wavelengths 384 and 307 nm. *i.e.*, the optimum excitation wavelengths for BaP and BkF. For curves for the solutions in toluene are shown in Figure 104:6. These are prepared by plotting the height of the peaks against the concentration. The fluorescence emission intensities of air sample extracts or

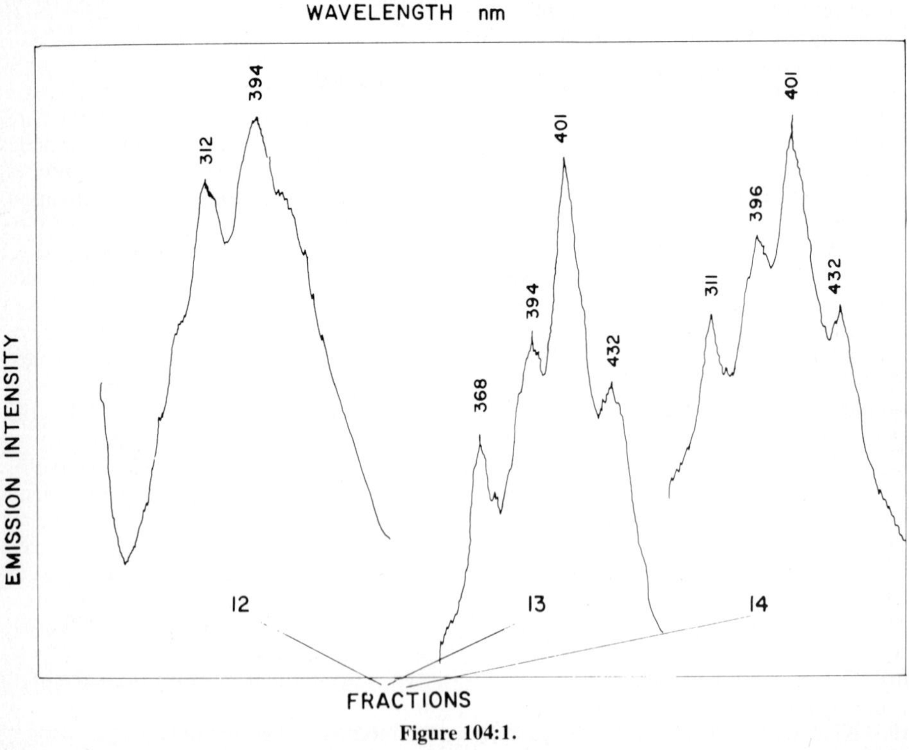

Figure 104:1.

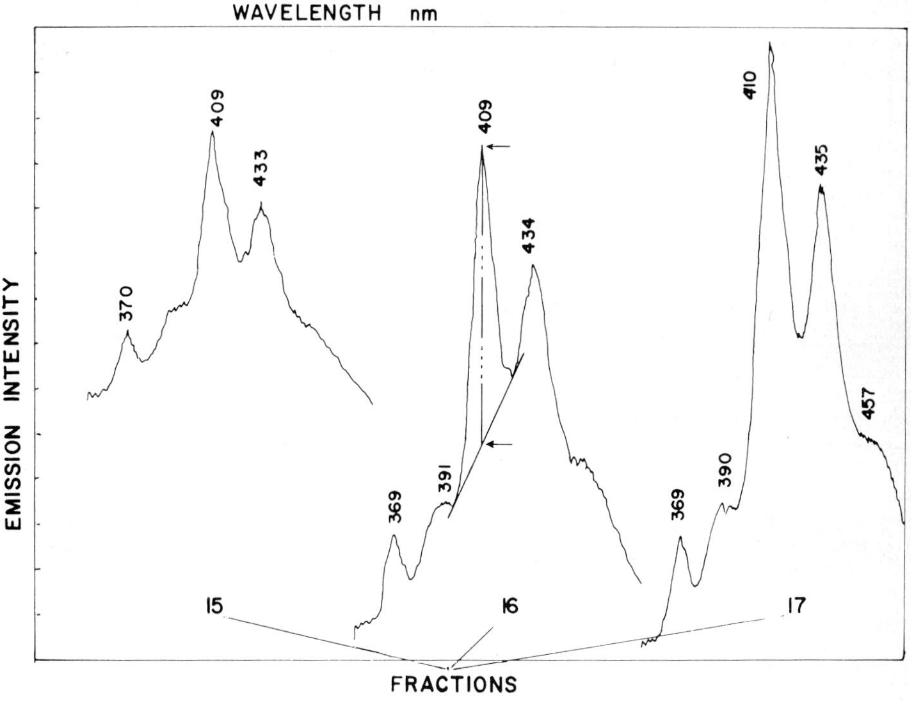

Figure 104:2.

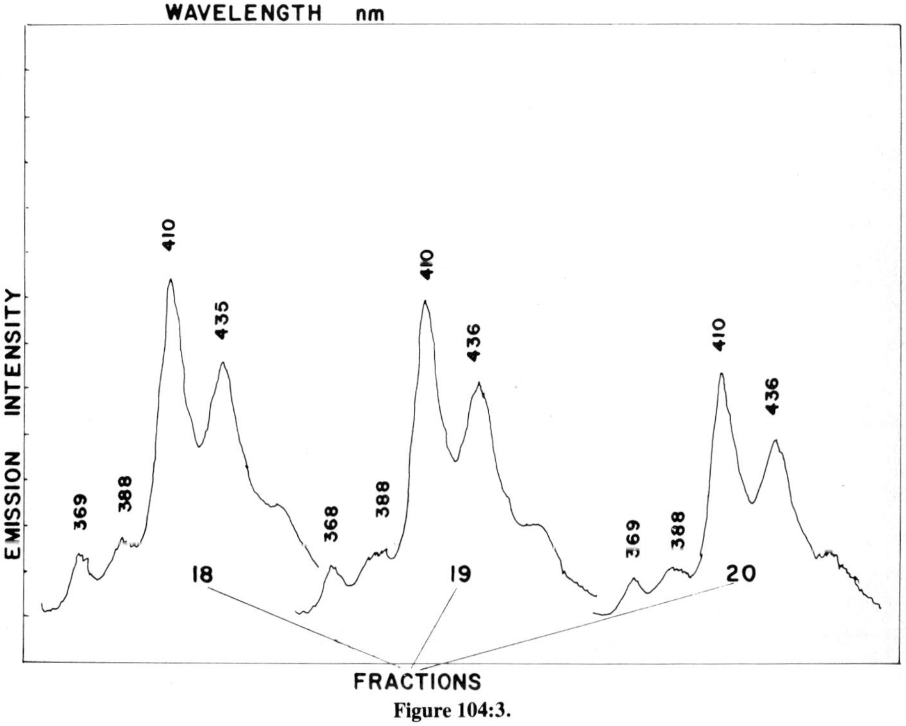

Figure 104:3.

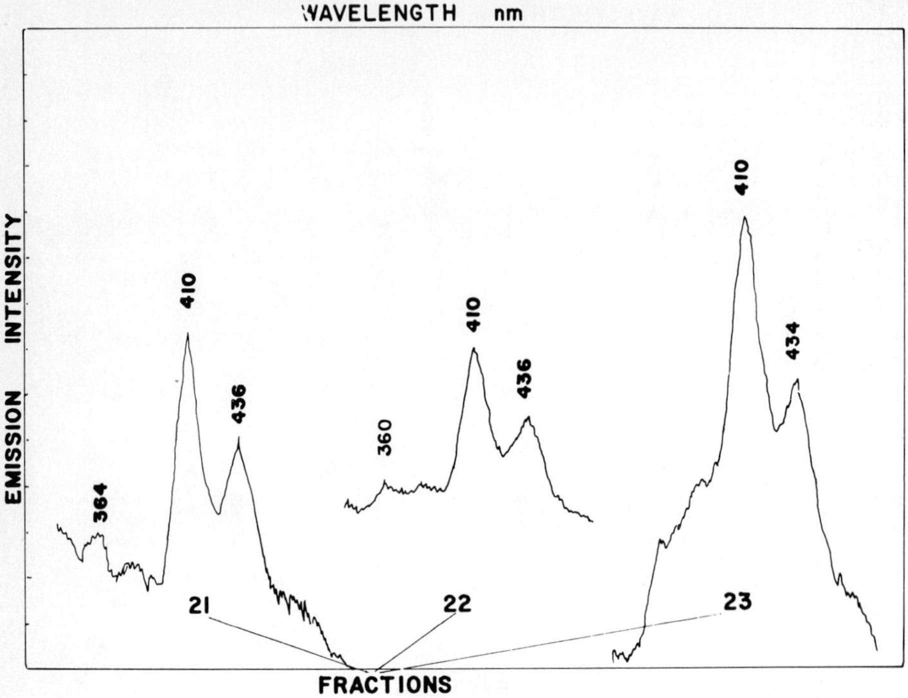

Figure 104:4.

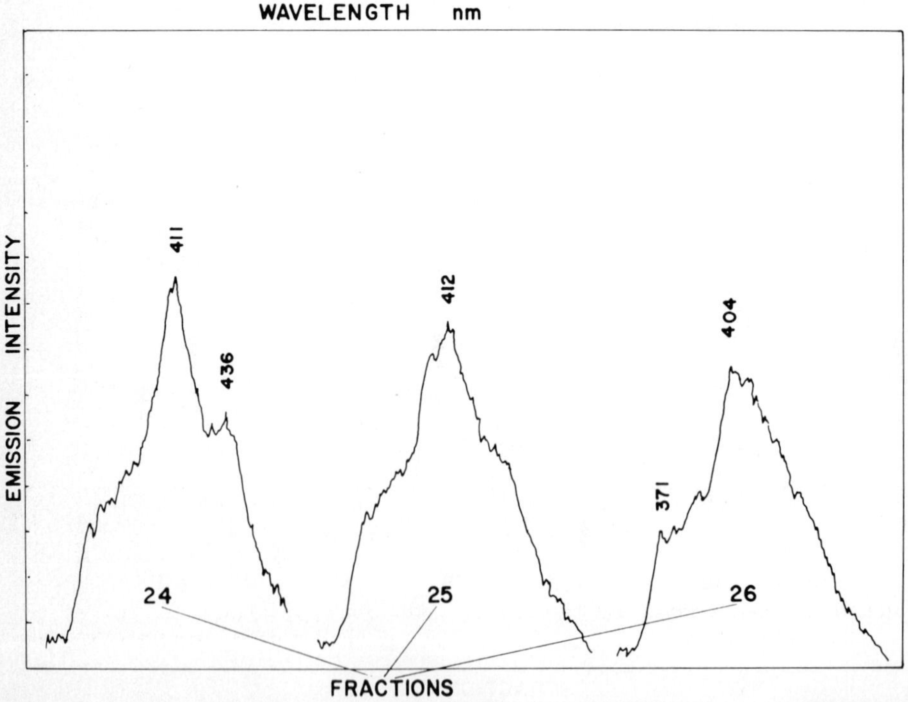

Figure 104:5.

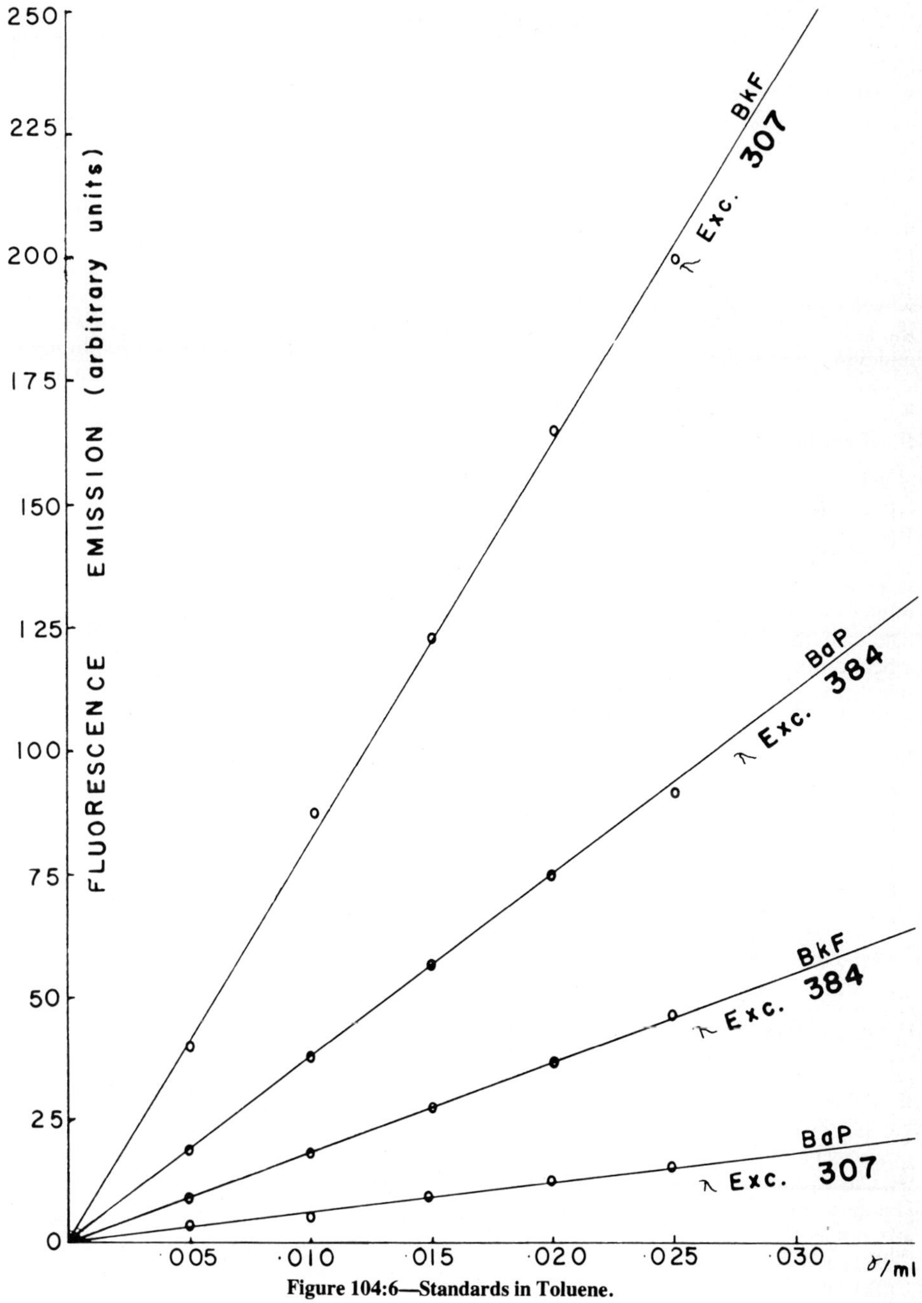

Figure 104:6—Standards in Toluene.

fractions are also measured using the same two exciting wavelengths.

8.2 Similar calibration curves are obtained when standard solutions are made up in cyclohexane. The sensitivity is noticeably less in toluene than in cyclohexane solution. However, though the sensitivity may be less, there is some advantage in making measurements in toluene. The emission of BaP with 384 excitation, is higher than the emission of BkF with 384 excitation, so that the toluene-based BaP measurement is somewhat better than a measurement made in cyclohexane. The optimum excitation wavelengths for BaP differ somehwat with the solvent, being 382 nm in cyclohexane and 384 nm in toluene.

9. Calculation

9.1 Since the fluorescence emission intensity of BkF is much greater than that of BaP when a mixture of the two is excited at 307 nm, the reading at this wavelength is essentially due to BkF. Having determined the concentration of BkF, one can calculate the effect of this hydrocarbon when a mixture is excited at 384 nm after which the BaP concentration may be calculated. Thus:—

$$\text{Conc. BkF } (\mu g/ml) = \frac{\text{Emission of sample at 307 nm excitation}}{\text{Slope of BkF Standard curve at 307 nm excitation}}$$

$$\text{Conc. BaP } (\mu g/ml) = \frac{\text{Emission of sample at 384 nm—(conc. BkF X Slope BkF standard curve at 384 nm excitation)}}{\text{Slope of BaP standard curve at 384 nm excitation}}$$

Example calculation—Suppose a sample yields a peak height of 14.8 arbitrary units at an excitation wavelength of 384 nm and 38.5 at λ_E 307 nm. By reference to the standard curves in Figure 104:6, the standard slopes are,

BkF at λ_{307} = 6610 (arbitrary units)
BkF at λ_{384} = 1890
BaP at λ_{384} = 1500
Conc. BkF (μg/ml) =

$$\frac{38.5}{6610} = 0.00582$$

Conc. BaP (μg/ml) =

$$\frac{14.8—(0.00582 \times 1890)}{1500}$$

$$= 0.00252$$

The peak height at a given emission wavelength is measured by the base line technique as illustrated in Figure 104:2.

The concentrations of the BaP and BkF in μg/g of particulate matter or per 100 m^3 of air sample are calculated as follows:—

To determine the concentration in μg/g, multiply the observed concentration in μg/ml (as above) by the total liquid volume (5 ml) and by the area factor of the glass fiber filter (10.5), then divide by the total weight of particulates collected on the filter (g).

$$\text{Conc. of hydrocarbon, } \mu g/g = \frac{\mu g/ml - 5 \times 10.5}{\text{wt. of particulates, g}}$$

$$\text{Conc. of hydrocarbon, } \mu g/1000M^3 = \frac{\mu g/ml \times 5 \times 10.5 \times 1000}{\text{Volume of air sample, } M^3}$$

The volume of the air sample should be corrected to standard conditions.

10. Effect of Storage

Polycyclic aromatic hydrocarbons in standard solutions or in sample extracts appear to remain constant for many months when such solutions are stored in the refrigerator in borosilicate glass bottles containing similar glass stoppers. The samples must be protected from light.

11. References

1. DUBOIS, L., A. ZDROJEWSKI, C. BAKER, and J.L. MONKMAN. 1967. Some Improvements in the Determination of Benzo[a]Pyrene in Air Samples. Air Poll. Control Assoc. J. 17:818–821.
2. AMERICAN PUBLIC HEALTH ASSOCIATION. 1977. Methods of Air Sampling and Analysis. 2nd ed. p. 235. Washington, D.C.

Subcommittee 5

E. SAWICKI, *Chairman*
R. C. COREY
A. E. DOOLEY
J. B. GISCLARD
J. L. MONKMAN
R. E. NELIGAN
L. A. RIPPERTON

105.

Tentative Method of Analysis for 7H-Benz(de)anthracen-7-one and Phenalen-1-one Content of the Atmosphere (Rapid Fluorimetric Method)

17502-01-70T

1. Principle of the Method

1.1 The benzene-soluble fraction of an urban airborne particulate sample is separated alongside pure 7H-benz(de)-anthracen-7-one (benzanthrone or BO) and pure phenalen-1-one (perinaphthenone or PO (Figure 105:1) on glass-fiber paper impregnated with silica gel (ITLC). The unknown and standard spots are eluted, their solutions are evaporated, and the residues are dissolved in concentrated sulfuric acid. Bright yellow and emerald-green fluorescent cationic salts are obtained for BO and PO, respectively. Readings of standard BO and test spot solutions are taken at F 370/560 on a spectrophotofluorimeter, using Aminco filter No. 4-7164, which peaks at 535 nm. Readings of standard PO and test spot solutions are taken at F 400/498 using Corning filter No. CS3-72, which excludes wavelengths below 435 nm. **(1, 2)**.

2. Range and Sensitivity

2.1 The lower limit of detection of BO with a spectrophotofluorimeter at its full potential of sensitivity is 10 ng; for PO it is 20 ng. Range of analysis is 10 to at least 1000 ng of BO and 20 to 1000 ng of PO. Adequate analysis of the average contaminated urban air sample **(3)** ($\sim$ 800 μg BO/g benzene-soluble fraction) requires more than 12.5 μg of benzene-soluble fraction or more than 1.3 m^3 of air. For samples of low contamination ($\sim$ 80 μg BO/g benzene-soluble fraction) at least 125 μg of benzene-soluble fraction or 12.5 m^3 of air are needed. In analysis for PO the amount of fraction or the volume of air needed is approximately doubled.

3. Interferences

3.1 This method is simpler and faster than all previous procedures. It is highly selective; no interferences in the determination of PO or BO are known. Although the natural mixtures contain many thousands of compounds, the spots are easily obtained and stand out vividly in the chromatogram, Figure 105:1. The fluorescence spectra of the unknown spot and the standard adjacent to it are identical.

4. Precision and Accuracy

4.1 Samples from urban atmospheres of 11 different cities were analyzed in triplicate, using 1 mg of the sample each time, for the presence of benzanthrone. In each case, the triplicate values were very close, an indication of the reproducibility of this method, Table 105:I. Similar precision was found in the determination of PO, Table 105:II.

4.2 Pure benzanthrone was added to an airborne particulate sample and recovered from the benzene extract in a 91% yield and from the methylene chloride extract in a 106% yield. In a similar manner, PO was recovered from the benzene extract in a 101% yield and from the methylene chloride extract in a 104% yield.

5. Apparatus

5.1 GLASS-FIBER PAPER IMPREGNATED WITH SILICA GEL, available from the Gel-

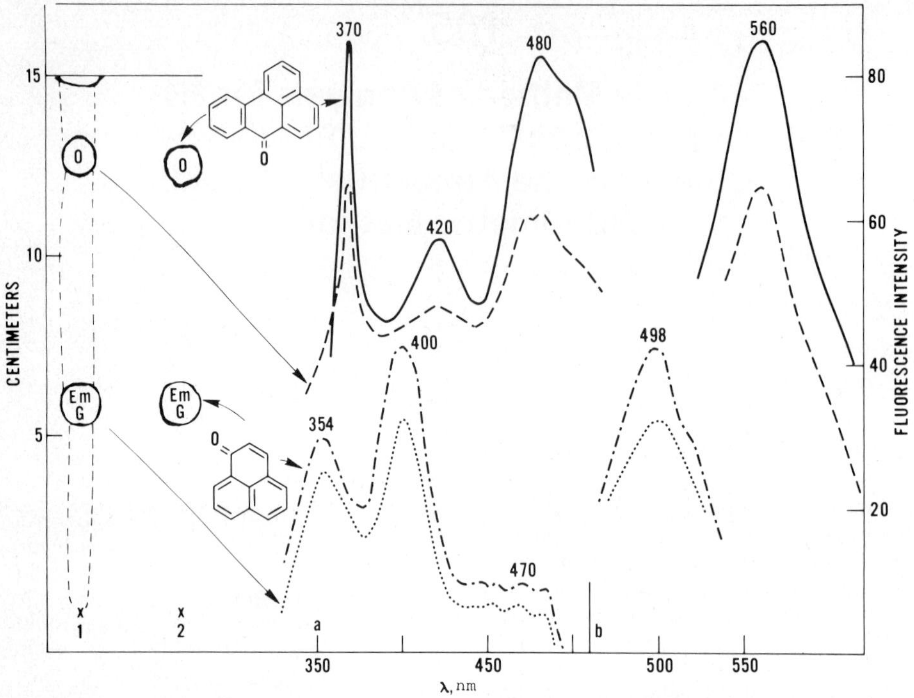

Figure 105:1—On the left side appears the separation on silica gel glass-fiber paper of (1) 1 mg of a benzene-soluble fraction of urban airborne particulate from Greenville, S.C., and (2) a mixture containing 0.4 μg of 7H-benz(de)anthracen-7-one and 0.05 μg of phenalen-1-one. The fluorescence colors shown appeared after TFA fuming. O = orange and EmG = emerald-green fluorescences. On the right side are shown the excitation (a) and emission (b) spectra of standard BO in sulfuric acid at F 370/560 (———), of the unknown spot in sulfuric acid at F 370/560 (-------), of standard PO in sulfuric acid at F 400/498 (–.–.–.–) and unknown spot in sulfuric acid at F 400/498 (. . . . . . .). Characterizations of the 'unknowns' are based on the Rf values and these spectra.

man Instrument Company, Ann Arbor, Michigan, is used without further preparation.

5.2 Thin-layer chromatographic chamber.

5.3 Chromato-Vue cabinet, or some other long wavelength ultraviolet source.

5.4 Spectrophotofluorimeter, with Aminco filter No. 4-7164 peaking at 535 nm (American Instrument Company, Silver Spring, Maryland) and a Corning filter No. CS3-72 used to exclude light of wavelengths below 435 nm (Corning Glass Works, Corning, New York).

6. Reagents

6.1 Purity. All reagents should be of analytic reagent grade.

6.2 Pentane, distilled before use.

6.3 Methylene chloride.

6.4 Conc sulfuric acid.

6.5 Trifluoroacetic acid.

6.6 Phenalen-1-one, recrystallized 2 times before use with hexane the first time and ethylcyclohexane the second time.

6.7 7H-Benz(de)anthracen-7-one: recrystallized before use with ethanol.

7. Procedure

7.1 The particulate matter collected from several cubic meters of urban atmosphere is extracted in a microsoxhlet extractor with 10 ml of either benzene or methylene chloride. The resulting extracts are evaporated to dryness, and the residue is dissolved in the minimum amount of

Table 105:I. Assay* for 7H-Benz(de)anthracen-7-one on ITLC (4)

Location†	mg BO/g Sample‡	Average Value
Harrisburg, Pa.	0.13, 0.20, 0.10	0.14
Greenville, S.C.	0.40, 0.43, 0.48	0.44
Asheville, N.C.	0.46, 0.45, 0.50	0.47
Mt. Vernon, N.Y.	0.22, 0.22, 0.22	0.22
Lynn, Mass.	0.20, 0.25, 0.28	0.24
New Rochelle, N.Y.	0.13, 0.14, 0.16	0.14
Memphis, Tenn.	0.20, 0.23, 0.25	0.23
Ft. Wayne, Ind.	0.23, 0.31, 0.29	0.28
Wilkes-Barre, Pa.	0.13, 0.14, 0.14	0.14
Glen Cove, N.Y.	0.14, 0.13, 0.13	0.13
Saginaw, Mich.	0.25, 0.34, 0.32	0.31

*Developer: Pentane-methylene chloride (3:1, v/v).

Adsorbent: Glass-fiber paper impregnated with silica gel., ITLC

Time for solvent from to travel 15 = 18 minutes.

†Sample size was 1 mg for each determination.

‡Corrected value based on recovery data.

chloroform. An aliquot of this solution, in the amount of 10 μl to the whole sample, is spotted on a glass-fiber paper impregnated with silica gel (ITLC) 1.5 cm from the bottom. The benzene-soluble fraction (1 mg) can also be used. Standard solutions of BO and PO (0.1 μg to a spot) are also spotted on the glass-fiber paper at the origin. The glass-fiber paper is then placed in a thin-layer tank containing approximately 150 ml of pentane—methylene chloride (3:1, v/v) as the developer. A separation time of 18 min is required for the solvent front to travel 15 cm and 6 min for the solvent front to travel 10 cm. After development, the paper is examined under ultraviolet light and the BO and PO spots, opposite the standards, are located with trifluoroacetic acid fumes. The spots are then cut out and extracted with 3 1-ml volumes of boiling acetone; the resulting extracts are evaporated to dryness in a boiling-water bath. The BO residue is dissolved in 0.5 ml of concentrated sulfuric acid. With a filter that peaks at 535 nm (Aminco filter No. 4-7164), a reading is taken at the emission wavelength of 560 nm with the instrument set at an excitation wavelength of 370 nm. The standard BO spot is treated in the same manner.

7.2 For the PO residue of the standard or unknown spot dissolved in 0.5 ml of sulfuric acid, readings are obtained at an emission wavelength of 498 nm with the spectrophotofluorimeter set at an excitation wavelength of 400 nm. Corning filter No. CS3-72 is placed in the instrument in the filter holder in front of the phototube.

8. Calibration and Standardization

8.1 A straight-line relation through the origin between the concentration and the fluorescence intensity is obtained for 10 (20) to at least 1000 ng of BO (PO). Conformance with Beer's Law is observed if the readings are made with the instrument set at a maximum sensitivity of 50 or at a sensitivity of 25.

8.2 Each unknown analysis is checked against standards when it is run, as discussed in the previous section. Knowing Beer's Law is obeyed, a calibration is run at the same time an analysis is made.

Table 105:II. Assay* for Phenalene-1-one on ITLC (4)

Location†	Sample mg PO/g	Value Average
Memphis, Tenn.	0.22, 0.17, 0.17	0.19
Saginaw, Mich.	0.33, 0.30, 0.33	0.32
Harrisburg, Pa.	0.22, 0.21, 0.22	0.22
Glen Cove, N.Y.	0.12, 0.09, 0.16	0.12
Wilkes-Barre, Pa.	0.13, 0.12, 0.14	0.13
Ft. Wayne, Ind.	0.43, 0.41, 0.43	0.42
New Rochelle, N.Y.	0.24, 0.25, 0.22	0.24
Mt. Vernon, N.Y.	0.12, 0.11, 0.12	0.12
Lynn, Mass.	0.23, 0.25, 0.23	0.24
Asheville, N.C.	0.52, 0.52, 0.51	0.52
Greenville, S.C.	0.38, 0.36, 0.40	0.38

*Developer: Pentane-TFA (50:1, v/v)

Adsorbent: Glass-fiber paper impregnated with silica gel., ITLC

Time for solvent front to travel 15 cm = 20 min.

Filter No. CS3-72 used in spectrophotofluorimeter

†Sample size in all determinations was 0.6 mg except for Asheville, N.C. and Greenville, S.C., which were 0.4 mg samples.

9. Calculations

9.1 Although the relationship between the concentration and the product of the meter multiplier and transmittance readings is linear, it is suggested that a standard be run each time with an unknown sample. On this basis, the amount of BO and PO in the benzene-soluble fraction or in airborne particulates is readily calculated by the following equations:

$$\mu g \text{ BO (or PO)/g organic sample} = \frac{W_s \cdot R_x}{W_t \cdot R_s}$$

and

$$\text{ng BO (or PO)/m}^3 \text{ air} = \frac{W_s \cdot R_x}{V \cdot R_s}$$

where:

W_s = Weight in nanograms of standard PO or BO placed on the glass-fiber paper.

W_t = Weight in milligrams of unknown organic sample placed at origin.

R_x = Product of meter multiplier and transmittance readings of unknown spot from glass-fiber paper following extraction, evaporation, and solution in 0.5 ml of conc sulfuric acid.

R_s = Product of meter multiplier and transmittance readings of standard BO or PO spot treated in the same manner.

V = Volume in cubic meters from which unknown organic material at origin was obtained.

10. Effects of Storage

10.1 The fluorescence intensity of PO and BO in sulfuric acid is stable for at least 2 hr in room light and decreases only slightly after 8 hr in room light.

10.2 BO is stable on an alumina plate for greater than 20 hr in a dark drawer, greater than 20 hr in room light, and greater than 0.25 hr under UV light in an ultraviolet box **(2)**. Since the hydrocarbons are not very stable on the alumina plate under room light for a prolonged period, are decomposed in minutes under UV light, but are stable in the refrigerated organic extract of airborne particulates for years, one would expect BO to be stable in the refrigerated organic extract for years.

11. References

1. SAWICKI, E., T.W. STANLEY, W.C. ELBERT. 1965. Mikrochimica Acta, 1110.
2. SAWICKI, E., H. JOHNSON. M. MORGAN. 1967. Mikrochimica Acta, 297.
3. SAWICKI, E. 1967. Arch. Environ. Health, 14:46.
4. ENGEL, C.R. and E.J. SAWICKI. 1968. Chromatog., 37:508.

Subcommittee 5

E. SAWICKI, *Chairman*
R. C. COREY
A. E. DOOLEY
J. L. MONKMAN
L. A. RIPPERTON
J. E. SIGSBY
L. D. WHITE

106.

Tentative Method of Analysis for Polynuclear Aromatic Hydrocarbon Content of Atmospheric Particulate Matter

11104-01-69T

1. Principle of the Method

1.1 The polynuclear aromatic hydrocarbon content of air particulate matter is determined by a lengthy process involving air sampling, Soxhlet extraction, column chromatography, and ultraviolet-visible spectrophotometry.

1.2 Air particulate samples are collected on glass fiber filters using a high volume air sampler. The samples are then extracted with benzene in a Soxhlet extractor to obtain a benzene soluble fraction of air particulate matter which is normally referred to as the "organic fraction". This organic fraction is then further separated into approximately 30 to 40 sub-fractions by means of column chromatography on alumina using increasing amounts of diethyl ether in pentane as the eluent. Each of the subfractions is dried, dissolved in pentane, and spectrophotometrically analyzed for polynuclear aromatic hydrocarbon content. The polynuclear aromatic hydrocarbons that can be effectively separated and analyzed by this technique if they are present in a particular air sample are anthracene, phenanthrene, fluoranthene, pyrene, benz [a] anthracene, chrysene, benzo [a] pyrene, benzo [e] pyrene, perylene, benzo [g,h,i] perylene, anthanthrene, and coronene.

1.3 A description of the separation, characterization, and analysis of polynuclear aromatic hydrocarbons in urban airborne particulate matter has been published by Sawicki, *et al.* (**7, 8, 10**). The various techniques involved are essentially a modification and application of the work of Falk (**3**) and the method has been successfully applied to a great number of air samples (**6, 10**) and to other sources such as auto exhaust, industrial and domestic heating plants, coal tar pitch, etc. (**9**). An excellent review article (**5**) which discusses the numerous techniques available for the analysis of polynuclear aromatic hydrocarbons has also been published.

2. Range and Sensitivity

2.1 The method can easily determine microgram quantities of the various hydrocarbons/1000 m^3 of air. The exact sensitivity is dependent upon volume of air sampled, the amount of particulate matter present, and the corresponding amount of organic fraction obtained. If the hydrocarbons are present in microgram quantities, they can be determined readily.

2.2 Since the Beer-Lambert Law holds true for varying concentrations of the hydrocarbons, the range of the method can be easily extended by simple dilution of the fraction or by increasing the weight of sample extracted.

3. Interferences

3.1 Any material eluted from the column along with a particular hydrocarbon

Warning: Some polynuclear aromatic hydrocarbons are carcinogenic. Consequently, care must be taken in handling laboratory equipment to avoid spilling solutions or PAH material on hands or other parts of the skin. Manipulations involving the PAH compounds should be carried out in a fume hood and must not be allowed to contaminate the laboratory air.

that exhibits a spectral band at the wavelength at which the hydrocarbon is measured will probably mask the peak to be measured and of course be an interference. Thus far this type of interference has been found to be minimal.

4. Precision and Accuracy

4.1 The precision of the method, can be considered from two viewpoints. If the entire method is taken into account, (*i.e.*,—collection of particular sample, Soxhlet extraction, and finally actual analysis by relatively inexperienced hands) the results for samples taken in the same vicinity at the same time will show a precision of + 15 to 20%. If duplicate analyses are performed on the same organic fraction, a precision of approximately + 10% will result.

5. Apparatus

5.1 The apparatus needed once the organic extracts are obtained are the following:

5.1.1 *Chromatographic Columns.* All borosilicate glass columns have an internal diameter of about 13 mm and an overall height of approximately 390 mm. They should be equipped at the top with a S/J 18/9 ball glass joint to accommodate a separatory funnel as a solvent reservoir and on the bottom with a drip tip. (See Figure 106:1). The glassware is cleaned before use by rinsing with dilute hydrochloric acid, then with distilled water, alcohol and finally with redistilled isoctane to remove all traces of organic matter such as oil, grease, and detergent residues etc. No grease shall be used on stopcocks or joints. Examine the glassware under ultraviolet light to detect any fluorescent contamination. Because some polynuclear hydrocarbons are very susceptible to photo-oxidation all operations should be carried out in subdued light or the apparatus should be covered with aluminum foil.

5.1.2 *Fraction Collector.* A fraction collector for the liquid from the chromatographic column capable of collecting 15 to 25 ml fractions in test tubes. A time flow collector is recommended rather than a drop counter collector because most drop counters employ a photo-electric cell in counting. The light intensity of the photoelectric cell may destroy some polynuclear hydrocarbons which are light-sensitive.

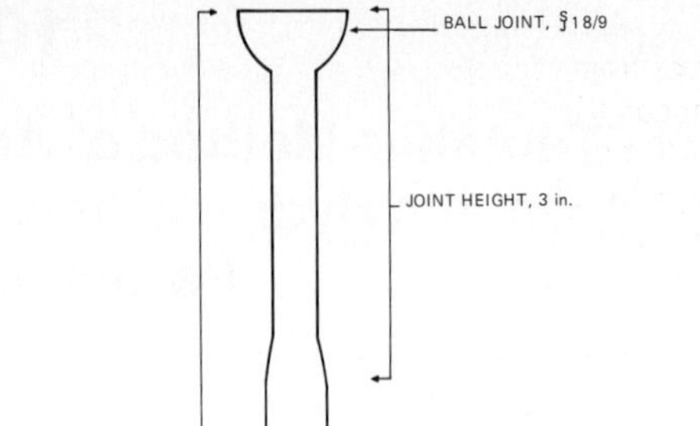

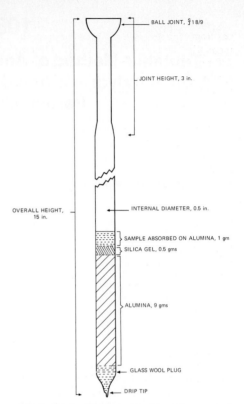

Figure 106:1—Chromatographic column used for separation.

5.1.3 A "Vacuum Oven" can be used for evaporation of the pentane-ether subfractions to prepare the samples at room temperatures in the dark before measuring their absorbances. If the evaporation is carried out at an elevated temperature, add 1 ml of PNA-free hexa-decane and discontinue the evaporation when 1 ml of liquid remains in the flask, to avoid loss of any benzene soluble sample by volatilization. In the interest of safety, nitrogen should be added above the heated liquid to avoid a possible explosion.

5.1.4 *Spectrophotometer*. A ratio recording spectrophotometer capable of measuring absorbancies between 450 and 220 nm is suitable for analysis. A manual spectrophotometer is highly impractical because of the many absorbance maxima and minima measured for each subfraction.

5.1.5 *Spectrophotometric Cells for Analysis*. Cells having a 10 mm light path and minimum transmittance of 80% at 200 nm are used for spectrophotometric analysis. For convenience, the cells are also used as miniature volumetric flasks and should be calibrated to contain a volume of 3 ml. The 10 mm cylindrical cells (Pyrocell Manufacturing Co. or Applied Physics Corp. or equivalent cells from any other suitable source) contain 3 ml when filled just to the bottom of the glass joint.

6. Reagents

6.1 PURITY. All chemicals should be analytical reagent grade unless otherwise specified.

6.2 ALUMINA. Merck Acid Washed Aluminum Oxide (100 to 200 mesh) or equivalent from another manufacturer is suitable for the chromatography. If a great number of samples are to be analyzed, a large batch of alumina can be prepared since the water content of the prepared alumina will remain relatively stable for at least 6 months if stored in a tightly stoppered glass container. The quantity of alumina desired is placed in a beaker, covered with diethyl ether and stirred for 5 min. The alumina is then allowed to settle and the supernatant ether is decanted and filtered by vacuum using a Buchner funnel. The process is repeated 4 times on the settled alumina and after the final washing the entire quantity of alumina is transferred to the funnel using diethyl ether as a wash. The alumina is left drying on the funnel until there is no noticeable odor of ether in the alumina. The alumina is then transferred to an evaporating dish and further dried in an oven at 130 C for 30 hr. The percentage of water (approximately 11.5 to 12%) remaining in the alumina is determined by heating an accurately weighed sample (1 to 2 g) in a crucible at red heat (800 C) for 10 min, cooling in a desicator over phosphorous pentoxide or anhydrone, and reweighing. The water content of the large batch of alumina is then adjusted to 13.7% by adding the required amount of water directly to the alumina. The adjusted alumina is then mixed well by shaking and allowed to stand 12 hr in a stoppered container to attain equilibrium.

6.3 SILICA GEL. Davison, or other suitable silica gel (100 to 200 mesh) is thoroughly washed and diethyl ether, dried in the same manner as the alumina and heated at 130 C for 1 hr. Silica gel should be stored in tightly stoppered glass bottles.

6.4 PENTANE. ACS reagent grade pentane is redistilled before use in an all glass apparatus. The fraction boiling at 35 to 36 C is used for analysis.

6.5 ETHER. The diethyl ether should be redistilled anhydrous ACS reagent grade.

6.6 BENZENE. ACS reagent grade benzene is redistilled before use in an all glass apparatus. The fraction boiling at 78 C is used for extraction.

6.7 CHLOROFORM. Distilled chloroform used should be ACS reagent grade.

6.8 POLYNUCLEAR AROMATIC HYDROCARBONS. The hydrocarbons needed for calibration are available from several sources *e.g.*, Aldrich, Eastman, K & K, also Burdick and Jackson Labs, Muskegon, Mich. (Distilled in glass; solvents PNA free). Most of the hydrocarbons must be purified by recrystallization from PNA-free solvents that have been redistilled in glass stills. The hydrocarbons are considered pure when their ultraviolet-visible spectra compare favorably with existing data **(11)**. The purification of hydrocarbons may not be a simple operation depending upon the impurities present. All of the hydrocarbons can be recrystallized from either the non-polar solvents heptane or ethylcyclohexane or from the polar solvents methanol and methanol containing increasing amounts of methyl cellosolve or water depending upon the melting point and molecular weight of the hydrocarbon. If the impurity is polar in nature, a non-polar solvent should be selected and *vice-versa*.

6.9 Sometimes the impurities cannot be removed by simple recrystallization and other techniques such as chromatography or destruction of the impurity by reaction may be necessary. These problems in purification will have to be individually coped with as they arise.

7. Procedure

7.1 SAMPLING. Air particulate matter is usually collected on flash fired glass fiber filter paper using a high volume air sampler **(12)** although other methods of collecting particulate matter may also be employed. Before use the filter should be thoroughly washed with pentane to make the blank as low as possible. The amount of particulate matter collected by a high-volume sampler depends upon many variables (*e.g.*, the particulate loading in the air, sampler location, volume and rate of air sampled, etc.). On the average, when a high volume sample is located in an urban area, the sampler will collect approximately 250 to 350 mg of particulate matter while sampling 2000 to 2400 m³ of air during a 24-hr period. Of the 250 to 350 mg of particulate matter approximately 10% will be soluble in benzene. Hence, the benzene soluble portion of air particulate matter (hereinafter referred to as the organic fraction) for an average high volume sample will amount to approximately 25 mg. A 50 to 150 mg quantity of the organic fraction is needed for the analysis of polynuclear aromatic hydrocarbons in air particulate matter. Therefore, to obtain this needed amount of organic fraction it will probably become necessary to pool together organic fractions of several individual high volume air samples from a single source.

7.1.1 To obtain the organic fraction, the collected particulate matter is extracted with benzene using a Soxhlet extractor for approximately 5 to 6 hr. The benzene extract is filtered through a sintered glass funnel to remove any glass fiber particles and dried in an oven at 60 C. The dry extract is weighed.

7.2 ANALYSIS. From 50 to 150 mg of the organic fraction is dissolved in enough chloroform (usually 5 ml or less) to dissolve the entire sample. One g of the pretreated alumina is added to the solution and the chloroform is evaporated so that the organic fraction is homogeneously dispersed on the alumina. This dry mixture is then transferred to the top of a chromatograph column that contains at the bottom a glass wool plug, a middle layer of 9.0 g of pretreated alumina, and an upper layer of 0.5 g of silica gel (See Figure 106:1). A new column is needed for each analysis. The column is then clamped into position on the fraction collector and the solvent reservoir is clamped to the ball joint at the top of the column. The column is then eluted with successive 100 ml volumes of pentane (added through the reservoir) containing 0, 3, 6, 9, 12, 15, 18, and 25% diethyl ether respectively. Finally, 100 ml of pure diethyl ether is passed through the column to ensure complete recovery. The column is protected from light during fractionation and each succeeding volume of pentane-ether is added when the previous volume has drained to a level about 1 cm above the top of the alumina so as not to disturb the top of the column.

7.2.1 Approximately 40 to 45 chromatographic fractions of 20 ml each are collected in test tubes using the fraction collector. Some of the solvent is lost through evaporation. The fractions are evaporated in a vacuum oven at room temperature in the dark.

7.2.2 The residue in each tube is dissolved in a small amount (0.5 to 1.0 ml) of pentane and transferred to a precalibrated 3 ml spectral cell of 1.0 cm light path. A disposable capillary pipet is very adequate for this transfer. This procedure is repeated 3 or 4 times so that the residue in the tube is quantitatively transferred to the cell. The final volume in the cell is adjusted to 3 ml with pentane and the ultraviolet-visible absorption spectra of each fraction is then determined from 220 to 450 nm against a pentane blank. Comparison of these spectra with predetermined standards enables the calculation of the polynuclear hydrocarbon content of the air particulate matter.

7.3 Notes on Analytical Procedures. The organic fraction to be analyzed is dispersed in the alumina prior to the fractionation because it is only slightly soluble in the primary eluting solvent.

7.3.1 Care must be taken in plugging the column with glass wool. If not, small particles of alumina will channel through the wool and into the subfractions. It is better to plug the column with two or three small pieces of glass wool rather than one large one to eliminate any channeling.

7.3.2 Upon the addition of the first 100 ml portion of 100% pentane, the interaction of the solvent and alumina produces a small amount of heat which sometimes causes the pentane to vaporize slightly. If this happens, some pentane vapor pockets may form in the sample area of the column which essentially results in column splitting. This can be easily controlled by cooling the column until the column becomes saturated with pentane. The cooling of the column can be simply achieved by rubbing the column with an ice cube wrapped in a towel or by a piece of toweling which is cooled by an evaporating solvent such as acetone.

7.4 Order of Elution. The hydrocarbons are eluted from the column in the following order: aliphatics, olefins, benzene derivatives, napthalene derivatives, dibenzofuran fraction, anthracene fraction, pyrene fraction, benzofluorene fraction, chrysene fraction, benzopyrene fraction, benzoperylene fraction, and coronene fraction. The smaller hydrocarbons (aliphatics, olefins, benzene, napthalene, and dibenzofuran) are not analyzed by this procedure because they come off the column together in the initial phases of chromatography. Anthracene and phenanthrene are volatile under the conditions of air sampling and therefore are not found in normal ambient air samples. However, they are found in air polluted by coal tar pitch fumes (**9**).

7.4.1 Three factors affect the chromatography of the organic fraction. These are: (a) the retardation volume, *i.e.* the volume of eluent passing through the column per g of absorbent before the substance in question leaves the column; (b) the volume spread of the eluted substance, *i.e.*, the volume of eluent in which the substance is found; and (c) the volume separation, *i.e.*, the volume of eluent separating one eluent substance from another.

7.4.2 These factors can be affected by three variables: (a) percentage of water in the alumina, (b) percentage of ether in the pentane, and (c) possible miscellaneous factors such as the composition and weight of the organic fraction. An increase in either the amount of water in the alumina or the percentage of ether in the eluent decreases the retardation volume. Conversely, a decrease in either the amount of water on the alumina or the percentage of ether in the eluent increases the retardation volume. The retardation volume, the volume spread, and the volume separation of aliphatic hydrocarbons, and mono-, di-, tri-, and tetracyclic aromatic hydrocarbons are optimal in alumina containing 12 to 13% water while alumina containing 14 to 15% water works best for penta-, hexa-, and heptacyclic aromatic hydrocarbons, when eluted with 100 ml volumes of pentane containing 0, 3, 6, 9, 12, 15, 18, and 25% ether respectively. Therefore alumina containing 13.7% water is used for the fractionation.

7.4.3 Although the relative location of the hydrocarbons in the eluted fractions is almost always the same, unknown variables (probably due to the nature of the sample or discrete differences in alumina) sometimes cause the hydrocarbons to be eluted sooner or later than expected. For this reason, the retardation volume, the volume spread, and the volume separation should be determined whenever, (a) the procedure is being initially performed, (b) a new sample source is to be evaluated, and (c) a new batch of alumina is prepared for analysis.

8. Calibration and Standards

8.1 In order to calibrate the analytical procedure it is essential to become familiar with the quantitative ultraviolet-visible spectra of the polynuclear hydrocarbons being analyzed. Although the spectra and absorptivity of these polynuclear hydrocarbons in pentane that have been published (**11**) can be used for comparison purposes, new spectra of each pure hydrocarbon in pentane should be determined in order to allow for any small operative or instrumental differences that may occur.

8.2 Next to be determined for calibration purposes are the column retention of the hydrocarbons (*i.e.*, the amount of each hydrocarbon eluted from the column as compared to the amount placed on the column) and the relative positions of elution of the hydrocarbons (*i.e.*, the percentage of ether in pentane necessary to elute the hydrocarbon) for each batch of treated alumina. For this reason a large batch of alumina should be prepared.

8.3 Solutions containing 100 μg of hydrocarbon/ml of chloroform are prepared. Individual solutions are prepared for each hydrocarbon to be tested. One ml (100 μg) of the pure hydrocarbon is dispersed in 1.0 g of pretreated alumina and the previously described analytical procedure is carried out using the individual hydrocarbons instead of a complex mixture of the hydrocarbons. The spectrophotometric analysis of the fractions obtained will show, (a) the amount of ether in the pentane necessary to elute the hydrocarbon from the column, and (b) the amount of hydrocarbon retained on the column as calculated by the base line technique. The amount of hydrocarbon retained on the column (about 10% to 30%) is fairly uniform for each hydrocarbon.

8.4 Next, a synthetic mixture containing 100 μg of all the hydrocarbons previously tested/ml of chloroform is prepared. One ml (containing 100 μg of each hydrocarbon) of the mixture is added to 1.0 g of alumina and the analytical procedure is again carried out. Examination of the ultra-violet-visible curves of the fractions obtained will demonstrate the effectiveness of the chromatographic separation.

8.5 Once an analyst becomes familiar with the procedure, it is better to use an actual organic fraction of air particulate matter as the standard. This is accomplished by pooling individual organic fractions into one quite large fraction and performing repeated analyses. In this manner, the standard is an actual air sample and will contain most of the interfering substances normally found in ambient atmospheres. This material can be used at a later time if recalibration becomes necessary since it will remain stable for at least a year when stored in the dark in a refrigerator.

9. Calculations

9.1 The base line technique of Commins (**1**) and Cooper (**2**) is used to calculate the amounts of polynuclear hydrocarbons in the various chromatographic fractions collected. Since the contents of the tubes to be analyzed for polynuclear hydrocarbons are dissolved in 3 ml of pentane, it is convenient for final calculations to determine the micrograms of each individual hydrocarbon necessary to give an absorbance of 1.00 when dissolved in 3 ml of pentane and measured at a specified wave length by the base line technique. The base line technique involves the drawing of a base line between two designated wave lengths beneath the specified wave length maximum to be measured for each hydrocarbon. A perpendicular line is dropped from the apex of the curve to the base line drawn beneath the curve. The absorbance covered by the perpendicular is directly proportional to the micrograms of the hydrocarbons present. Thus, the micrograms of hydrocarbons necessary to give an absorbance covered by the perpendicular of 1.00 can be easily calculated. An example of this calculation for pyrene is shown in Figure 106:2 and the wavelengths in nanometers at which these calculations are performed for each hydrocarbon along with typical calibration factors and abbreviations used in Figure 106:3–8 are listed in Table 106:I.

9.2 To demonstrate the spectral curves obtained and the calculations performed during analysis, the data obtained from an actual organic fraction of air particulate matter is presented. All spectral curves presented had a pentane blank of 0.00 from 450 to 220 nm.

Table 106:I. Illustrative data for measurement of polynuclear aromatic hydrocarbons*

Hydrocarbon In Order of Elution	Per cent Ether†	Base Line From nm	Drawn To nm	Measuring Wave Length nm	Calibration Factor‡ μg
Anthracene (ATC)	3	370	378	374	69
Phenanthrene (PHT)	3	290	297	293	43
Pyrene (PYR)	6	330	340	334	14.4
Fluoranthene (FLT)	6	284	290	287	14.8
Benz(a)anthracene (BAA)	9	282	290	287	14.4
Chrysene (CHY)	9	312	330	320	83.0
Benzo(e)pyrene (BEP)	12	328	335	331	31.0
Benzo(a)pyrene (BAP)	12	375	390	382	42.0
Perylene (PER)	12	430	440	434	45.0
Benzo(g,h,i,)perylene (GEE)	15	370	390	382	36.0
Anthranthrene (ANT)	15	418	423	420	29.6
Coronene (COR)	18–25	335	342	338	16.1
SUPPLEMENTARY WAVE LENGTHS					
Fluoranthene	. .	355	360	358	446
Fluoranthene	. .	354	361	358	257
Chrysene	. .	262	270	267	7.1
Anthanthrene	. .	424	432	428	14.0
Coronene	. .	298	310	301	5.14

*All wave lengths and calibration factors are listed here for comparison purposes and should be redetermined to account for minor operative and instrumental differences.

†Approximate per cent of ether in pentane needed to elute the hydrocarbon from the column.

‡μg of hydrocarbon in 3 ml pentane that will give an absorbance of 1.00 when calculated by the base line technique.

9.3 The first fraction to come off the column contained the aliphatic and small cyclic structures which are eluted using the 0 and 3% ether-pentane solutions. This fraction was collected in tubes #1 through #8 inclusive. An example of the spectral curve of the fraction appears in Figure 106:3, tube #7 of the chromatogram.

9.4 The pyrene was found in tubes #9 and #10 (the beginning of the 6% ether-pentane eluent) while the fluoranthene which normally comes out with the pyrene was found only in tube #10. The spectral curves obtained for tubes #8, #9, #10 and #11 are presented in Figures 106:3 and 106:4. Tubes #12 and #13, much like tube #11, do not show any discernible maxima.

9.5 The chrysene and benz[a]anthracene came off the column at the beginning of the 9% eluent with chrysene appearing in tubes #14, #15, and #16 while benz[a] anthracene was found only in tubes #14 and #15 (see Figures 106:4 and 106:5).

9.6 The benzo[a]pyrene, benzo[e]pyrene, benzo[k]fluoranthene and perylene came off the column in the 12% ether eluent with benzo[a]pyrene appearing in tubes #18, #19, #20, #21 and #22; benzo[e]pyrene in tubes #18, #19, #20, and #21; benzo[k]fluoranthene in tubes #18, #19, #20, #21 and #22; and perylene in tubes #19, #20, and #21. Benzo[k]fluoranthene is demonstrated by the spectral band at 401 nm, but since benzo[a]pyrene also has a band at this wave length, the contribution of benzo[a]pyrene must be subtracted from the overall absorbance at this wave length if benzo[k]fluoranthene is to be determined by this method. Figure 106:6 (tube #18) shows the beginning of the fraction; Figure 106:6 (tube #19) shows the middle of the fraction; and Figure 106:6 (tube #22) shows the end of the fraction. Tube #23 shows no distinguishable maxima. However, benzo[k]fluoranthene is not specifically identified on Figure 106:6, tubes #18, #19 and #22.

9.7 The benzo[g,h,i]perylene and anthanthrene came off the column in the 15%

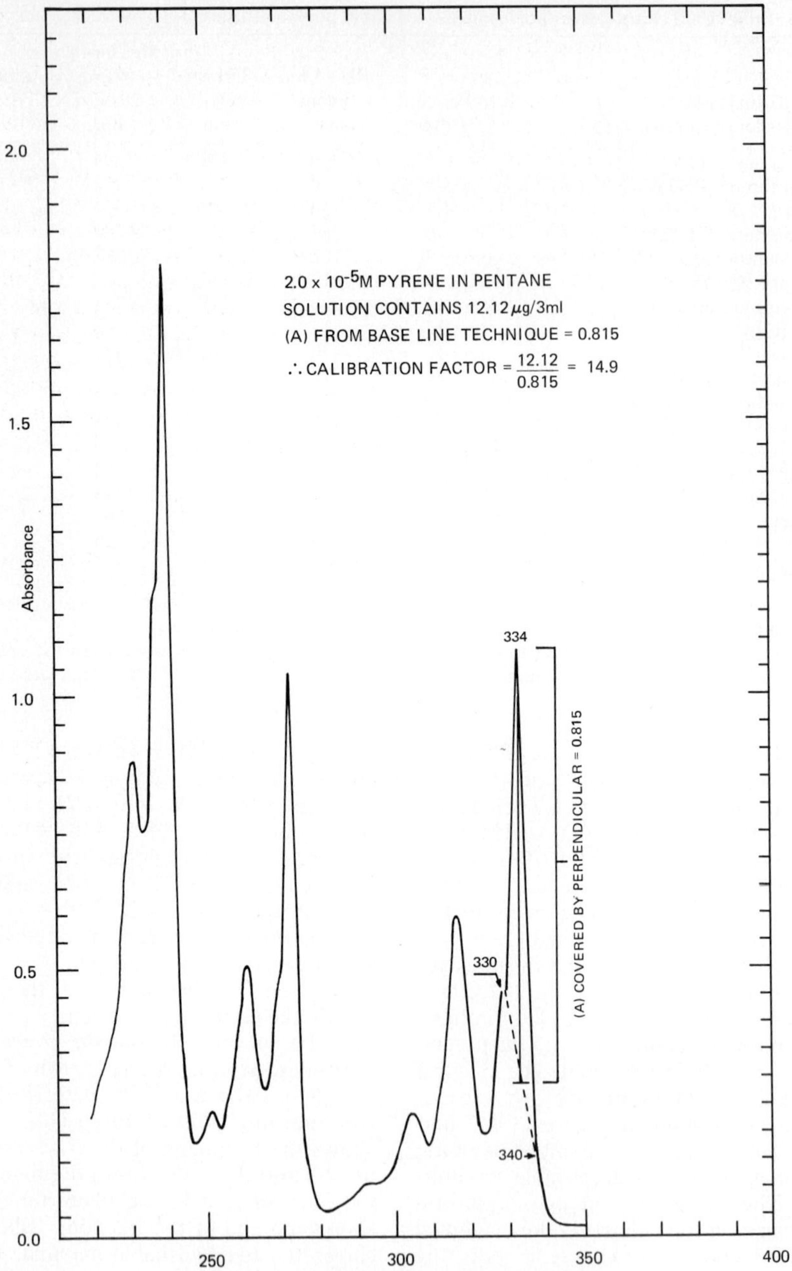

Figure 106:2—Application of the base line technique.

ether eluent with benzo[g,h,i]perylene appearing in tubes #24, #25, #26, #27, #28 and #29 while anthanthrene appeared in tubes #24, #25, #26, #27, and #28. Figure 106:7 (tube #24) shows the onset of the fraction; Figure 106:7 (tube #26) the middle of the fraction; and Figure 106:7 (tube #29) the cessation of the benzo[g,h,i]perylene and anthanthrene. In some fractionations, the benzo[a]pyrene and benzo[g,h,i]perylene may overlap but the end of the benzo[a]pyrene can be distinguished from the benzo[g,h,i]perylene by noting the increase in absorbance and the shape of the spectral curve in the 370 to 390 nm region. The benzo[e]pyrene is normally all eluted from the column before the benzo[g,h,i]perylene starts to come off the column.

9.8 The coronene started to come off the column in the end of the 18% ether eluent and was completely eluted in the 25% ether eluent. The coronene first appeared in tube #31 (Figure 106:8), reached an apex in tube #33 (Figure 106:8) and was last found in tube #35 (Figure 106:8). Tube #36 did not show any identifiable peaks.

9.9 Once the spectral curves are obtained and the presence of the hydrocarbons in the various tubes is established, the base line technique of calculation is employed. The base line is drawn between wave lengths designated in Table 106:I and the perpendicular is drawn from the specified measuring wave length (Table 106:I) to the base line. This is done for each hydrocarbon in every tube it appears. The total absorbance exhibited by any one hydrocarbon is then found by adding together the absorbancies found in the various tubes due to the hydrocarbon. This total absorbance is then multiplied by the calibration factor and the column factor to give the total μg of hydrocarbon present.

μg = A. × calibration factor × column factor

where μg = total μg of hydrocarbon found

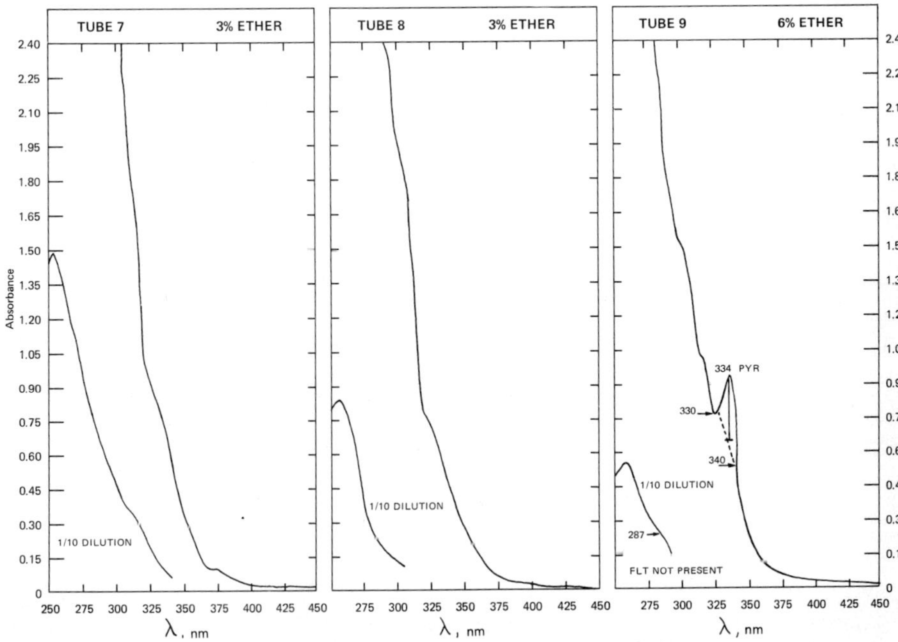

Figure 106:3—Ultraviolet-visible absorption spectra of chromatographic fractions obtained in pentane.

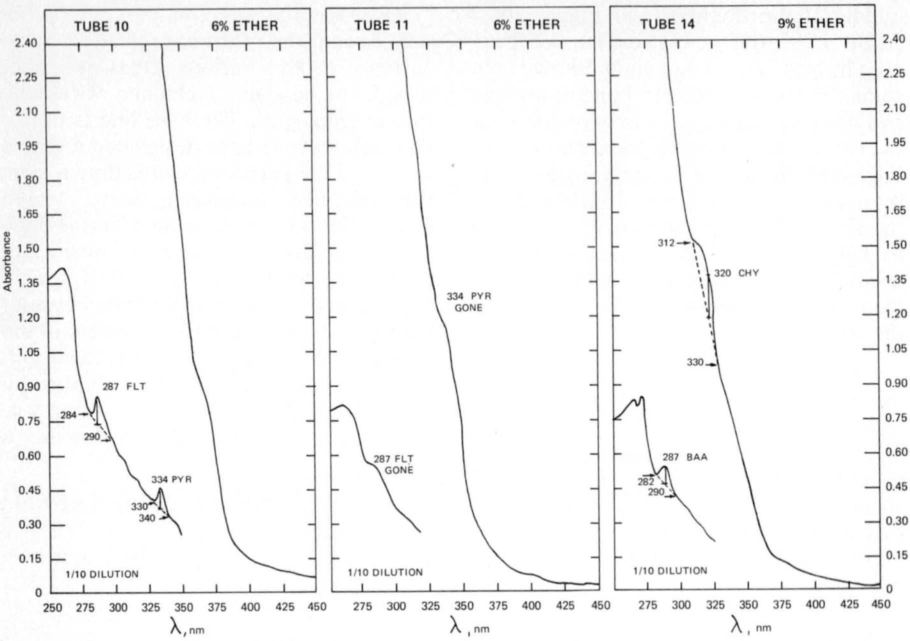

Figure 106:4—Ultraviolet-visible absorption spectra of chromatographic fractions obtained in pentane.

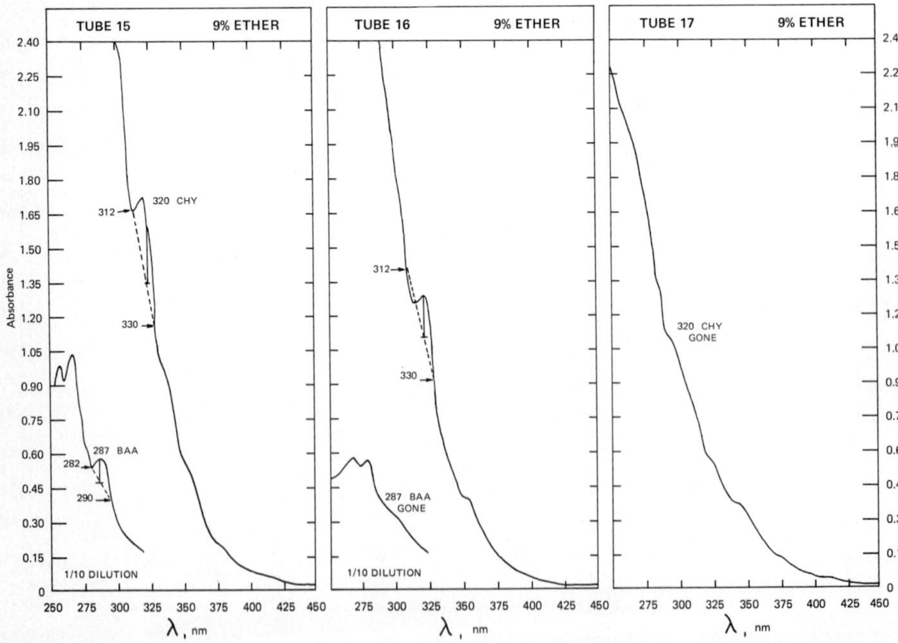

Figure 106:5—Ultraviolet-visible absorption spectra of chromatographic fractions obtained in pentane.

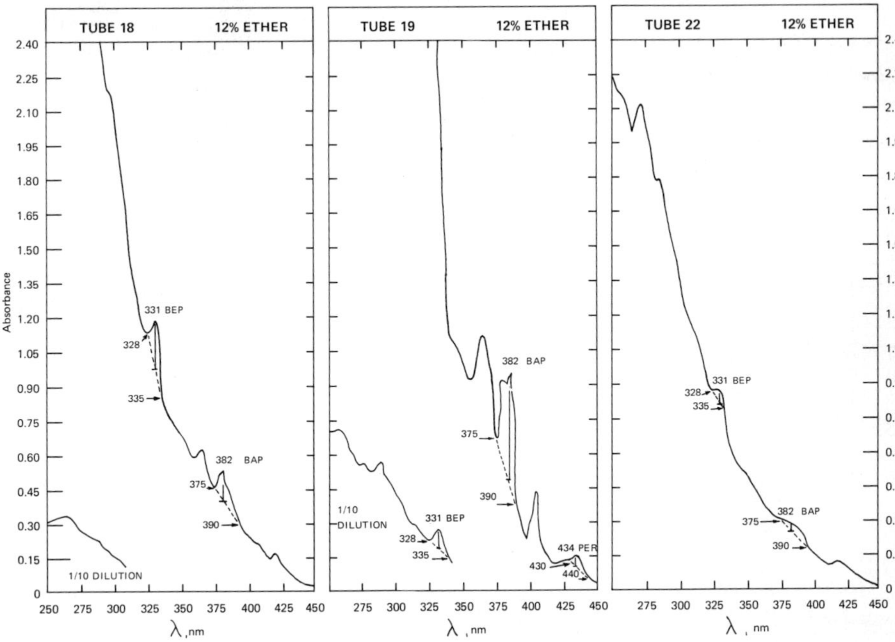

Figure 106:6—Ultraviolet-visible absorption spectra of chromatographic fractions obtained in pentane.

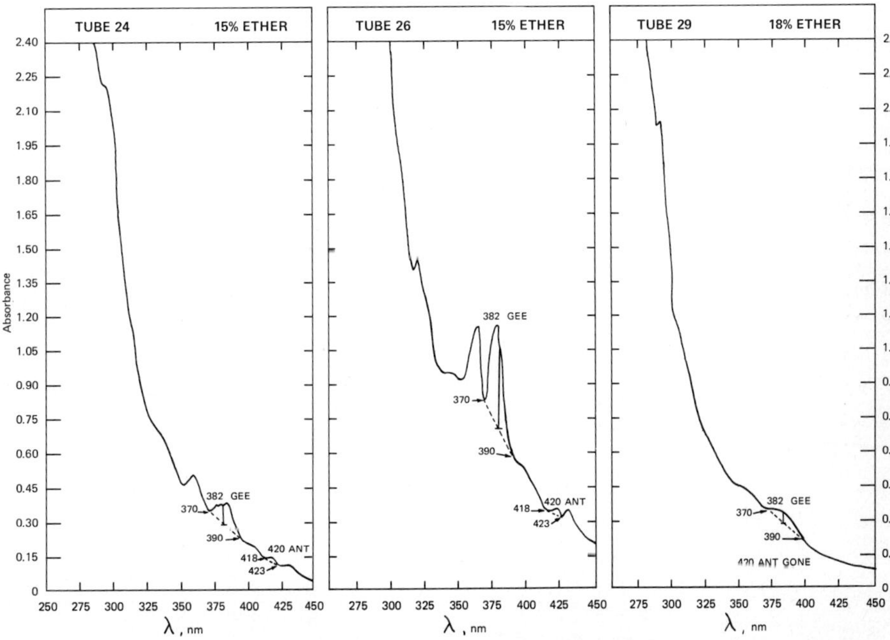

Figure 106:7—Ultraviolet-visible absorption spectra of chromatographic fractions obtained in pentane.

Table 106:II. Summary of methods of analysis for atmospheric arenes in particulate matter

Method	Separation	Photometric Method of Assay	Compounds Determined*	Range of Detm	Volume of urban air necessary M^3	Anal. Time for Org. fraction	Comments
5-1-ArH Polynuclear Aromatic Hydrocarbons	Column Chromatography (CC)	UV-Vis absorption spectrophotometry (SP)	P, FT, C, BaA, BaP, BeP, BkFT, PER, BghiPER, ANT, COR	Microgram Amounts	5000–10,000	1 ½ to 2 days	Most arenes determined by this method and method 5-2
5-2-ArH Routine Analysis	1. CC for arom. fraction 2. CC for sepn.	SP and Spectrophotofluorimetry (SPF)	as above	as above < 0.1 μg for BaP and BkFT	5000–10,000	2 days	Most man-hours of work. Lower concentrations of BaP and BkFT determined.
5-3-BaP Microanalysis	Thin-layer chromatography (TLC)	SPF or filter fluorimetry (FF)	BaP	3–200 ng (SPF) 10–300 ng (FF)	1–6	1 ½ hrs.	Highest cleanliness of lab air & glassware required. No smoking. Most sensitive method.
5-4-BaP + BkFT Chromatographic Analysis	CC	SPF	BaP + BkFT	0.01 to 0.1 μg	50–500	1 day	A 10% portion of a 24 to 96 hr air sample can be analysed.
5-5-BaP Spectrophotometric Analysis	TLC	SP	BaP	0.2–10 μg	40–200	2 hrs.	Probably the simplest method of assay for BaP.

*P = pyrene, FT = fluoranthene, C = chrysene, BaA = benz(a)anthracene, BaP = benzo(a)pyrene, BeP = benzo(e)pyrene, BkFT = benzo(k)fluoranthene, PER = perylene, Bgh.PER = benzo(ghi)perylene, ANT = anthanthrene, COR = coronene.

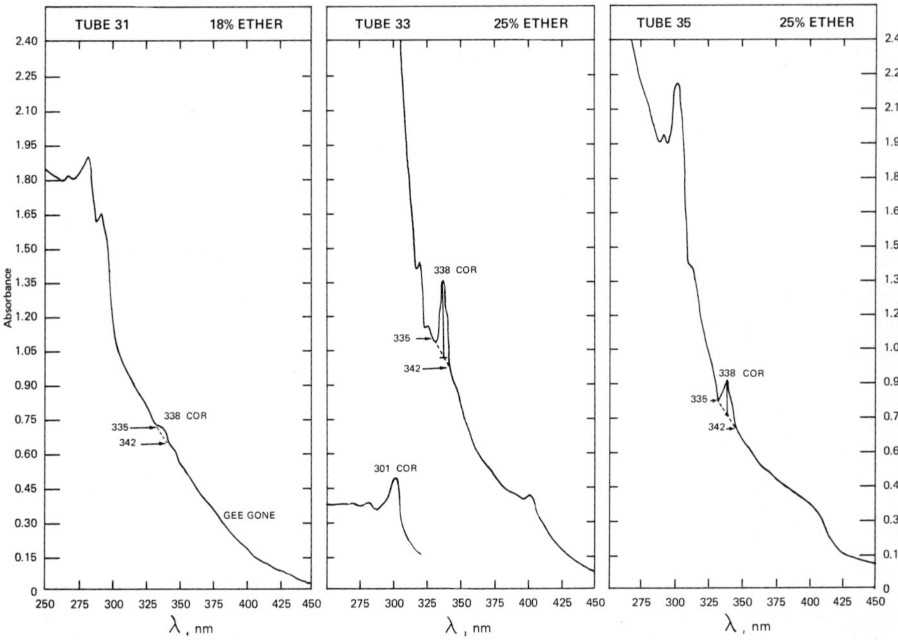

Figure 106:8—Ultraviolet-visible absorption spectra of chromatographic fractions obtained in pentane.

A. = total absorbance exhibited by the hydrocarbon on fractionation.

Calibration factor = the μg of hydrocarbon in 3 ml pentane necessary to give an absorbance of 1.00.

Column factor = factor to correct the amount eluted from the column to the amount placed on the column.

9.10 The following calculations are performed on air particulate samples:

9.10.1 μg of hydrocarbon/g benzene soluble (organic fraction) =

$$\frac{\text{Total } \mu\text{g found (from above)} \times 1000}{\text{Weight of the organic fraction analyzed in mg.}}$$

9.10.2 μg of hydrocarbon/g of air particulate matter =

$$\frac{\mu\text{g/g Benzene soluble (9.10.1 above)} \times \text{total wt. of organic fraction in g.}}{\text{Weight of particulate matter in g.}}$$

9.10.3 μg of hydrocarbon/1000 cm of air =

$$\frac{1000 \times \text{numerator of 9.10.2 above}}{\text{Total air sampled to obtain the organic fraction in m}^3}$$

10. Effects of Storage

The organic fractions of air particulate matter will remain fairly stable for a period of at least a year when they are stored in the dark in a refrigerator. It has been shown that strong light will destroy some of the hydrocarbons present especially if the hydrocarbons are absorbed on a substrate **(4)**.

11. References

1. COMMINS, B.T. 1958 *Analyst*, 83:386.
2. COOPER, R.L. 1954 *Analyst*, 79:573.
3. FALK, H.L., and P.E. STEINER. 1952. *Cancer Res.*, 12:30.
4. INSCOE, M.N. 1964. *Anal. Chem.*, 36:2505.
5. SAWICKI, E. 1964. *Chem. Anal.*, 53:24–26, 28–30, 56–62, 88–91.

6. SAWICKI, E., W.C. ELBERT, T.R. HAUSER, F.T. FOX, and T.W. STANLEY. 1960. *Amer. Ind. Hyg. Assoc. J.*, 21:443.
7. SAWICKI, E., W.C. ELBERT, T. W. STANLEY, T.R. HAUSER, and F.T. FOX. 1960. *Int. J. Air Poll.*, 2:273.
8. SAWICKI, E., W.C.ELBERT, T.W. STANLEY, T.R. HAUSER, and F.T. FOX. 1960. *Anal. Chem.*, 32:810.
9. SAWICKI, E., F.T. FOX, W.C. ELBERT, T. R. HAUSER, and J.E. MEEKER. 1962. *Amer. Ind. Hyg. Assoc. J.*, 23:482.
10. SAWICKI, E., T.R. HAUSER, W.C. ELBERT, F.T. FOX, and J.E. MEEKER. 1962. *Amer. Ind. Hyg. Assoc. J.*, 23:137.
11. SAWICKI, E., T.R. HAUSER, and T.W. STANLEY. 1960. *Int. J. Air Poll.*, 2:253.
12. TABOR, E.C., T.R. HAUSER, J.P. LODGE, and R.H. BURTTSCHELL. 1958. *A.M.A. Arch. of Ind. Health*, 17:58.

Subcommittee 5
E. SAWICKI, *Chairman*
R. C. COREY
A. E. DOOLEY
J. B. GISCLARD
J. L. MONKMAN
R. E. NELIGAN
L. A. RIPPERTON

107.

Tentative Method of Routine Analysis for Polynuclear Aromatic Hydrocarbon Content of Atmospheric Particulate Matter

11104-02-69T

1. Principle of the Method

1.1 Samples of the polycyclic hydrocarbons in the atmosphere are collected by "hi-vol" filtration through glass fiber filters. The particulate material is separated into 40 to 60 fractions by column chromatography which are subjected to spectrophotometric and fluorimetric analysis.

2. Range and Sensitivity

2.1 Separation and estimation of about 6 to 12 polycyclic hydrocarbons can be carried out by the procedure described in this method. This method can determine microgram quantities of the various hydrocarbons/1000 m^3 of air, depending upon the volume of air sampled, the amount of particulate matter present and the corresponding amount of organic fraction obtained.

3. Interferences

3.1 As the identification and estimation of the individual polycyclics is carried out by comparison of the unknown spectra with spectra of known pure compounds, every effort must be made to eliminate interfering compounds and maintain the systems free of background contamination which would modify the wavelength of the bands.

4. Precision and Accuracy

4.1 Since all identification and concentration measurements are made by comparison with standard curves prepared us-

Warning: Some polynuclear aromatic hydrocarbons, benzo[a]pyrene in particular, are carcinogenic. Consequently, care must be taken in handling laboratory equipment to avoid spilling solutions or PAH material on hands or other parts of the skin. Manipulations involving the PAH compounds should be carried out in a fume hood and must not be allowed to contaminate the laboratory air.

ing pure compounds, the greatest accuracy will be obtained from measurements made on spectra which resemble as closely as possible the standard curves for the pure substances. The greater the departure of the spectrum of the unknown sample from that of the standard the greater the inherent error of any quantitative measurement.

4.2 For the most accurate work only the very best instrumentation must be employed. The wavelength scale of the ultraviolet instrument usually used in this work should be accurate to 0.2 nm. The wavelength scale of the fluorimeter should also be accurate to 0.2 nm with the "instrumental" reproducibility equal to 0.1 or 0.2 nm. The complete analysis of a prepared air sample extract, including chromatographic measurements by ultraviolet or fluorescence and calculations requires two days.

5. Apparatus

5.1 Laboratory Ware. All glassware is cleaned before use by rinsing with dilute hydrochloric acid, then with distilled water, alcohol and finally with redistilled or spectrograde cyclohexane to remove all traces of organic matter such as oil, grease, and detergent residues, etc. No grease shall be used on stopcocks or joints. Examine the glassware under ultraviolet light to detect any fluorescent contamination. Some polynuclear hydrocarbons are very susceptible to photooxidation, consequently all operations should be carried out in subdued light or the apparatus should be covered with clean aluminum foil.

5.2 A Soxhlet extractor of 300 ml capacity is required.

5.3 Chromatographic columns are prepared by joining a stopcock with 1 mm bore to a 25 cm length of glass tubing of 1 cm ID.

5.4 A fraction collector may be used to collect chromatographic fractions overnight. However, it is quite convenient to run succeeding fractions on a recording spectrophotometer immediately after each fraction is eluted from the column. Evaporation of the sample extract or of an aliquot may be accomplished by using a vacuum oven at room temperature, or by the use of a current of pure nitrogen gas.

5.5 Spectrophotometer. A ratio recording spectrophotometer is used because of the large number of chromatographic fractions which have to be analysed. The useful wavelength range of the instrument used in this method is from 220 to 450 nm with a resolution of 0.1 to 0.2 nm. (Cary 14 of Applied Physics Corp. or equivalent instrument from any other source.)

5.6 Fluorimeter. The analysis of benzo[a]pyrene (BaP) in the presence of benzo[k]fluoranthene (BkF) is readily accomplished by using fluorescence techniques (**5**). An instrument capable of providing the necessary selection of excitation wavelengths is required. The instrument must also be able to scan the emission spectrum. An Aminco-Bowman SPF or any other instrument with a resolution accurate to 0.2 nm is suitable.

5.7 Cuvette. The cells used for ultraviolet absorption can be of the conventional rectangular fused silica design with a 1 cm light path. For fluorescence measurements similar cells, polished on all four sides, are used. These fluorescence cells may also be used for ultraviolet absorption measurements.

6. Reagents

6.1 Alumina. The alumina is type H, 100 to 200 mesh, of Peter Spence, England, or similar type from any other source. This alumina is slightly alkaline, but it is unnecessary to prewash this alumina before use. A quantity of alumina, perhaps 300 g, is activated by placing in an oven overnight at 140 C. About 100 g of such activated alumina is removed from the oven, cooled, weighed and placed in a flask having a ground glass stopper. The alumina is covered with sufficient cyclohexane to give a supernatant depth of perhaps 1.2 cm. Water equal in weight to about 1.8% of the alumina is added to the alumina-cyclohexane slowly and the mixture is shaken thoroughly. Because water and cyclohexane are immiscible, the water is evenly dispersed throughout the alumina assuring *uniform deactivation*. The alumina is then equilibrated by allowing it to stand overnight. Methods of determining the activity of deactivated alumina have been suggested by Commins and by Grimmer (**3, 6**). It is not possible to perform a practically useful chromatographic separation if the activity of the absorbent is too high (**1**).

6.2 Silica Gel. Material of 28 to 200 mesh (Davison Co., Baltimore or equivalent material) is kept in an oven at 140 C and removed for preliminary separations as required (**7**).

6.3 Cyclohexane. Fluorimetric or spectrograde cyclohexane must be free from benzene as impurity. Cyclohexane can be readily purified by using a suitable starting material a technical grade which is free from aromatics. This is purified by percolation through a glass column 6.5 cm in OD, 33 cm in length and provided with a Teflon plug stopcock. This is packed with activated carbon (Pittsburgh Chemical Co., 12 × 30 mesh, or equivalent material from another source) supported on a glass wool plug. This purified cyclohexane is free from ultraviolet absorbing and fluorescing impurities.

6.4 Iso-octane. The spectrograde solvent is purified in the same way as the cyclohexane.

6.5 Ethyl Ether and Benzene. Satisfactory spectro fluorimetric grades are available from Hartmann-Leddon, Philadelphia, or other source of equivalent grade. These solvents can be purified in the laboratory as directed under Section 6.3.

6.6 Polycyclic Aromatic Hydrocarbons. Standards of the various hydrocarbons of interest are available from a number of sources that supply purified organic compounds in North America and Europe. Zone refined quality is preferred if available. Benzo[k]fluoranthene is not available commercially but can be synthesized according to the methods described by Moreau *et al.* (**9**) and Buu-Hoi *et al.* (**10**). The purity of such standards must be confirmed before use by all available techniques such as chromatography, melting point, ultraviolet absorption, fluorescence and any other method which may be of help. Most impurities can be removed by several recrystallizations from appropriate solvents, if necessary.

6.7 The glass fiber sheets, on which the particulate matter from air samples are usually taken, may be extracted in toto or aliquot circles may be punched from the sheet. The whole sheets, or discs cut from them, are placed in a Soxhlet and extracted with purified cyclohexane for 24 hr. The total cyclohexane extract is filtered to remove carbon and glass fiber, made up to a standard volume and stored in the refrigerator in the dark until required.

7. Procedure

7.1 The cyclohexane extractable content of the total extract is established by carefully evaporating a 10 ml aliquot. A measured portion of the total cyclohexane extract, sufficient to contain 15 to 25 mg of cyclohexane-extractable material is measured into a weighed 5-ml beaker, carefully reduced in volume and then diluted with iso-octane. The solution is refluxed carefully in the beaker under a watch glass to remove the remaining cyclohexane and it is then reduced to 2 to 3 ml and placed on a column of activated silica gel (**7**). Elution with iso-octane is continued until 400 ml have passed through the column. The aromatic fraction left on the silica gel column is then removed with 100 ml of benzene. This "aromatic fraction" in benzene is evaporated carefully to dryness using a stream of nitrogen and taken up in 4 to 5 ml of cyclohexane. The concentrate is carefully poured onto the alumina column and allowed to pass almost entirely into the alumina. The column is then capped with 0.5 cm of deactivated alumina and elution with pure cyclohexane is begun. Approximately 70 chromatographic fractions of 8 ml each are collected in clean centrifuge or test tubes. Nothing appears in the first six fractions. Pyrene and fluoranthene may be seen in fractions 7 to 12 (Figure 107:1). 4-methyl pyrene is found in fractions 13, 14 and 15, in addition to pyrene and fluoranthene (Figure 107:2). In the fractions 16 to 34, benz[a]anthracene, and chrysene appear. The decline of chrysene structure at wavelengths 263.2 nm and 257.9 nm is the signal for changing the eluting solvent from pure cyclohexane to a mixture of 2% ether in cyclohexane. The first fraction #35, associated with this change is shown in Figure 107:3. This more polar mixed solvent will elute benzo[e]pyrene (BeP), benzo[a]pyrene (BaP), benzo[k]fluoranthene (BkF) and benzo[g,h,i]perylene (BghiP) in 20 to 25 fractions. The first three of this group are shown in fraction No. 45 Figure 107:4. After concentration by evaporation, two peaks for perylene may be seen in the same fraction (Figure 107:5). This illustrates one of the problems with ultraviolet absorption as a monitoring procedure, namely, the possibility of failing to see certain components originally present in low concentrations in the fractions, or whose absorbing effect is low, with accompanying low sensitivity. When the BghiP structure at peaks 362.5 nm and 384.0 nm begins to decline, the eluting solvent is changed to 4% ether in cyclohexane and anthanthrene and later coronene begins to show. Fraction 62 (Figure 107:6) shows four characteristic peaks for anthanthrene. Coronene is eluted last, over a group of 13 fractions, one of which, No. 69, is illustrated (Figure 107:7). Seven rather good peaks for coronene are shown.

7.2 Due to the variable composition of air samples, it is possible that there may be some variability in the optimum location for changing solvent polarity. For this reason, the operator should follow the elution process by observing the appearance and disappearance of characteristic peaks, changing solvent polarity when this is appropriate.

7.3 The fractions taken from the column may be weighed and the polycyclic concentrations measured fraction by fraction, or the fractions containing a particular polycyclic may be combined, concentrated and the total concentration of the polycyclic determined with one ultraviolet measurement. The second procedure is very useful in many cases where the concentration of aromatics is very low. The fractions containing BeP, BaP and BkF are grouped together but handled in a different manner. The BeP concentration is measured by ultraviolet absorption in the usual way. The solution is now diluted 10 to 100 fold to bring the concentration within the linear range for fluorescence. With the emission monochromator set at 403.0 nm the solution is excited at 307.0 nm and 385.0 nm and two fluorescence emission values are obtained from which the individual concentrations of BaP and BkF are determined **(5)**, (see also Intersociety Committee Method **(11)**).

7.4 In order to be of maximum value the spectra obtained must provide both qualitative and quantitative information. Whether a compound can be identified within a mixture of unknown composition will depend largely on two factors:

7.4.1 The identity, or otherwise, of the band wavelengths of the unknown with the bands of known standard solutions.

7.4.2 The shapes and ratios of the various peaks shown in the unknown and standard spectra.

7.5 Ideally, each of the bands displayed by the standard should also be identifiable in the unknown sample. Background and other interference frequently prevent the appearance of all bands. Consequently then, only the more intense bands are regularly available for identification comparisons. Much of the useful structure of airborne polycyclics lies between 240.0 nm and 300.0 nm. Unfortunately, some aliphatic hydrocarbons and other, so far unidentified, "background" materials also have absorption in this region. Particular attention must therefore be devoted at all times to minimizing interferences of this kind. Prior separations on silica gel are in tended to isolate the aromatics from the aliphatic groups. The use of ultra pure solvents is mandatory. The amount of the sample placed on the chromatographic column must, as a minimum, strike a balance between the sensitivities afforded by the analytical technique but the amount, as a maximum, must not overload the column.

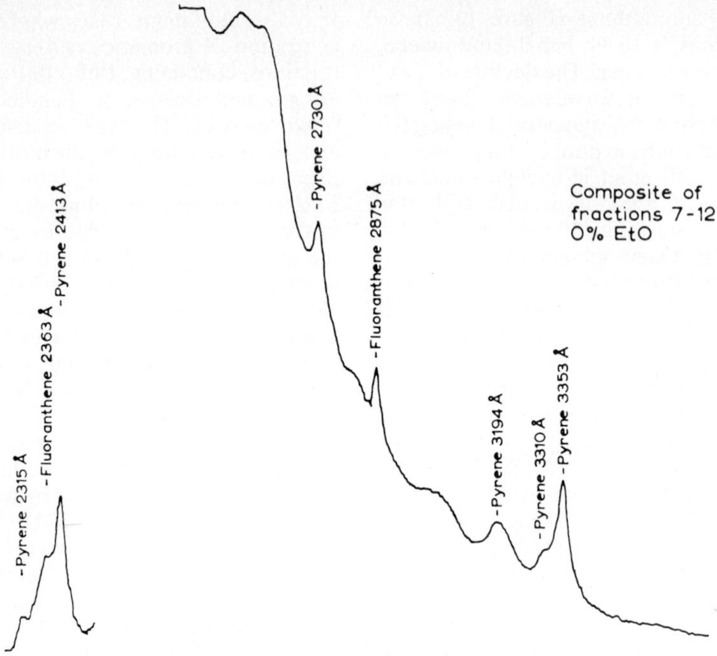

Figure 107:1—Pyrene and fluoranthene in combined fractions 7–12.

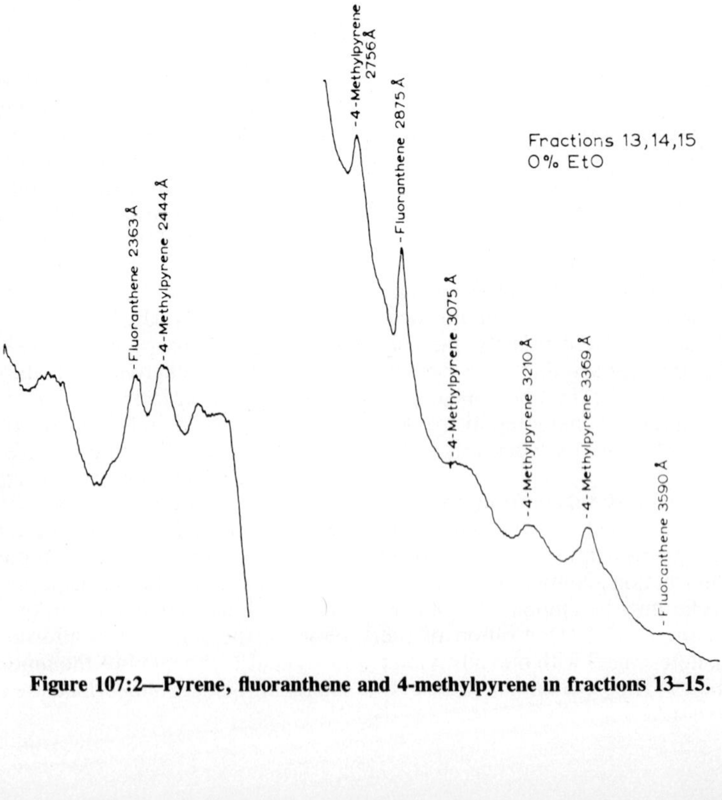

Figure 107:2—Pyrene, fluoranthene and 4-methylpyrene in fractions 13–15.

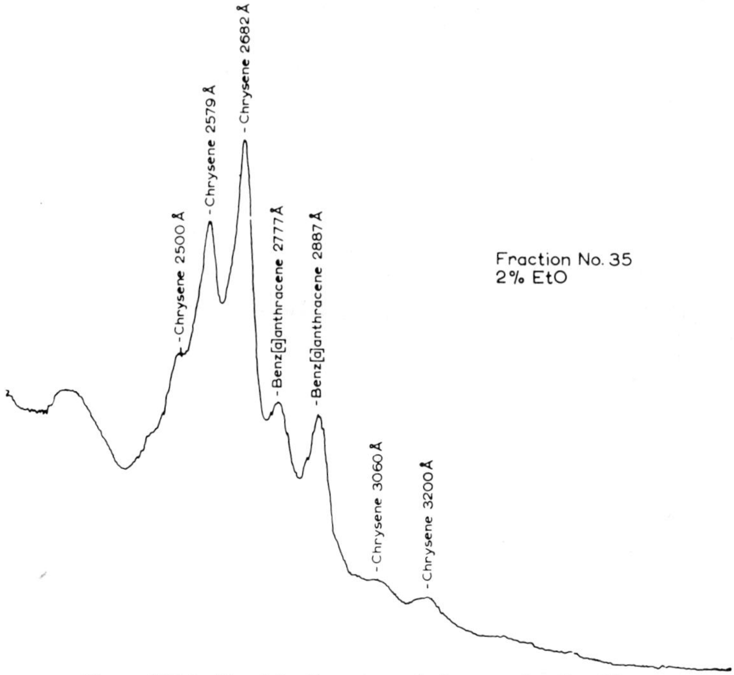

Figure 107:3—Benz[a]anthracene and chrysene, fraction 35.

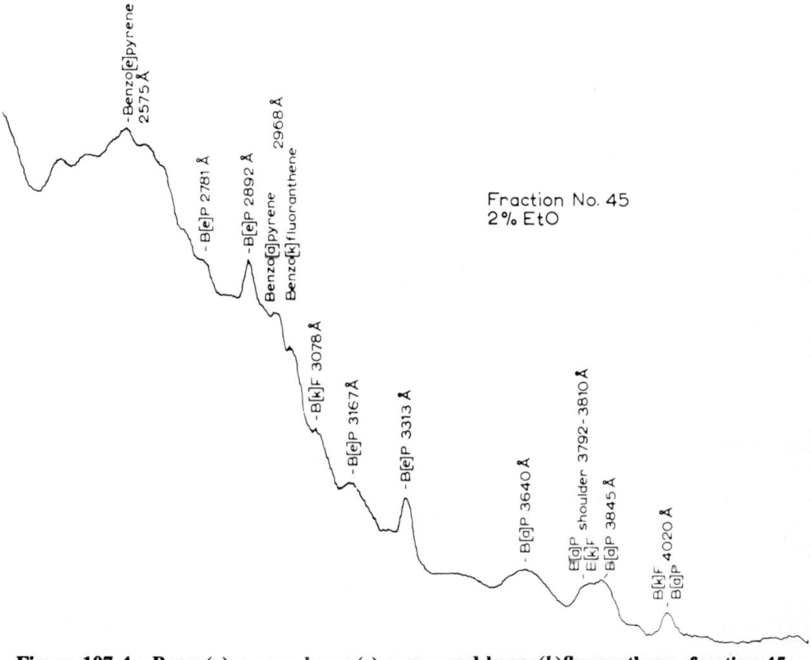

Figure 107:4—Benzo(*a*)pyrene, benzo(*e*)pyrene and benzo(*k*)fluoranthene, fraction 45.

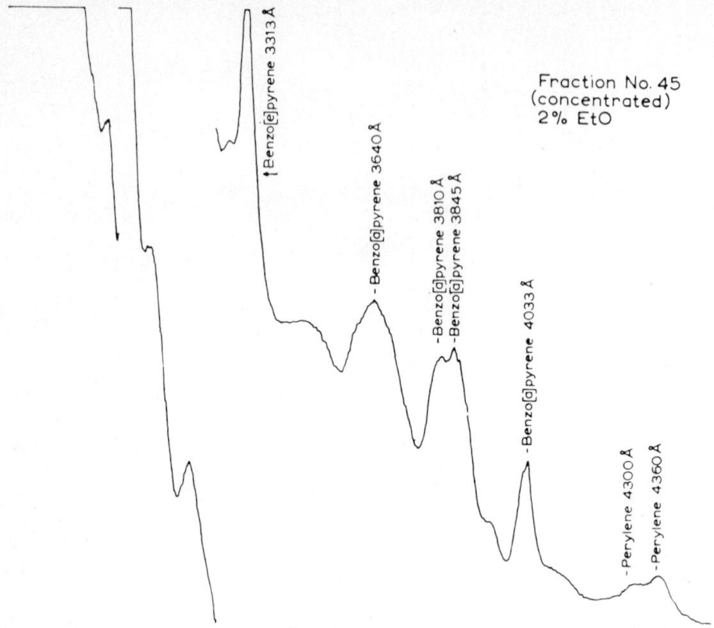

Figure 107:5—Perylene appearing in fraction 45 after concentration.

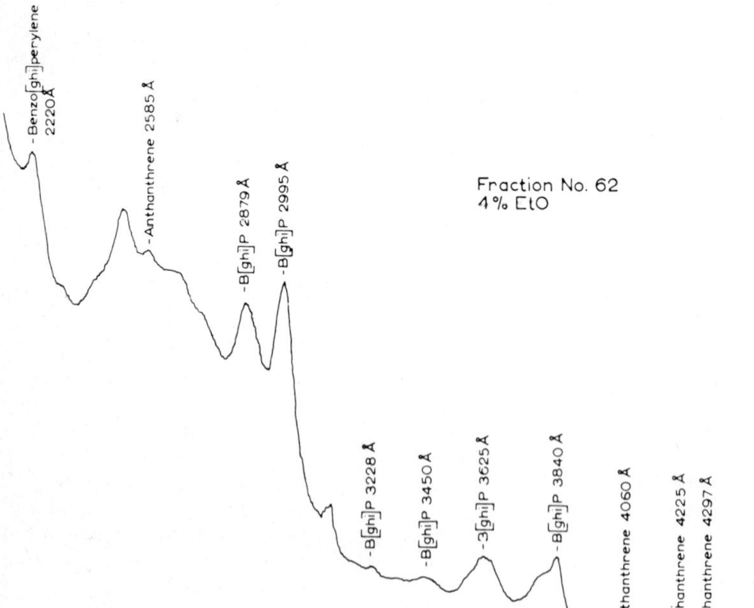

Figure 107:6—Benzo[ghi]perylene and anthanthrene, fraction 62.

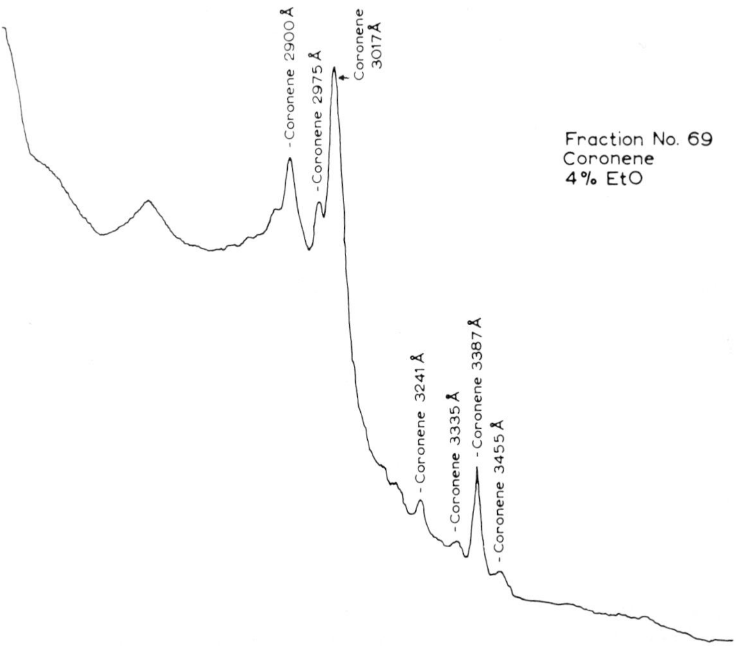

Figure 107:7—Coronene only, fraction 69.

7.6 As may be seen from the spectrum of the composite, formed from fractions 7 to 12 and that of 13 to 15, subtle changes occur in the structure of the absorption bands, which if not noticed, could lead to mistakes both in identification and measurement. In composite Figure 107:1 the pyrene bands at 335.3, 329.4, 275.0, 241.3, and 231.5 (Ave. 282.5) indicate unequivocally the presence of pyrene. In composite Figure 107:2 these bands have shifted to 336.9, 321.0, 307.5, 275.6 and 244.4 (Avc. 297.1). This is a difference of 14.6 nm based on a comparison of all 5 pairs of bands. If the 231.5 band is omitted from the first group and the 307.5 band from the second group the averages of the 4 closely agreeing pairs of bands are 295.2 and 294.5 or a difference of only 0.7 nm and therefore easily overlooked. The 4-methylpyrene identification in fractions 13 to 15 is a common and important interference in the accurate measurement of pyrene. In completely analogous fashion alkyl derivatives of fluoranthene and chrysene can and do display similar upscale shifts which can cause both qualitative and quantitative errors.

7.7 Although the method outlined is not the only way to analyze polycyclic hydrocarbons in air samples, it is, however, a method with which much experience has been obtained. The separations are sharper than that in most other procedures, the separation sequences are unvarying, and the solute elution positions are accurate to several ml.

8. Comparison with Standards.

8.1 Ultraviolet and fluorescence measurements are usually made by the "base line" method of Commins (**4**).

8.1.1 Establish the optimum instrumental parameters of the spectrophotometer (**12**). The ultraviolet absorption spectrum of the standard hydrocarbon is determined on at least 4 concentrations in spectrograde cyclohexane. The concentration range of the standard solutions should be between 0 to 10 μg/ml. Measurements arc made in 1.0-cm light path silica cells using cyclohexane as reference.

8.2 A baseline is drawn to subtend the particular absorption band in the reference spectrum which is to be used later as the analytical band (**13**). Alternatively, a horizontal baseline may be used after Dubois *et al* (**14**). The measurement of the peak height above the baseline is divided by the concentration giving the slope of S/μg/ml.

8.2.1 Example for estimation of slope: for pyrene, the analytical band chosen is 335.3 nm. Draw a baseline between 327.0 and 343.0 nm. Peak heights above the baseline are measured and tabulated as below for four concentrations.

Peak height H	Concentration μg/ml	S/μg/ml
0.683	2.600	0.263
0.333	1.250	0.266
0.237	0.886	0.267
0.115	0.425	0.271
	Mean	0.267

Note that the mean value for measurements of 4 different concentrations at the 335.3 nm wavelength of pyrene, is 0.267. A similar exercise must be completed for each of the compounds to be analyzed.

8.3 Benzo[a]pyrene and benzo[k]-fluoranthene are analyzed by fluorescence using the Intersociety Committee Method (**11**) or the method of Dubois and Monkman (**15**).

9. Calculation

9.1 The various composites of elution fractions are made up to a measured volume and run on the spectrophotometer. The diagnostic band for each compound is measured above the baseline as in 8.2.

9.2 The concentrations are calculated as follows:—

$$C_f\ (\mu g/ml) = \frac{\text{observed peak height}}{S/\mu g/ml}$$

$$\text{and } C\ (\mu g/m^3) = \frac{C_f \times V_f}{\frac{a}{A} \cdot V}$$

where

C_f = concentration of compound (PAH) in composite fraction, μg/ml.

C = concentration of compound (PAH) in air, μg/m³.

V_f = volume of composite fraction, ml.

V = volume of air sampled in m³ at standard conditions.

a = area aliquot of total sample area of extracted glass fiber filter, cm².

A = Total sample area of glass fiber filter, cm².

9.2.1 If L is the particulate loading of the filter in g/m³ of air sample, then C/L = Conc of PAH in μg/g of particulates.

10. Effects of Storage

10.1 The organic fractions of air particulate matter will remain fairly stable for a period of at least a year when they are stored in the dark in a refrigerator. It has been shown that strong light will destroy some of the hydrocarbons present especially if the hydrocarbons are absorbed on a substrate.

11. References

1. Cahnmann, H.J. 1957. Anal. Chem. 29:1307.
2. Cleary, G.J. 1962. J. Chromatog. 9:204.
3. Commins, B.T. 1965. Int. J. Air Wat. Poll. 9:10.
4. Commins, B.T. 1958. Analyst, (London) 83:386.
5. Dubois, L., and J.L. Monkman. 1965. Int. J. Air Wat. Poll. 9:131.
6. Grimmer, G. 1965. J. Chromatog. 20:89.
7. Rosen, A.A., and F.M. Middleton. 1955. Anal. Chem. 27:790.
8. Sawicki, E., W.C. Elbert, T.R. Hauser, F.T. Fox, and T.W. Stanley. 1960. Am. Ind. Hyg. Assoc. J. 21:443.
9. Moreau, H., P. Chovin, and G. Rivoal. 1946. Comp. rend. Acad. Sci. Paris. 223:951.
10. Buu-Hoi, Ng. Ph., D. Lavit, and J. Lamy. 1959. J. Chem. Soc. 1845.
11. American Public Health Association, Inc. 1977. Methods of Air Sampling and Analysis. 2nd ed. p. 224. Washington, D.C.
12. Cary Instruments. 1964. Applied Physics Corp. Monograph. "Optimum Spectrometer Parameters", Appl. Report AR 14-2.
13. Cooper, R.L. 1954. Analyst (London) 79:573.
14. Dubois, L., A. Zdrojewski, and J.L. Monkman. 1967. Mickrochim. Acta. 834.
15. Dubois, L., and J.L. Monkman. 1965. Int. J. Air Wat. Poll. 9:131.

Subcommittee 5

E. Sawicki, *Chairman*
R. C. Corey
A. E. Dooley
J. B. Gisclard
J. L. Monkman
R. E. Neligan
L. A. Ripperton

108.

Tentative Method For The Continuous Analysis Of Total Hydrocarbons In The Atmosphere (Flame Ionization Method)

43101-02-71T

1. Principle and Applicability

1.1 The sampling system is an integral part of most commercial total hydrocarbon analyzers designed to measure continuously the concentration of hydrocarbons (and other organic compounds) in the atmosphere. The sample line is attached to the inlet and the sample is pumped into a flame ionization detector. A sensitive electrometer coupled with a potentiometric recorder detects the increase in ion intensity resulting from the introduction into a hydrogen flame of a sample air containing organic compounds (*e.g.*, hydrocarbons, aldehydes, alcohols). The response is approximately proportional to the number of carbon atoms in the sample. The analyzer is calibrated using methane and the results reported as methane equivalents.

1.2 The sample introduction pump may be by-passed for analysis of gases under pressure as is done with calibration gases.

2. Range and Sensitivity

2.1 The range of the analyzer may be varied so that full scale may be 2.6 mg/m^3 (4 ppm) to 1960 mg/m^3 (3000 ppm) hydrocarbon as methane by varying the attenuation and the sample flow rate to the detector. The 13 mg/m^3 (20 ppm) range is normally used for atmospheric sampling.

2.2 Sensitivity is 1% of full scale recorder response.

3. Interferences

3.1 Carbon atoms bound to oxygen, nitrogen, or halogens give reduced or no response. There is no response to nitrogen, carbon monoxide, carbon dioxide, or water vapor.

4. Precision and Accuracy

4.1 Precision is approximately 0.5% full recorder scale on the 13 mg/m^3 (20 ppm) scale.

4.2 Accuracy is dependent on instrument linearity and absolute concentration of the calibration gases used. Generally, accuracy is ± 1% of full scale on the 0 to 13 mg/m^3 range.

4.3 Zero drift necessitates frequent calibration. The magnitude of the drift depends on the air flow rate, sample flow rate, fuel flow rate, ambient temperature changes, detector contamination, and electronic drift. Zero drift observations on various instruments indicate 2%/24 hr on the 13 mg/m^3 scale.

5. Apparatus

5.1 COMMERCIALLY AVAILABLE TOTAL HYDROCARBON ANALYZER. Instruments obtained must be installed on location and demonstrated by the manufacturer to meet or exceed manufacturer's specifications or those described in this method. Generally, hydrocarbon analyzers consist of a regulated fuel and air delivery system for the hydrocarbon burner, a regulated sample injection system, electrometer for measuring the flame ion current, meter readout with connections for a recorder, and a sample pump.

5.2 RECORDER. Potentiometric type, compatible with analyzer with an accuracy of 0.5% or better.

5.3 SAMPLE LINE, any tubing that is not a source of interference or an absorbent of hydrocarbons. Inert materials such as glass, stainless steel, and Teflon are recommended. All tubing should be clean prior to and during use. A particulate filter should be installed in the sample line.

5.4 MINIMUM PERFORMANCE SPECIFICATIONS, see Table 108:I.

6. Reagents

6.1 COMBUSTION AIR, high purity air containing less than 1.3 mg/m^3 (2 ppm) hydrocarbon as methane, purified by passage over copper oxide maintained at 500 C and through a clean 55 μ filter.

6.2 FUEL, hydrogen or a hydrogen-inert gas mixture; when ordering, specify hydrocarbon-free gas. A hydrogen generator is strongly advised for safety reasons. An automatic cut-off should be provided for the instrument as a safety measure in the event of a detector flameout.

6.3 ZERO GAS, less than 0.05 mg/m^3 (0.1 ppm) hydrocarbon as methane in air.

6.4 SPAN GAS, methane in air corresponding to 80% of full scale, 10.4 mg/m^3 (16.0 ppm) for 13 mg/m^3 range. A certified or guaranteed analysis is required.

7. Procedure

7.1 For specific operating instructions, refer to the manufacturer's manual (see also Section **5.1**).

8. Calibration and Efficiencies

8.1 Calibrate the instrument at the desired flow rate and attenuator setting. Introduce zero gas and set zero control to indicate proper value on the recorder. If a live zero recorder is not used, it is recommended that the zero setting be offset at least 5% of scale to allow for negative zero drift. In this case, the span setting must also be offset by an equal amount. Introduce span gas and adjust span control to indicate proper value on recorder scale, (*e.g.* 0 to 13 mg/m^3 (0 to 20 ppm) scale set at 10.4 mg/m^3 (16.0 ppm) standard to read 80% of recorder scale). Recheck zero and span until adjustments are no longer necessary. Since the scale is linear, the two-point calibration is valid.

8.1.1 If attenuation is varied, some discrepancy between the true attenuation and the nominal attenuation may exist. The instrument should be calibrated using appropriate standards at each attenuator setting used.

Table 108:I. Suggested Minimum Performance Specifications for Total Hydrocarbon Analyzers

Range (minimum)	0–13mg/m^3 (0–20 ppm)
Output (minimum)	0–10,100,1000,5000 mv full scale
Minimum Detectable Sensitivity	0.1 mg/m^3 (0.16 ppm)
Lag Time (maximum)	30 seconds
Time to 90% response (maximum)	60 seconds
Rise Time (90% maximum)	30 seconds
Fall Time (90%)	30 seconds
Zero Drift (maximum)	2%/24 Hours
Span Drift (maximum)	2%/24 Hours
Precision (maximum)	±1%
Operational Period (minimum)	3 days
Noise (maximum)	±0.5%
Interference Equivalent (maximum)	1% of full scale
Operating Temperature Range (minimum)	5–40 C
Operating Humidity Range (minimum)	10–100%
Linearity (maximum)	1%

9. Calculations

9.1 The recorder is read directly in terms of concentration expressed as ppm, methane by volume, (1 ppm = 0.65 mg/m^3).

10. Effects of Storage

10.1 Not applicable.

11. References

1. NEWSOME, J.R., V. NORMAN, and C.H. KEITH. 1966. *Tobacco Science*, page 102–110.
2. ETTRE, L.S.S. 1962. *Chromatog*. 8:525–530

Subcommittee 5
E. SAWICKI, *Chairman*
R. C. COREY
A. E. DOOLEY
J. L. MONKMAN
L. A. RIPPERTON
J. E. SIGSBY
L. D. WHITE

109.

Addendum to "Tentative Method for the Continuous Analysis of Total Hydrocarbons in the Atmosphere (Flame Ionization Method)": Flame Ionization Detector

43101-02-71T **(10)**

1. Principle of the Detector

1.1 A flame ionization detector (FID) is a device which incorporates regulated fuel, air and sample delivery systems, an internal burner, and associated electronics for measuring the ion current produced by species introduced into the flame. The FID is used to sense and measure small amounts of gaseous organic-type components present in the carrier gas stream leaving the column of a gas chromatograph (G.C.) or to monitor methane and/or total hydrocarbon concentrations in ambient air samples.

1.2 Ideal characteristics of a detector may be defined as high sensitivity, low noise level, a wide linearity of response, ruggedness, insensitivity to flow and temperature changes, response to all types of compounds or an advantageous specificity for a particular type of compound. The FID is applicable to a wide range of compounds and has been demonstrated to be highly sensitive, reasonably stable, moderately insensitive to flow and temperature changes, rugged, and reliable **(12, 13)**. The FID has been reported to have the best record for reliable performance among the ionization type detectors **(7)**.

1.3 A schematic diagram of a typical flame ionization burner (the sensing unit of the FID) is shown in Figure 109:1 **(13)**. The sample air or column effluent is mixed with fuel and burned at the tip of the metal jet in a diffusion flame with an excess of air.

1.4 The FID makes use of the principle that very few ions are present in the flame produced by burning pure hydrogen, or hydrogen diluted with an inert gas. However, the introduction of mere traces of organic matter into such a flame produces a large amount of ionization. The charged particles generated in the flame are collected at two terminals resulting in a small flow of current which is amplified by an electrometer for output to a suitable recording device.

1.5 The response of the detector is roughly proportional to the carbon content of the solute. The response to most organic compounds on a molar basis increases with molecular weight. For example, propane has three times the response that an equimolar quantity of methane has. Although the FID does not respond to atoms other than carbon, other atoms change instrument sensitivity to carbon by altering the chemical environment of the carbon atom. The sensitivity of the FID varies with structure of the oxygenated or nitrogenated molecule and decreases with increasing hetero-atom content (**13**). The relative response or effective carbon number of any particular type of carbon atom is defined as the ratio between the instrument response caused by an atom of this type and the instrument response caused by an aliphatic carbon atom. Tables of effective carbon numbers have been prepared by various investigators which permit reasonable estimates to be made from chromatograms (**4, 12, 17**). However, each detector must be calibrated for its response to each compound for accurate analysis.

2. Range and Sensitivity

2.1 The most extensive use of the FID in trace analysis has been in those cases where the matrix is selectively removed (traces concentrated) or where the detector does not respond to the matrix material. The FID does not respond generally to the compounds listed in Table 109:I (**4, 7**), and has relatively low sensitivity to perfluoro compounds. The lack of response to air and water enables the FID to be especially suitable for the analysis of many air and water pollutants.

2.2 The FID has the widest linear range of any detector in common use. Linear dynamic range is about 6 to 7 orders of magnitude which allows quantitation of a broad range of sample sizes (**11**).

2.3 The wide linear range, lack of response to major atmospheric constituents, and excellent stability of the FID are important advantages for ambient air analysis in the 0.001 to 1 ppm concentration range. The detection limit for the FID is near 0.001 ppm of sample gas concentration (about 2 mg/m^3 or 2 pg/ml) with optimized operating parameters (**3, 14**). It is accepted practice to use twice the peak-to-peak noise level for establishing a limit of detection (**7, 13**).

2.4 Concentration procedures have been described in the literature which can be utilized to estimate atmospheric concentrations far below the 1 ppb concentration level (**2, 13, 18**). The upper limit of measurement can be considerably greater than the concentrations normally encountered in the atmosphere.

3. Interferences

3.1 Any compound capable of ionizing in the burner flame is a potential interference. The FID is not specific for hydrocarbons, as other organic types such as alcohols, aldehydes, and ketones will respond to varying degrees. This should be taken into account in interpretation of data (**12**).

3.2 The FID is not a versatile qualitative tool since use of the FID without prior separation steps does not distinguish among individual compounds or hydrocarbon classes. Identification of components in G.C. atmospheric analysis usually is based on one or more retention times. However, the possibility always exists of more than one compound having the same retention time for any given set of instrumental parameters (**1**).

4. Precision and Accuracy

4.1 Precision is affected by instrumentation noise and drift and by variations in sample, air and fuel flow rates. Instrument stability will determine reproducibility and the need and frequency of recalibrations.

4.2 Accuracy is affected primarily by sampling error, instrument linearity, the species being determined and the standardization or calibration employed in the analysis. Non-linearity may be overcome by calibration, which normally is easy and rapid if span gases or the pure compounds are available.

4.3 Replicate quantitative analyses of standards by the FID should not deviate by more than 2% relative standard deviation at optimum operating conditions. For the most accurate work, only the best commercially available instrumentation should be utilized.

5. Apparatus

5.1 Commercially available instruments have varying modes of controlling the flow of sample air, carrier gas, fuel and combustion air streams.

5.1.1 Most of the continuous hydrocarbon analyzers utilize a pressure drop across a capillary to regulate the flow.

5.1.2 A constant differential-type low-flow controller in conjunction with a rotameter is used to regulate the carrier gas stream in most gas chromatographs. The controllers maintain a constant differential in the pressure across an external needle valve, making the flow a function of the needle valve. The use of a glass wool plug, sintered disk, or porous bronze filter is recommended to protect capillaries or needle valves against particles that might change flow characteristics.

5.1.3 The rotameters are sized for the appropriate flows to give immediate visible readout on the front panel. Flow rates may appear to fluctuate on short rotameters even though an independent measurement will show a well-regulated constant flow.

In any type of flow control system, the flow rates should be verified periodically by a flow calibration device.

5.2 A burner system is recommended which is linear over a current range of at least 10^6 to allow wide versatility in application of the equipment.

5.2.1 After the ionization process, recombination of the electrons and positive ions occurs to a degree that is determined by the ion concentrations and the electrode voltage. At low voltages the ion current is proportional to applied voltage, *i.e.*, the electrode gap acts as a fixed resistance, while at high voltages a maximum number of ions are accelerated to the electrodes as determined by the equilibrium between the rates of ionization, recombination, and discharge at the electrodes.

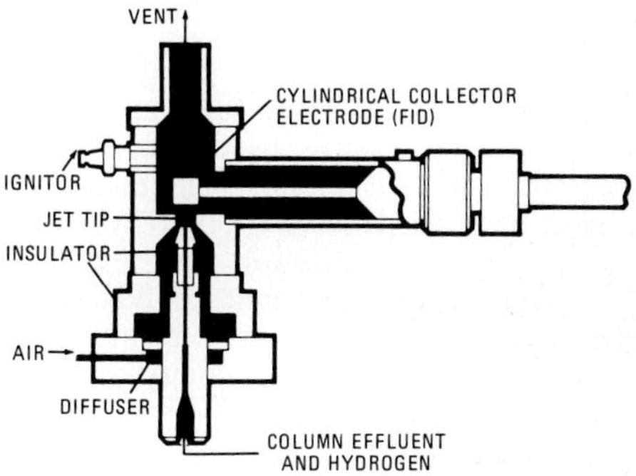

Figure 109:1—Flame Ionization Detector

If the jet is made the anode, higher voltages are required to obtain the saturation current. Presumably this is due to the low mobility and greater recombination rate of the positive ions, which must now move the increased distance from the inner zone of the flame to the electrode. Consequently, the best performance is obtained with the electrode over the flame as the electron collector (**13**). In general, electrode spacing and size are not critical; spacing determines only the minimum voltage required to obtain a saturation current. A fixed-position, cylindrical collector electrode is recommended to assure a

Table 109:I. Compounds giving little or no response in the FID

Ar	Ne
HCHO	NH_3
CO	NO
CO_2	N_2O
COS	NO_2
CS_2	O_2
He	$SiCl_4$
H_2O	SiF_4
H_2S	$SiHCl_3$
Kr	SO_2
N_2	Xe
HCOOH	

uniform spacing between the electrodes and provide long-term reproducibility.

5.2.2 The burner should be thermally isolated. Proportional temperature control up to at least 300 C is recommended to assure minimum baseline perturbations and long-term stability in G.C. isothermal and temperature programming applications. The FID must be maintained at a high enough temperature to prevent condensation of the sample, G.C. stationary phase, and the water and other by-products formed in the ionization process.

5.2.3 Corrosive samples may hasten the deterioration of the burner jet and electrodes. A FID with high-velocity air flow helps minimize the build-up of contaminants and causes SiO_2 dust, and other solids formed in the flame, to be blown out the detector's chimney. Suitable filters should be installed in the fuel, air and sample lines to protect the FID from particulate material.

5.2.4 A burner sensitivity of 0.01 coulombs/g (5×10^{-12} g/sec minimum detectability) or better is recommended. Sensitivity is a measure of the effectiveness of the detector as a transducer in converting the sample into a measurable electrical signal.

5.2.5 The burner should be equipped with a flame-sensing device to close the hydrogen supply solenoid valve in event of power failure or a flame out. This device is an important safety feature of methane and hydrocarbon analyzers which operate unattended for extended periods of time. The flame ionization detectors supplied on commercially available gas chromatographs are normally not equipped with such a device. Without exception, an exhaust hood must be provided above the instrument to remove all FID combustion products and any hydrogen which might be released to the atmosphere.

5.3 Electrometers furnished with FID's are usually of four types: (a) electron tube input, (b) vibrating capacitor input, (c) solid-state field effect, transistor input, and (d) solid-state parametric oscillator amplifier. The solid state devices are greatly improved in terms of ruggedness, long instrument life and freedom from calibration drift.

5.3.1 The ion current produced in the flame is on the order of 4×10^{-11} amperes/10 ppm carbon. An extremely stable electrometer is required to measure currents of this magnitude. Criteria for such an electrometer include negligible noise (less than 10^{-13} amperes), low time constant, high input impedance circuitry, minimum zero shift due to mechanical shock or temperature variations (less than 0.5% drift per C), overload protection, fast warm-up time, low background suppression current, and maximum reliability.

5.3.2 The current amplifier used to measure the signal should be capable of at least 5×10^{-12} amp/mV for full scale output, have appropriate attenuation settings to reduce this output, and have a linear dynamic range of over 10^6. FID amplifiers are commercially available which provide output for electronic integrator and computer as well as strip chart recorders.

5.4 Recording devices include meters, servo-recorders, electronic integrators, and complete data handling systems. The device should meet user requirements and be compatible with the output of the FID amplifier selected.

6. Reagents

6.1 PURITY. All organic compounds utilized to make G.C. standards should be of ACS analytical grade or spectrograde quality.

6.2 FUEL. Fuels range from 40% hydrogen in nitrogen to pure hydrogen depending on FID design requirements. Hydrogen in compressed gas cylinders should be prepared from water pumped sources.

High quality electrolytic generators which produce ultrapure hydrogen under high pressure are recommended. These provide some degree of safety when compared to the explosive hazard of bottled hydrogen. All connections should be made with thoroughly cleaned stainless steel tubing and properly tested for leaks.

6.3 AIR. Filtered compressed air in cylinders, low in hydrocarbon content. Because hydrocarbon-free air **(10)** is difficult to obtain, zero adjustment must be carried out for each new cylinder used if this air is also used as a calibration gas to set the zero point. Alternately, hydrocarbon-free synthetic air can be conveniently obtained by mixing oxygen and water-pumped nitrogen or other equivalent diluent. Compressed commercial grade oxygen in cylinders is suitable for detectors requiring oxygen.

6.4 CARRIER GAS. Compressed nitrogen, argon, or helium should be prepared from water pumped sources or be of guaranteed high purity.

6.5 SPAN GASES. At least three different concentrations of hydrocarbons in air (methane and/or propane are normally used) are required for calibration purposes. Commercial gases having a certified analysis should be used. If possible, an independent G.C. analysis should be made to verify the hydrocarbon concentration in the tank and particularly if the span gas has been calculated and prepared in the laboratory **(8, 10, 15, 16)**.

7. Operation

7.1 Instrument manuals are provided by manufacturers of commercially available flame ionization detectors and should be consulted for specific operating procedures.

7.2 FID performance depends on the proper choice of gas flow rates. Individual detectors should be calibrated to determine the hydrogen flow rate which gives maximum sensitivity and stability. At this point the signal is relatively insensitive to small variations in fuel rate **(11)**. Optimum performance is particularly dependent on the ratio of hydrogen to carrier or sample gas flow rate, since this ratio determines the flame temperature and therefore the efficiency of ionization. The optimum ratio of hydrogen to carrier gas has been determined to be about 1. The supply of air should be such that the ratio of air to hydrogen is about 10 **(4)**. Depending on the geometry of the detector chamber, an excessive supply of air may lead to turbulence in the flame zone with resultant noise and result in a loss of detectability of small sample concentrations.

7.3 Oxygen may be substituted for air in some FID's with a resulting increase in sensitivity but often with a decrease in the stability of the flame. All flow rates should be optimized for a particular instrument and maintained carefully.

7.4 When the burner is not operating, it is recommended that an inert gas (helium, argon, or nitrogen) be used to purge the FID. This will eliminate corrosion caused by the laboratory environment.

7.5 The FID is destructive in operation, since the sample stream is burned at the jet. However, this is not a serious handicap in G.C. applications, since the highly sensitive FID may be arranged in series with an effluent splitter which would allow a fraction of the components from the effluent stream to be collected for further analysis (mass spectrometry, nuclear magnetic resonance, etc.) **(5, 13)**.

7.6 While the FID has generally been touted for its insensitivity to flow, exception must be taken to this claim from a consideration of liquid phase volatility (G.C. applications). The signal represents a rate at which solute is being ionized in the detector and any background contributed by liquid phase vapor will vary with flow rate. The concentration of liquid phase in the detector is a function of its vapor pressure which is established by the flow rate of the carrier gas and the column temperature. Consequently, both carrier gas flow rate and column temperature can alter the performance of the FID, particularly when temperature programming is employed **(6)**.

7.7 After continued operation of the FID, the detector may become contaminated resulting in increased noise and decreased sensitivity. The FID should be carefully disassembled and cleaned with a soft brush and a suitable solvent such as methylene chloride. Clean tweezers should be used to avoid contamination from fingers when replacing the jets. The electrodes should be accurately repositioned as the FID is reassembled. It is recommended that the FID be cleaned every 3 to 6 months to assure proper maintenance and optimum performance.

7.8 A malfunction in the FID normally manifests itself as an erratic reading on a strip chart recorder. Most instrument manuals include a section on troubleshooting to determine whether the malfunction is in the electronic system, the detector or the flow system.

8. Calibration and Standards

8.1 Hydrocarbon analyzers should be zeroed with a hydrocarbon-free air mixture and calibrated with span gases in the 25 to several hundred ppm range to be analyzed on a regular basis, at least once daily. At optimum operation conditions, FID's require only a one point, one component calibration since response is linear with the carbon content of the sample being admitted to the burner. However, a three or more point calibration is recommended to assure a more accurate analysis (confirm linearity) over the range of expected sample concentrations.

8.2 The instrument sample regulator is adjusted to cause the recorder reading to equal the value of the concentration of hydrocarbon as indicated on the cylinder of span gas. It is recommended that the use of tanks at pressures lower than 500 psig be discontinued since calibration gases in the lower ranges could change markedly in concentration (because of desorption).

8.3 Care must be exercised in selecting the sample flow rate to the burner, since at some maximum flow rate the flow versus response relationship becomes nonlinear because of saturation of the electronics. Since the output of the FID is directly dependent on the amount of hydrocarbons in the flame, the same setting of the flow regulator is used when the atmospheric sample is introduced, as the calibration (for hydrocarbon analyzers) is flow dependent (**10, 12**).

8.4 Quantitative analysis requires adequate calibrations (three or more points) for individual compounds. The same G.C. operating parameters that are used for analyzing the standards must be used when analyzing the samples. The samples and standards should be analyzed in sequence to minimize analytical, analyst, and instrumentation errors. Calibration standards for G.C. applications may be prepared by use of procedures recommended by the Intersociety Committee (**9**).

Note: Preparation of Hydrocarbon Mixtures. In the preparation of hydrocarbon mixtures in air in tanks, experience has shown that the lowest concentration feasible on a routine basis is 10 ppm. Even at this concentration, careful checks must be made by independent means to determine the change in concentration with time. To prepare stable hydrocarbon mixtures at lower levels of concentration in tanks under pressure, down to less than 1 ppm, extreme care and effort must be taken to account for sorption and desorption effects that are influenced by metal composition of the tank, previous "chemical history," temperature and other factors.

Calibration gas mixtures at levels normally used, such as 15 ppm propane, change markedly in concentration, because of desorption, as the tank pressure drops below about 300 psig. It is recommended that tank use be discontinued at pressures below 500 psig.

9. Calculation

9.1 The meter or recorder recording of a calibrated FID (hydrocarbon analyzer) will normally represent the hydrocarbon concentration of the measured gas in parts per million carbon or methane.

9.2 The G.C. standard curves are drawn by plotting the concentration of the standards versus the respective peak area or integrator output. The concentration of any given organic contaminant in the sampled air is determined by referring its peak area or integrator output to the corresponding standard curve.

10. Effects of Storage

10.1 Not applicable.

11. References

1. ALTSHULLER, A.P. 1968. *Advances in Chromatography*, J.C. Giddings, and R.A. Keller, eds., Marcel Dekker, Inc., New York, N.Y., p. 236.
2. BELLAR, T.A., BROWN, M.F., and SIGSBY, J.E. 1963. *Anal. Chem.* 35:1924.
3. BELLAR, T., SIGSBY, J.E., CLEMONS, C.A., and ALTSHULLER. A.P. 1962. *Anal. Chem.* 34:763.
4. CONDON, R.D., SCHOLLY, P.R., and AVERILL, W. 1960. *Gas Chromatography*. R.P.W. Scott, ed., Butterworths, Washington, D.C., p. 30.

5. Cooper, C.V., White, L.D., and Kupel, R.E. 1974. Am. Ind. Hyg. Assoc. J. 32:383.
6. Gerrard, W., Hawkes, S.J., and Mooney, E.F. 1960. *Gas Chromatography*. R.P.W. Scott, ed., Butterworths, Washington, D.C., p. 199.
7. Hartmann, C.H. 1971. Anal. Chem. 43:113A.
8. American Public Health Association, Inc. 1976. *Methods of Air Sampling and Analysis*. 22nd ed. pp. 209. Washington, D.C.
9. Ibid., pp. 359.
10. Ibid., pp. 257.
11. McNair, H.M., and Bonelli, E.J. 1969. *Basic Gas Chromatography*. Varian Aerograph, Walnut Creek, Calif., pp. 99–105.
12. Morris, R.A., and Chapman, R.L. 1961. Air Poll. Control Assoc. J. 11:467.
13. Nogare, S.D., and Juvet, R.S. 1966. *Gas-Liquid Chromatography, Theory and Practice*. Interscience Publishers, New York, N.Y., pp. 180–239.
14. Rasmussen, R.A., and Uent, F.W. 1965. Proc. Natl. Acad. Sci. U.S. 53:215.
15. Saltzman, B.E. 1972. *The Industrial Environment . . . Its Evaluation and Control: Syllabus*. 3rd Ed., Chapter 12, Public Health Service Publication No. 614, U.S. Government Printing Office, Washington, D.C.
16. Saltzman, B.E., Burg, W.R., and Ramaswamy, G. 1971. *Environ. Sci. Technol.* 5:1121.
17. Stenberg, J.C., Gallaway, W.S., and Jones D.T.L. 1962. *Gas Chromatography*. N. Brenner, J.E. Callen, and M.D. Weiss, eds., Academic Press, New York, N.Y., p. 231.
18. White, L.D., Taylor, D.G., Mauer, P.A., and Kupel, R.E. 1970. Am. Ind. Hyg. Assoc. J. 31:225.

Subcommittee 5
E. Sawicki, *Chairman*
P. R. Atkins
T. Belsky
R. A. Friedel
D. L. Hyde
J. L. Monkman
R. A. Rasmussen
L. A. Ripperton
J. E. Sigsby
L. D. White

110.

Tentative Method of Fluorimetric Analysis for Aliphatic Hydrocarbons in Atmospheric Particulate Matter

16216-01-73T

1. Principle of the Method

1.1 The benzene-soluble fractions of airborne particulate extracts are separated along with standard solutions of n-docosane on silica gel TLC plates impregnated with rhodamine 6G. The saturated aliphatic hydrocarbons show as pink spots on the developed and dried plates. Under UV light they exhibit a yellow fluorescence against a lime-colored plate background. The difference in fluorescence of the alkane spot versus that of the rhodamine coated silica gel plate is derived from the solvent effect on the emission spectrum of rhodamine 6G. This dye has a cationic resonance structure. The absorption band obtained from such a structure shifts to longer wavelengths when its environment is changed from the polar silica gel type to the nonpolar alkane type. The fluorescence intensity is measured *in situ* in the TLC scanner of a spectrophotofluorimeter at an excitation wavelength of 560 nm and emission wavelength of 592 nm (F 560/592). Although this is a longer wavelength than that at which the dye emits (the dye in the alkane spot fluoresces maximally at F 525/554), the monochromator settings F 560/592 are used because the largest signal is obtained (**1**). Fluorescence and fluorescence quenching at or near the fluorescence excitation and emission wavelength maxima of the dye in the alkane spot can also be used but are less sensitive. The area under the peak of the recorded scan is measured by planimetry or triangulation and the alkane concentration in the benzene soluble air particulate fraction is estimated in terms of the docosane standard. This method is simpler, faster, more economical and more sensitive than previous methods (**2**).

2. Range and Sensitivity

2.1 The lower limit of detection of n-docosane with a spectrophotofluorimeter at its full potential of sensitivity is 0.1 μg. Range of analysis is 0.1 to at least 20 μg of docosane. Adequate analysis of the average contaminated air sample requires less than 30 μg of benzene-soluble organic fraction (BSO) or 3.5 m^3 of air for urban and approximately 50 μg of BSO or 20 m^3 of air for non-urban collection sites.

3. Interferences

3.1 Although rhodamine 6G is a reagent applicable to the analysis of most lipids (**3**), the long-chain saturated aliphatic hydrocarbons present in the air particulate BSO are distinguishable by their high Rf value obtained with a non-polar solvent on a silica gel plate.

3.2 Olefins, especially those containing one double bond in a long unbranched hydrocarbon chain, could interfere with the determination of the alkanes. However, unsaturated aliphatics will turn brown on exposure of the TLC plate to iodine vapor whereas alkanes do not react with iodine. Furthermore, analyzed air particulate extracts contain comparatively small amounts of olefins.

3.3 Aliphatic hydrocarbons separate well from polynuclear aromatic hydrocarbons present in the BSO fraction of air particulate extracts. The aliphatics travel at a higher rate than any of the other constituents, see Table 110:I. Pure benzo(a)pyrene added to a solution of docosane before chromatography and scanned at two different wavelength settings, showed a good separation and a negative response on the recorded plot, see Figure 110:1. Most of the aromatic ring compounds analyzed quenched the rhodamine 6G and were visible as dark spots or streaks on the developed TLC plate giving a negative response on the scan, see Figure 110:2. The more polar constituents remained at or near the origin.

3.4 The benzene soluble fraction from 3 urban and 3 non-urban air particulate samples were analyzed, see Figure 110:3. The scans of the urban samples show a large aliphatic peak and a shoulder of slightly lower Rf values. The spots yielding this shoulder gave a light brown color on exposure to iodine vapor indicating the presence of some unsaturation. However, compared with the very large peak given by the alkanes, the area under the shoulder is small and can be eliminated by drawing a perpendicular line to the baseline before integrating the area under the main peak, or the sample could be rechromatographed in a considerably lower concentration diluting out the spot responsible for the shoulder. Neither of the non-urban samples showed this shoulder nor did they give a noticeable color reaction when exposed to iodine vapor.

3.5 Contamination may cause considerable interference. Saturated aliphatic hydrocarbons are ubiquitous. They contaminate solvents, instruments, table tops, glassware, fingertips and laboratory air.

3.6 Aliphatic contamination of organic solvents can be determined using rhodamine 6G coated plates and two-dimensional chromatography. The plate is developed in both directions under identical conditions in the solvent to be tested. Development is interrupted when the solvent front is 8 to 10 cm above the starting line. If the solvent is contaminated, two bands are seen on the dried plate crossing each other, showing the characteristic pink color and yellow fluorescence. If the segment of the plate that has undergone development in both directions does not show such lines, the solvent is free of aliphatic contamination.

3.7 Room air contamination can be determined by exposing a clean rhodamine 6G coated silica gel plate to the air in a darkened room for a predetermined length of time. Subsequently the plate is developed in cyclohexane together with a clean control plate that was kept protected from the room air in a closed container for an equal length of time. After drying, both plates are scanned in the direction of the solvent flow (Figure 110:4). This simple but sensitive test does not require any collection devices or cleanup procedures. Generally speaking, the result of this study underlines the importance of clean-air rooms when employing very sensitive analytical methods for some of the more common air pollutants. In laboratories where hot or cool air is blown in from ducts in the ceiling, aliphatic air contaminants were found, especially during

Table 110:I. *Aliphatic and Aromatic Compounds Analyzed

Compound	100 Rf	# of Rings	Color†	UV Light‡
Normal alkanes, $C_{33}H_{68}$–$C_{36}H_{74}$	69	0	pk	UF
Normal alkanes, $C_{18}H_{38}$–$C_{32}H_{66}$	67	0	pk	YF
Pentatriacontene	67	0	pk	YF
Docoso-l-ene	65	0	pk	YF
Napthalene	65	2	pk	YF§
Squalene	62	0	pk	YF
Acenaphthene	40	3	None	Q
Pyrene	34	4	pu	Q
Thioxanthene	34	3	None	Q
Anthracene	33	3	pu	Q
2-Methylpyrene	32	4	pu	Q
2-Methylphenanthrene	31	3	pk	YF
Fluorene	31	3	pk	YF
Fluoranthene	30	4	pk	Pk-W F
Chrysene	25	4	pk	PK F, Center BF
Benzo(a)pyrene	24	5	pu	Q
Benzo(ghi)fluoranthene	23	5	pk	OF
Benz(a)anthracene	23	4	pu	Q
Perylene	22	5	y	Q
Benzo(k)fluoranthene	22	5	rpu	Q
Azulene	22	2	g	Q
Benzo(e)pyrene	22	5	pu	Q
Benzo(b)fluoranthene	21	5	pk	pk
Benzo(ghi)perylene	20	6	pu	Q
4H-Cyclopenta(def)-phenanthrene	20	3	None	Q
Coronene	19	7	pu	Q
Dibenzocoronene	15	9	pu	Q
9-Nitroanthracene	7	3	b	Q
Azobenzene	5	2	None	Q
Carbazole	3	3	pu	Q
11H-Benzo(a)carbazole	2	4	b-pu	Q
2-Nitrofluorene	2	3	pu	Q
1-Nitropyrene	2	4	y	Q
Acridine	0	3	pu	Q
2-Aminoanthracene	0	3	b	Q
7H-Dibenzo(bg)carbazole	0	5	r-b	Q
Pentacene	0	5	pu	Q

*Developer: Cyclohexane.
Absorbent: Silica gel TLC plates impregnated with rhodamine 6G.
†b-brown, g-gray, pk-pink, pu-purple, r-reddish, y-yellow.
‡F-Fluorescence, Q-Quenching, O-Orange, W-White, Bl-Blue, Y-Yellow.
§Unstable.

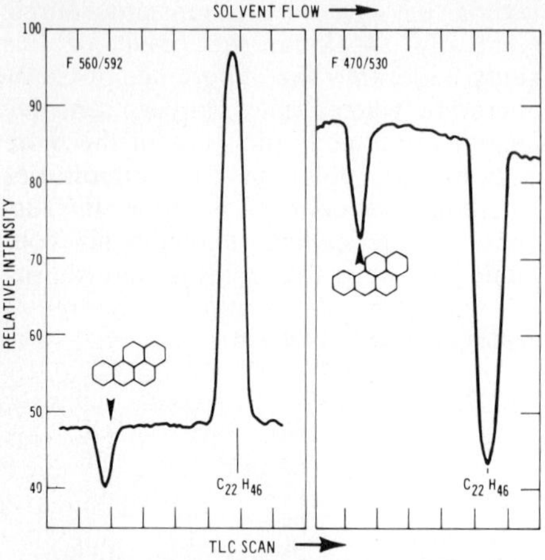

Figure 110:1—Chromatographic separation of a mixture of pure benzo(a)pyrene and n-docosane on a silica gel plate impregnated with rhodamine 6G developed 40 minutes in cyclohexane and scanned in direction of the solvent flow. At F 470/530 aliphatics quench the fluorescence; at F 560/592 they show a strong positive response. Benzo(a)pyrene gives a negative band at any wavelength setting.

the seasonal changeover from heating to air conditioning or vice versa.

4. Precision and Accuracy

4.1 Sixteen normal alkanes ranging from $C_{10}H_{22}$ to $C_{36}H_{74}$ were analyzed. The compounds containing 18 to 32 carbon atoms had identical Rf values. Dotriacontane, tetratriacontane and hexatriacontane consistently traveled at a slightly higher rate. Hexadecane and lower alkanes gave unstable colors; color and fluorescence faded within a short time after development. n-Docosane was chosen as a reference standard because it is the least expensive of the higher alkanes. Solutions of n-docosane of varying concentrations were analyzed in triplicate. The resulting scans were very similar indicating the reproducibility of the method (Figure 110:5). A straight-line relation of concentration and fluorescence peak area exists although the line does not go through the origin (Figure 110:6). Relative standard deviation at various concentration levels varied from 3 to 6%. At the 1 μg level, with n = 14, it was 5.4%.

4.2 The benzene soluble organic fraction from 3 urban and 3 non-urban sites were analyzed in triplicate for aliphatic hydrocarbons. In each case the triplicates were close, see Table 110:II. The scans of the thin-layer chromatograms show that the urban samples contained considerably higher amounts of aliphatics than the non-urban extracts, (Figure 110:3). The ratio of aliphatics to benzo(a)pyrene ranges from 315 to 850, see Table 110:III. The air volume and BaP values were supplied by

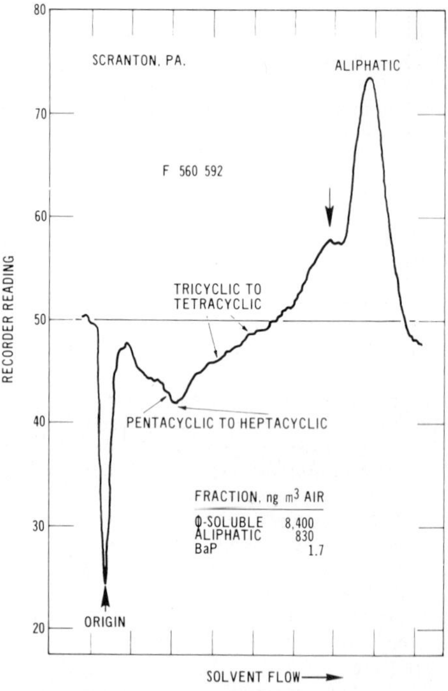

Figure 110:2—*In situ* TLC scan of benzene-soluble fraction from an urban air particulate extract separated on a silica gel plate impregnated with rhodamine 6G. Amount spotted: 70 μg of benzene-solubles in 1 μl chloroform. Developer, cyclohexane; developing time, 40 min. The positive response of the aliphatic hydrocarbons and the negative response of the aromatic hydrocarbons and of the compounds at the origin is evident. TLC spot giving band next to aliphatics (arrow) contains unsaturated aliphatics.

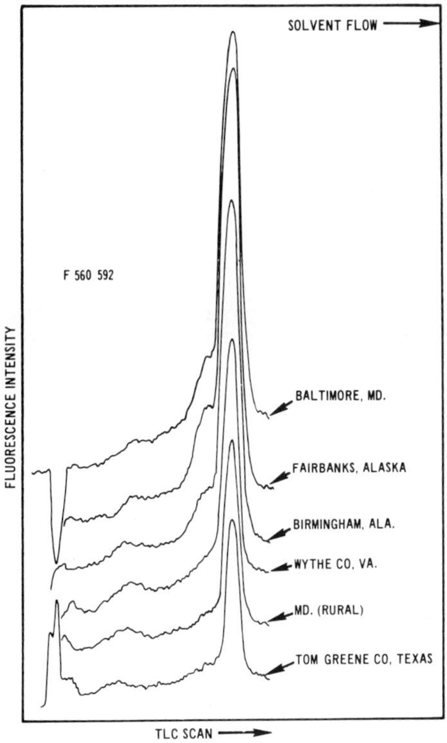

Figure 110:3—*In situ* TLC scans of benzene-soluble extracts from six different collection sites separated on rhodamine 6G-silica gel plates. Tall peaks at right are positive bands derived from the environmental effect of the aliphatic hydrocarbons. Negative band at left due to quenching compounds remaining at origin. The top five plots showed this negative peak, but four were omitted for reason of clarity. Bottom scan (T. Greene Co.) was the only one to show positive band instead. All samples adjusted so that each spot contained approximately 30 μg of benzene-soluble material.

EPA's National Air Surveillance Network.

4.3 Although a straight-line relation exists, (Figure 110:6), a standard in two different concentrations must be run on each plate alongside the unknown samples because (a) the line does not go through the origin and (b) variations between different batches of plates exist, and (c) Rf values in thin-layer chromatography are frequently not adequately reproducible.

5. Apparatus

5.1 LABORATORY WARE. All glassware must be free of organic matter particularly of oil and grease. It must be handled with utmost care so as not to cause recontamination. Glassware, including chromatography tanks and covers, that do not clear when rinsed with distilled water are usually covered with a thin film of oil and grease. This is best removed by applying a solution of 70% aqueous ethanol containing 25 g sodium hydroxide/100 ml. After rinsing with water, dilute hydrochloric acid, tap water, distilled water, and acetone, a final rinse with spectrograde cyclohexane is applied. In order to obtain an air tight seal between chromatography tank and lid, Teflon tape is used instead of a grease or lubricant.

5.2 Two thin-layer chromatographic chambers, with glass lids, one containing the rhodamine 6G dye solution.

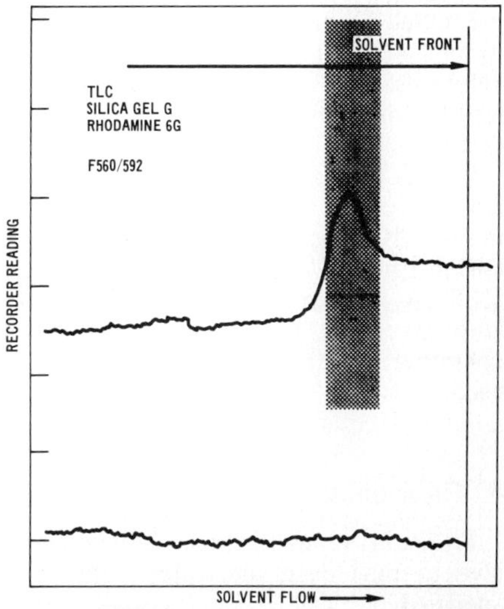

Figure 110:4—Fluorimetric scan of a rhodamine 6G-silica gel plate exposed to laboratory air for 36h and developed 35 min in cyclohexane. The shadowed region represents pink band observed after development giving the typical appearance of aliphatics. Higher baseline at right of peak showed portion of plate where hydrocarbon contamination is still present. Flat scan at bottom represents control plate.

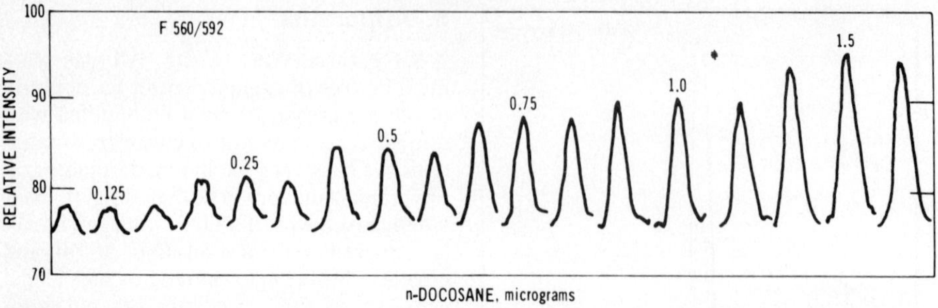

Figure 110:5—Chromatogram of several concentrations of n-docosane spotted in triplicates in 1 μl aliquots on rhodamine 6G impregnated silica gel plates and developed 40 min in cyclohexane. Area under the peak is concentration dependent.

5.3 Thin-layer chromatographic plates 20 × 20 cm, coated with a 250 μ layer of silica gel containing an inorganic binder such as calcium sulfate ($CaSO_4$). A binder is necessary to prevent the silica gel layer from being washed off the glass plate during the impregnation in the rhodamine 6G dye bath.

5.4 Self-filling micropipets with a capacity of 1 microliter. All pipets are checked before use, and only those that drain uniformly while spotting are used for quantitative work.

5.5 SPECTROPHOTOFLUORIMETER. A fluorimeter equipped with a scanner for thin-layer chromatograms is required. A Perkin-Elmer Spectrophotofluorimeter MPF-2A, a Farrand UV-Vis Chromatogram Analyzer, or any other instrument operating on the same principle, may be used.

6. Reagents

6.1 PURITY. All chemicals should be analytical reagent grade unless otherwise specified.

6.2 METHANOL.

6.3 RHODAMINE, 6G. This dye is available through regular suppliers. Its major producer is Allied Chemicals, National Biological Stains Department, Morristown, N.J. A stock solution is prepared by dissolving 200 mg rhodamine 6G in 100 ml methanol. Stored in dark bottles and protected from light, this solution keeps for many months.

6.4 CYCLOHEXANE. Each batch received is checked for long-chain aliphatic contaminants by two dimensional thin-layer chromatography on silica gel plates impregnated with rhodamine 6G. Contaminated cyclohexane is redistilled, collecting the 79 to 80 C cut.

6.5 CHLOROFORM. Spectrograde.

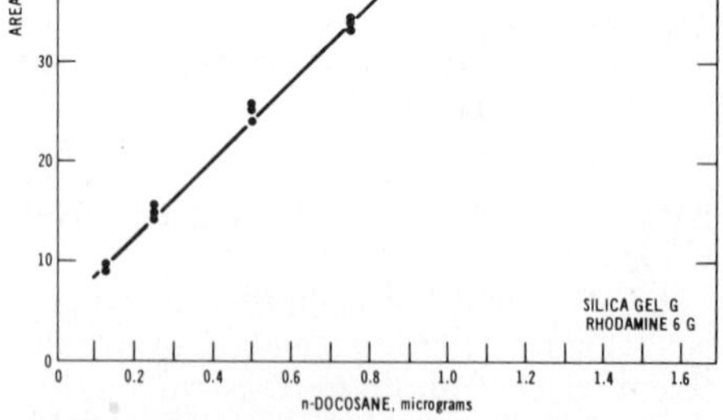

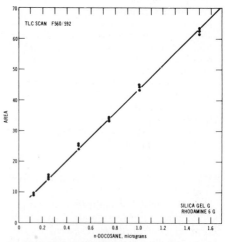

Figure 110:6—Peak areas of *in situ* TLC scans plotted against concentration of standard solutions of n-docosane spotted in triplicate on rhodamine 6G impregnated silica gel plates; developer, cyclohexane; and developing time, 40 min.

Table 110:II. Estimate* of Aliphatic Hydrocarbons

Location†	mg Aliph./g BSO			Average Value
Baltimore, Md.	280	270	260	270
Fairbanks, Alaska	281	290	280	280
Birmingham, Ala.	264	270	280	270
Wythe Co., Va.	150	150	140	150
Calvent Co., Md.	92	90	90	90
Tom Green Co., Tex.	90	80	90	90

*Developer: Cyclohexane.
Adsorbent: Silica gel coated glass plates impregnated with Rhodamine 6G.
Time for solvent front to travel 10 cm = 35 min.
†Sample size in all determinations was between 30 and 34 μg spotted in μl volume.

6.6 ETHER, ANHYDROUS.

6.7 N-DOCOSANE. The pure alkane is available from various manufacturers of research chemicals such as Analabs, Applied Sciences Laboratory, Supelco, Pfaltz & Bauer, etc.

6.8 Standard solutions of n-docosane are prepared ranging in concentration from 1 to 10 mg/l ml in spectrograde chloroform.

7. Procedure

7.1 An adequate volume (10 to 20 ml) of rhodamine 6G stock solution is diluted 200-fold with methanol and a glass TLC developing tank is filled to a height of 19 cm with this solution. When a 20 × 20 cm TLC plate is immersed in the solution, a 1 cm top edge of the plate is exposed which is sufficient to allow handling of the plate. Depending on the manufacturer, TLC tanks vary in size, the most popular contain approximately 3.5 to 5.0 l. Depending on the kind of tank used, 4 to 8 TLC plates can be impregnated simultaneously. The plates are placed in the dye solution where they remain for approximately 48 hr. (Although the impregnation time is not critical, it is advisable to follow the adopted procedure in order not to introduce an additional variable.) The tank is covered with a heavy glass lid using Teflon tape as a gasket to provide an airtight seal instead of greasing the rim of the tank. Tanks and impregnated plates are protected from light as much as possible during all phases of operation. After removing the plates from the tank, they are air-dried in a horizontal position and fully dried in a vacuum oven under reduced pressure at 50 C. The plates are stored in closed, clean containers until used. In order to eliminate possible contamination, the plates are pre-developed in ether before use by letting the ether run twice to the top of the plate.

7.2 The chromatographic developing tank is lined with a blotter type filter paper and cyclohexane is added to the tank so that it is at a height of 1 cm after the filter paper is saturated with the solvent. The tank is covered and the air in the tank is allowed to equilibrate with the solvent for 1 hr.

7.3 The benzene-soluble organic fraction (BSO) of airborne particulate samples, obtained and dried in the usual manner **(4)**, is redissolved in spectrograde chloroform to give a BSO concentration of 20 to 30 μg/μl. A cyclohexane-soluble organic fraction can also be used here. Seven to 9 aliquots of 1 μl each are spotted very carefully on the impregnated plate 2 cm from the bottom, 2 to 2.5 cm apart. Standard solutions of n-docosane in two concentrations are spotted on the same plate. In order to obtain reproducible results it is very important that samples and standards are spotted on the plate very uniformly. After the spots have dried, the plate is developed in cyclohexane until the solvent front has travelled 10 cm which requires a developing time of 35 to 45 min. After development, the plate is allowed to dry thoroughly for the pink aliphatic spots at an Rf of approximately 0.67 to become visible. The plate is then placed in the

Table 110:III. Amounts of Benzo(a)pyrene and Particulate Aliphatic Hydrocarbons in Ambient Atmospheres

Sampling Site*	Air Volume m^3	mg/air vol.BSO†	$\mu g/m^3$	BaP ng/m^3	Aliphatics $\mu g/m^3$	Ratio Aliphatics to BaP
Fairbanks, Alaska	3475	33.5	9.6	3.2	2.7	850
Birmingham, Ala.	3090	29.9	9.7	8.3	2.6	320
Baltimore, Md.	2484	14.8	6.0	2.5	1.6	650
Calvent Co., Md.	7972	13.1	1.6	0.65	0.16	250
Tom Green Co., Texas	9656	13.1	1.4	0.15	0.12	710
Wythe Co., Va.	9951	12.3	1.2	0.80	0.24	300

*1st Quarter 1970.
†BSO = Benzene-soluble organic fraction of airborne particulates.

scanning attachment of the spectrophotofluorimeter. The excitation monochromator is set at 560 wavelength, the emission monochromator at 592 nm. The spot that shows the densest pink color and brightest fluorescence under UV light, indicating the highest concentration of aliphatic hydrocarbons, is used to set the recorder pen at or near 100% using monochromator slits and sensitivity settings for the adjustment. The resulting baseline which will generally fall between 40% and 70% recorder reading is used for all sample and standard spots of that particular run. The high baseline is due to the fact that the intensity of the fluorescence of the spot is measured against a fluorescent background. Therefore a compromise in baseline adjustment has to be made when using a record of limited zero-suppression capability. When the monochromator slits are opened to increase the sensitivity, the plate background gives a high intensity reading and the intensity of the spot is off scale. On the other hand, when the sensitivity is reduced to suppress the fluorescence of the plate, the intensity of the sample spot is likewise reduced. A blank area close to the spots is also scanned using the same baseline as for the alkane spots.

7.4 The peak area of the recorded scan is measured using a planimeter or triangulation, and the concentration of the aliphatics in the sample is calculated in terms of n-docosane.

8. Calibration and Standardization

8.1 One μl aliquots of n-docosane solutions of increasing concentrations are spotted in triplicates, using two TLC plates from the same production batch and processed simultaneously under identical conditions. The concentrations used range from 0.125 to 1.5 μg. (Figure 110:5).

8.2 The curve obtained by plotting the concentration of the reference standard n-docosane versus the peak area in planimeter units or mm^2 is a straight line although the line does not go through the origin, (Figure 110:6). In order to increase the sensitivity of the method, lower concentration ranges are read at a higher sensitivity level. Samples and reference standards containing higher amounts of aliphatics are scanned at a lower baseline setting. Straight lines are obtained for any one of these ranges although the slopes for each set may vary. With a baseline at 70% relative intensity, standards containing 0.125 to 1.5 μg of n-docosane are scanned. Concentrations of 1 μg to 6 μg and 5 to 20 μg are scanned at 60% and 50% baseline setting, respectively.

8.3 Each unknown analysis is checked against two concentrations of docosane spotted on the same impregnated TLC plate in a comparable concentration range. Since the urban samples analyzed contained approximately 270 mg of aliphatic hydrocarbons/l g of BSO and samples from non-urban collection sites about 100 mg/l g BSO, approximately 8 μg and 3 μg respectively of aliphatic hydrocarbons were present in a 3 mg spot at the origin.

9. Calculations

9.1 Although a straight line relation exists between concentration and peak area, slight variations between different batchs of Rhodamine 6G impregnated plates or between plates prepared on different days cause the slope of the line to vary between different runs. A docosane standard must therefore be run alongside unknown samples, preferably in two concentrations in order to compensate for the fact that the line does not go through the origin. On this basis, the amount of aliphatics in the benzene-soluble fraction of airborne particulate material may be estimated by the following equations.

a) mg aliphatics/g BSO =

$$\frac{W_s \times A_u}{W_t \times A_s} \times 10^3$$

b) % aliphatics in BSO =

$$\frac{W_s \times A_u}{W_t \times A_s} \times 100$$

c) μg aliphatics/m³ air =

$$\frac{W_s \times A_u}{V \times A_s}$$

where:

W_s = weight of docosane standard spotted in μg

W_t = weight of benzene-soluble organic fraction spotted on TLC plate in μg

A_s = area of standard band of plotted scan in mm²

A_u = area of unknown aliphatic band of plotted scan on same plate in mm²

V = volume of air in m³ sampled to obtain weight of spotted BSO

10. Effects of Storage

10.1 The fluorescence intensity of n-alkanes and the corresponding spot in the unknown sample on the rhodamine 6G impregnated TLC plates is stable for at least 16 hr if protected from light. In fact, spots on plates that were left in the scanning attachment of the spectrofluorimeter in the dark and were rescanned in certain time intervals gave identical fluorescence intensity for as long as 48 hr. Alkanes having less than 18 carbon atoms are much less stable. Apparently their higher vapor pressure makes them evaporate faster on the plate.

10.2 Since the higher alkanes are stable compounds they are expected to be present in BSO extracts for a long time. This was found to be true when extracts that were stored for several years were analyzed by this method. A composite sample used in the analysis of air particulate benzene-soluble extract of 100 American cities collected during the years 1957 to 1959 gave a strong aliphatic band and corresponding large peak area (240 mg/g BSO).

10.3 Rhodamine 6G coated plates are stable for several weeks if protected from light and air. If faded, their color can be restored by renewed immersion in the dye bath for 16 hr.

10.4 The rhodamine 6G impregnating bath keeps for over 2 months if tightly covered and kept in the dark. It can be restored by adding 10 ml of the rhodamine 6G-methanol stock solution and stirring well.

11. References

1. WITTGENSTEIN, E., E. SAWICKI. 1972. *Intern. J. Environm. Anal. Chem.*, 1.
2. MCPHERSON, S.P., E. SAWICKI, F.T. FOX. 1966. *J. Gas Chrom.*, 156.
3. ROUSER, G., KRITCHEVSKY, G., HELLER, D. and E. LIEBER. 1963. *J. Am. Oil Chemists' Soc.*, 40:425.
4. SAWICKI, E., W.C. ELBERT, T.R. HAUSER, F.T. FOX, T.W. STANLEY, *Am. Indust. Hyg Assoc. J.*, 21:443.

Subcommittee 5
E. SAWICKI, *Chairman*
P. R. ATKINS
T. BELSKY
R. A. FRIEDEL
D. L. HYDE
J. L. MONKMAN
R. A. RASMUSSEN
L. A. RIPPERTON
J. E. SIGSBY
L. D. WHITE

111.

Tentative Method for Continuous Analysis of Methane in the Atmosphere (Flame Ionization Method)

43201-01-73T

1. Principle of the Method

1.1 The system for methane measurement operates by continuously passing atmospheric air through a treated charcoal column prior to introduction into an instrument equipped with a flame ionization detector **(1, 4, 6)**. The solid adsorbent selectively removes all hydrocarbons other than methane so that this procedure is a specific method for analysis of methane.

1.2 Commercially available total hydrocarbon analyzers may be modified for the continuous analysis of methane by connecting the activated charcoal column in the sample line upstream from the inlet to the instrument **(1, 5, 6)**.

2. Range and Sensitivity

2.1 The concentration of methane in the atmosphere has been determined to vary between 0.6 mg/m^3 (1 ppm) and 5.2 mg/m^3 (8 ppm) depending upon local sources of methane of atmospheric ventilation **(1)**. It is recommended that the sample flow rate to the detector and the attenuation be adjusted so that full scale response is 6.54 mg/m^3 (10 ppm) of methane concentration.

3. Interferences

3.1 Any compound capable of ionizing in the flame ionization detector (FID) is an interference.

3.2 Ethane is the hydrocarbon most likely to break through the carbon column when ambient air is sampled **(1, 6)**. The service life of the activated charcoal column has been measured by introducing known concentrations of ethane into the system **(6)**. It is recommended that the carbon column be changed twice weekly to ensure that nonmethane hydrocarbons do not break through and interfere with the analysis of methane **(1)**.

3.3 The charcoal column should be protected from sudden temperature increases which would cause positive error due to the release of hydrocarbons from the solid adsorbent.

3.4 The charcoal column should be protected from water mist (and rain) which would interfere with adsorption of the hydrocarbons by wetting the charcoal. Water vapor does not interfere.

4. Precision and Accuracy

4.1 Precision should be approximately ± 1% full recorder scale and is affected by instrumentation noise and drift as well as variations in sample, air and fuel gas flow rates.

4.2 Accuracy is dependent on the absolute concentration of methane in the gas mixtures used to calibrate the instrument. If instrument linearity is assumed, nonlinearity of instrument response can adversely affect accuracy. A three or more point calibration over the expected range of sample concentrations is recommended to improve accuracy and evaluate instrument linearity. It is believed that under ideal conditions accuracy differs less than ± 5% from the true value.

5. Apparatus

5.1 CHARCOAL COLUMN. A 5″ (12.5 cm) long threaded nipple of ¼″ (0.63 cm) pipe packed with 4.5 g of $^{6}/_{14}$-mesh activated charcoal (**1**). Round cloth discs are used to contain the activated charcoal. The freshly packed tube is blown out forcibly with helium for two min to remove fine dust, which otherwise would interfere with the operation of the FID.

5.2 The sample line should be made of an inert material such as glass, stainless steel, or Teflon. A particulate filter should be installed in the sample line between the charcoal column and the sample inlet to the analyzer.

5.3 Any commercially available hydrocarbon analyzer purchased for use must be installed on location and demonstrated by the manufacturer to meet or exceed manufacturer's specifications or those described in the Intersociety Committee Manual (**2**). Hydrocarbon analyzers (**5**) normally consist of a regulated fuel and air delivery system for the flame ionization detector (FID), a sample pump or regulated sample injection system, and an electrometer for measuring the ion current produced by compounds introduced into the flame.

5.4 Recording devices include meters or potentiometric-type recorders which should meet user requirements and be compatible with the output of the analyzer.

6. Reagents

6.1 FUEL GAS. Pure hydrogen or a 40% hydrogen/60% nitrogen mixture, depending on FID design requirements. Hydrogen in compressed gas cylinders should be prepared from water pumped sources. High quality electrolytic generators which produce ultrapure hydrogen under pressure may be used. These provide some degree of safety when compared to the explosive hazard of bottled hydrogen. All connections should be made with thoroughly cleaned stainless steel tubing and properly tested for leaks. An automatic cut-off should be provided for the instrument in the event of an FID flame-out.

6.2 COMPRESSED AIR. Water-pumped, filtered compressed air in cylinders, low in hydrocarbon content. Alternately, hydrocarbon-free synthetic air can be conveniently obtained by mixing oxygen and water-pumped nitrogen or other equivalent diluent.

6.3 AERO GAS. Less than 65 $\mu g/m^3$ (0.1 ppm) hydrocarbon as methane in air.

6.4 SPAN GASES. At least three different concentrations of methane in air are required for calibration purposes. Calibration gas corresponding to 80% of full scale (8 ppm or 5.2 mg/m^3) is recommended for use in spanning the instrument. Commercial gases having a certified or guaranteed analysis should be used. An independent gas chromatographic analysis should be made to verify the hydrocarbon concentration in the tank and particularly if the span gas has been prepared in the laboratory (**2, 4, 7**). Due to desorption effects, it is recommended that use of calibration gases be discontinued when the tank pressure is less than 500 psi.

6.5 The helium required for reconditioning the charcoal column should be prepared from water-pumped sources or be of guaranteed high purity.

7. Procedure

7.1 The volatile hydrocarbons are purged from the activated charcoal column by passing pure helium through the column for 30 min at flow rates of about 5 l per min (**1**). Columns can be rejuvenated as many as 8 times by this helium treatment. The carbon can be reactivated by placing the column in a furnace at 300 C and passing helium through the charcoal at a flow rate of about 1 liter/min for one hr.

7.2 The charcoal column is conditioned for use by passing a methane span-gas mixture through it for about 15 min or until the instrument indicates breakthrough of methane (**1**).

7.3 Instrument manuals are provided by manufacturers of commercially available hydrocarbon analyzers and should be consulted for specific operating procedures.

8. Calibration, Standards, and Efficiencies

8.1 The FID should be calibrated to determine the fuel, air and sample flow rates which give optimum sensitivity and stability (**3, 4**). All flow rates should be optimized for a particular instrument and maintained carefully by checking the flow rates with a calibrated bubble flow meter whenever the charcoal column is changed.

8.2 Calibrate the analyzer at the attenuator setting and sample flow rate required so that full scale response is 6.54 mg/m^3 (10 ppm) of methane concentration. Connect the zero-gas cylinder outlet to the carbon column so that the zero gas passes through the freshly purged and conditioned charcoal column prior to introduction into the analyzer. Operate the instrument on zero gas for 10 min and adjust the zero control to indicate the proper setting on the recorder (according to the analysis tag attached to the zero-gas cylinder). The calibration gas corresponding to 80% of full scale (8 ppm) is used to span the instrument. Connect the span-gas cylinder to the charcoal column and introduce span gas. Adjust the span control to bring the recorder pen to the indicated (analysis tag) methane concentration in ppm. Recheck zero and span settings until adjustments are no longer required.

8.3 At optimum operating conditions, a three or more point calibration is recommended to assure a more accurate analysis (**4**).

9. Calculations

9.1 The recorder is calibrated to be read directly in terms of concentration expressed as ppm (1 ppm = 0.65 mg/m^3) methane by volume.

10. Effects of Storage

Not applicable.

11. References

1. ALTSHULLER, A.P., G.C. ORTMAN, B.E. SALTZMAN, and R.E. NELLIGAN. 1966. *Air Poll. Control Assoc. J.*, 16:87.
2. AMERICAN PUBLIC HEALTH ASSOCIATION. 1977. Methods of Air Sampling and Analysis. 2nd ed. p. 257. Washington, D.C.
3. Ibid. p. 209.
4. Ibid. p. 257.
5. NADER, J.S. 1972. *Air Sampling Instruments for Evaluation of Atmospheric Contaminants*. 4th Ed., pp U54–U62, American Conference of Governmental Industrial Hygienists, Cincinnati, Ohio.
6. ORTMAN, G.C. 1966. *Anal. Chem.* 38:644.
7. SALTZMAN, B.E., W. R. BURG, and G. RAMASWAMY, 1971. *Environ. Sci. Technol.*, 5:1121.

Subcommittee 5

E. SAWICKI, *Chairman*
P. R. ATKINS
T. BELSKY
R. A. FRIEDEL
D. L. HYDE
J. L. MONKMAN
R. A. RASMUSSEN
L. A. RIPPERTON
J. E. SIGSBY
L. D. WHITE

112.

Tentative Method of Analysis for Polynuclear Aromatic Hydrocarbons in Coke Oven Effluents

11104-03-73T

1. Principle of the Method

1.1 Samples are prepared as a solution in a saturated hydrocarbon solvent such as cyclohexane. After the addition of 1,3,5-triphenylbenzene (TPB) as an internal standard to the sample, a concentrate is prepared by careful evaporation of the solvent. An aliquot of this concentrate is injected into a gas chromatograph to obtain fractions for measurement by ultraviolet absorption spectrophotometry (UV). The method is designated as the GC/UV procedure.

1.2 Effluent from the gas chromatograph is split 15% to a flame ionization detector for the GC measurement and 85% to an ice cooled trap. Selected peaks are trapped by condensing in metal tubes and the condensate is rinsed out with cyclohexane for UV measurement. Based on the UV spectra, microgram quantities of the various polynuclear aromatic (PNA) compounds are next calculated. The quantity of internal standard (TPB) is measured and compared with the amount originally added. This gives a factor by which to relate the microgram quantity of each PNA to its total weight in the sample. A more complete description of the method is given by Searl, *et al* (**1**).

1.3 Although the calculations are straightforward, it is a time-saver to handle them by computer.

2. Range and Sensitivity

2.1 From 0.1 to 10 mg of cyclohexane soluble material can be handled by the method. Individual PNA's are present at concentrations of 1 to 40 μg. The lower limit of detectability for each PNA is 0.1 μg.

3. Interferences

3.1 Plastic tubing is frequently used in laboratory work and a phthalate plasticizer is readily extractable and can cause contamination. This contaminant appears in the GC peak containing benz(a)anthracene (BaA), chrysene, and triphenylene and adversely affects the UV measurement.

3.2 The method utilizes the GC and UV measurements so as to minimize interferences. Pyrene and fluoranthene, for example, occur by themselves in separate GC fractions; BaÁ, chrysene, and triphenylene occur together in a GC fraction but their characteristic UV spectra permit them to be separately measured. Benzo(a)pyrene (BaP) and benzo(e)pyrene (BeP) occur together in a GC fraction but can be measured from the UV spectrum. All UV measurements employ the base line technique as first described by Wright (**2**) and amplified upon by Heigl, *et al* (**3**).

4. Precision and Accuracy

4.1 Little precision data are available. One sample was analyzed in duplicate in which the first analysis was done in March, 1969, and then reanalyzed in March, 1970. Microgram quantities were observed to be: fluoranthene—45, —; pyrene—20, 21; BaA—22, 26; chrysene—15, 15; BaP—20, 21; and BeP—10, 9. Duplicate filter samples were collected and analyzed in three instances for BaP and once for BaA. Results in micrograms were: BaP—1, 1 (sample 1); BaP—0.3, 0.3 (sample 2); BaP—3.7, 4.4; and BaA—3, 3 (sample 3).

4.2 Accuracy of the method was investigated only with regard to BaP. In a number of samples its GC peak was measured in terms of UV absorption and fluorescence (**4**). Typical results for a range of quantities are shown in Table 112:I.

5. Apparatus

5.1 Because of the sensitivity of the method, the possibility of error due to contamination is great. It is important that all glassware be scrupulously cleaned to remove all organic matter such as oil, grease, and detergent residues. Examine all glassware, including stopper and stopcocks, under ultraviolet light, to detect any residual fluorescent contamination. As a precautionary measure it is recommended practice to rinse all glassware with purified isooctane immediately before use. No grease is to be used on stopcocks or joints. Tygon tubing must *absolutely* be avoided to eliminate di-isooctyl phthalate contamination which interferes in both the GC and UV. Because some of the polynuclear hydrocarbons sought in this test are very susceptible to photooxidation (**5**), the entire procedure is to be carried out under subdued light. For laboratory lighting, it is desirable to use yellow lights in preference to the usual fluorescent lamps.

Table 112:I. Benzo(a)pyrene by UV Absorption and Fluorescence Measurement, μg

Sample No.	Absorption	Fluorescence
344	6	5
373	4	4
358	9	7
394	17	22
371	25	20

5.2 The gas chromatograph is equipped with a flame ionization detector, a linear temperature programmer, and a flow controller. The injection port consists of an aluminum block with a stainless steel tube insert (quartz sleeve). The chromatograph is modified by the addition of a bypass and trap so that 15% of the effluent of the column will go to the detector and 85% to the trap.

5.2.1 Figure 112:1 shows the gas chromatograph effluent splitter and trapping assembly. As shown in Figure 112:1, tubing of 0.25-mm ID connects the splitter block A to the detector, whereas a 0.5-mm ID tubing goes to the point of insertion of the trapping tube. The overall length of each set of tubes is approximately 22 cm. In this arrangement, 15% of the effluent passes to

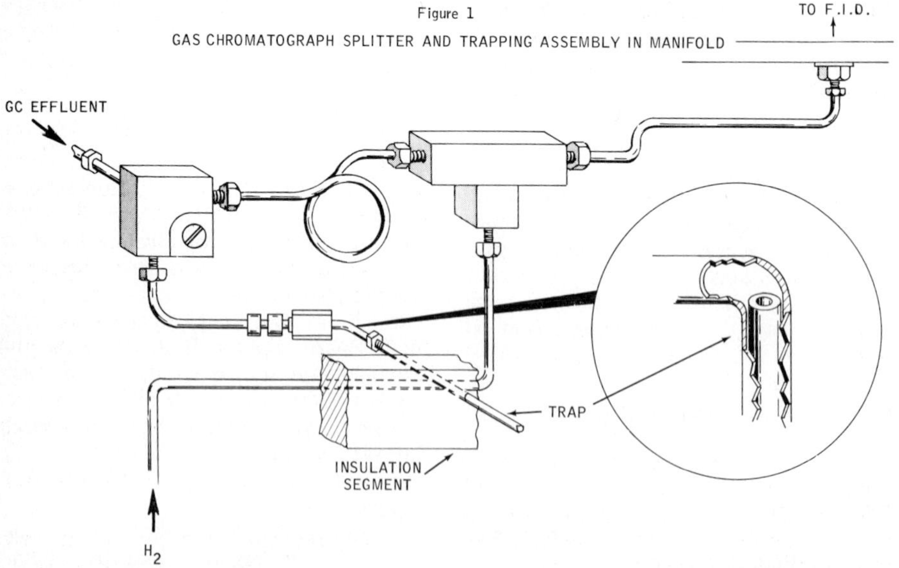

Figure 112:1—Gas Chromatograph Splitter and Trapping Assembly in Manifold.

the detector and 85% goes to the air cooled trap. To trap a fraction, the stainless steel tube (1/8″ tubing, 7″ long) is slid into position as shown. To assure condensation in the trap, it is also necessary to have the outer tube partially covered with ice. To accomplish this, a plastic cup is mounted on the tube by simply piercing the cup through its center and placing ice therein. To stop trapping, the tube is taken out.

5.2.2 *Recorder.* 0–1 mV, 2 min/inch chart speed.

5.2.3 *Trapping Tubes.* 30 or more stainless steel tubes 7″ long by 1/8″ outside diameter.

5.2.4 *Column.* Ten feet of 1/8″ OD stainless steel tubing packed with 2 per cent SE-30 (GC grade) on Chromosorb G (acid washed and DMCS treated), 80/100 mesh. The prepared packing is available from Supelco Inc., Bellefonte, Pennsylvania, and other suppliers. To condition a new column for use, the column is connected to the inlet of the chromatograph but not with the detector. With the column at 300 C, helium is then passed through at 30 ml per min for 1 hr. If a noticeable base line drift occurs upon trying to use, condition at 300 C for another hr.

5.3 SYRINGE—10 μL. The syringes, Cat. #701N, manufactured by the Hamilton Co., P.O. Box 307, Whittier, California 90608, and Model C of Precision Sampling Corp., P.O. Box 15119, Baton Rouge, La. 70821, are satisfactory.

5.4 SCANNING SPECTROPHOTOMETER. Spectral range 225 to 400 nm, with spectral slit width of 2 nm or less. Under instrument operating conditions for these absorbance measurements, the spectrophotometer shall also meet the following performance requirements: absorbance repeatability, ± 0.01 at 0.4 absorbance; absorbance accuracy, ± 0.05 at 0.4 absorbance; wavelength repeatability, ± 0.2 nm; wavelength accuracy, ± 1.0 nm.

5.4.1 *Spectrophotometric cells.* Fused quartz cells having optical path lengths of 1.000 ± 0.005 cm and 5.000 ± 0.005 cm. The 5-cm cells are microcells and hold about 3 ml of solution. They are available from Pyrocell Manufacturing Co., 91 Carver Ave., Westwood, N.J. 07675.

5.5 GRADUATE GLASS—5 ML.

5.6 VIALS—ONE-DRAM WITH CAP.

5.7 STEAM BATH. Equipped with nitrogen outlet for purging vials.

5.8 MEDICINE DROPPER. Pasteur Pipet—with tip drawn out.

5.9 SPECIAL LOW UV EMISSION ROOM LIGHTS. Westinghouse W-40-gold fluorescent lamps.

5.10 SOXHLET EXTRACTOR. Extraction tube 40 mm ID, and a 500-ml flask.

5.11 EXTRACTION THIMBLE. 33 mm diam, 80 mm height.

6. Reagents

6.1 PURITY. All solvents should be of highest purity.

6.2 CYCLOHEXANE. To check purity, evaporate 180 ml down to 5 ml. Run a UV scan on this residue in a one-cm cell from 400 to 225 nm. The absorbance should not exceed 0.01 units. To purify, percolate through activated silica gel, Grade 12, in a glass column, 90 cm long and 5 to 8 cm diam. Satisfactory solvent may be purchased from several sources including Burdick and Jackson Laboratories, Inc., Muskegon, Michigan 49442. Their product is called "Distilled in Glass Solvents."

6.3 TOLUENE.

6.4 ACETONE.

6.5 ISOPROPANOL.

6.6 BENZENE. Burdick and Jackson Laboratories, Inc.

6.7 BENZ(A)ANTHRACENE, EASTMAN 4672. Prior to use, this compound, as well as other PNA hydrocarbons, must be of acceptable purity as shown by gas chromatograph and UV.

6.8 BENZO(A)PYRENE, EASTMAN 4951. Repurify by elution through deactivated alumina column, if necessary.

6.9 PYRENE, EASTMAN 3627.

6.10 1,3,5-TRIPHENYLBENZENE. Aldrich Chemical, T8200-7.

6.11 FLUORANTHENE. Aldrich Chemical Co.

6.12 CHRYSENE. Matheson, Coleman & Bell Co.

6.13 TRIPHENYLENE. K & K Chemical Co.

6.14 BENZO(E)PYRENE. Source unknown.

6.15 TEST BLEND. Weigh exactly 25.0 mg of each of the four compounds listed as Items **6.7** through **6.10**. Place in a 25-ml volumetric flask, add 20 ml of toluene and swirl until compounds are in solution. Then make up to volume. This will contain 1 μg of each compound per μl of solution. Pour into a small, narrow neck brown bottle and keep in a cool, dark place, preferably a refrigerator.

6.16 INTERNAL STANDARD SOLUTION. Weigh exactly 250 mg of 1,3,5-triphenylbenzene, place in a 250-ml flask, dissolve and make up to volume with toluene.

7. Procedure

7.1 COLLECTION OF SAMPLE. Samples are collected on a silver membrane filter. A suitable procedure is described by Richards, *et al* (**6**).

7.2 The individual steps for the method of analysis are:

7.2.1 Extraction of organics from the silver membrane filter.

7.2.2 Concentration of organics.

7.2.3 Run in GC and "trap" peaks.

7.2.4 Measure UV spectra of "trapped" GC peaks.

7.2.5 Calculate micrograms of BaA, BaP, fluoranthene, pyrene, chrysene, triphenylene, BeP and TPB.

7.3 A conventional Soxhlet extraction is used as described below. Consideration can be given, however, to a recently described procedure which shortens the extraction from eight hours to one-half hour (**7**).

7.3.1 Before the filter is used to obtain the air sample, extract it in a Soxhlet extractor with benzene for about two hours. Remove from the Soxhlet, evaporate most of the solvent at room temperature, and finally dry in an oven at 105 C for 10 min. Cool in the air and weigh to the 5th decimal place.

7.3.2 After the sample has been collected, reweigh the filter to obtain the total weight of particulate matter collected. Since moisture is present, place the sample in a desiccator for 24 hr prior to weighing.

7.3.3 Place new Soxhlet thimbles in the extractor and extract with acetone for two hours. Remove the thimbles, draw off and discard the acetone but dry the thimbles.

7.3.4 Until assured that the blank is low, carry a blank through the whole procedure of extraction, gas chromatography, and UV measurement.

7.3.5 Place filter in Soxhlet thimble, pour 250 ml of cyclohexane into the flask. With a syringe or micro pipet, add exactly 20 μl of the TPB internal standard, add several glass beads and assemble the extractor. Completely cover the outside of the extractor with aluminum foil to keep out light and then extract for six hr at about 10 extraction cycles/hr.

7.3.6 Lift thimble containing the filter from the solvent and allow excess cyclohexane to draw off into Soxhlet apparatus. Drain all of the solvent into the flask.

7.4 Concentrate the PNA's by evaporating to a residual liquid volume:

7.4.1 Transfer some of the extract to a 150-ml beaker and place on the steam bath under a small jet of nitrogen.

7.4.2 As evaporation progresses, add portions of the remaining extract until completely transferred. Rinse out the flask and add rinsings to the beaker. Do not evaporate to less than 2 ml.

7.4.3 Transfer the concentrated solution to a one-dram vial.

7.4.4 Place the vial in a 30-ml beaker for support. The beaker is placed on a steam bath and a gentle jet of nitrogen directed into the vial.

7.4.5 Rinse the 150-ml beaker with 3 or 4 small portions of cyclohexane ($\sim$ 1 ml) and combine with the contents of the one-dram vial as evaporation proceeds.

7.4.6 Evaporate the solution to about 50 μl (0.05 ml). Add about 0.5 ml of toluene and evaporate to 50 μl.

7.4.7 It may be necessary to add 10 to

20 μl of toluene to lower the viscosity of the sample. Cap and save.

7.5 The following parameters are employed in the gas chromatographic separation with the Perkin-Elmer 900. Other instruments might use different conditions.

Carrier Gas—Helium.
Flow Rate—30 ml/min.
Hydrogen and Air—at manufacturers recommendation or optimum rates.
Injection Port—300 C.
Detector—340 C.
Program—175 C to 300 C at 4 per min. The temperature is held at 300 until all peaks are eluted, but in any case, for 20 min.

Dual columns are not used.

7.5.1 To inject the PNA concentrate to the chromatograph, flush the 10 μl syringe twice with the sample and then fill the syringe and hold vertically, point up. Stick the tip in a rubber septum and depress the plunger to compress gas bubbles which rise to the top.

7.5.2 Now advance the plunger to the desired amount, 5 to 10 μl; remove the rubber, and retract the plunger until the air-liquid meniscus enters the glass bore.

7.5.3 Note the volume from meniscus to plunger.

7.5.4 Insert the syringe up to the hilt in the GC injection port, depress the plunger. Start the recorder.

7.5.5 Remove the syringe from the port and retract the plunger.

7.5.6 Note the volume remaining (generally about 0.2 μl) and subtract from the previously noted volume to determine the net volume injected.

7.5.7 Have stainless steel trapping tubes and one-dram vials available. The vials may be in holes drilled in a board.

7.5.8 Based on retention times as established by the calibration procedure, tubes are inserted as the peak starts to rise and remove as the trace returns to the base line. Tubes are numbered to correspond with the peak and placed in a similarly labeled one-dram vial.

7.5.9 When the GC run is completed, take the first tube and place in a 5-ml graduate. With an elongated medicine dropper add enough cyclohexane through the top of the tube to fill the spectrophotometric cell to be used. Our cells required 3.8 ml.

7.5.10 Pour the solution into the original vial, cap and protect from light. Rinse out the graduate with cyclohexane and use for the next tube. Save the fractions for UV.

7.5.11 Measure and record the retention time in min from injection point for each peak.

7.6 Obtain the UV spectrum of each selected fraction and measure specified wavelength(s). A 1-cm or 5-cm cell should be used, depending upon the level of UV absorption.

8. Calibration and Standards

8.1 A performance test is used to measure GC retention times and demonstrate suitable recovery of individual PNA's. With the GC column conditioned and the parameters set as described in **7.5**, prepare to inject a sample of the test blend, **6.14**.

8.1.1 Inject a known volume of sample to the gas chromatograph. Because each PNA is present as one μg/μl, the known volume indicates the quantity of each injected.

8.1.2 As each PNA elutes, trap the peak and label properly. Also record the retention time.

8.1.3 Measure the UV absorption of each peak as required to allow calculation of the micrograms recovered for each component. Recoveries of ≥83% for pyrene, BaA and BaP, and TPB indicate satisfactory operation of the chromatograph. The observed retention times provide a guide

TABLE 112:II. Observed Retention Times

Compound	Retention Time, Minutes
Fluoranthene	13.1
Pyrene	13.6
Triphenylene	20.0
Benz(a)anthracene	20.1
Chrysene	20.4
Benzo(e)pyrene	27.2
Benzo(a)pyrene	27.3
1,3,5-Triphenylbenzene	28.5

TABLE 112:III. UV Calibration Data For Some PNA Compounds

Wavelength, nm									
Peak	Base Line	Fluor-Anthene	Pyrene	BaA	Chrysene	Tri-Phenyl-ene	BaP	BeP	1,3,5-Tri-Phenyl-Benzene
		Absorptivity (ml/μg/cm)							
288	285-293	*0.199*							
336	327-343		*0.240*						
289	283-295			*0.332*					
269	263-277			0.039	*0.464*				
259	252-263			0.015	0.107	*0.480*			
383	373-390						*0.092*		
333	325-338							*0.135*	
254	225-290								*0.142*

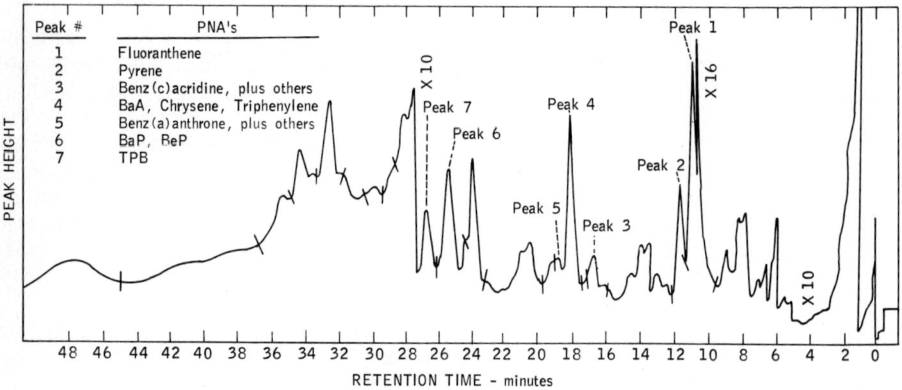

Figure 112:2—Chromatogram of coke oven effluent.

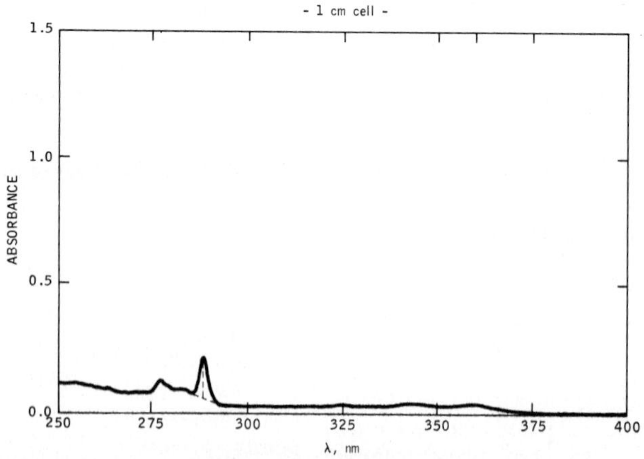

Figure 112:3—UV spectrum of GC fraction containing fluoranthene.

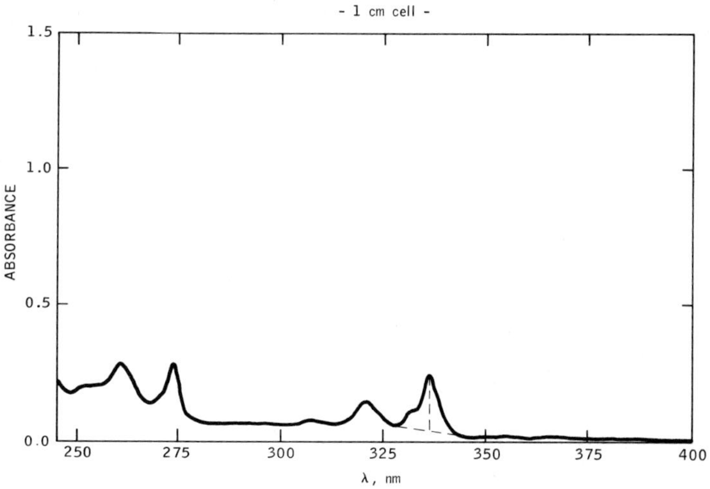

Figure 112:4—UV spectrum of GC fraction containing pyrene.

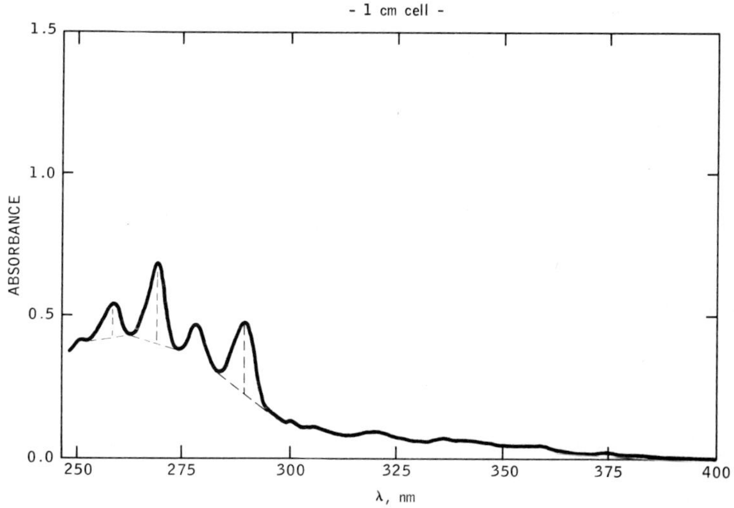

Figure 112:5—UV spectrum of GC fraction containing benz(a)anthracene, chrysene, and triphenylene.

TABLE 112:IV. Calibration Matrix for Calculation of BaA, Chrysene, and Triphenylene

	BaA	Chrysene	Tri-phenylene	
(1)	*0.332*	0	0	A289
(2)	0.039	*0.464*	0	A269
(3)	0.015	0.107	*0.480*	A259

A289, A269, A259 = UV absorbance of GC peak in a 1-cm cell. The equations can be solved directly by substitution.

$$\text{BaA} = \frac{\text{A289}}{0.332}$$

$$\text{Chrysene} = \frac{\text{A269} - \text{BaA} \times 0.039}{0.464}$$

$$\text{Triphenylene} = \frac{\text{A259} - \text{BaA} \times .015 - \text{Chrysene} \times 0.107}{0.480}$$

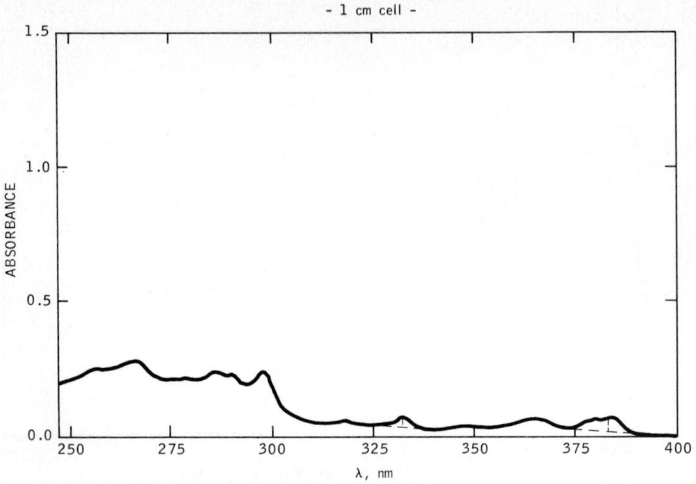

Figure 112:6—UV spectrum of GC fraction containing BaP and BeP.

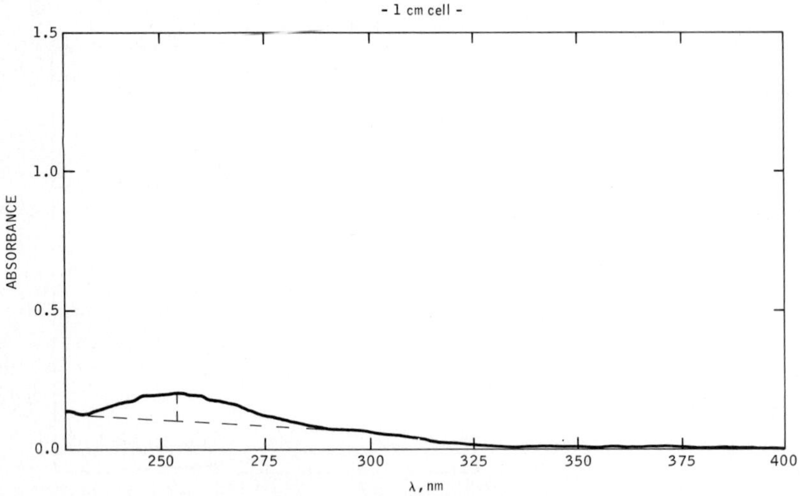

Figure 112:7—UV spectrum of GC fraction containing triphenylbenzene.

as to the appearance of compounds as they elute in a mixture. Typical retention times for the four compounds of the test blend, as well as other PNA's, are shown in Table 112:II.

8.2 Absorptivity coefficients are measured for fluoranthene, pyrene, BaA, chrysene, triphenylene, BaP, BeP, and TPB.

8.2.1 Prepare solutions of known concentrations of each PNA in cyclohexane. Concentrations of 1 to 2 $\mu g/\mu l$ are suitable.

8.2.2 With cyclohexane in the reference cell, obtain the spectrum of each solution from 400 to 225 nm.

8.2.3 Rinse the cell thoroughly with cyclohexane between spectra and remove the last traces of solvent by vacuum or a gentle jet of air.

8.2.4 From the spectra, obtain base line absorbancies of the various compounds at the wave lengths shown in Table 112:III. Calculate the absorptivity in each case from the known concentration. Typical calibration data are shown in Table 112:III.

9. Calculations

9.1 To measure the PNA hydrocarbons by ultraviolet spectrophotometry, it is necessary to determine whatever components occur in each GC peak. Thus, the peak containing BaA also includes chrysene and triphenylene, each of which must be determined based on its UV absorption. It is also required to resolve the mixture, BaP/BeP. Fluoranthene, pyrene, and TPB are measured directly from the UV spectrum of their respective GC peaks.

9.1.1 The concentration of each component is determined from the UV absorbance.

Equation 1. $D = \frac{A}{aC}$

where

D = concentration of each PNA (μg/ml)
A = baseline absorbance of GC peak solution
a = absorptivity (from Table 112:III)
C = cell length (cm)

9.2 The nature of a chromatogram in which these peaks occur is illustrated for an actual sample in Figure 112:2.

9.3 To demonstrate the spectral curves and the calculations, data from a sample are presented. The quantity of each PNA as measured in the total sample is shown in parentheses. Only an aliquot of the sample is analyzed in the GC/UV step. Thus, the quantities observed in the spectral measurements of Figures 112:3 to 7, inclusive, are approximately one-seventh of what is present in the total sample.

9.3.1 Fluoranthene (45 μg) is the first PNA of the method to elute and the UV spectrum observed for this GC peak is shown in Figure 112:3.

9.3.2 Figure 112:4 is the spectrum of pyrene (20 μg).

9.3.3 The next peak of interest is a mixture of BaA (22 μg), chrysene (15 μg), and triphenylene (3 μg). The measurement of base line absorptions is illustrated in Figure 112:5. To calculate the concentration of each compound, a set of three simultaneous equations is solved. The array of calibration coefficients from Table 112:III is shown below in Table 112:IV.

9.3.4 The peak comprised of BaP (20 μg) and BeP (10 μg) is next trapped to obtain the spectrum as shown in Figure 112:6. Base line measurements are indicated.

9.3.5 Measurement of the internal standard, TPB, is illustrated by the spectrum in Figure 112:7.

9.4 To calculate the amount of the various compounds in the entire sample, the aliquot factor must be determined based on the internal standard, TPB.

Equation 2. $F = \frac{I}{Dtpb}$

where

F = aliquot factor
I = micrograms of internal standard added to sample, usually 20
D_{tpb} = concentration of internal standard found in GC/UV step.

9.4.1 Micrograms of PNA in the total sample becomes:

Equation 3. $T = D \times F$

T = micrograms in total sample
D (from Equation 1)

10. Effect of Storage

10.1 In carrying out an analysis, sample solutions may stand for several hours or overnight. In those instances, they should be kept in a cold box. Aliquots of finished samples can also be safely stored for months and perhaps longer.

11. References

1. SEARL, T.D., F.J. CASSIDY, W.H. KING, and R.A. BROWN. 1970. Anal. Chem. 42, 954.
2. WRIGHT, N. 1941. Anal. Chem. 13, 1.
3. HEIGL, J.J., M.F. BELL, and J.U. WHITE. 1947. Anal. Chem. 19, 293.
4. SAWICKI, E., T.R. HAUSER, and T.R. STANLEY. 1960. Intern. J. Air Pollution, 2, 253.
5. KURATSUNE, M., T. KIROHATA. 1962. Nat. Cancer Inst. Monograph 9, 117.
6. RICHARDS, R.T., D.T. DONOVAN, and J.R. HALL. 1967. Amer. Ind. Hyg. Assn. J., 28, 590.
7. CHATOT, G. 1971. Anal. Chim. Acta, 53, 259.

Subcommittee 5
E. SAWICKI, *Chairman*
P. R. ATKINS
T. BELSKY
R. A. FRIEDEL
D. L. HYDE
J. L. MONKMAN
R. A. RASMUSSEN
L. A. RIPPERTON
J. E. SIGSBY
L. D. WHITE

113.

Tentative Method of Analysis for Polynuclear Aromatic Hydrocarbons in Automobile Exhaust

11104-04-73T

1. Principle of the Method

1.1 Auto exhaust tar samples are prepared as a solution in a saturated hydrocarbon such as cyclohexane. Carbon-14 labeled benz(a)anthracene (BaA) and benzo(a)pyrene (BaP) internal standards are added and the sample extracted with caustic. Chromatographic fractionation on deactivated alumina provides a concentrate of polynuclear aromatic (PNA) hydrocarbons. An aliquot is injected into a gas chromatograph and eluted fractions trapped for measurement of individual PNA by ultraviolet absorption spectrophotometer (UV). The method is designated as the GC/UV procedure. Additional descriptions of the method are available (**1, 2, 3**).

1.2 The effluent from the gas chromatograph is split 15% to a flame ionization detector for the GC measurement, and 85% to an ice cooled trap. Selected peaks are trapped by condensing in metal tubes and the condensate is rinsed out with cyclohexane for UV measurement. Based on the UV spectra, microgram quantities of the various PNA's are next calculated. Peaks containing BaA and BaP, respectively, are counted for carbon-14 radioactivity. These activities, compared with the amounts originally added, give factors by which to relate the microgram quantity of each PNA to its total weight in the sample. The carbon-14 BaA factor is used to calculate BaA, chrysene, and triphenylene which elute together. BaP and benzo(e)pyrene (BeP) are calculated based on the factor derived from carbon-14 BaP. Pyrene, methyl BaA, dimethyl BaA, methyl BaP, methyl BeP, and benzo(g,h,i)perylene are calculated using whichever radioactivity is lower.

1.3 Although the calculations are straightforward, it is a time saver to handle them by computer.

2. Range and Sensitivity

2.1 From 0.025 to 1.0 g of tar in 25 to 50 ml of solution can be handled by the method. The lower limit of determination for each PNA is 0.1 μg. The concentration of individual PNA's in tar is in the range 10 to 200 ppm.

3. Interferences

3.1 Plastic tubing, *e.g.*, polyvinyl chloride, contains phthalate plasticizers which are extracted by saturated hydrocarbons used to dissolve the tar sample. Such phthalates elute with PNA's from the alumina column and later also elute from the gas chromatograph with BaA, chrysene, and triphenylene. Phthalates absorb in the ultraviolet to interfere with the UV measurement of individual PNA's.

3.2 The method utilizes the GC and UV measurements so as to minimize interferences. Pyrene, for example, occurs by itself in a GC fraction; BaA, chrysene, and triphenylene occur together in a GC peak but their characteristic UV spectra permit them to be separately measured. Similarly, BaP and BeP occur together in a GC fraction but can be individually measured from the UV spectrum of the fraction. All UV measurements employ the base line technique as first described by Wright **(4)** and amplified by Heigl, *et al* **(5)**.

4. Precision and Accuracy

4.1 One sample was analyzed 6 times over a period of 8 months. Eleven PNA hydrocarbons were measured and results are summarized in Table 113:I. For 9 compounds the standard deviation was 3.8 to 11.8 relative %. The standard deviation for benzo(g,h,i)perylene was 18.2 relative %. Methyl BaP, at a level of 5 μg, had a standard deviation of 2.2 mcg (45.2 relative %). The poor precision for methyl BaP may be attributed to the small quantity present.

4.2 Analytical results by the GC/UV method were compared with results by a longer, better established procedure. In the latter method a preliminary separation is carried out in a deactivated alumina column. A composite of these fractions is next separated on a thin layer plate containing 20% acetylated cellulose. Fractions from the plate, rich in BaA or BaP are then examined by UV to obtain a quantitative measurement.

Table 113:I. Precision* of GC/UV Method as Applied to a Tar

Compound	Amount (μg)	σ (μg)	σ (%)
Pyrene	676	25.4	3.8
BaA	117	5.7	4.9
Chrysene	111	13.1	11.8
Triphenylene	25	1.4	5.6
Methyl BaA	58	4.0	6.9
Dimethyl/ethyl BaA	14	1.1	7.9
BaP	15	1.4	10.0
BeP	153	8.5	5.6
Benzo(g,h,i)-perylene	137	24.9	18.2
Methyl BaP	5	2.2	45.2
Methyl BeP	63	6.2	9.8

* = six analyses from March–October, 1971.

4.3 A comparison of the two methods is shown graphically in Figure 113:1. Shown are the ratios of GC to TLC measurements. BaP measurements by GC/UV are consistently higher, particularly for small quantities. An opposite relationship exists for BaA in that the measurement by TLC is higher than GC/UV. This is attributed to the presence of methyl BaA in TLC fractions so that the UV measurement reflects the presence of both BaA and methyl BaA **(6)**.

5. Apparatus

5.1 Because of the sensitivity of the method, the possibility of error due to contamination is great. It is important that all glassware be scrupulously cleaned. Examine all glassware, including stoppers and stopcocks, under ultraviolet light to detect any residual fluorescent contamination. As a precautionary measure it is recommended that all glassware be rinsed with purified isooctane immediately before use. No grease is to be used on stopcocks or joints. Tygon tubing must *absolutely* be avoided to eliminate di-isooctyl phthalate (DOP) contamination which interferes with UV measurement of PNA. Because some of the polynuclear hydrocarbons

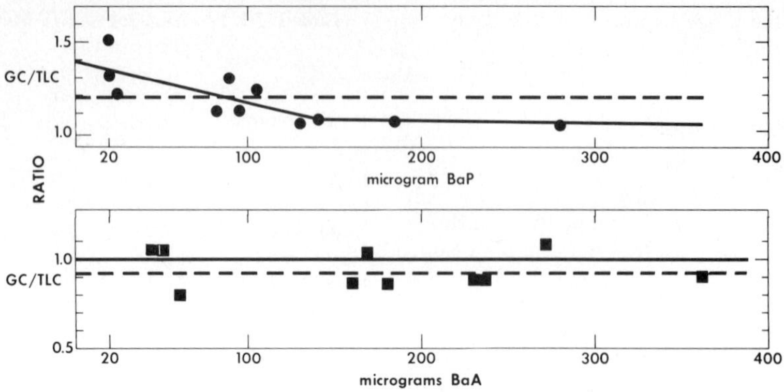

Figure 113:1—Comparison of Tar Analyses

sought in this test are susceptible to photooxidation (7), the entire procedure should be carried out under subdued light. For laboratory lighting it is desirable to use yellow lights in preference to the usual fluorescent lamps.

5.2 Radioactive materials must be handled and disposed of by accepted methods. Glassware used with radioactive species should be rinsed with chloroform-acetone solvent and washed in chromate-sulfuric acid cleaning solution.

5.3 The gas chromatograph is equipped with a flame ionization detector, a linear temperature programmer, and a flow controller. The injection port consists of an aluminum block with a stainless steel tube insert (quartz sleeve). The chromatograph is modified by the addition of a by-pass and trap so that 15% of the effluent of the column will go to the detector and 85% to the trap. A Perkin-Elmer 900 chromatograph was used but equivalent instruments should perform satisfactorily.

5.3.1 Figure 113:2 shows the gas chromatograph effluent splitter and trapping assembly. As shown in Figure 113:2, tubing of 0.25 mm ID connects the splitter block, **A**, to the detector, whereas a 0.5 mm ID tubing goes to the point of insertion of the trapping tube. The overall length of each set of tubes is approximately 22 cm. In this arrangement, 15% of the effluent passes to the detector and 85% goes to the air cooled trap. To trap a fraction, the stainless steel tube (⅛″ tubing, 7″ long) is slid into position as shown. To assure condensation in the trap, it is also necessary to have the outer tube partially covered with ice. To accomplish this, a plastic cup is mounted on the tube by simply piercing the cut through its center and placing ice therein. To stop trapping, the tube is taken out.

5.3.2 *Recorder.* 0–1 mV, 2 min/inch chart speed.

5.3.3 *Trapping Tubes.* 30 or more stainless steel tubes 7″ long by ⅛″ outside diam.

5.3.4 *Column.* Ten feet of ⅛″ OD stainless steel tubing packed with 2% SE-30 (GC grade) on Chromosorb G (acid washed and DMCS treated), 80/100 mesh. The prepared packing is available from Supelco Inc., Bellefonte, Pennsylvania, and other suppliers.

5.4 SYRINGE. 10 μl. The syringes manufactured by the Hamilton Company and Model C of Precision Sampling Corp. are satisfactory.

5.5 SPECTROPHOTOMETER. Spectral range 225 to 400 nm, with spectral slit width of 2 nm or less. Under instrument operating conditions for these absorbance measurements, the spectrophotometer shall also meet the following performance requirements: absorbance repeatability, ± 0.01 at 0.4 absorbance; absorbance accuracy, ± 0.05 at 0.4 absorbance;

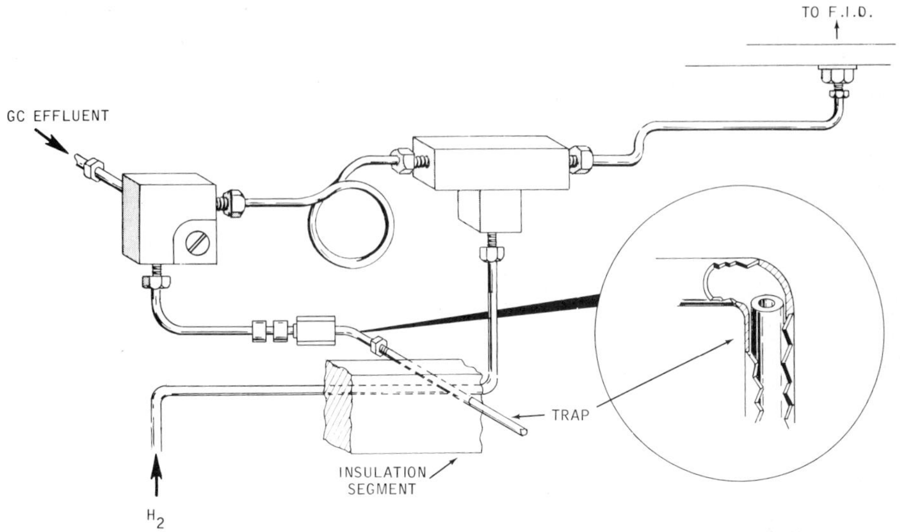

Figure 113:2—Gas Chromatograph Splitter and Trapping Assembly in Manifold

wavelength repeatability, ± 0.2 nm; wavelength accuracy, ± 1.0 nm.

5.5.1 *Spectrophotometric Cells.* Fused quartz cells having optical path lengths of 1.000 ± 0.005 cm and 5.000 ± 0.005 cm. The 5-cm cells are microcells and hold about 3 ml of solution.

5.6 Graduate Glass. 5-ml.

5.7 Vials. One- and three-dram with cap.

5.8 Steam Bath. Equipped with nitrogen outlet for purging vials.

5.9 Medicine Dropper. (Pasteur Pipet) with tip drawn out.

5.10 Special Low Emission Room Lights. Westinghouse W-40-gold fluorescent lamps.

5.11 Glass Chromatographic Column. 60-cm long, 1.25-cm OD, and 1.1 cm-ID, with a 200 ml receiving bulb on top and stopcock with Teflon plug on the bottom.

5.11.1 Seventy-five g of Woelm Neutral Alumina are dried at 150 C in an oven one hour. The alumina is transferred to a bottle and allowed to reach room temperature. 1.50 ml of water are added dropwise with shaking. The bottle is capped and placed on a paint shaker for 15 min to obtain a uniform mixture.

5.11.2 A glass wool plug is placed in the end of the column and the alumina is poured in while gently tapping the column to pack the alumina. The column is filled to within one inch of the top and another glass wool plug is placed on top.

5.12 Scintillation Counter.

5.13 Counting vials. 20-ml with cap.

6. Reagents

6.1 Cyclohexane. To check purity, evaporate approximately 180 ml down to 5 ml. Run a UV scan on this residue in a one cm cell from 400 to 280 nm. The absorbance should not exceed 0.01 units. To purify, percolate through activated silica gel, Grade 12, in a glass column, 90 cm long and 5 to 8 cm dia. Satisfactory solvent may be purchased from several sources including Burdick and Jackson Laboratories, Inc. "Distilled in Glass" purity is required.

6.2 Toluene.

6.3 Acetone.

6.4 Sodium hydroxide.

6.5 Hydrochloric acid.

6.6 Benzene. Burdick and Jackson Laboratories, Inc.

6.7 Methanol. Burdick and Jackson Laboratories, Inc.

6.8 Benz(a)anthracene, Eastman

4672. Prior to use, this compound as well as other PNA hydrocarbons must be of acceptable purity as shown by gas chromatography and UV.

6.9 Benzo(a)pyrene, Eastman 4951. Repurify by elution through deactivated alumina column.

6.10 Pyrene, Eastman 3627.

6.11 Benzo(g,h,i)perylene. Columbia Organic.

6.12 ^{14}C Benz(a)anthracene, ^{14}C Benzo(a)pyrene. Mallinckrodt Chemical Works.

6.13 Organic Counting Solution. 8 g of BBOT in a liter of toluene.

6.14 Chrysene. Matheson, Coleman and Bell Co.

6.15 Triphenylene. K & K Chemical Co.

6.16 Benzo(e)pyrene. Source unknown.

6.17 Test Blend. Weigh exactly 25.0 mg each of the four compounds listed as Items **6.8** through **6.11**. Place in a 25-ml volumetric flask, add 20 ml of toluene and swirl until compounds are in solution. Then make up to volume. This will contain 1 μg of each compound per μl of solution. Pour into a small, narrow-neck brown bottle and keep in a cool, dark place, preferably a refrigerator.

7. Procedure

7.1 Collection of Sample. To date, the method has been applied to samples that consist of several gallons of solvent containing the tar. The collection of the sample is described elsewhere **(8)**. Starting with several gallons of solvent, it is necessary to distill down to a residual volume of one liter. Before commencing the distillation of the seven-gallon sample, prepare the radioactive BaA and BaP spikes. Add 50 λ of BaA to a 25-ml volumetric flask and 50 λ of BaP to a second. Make up to volume with cyclohexane and remove duplicate 100 λ aliquots from each for counting. Pour the contents of the flasks into the still and rinse out well with cyclohexane and acetone. In some cases, advance knowledge of the sample may indicate that less than fifty μg of BaP is to be measured. For this sample, use a spike of 10 λ each for BaA and BaP.

7.2 It is impractical to frequently run blanks on several gallons of solvent. Prior to using a given batch of solvent, a blank of several gallons should be analyzed as a sample.

7.3 Preparation of Sample. After distillation, a liter sample solution of cyclohexane is on hand. One-half of this sample is used.

7.4 The individual steps for the method of analysis are:

7.4.1 Caustic extraction of tars to remove phenols and acids.

7.4.2 Obtain a PNA concentrate by separation of sample on alumina column.

7.4.3 Evaporate to a small residual liquid volume.

7.4.4 Run in GC and trap peaks.

7.4.5 Measure UV spectra of trapped GC peaks.

7.4.6 Measure ^{14}C activity of trapped BaA and BaP peaks.

7.4.7 Calculate micrograms of BaA, BaP, pyrene, chrysene, triphenylene, BeP, benzo(g,h,i)perylene, methyl BaA, dimethyl BaA, methyl BaP, and methyl BeP.

7.5 To start the caustic extraction, place 500 ml of the tar solution in a one-liter separatory funnel.

7.5.1 Add 50 ml of 0.5 N aqueous sodium hydroxide solution, shake one min, and remove lower aqueous phase.

7.5.2 Repeat the extractions with 30 and 20 ml portions of sodium hydroxide.

7.5.3 Wash with 50 ml of water, then 50 ml of 1/10 N HCl.

7.5.4 Finally, wash with 25 ml of water.

7.5.5 Evaporate cyclohexane to about 150 ml and filter, if necessary.

7.5.6 Evaporate filtrate to below 50 ml, place in a 50-ml volumetric flask and make up to volume with cyclohexane.

7.5.7 Place 25 ml on the alumina chromatographic column.

7.6 Steps in the liquid column separation are:

7.6.1 Ten ml of cyclohexane is placed on top of the column and forced into the column under nitrogen pressure of 2 pounds. The remainder of the cyclohexane solution of sample is then poured into the 200-ml bulb on top and allowed to run into the column by gravity. 100 ml cyclohexane is added to the bulb. From this point on to the completion of the column separation, flow should not be stopped or absorbant allowed to run dry between solvent solutions.

7.6.2 When the cyclohexane has run into the column below the top glass wool plug, 100 ml of cyclohexane/benzene (4 : 1) are added and run into the column. The first one hundred ml of cyclohexane is collected and set aside. Following this, 10 ml fractions are collected in 3-dram vials and capped.

7.6.3 After the cyclohexane/benzene (4/1) is below the top of the glass wool plug, 100 ml of benzene are added to the column. Collection of 10 ml fractions is continued.

7.6.4 After all the benzene has passed the top glass wool plug, 100 ml of benzene/methanol (1 : 1) is added to the column. The benzene/methanol front on the column is followed by the movement of an orange colored band. In addition, the methanol front can be detected by UV lamp or as a warm band moving down the column, apparent to the touch. When the front reaches the end of the column, a two-phase system becomes visible in the collecting vial. At this point, the remaining solution is collected in one large bottle and allowed to run off the column.

7.6.5 The solutions in the individual 3-dram vials are scanned in order of elution on a UV spectrophotometer. The peak at 340 nm is used as a guide in detecting the start of elution for tetracyclic PNA compounds, and all remaining vials ahead of the methanol front are combined as the PNA fraction in a 150-ml beaker. Rinse each vial twice with 1 to 2 ml of cyclohexane and add to the beaker.

7.7 Concentrate the PNA's by evaporating to a residual liquid volume:

7.7.1 Place the 150-ml beaker on the steam bath under a small jet of nitrogen.

7.7.2 As evaporation progresses, add the contents of the remaining vials with similar rinsing to the beaker. Do not evaporate to less than 2 ml.

7.7.3 Transfer the concentrated solution to a one-dram vial. Wash beaker with cyclohexane and add to vial.

7.7.4 Place the vial in a 30-ml beaker for support. The beaker is placed on a steam bath and a gentle jet of nitrogen directed into the vial.

7.7.5 Rinse the 150-ml beaker with 3 or 4 small portions of cyclohexane (~ 1 ml) and combine with the contents of the one-dram vial as evaporation proceeds.

7.7.6 Evaporate the solution to about 50 μl (0.05 ml). Add about 0.5 ml of toluene and evaporate to 50 μl.

7.7.7 It may be necessary to add 10 to 20 μl of toluene to lower the viscosity of the sample. Cap and save.

7.8 The following parameters are employed in the gas chromatographic separation with the Perkin-Elmer 900. A slight column bleed may occur but not enough to affect subsequent measurements. Other instruments might use different conditions.

Carrier Gas—Helium.
Flow Rate—30 ml/min.
Hydrogen and Air—at manufacturers recommendation or optimum rates.
Injection Port—300 C.
Detector—340 C.
Program—175 C to 300 C at 4 per min. The temperature is held at 300 until all peaks are eluted, but in any case, for 20 min.

Dual columns are not used.

7.8.1 To inject the PNA concentrate to the chromatograph, flush the 10-μl syringe twice, and then fill the syringe and hold vertically, point up. Stick the tip in a rubber septum and depress the plunger to compress air bubbles which rise to the top. Advance the plunger to the desired volume, 5 to 10 μl; remove the rubber, and retract the plunger until the air-liquid meniscus enters the glass bore. Note the volume from meniscus to plunger.

7.8.2 Insert the syringe up to the hilt in the GC injection port, depress the plunger. Start the recorder.

7.8.3 Remove the syringe from the port and retract the plunger.

7.8.4 Note the volume remaining

(generally about 0.2 μl) and subtract from the previously noted volume to determine the net volume injected.

7.8.5 Have stainless steel trapping tubes and one-dram vials available. The vials may be in holes drilled in a board.

7.8.6 Based on retention times as established by the calibration procedure, tubes are inserted as the peak starts to rise and removed as the trace returns to the baseline. Tubes are numbered to correspond with the peak and placed in a similarly labeled one-dram vial.

7.8.7 When the GC run is completed, measure and record the retention time from injection point for each peak. Place the first tube in a 5-ml graduate. With an elongated medicine dropper add enough cyclohexane through the top of the tube to fill the spectrophotometric cell to be used. Our cells required 3.8 ml.

7.8.8 Pour the solution into the original vial, cap and protect from light. Rinse out the graduate with cyclohexane and use for the next tube. Save the fractions for UV.

7.9 Obtain the UV spectrum of each selected fraction and measure specified wavelength(s). A 1-cm or 5-cm cell should be used, depending upon the level of UV absorption.

7.10 Measure recovery of ^{14}C isotopes. To do this, 1 ml of the separate fractions containing BaA and BaP is placed in a counting vial along with 9 ml of toluene and 10 ml of counting solution. Counts per min are measured using a scintillation counter.

8. Calibration and Standards

8.1 A performance test is used to measure GC retention times and demonstrate suitable recovery of individual PNA's. With the GC column conditioned and the parameters set as described in **7.8**, prepare to inject a sample of the test blend—see **6.17**.

8.1.1 Inject a known volume of sample to the gas chromatograph. Because each PNA is present as one μg/μl, the known volume indicates the quantity of each injected.

8.1.2 As each PNA elutes, trap the peak and label properly. Also record the retention time.

Table 113:II. Observed Retention Times

Compound	Retention Time, Min.
Pyrene	13.6
Triphenylene	20.0
Benz(a)anthracene	20.1
Chrysene	20.4
Methylbenz(a)anthracenes	23.3
Dimethyl/ethylbenz(a)anthracenes	25.0
Benzo(e)pyrene	27.2
Benzo(a)pyrene	27.3
Methylbenzo(e)pyrenes* / Methylbenzo(a)pyrenes*	29.0, 29.6, 30.7
Benzo(g,h,i)perylene	33.3

*Measured in three different peaks.

8.1.3 Measure the UV absorption of each peak as required to allow calculation of the micrograms recovered for each component. Recoveries of ≥ 83% for pyrene, BaA and BaP, and for benzo(g,h,i)perylene indicate satisfactory operation of the chromatograph. The observed retention times provide a guide as to the appearance of compounds as they elute in a mixture. Typical retention times for the four compounds of the test blend as well as other PNA's are shown in Table 113:II.

8.2 Absorptivity coefficients are measured for pyrene, BaA, chrysene, triphenylene, BaP, BeP, and benzo(g,h,i)perylene.

8.2.1 Prepare solutions of known concentrations of each PNA in cyclohexane, e.g., 2.4 μg/ml.

8.2.2 With cyclohexane in the reference cell, obtain the spectrum of each solution from 400 to 225 nm.

8.2.3 Rinse the cell thoroughly with cyclohexane between spectra and remove all solvent.

8.2.4 From the spectra obtain the absorbances of the various compounds at the wavelength shown in Table 113:III. Calculate the absorptivity from the known concentrations. Absorptivity of methyl and dimethyl/ethyl derivatives are estimated as being 0.815 of that for the parent compound. This factor is based on the comparison between BaA and a number of methyl- and dimethylbenz(a)anthracenes. Typical calibration data are shown in the table.

9. Calculations

9.1 To measure the PNA hydrocarbons by ultraviolet spectrophotometry, it is necessary to resolve whatever components occur in each GC fraction. Thus, the peak containing BaA also includes chrysene and triphenylene, each of which must be determined based on its UV absorption. The methyl and dimethyl isomers of BaA occur in peaks which include methyl substituted chrysenes and triphenylenes. In these two cases, only BaA compounds are measured and they are calculated as though the chrysenes and triphenylenes were absent. It is also required to resolve the mixtures BaP/BeP and methyl BaP/methyl BeP. Pyrene and benzo(g,h,i)perylene are measured directly from the UV spectrum of their respective GC fractions. In all cases, the concentration (μg/ml) of each PNA is calculated from the UV spectra and these are finally converted to total micrograms by equation 4.

9.2 A chromatogram for an actual sample is shown in Figure 113:3.

9.3 The spectral curves and base line measurements are demonstrated for an actual sample.

9.3.1 Pyrene is the first PNA of the method to elute and the spectrum observed for this GC peak is shown in Figure 113:4.

9.3.2 The next peak of interest is a mixture of BaA, chrysene, and triphenylene. The measurement of base line absorptions is illustrated in Figure 113:5. Since BaA contributes to the absorption of chrysene and triphenylene, a set of three simultaneous equations is solved to obtain concentrations for each compound. The array of calibration coefficients from Table 113:IV is shown below.

9.3.3 Spectra for the next two peaks of interest are shown separately in Figures 113:6 and 113:7. This includes a mixture of methyl BaA, methyl chrysene, and methyl triphenylene in one spectrum, and their dimethyl and/or ethyl substituted isomers in the other. The BaA compounds can be measured as though the other compounds were absent. Methyl chrysenes and methyl triphenylenes cannot be calculated, however, due to lack of calibration data.

9.3.4 The fraction comprised of BaP and BeP is illustrated in Figure 113:8.

9.3.5 Methyl BaP and methyl BeP occur in three successive peaks whose spectra are depicted in Figure 113:9. To calculate each compound, the sum of base line measurements in the three spectra is used. Small differences in the position of the maxima are evident. This is due in part to the differences in structure among the methyl substituted isomers and, in part, to the presence of some dimethyl or ethyl isomers **(3)**.

TABLE 113:IV. Calibration Matrix for Calculation of BaA, Chrysene, and Triphenylene

	BaA	Chrysene	Triphenylene	
(1)	*0.332*	0	0	A289
(2)	0.039	*0.464*	0	A269
(3)	0.015	0.107	*0.480*	A259

A289, A269, A259 = UV absorbance of GC peak in a 1 cm cell. The equations can be solved directly by substitution.

$$\text{BaA} = \frac{\text{A289}}{0.332}$$

$$\text{Chrysene} = \frac{\text{A269} - \text{BaA} \times 0.039}{0.464}$$

$$\text{Triphenylene} = \frac{\text{A259} - \text{BaA} \times .015 - \text{Chrysene} \times 0.107}{0.480}$$

9.3.6 Figure 113:10 shows the spectrum of benzo(g, h, i)perylene.

9.4 To calculate the amount of the various compounds in the entire sample, the aliquot factor (activity ratio) must be determined based on radioactivity per ml of the cyclohexane solutions of GC fractions containing ^{14}C labeled BaA and BaP, respectively.

Table 113:III. UV Calibration Data for some PNA Hydrocarbons

Wavelength nm												
Peak	Base Line	Pyrene	BaA	Chrysene	Triphenylene	Methyl BaA	Dimethyl /ethyl BaA	BaP	BeP	Methyl BaP	Methyl BeP	Benzo-(g,h,i) perylene
	Absorptivity (ml/μg cm)											
336	327–343	*0.240*										
289	283–295		*0.332*									
269	263–277		0.039	*0.464*								
259	252–263		0.015	0.107	*0.480*							
293*	285–300					*0.27*						
293*	285–300						*0.27*					
383	373–390							*0.092*				
333	325–338								*0.135*			
384†	373–390									*0.075*		
335	325–340										*0.110*	
382	370–390											*0.074*

*Occurs in range, 290–293 nm.
†Occurs in range, 383–385 nm.

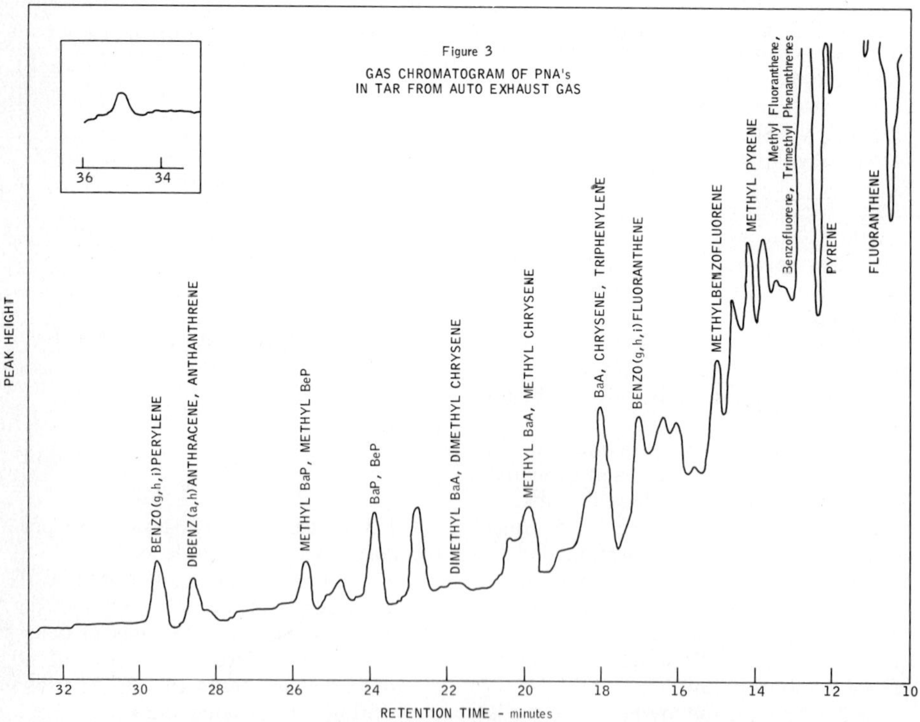

Figure 113:3—Gas Chromatogram of PNA's in Tar from Auto Exhaust Gas.

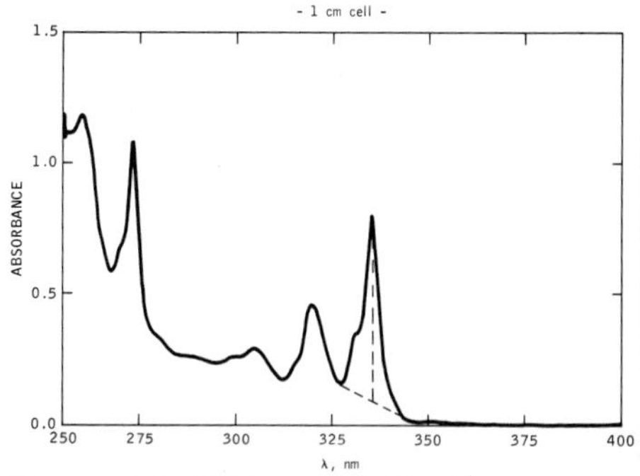

Figure 113:4—UV Spectrum of GC Fraction Containing Pyrene

Equation 1. $M = \dfrac{QN}{(K - L)}$

Q = activity in DPM added to sample
M = activity ratio
K = average CPM per ml of solution containing GC fraction
L = Background in CPM
N = counting efficiency CPM per DPM

CPM is counts per minute
DPM is disintegrations per minute.

9.4.1 The concentration of each component is determined from the UV absorbance.

Equation 2. $D = \dfrac{A}{aC}$

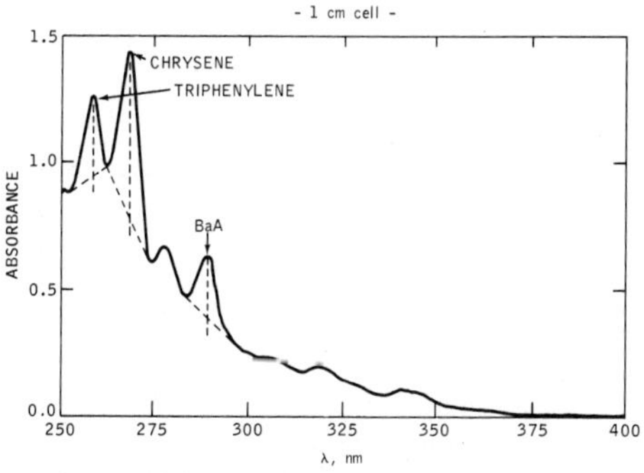

Figure 113:5—UV Spectrum of GC Fraction Containing BaA, Chrysene and Triphenylene.

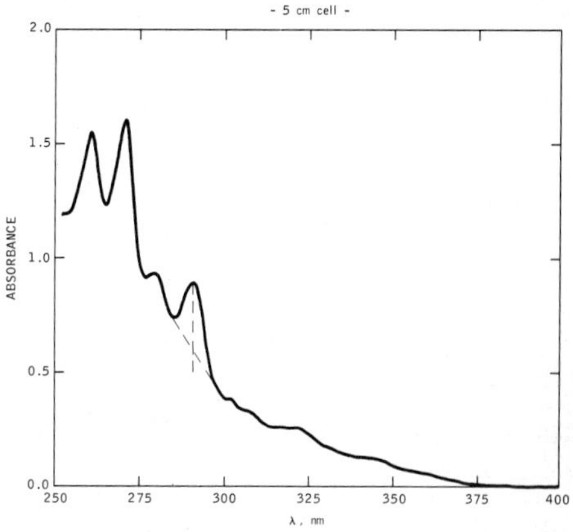

Figure 113:6—UV Spectrum of GC Fraction Containing Methyl BaA.

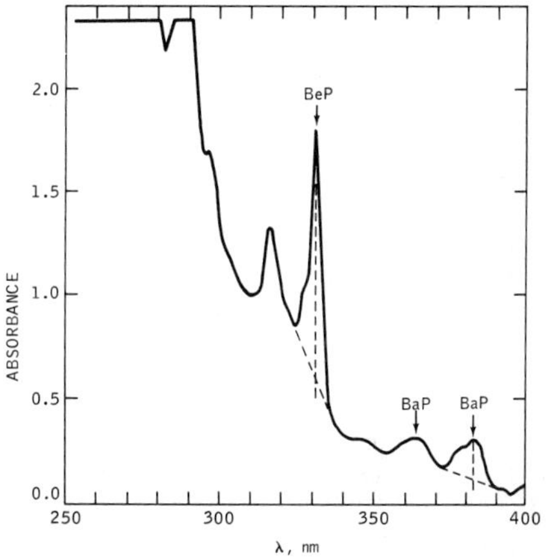

Figure 113:7—UV Spectrum of GC Fraction Containing Dimethyl/Ethyl BaA.

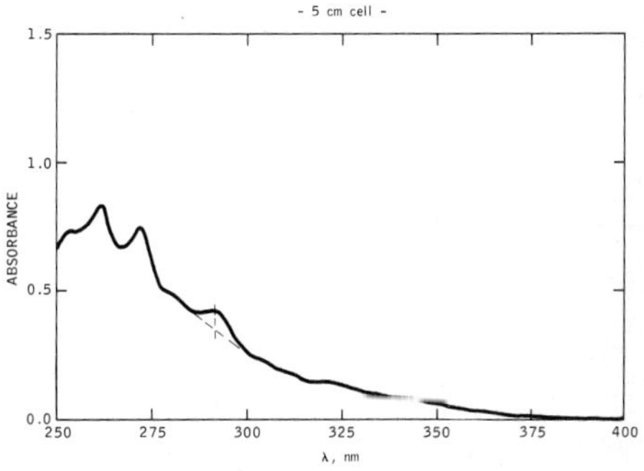

Figure 113:8—UV Spectrum of GC Fraction Containing BaP and BeP.

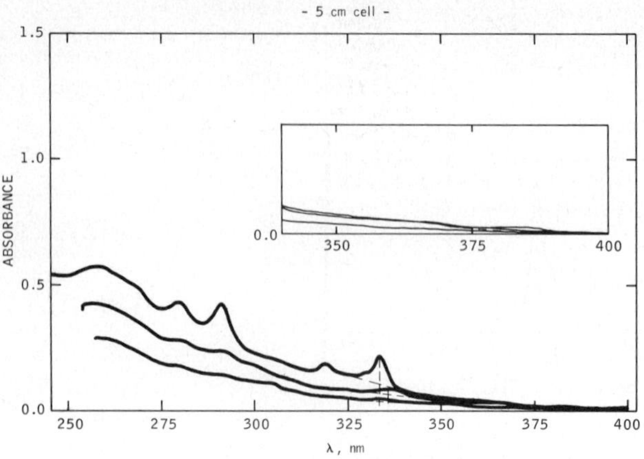

Figure 113:9—UV Spectra of GC Fractions Containing Methyl BaP and Methyl BeP

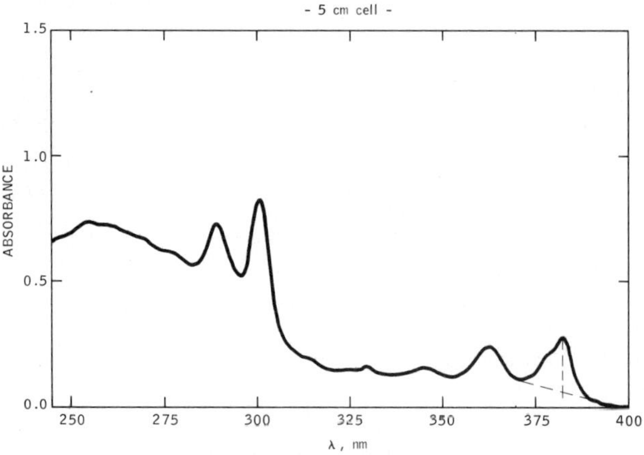

Figure 113:10—UV Spectrum of GC Fraction Containing Benzo(g,h,i)perylene

D = concentration of each PNA (μg/ml)
A = absorbance of GC peak.
a = absorptivity of each PNA (ml/μg/cm)
C = cell length (cm)

The weight of added radio tracer is determined.

Equation 3. $X = \frac{Q}{S}$

X = weight of radio tracer added, *e.g.*, 1 μg of ^{14}C-BaP.
S = specific activity (DPM/μg)
Q = activity in DPM added to sample.

Calculate weight of PNA in the sample

Equation 4. $T = MDV - X$

T = weight of PNA in the sample.
V = volume in ml of the solution containing the GC fraction taken for counting.

The subtraction of X (equation 4) is done only for BaA and BaP.

9.4.2 Where more than one internal standard is used, all components eluting in the same peak with a standard are calculated based on that standard's activity ratio. All other components are calculated using the smallest activity ratio since this represents the maximum recovery of standard.

10. Effect of Storage

10.1 In carrying out an analysis, sample solutions may be standing for several hours or overnight. In those instances, they should be kept in a cold box. Aliquots of finished samples can also be safely stored for months and perhaps longer.

11. References

1. Searl, T.D., F.J. Cassidy, W.H. King, R.A. Brown. 1970. Anal. Chem. 42, 954.
2. Brown, R.A., J.M. Kelliher, W.H. King. First Annual Report, Progress in Development of Rapid Methods of Analysis for Trace Quantities of Polynuclear Aromatic Hydrocarbons in Automobile Exhaust. Report #PB-196-808, National Technical Information Service, Springfield, Va. 22151.
3. Brown, R.A., T.D. Searl, W.H. King, Jr., W.A. Dietz, and J.M. Kelliher, Final Report. Rapid Methods of Analysis for Trace Quantities of Polynuclear Aromatic Hydrocarbons and Phenols in Automobile Exhaust, Gasoline, and Crankcase Oil. CRC-APRAC Project CAPE-12-68. Report # (to be assigned), National Technical Information Service, Springfield, Va. 22151.
4. Wright, N. 1941. Anal. Chem. 13, 1.
5. Heigl, J.J., M.F. Bell, and J.U. White. 1947. Anal. Chem. 19, 293.
6. Brown, R.A., and J.M. Kelliher. 1971. Amer. Petrol. Inst. Proc., Div. Ref. 51, 349.
7. Kuratsune, M., and T. Hirohata. 1962. Nat. Cancer Inst. Monograph 9, 117.
8. Gross, G.P. 1971. Gasoline Composition and Vehicle Exhaust Gas Polynuclear Aromatic Content. U.S. Clearinghouse Federal Science Technology Information, PB Rep. Issue No. 200266 124 pp.

Subcommittee 5

E. Sawicki, *Chairman*
P. R. Atkins
T. Belsky
F. A. Friedel
D. L. Hyde
J. L. Monkman
R. A. Rasmussen
L. A. Ripperton
J. E. Sigsby
L. D. White

114.

Tentative Method of Analysis for Acrolein Content of the Atmosphere (Colorimetric)

43505-01-70T

1. Principle of Method

1.1 The reaction of acrolein with 4-hexylresorcinol in an ethyl alcohol-trichloroacetic acid solvent medium in the presence of mercuric chloride results in a blue colored product with strong absorption maximum at 605 nm **(1, 2)**.

2. Range and Sensitivity

2.1 The absorbances at 605 nm are linear for at least 1 to 30 μg of acrolein in the 10 ml portions of mixed reagent **(3)**.

A concentration of 0.01 ppm of acrolein can be determined in a 50-l air sample based on a difference of 0.05 absorbance unit from the blank using a 1-cm cell. Greater sensitivity could be obtained by use of a longer path length cell.

3. Interferences

3.1 There is no interference from ordinary quantities of sulfur dioxide, nitrogen dioxide, ozone and most organic air pollutants. A slight interference occurs from dienes: 1.5% for 1,3-butadiene and 2% for 1,3-pentadiene. The red color produced by some other aldehydes and undetermined materials does not interfere in spectrophotometric measurement **(3)**.

4. Precision and Accuracy

4.1 The available information is insufficient to determine the precision and accuracy of this tentative method.

5. Apparatus

5.1 ABSORBERS. All glass standard midget impingers with fritted glass inlets are acceptable. The fritted end should have a porosity approximately equal to that of Corning EC (170 to 220 μ maximum pore diameter). A train of 2 bubblers is needed **(3)**.

5.2 WATER BATH. Any bath capable of maintaining a temperature of 58 to 60 C is acceptable.

5.3 AIR PUMP. A pump capable of drawing at least 2 l of air/min for 60 min through the sampling train is required. A trap at the inlet to protect the pump from the corrosive reagent is recommended.

5.4 AIR-METERING DEVICE. Either a limiting orifice of approximately 1 or 2 l min capacity or a glass flow meter can be used. If a limiting orifice is used, regular and frequent calibration is required.

5.5 SPECTROPHOTOMETER. This instrument should be capable of measuring the developed color at 605 nm. The absorption band is rather narrow, and thus a lower absorptivity may be expected in a broad-band instrument **(3)**.

6. Reagents

6.1 PURITY. All reagents must be analytical reagent grade. An analytical grade of distilled water must be used.

6.2 ETHANOL (96%).

6.3 TRICHLOROACETIC ACID SOLUTION, SATURATED.* Dissolve 100 g of the acid (reagent grade) in 10 ml of water by heating on a water bath. The resulting solution has a volume of approximately 70 ml. Even reagent grade trichloroacetic acid has an impurity which affects product intensities. Every new batch of solution should be standardized with acrolein. It is convenient to prepare a large quantity of solution from a single batch of trichloroacetic acid to maintain a uniformity of response **(1)**.

6.4 MERCURIC CHLORIDE SOLUTION (3%). Dissolve 3 g of mercuric chloride in 100 ml of ethanol.

6.5 4-HEXYLRESORCINOL SOLUTION. Dissolve 5 g of 4-hexylresorcinol (MP 68 to 70 C) in 5.5 ml of ethanol. This makes about 10 ml of solution.

6.6 MIXED SAMPLING REAGENT. Mix, in order, reagents in the following proportions: 5 ml ethanol, 0.1 ml 4-hexylresorcinol solution, 0.2 ml mercuric chloride solution, and 5 ml saturated trichloroacetic acid solution. The mixed reagent may be stored for a day at room temperature. Prepare the needed quantity by selecting an appropriate multiple of these amounts. Protect from direct sunlight.

6.7 ACROLEIN, PURIFIED. Freshly prepare a small quantity (less than 1 ml is sufficient) by distilling 10 ml of the purest grade of acrolein commercially available. Reject the first 2 ml of distillate. (The acrolein should be stored in a refrigerator to retard polymerization.) The distillation should be done in a hood because the vapors are irritating to the eyes.

6.8 ACROLEIN, STANDARD SOLUTION "A" (1 MG/ML). Weigh 0.1 g (approximately 0.12 ml) of freshly prepared, purified acrolein into a 100-ml volumetric flask and dilute to volume with ethanol. This solution may be kept for as long as a month if properly refrigerated.

***CAUTION**—*Both the solid acid and solution are corrosive to the skin. Breathing of the fumes evolved during preparation of the solution.*

†CAUTION—*Mercuric chloride is highly toxic and direct contact should be avoided.*

6.9 ACROLEIN, STANDARD SOLUTION "B" (10 μG/ML). Dilute 1 ml of standard solution "A" to 100 ml with ethanol. This solution may be kept for as long as a month if properly refrigerated.

7. Procedure

7.1 AIR SAMPLING. Draw measured volumes of the vapor laden air at a rate of either 1 l/min for 60 min or 2 l/min for 30 min through 2 bubblers in series, each containing 10 ml of mixed sampling reagent. An extra bubbler containing water may be added as a trap to protect the pump. A maximum of 60 l of air can be sampled before possible reagent decomposition may occur.

This sampling system collects 70 to 80% of the acrolein in the first bubbler and 95% of the acrolein in the first 2 bubblers (1), using absorbers with EC fritted glass inlets. The absorption efficiency might be increased by use of a C porosity frit (60 μ maximum pore diameter).

7.2 ANALYSIS.

7.2.1 If evaporation has occurred during sampling the sampling solution is diluted to its original 10 ml volume with ethanol.

7.2.2 Transfer the samples to glass stoppered test tubes. Immerse the tubes in a 60 C water bath for 15 min to develop the colors. A test tube containing only 10 ml of mixed sampling reagent must be run similarly and simultaneously. This serves as the reagent blank.

7.2.3 Cool the test tubes in running water immediately upon removal from the water bath.

7.2.4 After 15 min read the absorbances at 605 nm in a suitable spectrophotometer using 1-cm cells. There is no appreciable loss in accuracy if the samples are allowed to stand up to 2 hr before reading the absorbances. Determine the acrolein content of the sampling solution from a curve previously prepared from the standard acrolein solution.

For very low acrolein concentrations it may be convenient to use a longer path length cell.

8. Calibration

8.1 PREPARATION OF STANDARD CURVE.

8.1.1 Pipet 0, 0.5, 1.0, 2.0, and 3.0 ml of standard solution "B" into glass stoppered test tubes.

8.1.2 Dilute each standard to exactly 5 ml with ethanol.

8.1.3 Add in order, to each tube, exactly 0.1 ml of 4-hexylresorcinol solution, 0.2 ml of mercuric chloride solution, and 5 ml of trichloroacetic acid solution.

8.1.4 Mix, develop and read the colors as described in the analytical procedure.

8.1.5 Plot absorbance against micrograms of acrolein in the color developed solution.

9. Calculation

9.1 The concentration of acrolein in the sampled atmosphere may be calculated by using the following equation.

$$\text{PPM (Vol.)} = \frac{C \times 24.45}{V \times M.W. \times E}$$

E = Correction factor for sampling efficiency
V = liters of air sampled
C = μg of acrolein in sampling solution
M.W. = molecular weight of acrolein (56.06)
24.45 = ml. of acrolein vapor in one millimole at 760 Torr and 25 C.

10. Effect of Storage

10.1 The color forms in the sampling solution and is fully developed in 2 hr at room temperature. The solution starts fading after about 3 hr. Therefore it is probably best to analyze the samples almost immediately after completion of sampling.

11. References

1. COHEN, I.R., and A.P. ALTSHULLER. 1961. A New Spectrophotometric Method for the Determination of Acrolein in Combustion Gases and the Atmosphere. *Anal. Chem.* 33:726.
2. ALTSHULLER, A.P., and S.P. MCPHERSON. 1963.

Spectrophotometric Analysis of Aldehydes in the Los Angeles Atmosphere. *J. Air Poll. Control Assoc.* 13:109.
3. COHEN, ISRAEL, R., and BERNARD E. SALTZMAN. 1965. Determination of Acrolein:4-Hexylresorcinol Method. Selected Methods for the Measurement of Air Pollutants, Public Health Service Publication No. 999-AP-11, Page G-1.

Subcommittee 4
R. G. SMITH, *Chairman*
R. J. BRYAN
M. FELDSTEIN
B. LEVADIE
F. A. MILLER
E. R. STEPHENS
N. G. WHITE

115.

Tentative Method of Analysis for Low Molecular Weight Aliphatic Aldehydes in the Atmosphere

43501-01-71T

1. Principle

1.1 Formaldehyde, acrolein and low molecular weight aldehydes are collected in 1% $NaHSO_3$ solution in midget impingers. Formaldehyde is measured in an aliquot of the collection medium by the chromotropic acid procedure, acrolein by a modified mercuric-chloride-hexylresorcinol procedure, and C_2-C_5 aldehydes by a gas chromatographic procedure. The method permits the analysis of all C_1–C_5 aldehydes in a sample **(1)**.

The sampling procedure is not applicable for the determination of alcohols, esters or ketones in atmospheric samples, since bisulfite does not efficiently collect these materials. However, should some of these compounds be present in the atmosphere, their presence may be indicated by the appearance of peaks corresponding to their retention times in the chromatograms. The retention times for several of these compounds are shown along with the aldehydes in Table 115:I.

2. Range and Sensitivity

2.1 At sampling rates of 2 l/min over a 1 hr period, the following minimum concentrations can be determined:

CH_2O:	0.02 ppm
CH_3CHO:	0.02 ppm
CH_3CH_2CHO:	0.03 ppm
$(CH_3)_2CHCHO$:	0.03 ppm
$CH_2 = CHCHO$:	0.01 ppm

Shorter sampling periods are permissible for higher concentrations.

3. Interferences

3.1 FORMALDEHYDE.

3.1.1 The chromotropic acid procedure has very little interference from other aldehydes. Saturated aldehydes give less than 0.01% positive interference, and the unsaturated aldehyde acrolein results in a few per cent positive interference. Ethanol and higher molecular weight alcohols and olefins in mixtures with formaldehyde are negative interferences. However, concentrations of alcohols in air are usually much lower than formaldehyde concentrations and, therefore, are not a serious interference.

3.1.2 Phenols result in a 10 to 20% negative interference when present at an 8:1 excess over formaldehyde. They are, however, ordinarily present in the atmosphere at lesser concentrations than formaldehyde and, therefore, are not a serious interference.

3.1.3 Ethylene and propylene in a 10:1 excess over formaldehyde result in a

5 to 10% negative interference and 2-methyl-1, 3-butadiene in a 15:1 excess over formaldehyde showed a 15% negative interference. Aromatic hydrocarbons also constitute a negative interference. It has recently been found that cyclohexanone causes a bleaching of the final color.

3.2 ACROLEIN.

3.2.1 There is no interference in the acrolein determination from ordinary quantities of sulfur dioxide, nitrogen dioxide, ozone and most organic air pollutants. A slight interference occurs from dienes: 1.5% for 1,3-butadiene and 2% for 1,3-pentadiene. The red color produced by some other aldehydes and undetermined materials does not interfere in spectrophotometric measurement.

4. Precision and Accuracy

4.1 Known standards can be determined to within ± 5% of the true value.

Table 115:I. Retention Times for Aldehydes, Ketones, Alcohols and Esters*

Compound	Time, Retention minutes
Acetaldehyde	3.5
Propionaldehyde	4.6
Acetone	5.1
Isobutylraldehyde	5.5
Methyl alcohol	6.1
Ethyl alcohol	6.7
Isopropyl alcohol	6.7
Ethyl acetate	7.0
n-Butyraldehyde	7.1
Methyl-ethyl ketone	7.7
Isopentanal	12.0
Crotonaldehyde	14.0

*Flow rate, temperature and conditions described in text.

No data are available on precision and accuracy for atmospheric samples.

5. Apparatus

5.1 ABSORBERS. All glass standard midget impingers are acceptable. A train of 2 bubblers in series is used.

5.2 AIR PUMP. A pump capable of drawing at least 2 l of air/min for 60 min through the sampling train is required.

5.3 AIR METERING DEVICE. Either a limiting orifice of approximately 2 l/min capacity or a glass flow meter can be used. Cleaning and frequent calibration are required if a limiting orifice is used.

5.4 SPECTROPHOTOMETER. This instrument should be capable of measuring the developed colors at 605 nm and 580 nm. The absorption bands are rather narrow, and thus a lower absorptivity may be expected in a broad-band instrument.

5.5 GAS CHROMATOGRAPH, with hydrogen flame detector and injection port sleeve (Varian 1200 or equivalent).

5.6 BOILING WATER BATH.

6. Reagents

6.1 DETERMINATION OF FORMALDEHYDE.

6.1.2 *Sodium formaldehyde bisulfite* (E.K. P6450)

6.1.3 *Chromotropic acid sodium salt, EK P230. 0.5% in water.* Filter just before using. Stable for one week if kept refrigerated.

6.1.4 *Sulfuric acid,* Conc reagent grade.

6.2 DETERMINATION OF ACROLEIN.

6.2.1 *$HgCl_2$-4-hexylresorcinol.* 0.30 g $HgCl_2$ and 2.5 g 4 hexylresorcinol are dissolved in 50 ml 95% ethanol. (Stable at least 3 weeks if kept refrigerated.)

6.2.2 *TCAA.* To a 1 lb bottle of trichloracetic acid add 23 ml distilled water and 25 ml 95% ethanol. Mix until all the TCAA has dissolved.

6.3 COLLECTION MEDIUM. Sodium bisulfite, 1% in water.

7. Procedure

7.1 COLLECTION OF SAMPLES. Two midget impingers, each containing 10 ml of 1% $NaHSO_3$ are connected in series with Tygon tubing. These are followed by and connected to an empty impinger (for meter protection) and a dry test meter and a source of suction. During sampling the impingers are immersed in an ice bath. Sampling rate of 2 l/min should be maintained.

Sampling duration will depend on the concentration of aldehydes in the air. One hr sampling time at 2 l/min is adequate for ambient concentrations.

After sampling is complete, the impingers are disconnected from the train, the inlet and outlet tubes are capped, and the impingers stored in an ice bath or at 6 C in a refrigerator until analyses are performed. Cold storage is necessary only if the acrolein determination cannot be performed within 4 hr of sampling.

7.2 Analysis of Samples. (Each impinger is analyzed separately).

7.2.1 *Formaldehyde* **(1) (2)**. Transfer a 2-ml aliquot of the absorbing solution to a 25-ml graduated tube. Add 0.2 ml chromotropic acid, and then, cautiously, 5.0 ml conc sulfuric acid. Mix well. Transfer to a boiling water bath and heat for 15 min. Cool the samples and add distilled water to the 10-ml mark. Cool, mix and transfer to a 16-mm cuvette, reading the transmittance at 580 nm. A blank containing 2 ml of 1% sodium bisulfite should be run along with the samples and used for 100% T setting. From a standard curve read μg of formaldehyde.

7.2.2 *Acrolein* **(1) (3)**. To a 25-ml graduated tube add an aliquot of the collected sample in bisulfite containing no more than 30 μg acrolein. Add 1% sodium bisulfite (if necessary) to a volume of 4.0 ml. Add 1.0 ml of the $HgCl_2$-4-hexylresorcinol reagent and mix. Add 5.0 ml of TCAA reagent and mix again. Insert in a boiling water bath for 5 to 6 min, remove, and set aside until tubes reach room temperature. Centrifuge samples at 1500 rpm for 5 min to clear slight turbidity. One hr after heating, read in a spectrophotometer at 605 nm against a bisulfite blank prepared in the same fashion as the samples.

7.2.3 *C_2-C_5 Aldehydes* **(1)**.

a. Analytical column 12′ × ⅛″ stainless steel packed with 15% w/w Carbowax 20 M on Chromosorb, 60 to 80 mesh, followed by 5′ × ⅛″ stainless steel Uncondinonylphthalate on firebrick, 100 to 200 mesh, prepared as follows: Ucon 50-HB-200, 1.5 g, and 1.4 g of dinonylphthalate are dissolved in chloroform and added to 13 g of firebrick. The solvent is evaporated at room temperature and the column packed in the usual manner.

b. Injection port sleeve: The inlet of the injection port contains a glass sleeve packed with solid Na_2CO_3. The Na_2CO_3 is held in place with glass wool plugs.

c. Conditions:

Injection port temperature, 160 to 170 C
Column temperature, 105 C
Detector temperature, 200 C
Nitrogen carrier gas flow rate, 14 ml/min
Hydrogen flow rate, 20 ml/min
Combustion air flow rate, 400/min

d. Procedure: A 4 μl sample of the bisulfite collection solution is injected into the packed sleeve at the injection port and the chromatogram is recorded. Table 114:I shows the relative retention times for a series of aldehydes and ketones in the C_2-C_5 range.

8. Calibration

8.1 Formaldehyde.

8.1.1 *Preparation of standard curve*. To a 1-l volumetric flask add 0.4466 g sodium formaldehyde bisulfite and dilute to volume. This solution contains 0.1 mg formaldehyde per ml. Dilute to obtain standard solutions containing 1, 3, 5 and 7 μg formaldehyde per ml. Treat 2-ml aliquots as described in the procedure for color development. Read each at 580 nm after setting instrument at 100% T with the blank. Using semilog paper, graph the respective concentrations *vs*. transmittance.

8.2 Acrolein.

8.2.1 *Preparation of standard curve*. To 250 ml of 1% sodium bisulfite add 4.0 μl freshly distilled acrolein. This yields a standard containing 13.4 μg/ml. To a series of tubes add 0.5, 1.0, 1.5, and 2.0 ml of standard. Adjust the volumes to 4.0 ml with 1% bisulfite and develop color as described above. Plot data on semi-log paper.

8.3 C_2—C_5 Aldehydes.

8.3.1 *Calibration*. A mixed standard of C_2–C_5 aldehydes and ketones is prepared as follows:

a. Acetaldehyde-bisulfite solution: 0.336 g CH_3 CHO · $NaHSO_3$ (EK 791) is dissolved in 1 l of 1% $NaHSO_3$. This gives a solution containing 100 μg/ml acetaldehyde.

b. To 10.0 ml of the above solution are added 40.0 ml of 1% $NaHSO_3$, and 8 μl of a mixture of equal volumes of propanal, isobutanal, butanal, isopentanal, pentanal crotonaldehyde, acetone and butanone.

The final solution contains 20 μg/ml acetaldehyde and 0.02 μl of each of the C_2–C_5 aldehydes and ketones per ml. Four μl of the standard are injected into the glass sleeve in the injection port of the chromatograph as described in the procedure, and the chromatogram is recorded.

9. Calculations

(1.23 μg formaldehyde = 1 μl (vol) at 25 C and 760 Torr)

9.1 FORMALDEHYDE, ppm formaldehyde (CH_2O) =

$$\frac{\text{total micrograms of } CH_2O \text{ in sample}}{1.23 \times \text{sample volume in liters}}$$

9.2 ACROLEIN.

(2.3 μg acrolein = 1.0 μl (vol) acrolein)

$$\text{ppm} = \frac{\text{total } \mu\text{g of acrolein in sample}}{2.3 \times \text{sample volume in liters}}$$

9.3 ALDEHYDES. Calculation of unknown sample concentration is made on the basis of comparative peak heights between standards and unknowns.

10. Effect of Storage

10.1 After sampling is complete, collection media are stored in an ice bath or refrigerator at 6 C. Cold storage is necessary only if acrolein is to be determined. Under cold storage conditions, analyses can be performed within 48 hr with no deterioration of collected samples.

11. References

1. LEVAGGI, D.A., and M. FELDSTEIN. 1970. The Determination of Formaldehyde, Acrolein and Low Molecular Weight Aldehydes in Industrial Emissions on a Single Collected Sample. *JAPCA*, 20:312.
2. AMERICAN PUBLIC HEALTH ASSOCIATION. 1977. Methods of Air Sampling and Analysis. 2nd ed. p. 297. Washington, D.C.
3. Ibid. p. 297.

Subcommittee 4
R. G. SMITH, *Chairman*
R. J. BRYAN
M. FELDSTEIN
B. LEVADIE
F. A. MILLER
E. R. STEPHENS
N. G. WHITE

116.

Tentative Method of Analysis for Formaldehyde Content of the Atmosphere (Colorimetric Method)

43502-01-69T

1. Principle of the Method

1.1 Formaldehyde reacts with chromotropic acid-sulfuric acid solution to form a purple monocationic chromogen. The absorbance of the colored solution is read in a spectrophotometer at 580 nm and is proportional to the quantity of formaldehyde in the solution (**2, 6**).

1.2 Feigl (**3**), though stating "The chemistry of this color reaction is not known with certainty.", proposes "the following reactions:

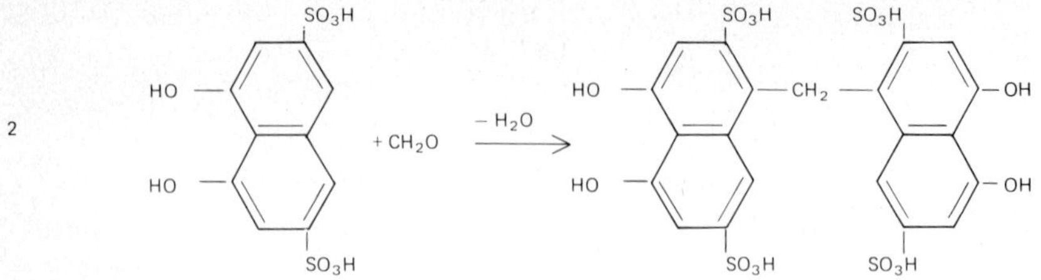

(1)

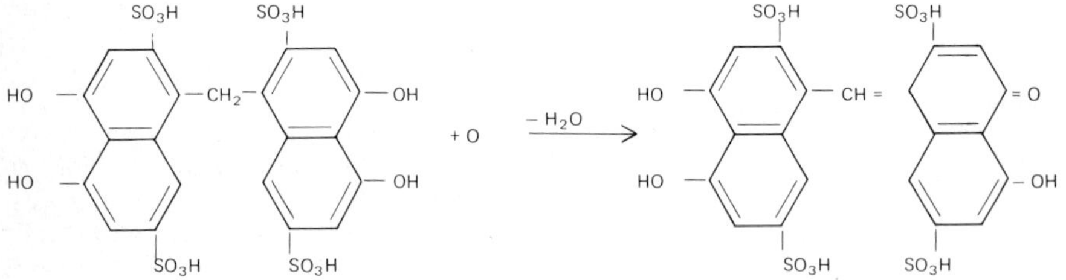

(2)

2. Range and Sensitivity

2.1 From 0.1 μg/ml to 2.0 μg/ml of formaldehyde can be measured in the color developed solution (10.1 ml).

2.2 A concentration of 0.1 ppm of formaldehyde can be determined in a 25-l air sample based on an aliquot of 4 ml from 20 ml of absorbing solution and a difference of 0.05 absorbance unit from the blank.

3. Interferences

3.1 The chromotropic acid procedure has very little interference from other aldehydes. Saturated aldehydes give less than 0.01% positive interference, and the unsaturated aldehyde acrolein results in a few per cent positive interference. Ethanol and higher molecular weight alcohols and olefins in mixtures with formaldehyde are negative interferences. However, concentrations of alcohols in air are usually much lower than formaldehyde concentrations and, therefore, are not a serious interference.

3.2 Phenols result in a 10 to 20% negative interference when present at an 8:1 excess over formaldehyde. They are, however, ordinarily present in the atmosphere at lesser concentrations than formaldehyde and, therefore, are not a serious interference.

3.3 Ethylene and propylene in a 10:1 excess over formaldehyde result in a 5 to 10% negative interference and 2-methyl-1,3-butadiene in a 15:1 excess over formaldehyde showed a 15% negative interference. Aromatic hydrocarbons also constitute a negative interference **(6)**. It has recently been found that cyclohexanone causes a bleaching of the final color **(4)**.

4. Precision and Accuracy

The method was checked for reproducibility by having three different analysts in three different laboratories analyze formaldehyde samples. The results listed in Table 116:I agreed within ± 5%.

5. Apparatus

5.1 ABSORBERS. All glass samplers with extra coarse fritted tube inlet. (The following sketch, Figure 116:1 shows an acceptable absorber).

5.2 AIR PUMP. A pump capable of drawing at least 1 l of air per min for 24 hr through the sampling train is required.

5.3 AIR METERING DEVICE. Either a limiting orifice of approximately 1 lpm capacity or a wet test meter can be used. If a limiting orifice is used regular and frequent calibration is required.

Table 116:I—Comparison of Formaldehyde Results From Three Laboratories

Micrograms Formaldehyde	Absorbance		
	Lab. 1	Lab. 2	Lab. 3
1	0.057	0.063	0.061
3	0.183	0.175	0.189
5	0.269	0.279	0.262
7	0.398	0.381	0.392
10	0.566	0.547	0.537
20	1.02	0.980	1.07

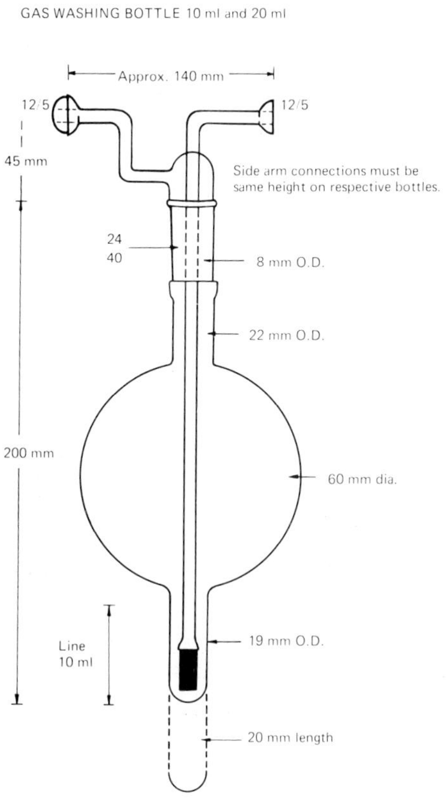

Figure 116:1—Gas washing bottle.

5.4 SPECTROPHOTOMETER OR COLORIMETER. An instrument capable of measuring the absorbance of the color developed solution at 580 nm.

6. Reagents

6.1 CHROMOTROPIC ACID REAGENT. Dissolve 0.10 g of 4,5-dihydroxy-2,7-naphthalenedisulfonic acid disodium salt (Eastman Kodak Company, Rochester, New York, Cat. No. P230) in water and dilute to 10 ml. Filter if necessary and store in a brown bottle. Make up fresh weekly.

6.2 CONCENTRATED SULFURIC ACID.

6.3 FORMALDEHYDE STANDARD SOLUTION "A" (1 MG/ML). Dilute 2.7 ml of 37% formalin solution to 1 l with distilled water. This solution must be standardized as described in section 8.1. The solution is stable for at least a 3-month period. Alternatively sodium formaldehyde bisulfite (Eastman Kodak Company, Cat. No. P6450) can be used as a primary standard (4). Dissolve 4.4703 g in distilled water and dilute to 1 l.

6.4 FORMALDEHYDE STANDARD SOLUTION "B" (10 μG/ML). Dilute 1 ml of standard solution "A" to 100 ml with distilled water. Make up fresh daily.

6.5 IODINE, 0.1 N (APPROXIMATE). Dissolve 25 g of potassium iodide in about 25 ml of water, add 12.7 g of iodine and dilute to 1 l.

6.6 IODINE, 0.01 N. Dilute 100 ml of the 0.1 *N* iodine solution to 1 l. Standardize against sodium thiosulfate.

6.7 STARCH SOLUTION, 1%. Make a paste of 1 g of soluble starch and 2 ml of water and slowly add the paste to 100 ml of boiling water. Cool, add several ml of chloroform as a preservative, and store in a stoppered bottle. Discard when a mold growth is noticeable.

6.8 SODIUM CARBONATE BUFFER SOLUTION. Dissolve 80 g of anhydrous sodium carbonate in about 500 ml of water. Slowly add 20 ml of glacial acetic acid and dilute to 1 l.

6.9 SODIUM BISULFITE, 1%. Dissolve 1

g of sodium bisulfite in 100 ml of water. It is best to prepare a fresh solution weekly.

7. Procedure

7.1 Air sampling.

7.1.1 Draw measured volumes of ambient air at a rate of 1 lpm for 24 hr through 20 ml of distilled water contained in the absorber (**1, 5**). However, a shorter sampling time can be used providing enough formaldehyde is collected to be above the lower limit of sensitivity of the method. Two absorbers must be used in series because under conditions of sampling, collection efficiency of one absorber is approximately 80%. With two absorbers in series the total collection efficiency is approximately 95%.

7.1.2 Note that some loss of sampling solution due to evaporation will take place over a 24 hr period so that it is either necessary to add water during the sampling period or else start out with a larger volume than 20 ml. Tests have shown that 35 ml in the first bubbler and 25 ml in the second bubbler is satisfactory.

7.2 Analysis.

7.2.1 Transfer the sample from each absorber to either a 25 ml- or 50 ml-graduate. Note the volume of each solution.

7.2.2 Pipet a 4 ml aliquot from each of the sampling solutions into glass stoppered test tubes. A blank containing 4 ml of distilled water must also be run. If the formaldehyde content of the aliquot exceeds the limit of the method a smaller aliquot diluted to 4 ml with distilled water is used.

7.2.3 Add 0.1 ml of 1% chromotropic acid reagent to the solution and mix.

7.2.4 To the solution, pipet slowly and cautiously 6 ml of concentrated sulfuric acid. The solution becomes extremely hot during the addition of the sulfuric acid. If the acid is not added slowly, some loss of sample could occur due to spattering.

7.2.5 Allow to cool to room temperature. Read at 580 nm in a suitable spectrophotometer using a 1-cm cell. No change in absorbance was noted over a 3 hr period after color development. Determine the formaldehyde content of the sampling solution from a curve previously prepared from standard formaldehyde solutions.

8. Calibration and Standards

8.1 Standardization of formaldehyde solution.

8.1.1 Pipet 1 ml of formaldehyde standard solution "A" into an iodine flask. Into another flask pipet 1 ml of distilled water. This solution serves as the blank.

8.1.2 Add 10 ml of 1% sodium bisulfite and 1 ml of 1% starch solution.

8.1.3 Titrate with 0.1 *N* iodine to a dark blue color.

8.1.4 Destroy the excess iodine with 0.05 *N* sodium thiosulfate.

8.1.5 Add 0.01 *N* iodine until a faint blue end point is reached.

8.1.6 The excess inorganic bisulfite is now completely oxidized to sulfate, and the solution is ready for the assay of the formaldehyde bisulfite addition product.

8.1.7 Chill the flask in an ice bath and add 25 ml of chilled sodium carbonate buffer. Titrate the liberated sulfite with 0.01 *N* iodine, using a microburet, to a faint blue end point. The amount of iodine added in this step must be accurately measured and recorded.

8.1.8 One ml of 0.0100 *N* iodine is equivalent to 0.15 mg of formaldehyde. Therefore, since 1 ml of formaldehyde standard solution was titrated, the ml of 0.01 *N* iodine in the final titration multiplied by 0.15 mg gives the formaldehyde concentration of the standard solution in mg/ml.

8.2 Preparation of standard curve.

8.2.1 Pipet 0, 0.1, 0.3, 0.5, 0.7, 1.0, and 2.0 ml of standard solution "B" into glass stoppered test tubes.

8.2.2 Dilute each standard to 4 ml with distilled water.

8.2.3 Develop the color as described in the analytical procedure (**7, 2**).

8.2.4 Plot absorbance against micrograms of formaldehyde in the color developed solution.

9. Calculation

The concentration of formaldehyde in the sampled atmosphere may be calculated by using the following equation, assuming standard conditions are taken as 760 Torr and 25 C:

$$\text{ppm (volume)} = \frac{C \times A \times 24.47}{V \times M.W.}$$

where:

V = liters of air sampled
C = μg of formaldehyde in aliquot

$$A = \text{aliquot factor} \; \frac{\text{sampling solution volume in ml}}{\text{ml in aliquot}}$$

M.W. = molecular weight of formaldehyde (30.03)
24.47 = ml of formaldehyde gas in one millimole at 760 Torr and 25 C.

10. Effect of Storage

10.1 The absorbance of the reaction product increases slowly on standing. An increase of 3% in absorbance was noted after 1 day standing and an increase of 10% after 8 days standing **(6)**.

10.2 No information is available on the effect of storage on the collected air sample.

11. References

1. ALTSHULLER, A.P., L.J. LENG, and A.F. WARTBURG. 1962. Source and Atmospheric Analyses for Formaldehyde by Chromotropic Acid Procedures. *Int. J. Air Wat. Poll.*, 6:381.
2. EEGRIWE, EDWIN. 1937. Reaktionen und Reagenzien zum Nachweis Organischer Verbindungen IV. *Z. Anal. Chem.*, 110:22.
3. FEIGL, FRITZ. 1966. Spot Tests in Organic Analysis. Seventh Edition, American Elsevier Publishing Company, 434, New York.
4. FELDSTEIN, MILTON. March, 1968. (Bay Area Air Pollution Control District) Personal communication.
5. MACDONALD, WILLIAM W. 1954. Formaldehyde in Air—A Specific Field Test. *Amer. Ind. Hyg. Assoc. Quarterly*, 15:217.
6. SLEVA, STANLEY F. 1965. Determination of Formaldehyde: Chromotropic Acid Method. Selected Methods for the Measurement of Air Pollutants. Public Health Service Publication No. 999-AP-11, H-1.
7. TREADWELL AND HALL. 1951. Analytical Chemistry. Vol. II. p. 590. Ninth English Edition. John Wiley & Sons, Inc. New York.
8. TREADWELL AND HALL. 1951. Analytical Chemistry. Vol. II. p. 588. Ninth English Edition. John Wiley & Sons. Inc. New York.

Subcommittee 4
R. G. SMITH, *Chairman*
R. J. BRYAN
M. FELDSTEIN
B. LEVADIE
F. A. MILLER
E. R. STEPHENS
N. G. WHITE

117.

TENTATIVE METHOD OF ANALYSIS FOR FORMALDEHYDE CONTENT OF THE ATMOSPHERE (MBTH—COLORIMETRIC METHOD—APPLICATIONS TO OTHER ALDEHYDES)

43502-02-70T

1. Principle of Method

1.1 The aldehydes in ambient air are collected in a 0.05% aqueous 3-methyl-2-benzothiazolone hydrazone hydrochloride (MBTH) solution. The resulting azine is then oxidized by a ferric chloride-sulfamic acid solution to form a blue cationic dye in acid media, which can be measured at 628 nm (**1, 2, 3**).

1.2 The mechanism of the present procedure as applied to formaldehyde includes the following steps: reaction of the aldehyde with 3-methyl-2-benzothiazolone hydrazone, A, to form the azine, B; oxidation of A to a reactive cation, C; and formation of the blue cation, D (**1**).

2. Range and Sensitivity

2.1 From 0.03 μg/ml to 0.7 μg/ml of formaldehyde can be measured in the color developed solution (12 ml). A concentration of 0.03 ppm of aldehyde (as formaldehyde) can be determined in a 25 l air sample based on an aliquot of 10 ml from 35 ml of absorbing solution and a difference of 0.05 absorbance unit from the blank.

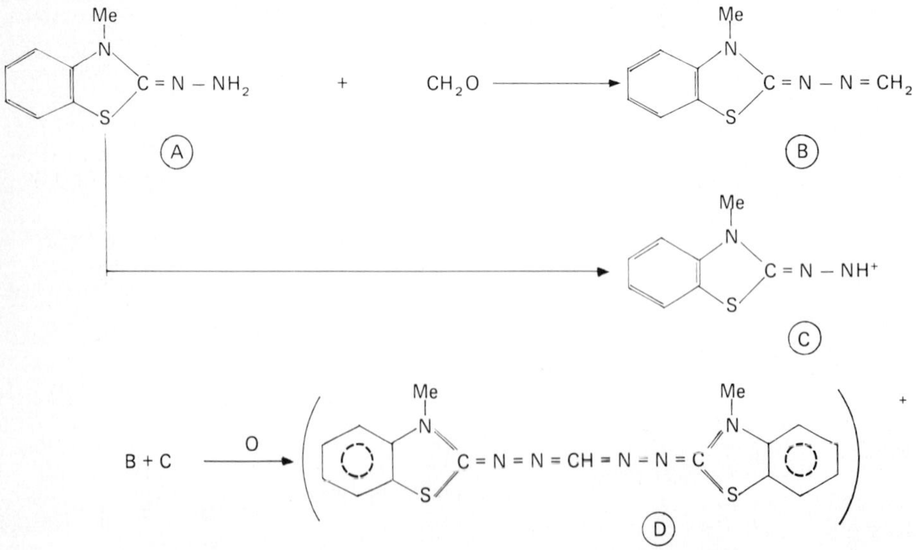

3. Interferences

3.1 The following classes of compounds react with MBTH to produce colored products. These are aromatic amines, imino heterocyclics, carbazoles, azo dyes, stilbenes, Schiff bases, the aliphatic aldehyde 2,4-dinitrophenyl hydrazones, and compounds containing the p-hydroxy styryl group. Most of these compounds are not gaseous or water soluble and, consequently, should not interfere with the analysis of water soluble aliphatic aldehydes in the atmosphere (**3**).

4. Precision and Accuracy

4.1 The method was checked for reproducibility by having three different analysts in three different laboratories analyze standard formaldehyde samples. The results listed in Table 117:I agreed within ± 5%.

5. Apparatus

5.1 ABSORBERS. (All glass samplers with coarse fritted tube inlet. Figure 117:1 shows an acceptable absorber.)

5.2 AIR METERING DEVICE. Either a limiting orifice of approximately 0.5 lpm capacity or a wet test meter can be used. If a limiting orifice is used, regular and frequent calibration is required.

5.3 AIR PUMP. A pump capable of drawing at least 0.5 l of air/min for 24 hr through the sampling train is required.

5.4 SPECTROPHOTOMETER. An instrument capable of measuring accurately the developed color at the narrow absorption band of 628 nm.

6. Reagents

6.1 PURITY. All reagents must be analytical reagent grade.

6.2 3-METHYL-2-BENZOTHIAZOLONE HYDRAZONE HYDROCHLORIDE ABSORBING SOLUTION (0.05%). Dissolve 0.5 g of MBTH in distilled water and dilute to 1 l. This colorless solution is filtered by gravity, if slightly turbid, and is stable for at least 1 week after which it becomes pale yellow. Stability may be increased by storing in a dark bottle in the cold.

6.3 OXIDIZING REAGENT. Dissolve 1.6 g of sulfamic acid and 1.0 g of ferric chloride in distilled water and dilute to 100 ml.

6.4 FORMALDEHYDE STANDARD SOLUTION "A" (1 MG/ML). Dilute 2.7 ml of 37% formalin solution to 1 l with distilled water. This solution must be standardized as described in "Calibration" section. This solution is stable for at least a 3-month period.

6.5 FORMALDEHYDE STANDARD SOLUTION "B" (10 μG/ML). Dilute 1 ml of standard solution "A" to 100 ml with 0.05% MBTH solution. Make up fresh daily.

6.6 IODINE 0.1 N (APPROXIMATE). Dissolve 25 g of potassium iodide in about 25 ml of water, add 12.7 g of iodine and dilute to 1 l.

6.7 IODINE 0.01 N. Dilute 100 ml of the 0.1 *N* iodine solution to 1 l. Standardize against sodium thiosulfate.

6.8 STARCH SOLUTION 1%. Make a paste of 1 g of soluble starch in 2 ml of water and slowly add the paste to 100 ml of boiling water. Cool, add several mls of chloroform as a preservative, and store in a stoppered bottle. Discard when a mold growth is noticeable.

6.9 SODIUM CARBONATE BUFFER SOLUTION. Dissolve 80 g of anhydrous sodium carbonate in about 500 ml of water. Slowly add 20 ml of glacial acetic acid and dilute to 1 l.

6.10 SODIUM BISULFITE 1%. Dissolve 1 g of sodium bisulfite in 100 ml of water. It is best to prepare a fresh solution weekly.

Table 117:I. Comparison of Formaldehyde Results from Three Laboratories (Analysis of Standard Formaldehyde Samples)

	Absorbance		
Micrograms/ml Formaldehyde	Laboratory 1	Laboratory 2	Laboratory 3
0.05	0.078	0.077	0.082
0.10	0.151	0.156	0.146
0.30	0.430	0.457	0.445
0.50	0.720	0.700	0.728
0.70	0.990	1.04	1.02

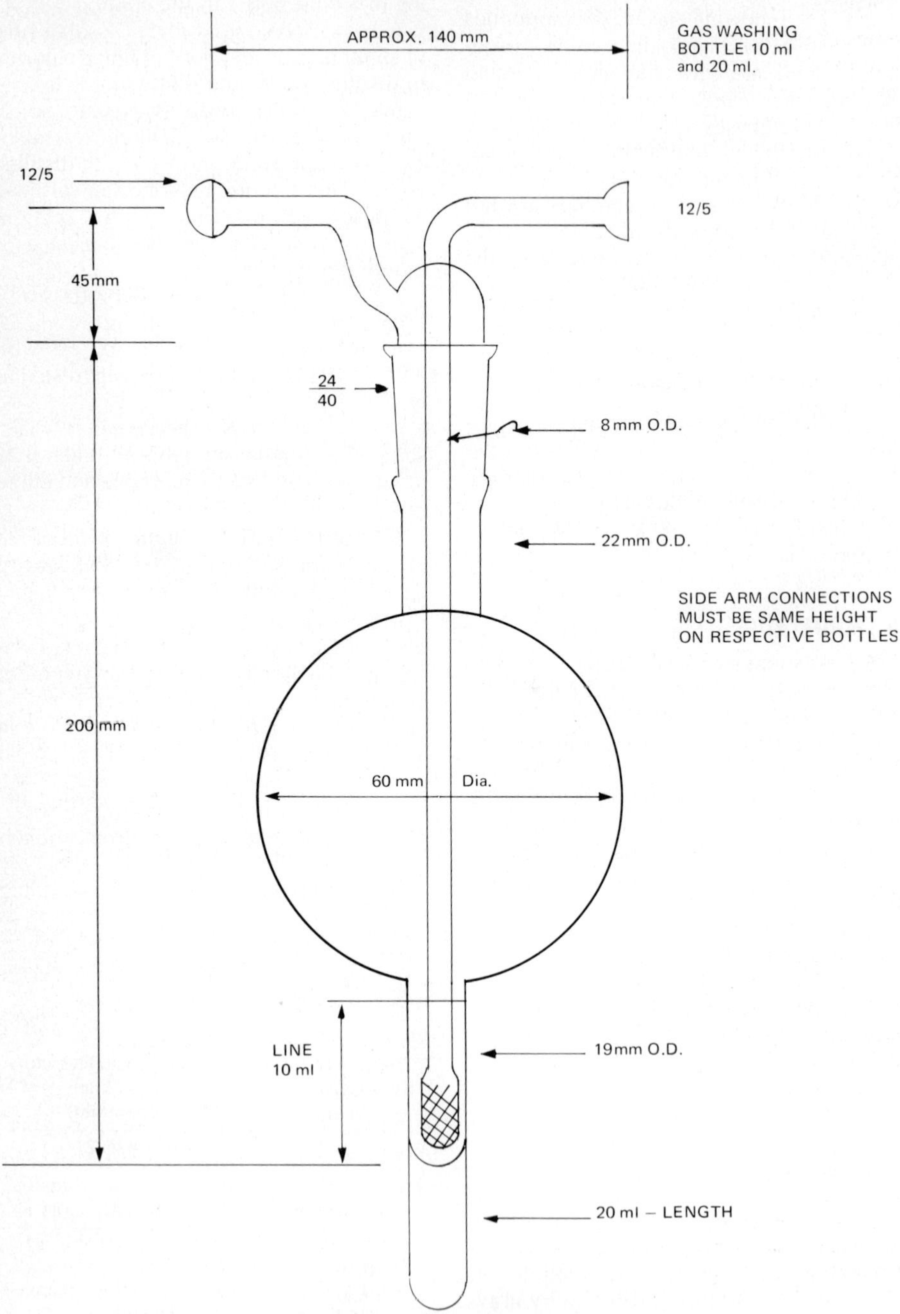

Figure 117:1—Absorber.

7. Procedure

7.1 AIR SAMPLING. Draw measured volumes of the vapor laden air at a rate of 0.5 lpm for 24 hr through 35 ml of MBTH absorbing solution contained in the absorber. A shorter sampling time can be used providing enough formaldehyde is collected to be above the lower limit of sensitivity of the method.

The average collection efficiency of formaldehyde in air has been determined to be 84% when air was sampled at a rate of 0.5 lpm over a 24-hr period in 35 ml of collecting reagent (**3**) in an absorber equipped with an extra coarse (EC) fritted tube inlet. Absorption efficiency may be improved by using a coarse (C) frit although data are lacking on this likelihood.

7.2 ANALYSIS.

7.2.1 Transfer the samples from the sampling bottles to 50-ml graduates, dilute to 35 ml with distilled water and allow to stand for 1 hr.

7.2.2 Pipet a 10 ml aliquot of the sampling solution into a glass stoppered test tube. A blank containing 10 ml of MBTH solution must also be run. If the aldehyde content of the aliquot exceeds the limits of the method, a smaller aliquot diluted to 10 ml with MBTH solution is used.

7.2.3 Add 2 ml of oxidizing solution and mix thoroughly.

7.2.4 After standing for at least 12 min, read at 628 nm on a suitable spectrophotometer using a 1-cm cell. No significant change in absorbance was noted over a 3 hr period after color development. Determine the aldehyde content of the sampling solution from a curve previously prepared from the standard formaldehyde solution. This will give total aldehyde calculated as formaldehyde.*

*__Note:__ The final colored solution tends to form bubbles that cling to the sides of the cuvettes. In order to eliminate this, the solution should be thoroughly shaken periodically during the 12 min standing time waiting for full color development. It has been found that this thorough shaking will eliminate bubble formation.

8. Calibration

8.1 Pipet 1 ml of formaldehyde standard solution "A" into an iodine flask. Into another flask pipet 1 ml of distilled water. This solution serves as the blank.

8.2 Add 10 ml of 1% sodium bisulfite and 1 ml of 1% starch solution.

8.3 Titrate with 0.1 *N* iodine to a dark blue color.

8.4 Destroy the excess iodine with 0.05 *N* sodium thiosulfate.

8.5 Add 0.01 *N* iodine until a faint blue end point is reached.

8.6 The excess inorganic bisulfite is now completely oxidized to sulfate and the solution is ready for the assay of the formaldehyde bisulfite addition product.

8.7 Chill the flask in an ice bath and add 25 ml of chilled sodium carbonate buffer. Titrate the liberated sulfite with 0.01 *N* iodine using a microburet, to a faint blue end point. The amount of iodine added in this step must be accurately measured and recorded.

8.8 One ml of 0.0100 *N* iodine is equivalent to 0.15 mg of formaldehyde. Therefore, since 1 ml of formaldehyde standard solution was titrated, the milliliters of 0.01 *N* iodine used in the final titration multiplied by 0.15 mg gives the formaldehyde concentration of the standard solution in mg/ml.

8.9 PREPARATION OF STANDARD CURVE.

8.9.1 Pipet 0, 0.5, 1.0, 3.0, 5.0, and 7.0 ml of standard formaldehyde solution "B" into 100-ml volumetric flasks. Dilute to volume with 0.05% MBTH solution. These solutions contain 0, 0.05, 0.1, 0.3, 0.5, and 0.7 μg of formaldehyde/ml.

8.9.2 After final dilution let stand for 1 hr.

8.9.3 Transfer 10 ml of each solution to a glass stoppered test tube and add 2 ml of oxidizing reagent and mix.

8.9.4 After 12 min read the absorbance at 628 nm in a suitable spectrophotometer using 1-cm cells.

8.9.5 Plot absorbance against micrograms of formaldehyde/ml of solution.

9. Calculation

9.1 The concentration of total aliphatic aldehyde (as formaldehyde) in the sampled atmosphere may be calculated by using the following equation:

$$\text{PPM (Vol.)} = \frac{C \times 35 \times 24.45}{V \times M.W. \times E}$$

E = correction factor for sampling efficiency (0.84 may be used if absorber contains an EC frit)
V = liters of air sampled.
C = μg/ml of formaldehyde in sampling solution. (Since each sample is diluted to 35 ml, this figure must be multiplied by 35 to give total micrograms in sampling solution.)
M.W. = molecular weight of formaldehyde (30.03)
24.45 = ml of formaldehyde gas in one millimole at 760 Torr and 25 C.

10. Effect of Storage

10.1 The time study of the reaction of microgram quantities of formaldehyde with 0.05% MBTH shows that the reaction is complete in approximately 45 min; therefore, a reaction time of 1 hr is selected for this procedure. Formaldehyde is fairly stable in 0.05% MBTH since only approximately 5% of the formaldehyde is lost after standing in the MBTH for 13 days. The samples are, therefore, stable enough for later analysis **(3)**.

11. References

1. SAWICKI, E., T.R. HAUSER, T.W. STANLEY, and W. ELBERT. 1961. The 3-Methyl-2-Benzothiazolone Hydrazone Test. *Anal. Chem.* 33:93.
2. HAUSER, T.R., and R.L. CUMMINS. 1964. Increasing the Sensitivity of 3-Methyl-2-Benzothiazolone Hydrazone Test for Analysis of Aliphatic Aldehydes in Air. *Anal. Chem.* 37:679.
3. HAUSER, THOMAS R. 1965. Determination of Aliphatic Aldehydes: 3-Methyl-2-Benzothiazolone Hydrazone Hydrochloride (MBTH) Method. Selected Methods for the Measurement of Air Pollutants. *Public Health Service Publication No. 999-AP-11, Page F-1.*

ADDENDUM

Applications to Other Aldehydes

Acetaldehyde and propionic aldehyde both yield a blue dye after reaction with 3-methyl-2-benzothiazolone hydrazone hydrochloride and a ferric chloride-sulfamic acid solution. It has been found that as the length of chain increases, the sensitivity decreases. Therefore when measuring total aldehydes as formaldehyde this method would give low results if any aldehyde other than formaldehyde is present.

From 0.05 μg/ml to 1.0 μg/ml of both acetaldehyde and propionic aldehyde can be measured in the color developed solution (12 ml). For the lower concentrations the method has poor reproducibility. However, at higher concentrations (0.30 μg/ml and above) reproducibility was very good. These data are summarized in Tables 117:II and III.

Table 117:II. Acetaldehyde

μg/ml	Number of Samples	Average Absorbance	Range	% Variance From Avg.
0.05	29	0.063	0.050–0.074	±20
0.10	29	0.125	0.106–0.144	±15
0.30	29	0.339	0.316–0.355	± 7
0.50	29	0.519	0.495–0.538	± 4
0.70	29	0.685	0.660–0.710	± 3
1.00	15	0.900	0.890–0.910	± 1

Table 117:III. Propionic Aldehyde

μg/ml	Number of Samples	Average Absorbance	Range	% Variance From Avg.
0.05	29	0.046	0.032–0.057	±27
0.10	29	0.082	0.063–0.095	±20
0.30	29	0.243	0.225–0.250	± 5
0.50	29	0.399	0.380–0.422	± 5
0.70	29	0.538	0.515–0.568	± 5
1.00	15	0.732	0.710–0.750	± 2

Acetaldehyde (Eastman Kodak Company, Cat. No. 468) and propionic aldehyde (Eastman Kodak Company, Cat. No. 653) were considered to be primary standards when preparing solutions of known concentration. Exactly 1.28 ml of acetaldehyde was diluted to 1 l with distilled water and then 1 ml of this solution was diluted to 100 ml with MBTH solution giving a final concentration of 10 μg/ml. Exactly 1.24 ml of propionic aldehyde was diluted to 1 l with distilled water and then 1 ml of this solution was diluted to 100 ml with MBTH solution giving a final concentration of 10 μg/ml. The strong standard solutions have a 2 month shelf life. The dilute standard solutions must be prepared fresh daily.

A series of 34 ambient air samples were collected in 35 ml of MBTH solution contained in each of two absorbers in series. The sampling time was 24 hr and the sampling rate was 1 l/min. Collection efficiencies varied from 69% to 100% with the average for the 34 samples being 82%.

Subcommittee 4
R. G. Smith, *Chairman*
R. J. Bryan
M. Feldstein
B. Levadie
F. A. Miller
E. R. Stephens
N. G. White

118.

Tentative Method of Analysis for Mercaptan Content of the Atmosphere

43901-01-70T

1. Principle of the Method

1.1 Mercaptans (organic thiols) are collected by aspirating a measured volume of air through an aqueous solution of mercuric acetate-acetic acid. The collected mercaptans are subsequently determined by spectrophotometric measurement of the red complex produced by the reaction between mercaptans and a strongly acid solution of N,N-dimethyl-p-phenylenediamine and ferric chloride (**3–5**). The method determines total mercaptans and does not differentiate among individual mercaptans although it is most sensitive to lower molecular weight alkanethiols.

2. Sensitivity and Range

2.1 This method is intended to provide a measure of mercaptans in the range below 200 $\mu g/m^3$ (102 ppb). For concentrations above 100 ppb, the sampling period can be reduced or the liquid volume increased either before or after aspirating. The minimum detectable amount of methyl mercaptan is 0.04 $\mu g/ml$ (**4**) in a final liquid volume of 25 ml. When sampling air at the maximum recommended rate of 1 l/min for 2 hr, the minimum detectable mercaptan concentration is 3.9 $\mu g/m^3$ (2.0 ppb methyl mercaptan at 760 Torr and 25 C).

3. Interferences

3.1 The N,N-dimethyl-p-phenylenediamine reaction is also suitable for the determination of other sulfur-containing compounds including hydrogen sulfide (**3**) and dimethyl disulfide. The potential for interference from these latter compounds is especially important since all of these compounds commonly coexist in certain industrial emissions. Appropriate selection of the color formation conditions minimizes the interference from hydrogen sulfide and dimethyl disulfide.

3.2 Hydrogen sulfide, if present in the sampled air, may cause a turbidity in the sample absorbing solution. This precipitate must be filtered before proceeding with the analysis. One study showed that 100 μg H_2S gave a mercaptan color equivalent to 1.5 to 2.0 μg methyl mercaptan (**6**). Other studies reported no absorption at 500 nm in the presence of 150 μg of hydrogen sulfide (**1, 2, 4**).

3.3 An unexplained yellow tinge has been randomly observed in a few impingers after sampling. In these instances, the absorbing solution subsequently turned pink. A black precipitate then formed when the color developing reagent was added. Although the precipitate was removed by filtration just before the absorption was measured on the colorimeter, it is not known if this condition changed the apparent mercaptan concentration (**7**).

3.4 Approximately equimolar response is obtained from the hydrolysis products of dimethyl disulfide. The molar absorptivity for the amine-mercaptan reaction product is 4.4×10^3 and is 5.16×10^3 for the amine-dimethyl disulfide reaction product (**2**). In practice, however, the collection efficiency for dimethyl disulfide in aqueous mercuric acetate is inefficient. Thus the actual interference is negligible.

3.5 Sulfur dioxide up to 250 μg does not influence the color development even when sampling a test atmosphere containing 300 ppm SO_2.

3.6 Nitrogen dioxide does not interfere up to 700 μg NO_2 when sampling a test atmosphere containing 6 ppm NO_2. Higher concentrations of NO_2 caused a positive interference when mercaptans were present but no interference in the absence of mercaptans.

3.7 The supply of mercuric acetate must be free of mercurous ion. If mercurous ion is present turbidity will result when the chloride ion-containing reagents are added to the last step of the analytical procedure.

4. Precision and Accuracy

4.1 The relative standard deviation for 4 mercaptans from methyl to hexyl mercaptan ranged from 0.0 to ±2.6% (**4**). The relative standard deviation increased with increasing molecular weight of the mercaptans.

5. Apparatus

5.1 ABSORBER. Midget impinger fitted with coarse porosity frit.

5.2 AIR PUMP, with flow meter and/or gas meter having minimum capacity of 2 l/min of air through a midget impinger.

5.3 COLORIMETER, OR SPECTROPHOTOMETER AT 500 NM. Use 2.5 or 5.0 cm path length to obtain adequate sensitivity.

5.4 AIR VOLUME MEASUREMENT. The air meter must be capable of measuring the air flow with ±2%. A wet or dry gas meter, with contacts on the 1-ft^3 or 10-l dial to record air volume, or a specially calibrated rotameter, is satisfactory. Instead of these, calibrated hypodermic needles may be used as critical orifices if the pump is capable of maintaining greater than 0.7 atmosphere pressure differential across the needle.

6. Reagents

6.1 PURITY. Reagents must be ACS analytical reagent quality. Distilled water should conform to the ASTM Standards for Referee Reagent Water.

6.2 Solutions should be refrigerated when not in use.

6.3 AMINE-HYDROCHLORIC ACID STOCK SOLUTION. Dissolve 5.0 g N,N-dimethyl-p-phenylenediamine hydrochloride (p-aminodimethylaniline hydrochloride) in 1 l of conc hydrochloric acid. Refrigerate and protect from light. This solution is stable for at least 6 months.

6.4 REISSNER SOLUTION. Dissolve 67.6 g ferric chloride hexahydrate in distilled water, dilute to 500 ml and mix with 500 ml nitric acid solution containing 72 ml boiled conc nitric acid (sp gr 1.42). This solution is stable.

6.5 COLOR DEVELOPING REAGENT. Mix 3 volumes of amine solution and 1 volume of Reissner solution. Prepare this solution freshly for each set of determinations.

6.6 ABSORBING SOLUTION*. Dissolve 50 g of mercuric acetate in 400 ml distilled water and add 25 ml glacial acetic acid. Dilute to 1 l. The mercuric acetate must be free of mercurous salts to prevent precipitation of mercurous chloride during color development. Reagent grade mercuric acetate sometimes contains mercurous mercury. Determine the acceptability of each new bottle of mercuric acetate by adding 3 ml of conc hydrochloric acid to 3 ml of the 5 per cent mercuric acetate. If the solution becomes cloudy, the mercuric acetate is not acceptable.

6.7 LEAD METHYL MERCAPTIDE. Bubble tank methyl mercaptan gas into 10% lead acetate solution in an adequate fume hood (**5**). Collect the yellow crystals by vacuum filtration, wash with distilled water and dry overnight in a vacuum oven at 45 C. Store crystals in a vacuum-sealed container in the dark. One mole of this mercaptide is equivalent to 2 moles of a mercaptan. Lead mercaptide may be purchased from commercial sources, if desired.†

6.8 CONCENTRATED, STANDARD LEAD MERCAPTIDE SOLUTION. Weigh out 156.6 mg of the crystalline lead mercaptide and make up to 100 ml with the 5% mercuric acetate absorbing solution. This solution contains the equivalent of 500 μg of methyl mercaptan/ml.

6.9 DILUTED STANDARD MERCAPTAN SOLUTION. Dilute 2 ml of the concentrated standard solution to 100 ml with the 5% mercuric acetate absorbing solution. This solution contains the equivalent of 10 μg CH_3SH/ml.

6.10 METHYL MERCAPTAN PERMEATION TUBE. Prepare (or purchase‡ if available) a triple-walled or thick-walled Teflon permeation tube (**8, 9**) which delivers methyl mercaptan at a maximum rate of approximately 1 μg/min at 25 C. This loss rate will produce a standard atmosphere containing 500 $\mu g/M^3$ (255 ppb) CH_3SH when the tube is swept with a 2 l/min air flow. This concentration, however, is not realistic. The desired permeation rate would be in the range of 0.004 to 0.1 μg/min. At these permeation rates, a 2 l/min air flow over the tube would produce standard air concentrations in the realistic range of 2 to 50 $\mu g/M^3$ (1 to 25 ppb) CH_3SH.

The permeation tubes must be stored in a wide-mouth glass bottle containing silica gel and solid sodium hydroxide to remove

*CAUTION: Mercuric acetate is highly toxic. If spilled on skin, wash off immediately with water.

†Distillation Products Industries, Eastman Organic Chemicals Department, Rochester, N.Y. 14603.

‡Available from Metronics, Inc., 3201 Porter Avenue, Palo Alto, Calif. 94304 or Polyscience Corp., 909 Pitner Avenue, Evanston, Ill. 60202.

moisture and methyl mercaptan. The storage bottle is immersed to two-thirds its depths in a constant temperature water bath in which the water is controlled at 25 C ± 0.1 C.

7. Procedure

7.1 Collection of Sample. Aspirate the air sample through 15 ml of the absorbing solution in a midget impinger at 1.0 to 1.5 l/min for a selected period up to 2 hr.

7.2 Analysis. Quantitatively transfer the sample from the impinger to a 25 ml volumetric flask and dilute to approximately 22 ml with distilled water that has been used to rinse the fritted bubbler and flask. Add 2.0 ml of freshly prepared color developing reagent, dilute to volume with distilled water and mix well. Prepare a reference blank in the same manner using 15 ml of unaspirated 5% mercuric acetate, 2 ml color developing reagent and dilute to 25 ml. After 30 min, measure the absorbance at 500 nm with spectrophotometer against the mercaptan-free reference blank.

8. Calibration

8.1 Aqueous Mercaptide. Prepare a calibration curve by pipetting appropriate aliquots of the diluted standard lead mercaptide into a series of 25-ml volumetric flasks, diluting each with 15 ml of 5% mercuric acetate absorbing solution and developing the color in the same manner as the samples. Prepare a reference blank in the same manner without lead mercaptide. Determine the absorbance at 500 nm against the mercaptan-free reference blank. Prepare a standard curve of absorbance vs μg methyl mercaptan/ml.

8.2 Gaseous Mercaptan. Commercially available permeation tubes containing liquefied methyl mercaptan may be used to prepare calibration curves for use at the upper range of atmospheric concentration. Triple-walled and drilled rod tubes which deliver methyl mercaptan within a minimum range of 0.1 to 1.2 μg/min at 25 C have been prepared by Bamesberger and Adams (**2**). Preferably the tubes should deliver methyl mercaptan within a loss rate range of μg/min to provide realistic concentrations of CH_3SH (2 to 50 μg/M^3 or 1 to 25 ppb) without having to resort to a dilution system to prepare the concentration range required for determining the collection efficiency of midget impingers.

Analyses of these known concentrations give calibration curves which simulate all of the operational conditions performed during the sampling and chemical procedure. This calibration curve includes the important correction for collection efficiency at various concentrations of methyl mercaptan.

8.2.1 Prepare or obtain a Teflon permeation tube that emits methyl mercaptan at a rate of 0.1 to 0.2 μg/min (0.05 to 0.10 μl/min at standard conditions of 25 C and 1 atmosphere). A permeation tube with an effective length of 2 to 3 cm and a wall thickness of 0.125″ (0.318 cm), will yield the desired permeation rate if held at a constant temperature of 20 to 25 C. Permeation tubes containing methyl mercaptan are stored and calibrated under a stream of dry nitrogen to prevent precipitation of sulfur in the walls of the tube.

8.2.2 To prepare standard concentrations of methyl mercaptan,§ assemble the apparatus consisting of a water-cooled condenser; constant temperature bath maintained at 20 to 25 C; cylinders containing pure dry nitrogen and pure dry air with appropriate pressure regulators; needle valves and flow meters for the nitrogen and dry air, diluent-streams. The diluent gases are brought to temperature by passage through a 2-meter-long copper coil immersed in the water bath. Insert a calibrated permeation tube into the central tube of the condenser maintained at the selected constant temperature by circulating water from the constant temperature bath, and pass a stream of nitrogen over the tube at a fixed rate of approximately 50 ml/

§*Note:* Select either the system designed for laboratory or field use as illustrated in Figures 704:1 and 704:2, respectively, of Intersociety Committee Tentative Method of Analysis for Sulfur Dioxide Content of the Atmosphere, pp. 00, 00.

Table 118:I. Typical Calibration Data

Concentration CH_3SH, ppb	Amount of CH_3SH in μl per 120 liters air sample	Absorbance of sample
21	2.5	0.027
42	5.0	0.056
83	10.0	0.085
125	15.0	0.120
208	25.0	0.209
417	50.0	0.417

min. Dilute this gas stream to the desired concentration by varying the flow rate of the clean, dry air. This flow can normally be varied from 0.2 to 15 l/min. The flow rate of the sampling system determines the lower limit for the flow rate of the diluent gases. The flow rates of the nitrogen and the diluent air must be measured to an accuracy of 1 to 2%. With a tube permeating methyl mercaptan at a rate of 0.1 μl/min, the range of concentration of methyl mercaptan will be between 3 to 200 $\mu g/M^3$ (1.5 to 100 ppb) a generally satisfactory range for ambient air conditions. When higher concentrations are desired, calibrate and use longer permeation tubes.

8.2.3 Procedure for preparing simulated calibration curves. Obviously one can prepare a multitude of curves by selecting different combinations of sampling rate and sampling time. The following description represents a typical procedure for ambient air sampling of short duration, with a brief mention of a modification for 24 hr sampling.

The system is designed to provide an accurate measure of methyl mercaptan in the 3 to 200 $\mu g/M^3$ (approximately 1.5 to 100 ppb) range. It can be easily modified to meet special needs.

The dynamic range of the colorimetric procedure fixes the total volume of the sample at 120 l; then, to obtain linearity between the absorbance of the solution and the concentration of methyl mercaptan in ppb, select a constant sampling time. This fixing of sampling time is desirable also from a practical standpoint: in this case, select a sampling time of 120 min. Then to obtain a 120 l sample of air requires a flow rate of 1.0 l min. The concentration of standard CH_3SH in air is computed as follows:

Where

$$C = \frac{Pr \times M}{R + r}$$

C = Concentration of CH_3SH in ppm,
Pr = Permeation rate in μg/min,
M = Reciprocal of vapor density, 0.50 $\mu l/\mu g$
R = Flow rate of diluent air, l/min
r = Flow rate of diluent nitrogen, l/min

Data for a typical calibration curve are listed in Table 118:I.

A plot of the concentration of methyl mercaptan in ppb (x-axis) against absorbance of the final solution (g-axis) will yield a straight line, the reciprocal of the slope of which is the factor for conversion of absorbance to ppb. This factor includes the correction for collection efficiency. Any deviation from the linearity at the lower concentration range indicates a change in collection efficiency of the sampling system. If the range of interest is below the dynamic range of the method, the total volume of air collected should be increased to obtain sufficient color within the dynamic range of the colorimetric procedure. Also, once the calibration factor has been established under simulated conditions, the conditions can be modified so that the concentration of CH_3SH is a simple multiple of the absorbance of the colored solution.

For long-term sampling of 24-hr duration, the conditions can be fixed to collect 1200 l of sample in a larger volume of mercuric acetate-acetic acid. For example, for 24 hr at 0.83 l/min, approximately 1200 l of air are collected. An aliquot representing 0.1 of the entire amount of sample is taken for the analyses.

The remainder of the analytical procedure is the same as described in the previous paragraph.

8.2.4 The permeation tubes must be stored in a wide-mouth glass bottle containing silica gel and solid sodium hydroxide to remove moisture and methyl mercaptan. The storage bottle is immersed to two-thirds its depth in a constant temperature water bath in which the water is controlled at 25 C ± 0.1 C.

Periodically (every week or 2 weeks) the permeation tubes are removed and rapidly weighed on a semi-micro balance (sensitivity ± 0.01 mg) and then returned to the storage bottle. The weight loss is recorded. The tubes are ready for use when the rate of weight loss becomes constant (within ± 2%).

9. Calculation

9.1 Determine the sample volume in liters from the gas meter or flow meter readings and time of sampling. Adjust volume to 760 Torr and 25 C (V_s). Compute the concentration of methyl mercaptan in the sample by one of the following formulas:

$$\text{ppb} = \frac{(A - A_0)\ 0.510B}{V_s}$$

$$\mu g/m^3 = \frac{(A - A_0)\ B}{V_s}$$

Where: A is the sample absorbance

A_0 is the reagent blank

0.510 is the volume (μl) of 1 μg CH_3SH at 25 C, 760 Torr

B is the calibration factor, μg/absorbance unit

V_s is the sample volume in cubic meters corrected to 25 C, 760 Torr (by PV = nRT).

10. Effect of Storage

10.1 No information available.

11. References

1. Adams, D.F. 1969. Analysis of malodorous sulfur-containing gases. TAPPI 52:53.
2. Bamesberger, W.L. and D.F. Adams. 1968. Unpublished information.
3. Marbach, E.P. and D.M. Doty. 1956. Sulfides released from gamma-irradiated meat as estimated by condensation with N,N-Dimethyl-p-phenylenediamine. *J. Agr. Food Chem.* 4:881.
4. Moore, H., H.L. Helwig, and R.J. Graul. 1960. A spectrophotometric method for the determination of mercaptans in air. *Ind. Hyg. J.* 21:466.
5. Silwinski, R.A., and D.M. Doty. 1958. Determination of micro quantities of methyl mercaptan in gamma-irradiated meat. *J. Agr. Food Chem.* 6:41.
6. ACGIH Recommended Method. 1964. Determination of total mercaptans in air. July.
7. Hendrickson, E.R. and D.A. Falgout. 1969. Personal communication.
8. O'Keeffe, A.E., and G.C. Ortman. 1966. 1969. Primary standards for trace gas analysis. *Anal. Chem.* 38:760. Ibid., 44, 1598.
9. Scaringelli, F.P., S.A. Frey, and B.F. Saltzman. 1967. Evaluation of permeation tubes for use with sulfur dioxide. *Am. Ind. Hyg. Assoc. J.* 28.

Subcommittee 1

D. F. Adams, *Chairman*
D. Falgout
J. O. Frohliger
J. B. Pate
A. L. Plumley
F. P. Scaringelli
P. Urone

119.

Tentative Method of Analysis for Peroxyacetyl Nitrate (PAN) in the Atmosphere (Gas Chromatographic Method)

44301-01-70T

1. Principle of the Method

1.1 Gas chromatography, which is used in this method for measuring peroxyacetyl nitrate (PAN) **(4, 6, 7, 8, 9)** in ambient air, is a very powerful procedure for separating mixtures into their chemical components after separation **(3, 13, 14)**. The present method exploits the exceptional ability of PAN to capture free electrons to provide a method for measuring ppb concentrations of PAN in a 2 or 3 ml sample of ambient air. Higher homologues in the series such as peroxypropionyl nitrate can be detected but procedures for calibration have not been worked out since the amounts present are much smaller than the amounts of PAN found. The recently reported peroxybenzoyl nitrate **(2)** might be measured by an adaptation of the present technique.

1.2 Three major components are required: (a) a chromatographic column to separate PAN from the many other compounds present; (b) a sample introduction system, in this case a gas sample valve activated by a timer and solenoid; (c) an electron capture detector with associated amplifier and recorder. The instrument injects a sample every 15 min and about 1.5 min are required to develop the chromatogram. The minimum detectable quantity is about one ppb although this could be lowered if there were need.

2. Range and Sensitivity

2.1 Since the detector used in this method is based on the capture of electrons, large amounts of substances which have a high affinity for electrons can saturate the detector by capturing all the electrons. As saturation is approached, the response becomes very non-linear. To maintain linear response it is necessary to restrict the sample so that the steady current through the detector is reduced by no more than about 25%. This limits the upper end of the concentration range. The maximum depends on sample size, column length, temperature, and carrier gas flow rate, and is usually less than 1 ppm by volume.

2.2 On the other hand maximum sensitivity permits detection of concentrations of less than one ppb (10^{-9} by volume or 4.95 $\mu g/M^3$ at 25 C and 760 Torr).

3. Interferences

3.1 There are no known interferences with this PAN method. An interfering substance would have to meet three conditions: (a) it must have high electron affinity; (b) it must have an elution time very close to that of PAN; and (c) it must be present in the sample at a concentration detectable by this procedure. These stringent conditions eliminate virtually everything.

4. Precision and Accuracy

4.1 Repetitive sampling of standard samples containing about 60 ppb of PAN gave a relative standard deviation of less than 2% for the amount present. Accuracy of course is dependent on the validity of the calibration which in turn depends on the accuracy of the infrared method (section 8). This has been compared with a colorimetric method based on hydrolysis to nitrite ion with good agreement (**12**). It is probable that the overall accuracy of the method is within 5%.

5. Apparatus

5.1 The procedures for measuring PAN described by Darley *et al.* (**1**) have been modified and used in the automated system. An Aerograph Model 681 gas chromatograph* with an electron capture detector or any similar instrument is equipped with an automatic system which injects a sample of ambient air into the following column every 15 min. A 9″ long column of ⅛″ ID Teflon tubing packed with 5% carbowax E400 on 100 to 120 mesh HMDS treated chromosorb-W is used. It is operated at 25 C oven temperature with a nitrogen carrier gas flow of 40 ml/min. Under these conditions the retention time for PAN is 60 sec.

5.2 A stainless steel hexaport valve (similar to the valve supplied by Varian Aerograph, Walnut Creek, California) with an external 2-ml stainless steel sample loop is used instead of the glass sample valve described by Darley *et al.* (**1**). With this system there is no detectable decomposition of PAN.

5.3 The automatic sampling mechanism attached to the outside of the chromatograph case, Figure 119:1, consists of: (a) a timing mechanism; (b) sample valve; (c) solenoid activator; and (d) time delay relay. The timing mechanism consists of a micro switch, a cam to operate the microswitch, and a standard Synchron 4 RPH motor. The cam is designed to energize the solenoid for 90 sec during each 15 min period. The shaft of the hexaport valve is attached rigidly to the solenoid shaft and positioned so that the air sample flows through the 2 ml sample loop during the time the solenoid is energized. A coil spring of proper tension installed between the solenoid and sample valve returns the valve to injection position when the solenoid is de-energized.

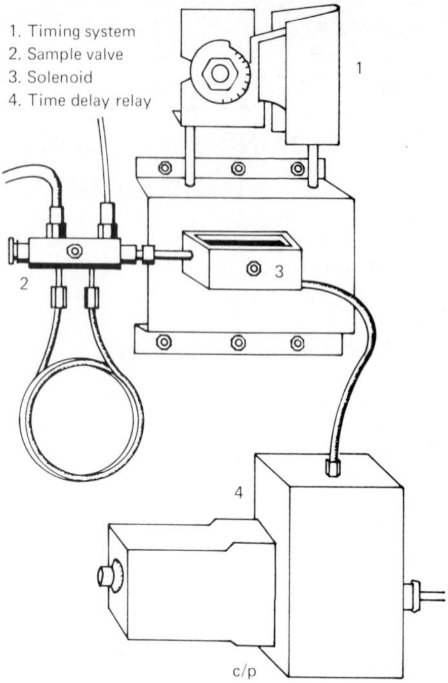

Figure 119:1—Automatic sampling system consisting of: (a) standard timing system; (b) hexaport valve and 2 ml sample loop; (c) solenoid to activate valve; (d) time delay relay to control recorder chart drive motor.

5.4 Calibration requires reference to a known PAN mixture which is most easily measured using a 10-cm path length gas cell and a spectrometer capable of scanning the 1161 cm^{-1} band. Since this system cannot measure concentrations less than a few hundred ppm, provision must be made for quantitative dilution by a factor of about 1,000.

6. Reagents

6.1 The only reagents required for this determination are the column packing, nitrogen carrier gas, and PAN for calibration. The column packing is made by coating 100 to 120 mesh HMDS treated

*Varian Aerograph, 2700 Mitchell Drive, Walnut Creek, California. This model is no longer available but other chromatographs can be adapted to this purpose.

chromosorb-W with 5% by weight of carbowax E400. Both these materials can be obtained from Varian Aerograph or other supply sources. Dry "prepurified" nitrogen can be obtained in 220 ft^3 cylinders from various suppliers.

6.2 PAN is synthesized, purified as described by Stephens *et al.*, **(11)** and stored in 34-l stainless steel cylinders for use in the dynamic calibration system for the chromatograph **(5)**. The cylinders, pressurized with nitrogen to 100 psig, contain 500 to 1000 ppm PAN and are stored at 60 F (16 C). An infrared spectrophotometer with a 10-cm cell is employed to determine the concentration of PAN in the cylinders, using the absorptivities of 13.9×10^{-4} ppm^{-1} m^{-1} at 8.6μ (1161 cm^{-1}) **(9)**.

6.3 A simplified alternate method which requires much less apparatus may also be used. Small quantities of PAN are prepared directly in a 10-cm infrared cell with a pyrex glass body by photolyzing traces of ethyl nitrite, CH_3CH_2ONO, in pure oxygen. The PAN formed is then measured in the cell and is not separated from unreacted nitrite or from by-products. But it must be diluted by a factor of several thousand to reach the electron capture concentration range. Blacklight fluorescent lamps provide a good energy source for the necessary 300 to 400 nm radiation but it might well be possible to use the sun for this photolysis. The cell is first flushed with pure oxygen and then small amounts (50μl) of vapor from the space above the liquid in the ethyl nitrite bottle are added at 15 min intervals while irradiating. The ethyl nitrite concentration must be kept low to maximize yield.

7. Procedure

7.1 Dry nitrogen carrier gas is flushed continuously through the 2-ml sample loop of the hexaport valve for 13.5 min of each period when the solenoid is de-energized. When the solenoid is energized the sample loop is open to the atmosphere to be sampled and a Neptune pump Model 4-K or similar pump draws a continuous flow of air through the loop for 90 sec. When the solenoid is again de-energized, the carrier gas flushes the 2 ml sample into the column.

7.2 The time delay relay, Figure 119:1, energizes the chart drive motor on the strip chart recorder at the end of the 90 sec sample period when the solenoid is de-energized. The chart drive motor is allowed to run for 90 sec which is ample time to record the chromatogram. The column is too short to separate peroxypropionyl nitrate (PPN) and peroxybutyryl nitrate (PBN), although a shoulder on the PAN peak, assumed to be PPN, appears occasionally.

7.3 A typical chromatogram for the period between 8:00 and 9:00 p.m. on March 27, 1968, Figure 119:2, shows that significant concentrations may be measured long after sundown.

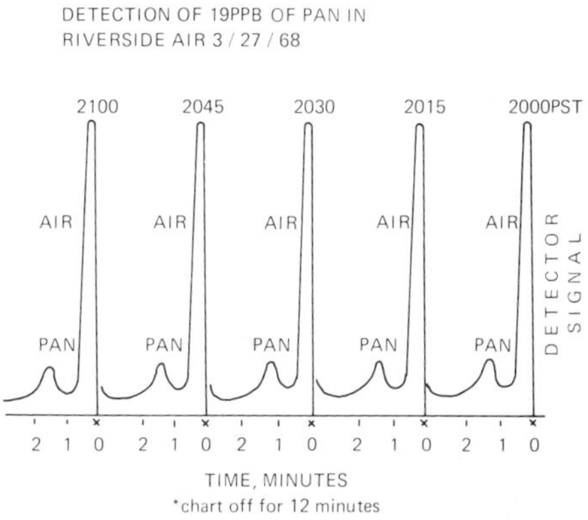

Figure 119:2—A chromatogram produced by the automated PAN analyzer from ambient air samples during the period 8:00 to 9:00 p.m. on March 27, 1968. The large air peak is followed by a PAN peak in each of the 15 minute sample periods.

8. Calibration and Standards

8.1 The chromatograph is calibrated with a flow dilution panel (Figure 119:3) from a supply containing PAN at a concentration high enough to be measured in a 10 cm infrared cell (~ 1000 ppm).

8.2 In the first dilution of the dynamic system **(5)** (Figure 119:3), 1 part PAN from the storage cylinder is diluted with 100

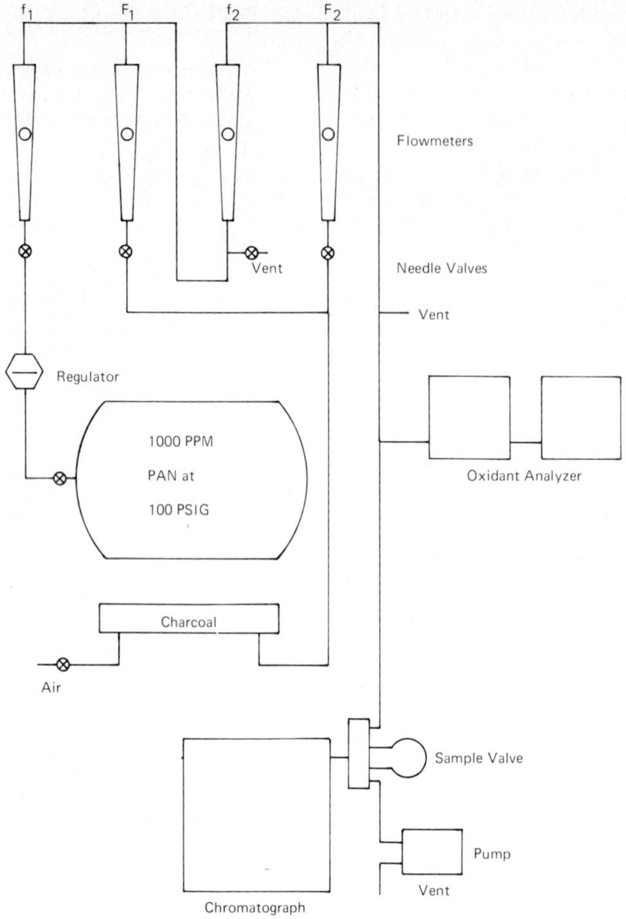

Figure 119:3—Dynamic calibration system.

parts activated-charcoal-filtered air. A similar 1 to 100 dilution with filtered air in the second step of the dilution system reduces the concentration of PAN in a constant flow of air to the ppb range. Calibration curves are plotted from calculated concentrations of PAN and peak height. The oxidant recorder shown in Figure 119:3 is not essential.

9. Calculation

9.1 The tank concentration is calculated from the infrared spectrum **(10)** by applying the Beer-Lambert equation to the absorbance at 8.6μ (8.6×10^{-6}m).

$$C\ (\text{ppm}) = A/(0.1\ \text{m}) \times (13.9 \times 10^{-4}\ \text{ppm}^{-1}\ \text{m}^{-1})$$

where C is conc. in ppm

A is absorbance $= \log_{10}(^{Io}/_{I})$

For accurate measurement this concentration must be of the order of a 1000 ppm. This is about 10,000 times higher than the concentration required for calibration so dilution is carried out in two steps of about one hundredfold using the dilution panel shown in Figure 119:3. The final concentration is then calculated by the formula:

$$C_e = C\,[f_1/(F_1 + f_1)]\,[f/_2(F_2 + f_2)]$$

In which f_1 and f_2 are rates in the small

streams and F_1 and F_2 are flow rates in the large streams.

9.2 Calculated concentration of PAN and peak height show a linear relationship in the range from 1 to about 50 ppb; therefore, concentration within this range can be calculated by multiplying the peak height by a constant. The constant varies from one instrument to another and varies inversely with changes in standing current of a particular instrument.

9.3 The automatic system has been operated continuously 24 hr a day for 11 months with only brief interruptions to clean the electron capture detector. Contamination of the detector causes a slow decline in standing current and a concomitant reduction in sensitivity. It is desirable to calibrate the instrument about once a week to compensate for the gradual reduction in sensitivity. When the detector is thoroughly cleaned the original sensitivity is regained.

10. Effect of Storage

10.1 No reliable means of retaining PAN-containing air samples is known at this time.

11. References

1. Darley, E.F., K.A. Kettner, and E.R. Stephens. 1963. Analysis of peroxyacetyl nitrates by gas chromatography with electron capture detection. *Anal. Chem.* 35:589–591.
2. Heuss, J.M., and W.A. Glasson. 1968. Hydrocarbon reactivity and eye irritation. *Env. Sci. & Tech.* 2:1109.
3. Mayrsohn, H., and C. Brooks. Nov. 18, 1965. The analysis of PAN by electron capture gas chromatography. Western Regional meeting of the Amer. Chem. Soc.
4. Middleton, J.T. 1961. Photochemical air pollution damage to plants. *Ann. Rev. Plant Physiol.* 12:431.
5. Plata, R.L. 1968. Calibration and comparison of coulometric and flame ionization for monitoring PAN in experimental atmospheres. Ninth Conference on Methods in Air Pollution and Industrial Hygiene Studies, Huntington-Sheraton Hotel, Pasadena, California.
6. Stephens, E.R., P.L. Hanst, R.C. Doerr, and W.E. Scott. 1956. Reactions of nitrogen dioxide and organic compounds in air. *Ind. Eng. Chem.* 48:1498–1504.
7. Stephens, E.R., W.E. Scott, P.L. Hanst, and R.C. Doerr. 1956. Recent developments in the study of the organic chemistry of the atmosphere. *J. Air Poll. Cont. Assoc.* 6:159–165.
8. Stephens, E.R., E.F. Darley, O.C. Taylor, and W.E. Scott. 1961. Photochemical reaction products in air pollution, *Proc. API 40* III:325–338, 1960; *Int. J. Air and Water Poll.* 4:79–100.
9. Stephens, E.R. 1961. The photochemical olefin-nitrogen oxides reaction in the lower and upper atmosphere. *Interscience,* New York.
10. Stephens, E.R. 1964. Absorptivities for infrared determination of peroxyacyl nitrates. *Anal. Chem.* 36:928–929.
11. Stephens, E.R., F.R. Burleson, and E.A. Cardiff. 1965. The production of pure peroxyacyl nitrates. *J. Air Poll. Control Assoc.* 15:87–89.
12. Stephens, E.R., and M.A. Price. 1967. Colorimetric analysis of PAN. Eighth Conference on Methods in Air Pollution and Industrial Hygiene Studies. Oakland, California.
13. Taylor, O.C., E.R. Stephens, and E.A. Cardiff. 1968. Automatic chromatographic measurement of PAN. 61st Annual Meeting of the Air. Poll. Cont. Assoc. St. Paul, Minnesota.
14. Tingey, D.T., and A.C. Hill. 1968. The occurence of photochemical phytotoxicants in the Salt Lake Valley, Utah Academy Proceedings 44:387–395.

Subcommittee 4
R. G. Smith, *Chairman*
R. J. Bryan
M. Feldstein
B. Levadie
F. A. Miller
E. R. Stephens

120.

Tentative Method of Analysis for Determination of Phenolic Compounds in the Atmosphere (4-Aminoantipyrine Method)

17320-01-70T

1. Principle of the Method

1.1 Air is scrubbed with an alkaline solution in a standard impinger. Particulate phenolic substances are collected by passing air through a fiberglass filter. Phenolic compounds are separated from other compounds by distillation from an acidified system. Phenols are determined by coupling them with 4-aminoantipyrine in an alkaline medium containing an oxidant.

1.2 The method is based on a reaction discovered by Emerson (**1**). This procedure is essentially that of Smith *et al* (**5**). Discussions of theory and efficiency are given by Ettinger (**2**) and Mohler (**4**).

1.3 In the presence of a strong alkaline oxidizing reagent this coupling reaction will proceed as shown in 1.4 below. If the system is not sufficiently alkaline dimerization of 4-aminoantipyrine to antipyrine red will take place, as in 1.5 below. It is important, therefore, to have a high pH when the coupling reaction is induced.

1.4 Coupling reaction of 4-aminoantipyrine and phenol.

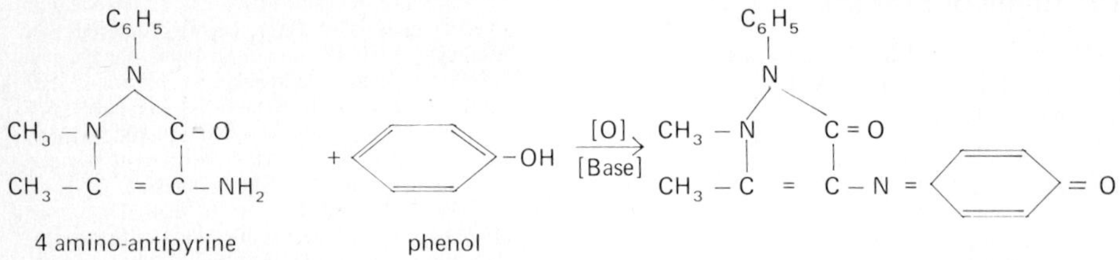

1.5 Dimerization of 4-aminoantipyrine.

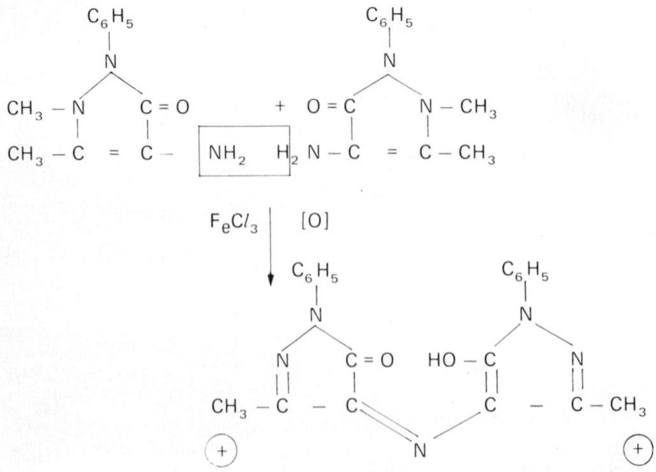

Antipyrine Red.

2. Range and Sensitivity

2.1 One cubic meter of air containing 1.3 ppb of phenol will produce sufficient sample to give a coupling product absorbance of approximately 0.2 units in a 20 mm cuvet when measured at 460 nm wavelength in a spectrophotometer.

3. Interferences

3.1 Any color, other than that due to the reagents used, interferes with the method. Turbidity, sulfur compounds and certain metallic ions interfere (**4**). However, the distillation procedure described by Smith eliminates these interferences.

4. Precision and Accuracy

4.1 In the range of 0.49 to 1.93 ppm the standard deviation is 0.022 and at the 95% confidence level 0.065. At 18 to 75 ppb the standard deviation is 3.3 and at the 95% confidence level 10.7 ppb. Accuracy is ±2% (**4**).

5. Apparatus

5.1 SOXHLET EXTRACTORS.

5.2 DISTILLATION APPARATUS, ALL GLASS.

5.3 IODINE BOTTLES, 500 ML SIZE.

5.4 IMPINGERS. Standard, midget or equipped with fritted absorbers (extra coarse porosity).

5.5 FIBER GLASS FILTER SHEETS, FLASH FIRED.

5.6 SPECTROPHOTOMETER. Any spectrophotometer capable of measuring the absorbance of the solution complex and 460 to 510 nm, as required.

5.7 HIGH VOLUME SAMPLER, or other air sampling device for collection of particulate samples, equipped with a calibrated gauge or flow meter to measure air volume flow accurately.

6. Reagents

6.1 PURITY. All reagents should be ACS analytical grade.

6.2 4-AMINOANTIPYRINE SOLUTION. Dissolve 2 g of 4-aminoantipyrine in distilled water and make up to 100 ml. This solution should not be kept longer than 1 week.

6.3 POTASSIUM FERRICYANIDE SOLUTION. Dissolve 8 g of analytical reagent grade potassium ferricyanide in distilled water and make up to 100 ml. Discard when the solution becomes darkened.

6.4 AMMONIUM CHLORIDE SOLUTION. Dissolve 50 g of analytical reagent grade salt in distilled water and make up to 1 l.

6.5 COPPER SULFATE SOLUTION. Prepare a 10% solution of the pentahydrate.

6.6 SODIUM HYDROXIDE SAMPLING SOLUTION. Prepare a 1 *N* solution.

6.7 BROMIDE/BROMATE SOLUTION. Dissolve 2.784 g of analytical reagent grade potassium bromate in distilled water; add 10 g of analytical reagent grade potassium bromide and make up to 1 l.

6.8 AMMONIUM HYDROXIDE.

6.9 HYDROCHLORIC ACID.

6.10 PHOSPHORIC ACID SOLUTION. Prepare a 10% solution of orthophosphoric acid.

6.11 POTASSIUM IODIDE.

6.12 SODIUM THIOSULFATE SOLUTION. Prepare a 0.1 *N* solution of the salt and standardize according to classical laboratory procedures. Dilute to make an exactly 0.025 *N* solution.

6.13 STARCH SOLUTION. Dissolve 1 g of soluble starch in 100 ml of distilled water. Prepare a fresh solution daily.

6.14 PHENOL.

6.15 BENZENE.

6.16 CHLOROFORM.

6.17 METHYL ORANGE INDICATOR. Prepare according to classical laboratory method.

7. Procedure

7.1 AIR SAMPLING: PARTICULATES. Draw a 24-hr air sample at a measured flow rate through a flash-fired fiberglass filter using a high volume sampler.*

*Proper methods for calibrating the high volume sample or other sampling devices should be provided by the manual supplied by the manufacturer.

7.2 AIR SAMPLING: VAPORS AND PARTICULATES. WET METHOD. Draw a 30 min sample (or larger if desired) of air through a 0.1 *N* solution of sodium hydroxide in distilled water, at a standard impinger flow rate of 1 ft^3/min. If only vapor phenolics are required, use a membrane filter in the sampling train to remove particulates.

7.3 ANALYSIS (**CAUTION:** DO NOT USE STOPCOCK GREASE IN ANY APPARATUS).

7.4 FILTER SAMPLES. Extract the filter or any desired portion of it in a Soxhlet extractor by refluxing with benzene for 3 hr. Transfer the benzene extract to a separatory funnel, filtering it through a close (Whatman 42 or equivalent) paper. Extract 3 times with 10 ml portions of 1.0 *N* NaOH. Treat according to Section **7.6** DETERMINATION OF PHENOLS.

7.5 SAMPLES COLLECTED IN 0.1 *N* NAOH.

7.5.2 *Exhaust gases or process effluents.* Use the whole sample. Add 1 ml of 10% copper sulfate solution. Acidify, using methyl orange as indicator and 10% phosphoric acid solution. Transfer to an all glass distillation apparatus and distill, collecting 90 ml of the distillate. Cool the distillation flask and add 10 ml of distilled water. Continue the distillation until exactly 100 ml of distillate has been collected. Acidify with 0.5 ml of 10% phosphoric acid solution, add 1.0 ml of 10% copper sulfate solution and transfer to a separatory funnel. Add 30 g of reagent grade sodium chloride and extract with 3 to 10 ml portions, of chloroform. Discard the aqueous phase. Shake the chloroform extract with 2 to 15 ml portions of 0.1 *N* NaOH. Discard the chloroform phase. Heat the alkali extracts until the traces of chloroform have been removed, dilute the alkaline extract to 100 ml volume with distilled water and treat according to Section **7.6.2.**

7.6 DETERMINATION OF PHENOLS.

7.6.1 Adjust the alkaline extracts to volume of 100 ml, either by aliquoting or diluting to volume with distilled water. Add 1 ml of 10% copper sulfate solution. Acidify with 10% phosphoric acid solution using methyl orange indicator. Distill from an all glass distillation apparatus until 90 ml have been collected. Add 10 ml of distilled water to the cooled distillation flask and continue the distillation until a total volume of 100 ml of distillate has been collected.

7.6.2 Take a 50 ml aliquot of the distillate. Prepare standards containing 0.5, 1.0, 5.0, 10.0 and 20.0 μg of phenol. Adjust the volumes of sample and standards to 100 ml with distilled water. Add 2 ml of ammonium chloride solution. Using a pH meter adjust to pH 10.0 $\pm$ 0.2, with conc ammonium hydroxide. Add 1 ml of 4-aminoantipyrine solution and mix. Add 1 ml of potassium ferricyanide solution, transfer to a separatory funnel and wait 3 minutes. Extract with 3 to 5 ml portions of chloroform and discard the aqueous phase. Make up the chloroform extract to a known volume with chloroform. Using a blank as reference, record the absorbance at 460 nm. For higher concentrations of phenol the chloroform extraction may be omitted and the absorbance measured directly at 510 nm.

8. Standardization of Phenol Solution

8.1 STOCK STANDARD PHENOL SOLUTION. Prepare a 0.1% solution of phenol in distilled water. Into a 500-ml iodine flask transfer 50 ml of stock standard and add 100 ml of distilled water. Add exactly 10 ml of bromide-bromate solution. Add carefully 0.5 ml of conc hydrochloric acid. Swirl the flask gently, making certain that the stopper is seated. If, at this point, the color of bromine does not persist, continue to add exactly 10 ml portions of bromide/bromate solution until the reddish-brown bromine color does persist. If the stock solution is made up to contain 1000 mg of phenol/l, 4 to 10 ml portions of bromide/bromate solution will be required. With the stopper in position let the reaction flask sit for 10 min. Add quickly 1 g of potassium iodide. Prepare a blank in exactly the same manner, using 10 ml of bromide/bromate solution and distilled water. Titrate both blank and sample with 0.025 *N* sodium thiosulfate, using starch as indicator. Calculate the concentration of phenol solution as follows:

Milligrams of phenol per liter = [(A × B) − C] × 7.835

A = ml of 0.025 *N* thiosulfate used for blank.
B = ml of bromide-bromate solution used for sample, divided by 10.
C = ml of 0.025 *N* thiosulfate used for sample. The factor, 7.835, is based on the use of an exactly 0.025 *N* thiosulfate solution in the titration.

8.2 PRIMARY STANDARD PHENOL SOLUTION. Dilute the stock standard so that 1 ml is equal to 10 μg of phenol.

8.3 WORKING STANDARD. Dilute the primary standard 1:10 with distilled water. This solution contains 1 μg of phenol in 1 ml.

9. Calculation

9.1 Working standards are used to prepare a concentration *vs* absorbance curve from which the concentration of phenol in samples is determined. One μg of phenol per l of air is equal to 0.26 ppm and one ppm is equal to 3.84 μg per l at 25 C and standard pressure.

10. Effects of Storage

10.1 The addition of 5 ml of copper sulfate solution to the alkaline solution of phenols will serve to stabilize the sample.

10.2 CAUTIONS.

10.2.1 Equipment which has been lubricated with stopcock grease should not be used.

10.2.2 Temperature variations will affect the blank.

10.2.3. Filtration of the chloroform extract before reading it in the spectrophotometer will remove possible turbidity due to presence of water dispersion.

10.2.4 The chloroform extract of the dye will fade on standing.

10.2.5 It is advisable to work quickly when serial readings are made.

11. References

1. EMERSON, E.J. 1943. The Condensation of Aminoantipyrine: A New Test for Phenolic Compounds. *J. Organic Chem.*, 8:417.
2. ETTINGER, M.D., C.C. RUCHHOFT, and R.J. LISHKA. 1951. Sensitive 4-Aminoantipyrine Method for Phenolic Compounds. *Anal. Chem.* 23:1783–1788.
3. JACOBS, M.B. The Analytical Chemistry of Industrial Poisons, Hazards and Solvents. Interscience Publishers, 2nd Edition. (pp 705).
4. MOHLER, E.F., and L.N. JACOB. 1952. Determination of Phenolic-Type Compounds in Water and Industrial Waste Waters. *Anal. Chem.* 29:1369–1374.
5. SMITH, R.G., J.D. MACEWEN, and R.E. BARROW. 1959. Sampling and Analysis of Phenols in Air. *Amer. Indust. Hyg. Assoc. Jour.* April, 142–148.

Subcommittee 4
R. G. SMITH, *Chairman*
R. J. BRYAN
M. FELDSTEIN
B. LEVADIE
F. A. MILLER
E. R. STEPHENS
N. G. WHITE

121.

Tentative Method of Analysis for Phenols in the Atmosphere (Gas Chromatographic Method)

17320-02-72T

1. Principle of the Method

1.1 Phenolic compounds react with sodium hydroxide to form "phenates". These compounds are hydrolyzed by acid. The released phenols are separated from the acid aqueous system by steam distillation. The aqueous solution of the phenols is analyzed by gas chromatography, using a flame ionization detector and a short column. Total phenolic content is expressed as phenol. Particulate phenols are collected by passing the air sample through a fiber glass filter; gaseous phenols by scrubbing with an alkaline solution in a standard impinger.

2. Range and Sensitivity

2.1 Using phenol as a standard test substance, and under the operating conditions given below where bubbler recovery is in the range of 88 to 95%, the sensitivity of the method is 1 ppm in solution (μg of phenol in 1 ml of final distillate) or approximately 40 $\mu g/m^3$, based upon sampling as described in Section **7.1**, sampling time 1.5 hr.

3. Interferences

3.1 This method is not subject to interferences by other organic compounds. It involves the actual physical separation of total phenolic compounds on a highly polar substrate placed upon an inert absorptive packing. If the chromatographic column is changed by increasing the absorptive area without increasing the cross-sectional area of the column it is possible to separate the phenolic compounds and to determine them individually.

4. Precision and Accuracy

4.1 Generally, the precision and accuracy of gas chromatographic systems are governed by the care and technical competence of the individuals operating them. With phenol as the standard, and applying the operating conditions given above, the sensitivity of this system is 1 ppm in solution.

For a single operator, a 10 ppm standard gave a mean area under the peak of 10.82 cm^2 measured by the triangulation method. Over a period of several days, for a 10 ppm standard the following areas under the peak were obtained:

9.01 cm^2
9.52 cm^2
11.30 cm^2
11.34 cm^2
11.31 cm^2
12.42 cm^2

Mean area = 10.82 cm^2
Standard error of the mean = 1.29
Coefficient of variation = 12%, or 4.8 $\mu g/m^3$ based on extrapolation to the lower limit of detection.

For any one day the standard error of the mean was found to be 0.028 and the coefficient of variation 1% (**1**).

5. Apparatus

5.1 A gas chromatographic system consisting of the following components:

5.1.1 An injection port, the operating temperature of which can be controlled and which will deliver at least 300 C.

5.1.2 A column heating chamber capable of precise temperature control (to within 0.05 C).

5.1.3 A flame ionization detector (FID) having a temperature control system.

5.1.4 A stainless steel gas chromatographic column consisting of 5% Carbowax 20M, terephthalic acid liquid phase on 60 to 80 mesh acid washed Chromosorb W. For determination of total phenols a 2 foot stainless steel column is used.

5.1.5 A recorder capable of response of 5 mV full scale in one second, and of chart drive speed of ½″/min.

5.1.6 A gas chromatographic sample injection syringe capable of delivering from 1 to 10 μl with a reproducibility of 95% or better.

5.2 An air sampling system consisting of:

5.2.1 An air scrubber such as the Greenburg-Smith Impinger or its equivalent, capable of sampling at a rate of 1 ft^3/min, (28.3 l/min).

5.2.2 An air pump capable of drawing one cubic foot of air through an orifice having a diameter of 3 mm, through 100 ml of solution in not less than 1 min, and being equipped with a means of measuring the static pressure of and controlling the air flow.

5.2.3 A wet test meter and a rotameter.

5.3 All glass distillation apparatus for steam distillation.

5.4 A pH meter or pH indicator paper capable of reading units of pH.

6. Reagents

6.1 PURITY. All chemicals should be ACS analytical reagent grade.

6.2 COPPER SULFATE SOLUTION. 10% aqueous prepared in double distilled water, organic-matter-free.

6.3 SODIUM HYDROXIDE SOLUTION, (one molar), prepared in double distilled water, organic-matter-free.

6.4 5% CARBOWAX 20 M. Terephthalic acid liquid phase on 60 to 80 mesh acid washed Chromosorb W.

6.5 CHLOROFORM.

6.6 CHROMOSORB W: 60 TO 80 MESH. Acid washed, commercially available.

6.7 PHENOL.

6.8 Preparation of the chromatographic column: (**Note:** A commercially prepared column may be used).

6.8.1 There are several methods for applying the substrate to the chromatographic support, to prepare the column packing. One simple approach is to immerse the appropriate weight of substrate in chloroform. Sufficient solution should be prepared to just cover the solid phase when the two are mixed. This helps to avoid loss of substrate to the surface of the mixing vessel. As a rule it may be assumed that a firebrick type of solid (chromosorb) will absorb about 60% of its own volume of solution.

6.8.2 It is useful to stir the mixture mildly with a spatula or long haired "camel's hair" brush about one inch wide. Gentle heat from a boiling water bath helps to accelerate the evaporation. Any physical treatment of the impregnated packing should be careful to avoid comminution of the particles.

6.8.3 When the packing is sufficiently dry, those particles less than 80 mesh should be removed by sieving carefully.

6.9 The prepared packing now may be introduced to the stainless steel column. This is done in the following manner.

6.9.1 Connect one end, plugged with fine glass wool for a depth of about ¼″, to a vacuum system fitted with a bleeder control.

6.9.2 A small funnel is mounted to the open end of the column which should be adequately supported.

6.9.3 Commencing with mild suction, packing is introduced to the column. The agitation of a vibrating type of tool, or gentle tapping with the handle of a 6″ spatula assists in this operation.

6.9.4 Continue adding packing to the funnel until further vacuum and vibration do not cause the packing to disappear down the funnel stem.

6.9.5 Once the column has been fully packed, about a quarter of an inch of the packing should be shaken out and the open end plugged with glass wool.

6.10 If enough extra packing has been

made to warrant storage, this may be achieved by sealing the excess in a polyethylene bag after pressing any residual air from the container. The packing may be stored refrigerated or at room temperature.

7. Procedure

7.1 Air Sampling. Particulates—Draw a 24 hr air sample at a measured flow rate through a flash-fired glass fiber filter using a high volume sampler.*

7.1.1 *Air Sampling Vapors and Particulates.* Draw air at 1 ft^3/min. into 100 ml of 0.1 N sodium hydroxide in a standard impinger for a sufficient period to obtain a minimum of one microgram of phenol in 1 ml of sampling solution. If vapor phenolics only are required, use a membrane filter in the sampling train to remove particulate matter. Normal air levels require sampling times of 10 min to 15 hr.

7.2 Analyis. (**Caution**—do not use stopcock grease in any apparatus.)

7.2.1 *Extraction of phenols.*

7.2.2 *Filter samples.* Extract the filter or any desired portion of it in a Soxhlet extractor by refluxing with benzene for 3 hr. Transfer the benzene extract to a separatory funnel, filtering it through a close (Whatman 42 or equivalent) paper. Extract 3 times with 10 ml portions of 1.0 N NaOH. Treat according to Section **7.2.5** Determination of Phenols.

7.2.3 Samples collected in 0.1 N NaOH.

7.2.4 *Air samples.* Proceed to Section 7.2.5 Determination of Phenols.

7.2.5 Determination of Phenols.

Adjust the alkaline extracts to volume of 100 ml, either by aliquoting or diluting to volume with distilled water. Add 1 ml of 10% copper sulfate solution. Acidify with 10% phosphoric acid solution using methyl orange indicator. Distill from an all glass distillation apparatus until 90 ml have been collected. Add 10 ml of distilled water to the cooled distillation flask and continue the distillation until a total volume of 100 ml of distillate has been collected.

*Proper methods for calibrating the high volume sampler or other sampling devices should be provided by the manual supplied by the manufacturer.

7.3 Preparation of the Chromatograph. (**4**).

7.3.1 With suitable fittings, the column should be installed, using an antigalling thread lubricant if desirable.

7.3.2 A leak test may be performed in the following manner:

a. Shut off the downstream end of the system.

b. Pressurize the carrier gas supply in the system to 15 psi above the operating pressure conditions selected.

c. Observe the pressure gage reading of the chromatograph.

d. Observe the pressure gage reading after 10 to 15 min. If no drop in pressure is noted at this time, the system is tight.

e. If pressure drop is observed, a leak detecting solution, such as "snoop" may be used to locate minor leaks. This method should *not* be used to detect leaks near the ionization detector. Generally, caution should be used to prevent the entry of leak detecting solution into the system. Such a leakage will produce extraneous peaks during analysis.

7.4 Column Conditioning.

7.4.1 Column conditioning should continue for at least 24 hr, and at a temperature 50° above the expected operating temperature before use. The maximum allowable temperature for both packing and substrate must not be exceeded.

7.4.2 Before beginning the conditioning operation, disconnect the column from the detector to avoid contaminating the detector.

7.4.3 Adjust the carrier gas flow to from 20 to 40 ml per min.

7.4.4 Occasional injection of a few μl of water into the column during this procedure will help to clear the column of impurities.

7.4.5 At the end of the conditioning period connect the column to the detector block.

7.5 Chromatography of phenols.

7.5.1 For the determination of "total phenols" a short, 2-ft long ¼″ stainless steel column packed as above described is satisfactory. With such a column the operating conditions of the system need not be restricted rigidly between series of analyses. Generally the temperatures should be maintained within 0.2 C. Recommended (**1**) temperature criteria are: Injection port: 300 C, column temperature: 200 C, detector block: 260 C. For each day and for each series of analyses a set of standards should be run.

7.5.2 The setting of the gas flow rate for any series of analyses done at one time is a matter of choice for such a short column. Once a convenient flow rate has been selected after some preliminary work with the column, a closer control of the flow rate may be maintained by the use of a soap-film flowmeter.

7.5.3 The chromatographic apparatus has reached stability when there is no baseline drift observable. The time for this will vary with individual instruments.

7.5.4 A preliminary test of the instrument's performance may be run by injecting one of the standards into the system. A 1 μl portion of the sample may also be run as a further check.

7.5.5 Generally, once the operating parameters for a given system have been established for this analysis it will be necessary only to run a series of standards prior to an analysis session using this column.

7.5.6 Size of the sample injected is governed by the concentration of total phenolic content of the sample. Generally from 1 to 5 μl of distillate containing the phenols may be used.

8. Calibration and Standards

8.1 Standardization of Phenol Solution (**2**).

8.1.1 *Stock standard phenol solution.* Prepare a 0.1% solution of phenol in distilled water. Into a 500-ml iodine flask transfer 50 ml of stock standard and add 100 ml of distilled water. Add exactly 10 ml of bromide-bromate solution.† Add carefully 0.5 ml of conc hydrochloric acid. Swirl the flask gently, making certain that the stopper is seated. If, at this point, the color of bromine does not persist, continue to add exactly 10 ml portions of bromide/bromate solution until the reddish-brown bromine color does persist. If the stock solution is made up to contain 1000 mg of phenol/1, 4 to 10 ml portions of bromide/bromate solution will be required. With the stopper in position let the reaction flask sit for 10 min. Add quickly 1 g of potassium iodide. Prepare a blank in exactly the same manner, using 10 ml of bromide/bromate solution and distilled water. Titrate both blank and sample with 0.025 N sodium thiosulfate, using starch as indicator. Calculate the concentration of phenol solution as follows:

$$\text{Milligrams of phenol per liter} = [(A \times B) - C] \times 7.835$$

A = ml of 0.025 N thiosulfate used for blank.

B = ml of bromide-bromate solution used for sample, divided by 10.

C = ml of 0.025 N thiosulfate used for sample. The factor, 7.835, is based on the use of an exactly 0.025 N thiosulfate solution in the titration.

8.1.2 *Primary standard phenol solution.* Dilute the stock standard so that 1 ml is equal to 10 μg of phenol.

8.1.3 *Working Standard.* Dilute the primary standard 1:10 with distilled water. This solution contains one μg of phenol in 1 ml.

8.1.4 Since the readout of a flame ionization detector is based upon the actual mass engaged by the detector, data produced by the recorder will be equivalent to micrograms of phenol. The relationship be-

†Bromide-bromate solution—dissolve 2.784 g of analytical reagent grade potassium bromate in distilled water; add 10 g of analytical reagent grade potassium bromide and make up to 1 liter with distilled water.

tween peak height or area under the peak at half height and concentraton of phenol per volume of injected sample may be considered linear for practical purposes. If concentration is plotted against peak height to make a standard curve, when a series of standards are run, this curve may be used to determine the total phenols in the injected sample.

Quantitative information may be obtained by using the integrated area under the phenolic peak as produced by planimetry, or be recording devices on the chromatograph. Regardless of the method of quantitation, a series of standards should be run for each day the analysis is performed.

9. Calculations

Let

V_d = total final volume of acid distillate in milliliters

V_a = volume of air sampled, in cubic meters at 25 C and 760 Torr

V_s = volume of sample injected for gas chromatography, in microliters

W = mass of total phenolics in the injected sample, expressed in micrograms

then:

$$\frac{V_d \times 1000 \times W}{V_s \times V_a} = \text{micrograms}$$

of total phenolics per cubic meter of air.

10. Effects of Storage

10.1 The addition of 5 ml of copper sulfate solution to the alkaline solution of phenols will serve to stabilize the sample.

10.2 CAUTION.

10.2.1 Equipment which has been lubricated with stopcock grease should not be used.

11. References

1. ROBERT G. KEENAN, of George D. Clayton & Associates, personal communication.
2. AMERICAN PUBLIC HEALTH ASSOCIATION, 1977. Methods of Air Sampling and Analysis, 2nd ed. p. 324. Washington, D.C. 20036.
3. Ibid. p. 327.
4. PHENOLS IN WATER by Gas Liquid Chromatography: ASTM Standard Method of Test D2580-68.

Subcommittee 4

R. G. SMITH, *Chairman*
R. J. BRYAN
M. B. DENTON
M. FELDSTEIN
B. LEVADIE
D. C. LOCKE
F. A. MILLER
E. R. STEPHENS
P. O. WARNER

122.

Tentative Method of Analysis for Formaldehyde, Acrolein and Low Molecular Weight Aldehydes in Industrial Emissions on a Single Collection Sample

43501-02-74T

1. Principle of the Method

1.1 Formaldehyde, acrolein, and low molecular weight aldehydes are collected in 1% $NaHSO_3$ solution in Greenberg-Smith impingers from industrial effluents. Efficiency of collection is approx. 95% when two impingers are used in series. Formaldehyde is measured in an aliquot of the collection medium by the chromotropic acid procedure, acrolein by a modified mercuric-chloride-hexylresorcinol procedure, and $C_2 - C_5$ aldehydes by a gas chromatographic procedure. The method permits the analysis of $C_1 - C_5$ aldehydes on a single collection sample. **(1)**.

2. Range and Sensitivity

2.1 At a sampling rate of 0.5 to 1.0 CFM over a 15-min period, the following minimum concentrations of each substance can be determined:

Formaldehyde:	0.1 ppm
Acrolein:	0.1 ppm
$C_2 - C_5$ aldehydes:	0.1 ppm

Shorter sampling periods are permissible for higher concentrations.

3. Interferences

3.1 FORMALDEHYDE.

3.1.1 The chromotropic acid procedure has very little interference from other aldehydes. Saturated aldehydes give less than 0.01% positive interference, and the unsaturated aldehyde acrolein results in a few per cent positive interference. Ethanol and higher molecular weight alcohols and olefins in mixtures with formaldehyde are negative interferences.

3.1.2 Phenols result in a 10 to 20% negative interference when present at an 8:1 excess over formaldehyde.

3.1.3 Ethylene and propylene in a 10:1 excess over formaldehyde result in a 5 to 10% negative interference and 2-methyl-1,3-butadiene in a 15:1 excess over formaldehyde showed a 15% negative interference. Aromatic hydrocarbons also constitute a negative interference. It has recently been found that cyclohexanone causes a bleaching of the final color.

3.2 ACROLEIN.

3.2.1 There is no interference in the acrolein determination from ordinary quantities of sulfur dioxide, nitrogen dioxide, ozone and most organic air pollutants. A slight interference occurs from dienes: 1.5% for 1,3-butadiene and 2% for 1,3-pentadiene. The red color produced by some other aldehydes and undetermined materials does not interfere in spectrophotometric measurement.

4. Precision and Accuracy

4.1 Known standards can be determined to within ±5% of the true value.

5. Apparatus

5.1 ABSORBERS. All glass standard Greenberg-Smith impingers are acceptable. A train of two bubblers in series is used.

5.2 AIR PUMP. A pump capable of drawing at least 1 CFM of air/min for 15 min through the sampling train is required.

5.3 AIR METERING DEVICE. Dry test meter, or other suitable device.

5.4 SPECTROPHOTOMETER. This instrument should be capable of measuring the developed colors at 605 nm and 580 nm. The absorption bands are rather narrow, and thus a lower absorptivity may be expected in a broad-band instrument.

5.5 GAS CHROMATOGRAPH. With hydrogen flame detector and injection port sleeve (Varian 1200 or equivalent).

5.6 BOILING WATER BATH.

5.7 ICE BATH.

5.8 CENTRIFUGE.

5.9 GRADUATED 25-ML G.S. TUBES.

6. Reagents

6.1 All reagents must conform to ACS Reagent Grades or equivalent.

6.2 DETERMINATION OF FORMALDEHYDE.

6.2.1 *Sodium formaldehyde bisulfite*. (E.K. P6450).

6.1.2 *Chromotropic acid sodium salt, EK P230*. 0.5% in water. Filter just before using. Stable for one week if kept refrigerated.

6.1.3 *Sulfuric acid*. Conc.

6.3 DETERMINATION OF ACROLEIN.

6.3.1 *$HgCl_2$ 4-hexylresorcinol*. 0.30 g $HgCl_2$ and 2.5 g 4-hexylresorcinol are dissolved in 50 ml of 95% ethanol. (Stable at least 3 weeks if kept refrigerated.)

6.3.2 *TCAA*. To a 1 lb bottle of trichloracetic acid add 23 ml distilled water and 25 ml 95% ethanol. Mix until all the TCAA has dissolved.

6.4 COLLECTION MEDIUM. Sodium bisulfite, 1% in water.

7. Procedure

7.1 COLLECTION OF SAMPLES. Two Greenberg-Smith impingers, each containing 10 ml of 1% of $NaHSO_3$ are connected in series with Tygon tubing. These are following by and connected to an empty impinger (for meter protection) and a dry test meter and a source of suction. During sampling the impingers are immersed in an ice bath. Sampling rate of 0.5 to 1.0 CFM should be maintained. Sampling duration will depend on the concentration of aldehydes in the sample stream; 15 min is adequate for most industrial effluents. After sampling is complete, the impingers are disconnected from the train, the inlet and outlet tubes are capped, and the impingers stored in an ice bath or at 6 C in a refrigerator until analyses are performed. Cold storage is necessary only if the acrolein determination cannot be performed within 4 hr of sampling.

7.2 ANALYSIS OF SAMPLES (each impinger is analyzed separately). If the second impinger contains more than 10% of the amount found in the first impinger, sample collection should be repeated for a shorter time period.

7.2.1 *Formaldehyde* (**1, 2**). Transfer a 2-ml aliquot of the absorbing solution to a 25-ml graduated tube. Add 0.2 ml chromotropic acid, and then, cautiously, 5.0 ml conc sulfuric acid. Mix well. Transfer to a boiling water bath and heat for 15 min. Cool the samples and add distilled water to the 10 ml mark. Cool, mix and transfer to a 16 mm cuvette, reading the transmittance at 580 nm. A blank containing 2 ml of 1% sodium bisulfite should be run along with the samples and used for 100% T setting. From a standard curve read micrograms of formaldehyde.

7.2.2 *Acrolein* (**1, 3**). To a 25 ml graduated tube add an aliquot of the collected sample in bisulfite containing no more than 30 μg acrolein. Add 1% sodium bisulfite (if necessary) to a volume of 4.0 ml. Add 1.0 ml of the $HgCl_2$ 4-hexylresorcinol reagent and mix. Add 5.0 ml of TCAA reagent and mix again. Insert in a boiling water bath for 5 to 6 min, remove, and set aside until tubes reach room temperature. Centrifuge samples at 1500 rpm for 5 min to clear slight turbidity. One hr after heating, read in a spectrophotometer at 605 nm against a bisulfite blank prepared in the same fashion as the samples.

7.3 $C_2 - C_5$ ALDEHYDES (**1**).

7.3.1 *Analytical column.* 12′ × 1/8″ stainless steel packed with 15% w/w Carbowax 20 M on Chromosorb, 60 to 80 mesh, followed by 5′ × 1/8″ stainless steel tube of Ucon-dinonylphthalate on firebrick, 100 to 200 mesh, prepared as follows: Ucon 50-HB-200, 1.5 g, and 1.4 g of dinonylphthalate are dissolved in chloroform and added to 13 g of firebrick. The solvent is evaporated at room temperature and the column packed in the usual manner.

7.3.2 *Injection port sleeve.* The inlet of the injection port contains a glass sleeve packed with solid Na_2CO_3. The Na_2CO_3 is held in place with glass wool plugs.

7.3.3 *Conditions.*

Injection port temperature, 160 to 170 C
Column temperature, 105 C
Detector temperature, 200 C
Nitrogen carrier gas flow rate, 14 ml/min
Hydrogen flow rate, 20 ml/min
Combustion air flow rate, 400 ml/min

7.3.4 *Procedure.* A 4 μl sample of the bisulfite collection solution is injected into the packed sleeve at the injection port and the chromatogram is recorded. Table 122:I shows the retention times for a series of aldehydes and ketones in the $C_2 - C_5$ range.

8. Calibration

8.1 FORMALDEHYDE.

8.1.1 *Preparation of standard curve.* To a 1-liter volumetric flask add 0.4466 g sodium formaldehyde bisulfite and dilute to volume. This solution contains 0.1 mg formaldehyde per ml. Dilute to obtain standard solutions containing 1, 3, 5 and 7 μg formaldehyde per ml. Treat 2-ml aliquots as described in the procedure for color development. Read each at 580 nm after setting instrument at 100% T with the blank. Using semilog paper, graph the respective concentrations *vs*. transmittance.

8.2 ACROLEIN.

8.2.1 *Preparation of standard curve.* To 250 ml of 1% sodium bisulfite add 4.0 μl freshly distilled acrolein. This yields a standard containing 13.4 μg/ml. To a series of tubes add 0.5, 1.0, 1.5, and 2.0 ml of standard. Adjust the volumes to 4.0 ml with 1% bisulfite and develop color as described above. Plot data on semi-log paper.

TABLE 122:I. Retention Times for Aldehydes, Ketones, Alcohols, and Esters*

Compound	Retention Time, min
Acetaldehyde	3.5
Propionaldehyde	4.6
Acetone	5.1
Isobutyraldehyde	5.5
Methyl alcohol	6.1
Ethyl alcohol	6.7
Isopropyl alcohol	6.7
Ethyl acetate	7.0
n-Butyraldehyde	7.1
Butanone	7.7
Isopentanal	9.2
n-Pentanal	12.0
Crotonaldehyde	14.0

*Flow rate, temperature and conditions described in text.

8.3 $C_2 - C_5$ ALDEHYDES.

8.3.1 *Calibration.* A mixed standard of $C_2 - C_5$ aldehydes and ketones is prepared as follows:

a. Acetaldehyde-bisulfite solution: 0.336 g $CH_3CHO \cdot NaHSO_3$ (EK 791) is dissolved in 1 liter of 1% $NaHSO_3$. This gives a solution containing 100 μg/ml acetaldehyde.

b. To 10.0 ml of the above solution are added 40.0 ml of 1% $NaHSO_3$, and 8 μl of a mixture of equal volumes of propanal, isobutanal, butanal, isopentanal, pentanal, crotonaldehyde, acetone and butanone.

The final solution contains 20 μg/ml acetaldehyde and 0.02 μl of each of the $C_2 - C_5$ aldehydes and ketones per ml. Four μl of the standard are injected into the glass sleeve in the injection port of the chromatograph as described in the procedure, and the chromatogram is recorded.

9. Calculations

(1.23 μg formaldehyde = 1 μl (vol) at 25 C and 760 Torr)

9.1 FORMALDEHYDE.

ppm formaldehyde (CH_2O)

$$= \frac{\text{total micrograms of } CH_2O \text{ in sample}}{1.23 \times \text{sample volume in liters}}$$

9.2 ACROLEIN.

(2.3 μg acrolein = 1.0 μl (vol acrolein)

$$\text{ppm} = \frac{\text{total } \mu g \text{ of acrolein in sample}}{2.3 = \text{sample volume in liters}}$$

9.3 ALDEHYDES. Calculation of unknown sample concentration is made on the basis of comparative peak heights between standards and unknowns.

10. Effect of Storage

10.1 After sampling is complete, collection media are stored in an ice bath or refrigerator at 6 C. Cold storage is necessary only if acrolein is to be determined. Under cold storage conditions, analyses can be performed within 48 hr with no deterioration of collected samples.

11. References

1. LEVAGGI, D.A., and M. FELDSTEIN. 1970. The Determination of Formaldehyde, Acrolein and Low Molecular Weight Aldehydes in Industrial Emissions on a Single Collection Sample. *JAPCA* 20:312.
2. AMERICAN PUBLIC HEALTH ASSOCIATION. 1977. Methods of Air Sampling and Analysis. 2nd ed. p. 297.
3. Ibid. p. 297.

Subcommittee 4

R. G. SMITH, *Chairman*
M. FELDSTEIN
D. C. LOCKE
P. O. WARNER

123.

Tentative Method for Determination of Benzaldehyde (Aromatic Aldehydes) in Auto Exhaust or Other Sources

1. Principle of the Method

1.1 Benzaldehyde, together with other aromatic aldehydes, is collected in a hydrochloric acid solution of 2,4-dinitrophenyl hydrazine. Collection efficiency is about 100% with the use of two impingers in series and an icewater bath to reduce the temperature of collection. Precipitate formed is dried under a stream of nitrogen, dissolved in benzene and injected into a gas chromatograph.

1.2 The chromatograph is calibrated using known concentrations of benzaldehyde (or other aromatic aldehydes) in benzene.

2. Range and Sensitivity

2.1 The estimated detection limit is 10 ppm of benzaldehyde or other aromatic aldehydes, with a range of 10 to 1000 ppm.

3. Interferences

3.1 Hydrocarbons ordinarily found in combustion sources do not interfere.

3.2 Oxygenated compounds ordinarily present in ambient air or auto exhaust have not been found to interfere.

4. Precision and Accuracy

4.1 Ten % relative deviation has been found between replicate analysis on standards where peak height is used as a measure of benzaldehyde present.

4.2 Variation in collection temperature and reentrainment losses can contribute to sampling error.

5. Apparatus

5.1 GAS CHROMATOGRAPHY. A gas chromatograph equipped with dual columns and flame ionization detectors is suitable.

5.1.1 *GC columns.* A 4-meter, 0.32 cm OD, 0.24 cm ID, stainless steel tube packed with 75% V-17 (methyl phenyl silicone) on 100 to 120 mesh Gas Chrom Q. (Column is prepared according to methods of Air Sampling and Analysis, Intersociety Committee, p. 92, American Public Health Assoc., 1977)

5.1.2 GASES. F.I.D. gases—air pressure and H_2 pressure optimized for each detector. Carrier gas—Helium 30 cc/mm.

5.1.3 TEMPERATURES. Oven temperature 140 to 270 C programmed to rise at 8 C/min. Detector manifold temperature 300 C.

5.2 SAMPLING.

5.2.1 *Pump.* Any vacuum pump capable of providing a constant flow of 0.5 to 2.5 liters per min.

5.2.2 IMPINGERS **(2)**, Greenberg-Smith with orifice type sparging tubes.

6. Reagents

6.1 All reagents must be ACS Reagent Grade or better.

6.2 CHROMATOGRAPHIC.

6.2.1 Calibration standards are prepared from reagent grade benzaldehyde and purified air.

6.2.2 Carrier gas, helium should be best grade available, passed through molecular sieve to remove traces of water and organic vapors.

6.3 ABSORPTION.

6.3.1 *Reagent solution.* 2,4-dinitrophenyl hydrazine (DNPH), saturated at 0 C.

6.3.2 *Preparation.* To a 1-l volumetric flask, add 163 ml conc HCl and 2.5 g DNPH crystals; this amount of DNPH yields 1 l of saturated DNPH at 0 C. When conversion to the hydrochloride is complete, orange color changes to yellow; add distilled water to 1-l mark. Stopper flask and agitate by ultrasonic generation to speed solution of solids. Store 5 to 10 C up to 1 week.

6.4 SOLVENT.

6.4.1 *Preparation*. Traces of carbonyls must be removed from pentane and methylene chloride which are used as hydrazone solvents.

6.4.2 Pentane is purified by vigorous agitation with conc sulfuric acid for 3 to 5 min. Immediately after separation of the acid portion, shake pentane 3 to 5 min with solid sodium bicarbonate and decant.

6.4.3 If methylene chloride other than C.P. is used, purify this solvent of carbonyls by the method of Lappin and Clark **(1)**.

7. Procedure

7.1 SAMPLE COLLECTION PROCEDURE.

7.1.1 Pipet 40 ml of reagent solution into 1st scrubber and 10 ml into 2nd scrubber.

7.1.2 Assemble the scrubbers and place each into a vacuum insulated flask filled with ice and water.

7.1.3 Connect the scrubbers in series to the manifold positioning them in such a manner that the exhaust gas or source gas mixture to be sampled passes first through the scrubber containing 40 ml of reagent solution.

7.2 RECOVERY OF CARBONYL-DNPH DERIVATIVES.

7.2.1 Attach the side arm of an assembled vacuum filter flask and funnel to the vacuum line and begin evacuation.

7.2.2 Remove the scrubbers from the ice bath. Partially remove the sparging tube assembly until the stem is above the liquid and wash the loose precipitate and reagent from both the internal and external surfaces of the stem with a few milliliters of distilled water. Allow the excess water to drip from the stem and remove the sparging tube assembly from the scrubber.

7.2.3 Transfer the contents of the scrubber to the Buchner funnel; wash the sides of the scrubber with a few milliliters of distilled water; transfer the wash to the filter and replace the sparging tube assembly.

7.2.4 The second scrubber is treated in the same manner as the first; the solids of both scrubbers are recovered on the same filter.

7.2.5 Repeat the washing in steps 2 and 3 until the filtrate is nearly colorless (three washes are generally sufficient), then drain as much of the water as possible from the bubblers.

7.2.6 When the water wash has completely passed through the filter, release the vacuum and remove the funnel from the flask.

7.2.7 Wash the sparging tubes and scrubbers thoroughly (until the wash is colorless) but with minimal volumes of carbonyl-free methylene chloride, and quantitatively transfer the washings to the filter.

7.2.8 Swirl the solvent in the funnel to aid solution, then, turn the vacuum on just enough to draw the solution into the flask (a hard vacuum may freeze the residual water in the fritted glass disc forming a plug).

7.2.9 Wash the sides of the funnel and the glass disc with small volumes of methylene choloride until the solvent passing through the filter is colorless and all traces of color are gone from the filter funnel, then, turn off the vacuum and remove the funnel while washing the stem.

7.2.10 Transfer the contents of the filter flask to the 50-ml Erlenmeyer flask, completing the transfer with incremental methylene chloride washes

7.2.11 Label the Erlenmeyer flask and place it under a stream of nitrogen from the manifold to evaporate the solvent.

7.2.12 For chromatographic analysis, dissolve the carbonyl-DNPH derivatives from step 7.2.11 in a measured volume—as small as possible—of benzene.

7.3 GAS CHROMATOGRAPHIC PROCEDURE.

7.3.1 Injection is accomplished by turning off oven power and cooling the oven to below the initial program starting temperature of 130 C. The chromatographic column is disconnected and the solution sample is injected directly into the column approximately 3/4 of an inch from the entrance point. A sample size of 4 μl is normally used for analysis. Upon connecting the column, oven power is turned on and column temperature is programmed from 130 C to 290 C at 8 C/min.

7.3.2 A detector temperature of 200 C is used with flows of approx. 50 cc/min. of hydrogen and 50 cc/min nitrogen diluent.

8. Calibration and Standards

8.1 BENZALDEHYDE.

8.1.1 A standard curve of peak area or number of integrator counts versus ppm benzaldehyde concentration must be prepared using CP benzaldehyde.

8.1.2 Synthesize pure DNPH derivative of benzaldehyde using MP 242-243 C to check purity of compound.

8.2 DETERMINATION OF OTHER CARBONYLS.

8.2.1 Identification of individual carbonyl hydrazones in air samples is based on retention times alone.

8.2.2 Use anthracene as internal standard and prepare standard solution by dissolving anthracene in benzene to a concentration of 0.25 mg/ml. Use standard solution as a solvent for chromatographic analysis of carbonyl-DNPH derivatives.

8.2.3 Synthesize pure DNPH derivatives of carbonyls, such as those listed in Table 123:I, and check purity from melting points (**2**). Prepare solutions containing mixtures of carbonyl-DNPH derivatives at concentrations which approximate those resulting from normal size samples (**3**). For calibration purposes, prepare one standard mixture with total hydrazone concentration equal to from 4 to 6 mg/ml. A typical synthetic derivative mixture can be prepared by the data of Table 123:II.

8.2.4 Prior to chromatographic analysis, condition the column with a single 4 μl sample of a hydrazone calibration blend. Repeat conditioning process whenever samples are not analyzed for a period of an hour or more.

8.2.5 Chromatograph blends using standard conditions and from digitized areas calculate relative response factors (F) for the individual hydrazones.

TABLE 123:I Melting Points of Carbonyl-DNPH Derivatives

Carbonyl Compound	Melting Point °C Literature (1)	Melting Point °C Observed
Formaldehyde	166	163.2–165.2
Acetaldehyde	147–168	149.1–150.8
Acetone	126	124.7–126.2
Propionaldehyde	154	146.7–149.2
Acrolein	165	162.1–163.3
Isobutyraldehyde	182	178–179
n-Butyraldehyde	122	120.3–121.8
n-Valeraldehyde	106	104.8–105.8
Crotonaldehyde	190	187.5–188.5
n-Hexanaldehyde	104	104.0–105.0
Benzaldehyde	237	242–243
m-Tolualdehyde	194	209–210
Salicylaldehyde	252	

TABLE 123:II. Composition of Standard Mixture of DNPH Derivatives Appropriate for Calibration of GC System

DNPH Derivative	No. of Determinations	Conc mg/m	F Factors
Formaldehyde	11	2.238	2.79
Acetaldehyde	11	.265	2.38
Acetone	11	.105	1.96
Acrolein	11	.110	2.82
n-Butyraldehyde	11	.190	2.29
Crotonaldehyde	11	.038	3.08
n-Valeraldehyde	11	.033	
Benzaldehyde	11	.265	2.16
o-Tolualdehyde	11	.155	

$$F = \frac{\text{Peak area anthracene}}{\text{Weight anthracene (mg/ml)}} \times \frac{\text{Weight hydrazone (mg/ml)}}{\text{Peak area hydrazone}}$$

8.2.6 Typical values of response factors (F) for a number of hydrazone derivatives are listed in Table 123:II.

9. Calculation

9.1 Peak areas are measured by peak height multiplied by peak width at one-half the peak height or by summation of integrator counts. Either of these values is plotted versus ppm of the individual component to establish the calibration.

9.2 Benzaldehyde calculation.

ppm benzaldehyde

$$= \frac{\text{weight of DNPH of benzaldehyde} \times 10^3 \times T}{\text{MW of DNPH of benzaldehyde} \times V_s \times P \times 0.01604} \quad (2)$$

where T = ° K and V_s = volume in liters of air sample P = barometric pressure in mm Hg. 0.01604 = gravimetric constant. Calculate individual aromatic aldehyde. Component concentrations in the chromatographed sample are obtained from peak area data using equation (3)

Concentration DNPH (mg/ml)

$$= \frac{F \times \text{weight anthracene (mg/ml)} \times \text{peak area DNPH}}{\text{Peak area anthracene}} \quad (3)$$

Calculate weights for each of the identified derivatives in the air sample using the predetermined (F) values and total solvent volume.

Calculate individual carbonyl concentration in the gaseous samples using equation (2).

10. Effect of Storage

10.1 The reagent solution of DNPH may be stored at 5 to 10 C for not longer than one week.

11. References

1. LAPPIN, S.R. and L.C. CLARK. 1951. Colorimetric Method for Determination of Traces of Carbonyl Compounds, *Anal. Chem.*, 23:541.
2. SHRINER, R.L. and R.C. FUSON. 1948. Identification of Organic Compounds, John Wiley and Sons, Third Edition, New York, N.Y.
3. OBERDORFER, P.E. 1967. The Determination of Aldehydes in Automobile Exhaust Gas, Society of Automotive Engineers Paper No. 670123, Jan.

Subcommittee 4

R. G. SMITH, *Chairman*
R. J. BRYAN
M. FELDSTEIN
D. C. LOCKE
P. O. WARNER

124.

Tentative Method of Analysis for Primary and Secondary Amines in the Atmosphere (Ninhydrin Method)

43724-01-73T

1. Principle of the Method

1.1 Amines in the atmosphere are collected by bubbling a measured volume of air through an acidified isopropanol mixture.

1.2 The absorbing solution is reacted with ninhydrin (1,2,3-triketohydrindene). Primary and secondary amines produce "Ruhemann's purple" (**1**). Tertiary amines do not react. The following reaction occurs, omitting the intermediate steps (**2**):

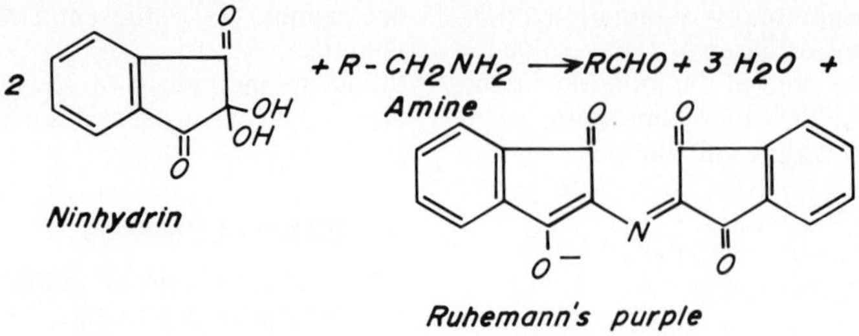

1.3 The amines collected are expressed as an equivalent molar amount of n-butylamine.

2. Range and Sensitivity

2.1 When using 10 ml of absorbing solution, a 30 liter sample will provide a limit of detection equivalent to 300 $\mu g/m^3$ (0.1 ppm) of butylamine vapor (**1**).

2.2 The working range is 0.15 to 2 μMol/10 ml absorbing solution, which corresponds to absorbances between 0.1 and 1.0 when using a one-cm cell at 575 nm.

3. Interferences

Under conditions of this test, ammonia and alkaline substances other than primary and secondary amines do not interfere. Some aminoacids interfere on a molar basis.

4. Precision and Accuracy

4.1 Precision and accuracy data are not available.

4.2 ABSORPTION EFFICIENCY. Using 30 ml of absorbing solution in each of two gas washing bottles in series, recovery of butylamine was quantitative in the first bottle in concentrations of 16 and 300 mg/m^3 in air (**1**). No data is available for lower concentrations in air.

5. Appratus

5.1 VACUUM PUMP. Any explosion-proof pump that will maintain an air flow of 2 l/min during the required sampling time is suitable. If a critical orifice is used as a flow control device, the pump must be also capable of maintaining a minimum vacuum of 500 Torr across the orifice at the required flow rate of 2 l per min.

5.2 FLOW MEASURING DEVICE. A calibrated flow meter or critical orifice to measure or control the air flow from 1 to 2 l per min. Either measuring devices should be calibrated under conditions of use.

5.3 ABSORBER. A gas washing bottle or midget impinger with fritted gas dispersing tube (**3, 4**) can be used if it is designed for operating at a flow rate of 1 to 2 l per min with 10 to 50 ml of absorbing solution.

5.4 SPECTROPHOTOMETER. The instrument used must be capable of measuring the absorbance at 575 nm. One-cm cells are suitable, although longer path lengths may be used.

5.5 WATER BATH. The bath must be capable of holding the temperature of the stoppered test tubes at 85 ± 1 C.

5.6 TEST TUBES. The tubes for color development should have a ground glass closure and a minimum capacity of 20 ml. Glass stoppered 25-ml graduated cylinders can also be used.

6. Reagents

6.1 All reagents should conform to the ACS specifications (**5**). Reagent water should conform to ISC standards (**6**).

6.2 NINHYDRIN REAGENT. Prepare a 0.2% (w/w) solution of ninhydrin (1,2,3 triketohydrindene) in isopropanol (0.16 g dissolved in 100 ml isopropanol). This reagent is stable for 2 weeks if kept in a brown bottle.

6.3 ABSORBING SOLUTION. Dilute one

volume of conc hydrochloric acid to 100 volumes with isopropanol.

6.4 REAGENT GRADE PYRIDINE.

6.5 N-BUTYLAMINE STOCK SOLUTION (500 μG/ML). Weigh 0.250 g of n-butylamine in a weighing bottle and dilute to 500 ml with absorbing solution.

7. Procedure

7.1 SAMPLE COLLECTION. 10 ml or more of the absorbing solution are placed in each gas scrubber for samples and field blank. Cap the scrubbers for transport. At the site assemble in order: gas scrubber, membrane filter, flow meter, needle valve or critical orifice and pump. Sample at the rate of 1 to 2 l per min for about 30 min. Record sampling time and flow rate. After sample collection, recap the gas scrubbers. Longer sampling time may be necessary for increased accuracy and can be established by a previous trial.

7.2 ANALYTICAL PROCEDURE. Make up the volume of sampling solution with isopropanol to compensate for evaporative losses. Transfer a measured portion from each sample (usually 3.0 ml or less) to a numbered glass-stoppered test tube. Run a blank for each series of samples by using the same volume of the absorbing solution.

7.3 To each test tube add 5 ml pyridine, 2 ml ninhydrin solution, and make up to 10 ml with absorbing solution. Stopper the test tubes, thoroughly mix and place in the water bath at 85 C for 7 min. **Caution!** Avoid contact of the solution with the skin and operate under a hood. Remove the tubes from the water bath and immerse them in cold water (<25 C) for 10 min. Transfer the solution to measuring cells and read the absorbance against that of the blank in a spectrophotometer set at 575 nm.

8. Calibration and Standardization

8.1 N-BUTYLAMINE CALIBRATING SOLUTIONS. Prepare a series of standards containing 2 to 100 μg/ml n-butylamine by diluting 0.4 to 20 ml of n-butylamine stock solution to 100 ml with absorbing solution.

8.2 Prepare a standard curve by using 1 ml portions of the calibrating solutions in the analytical procedure starting at 7.3.

8.3 Plot the absorbance of the calibrating solution readings as the ordinate versus the concentration as the abscissa on linear graph paper.

8.4 Absorbance values of reaction products of other alkylamines with ninhydrin follow Beer's law approximately on a molar basis.

9. Calculations

9.1 The concentration of amine in the original air sample is calculated as follows:

$$\text{Concentration (mg/m}^3) = \frac{M(\text{mg RNH}_2)\ F(\text{ml/ml})}{Q(\text{l/min})\ (10^{-3}\text{m}^3/\text{l})\ t(\text{min})}$$

where:

M = total mg of amine in the 10 ml colored solution

F = the ratio of the total volume of the sampling solution to the volume of the portion used for analysis, in milliliters. (The inverse of the aliquot, $F \geq 1$).

Q = air sampling rate in liters per min corrected to 25 C and 760 Torr.

t = sampling time in min.

9.2 Conversion of mg/m³ to ppm (v/v).

$$\text{ppm (v/v)} = \text{Concentration (mg/m}^3) \times \frac{24.45}{MW}$$

where:

24.45 = molar volume of an ideal gas at 25 C and 760 Torr. (liters)

MW = molecular weight of amine (n-butylamine = 73).

9.3 These equations are applicable primarily where the amine identity is known. Where the identity is unknown, the values are most conveniently expressed as an equivalent molar amount of n-butylamine.

10. Effect of Storage

10.1 The amine-hydrochloride solution is stable over a period of 3 weeks **(1)**.

11. References

1. SCHERBERGER, R.F., F.A. MILLER, and D.W. FASSETT, 1960. The determination of butylamine in air. *Am. Ind. Hyg. Assoc. J.* 21:471.
2. MCCALDIN, D.J. 1960. The chemistry of ninhydrin. *Chem. Reviews* 60:39.
3. FEDERAL REGISTER 1971. 36(84): part II, 8201.
4. AMERICAN PUBLIC HEALTH ASSOCIATION. 1977. Methods of Air Sampling and Analysis. 2nd ed. p. 527. Washington, D.C. 20036.
5. REAGENT CHEMICALS, American Chemical Society Specifications 1966. American Chemical Society, Washington, D.C.
6. AMERICAN PUBLIC HEALTH ASSOCIATION. 1977. Methods of Air Sampling and Analysis. 2nd ed. p. 53. Washington, D.C. 20036.

Subcommittee 3
E. L. KOTHNY, *Chairman*
W. A. COOK
J. E. CUDDEBACK
B. DIMITRIADES
E. F. FERRAND
R. G. KLING
P. W. MCDANIEL
G. D. NIFONG
F. T. WEISS

125.

Tentative Method for Preparation of Carbon Monoxide Standard Mixtures

42101-01-69T

1. Principle of the Method

1.1 Carbon monoxide standards are prepared by the volumetric dilution technique and calibrated by oxidation to carbon dioxide by the catalyst Hopcalite. The CO_2 formed is measured gravimetrically by absorption on Ascarite. **(1, 2)**

2. Range and Sensitivity

2.1 A minimum of 10 mg of CO_2 is required to achieve adequate weighing accuracy. Greater sensitivity is achieved by weighing greater quantities of CO_2 derived from larger sample volumes. As a matter of practice, it is generally impracticable to pass more than 60 liters of sample through the train. At 500 ml/min, this would require 2 hr for completion of the analysis. Using a 60-l sample as the required volume, a 100 ppm sample would yield 10.8 mg CO_2. For higher CO concentrations, smaller volumes of sample can be taken.

3. Interferences

3.1 Using pure gases and CO-free dilution air there are no interfering chemical compounds in the calibration procedure for CO.

4. Precision and Accuracy

4.1 Replicate analyses of CO standards which yield a minimum of 10 mg CO_2 on oxidation agree to within ± 2.5%. Sufficient volume of sample to yield a minimum of 10 mg CO_2 should be taken for analysis.

5. Apparatus

5.1 STAINLESS STEEL TANKS, 34-liter capacity (calibrated).

Note: 10 ppm CO in air by volume = 11.45 mg/M^3 at 25 C and standard pressure.

5.2 COMPRESSOR.

5.3 HYPODERMIC NEEDLES, 24-gauge.

5.4 RUBBER SERUM BOTTLE CAPS. #2330 A. H. Thomas Co., Phila., Pa., or similar material from any other suitable source.

5.5 STAINLESS STEEL STOPCOCKS. Tomac #44251, American Hospital Supply Corp., Evanston, Illinois, or similar material from any other suitable source.

5.6 PRESSURE GAUGES, 0 to 100 Psig., 3½″ dial. #63-4112, Matheson Co. or similar material from any other suitable source.

5.7 PRESSURE GAUGE, 0 to 100 Psig. High accuracy, #63-5212, Matheson Co. or similar material from any other suitable source.

6. Reagents

6.1 HOPCALITE, ACTIVE. Mine Safety Appliances, 14 to 20 mesh, or similar material from any other suitable source.

6.2 ASCARITE. A. H. Thomas Co., 8 to 20 mesh, or similar material from any other suitable source.

6.3 CARBON MONOXIDE, C. P.

6.4 NITROGEN WATER PUMPED, C. P.

6.5 AIR, free from CO.

7. Procedure

7.1 An aliquot of pure CO is delivered to the inlet of an evacuated, stainless steel tank of 34-l capacity by means of standard Luerlok syringe and 24-gauge needle. The gas is permitted to be drawn into the tank from the syringe via a rubber serum bottle cap connected to the outlet of the tank. After the desired quantity of gas has entered the tank, it is filled with air free of CO to atmospheric pressure and pressurized to a final measured pressure, depending on the dilution required. For example, a 34-l tank containing 100 ml CO and pressurized to four atmospheres with air has a calculated CO concentration of 588 ppm (100 ml CO in 170 l of air).

7.2 A few precautions should be observed in the preparation of standards by this technique. A minimum of 4 to 5 hr should elapse before the standard is used, in order to allow complete mixing of the CO to take place in the tank. Care should be taken to determine that the syringes used are airtight, and that the volume is accurate. Finally, the pressure in the tank should be measured with an accurate pressure gauge.

7.3 Samples of the standard gas mixture are easily removed from the tank by means of a syringe and needle inserted through the serum bottle cap on the fitting of the tank. The valve is opened slightly, and the sample is forced into the syringe by the pressure in the tank. It is an added convenience to have the needle mounted on the syringe via a stainless steel stopcock, so that the sample can be "locked" in the syringe until it is ready for use.

8. Calibration and Standards

8.1 Standards in the range of a few ppm up to a few per cent can be prepared in this manner. Experience has shown that, using care, standards can be prepared with an error of less than 5%. Absolute analysis of such standards can be accomplished by oxidation to CO_2, with Hopcalite and absorption, and weighing of the CO_2 on Ascarite.

8.2 A sampling train is set up as shown in Figure 125:1. The train consists of a Drierite trap to remove moisture, an Ascarite trap to remove CO_2, a heated (100 C) Hopcalite trap to oxidize CO, a Drierite trap to remove moisture, a final Ascarite trap to collect and weigh the CO_2 formed in the oxidation, and a wet test meter. The final Ascarite trap is made in a borosilicate tube with spherical glass joints for ease in connecting and removing from the sampling train.

8.3 Traps are constructed of borosilicate glass of 7 to 10 mm ID and approximately 15 cm in length. The traps are filled with appropriate material and plugged at each end with glass wool. For convenience, all traps should be fitted with spherical joints 19/38 for ease in assembling and disassembling.

8.4 Before an analysis, the final Ascarite trap is removed and the system flushed out with several hundred liters of

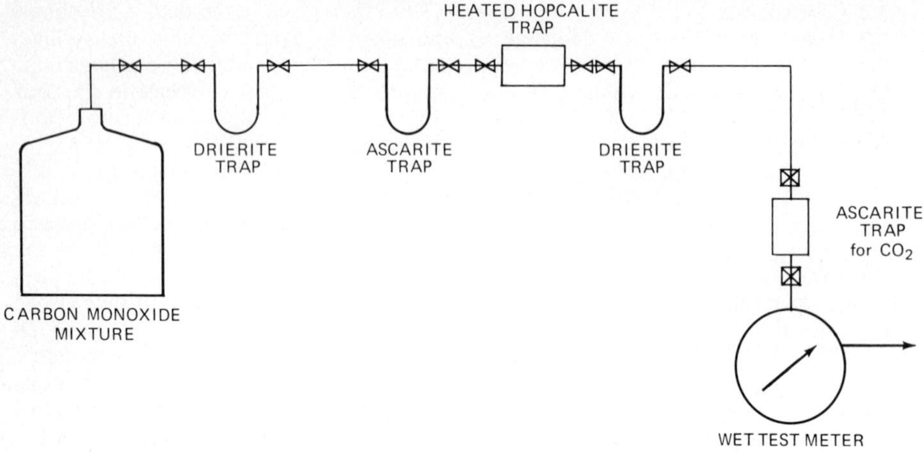

Figure 125:1.—Gravimetric determination of carbon monoxide.

nitrogen, with the Hopcalite at 100 C. It is important that all moisture be removed from the Hopcalite for efficient oxidation of CO. The Ascarite tube is carefully weighed and connected to the train. The sample is sent through the system at a rate of 500 ml/min. The volume of sample required depends on the concentration of CO in the sample. Best results are obtained when the weight gain on the Ascarite tube is 10 mg or more CO_2. Thus for a 100-ppm CO sample about 60 liters are required.

8.5 After the sample has been collected, the Ascarite tube is again carefully weighed, and the CO concentration calculated from the weight gained and volume of sample used.

9. Calculation

(1 mg/liter at 25 C and 760 Torr pressure = 873 ppm CO)

$$\text{ppm CO} = \frac{\text{mg } CO_2 = \dfrac{28}{44}}{\text{Volume of sample (liters) (25 C, 760 Torr)}} \times 873$$

10. Effect of Storage

10.1 CO standards prepared in pressurized tanks, as described, show no change in concentration on standing.

11. References

1. Lamb, A.B., W.C. Gray, and J.C. Frazier. 1920 *Ind. Eng. Chem.* 12:213.
2. Feldstein, M. 1967. Progress in Chemical Toxicology. Vol. 3. p. 105–106. Ed. A. Stolman. *Academic Press*, New York.

Subcommittee 4
R. G. Smith, *Chairman*
R. J. Bryan
M. Feldstein
B. Levadie
F. A. Miller
E. R. Stephens
N. G. White

126.

Tentative Method of Analysis for Carbon Monoxide Content of the Atmosphere (Manual—Colorimetric Method)

42101-02-69T

1. Principle of the Method

1.1 Carbon monoxide reacts with an alkaline solution of the silver salt of p-sulfaminobenzoic acid to form a colloidal suspension of silver. The absorbance of the resulting suspension is read in a spectrophotometer, and is proportional to the carbon monoxide concentration (**1, 2**).

2. Range and Sensitivity

2.1 Range of concentration which can be measured is dependent on the wavelength and sample concentration used to measure absorbance. A low range of 0 to 20 μl CO/50 ml sample corresponding to 0 to 400 ppm is covered by the 425 nm wavelength, and 20 to 90 μl CO corresponding to 400 to 1800 ppm, by the 600 nm wavelength.

2.2 For the low range, the minimum detectable concentration is 5 ppm using a 50 ml sample; using a flask volume of 125 ml as the sample the minimum concentration is 2 ppm. Concentrations greater than 400 ppm should be read at the 600 nm wavelength. Higher concentrations than 1800 ppm can be measured by using smaller samples.

3. Interferences

3.1 Acetylene, olefins, H_2S and aldehydes in high concentration interfere with the procedure either by reducing the silver salt or precipitating Ag_2S. These are effectively removed by passing the sample through a silica gel filter containing $HgSO_4$.

4. Precision and Accuracy

4.1 When standard samples are analyzed with each series of unknowns, the method gives a repeatability of ± 5% of the mean.

5. Apparatus

5.1 WRIST ACTION SHAKER.

5.2 SPECTROPHOTOMETER. 400 to 625 nm wavelength range, nominal band width 20 nm or less.

5.3 1-CM CELLS, CALIBRATED.

5.4 5-ml, 10-ml, or 20-ml-gastight syringes, fitted with either fixed or luerlock needles (available from Hamilton Co., Whittier, Calif.).

5.5 125-ML ERLENMEYER FLASKS, 24/40 joint.

5.6 24/40 JOINT STOPCOCK ADAPTERS, Corning 9120.

5.7 RUBBER SERUM BOTTLE CAPS. #2330, A. H. Thomas Co., Phila., Pa.

5.8 The *filter* consists of a 7.5 cm × 7 mm OD glass tube containing $HgSO_4$ absorbed on silica gel. The tube is fitted with a 24-gauge hypodermic needle on one end, and a serum bottle rubber cap on the other for convenience in adding samples to the reaction flask. The absorbant is prepared as follows:

Two g HgO are dissolved in 50 ml H_2O containing 2 ml conc H_2SO_4. After solution is complete, 35 g Davison silica gel, 40 to 60 mesh, are added and the solution stirred for 10 min. The excess solution is decanted and the silica gel dried at 110 C in an oven for 5 to 6 hr. The absorbant is stored in a tightly closed bottle prior to use, and is stable for at least a year.

6. Reagents

6.1 P-SULFAMINOBENZOIC ACID ($C_6H_4(COOH)SO_2NH_2$). Available K & K Laboratories, 177-10 93rd Ave., Jamaica 33, New York, 0.1 M, 2 g reagent dissolved in 100 ml 0.1 M NaOH.

6.2 $AgNO_3$, C.P. 0.1 M.

6.3 NaOH, C.P. 1.0 M.

6.4 SILICA GEL, Davison, PA 400, 40 to 60 mesh.

6.5 HgO, C.P.

7. Procedure

7.1 To a 125-ml Erlenmeyer flask add 20 ml of the p-sulfaminobenzoic acid reagent and 20 ml of the $AgNO_3$ solution. Mix and add 10 ml 1.0 M NaOH with shaking. The resultant solution should be clear and colorless.

7.1.1 Alternatively, where many samples are to be analyzed, it is more convenient to make up the absorbing solution in ½ to 1 liter quantities, using 10 or 20 times the quantities mentioned above, in the same order. Store the absorbing solution in an amber bottle, out of direct light.

7.2 Place 10 ml of the solution in a 125-ml, 24/40 joint Erlenmeyer flask. Seal the flask with the stopcock adapter and evacuate until the solution begins to bubble. Close the stopcock. Fit the rubber serum bottle cap over the exit tube of the stopcock and introduce the gas sample by means of a gastight syringe with the needle inserted through the cap. The sample is drawn into the evacuated flask by simply opening the stopcock. (If interferences are present, introduce the sample via the $HgSO_4$-silica gel trap.)

7.3 Bring the pressure in the flask to atmospheric by opening the stopcock in a well-ventilated hood free from CO. Immediately close the stopcock and shake the flask on a wrist shaker for 2 hr.

7.4 Air samples are collected by opening the evacuated flask to the atmosphere to be tested, and filling to atmospheric pressure, or by transferring aliquots to the flask in gastight syringes as described above. If the entire flask is to be used to collect the air sample, the volume of the flask and its matched stopcock adapter must be determined by filling it with water and measuring the total water volume to fill it. In general this volume will not be the same for different flasks. If the flask is filled with room air sample, care should be taken to obtain the sample over a few seconds of time in order to have a representative sample.

7.5 A reagent blank and a series of standards spanning the expected range are prepared at the same time as the unknowns.

7.6 At the end of the 2 hr shaking period, which should be the same for all analyses to within ± 5 min, the samples and standards are read in a spectrophotometer, using 1-cm calibrated cells against the blank set at zero absorbance. The choice of wavelength is dependent on the concentration of CO as described in the calibration procedure.

7.7 Since the method is empirical, it is important that all conditions be the same for analysis of samples and standards. This includes time of shaking, temperature and interval of time between completion of shaking and measurement of absorbance. Samples should be measured as soon after shaking is completed as possible. However, since the reaction proceeds extremely slowly after the 2 hr of shaking, delays of 1 hr in reading absorbance can be tolerated, but the delay interval should be held constant to within ± 15 min. The analyses are frequently carried out in the open laboratory where considerable fluctuation of temperature can occur. Thermostating of the laboratory is apparently unnecessary, but large fluctuations, greater than 5 to 10 F, should be avoided, even when standards and blanks are run simultaneously with the samples Laboratory temperature should be 70 F ± 2.

7.8 Glassware must be scrupulously clean for best results. After use, or before absorbing solution is added to the flasks they should be rinsed in 1:3 nitric acid, followed by 1:1 NH_4OH and distilled water.

8. Calibration and Standards

8.1 Carbon monoxide standards of known concentration are prepared in 8- or 34-l stainless steel tanks. The tanks are evacuated, the appropriate volume of carbon monoxide added by means of a syringe, and the tank brought to atmospheric pressure with CO-free air or nitrogen. Experience has shown that with careful technique in filling syringes with pure CO and transferring to evacuated tanks, standards can be prepared to within ± 2% of the calculated value.

8.2 Five to 50 ml aliquots of the CO standard are removed from the tank by means of gastight syringes and analyzed as described above. Since the entire volume of gas in the syringes is drawn into the vacuum in the flasks, including the dead volume in the needles, the syringes must be calibrated by filling with water and measuring the volume of the water held. The solutions are read in the spectrophotometer. Absorbance is plotted against total μl CO. The resulting calibration curves are linear in the ranges indicated. Repeated calibrations show less than 3% variation between runs.

9. Calculation

9.1 The absorbances of the unknown solutions are read in the spectrophotometer and concentrations of CO calculated by reference to the calibration curve. Values are based on the volume of the CO standards or samples taken for analysis and the μl CO needed for the calibration curve:

$$\text{ppm CO} = \frac{\mu\text{l CO from curve}}{\text{Vol of sample (liters) (25 C, 760 Torr)}}$$

10. Effects of Storage

10.1 After samples have been added to the reaction flask, shaking should begin within 30 min. Since the reaction is relatively slow, shaking is necessary to increase surface contact between reagent and gas sample. Shaking time should be controlled accurately to precisely 2 hr, and absorbance measurements should be made within 1 hr after shaking time is completed.

11. References

1. CIUHANDU, G. 1957. Colorimetric Determination of Carbon Monoxide in Air. *Z. Anal. Chem.*, 155:321–327.
2. LEVAGGI, D.A., and M. FELDSTEIN. 1964. The Colorimetric Determination of Low Concentrations of Carbon Monoxide. *Am. Ind. Hyg. Assoc. J.*, 25:64–66.

Subcommittee 4

R. G. SMITH, *Chairman*
R. J. BRYAN
M. FELDSTEIN
B. LEVADIE
F. A. MILLER
E. R. STEPHENS
N. G. WHITE

127.

Tentative Method of Analysis for Carbon Monoxide Content of the Atmosphere (Infrared Absorption Method)

42101-03-69T

1. Principle of the Method

1.1 Carbon monoxide has absorption bands at 4.67μ (2143 cm^{-1}) and 4.72μ (2119 cm^{-1}) in the infrared region of the spectrum (**3**). The absorbance is measured at 4.67 μ in a 10-meter infrared gas cell (**1**). Absorbance is proportional to concentration of CO present.

2. Range and Sensitivity

2.1 The 10-meter path length infrared cell can measure accurately concentrations of CO in excess of 10 ppm at high energy and slow scan speeds at the 4.67 μ wavelength. The upper limit is dependent on the actual volume of sample introduced into the cell. For concentrations near 10 to 500 ppm a sample volume of 3.85 liters is required (the volume of the cell). For higher concentrations, proportionately smaller volumes are used and the cell brought to atmospheric pressure with CO-free air. (CO-free air can be obtained by using cylinder air known to be free of CO, or by allowing dilution air to enter the cell via a Hopcalite absorption trap connected to the inlet of the cell).

3. Interferences

3.1 There are a few known gases which absorb in the 4.6 to 4.7 μ region of the infrared spectrum. These gases include acetylene, aldehyde, cyanogen, diazomethane, hydrogen sulfide, nitrosylchloride, nitrous oxide, olefines and propyne. These gases are not usually present in significant concentration in normal atmospheres, but they can readily be ruled out by reference to the complete spectrum of the gas sample scanned from 2 to 14 microns. Each of the interfering gases has additional absorption lines in other areas of the spectrum which readily denote their presence. Carbon dioxide in concentrations as high as 3% does not interfere. Water vapor is eliminated by the use of drying agent traps on the inlet of the IR cell.

4. Precision and Accuracy

4.1 Replicate samples greater than 10 ppm can be analyzed with an error of ± 10 per cent. Care must be taken in measurement of sample volumes, pressures and volume of the infrared cell. For preparing known concentrations of carbon monoxide, refer to Intersociety Committee Method of CO-Standardization (**4**).

5. Apparatus

5.1 Infrared spectrometer. 2 to 15 micron range, wavelength accuracy ± 0.015μ, resolving power 0.01μ, wavelength reproducibility 0.005μ, % transmission accuracy ± 0.5%.

5.2 10-meter path length gas cell.

5.3 Manometer.

5.4 Vacuum pump. (capable of reducing pressure to 1 mm).

5.5 Rubber serum bottle caps. #2330, A. H. Thomas Co., Phila., Pa., or similar material from any other suitable source.

5.6 Glass tubes. 10 cm diam × 15 cm length.

Note: 10 ppm CO in air by volume = 11.45 mg/M^3 at 25 C and standard pressure.

5.7 NEEDLE VALUES.
5.8 SYRINGES.
5.9 HYPODERMIC NEEDLES. 24 gauge.

6. Reagents

6.1 ANHYDROUS CALCIUM CHLORIDE. 10–20 mesh, C.P.

6.2 HOPCALITE. Active MSA 8 to 14 mesh, or similar material from any other suitable source.

6.3 CARBON MONOXIDE, C.P.

7. Procedure

7.1 The 10-meter path length gas cell is connected to a manometer via a T connection, as shown in Figure 127:1. The cell is then evacuated to approximately 1.0 Torr, and the sample is transferred to the cell *via* a calcium chloride drying tube. When the sample container and the 10-meter cell are at equilibrium, the pressure is noted, (manometer equilibrium pressure, Pe_1), and the cell is then filled to atmospheric pressure with CO-free air. Samples may also be introduced into the evacuated cell by means of syringes and needles through a rubber serum bottle cap attached to the inlet of the cell. The pressure in the cell is then brought to atmospheric with dry air free of carbon monoxide and 15 min are allowed for equilibrium to be established.

7.2 The spectrum is scanned from 4 to 6 μ, and the absorbance at 4.67 μ is measured by the base-line technique **(2)**.

8. Calibration and Standards

8.1 The volume of the 10-meter cell is determined by standard techniques. The simplest procedure is to evacuate the cell, and then bring to atmospheric pressure by permitting air to enter via a wet test meter. The volume of the cell is the volume of air shown on the meter. After the volume has been ascertained, the cell is again evacuated, and known volumes of CO added with the aid of standard syringes, through a rub-

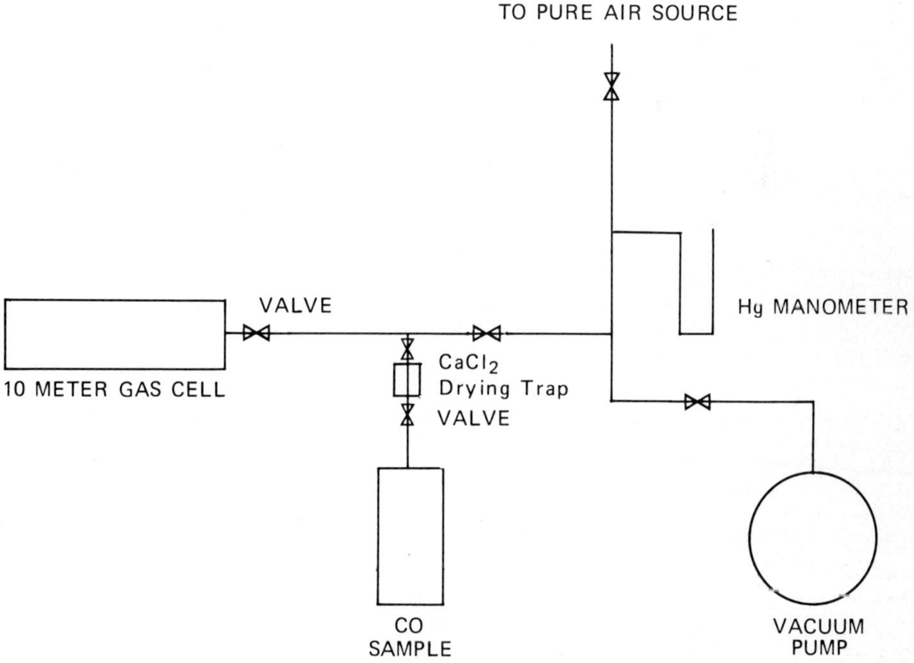

Figure 127:2—Introduction of gas samples into IR cells.

ber serum bottle cap attached to the inlet of the cell. The pressure in the cell is then brought to atmospheric with CO-free dry air. The absorbance at 4.67 μ is measured as described in the procedure. The CO concentration is calculated from the quantity of CO added, and the volume of the cell. For example, 1.0 ml CO in a 3.85-l volume cell gives a concentration of 260 ppm CO. A calibration curve relating absorbance and concentration is prepared for a series of known volumes of CO introduced into the TR cell.

9. Calculation

9.1 The concentration of the unknown is read from the calibration curve. The value obtained is corrected for the volume of sample actually introduced into the cell. For samples introduced into the cell with syringes, the volume is readily known and the correction applied. For samples introduced by pressure measurements, the volume is calculated as follows:

9.1.1 Volume of sample = volume of

$$\text{infrared cell} \times \frac{Pa - Pe}{Pa} \quad \text{where:}$$

Pe = equilibrium pressure after connecting the sample container to the IR cell.

Pa = atmospheric pressure

9.2 Sample concentration (ppm) = observed concentration from

$$\text{calibration curve} \times \frac{Vs}{Vc}$$

where

Vs = volume of sample

Vc = volume of IR cell

10. Effect of Storage

10.1 CO samples do not deteriorate on standing. If samples collected for analysis are under reduced pressure, care should be taken to see that dilution does not occur because of leaks in the container. Atmospheric samples can generally be stored at atmospheric pressure to avoid dilution.

11. References

1. AIR POLLUTION FOUNDATION. 1956. Analysis of Air Near Heavy Traffic Arteries. Los Angeles, California. Report No. 16.
2. HEIGL, J.J., M.F. BELL, and J.U. WHITE. 1947. *Anal. Chem.* 19:293.
3. HERTZBERG, W. 1950. Molecular Spectra and Molecular Structure. Vol. I., Daitomic Molecules. Van Nostrand, Princeton, N.J.
4. AMERICAN PUBLIC HEALTH ASSOCIATION, INC. 1977. *Methods of Air Sampling and Analysis.* 2nd ed. p. 342. Washington, D.C.

Subcommittee 4

R. G. SMITH, *Chairman*

R. J. BRYAN

M. FELDSTEIN

B. LEVADIE

F. A. MILLER

E. R. STEPHENS

N. G. WHITE

128.

Recommended Method of Continuous Analysis for Carbon Monoxide Content of the Atmosphere (Nondispersive Infrared Method)

42101-04-69T

1. Principle of the Method

1.1 Nondispersive infrared photometry provides a method for utilizing the integrated absorption of infrared energy over most of the spectrum for a given compound, to provide a quantitative determination of the concentration of that compound in a gas mixture. Specifically, the technique involves determining the difference in infrared energy absorption over all wave lengths passed by the optical system between a gas sample containing the compound of interest and a sealed reference sample consisting of an infrared transparent gas. The assumption is made that this difference in energy absorption is directly proportional to the concentration of the subject compound in the sample gas.

1.2 Detection of the energy difference is accomplished through absorption of the residual infrared energy by a mixture of the subject compound with an inert gas in a sealed detector cell. Only that energy is absorbed which is defined by the absorption spectrum of the compound. The energy absorbed is transformed to heat causing an alternate expansion and contraction of the gas in the detector cell. Means are provided for obtaining an electrical signal from this action (**1, 2, 3, 4**).

2. Range and Sensitivity

2.1 The range and sensitivity of the nondispersive infrared method for carbon monoxide analysis is dependent principally upon analyzer design. Even so, considerable flexibility in range can be obtained with commercially available analyzers. The range is principally determined by the sample cell length and the operating pressure used. This derives from the application of Beer's law governing the absorption of radiation by a solute in a non-radiation-absorbing solvent (for gases, the compound of interest is the solute, and the carrier gas is the solvent). In the integrated form, Beer's law is expressed as:

$I = I_0 \exp - KCl$, where

- I = transmitted radiation
- I_0 = incident radiation
- K = absorption coefficient
- C = molar concentration (for gases, volume concentration is equivalent to molar concentration)
- l = path length

2.2 Since the available signal from the detector is directly related to the transmitted radiation, I, it can be seen that change in range can be obtained either by a change in sample pressure or in sample cell length. Most analyzers for atmospheric range service are designed to provide full scale deflection at about 100 ppm (vol).

2.3 Sensitivity of the nondispersive infrared technique is determined principally by electrical and optical noise and, in addition, by characteristics and performance of the signal processing components. In most cases this is dependent upon amplifier gain and servo-loop performance when a null-balancing potentiometric recorder is used. Generally, the smallest concentration change which will result in a change in output display is 0.5% of full-

---

Note: 10 ppm CO in air by volume = 11.45 mg/M^3 at 25 C and standard pressure.

scale concentration (0.5 ppm when scale is 0 to 100 ppm).

3. Interferences

3.1 Interferences may arise from gases that absorb infrared radiation in bands that overlap that of carbon monoxide. Fortunately, this potential chance for error is relatively small for carbon monoxide. Being a heteronuclear diatomic gas, it absorbs infrared radiation but the pattern is simple. The single band involved has its absorption peak at 2165 cm^{-1} (4.6 μ). At this point the principal interference would be from water and carbon dioxide. Limited interference from methane and ethane might also be expected.

3.2 The major interference possibilities should be eliminated by use of an interference cell in line with the sample cell, containing the principal interferents in a concentration sufficient to block the radiation from the overlapping portion of this absorption spectra. Commercially available instruments incorporate this feature routinely. For such analyzers the ranges of discrimination ratios (discrimination ratio equals concentration of interferents required to give response equivalent to one ppm of carbon monoxide) for various possible interfering compounds are as follows:

CO_2	2000–5000:1
H_2O	4000–7000:1
CH_4	1000–5000:1
C_2H_6	150– 500:1

3.3 Based upon the above discrimination ratios, water interference is the principal concern. Application of climatological data from the U.S. Weather Bureau indicates that, without further precautions, carbon monoxide readings obtained in tropical air bordering the Gulf of Mexico might be as much as 10 ppm too high due to water interference. Over much of the United States, however, positive interference from water vapor would more likely be in the range of 3 to 5 ppm. Even less interference would be expected during the winter when air of polar land mass origin prevails.

3.4 A statement from the manufacturer should be obtained as to the specifications for discrimination against expected interferents. This can change should leaks occur in either the reference or detector cells.

3.5 Additional measures can be taken to minimize water interference by providing a means of drying the sample air or by saturating both the zero, span, and sample gas streams with water vapor at the same controlled temperature. The preferred procedure is to remove water by refrigerated dehumidification or by an indicating drying agent. Based upon an air sample rate of 0.5 l/min and an average dew point of 60 F, a drying column containing 400 g of silica gel will last about a week before regenerating or changing is required. Not more than 20% of the weight of silica gel should be used to judge the total water holding capacity. The column diameter should not exceed about two inches for best performance.

4. Precision and Accuracy

4.1 A great many factors can effect the expected precision and accuracy of measurements with this method. Certain of these are related to analysis uncertainties and others to instrument characteristics.

4.2 The principal analysis uncertainties affecting accuracy of measurement are the presence of unknown amounts of interfering gases in the sample and the accuracy of carbon monoxide assay in the span and zero gases. With respect to interferences it has been shown that the extent of water vapor interference may range from a negligible ($<$ 1.0 ppm) amount to as much as 10 ppm. This latter would be an extreme case. For most of the United States, the long-term mean interference would more likely be + 3 ppm for an analyzer not having water vapor removal. Interference from water vapor for an analyzer incorporating water removal or saturation should be $<$ 1 ppm.

4.3 Analysis errors due to inaccurate zero and span gas assay are more difficult to assess. Errors of ± 2 to 3 ppm are not uncommon with certified cylinder sources. Error in zero gas assay leads to greater relative inaccuracies in low concentration samples than does error in span gas assay. If possible, a true carbon-monoxide-free zero gas should be used.

4.4 Leaks in either the reference or detector cells will result in an unknown amount of absorption by carbon monoxide or interferents.

4.5 The assumption of linearity in scale preparation introduces an error because of the logarithmic nature of the absorption equation. For greatest accuracy a calibration curve should be prepared from experimental data using varying span gas concentrations. The nonlinearity effect will be accentuated if nonoptimum cell lengths are used considering the range of concentrations to be encountered. If a single concentration span gas is used for calibration, the value should be approximately two-thirds the maximum expected concentration. This will minimize the mean error for the concentration frequency distribution expected in ambient air.

4.6 Several factors affect both precision and accuracy. Of most concern is span and zero drift. These are instrument dependent. The combined effect may be ± 1% over an 8 hr period. Dirty cell windows or loose dirt in the sample cell may affect precision and accuracy. Out of phase signals due to optical system misalignment as well as those due to scattered radiation will contribute to random noise and loss in accuracy. Sampling at higher than design rate may increase sample cell pressure to the extent that high values result.

4.7 Signal noise from optical or electrical system may result from vibration, poor voltage regulation, or operation at too high amplifier gain. These factors contribute to loss in precision.

4.8 A formal analysis of errors involves too many assumptions for the general case in which all the above factors might operate. Under optimum conditions, however, for analyzers spanned at 100 ppm full scale, the accuracy in terms of reproducibility would not be expected to be better than ± 3 ppm.

4.9 Short term precision would be expected to be better, about ± 1 ppm.

5. Apparatus

5.1 A general description of instrument components is given because more than one make of analyzer is available commercially which will meet the method criteria.

5.2 The basic components of an analyzer include:

5.2.1 Sample handling system including flow regulator, particulate filter and/or water removal system, and flow meter. These components are generally quoted separately to meet user requirements.

5.2.2 Analyzer section, containing infrared sources, beam chopper assembly, sample, comparison and interference cells, and detector cell.

5.2.3 Amplifier and control section containing amplifier, demodulator, output filter, zero and span controls, and other internal controls used less frequently.

5.2.4 Signal display usually is a servo-recorder, but may be meter output.

5.3 In some cases all components are mounted upon a single chassis; in others each of the four major assemblies listed above are mounted separately. Figure 128:1 gives a schematic layout of all major components.

6. Reagents

6.1 Calibration Gas. A pressure cylinder equipped with a two-stage regulator and containing a mixture of carbon monoxide in nitrogen at a concentration approximately two thirds the full-scale range selected. The concentration should be known within ± 2 ppm (when full-scale = 100 ppm).

6.2 Zero Gas. A pressure cylinder equipped with a two-stage regulator and containing, if possible, carbon monoxide-free nitrogen. In any case, the carbon monoxide content should be no higher than 5 ppm, and known within ± 1 ppm (when full-scale = 100 ppm).

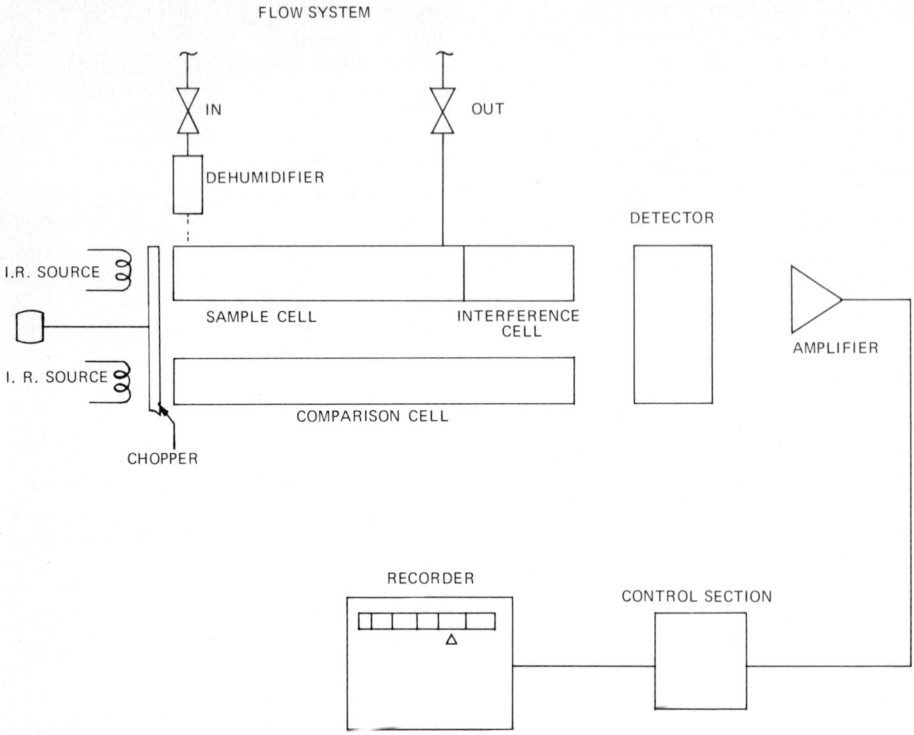

Figure 128:1—NDIR Schematic Diagram.

7. Procedure

7.1 Specific operating procedures are supplied by manufacturers of the commercially available analyzers. Certain general operating steps only, therefore, are given. In particular, a distinction is given between zeroing and standardizing operations as contrasted to actual calibration.

7.2 The first step common to all analyzers is to turn on the power supply and allow the recommended (or necessary) warm-up time before actual operation is begun using standard or sample gases. This time varies among analyzers and the length of time the analyzer has spent without power applied. It can range between 30 min and 3 hr. Under any circumstances warm-up time should be long enough to stabilize drift.

7.3 Several classifications of procedures are actually involved. First is preliminary or initial setup; second, routine operation; and third, trouble shooting. The first and third classifications should be carried out by a competent technician or engineer. They involve the use of internal adjustments, and often require special electronic instrumentation. Routine operations involve the use of external controls only and require relatively little training.

7.4 Typical initial or start-up procedures are setting of electrical zero, making coarse adjustments of infrared sources, oscillator tuning (if used), signal phase adjusting, amplifier gain adjusting, and initial span and zero adjusting.

7.5 Routine operations that might ordinarily be involved are described in the sequence below.

7.5.1 Turn power on.

7.5.2 Allow analyzer to warm up for time period earlier found to be necessary.

7.5.3 Introduce "zero" gas and adjust zero control to proper reading.

7.5.4 Introduce "span" gas and just span control to proper reading.

7.5.5 Check sample flow rate. This must not be in excess of rate found to cause pressure effects. Minimum rate is adjusted on the requirement for the resolution of fine sample structure.

7.5.6 Log all operations, control settings, adjustments, and the time of these operations.

7.6 Trouble shooting and maintenance are carried out both routinely and on demand. The principal routine servicing includes replacement of particle filter, water removal check and recorder servicing. Other servicing that can be expected includes sensitivity checks, analyzer signal output determination, infrared source checks, amplifer servicing, sample cell cleaning, phase adjustments, and servicing or changing detector. Recorder service requirements are the same as for any similar null-balancing instrument used for signal recording.

8. Calibration and Standards

8.1 As contrasted with standardization, calibration is defined as the verification by independent means of the carbon monoxide concentrations present in the "zero" and "span" gases. For convenience, interference checks are also included under this heading. Absolute reliance should not be placed in "certified analyses" of commercially available gases, unless there is opportunity for interlaboratory checking where at least one has the capability for conducting primary calibration checks. Calibration standards may be prepared by use of Intersociety Method **(5)** for CO-Standardization.

8.2 CALIBRATION PROCEDURE.

8.2.1 Using established operating procedure, introduce "zero" gas and adjust zero control to proper setting so recorder gives proper value. A pressure check is recommended to verify the concentration of carbon monoxide in the "zero gas". With the sample cell discharge valve closed, introduce the "zero gas" at atmospheric pressure. Set zero control at indicated carbon monoxide concentration on cylinder. Increase the pressure of the "zero gas" in the sample cell to twice atmospheric. If no carbon monoxide is present, there will be no indicated change in the recorder reading (assuming no significant interferences). If carbon monoxide is present the net increase in scale reading should be approximately equal to the carbon monoxide concentration in the "span gas".

8.2.2 Introduce "span gas" having a concentration of about two-thirds the maximum expected concentration. Adjust span control to obtain correct recorder value.

8.2.3 Introduce several (at least three) other known concentrations of carbon monoxide in nitrogen covering the full-scale range and note recorder values.

8.2.4 Prepare calibration chart, plotting value versus true value.

8.3 INTERFERENCE CHECK. A test should be made for determining the response factor of the analyzer to water vapor. This is accomplished by sampling an atmosphere containing a known amount of water vapor with and without the water removal system. The ambient atmosphere may be used by determining water content with a psychrometer or calibrated hygrothermograph. When so doing, it is suggested that the sample atmosphere be drawn through a mixing container sufficient to smooth out high frequency concentration variations (a 50-liter glass flask is acceptable.).

8.4 Another approach involves the injection of liquid water into an inert plastic sample bag, filling with air and allowing the mixture to equilibrate at room temperature for one-half hour. Sufficient water is added to insure complete saturation. As is the case of the atmospheric air sample procedure, the samples is drawn to the analyzer with and without water removal system connected. In either case, analyzer response to water is determined by calculating the volumetric water content necessary to produce one ppm response as carbon monoxide.

8.5 A record should be kept of this factor (earlier defined as discrimination ratio). Any abrupt increase in water interference should be investigated to check on effectiveness of the water removal system and for possible instrument malfunction.

9. Calculation

9.1 For greatest accuracy the calibration chart should be used to reduce recorder values. Oftentimes, however, the assumption is made that, over the range used in atmospheric monitoring (most often 0 to 100 ppm), a linear calibration can be used. The procedure selected affects the accuracy of reported measurements. To minimize error when assuming linearity together with a single point standardization, the standardization gas concentration should be about two-thirds the expected maximum value.

9.2 Data reduction procedures used are dependent upon the parameters to be reported. For determination of mean values, graphical, mechanical or electronic integration techniques have been used. Mean values may also be obtained from a sufficient number of discrete values to give the confidence interval desired. Instantaneous peak values should be defined as having a minimum baseline time of 5 min. The value at the base line is reported.

9.3 Concentrations in the atmosphere should be reported to the nearest ppm with the confidence level stated.

10. Effects of Storage

10.1 Not applicable.

11. References

1. Beckman Instructions 1307-C. 1966. Models IR 215, IR 315, and IR 415 Infrared Analyzers. Beckman Inst., Inc., Sci. and Proc. Inst. Div., Fullerton, California.
2. Bryan, R.J. 1966. In Standard Methods of Chemical Analysis. Sixth edition. F.J. Welcher, Ed., Van Nostrand. Princeton, N.J. Chapter 42:850.
3. Hartz, N.W., and J.L. Waters. 1952. An improved Luft infrared analyzer. Instruments, 25:622.
4. MSA LIRA Infrared Gas and Liquid Analyzer Instruction Book, Mine Safety Appliances Co., Pittsburgh, Pa.
5. American Public Health Association, Inc. 1977. Methods of Air Sampling and Analysis. 2nd ed. p. 342. Washington, D.C.

Subcommittee 4
R. G. Smith, *Chairman*
R. J. Bryan
M. Feldstein
B. Levadie
F. A. Miller
E. R. Stephens
N. G. White

129.

Tentative Method of Analysis for Carbon Monoxide Content of the Atmosphere (Hopcalite Method)

42101-05-71T

1. Principle of Method

1.1 Air containing carbon monoxide is passed over an oxidation catalyst (Hopcalite) where the carbon monoxide is oxidized to carbon dioxide. The heat of oxidation results in a temperature increase in the air stream which is proportional to the carbon monoxide concentration.

1.2 Earliest use of this method was for removal of carbon monoxide from air (**1**).

The method was later adapted for carbon monoxide analysis and was suggested as being one of the most reliable of seven methods studied early in World War II (**2**). Reaction conditions, sample treatment, and catalyst conditioning have been described by several investigators (**3, 4, 5**). The method is used commonly in garages and other indoor situations.

1.3 Hopcalite, the oxidation catalyst, has sometimes been described as consisting of more than two metal oxides (**6**), but for purposes of this method is assumed to consist of a mixture of MnO_2 and CoU (**7**).

2. Range and Sensitivity

2.1 This method is intended for intermittent sampling and analysis in carbon monoxide concentrations normally exceeding 50 ppm in air. Commercially available analyzers usually provide a scale of 0 to 500 ppm with a quoted accuracy under ideal conditions of ± 5 ppm.

2.2 For carbon monoxide concentrations in air lower than this range or for continuous analysis, the nondispersive infrared (**8**) or the gas chromatographic (**9**) methods should be used.

3. Interferences

3.1 Low results may be expected if the Hopcalite catalyst is deactivated by water or otherwise poisoned. Drying agents, ascarite, and activated carbon have been used to protect the catalyst from deactivation (**2, 3, 4**). As a minimum some form of water removal is required.

3.2 Other combustible substances in the sample air may produce a positive interference. The effect of these possible interferences may be minimized by selection of an optimum catalyst temperature and by sample pretreatment. Methane, ethane, and propane are resistant to oxidation at temperatures below 150 C (**4, 7**). Aromatic hydrocarbons are refractory at temperatures below 200 C (**10**), but higher molecular weight aliphatic hydrocarbons, alcohols and ketones will be partially oxidized at temperatures near 100 C (**7, 10**). When carbon monoxide is present in excess over such compounds an activated carbon adsorption unit will minimize such interferences.

4. Precision and Accuracy

4.1 Insufficient information is available to make definitive statements about the accuracy of this method over a wide range of operating conditions. For one commercial analyzer having a full scale range of 0 to 500 ppm, the manufacturer states that reproducibility is ± 1% of full scale and that maximum variation in linearity is ± 2% of full scale (**11**).

5. Apparatus

5.1 This method is intended for use in the manual sampling and analysis of carbon monoxide by the measurement of its heat of combustion. It is most likely that a commercially available analyzer will be used in the application of this method. A general description only is given of the basic components of such an analyzer.

5.2 SAMPLE HANDLING SYSTEM. A particulate filter and dryer unit are used for pre-treatment of the sample air. A sample pump, flow meter(s), and means for flow regulation are used to draw the sample through the pre-treatment devices and thence the Hopcalite cell at a controlled rate.

5.3 HOPCALITE OXIDATION CELL. The cell is an insulated container for one bed each of inactive and active Hopcalite. Cell temperature is maintained at 100 C by a heater regulated by a proportioning controller. Higher temperatures may result in increased positive interference by any organic gases present.

5.4 MEASUREMENT CIRCUIT AND CONTROLS. Two thermistors, one each in the inactive and active catalyst beds, are connected in the legs of a Wheatstone bridge circuit. Any carbon monoxide passing through the cell is oxidized in the active Hopcalite bed, thus increasing temperature and changing the resistance of the thermistor. The resulting unbalance of the bridge is detected by a meter circuit or an external recorder. A zero control is provided to adjust electrical zero while carbon monoxide-free air is being passed through the analyzer. A span control is provided to adjust the meter or recorder reading when a known concentration of carbon monoxide in air is being introduced to the sample system. A cell temperature adjustment control should be provided. It is also possible to provide audible and visual alarm circuitry for an analyzer of this type.

6. Reagents

6.1 Span Gas. A known concentration of carbon monoxide in air prepared in a pressurized cylinder. Such gases are available from a variety of commercial sources. For use with this method it is important that the span gas diluent be hydrocarbon and carbon monoxide-free air. A number of unofficial terms such as "primary standard", certified standard, "custom mixture", certified calibration standard, "high accuracy mixture", and "custom accuracy mixture" are used to describe commercial calibration gases. Each is associated with a preparation tolerance and accuracy as stated by the supplier. Standards must be of equivalent quality to ISC procedure (**12**).

6.2 Zero Gas. Hydrocarbon and carbon monoxide-free air in a pressurized cylinder. Air with very low hydrocarbon and carbon monoxide content may be used if the concentrations are known. Such air is available from commercial sources.

6.3 Active Hopcalite, 14 to 20 Mesh. Available from Mine Safety Appliances Co., Pittsburgh, Pa.

6.4 Inactive Hopcalite, 14 to 20 Mesh. Available from Mine Safety Appliances Co., Pittsburgh, Pa.

6.5 Particulate Filter. Available from Mine Safety Appliances Co., Pittsburgh, Pa.

6.6 Dryer: Molecular Sieve Type. Available from Mine Safety Appliances Co., Pittsburgh, Pa., as refill.

7. Procedure

7.1 Specific operating procedures will vary with the particular analyzer used. Only generalized steps are given here.

7.2 Install analyzer as specified by manufacturer. The proper particulate filter, fresh drying agent and activated charcoal should be inserted in the sample line.

7.3 Active and inactive Hopcalite are added to the cell as directed.

7.4 The analyzer is turned on and allowed to warm up for at least one half hour to bring the cell to thermal equilibrium. Adjust sample flow rate to manufacturer's specification.

7.5 After warm-up switch the sample inlet to the zero air source and turn the zero control to adjust the meter pointer to electrical zero.

7.6 Switch sample inlet to source of span gas. Check sample flow rate. Adjust span control to give proper meter and recorder indication.

7.7 Switch sample inlet to atmosphere to be sampled. Adjust to proper sampling rate.

7.8 Check sample flow rate, zero and span values periodically. Frequency should be determined by field experience with time stability of these parameters.

7.9 The particulate filter should be changed whenever it is difficult to adjust the air flow to the proper rate. Dryer life is estimated to be approximately two months. However, should span check operations indicate necessity for frequent change of Hopcalite, more frequent changes of the dryer may be necessary. In some cases it may be desirable to use a supplemental in-line indicating dryer unit. The active Hopcalite should be changed whenever the span check operation indicates lack of sensitivity to carbon monoxide concentrations in the instrument range.

8. Calibration and Standardization

8.1 The signal output of the analyzer described in this method should be linear with carbon monoxide concentration. Thus, when zeroed and spanned with accurately known gases the previously described operating procedures include the normal calibration operations.

8.2 Under normal conditions the span setting should require little adjustment. Should water or other impurities penetrate the sample system protective units the catalyst may become deactivated. The periodic introduction of span gas is desirable as a check on catalyst activity.

8.3 Should unusually high concentrations of higher molecular weight (C_6 and higher) aliphatic hydrocarbons, alcohols, or ketones be expected, the response of the analyzer to these substances may be checked with cylinder sources at appropriate concentrations.

9. Calculations

9.1 Carbon monoxide concentrations in ppm are obtained directly by this method. Consequently, no calculations as such are required. 1 ppm = 1.14 mg/m^3 at 25 C and 760 Torr.

10. Effects of Storage

10.1 Not applicable.

11. References

1. LAMB, A.B., W.C. BRAY, and J.C.W. FRASER. 1920. The Removal of Carbon Monoxide from Air, *Ind. Engr. Chem.* 12:213.
2. GOLDMAN, F.H. and A.D. BRANDT. 1942. A Comparison of the Methods for Carbon Monoxide. *AJPH* 32:475–480.
3. SALSBURY, J.M., J.W. COLE, and J.H. YOE. 1947. Determination of Carbon Monoxide. *Anal. Chem.* 19:66–68.
4. LINDSLEY, C.H., and J.H. YOE. 1948. Simple Thermometric Apparatus for the Estimation of Carbon Monoxide in Air. *Anal. Chim. Acta.* 2:127–132.
5. LYSYJ, I., J.E. ZAREMBO, and A. HANLEY. 1959. Rapid Method for Determination of Small Amounts of Carbon Monoxide in Gas Mixtures. *Anal. Chem.* 31:902–904.
6. RIENAECKER, G., and J. SCHEVE. 1964. *Anorg. Allgem. Chem.,* 330:18–26.
7. CHRISTIAN, J.G., and J.E. JOHNSON. 1965. Catalytic Combustion of Atmospheric Contaminants over Hopcalite. *Int. J. Air and Water Poll.* 9:1–10.
8. AMERICAN PUBLIC HEALTH ASSOCIATION, INC. 1977. *Methods of Air Sampling and Analysis.* 2nd. ed. p. 342. Washington, D.C.
9. *Ibid.* p. 359.
10. JOHNSON, J.E., J.G. CHRISTIAN, and H.W. CARHART. 1961. Hopcalite Catalyzed Combustion of Hydrocarbon Vapors at Low Concentrations. *Ind. Engr. Chem.* 53:900–902.
11. MINE SAFETY APPLIANCES CO. Bulletin No. 0802-2.
12. AMERICAN PUBLIC HEALTH ASSOCIATION, INC. 1977. Methods of Air Sampling and Analysis. 2nd ed. p. 342. Washington, D.C.

Subcommittee 4
R. G. SMITH, *Chairman*
R. J. BRYAN
M. FELDSTEIN
B. LEVADIE
F. A. MILLER
E. R. STEPHENS
N. G. WHITE

130.

Tentative Method of Analysis for Methane and Carbon Monoxide Content of the Atmosphere (Gas Chromatographic Method by Reduction to Methane)

43201-01-71T

1. Principle of the Method

1.1 An air sample is introduced onto a prechromatographic or stripper column, which passes the methane and carbon monoxide quantitatively to the gas chromatograph. The gas chromatographic column separates the methane from the carbon monoxide.

1.2 The methane is eluted first from the gas chromatographic column through a catalytic reduction tube into a flame ionization detector. The carbon monoxide is eluted next into the catalytic reduction tube and is reduced to methane before entering the detector.

1.3 The response of the detector is directly proportional to the weight of hydrocarbon in the carrier gas stream. The analysis has no interferences.

2. Range and Sensitivity

2.1 The linear range of the gas chromatographic system for carbon monoxide described here is from 0 to 1 ppm to 0 to 1000 ppm full scale, depending on the attenuator setting of the electrometer. In the 0 to 1 ppm range the sensitivity is 50 ppb (thousand million). For ambient air analysis a logarithmic amplifier system can be used to obtain high sensitivity for low concentrations and still contain the tracings of peak concentrations.

(1140 $\mu g/m_3$ = 1 ppm CO at 25 C and 760 Torr).
(655 $\mu m/m_3$ = 1 ppm methane at 25 C and 760 Torr).

3. Interference

3.1 The stripper column used with the instrument is designed to prevent hydrocarbons other than methane, and carbon monoxide from reaching the analytical column. As long as this stripper column is effective, interferences with the methane and carbon monoxide measurements will not occur. The stripper column must be checked frequently with known gas mixtures to determine efficiency.

4. Precision and Accuracy

4.1 Repeatability of the measurement of carbon monoxide and methane in a sample introduced into the gas chromatographic system is primarily a function of the carrier gas and hydrogen flow. A change in the carrier or hydrogen flow of 10 to 15% can vary the detector response as much as 15 to 20%. However, variations in the carrier and hydrogen flow are so infrequent that weekly checks on these parameters are sufficient to maintain a steady flow.

4.2 The accuracy of the carbon monoxide and the methane measurements has been established at ± 2% of the absolute value based on a known standard.

4.3 The system is stable to the extent that flow rates are maintained constant. In practice day-to-day variation is about 2%. The baseline drift due to temperature and flow fluctuations is rarely more than 1% per 24 hr.

5. Apparatus

5.1 The analytical system (Figure 130:1) consists of the following:*

5.1.1 *Automatic Gas Sampling Valve.* With two 15-ml sample loops (Carle Valve Co., No. 2014 and No. 2050 automatic actuator).

5.1.2 *Automatic Column Switching Valve.* (Carle Valve Co., No. 2014 and No. 2050 automatic actuator).

5.1.3 *Time Sequence Programmer.* (Carle Valve Co., and Merco Valve Programmer No. 2100).

5.1.4 *Stripper Column.* A ¼″-OD, 12″-long stainless steel tube packed with 5″ of 10% Carbowax 400 on 60/80 mesh Chromosorb-W.H.P., 5″ of 60/80 mesh silica gel, and 2″ of Malcosorb (Mallinckrodt).

5.1.5 *Gas Chromatographic Oven.* Capable of maintaining 115 C.

5.1.6 *Gas Chromatographic Column.* 12′ of ¼″-OD stainless steel tubing packed with 5A molecular sieve, 60/80 mesh.

5.1.7 *Catalytic Reactor.* 6″ of ¼″-OD stainless steel tube packed with 10 per cent Ni on 42/60 mesh C-22 firebrick. Add 24 ml of nickel nitrate solution (see **6.4**) to 10 g 42/60 mesh C-22 firebrick. Dry the mixture slowly in a fluidizer at 100 C while purging with a stream of dry nitrogen flowing at 300 ml/min. Break up the dried coated firebrick lumps formed during the drying process, sieve to 42/60 mesh size and pack the material into a 6″ length of

*Equivalent supplies from any other source may be used.

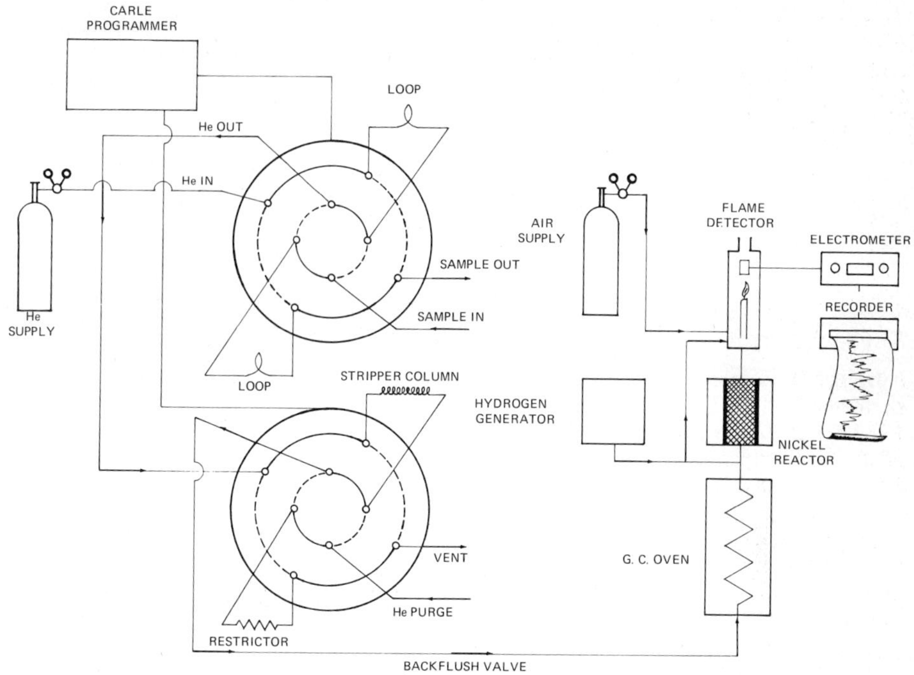

Figure 130:1—Analytical sampling system for methane and carbon monoxide by reduction to methane.

¼″-OD stainless tube. Heat the tube to 600 C for 1 hr while purging it with oxygen at 100 ml/min. Cool the reactor, then install it downstream of the molecular sieve column (Section **5.1.6**) and slowly heat to 360 C while purging with a mixture of 200 ml/min helium and 30 ml/min of hydrogen for 2 hr. For optimum results, maintain the reactor at 360 C with the prescribed ratio of helium-hydrogen gas flowing through the reactor.

5.1.8 *Flame Ionization Detector*. A flame tip with an inside diameter of 0.508 mm.

5.1.9 *Electrometer*. An amplification range $1 \times 10^{-12} - 1 \times 10^{-7}$ amperes is recommended. For ambient air a logarithmic amplifier system set to amplify signals between 1×10^{-11} and 1×10^{-9} would normally cover variations in concentration that occur in densely populated urban areas.

5.1.10 *Recorder*. The recorder input must be compatible with the electrometer output.

5.1.11 *Pump*. A noncontaminating diaphragm pump with a pumping rate of 5 liters/hr.

5.2 Calibrated Steel Cylinders. A standard 44-l cylinder whose volume is known within ± 10 ml.

5.3 Transfer Pipets. 1, 5, 10-ml. Calibrated by weighing with mercury to determine absolute volume.

5.4 Pressure Guage. Capable of measuring pressure within 1% or less.

5.5 High Pressure Transfer Line. For pressurizing cylinder.

6. Reagents

6.1 Helium. Bureau of Mines Grade.

6.2 Air. Water pumped.

6.3 Hydrogen. Ultra pure hydrogen or from a hydrogen generator.

6.4 Nickel Nitrate Solution. Dissolve 238.5 g nickel nitrate hexahydrate ($Ni(NO_3)_2 \cdot 6H_2O$) in 100 ml distilled water.

6.5 METHANE. Ultra high purity (99.5%).

6.6 CARBON MONOXIDE. Ultra high purity (99.5%).

6.7 ULTRA PURE AIR. Containing less than 0.1 ppm CO and CH_4. Scott Laboratories can supply air to meet these specifications.

7. Procedure

7.1 RECOMMENDED OPERATING PARAMETERS.

7.1.1 *Temperatures.*

Stripper column	25 C ± 5 C
Molecular column	115 C
Detector	150 C
Reactor	360 C

7.1.2 *Gas Flows.*

Carrier	200 ml/min helium
Hydrogen to reactor	30 ml/min.
Hydrogen to flame	60 ml/min.
Air to flame	400 ml/min.

7.2 PROCEDURE I. Sample air is pulled through the sample loop at a rate of 100 ml/min with the pump positioned after the sample loop. Once every 10 min a sample is injected into the analyzer. The sample flows through the loop into the stripper column before entering the gas chromatographic oven and molecular sieve (M.S.) column. After 30 sec the backflush actuates, reversing the carrier flow in the stripper column to vent while maintaining the carrier flow through the M.S. column. Oxygen and nitrogen are eluted first from the M.S. column into the reactor and flame, causing fluctuations in the signal from the detector. Following the oxygen and nitrogen is the methane, followed in turn by carbon monoxide. The carbon monoxide is quantitatively converted by the catalytic reactor to methane before it is detected by the flame ionization detector.

7.3 PROCEDURE II. Instead of being pumped directly into the sample loop from the air source, the sample is first pulled through an integrating vessel. The dimension of the vessel and the flow through the vessel are adjusted so that the sample pulled into the gas chromatographic system represents the concentration averaged over the sample residence time in the vessel, which in turn is arranged to correspond to the sampling interval. This sampling procedure gives an average of the concentrations of CO and CH_4 in the atmosphere that prevail between sample injections to the chromatograph.

7.4 PROCEDURE III. Manual samples can be analyzed by directly injecting 15 ml samples into the sample loop. Samples of ambient air may be collected by filling evacuated stainless steel cylinders in the field. For convenience in removing samples, the cylinders may be pressurized to 860 Torr with nitrogen, and syringe samples withdrawn through a rubber septum. Results are corrected for dilution.

8. Calibration and Standards

8.1 CALIBRATION STANDARDS. To calibrate the analyzer and determine the efficiency of the conversion of carbon monoxide to methane by the reactor over the concentration ranges of interest, prepare calibration standards of carbon monoxide and methane. Evacuate the calibrated steel cylinder to approximately 1 Torr. Attach a rubber septum to allow introduction of the gases from the transfer pipet to the tank. Flush the transfer pipet with methane, seal rapidly by holding between rubber-gloved hands, and introduce to the tank. Allow the contents of the pipet plus a small rinse of room air to be drawn into the tank. Add an identical volume of carbon monoxide in the same manner. Pressurize the cylinder with ultra pure air to give the desired concentration. Prepare at least four cylinders of different concentrations over the range of interest. Construct a calibration curve from the chromatographic analysis of the calibration standards. (**Caution:** This calibration procedure is a hazardous operation and should be performed only with armour plate protection).

8.2 A plot of the area of the methane peak should be identical to that of the area of the carbon monoxide peak if the reactor is converting the CO and CH_4 quantitatively. If the area of the CO peak is smaller than that of methane, the concentration of hydrogen in carrier gas entering the reactor may be low, or the reactor temperature may be too low to effect quantitative conversion. Usually the conversion efficiency is a function of hydrogen concentration entering the reactor.

9. Calculations

9.1 For most applications the peak heights of the CH_4 and CO are adequate to quantitate the concentrations of these gases in an unknown air sample. An automatic electronic integrator or a ball and disc integrator can be used for quantitation.

10. Effects of Storage

10.1 None.

11. References

1. STEVENS, R.K., A.E. O'KEEFFE, and G.C. ORTMAN. A Gas Chromatographic Approach to the Semi-Continuous Monitoring of Atmospheric Carbon Monoxide and Methane. *U.S. Dept. of Health, Education, and Welfare, Public Health Services.*
2. ORTMAN, G.C. 1966. *Anal. Chem.* 38:644–646.

Subcommittee 4
R. G. SMITH, *Chairman*
R. J. BRYAN
M. FELDSTEIN
B. LEVADIE
F. A. MILLER
E. R. STEPHENS
N. G. WHITE

131.

Tentative Method of Analysis for Carbon Monoxide—Mercury Replacement Method

42101-06-72T

1. Principle of the Method

1.1 The air sample, after removal of particulate matter with a filter, is passed through an adsorption unit to remove water vapor and interfering hydrocarbons. Subsequent reaction of carbon monoxide (CO) in the sample with hot mercuric oxide produces mercury vapor (Hg) which is determined photometrically, using the 253.7 nm Hg emission line (**1, 2**). The reaction is as follows:

$$CO + HgO(S) \xrightarrow{\Delta} CO_2 + Hg(V).$$

2. Range and Sensitivity

2.1 RANGE. The range of this method is particularly suited for use in low ambient and natural background carbon monoxide concentrations. Full scale detection over a range of 0.05 to 10 ppm may easily be attained with an analyzer specifically designed to be used in a study of remote locations (**2**). A commercial version of an instrument using this approach has a nominal range of 0 to 50 ppm (**3**). The instrument design features having major influence on range are the optical path length of the mercury vapor detector cell and whether a stoichiometric or low yield CO − HgO reaction is utilized (1 ppm CO = 1.14 mg/m^3 at 25 C and 760 Torr).

2.2 SENSITIVITY. The minimum detectable concentration observable after a step input to the analyzer is directly related to the operating range of the analyzer. Concentrations as low as 30 ppb (34 $\mu g/m^3$) have been reported with a stoichiometric analyzer having an 18-cm optical cell (**2**). Another study reports a sensitivity of 5 ppb (**4**). A commercial low yield instrument with a full scale range of 50 ppm has a minimum detectable concentration of 0.05 ppm (short range) and 0.5 ppm (long range). Under field conditions where a null balancing type potentiometric recorder is used, a sensitivity (minimum detectable concentration) of 1% of full scale range should easily be attainable.

3. Interferences

3.1 Any compound or condition which will result in the release of mercury vapor from mercuric oxide will interfere with this analysis. Reactive hydrocarbons, some organics, and hydrogen have been evaluated. Most investigators have found that methane does not interfere, that acetylene, ethylene, propylene, and formaldehyde interfere quantitatively in a positive direction, and that acetone and hydrogen interfere, but not quantitatively. Hydrogen interference is influenced by the heat of the mercuric oxide bed, being less at lower temperatures. At 200 C, Palanos found that hydrogen responds at 5.8% of its volumetric concentration (**5**). At a nominal concentration of 0.5 ppm hydrogen in the atmosphere interference due to this gas would be approximately 0.03 ppm. There is also an interference due to the thermal dissociation of mercuric oxide to produce mercury vapor. This reaction is also temperature dependent. At 200 C the thermal emission level has been reported at 0.4 ppm carbon monoxide equivalent (**5**). Interference by water vapor apparently involves a complex catalytic effect. Positive interference was found by Robbins, *et al* (**2**) in a stoichiometric (quantitative yield) instrument while Palanos (**5**) found a negative interference in a low yield instrument.

3.2 Several approaches have been used to eliminate or minimize interferences. Precise temperature and flow control are necessary to stabilize the thermal dissociation background. A drier unit can be used to remove water vapor. One approach used to compensate for atmospheric hydrogen involves the addition of a silver oxide oxidation catalyst which when operated at room temperature oxidizes carbon monoxide but not hydrogen. The periodic passing of sample air through this catalyst provides data on the hydrogen blank. Operation of the mercuric oxide bed at reduced temperature such as is done in the low yield type analyzer minimizes the mercuric oxide—hydrogen reaction so as to result in hydrogen interference at normal atmospheric concentrations of less than 0.05 ppm CO equivalent. Acetylene and C_2–C_3 olefins will interfere unless removed. This may be accomplished by selective adsorbents such as activated carbon or 5A molecular sieves.

4. Precision and Accuracy

4.1 Only limited data is available on precision and accuracy for this method. Depending upon mode of operation, and the extent and rate of variation of hydrogen concentration in the test atmosphere, accuracy can be hydrogen concentration limited. Using an analyzer equipped with silver oxide catalyst for obtaining a hydrogen blank, Robbins, *et al* reported field trials showing very constant hydrogen concentrations in the atmosphere (**2**). In this study, blanking operations were conducted at 2 hr intervals, permitting nearly total elimination of hydrogen interference. Palanos reports that in an analyzer using a non-stoichiometric (low yield) carbon monoxide—mercuric oxide reaction, very close control of sample flow rate and mercuric oxide cell temperature permits attainment of span and zero drifts as low as 1% of full scale over a 24 hr period. A reproducibility of better than 2% is claimed (**5**). The precision of this type of analyzer should be ± 0.5% of full scale.

4.2 The effect on accuracy of the optical and electronics components are expected to be minimal in a conservatively designed analyzer incorporating an optical reference cell.

5. Apparatus

5.1 Sampling

5.1.1 *Particulate Filter.* A high efficiency filter is used to remove particulates from the sample air.

5.1.2 *Sample Pump.* Use a sample pump of the non-lubricated type, with the capacity to deliver the specified rate of air sample at the total system back-pressure. In a commercial analyzer this rate will have been determined on the basis of consumption of HgO, dimensions of optical cell and rise time specifications. A typical rate is one liter/min.

5.1.3 *Adsorption Unit.* A two column cycling molecular sieve adsorption unit is used to remove water vapor and interfering hydrocarbons. The 5A molecular sieve material, size of column and cycle rate are selected to pass only carbon monoxide, methane, hydrogen, and no more than 1 ppm water vapor. Heated purge air is passed through each column on alternate cycles.

5.1.4 *Flow regulator.* A pressure feedback type flow regulator is used to hold sample air flow to ± 2% of the nominal rate.

5.1.5 *Assembly.* The components of the sampling unit are assembled as shown in Figure 131:1A.

5.2 Analyzer Unit

5.2.1 *Heat Exchanger and Reactor.* An integral unit in which the air sample is pre-heated to the desired reactor temperature and then is passed over a mercuric oxide pellet having a dimensional configuration such that the proper volumetric flow rate per unit surface area of HgO is maintained. A proportional temperature controller capable of maintaining the temperature to within ± 0.5 C is to be used. The reactor volume is within a metal block of sufficient mass to aid in stabilization of the reactor temperature.

5.2.2 *Detector.* A double beam photometer using the 253.7 nm Hg emission line is employed as the mercury vapor detector. The line widths of both the emission and the absorption wavelengths in the sample cell are stabilized by temperature regulation of the mercury vapor source lamp and the detectors. UV filters are used between the sample and reference cells, and the detectors to pass the desired 253.7 nm signals.

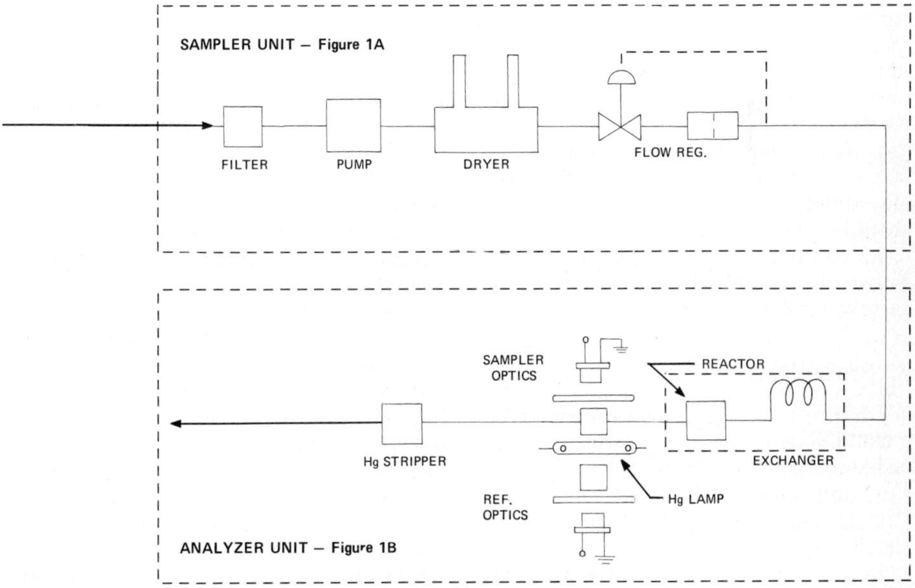

Figure 131:1A—Sampler unit
Figure 131:1B—Mercury replacement analyzer for CO

5.2.3 *Mercury Stripper.* An activated charcoal adsorption unit is used after the detector to remove mercury vapor from the air stream discharged from the instrument. This unit should be as specified by the analyzer manufacturer.

5.2.4 *Assembly.* The analyzer unit is assembled as shown in Figure 131:1B.

6. Reagents

6.1 Red mercuric oxide (HgO) (ACS analytical reagent grade) in the form of standard pellets of specified shape and mass. This material would in general be supplied by the manufacturer of the analyzer.

6.2 CALIBRATION AND ZERO GASES.

6.2.1 *Calibration Gases.* A series of accurately known mixtures of pure carbon monoxide in pre-purified nitrogen. Concentrations of 20, 40, 60, and 80% of the full scale range of the analyzer are employed. Suppliers of specialty gases offer a variety of mixture grades, but there is no standardization as to terms such as—high accuracy, custom mixture, certified standard, etc. An initial series of reference standards should be obtained which are known to within ± 2% of the true concentration. Alternatively, standard mixtures may be prepared by the Intersociety Committee Method (**6**).

6.2.2 *Zero Gas.* Helium or pre-purified nitrogen may be used as a zero gas. Helium may be presumed to be carbon monoxide free, but the carbon monoxide content of any high grade nitrogen should be verified.

6.3 ACTIVATED CARBON. Iodine treated activated charcoal is used to strip the mercury vapor from the discharge air.

7. Procedure

7.1 SPECIFIC. Operating procedures for a continuous monitoring instrument must be based upon the specific analyzer in use. For any commercially available instrument the manufacturers instructions should be followed.

7.2 GENERAL. The following operating procedures will in general apply to any mercury displacement CO analyzer.

7.2.1 *Start Up.* The analyzer should be turned on and allowed to warm up to operating temperature. This may be determined by the observation of a stable output when sampling from a calibration gas source.

7.2.2 *Flow Rate Check.* The air sample rate with the type of flow regulation used should be very stable. Upon start up and periodically thereafter (once per month) the flow rate should be independently verified with a calibrated rotameter or a volumetric flow metering device.

7.2.3 *Moisture Test.* The performance of the cycling dryer unit should be checked by bubbling calibration gas through water at room temperature. No change in output should be noticeable.

7.2.4 *Standardization and Zeroing.* The calibration gas used for routine standardization is introduced in place of the air sample. The span control is adjusted to give the proper recorder output. Zero gas is then introduced and the zero control adjusted to give a zero scale reading (if CO free zero gas is used) or the proper low scale reading (if a small known CO concentration is present). Repeat span and zeroing operations in the same sequence until no further adjustments are necessary.

7.2.5 *HgO Activity Check.* HgO pellet will gradually be depleted and should be replaced when the rate of span drift exceeds specifications or when total span drift exceeds the span control range. This latter point is reached when it is no longer possible to adjust the span control so as to obtain the proper recorder output when a span gas of known concentration is introduced to the analyzer.

7.2.6 *Hg Vapor Stripper Check.* The activated charcoal unit used to strip the Hg vapor from instrument exhaust air should be replaced on a schedule recommended by the manufacturer. This schedule should be such that the capacity of the stripper will not be exceeded even under the condition that carbon monoxide at the concentration of the full scale detection is continuously present in the sample air. This procedure should be followed because of the toxicity of mercury vapor. An alternate procedure would call for the exhausting of instrument discharge air through a vent to the atmosphere.

8. Calibration

8.1 INITIAL. A full scale calibration should be performed upon start up and after each change of the HgO pellet. Because a non-stoichiometric analyzer has non-linear response characteristics, at least four different concentrations of carbon monoxide covering the range of interest should be used. Analyzed mixtures, certified in accuracy to within ± 2% of the analyzed value may be used if from an acceptable source.

Calibration mixtures may also be prepared in the laboratory by the precision dilution technique and analyzed by the gravimetric method described in Intersociety Method 42101-01-69T, "Tentative Method for Preparation of Carbon Monoxide Standard Mixtures (**6**).

8.1.1 The multiple gas calibration is performed by selecting a concentration at approximately mid-range and introducing to the analyzer. After a stable reading is attained the span control is adjusted so that the meter (or recorder output) output on a 0 to 100% scale reads at that value representing the per cent full scale range equivalent to the known calibration concentration. As an example, suppose full scale deflection is to be 50 ppm and the calibration gas is 24 ppm. The span control will be adjusted until meter output is

$$\frac{24}{50} \times 100 = 48$$

Carbon monoxide-free zero gas is then introduced and the zero control adjusted until meter output reads zero. This sequence of operations is repeated until no further adjustment of controls is required. The other span gas mixtures are then introduced serially and a plot made of carbon monoxide concentration vs. meter (or recorder) output.

8.2 ROUTINE. Routine calibrations are performed as described under **7.2.4** "Standardization and Zeroing."

9. Calculations

9.1 No calculations are required except as described in Section **8.0** "Calibration." Recorder values are read on the calibration plot to obtain true carbon monoxide concentrations.

10. Effects of Storage

10.1 The only effects of storage would pertain to calibration gas standards. There is no data reported indicating any deterioration of CO in dry CO-free air when stored in clean dry steel cylinders.

11. References

1. MULLER, R.A. 1954. *Anal. Chem.* 26, (9), 39A.
2. ROBBINS, R.C., K.M. BORG, and E. ROBINSON. 1968. *J. Air Poll. Control Assoc.*, 18, 106–110.
3. BACHRACH INSTRUMENT CO. Performance Specifications, Model US400L Carbon Monoxide Analyzer.
4. JUNGE, C.E. 1963. *Air Chemistry and Radioactivity*, Academic Press, New York.
5. PALANOS, P.N. 1971."Selective Detection of Ambient CO with a Mercury Replacement Analyzer", paper No. 71-1128, AIAA Joint Conf. on Sensing the Environment, Palo Alto, Calif., November 8–10.
6. AMERICAN PUBLIC HEALTH ASSOCIATION, INC. 1977. *Methods of Air Sampling and Analysis*. 2nd ed. pp. 342. Washington, D.C. 20036.

Subcommittee 4
R. G. SMITH, *Chairman*
R. J. BRYAN
M. FELDSTEIN
B. LEVADIE
D. C. LOCKE
F. A. MILLER
E. R. STEPHENS
P. O. WARNER

132.

Tentative Method for the Determination of Carbon Monoxide (Detector Tube Method)

42101-07-74T

1. Principle of Method

1.1 A known volume of air at a standard flow rate is drawn through a glass tube containing a solid adsorbant impregnated with a reagent that reacts with carbon monoxide to form a colored stain. The length of the stain or depth of the color is proportional to the quantity of carbon monoxide in the air being tested.

Usually the air is sampled by means of a portable hand pump which is factory calibrated to deliver a known amount of air at a standard flow rate.

Pumps and tubes are available from several different manufacturers.

2. Range and Sensitivity

2.1 The range of concentrations that can be determined will vary with the detector tubes from the different manufacturers. In general, the effective range of a tube is from 10 ppm to 1,000 ppm.

3. Interferences

3.1 Depending on the manufacturer's tube used, various chemicals such as acetylene, hydrogen sulfide, nitrogen dioxide, sulfur dioxide and ethylene will interfere. Before use, the list of possible interferences listed by the manufacturer should be consulted.

4. Precision and Accuracy

4.1 A study of the precision and accuracy of tubes for the measurement of carbon monoxide at concentrations near the Threshold Limit Value was published in 1970 (**1**). None of the tubes tested met the approval criterion (± 25%) in the range considered (25 to 100 ppm), although eight tubes were within the alternate performance level limits of ± 50%. Currently, however, the tubes from four major manufacturers meet the approval criterion of ± 25%, as specified by the National Institute for Occupational Safety and Health.

5. Apparatus

5.1 A portable hand pump delivering, according to manufacturer's specifications, a known amount of air in a specified time period.

5.2 Carbon monoxide detector tubes supplied by the same manufacturer as the pump specified in **5.1**.

6. Reagents

6.1 None

7. Procedure

7.1 Break off both ends of the sealed carbon monoxide detector tube.

7.2 Insert tube into tube holder on the pump and apply slight pressure to be sure a snug fit is obtained.

7.3 Draw a known amount of air through the tube by means of the pump.

7.4 Either match color in indicating tube with a manufacturer's supplied color chart or measure length of stain. Determine carbon monoxide concentration from the calibration supplied by the manufacturer.

7.5 Periodic routine checks of the pump according to manufacturer's directions should be made to be sure the pump is operating properly.

8. Calibration

8.1 It is good operating procedure to routinely check a tube from a new shipment that has a lot number different from the previous tubes used and supplied by the same manufacturer.

8.2 Known carbon monoxide concentrations may be prepared according to the Intersociety Committee method (**2**).

8.3 Temperature corrections may be necessary for the calibrations. In the case of color tint tubes, these corrections may be as high as a factor of two for each 10 C deviation from standard calibration conditions.

9. Calculations

9.1 Not applicable

10. Effect of Storage

10.1 Tubes should be stored under refrigeration.

11. References

1. MORGENSTEIN, A.S., R.M. ASH, and J.R. LYNCH. 1970. The Evaluation of Detector Tube Systems: 1. Carbon Monoxide. *Amer. Ind. Hyg. Assoc. J.* 31:630.
2. AMERICAN PUBLIC HEALTH ASSOCIATION, INC. 1977. *Methods of Air Sampling and Analysis*, 2nd ed. p. 342. Washington, D.C. 20036.

Subcommittee 4
R. G. SMITH, *Chairman*
R. J. BRYAN
M. FELDSTEIN
D. C. LOCKE
P. O. WARNER

133.

Tentative Method for Gas Chromatographic Analysis of O_2, N_2, CO, CO_2, and CH_4

1. Principle of the Method

1.1 A dual column/dual thermal conductivity detector gas chromatograph is used to separate and quantitate O_2, N_2, CO, CO_2, and CH_4 in gas samples (**1**). The sample is introduced as a plug into the carrier gas, and after drying in a desiccant tube it passes successively through two carefully matched gas chromatography columns. The first column contains a very polar stationary liquid phase while the second is packed with molecular sieve 13-X. Detectors are placed at each end of the column. The first column retains only CO_2, which is eluted after passage of the rest of the mixture (the composite peak). The first detector thus records two peaks, one corresponding to the unresolved O_2, N_2, CH_4, and/or CO, and the second to CO_2. The gases are swept into the molecular sieve column which separates all the components. The second detector records the elution of O_2, N_2, CO, and/or CH_4. The CO_2 is irreversibly adsorbed on molecular sieve 13-X and does not elute. As shown in Figure 133:1, the retention time of O_2 is sufficiently long to allow CO_2 to elute from the polar column before the O_2 elutes from the second column.

1.2 Peak heights are used in conjunction with calibration plots for quantitative measurements. Alternatively, electronic integration of peak areas may be used.

1.3 The separation is complete in 8.5 minutes.

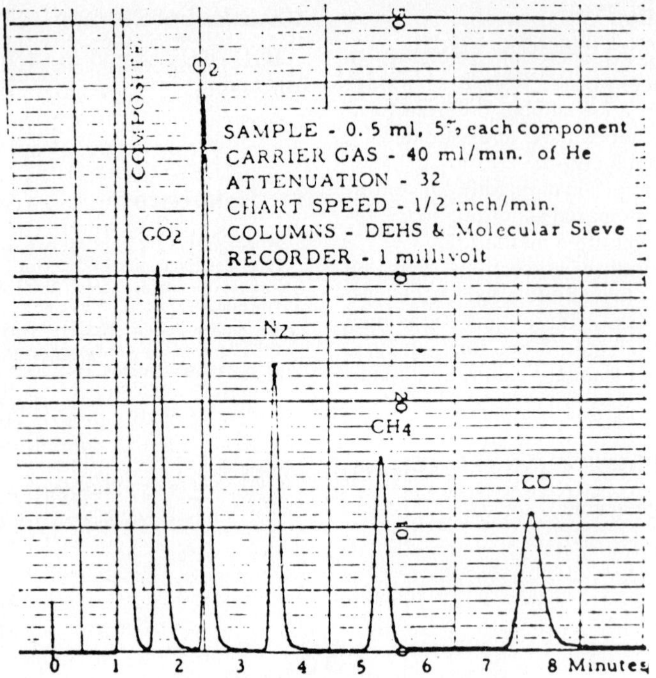

Figure 133:1—Typical chromatogram

2. Range and Sensitivity

2.1 The limits of detection with hot wire thermal conductivity detectors and helium carrier gas, expressed as ppm of a gas in a 1-ml sample that produces a 0.01 mV signal on a 1-mV recorder, are given below:

Gas	Limits of Detection
CO_2	250 ppm
O_2	300
N_2	300
CO	500
CH_4	300

2.2 Full scale sensitivity can be varied with an attenuator from 1 mV to 256 mV to accomodate diverse levels of different components in a sample.

3. Interferences

3.1 Argon is not separated from O_2, but is present in natural air at 0.9 volume percent. For samples with low oxygen concentrations, a correction may be necessary depending upon the preparation of the calibration standards; see Section **8.2**.

3.2 Any compound present in a sample at a detectable level which elutes from either column at a time close to that of a component of interest is a potential interference. Polar compounds including acid gases are strongly retained on both columns at ambient temperatures and will not interfere. Heavier hydrocarbons than methane are retained somewhat by the polar column and elute in order of increasing molecular weight.

3.3 Hydrogen is not detected using helium carrier gas, but can be measured using argon carrier gas.

4. Precision and Accuracy

4.1 Accuracy depends upon the availability of accurate calibration standards.

These may be obtained with a certificate of analysis from commercial suppliers.

4.2 Precision is controlled by the mode of sample introduction, primarily. Gas sampling valves provide precision of ± 0.3%. Reproducibility with a 1 ml gas-tight syringe is about ± 1.5%.

4.3 Precision is also affected by detector drift, which in turn depends upon the control of carrier gas flow rate and system temperature. Standard commercial gas chromatographic equipment capable of detector temperature control to ± 0.5 C and flow rate control to ± 1% is adequate.

5. Apparatus

5.1 GAS CHROMATOGRAPH. Any commercial gas chromatograph equipped with dual column fittings, a four-channel thermal conductivity detector, and both a six-port gas sampling valve and a syringe injection port, can be adapted to this analysis. Commercially available models **(2, 3)** designed specifically for this analysis are recommended. A schematic diagram of one such apparatus **(2)** is given in Figure 133:2.

5.1.1 *Detector.* Either tungsten filament or thermistor thermal conductivity detector elements are acceptable. Gold plated filaments are resistant to oxidation by O_2. Greatest detector stability results when the detector is thermostatted and controlled at a temperature slightly above ambient.

5.1.2 *Carrier gas.* A cylinder of purified helium with a two-stage regulator is required. Flow rate is measured at the exit of the second detector with a soap film flow meter.

5.1.3 *Sample introduction.* A six-port gas sampling valve with a 1-ml sample loop provides the better precision. Alternatively, a 1-ml precision gas-tight syringe with needle may be used.

5.1.4 *Drying tube.* A tube of 200-ml capacity with gas-tight fittings at either end is filled with 10/20 mesh Indicating Drierite. The tube is installed between the sample introduction system and the first gas chromatography column. The desiccant must be replaced when the indicator color changes from blue to pink.

5.1.5 *Gas chromatography columns.* Column number 1 is a 6 ft. × ¼ in. column packed with 30% by weight hexamethylphosphoramide (HMPA) on 60/80 mesh Chromosorb P. Alternatively the column may be packed with 30% by weight di-2-ethyl-hexylsebacate (DEHS) on 60/80 mesh Chromosorb P. The DEHS column

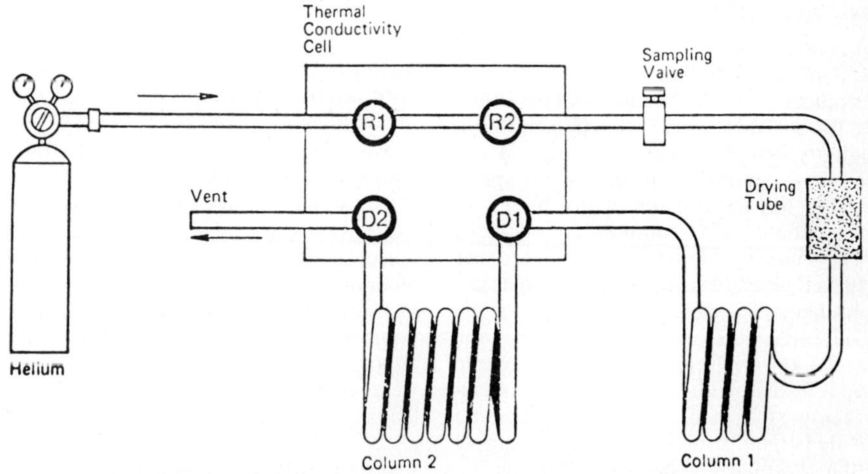

Figure 133:2—Dual Column/Dual Detector GC Schematic

has a longer lifetime than the HMPA column, but DEHS does not separate ethane and ethylene from CO_2. Column number 2 is a 6.5 ft. × 3/16 in. column packed with 40/60 mesh molecular sieve 13-X. The columns must be carefully matched to ensure that the retention times of the components allow separation of the CO_2 from the O_2.

5.1.6 *Temperature.* The columns are operated at room temperature. As noted above, best precision results when the detector is operated slightly above ambient temperature.

5.1.7 *Recorder.* Any 1 mV potentiometric strip-chart recorder with a chart speed of 1 in./min. is suitable.

5.1.8 *Electronic integrator.* Any suitable electronic integrator compatible with the chromatograph may be used to measure peak areas for quantitation.

6. Reagents

6.1 Helium. High purity grade helium (99.995%) is required.

6.2 Calibration standards. Standard blends encompassing the concentration range of components in the samples can be obtained from commercial suppliers.

7. Procedure

7.1 Gas Chromatograph. The carrier gas is turned on and the flow rate adjusted to 50 ml/min. The flow should be checked periodically. After the gas has been flowing for at least 3 min, the thermal conductivity detectors may be turned on and the currents adjusted to the values specified for the instrument by its manufacturer. About thirty minutes are required for instrument stabilization. The recorder is turned on and zeroed before samples are introduced.

7.2 Injection of sample.

7.2.1 *Sampling valve.* The sample loop is flushed with several volumes of calibration standard or sample gas. The handle is then turned to divert the sample to the chromatograph.

7.2.2 *Syringe.* Sample is withdrawn from the sample vessel and quickly injected, guarding against blow-back of the plunger.

7.3 Repetitive analyses. A new sample may be analyzed immediately after the last peak in the sample has emerged. Samples should be analyzed in duplicate.

8. Calibration and Standards

8.1 A standard curve of peak height or peak area *vs.* volume per cent is prepared for each constituent of interest, by analyzing the calibration standards. The calibration plots should bracket the sample concentrations. Linear plots should result. However, in the presence of 5 to 7% CO_2, the calibration for O_2 is not linear up to 20% (**4**), but the calibration plot may still be used.

8.2 If the sample source is natural air, the result for O_2 may need correction for argon present in the sample but not separated from the O_2.

8.2.1 If the O_2 calibration standard mixtures contain pure O_2 diluted with pure N_2, the apparent volume percent of O_2 in the samples must be reduced by 0.93%. The thermal conductivity detector molar response factors for Ar and O_2 are sufficiently close (**5**) that no appreciable error will result from assuming identical relative responses.

8.2.2 If the O_2 calibration standard mixtures contain natural air diluted with pure N_2, both samples and standards contain argon and no correction is necessary.

8.3 The standard curves should be checked periodically.

8.4 A severe loss in resolution of the CO_2/composite peaks and/or of the N_2/O_2 peaks indicates the need for replacement of columns. The polar gas chromatography column continuously loses stationary liquid phase through volatilization. These vapors are adsorbed on the molecular sieve column along with the CO_2, which leads to slow deterioration of the performance of that column. Normally this will happen slowly over a long period of time.

9. Calculations

9.1 Concentrations are determined directly from the calibration plots. The following conversion factors apply at 1 atm and 25 C.

Gas	(mg/m³)/ppm
CO_2	1.80
O_2	1.31
N_2	1.14
CH_4	0.654
CO	1.14

10. Effects of Storage

10.1 Not applicable.

11. References

1. JEFFERY, P.G., and P.J. KIPPING. 1972. *Gas Analysis by Gas Chromatography*, 2nd ed. Pergamon Press, Oxford.
2. FISHER SCIENTIFIC COMPANY, Springfield, New Jersey. Fisher-Hamilton Gas Partitioner, Models 25A and 29.
3. CARLE INSTRUMENTS, INC., Fullerton, California. Gas Chromatograph Model 8000.
4. M. FELDSTEIN, Bay Area Air Pollution Control District, San Francisco, California, private communication.
5. DIETZ, W.A. 1967. *J. Gas Chromatog.*, 5:68.

Subcommittee 4

R. G. SMITH, *Chairman*
R. J. BRYAN
M. FELDSTEIN
D. C. LOCKE
P. O. WARNER

134.

Tentative Method for Constant Pressure Volumetric Gas Analysis for O_2, CO_2, CO, N_2, Hydrocarbons (ORSAT)

1. Principle of the Method

1.1 Sample gas is contacted successively by a series of chemically reactive solutions. Each solution removes a specific constituent of the sample gas mixture with the corresponding decrease in gas volume at each step representative of the volume of the specific gas removed. A leveling bulb is used to adjust all gas volume measurements to atmospheric pressure. Ordinarily, the analysis is applied in the field using the portable Orsat apparatus to determine the volume composition of carbon monoxide, carbon dioxide, oxygen, and unsaturated hydrocarbons in the gaseous effluent from combustion processes.

Results are usually expressed in volume per cent of each component gas.

Methane and ethane may be determined by fractional combustion, and nitrogen is calculated by difference.

2. Range and Sensitivity

2.1 The limit of detection for each component is given as 0.1% of the total volume based on a 100.0 ml sample.

3. Interferences

3.1 Errors due to physical absorption can be minimized by proper air-solution contact allowing at least 3 min per contact for proper equilibration. Otherwise, no interference is observed from components of ordinary combustion air at levels normally encountered.

Negligible interference results from the

presence of H_2S, SO_2, and acid gases which are absorbed by the caustic solution and reported as CO_2.

4. Precision and Accuracy

4.1 HYDROCARBONS. Since no more than two hydrocarbons may be determined at the same time by this method, errors may be caused by the presence of other hydrocarbons depending upon kind and concentration. Consequently, it is not proper to express accuracy for individual hydrocarbons, although relative precision has been determined for the most common combustion related hydrocarbons exclusive of other hydrocarbon interferences.

Gas	Reproducibility (%) (**2**) Different Laboratories and Apparatus	One Laboratory and Apparatus
Unsaturated hydrocarbons as a group	—	0.01
Methane	1.0	0.2
Ethane	1.0	0.2

4.2 CARBON OXIDES. Oxygen and Nitrogen

Gas	Probable Accuracy	Reproducibility (%) (**2**) Different Laboratories and Apparatus	One Laboratory and Apparatus
Carbon dioxide	0.05	0.05	0.02
Carbon monoxide	0.1	—	—
Nitrogen	0.6	0.6	0.1
Oxygen	0.1 to 0.2	0.1	0.03

5. Apparatus (3, 4)

The apparatus should consist of the conventional Orsat type in which volumes are made comparable by a pressure-temperature compensator, with a manometer interposed between the compensating tube and buret.

5.1 BURET. The buret employed must have a 600 mm length of the graduated section which should contain a volume of 100 ml graduated at 0.2 ml intervals, each graduation to be separated by a distance of 1.2 mm. The buret should be calibrated by weighed volumes of mercury and must be accurate to within 0.1 ml/100 ml delivery and to 0.02 ml/each 10 ml interval (**1**). A glass leveling bulb is connected with rubber tubing to the buret.

5.2 PIPET. Gas absorption pipets are placed following the gas measuring buret, 5.1, in the following order:

5.2.1 A bubbling or contact pipet containing KOH solution (**Caution:** avoid contact to skin and eyes).

5.2.2 A bubbling pipet containing fuming sulfuric acid. (**Caution:** Avoid contact with skin and eyes).

5.2.3 A distributing tip bubbling pipet containing alkaline pyrogallol solution.

5.2.4 A slow combustion pipet with platinum spiral.

5.2.5 A bubbling pipet containing KOH solution. (Duplicate of **5.2.1**) (**Caution:** Avoid contact to skin and eyes).

5.2.6 A distributing tip pipet containing alkaline pyrogallol solution. (Duplicate of **5.2.3**).

5.2.7 These pipets must possess smooth surfaces which will not entrap gas bubbles, and must be so efficient as to absorb the following gases from the sample after the required number of contacts with reagent.

Gas	Number of contacts with Reagent
Oxygen	4–5
Carbon dioxide	3
Unsaturated hydrocarbons	3

5.3 MANOMETER. This apparatus must demonstrate reproducibility of measurement of 0.02 ml per single contact or 0.05 ml on three successive contacts using the same reference gas or air.

6. Reagents

6.1 All reagents must conform to ACS Standards of Purity, ACS Reagent Grade or Equivalent.

6.2 POTASSIUM HYDROXIDE SOLUTION (SATURATED). In 200 ml of distilled water, dissolve solid potassium hydroxide until excess KOH remains. Cool the saturated solution to at least 3 C below lowest expected temperature at which analysis will be carried out. Decant and store the supernate liquid.

6.3 FUMING SULFURIC ACID. Sulfuric acid which, when fresh, contains 15% free SO_3.

6.4 ALKALINE PYROGALLOL SOLUTION. Dissolve 17 g of pyrogallol crystals in 100 ml of potassium hydroxide solution, **6.1**. Store under refrigeration in a glass-stoppered bottle.

6.5 ACIDIC, COPPER (II) CHLORIDE SOLUTION. Dissolve 450 g of copper (II) chloride in 2500 ml of ACS reagent, special grade, 1.18 hydrochloric acid. If this solution appears greenish or black in color after preparation, strips or turnings of copper must be added to the solution until a straw-yellow colored liquid is produced on standing. Store solution over copper turnings or wire.

6.6 Ascarite may be used when an unusually accurate measure of carbon dioxide is required. This absorber may also be used to remove sulfur dioxide for a more accurate determination of unsaturated hydrocarbons **(1)**.

6.7 Seventy-five % saturated salt solution contains 30 g NaCl or Na_2SO_4, or both, 5 ml conc HCl, 2 drops methyl red, per 100 ml distilled water.

7. Analysis Procedure

7.1 Analysis with portable Orsat apparatus for carbon dioxide, oxygen, and carbon monoxide.

7.1.1 *Procedure.* The portable Orsat apparatus is fitted with a metal or wooden carrying case and uses a shortened form of buret with three gas absorbing pipets. In order, starting from the buret, the pipets are filled with potassium hydroxide, pyrogallol and cuprous chloride solutions respectively.

7.1.2 After filling the above pipets to the engraved mark with the above solutions, and before making an analysis, adjust the level of each to atmospheric pressure using the leveling bulb.

7.1.3 Open the stopcock of the buret to the atmosphere. Raise the leveling bulb until the buret fills to the stopcock with salt water (saturated). Connect the stopcock of the buret to the atmosphere to be sampled or to a sample container and fill the buret with sample gas by lowering the leveling bulb until the meniscus of the water level reads the desired volume in the buret.

7.1.4 Open the stopcock connecting the buret to the manifold of the absorbing system and also open the stopcock of the potassium hydroxide pipet. Pass the gas contained in the buret into the potassium hydroxide pipet by first raising, then lowering the leveling bottle. Repeat until three to five full contacts have been made. Return the remainder of the gas sample to the buret using the leveling bulb until the level of potassium hydroxide solution returns to the engraved mark, and with the pipet stopcock closed, again, adjust the water level in the buret to atmospheric pressure using the leveling bulb. Measure the volume V_2 of the remaining gas, and record the percent carbon dioxide as follows:

$$\text{per cent } CO_2 = \frac{100\,(V_1 - V_2)}{V_1}$$

7.1.5 Similarly, oxygen is removed from the remaining gas volume V_2 by passing this gas into the pyrogallol solution in the second pipet. Measurement of the remaining volume, V_3, is used to calculate the per cent oxygen by:

$$\text{per cent } O_2 = \frac{100\,(V_2 - V_3)}{V_1}$$

7.1.6 Carbon monoxide is measured by manipulating the remaining gas volume

V_3, as previously, to admit this sample into the pipet containing copper II chloride. However, before returning the gas volume V_4 to the buret for measurement, volume V_4 is passed once into the potassium hydroxide pipet to remove any HCl vapors evolved from copper II chloride.

$$\text{per cent CO} = \frac{100\,(V_3 - V_4)}{V_1}$$

7.2 LABORATORY ANALYSIS BY CONSTANT PRESSURE VOLUMETRY. In the laboratory bench apparatus, the buret is filled with mercury and enclosed in a water jacket, the pipets are connected to the buret by a manifold and a bubbling pipet containing sulfuric acid **(5.2.2)** is used together with a slow combustion pipet **(5.2.4)** which is equipped with a separate leveling bulb. Pipets **(5.2.5)** and **(5.2.6)** are added to the system. Results of both absorption and combustion analyses are reported.

7.2.1 *Removal of Gases by Absorption.*

7.2.2 *Procedure.* Transfer a 95 to 100 ml volume of sample gas to the buret allowing 2 to 3 min for temperature and humidity equilibration. Using the leveling bulb to bring the sample volume to atmospheric pressure, read the exact volume V_1 to the nearest 0.02 ml.

7.2.3 *Removal of Carbon Dioxide (or Acid Gases).* Displace the gas sample into the manifold and then transfer into the KOH pipet. Return the sample gas to the buret. Then contact the KOH pipet twice, finally returning the sample to the buret and allowing 2 to 3 min before equilibrating the sample to atmospheric pressure with the leveling bulb. Then read the volume V_2 to the nearest 0.02 ml.

7.2.4 *Removal of Unsaturated Hydrocarbons.* Displace the gas sample from the manometer arm, and pass twice in and out of the pipet containing fuming sulfuric acid. Transfer the sample to the KOH pipet and return to the sulfuric acid pipet for two successive contacts. Finally, return to the buret for measurement, V_3, of the gas after standing 3 min in the buret.

7.2.5 *Removal of Oxygen.* Displace the gas sample from the manometer and transfer twice to the pipet containing alkaline pyrogallol, then, transfer to the KOH pipet, and the sulfuric acid pipet in sequence. Finally, transfer twice to the alkaline pyrogallol pipet and return to the buret for measurement of the residual volume (V_4).*

7.3 COMBUSTION ANALYSIS.

7.3.1 Prepare nitrogen to be used as a transfer gas by absorption of oxygen from uncontaminated air by contact with alkaline pyrogallol pipet. Flush the manifold with this nitrogen, then, transfer approximately 40 ml of the pure nitrogen to the duplicate KOH pipet **(5.2.5)** for storage, V_5.

7.3.2 Transfer approximately 95 ml of pure cylinder oxygen, V_6, to the buret; measure, and transfer to the slow combustion pipet for storage. Let the inert impurities shown by the published analysis of this cylinder oxygen be represented as V_7.

7.3.3 Measure a fresh 30 to 35 ml of sample gas, V_8, for combustion analysis. If prior absorption analysis shows a level of unsaturated hydrocarbons greater than 0.2%, pass sample V_8 through the fuming sulfuric acid pipet prior to combustion analysis.

With the combustion gas sample contained in the buret, adjust the pressure in the combustion pipet, and the buret to atmospheric, and with the platinum wire glowing dull red, open the combustion pipet and slowly admit the gas sample over the hot platinum wire. Allow a full 15 min for the first pass of sample into the combustion tube. When all of the gas sample has been transferred to the combustion pipet, displace the gas contained in the manometer arm through the distributer into the combustion pipet. Over a period of about 5 min, return the contents of the combustion pipet to the buret until the mercury level is just below the platinum

*When acid gases, oxygen and unsaturated hydrocarbons occur at levels below 0.5%, they are best determined using reaction tubes (2), rather than pipets.

spiral, then return the gas slowly to the combustion pipet and repeat the slow combustion three times. Allow the final pass of sample gas in the combustion pipet to cool before returning to the buret. Measure this residue and record as V_9.

7.3.4 *Removal of CO_2 produced by combustion.* Displace the gas from the manometer and the contents of sample residue in the buret into the duplicate KOH pipet (**5.2.5**) three times. Return this sample volume to the combustion pipet and then repeat contact with the KOH solution before returning to the buret for measurement as V_{10}.

7.3.5 *Removal of excess oxygen after combustion.* Displace the sample gas from the manometer and contact 4 times the duplicate alkaline pyrogallol pipet (**5.2.6**). Then, contact once the duplicate KOH pipet (**5.2.5**) and pass once through the slow combustion pipet before returning to the alkaline pyrogallol pipet. Transfer this residue to the buret and measure, V_{11}.

8. Calibration and Standards

8.1 Calibration may be performed using commercially purchased oxygen, nitrogen, carbon monoxide, and specific hydrocarbon gases, as the application of the method dictates, to prepare synthetic mixtures for admission into the apparatus as calibration gas. Such calibration gases may be standardized, if desired, by gas chromatography and/or gravimetric methods

9. Calculations (5)

9.1 Absorption Analysis.

$$\text{Per cent } CO_2 = \frac{(V_1 - V_2)}{V_1} \times 100$$

$$\text{Per cent unsaturated hydrocarbons} = \frac{(V_2 - V_3)}{V_1} \times 100$$

$$\text{Per cent } O_2 = \frac{(V_3 - V_4)}{V_1} \times 100$$

Volume designations refer to steps indicated in **7.2.2** to **7.2.5** where these designations are:

V_1 = initial volume of sample for absorption analysis
V_2 = volume of sample after removal of CO_2 (and the acid gases)
V_3 = volume of sample after removal of unsaturated hydrocarbons
V_4 = volume of sample after removal of oxygen

9.2 Combustion Analysis.

Volume designations refer to steps indicated in **7.2** and **7.3** where these designations are:

V_1 = initial volume of sample
V_2 = volume of sample after removal of CO_2 (and the acid gases)
V_3 = volume of sample after removal of unsaturated hydrocarbons
V_4 = volume of residual gas after removal of oxygen

Per cent methane

$$= \tfrac{1}{3}\,[7(TC + CO_2) - 9\,O_2]\,\frac{100}{V_8}$$

Per cent ethane

$$= \tfrac{1}{3}\,[6\,O_2 - 4\,(TC + CO_2)]\,\frac{100}{V_8}$$

where,

TC $= V_6 + V_8 - V_9$, total sample contraction after combustion.
$CO_2 = V_5 + V_9 - V_{10} - V_{12}$, carbon dioxide produced upon sample combustion.
$O_2 = V_6 - V_7 + V_{13} - (V_{10} - V_{11})$, oxygen consumed during combustion.
$N_2 = V_{11} - V_5 - V_7$, balance nitrogen after subtraction of transfer nitrogen and inert impurities in oxygen taken for combustion.

Other volumes represented:

$$V_{12} = (V_1 - V_2)\left(\frac{V_8}{V_1}\right)$$

$$V_{13} = (V_3 - V_4)\left(\frac{V_8}{V_1}\right)$$

V_5 = Volume of nitrogen taken as transfer gas.

10. Effects of Storage

10.1 Not applicable.

11. References

1. STANDARD METHOD FOR ANALYSIS OF NATURAL GASES BY THE CHEMICAL VOLUMETRIC METHOD, A.S.T.M. Designation D1136-53 (reapproved 1970).
2. SHEPHERD, M. 1947. Analysis of a Standard Sample of Natural Gas by Laboratories Cooperating with the American Society for Testing Materials. Journal of Research, National Bureau of Standards (JRNBS) **38**, (1) p. 19, January. (Research Paper RP 1759).
3. MATUSZAK, M.P. Fisher Gas Analysis Manual, Fisher Scientific Co., Pittsburgh, Pa.
4. SERIES 621 GAS ANALYZERS. 1967. Hand Orsat Type, Hays Corporation, Michigan City, Indiana.
5. STANDARD METHODS OF CHEMICAL ANALYSIS, Vol. II, Part B, F.J. Welcher ed., Princeton, N.J., D. Van Nostrand Company, Inc., p. 1521.

Subcommittee 4

R. G. SMITH, *Chairman*
R. J. BRYAN
M. FELDSTEIN
D. C. LOCKE
P. O. WARNER

201.

Tentative Method of Analysis for Chloride Content of the Atmosphere (Manual Method)

12203-01-68T

1. Principle of the Method

1.1 Dissolved chlorides are titrated with a dilute solution of mercuric nitrate in the presence of a mixed diphenylcarbazone-bromphenol blue indicator. The chloride combines with the mercuric ions to form essentially un-ionized $HgCl_2$. When all of the chloride has been titrated, the excess Hg^{++} reacts with the diphenylcarbazone to form a blue-violet complex (**3**).

1.2 This procedure is applicable to aqueous solutions containing at least 2 ppm of chloride and free of the interferences listed below. Gaseous chlorides may be collected by fritted glass absorbers or impingers. Particulate chlorides may be collected by filtration, impingement, or electrostatic precipitation.

2. Range and Sensitivity

2.1 This procedure will determine 2 ppm chloride in solution (0.1 mg Cl^- in 50 ml) with reasonable accuracy (**3**).

2.2 As written, the method applies to solutions containing 0.01 to 12 mg of chloride in an appropriate volume. The upper limit is set by the conc of the $Hg(NO_3)_2$ reagent solution and the assumption that a 25-ml buret is used. This limit can be extended indefinitely by using more concd or larger volumes of titrant or by appropriate aliquotting.

2.3 The lower limit is set by the indistinctness of the endpoint in very dilute solutions. For small volumes of titrant, a microburet is recommended. The limit can be extended somewhat by detecting the endpoint photometrically at 525 nm (**1**) or by concentrating the chloride by evaporation or ion exchange.

3. Interferences

3.1 Heavy metal ions can change the color of the solution at the neutral endpoint and at the chloride endpoint without affecting the accuracy of the titration. In such cases the analyst should familiarize himself with the proper endpoint colors by titrating standard solutions approximating the composition of the samples. These interferences can also be removed by ion exchange on a cation exchanger in the hydrogen or sodium cycle (**4**).

3.2 If chromates or ferric ion are present, limit the sample to a volume containing no more than 2.5 mg of ferric ion plus chromate ion. Add 2 ml of freshly prepared hydroquinone solution and proceed as outlined in steps **7.2** to **7.5**.

3.3 If sulfites or sulfides are present, dilute the sample to 50 ml, add 0.5 ml of 30% hydrogen peroxide and allow to stand for one minute. Continue as outlined in steps **7.2** to **7.5**.

3.4 Iodide, bromide, cyanide and thiocyanate are titrated and reported as chlordide.

4. Precision and Accuracy

4.1 Precision is 0.1 ppm or 2% relative, whichever is greater (**2**).

5. Apparatus

5.1 Only conventional laboratory glassware is required.

6. Reagents

6.1 MERCURIC NITRATE SOLUTION, 0.014N. Dissolve 2.42 g of mercuric nitrate monohydrate in 25 ml of water containing 0.25 ml of conc nitric acid. Dilute the solution to 1 l and mix. One ml of this solution is equivalent to about 0.5 mg of Cl^-. Standardize the solution by titrating 10.00 ml of standard 0.025N sodium chloride solution as described in the procedure. The exact normality of the mercuric nitrate solution is then:

$$N = \frac{\text{ml} \cdot \text{NaCl} \times \text{NaCl normality}}{\text{ml} \cdot \text{Hg } (NO_3)_2}$$

6.2 STANDARD SODIUM CHLORIDE SOLUTION, 0.0250N. Dry 2 to 3 g of sodium chloride in an oven at 110 C for 2 hr. Dissolve 1.4613 g of the dried material in water, dilute to 1 l and mix.

6.3 MIXED INDICATOR. Dissolve 0.50 g of diphenylcarbazone and 0.50 g of bromphenol blue powder (tetrabromophenolsulfonephthalein) in 100 ml of methanol. Store in a brown glass bottle and keep under refrigeration. If properly stored, the indicator is stable for at least 6 months.*

6.4 NITRIC ACID, DILUTE SOLUTION. Dilute 3 ml of nitric acid to 1 l and mix.

6.5 SODIUM HYDROXIDE, DILUTE SOLUTION. Dissolve 10 g of sodium hydroxide in a liter of water.

6.6 HYDROQUINONE SOLUTION. Dissolve 1 g of hydroquinone in 100 ml of water.

6.7 HYDROGEN PEROXIDE. 30% solution.

7. Procedure

7.1 Transfer the sample or a suitable aliquot into a 250-ml Erlenmeyer flask and dilute, if necessary, to about 50 ml. The sample volume should not exceed 50 ml or contain more than 12 mg of chloride ion.

*This mixed indicator is covered by U.S. Patent 2,784,064. In publishing this procedure, the Intersociety Committee does not undertake to ensure any user of the procedure against liability for patent infringement.

7.2 Add 5 drops of mixed indicator and mix by swirling the flask.

7.3 If the solution is red or blue-violet, add dilute nitric acid solution until the color changes to yellow and then 1.0 ml in excess.†

7.4 If the solution is yellow or orange, add dilute sodium hydroxide solution dropwise until the color changes to blue-violet. Then add dilute nitric acid solution dropwise until the color changes to yellow. Add 1.0 ml of dilute nitric acid in excess.†

7.5 Titrate the solution with standard mercuric nitrate solution until the color, viewed by transmitted light, is blue-violet. Let the ml of $Hg(NO_3)_2$ solution required equal A. Run a blank on all the reagents substituting chloride-free water for the sample. Let this titration equal B.

8. Calibration and Standards

9. Calculation

9.1

$$\frac{62400 \times (A\text{-}B) \times N \times M \times (273 + t)}{V \times P \times m} = \text{ppm HCl by volume}$$

$$\frac{90500 \times (A\text{-}B) \times N \times M \times (273 + t)}{V \times P \times m} = \text{mg HCl/m}^3 \text{ at 25 C}$$

Where:

(A-B) is the net titration of the sample or aliquot,

N is the normality of the Hg $(NO_3)_2$ solution.

M is the total volume of sample solution.

m is the aliquot of volume. (M/m = aliquot factor.)

†The pH at this point should be 3.0 to 3.5 Potentiometric determination of the pH should not be attempted with a calomel electrode, however, as the electrode may contribute chloride ion to the solution. A KNO_3-agar salt bridge may be used or the calomel electrode may be used on a separate aliquot treated in the same manner as the aliquot to be titrated. Another alternative is to apply a drop of the solution to a narrow range pH paper. Do not dip the paper into the solution.

t is the sampling in degrees Celsius
V is the volume of air sampled in liters.
P is the sampling pressure in Torr.

10. Effect of Storage

10.1 Negligible in neutral or slightly basic solutions.

11. References

1. ALLIED CHEMICAL CORPORATION. Morristown, New Jersey. *Methods of Analysis*. Unpublished.
2. AMERICAN SOCIETY FOR TESTING AND MATERIALS. 1964. Method D 512. *Test for Chlorides in Industrial Water:* Part 23, 22–24.
3. CLARKE, FRANK E. 1950. Determination of Chloride in Water. *Analytical Chemistry*. 22:553–555 and 1458.
4. SAMUELSON, OLOF. 1963. *Ion Exchange Separations in Analytical Chemistry*. John Wiley & Sons, New York, pp. 257–258.

Subcommittee 2
L. V. CRALLEY, *Chairman*
L. V. HAFF
A. W. HOOK
E. J. SCHNEIDER
J. D. STRAUTHER
C. R. THOMPSON
L. H. WEINSTEIN

202.

Tentative Method of Analysis for Free Chlorine Content of the Atmosphere (Methyl Orange Method)

42215-01-70T

1. Principle of the Method

1.1 Near a pH of 3.0 the color of a methyl orange solution ceases to vary with acidity. The dye is quantitatively bleached by free chlorine, and the extent of bleaching can be determined colorimetrically. The optimum concentration range is 0.05 to 1.0 ppm in ambient air (145 μg to 2900 μg per M^3 at 25 C and 760 Torr).

2. Sensitivity and Range

2.1 The procedure given is designed to cover the range of 5 to 100 μg of free chlorine/100 ml of sampling solution. For a 30-l air sample, this corresponds to approximately 0.05 to 1.0 ppm in air, which is the optimum range.

Increasing the volume of air sampled will extend the range at the lower end, but only within limits, since 50 liters of chlorine-free air produce the same effect as about 0.01 ppm of chlorine.

By using a sampling solution more dilute in methyl orange, a concentration of 1 μg/100 ml of solution may be measured, but beyond this, problems are encountered because of the absorption of ammonia and other gases from the air and by the presence of minute amounts of chlorine-consuming materials even in distilled water.

3. Interferences

3.1 Free bromine, which gives the same reaction, interferes in a positive direction. Manganese (III, IV) in concentrations of 0.1 ppm or above also interferes positively. In the gaseous state, interference from SO_2 is minimal, but in solution, negative interference from SO_2 is significant. Nitrites impart an off-color orange to the methyl orange reagent. NO_2 interferes positively, reacting as 20% chlorine. SO_2 interferes negatively, decreasing the chlorine by an amount equal to one-third the SO_2 concentration.

4. Precision and Accuracy

4.1 The data available **(5)** indicate that 26 chlorine concentrations produced by two different methods (flowmeter calibrated by KI absorption, and gas-tight syringe) were measured by this procedure with an average error of less than ± 5% of the amount present.

5. Apparatus

5.1 Spectrophotometer. Suitable for measurement at 505 nm, preferably accommodating 5-cm cells.

5.2 Fritted bubbler. Coarse porosity, of 250 to 350 ml capacity (Figure 202:1).

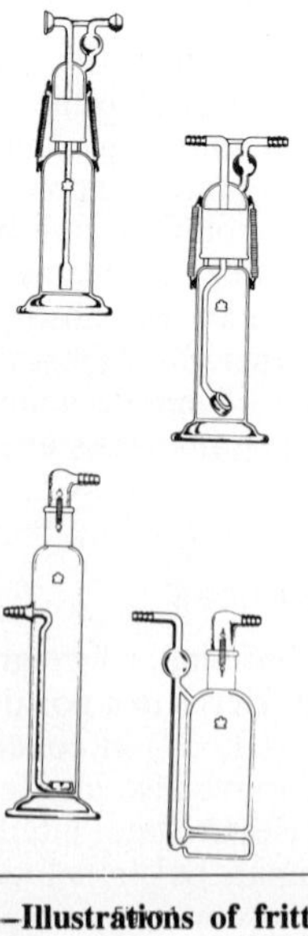

Figure 202:1—Illustrations of fritted bubblers, coarse porosity, of 250 to 350 ml capacity.

6. Reagents

6.1 Reagents must be ACS analytical grade quality. Distilled water should conform to ASTM Standard for Referee Reagent Water.

6.2 Methyl orange stock solution, 0.05%. Dissolve 0.500 g reagent grade methyl orange in distilled water and dilute to one liter. This solution is stable indefinitely if freshly boiled and cooled distilled water is used.

6.3 Methyl orange reagent, 0.005%. Dilute 100 ml of stock solution to 1 l with distilled water. Prepare fresh for use.

6.4 Sampling solution. 6 ml of 0.005% methyl orange reagent is diluted to 100 ml with distilled water and 3 drops (0.15 to 0.20 ml) of 5.0 N HCl added. One drop of butanol may be added to induce foaming and increase collection efficiency.

6.5 Acidified Water. To 100 ml of distilled water, add 3 drops (0.15 to 0.20 ml) of 5 N HCl.

6.6 Potassium Dichromate Solution, 0.1000 N. Dissolve 4.904 g anhydrous $K_2Cr_2O_7$, primary standard grade, in distilled water and dilute to 1 l.

6.7 Starch Indicator Solution. Prepare a thin paste of 1 g of soluble starch in a few ml of distilled water. Bring 200 ml of distilled water to a boil, remove from heat, and stir in the starch paste. Prepare fresh before use.

6.8 Potassium iodide, reagent grade.

6.9 Sodium thiosulfate solution, 0.1 N. Dissolve 25 g of $Na_2S_2O_3 \cdot 5H_2O$ in freshly boiled and cooled distilled water and dilute to 1 l. Add 5 ml chloroform as preservative and allow to age for 2 weeks before standardizing as follows: to 80 ml of distilled water add with constant stirring, 1 ml conc H_2SO_4, 10.00 ml 0.1000 N $K_2Cr_2O_7$, and approximately 1 g of KI. Allow to stand in the dark for 6 min. Titrate with 0.1 N thiosulfate solution. Upon approaching the end-point (brown color changing to yellowish green), add 1 ml of starch indicator solution and continue titrating to the end-point (blue to light green).

Normality $Na_2S_2O_3 \doteq \frac{1.000}{\text{mls of } Na_2S_2O_3 \text{ used}}$

6.10 Sodium Thiosulfate Solution, 0.01 N. Dilute 100 ml of the aged and standardized 0.1 N $Na_2S_2O_3$ solution to 1 l with freshly boiled and cooled distilled water. Add 5 ml chloroform as preservative and store in a glass-stoppered bottle. Standardize frequently with 0.0100 N $K_2Cr_2O_7$.

6.11 Chlorine Solution, Approximately 10 ppm. Prepare by serial dilution of household bleach (approx. 50,000 ppm), or by dilution of strong chlorine water made by bubbling chlorine gas through cold distilled water. The diluted solution should contain approximately 10 ppm of free (available) chlorine. Prepare 1 liter.

7. Procedure

7.1 100 ml of sampling solution is placed in the fritted bubbler, and a measured volume of air drawn through at a rate of 1 to 2 l/min for a period of time appropriate to the estimated chlorine concentration. Transfer the solution to a 100-ml volumetric flask and make to volume, if necessary, with acidified water. Measure absorbance at 505 nm in 5-cm cells against distilled water as reference.

7.2 The volume of sampling solution, the concentration of methyl orange in the sampling solution, the amount of air sampled, the size of the absorbing vessel, and the length of the photometer cell can be varied to suit the needs of the situation as long as proper attention is paid to the corresponding changes necessary in the calibration procedure.

8. Calibration

8.1 Prepare a series of six 100-ml volumetric flasks containing 6 ml of 0.005% methyl orange reagent, 75 ml distilled water, and 3 drops (0.15 to 0.20 ml) of 5.0 N HCl. Carefully and accurately pipet 0, 0.5, 1.0, 5.0 and 9.0 ml of chlorine solution (approximately 10 ppm) into the respective flasks, holding the pipet tip beneath the surface. Quickly mix and make to volume with distilled water.

8.2 Immediately standardize the 10 ppm chlorine solution as follows: To a flask containing 1 gm KI and 5 ml glacial acetic acid, add 400 ml of chlorine solution, swirling to mix. Titrate with 0.01 N $Na_2S_2O_3$ until the iodine color becomes a faint yellow. Add 1 ml of starch indicator solution and continue the titration to the endpoint (blue to colorless). One ml of 0.0100 N $Na_2S_2O_3$ = 0.3546 mg of free chlorine. Compute the amounts of free chlorine added to each flask in **8.1**.

8.3 Transfer the standards prepared in **8.1** to absorption cells and measure absorbance *vs.* micrograms of chlorine to draw the standard curve.

9. Calculation

9.1

$$\text{ppm } Cl_2 = \frac{\text{mg } Cl_2 \text{ found}}{\text{liters of air sampled (25 C, 760 Torr)}} \times \frac{24{,}450}{71}$$

For different temperatures and atmospheric pressures, proper correction for air volume should be made.

10. Effect of Storage

10.1 The color of sampled solutions is stable for at least 24 hr if protected from direct sunlight, although the presence of certain interferences (Fe III) may cause slow color change.

11. References

1. Taras, M. 1947. Colorimetric Determination of Free Chlorine with Methyl Orange. *Anal. Chem.* 19:342.
2. Boltz, D.F. 1958. Colorimetric Determination of Nonmetals, p. 163. Interscience Publishers, New York.
3. American Public Health Association. 1976. Standard Methods for the Examination of Water and Waste Water. 14th ed. Washington, D.C. 20036.
4. Traylor, P.A., and S.A. Shrader. Determination of Small Amounts of Free Bromine in Air. Dow

Chemical Co., Main Laboratory Reference MR4N, Midland, Michigan.
5. THOMAS, M.D., and R. AMTOWER. Unpublished work.

Subcommittee 2
C. R. THOMPSON, *Chairman*
L. V. CRALLEY
L. V. HAFF
A. W. HOOK
E. J. SCHNEIDER
J. D. STRAUTHER
L. H. WEINSTEIN

203.

Tentative Method of Analysis for Fluoride Content of the Atmosphere and Plant Tissues (Manual Methods)

12202-01-68T

PART A: *Introduction (General Precautions and Sample Preparation)*

PART B: *Isolation of Fluoride (Willard-Winter Distillation)*

PART C: *Isolation of Fluoride (Ion Exchange)*

PART D: *Isolation of Fluoride (Diffusion)*

PART E: *Determination of Fluoride (Titrimetric Method)*

PART F: *Determination of Fluoride (Spectrophotometric Method)*

PART A: *Introduction (General Precautions and Sample Preparation)*

1. General Precautions

1.1 Fluorine is one of the more common elements and occurs in at least trace amts in virtually all natural and manufactured materials. Contamination by extraneous fluoride may, therefore, come from such sources as sampling and laboratory apparatus, reagents, and from exposure to laboratory dust and fume. Care must be exercised in the selection, purification and testing of reagents and apparatus, and only minimal exposures of samples should be permitted.

1.2 Vessels used for evaporation, ashing, or caustic fusion of samples are first rinsed with warm, dilute acid (hydrochloric or nitric) solution, then with distilled water and air dried under clean toweling. Inconel crucibles used for fusion of ash may require additional cleaning by boiling in 10% (w/v) NaOH for 1 hr. Glassware is washed with hot detergent solution followed by a rinse in warm, dilute acid; it is finally rinsed with distilled water and dried (see Footnote 3 on the cleaning of distilling flasks). All sampling devices, containers, volumetric glassware, reagent solutions, etc., are stored under suitable conditions of protection from airborne dusts and fumes, and are reserved for exclusive use in low-fluoride analysis.

1.3 Before proceeding with analysis of samples, blank determinations are repeated until satisfactorily low values (5 μg, or less, total fluoride per determination) are consistently obtained. Calibration standards are analyzed whenever new batches of reagent solutions are prepared. In addition, one blank and one standard determination are carried through the entire analytic procedure with each set of 10 or fewer samples. If samples are handled in larger sets, the ratio of one blank and one standard per 10 samples should be maintained.

2. Sample Preparation

2.1 The techniques of sample recovery and preparation will vary, as described below, with the sampling method and equipment, and will also depend upon the procedures selected for isolation and measurement of fluoride. Until proof to the contrary is established, samples are assumed to contain fluoride in refractory forms in addition to the commonly encountered interfering materials. Many details involved in the determination of gaseous and particulate fluorides in the atmosphere and in vegetation are discussed by Pack, *et al,* **(13)** and automatic apparatus for the determination of ambient atmospheric hydrogen fluoride concn down to 0.1 ppb has also been described **(21)**.

2.2 PARTICULATE FLUORIDES.

2.2.1 Particulate matter collected in air sampling generally requires fusion with sodium hydroxide for conversion into soluble form prior to separation of fluoride by Willard-Winter distillation **(22)**. This treatment is also necessary for materials containing fluoride associated with aluminum, for materials high in silica, and for many minerals.

2.2.2 Transfer the sample-bearing paper filter to a resistant crucible, *e.g.*, platinum, nickel, or Inconel, moisten with water and make alkaline to phenolphthalein with special low-fluoride calcium oxide.* After evaporation to dryness, ignite the paper in a muffle furnace at a temp of 550 to 600 C until all carbonaceous material has been oxidized. Control combustion of filters of the membrane (cellulose ester) type by drenching with ethanolic sodium hydroxide and igniting with a small gas flame.

2.2.3 Remove particulate matter, collected by electrostatic precipitation, from the surfaces of both electrodes with the aid of a rubber policeman and distilled water. Make the resulting suspension alkaline and evaporate to dryness.

2.2.4 Integrated samples, *i.e.*, those containing both gaseous and particulate fluorides, which have been collected on glass fiber filters, are not amenable to fusion; filters are transferred directly to the distillation flask. Integrated samples collected in impingers are transferred to beakers made of nickel, platinum, or other resistant materials, evaporated to dryness in the alkaline condition, and the residue ashed if organic matter is present.

2.2.5 Fuse the impinger sample, residue from ashing of a filter, or electrostatic precipitator catch, with 2 g of sodium hydroxide. Dissolve the cold melt in a few ml of water, add a few drops of 30% hydrogen peroxide to oxidize sulfites to sulfates, and boil the solution to destroy excess peroxide. The sample solution is then ready for isolation of fluoride.

2.3 GASEOUS FLUORIDES.

2.3.1 *Dry Collectors.*

a. Treat filter papers impregnated with calcium-based fixative agents as described above for particulate fluorides, except that

*Available on special order from G. Frederick Smith Chemical Co., P.O. Box 23344, Columbus, Ohio 43223.

caustic fusion of the ashed residue is not required.

b. Filters impregnated with soluble alkalies are leached with water, as are fixative-coated beads or tubes. Evaporate the washings in a suitable vessel and maintain in an alkaline condition during reduction to a volume convenient for the subsequent fluoride separation procedure. Add a few drops of 30% hydrogen peroxide and boil the solution to destroy excess peroxide.

c. Gaseous fluorides collected on glass fiber filters cannot be quantitatively removed by leaching with water. Transfer such filters directly to the flasks in which a Willard-Winter distillation is to be conducted.

2.3.2 *Wet Collectors.*

a. Transfer a sample collected in water or alkaline solution to a suitably sized vessel, make alkaline to phenolphthalein with sodium hydroxide, and evaporate to the desired volume. Treat the solution with 30% hydrogen peroxide and destroy the excess peroxide by boiling before proceeding with the isolation and determination of fluoride.

2.4 VEGETATION.

2.4.1 Reduce the gross specimen to manageable size for mixing by use of hand shears or, in the case of dried materials, a Wiley cutting mill. Take a small portion (10 to 25 g) of the mixed specimen for determination of moisture by oven-drying at 80 C for 24 to 48 hr.

2.4.2 Adjust the weight of material taken for fluoride determination according to condition of the specimen: 100 to 105 g of fresh or frozen vegetation is satisfactory, while for dried materials, such as cured hay, dried leaves, straw, etc., a 50 g portion is adequate.

2.4.3 Weigh the sample into a resistant vessel,† bake alkaline to phenolphthalein by addition of low-fluoride calcium oxide slurry, and maintain alkalinity during evaporation to dryness on a hot plate. Raise the temp of the hot plate until charring and partial ashing have occurred. Complete the ashing at 500 to 600 C in an electric furnace reserved for the ignition of low-fluoride materials. The ash should be white or gray, indicating removal of organic matter.

2.4.4 When the ash has cooled, pulverize it and scrape all material from the dish; mix and determine the net weight. Store in a tightly stoppered bottle.

2.4.5 In order to effect quantitative release of fluoride combined with silica in many varieties of vegetation, fusion of the limed-ash with sodium hydroxide is required and is routinely performed on all vegetation specimens (**15, 18**). Transfer approximately one g of ash into a tared nickel, Inconel, or platinum crucible and weigh accurately. Add about 5 g of sodium hydroxide pellets, cover the vessel and fuse the contents for a few min over a gas burner. After cooling the melt note its color; a blue-green color indicates the presence of manganese and treatment with hydrogen peroxide is required as described under *Procedure for Single Distillation, Vegetation Ash.* Disintegrate the melt with hot water, washing down the lid and walls of the crucible. Reserve the resulting material for isolation of fluoride.

†Inconel dishes 14 cm in diam and 8 cm high, obtainable from Precision Metal Spinning Company, 9825 Dixie Highway, Clarkston, Michigan, are satisfactory.

PART B: *Isolation of Fluoride (Willard-Winter Distillation)*

1. Principle of the Method

1.1 The prepared sample is distilled from a strong acid such as sulfuric or perchloric, in the presence of a source of silica. Fluoride is steam-distilled as the fluosilicic acid under conditions permitting a minimum of volatilization and entainment of the liberating acid (**1**).

2. Range and Sensitivity

2.1 The Willard-Winter distillation method, on the macroscale, can accommodate quantities of fluoride ranging from 100 mg down to a few μg.

3. Interferences

3.1 Samples relatively free of interfering materials, and containing fluoride in forms from which it is easily liberated, may be subjected to a single distillation from perchlorate acid at 135 C. Samples containing appreciable amts of aluminum, boron, or silica require a higher temp and larger volume of distillate for quantitative recovery. In this case a preliminary distillation from sulfuric acid at 165 C is commonly used. Large amts of chloride are separated by precipitation with silver perchlorate following the first distillation. Small amts are held back in the second distillation from perchloric acid by addition of silver perchlorate solution to the distilling flask.

4. Precision and Accuracy

4.1 Recovery data for the Willard-Winter distillation, as given in the literature, are difficult to dissociate from inaccuracies inherent in various methods of sample preparation and final evaluation of fluoride. Recovery data from field samples are further complicated by variability of interfering substances and ranges of fluoride contained. In general, recoveries should be within ± 10% of the amt of fluoride present. Under favorable circumstances of sample composition and fluoride range, mean recoveries of approximately 99% with standard deviation of about 2.5%, have been reported (**20**).

5. Apparatus

5.1 Steam Generator. (Figure 203:1) 2000-ml Florence flask made of heat-resistant glass. Each flask is fitted with a stopper having at least 3 holes for inserting 6-mm OD heat-resistant glass tubing. Through one of the glass tubes, bent at right angles, steam is introduced into the distilling flask. The second tube is a steam release tube (Figure 203:1,D) which controls the steam pressure. The small piece of rubber tubing which is slipped over the end of the steam release tube is clamped shut during sample distillation. The third tube is a safety tube. If desired, other tubes may be added to permit the steam generator to supply a max of 3 distilling flasks. Any suitable heating device may be used.

5.2 Distilling Flask. (Figure 203:1,B) A 250-ml modified Claisen flask made of heat-resistant glass. The auxiliary neck of this flask is sealed and the outer end of the side tube is bent downward so that it may be attached to an upright condenser. The side tube is fitted with a one-hole rubber stopper to fit the condenser and main neck with a two-hole stopper through which passes a thermometer and a 6-mm OD heat-resistant glass inlet tube for admitting the steam. Any suitable heating device may be used.

5.3 Liebig Condenser. (Figure 203:1,C) heat-resistant glass, 300-mm jacket.

5.4 Steam Release Tube. (Figure 203:1,D) (See Steam Generator, **Section 5.1.**)

5.5 Thermometer. (Figure 203:1,E) Partial Immersion Thermometer having a range of 0 to 200 C.

5.6 Support Plate. (Figure 203:1,F) metal, ceramic, or hard asbestos board. The plate shall have a perfectly round 5-cm hole in which the distilling flask is placed as shown in Figure 203:1. The Claisen flask must fit well in the 5-cm hole so that the flask wall, above the liquid level, is not subjected to direct heat. Excessive heat on the wall of the flask causes the liberating acid to be distilled.

5.7 Receiver. (Figure 203:1,G) 250- or 500-ml volumetric flask, or 400-ml beaker.

5.8 Safety Tube. (Figure 203:1,H) A 6-mm OD heat-resistant glass tubing, 60-cm long, one end of which is 1 cm from the bottom of the steam generator flask.

5.9 Rubber Tubing. (Figure 203:1,I) for flask connections, made from natural

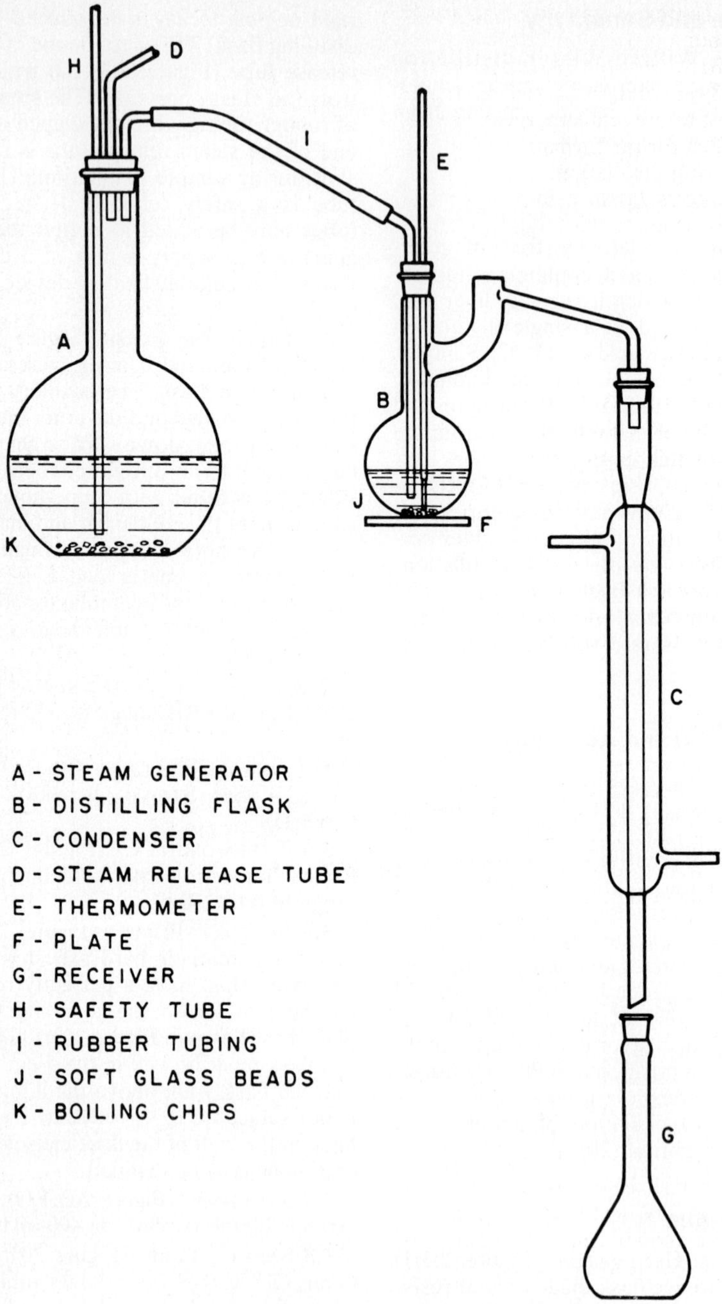

Figure 203:1—Apparatus for distillation of fluoride.

rubber. Lengths of rubber tubing shall be kept as short as possible.

5.10 SOFT GLASS BEADS. (Figure 203:1,J) 3-mm diam, for use in the distilling flask to prevent superheating and to supply silica for the formation of fluosilicic acid during distillation.

5.11 POROUS PUMICE STONES or BOILING CHIPS. (Figure 203:1,K).

5.12 PINCHCOCK. (Figure 203:1,L) to control steam supply from the generator.

6. Reagents

6.1 PERCHLORIC ACID. (70 to 72% by wt) Concentrated perchloric acid ($HClO_4$).‡

6.2 SILVER PERCHLORATE SOLUTION. 50% w/v)—Dissolve 100 g of silver perchlorate ($AgClO_4$) in 100 ml of water.

6.3 SULFURIC ACID. (96% by wt)—Concentrated sulfuric acid (H_2SO_4).‡

6.4 WATER. All references to water shall be understood to mean distilled or deionized water of reagent purity.

7. Procedure

7.1 PROCEDURE FOR SINGLE DISTILLATION, MISCELLANEOUS MATERIALS.

7.1.1 Fill a steam generator about two-thirds full of water. Add to it a pellet of sodium hydroxide and a few drops of phenolphthalein indicator solution to insure that the water remains alkaline at all times. Add a piece of pumice to permit free boiling, and heat the water to boiling. Keep the steam release tube open at this time and place a pinchcock on the steam supply tubing.

7.1.2 Introduce the sample into a Claisen distilling flask containing five or six glass beads. Wash down the sides of the flask with water and bring the volume to 50 to 75 ml, the lesser volume being more desirable. Insert in the main neck of the flask the rubber stopper that contains the thermometer and steam inlet tube. Set the flask in the 5-cm diam hole in the plate and connect the outlet to a condenser.

7.1.3 Rinse the sides of the beaker or crucible which contained the sample with 50 ml of perchloric acid (70 to 72%)§ and add 1 ml of silver perchlorate. Transfer the rinsings to the distilling flask by means of a small funnel attached to the steam inlet tube. Rinse the beaker or crucible with water and add the rinsings to the flask. Mix the contents of the flask by gentle shaking and attach the flask to the steam generator. Place a 250-ml volumetric flask under the condenser to receive the distillate and begin heating the solution in the flask. Keep the pinchcock in place on the steam inlet tube until the contents of the distilling flask reach 135 C.

7.1.4 Remove the pinchcock on the steam inlet tube and place it on the steam release tube of the steam generator. Maintain the distillation temp at 135 ± 2 C. Swirl the contents of the distilling flask frequently to minimize deposition on the flask wall of any siliceous residues that might retain fluoride. After collecting 250 ml of distillate during a period of about 1 hr, remove the pinchcock from the steam release tube and place it on the steam inlet tube. Disconnect the rubber tubing from the steam inlet tube, and discontinue heating.‖

7.2 PROCEDURE FOR SINGLE DISTILLATION, VEGETATION ASH.

7.2.1 Transfer the disintegrated melt

‡Acid giving excessively high fluoride blanks requires preboiling at 135 C, with admission of steam, prior to addition of samples.

§*Caution.* When using perchloric acid, the usual precautions should be taken. Hot concd perchloric acid may react explosively with reducing substances, such as organic matter. Therefore, it is wise to see that any organic matter in the sample is destroyed in the ashing process prior to distillation. Precautions for the use of perchloric acid are available in "Chemical Safety Data Sheet SD-11, Perchlorid Acid Solution," published by the Manufacturing Chemists' Association of the United States (**14**).

‖*Caution.* The distilling flasks should be cleaned using only a brush and distilled water. Repeated use of alkaline cleaning solution produces an etched surface that is difficult to clean and tends to retain fluoride.

to a Claisen distilling flask, as described in section **7.1** for Miscellaneous Materials.

7.2.2 Rinse the sides of the crucible in which the fusion was made with 50 ml of perchloric acid (70 to 72%) and add 1 ml of silver perchlorate solution.

7.2.3 If the sample contains manganese, add sufficient (2 to 10 drops) 3% hydrogen peroxide solution to the contents of the distilling flask to reduce manganese dioxide and permanganates (**6**).

7.2.4 Carry out the distillation as previously described, except that a 500-ml volumetric flask is used as receiver and filled with distillate during a period of about 2 hr.

7.3 Procedure for Double Distillation.

7.3.1 Fill a steam generator, as directed under Procedure for Single Distillation. Transfer the sample solution to a Claisen flask and rinse the sides of the beaker or crucible which contained the sample with 50 ml of conc sulfuric acid. Transfer the rinsings to the distilling flask through a small funnel attached to the steam inlet tube. Mix the contents of the flask by swirling, rinse and remove the funnel, and connect the distilling flask to the steam generator. Place a 400-ml beaker under the condenser and begin heating the distilling flask, and steam generator. Keep the pinchcock in place on the steam inlet tube until the contents of the distilling flask reach 165 ± 5 C. Swirl contents of the flask as required to prevent accumulation of insoluble material on the walls of the flask above the liquid level. Collect about 375 ml of distillate during a period of about 1½ to 2 hr.

7.3.2 Add sodium hydroxide solution (10 g/l) to the distillate until alkaline by phenolphthalein indicator. Evaporate the distillate to 10 to 15 ml by heating below the boiling point.

7.3.3 The concentrated distillate is redistilled from perchloric acid as directed under Procedure for Single Distillation. Small quantities of chloride are fixed in the distilling flask by the addition of 1 ml of silver perchlorate solution. A 250-ml quantity of distillate is collected in a volumetric flask.

8. Calibration and Standards

9. Calculation

10. Effect of Storage

11. References (See Part F, Section 11.)

PART C: *Isolation of Fluoride (Ion Exchange)*

1. Principle of the Method

1.1 The sample is freed of interferences by preferential sorption on an ion exchange resin, followed by desorption of fluoride in a small volume of eluting solution. Thus, concentration of fluoride from impinger- or bubbler-collection media may be achieved without the attendant danger of contamination on prolonged exposure of solutions during evaporation (**12**).

2. Range of Sensitivity

2.2 The ion exchange procedure can be adapted to quantities of fluoride in the low-mg to μg range.

3. Interferences

3.1 Interfering cations may be eliminated by sorption of fluoride on an anion exchange resin. Fluoride is then eluted with sodium hydroxide solution.**

When interferences are present in cationic as well as anionic forms, both may be removed by use of a strongly basic anion exchange resin. This is accomplished by conversion of cations into strongly-held complex anions, while the weakly-held fluoride ions are quantitatively eluted from the column (10**).

4. Precision and Accuracy

4.1 Net recoveries for quantities of fluoride of 20 μg or more should be within ±5% of the quantity present. Low recovery indicates incomplete preconditioning of a new column, while high recovery may be due to contamination or failure to elute completely the previous sample.

5. Apparatus

5.1 CHROMATOGRAPHIC COLUMN. Dimensions of the column are not critical and many types, available from suppliers' stocks, are usable. A column made of borosilicate glass tubing 10-mm ID and 16 cm long, having a fritted glass disc fused into the constricted base, and a reservoir of about 100 ml capacity at the top, is satisfactory. A short piece of polyvinyl chloride tubing attached to the bottom and closable with a screw hose clamp permits adjustment of flow rates and prevents complete drainage of liquid from the column.

5.2 QUARTZ SAND, WHITE. Sand of −60 +120 mesh, purified by hot extraction with 20% sodium hydroxide solution, followed by hot 10% hydrochloric acid solution, is used as a protective layer at the top of the resin bed.

6. Reagents

6.1 ANION EXCHANGE RESIN. Intermediate-Base of the granular aliphatic polyamine type:

Duolite A-41, A-43 (Diamond Alkali Co.)

Ionac A 302 (Ionac Chem. Co., Div. of Ritter Pfaudler Corp.)

Permutit A (Permutit Co., Ltd., Div. of Ritter Pfaudler Corp.)

Rexyn 205 (OH) (Fisher Scientific Co.)

Mesh size is not critical but, along with column diam and height, is a factor in controlling flow rate of solutions. Mesh sizes of 60 +100 or −100 +200 are usable.††

6.2 HYDROCHLORIC ACID, 2.0 N AND 1.0 N SOLUTIONS.

6.3 SODIUM HYDROXIDE, 2.0 N, 0.1 N, AND 0.01 N SOLUTIONS.

7. Procedure

7.1 Prepare the resin column by adding a few ml of water to the chromatographic tube, then a slurry of resin (1:1) in water. Add sufficient slurry so that when the resin has settled, a layer 10 to 12 cm in height will result. Level the resin bed by twirling the tube, and before the water level has dropped below the surface of the resin, add a 2-cm layer of quartz sand. Wash the resin with 200 ml of 2.0 N hydrochloric acid solution, rinse with water; wash with 200 ml of 2.0 N sodium hydroxide solution, and, finally, rinse with 200 ml water.

7.2 Precondition the resin by passing 400 ml of a solution containing about 1 ppm hydrofluoric acid and an equal volume containing 1 ppm sodium fluoride through the column at a rate of about 5 ml/min. Follow this with 50 ml of 0.1 N sodium hydroxide solution, then 25 ml of 0.01 N sodium hydroxide solution. Discard the eluate. The resin is now ready for use.

7.3 Acidify the sample solution by addition of 0.5 ml of 1 N hydrochloric acid per 100 ml, but add no more than 3 ml of 1 N hydrochloric acid per sample. Remove, by filtration, any solids remaining in the sample after acidification. Pass the sample solution through the resin column at a rate of about 10 ml/min, followed by a water rinse of a few ml. Elute fluoride with a 25-ml portion of 0.1 N sodium hydroxide solution, followed by a 25-ml portion of 0.01 N sodium hydroxide solution.

8. Calibration and Standards

9. Calculation

††Separation of fluoride may be achieved with other resins and appropriate eluting solutions; for example, Dowex 1-X8 in the acetate form may be used with elution by sodium acetate solution (**11**).

10. Effect of Storage

10.1 An ion exchange column may be preserved indefinitely if the resin is covered with water. Before the column is reused, a recovery test should be made by passage of a measured quantity of standard fluoride solution through the column; 200 ml of neutral sodium fluoride solution, at a rate of 10 ml/min, is suggested. The quantity of fluoride added should approximate that expected in the sample. Fluoride is eluted, as described under **Procedure**, and determined by the method selected for evaluation of samples.

11. References (See Part F, Section 11.)

PART D: *Isolation of Fluoride (Diffusion)*

1. Principle of the Method

1.1 An aliquot of the prepared sample is mixed with a strong acid, gently heated in a sealed container, and the liberated hydrogen fluoride is absorbed by an alkali **(7, 17, 19)**.

2. Range and Sensitivity

2.1 Quantities of fluoride from about 30 μg to a few tenths of a μg may be used. In routine work, blanks range from 0.5 μg to 0.0 μg.

3. Interferences

3.1 Interfering materials that volatilize from acid medium must be eliminated. Sulfites are oxidized to sulfate by preliminary treatment with 30% hydrogen peroxide solution. Relatively large amts of chloride may be fixed in the diffusion vessel as silver chloride, by addition of 0.1 to 0.2 g silver perchlorate to the sample aliquot prior to diffusion. Samples high in carbonates require caution upon acidification, to control effervescence.

4. Precision and Accuracy

4.1 Recoveries from five sodium fluoride standards, covering the range 4 μg to 20 μg F^-, varied from 97.5% to 102.5%; average 99.4% **(17)**. By a slightly modified technique, standards containing 0.2 μg, 0.5 μg, and 1.0 μg F^- (five replicates of each) yielded recoveries of 94% to 101%; average 98.1% **(7)**.

5. Apparatus

5.1 MICRODIFFUSION DISH. Disposable plastic petri dish, 48 mm ID by 8 mm deep. (Obtainable from Millipore Filter Corp., Bedford, Mass. 01730).#

5.2 OVEN. A thermostatically controlled oven capable of maintaining temp with ±1 C in the 50 to 60 C range.

5.3 PIPET, MOHR. Capacity 0.1 ml, 0.01 ml subdivisions.

6. Reagents

6.1 PERCHLORIC ACID, 70 to 72%.§§

6.2 SILVER PERCHLORATE. ANHYDROUS, C.P. ($AgC1O_4$).

6.3 SODIUM HYDROXIDE. 1 N alcoholic solution. Dissolve 4 g sodium hydroxide (NaOH) in 5 ml of water and dilute to 100 ml with ethyl-, methyl-, or Formula 30 denatured alcohol.

7. Procedure

7.1 Place 0.05 ml of 1 N alcoholic sodium hydroxide solution on the center of the inside top of the plastic petri dish. Use the tip of the 0.1-ml Mohr pipet to spread the droplet into a circular spot of about 3 to 4 cm diam. Dry the top for about 1 hr,

#The Conway Microdiffusion Dish, with Obrink modification, made of methyl methacrylate resin or similar plastic capable of withstanding temps up to 60 C, may also be used.

§§See footnote §, page 00.

under slightly reduced pressure, in a desiccator containing activated alumina.

7.2 Transfer a 1.0 ml aliquot of prepared sample solution to the diffusion unit. Add 2.0 ml of perchloric acid and immediately close the dish with a prepared lid. Place the unit in an oven maintained at the selected temp (50 to 60 C) and allow to remain for 16 to 20 hr.

7.3 Carefully remove the diffusion vessel from the oven and take off the lid. Wash the alkaline absorbent into a 10- or 25-ml volumetric flask (a small funnel is helpful), the size of the flask depending upon amt of fluoride expected and method of measurement chosen.

8. Calibration and Standards

9. Calculation

10. Effect of Storage

11. References (See Part F, Section 11.)

PART E: *Determination of Fluoride (Titrimetric Methods)*

1. Principle of the Method

1.1 In the direct titration of fluoride with standard thorium nitrate solution, the sample solution or distillate containing sodium alizarinsulfonate is buffered at pH 3.0. Upon addition of thorium nitrate, insoluble thorium fluoride is formed. When the endpoint is reached, and all fluoride has reacted, the addition of another increment of thorium nitrate causes formation of a pink "lake" **(16)**.

1.2 In the back titration procedure, the pink lake is first formed by addition of sodium alizarinsulfonate and a slight excess of thorium nitrate to the sample. Equal amounts of dye and thorium solution are added to a fluoride-free reference. The reference solution is then titrated with standard sodium fluoride solution until a color match is achieved with the unknown sample **(16)**.

2. Range and Sensitivity

2.2 The direct titration procedure can accommodate 10 to 0.05 mg fluoride in the total sample. The back titration modifications can measure 50 to about 5 μg fluoride in the total sample. With photometric endpoint detection, direct titration can also be used for the lower ranges.

3. Interferences

3.1 Ions capable of forming insoluble or undissociated compounds with fluorine or with thorium interfere with these titrimetric methods and must be separated by an appropriate technique (*e.g.*, distillation, diffusion or ion exchange). Among the more common of the interfering cations are Al^{+3}, Ba^{+2}, Ca^{+2}, Fe^{+3}, Th^{+4}, TiO_2^{+2}, VO^{+2}, and Zr^{+4}. The principal interfering anions are PO_4^{-3} and SO_4^{-2}. However, any material which constitutes an appreciable change in total ionic strength of the sample solution will affect the endpoint color as well as stoichiometry of the reaction. Thus, excessive acidity in the distillate from a Willard-Winter distillation, as from the liberating acid or chloride content of the sample, will interfere. This effect may be reduced by careful control of temp and rate of admission of steam, and by separation of chloride. Similarly, acidity or alkalinity of eluates from ion exchange separations must be matched with that of standards used in calibration, and with the requirements of the method of evaluation.

3.2 Sulfide and sulfite interferences are prevented by preliminary oxidation with 30% hydrogen peroxide in boiling solution, as described under **Sample Prepara-**

tion. Interference by free chlorine is eliminated by addition of hydroxylamine hydrochloride solution.

4. Precision and Accuracy

5. Apparatus

5.1 FLUORESCENT LAMP. To provide illumination for titrating.

5.2 MICROBURET. 5-ml capacity, 0.01-ml divisions, and a reservoir holding about 50 ml.

5.3 NESSLER TUBES. Matched set of 50-ml, tall-form tubes with shadowless bottoms. Tubes may be fitted with either ground glass or rubber stoppers. The set should be checked for optical similarity as follows: Add to the tubes 40 ml of water, 1 ml of sodium alizarinsulfonate solution, and 2 ml of 0.05 N hydrochloric acid. Add thorium nitrate solution from a buret until the color of the solution just changes to pink. Close the top of the tube and invert several times. Add the same quantity of thorium nitrate solution to the remaining tubes. Fill all the tubes to the 50 ml mark with water and mix. Compare the colors and reject any tubes showing differences in shade or intensity.

5.4 NESSLER TUBES. Matched set of 100-ml, tall-form tubes with shadowless bottoms. The set should be checked for optical similarity, using the same technique as with the 50-ml tubes, except that the quantities of reagents shall be doubled.

5.5 NESSLER TUBE RACK OR COMPARATOR.

5.6 PHOTOMETRIC TITRATOR. A Beckman Model B Spectrophotometer, equipped with an Alcoa Research Laboratories' titration attachment (Figure 203:2) or equivalent (**8**). Light from the monochromator passes through a 20.3-cm (8″) sample cell to the blue-sensitive phototube mounted at the outboard end of the cell housing. A magnetic stirrer is attached under the cell compartment. The tip of a semimicroburet passes through the cell housing and is immersed in the solution to be titrated. A ball-and-socket joint connects the tip to the buret, facilitating removal of the sample cell. The titration cell is 5.1 cm (2″) wide, 7.6 cm (3″) deep, and 20.3 cm (8″) long.

6. Reagents

6.1 BUFFER-INDICATOR SOLUTION. Dissolve 0.40 g of sodium alizarin-sulfonate in about 200 ml of water. Weight 47.25 g of monochloracetic acid into a 600-ml beaker and dissolve in 200 ml of water. Add indicator solution with stirring. Dissolve 10 g of sodium hydroxide pellets in 50 ml of water, cool to approximately 15 to 20 C, and add to the above solution slowly with stirring. Filter and make to 500 ml. Prepare fresh weekly.

6.2 CHLOROACETATE BUFFER SOLUTION. Dissolve 9.45 g of monochloroacetic acid and 2.0 g of sodium hydroxide (NaOH) in 100 ml of water. This solution is stable for more than 2 weeks if stored under refrigeration.

6.3 HYDROCHLORIC ACID, STANDARD SOLUTION (0.05 N). Dilute 4.28 ml of hydrochloric acid (HCl, sp gr 1.19) to 1 l. The normality of this solution should be exactly equal to that of the sodium hydroxide (NaOH) solution (0.05 N).

6.4 HYDROXYLAMINE HYDROCHLORIDE SOLUTION. 1 g of $NH_2OH.HCl$/100 ml of water.

6.5 PHENOLPHTHALEIN INDICATOR SOLUTION (0.5 G/L). Dissolve 0.5 g of phenolphthalein in 60 ml of ethyl alcohol and dilute to 1 l with water.

6.6 SODIUM ALIZARINSULFONATE SOLUTION (0.80 G/L). Dissolve 0.40 g of sodium alizarinsulfonate in 500 ml of water.‖ ‖

‖ ‖In the literature, this reagent is also known as alizarin Red S, alizarin Red, alizarin-S, alizarin carmine, alizarin, sodium alizarin sulfonate, sodium alizarin monosulfonate, monosodium alizarin sulfonate, and 3-alizarinsulfonic acid sodium salt. The dye is identified by Color Index No. 58005.

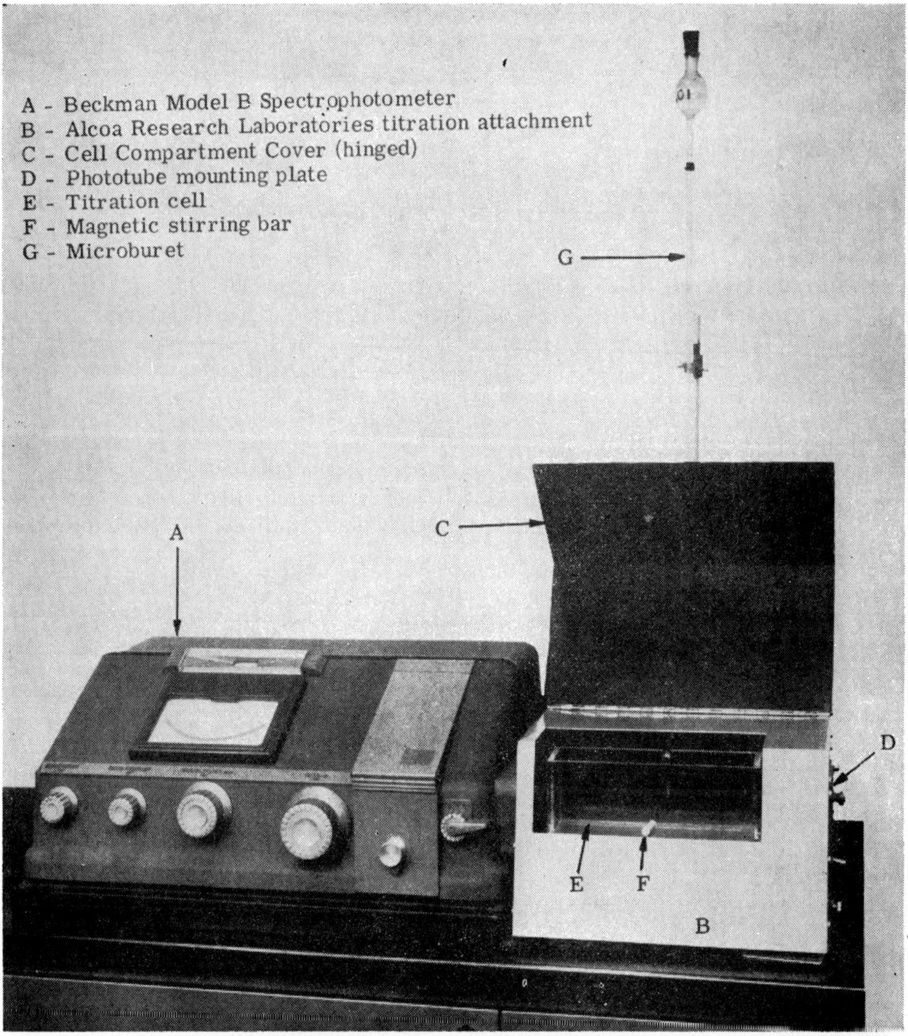

Figure 203:2—Photometric Titrator.

6.7 SODIUM ALIZARINSULFONATE SOLUTION (0.01 G/L). Dissolve 0.01 g of sodium alizarinsulfonate in 1000 ml of water.

6.8 SODIUM FLUORIDE, 100%.

6.9 SODIUM FLUORIDE, STANDARD SOLUTION (1 ML = 1.00 MG F). Dissolve 2.2105 g of sodium fluoride (NaF, 100%) in water and dilute to 1 l in a volumetric flask, mix, and transfer to a polyethylene bottle for storage.

6.10 SODIUM FLUORIDE, STANADRD SOLUTION (1 ML = 0.01 MG F). Dilute 10 ml of NaF solution (1 ml = 1.00 mg F) to 1 l with water in a volumetric flask, mix, and transfer to a polyethylene bottle for storage.

6.11 SODIUM HYDROXIDE SOLUTION (10 G/L). Dissolve 10 g of NaOH in water, dilute to 1 l and mix. Store in a polyethylene bottle.

6.12 Sodium Hydroxide, Standard Solution (0.05 N). Dissolve 2.00 g of NaOH in water and dilute to 1 l. The normality of this solution should be exactly equal to that of the standard HCl (0.05 N). Store in a polyethylene bottle.

6.13 Thorium Nitrate, Standard Stock Solution (1 ml = 1.9 mg F). Dissolve 13.80 g of thorium nitrate tetrahydrate ($Th(NO_3)_4 \cdot 4H_2O$) in water and dilute to 1 l.

6.14 Thorium Nitrate Solution (0.25 g/l). Dissolve 0.25 g of thorium nitrate tetrahydrate ($Th(NO_3)_4 \cdot 4H_2O$) in water, dilute to 1 l and mix. Store in a polyethylene bottle.

6.15 Thorium Nitrate Solution, 0.01 N (0.19 g F/l). Dilute a 100-ml aliquot of the stock solution (**6.13**) to 1 l and store in polyethylene bottle.

7. Procedure

7.1 Procedure for Direct Titration, High Concentrations. (10 to 0.05 mg F in Total Sample).

7.1.1 Pipet an aliquot of the distillate into a 400-ml beaker and dilute to 100 ml. Add 1 ml of sodium alizarinsulfonate solution (0.80 g/l), and then sodium hydroxide solution (10 g/l) dropwise until a pink color is obtained. Discharge the pink color by adding 0.05 N hydrochloric acid dropwise. Add 1 ml of chloroacetate buffer solution dropwise, and titrate with thorium nitrate solution (1 ml = 1.9 mg F) to a faint, persistent, pink endpoint. Determine a blank obtained by carrying the same amt of all reagents through the entire procedure.

7.2 Procedure for Back Titration, Medium Concentration. (0.05 to 0.01 mg F in Total Sample).

7.2.1 Transfer 50 ml of the distillate into a 50-ml Nessler tube, add 1 ml of sodium alizarinsulfonate solution (0.01 g/l) and sufficient 0.05 N sodium hydroxide solution to produce a pink color. Note precisely the volume of 0.05 N sodium hydroxide solution required for neutralization; then discard the titrated solution. If more than 4 ml of 0.05 N sodium hydroxide solution is required, make the remaining distillate alkaline, evaporate to 10 to 15 ml, and transfer it to a distilling flask. Repeat the distillation, precautions being taken to reduce the amt of perchloric acid distilled over.

7.2.2 Transfer another 50-ml portion of distillate into a 50-ml Nessler tube (sample tube) and add 1 ml of sodium alizarinsulfonate solution (0.01 g/l). Adjust the acidity with 0.05 N hydrochloric acid until the equivalent of exactly 2 ml of acid is present; that is, 2 ml minus the number of ml of 0.05 N sodium hydroxide solution required for neutralization as described. If between 2 ml and 4 ml of 0.05 N sodium hydroxide solution were required for neutralization, omit the addition of hydrochloric acid to the distillate. Add thorium nitrate solution (0.25 g/l) from a microburet until a faint pink color appears. Note the volume of thorium nitrate solution required, and save the Nessler tube for comparison with the standard.

7.2.3 Pour 50 ml of water into a 50-ml Nessler tube (standard tube) and add 1 ml of sodium alizarinsulfonate solution of (0.01 g/l). If neutralization of the sample requires 2 ml or less of 0.05 N sodium hydroxide solution, pipet exactly 2 ml of 0.05 N hydrochloric acid into the standard tube. If the 50-ml aliquot of the distillate requires more than 2 ml of 0.05 N sodium hydroxide solution for neutralization, no further acidification of the distillate is necessary, but add to the standard tube a quantity of acid equivalent to that found in the sample distillate.

7.2.4 From a microburet add sodium fluoride solution (1 ml = 0.01 mg F) equivalent to about 80% of the fluoride present in the sample aliquot, as indicated by the thorium nitrate solution required. Mix thoroughly, add the same volume of thorium nitrate solution as that required for titration of the sample aliquot as described, and again mix thoroughly. The color in the standard tube will be deeper than that in the sample tube.

7.2.5 From the microburet continue

to add sodium fluoride solution (1 ml = 0.01 mg F) to the standard tube until its color matches that of the sample tube. (If the colors cannot be matched, repeat the distillation.) Equalize the volumes in the sample and standard tubes by adding water. After the addition of water, mix thoroughly, then allow all bubbles to escape before making the final color comparison. Check the end point by adding 1 or 2 drops of sodium fluoride solution (1 ml = 0.01 mg F) to the standard tube. If the colors were originally matched, the color in the standard tube will be distinctly lighter in shade than that in the sample tube.

7.2.6 Determine a blank by carrying the same amt of all reagents through the procedure described. With proper attention to details, blanks of 5 μg of fluoride, or less, can be obtained.

7.3 PROCEDURE FOR BACK TITRATION, LOW CONCENTRATIONS (LESS THAN 0.01 MG F IN TOTAL SAMPLE)

7.3.1 Distill successive 85 to 90-ml portions of distillate directly into three or four 100-ml Nessler tubes. Take care to keep the amt of perchloric acid distilling over as small as possible, because the entire distillate is titrated and there is no aliquot available for a separate acidity determination. Analyze each of the distillate portions in the 100-ml Nessler tubes separately as follows:

a. Add 2 ml of sodium alizarinsulfonate solution (0.01 g/l) and neutralize the acid by adding 0.05 N sodium hydroxide solution until a pink color is produced. Add 4 ml of 0.05 N hydrochloric acid and sufficient thorium nitrate solution (0.25 g/l) to provide a faint pink color. Compare the treated distillate portion with a standard of equal total volume containing 2 ml of sodium alizarinsulfonate solution (0.01 g/l), 4 ml of hydrochloric acid and the same volume of thorium nitrate solution (0.25 g/l) as is required to produce the pink color in the sample tube. Add sodium fluoride solution to the standard tube until the color matches that of the sample tube. The sum of all significant amts of fluoride found in each successive portion of distillate is the total amt of fluoride in the sample.

7.4 PROCEDURE FOR PHOTOMETRIC TITRATION.

7.4.1 Transfer the distillate to a 20.3 cm (8″) titration cell and add 5 ml of hydroxylamine hydrochloride solution. Adjust, if necessary, to pH 3.6 with 0.05 N perchloric acid, and then add 5 ml of buffer-indicator solution.***

7.4.2 Place the cell in the titrating attachment, immerse the buret tip, and start the stirring motor. Close the titrator lid, set the wavelength to 525 nm, and the sensitivity knob to the proper position (usually 1). Close the shutter and adjust the slit width to give a transmittance reading of 100.

7.4.3 Titrate with standard thorium nitrate (0.01 N solution) to a transmittance of 75%. Record the volume to the nearest 0.005 ml.

7.4.4 Deduct a blank obtained by carrying the same amt of all reagents through the entire procedure, including the preliminary distillation and titration steps. Determine the amt of fluoride present from the calibration chart.

8. Calibration and Standards

8.1 THORIUM NITRATE, STANDARD STOCK SOLUTION (1 ML = 1.9 MG F). Standardize this solution as follows:

***The addition of buffer-indicator solution should adjust the pH to 3.0. For amts of fluoride ordinarily encountered, the pH of the distillate should be 3.5 to 3.7 if the distillation is properly controlled. The addition of buffer-indicator solution will maintain a pH of 3.0 under these conditions. For extreme cases, where acidity of the distillate is less than pH 3.5, 0.05 N, sodium hydroxide may be used to raise the pH to the proper level. However, it has been found to be the rule that distillations properly conducted will have a pH greater than 3.5. The use of sodium hydroxide for neutralization produces a slight change in the factor due to the sodium perchlorate formed.

8.1.1 Weigh 0.100 g of sodium fluoride into a distilling flask and collect 250 ml of distillate as previously described. Titrate 50 ml of the distillate (20 mg of NaF) with the solution being standardized. Carry a blank through the same procedure. Calculate the strength of the thorium nitrate solution in terms of mg of fluoride ion/ml of solution as follows:

$$\text{Fluoride ion, mg/ml} = \frac{C \times 20}{A - B}$$

Where:

A = ml of $Th(NO_3)_4 \cdot 4H_2O$ solution required for titration of the fluoride,
B = ml of $Th(NO_3)_4 \cdot 4H_2O$ solution required for titration of the blank, and
C = 0.4524 when titrating sodium fluoride (NaF).

8.2 THORIUM NITRATE SOLUTION, 0.01 N (0.19 G F^-/L). Prepare a calibration curve for this solution from data obtained in the following way:

8.2.1 Pipet aliquots of standard sodium fluoride solution covering the range 10 to 1000 μg of fluoride into 500-ml volumetric flasks and dilute to volume. Transfer to a 20.3-cm titration cell, add 5 ml of hydroxylamine hydrochloride solution, and adjust to pH 3.6 with 0.05 N perchloric acid. Add 5 ml of buffer-indicator solution and titrate as described in the **Procedure for Photometric Titration.**

9. Calculation

9.1 Calculate the fluoride ion content of the total distillate‡‡‡ in mg as follows:

$$F = \frac{(A - B)\,CD}{E}$$

Where:

F = mg of fluoride ion in total distillate,
A = ml of titrating solution††† used in titration of sample aliquot,
B = ml of titrating solution††† used in titrating the blank,
C = fluoride equivalent of titrating solution§§§ in mg of fluoride ion/ml of solution.
D = ml of total distillate collected,††† and,
E = ml of distillate titrated.†††

9.2 Calculate the fluoride concn in the atmosphere at 25 C and 760 Torr, in terms of ppm of hydrogen fluoride (HF) or fluorine (F_2), or mg of particulate fluoride/m_3, as follows:

$$\text{Hydrogen fluoride, ppm} = \frac{3450 \times F \times (273 + t)}{PV}$$

$$\text{Fluorine, ppm} = \frac{1640 \times F \times (273 + t)}{PV}$$

$$\text{Particulate fluoride, mg/m}^3 = \frac{2550 \times F \times (273 + t)}{PV}$$

Where:

F = mg of fluoride ion in total sample
P = sampling pressure in Torr (mm of mercury),

‡‡‡The volume of total distillate collected normally is 250 ml. However, if any other volume of total distillate is collected, this volume shall be substituted for 250. The volume of the distillate titrated normally is 50 ml but may vary as described in *Procedure for Back Titration, Low Concentrations*. If this procedure applies, for each portion of distillate titrated, the value of E is equal to the value of D.

†††The term "titrating solution" refers to either the $Th(NO_3)_4$ solution used in accordance with the Procedure for Direct Titration, or the NaF solution (1 ml = 0.01 mg F) used in Procedure for Back Titration, Medium Concentration and in Procedure for Back Titration, Low Concentrations.

t = sampling temp in degree Celsius, and
V = sample volume in liters.

9.3 Calculate the fluoride concn in vegetation on the oven-dry basis, as follows:

$$\text{Fluoride, ppm (dry basis)} = \frac{F \times A \times 1000}{W \times S \times [1 - (M/100)]}$$

Where:

F = mg of fluoride ion in total distillate,
A = g of total ash
W = g of ash distilled
S = g of fresh sample
M = percentage of moisture in fresh sample.

10 Effects of Storage

10.1 All reagent solutions listed are stable at room temp except as individually noted.

11. References (See Part F, Section 11).

PART F: *Determination of Fluoride (Spectrophotometric Methods)*

1. Principle of the Method

1.1 Reaction of fluoride with the metal ion moiety of a metal-dye complex results in fading [Zirconium-Eriochrome Cyanine R **(9)** and Zirconium-SPADNS **(4)** reagents) or increase (Lanthanum-Alizarin Complexone **(3)** reagent] in the absorbance of the solution.

2. Sensitivity and Range

2.1 Both Zirconium-Eriochrome Cyanine R and Zirconium-SPADNS reagents obey Beer's Law over the range of 0.00 μg to 1.40 μg fluoride/ml with a detection limit of the order of 0.02 μg/ml. The procedure given for the lower-range, Lanthanum-Alizarin Complexone, covers the range 0.00 μg to 0.5 μg fluoride/ml with a detection limit of approximately 0.015 μg/ml.

2.2 In common with other spectrophotometric methods, these are temp-sensitive and absorbances must be read within ± 2 C of the temp at which the respective calibration curve was established.

3. Interferences

3.1 Moderate variations in acidity of sample solutions will not interfere with the Zirconium-Eriochrome Cyanine R or Zirconium-SPADNS reagents. The Lanthanum-Alizarin Complexone reagent has greater pH sensitivity and solutions must not exceed the capacity of the buffer system to maintain an apparent pH 4.50 ± 0.02.

3.2 Many ions interfere with these fluoride reagents, but those most likely to be encountered in analysis of ambient air are aluminum, iron, phosphate and sulfate. If these are present above the trace level their effects must be eliminated. Distillation, diffusion, or ion exchange may be employed but, in certain cases, complexation-extraction **(2)** may be advantageous.

3.3 In vegetation analysis, ashing and distillation by the Willard-Winter technique generally assure a sample solution sufficiently free of interfering ions for direct colorimetric evaluation. Traces of free chlorine in the distillate, if present, must be reduced with hydroxylamine hydrochloride.

4. Precision and Accuracy

4.1 Because of the wide variability in composition of samples, and in methods and conditions of sampling, no general statements of precision and accuracy for field samples can be given. Precision studies of pure sodium fluoride standards indicate that, within the concn ranges for which the reagents follow Beer's Law, standard deviation of ± 0.015 to 0.020 μg of fluoride/ml should be expected.

5. Apparatus

5.1 SPECTROPHOTOMETER. An instrument capable of accepting sample cells of 1-cm to 2.5-cm optical path, and which is adjustable throughout the visible wavelength region, is required. Each spectrophotometer sample cell is given an identification mark and calibrated by reading a portion of the same reagent blank solution at the designated wavelength. The determined cell correction is subsequently applied to all absorbance readings made in that cell.

6. Reagents

6.1 ACETIC ACID, GLACIAL.

6.2 ACETONE, REAGENT GRADE.

6.3 ALIZARIN COMPLEXONE. (1,2-dihydroxy-3-anthraquinonylamine-N, N-diacetic acid, available from Hopkins & Williams, Ltd., Chadwell Heath, Essex, England, cat. no. 1369.4)§§§

6.4 AMMONIUM ACETATE SOLUTION (20% w/v). Dissolve 20.0 g ammonium acetate in water and dilute to 100 ml.

6.5 AMMONIUM HYDROXIDE, SP GR 0.880.

6.6 ERIOCHROME CYANINE R SOLUTION. Dissolve 1.800 g of Eriochrome Cyanine R (Mordant Blue 3, Color Index No. 43820) in water to make 1 l. Solution is stable for more than a year when protected from light.

6.7 LANTHANUM CHLORIDE, 99.9% ASSAY. (Available from Kleber Laboratories, Burbank, Calif.)

6.8 LANTHANUM-ALIZARIN COMPLEXONE REAGENT. Dissolve 8.2 g sodium acetate in 6 ml of glacial acetic acid, and sufficient water to permit solution, and transfer to a 200-ml volumetric flask. Dissolve 0.0479 g alizarin complexone in 1.0 ml of 20% ammonium acetate solution, 0.1 ml ammonium hydroxide and 5 ml water. Filter this solution through a Whatman #1 paper into the 200-ml volumetric flask. Wash the filter with a few drops of water and discard the residue. Add 100 ml of acetone, slowly and with mixing, to the flask. Dissolve separately 0.612 g of lanthanum chloride in 2.5 ml 2 N hydrochloric acid solution, warming slightly to promote solution, and combine this with the flask contents.‖‖‖ Dilute, mix well, cool the solution to room temp and adjust the volume to the mark. The reagent solution is stable for about 1 week if kept under refrigeration.

6.9 SODIUM FLUORIDE STOCK SOLUTION. (1 ml = 1.0 mg F^-). Dissolve 2.2105 g of 100% sodium fluoride, or the equivalent weight of reagent grade sodium fluoride, in water and dilute to 1 l. Store in a polyethylene bottle.

6.10 SODIUM FLUORIDE WORKING STANDARD SOLUTION (1 ML = 10 μG F^-). Dilute 5.0 ml of the stock solution to 500 ml. Store in a polyethylene bottle.

6.11 SPADNS SOLUTION. (4,5-dihydroxy-3-(p-sulfophenylazo)-2,7-napthalene disulfonic acid trisodium salt, available from Eastman Organic Chemicals, Rochester, N.Y., cat. no. 7309)—Dissolve 0.985 g SPADNS dye in water and dilute to 500 ml.

6.12 SPADNS REFERENCE SOLUTION. Add 10 ml SPADNS Solution to 100 ml of water and acidify with a solution prepared by diluting 7 ml concd hydrochloric acid to 10 ml. This solution may be stored and reused repeatedly.

6.13 ZIRCONIUM SOLUTION. Dissolve 0.265 g of zirconyl chloride octahydrate ($ZrOCl_2 \cdot 8H_2O$) in 50 ml of water, add 700 ml of conc hydrochloric acid, and dilute to 1 l with water.

6.14 ZIRCONIUM-ERIOCHROME CYANINE R REAGENT. Mix equal volumes of the Eriochrome Cyanine R and the zirconium solutions. Cool to room temp before use. Prepare fresh daily.

§§§This reagent has been variously named in the literature, *e.g.*, alizarin complexan, alizarin complexon, alizarin complexone, and alizarin Fluorine Blue.

‖‖‖An equimolar conc of lanthanum nitrate may be substituted for the chloride.

6.15 ZIRCONIUM-SPADNS REAGENT. Mix equal volumes of the SPADNS and the zirconium solutions. Cool to room temp before use. This reagent may be stored for several months, at room temp, in a polyethylene bottle.

7. Procedure

7.1 PROCEDURE FOR INTERMEDIATE RANGE.

7.1.1 *Zirconium-Eriochrome Cyanine R Reagent.* Transfer an aliquot of the prepared sample, standard, or blank solution to a 25-ml volumetric flask containing 4 ml of the Zirconium-Eriochrome Cyanine R Reagent. Dilute the solution to the mark, mix well, and allow to stand for 30 min for temp equilibration. Transfer the solution to a calibrated spectrophotometer cell of about 2.5-cm light patch. The spectrophotometer is set on a wavelength of 536 nm and the light control adjusted to an absorbance value of 0.500 on a Reagent Blank similarly prepared.

7.1.1.1 Spectrophotometer cells of 1-cm light path may be used by making the following adjustments of volumes: transfer the aliquot of sample, standard or blank solution to a 10-ml volumetric flask, and add 3 ml of Zirconium-Eriochrome Cyanine R Reagent. Dilute the mixture to the mark, mix well and allow to stand for 30 min; then read absorbance as previously described.

7.1.2 *Zirconium-SPADNS Reagent.* Dilute a suitable aliquot of the sample solution to 25 ml and add 5.0 ml of Zirconium-SPADNS Reagent. Mix and allow to stand for 30 min to establish temp equilibrium before transferring the solution to a spectrophotometer cell (cells of 1 cm to 2.5 cm optical path may be used). Measure the absorbance value at 570 nm with the spectrophotometer adjusted to read zero absorbance on the SPADNS Reference Solution.

7.2 PROCEDURE, LOWER RANGE. Lanthanum-Alizarin Complexone Reagent: Transfer a suitable aliquot of sample solution, containing no more than 4 μg of fluoride, to a 10-ml volumetric flask. Add 3 ml of Lanthanum-Alizarin Complexone Reagent, dilute to the mark and mix well. Allow to stand for 30 min. Measure the absorbance at 622 nm, in a calibrated 1-cm cell, using a reagent blank as reference.

8. Calibration and Standards

8.1 ZIRCONIUM-ERIOCHROME CYANINE R REAGENT. Prepare a standard series, spanning the range of zero to 20 μg of fluoride, by pipetting aliquots of the standard sodium fluoride solution (10 μg F^-/ml) into 25-ml volumetric flasks. Add 4 ml of Zirconium-Eriochrome Cyanine R Reagent to each flask, dilute to the mark and mix thoroughly. Allow the standards to stand until solution temp has equilibrated at the desired value. Measure absorbances at 536 nm, in 2.5-cm cells, against a reagent blank for which the spectrophotometer is adjusted to read 0.500 absorbance unit. Plot a calibration curve relating fluoride concn, in μg, to absorbance values at the selected working temp.

8.1.1 If 1-cm cells are to be used, a similar standard series is prepared in 10-ml volumetric flasks, adding 3 ml of Zirconium-Eriochrome Cyanine R Reagent to each flask, and reading as described above.

8.2 ZIRCONIUM-SPADNS REAGENT. Prepare a standard series containing from zero to 35 μg of fluoride by pipetting aliquots of the standard sodium fluoride solution (10 μg F^- per ml) into 25-ml volumetric flasks. Add 5 ml of Zirconium-SPADNS Reagent to each flask, dilute to the mark and mix well. Allow the standards to stand 30 min to reach temp equilibrium at the desired value. Measure absorbances at 570 nm after zeroing the spectrophotometer on the SPADNS Reference Solution. Prepare a calibration curve relating fluoride concn, in μg, to absorbance values at the selected working temp.

8.3 LANTHANUM-ALIZARIN COMPLEXONE REAGENT. Prepare a standard series containing zero to 4 μg of fluoride by measuring portions of the standard sodium fluoride solution (10 μg F^-/ml) into 10-ml volumetric flasks. Add 3 ml of Lanthanum-Alizarin Complexone Reagent to each flask, dilute to the mark, mix and let stand for 30 min at the selected temp. Measure the absorbances at 622 nm in 1-cm cells, using a reagent blank as reference.

9. Calculation

9.1 Concentrations of fluoride in air or vegetation samples are calculated by use of the formulas given under *Part E: Titrimetric Methods*, Calculations **9.2** and **9.3**

10. Effects of Storage

10.1 Reagents used in these procedures are stable, except as individually noted.

11. References

1. ASTM STANDARDS ON METHODS OF ATMOSPHERIC SAMPLING AND ANALYSIS, 1962. 2nd Ed, Oct.
2. BELCHER, R. and T.S. WEST. 1961. A study of the cerium^{III}-alizarin complexan-fluoride reaction. Talanta 8:853.
3. BELCHER, R., and T.S. WEST. 1961. A comparative study of some lanthanum chelates of alizarin complexan as reagents for fluoride. Talanta 8:863.
4. BELLACK, ERVIN and P.J. SCHOUBOE. 1958. Rapid photometric determination of fluoride in water. *Anal. Chem.* 30:2032.
5. DAHLE, DAN, R.U. BONNAR, and H.J. WICHMANN. 1938. Titration of small quantities of fluorides with thorium nitrate. I. Effect of changes in the amount of indicator and acidity. *J. Assoc. Official Agr. Chem.* 21:459. II. Effects of chlorides and perchlorates. Ibid. 21:468.
6. DEUTSCH, SAMUEL. 1955. Overcoming the effect of manganese dioxide in fluoride determinations. *Anal. Chem.* 27:1154.
7. HALL, R.J. 1963. The spectrophotometric determination of sub-microgram amounts of fluorine in biological specimens. *Analyst.* 88:76.
8. MAVRODINEANU, R., and J. GWIRTSMAN. 1955. Photoelectric end-point determination in the titration of fluorides with thorium nitrate. Contrib. Boyce Thompson Inst. 18, 3:181.
9. MEGREGIAN, STEPHEN. 1954. Rapid spectrophotometric determination of fluoride with Zirconium-Eriochrome Cyanine R lake. *Anal. Chem.* 26:1161.
10. NEWMAN, A.C.D. 1958. The separation of fluoride ions from interfering anions and cations by anion exchange chromatography. *Anal. Chem. Acta.* 19:471.
11. NIELSEN, H.M. 1958. Determination of microgram quantities of fluoride. *Anal. Chem.* 30:1009.
12. NIELSEN, J.P., and A.D. DANGERFIELD. 1955. Use of ion exchange resins for determination of atmospheric fluorides. *A.M.A. Arch. Ind. Health* 11:61.
13. PACK, M.R., A.C. HILL, M.D. THOMAS, and L.G. TRANSTRUM. 1959. Determination of gaseous and particulate inorganic fluoride in the atmosphere. *A.S.T.M. Special Publication #281.*
14. MANUFACTURING CHEMISTS' ASSOC., 1947. Perchloric Acid Solution. *Chemical Safety Data Sheet SD-11.* 1825 Connecticut Ave., N.W. Washington, D.C. 20009.
15. REMMERT, L.F., T.D. PARKS, A.M. LAWRENCE, and E.H. MCBURNEY. 1953. Determination of fluorine in plant materials. *Anal. Chem.* 25:450.
16. ROWLEY, R.J., and H.V. CHURCHILL. 1937. Titration of fluorine in aqueous solutions. *Ind. Eng. Chem., Anal.* Ed. 9:551.
17. ROWLEY, R.J., and G.H. FARRAH. 1962. Diffusion method for determination of urinary fluoride. *Amer. Ind. Hyg. Assoc. J.* 23:314.
18. ROWLEY, R.J. J.G. GRIER, and R.L. PARSONS. 1953. Determination of fluoride in vegetation. *Anal. Chem.* 25:1061.
19. SINGER, LEON and W.D. ARMSTRONG. 1954. Determination of fluoride. Procedure based upon diffusion of hydrogen fluoride. *Anal. Chem.* 26:904.
20. SMITH, F.A. and D.E. GARDNER. 1955. The determination of fluoride in urine. *Amer. Ind. Hyg. Assoc. Quarterly* 16:215.
21. THOMAS, M.D, G.A. ST. JOHN, and S.W. CHAIKEN. 1958. An atmosphere fluoride recorder. *A.S.T.M. Special Publication #250.*
22. WILLARD, H.H. and O.B. WINTER. 1933. Volumetric method for determination of fluorine. *Ind. Eng. Chem., Anal.* Ed. 5:7.

Subcommittee 2

L. V. CRALLEY, *Chairman*
L. V. HAFF
A. W. HOOK
E. J. SCHNEIDER
J. D. STRAUTHER
C. R. THOMPSON
L. H. WEINSTEIN

204.

Tentative Method of Analysis for Fluoride Content of the Atmosphere and Plant Tissues (Semiautomated Method)

12202-02-68T

1. Principle of the Method

1.1 GENERAL. The plant material including leaf samples, washed or unwashed, is dried and ground, then ashed, alkali fused and diluted with water to 25 ml. In the case of leaf samples, an appreciable amt of fluoride may be deposited on the external leaf surfaces. This fluoride behaves differently physiologically from fluoride absorbed into the leaf and it is often desirable to wash it from the surface as a preliminary step in the analysis. Details of a leaf-washing process are given in **Section 7.1.** The suspended digest and sulfuric acid are pumped into the Teflon coil of a microdistillation device maintained at 170 C. A stream of air carries the acidified sample swiftly through a coil of Teflon tubing to a fractionation column. The fluoride and water vapor distilled from the sample are swept up the fractionation column into a condenser, and the condensate passes into a small collector. The distillate is pumped continuously from the sample collector. Acid and solids are removed from the bottom of the fractionation column and are drawn to waste. The distillate is mixed continuously with alizarin fluorine blue-lanthanum reagent, the colored stream passes through a 15-mm tubular flow cell of a colorimeter, and the absorbance is measured at 624 nm. The impulse is transmitted to a recorder. All major pieces of the apparatus are components of the Technicon AutoAnalyzer (**15**). Details of construction of the microdistillation device are given in **Section 5.9.** Earlier versions of this method have been published (**10, 19, 20**).

1.2 PRINCIPLE OF OPERATION.

1.2.1 *Colorimetric System.* The absorbance of an alizarin fluoride blue-lanthanum solution is changed by very small amounts of inorganic fluoride. In addition, a number of other materials also cause changes in absorbance at 624 nm. Potential interfering substances commonly found in plant tissues are metal cations such as iron and aluminum, inorganic anions such as phosphate, chloride, nitrate, and sulfate, and organic anions such as formate and oxalate. Fortunately, metal cations and inorganic phosphate are not distilled in this system, and organic substances are destroyed by preliminary ashing. The remaining volatile inorganic anions may interfere if present in a sufficiently high concn because they are distilled as acids. Their hydrogen ions bleach the reagent which, in addition to being an excellent complexing agent, is also an acid-base indicator. To reduce the danger of acidic interferences, a relatively high concn of acetate buffer is used in the reagent solution despite some reduction in sensitivity. All plant tissues growing with normal nutrient supplies thus far tested have not contained concns of interfering substances sufficient to cause problems (Table 204:I, Section **3**).

1.2.2 *Distillation System.* Since HF has a high vapor pressure, it is more efficiently distilled than the other acids previously mentioned (**1.2.1**). The factors controlling efficiency of distillation are temp, concn of acid in the distillation coil, and vacuum in the system. Very large amounts of solid matter, particularly silicates, will also retard distillation. Accordingly, the smallest sample of vegetation consistent

Table 204:I. Maximum concentration of several anions present in samples at which there is no analytical interference

Compound Tested	Interfering Anion	Molarity Tolerated
Na_2SO_4	SO_4 =	2×10^{-2}
Na_2SiO_3	SiO_3 =	5×10^{-3}
NaCl	Cl^-	1×10^{-3}
NaH_2PO_4	PO_4 =	3.8
$NaNO_3$	NO_3^-	5×10^{-3}

with obtaining a suitable amount of F should be analyzed. The aforementioned conditions must be carefully controlled, since accurate results depend on obtaining the same degree of efficiency of distillation from samples as from the standard F solutions used for calibration.

1.2.3. Temp control is maintained within ± 2 C by the thermoregulator and by efficient stirring of the silicone oil. Acid concentration during distillation is regulated by taking plant samples in the range of 0.1 to 2.0 g and by using 100 ± 10 mg of calcium oxide and 3.0 ± 0.1 g of sodium hydroxide for ashing and fusion of each sample. Vaccum in the system is controlled with flowmeters and a vacuum gauge. Any marked change in vacuum (greater than 0.2 inch Hg or 5 Torr) over a short time period indicates either a leak or a block in the system. Distillation should take place at the same vacuum each day unless some other change in the system has been made. It is also essential to maintain the proper ratio between the air flow on the line drawing liquid and solid wastes from the distillation coil and on the line drawing HF and water vapor (Figure 204:1.) from the distillation unit. Occasional adjustments on the two flowmeters should be made to keep this ratio constant and to maintain higher vacuum on the line drawing HF vapor so that little or no HF is diverted into the liquid and solid waste line. (See Section **7.3.2** for description of air flow system.)

2. Range and Sensitivity

2.1 In normal use, the procedure can detect 0.1 μg F/ml. The normal range of analysis is from 0.1 to 4.0 μg F/ml. Higher concentrations can be analyzed by careful dilution of samples with 3N sodium hydroxide. If digested samples to be analyzed routinely exceed 4.0 μg F/ml, the analytical portion of the pump manifold can be modified to reduce sensitivity. However, the best procedure is to analyze a smaller aliquot of the sample. Most accurate results are obtained with the F concentration falls in the middle or upper part of the calibration curve. It is not necessary that the total of sample and water pumped be exactly equal to the original amount of distillate used (2.50 ml/min). For example, a decrease in sensitivity of about one-half can be achieved by using a distillate sampling tube of 0.081″ (2.50 ml/min) and a water diluent tube of 0.065 inch (1.6 ml/min). The total volume of the distillate and water diluent should approximate the original volume of distillate used (4.10 ml/min) to maintain the correct proportion of sample with alizarin fluorine blue-lanthanum reagent (1.69 ml/min). When changes of the type described are made in the analytical system, an appropriate fluoride standard curve must be run.

3. Interferences

3.1 Since the air which is swept through the microdistillation unit is taken from the ambient atmosphere, airborne contaminants in the laboratory may contaminate samples. If this is a problem, a small drying bulb filled with calcium carbonate granules can be attached to the air inlet tube of the microdistillation unit.

3.2 If the Teflon distillation coil in the microdistillation unit is not cleaned period-

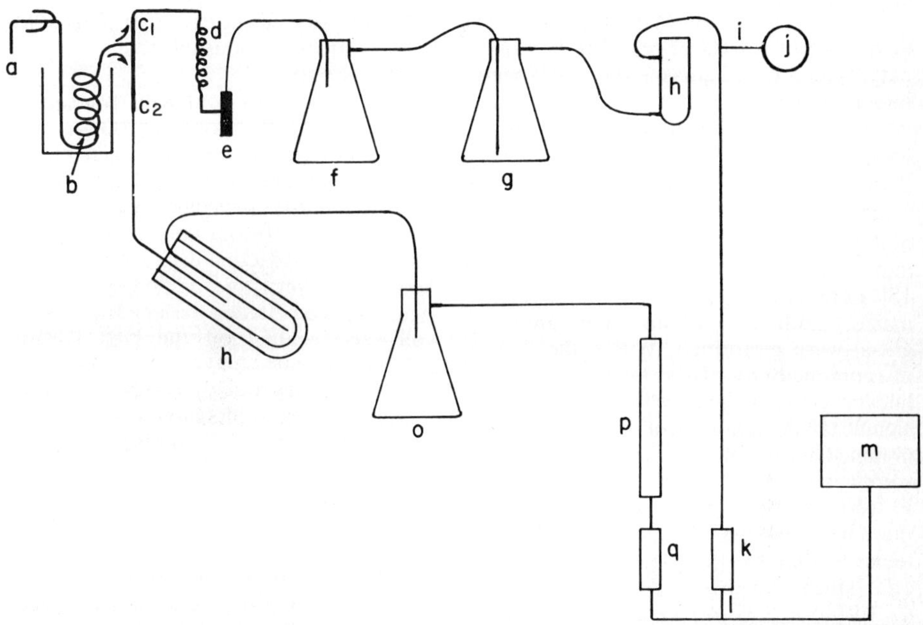

Figure 204:1—Schematic drawing of air flow system used in semi-automated analysis of fluoride in vegetation.

ically, particulate matter will accumulate and will reduce sensitivity.

3.3 Silicate, chloride, nitrate, and sulfate ions in high concn can be distilled with fluoride ion and will interfere with the analysis by bleaching the alizarin fluorine blue-lanthanum reagent. Phosphate ion is not distilled and therefore does not interfere. Metals such as iron and aluminum are not distilled and will also not interfere with the analysis. Max concentrations of several common anions at which there was no interference are given in Table 204:I. The sulfate concn shown is the amt tolerated above the normal amt of sulfuric acid used in microdistillation.

4. Precision and Accuracy

4.1 General. It is essential that each laboratory occasionally perform the tests outlined below (**7.3.6a, 7.3.6b, 7.3.6.c**) to insure good results. The degree of accuracy and precision obtained in one laboratory may not be a fair representation for another laboratory, since much depends on the thoroughness with which any multistep analytical procedure is carried out. The wide variation in results obtained in a recent cooperative study among 31 laboratories (**6**) emphasizes the importance of proper performance of methods. The following information on accuracy and precision is meant, therefore, as a guide and is taken from a published report (**7**) in which results of analyses of tissues by the semi-automated and Willard-Winter (**11, 21**) methods were compared. The paper should be consulted for details.

4.2 Precision. The standard error of a single determination is between 2 and 8 μg, depending on the kind of plant tissue used and the level of F present. With higher amounts of F, the standard deviation increases, although the coefficient of variation (standard deviation expressed as per cent of F content) decreases. The coefficients of variation between 20 and 100 ppm F are generally 10% or less. Analysis of high silicate tissues (grasses) for F has always been a problem and standard deviations of results of replicate determinations on orchard grass tissues have been larger than when other tissues are analyzed. Recent modifications in procedures for transferring and distilling samples have improved analysis of high silicate samples.

4.3 Accuracy. Since no direct means of determining accuracy are yet available, indirect means have been used. A total of 180 determinations of 4 tissues (Milo maize, gladiolus, alfalfa, and orchard grass) were performed. Within the limits of reproducibility of the results, multiple linear regression analysis indicated that no significant deviations from linearity were obtained when different amts of a tissue were analyzed and when different amts of F were added to a tissue. In addition, there were no significant second-order effects. Systematic errors were not significant since the intercept values found ($-$ 3.97 to + 2.17 μg) were not significantly different from zero.

5. Apparatus (See Figure 204:2.)

5.1 Multichannel Proportioning Pump, with assorted pump tubes, nipple connectors, glass connectors, and manifold platter.

5.2 Pulse Suppressors, for the sample and alizarin fluorine blue-lanthanum reagent streams are each made from 10-ft lengths of 0.035-inch ID Teflon standard wall tubing. The tubes are coiled around a mailing tube or other suitable object about 2½″ in diam. The outlet ends of the suppressor tubes are forced into short lengths of 0.045″ ID Tygon tubing, which is then sleeved with a piece of 0.081″ Tygon tubing. The sleeved ends are then slipped over the "h" fitting which joins the sample and reagent streams.

5.3 Automatic Sampler. Technicon Sampler II with rotary stirrer and with 8.5 ml plastic sample cups.

5.4 Voltage Stabilizer.

5.5 Colorimeter, with 15-mm tubular flow cell and 624 nm interference filters.

5.6 Rotary Vacuum and Pressure Pump, with continuous oiler.*

5.7 Recorder.

5.8 Range Expander (optional).

5.9 Microdistillation Apparatus. A schematic drawing is shown in Figure 204:2. Major components of the microdistillation apparatus include the following:

5.9.1 The bottom only of a jacketed 1,000-ml resin reaction flask with a conical flange.† The kettle is modified by evacuating the space between the inner and outer walls and sealing off the port (Figure 204:2,a).

5.9.2 Resin reaction flask top with conical flange, modified to have one 29/42 center joint and four 24/40 side joints (Figure 204:2,a).‡

5.9.3 Resin reaction flask clamp.§

5.9.4 Variable high speed stirrer (Figure 204:2,d).‖

5.9.5 Stainless steel, heavy duty stirrer stuffing box with a 29/42 standard taper and shredded Teflon packing.¶

5.9.6 Ten mm diam stainless steel stirrer rod with propeller to fit the stuffing box.

5.9.7 Thermometer-thermoregulator, range 0–200 C (Figure 204:2,c).**

5.9.8 *Electronic relay control box.*††

5.9.9 *Low drift immersion heater, 750 watts* (Figure 204:2,b).§§

---

*Gast #0211-V36-G 10 pump with AA 930 oiler.

†Scientific Glass Apparatus Co., Inc., Bloomfield, N.J., Catalog #JR-5130.

‡Scientific Glass Apparatus Co., Inc., Bloomfield, N.J., Catalog #JR-7935.

§Scientific Glass Apparatus Co., Inc., Bloomfield, N.J., Catalog #JR-9210.

‖Scientific Glass Apparatus Co., Inc., Bloomfield, N.J., Catalog #S-6362.

¶Scientific Glass Apparatus Co., Inc., Bloomfield, N.J., catalog #JS-1160 and #JS-3050.

**Scientific Glass Apparatus Co., Inc., Bloomfield, N.J., Catalog #T-5715.

††Scientific Glass Apparatus Co., Inc., Bloomfield, N.J., Catalog #T-5905.

§§Scientific Glass Apparatus Co., Inc., Bloomfield, N.J., Catalog #H-1265.

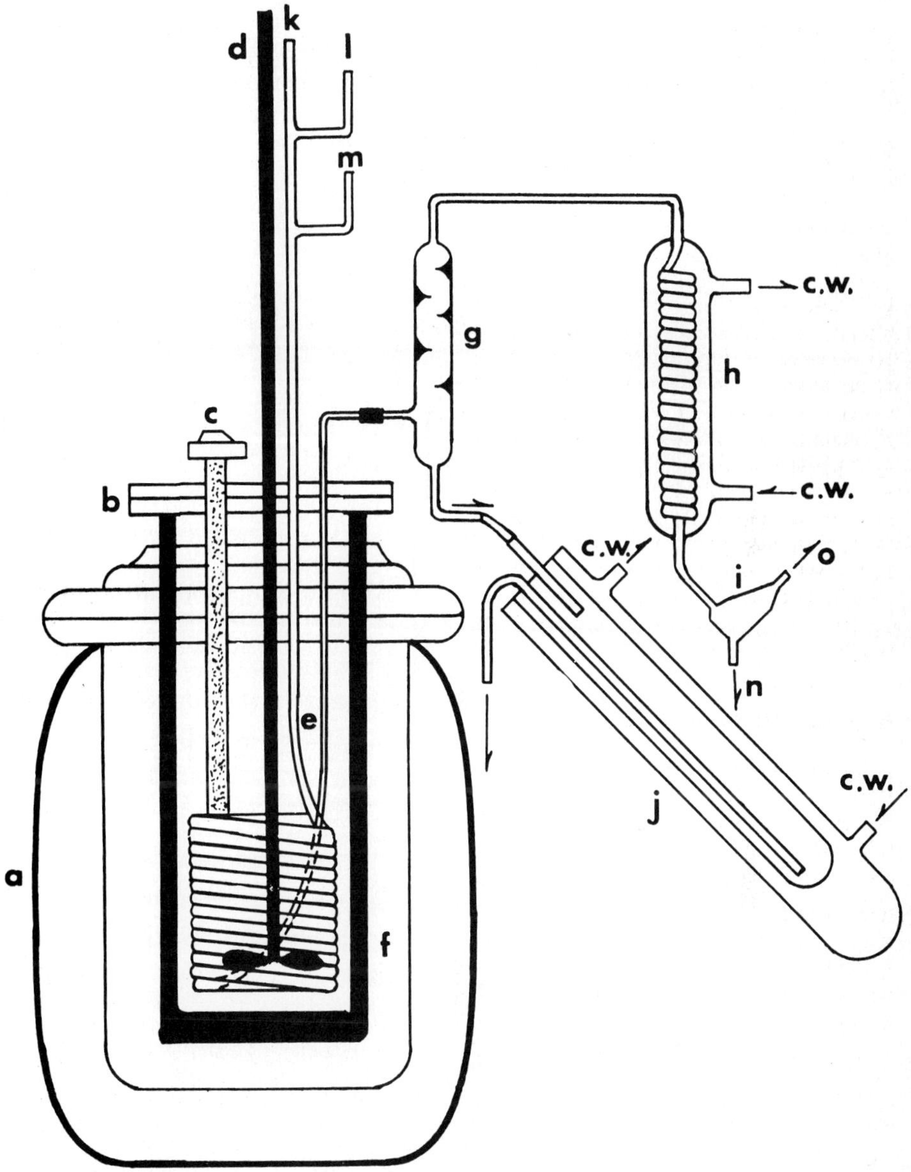

Figure 204:2—Schematic drawing of microdistillation apparatus. See text.

5.9.10 *Flexible "Teflon"*. TFE tubing, ⅛″ ID, $^{3}/_{16}$″ OD, 0.030″ wall. A 30-ft length is coiled on a rigid support of such a diam that the completed coil will fit into the resin reaction flask (**5.8.1**). Care must be taken to prevent kinking of the tubing (Figure 204:2,e).

5.9.11 *Two flowmeters*, with ranges

of 0 to 1 and 0 to 5 l/min, both with needle valve controls.

5.9.12 *Vacuum gauge,* with a range of 0 to 10 inches Hg or 0 to 254 Torr.

5.9.13 *Fractionation column*, of boro-silicate glass (Figure 204:2,g; also see Figure 204:3).

5.9.14 *Distillate Collector*.[‖]

5.9.15 *Water-jacketed condenser* (Figure 204:2,h). ##

5.9.16 *Dow-Corning 200 Fluid* (100 cs at 25 C) (Figure 204:2,f).

6. Reagents

6.1 REAGENTS FOR AUTOMATED FLUORIDE DISTILLATION.

6.1.1 *Sulfuric acid,* analytical grade, 50%, v/v. Mix 500 ml of concd sulfuric acid with 500 ml of deionized water. Cool before use.

6.1.2 *Acetate buffer, pH 4.0.* Dissolve 60 g of sodium acetate trihydrate in

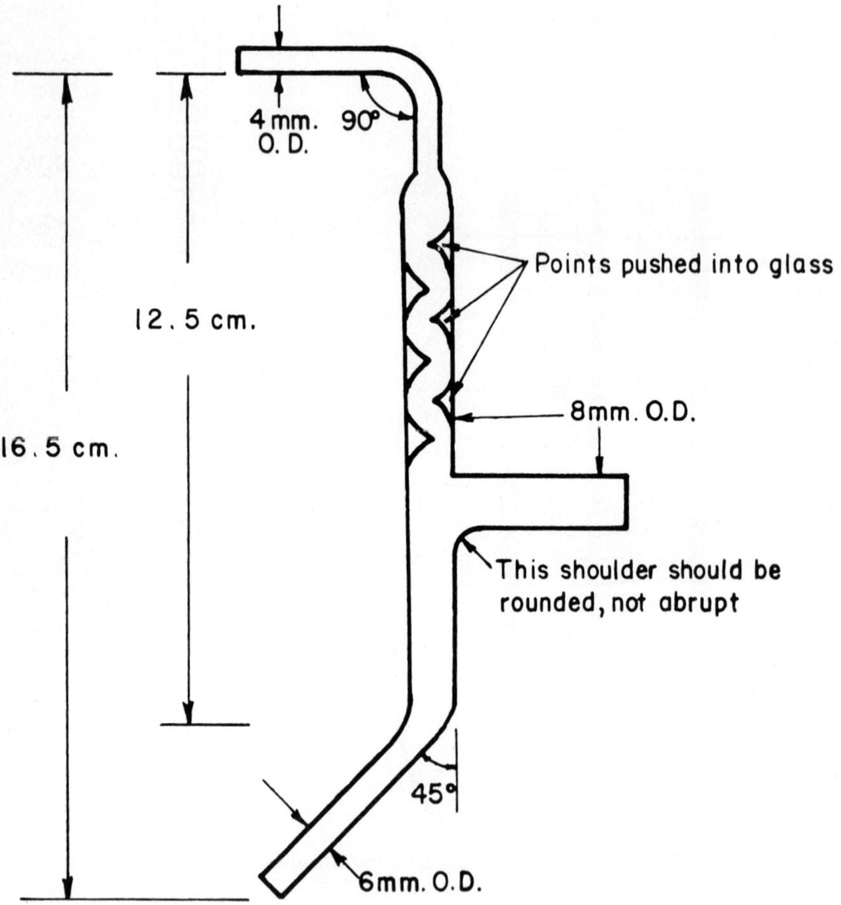

Figure 204:3—Microdistillation column.

‖Technicon Instrument Co., Chauncey, N.Y., BO Electrolyte Trap.

##Technicon Instrument Co., Chauncey, N.Y., Catalog #114-209.

500 ml of deionized water. Add 100 ml of analytical grade glacial acetic acid and dilute to 1 l with deionized water.

6.1.3 *Acetone, analytical grade.*

6.1.4 *Alizarin fluorine blue (alizarin complexone, 3-amino-ethylalizarin-N, N-diacetic acid) stock solution, 0.01M.* Suspend 0.9634 g of alizarin fluorine blue*** in about 100 ml of deionized water in a clean 250-ml volumetric flask. Add 2 ml volumetric flask. Add 2 ml of conc ammonium hydroxide and shake until the dye has completely dissolved. Add 2 ml of glacial acetic acid. Dilute the solution to 250 ml volume with deionized water and store at 4 C.

6.1.5 *Lanthanum nitrate stock solution, 0.01M.* Dissolve 2.1652 g of lanthanum nitrate hexahydrate [$La(NO_3)_3 \cdot 6H_2O$] in deionized water in a 250-ml volumetric flask and dilute to volume with deionized water.

6.1.6 *Brij 35*** wetting agent, 30% w/v.* Dissolve 50 g of Brij 35 in 100 ml of deionized water by heating over a hot plate.

6.1.7 *Working reagent.* The reagent is a modification of that reported by Yamamura, *et al* (**23**). Mix the following quantities of solutions in the order listed to make 1 l of working reagent: 300 ml of acetate buffer, 244 ml of deionized water, 300 ml of acetone, 100 ml t-butanol, 36 ml of alizarin fluorine blue, 20 ml of lanthanum nitrate, and 40 drops of Brij 35. Unused Working reagent is stable at 4 C for at least 7 days. Just prior to using reagent, place under vacuum for 10 min to remove air from solution.

6.1.8 *Standard fluoride solutions.* Dissolve 0.2207 g of dry reagent grade sodium fluoride in deionized water and dilute to 1 l. The stock solution will contain 100 μg F per ml. Prepare working standards by taking suitable aliquots to seven final concentrations of 0.2, 0.4, 0.8, 1.6, 3.2, and 4.0 μg F per ml. Diluted working standards for plant analysis should contain 6 g of reagent grade sodium hydroxide and 20 ml conc perchloric acid (70%) for each 100 ml of solution in order to compensate for the amounts of these substances used in alkali fusion of the ashed plant samples. Standard solutions for fluoride analysis of water samples or of air samples absorbed in water are made up in deionized water. Store stock solutions in clean polyethylene bottles in the cold. Since, as will be seen later, plant tissue samples are diluted to a 50-ml volume before analysis, the standard containing 0.20 μg fluoride/ml is equivalent to a sample of plant material containing 10.0 μg fluoride (0.2 μg/ml × 50 ml).

6.1.9 *Tetrasodium ethylenediamine-tetraacetate (Na_4EDTA), 1% (w/v).* Dissolve 1 g of Na_4EDTA in 99 ml of deionized water.

6.2 Reagents for Ashing and Alkali Fusion of Plant Samples.

6.2.1 *Calcium oxide,* low in fluorine.

6.2.2 *Sodium hydroxide pellets,* analytical grade, low in fluorine.

6.2.3 *Phenolphthalein solution, 1%.* Dissolve 1 g phenolphthalein in 50 ml of absolute ethanol or isopropanol. Add 50 ml of deionized water.

6.2.4 *Nickel, Inconel, or platinum crucibles*, 40-ml capacity.

6.3 Reagents for Transfer of Ashed and Fused Samples.

6.3.1 *Deionized water.*

6.3.2 *1:1 (v/v) perchloric acid-water solution.* Using conc perchloric acid (70%).

7. Procedure

7.1 Preparation of Plant Tissues for Analysis.

7.1.1 Wash leaf samples prior to analysis as already discussed in Section **1.1**. Fluorides may accumulate on plant surfaces by deposition of particulate matter and by absorption of gaseous fluoride onto components of the plant surface. Washing

***Hopkins and Williams, Ltd., Chadwell Heath, Essex, England. Reagent also known as alizarin complexone, alizarine complexone, alizarin complexon, alizarin complexan.

‡‡‡Atlas Powder Co., Chemicals Division, Wilmington, Delaware.

of plant samples before analysis is therefore necessary in order to have a meaningful estimate of the amt of F which has penetrated the tissue and which may be potentially harmful.

a. A standard washing procedure should meet several criteria: (1) it should be simple and gentle; (2) it should remove surface fluoride quantitatively with a min of leaching of internal fluoride; (3) it should not leave residues which might interfere with subsequent analysis; and (4) in the event that the tissue is to be analyzed also for nutrient status, it should not leach other internal elements.

b. Methods proposed in the literature may be summarized as follows:

(1) Parberry **(14)** and Thorne and Wallace **(18)** have recommended washing fresh leaves with distilled water prior to analysis of nutrient elements. Boynton, Cain and Compton **(2)**, Chapman and Brown **(4)** and Cameron, Mueller, Wallace and Sartori **(3)** have prepared leaves for analysis by wiping the surface with a piece of damp cheesecloth or diaper cloth.

(2) Jacobson **(7)** reported that a short tinse in 0.3*N* HCl followed by distilled water rinses did not leach iron from leaf cells. Arkley, Munns, and Johnson **(1)** found that rinsing with 0.1*N* HCl was as effective as 0.1*N* HCl detergent. Jacobson and Oertli **(9)** prepared leaves for iron analysis by washing with 2% Dreft followed by distilled water washes.

(3) Taylor **(17)** reported on a comprehensive study of the effectiveness of five cleaning procedures in the preparation of apple leaves for nutrient element analysis, various combinations of sulfur, lead arsenate, fermate, DDT, and phenylmercuric acetate. A solution containing Alconox and tetrasodium ethylenediaminetetraacetate (Na_4EDTA) was found to be excellent for the removal of iron and lead deposits on the leaf surfaces. Analyses showed that soluble elements were not leached from leaves.

(4) Nicholas, Lloyd-Jones, and Fisher **(12, 13)** studied the use of Teepol (sodium higher alkyl sulfate) for removal of surface iron contamination from leaves. Tissues were immersed in 0.3% Teepol in distilled water and agitated for 30 sec. This was followed by a 30-sec rinse in distilled water. Iron as a surface contaminant was removed quantitatively, while there was no loss of ^{59}Fe from within. Steyn **(16)** recommended that both sides of leaves be sponged with cotton in a 0.1% Teepol solution, followed by several rinses in distilled water.

(5) Vanselow and Bradford **(22)** have reported a procedure for removal of surface contaminants from leaves prior to spectrographic analysis. Washing efficiency was tested by the absence of titanium lines on the spectrogram. Leaf samples were washed individually by hand in a small stream of tap water with a small amt of Ivory soap on the fingertips. All traces of soap were then rinsed off, and the leaves were rinsed with distilled water. This method would appear to be satisfactory for leaves such as citrus, but its usefulness for more delicate leaves is doubtful.

(6) Grant **(5)** has also reported a method satisfactory for spectrographic analysis of leaves. Leaves were agitated for 30 sec in 0.05% Alconox in a polyethylene container, followed immediately by several distilled water rinses.

(7) On the basis of the methods reviewed, the following procedure has been found to meet the criteria outlined earlier for removal of surface contamination from vegetation: freshly harvested tissues are washed in a solution of 0.05% Alconox‡‡‡ and 0.05% Na_4EDTA in a polyethylene container for 30 sec with gentle agitation. The tissue is removed and allowed to drain for a few sec and is then rinsed for 10 sec in each of three beakers of deionized water.

7.1.2 Dry fresh plant tissue in a forced-draft oven at 80 C for 24 to 48 hr.

7.1.3 Grind dried tissues in a semimicro Wiley mill to pass a 40-mesh sieve. Place ground tissue in a clean bottle and cap.

7.2 Ashing and Alkali Fusion of Plant Tissues.

‡‡‡Alconox, Inc., New York, New York.

7.2.1 Mix the dried sample thoroughly and carefully weigh from 0.1 to 2.0 g of plant tissue, depending on the fluoride content, into a clean crucible.

7.2.2 Add 100 ± 10 mg of calcium oxide, sufficient deionized water to make a loose slurry, and 2 drops of phenolphthalein. Mix thoroughly with a polyethylene policeman. The final mixture will be uniformly red in color and will remain red during evaporation to dryness.

7.2.3 Place crucibles on a hot plate and under infrared lamps. Turn on infrared lamps (do not turn on hot plate) until all liquid is evaporated. Turn on hot plate and char samples for 1 hr.

7.2.4 Transfer crucibles to a muffle furnace at 600 C and ash for 2 hr.(**Caution**: To avoid flaming, place crucibles at front of muffle furnace with door open for about 5 min to further char samples. Crucibles may then be positioned in the furnace.)

7.2.5 After ashing, remove crucibles (not more than eight at one time), add 3 ± 0.1 g of sodium hydroxide pellets, and replace in the furnace with door closed for 3 min. (**Caution**: Watch out for "creeping" of the molten NaOH.) Remove crucibles one at a time and swirl to suspend all particulate matter until the melt is partially solidified. Allow crucible to cool until addition of small amt of water does not cause spattering. Wash down inner walls of crucible with 10 to 15 ml of deionized water.

7.2.6 After crucibles have cooled to room temp suspend the melt with a polyethylene policeman and transfer to a plastic tube graduated at 50.0 ml with deionized water. Rinse crucible with 20.0 ml of a 1:1 (v/v) 70% perchloric acid-deionized water solution and add to the tube. Make sample to 50.0 ml/volume with water.

7.2.7 Run several blank crucibles (about one blank for each 10 samples) containing all reagents through the entire procedure.

7.2.8 *Cleaning of crucibles*. Crucibles should be cleaned as soon as possible after use. Inconel crucibles are boiled in 10% (w/v) sodium hydroxide solution for 1 hr. Follow by washing with hot water and detergent and rinse thoroughly with distilled water. Crucibles which held samples containing more than 100 μgrams of fluoride are immersed in 4*N* HCl for 45 min before boiling in NaOH. Blank analyses are performed on these crucibles before use to check on contamination.

7.3 AUTOMATED ANALYSIS OF SAMPLES.

7.3.1 *Description of Analytical System* (Refer to Figure 204:2). All flow rates given are nominal values. Standard fluoride solutions, ashed and alkali-fused samples, or impinged air samples are placed in 8.5-ml plastic cups in the sampler module. The sampler is actuated, and the sample is pumped from the cup at a net rate of 2.48 ml/min with air segmentation of 0.42 ml/min after the sampler crook (2.90 ml − 0.42 = 2.48 ml) and is pumped into the microdistillation device through the sample inlet 1, (Figure 204:2). Teflon tubing of 0.053-inch ID is employed for sample transmission. Sulfuric acid is pumped at 2.50 ml/min through the acid inlet m (Figure 204:2). Acid and ashed solids are cooled and discarded. Distillate is pumped from the sample trap at 2.50 ml/min through 0.053-inch ID Teflon tubing, alizarin fluorine blue-lanthanum reagent is added at 1.69 ml/min, and the two liquid streams are combined for color development. The mixing device consists of a 4″ piece of ⅛″ ID glass tubing packed with pieces of 20-mesh broken borosilicate glass. The colored stream than passes through a time delay coil consisting of 15 ft of 0.035″ Teflon spaghetti tubing where color development is enhanced. The reagent stream then passes through a debubbler fitting where a small portion of the sample stream is removed (along with any bubbles) at a rate of 0.70 ml/min and passes into a waste bottle. The remainder of the sample stream passes through a 15-mm tubular flow cell of the colorimeter, and the absorbance is measured at 624 nm. Results are plotted on a chart recorder. The lag time from sampling to the appearance of a peak on the chart recorder is about 5 min. The time between samples is 6 min when the sampling rate is 10 per hr. A flow diagram of this automated procedure is shown in Figure 204:4.

7.3.2 *Description of Air Flow Sys-*

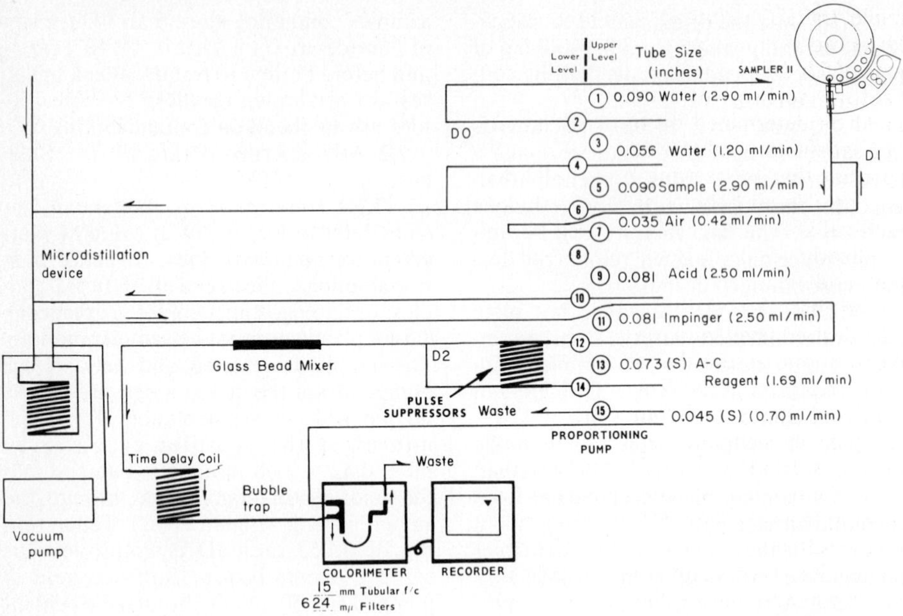

Figure 204:4—Flow diagram of semi-automated procedure for fluoride in vegetation.

tem. (Refer to Figure 204:1.) Air is drawn through the air inlet tube (a) before the Teflon microdistillation coil (b). Air sweeps through (b) to the fractionation column (c) where the air stream is diverted into two channels (c_1 and c_2). Air diverted through one channel (c_1) passes through the water-jacketed condensor (d) and sample trap (c) to waste bottle (f). The air then passes through a ⅛″ ID glass tube directed against the surface of conc sulfuric acid contained in waste bottle (g). The partially dehydrated air passes through a gas-drying tower (h) containing one lb of indicating silica gel. Air leaving the outlet of the drying tower passes through a T-tube (i) to which a vacuum gauge (j) (0 to 10″ Hg or 0 to 254 Torr) is connected, through a flowmeter (k) (0 to 5 l/min), through a T-tube (l), and then to the vacuum pump (m). Air diverted through the second channel passes through a water-jacketed waste trap (n) and to a waste bottle (o). Air leaving the waste bottle flows through a drying bulb (p) filled with indicating silica gel, and the dry air then passes through a flowmeter (q) (0 to 1 l/min). The air stream then connects through T-tube (l) with air flowing from the first channel (c_1). Flowmeter and vacuum gauge settings are described in **7.3.3**.

7.3.3 *Start-Up Procedures.* Turn on water to condenser and cooling jacket. Turn on colorimeter. Engage the manifold on the proportioning pump and start pump. Turn on the stirring motor on the microdistillation unit, the vacuum pump (adjusted for full vacuum), and the heater in the microdistillation unit. Connect the lines to the sulfuric acid, to the alizarin fluorine blue-lanthanum reagent, and to the deionized water bottles. The sampling tube of the sampler unit should be in the water reservoir. Allow the apparatus to equilibrate until the silicone oil in the microdistillation unit has reached 170 C. Be sure that all tubing connections are secure. Adjust flowmeter which controls distillation (Figure 204:1,k) to 2.5 to 3 l/min. Adjust flowmeter which carries the flow of waste from the distillation (Figure 204:1,g) to 0.3 l/min. Distillate should now fill the sample trap. Readjust flowmeter (Figure 204:1,k) to give a reading on the vacuum gauge of 5 to 5.9″ Hg (127 to 150 Torr). Satisfactory setting for each instrument should be determined by trial and error. Once satisfactory value is determined, it is important that this setting be maintained each day. No air bubbles should be in the analytical system beyond the point where the alizarin fluorine blue-lanthanum reagent and distillate streams are joined. Turn on chart recorder, adjust the baseline to the desired level, and run a baseline for several min to assure that all components are operating properly. Transfer standard fluoride solutions to 8.5-ml plastic cups and place in sampler. Separate the last standard sample from unknown samples with one cup containing deionized water. Program the sampler for 10 samples/hr with 1:2 sample to wash ratio or to other desired rate. Turn sampler on. Determine a standard curve before and after each set of 40 to 80 samples or even when fewer samples are analyzed. Analyze calibration solutions before and after each day's set of samples. A net absorbance of 0.7 to 0.9 should be obtained with a solution containing 4 μg fluoride/ml. The absorbance of each standard solution should be reproducible within 10% from day to day.

7.3.4 *Shut-Down Procedure.* Turn off chart recorder. Disconnect the sulfuric acid line and place in deionized water. Disconnect alizarin fluorine blue-lanthanum reagent line and place in 0.01% EDTA solution for about 1 min. Transfer the line to deionized water and allow water to pass through the analytic system for about 5 min. Clean out Teflon distillation coil with Na_4EDTA solution (see **7.3.5.a.**) and follow with several water washes. Turn off the heater and stirring motor in the microdistillation unit. Turn off the vacuum pump. Release pump tube manifold. Turn off water to the condenser and cooling trap.

7.3.5 *Maintenance.*

a. After use with samples containing particulate matter, the Teflon distillation coil should be cleaned out. This is easily accomplished by inserting the Tygon tube connected to the air inlet line of the microdistillation device (Figure 204:2,c) briefly into 0.01% Na_4EDTA solution. A quantity of the wash fluid will be swept through the Teflon coil and will remove deposited material. When all particulate matter is removed from the coil, wash briefly with deionized water several times.

b. Pump tubes should be replaced after 200 working hr or prior to that if they become hard and inflexible or flattened.

c. Indicating silica gel should be regenerated when two-thirds of the amount in each container loses its normal blue color.

d. Proportioning pumps should be oiled once a month and gain on the recorder checked and adjusted monthly.

e. All tubing containing reagent should be cleaned after each daily run with Na_4EDTA solution followed by deionized water.

f. Pump tubes are always left in the relaxed position when not in use.

7.3.6 *Check Procedures.*

a. An estimate of the degree of F contamination from reagents and equipment should always be made. Due to the ubiquity of F, crucibles without sample but with all reagents should be carried through the entire procedure. Contamination from previously analyzed samples, from a contaminated muffle furnace, and from reagents can then be detected and corrective efforts made. Blank values over 5 μg F are considered evidence of contamination. Two blank determinations should be made with every batch of samples analyzed (approximately 2/40 samples). Blank values are usually found to range from 1 to 3 μg.

b. Occasionally a special calibration curve should be made by adding known amounts of sodium fluoride solution from a microburet to aliquots of a low-F tissue. Recovery of added F should be 100 ± 10%. Low values indicate loss of F, possibly during pretreatment, and high values indicate contamination.

c. Different amounts (0.1 to 2.0 g) of a plant sample containing 50 to 65 ppm F should be analyzed occasionally. A linear relationship should exist between F found and amount of tissue taken. A nonlinear relationship may indicate that some component of the tissue is retarding distillation or interfering with color development.

d. Calibration curves should be repeated at least twice daily to correct for any small changes in distillation efficiency that might occur.

7.3.7 *Trouble Shooting.* Since no method for determining μg quantities of an element in an overwhelming excess of other materials is free of occasional problems, suggestions on how to recognize the difficulties and systematically locate the problems may be of value. Fortunately, most of the potential problems in distillation and F analysis are manifested by obvious irregularities on the chart recorder.

a. Irregular fluctuations in the baseline may result from the following: (1) excessive surge pressures in the liquid streams; (2) air bubbles passing through the flow cell in the photometer; or (3) bleaching of the reagent by excess sulfuric acid carryover during distillation or insufficient buffer in the reagent. Excessive pulse pressures may be due to faulty pump tubes, the absence of surge suppressors, or the presence of pulse suppressors that are improperly made or placed. Air bubbles in the photometer flow cell may be due to the absence of a debubbler bypass, a blockage in the reagent pump tube, or a periodic emptying of the sample trap. The last condition will result if the air flow to the distillation trap becomes too great. Excessive sulfuric acid carryover can be caused by too high a temp in the oil bath, improper sulfuric acid concentration, or too high a vacuum on the system. Large fluctuations or imbalances in vacuum or air flow rates in the distillation or waste systems will also produce baseline irregularities. An improper flowmeter setting, trapped air in the tubing, or a leak or block in the system should be sought as the probable cause of this type of difficulty.

b. Asymmetrical peaks, double peaks, or peaks with shoulders may result from: (1) baseline irregularities; (2) interfering substances from the sample or impure reagents; (3) inadequate buffer concentration; or (4) excessive amts of solid material in the Teflon distillation coil. The presence or accumulation of excessive solids may be due to insufficient flow of sulfuric acid, too large a sample, excessive amounts of calcium oxide or sodium hydroxide in the sample, inadequate suspension of particles in the samples, or lack of proper air segmentation in the sample tubing.

c. Poor reproducibility can be caused by improper sample pickup; by faulty pump tubes; by inadequate washout of Teflon distillation coils between samples; by large deviations in acid concn, temp, or air flow in the distillation coil; or by changes in vacuum on the waste systems.

8. Calibration and Standards

8.1 Transfer aliquots of each calibration fluoride solution (0.2, 0.4, 0.8, 1.6, 2.4, 3.2, and 4.0 μg F/ml) to 8.5-ml sample cups. Proceed with analysis as described in **7.3**. A calibration curve should precede and follow each 40 samples.

8.2 After the calibration solutions have been anaylzed and the peaks plotted by the chart recorder, draw a straight line connecting the baseline before and after the analysis. Record the absorbance of each peak and subtract the absorbance of the baseline. Plot net absorbance readings (abscissa) against μg F/ml on graph paper. Draw a line connecting the points plotted. Typical chart results at an analysis rate of 20 samples/hr are shown in Figure 204:5A. Figure 204:5B shows results for an analysis rate of 30 samples/hr and using a slightly different manifold. The standard curves derived from data of Figure 204:5 are shown in Figure 204:6.

9. Calculation

9.1 The fluoride content of the sample in ppm is calculated as follows:

$$F_T = \frac{F_s V_s D}{W_s}$$

Where:

F_T = ppm of fluoride in sample
F_s = μg of fluoride/ml of unknown

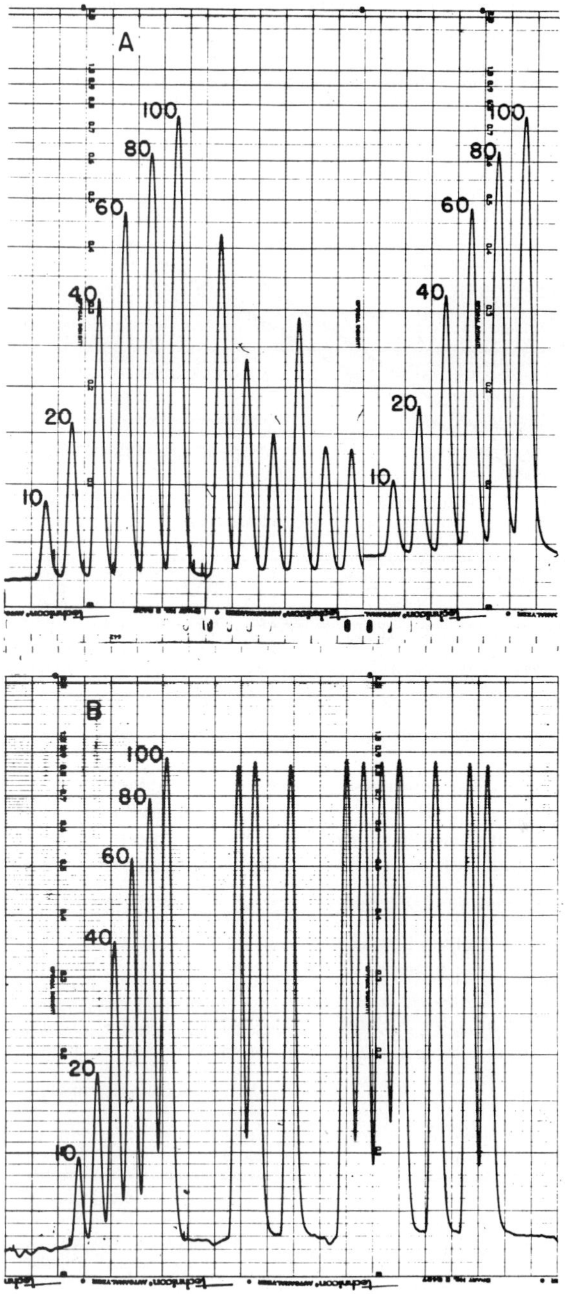

Figure 204:5—A. Chart record of automated fluoride analysis of calibration solutions at rate of 20 samples/hour.

B. Similar chart for an analysis rate of 30 samples/hr., using a modified manifold.

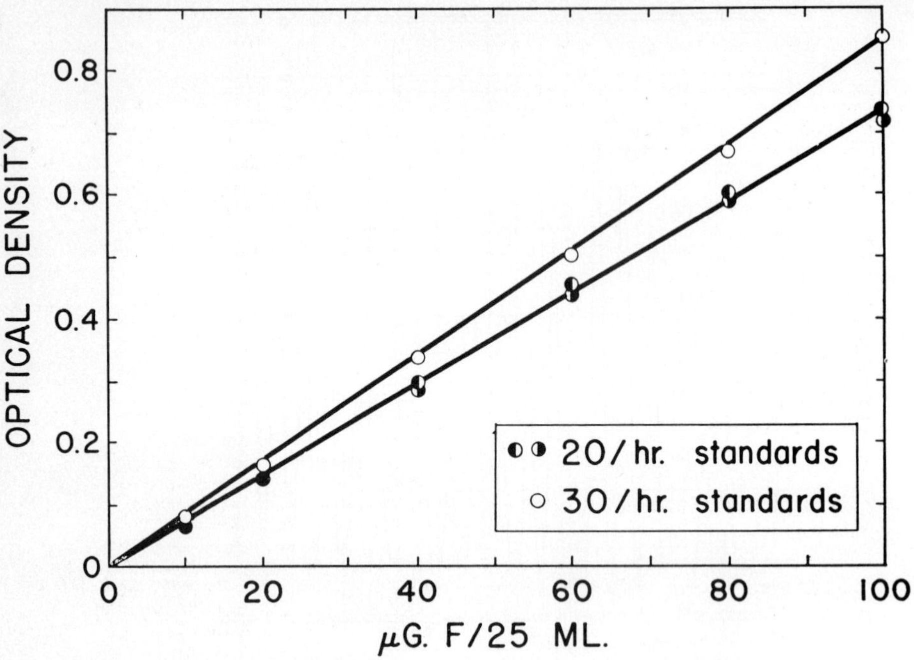

Figure 204:6—Standard curves derived from data of Figure 204:5.

sample as taken from the calibration curve.

V_s = volume in ml of the unknown sample (usually 50 ml).

D = dilution factor used when F in unknown sample exceeds the standard curve. For example, if the original sample is diluted from 50 ml to 100 ml, D will equal 2. (Note: All dilutions of plant samples should be made with 3N sodium hydroxide.) If the unknown sample is not diluted, D should be dropped from calculations.

W_s = weight of sample taken for analysis in grams.

10. Effect of Storage

10.1 The acetate buffer (**6.1.2**) and Brij 35 (**6.1.6**) solutions are stable at room temp. Stock solutions of alizarin fluorine blue (**6.1.4**) and lanthanum nitrate (**6.1.5**) are stable indefinitely at 4 C. The alizarin fluorine blue-lanthanum working reagent (**6.1.7**) is stable at 4 C for at least 7 days. Dilute NaF solutions should be stored in the cold in polyethylene bottles and are stable in the presence of sodium hydroxide. Tightly covered ashed and fused plant samples appear to be stable indefinitely.

11. References

1. ARKLEY, T.H, D.H. MUNNS, and C.M. JOHNSON. 1960. Preparation of plant tissues for micronutrient analysis. Removal of dust and spray contaminants. *J. Agr. Food Chem.* 8:318–321.
2. BOYNTON, DAMON, J. CARLTON CAIN, and O.C. COMPTON. 1944. Soil and season influences on the chemical composition of McIntosh apple leaves in New York. Proc. Am. Soc. Hort. Sci. 44:15–24.
3. CAMERON, S.H., R.T. MUELLER, A. WALLACE, and E. SARTORI. 1952. Influence of age of leaf, season of growth, and fruit production on the size and inorganic composition of Valencia orange leaves. Proc. Am. Soc. Hort. Sci. 60:42–50.
4. CHAPMAN, H.D.; and S.M. BROWN. 1950. Analysis of orange leaves for diagnosing nutrient status with reference to potassium. Hilgardia 19:501–539.
5. GRANT, CLARENCE L. 1961. Personal communication.
6. JACOBSON, J.S., and D.C. MCCUNE. An interlaboratory study of analytical techniques for fluo-

ride in vegetation. *J. Assoc. Offic. Anal. Chemists* (In press).
7. JACOBSON, J.S., D.C. MCCUNE, L. H. WEINSTEIN, R.H. MANDL, and A.E. HITCHCOCK. 1966. Studies on the measurement of F in air and plant tissues by the Willard-Winter and semi-automated methods. *J. Air Pollut. Contr. Assoc.* 16(7):367–371.
8. JACOBSON, L. 1945. Iron in the leaves and chloroplasts of some plants in relation to their chlorophyll content. Plant Physiol. 20:233–245.
9. JACOBSON, LOUIS, and J.J. OERTLI. 1956. The relation between iron and chlorophyll contents in clorotic sunflower leaves. Plant Physiol. 31:199–204.
10. MANDL, R.H., L.H. WEINSTEIN, J.S. JACOBSON, D.C. MCCUNE, and A.E. HITCHCOCK. 1966. Simplified semi-automated analysis of fluoride. Proc. Technicon Symposium "Automation in Analytical Chemistry." New York, Sept. 8, 1965, pp. 270–273.
11. MAVRODINEANU, R., J. GWIRTSMAN, D.C. MCCUNE, and C.A. PORTER. 1962. Summary of procedures used in the controlled fumigation of plants with volatile fluorides and in the determination of fluorides in air, water, and plant tissues. Contrib. Boyce Thompson Inst. 21:453–464.
12. NICOLAS, D.J.D., C.P. LLOYD-JONES, and D.J. FISHER. 1956. Some problems associated with determining iron in plants. Nature 177:336–337, 1956.
13. ______. 1957. Some problems associated with determining iron in plants. Plant Soil 8:367–377.
14. PARBERRY, N.H. 1935. Mineral constituents in relation to chlorosis of orange leaves. Soil Science 39:35–45.
15. SKEGGS, L.T., JR. 1957. An automatic method for colorimetric analyses. *Am. J. Clin. Pathol.* 28:311–322.
16. STEYN, W.J.A. 1959. Leaf analysis. Errors involved in the preparative phase. J. Agr. Food Chem. 7:344–348.
17. TAYLOR, G.A. 1956. The effectiveness of five cleaning procedures in the preparation of apple leaf samples for analysis. Proc. Am. Soc. Hort. Sci. 67:5–9.
18. THORNE, D.W., and ARTHUR WALLACE. 1944. Some factors affecting chlorosis on high-lime soils: 1. Ferrous and ferric iron. Soil Science 57:299–312.
19. WEINSTEIN, L.H., R.H. MANDL, D.C. MCCUNE, J.S. JACOBSON, and A.E. HITCHCOCK. 1963. A semiautomated method for the determination of fluorine in air and plant tissues. Contrib. Boyce Thompson Inst. 22:207–220.
20. WEINSTEIN, L.H., R.H. MANDL, D.C. MCCUNE, J.S. JACOBSON, and A.E. HITCHCOCK. 1965. Semi-automated analysis of fluoride in biological materials. *J. Air Pollut. Contr. Assoc.* 15:222–225.
21. WILLARD, H.H., and O.B. WINTER. 1933. Volumetric method for determination of fluorine. Ind. Eng. Chem. Anal. Ed. 5:7–10.
22. VANSELOW, A.P., and G.R. BRADFORD. 1961. Spectrographic Techniques. p. 212 In: Chapman, H.D. and P.F. Pratt. *Methods of Analysis for Soils, Plants and Water*. Univ. Calif. Div. of Agr. Sciences.
23. YAMAMURA, S.S., M.A. WADE, and J.H. SIKES. 1962. Direct spectrophotometric fluoride determination. Anal. Chem. 34:1308–1312.

Subcommittee 2
L. V. CRALLEY, *Chairman*
L. V. HAFF
A. W. HOOK
E. J. SCHNEIDER
J. D. STRAUTHER
C. R. THOMPSON
L. H. WEINSTEIN

205.

Tentative Method of Analysis for Fluoride Content of the Atmosphere and Plant Tissues

12202-01-72T

PART G. *Determination of Fluoride (Potentiometric Method)**

1. Principle of the Method

1.1 This method can be used for the determination of inorganic fluoride in aqueous solution obtained from impingers, bubblers, impregnated filter papers, coated tubes or other air sampling devices. (**1, 2, 3**). The alkaline sample solution containing atmospheric fluoride is diluted 1:1 with a combined buffer, ionic strength ad-

*Reference **203:10**

justed and complexing agent solution, adjusted to pH 5.9 to 6.1, prior to analysis with a fluoride ion selective electrode. The potential corresponding to the fluoride ion concentration is measured in millivolts.

1.2 Aqueous solutions containing inorganic fluoride are prepared for analysis by adding buffer to control pH, salts to provide constant ionic strength, and by dilution and addition of complexing agents to avoid interference with fluoride analysis by substances enumerated in Section **3**.

1.3 A combination fluoride ion electrode or fluoride ion and separate reference electrodes are inserted into liquid samples contained in polyethylene, Teflon, or polypropylene beakers. Temperature of both samples and solutions of known fluoride concentration (for calibration) is controlled to $\pm$ 2 C and all solutions are stirred during analysis.

1.4 Potential in mV is recorded and converted to μg of fluoride per ml of solution, using a calibration curve.

2. Range and Sensitivity

2.1 The range of measurement of the electrode is 0.019 to 19,000 μg fluoride/ml of solution (10^{-6} to 1*M*), (**4**); however, the recommended range of analysis for air samples is between 0.1 and 10 μg/ml F in solution. Slow response time and non-linearity of the calibration curve make measurements below 0.1 μg/ml less desirable. The minimum concentration that can be measured in the recommended range is 1 μg F/m^3 if the air sample volume is 10 m^3.

3. Interferences

3.1 The electrode measures the free fluoride ion (fluoride activity); therefore, any substances which complex the fluoride or conditions which reduce the dissociation of fluoride will produce errors. For example, fluoride in particulate matter will not be measured unless fluoride is solubilized. Fluoride in aluminum fluoride will not be measured unless it is released from combination with aluminum. Any substance which coats or reacts with the electrode crystal or otherwise slows the response time to fluoride will produce analytical errors. Hydroxyl ions are measured by the electrode when the OH^- to F^- concentration ratio is in excess of ten to one. A high concentration of dissolved salts depresses fluoride ion activity; therefore, differences in total ionic strength between samples and standard fluoride solutions can be a source of error. Finally, differences in temperature greater than 2 C between samples and standards can cause errors, particularly at low fluoride concentrations.

3.2 Pretreatment steps and conditions of analysis should be carefully considered with respect to the objective of the air sampling program. For example, electrode analysis of an impinged sample from air containing HF, SiF_4 and AlF_3 (as a dust) will give a positive error for gaseous fluoride when aluminum-complexing agents (CDTA or citrate) are used. However, if nonfluoride-containing aluminum salts are present, it is necessary to use complexing agents to obtain accurate measurement of gaseous fluoride. A more suitable approach in these situations would be to separate gaseous and particulate matter.

3.3 Other interferences in measurement of fluoride concentration by this method in addition to aluminum are iron, silicate, and rare earth elements. Metallic ions such as lead, manganese, magnesium, and calcium can interfere particularly if the counter-ion to fluoride ratio is very high. Usually, water-soluble forms of these substances are present in low concentrations in ambient air. Dilution and use of complexing agents (**5**) reduces the possibility of interference by these ions.

3.4 When high concentrations of silicates are expected, adjustment to pH 9 with sodium hydroxide can be employed to hydrolyze fluosilicates (**6**). The solutions must then be readjusted to the acceptable pH range for electrode measurement (pH 5 to 8).

3.5 For accurate results, the pH of both calibration and sample solutions should be approximately the same. Use of acetate buffer achieves this control and avoids interference by hydroxyl ions or by formation of hydrofluoric acid (not measured by the electrode).

3.6 Large amounts of salts or other dissolved substances can produce errors by depressing fluoride ion activity or by slowing response time. Every effort should be made to either have similar types and concentrations of substances present in both calibration and sample solutions or to swamp out the effects of unknown substances by adding large but constant amounts of sodium chloride to samples and standards.

3.7 The electrode responds more slowly at lower concentrations consequently it is advisable to arrange the sequence of measurements to avoid analysis of a sample low in fluoride following one containing a high fluoride concentration.

3.8 Differences in temperature greater than 2 C between calibration and sample solutions can cause errors, particularly when less than 0.2 μg/ml fluoride are measured, a water bath with ± 1 C temperature control should be used.

3.9 Analysis should be made in acid-washed polyethylene, Teflon or polypropylene beakers. Glass surfaces should be avoided especially with fluoride concentrations below 20 ppm.

4. Precision and Accuracy

4.1 The electrode method of fluoride analysis of water samples was found to compare favorably with an ion-exchange-colorimetric method in a colaborative study **(7)**. The relative standard deviation ranged from 2.9 to 5.8% and relative errors of 0.2 to 4.9% were found. Use of a complexing agent significantly reduced the negative errors caused by the presence of aluminum.

5. Apparatus

5.1 A combination fluoride ion electrode or separate fluoride ion and reference electrodes.

5.2 An electrometer or an expanded scale pH meter with a mV scale for measurement of potential.

5.3 A compatible recorder can be attached to the electrometer for obtaining a permanent record of analytical results.

5.4 Teflon-coated magnetic stirring bars and air-driven magnetic stirrers. The latter device avoids the heating effect of motor-driven stirrers.

6. Reagents

6.1 All reagents should be ACS analytical grade.

6.2 A combined buffer, ionic strength adjuster and complexing agent solution (Total Ionic Strength Adjustment Buffer, TISAB) **(8, 9)** is mixed with samples and standards prior to electrode analysis, as follows: To approximately 500 ml of distilled water in a 1-l beaker add 57.0 ml glacial acetic acid, 58.0 g sodium chloride (reagent grade), and 4.0 g of cyclohexylene-dinitrilo-tetraacetic acid, (CDTA). CDTA is available as either 1,2-diaminocyclohexane N, N, N′, N′-tetraacetic acid or (1,2-cyclohexylene-dinitrilo)-tetraacetic acid. Place the beaker in a water bath at room temperature and adjust the pH to 5.9 to 6.1 with 5 *M* sodium hydroxide. Pour into a 1-l flask and make to volume with distilled water.

6.3 A stock solution containing 100 μg/ml F is made by dissolving 0.222 g of dry reagent-grade sodium fluoride in 500 ml distilled water and diluting to 1-l with TISAB (Section **6.1**). The stock solution is stored in a plastic bottle in the cold and allowed to come to room temperature before use.

7. Procedure

7.1 For measurement of dissolved fluoride, samples are diluted 1:1 with TISAB in labelled plastic beakers. The fluoride electrode is inserted into each beaker and the potential is recorded with constant stirring after 2 min.

7.2 Measurement of fluoride in particulate matter may be accomplished by treating suspensions to dissolve fluoride and yield the free ion prior to analysis. If addition of acid is used, samples must be adjusted in pH to 5.9 to 6.1 with NaOH and/or TISAB prior to analysis.

7.3 PRECAUTIONS.

7.3.1 Proper electrode reponse is essential to accurate measurements. Nonlinearity of the calibration curve, poor reproducibility of replicate analyses or slow response time are indications that the electrode may not be operating properly. To check the electrometer, substitute a pH electrode for the fluoride electrode. If pH measurements (on expanded scale) of sample solutions are satisfactory, faulty operation of the fluoride electrode is indicated.

7.3.2 If response of the electrode is slow and a combination electrode is used, the reference solution should be drained and replaced. If operation is not improved, the end of the electrode can be dipped briefly in absolute methanol followed by 0.1 *N* HCl. After rinsing with deionized water the electrode is ready for use. This procedure should be used infrequently to avoid an effect of solvents on the plastic housing of the electrode.

8. Calibration and Standards

8.1 For calibration of the electrode, 50.0 ml TISAB and 50.0 ml distilled water are placed into a plastic beaker containing a Teflon-coated stirring bar. 0.10 ml of stock fluoride solution is pipetted into the stirred solution and the potential recorded. This procedure is repeated adding 1.0 ml and 10.0 ml stock solution successively to the same beaker. One set of standards at 0.1, 1.09 and 9.99 μg/ml should be analyzed before and after analyzing each set of samples.

8.2 A calibration curve is prepared on three-cycle semilog paper relating potential in mV *vs*. logarithm of concentration of fluoride standard solutions (0.1, 1.09 and 9.99 μg/ml.). Reproducibility of each point should be ± 1 mV. A linear calibration curve is obtained in the 0.1 to 10 μg/ml range with a slope of between 57 and 59 millivolts/tenfold change in fluoride concentration. If solutions containing less than 0.1 μg/ml are measured, additional standards should be used since the calibration curve is not linear at low fluoride concentrations.

9. Calculation of Results

9.1 Concentration of fluoride in the solution is read from the calibration curve in μg ml using millivolt values given by analysis. The final atmospheric fluoride concentration is determined by calculating total μg of fluoride in the original air sample and dividing by the total volume of air sampled as indicated by the following equation:

$$\text{Microgram F per cubic meter} = \frac{F \times V}{A}$$

where F = fluoride concentration found in μg/ml, V = volume of sample solution (ml) and A = volume of air sample (cubic meters at 760 Torr and 25 C).

10. Effects of Storage

10.1 The TISAB buffer and stock solutions of fluoride are stable under refrigeration.

10.2 If sample solutions are to be stored even for a few hours before analysis they should be held in plastic rather than glass containers.

10.3 For overnight storage place electrode in a solution containing 1 μg/ml fluoride and no salts or buffer. For long periods of storage, electrode may be kept dry after rinsing exterior with distilled water.

11. References

1. DORSEY, J.A., and KEMNITZ, D.A., JR. 1968. A source sampling technique for particulate and gaseous fluorides. *Air Pollution Control Assoc.* 18:12–14.
2. ELFERS, L.A., and DECKER, C.E. 1963. Determination of fluoride in air and stack gas samples by use of an ion specific electrode. *Anal. Chem.* 40:1658–1661.
3. MORGAN, G.B., FEUSTERSTOCK, J.C., and MCMULLEN, T.B. Personal communication from National Air Pollution Control Administration. The determination of atmospheric fluorides using the fluoride ion electrode.
4. FRANT, M.S., and ROSS, J.W., JR. 1966. Electrode for sensing fluoride ion activity in solution. *Science* 514:1553–1555.
5. HARWOOD, J.E. 1969. The use of an ion selective electrode for routine fluoride analyses on water samples. *Water Research* 3:273–280.
6. JORDAN, D.F. 1970. Determination of total fluo-

ride and/or fluosilicic acid concentrates by specific ion electrode potentiometry. *Journal of the AOAC* 53:447–450.
7. McFarren, E.F., Moorman, B.J., and Parker, J.H. 1969. Water Fluoride No. 3, Study No. 33, Analytical Reference Service, U.S. Dept. Health, Education and Welfare, Environmental Control Administration. Cincinnati, Ohio.
8. Frant, M.S., and Ross, J.W., Jr. 1968. Use of a total ionic strength adjustment buffer for electrode determination of fluoride in water supplies. *Anal. Chem.* 40:1169–1171.
9. Instruction Manual Fluoride Electrode. 1970. Method 94-09, Orion Research, Inc., p. 15.
10. American Public Health Association. 1977. Methods of Air Sampling and Analysis. 2nd ed. p. 403. Washington, D.C.

Subcommittee 2
C. R. Thompson, *Chairman*
G. H. Farrah
L. V. Haff
W. S. Hillman
A. W. Hook
E. J. Schneider
J. D. Strauther
L. H. Weinstein

206.

Tentative Method of Analysis for Gaseous and Particulate Fluorides in the Atmosphere (Separation and Collection with Sodium Bicarbonate Coated Glass Tube and Particulate Filter)

42222-01-72T; 12202-03-72T

1. Principle of the Method

1.1 Gaseous fluorides are removed from the air stream by reaction with sodium bicarbonate coated on the inside wall of a pyrex glass tube. Particulate fluorides are collected on a filter following the tube. The fluoride collected by the primer coating on the inside of the tube is removed with water or buffer and analyzed for fluoride. The particulates collected by the filter are eluted with acid and analyzed for fluoride. The results are reported as μg of gaseous fluoride and μg of particulate fluoride/m³ of air at 25 C and 760 Torr, (**1, 2**).

1.2 By collecting the samples on the dry tube and filter, the fluoride may be eluted with a small volume of eluant, thus concentrating the collected fluoride. This allows the analysis of the collected fluoride into fractional parts of a μg/m³.

2. Range and Sensitivity

2.1 The bicarbonate-coated tube method can be used to collect from 1 to 500 μg gaseous fluoride at a sampling rate of 0.5 CFM (14.3 liters/min). The length of the sampling period can therefore be adjusted so that the amount of fluoride collected will fall within this range. For a 12-hr sampling period, the detectable atmospheric concentration would be from about 0.1 to 50 μg F/m³. The actual lower limit of the method will depend upon the sensitivity of the analytical method employed and the quality of reagents used in tube preparation and analysis. It is recommended that the lower limit of detection should be considered as 2 times the standard deviation of the monthly mean blank value. Any values greater than the blank by less than this amount should be reported as "blank value".

3. Interferences

3.1 Significant amounts of acid aerosols or gases might neutralize the sodium bicarbonate coating and prevent quantitative uptake of gaseous fluoride from the atmosphere.

3.2 The presence of large amounts of aluminum or certain other metals, or of phosphates can interfere with subsequent analyses by colorimetric or electrometric methods. This is a problem inherent with any collection method for fluoride.

4. Precision and Accuracy

4.1 PRECISION. The root mean square difference of duplicate bicarbonate-coated tubes within the range of 0.5 to 3.3 μg F/m^3 is 0.051 μg F/m^3.

4.2 ACCURACY. Recovery of known amounts of gaseous HF was better than 95% with amounts of F up to about 40 μg and at sampling periods of 15 to 120 min. Data on particulate F are not sufficient to establish recovery under field conditions.

5. Apparatus

5.1 GLASS TUBING. Four foot lengths of 7 mm ID pyrex tubing. The tubing will be coated with sodium bicarbonate according to the requirements outlined in **6.8**.

5.2 FILTER AND HOLDER. The tubing will be followed by a filter holder and filter for the collection of particulates for particulate fluoride analysis (Figure 206:1). Nuclepore (one μ pore size), Acropor AN-800, and Whatman Nos. 42 and 52 (47 mm filter) have been tested and found satisfactory for this application.

5.3 The tube and filter are followed by an air sampling system which is capable of sampling at a rate of 14.3 l/min (0.5 CFM) and measuring the total air sampled either on a time rate basis or with a totalizing meter.

5.4 The system should be equipped so that pressure and temperature of the gas at the point of metering also are known for correcting sample volumes to standard conditions of 760 Torr and 25 C. Relative humidity of the gases sampled has not been considered as a required part of the data.

5.5 The sampling system should be assembled so that the inlet of the tube is 12 to 20 ft above ground level and protected from rain in such a manner as not to interfere with the free passage of aerosol fluorides.

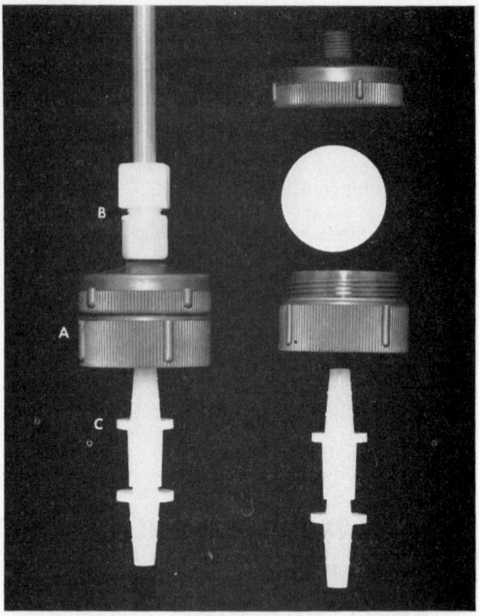

Figure 206:1—Details of attachment of the filter assembly and limiting orifice to a bicarbonate-coated tube (7 mm. I.D.). (A) Polypropyene filter holder, (B) plastic female connector, (C) limiting orifice.

5.6 A 25- or 40-watt light bulb should be installed to condition the gases to a point where condensation will not occur.

5.7 CONFIGURATION OF SAMPLING EQUIPMENT.

5.7.1 Figure 206:2 is a sketch of a suggested sampling system.

5.7.2 Other equivalent systems which meet the requirements outlined also would be satisfactory.

5.8 CRITERIA FOR COATING OF THE PYREX TUBES. Apparatus for cleaning, coating and drying the tubes may vary but the following requirements should be met.

5.8.1 The coating should be a visible uniform coating on the full length of the tube.

5.8.2 The coating should not contain any large crystals or heavy local deposits

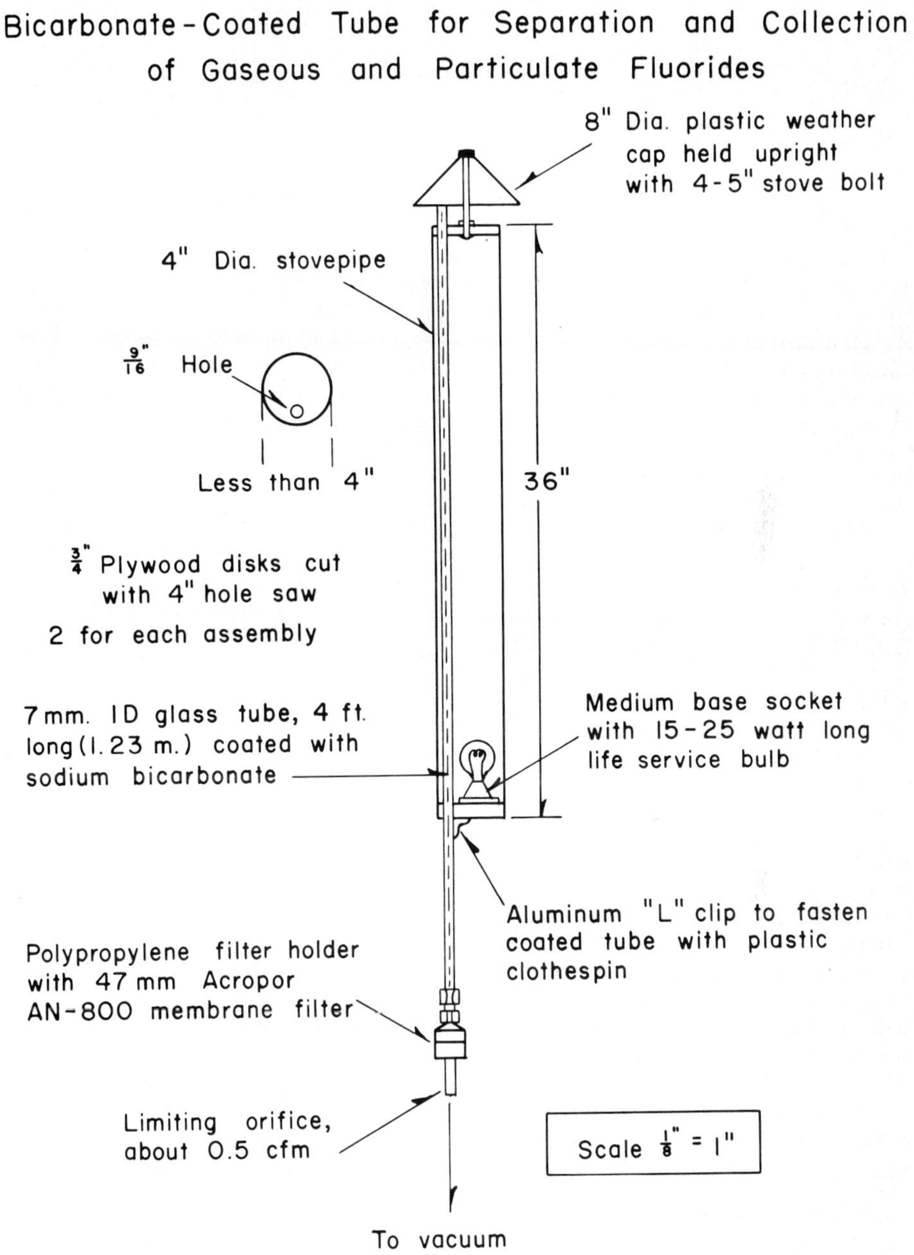

Figure 206:2—Sodium bicarbonate-coated glass tube illustrating simple heating device.

which could flake off and be collected with the aerosol fluorides.

5.8.3 The total coating should contain less than 1 μg of fluoride when analyzed without exposure, including all the reagents used in the procedure.

5.8.4 Prepared tubes should be sealed until time of use. Parafilm, serum tube caps, or other materials may be used for this purpose.

5.8.5 Any materials which may come into contact with the tubes should be checked for freedom from contamination with fluoride.

6. Reagents for Cleaning and Coating the Pyrex Tubes

6.1 All reagents should be ACS analytical grade.

6.2 A detergent solution low in fluoride and phosphate is used for initial cleaning of the tubes.

6.3 Alcoholic—KOH, 10% W/V. Prepare a solution of 10% by weight of KOH in methanol by dissolving 100 g of KOH in methanol and making the volume to 1 l. Mix thoroughly.

6.4 Sodium Bicarbonate, 5% W/V. Prepare a solution of 5% $NaHCO_3$ by dissolving 50 g of $NaHCO_3$ in water and making the volume to 1 l. Mix thoroughly.

6.5 A wetting agent should be used in the sodium bicarbonate solution to promote even wetting of the tube.*

6.6 Total Ionic Strength Adjustment Buffer (TISAB). Add 57 ml glacial acetic acid, 58 g sodium chloride and 4.0 g CDTA ((1,2-cyclohexylenedinitrilo)—tetraacetic acid), to 500 ml distilled water. Stir and add 5 *M* sodium hydroxide slowly until pH is between 5.0 and 5.5. Cool and dilute to 1 l.

6.7 1.0*N* sulfuric acid in distilled water.

6.8 1.0*N* sodium hydroxide in distilled water.

*1 ml of Brij 35 (1 : 1 dilution, Atlas Powder Co., Chemicals Division, Wilmington, Delaware) may be used for each 100 ml of coating solution, or equivalent wetting agent from any other source.

7. Procedure

7.1 Coating the pyrex tubes.

7.1.1 Clean the tubes with detergent, alcoholic KOH solution and deionized or distilled water.

7.1.2 While still wet from the cleaning, wet the internal surface of the tube with the 5% $NaHCO_3$ solution.

7.1.3 Allow the tube to drain for about 10 sec and dry the coating rapidly by passing hot, dry fluoride-free air downward through the tube, hanging in a vertical position.

7.1.4 The hot fluoride-free air stream can be provided by blowing air through a 4 to 8 mesh soda lime trap and then through four feet of coiled copper tubing heated by a small gas burner or heating tape. To simplify the drying, the hot air stream can be run through a manifold terminating in several outlet ports. The flow rate through the system should be in the order of about 3 l/min/tube, and the drying should be complete in about 1 min.

7.1.5 After the tubes are dry, seal the ends and store in a clean area until used.

7.2 Preparation of filters.

7.2.1 The filters and filter holders should be assembled in the laboratory and sealed if not used immediately.

7.2.2 Prior to the taking of the tubes and filter holders to the sampling site, the tubes and filter holders should be assembled into a sampling unit. The ends should be kept sealed until installed at the sampling site.

7.3 Air Sampling Procedure.

7.3.1 Using this method, samples should be taken for 12-hr periods to provide sufficient fluoride for accurate measurement.

7.3.2 It is recommended that sampling cover day and night conditions.

7.3.3 The sample should be taken at abost 14.3 l/min (0.5 CFM) using a calibrated limiting orifice or other suitable device to control the flow at this rate (Figure 206:3). A totalizing gas meter is recommended to measure total sample volume.

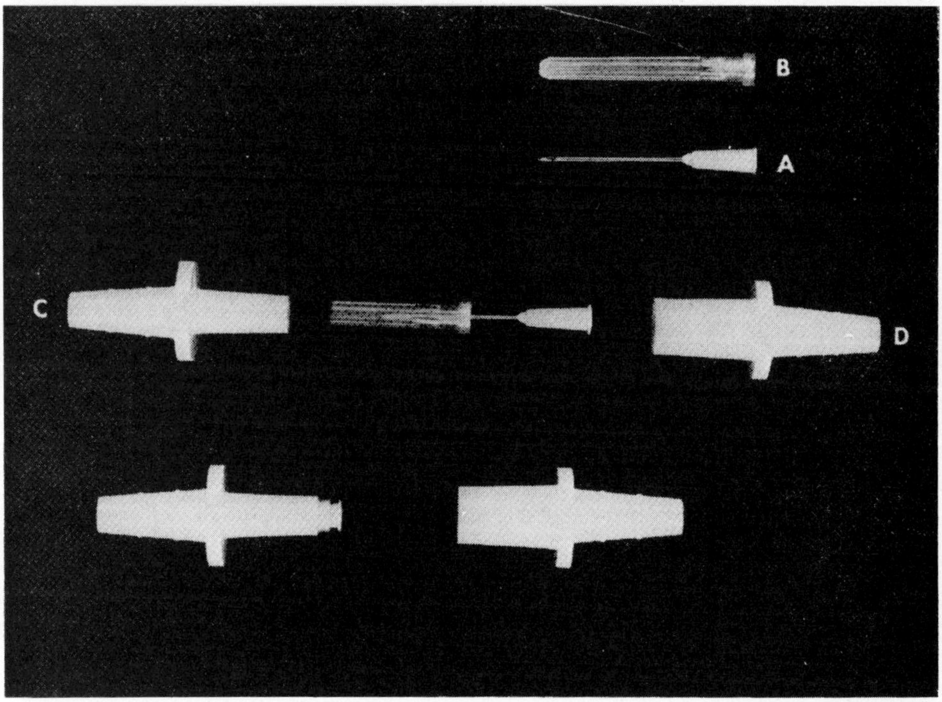

Figure 206:3—Details of construction of a limiting orifice with "quick disconnects" and disposable hypodermic needles. (A) Disposable needle, (B) needle sheath, (C) male end of "quick-disconnect", (D) female end of "quick-disconnect".

Pressure drop and temperature at the meter should be recorded at the beginning and end of each sampling period.

7.3.4 At the end of the sampling period, the tube and filter assemblies are capped and returned to the laboratory for analysis.

8. Preparation of Samples for F Analysis

8.1 In a laboratory free of contamination by process-generated fluorides, the tubes and filter assemblies are separated and each analyzed for total fluoride content.

8.2 PREPARATION OF TUBES FOR F ANALYSIS.

8.2.1 *For potentiometric analysis.*

a. With the tube in a vertical position and the lower end capped, pipet in 5.0 ml of ½-strength TISAB buffer.

b. Gently agitate the tube to wet all surfaces and empty the tube into a clean plastic beaker.

c. For details of potentiometric analyses refer to 12202-01-72T, Part G, **(3)**.

8.3 PREPARATION OF PARTICULATE FILTERS FOR F ANALYSIS.

8.3.1 *For potentiometric analysis.*

a. Place particulate filters in clean 15 × 150 mm test tubes. Add 5.0 ml 1*N* H_2SO_4 and mix for several seconds with a vortex mixer and allow to stand for 5 min.

b. Add 5.0 ml 1*N* NaOH and 10.0 ml TISAB.

c. Analyze using potentiometric method, 12202-01-72T, Part G, **(3)**.

8.3.2 *For semiautomated analysis.* **(4).**

a. Place the particulate filter in a clean test tube, pipet in 5.0 ml of 1*N* H_2SO_4, mix for several sec with a vortex mixer and allow to stand for 5 min. Filter samples to remove cellulose fibers into 8.5 ml sample cups.

b. Analyze the sample using the semiautomated method as described in **(4)** 12202-02-68T, Section **7.3**. Standards for semiautomated analysis should be made up in 1*N* H_2SO_4.

9. Calculation of Results

9.1 The fluoride concentration of the atmosphere is calculated by the following relationship:

$$\text{Fluoride, } \mu g/m^3 = \frac{(A - B)}{C}$$

Where:

A = number of micrograms of F in the sample.
B = number of micrograms of F in the blank.
C = volume of air sample in cubic meters at 760 Torr and 25 C.

10. Effect of Storage

10.1 Coated tubes and filter holders can be sealed and stored indefinitely in a clean, fluoride-free area, prior to analysis.

11. References

1. MANDL, R.H., WEINSTEIN, L.H., WEISKOPF, G.J., and MAJOR, J.L. 1970. The separation and collection of gaseous and particulate fluorides. Paper CP-25A. 2D Internat. Clean Air Congress, Washington, D.C., December 6–11.
2. WEINSTEIN, L.H., and MANDL, R.H. 1971. The separation and collection of gaseous and particulate fluorides. VD1 Berichte Nr. 164:53–63.
3. AMERICAN PUBLIC HEALTH ASSOCIATION. 1977. Methods of Air Sampling and Analysis. 2nd ed. p. 417. Washington, D.C.
4. Ibid. p. 405.

Subcommittee 2
C. R. THOMPSON, *Chairman*
G. H. FARRAH
L. V. HAFF
W. S. HILLMAN
A. W. HOOK
E. J. SCHNEIDER
J. D. STRAUTHER
L. H. WEINSTEIN

207.

Tentative Method of Analysis for Gaseous and Particulate Fluorides in the Atmosphere (Separation and Collection with a Double Paper Tape Sampler)

42222-02-72T; 12202-04-72T

1. Principle of the Method

1.1 The double paper tape sampler provides a method for the automatic separation and collection on chemically-treated paper tapes of inorganic particulate and gaseous atmospheric forms of fluoride. The device may be programmed to collect and store particulate and gaseous fluoride from individual air samples obtained over time periods ranging from several minutes to several hours. A sufficient quantity of paper tapes will allow unattended operation for the automatic collection of up to 600 samples.

1.2 Air is drawn through an air inlet tube and is first passed through an acid-treated prefilter paper tape to remove particulate fluoride and then through an alkali-treated paper tape to remove gaseous fluorides.

1.3 The exhaust air is filtered through soda lime and glass wool and the fluoride-free air is used to pressurize the front compartment, preventing fluoride contamination of the paper tapes from the ambient air.

1.4 At the end of the preset sampling period, the vacuum pump is turned off and the tapes are indexed. After indexing, the vacuum pump is again activated. Indexing results in a "dead time" of several seconds.

1.5 The individual sample spots are cut out, treated with an ionic strength adjustment buffer solution to dissolve fluoride, and the sample solution is analyzed by potentiometric or colorimetric methods (**1, 2, 3, 4**).

2. Range and Sensitivity

2.1 The lower limit of detection of either gaseous or particulate fluoride will depend upon the length of the sampling period selected. For a 1-hr period, however, an atmospheric concentration of as low as about 0.1 μg F/m^3 (gaseous or particulate) can be detected. The upper limit is about 100 μg F/m^3.

3. Interferences

3.1 The presence of significant amounts of particulate metallic salts such as those of aluminum, iron, calcium, magnesium or rare-earth elements may react with gaseous fluoride and remove some or all of it on the prefilter.

3.2 Significant amounts of acid aerosols or gases may neutralize or acidify the alkali-treated tape and prevent quantitative uptake of gaseous fluoride from the atmosphere.

3.3 The presence of large amounts of aluminum or certain other metals, or of phosphates, can interfere with subsequent analysis of the tapes by colorimetric or electrometric methods. This is a problem inherent with any collection method for fluoride.

3.4 There are several other possible limitations of the method:

3.4.1 Although the acid-treated Whatman No. 52 prefilter has been shown to allow passage of HF, it will restrict passage of particulate F only as small as about 1 μm. Thus, smaller particulates may pass through the prefilter and impinge on or pass through the alkali-treated second tape. If prefilters of higher retentivity are required for removal of submicron-size particles, citric acid-impregnated tapes of Whatman No. 42 filter paper or Acropor AN-800 may be used.

3.4.2 Sampling of the atmosphere for long periods of time may result in the collection of a sufficient amount of particulate matter to result in sorption of gaseous fluoride on the particulate matter or a change in the air sampling rate.

4. Precision and Accuracy

4.1 PRECISION.

4.1.1 *Gaseous fluoride.* A relative standard deviation of about 16% was found in the range of 1 to 3 μg F/m^3; in the range of 12 to 45 μg F/m^3, it was about 5%. Relative humidities between 40 to 85% had no effect on precision.

4.1.2 *Particulate fluoride.* No information is available.

4.2 ACCURACY.

4.2.1 *Gaseous fluoride.* Recovery of known amounts of gaseous HF was better than 95% with amounts of fluoride of up to 40 μg and at sampling periods of 15 to 120 min. Removal of gaseous fluoride on the prefilter was negligible at relative humidities of 85% or less. No information is available above 85% relative humidity or in the presence of very dusty atmospheres.

4.2.2 *Particulate fluoride.* No information is available.

5. Apparatus

5.1 The double paper tape sampler is a modification of and utilizes the basic principles of the sequential paper tape sampler used for dust collection.

5.2 DESCRIPTION OF SAMPLER. (see Figure 207:1)

5.2.1 The sampler has two supply reels and two take-up reels with appropriate capstans to guide the tapes through the sampling block.

5.2.2 The sampling block and sample inlet tube are constructed of Teflon or stainless steel to minimize reactivity with gaseous F. The upper part of the sampling block has a cylindrical cavity one″ in diameter and the inlet tube is perpendicular to the paper tapes. The lower part of the sampling block is constructed of stainless steel with a one-inch diameter cylindrical cavity and with an outlet tube which passes at a right angle into the pump compartment. The lower block is spring-loaded with a total pressure of 3 pounds against the lower surface of the upper block. The surfaces of the two blocks are machined flat to insure a tight seal. The lower block is lowered by means of an electric solenoid which counteracts the spring pressure.

5.2.3 The paper tapes are 1 ½″ wide.

5.2.4 Capstans are positioned to

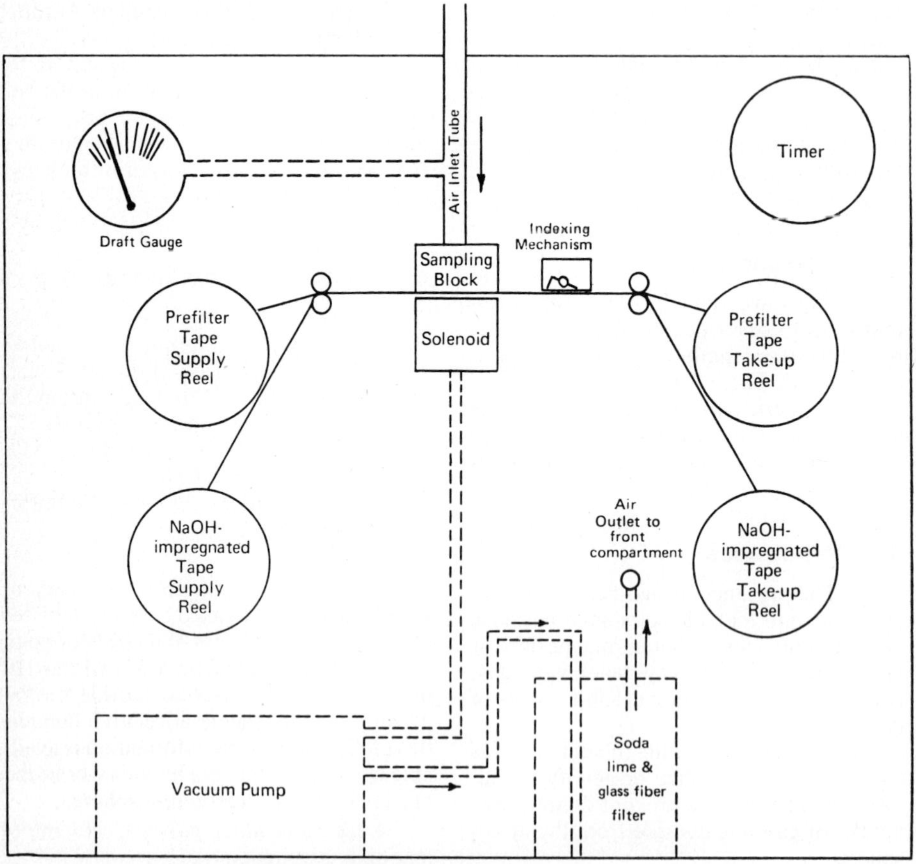

Figure 207:1—Schematic drawing of double paper tape sampler.

guide the paper tapes through the sampling block and to the take-up reel.

5.2.5 The paper tapes are drawn through the sample block and wound on the take-up reels by 2 rpm synchronous motors. Indexing is accomplished either by mechanical or photoelectric means to provide even spacing between samples. Provision is made by the use of tape perforated at regular intervals, or by some other means, to locate the collected sample spots for subsequent analysis. A relay is wired in series with the indexing mechanism to turn off the vacuum pump during tape transport.

5.2.6 An interval timer is used to provide desired sampling times.

5.2.7 Air is sampled with a carbon-vane vacuum pump of nominal 1 cfm (28.3 l/min) free-air capacity. This should provide a sampling rate through the two tapes of about 0.5 cfm. (14.1 l/min).

5.2.8 Exhaust air from the pump passes through a soda lime-glass wool filter and the filtered air is used to pressurize the front compartment and prevent contamination by F from the ambient air.

5.2.9 Air flow through the sample inlet tube is measured by means of a calibrated draft tube connected to a static pressure meter. The distance from the top of the inlet tube to the point of attachment of the draft gauge should be constant for all measurements.

5.2.10 Provision is made for manual override of the tape transport mechanism.

5.2.11 Precautions. Strain on the sample probe (Air inlet tube, Figure 207:1) should be avoided since this may displace the head (Sampling block, Figure 207:1) and allow air to enter the seal (between the sampling block and solenoid, Figure 207:1). The probe should be dismantled and both the probe and the head (Air inlet tube and sampling block, Figure 207:1) cleaned periodically in case dust has accumulated.

6. Reagents

6.1 PURITY. All reagents should be ACS analytical grade.

6.2 Whatman No. 52 filter paper tape, 1 ½″ wide, is to be used as the prefilter. If prefilters of higher particle retention are required, Whatman No. 42 or Acropor AN-800 may be used.

6.3 Whatman No. 4 filter paper tape, 1 ½″ wide, is to be used to remove gaseous F.

6.4 0.1*M* citric acid in 95% ethyl alcohol.

6.5 0.5*N* NaOH in 95% ethyl alcohol containing 5% glycerol.

6.6 TOTAL IONIC STRENGTH ADJUSTMENT BUFFER (TISAB). Add 57 ml glacial acetic acid, 58 g sodium chloride and 4.0 g CDTA ((1,2-cyclohexylenedinitrilo)-tetraacetic acid) to 500 ml distilled water. Stir and add 5*M* sodium hydroxide slowly until pH is between 5.0 and 5.5. Cool and dilute to 1 l.

6.7 1.0*N* sulfuric acid in distilled water.

6.8 1.0*N* sodium hydroxide in distilled water.

7. Procedure

7.1 Treat the Whatman No. 52 paper tape with the citric acid solution in ethyl alcohol and dry.

7.2 Treat the Whatman No. 4 paper tape with the NaOH-glycerol solution in ethyl alcohol and dry.

7.3 Place the citric acid-treated tape on the upper supply reel and the NaOH-treated tape on the lower supply reel. Thread the tapes through the sampling block and to their respective take-up reels.

7.4 Set the interval timer to the desired sampling period.

7.5 Turn the tape sampler to the "on" position.

7.6 Record the rate of air flow through the inlet tube on the static pressure meter.

7.7 At convenient intervals remove the paper tapes and place each in separate clean containers.

7.8 Analysis of the individual samples is carried out in a laboratory by colorimetric or electrometric methods.

7.9 Portions of the unused tapes are also analyzed periodically to provide a blank correction.

8. Preparation of Samples for Fluoride Analysis

8.1 Analysis is carried out in a laboratory free of contamination by process-generated fluorides.

8.2 PREPARATION OF ALKALI-TREATED TAPES FOR FLUORIDE ANALYSIS.

8.2.1 *For potentiometric analysis:*

a. Cut out individual sample spots and place in clean 15 × 150 mm test tubes. Add 5.0 ml of ½-strength TISAB and mix for several seconds with a vortex mixer. Decant into a clean plastic beaker.

b. Analyze using potentiometric method, (12202-01-72T, Part G), (**3**).

8.2.2 *For semiautomated analysis:*

a. Cut out individual sample spots and place in clean 15 × 150 mm test tubes. Add 5.0 ml of deionized water and mix for several seconds with a vortex mixer. Filter samples (to remove cellulose fibers) into 8.5 ml sample cups by a semi-micro filtration method.

b. Analyze the sample using the semiautomated method (12202-02-68T, Section **7.3**), (**4**).

8.3 PREPARATION OF CITRIC ACID-TREATED (PREFILTER) TAPES FOR FLUORIDE ANALYSIS:

8.3.1 *For potentiometric analysis:*

a. Cut out individual sample spots and place in clean 15 × 150 mm test tubes. Add 5.0 ml 1*N* H_2SO_4, mix for several seconds with a vortex mixer, and allow to stand for 5 min.

b. Add an equal volume of 1*N* NaOH. Decant into a clean plastic beaker and add 10.0 ml TISAB.

c. Analyze using potentiometric method (12202-01-72T, Part G), (**3**).

8.3.2 *For semiautomated analysis:*

a. Cut out individual sample spots and place in clean 15 × 150 mm test tubes. Add 5.0 ml 1*N* H_2SO_4, mix for several seconds with a vortex mixer, and allow to stand for 5 min. Filter samples (to remove cellulose fibers) into 8.5-ml sample cups by a semi-micro filtration method.

b. Analyze the sample using the semiautomated method (12202-02-68T, Section **7.3**), (**4**).

8.4 CALIBRATION AND STANDARDS.

8.4.1 None required.

9. Calculation

9.1 The fluoride concentration of the atmosphere is calculated by the following relationship.

$$\text{Fluoride, } \mu g/m^3 = \frac{(A - B)}{C}$$

Where:

A = number of micrograms of F in sample

B = number of micrograms of F in blank

C = volume of air sample in cubic meters at 760 Torr and 25 C.

10. Effect of Storage

10.1 Paper tape samples prior to analysis can be stored indefinitely under dry, fluoride-free conditions.

11. References

1. MANDL, R.H., WEINSTEIN, L.H., WEISKOPF, G.J., and MAJOR, J.L. 1970. The Separation and Collection of Gaseous and Particulate Fluorides. Paper CP-25A. 2D Internat. Clean Air Congress, Washington, D.C., December 6–11.
2. WEINSTEIN, L.H., and MANDL, R.H. 1971. The Separation and Collection of Gaseous and Particulate Fluorides. VDI Berichte Nr. 164:53–63.
3. AMERICAN PUBLIC HEALTH ASSOCIATION. 1977. Methods of Air Sampling and Analysis. 2nd ed. p. 417. Washington, D.C.
4. Ibid. p. 403.

Subcommittee 2
C. R. THOMPSON, *Chairman*
G. H. FARRAH
L. V. HAFF
W. S. HILLMAN
A. W. HOOK
D. J. SCHNEIDER
J. D. STRAUTHER
L. H. WEINSTEIN

301.

Tentative Method of Analysis for Antimony Content of the Atmosphere

12102-01-69T

1. Principle of the Method

1.1 Pentavalent antimony in the presence of a large excess of chloride ion **(2)** reacts with Rhodamine B to form a colored complex, which may be extracted with organic solvents such as benzene, toluene, xylene, or isopropyl ether **(2, 3, 4, 5, 6, 7, 8, 9)**. Trivalent antimony will not react with Rhodamine B and must be oxidized to the pentavalent state, (i.e., Sb_2O_5) by digestion in sulfuric, nitric, and perchloric acids. Phosphoric acid is used to minimize interference by iron. The absorbance of the pink-colored extract is measured in a spectrophotometer at 565 nm, or in a colorimeter using a green filter, and the amount of antimony present determined by reference to a calibration curve prepared from known amounts of antimony.

2. Range and Sensitivity

2.1 With a 1-cm light path cell, the method covers a range from 0 to 10 μg of antimony in 10 ml of solvent.

2.2 The sensitivity being 1.0 μg antimony (Sb)/determination (0.05 absorbance units), a concentration of 0.05 μg of antimony/m^3 can be measured if a 20 m^3 air sample is taken for analysis.

3. Interferences

3.1 According to Maren **(6)**, of the commonly encountered ions, only iron is likely to interfere. Maren suggests iron interference can be eliminated by extracting with isopropyl ether instead of benzene. For 750 μg of iron/ml, results obtained by the ACGIH, show interference from iron to be insignificant **(1)**.

4. Precision and Accuracy

4.1 Three samples containing antimony

Sample No.	Antimony present μg/ml	Iron present μg/ml
1	3.0	0.0
2	5.0	750.0
3	8.0	0.0

were analysed in triplicate by 10 collaborating laboratories in addition to the referee, and the standard deviation calculated

	Sample 1	Sample 2	Sample 3
Standard deviation	0.113	0.237	0.245

4.2 Seven tests were made to determine the recovery of antimony added in solution by pipet to 9 cm Whatman #1 filters which were then oven dried. The table shows that recovery of the antimony ranged from 95 to 102.5% of the 2 to 10 μg added.

Sb Added	Sb Found	Deviation Per cent
2.0 μg	1.90 μg	− 5.0
4.0 μg	4.08 μg	+ 2.0
6.0 μg	6.15 μg	+ 2.5
6.0 μg	6.06 μg	+ 1.0
8.0 μg	8.0 μg	0
10.0 μg	10.03 μg	+ 0.03
10.0 μg	9.94 μg	− 0.06

5. Apparatus

5.1 A SPECTROPHOTOMETER OR A COLORIMETER, with a green filter having maximum transmittance in the 565 nm range.

5.2 125-ML ERLENMEYER FLASKS, with ground glass stoppers.

5.3 125-ML SQUIBB SEPARATORY FUN-

NELS, with Teflon stopcocks (prior to use, these should be cooled in 5 C in a refrigerator).

5.4 BOILING AIDS. Perforated pyrex beads, platinum tetrahedra or other materials may be used if necessary, to promote smooth boiling after ascertaining that no measurable contribution to the blank results from their use.

6. Reagents

6.1 SULFURIC ACID C.P., SP GR 1.84.

6.2 NITRIC ACID C.P., SP GR 1.42.

6.3 PERCHLORIC ACID C.P., REDISTILLED 72%.

6.4 BENZENE C.P., OR TOLUENE C.P., OR XYLENE C.P., OR ISOPROPYL ETHER C.P.

6.5 HYDROCHLORIC ACID 6 N. Prepare from C.P. hydrochloric acid (S.G. 1.19) by mixing with an equal volume of distilled water.

6.6 ORTHOPHOSPHORIC ACID 3 N. Dilute 70 ml C.P. phosphoric acid to 1 l with distilled water.

6.7 RHODAMINE B, 0.02%. Prepare solution 0.02% w/v in distilled water.

6.8 ANTIMONY STANDARD SOLUTION. Place 0.1000 g C.P. antimony in 25 ml conc sulfuric acid and heat till dissolved; dilute to 1 l with distilled water. This solution contains 100 μg trivalent antimony/ml, and from this more dilute working standards may be prepared to cover the range 1 to 10 μg/ml.

6.8.1 *Pentavalent antimony* in the form Sb_2O_5 is available commercially from K & K Laboratories, and probably from other sources. The powder is soluble in HCl, HI, and KOH. Possibly it would facilitate calibration if the digestion step could be eliminated by using antimony in pentavalent form. However, in the process of dissolving the Sb_2O_5 the antimony was reduced so that it could not be used directly for standardization.

6.9 DISTILLED WATER. Where distilled water is mentioned, this should be double distilled, with the final distillation being from glass.

6.10 CAPRYL ALCOHOL C.P. This may be used if excessive frothing occurs during digestion.

6.11 All the foregoing reagents and solutions should be cooled to 5 C in a refrigerator before use.

7. Procedure

7.1 Antimony collected from the atmosphere on membrane, cellulose or glass fiber filters must be oxidized to the pentavalent state before analysis with Rhodamine B is possible. The filter or an aliquot is placed in a 125-ml Erlenmeyer flask; 5 ml conc sulfuric acid is added, followed by 5 ml conc nitric acid. The flask is then heated on a hot plate or over a gas flame till brownish-red fumes of NO_2 are driven off, the solution blackens (for organic materials) and dense white fumes of sulfur trioxide are driven off. The solution is allowed to cool slightly (2 to 5 min) and 10 drops of 72% perchloric acid are added. The solution is then heated once more till white fumes of SO_3 are driven off, the solution clears and becomes colorless, and boiling stops. The resulting digest is placed in an ice bath to cool.

7.2 After temperature equilibrium is established (at least 30 min) 5 ml precooled 6 N hydrochloric acid is added slowly (taking 1 to 2 min) by pipet to minimize temperature increase in the digest. The solution is cooled in the ice bath for at least 15 min, after which 8 ml of precooled 3 N phosphoric acid is added, followed immediately by the addition of 5 ml precooled 0.02% Rhodamine B solution. Without delay, the flask is stoppered, shaken vigorously, and the contents transferred to a precooled separatory funnel.

7.3 Ten ml of precooled benzene, or other solvent is added to the separatory funnel, the contents are shaken vigorously for 1 min, and after allowing the contents to separate, the aqueous layer is discarded. The benzene etc., phase (colored pinkish-red if antimony is present) is collected in a centrifuge tube and allowed to stand a few min for water to settle out or it may be centrifuged. A 1-cm light path cuvette is rinsed with several ml of the extract, then filled and read at 565 nm against

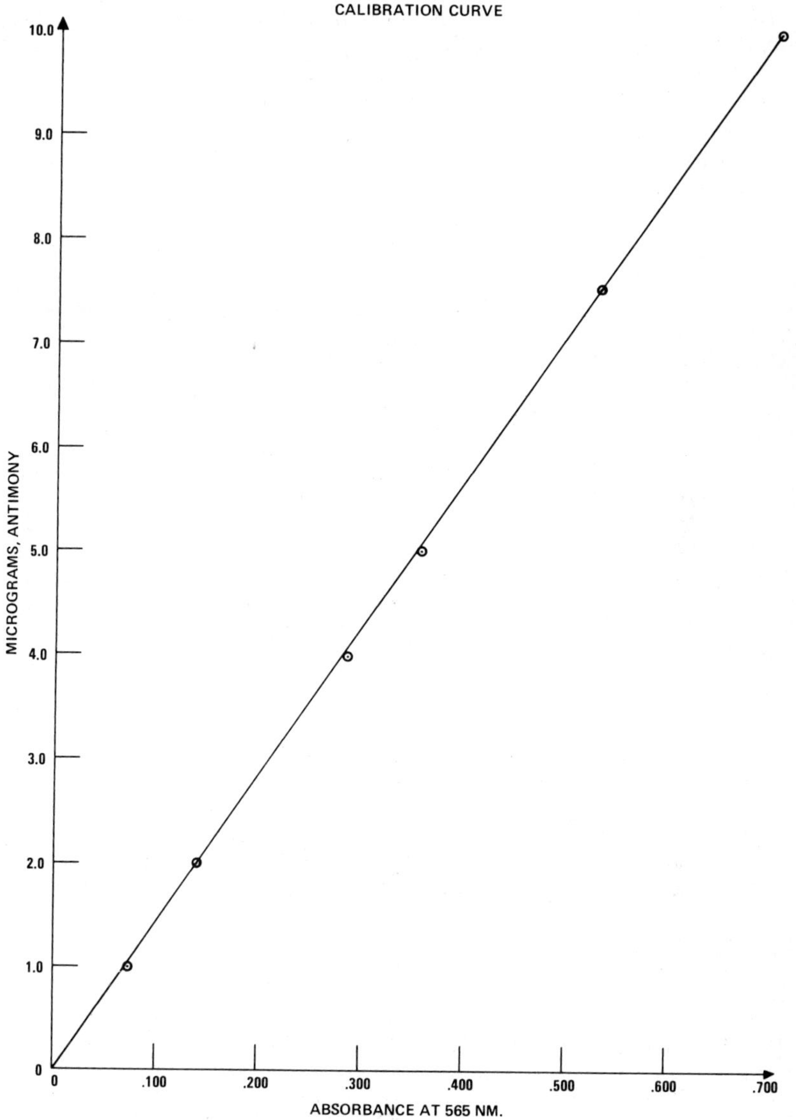

Figure 301:1—A typical plot of μg Sb/ml versus absorbance.

a benzene etc., blank which has been taken through the entire procedure.

7.4 It is convenient to cool separately funnels, hydrochloric acid, phosphoric acid, benzene or other solvent used, and Rhodamine B solution in a refrigerator before use. The analyst should aim at maintaining a temperature of 5 to 10 C during the analysis. Since phosphoric acid will lower the results if allowed to stand, the extraction should be completed within 5 min after the phosphoric acid is added. Color development and extraction should be carried out in subdued light.

7.5 If the color developed is too intense to measure, the extract may be diluted with the solvent used for extraction, and the antimony present found by multiplying by the aliquot factor.

8. Calibration

8.1 Dilute the antimony stock solution to prepare standards containing 1 to 10 μg Sb/ml. One to 10 μg Sb are analyzed by carrying not more than 1 ml of the appropriate reference solutions through the above digestion and analysis. Since NO_2 fumes may not be driven off visibly when no organic material is present, the solutions should be heated till white fumes of SO_3 are driven off strongly and the solution has stopped bubbling. The analysis will fail if any nitric acid remains in the digest. Plot μg Sb versus absorbance at 565 nm. The plot should be linear, as in Figure 301:1.

9. Calculations

9.1 If an air sample of V cubic meters (at standard conditions) is found to contain Y μg of antimony, then the concentration of antimony in the air is

$$\mu g\ Sb/m^3\ air = \frac{Y}{1000\ V}$$; if an aliquot

x is taken from the total sample X, then

the result must be multiplied by $\frac{X}{x}$.

10. Effect of Storage

10.1 Presumably particulate samples collected on filters may also be stored indefinitely. Question: How long may the digest be stored before completing analysis? It would be useful to be able to digest the samples en masse and store them for analysis at one time.

11. References

1. AMERICAN CONFERENCE OF GOVERNMENTAL INDUSTRIAL HYGIENISTS. 1963. Manual of Recommended Analytical Methods.
2. EDWARDS, F.C., and A.F. VOIGT. 1945. Separation of Antimonic Chloride from Antimonixous Chloride by Extraction into Isopropyl Ether. *Anal. Chem.* 21:1204–1205.
3. FREEDMAN, L.D. 1947. Rhodamine Method for Microdetermination of Antimony. *Anal. Chem.* 19:502.
4. LUKE, C.L. 1953. Photometric Determination of Antimony in Lead Using the Rhodamine B Method. *Anal. Chem.* 25:674–675.
5. MAREN, T.H. 1945. The Microdetermination of Antimony. Bull. Johns Hopkins University, 77:338–344.
6. MAREN, T.H. 1947. Colorimetric Microdetermination of Antimony with Rhodamine B. *Anal. Chem.* 19:487–491.
7. NIELSON, W., and G. BOLTS. 1954. Extraction and Photometric Determination of Antimony by Means of Rhodamine. *Z. Anal. Chem.* 143:(4) 264–272.
8. RAMETTE, R.W., and E.B. SANDELL. 1955. Rationalization of the Rhodamine-B Method for Antimony. Anal. Chem. Acta, 13:(5) 455–458.
9. WARD, F.N., and H.W. LAKIN. 1954. Determination of Traces of Antimony in Soils and Rocks. *Anal. Chem.* 26:(7) 1168–1173.

Subcommittee 7

E. C. TABOR, *Chairman*
M. M. BRAVERMAN
H. E. BUMSTEAD
A. CAROTTI
H. M. DONALDSON
L. DUBOIS
R. E. KUPEL

302.

Tentative Method of Analysis for Arsenic Content of Atmospheric Particulate Matter

12103-01-68T

1. Principle of the Method (5)

1.1 Arsenic exists in atmospheric particulates primarily in the form of inorganic oxides and arsenates. The arsenic compounds are dissolved from the sample of particulate matter with hydrochloric acid followed by reduction of arsenic to the trivalent state with KI and $SnCl_2$. The trivalent arsenic is further reduced to arsine, AsH_3, by zinc in acid solution in a Gutzeit generator. The evolved arsine is then passed through an H_2S scrubber which consists of glass wool impregnated with lead acetate and then into an absorber containing silver diethyldithiocarbamate dissolved in pyridine. In this solution, the arsine reacts with silver diethyldithiocarbamate forming a soluble red complex which is suitable for photometric measurement with a max absorbance at 535 nm **(2)**. This method replaces a Gutzeit treatment and a titrimetric method involving the absorption of arsine in mercuric chloride and subsequent iodometric determination **(3)**.

1.2 This method is suitable for the analysis of particulate samples collected either on membrane or glass fiber filters. Other techniques such as impingers or electrostatic precipitation will not provide adequate samples within a reasonable sampling period because of the low concentrations of arsenic in the ambient air.

1.3 Since the level of arsenic in ambient air usually lies between 0.1 and 1.0 $\mu g/m^3$, an aliquot representing at least 10m^3 of air is required if the sample absorbance is to fall on the calibration curve. The amt of filter used must be limited, or else the filter itself may physically and/or chemically interfere in the analysis. No difficulties will be encountered if the filter area used does not exceed 2 sq inches. Glass fiber filters gave better recovery of arsenic than Whatman #40 or #541 paper. In one series of tests 7 samples of the stock solution in the range from 0.5 to 8.0 μg As_2O_3 were added to 5-cm Whatman #40 filter paper and similar additions were made to glass fiber filters. The latter were digested with hot dilute hydrochloric acid as in **7.1** and **7.2**. Average recovery of the arsenic from the paper was 86 ± 4% and from the glass fiber 103 ± 5%.

2. Range and Sensitivity

2.1 If suitable samples (10 m^3 of air) are available, concentrations as low as 0.1 μg As/m^3 can be measured by this procedure. The max measurable concentration with a comparable sample is 1.5 μg As/m^3. Higher concentrations can be measured if smaller samples are used.

3. Interferences

3.1 Antimony present in the sample will form stibine, SbH_3, which will interfere slightly by forming a colored complex with silver diethyldithiocarbamate which has a max absorbance at 510 nm. The amt of antimony found in air is so small that any interference would be insignificant.* High concentrations of nickel, copper, chromium, and cobalt interfere in the generation of arsine. In addition certain combinations may produce a similar situation.

*By actual test, it has been demonstrated that antimony does not interfere to any appreciable extent **(4)**.

The presence of interferences can be determined by the internal standard technique where known amounts of arsenic are added to a sample and the recovery determined. Hydrogen sulfide would interfere, but it is removed by the lead acetate scrubber.

3.2 The filters used to collect atmospheric particulates may contain measurable quantities of arsenic. Portions of unused filters should be analyzed and corrections made if necessary. This can be done separately or as part of the reagent blank procedure.

4. Precision and Accuracy

4.1 Spiked samples containing 0.10, 1, 5, and 10 μg arsenic were analyzed with an accuracy of ± 0.04 μg based upon 7 replicate determinations at each concn. **(4)**.

4.2 Four samples containing known amts of arsenic were analyzed by 8 laboratories with the following results **(1)**.

Table 302:I. Analysis of Samples Containing Known Amounts of Arsenic

Sample	μg As/ml	Per cent Average Deviation
1	0.05	9.1
2	0.50	3.5
3*	1.00	6.1
4	1.50	4.0

*Contained 0.5 μg Sb/ml.

5. Apparatus

5.1 ARSINE GENERATOR, SCRUBBER, AND ABSORBER. Fisher arsine generator, #1–405. See Figure 302:1.

5.2 SPECTROPHOTOMETER, for use at 535 nm with 1-cm cells, *or*

5:3 FILTER PHOTOMETER, with blue filter having a max transmittance at or in the range of 530 to 540 nm with 1-cm cells.

6. Reagents

6.1 PURITY. All reagents must be prepared from analytic grade chemicals low in arsenic using double distilled water.

6.2 SILVER DIETHYLDITHIOCARBAMATE AG SCSN $(C_2H_5)_2$ Reagent. Dissolve 4.0 g of silver diethyldithiocarbamate (Fisher #S–666 or equivalent) in 800 ml of pyridine. The useful life of this reagent can be extended to at least 2 months by storing in a dark brown bottle or in the dark.

6.3 STANNOUS CHLORIDE REAGENT. Dissolve 10.0 g of fresh supply of stannous chloride dihydrate in 25 ml of 12 *N* (sp gr 1.19) hydrochloric acid. Place in a separatory funnel with a layer of pure mineral oil 5 mm thick on top to minimize oxidation. Drain a small quantity of the solution out of the stopcock before use. This solution is stable for 2 weeks.

6.4 LEAD ACETATE SOLUTION. Dissolve 10 g of $PbC_2H_3O_2 \cdot 3H_2O$ crystals in 100 ml of water. The solution will be slightly turbid as a small amt of the basic salt is formed, but this will not affect its usefulness.

6.5 POTASSIUM IODIDE SOLUTION. Dissolve 15 g of KI in 100 ml of water. The solution should be stored in a brown glass bottle.

6.6 STOCK ARSENIC STANDARD SOLUTION. Dissolve 0.132 g of As_2O_3 in 10 ml of 10 *N* NaOH and dilute with distilled water to 1000 ml. This solution contains 100 μg As/ml.

6.7 WORKING ARSENIC STANDARD. Dilute 10 ml of stock arsenic standard solution to 1000 ml with distilled water. This solution contains 1 μg As/ml. Prepare fresh weekly.

6.8 6N SODIUM HYDROXIDE. Dissolve 240 g of NaOH pellets in 200 ml of freshly boiled and cooled distilled water and dilute to 1 l.

6.9 6N HYDROCHLORIC ACID. Dilute 500 ml of 12 *N* HCl to 1 l with distilled water.

6.10 ZINC. Baker's analyzed; granular 20 mesh.

6.11 PYRIDINE. Some pyridine may contain colored materials which can be removed by passing the pyridine through a 2.5-cm diam by 15-cm depth alumina column at 150 to 200 ml/hr. Other pyridine is sufficiently pure so that repurification is not necessary.

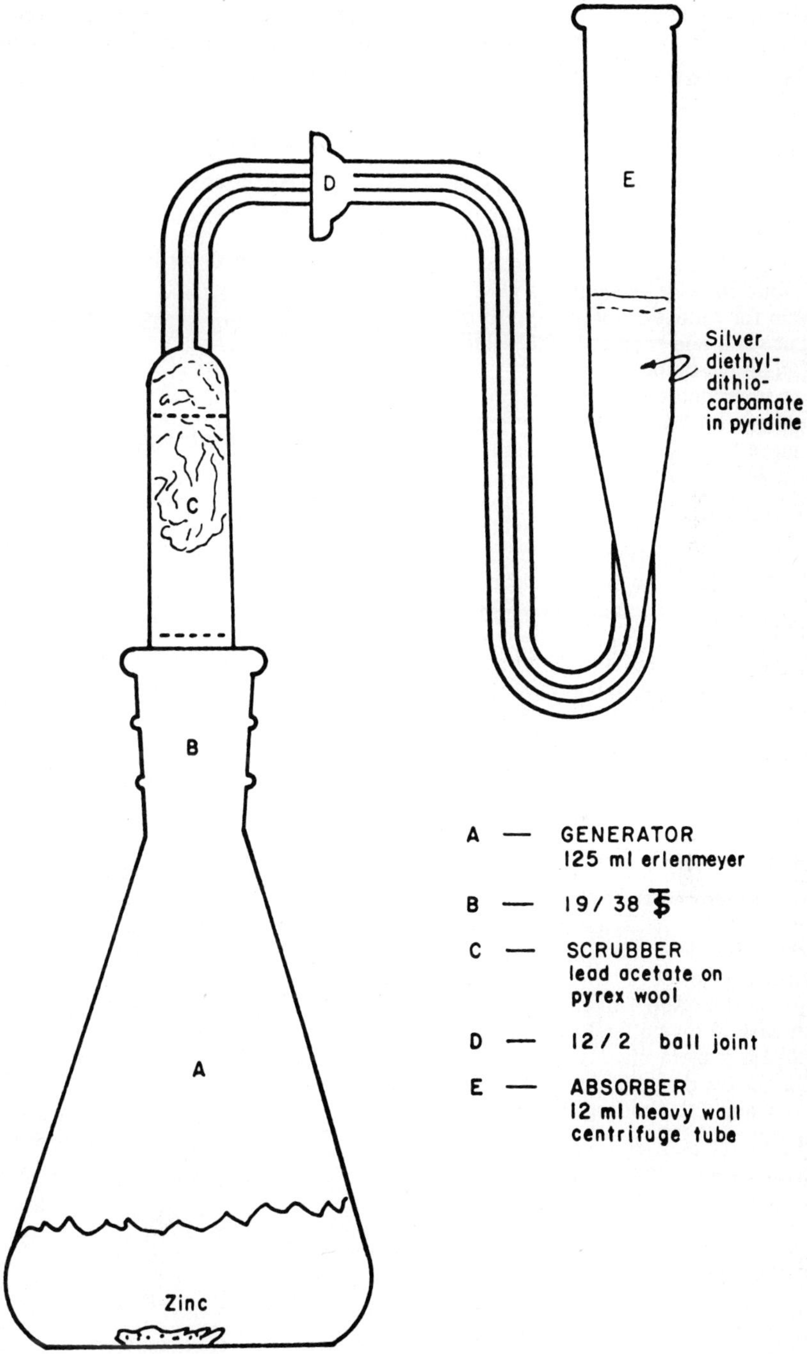

Figure 302:1—Arsine generator.

7. Procedure

7.1 Place a sample or aliquot of a sample representing from 0.1 to 15 μg of arsenic in a 125-ml flask (Figure 302:1) and add 30 ml of 6 *N* HCl. Digest at 90 to 95 C (boiling will result in loss of As) at least 1 hr. Cool and add 20 ml of 6 *N* NaOH. After cooling the solution to room temp, add 2 ml of potassium iodide solution, mix thoroughly, and then add 8 drops of the stannous chloride reagent and mix. Allow 15 min for reduction of arsenic to the trivalent state and then cool the solution to 4 C in an ice bath.

7.2 Impregnate the glass wool in the scrubber with lead acetate solution. About 1.5 ml of lead acetate is sufficient to saturate a 4-cm length of glass wool packing. Drain off any excess of solution. Pipet exactly 3 ml of the silver diethyldithiocarbamate (4 C) solution into the absorber tube. Add 3 g of zinc to the flask and immediately connect the scrubber-absorber assembly, making certain that all connections are tight. Allow 90 min for complete evolution of the arsine. Pour the solution directly from the absorber into the spectrophotometer cell and determine the absorbance of the solution using the reagent blank flask as the reference.

8. Calibration

8.1 Prepare a set of standard solutions by pipetting 1, 3, 5, 7, 10, and 15 ml of the working standard arsenic solution (1 μg/ml) into a series of 125-ml Erlenmeyer generator flasks. Add sufficient 6 *N* HCl to each flask to give a final volume of 30 ml. Include a reagent blank.

8.2 Follow the procedure as outlined under **7.1** and **7.2** omitting the initial digestion at 90 to 95 C.

8.3 Construct a calibration curve by plotting absorbance value *vs* μg arsenic in the standard.

9. Calculations

9.1 Use the absorbance of the sample to obtain the arsenic content from the calibration curve.

9.2 Concn as As in

$$\mu g/m^3 = \frac{\mu g \text{ As in sample analyzed}}{m^3 \text{ of Air represented by sample}}$$

10. Effect of Storage

10.1 Air particulate samples to be analyzed may be stored indefinitely at room temp without loss of arsenic.

11. References

1. AMERICAN CONFERENCE OF GOVERNMENT INDUSTRIAL HYGIENISTS. 1955. Manual. Determination of Arsenic in Air.
2. AMERICAN PUBLIC HEALTH ASSOCIATION. 1976. *Standard Methods for the Examination of Water and Wastewater*. 14th ed. Washington, D.C. 20036.
3. CASSIL, C.C., and H.J. WICHMANN. 1939. A rapid micromethod for determining arsenic. *J. Assoc. of Agric. Chemists*. 22:436.
4. MORGAN, GEORGE B. Personal Communication. Air Quality and Emission Data Program. National Center for Air Pollution Control, Public Health Service.
5. VASAK, V., and V. SEDIVEC. 1952. Colorimetric determination of arsenic. *Chem. Listy*. 46:341–344.

Subcommittee 7

E. C. TABOR, *Chairman*
M. M. BRAVERMAN
H. E. BUMSTEAD
A. CAROTTI
H. M. DONALDSON
L. DUBOIS
R. E. KUPEL

303.

Tentative Method of Analysis for Beryllium Content of Atmospheric Particulate Matter

12105-01-70T

1. Principle of the Method

1.1 The beryllium content of ambient air has been found to range from undetectable to as high as 0.01 $\mu g/m^3$ (**1**). The analytical method described herein involves measuring the fluorescence of an aqueous beryllium solution with ultra violet light in the presence of morin. It is well suited to the analysis of air particulate samples in this range.

2. Range and Sensitivity

2.1 The recommended conc range for analysis is 0.01 μg to 1.0 μg of beryllium in 10 ml of solution. Greater sensitivity can be achieved depending on instrumentation. Detection of as little as 0.0004 μg Be has been reported (**2**). Theoretically, a minimum concentration of 0.00002 $\mu g/m^3$ could be detected if a 20 m^3 sample of air particulates is taken for analysis; but from a more realistic point of view, the minimum detectable concentration is probably one order of magnitude greater. Higher concentrations can be measured by making proper adjustments in the procedure.

3. Interferences

3.1 The method is practically specific for beryllium (**3**) since most heavy metals which would interfere are removed. Zn, however, may interfere slightly even though it is complexed with KCN. According to Walkley (**4**), in a 10 ml solution containing 0.01 μg Be, 100 μg of Zn will give a blank of 0.004 μg Be though 75 μg Zn gave no detectable blank.

4. Precision and Accuracy

4.1 No good quantitative data are available. On 200 spiked samples on filter paper Welford and Harley (**5**) reported 92% recovery. Walkley (**4**) reported average recovery of 110% on 10 spiked samples on filter paper.

5. Apparatus

5.1 Sample Collection Equipment.

5.1.1 *Hi-Vol Sampler.*

a. Circular head 11-cm.

b. Rectangular head 8″ × 10″.

5.1.2 *Filter Media.*

a. Fluted pad, MSA #48570.

b. Whatman #41, 8″ × 10″. Membrane filter. Approximately 3 μ pore size (requires positive displacement pump).

c. Glass fiber filter (MSA 1106B or Gelman Type A).

5.2 Analytical Equipment—General.

5.2.1 *A photofluorimeter.* Equipped with a filter transmitting primary radiation of approximately 436 nm (Coleman B-2, Turner 2A) between lamp and sample and filter between sample and photocell which transmits a maximum fluorescent emission of 550 nm. (Coleman PC-2, Turner #58).

5.2.2 *Centrifuge.* Capable of 2000 rpm.

5.2.3 *Miscellaneous glassware.* Pipets, volumetric flasks, graduated cylinders, centrifuge tubes. (Vycor dishes (100-ml) if cellulose or membrane filter samples are to be analyzed.)

5.3 Analytical equipment for glass fiber filter samples.

5.3.1 *Low Temperature Asher.* (Tracer Lab Model 600) or Muffle Furnace.

5.3.2 *Sample Holder*. A pyrex cup 25-mm OD and 60-mm long which has a flat bottom with a hole in it (approximately 5 mm dia) is used. (See Figure 303:2—these are made from pyrex tubes. Fisher #149571).

5.3.3 *Thermometer Adapter*. With Male 24/40 standard taper joint. (Corning #8820).

5.3.4 *Extraction Tube*. See Figure 303:2. Indentions are made using a pointed carbon rod at a small spot softened with a torch followed by annealing.

5.3.5 *Erlenmeyer Flask*. 125-ml 24/40 standard taper joint.

5.3.6 *Condenser*. A short Ahlin condenser equipped with a male 34/45 standard taper joint.

5.3.7 *Crucibles*. Tracerlab ignition boats or 30-ml porcelain crucibles.

6. Reagents

6.1 PURITY. All reagents must be analytical reagent grade, and all water double distilled in glass.

6.2 MORIN STOCK SOLUTION. Dissolve 20 mg of morin in 5 ml of ethylene glycol and dilute to 100 ml with H_2O. (Eastman Morin T–4475 is used)—a purer grade of morin may be obtained from L. Light Co. Ltd., Coyle, Coinbrook, Bucks, England; or if desired, the technical grade may be purified by method of Bonner **(6)**. This stock solution is stable for approximately 1 month if stored in dark bottle in refrigerator.

6.3 MORIN WORKING SOLUTION. Dilute 12 ml of the stock solution to 200 ml with water. The solution must stand at least 8 hr previous to use. Store in dark bottle in refrigerator.

6.4 ALUMINUM NITRATE. 0.05 M. Dissolve 18.76 g of Al $(NO_3)_3$ in water and dilute to 1 l.

6.5 AMMONIUM CHLORIDE 25%. Dissolve 250 g NH_4Cl in water and dilute to 1 l.

6.6 PHENOL RED. Dissolve in water 0.1 g of sodium salt of phenol sulfophthalein and dilute to 250 ml. Eastman #6131.

6.7 AMMONIUM HYDROXIDE 3N. Dilute 200 ml of conc NH_4OH to 1 l with water.

6.8 SODIUM HYDROXIDE 4N. Dissolve 160 g NaOH in water and dilute to 1 l.

6.9 POTASSIUM CYANIDE 25 %. Dissolve 25 g KCN in water and dilute to 100 ml.

6.10 CONCENTRATED NITRIC ACID.

6.11 MIXED ACID. 50% conc HNO_3—50% conc H_2SO_4.

6.12 SULFURIC ACID. 1:1 and 1:5 dilutions.

6.13 QUININE SULFATE SOLUTION. 4 g quinine in a liter of 0.1N H_2SO_4.

6.14 BERYLLIUM STOCK SOLUTION. Dissolve the equivalent of 0.1 g of pure metal Be power in 15 ml of 1:1 H_2SO_4. (Be metal powder is usually 98 to 99% pure). Cool and dilute to 1 l with water. Stock solution contains 100 μg Be/ml.

6.15 BERYLLIUM STANDARD SOLUTION #1. Dilute 10 ml of stock solution to 100 ml. This solution contains 10 μg Be/ml. Prepare fresh daily.

6.16 BERYLLIUM STANDARD SOLUTION #2. Dilute 10 ml of standard solution #1 to 100 ml. This solution contains 1 μg Be/ml. Prepare fresh daily.

6.17 BERYLLIUM STANDARD SOLUTION #3. Dilute 5 ml of standard solution #2 to 100 ml. This solution contains 0.05 μg Be/ml. Prepare fresh daily.

7. Procedure

7.1 SAMPLE COLLECTION.

7.1.1 Using an acceptable sampling device with a suitable filter, sample the ambient air for a 24 hr period at the selected sampling site. Follow standard procedure for operating sampler, measuring volume of air sampled, and handling filter and sample **(8)**.

7.2 SAMPLE PREPARATION. Cellulose or membrane filter sample.

7.2.1 Transfer filter to Vycor dish and wet with 1:5 H_2SO_4. Heat on hot plate until charring occurs. Add a few drops of conc HNO_3 and take to dryness. Next fire with Meeker burner until carbon free. (Muffle furnace may be used). Allow dish to cool, wet the ash with mixed acid and take to fumes on hot plate or burner. Repeat if not carbon free. Cool, add a few drops of 1:1 H_2SO_4 and heat to strong SO_3

fumes. Remove from hot plate, cool, add enough water to dissolve salts. Transfer to 50-ml volumetric flask and make up to volume. **NOTE:** Samples suspected of containing beryl ore or high fired BeO which is difficult to put in solution should be decomposed using a potassium fluoride, sodium pyrosulfate fusion (**7**).

7.3 SAMPLE PREPARATION. Glass Fiber Filter Sample. From the exposed portion of an 8″ × 10″ glass fiber filter sample, cut a 0.5″ wide strip for urban samples and a 2″ wide strip for nonurban samples across the shorter dimension of the filter. This is based on a minimum sampling time of 24 hr using the Hi-Vol sampler.

7.3.1 *Low Temperature Ashing of Sample.* Cut off the unexposed border areas of filter and discard. Fold the exposed portion toward the center and crease. Bring the ends together and crease the strip again. Place the strip in a pyrex low temperature asher boat and place the loaded boat in the combustion compartment of the Low Temperature Asher (LTA). Ash the strips at about 150 C for one hr at 250 W and 1 mm Hg chamber pressure with an oxygen flow of 50 ml/min.

7.3.2 *Muffle Furnace Ashing of Sample.* Cut off the unexposed border area of filter and discard. Fold the remaining filter with the sample in the center. Place the folded sample in a clean 50-ml pyrex beaker, then place the beaker in the Muffle Furnace and ash at 500 C for one hr.

7.3.3 *Extraction of the Ashed Sample.* Place the ashed filter in a glass thimble (Figure 303:1) which is then placed in an extraction tube (Figure 303:2). To a 24/40 standard taper 125-ml Erlenmeyer flask add 8 ml of 20% hydrochloric acid and 32 ml of 40% nitric acid. Attach the flask to the extraction tube, and fit the extraction tube with the Ahlin condenser supplied with water. Reflux the acid through the sample for 3 hr. Remove the extraction tube and condenser from the Erlenmeyer flask and fit the flask with the thermometer adapter, which serves as a spray retainer. Concentrate the extracted liquid to 1 to 2 ml on a hot plate, and allow to cool and stand overnight. Decant the concentrated material into a 15-ml centrifuge tube and quantitatively transfer by washing the extract from the glass fibers in the Erlenmeyer flask by three washings of 1 ml each of distilled water and decanting the washings into the centrifuge tube. Centrifuge at 2000 rpm for 30 min and decant

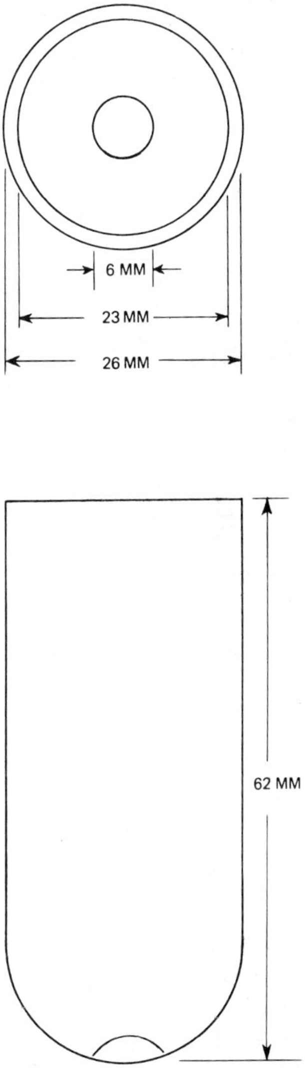

Figure 303:1—Sample holder for acid extraction of glass fiber filter particulate samples.

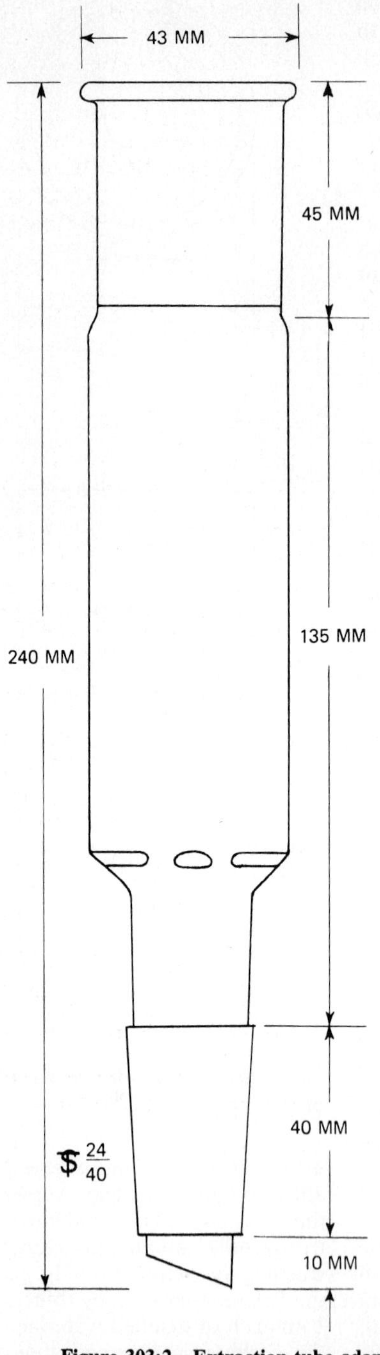

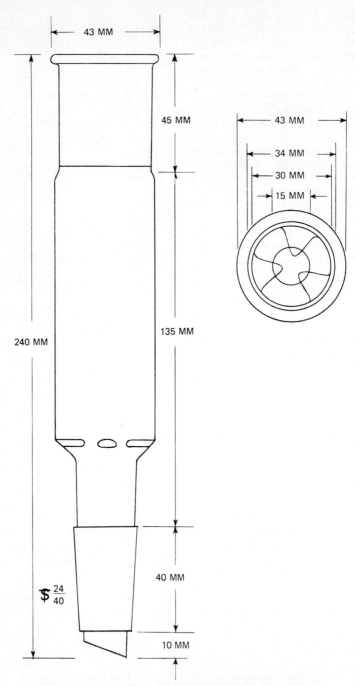

Figure 303:2—Extraction tube adapted for the sample holder.

the supernatant liquid into a 10-ml volumetric flask. Wash precipitate in tube with 1 to 2 ml distilled water, centrifuge again and decant into the volumetric flask. Dilute to volume with distilled water, mix.

7.4 Analysis of Sample Solution. Pipet 1 ml into 15-ml graduated centrifuge tube. Add 1 ml of Al $(NO_3)_3$ solution (0.05M) 1 ml of NH_4Cl solution (25%), and 1 drop of phenol red. Then add enough NH_4OH (3N) to make solution alkaline. Let stand 10 min. Centrigure at 2,000 rpm for 5 min, rotate tube 180° in centrifuge and centrifuge for 5 more min. Pour off and discard supernatant liquor. Dissolve precipitate in 1 ml of NaOH (4N). Add 1 drop of KCN (25%) and make up to 10 ml volume with water. Centrifuge 10 min at 2,000 rpm and pour liquid into Coleman cuvette, (19 mm dia × 100 mm long). Read in photofluorimeter set at 100% fluorescence with solution of 4 g/l of quinine in 0.1 N H_2SO_4. Add 1 ml morin reagent to sample, mix and read fluorescence immediately. Morin reading, minus sample reading without morin equals net instrument reading. Standards in the estimated range of the sample are run through the procedure with samples. These standards are prepared by pipetting the appropriate quantity of beryllium standard solution, usually solution #2 or #3 (**6.16** or **6.17**) upon the type filter collection media used.

8. Calibration

8.1 Add 2, 5, and 10 ml portions of Beryllium Standard Solution #3 and 1 ml Beryllium Standard Solution #2 to suitable portions of the filter medium used. Run the impregnated filter through the appropriate procedure.

9. Calculations

9.1 Calculation for Total Be in Sample:

$$\frac{\text{ml sample}}{\text{ml aliquot}} \times \frac{\text{net sample IR}}{\text{net standard IR}} \times \mu\text{g Be in standard} = \text{Total } \mu\text{g Be}$$

IR = Instrument Reading

ml sample = Volume of dilution of total sample

ml of aliquot = Portion of sample used for analysis

Net sample IR = IR sample with morin—IR of sample without morin

Net standard IR = IR with standard—IR of standard (without morin)

9.2 Calculation of Beryllium Concentration:

$$\frac{\text{Total } \mu\text{g Be}}{\text{m}^3 \text{ of air sample}} = \mu\text{g Be/m}^3 \text{ of ambient air*}$$

10. Effect of Storage

10.1 Samples can be stored indefinitely without loss of beryllium.

11. References

1. Air Quality Data from the National Air Sampling Networks and Contributing State and Local Networks, 1964–65; U.S. Department of Health, Education and Welfare, Public Health Service.
2. Sill, C.W., and C.P. Willis. 1959. *Analytical Chemistry* 31:598–608.
3. Sandell, E.B. 1940. *Industrial & Engineering Chemistry*. Analytical Edition, pp. 762–764, December 5.
4. Walkley, J. 1950. A Study of the Morin Method for the Determination of Beryllium in Air Samples. *A.I.H.A. Journal,* 20, 241–245.
5. Welford, G.A., and J. Harley. 1952. Fluorimetric Determination of Trace Amounts of Beryllium. *American Industrial Hygiene Association* (Quarterly) 13, 232.
6. Bonner, J.F. 1959. The Isolation of Morin as a Reagent for Beryllium Analysis. *U.S. Atomic Energy Commission Report* UR–111, April 1959.
7. Sill, C.W. 1961. Decomposition of Refractory Silicates in Ultramicro Analysis. *Analytical Chemistry* 33:1684.
8. American Public Health Association. 1977. Methods of Air Sampling and Analysis. 2nd ed. p. 578. Washington, D.C.

Subcommittee 7

R. E. Kupel, *Chairman*
M. M. Braverman
J. M. Bryant
A. Carotti
H. M. Donaldson
L. Dubois
E. C. Tabor

*Volume of air sample corrected to 760 Torr and 25 C.

304.

Tentative Method of Analysis for Cadmium Content of Atmospheric Particulate Matter

12110-01-73T

1. Principle of the Method

1.1 Cadmium ion is reacted with an excess of diphenylthiocarbazone (dithizone) in alkaline solution to form cadmium dithizonate which is extracted by chloroform. The intensity of the color is proportional to the amount of cadmium present and is determined spectrophotometrically at a wavelength of 518 nm (**1, 2**).

2. Range and Sensitivity

2.1 The method is applicable to cadmium concentrations of 1 to 20 μg in the aliquot taken for analysis. For an atmospheric concentration of 0.01 $\mu g/m^3$ Cd, an air sample volume of 200 m^3 is appropriate. The molar absorptivity of cadmium dithizonate is 75,000 l/mole-cm (**3**).

3. Interferences

3.1 The method, as described prevents interference from usual amounts of all other metals except thallium. Thallium can be removed by extraction with diethyl ether from 0.3 N hydrobromic acid containing bromine (**3**).

4. Precision and Accuracy

4.1 When the thallium separation is unnecessary, recoveries of 97% to 98% have been reported. If thallium must be separated, average recoveries of 81% are obtained (**1**). No detailed studies on the precision and accuracy of this method are available.

5. Apparatus

5.1 Sample Collection. Membrane filters such as Millipore, appropriately sized for sample volume and rate, pore size of 0.8μ, are satisfactory for usual air samples. Collection efficiencies greater than 99% are obtained from a 0.3μ DOP smoke at face velocities from 7.2 to 53 cm per sec (**4**). For membrane filters of 47mm, 4″ dia and 8″ × 10″ (10 cm and 20 × 25 cm) rectangular size, a face velocity of 25 cm/sec will result in flow rates of about 26, 120 and 770 l/min, respectively (see Section **2.1** for desired air sample volume).

5.2 Borosilicate glassware should be used throughout. Prior to use, rinse glassware several times with 1:1 hydrochloric acid, followed by distilled water.

5.3 Separatory Funnel. Squibb or similar type 125-ml capacity with Teflon stopcocks and stoppers.

5.4 A spectrophotometer capable of isolating wavelengths at 518 nm ± 10 nm is suitable. A 1-cm cell is used for the analysis.

5.5 Buret

6. Reagents

6.1 Purity. American Chemical Society reagent grade chemicals or equivalent shall be used in all tests. References to water should be understood to mean double distilled or equivalent. Prepared solutions of indicators must be free of chelating agents. Care in selection of reagents and in following the above precautions is essential if low blank values are to be obtained.

6.2 Chloroform. American Chemical Society reagent grade suitable for use in dithizone procedures.

6.3 Dithizone Extraction Solution. Dissolve 80.0 mg dithizone in chloroform and dilute to a liter in a volumetric flask. Store in a brown bottle in the refrigerator. This solution should be used while cold.

6.4 Dithizone Standard Solution. Dissolve 8 mg dithizone in chloroform as above and dilute to 1l in a volumetric flask. If a high blank is obtained, the dithizone can be purified according to Sandell **(3)**. Store the solution in a brown bottle in the refrigerator. This solution must be at room temperature when used. The fresh solution should be aged a day or two before standardizing and standardization should be checked at intervals of two weeks.

6.5 Sodium Potassium Tartrate. Dissolve 25 g of sodium potassium tartrate in water and dilute to 100 ml.

6.6 Sodium hydroxide. Dissolve 40 g sodium hydroxide in water and dilute to 100 ml. Store in polyethylene bottle.

6.7 Sodium Hydroxide, 40% Potassium Cyanide, 1%. Dissolve 200 g of sodium hydroxide and 5 g of potassium cyanide in water and dilute to 500 ml. Store in a polyethylene bottle. Discard this solution after 6 weeks.

6.8 Sodium Hydroxide, 40%—Potassium Cyanide, 0.05%. Dissolve 200 g of sodium hydroxide and 0.25 g potassium cyanide in distilled water and dilute to 500 ml. Store in a polyethylene bottle. Discard after two weeks.

6.9 Hydroxylamine Hydrochloride. Dissolve 20 g of hydroxylamine hydrochloride in distilled water and dilute to 100 ml.

6.10 Cadmium Metal. (stock solution)—Dissolve 10.0 mg of cadmium metal in dilute nitric acid (see Section **6.14**) and make up to volume of one liter with this acid.

6.11 Tartaric Acid, 2%. Dissolve 20 g of tartaric acid in water and dilute to 1l. The solution should be refrigerated and used cold.

6.12 Thymol Blue Indicator, Sodium Salt (free of chelating agents). Dissolve 0.1 g of thymol blue in 100 ml of water.

6.13 Nitric Acid, 70–71% (sp, gr 1.42)

6.14 Nitric Acid (dilute). Dissolve 10 ml of 70% HNO_3 and dilute to 1 l with water.

6.15 Hydrochloric Acid, 36–38% (sp gr 1.64)

6.16 Hydrochloric Acid (1:1). Add 50 ml HCl (sp gr 1.64) to 50 ml of water (for cleaning glassware).

6.17 Perchloric Acid, 72% (sp gr 1.77).

6.18 Nitric-Perchloric Acid (3:2). Add 60 ml HNO_3 (sp gr 1.42) to 40 ml $HC10_4$ (sp gr 1.77).

7. Procedure

7.1 Preparation of Samples

7.1.1 *Membrane Filter*. Place the filter in a 150-ml beaker with the exposed side down and add 15 ml of (3:2) nitric-perchloric acid. Digest on a hot plate to fumes of perchloric acid. Make sure that all of the carbonaceous material has been oxidized* before allowing the sample to cool.

7.2 Analysis

7.2.1 Rinse the sample into a separatory funnel with two 5-ml portions of water.

7.2.2 Add a few drops of thymol blue indicator and if yellow add 40% sodium hydroxide dropwise to a blue color.

Add the following reagents in the order listed and mix after each addition:

a. 1 ml sodium potassium tartrate

b. 5 ml sodium hydroxide (40%)-potassium cyanide (1%)

c. 1 ml hydroxylamine hydrochloride.

Add 15 ml dithizone extraction solution from a buret, shake gently, invert and release the pressure through the stopcock. Shake vigorously for 1 min, allow the layers to separate and drain the chloroform layer (to the top of the stopcock bore) into a second funnel containing 25 ml cold tartaric acid. *Do not allow any aqueous layer to enter the second funnel*. The time of contact with strong alkali must be kept to a minimum.

*If the carbonaceous residue remains, one or more additions of 10 ml of nitric acid (sp. gr. 1.42) with subsequent digestion may be required. All safety precautions must be observed in the use of hot conc perchloric acid.

Add 10 ml chloroform to the first funnel, shake one min, allow the layers to separate and drain the chloroform layer (to the bottom of the bore) into the second funnel. Do not attempt a third shakeout since lead and zinc, which are extracted to a slight extent in this step, would, if present in large amounts, give the misleading impression that all the cadmium had not been extracted. If the orange color of excess dithizone is absent in the aqueous layer (indicative of excess cadmium in the aliquot or the presence of excess amounts of other metals which react with dithizone), use a smaller sample aliquot. (The estimate of cadmium present is based on the difference in intensity of the pink color of the first extract and of the chloroform wash extract. Most of the cadmium is present in the first extract; the interfering metals give about the same color in both extracts.)

7.2.3 Shake the second funnel containing the tartaric acid-chloroform extract for 2 min and allow time for the layers to separate (about 2 min). Discard the chloroform layer (to the top of the bore).

7.2.4 Add 5 ml chloroform, shake one min, allow the layers to separate (approximately 2 min) and discard the chloroform layer. Make as clean a separation as possible.

7.2.5 Add 0.25 ml hydroxylamine hydrochloride, exactly 15 ml dithizone standard solution and 5 ml of 40% sodium hydroxide—0.05% potassium cyanide solution and immediately shake for one minute.

7.2.6 Insert a loose plug of cotton in the funnel stem and drain the chloroform layer into a dry spectrophotometer cell. Read the absorbance at 518 nm, with the blank as a reference.

7.2.7 A blank must be run through the entire procedure. Saltzman **(2)** obtained very low blanks when fresh dithizone was used.

8. Standardization

8.1 Standard Cadmium Solution. A 1 μg/ml Cd working solution in dilute nitric acid from the stock solution, previously prepared as in Section **6.10**.

8.2 Prepare a series of separatory funnels containing 0, 1, 2, 5, 10, 20 μg cadmium. Add tartaric acid to make a total volume of 25 ml in each. Follow the regular procedure beginning with the wash with 5 ml chloroform (Section **7.2.4**).

8.3 Prepare a calibration curve from the readings obtained.

9. Calculation

9.1 Micrograms Cd/m³ =

$$\frac{(\mu g \text{ of Cd found})}{\text{Volume of air sampled } (m^3)\dagger}$$

10. Effects of Storage

10.1 Filter samples may be stored indefinitely before analysis, if protected from contamination.

11. References

1. ACGIH. 1957. Recommended Method. Determination of Cadmium in Air, Monocolor Dithizone Method. August.
2. Saltzman, B.E. 1953. Colorimetric Microdetermination of Cadmium with Dithizone: *Anal. Chem.*, **25**:493.
3. Sandell, E.B. 1959. *Colorimetric Determination of Traces of Metals*, 3rd edition p. 353. Interscience Publishers, Inc., New York.
4. Lippmann, M. 1972. Filter Media for Air Sampling, in *Air Sampling Instruments for Evaluation of Atmospheric Contaminants*, 4th edition, p. N–4, Am. Conf. of Governmental Industrial Hygienists, Cincinnati, Ohio.

Subcommittee 6
T. J. Kneip, *Chairman*
R. S. Ademian
J. R. Carlberg
J. Driscoll
J. L. Moyers
L. Kornreich
J. W. Loveland
R. J. Thompson

†*Note*: Volume of air sample corrected to 25 C. and 76 C. Torr.

305.

Tentative Method of Analysis for Iron Content of Atmospheric Particulate Matter (Bathophenanthroline Method)*

12126-01-70T

1. Principle of the Method

1.1 Bathophenanthroline (4,7-diphenyl-1,10-phenanthroline) reacts with ferrous ion to form a red-colored complex, which may be extracted by chloroform. The intensity of the color is proportional to the amount of iron present, and is determined spectrophotometrically at a wavelength centered around 538 nm.

2. Range and Sensitivity

2.1 The method is applicable to iron concentrations of 1.0 to 15.0 μg in the aliquot taken for analysis.

2.2 The molar absorptivity of the bathophenanthroline-iron complex is 22,350. An absorbance of approximately 0.1 is obtained in a 1-cm cell for 0.25 μg iron/per ml.

3. Interferences

3.1 The method as described prevents interference from usual amounts of all other metals and anions found in the atmosphere (**1**).

4. Precision and Accuracy

4.1 The precision is high, and the accuracy of the procedure as described is ± 2.0% (**2**).

*Adapted from "ACGIH Recommended Method for Determination of Iron in Air and Biological Materials" (1964) and "The Determination of Iron in Dustfall Samples" (R. G. Reynolds and J. L. Monkman, AIHA Journal 23:415–418, 1962).

5. Apparatus

5.1 For collection of samples.

5.1.1 Analytical grade membrane filters or cellulose filter papers are satisfactory for usual air samples; however, blanks should be checked before use.

5.1.2 The electrostatic precipitator may be used at a nominal sampling rate when iron is present as fume or dust.

5.1.3 The midget impinger, containing 10 ml of 1 to 5% nitric acid as the collecting medium, may also be employed at a sampling rate of approximately 0.1 cfm, when iron is present as dust.

5.2 Any spectrophotometer capable of isolating wavelengths centered around 538 nm is suitable.

5.3 Borosilicate glassware should be used throughout the analysis. Prior to use, rinse glassware with 1:1 hydrochloric acid, followed by iron-free distilled water to remove traces of iron. Glassware should again be cleaned immediately after use.

6. Reagents

6.1 Purity. Reagent grade chemicals shall be used in all tests. Unless otherwise indicated, it is intended that all reagents shall conform to the specifications of the Committee on Analytical Reagents of the American Chemical Society where such specifications are available. Other grades may be used provided it is first ascertained that the reagent is of sufficiently high purity to permit its use without lessening the accuracy of the determination. Unless otherwise indicated, references to water shall be understood to mean iron-free distilled water. Prepared solutions of indicators should be free of chelating agents.

6.1.1 *Ethanol, 95%.*

6.1.2 *Bathophenanthroline, 0.001 M solution in ethanol.* Dissolve 33.2 mg in a small volume of 95% ethanol and dilute to 100 ml with ethanol. This solution must be refrigerated when not in use. Fresh reagent should be prepared every 30 days.

6.1.3 *Hydroxylamine Hydrochloride, 10%.* Dissolve 20 g in 200 ml of distilled water. To remove traces of iron, place this solution in a 250-ml conical separatory funnel, add 20 to 40 ml of chloroform, and 10 ml of bathophenanthroline reagent. Shake the contents of the funnel and allow 5 min for separation of aqueous and organic layers. Draw off the red-colored lower layer and discard. Repeat this process until the chloroform-bathophenanthroline layer is colorless. Discard the colorless lower layer. The aqueous solution is stable.

6.1.4 *Sodium Acetate Buffer.* Prepare 200 ml of a solution 2 M in acetic acid and 2 M in sodium acetate by adding 32.8 g of sodium acetate to 23 ml of conc acetic acid (17.4 N) and diluting to 200 ml with distilled water. As before, traces of iron are removed by extracting with chloroform in the presence of 10 ml of added bathophenanthroline reagent.

6.1.5 *Nitric Acid, Concentrated (Specific Gravity, 1.50).*

6.1.6 *Perchloric Acid, 70%.*

6.1.7 *Ammonium Hydroxide, Conc (sp gr, 0.90).*

6.1.8 *Chloroform Extraction Solution.* For extractions, use a mixture of chloroform and absolute ethanol, prepared in a volume ratio of 9:1.

6.1.9 *Congo Red Paper.*

6.1.10 *Standard Iron Solution A, Stock.* Dissolve 0.4 g of high purity (99.9%) iron wire in 1:1 hydrochloric acid (sp gr, 1.8–1.9) and dilute to 1 l with water.

6.1.11 *Standard Iron Solution B.* Transfer 50 ml of A to a 1-l volumetric flask together with 10 ml of concentrated hydrochloric acid, and make up to a volume of 1 l.

6.1.12 *Standard Iron Solution C.* Transfer 21 ml of B and 10 ml of conc hydrochloric acid and make up to 1 l with water. Solution C contains 0.5 μg iron/ml and is the working standard. Solution C should be prepared daily.

7. Procedure (2,3)

7.1 Preparation of Samples.

7.1.1 *Membrane or Cellulose Filters.* Transfer the membrane or cellulose filter to a 125-ml Erlenmeyer flask and wet ash by repeated addition of 5 ml conc nitric acid until a clear solution is obtained. Evaporate to dryness. The addition of 0.5 ml perchloric acid will expedite the ashing process.

7.1.2 *Impinger.* Transfer the sample to a 125-ml Erlenmeyer and take to dryness. The residue is then treated with conc nitric acid as described above.

7.1.3 *Electrostatic Precipitator.* Rinse the contents of the precipitator tube with the aid of a policeman and a stream of 5% nitric acid into a 150-ml beaker. Transfer the rinsings to a 125-ml Erlenmeyer, take to dryness, and treat like the previous samples.

7.2 Treatment of Digested Samples.

7.2.1 Since the reaction requires the iron to be in the ferrous state, any oxidizing material or acid, such as nitric, must be removed from the dry sample before color development. Once the digestion is complete, the residue is dissolved with approximately 15 ml of conc hydrochloric acid. The acid is added in installments, as gentle heat is applied to the sample. This process must be watched carefully to be certain that solution is proceeding efficiently. 10 to 20 ml of distilled water is now added to the acid solution and the volume of the solution is reduced to 5 to 10 ml by heating on a hot plate until boiling occurs. It may be necessary to filter the solution if any undissolved residue now remains. Immediately before analysis, the filtrate is brought almost to neutrality with ammonium hydroxide using congo red paper, and is made up to 100 or 200 ml in a volumetric flask with distilled water.

7.3 Analysis.

7.3.1 Prepare a standard curve (See **8.1**).

7.3.2 A suitable aliquot† of the sample solution is pipetted into a 125-ml separatory funnel, and the volume is made up to 25 ml with distilled water. 25 ml of distilled water is added to another funnel for a blank determination. 2.0 ml of hydroxylamine hydrochloride solution is added to each funnel with shaking. 5.0 ml of bathophenanthroline reagent is then added and the contents are shaken again. The pH is adjusted to 4 to 6 with congo red paper by adding ammonium hydroxide dropwise, and 5.0 ml of buffer solution is added with shaking. 8.0 ml of chloroform extraction solution is added, and the funnel is shaken thoroughly. After adding 5.0 ml of distilled water, the contents are again shaken gently and allowed to stand for at least 15 min. Fill a 1-cm light path cell with a portion of the dry chloroform layer. Measure the absorbance at 538 nm on a spectrophotometer, against a chlorofrom reference.

8. Calibration and Standards

8.1 Pipet 1, 5, 10, 15, 20, and 25 ml of the working standard iron solution (0.5 μg/ml) into a series of 125-ml separatory funnels. The volume is made up to 25 ml with water and the standards are treated and analyzed as outlined under **7.3.2**. A reagent blank must be included.

8.2 Construct a calibration curve by plotting absorbance value vs. μg iron in the standard, *i.e.*, 0.5, 2.5, 5.0, 7.5, 10.0, and 12.5 μg of iron respectively.

†If the entire sample contains 0.2 milligram of iron in a total solution volume of 200 ml, a 10-ml aliquot taken for analysis will provide 10.0 μg. Since Beer's Law is obeyed over the range 0 to 15 μg iron (Fe^{++}), this quantity of iron is more than adequate for an accurate determination. Aliquots, whose iron content is below the range of the standard curve, can be handled by extraction concentration if more sample is available **(3)**.

9. Calculations (2)

9.1 Use the absorbance of the sample to determine the iron content from the calibration curve.

9.2 If an aliquot of v ml is taken from V ml of the prepared sample, Va is the volume of air sampled in cubic meters at 760 Torr and 25 C and Y is the micrograms of iron found in v, then

$$\mu g\ Fe/m^3\ air = \frac{V \cdot Y}{\cdot v \cdot Va}$$

10. Effect of Storage

10.1 Untreated samples, protected against contamination, may be stored indefinitely.

11. References

1. SMITH, G.F., MCCURDY, W.H., and H. DIEHL. 1952. The Colorimetric Determination of Iron in Raw and Treated Municipal Water Supplies by the Use of 4,7-Diphenyl-1,10-Phenanthroline. *Analyst* 77:418.
2. AMERICAN CONFERENCE OF GOVERNMENTAL INDUSTRIAL HYGIENISTS. 1964. Manual. *Determination of Iron in Air and Biological Materials*.
3. REYNOLDS, R.G., and J.L. MONKMAN. 1962. The Determination of Iron in Dustfall Samples. *A.I.H.A. Journal* 23:415–418.
4. LEE, C.F., and W.J. STUMM. 1960. Determination of Ferrous Iron in the Presence of Ferric Ion with Bathophenanthroline. *J.A.W.W.A.* 52:1567.

Subcommittee 6

W. G. FREDRICK, *Chairman*
J. CHOLAK
A. MOFFIT
K. KUMLER
G. B. MORGAN
W. M. SMITH
L. G. VON LOSSBERG

306.

Tentative Method of Analysis for Iron Content of Atmospheric Particulate Matter (Volumetric Method)

12126-02-70T

1. Principle of the Method

1.1 Samples containing more than 1 mg of iron may be conveniently analyzed by volumetric procedures employing either ceric sulfate or potassium dichromate. Due to endpoint characteristics, samples, containing up to 40 mg of iron as Fe_2O_3 should be titrated with 0.01 N ceric sulfate and samples containing greater amounts with 0.1 N potassium dichromate.

1.2 Acid-soluble inorganic iron compounds or the acid-soluble ash of organic materials containing iron are dissolved by treatment of the sample with strong hydrochloric acid, the insoluble material separated, the volume adjusted to about 10 ml and reduced to ferrous iron by treatment of the hot solution with stannous chloride. The excess of stannous chloride is removed with mercuric chloride and the iron is determined volumetrically to a color indicator endpoint.

2. Range and Sensitivity

2.1 1 ml of 0.1 N potassium dichromate is equivalent to 7.99 mg Fe_2O_3; 1 ml of 0.01 N ceric sulfate is equivalent to 0.799 mg Fe_2O_3. (By use of a microburet, the lower limit can be reduced to about 100 μg.)

2.2 The working range extends from 100 μg. to 400 mg. in the sample or aliquot to be titrated.

3. Interferences

3.1 Significant interference from other materials in this type of air pollution sample is unlikely. However, in sufficient amount, fluoride ion forms complexes with Ce^{IV} and phosphates form a precipitate.

4. Precision and Accuracy

4.1 The relative standard deviation is less than ±5%.

5. Apparatus

5.1 In addition to membrane filter, impingement and precipitator collectors, fallout jars are frequently used. All glassware should be of borosilicate glass.

6. Reagents

6.1 PURITY. All reagents shall conform to the specifications of the Committee on Analytical Reagents of the American Chemical Society where such specifications are available. Otherwise high purity reagents must be used. Distilled water must conform to the ASTM standard for Referee Reagent Water.

6.2 HYDROCHLORIC ACID SP GR 1.18–1.19.

6.3 SULFURIC ACID SP GR 1.84.

6.4 STANNOUS CHLORIDE, APPROXIMATELY 0.5 N. Dissolve 15 gm $SnCl_2 \cdot 2H_2O$ in 100 ml of 1 part HCl in 2 parts distilled water. Make up fresh daily and protect from air.

6.5 MERCURIC CHLORIDE. Prepare a saturated solution in distilled water. (*Approximately* 7 g in 100 ml H_2O).

6.6 CERIC SULFATE METHOD.

6.6.1 *Ceric sulfate 0.1 N (approximately).* Add slowly 28 ml conc H_2SO_4 to 500 ml distilled water in a 1500-ml beaker. Dissolve in this dilute acid 35 to 40 g anhydrous ceric sulfate, add 500 ml distilled water, mix, place in a glass stoppered storage bottle and standardize. The solution is stable for at least 1 year.

6.6.2 *Ceric sulfate 0.01N.* Dilute the approx. 0.1 N solution with 10 to 15% sulfuric acid. 0.01 N ceric sulfate is not stable on long storage and should be freshly prepared and standardized before each use.

6.6.3 *1, 10- phenanthroline ferrous complex—0.025 N.* Dissolve appropriate amount of ortho-phenanthroline in a 0.25 M aqueous solution of ferrous sulfate. (*e.g.*, 1.5 g in 100 ml of a solution containing 7 g $FeSO_4 \cdot 7H_2O$).

6.6.4 *Iron wire.* 99.9+% pure.

6.7 POTASSIUM DICHROMATE METHOD.

6.7.1 *Potassium dichromate 0.1 N.* Dissolve 4.904 g anhydrous potassium dichromate, primary standard grade, in distilled water and dilute to 1 l in a volumetric flask. Transfer to a glass stoppered borosilicate bottle and store at room temperature in diffuse light.

6.7.2 *Barium diphenylamine sulfonate indicator solution.* Dissolve 0.4 g in 100 ml distilled water.

6.7.3 *Phosphoric acid solution.* Dissolve 125 ml phosphoric acid sp gr 1.37 in 375 ml distilled water and store in iron-free container.

7. Procedure

7.1 SAMPLE PREPARATION

7.1.1 *Fall-out Jars.* Treat dry residue after loss on ignition in a 250- to 400-ml beaker with 10 ml conc HCl and 2 ml distilled water at point of incipient boiling (use watch glass) until solution is complete (15 to 30 min). If the ignited sample is dispersed in water, add 10 ml conc HCl and evaporate down to a volume of 10 ml. If insoluble material is present, dilute somewhat, remove by filtration or centrifuge, and evaporate down to a volume of 10 ml.

7.1.2 *High volume filter samples.* Treat an appropriate area aliquot or entire fiter with 10 + 2 conc HCl-water mixture until iron is in solution. Dilute somewhat and separate solution from insoluble matter by filtration or centrifuge, transfer to a 250- to 400-ml beaker and reduce volume to approximately 10 ml.

7.1.3 *Impingement devices.* Transfer to 250- to 400-ml beaker with water and policeman, add 10 ml conc HCl and reduce volume to approx 10 ml by evaporation at incipient boiling.

7.1.4 *Important Note.* If alcohol or other non-aqueous solvents or antifreezes are present in the sample after transfer to the beaker, evaporate to dryness, cool, add 10 ml conc HCl and 2 ml H_2O and treat as residue from fallout jar samples; if insoluble debris is present, proceed as for high volume filter samples **7.1.2**.

7.2 POTASSIUM DICHROMATE TITRATION. With the prepared sample near the boiling point, add stannous chloride solution dropwise, swirling between additions, until the solution turns pale green or colorless. Add one drop excess stannous chloride, dilute immediately with approx. 100 ml cold water, add 3 to 5 ml mercuric chloride solution, swirl and let stand for up to five min. A scant precipitate or faint opalescence should form. If none, or if a heavy white or gray precipitate forms, too little or too much stannous chloride has been added and the sample must be discarded.

Without delay, add 3 to 4 ml of (1 + 3) phosphoric acid, 2 to 3 ml of (1 + 1) H_2SO_4 and 3 to 4 drops of barium diphenylamine sulfonate solution. Titrate with 0.10 N dichromate solution to a permanent purple or violet end point. The approach of the end point is signalled by the appearance of the purple violet color which disappears on stirring.

7.3 CERIC SULFATE TITRATION. With the prepared sample near the boiling point, add fresh 0.5 N stannous chloride solution dropwise until the color turns pale green or colorless, then add one drop excess. Dilute at once to approx 125 ml with cold distilled water, add 2 ml of saturated mercuric chloride solution and mix. If none or more than a faint precipitate appears within one minute, too little or too much stannous chloride has been added and the sample must be discarded, otherwise, without further delay, add 2 ml conc HCl, 2 drops orthophenanthroline indicator and titrate with the approx 0.01 N ceric sulfate solution. Replicate samples must be carried through the reduction and titration steps one at a time and without delay or atmospheric oxidation of the ferrous ion will result.

8. Calibration and Standards

8.1 0.01 N CERIC SULFATE. Although the 0.1 N master standard is stable, it is best to standardize the 0.01 N ceric sulfate standard just before use against 99.9% pure iron wire. Place 20 to 30 mg of iron wire in a 250-ml covered beaker, add 10 ml conc HCl and 2 ml H_2O, and keep at incipient boiling until the iron is dissolved. With the solution at this temperature, treat as directed in **7.3** for the prepared sample.

9. Calculations

9.1 One ml 0.01 N oxidant solution used is equivalent to 0.799 mg Fe_2O_3 in the aliquot titrated.

9.1.1 Mg Fe_2O_3 in aliquot = volume of 0.1 N potassium dichromate used in titration × 7.99 or volume of 0.01 N ceric sulfate × 0.799.

9.2 Mg Fe_2O_3/m^3 air (high volume filter samples) =

$$\frac{\text{Mg } Fe_2O_3 \text{ in aliquot} \times \text{vol. of original sample}}{\text{Vol. of air sample in } m^3 \text{ (760 Torr and 25 C)} \times \text{vol. of aliquot}}$$

9.3 Fall-out jar samples are usually reported in tons of Fe_2O_3 per square mile = 36.4 (w/D^2); where w = mg Fe_2O_3 and D = diameter of fall-out jar in cm.

10. Effect of Storage

10.1 Samples may be stored indefinitely before analysis when protected from contamination. Both 0.1 N potassium dichromate and ceric sulfate standard solutions are stable for at least one year when stored in stoppered iron-free glass containers at room temperature and in diffuse daylight. Solutions of 0.01 N strength may not be stable for more than 24 hr after preparation.

11. References

1. WILLIAMS, H.H., and P. YOUNG. 1928. Ceric Sulfate as a Volumetric Oxidizing Agent. II. The Determination of Iron. *J. Am. Chem. Soc.* 50:1368–71.
2. SMITH, G. FREDRICK. 1942, 1964. Cerate Oxidimetry. 1st ed., 2nd ed., Fredrick Smith Chemical Co., Columbus, Ohio.
3. WILLARD, H.H., and N. HOWELL FURMAN. 1940. Elementary Quantitative Analysis. Chapter 11. Oxidations with Ceric Sulfate and with Potassium Bichromate. D. Van Nostrand Co., N.Y.

Subcommittee 6

W. G. FREDRICK, *Chairman*
J. CHOLAK
A. MOFFITT
K. KUMLER
G. B. MORGAN
W. M. SMITH
L. G. VON LOSSBERG

307.

Tentative Method of Analysis for Inorganic Lead Content of the Atmosphere

12128-01-71T

1. Principle of the Method

1.1 Lead ion reacts quantitatively with diphenylthiocarbazone (dithizone), in the absence of chelating agents (**2, 7**), in the 8.5 to 11.5 pH range to form read lead dithiozonate which is extractable with organic solvents, *e.g.*, chloroform. The red color, which is proportional to the amount of lead present, is measured spectrophotometrically at 510 nm. In the presence of an ammoniacal cyanide solution the interfering elements are limited to stannous tin, bismuth, and thallium. A sufficient known amount of dithizone is added to provide an excess of reagent and form a mixed color of dithizone and dithizonate (mixed color method).

2. Range and Sensitivity

2.1 This method is applicable to the determination of lead in quantities of 0.5 to 50 μg in the aliquot taken for analysis. For atmospheric concentrations ranging from 0.1 to 100 $\mu g/m^3$, air sample volumes from 100 to 1 m^3 respectively are appropriate. An air sample of 20 m^3 is satisfactory for most situations.

3. Interferences

3.1 The method as described is free from interferences from metals other than thallium, bismuth, and tin (II). Although bismuth, thallium, and tin (II) have not been shown to be present in ambient air samples in sufficient amount to interfere, special procedures must be used should this occur. (**3, 7, 8**). Small amounts of oxidizing agents produce a colored oxidation product of dithizone. Hydroxlamine hydrochloride inhibits this effect. This method describes two alternative procedures, including a special procedure, given in Section **7.3**, that is necessary for very low lead levels where sufficient extraneous material may be present in the large air sample required.

4. Precision and Accuracy

4.1 No specific studies are known on the overall precision and accuracy of the method for air samples. A careful study of the method as applied to the determination of lead in blood (**4**) showed a mean recovery of 97.1% for added lead, and a coefficient of variation of 4.0 to 9.5% based on groups of 10 repeated analyses). A more recent interlaboratory study of blood lead determinations has shown much greater interlaboratory variation (**5**). This study suggested that the use of a "standardized procedure" might reduce the variation. The results from the earlier study are most likely representative of the capability of the method excluding the sampling error for air.

5. Apparatus

5.1 For collection of samples. Membrane filters such as Millipore, appropriately sized for sample volume and rate, pore size of 0.8 μm, are satisfactory for usual air samples. Collection efficiencies greater than 99% are obtained for a 0.3 μ DOP smoke at face velocities from 7.2 to 53 cm/sec (**6**). For membrane filters of 47 mm, 4″ diam and 8″ × 10″ rectangular size, a face velocity of 25 cm per sec will result in flow rates of about 26, 120 and 770 liters per min respectively (see Section **2.1** for desired air sample volume).

5.2 Borosilicate and Vycor glassware should be used throughout. Glassware which has been used for the analysis should be rinsed with tap water, 10% nitric acid, and with distilled water, immediately before and immediately after use.

5.3 Squibb form separatory funnels 60- or 125-ml capacity with Teflon stopcocks and stoppers.

5.4 Buret.

5.5 A spectrophotometer capable of isolating wavelengths centered around 510 ± 10 nm. One-cm matched cells.

6. Reagents

6.1 Purity. ACS reagent grade chemicals or equivalent shall be used in all tests. References to water shall be understood to mean lead-free double distilled water or equivalent. Prepared solutions of indicators must be free of chelating agents. Care in selection of reagents and in following the above precautions is essential if low blank values are to be obtained.

The chloroform layer from a blank solution run through the procedure, **7.2.1**, *et seq.*, should be green in color. If not, there is too much lead in the water or reagents being used. Aqueous solutions of reagents can be deleaded by extracting with a 100 mg/liter dithizone solution in chloroform and then washing with chloroform until dithizone free. Nitric acid and ammonium hydroxide can be purified by distillation from all borosilicate glass apparatus.

6.2 CHLOROFORM. Suitable for use in Dithizone Procedures.

6.3 HYDROCHLORIC ACID (36.5–38.0%)—Heavy metals as lead less than 0.0001%.

6.4 NITRIC ACID (69.0–71.0%)—Heavy metals as lead less than 0.00002%.

6.5 DILUTE NITRIC ACID (1.10). Dilute 100 ml of conc acid to 1 liter with water.

6.6 DILUTE NITRIC ACID (1:30). Dilute 100 ml of dilute nitric acid (1:10) to 300 ml with water.

6.7 AMMONIUM HYDROXIDE (28.0–30.0% AS NH_3)—Heavy metals as lead less than 0.00005%.

6.8 DITHIZONE. Specifications as given in *Reagent Chemicals*, 1968. 4th Edition, American Chemical Society, Washington, D.C. For purification see Sandell (**7**).

6.9 STRONG DITHIZONE SOLUTION FOR DOUBLE DITHIZONE EXTRACTION PROCEDURE OR DELEADING OF REAGENTS. Dissolve 50 mg dithizone in 500 ml of chloroform.

6.10 DITHIZONE SOLUTION FOR ANALYTICAL COLOR DEVELOPMENT. Dissolve 10 mg dithizone in 1 liter of chloroform. Store in a refrigerator. This solution should be at room temperature when used.

6.11 PHENOL RED INDICATOR SOLUTION—(002% IN DISTILLED WATER). Dissolve 0.02 g phenol red in 100 ml of water.

6.12 AMMONIUM CITRATE SOLUTION (5%). Dissolve 25 g of citric acid in 250 ml of water, neutralize with ammonium hydroxide to a red phenol red end point, dilute to 500 ml and extract with strong dithizone solution until the last extract remains gray, and with chloroform until the final extract is free of color.

6.13 THYMOL BLUE INDICATOR. Dissolve 0.1 g of thymolsulfonphthalein in 20 ml warm ethanol (95% U.S.P. grade) and dilute with water to 100 ml.

6.14 POTASSIUM CYANIDE SOLUTION. Dissolve 10 g of potassium cyanide in water and dilute to 100 ml. Since this solution does not keep well, only 100 ml should be prepared at a time. De-lead, as in **6.1**, if a blank run with the new solution is not green in color.

6.15 LEAD EXTRACTIVE SOLUTION. Mix 30 ml of the potassium cyanide solution and 40 ml of the ammonium citrate solution with 106 ml of conc ammonium hydroxide, and add 900 ml of water.

6.16 HYDROXYLAMINE HYDROCHLORIDE SOLUTION. Prepare a saturated solution by adding 85 g of hydroxylamine hydrochloride to 100 ml of water.

7. Procedure

7.1 PREPARATION OF SAMPLES.

7.1.1 *Membrane filter.* Place in a 150 ml Vycor beaker, moisten with conc nitric acid, evaporate to dryness and ash in a muffle furnace at 500 C ± 10 C for 30 min. The ash is put in solution by warming with 10 ml conc nitric acid and a little water. Transfer to a 100-ml volumetric flask and dilute to volume with water. As lead may be volatilized or fused by prolonged ashing or use of higher temperatures, care is necessary in the control of both time and temperature.

7.2 PROCEDURE.

7.2.1 Into a separatory funnel of suitable size, place 15.0 ml of lead extractive solution dispensed from a buret. Pipet a 5.00 ml aliquot of the prepared sample into the funnel, and add two drops of hydroxylamine hydrochloride solution. The solution should have a pH of 9 to 9.5.

7.2.2 To each sample, blank and standard add 5 ml of dithizone solution (10 mg/liter), and shake vigorously for 45 sec. The color of the dithizone layer will range from green through blue and violet to cherry red. If the sample or standard is red in color, add a second portion of dithizone and again shake vigorously for 45 sec. If the red color persists, suitable dilutions of the original sample should be made in 1:10 nitric acid and 5.00 ml of the diluted solution used for analysis as in **7.2.1**.

7.2.3 Determine the wavelength of maximum absorbance in a spectrophotometer at a wavelength near 510 nm (use chloroform as a reference). Extracted samples must be kept out of sunlight. The colors are stable in the extraction funnels for about 2 hr, but must be read immediately after dry transfer to a spectrophotometer cell. If the transferred chloroform layer is cloudy or contains water droplets, filter the solution from the funnel through a small plug of glass wool inserted in the stem.

7.3 DOUBLE DITHIZONE EXTRACTION.

7.3.1 For very low atmospheric lead levels an air sample greater than 20 m^3 must be taken. If sufficient extraneous material is present to interfere with the simple direct dithizone procedure described, proceed as follows:

a. Completely ash the sample in a muffle furnace at 500 ± 10 C. Cool the beaker and wash down the sides with approximately 3 ml of conc hydrochloric acid, cover with a tight watch glass and heat for a few minutes to dissolve the acid soluble residue, but do not evaporate to dryness. Cool the acid mixture. Add 20 ml of lead free ammonium citrate solution, and adjust to pH 9 to 9.5 with concentrated ammonium hydroxide using thymol blue indicator (color change from orange to green-blue to blue).

b. Transfer the sample to a Squibb separatory funnel, add 2 drops of saturated hydroxylamine hydrochloride, and extract with 10 ml portions of strong dithizone solution (100 mg/l), transfer the dithizone extracts to a second funnel containing 15 ml of the dilute nitric acid (1:30) and 5 drops of hydroxylamine hydrochloride, being careful not to allow any aqueous ammoniacal solution to flow into the acid. Continue the extraction in the original funnel with additional portions of the strong dithizone solution until the dithizone appears green in the stem of the funnel during the transfer.

c. Shake the second funnel for at least 1 min to extract the heavy metals into the aqueous acid solution (the dithizone layer turns green). Allow the layers to separate, and discard the dithizone layer. Wash the remaining dithizone from the acid layer by shaking with 5 ml of chloroform. Draw off and discard the chloroform layer.

d. Add 15 ml of lead extractive solution, 2 drops of hydroxylamine hydrochloride solution. Mix thoroughly then proceed as in section **7.2.2**.

8. Standards and Calibrations

8.1 STANDARD LEAD SOLUTIONS.

8.1.1 *Master Standard (1 mg lead per ml).* Dry a portion of lead nitrate at 100 C for 2 hr, cool, weigh out exactly 1.5980 g and dissolve in 1:10 nitric acid in a one liter volumetric flask. Dilute to volume with 1:10 nitric acid. Transfer to glass stoppered pyrex bottle. This laboratory master lead standard is stable for several years. One ml of this solution contains 1.000 mg lead.

8.1.2 *Dilute Standard (0.010 mg lead per ml).* Pipet 100 ml of the Master Standard to a 1-1 volumetric flask and dilute to volume with 1:10 nitric acid.

8.1.3 *Working standards.* Prepare 100 ml portions of working standards ranging from 0.2, 0.4, ... 6, 8 μg/1 ml volume by adding the appropriate volumes of Dilute Standard to 100 ml volumetric flasks, then diluting to volume with 1:10 nitric acid. The working standards are stored in 125-ml screw top Erlenmeyer flasks with Teflon liners in the caps. The working standards should be prepared fresh weekly.

8.1.4 A minimum of 5 known amounts of lead from 0 to 50 μg, added to the blank membrane filter, are carried through the procedure to establish a calibration graph for the spectrophotometer. The calibration should be performed with each group of samples analyzed. The calibration graph must be a straight line.

8.2 A blank of the membrane filter should be carried through the procedure. The blank should not exceed 1.0 μg Pb.

9. Calculations

9.1

$$\mu g\,Pb/m^3 \text{ air} = \frac{\mu g \text{ Pb per ml in aliquot} \times (\text{vol. of original sample})}{\text{total volume of air in } m^3}$$

(**Note:** volume of air sample corrected to 25 C and 760 Torr.)

10. Effects of Storage

10.1 Solid dithizone is stable indefinitely. Dithizone solutions in brown bottles and solutions containing cyanide and/or ammonium hydroxide should be stored in the refrigerator. For best results it is recommended that these solutions be made up fresh each week. Untreated samples, protected against contamination, may be stored indefinitely.

APPENDIX I

A comparison of the ISC method of analysis for inorganic lead content of the atmosphere with the new ASTM "Tentative Method of Test for Lead in the Atmosphere" reveals certain similarities and also differences. Both are colorimetric methods using the dithizone procedure. However, the ISC method is applicable only to particulate lead in the range of 0.1–100 $\mu g/m^3$. The ISC method gives more attention to removing traces of lead from the standard solutions, and the ashing at 500 C may prevent the interference from tin (II) although this has not been tested. In the presence of large amounts of tin (II), thallium or bismuth (and possibly indium) special separation methods are recommended by Sandell (7). These elements have not been shown to be present in ambient air samples in sufficient amount to cause interference.

The ISC method employs a 0.8 μm membrane filter to collect particulate lead while the ASTM method uses a 0.45 μm filter. On theoretical grounds there should be no essential difference in collection efficiency. The 0.45 μm filter may cause higher resistance to flow and involve a shorter sampling period.

In the ISC method, the particulate filter sample is ashed at 500 C for 30 min. The ASTM method employs a wet ashing procedure by digestion of the particulate lead filter with nitric and perchloric acids. The wet ashing method may be preferable, although tests of the 500 C ashing for only 30 min have indicated no measurable loss of particulate lead.

ADDENDUM

Precaution: Millipore filters should not be heated directly in a muffle furnace, as complete loss of sample may result due to the sharp ignition point. They are, therefore, first treated with acetone, then dried and slightly charred on the hot plate, then treated with concentrated nitric acid and again dried and charred more strongly, before ignition in the muffle furnace. (Walkley, J. 1959, AIHA Jour. 20:241–245.)

11. References

1. American Conference of Governmental Industrial Hygienists. 1958. *Manual of Analytical Methods. Lead (USPHS).*
2. Fischer, H. and G. Leopoldi. 1940. *Z. Anal. Chem.*, 119:182–184.
3. Bambach, K. and R.E. Burkey. 1942. *Ind. Eng. Chem., Anal. Ed.*, 14:904.
4. Kennan, R.G., D.H. Byers, E.E. Saltzman, and F.L. Hyslop. 1963. The "USPHS" Method for Determining Lead in Air and in Biological Materials. *Am. Ind. Hyg. Assoc. J.*, 24:481–491
5. Keppler, J.F., M.E. Maxfield, W.D. Moss, G. Tietjen, and A.L. Linch. 1970. Interlaboratory Evaluation of the Reliability of Blood Lead Analyses. *Am. Ind. Hyg. Assoc. J.*, 31:412–29.
6. Lockhart, L.B., Jr., R.L. Patterson, Jr., and W.L. Anderson. 1963. Characteristics of Air Filter Media Used for Monitoring Airborne Particulates. *NRL Report 6054*, Dec. 1963.
7. Sandell, E.B. 1965. *Colorimetric Determination of Traces of Metals*, 3rd Edition Revised and Enlarged. Interscience Publishers, Inc., New York, N.Y.
8. Woessner, W. W. and J. Cholak. 1953. Improvements in the Rapid Screening Method for Lead in Urine. *AMA Archives Ind. Hyg. and Occ. Medicine*, 7:249–254.

Subcommittee 6

T. J. Kneip, *Chairman*
R. S. Ajemian
R. R. Carlberg
J. Driscoll
H. Freiser
L. Kornreich
K. Kumler
R. J. Thompson

307.

Tentative Method of Analysis for Manganese Content of Atmospheric Particulate Matter

12132-01-70T

1. Principle of the Method

1.1 Manganese may be present in atmospheric particulate matter in the range 0.01 to 3.0 $\mu g/m^3$ (**1**). For estimating microquantities of manganese, the periodate method is one of the most suitable wet chemical method (**2, 3, 4**). By this method the collected manganese is dissolved in nitric and sulfuric acids and the resulting acid solution, free of chlorides, is oxidized with potassium periodate to convert the manganese to permanganate. The latter is determined in a spectrophotometer at 525 nm (**5**).

2. Range and Sensitivity

2.1 Using a light path of 2 cm in the spectrophotometer, concentrations of 6 to 120 μg of manganese in 10 ml of solution represent the limiting range of the instrument (**6**).

2.2 A sample or an aliquot representing 20 m^3 of air will contain sufficient manganese for analysis if the air contains a minimum of 3.0 $\mu g/m^3$.

2.3 A longer light path or a larger sample or both may be used for lower concentrations.

3. Interferences

3.1 The maximum amounts of interfering elements which can be present in the sample without significantly altering results are as follows: (**6**)

Beryllium	0.0001%
Bismuth	0.01%
Boron	0.005%
Chromium	0.01%
Copper	0.01%
Lead	0.03%
Iron	0.01%
Magnesium	0.01%
Nickel	0.02%
Silicon	0.05%
Tin	0.01%
Zinc	0.30%

3.2 Ferrous salts or other strong reducing agents will reduce the periodate and liberate iodine which will invalidate the results. The use of more HNO_3 in the digestion process should eliminate this interference (**2**).

3.3 If HCl is added to dissolve the sample, it must be completely removed by fuming with H_2SO_4 before the color is developed.

4. Precision and Accuracy

4.1 No data are available in the literature for the range required. On the basis of mg/m_3, an error of ± 7.7% was reported by the American Conference of Governmental Industrial Hygienists (**2**).

5. Apparatus

5.1 High Volume Sampler. 8″ × 10″ filter holder.

5.2 Glass Fiber Filters. 8″ × 10″—MSA 1160 BH or Gelman Type A.

5.3 Spectrophotometer. For use at 525 nm. Absorption cells—10-ml capacity or greater with 2-cm light path. Cells with light path greater than 2-cm optional.

6. Reagents

6.1 Reagents must be ACS analytical reagent quality. Distilled water should conform to ASTM Standards for Referee Reagent Water.

6.2 STANDARD MANGANESE SOLUTION (1 Ml = 1 μg of Mn)—Dissolve 0.288 g of $KMnO_4$ (dried to constant weight at 100 C) in 100 ml of H_2O, add 20 ml of H_2SO_4 (1:1) and reduce permanganate color by additions of small quantities of Na_2SO_3 or H_2O_2. Boil to remove excess SO_2 or H_2O_2. Cool, dilute to 1 l and mix. Dilute 10 ml of this solution to 1 l. One ml of this solution contains 1 μg of manganese.

6.2 POTASSIUM PERIODATE SOLUTION (10 g/l). Dissolve 10 g of KIO_4 in H_2O, and dilute to 1 l.

6.3 SODIUM NITRITE SOLUTION (20 g/l). Dissolve 0.2 of $NaNO_2$ in 10 ml H_2O. Prepare solution fresh before use.

7. Procedure

7.1 COLLECTION OF SAMPLE. High Volume sampler—pass air through a glass fiber filter 8″ × 10″ in a high volume sampler at approximately 1.5 m^3/min for 24 hr.

7.2 ANALYSIS OF SAMPLE

7.2.1 Place an aliquot of filter which represents at least 60 μg of manganese in a 250-ml beaker. Add 10 ml of H_2SO_4 (1:1), 15 ml of conc HNO_3, 10 ml conc HCl and evaporate to heavy white fumes of H_2SO_4. Cool, add 15 ml of H_2O and refume. Cool, add 50 ml of H_2O, 3 ml (85%) H_3PO_4, heat to boiling and boil gently for 1 min. Filter through a sintered glass funnel using suction. (Do not use filter paper.) Wash with 50 ml hot water and add to filtrate. Transfer quantitatively to a 250-ml beaker and reduce volume to about 70 ml by gentle boiling. Cool and treat as in **8.2**.

7.2.2 Run a blank on a section of filter and reagents following procedure **7.2.1**. The filters used to collect the air borne particulates may contain measurable quantities of manganese. Portions of unused filters should be analyzed and corrections made if necessary. This can be done separately or as part of the reagent blank procedure.

8. Calibrations

8.1 CALIBRATION SOLUTIONS. Transfer 1, 5, 10, 20, 40, 60 ml of standard manganese solution to separate 150-ml beakers. Add to each beaker 10 ml of H_2O, 10 ml conc HNO_3, 5 ml of conc (85%) H_3PO_4 and mix.

8.2 COLOR DEVELOPMENT. Add 10 ml of KIO_4 solution to each beaker. Heat to boiling, boil for 2 min and digest just below boiling for 10 min to develop full intensity of color. Cool, transfer to 100-ml volumetric flask, dilute to mark and mix.

8.3 PHOTOMETRY.

8.3.1 The reference solution for samples should be prepared from a 10 ml aliquot of prepared sample solution in which the color has been discharged by 1 drop of $NaNO_2$ solution.

8.3.2 Fill one absorption cell with solution containing reference solution (See **8.3.1**) and using a wave-length of 525 nm set photometer at zero optical density. While maintaining the zero setting of the photometer with this reference solution, fill other similar absorption cells with the calibrating (or sample) solutions and take absorbance readings.

8.4 CALIBRATION CURVE. Plot the photometric readings of the calibrating solution against μg of manganese per 100 ml of solution on linear paper. If the lower calibration points show insignificant changes of photometric readings, use a longer light path (5 cm) and repeat sections **8.3** and **8.4**. Or dilute to 50 ml instead of 100 ml for the lower points and prepare a different curve for μg of manganese in 50 ml of solution.

9. Calculations

9.1 Convert the photometric readings of the sample and blank to μg of manganese by means of the calibration curves.

9.2 Calculate to $\mu g/m^3$ as follows:

$$\mu g Mn/m^3 = \frac{(A - B)}{C}$$

A = Micrograms of manganese in sample.
B = Micrograms of manganese in blank.
C = Cubic meters of air represented by sample or aliquot at 760 Torr and 25 C.

10. Effect of Storage

10.1 Samples on glass fiber filters can be stored indefinitely. The permanganate color is stable for hours.

11. References

1. Air Quality Data from the National Air Sampling Networks and Contributing State and Local Networks 1964–1965. U.S. Department of Health, Education, and Welfare, Public Health Service, Cincinnati, Ohio, 1966.
2. Recommended Method for Manganese. 1958. Manual of Analytical Methods Recommended for Sampling and Analysis of Atmospheric Contaminants. American Conference of Governmental Industrial Hygienists. Cincinnati, Ohio.
3. Fairhall, L.T. 1940. U.S. Public Health Service, Bull. 247:31.
4. Richards, M.B. 1930. *Analyst*, 55:554.
5. Willard, H.H. and T.H. Greathouse. 1917. *J. Am. Chem. Soc.*, 39:2366.
6. Methods of Chemical Analysis of Metals, 391–3, 1956, Philadelphia, Pa. American Society for Testing and Materials.

Subcommittee 7
E. C. Tabor, *Chairman*
M. M. Braverman
H. E. Bumstead
A. Carotti
H. M. Donaldson
L. Dubois
R. E. Kupel

309.

Tentative Method of Analysis for Molybdenum Content of Atmospheric Particulate Matter

12134-01-70T

1. Principle of the Method

1.1 Molybdenum may exist in the atmosphere as the element, the oxide or as molybdates. Concentrations of molybdenum in ambient air range from below detectable amounts to maximum values in the order of 0.34 $\mu g/m^3$ of air (**1**). The satisfactory determination of molybdenum depends largely on the quantitative formation of a stable molybdenum (V) thiocyanate complex. This complex is formed, in the presence of iron and copper, by reduction of molybdenum VI with stannous chloride in a solution containing potassium thiocyanate. The thiocyanate complex is extracted with methyl isobutyl ketone. The absorbance of the extract is measured in a 10 cm cell at 500 nm using methyl isobutyl ketone in the reference cell. The absorbance is a linear function of molybdenum concentration, at least between 0 and 15 μg of molybdenum/25 ml of solvent.

2. Range and Sensitivity

2.1 With a 10-cm light path cell the method covers the range from 0 to 15 μg molybdenum/25 ml of solvent.

2.2 The sensitivity being 1.1 μg of molybdenum/0.05 absorbance units, a concentration of 0.055 μg of molybdenum/m^3 of air can be measured if a 20 m^3 sample is taken for analysis.

3. Interferences

3.1 The only elements which cause interferences are rhenium, platinum, palladium, rhodium, selenium and tellurium (**2**).

3.2 The interferences caused by these elements are negligible in concentrations found in ambient air.

4. Precision and Accuracy

4.1 Twenty solutions each containing 75 μg of molybdenum were extracted and the absorbances measured using a 2-cm cell. The absorbances of all 20 solutions were between 0.670 and 0.676. The method is accurate ± 4% (**2**).

5. Apparatus

5.1 A SPECTROPHOTOMETER, for use at 500 nm with 10-cm cells.

5.2 15-ML GLASS-STOPPERED CENTRIFUGE TUBES.

5.3 250-ML SEPARATORY FUNNELS, VOLUMETRIC AND GENERAL LABORATORY GLASSWARE.

5.4 CENTRIFUGE.

6. Reagents

6.1 PURITY. Reagents must be ACS analytical reagent quality. Distilled water should conform to ASTM Standards For Referee Reagent Water.

6.2 STANDARD MOLYBDENUM SOLUTION. Dissolve 3 g of molybdenum trioxide in 200 ml of 10% sodium hydroxide and dilute to 2 l with distilled water. Store in a polyethylene container. (This solution is stable indefinitely.) Standardize this solution by titrating with standard ceric sulfate using 0-phenanthroline indicator (2 drops of 0.025 M indicator are used) (**3, 4**). Dilute an aliquot with distilled water to obtain a solution containing approximately 25 μg of molybdenum/ml.

Molybdenum trioxide of over 99.5% purity is available. Consequently sufficiently accurate standards may be prepared without standardization.

6.2 TARTARIC ACID. 50% solution in distilled water—Filter if necessary.

6.3 SULFURIC ACID, SP GR 1.84.

6.4 HYDROCHLORIC ACID, SP GR 1.19.

6.5 IRON SOLUTION. 10 mg Fe (III)/ml as ferric chloride in distilled water.

6.6 COPPER SOLUTION. 50 μg Cu (II)/ml as copper sulfate in distilled water.

6.7 OXALIC ACID. 2% (W/V) in 1.2 N hydrochloric acid.

6.8 METHYL ISOBUTYL KETONE. (4-methyl-2 pentanone), bp 114 to 116 C.

6.9 POTASSIUM THIOCYANATE. 25% (W/V) in distilled water.

6.10 STANNOUS CHLORIDE. 25% (W/V) in 12 N hydrochloric acid.

6.11 WASH SOLUTION NO. 1. Add 4 ml of the stannous chloride solution to 50 ml of the oxalic acid solution. For convenience prepare this wash solution as needed.

6.12 WASH SOLUTION NO. 2. 2.4 N hydrochloric acid.

7. Procedure

7.1 Wet ash samples collected on membrane or cellulose filters or extract glass fiber filter samples in a Phillips beaker using enough concentrated nitric acid to cover the filter (take whole sample or suitable aliquot to give about 3 μg of molybdenum/ml). In the case of glass fiber filters heat the beaker below the boiling point of the acid to extract the molybdenum, decant the liquid into a second beaker and wash the Phillips beaker and filter with nitric acid and distilled water. Repeat washing 3 times. (Add all washings to the second beaker).

7.2 Place the second beaker on a low temperature hot plate (125 C and heat until the brown fumes of oxides of nitrogen appear then transfer the beaker to a high temperature hot plate (400 C) and heat until the fumes of nitrogen oxides disappear. Repeat the ashing treatment with further additions of 2 ml of concentrated nitric acid until all traces of organic material have disappeared.

7.3 Take up the residue with 5 ml of hydrochloric acid, add 10 ml of sulfuric acid and 5 g of tartaric acid and dilute to volume of about 90 ml with water. Transfer to a 250-ml separatory funnel and add 1 ml of the iron solution, and 1 ml of the copper solution. Add 10 ml of potassium thiocyanate solution and 9 ml of stannous chloride solution, mixing well after each addition.

7.4 Add exactly 25 ml of methyl isobutyl ketone and extract the molybdenum thiocyanate complex by shaking for 90 seconds. Discard the aqueous phase. Add 50 ml of wash solution No. 1 and shake for 1 min. Discard the acid wash. Transfer the extract to a glass-stoppered centrifuge tube and allow to stand for 18 hr for maximum color development. Centrifuge, transfer extract to a 10-cm cell, and then measure the absorbance of the extract at 500 nm using methyl isobutyl ketone in the reference cell. An absorbance of approximately 0.67 is obtained from 15 μg of molybdenum carried through the procedure.

8. Calibration

8.1 Prepare a calibration curve using standards containing 0 to 15 μg of molybdenum and carry them through the complete procedure, section 7. Plot absorbance *vs* concentration.

9. Calculations

9.1 Use the absorbance of the samples to obtain the molybdenum concentration from the calibration curve.

9.2 Concentration of molybdenum in μg/m³ =

$$\frac{\mu g \text{ Mo in sample analyzed}}{m^3 \text{ of air represented by sample at 760 Torr and 25 C.}}$$

10. Effect of Storage

10.1 Molybdenum collected on solid filter media can be stored indefinitely.

11. References

1. AIR QUALITY DATA FROM THE NATIONAL AIR SAMPLING NETWORK AND CONTRIBUTING STATE AND LOCAL NETWORKS. 1966. U.S. Department of Health, Education, and Welfare, Public Health Service, National Air Pollution Control Administration.
2. HIBBITS, J.O., and R.T. WILLIAMS. 1962. The Extraction and Determination of Molybdenum as the Thiocyanate Complex. *Anal. Chem. Acta*, 26, 363–370.
3. FURMAN, N.H., and W.M. MURRAY, JR. 1936. *J. Am. Chem. Soc.* 58, 1689.
4. SCOTTS STANDARD METHODS OF CHEMICAL ANALYSES. 1969. Sixth Edition, 1, 1205. D. Van Nostrand. N.Y.

Subcommittee 7
E. C. TABOR, *Chairman*
M. M. BRAVERMAN
H. E. BUMSTEAD
A. CAROTTI
H. M. DONALDSON
L. DUBOIS
R. E. KUPEL

310.

Tentative Method of Analysis for Selenium Content of Atmospheric Particulate Matter

12154-01-69T

1. Principle of the Method

1.1 Selenium may exist in the atmosphere most probably as the element, the oxide or as the selenite.* Sampling of the particulates is followed by oxidation of the sample with concentrated nitric acid. The Se(IV) reacts with 2,3-diaminonaphthalene (DAN); in acid solution to form the red-colored and strongly fluorescent 4,5-benzopiazselenol. The latter is extracted with toluene. The fluorescence intensity of the solution sample is measured at an exciting wavelength of 390 nm and a fluorescent wavelength of 590 nm, using a sample

*Hydrogen selenide, H_2Se, would be expected to have a short life in the atmosphere. Air oxidation of this compound will yield the element or oxide.

containing 1.0 μg of selenium as the reference standard.

2. Range and Sensitivity†

2.1 In the fluorimetric procedure, a linear calibration curve is observed over the range 0 to 1 μg of selenium in 10 ml of solvent (1-cm light path cell) **(1, 2)**. In the spectrophotometric procedure, the calibration curves follow Beer's law over the range 0 to 20 μg of selenium per 5 ml of solvent (1-cm light path cell) at a wavelength of 390 nm using a reagent blank **(1)**.

2.2 The sensitivity of the spectrophotometric procedure is 0.0032 μg/sq cm for $\log I_0/I = 0.001$, but the fluorimetric procedure with DAN is over 10 times as sensitive as the spectrophotometric procedure. The concentration of selenium in the atmosphere over an urban-industrial area may be expected to average 0.2 μg/200 m^3 of air **(3)**.

3. Interferences (1)

3.1 In the presence of masking agents, the primary interference which is observed at a 2000-fold excess of foreign ion to selenium is due to substances like hypochlorite, which oxidize the reagent, or reducing agents like Sn(II) which reduce selenium to the elemental state. Some ions like zinc, aluminum or sodium can be present at a million-fold excess without causing any interference. Macro amounts of ions like Al(III), Zn(II), Cu(II), Ca(II), Cd(II), Mn(II), Ni(II), Mg(II), Ba(II) and Sr(II) are separated from selenium (IV) by ion exchange (by passing the sample in solution through the Dowex 50 WX 8 resin). The presence of about 1200 μg of tellurium does not interfere in the fluorometric analysis of a sample containing 1 μg of selenium. The presence of about 2500 μg of tellurium causes little interference in the spectrophotometric analysis of a sample containing 10 μg of selenium. Such high concentrations of tellurium are normally not to be expected in the ambient air. Nitrite ion interferes in both methods. It is therefore important, in preparing the sample for analysis, (see section **11**) to use the minimum amount of nitric acid in dissolving the sample and to remove the oxides of nitrogen by boiling. Nitrate ion does not appreciably interfere with the spectrophotometric method and not at all with the fluorimetric method. Nevertheless, since both the sample solution to be analyzed and the standard stock selenium solution are prepared via the dissolution of the particulates or elemental selenium in nitric acid, it is important to establish a blank reading by running through the whole analytical procedure without a sample but with the unused paper or other type of filter.

3.2 Varying known amounts (0 to 10 μg) of selenium were added to 0.5 or 1.0 g copper samples. The samples were dissolved using 8 ml of 12 M HNO_3 per gram of cop-

---

†The control of variables in any fluorimetric procedure is extremely important. Because of the high sensitivity of the method, many errors may be introduced if blanks, standards and samples are not measured identically. Energy re-emitted from a fluorescent material is not directly proportional to concentration over all ranges, due to self-quenching. This may occur whenever the emitted light can be reabsorbed by the fluorescent material. Some self-absorption can be avoided by working at high dilution. However, standards should bracket the unknown in concentration in order that compensation exists for any self-absorption that may take place. Other substances present in the sample to be read, such as solvents and impurities, may behave in a similar manner. This phenomenon, known as quenching, is quite common. On the other hand, the presence of foreign material may, in some cases, enhance the fluorescence. Extreme care must therefore be taken to maintain identical conditions throughout the measurements. When the concentration of one unknown quenches or enhances the fluorescence of another unknown substance, families of concentration curves may be used to determine correction factors via an evaluation of the recorded deviations from the respective expected readings. Several excellent reviews on this subject can be found in the literature *e.g.*, Bowen, E. J. 1947. *Analyst,* 72:377, and White, C. E., *et al.* 1952. *Anal. Chem.*, 24:1965.

per and the resulting solutions boiled to remove oxides of nitrogen, diluted with 25 ml of water and reboiled. This final solution, containing a large excess of Cu^{+2} and NO_3^- was analyzed for selenium spectrophotometrically, (Section **7.2**) but with larger ion exchange column. Representative results were as follows:

Sample	Added, Se μg	Found, Se μg
Reagent Grade Copper (0.500 g, as nitrate)	none	4.9, 4.9
	2.0	7.0
	5.0	10.0
	10.0	14.1

Varying known amounts (0 to 2 μg) of selenium were added to 0.25 or 0.5 g samples of copper. The samples were dissolved in 5 ml of 12 M HNO_3 and the resulting solutions boiled to remove oxides of nitrogen, diluted with 25 ml of water and reboiled. This final solution, containing a large excess of Cu^{+2} and NO_3^-, was analyzed for selenium fluorimetrically, (section **7.1**) but with larger ion exchange column. Representative results were as follows: **(1)**

Sample	Added, Se μg	Found, Se μg
Spectrographic Copper, 0.500 g sample	none	0.16, 0.16
	0.25	0.39
Electrolytic Copper Foil, 0.500 g sample	none	0.46, 0.47
	0.25	0.71
	0.50	0.92

4. Precision and Accuracy (1)

4.1 Both the spectrophotometric and the fluorimetric procedures showed a relative standard deviation of 9.9% for the determination of selenium in copper samples.

5. Apparatus

5.1 Beckman Du, Perkin-Elmer Model 202 or Bausch and Lomb spectronic 20 Spectrophotometer with 1-cm cells, and an Aminco-Bowman Spectrofluorimeter, or equivalent instruments from other sources of supply.

5.2 pH METER, with micro electrodes (glass-calomel).

5.3 25 ML BURET. Filled to the 10 ml mark with Dowex 50 WX 8, 50- to 100-mesh regenerated resin.

5.4 STIRRER, SEPARATORY FUNNEL, VOLUMETRIC AND LABORATORY GLASSWARE. Thoroughly cleaned.

6. Reagents

6.1 PURITY. All reagents should be prepared from analytical grade chemicals low in selenium, using distilled water.

6.2 2,3-DIAMINONAPHTHALENE (DAN), $C_{10}H_0(NH_2)_2$, REAGENT. Dissolve 1.00 g of 2,3-diaminonaphthalene (Aldrich Chemical Co., Milwaukee, Wis. or equivalent) in 1 l of 0.10 N hydrochloric acid, using a stirrer. Pure DAN is a white crystalline solid. It is slowly oxidized in air to form a yellow-brown material. An aqueous solution is more readily oxidized but can be stored for three days under refrigeration without being sufficiently altered to affect its use for the intended purpose. Under the conditions described there was no significant difference in the results of analyses when either the commercial or purified reagents were used.

6.3 0.1 M ETHYLENEDIAMINE TETRAACETIC ACID DISODIUM SALT (EDTA) SOLUTION. Dissolve 33.6 g of the disodium salt of ethylenediamine tetraacetic acid in a volume of distilled water and dilute to 1000 ml in a volumetric flask.

6.4 0.1 M SODIUM FLUORIDE, NaF, SOLUTION. Dissolve 0.42 g of NaF in a volume of distilled water and dilute to 100 ml in a volumetric flask.

6.5 0.1 M SODIUM OXALATE, $Na_2C_2O_4$ SOLUTION. Dissolve 1.34 g of $Na_2C_2O_4$ (Sorensen, Fisher Certified Reagent) in a volume of distilled water and dilute to 100 ml in a volumetric flask.

6.6 TOLUENE, $C_6H_5CH_3$. Fisher Certified Reagent or equivalent. Fisher's toluene, purified, may also be used.

6.7 STANDARD STOCK SELENIUM SOLUTION (SELENIOUS ACID, H_2SeO_3). **(4)**—A solution containing 0.50 mg of selenium/ml is prepared by dissolving 50 mg of pure selenium metal (Fisher Certified Reagent, specially purified) in a few drops (minimum necessary) of concentrated nitric acid, boiling gently to expel brown fumes, and making up to 100 ml with distilled water.

6.8 STANDARD WORKING SELENIUM SOLUTIONS. The standard stock solution, as prepared under **6.7**, is diluted as necessary for preparing standard working solutions. Dilute 1 ml of stock selenium solution to 100 ml with distilled water. This solution contains 5 μg Se/ml. Dilute 1 ml of stock selenium solution to 500 ml with distilled water. This solution contains 1 μg Se/ml.

6.9 1 N AND 0.1 N SODIUM HYDROXIDE, NAOH. Dissolve 40 g of NaOH pellets in a volume of distilled water and dilute to 1000 ml in a volumetric flask (1 N NaOH). Dissolve 4.0 g of NaOH pellets in a volume of distilled water and dilute to 1000 ml in a volumetric flask (0.1 N NaOH).

6.10 CONC NITRIC ACID, HNO_3. Fisher Reagent, 69 to 71% by wt, 15.8 N, or equivalent analytical reagent grade.

6.11 60% PERCHLORIC ACID, $HClO_4$. Fisher Reagent, 60 to 62% by wt or equivalent analytical grade.

7. Procedures (1)

7.1 FLUORIMETRIC PROCEDURE.

7.1.1 After appropriate treatment to dissolve the sample (see section **11**) containing from 0 to 1 μg of selenium, adjust the solution to pH 2 with 1 N and 0.1 N NaOH and pass it through a 25-ml buret filled to the 10-ml mark with Dowex 50 WX 8, 50- to 100-mesh regenerated resin at a flow rate of 0.5 ml/min. Collect the effluent and any remaining traces of selenium washed from the column with 20 ml of distilled water. To this solution add 0.5 ml of 0.1 M EDTA, 0.5 ml of 0.1 M sodium fluoride, and 5 ml of DAN solution, and readjust to pH 2. Add 5 ml of 0.1% DAN solution, allow to stand for 2 hr.

7.1.2 Transfer to a separatory funnel, add exactly 10 ml of toluene and extract the piazselenol by shaking for ½ min. Separate. To remove droplets of water the toluene layer is filtered into the cuvette through a small filter paper plug placed in the stem of the separatory funnel.

7.1.3 Measure the fluorescence intensity of the sample at an exciting wavelength of 390 nm and a fluorescent wavelength of 590 nm, using a solution containing 1.0 μg of selenium as the reference standard. A linear calibration curve is observed over the range 0 to 1 μg of selenium per 10 ml of toluene. The zero sample gives a scale reading of 0.10. This blank reading, which is reproducible within 2%, would correspond to 0.16 μg of selenium per 10 ml of toluene.

7.2 SPECTROPHOTOMETRIC PROCEDURE.

7.2.1 After appropriate treatment to dissolve the sample (see section **11**) containing approximately 10 μg of selenium, adjust the solution to pH 2 with 1 N and 0.1 N NaOH and pass it through a 25-ml buret filled to the 10-ml mark with Dowex 50 WX 8, 50- to 100-mesh regenerated resin at a flow rate of 0.5 ml/min. Collect the effluent and any remaining traces of selenium washed from the column with 20 ml of distilled water. To this solution add 5 ml of 0.1 M EDTA, 1 ml of 0.1 M sodium fluoride and 1 ml of 0.1 M sodium oxalate solution, and readjust to pH 1.5 to 2.5. Add 5 ml of 0.1% DAN solution and allow to stand for 2 hr.

7.2.2 Transfer to a separatory funnel, add exactly 5 ml of toluene, and extract the piazselenol by shaking for ½ min. Separate. To remove droplets of water, the toluene layer is filtered into the cuvette through a small filter paper plug placed in the stem of the separatory funnel.

7.2.3 The absorbance is determined at 390 nm using a reagent blank. The calibration curves follow Beer's law over the range 0 to 20 μg of selenium per 5 ml of toluene in a 1-cm cell at this wavelength.

8. Calibration

8.1 Prepare a set of standard solutions by appropriately diluting, with distilled water, accurately measured volumes of the standard stock selenium solution (section **6.7**), and the standard working selenium solutions (section **6.8**). Representative concentrations should range from 0 to 1 μg of selenium for the fluorimetric procedure (use the 1 μg Se/ml standard working selenium solution, section **6.8**), and 0 to 20 μg of selenium for the spectrophotometric procedure (use the 500 and the 5 μg Se/ml standard stock resp. working selenium solutions, sections **6.7** resp. **6.8**). Include a reagent blank.

8.2 Follow the procedures as outlined under sections **7.1** and **7.2**.

8.3 Construct calibration curves by plotting fluorescence intensity values vs. μg selenium in the standard (fluorimetric procedure) and absorbance values vs. μg selenium in the standard (spectrophotometric procedure). A linear calibration curve is observed over the range 0 to 1 μg of selenium per 10 ml of toluene (fluorimetric). The zero sample gives a scale reading of 0.10. This blank reading, which is reproducible within 2%, corresponds to 0.16 μg of selenium per 10 ml of toluene. In the spectrophotometric procedure the calibration curves follow Beer's law over the range 0 to 20 μg selenium per 5 ml of toluene in a 1-cm-cell at the prescribed wavelength.

8.4 The filters used to collect the air borne particulates *may* contain measurable quantities of selenium. Portions of unused filters should be analyzed, going through sections **11** and **7**, and corrections made if necessary. This can be done separately or as part of the reagent blank procedure, see also section **3** above.

9. Calculations

9.1 Use the fluorescence intensity resp. the absorbance of the samples to obtain the selenium content from the calibration curves.

9.2 Concentration of selenium in

$$\mu g/m^3 = \frac{\mu g \text{ Se in sample analyzed}}{m^3 \text{ of air represented by sample}}$$

10. Effects of Storage

10.1 Solid, pure DAN is slowly oxidized in air to form a yellow-brown material. An aqueous solution is more readily oxidized but can be stored for 3 days under refrigeration without being sufficiently altered to affect its use for the intended purpose. Under the conditions described there was no significant difference in the result of an analysis when either the commercial or purified reagent was used.

10.2 All work can be performed under normal laboratory conditions *ie.*, in daylight, without deaerating solutions or purifying reagents **(1)**.

11. Collection and Preparation of Samples

11.1 Both methods are suitable for the analysis of particulate samples collected either on filters (cellulosic, membrane or glass fiber) or by an electrostatic precipitator **(5)**. Tape samplers marketed by Instrument Development Company and modified AISI smoke samplers are reportedly satisfactory for inplant sampling and for sampling atmospheres having high concentrations of selenium-containing particulate matter. Cellulosic, membrane (Millipore or Gelman) and glass fiber (Mine safety 1106 BH or Gelman Type A) filters should be satisfactory for ambient air sampling. (Analysts at American Smelting and Refining Company have used personal samplers (MSA Monitaire or similar) with Millipore AA filters (or glass fiber filters) to assay individual exposures in working atmospheres).

11.2 The section of filter containing the particulate sample is dissolved in or quantitatively washed with the minimum amount of a solution containing concentrated nitric acid and 60% perchloric acid (10 HNO_3:1 $HClO_4$). Allow the mixture (paper + sample + acid solution) to stand at room temperature with occasional shaking until the foaming ceases. Apply heat from a burner *very cautiously* until the initial rapid oxidation begins. Remove the heat and allow the oxidation to continue until reaction has subsided **(7)**. This treatment destroys the filter paper and oxidizes any elemental selenium to Se(IV). After dissolution,‡ boil to remove oxides of nitrogen and some excess nitric acid. Add 25 ml of distilled water and reboil. Boiling to remove oxides of nitrogen is necessary to prevent nitrite ion interference. Analyze the resulting solution for selenium following the prescribed procedures in section **7**.

11.3 The sample collected by an electrostatic precipitator (in sampling atmospheres having high concentrations of selenium-containing particulate matter) is transferred into and digested in the minimum amount of a solution containing conc nitric acid and 60% perchloric acid (10 HNO_3:$HClO_4$). **Cautiously** heat the mixture of solids in acid solution until all reaction has subsided. This treatment destroys the small amount of organic matter present and oxidizes any elemental selenium to Se(IV). After dissolution, follow the procedure in section **11.2** (including footnote).

‡Insolubles may be removed by quantitative filtration using a fritted-glass filter. As much as 25 ml of distilled water may be used for washing. Additional water need *not* then be added before the filtrate is reboiled, see following sentence.

12. References

1. Lott, P.F., P. Cukor, G. Moriber, and J. Solga. 1963. *Anal. Chem.* 35:1159.
2. Parker, C.A., and L.G. Harvey. 1962. *Analyst* 87:558.
3. Hashimoto, Y., and J.W. Winchester. 1967. *Environ. Sci. Technol.* 1:338–340.
4. West, P.W., and Ch. Cimerman. 1964. *Anal. Chem.* 36:2013.
5. Snell, F.D., C.T. Snell, and C.A. Snell. 1959. Colorimetric Methods of Analysis. Volume 11A, p. 678; D. Van Nostrand Company, Inc., New York.
6. Setterlind, A.N. 1947. What's New in Industrial Hygiene. *Am. Ind. Hyg. Assoc. J.* 4:11–14.
7. Kelleher, W.J. and M.J. Johnson. 1961. *Anal. Chem.* 33:1429.

Subcommittee 7
E. C. Tabor, *Chairman*
M. M. Braverman
H. E. Bumstead
A. Carotti
H. M. Donaldson
L. Dubois
R. E. Kupel

311.

Tentative Method of Analysis for Cadmium Content of Atmospheric Particulate Matter by Atomic Absorption Spectroscopy

12110-02-73T

1. Principle of the Method

1.1 Samples can be collected by drawing a known volume of air through membrane or glass fiber filters. The samples are ashed and extracted with a mixture of nitric and hydrochloric acids. The analysis is subsequently made by atomic absorption spectroscopy using the 228.8 nm cadmium line (**1**).

2. Range and Sensitivity

2.1 This method is applicable to the determination of cadmium in an optimum range of 0.05 to 20.0 μg of cadmium/ml of solution. An atmospheric concentration of 0.005 μg of cadmium/m^3 can be detected using the method as described. For this concentration, a minimum air sample volume of 2000 m^3 is recommended for samples taken on 8″ × 10″ filters with a Hi-Vol sampler (see Section **7.1**).

3. Interferences

3.1 No serious interferences have been reported for cadmium **(2)**. Any possible interference due to silica from the glass fiber filters or silicates from the samples can be overcome by allowing the acid extracts to stand overnight and centrifuging at about 2000 RPM for 30 min **(1)**.

4. Precision and Accuracy

4.1 The precision of the method has not been reported using air samples; however, in the determination of cadmium by a nearly identical method, a relative standard deviation of 16% was obtained for urban concentrations **(3)**.

4.2 The recovery by this method is 90% provided the matrix of the sample is not appreciably different from that of the standards **(1)**.

5. Apparatus

5.1 SAMPLE COLLECTION

5.1.1 *Membrane filters* such as Millipore, appropriately sized for sample volume and rate, with a pore size of 0.8 μ, are satisfactory for air sampling.*

5.1.2 *Glass fiber filters* can also be used with a high volume air sampler. The filter blank should be investigated to establish the extent of cadmium contamination and the potential matrix effect.

5.2 GLASSWARE. Borosilicate glassware should be used throughout the analysis. The glassware must be cleaned with 1:1 HCl and rinsed with distilled water. Beakers, pipets and auxiliary glassware will be required.

5.3 EQUIPMENT

5.3.1 *Acetylene Gas (Cylinder)*

5.3.2 *Air Supply*

5.3.3 *Atomic Absorption Spectrophotometer*

5.3.4 *Centrifuge*

5.3.5 *Cadmium Hollow Cathode Lamp*

5.3.6 *Low Temperature Asher-Tracerlab LTA-600 or equivalent*

6. Reagents

6.1 PURITY. ACS reagent grade chemicals or equivalent shall be used in all tests. References to water shall be understood to mean double distilled water or equivalent. Care in selection of reagents and in following the above precautions is essential if low blank values are to be obtained.

6.2 CONC HYDROCHLORIC ACID (36.5–38.0%). sp gr—1.18.

6.3 CONC NITRIC ACID (69.0–71%). sp gr 1.42.

6.4 CADMIUM METAL.

6.5 PERCHLORIC ACID 72%.

6.6 HYDROCHLORIC ACID 1:10. Dilute 100 ml of concentrated acid to 1 l with water.

6.7 PERCHLORIC-NITRIC ACID MIXTURE. Add 10 ml of conc perchloric to 90 ml of conc nitric.

7. Procedure

7.1 SAMPLE PREPARATION

7.1.1 An 8″ × 10″ glass fiber filter, of which a 7″ × 9″ (63 square inches) section is exposed on the high-volume sampler, is divided into sections. The amount of a filter used for the analysis of metals depends upon the type of sample being prepared, urban or non-urban, individual or composite. Table 311:I lists the amount of fil-

Table 311:I. Amount of filter used for each type of sample

Sample	Amount of filter
Urban individual	2.5 × 17.5 cm strip
Urban composite	1.25 × 17.5 cm strip
Non-urban individual	5.0 × 17.5 cm strip
Non-urban composite	2.5 × 17.5 cm strip

*Collection efficiencies greater than 99% are obtained for a 0.3 μ DOP smoke at face velocities from 7.2–53 cm/sec **(4)**.

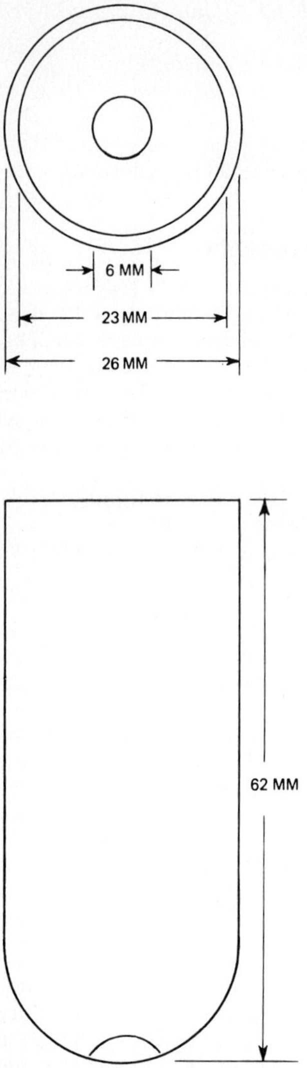

Figure 311:1—Sampler holder for acid extraction of glass fiber filters.

ter used for each type of sample. The strips for metal analysis are ashed in a low-temperature asher. In this procedure, an oxygen plasma is formed by passing oxygen through a high-frequency electromagnetic field and the activated oxygen reacts with the sample (**5**). Slow burning takes place at temperatures ranging from 50 to 250 C. A pyrex boat is placed into each of five compartments of the low-temperature asher and the strips are ashed at approximately 150 C for 1 hr at 250 W at 1 mm chamber pressure with an oxygen flow of 50 cc/min.

The ashed filter is placed in a glass thimble (Figure 311:1), which is then placed in an extraction tube (Figure 311:2). A 125-ml Erlenmeyer flask with a 24/40 female ₸ joint is charged with 8 ml of constant boiling (about 19%) hydrochloric acid and 32 ml of 40% nitric acid. The flask is attached to the extraction tube, and the extraction tube is fitted with an Allihn condenser. The acid is refluxed over the sample for 3 hr. The sample and extraction thimble remain at the temperature of the boiling acid throughout the extraction.

The extraction tube and condenser are removed from the Erlenmeyer flask and the flask is fitted with a thermometer adapter, which serves as a spray retainer. The extracted liquid is concentrated to 1 to 2 ml on a hot plate, and allowed to cool and stand overnight. The concentrated material is quantitatively transferred to a graduated 15-ml centrifuge tube with three washings of 5 to 10 drops 1:10 HCl. Urban samples are diluted to 4.4 ml/1″ strip (40 ml per 63 sq. in. of filter). Non-urban samples are diluted to 3.0 ml/2″ strip (13.3 ml per 63 sq. in. of filter). Following dilution, the samples are centrifuged at 2000 RPM for about 30 min and the supernatant liquid decanted into polypropylene tubes that are then capped and stored pending analysis. One ml from each solution is diluted with 1:10 HCl to 10 ml for atomic absorption analysis.

7.1.2 The membrane filters are wet ashed with 10 to 20 ml of the mixture of nitric and perchloric acid in a suitably sized beaker. In the case of difficulty in oxidizing the organic matter, a first oxidation may be performed with a mixture of 20 ml of nitric acid and 5 ml of perchloric acid, followed by addition of a second 20 ml volume of nitric acid when the volume is first reduced to nearly 5 ml. (**Note:** All safety precautions necessary to the use of perchloric acid must be followed.) After the filter has been completely ashed, the acid is boiled down to the appearance of fumes

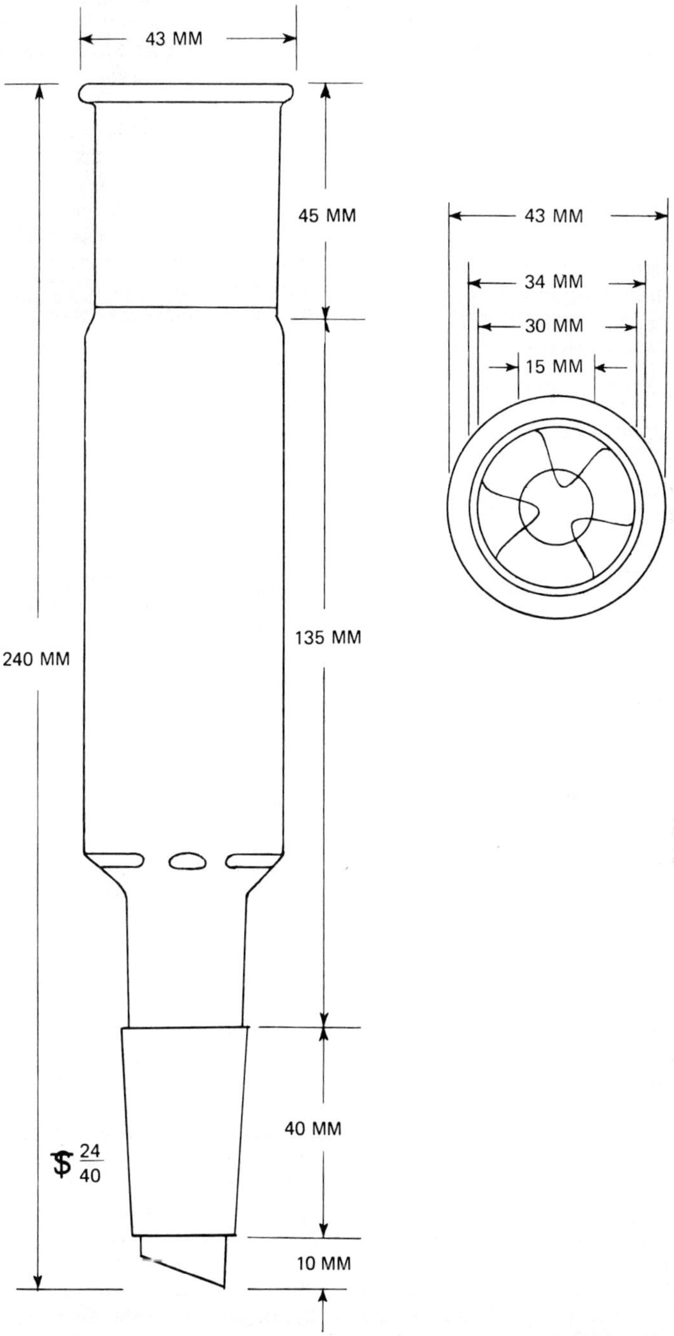

Figure 311:2—Extraction tube adapted for the sample holder.

of perchloric acid. All blanks and standards must be carried through with the same volumes of acid. Transfer to a graduated polypropylene tube with 1:10 HCl, add 1.2 ml of 1:1 HCl, and dilute to 25 ml with 1:10 HCl.

7.2 ANALYSIS

7.2.1 Set the instrument operating conditions as recommended by the manufacturer. The instrument should be set to the wavelength of maximum intensity for the 228.8 nm line from the hollow cathode lamp.

7.2.2 Aspirate the standards prepared fresh daily. Standards must be in the same matrix as the samples. Record the absorbance and prepare a calibration curve as described in section 8.1.†

7.2.3 Aspirate the samples directly into the instrument and record the absorbance for comparison with the standards. Should the absorbance be off scale, dilute an appropriate aliquot to 10 ml. Be certain to aspirate distilled water between each sample. A blank and standard must be aspirated with sufficient frequency to assure the accuracy of the sample determinations.

8. Standards and Calibrations

8.1 STANDARD CADMIUM SOLUTION

8.1.1 *Master Standard (1.000 mg cadmium per ml).* Dissolve 1.000 g of reagent grade cadmium in a minimum volume of a 1:1 nitric-hydrochloric mixture (10 to 25 ml). Upon complete dissolution, evaporate to 5 ml and dilute to 1 liter with 1:1 HCl in a stoppered volumetric flask. This master standard is stable for at least one year stored in a polyethylene bottle.

†All combustion products from the flame must be removed by direct exhaustion through the use of a good, separate ventilation system. This is of particular importance when using a nitrous oxide flame and when aspirating a sample containing a toxic element or a material with toxic combustion products.

8.1.2 *Dilute Standard (0.10 mg cadmium per ml.)* Pipet 10.0 ml of the master standard into a 100-ml volumetric flask and dilute to volume with 1:1 HCl.

8.1.3 *Working Cadmium Standards.* Prepare a 1.00 μg/ml working standard solution by pipetting 1.00 ml of the dilute standard into a 100-ml volumetric flask, adding 4.00 ml of 1:1 HCl and diluting to the mark with 1:10 HCl. Repeat the procedure with the appropriate amounts of the dilute standard and 1:1 HCl (total volume of 5.00 ml) to prepare standards containing 2.0 μg/ml, 3.0 μg/ml, 4.0 μg/ml and 5.0 μg/ml of cadmium. Working standards must be prepared daily.

8.1.4 *Calibrating Sample Preparation Procedure.* Five known amounts of cadmium are added to one inch strips of glass, or an equivalent portion of a membrane filter, and carried through the procedure to establish a calibration graph for the entire sample preparation procedure. If this calibration curve is in agreement with that obtained using the working standards, one need only perform this task periodically to insure that the sample preparation procedure is being carried out properly. If the calibration graph for the sample preparation differs significantly from that obtained with the working standards, then the sample preparation calibration graph should be used to determine the cadmium content of the samples and this calibration should be prepared with each set of samples analyzed.

8.2 BLANK. A blank filter must be carried through the entire procedure.

9. Calculations

9.1 GLASS FIBER FILTER. After a correction for the blank has been made, concentrations are calculated by multiplying the micrograms of cadmium per ml in the sample aliquot by ten and by the total number of ml in the urban or non-urban sample (See **7.1.1**). The result is divided by the number of cubic meters of air represented by the portion of the filter taken for analysis.

9.2 $\mu g\ Cd/m^3 = \dfrac{\mu g\ Cd/ml \text{ in aliquot} \times 10 \times (\text{total ml in urban or non-urban sample})}{(\text{total corrected } m^3 \text{ of air})^{\ddagger}}$

9.3 MEMBRANE FILTER. Make correction for blank. Calculate the concentration as follows (see **7.1.2**).

$$\mu g\ Cd/m^3 = \frac{\mu g\ Cd/ml \text{ in aliquot} \times 25}{(\text{corrected } m^3 \text{ of air for sample})}$$

10. Storage

10.1 Untreated filter samples may be stored indefinitely. Acid extracts may be stored for a month in polypropylene containers.

‡This is that volume of air in m³ represented by the fraction of filter taken for analysis. This volume is corrected to 25 C and 760 torr.

11. References

1. THOMPSON, R.J., G.B. MORGAN and L.J. PURDUE. 1970. Analysis of Selected Elements in Atmospheric Particulate Matter by Atomic Absorption. *Atomic Absorption Newsletter*, 9:53–57.
2. SLAVIN, W. 1968. *Atomic Absorption Spectroscopy*, Interscience Publishers, New York, New York.
3. KNEIP, T.J., M. EISENBUD, C.D. STREHLOW and P.C. FREUDENTHAL. 1970. Airborne Particulates in New York City, *APCA Journal*, 20:144–149.
4. LOCKHART, L.B., JR., R.L. PATTERSON, JR. and W.L. ANDERSON. 1963. Characteristics of Air Filter Media Used for Monitoring Airborne Particulates *NRL Report 6054*.
5. GLEIT, C.E. 1963. *Am. J. of Med. Electronics*, 2:112.

Subcommittee 6
T. J. KNEIP, *Chairman*
R. S. AJEMIAN
J. R. CARLBERG
J. DRISCOLL
J. L. MOYERS
L. KORNREICH
J. W. LOVELAND
R. J. THOMPSON

312.

Tentative Method of Analysis for Chromium Content of Atmospheric Particulate Matter by Atomic Absorption Spectroscopy

12112-01-73T

1. Principle of the Method

1.1 Samples can be collected by drawing a known volume of air through membrane or glass fiber filters. The samples are ashed and extracted with a mixture of nitric and hydrochloric acids. The analysis is subsequently made by atomic absorption spectroscopy using the 357.9 nm chromium line (**1**).

2. Range and Sensitivity

2.1 This method is applicable to the determination of chromium in an optimum range of 0.1 to 20.0 μg of chromium/ml of solution. An atmospheric concentration of 0.005 μg of chromium/m³ can be detected using the method as described. For this concentration, a minimum air sample volume of 2000 m³ is recommended for samples taken on 8″ × 10″ filters with a Hi-Vol sampler (see **7.1**).

3. Interferences

3.1 Silica extracted from the glass fiber filter and from the collected particulate matter may cause a significant interference with the measurement of chromium. This interference can be overcome by allowing the acid extracts to stand overnight and certrifuging at about 2000 RPM for 30 min.

3.2 The absorption of light by chromium may be suppressed by the presence of excessive amounts of iron and nickle **(2)**.

4. Precision and Accuracy

4.1 The precision of the method has not been reported using air samples; however, in the determination of chromium by a nearly identical method, a relative standard deviation of 22% was obtained for urban concentrations **(3)**.

4.2 The recovery by this method is 88% provided the matrix of the sample is not appreciably different from that of the standards **(1)**. Interferences due to the presence of other metals can reduce the accuracy of the method **(2)**.

5. Apparatus

5.1 SAMPLE COLLECTION

5.1.1 *Membrane filters.* Filters such as Millipore, appropriately sized for sample volume and rate, with a pore size of 0.8μ, are satisfactory for air sampling.*

5.1.2 *Glass fiber filters.* These may also be used with a high volume air sampler. The filter blank should be investigated to establish the extent of chromium contamination and the potential matrix effect.

5.2 GLASSWARE. Borosilicate glassware should be used throughout the analysis. The glassware must be cleaned with 1:1 HCl and rinsed with distilled water. Beakers, pipets and auxiliary glassware will be required.

5.3 EQUIPMENT

5.3.1 *Acetylene Gas (Cylinder).*

5.3.2 *Air Supply.*

5.3.3 *Atomic Absorption Spectrophotometer.*

5.3.4 *Centrifuge.*

5.3.5 *Chromium Hollow Cathode Lamp.*

5.3.6 *Low Temperature Asher-Tracerlab LTA-600* (or equivalent).

6. Reagents

6.1 PURITY OF CHEMICALS. ACS reagent grade chemicals or equivalent shall be used in all tests. References to water shall be understood to mean double distilled water or equivalent. Care in selection of reagents and in following the above precautions is essential if low blank values are to be obtained.

6.2 CONC HYDROCHLORIC ACID (36.5–38.0%). sp gr—1.18.

6.3 CONC NITRIC ACID (69.0–71.0%). sp gr—1.42.

6.4 CHROMIUM METAL.

6.5 PERCHLORIC ACID 72%.

6.6 HYDROCHLORIC ACID 1–10. Dilute 100 ml of conc acid to 1 l with water.

6.7 PERCHLORIC-NITRIC ACID MIXTURE. Add 10 ml of conc perchloric to 90 ml of conc nitric.

6.8 CONSTANT BOILING HYDROCHLORIC ACID. About 19% strength.

6.9 NITRIC ACID. 40% strength.

7. Procedure

7.1 SAMPLE PREPARATION.

7.1.1 An 8 × 10″ glass fiber filter (20 × 25 cm), of which a 7 × 9″ (17.5 × 22.5 cm) section is exposed on the high-volume sampler is divided into sections. The amount of a filter used for the analysis of metals depends upon the type of sample being prepared, urban or non-urban, individual or composite. Table 311:I lists the amount of filter used for each type of sample.

*Collection efficiencies greater than 99% are obtained for a 0.3 μ DOP smoke at face velocities from 7.2–53 cm/sec. **(4)**.

The strips for metal analysis are ashed in a low-temperature asher. In this procedure, an oxygen plasma is formed by passing oxygen through a high-frequency electro-magnetic field and the activated oxygen reacts with the sample (**5**). Slow burning takes place at temperatures ranging from 50 to 250 C. A pyrex boat is placed into each of five compartments of the low-temperature asher and the strips are ashed at approximately 150 C for 1 hr at 250 watts at 1 mm chamber pressure with an oxygen flow of 50 cc/min.

The ashed filter is placed in a glass thimble (Figure 211:1), which is then placed in an extraction tube (Figure 311:2). A 125-ml Erlenmeyer flask with a 24/40 female joint is charged with 8 ml of constant boiling (about 19%) hydrochloric acid and 32 ml of 40% nitric acid. The flask is attached to the extraction tube, and the extraction tube is fitted with an Allihn condenser. The acid is refluxed over the sample for 3 hr. The sample and extraction thimble remain at the temperature of the boiling acid throughout the extraction.

The extraction tube and condenser are removed from the Erlenmeyer flask and the flask is fitted with a thermometer adapter, which serves as a spray retainer. The extracted liquid is concentrated to 1 to 2 ml on a hot plate, and allowed to cool and stand overnight. The concentrated material is quantitatively transferred to a graduated 15-ml centrifuge tube with three washings of 5 to 10 drops 1:10 HCl. Urban samples are diluted to 4.4 ml/1″ strip (40 ml/63 sq in. of filter). Non-urban samples are diluted to 3.0 ml/2″ strip (13.3 ml/63 sq. in. of filter). Following dilution, the samples are centrifuged at 2000 RPM for about 30 min and the supernatant liquid decanted into polypropylene tubes that are then capped and stored pending analysis. One ml from each solution is diluted with 1:10 HCl to 10 ml for atomic absorption analysis.

7.1.2 The Millipore filters are wet ashed with 10 to 20 ml of the mixture of nitric and perchloric acid in a suitably sized beaker. In the case of difficulty in oxidizing the organic matter, a first oxidation may be performed with a mixture of 20 ml of nitric acid and 5 ml of perchloric acid, followed by addition of a second 20 ml volume of nitric acid when the volume is first reduced to nearly 5 ml. (**Note:** All safety precautions necessary to the use of perchloric acid must be followed.) After the filter has been completely ashed, the acid is boiled down to the appearance of fumes of perchloric acid. All blanks and standards must be carried through with the same volumes of acid. Transfer to a graduated polypropylene tube with 1:10 HCl, add 1.2 ml of 1:1 HCl, and dilute to 25 ml with 1:10 HCl.

7.2 ANALYSIS.

7.2.1 Set the instrument operating conditions as recommended by the manufacturer. The instrument should be set to the wavelength of maximum intensity for the 357.9 nm line from the hollow cathode lamp.

7.2.2 Aspirate the standards prepared fresh daily. Standards must be in the same matrix as the samples. Record the absorbance and prepare a calibration curve as described in Section **8.1.4**.
(**Note:** All combustion products from the flame must be removed by direct exhaustion through the use of a good, separate ventilation system. This is of particular importance when using a nitrous oxide flame and when aspirating a sample containing a toxic element or a material with toxic combustion products.)

7.2.3 Aspirate the samples directly into the instrument and record the absorbance for comparison with the standards. Should the absorbance be off scale, dilute an appropriate aliquot to 10 ml. Be certain to aspirate distilled water between each sample. A blank and standard must be aspirated with sufficient frequency to assure the accuracy of the sample determinations.

8. Standards and Calibrations

8.1 STANDARD CHROMIUM SOLUTION.

8.1.1 *Master Standard.* (1.0000 mg chromium/ml). Dissolve 1.0000 g of reagent grade chromium in a minimum volume of a 1:1 nitric-hydrochloric mixture (10 to 25 ml). Upon complete dissolution, evaporate to 5 ml and dilute to 1 l with 1:1 HCl in a stoppered volumetric flask. This master standard is stable for at least one year stored in a polyethylene bottle.

8.1.2 *Dilute Standard* (0.100 mg chromium/ml). Pipet 10.0 ml of the master standard into a 100-ml volumetric flask and dilute to volume with 1:1 HCl.

8.1.3 *Working Chromium Standards.* Prepare a 1.00 μg/ml working standard solution by pipetting 1.00 ml of the dilute standard into a 100-ml volumetric flask, adding 4.00 ml of 1:1 HCl and diluting to the mark with 1:10 HCl. Repeat the procedure with the appropriate amounts of the dilute standard and 1:1 HCl (total volume of 5.00 ml) to prepare standards containing 2.0 μg/ml, 3.0 μg/ml, 4.0 μg/ml and 5.0 μg/ml of chromium. Working standards must be prepared daily.

8.1.4 *Calibrating Sample Preparation Procedure.* Five known amounts of chromium are added to one inch strips of glass, or an equivalent portion of a membrane filter, and carried through the procedure to establish a calibration graph for the entire sample preparation procedure. If this calibration curve is in agreement with that obtained using the working standards, one need only perform this task periodically to insure that the sample preparation procedure is being carried out properly. If the calibration graph for the sample preparation differs significantly from that obtained with the working standards, then the sample preparation calibration graph should be used to determine the chromium content of the prepared with each set of samples analyzed.

8.2 BLANK. A blank filter must be carried through the entire procedure.

9. Calculations

9.1 GLASS FIBER FILTER. After a correction for the blank has been made, concentrations are calculated by multiplying the μg of chromium per ml in the sample aliquot by ten and by the total number of ml in the **7.1.1**). The result is divided by the number of m³ of air represented by the portion of the filter taken for analysis.

9.2
$$\mu g\ Cr/m^3 = \frac{\mu g\ Cr/ml \text{ in aliquot} \times 10 \times (\text{total ml in urban or non-urban sample})}{(\text{total corrected } m^3 \text{ of air})\dagger}$$

9.3 MEMBRANE FILTER. Make correction for blank. Calculate the concentration as follows (see **7.1.2**).

$$\mu g\ Cr/m^3 = \frac{\mu g\ Cr/ml \text{ in aliquot} \times 25}{(\text{corrected } m^3 \text{ of air for sample})}$$

10. Storage

10.1 Untreated filter samples may be stored indefinitely. Acid extracts may be stored for a month in polypropylene containers.

11. References

1. THOMPSON, R.J., G.B. MORGAN, and L.J. PURDUE. 1970. Analysis of Selected Elements in Atmospheric Particulate Matter by Atomic Absorption. *Atomic Absorption Newsletter*, 9:53–57.
2. SLAVIN, W. 1968. *Atomic Absorption Spectroscopy*, Interscience Publishers, New York, New York.
3. KNEIP, T.J., M. EISENBUD, C.D. STREHLOW, and P.C. FREUDENTHAL. 1970. Airborne Particulates in New York City. *APCA Journal*, 20:144–149.
4. LOCKHART, L.B., JR., R.L. PATTERSON, JR., and W.L. ANDERSON. 1963. Characteristics of Air Filter Media Used for Monitoring Airborne Particulates. *NRL Report 6054*, Dec.
5. CLEIT, C.E. 1963. *Am. J. of Med. Electronics*, 2:112.

Subcommittee 6
T. J. KNEIP, *Chairman*
R. S. AJEMIAN
J. R. CARLBERG
J. DRISCOLL
L. KORNREICH
J. W. LOVELAND
J. L. MOYERS
R. J. THOMPSON

†This is that volume of air in m³ represented by the fraction of filter taken for analysis. This volume is corrected to 25 C and 760 torr.

313.

Tentative Method of Analysis for Copper Content of Atmospheric Particulate Matter by Atomic Absorption Spectroscopy

12114-01-73T

1. Principle of the Method

1.1 Samples can be collected by drawing a known volume of air through membrane or glass fiber filters. The samples are ashed and extracted with a mixture of nitric and hydrochloric acids. The analysis is subsequently made by atomic absorption spectroscopy using the 324.7 nm copper line **(1)**.

2. Range and Sensitivity

2.1 This method is applicable to the determination of copper in quantities of 0.05 to 20.0 μg of copper/ml of solution. An atmospheric concentration of 0.005 $\mu g/m^3$ can be detected using the method as described. For this concentration, a minimum sample volume of 2000 m^3 of air is recommended. (See **7.1**)

3. Interferences

3.1 Silica extracted from the glass fiber filter and from the collected particulate matter may cause a significant interference with the measurement of copper. This interference can be overcome by centrifuging the acid extracts. Standards should be prepared in an acid extract of a blank glass fiber filter to minimize the matrix effect of silica.

3.2 A significant copper blank has been observed when hi-vol motors using brushes and copper comutators are used and the exhaust from the system is not prevented from reentering the filter **(5)**.

4. Precision and Accuracy

4.1 The precision of the method has not been reported using air samples; however, in the determination of copper by a nearly identical method, a relative standard deviation of 9% was obtained at normal urban concentrations **(2)**.

4.2 The recovery by this method is 92% provided the matrix of the sample is not appreciably different from that of the standards **(1)**. Interference due to the presence of other metals can reduce the accuracy of the method **(3)**.

5. Apparatus

5.1 Sample Collection. The fraction of the filter used in the analysis should be equivalent to the minimum air volume indicated as required in section **2.1**.

5.1.1 *Membrane filters*, such as Millipore, appropriately sized for sample volume and rate, with a pore size of 0.8 μ, are satisfactory for air sampling.

5.1.2 *Glass fiber filters*. These may also be used with a high volume air sampler. The filter blank should be investigated to establish the extent of copper contamination and the potential matrix effect.

5.2 Glassware. Borosilicate glassware should be used throughout the analysis. The glassware must be cleaned with 1:1 HCl and rinsed with distilled water. Beakers, pipettes and auxiliary glassware will be required.

5.3 Equipment.

5.3.1 *Acetylene Gas (Cylinder)*

5.3.2 *Air Supply*

5.3.3 *Atomic Absorption Spectrophotometer*

5.3.4 *Centrifuge*

5.3.5 *Copper Hollow Cathode Lamp*

5.3.6 *Low Temperature Asher-Tracerlab LTA-600 or equivalent*

6. Reagents

6.1 PURITY. ACS reagent grade chemicals or equivalent shall be used in all tests. References to water shall be understood to mean double distilled water or equivalent. Care in selection of reagents and in following the above precautions is essential if low blank values are to be obtained.

6.2 CONCENTRATED HYDROCHLORIC ACID (36.5–38.0%). sp gr—1.18.

6.3 CONCENTRATED NITRIC ACID (69.0–71.0%). sp gr—1.42.

6.4 COPPER.

6.5 PERCHLORIC ACID. 72%

6.6 HYDROCHLORIC ACID 1:10. Dilute 100 ml of conc acid to one liter with water.

6.7 PERCHLORIC-NITRIC ACID MIXTURE. Add 10 ml of conc perchloric to 90 ml of conc nitric.

7. Procedure

7.1 SAMPLE PREPARATION.

7.1.1 An 8″ × 10″ glass fiber, of which a 7″ × 9″ (63 square inches) section is exposed on the high-volume sampler, is divided into sections. The amount of a filter used for the analysis of metals depends upon the type of sample being prepared, urban or nonurban, individual or composite. Table 311:I lists the amount of filter used for each type of sample.

The strips for metal analysis are ashed in a low-temperature asher. In this procedure, an oxygen plasma is formed by passing oxygen through a high-frequency electro-magnetic field and the activated oxygen reacts with the sample (**4**). Slow burning takes place at temperatures ranging from 50 to 250 C. A pyrex boat holding the filter section is placed into each of five compartments of the low temperature asher, and the strips are ashed at approximately 150 C for 1 hr at 250 W at 1 mm chamber pressure with an oxygen flow of 50 cc per min.

The ashed filter is placed in a glass thimble (Figure 311:1), which is then placed in an extraction tube (Figure 311:2). A 125-ml Erlenmeyer flask with a 24/40 female T joint is charged with 8 ml of constant boiling (about 19%) hydrochloric acid and 32 ml of 40% nitric acid. The flask is attached to the extraction tube, and the extraction tube is fitted with an Alihn condenser. The acid is refluxed over the sample for 3 hr. The sample and extraction thimble remain at the temperature of the boiling acid throughout the extraction.

The extraction tube and condenser are removed from the Erlenmeyer flask, and the flask is fitted with a thermometer adapter, which serves as a spray retainer. The extracted liquid is concentrated to 1 to 2 ml on a hot plate, and allowed to cool and stand overnight. The concentrated material is quantitatively transferred to a graduated 15-ml centrifuge tube with three washings of 5 to 10 drops of 1:10 HCl. Urban samples are diluted to 4.4 ml/1″ strip (40 ml per 63 sq. in. of filter). Non-urban samples are diluted to 3.0 ml/2″ strip (13.3 ml per 63 sq. in. of filter). Following dilution, the samples are centrifuged at 2000 rpm for 30 min and the supernatant liquid decanted into polypropylene tubes that are then capped and stored pending analysis. One ml from each solution is diluted with 1:10 HCl to 10 ml for atomic absorption analysis.

7.1.2 The Millipore filters are wet ashed with 10 to 20 ml of the mixture of nitric and perchloric acid in a suitably sized beaker. In the case of difficulty in oxidizing the organic matter, a first oxidation may be performed with a mixture of 20 ml of nitric acid and 5 ml of perchloric acid, followed by addition of a second 20 ml volume of nitric acid when the volume is first reduced to nearly 5 ml. (**Note:** All safety precautions necessary to the use of perchloric acid must be followed.) After the filter has been completely ashed, the acid is boiled down to the appearance of fumes of perchloric acid. All blanks and standards must be carried through with the same volumes of acid. Transfer to a graduated polypropylene tube with 1:10 HCl, add 1.2 of 1:1 HCl, and dilute to 25 ml with 1:10 HCl.

7.2 ANALYSIS.

7.2.1 Set the instrument operating conditions as recommended by the manufacturer. The instrument should be set to the wavelength of maximum intensity for the 324.7 nm line from the hollow cathode lamp.

7.2.2 Aspirate the standards prepared fresh daily. Standards must be in the same matrix as the samples. Record the absorbance and prepare a calibration curve as described in section **8.1**.

(**Note:** All combustion products from the flame must be removed by direct exhaustion through the use of a good, separate ventilation system. This is of particular importance when using a nitrous oxide flame and when aspirating a sample containing a toxic element or a material with toxic combustion products.)

7.2.3 Aspirate the samples directly into the instrument and record the absorbance for comparison with the standards. Should the absorbance be off scale, dilute an appropriate aliquot to 10 ml. Be certain to aspirate distilled water between each sample. A blank and standard must be aspirated with sufficient frequency to assure the accuracy of the sample determinations.

8. Standards and Calibration

8.1 STANDARD COPPER SOLUTION

8.1.1 *Master Standard (1.000 mg copper per ml).* Dissolve 1.000 g of reagent grade copper in a 1:1 nitric-hydrochloric mixture (10 to 25 ml). Upon complete dissolution evaporate to 5 ml and dilute to 1 l with 1:1 HCl in a stoppered volumetric flask. This master standard is stable for one year and can be stored in a polyethylene bottle.

8.1.2 *Dilute Standard (0.010 mg copper per ml).* Pipet 10 ml of the master standard into a 1-l volumetric flask and dilute to volume with 1:10 HCl.

8.1.3 *Working Copper Standards.* Prepare a 1.00 μg/ml working standard solution by pipetting 10.0 ml of the dilute standard into a 100-ml volumetric flask and diluting to the mark with 1:10 hydrochloric acid. Five known amounts of copper ranging from 1.0 to 5.0 μg are added to the filters (membrane or glass) and carried through the procedure to establish a calibration graph. The calibration should be performed with each group of samples analyzed. The calibration graph must be a straight line. The working standard should be prepared fresh daily.

8.2 BLANK. A blank filter must be carried through the entire procedure.

9. Calculations

9.1 GLASS FIBER FILTERS. After a correction for the blank has been made, concentrations are calculated by multiplying the μg of copper per ml in the sample aliquot by 10 and by the total number of ml in the urban or non-urban sample (see **7.1.1**). The result is divided by the number of m^3 of air represented by the portion of the filter taken for analysis.

9.2 $$\mu g\ Cu/m^3 = \frac{\mu g\ Cu/ml\ in\ aliquot \times 10 \times (total\ ml\ in\ urban\ or\ non\text{-}urban\ sample)}{(total\ corrected\ m^3\ of\ air)^*}$$

9.3 MEMBRANE FILTER. Make correction for blank. Calculate the conc as follows (see **7.1.2**).

$$\mu g\ Cu/m^3 = \frac{\mu g\ Cu/ml\ in\ aliquot \times 25}{(corrected\ m^3\ of\ air\ for\ sample)}$$

10. Storage

10.1 Untreated filter samples may be stored indefinitely in polyethylene bottles. Acid extracts may be stored for a month in polypropylene containers.

11. References

1. THOMPSON, R.J., G. MORGAN and L. PURDUE. 1970. Analysis of Selected Elements in Atmospheric Particulate Matter by Atomic Absorption. *Atomic Absorption Newsletter*, 9:53–57.

---

*25 C and 760 torr. The total volume of air in cubic meters is for that fraction of the filter taken.

2. Kneip, T.J., M. Eisenbud, C.D. Strehlow and P.C. Freudenthal. 1970. Airborne Particulates in New York City. *APCA Journal*, 20:144–149.

3. Slavin, W. 1968. *Atomic Absorption Spectroscopy. Interscience Publishers*, New York, New York.

4. Gleit, C.E. 1963. *Am. J. of Med. Electronics*, 2:112.

5. Hoffman, G.L., and R.A. Duce. 1971. Copper Contamination of Atmospheric Particulate Samples Collected with Gelman Hurricane Air Samplers. *Environ. Sci. & Technology*, 5:1134–1136.

Subcommittee 6
T. J. Kneip, *Chairman*
R. S. Ajemian
J. R. Carlberg
J. Driscoll
J. L. Moyers
L. Kornreich
J. W. Loveland
R. J. Thompson

314.

Tentative Method of Analysis for Iron Content of Atmospheric Particulate Matter by Atomic Absorption Spectroscopy

12126-03-73T

1. Principle of the Method

1.1 Samples can be collected by drawing a known volume of air through membrane or glass fiber filters. The samples are ashed and extracted with a mixture of nitric and hydrochloric acids. The analysis is subsequently made by atomic absorption spectroscopy using the 248.3 nm iron line (**1, 2**).

2. Range and Sensitivity

2.1 This method is applicable to the determination of iron in an optimum range of 0.1 to 20.0 μg of iron/ml of solution. An atmospheric concentration of 0.005 μg of iron/cubic meter can be detected using the method as described. For this concentration, a minimum air sample volume of 2000 m^3 is recommended for samples taken on 8″ × 10″ filters with a Hi-Vol sampler (see Section **7.1**).

3. Interferences

3.1 Silica extracted from the glass fiber filter and from the collected particulate matter may cause a significant interference with the measurement of iron. This interference can be overcome by allowing the acid extracts to stand overnight and centrifuging at about 2000 RPM for 30 min (**1**).

4. Precision and Accuracy

4.1 The precision of the method has not been reported using air samples.

4.2 The recovery of this method is 89% provided the matrix of the sample is not appreciably different from that of the standards (**1**).

5. Apparatus

5.1 Sample Collection.

5.1.1 *Membrane filters* such as Millipore, appropriately sized for sample volume and rate, with a pore size of 0.8 μ, are satisfactory for air sampling.*

5.1.2 *Glass fiber filters* can also be used with a high volume air sampler. The filter blank should be investigated to es-

*Collection efficiencies greater than 99% are obtained for a 0.3 μ DOP smoke at face velocities from 7.2 to 53 cm/sec. (**4**).

tablish the extent of iron contamination and the potential matrix effect.

5.2 GLASSWARE. Borosilicate glassware should be used throughout the analysis. The glassware must be cleaned with 1:1 HCl and rinsed with distilled water. Beakers, pipets and auxiliary glassware will be required.

5.3 EQUIPMENT.

5.3.1 *Acetylene Gas (Cylinder).*

5.3.2 *Air Supply.*

5.3.3 *Atomic Absorption Spectrophotometer.*

5.3.4 *Centrifuge.*

5.3.5 *Iron Hollow Cathode Lamp.*

5.3.6 *Low Temperature Asher-Tracerlab LTA-600 or equivalent.*

6. Reagents

6.1 PURITY. ACS reagent grade chemicals or equivalent shall be used in all tests. References to water shall be understood to mean double distilled water or equivalent. Care in selection of reagents and in following the above precautions is essential if low blank values are to be obtained.

6.2 CONC HYDROCHLORIC ACID (36.5–38.0%) SP GR 1.18.

6.3 CONC NITRIC ACID (69.0–71.0%). SP GR 1.42.

6.4 IRON WIRE.

6.5 PERCHLORIC ACID. 72%.

6.6 HYDROCHLORIC ACID 1:10. Dilute 100 ml of conc acid to 1 l with water.

6.7 PERCHLORIC-NITRIC ACID MIXTURE. Add 10 ml of conc perchloric to 90 ml of conc nitric.

7. Procedure

7.1 SAMPLE PREPARATION.

7.1.1 An 8″ × 10″ glass fiber filter, of which a 7″ × 9″ (63 square inches) section is exposed on the high-volume sampler, is divided into sections. The amount of a filter used for the analysis of metals depends upon the type of sample being prepared, urban or non-urban, individual or composite. Table 311:I lists the amount of filter used for each type of sample.

The strips for metal analysis are ashed in a low-temperature asher. In this procedure, an oxygen plasma is formed by passing oxygen through a high-frequency electro-magnetic field and the activated oxygen reacts with the sample (**5**). Slow burning takes place at temperatures ranging from 50 to 250 C. A pyrex boat is placed into each of five compartments of the low-temperature asher and the strips are ashed at approximately 150 C for 1 hr at 250 W at 1 mm chamber pressure with an oxygen flow of 50 cc per min.

The ashed filter is placed in a glass thimble (Figure 311:1), which is then placed in an extraction tube (Figure 311:2). A 125-ml Erlenmeyer flask with a $^{24}/_{40}$ female ₮ joint is charged with 8 ml of constant boiling (about 19%) hydrochloric acid and 32 ml of 40% nitric acid. The flask is attached to the extraction tube, and the extraction tube is fitted with an Allihn condenser. The acid is refluxed over the sample for 3 hours. The sample and extraction thimble remain at the temperature of the boiling acid throughout the extraction.

The extraction tube and condenser are removed from the Erlenmeyer flask and the flask is fitted with a thermometer adapter, which serves as a spray retainer. The extracted liquid is concentrated to 1 to 2 ml on a hot plate, and allowed to cool and stand overnight. The concentrated material is quantitatively transferred to a graduated 15-ml centrifuge tube with three washings of 5 to 10 drops 1:10 HCl. Urban samples are diluted to 4.4 ml per 1″ strip (40 ml/63 sq. in of filter). Non-urban samples are diluted to 3.0 ml/2″ strip (13.3 ml per 63 sq. in of filter). Following dilution, the samples are centrifuged at 2000 RPM for about 30 min and the supernatant liquid decanted into polypropylene tubes that are then capped and stored pending analysis. One ml from each solution is diluted with 1:10 HCl to 10 ml for atomic absorption analysis.

7.1.2 The membrane filters are wet ashed with 10 to 20 ml of the mixture of nitric and perchloric acid in a suitably sized beaker. In the case of difficulty in oxidizing the organic matter, a first oxidation may be performed with a mixture of 20 ml of nitric acid and 5 ml of perchloric acid, followed by addition of a second 20 ml volume of nitric acid when the volume is first reduced to nearly 5 ml. (**Note**: All safety precautions necessary to the use of perchloric acid must be followed.) After the filter has been completely ashed, the acid is boiled down to the appearance of fumes of perchloric acid. All blanks and standards must be carried through with the same volumes of acid. Transfer to a graduated polypropylene tube with 1:10 HCl, add 1.2 ml of 1:1 HCl, and dilute to 25 ml with 1:10 HCl.

7.2 ANALYSIS.

7.2.1 Set the instrument operating conditions as recommended by the manufacturer. The instrument should be set to the wavelength of maximum intensity for the 248.3 nm line from the hollow cathode lamp.

7.2.2 Aspirate the standards prepared fresh daily. Standards must be in the same matrix as the samples. Record the absorbance and prepare a calibration curve as described in section **8.1**.†

7.2.3 Aspirate the samples directly into the instrument and record the absorbance for comparison with the standards. Should the absorbance be off scale, dilute an appropriate aliquot to 10 ml. Be certain to aspirate distilled water between each sample. A blank and standard must be aspirated with sufficient frequency to assure the accuracy of the sample determinations.

8. Standards and Calibration

8.1 STANDARD IRON SOLUTION.

8.1.1 *Master Standard (1.0000 mg iron per ml.)* Dissolve 1.0000 g of reagent grade iron wire in a minimum volume of a 1:1 nitric-hydrochloric mixture (10 to 25 ml). Upon complete dissolution, evaporate to 5 ml and dilute to 1 l with 1:1 HCl in a stoppered volumetric flask. This master standard is stable for at least one year stored in a polyethylene bottle.

8.1.2 *Dilute Standard (0.100 mg iron per ml.)* Pipet 10.0 ml of the master standard into a 100-ml volumetric flask and dilute to volume with 1:1 HCl.

8.1.3 *Working Iron Standards.* Prepare a 1.00 μg/ml working standard solution by pipetting 1.00 ml of the dilute standard into a 100-ml volumetric flask, adding 4.00 ml of 1:1 HCl and diluting to the mark with 1:10 HCl. Repeat the procedure with the appropriate amounts of the dilute standard and 1:1 HCl (total volume of 5.00 ml) to prepare standards containing 2.0 μg per ml, 3.0 μg per ml, 4.0 μg per ml and 5.0 μg per ml of iron. Working standards must be prepared daily.

8.1.4 *Calibrating Sample Preparation Procedure.* Five known amounts of iron are added to 1″ strips of glass, or an equivalent portion of a membrane filter, and carried through the procedure to establish a calibration graph for the entire sample preparation procedure. If this calibration curve is in agreement with that obtained using the working standards, one need only perform this task periodically to insure that the sample preparation procedure is being carried out properly. If the calibration graph for the sample preparation differs significantly from that obtained with the working standards, then the sample preparation calibration graph should be used to determine the iron content of the samples and this calibration should be prepared with each set of samples analyzed.

8.2 BLANK. A blank filter must be carried through the entire procedure.

9. Calculation

9.1 GLASS FIBER FILTER. After a correction for the blank has been made, concentrations are calculated by multiplying the μg of iron per ml in the sample aliquot by 10 and by the total number of ml in the ur-

†All combustion products from the flame must be removed by direct exhaustion through the use of a good, separate ventilation system. This is of particular importance when using a nitrous oxide flame and when aspirating a sample containing a toxic element or a material with toxic combustion products.

ban or non-urban sample (See **7.1.1**). The result is divided by the number of m^3 of air represented by the portion of the filter taken for analysis.

9.2 $\mu g\ Fe/m^3 =$

$$\frac{\mu g\ Fe/ml \text{ in aliquot} \times 10 \times (\text{total ml in urban or non-urban sample})}{(\text{total corrected } m^3 \text{ of air})‡}$$

9.3 MEMBRANE FILTER. Make correction for blank. Calculate the concentration as follows (see **7.1.2**).

$\mu g\ Fe/m^3 =$

$$\frac{\mu g\ Fe/ml \text{ in aliquot} \times 25}{(\text{corrected } m^3 \text{ of air for sample})}$$

10. Storage

10.1 Untreated filter samples may be stored indefinitely. Acid extracts may be stored for a month in polypropylene containers.

‡This is that volume of air in m^3 represented by the fraction of filter taken for analysis. This volume is corrected to 25 C and 760 Torr.

11. References

1. THOMPSON, R.J., G.B. MORGAN and L.J. PURDUE. 1970. Analysis of Selected Elements in Atmospheric Particulate Matter by Atomic Absorption. *Atomic Absorption Newsletter*, 9:53–57.
2. SLAVIN, W. 1968. Atomic Absorption Spectroscopy. Inter-Science Publishers, New York, New York.
3. KNEIP, T.J., M. EISENBUD, C.D. STREHLOW and P.C. FREUDENTHAL. 1970. Airborne Particulates in New York City. *APCA Journal*, 20:144–149.
4. LOCKHART, L.B., JR., R.L. PATTERSON, JR. and W.L. ANDERSON. Characteristics of Air Filter Media Used for Monitoring Airborne Particulates. NRL Report 6054.
5. GLEIT, C.E. 1963. *Am. J. of Med. Electronics*, 2:112.

Subcommittee 6
T. J. KNEIP, *Chairman*
R. S. AJEMIAN
J. R. CARLBERG
J. DRISCOLL
L. KORNREICH
J. W. LOVELAND
J. L. MOYERS
R. J. THOMPSON

315.

Tentative Method of Analysis for Lead Content of Atmospheric Particulate Matter by Atomic Absorption Spectroscopy

12128-02-73T

1. Principle of the Method

1.1 Samples can be collected by drawing a known volume of air through membrane or glass fiber filters. The samples are ashed and extracted with a mixture of nitric and hydrochloric acids. The analysis is subsequently made by atomic absorption spectroscopy using the 217.0 nm lead line (**1, 2**).

2. Range and Sensitivity

2.1 This method is applicable to the determination of lead in an optimum range of 0.1 to 20.0 μg of lead/ml of solution. An atmospheric concentration of 0.01 μg of lead/m^3 can be determined using the method as described. For this concentration, a minimum sample volume of 2000 m^3 is recommended for samples taken with a Hi-

Vol Sampler on 8″ × 10″ filters (see Section **7.1**).

3. Interferences

3.1 No serious interferences have been reported for lead (**2**). Any possible interference due to silica from the glass fiber filters or silicates from the samples can be overcome by allowing the acid extracts to stand overnight and centrifuging at about 2000 RPM for 30 min.

3.2 A sulfate interference has been reported (**3**). As sulfate is a potential component of the samples and may vary with location, the presence of this interference must be determined and accounted for if found.

4. Precision and Accuracy

4.1 The precision of the method has not been reported using air samples; however, in the determination of lead by a nearly identical method, a relative standard deviation of 10% was obtained for urban concentrations (**4**).

4.2 The recovery by this method is 82% provided the matrix of the sample is not appreciably different from that of the standards (**1**).

5. Apparatus

5.1 SAMPLE COLLECTION.

5.1.1 *Membrane filters* such as Millipore, appropriately sized for sample volume and rate, with a pore size of 0.8 μ, are satisfactory for air sampling.*

5.1.2 *Glass fiber filters* can also be used with a high volume air sampler. The filter blank should be investigated to establish the extent of lead contamination and the potential matrix effect.

5.2 GLASSWARE. Borosilicate glassware should be used throughout the analysis. The glassware must be cleaned with 1:1 HCl and rinsed with distilled water. Beakers, pipet and auxiliary glassware will be required.

5.3 EQUIPMENT

5.3.1 *Acetylene Gas (Cylinder).*

5.3.2 *Air Supply.*

5.3.3 *Atomic Absorption Spectrophotometer.*

5.3.4 *Centrifuge.*

5.3.5 *Lead Hollow Cathode Lamp.*

5.3.6 *Low Temperature Asher-Tracer lab LTA-600 or equivalent.*

6. Reagents

6.1 PURITY. ACS reagent grade chemicals or equivalent shall be used in all tests. References to water shall be understood to mean double distilled water or equivalent. Care in selection of reagents and in following the above precautions is essential if low blank values are to be obtained.

6.2 CONC HYDROCHLORIC ACID (36.5–38.0%). SP GR 1.18.

6.3 CONC NITRIC ACID (69.0–71.0%). SP GR 1.42.

6.4 LEAD NITRATE [$Pb(NO_3)_2$].

6.5 HYDROCHLORIC ACID 1–10. Dilute 100 ml of conc acid to one liter with water.

6.6 PERCHLORIC-NITRIC ACID MIXTURE. Add 10 ml of conc perchloric to 90 ml of conc nitric acid.

7. Procedure

7.1 SAMPLE PREPARATION.

7.1.1 An 8- by 10-″ glass fiber filter, of which a 7- by 9-″ (63 square inches) section is exposed on the high-volume sampler is divided into sections. The amount of a filter used for the analysis of metals depends upon the type of sample being prepared, urban or non-urban, individual or composite, Table 311:I lists the amount of filter used for each type of sample.

The strips for metal analysis are ashed in a low-temperature asher. In this procedure, an oxygen plasma is formed by passing oxygen through a high-frequency electro-magnetic field and the activated oxygen reacts with the sample (**5**). Slow

*Collection efficiencies greater than 99% are obtained for a 0.3 μ DOP smoke at face velocities from 7.2 to 53 cm/sec. (**6**)

burning takes place at temperatures ranging from 50 to 250 C. A pyrex boat is placed into each of five compartments of the low-temperature asher and the strips are ashed at approximately 150 C or 1 hr at 250 watts at 1 mm chamber pressure with an oxygen flow of 50 cc per min.

The ashed filter is placed in a glass thimble (Figure 311:1), which is then placed in an extraction tube (Figure 311:2). A 125-ml Erlenmeyer flask with a $^{24}/_{40}$ female ₸ joint is charged with 8 ml of constant boiling (about 19%) hydrochloric acid and 32 ml of 40% nitric acid. The flask is attached to the extraction tube, and the extraction tube is fitted with an Allihn condenser. The acid is refluxed over the sample for 3 hr. The sample and extraction thimble remain at the temperature of the boiling acid throughout the extraction.

The extraction tube and condenser are removed from the Erlenmeyer flask and the flask is fitted with a thermometer adapter, which serves as a spray retainer. The extracted liquid is concentrated to 1 to 2 ml on a hot plate, and allowed to cool and stand overnight. The concentrated material is quantitatively transferred to a graduated 15-ml centrifuge tube with three washings of 5-10 drops 1:10 HCl. Urban samples are diluted to 4.4 ml per 1″ strip (40 ml per 63 sq. in. of filter). Non-urban samples are diluted to 3.0 ml per 2″ strip (13.3 ml per 63 sq. in. of filter). Following dilution, the samples are centrifuged at 2000 RPM for about 30 min and the supernatant liquid decanted into polypropylene tubes that are then capped and stored pending analysis. One ml from each solution is diluted with 1:10 HCl to 10 ml for atomic absorption analysis.

7.1.2 The membrane filters are wet ashed with 10 to 20 ml of the mixture of nitric and perchloric acid in a suitably sized beaker. In the case of difficulty in oxidizing the organic matter, a first oxidation may be performed with a mixture of 20 ml of nitric acid and 5 ml of perchloric acid, followed by addition of a second 20 ml volume of nitric acid when the volume is first reduced to nearly 5 ml. (**Note:** All safety precautions necessary to the use of perchloric acid must be followed.) After the filter has been completely ashed, the acid is boiled down to the appearance of fumes of perchloric acid. All blanks and standards must be carried through with the same volumes of acid. Transfer to a graduated polypropylene tube with 1:10 HCl, add 1.2 ml of 1:1 HCl, and dilute to 25 ml with 1:10 HCl.

7.2 ANALYSIS.

7.2.1 Set the instrument operating conditions as recommended by the manufacturer. The instrument should be set to the wavelength of maximum intensity for the 217.0 nm line from the hollow cathode lamp.

7.2.2 Aspirate the standards prepared fresh daily. Standards must be in the same matrix as the samples. Record the absorbance and prepare a calibration curve as described in section **8.1**.†

7.2.3 Aspirate the samples directly into the instrument and record the absorbance for comparison with the standards. Should the absorbance be off scale, dilute an appropriate aliquot to 10 ml. Be certain to aspirate distilled water between each sample. A blank and standard must be aspirated with sufficient frequency to assure the accuracy of the sample determinations.

8. Standards and Calibrations

8.1 STANDARD LEAD SOLUTION.

8.1.1 *Master Standard (1.000 mg lead per ml).* Dissolve 1.598 g of reagent grade lead nitrate [$Pb(NO_3)_2$] in a minimum volume of a 1:1 nitric-hydrochloric mixture (10 to 25 ml). Upon complete dissolution, evaporate to 5 ml and dilute to 1 l with 1:1 HCl in a stoppered volumetric flask. This master standard is stable for at least one year stored in a polyethylene bottle.

†All combustion products from the flame must be removed by direct exhaustion through the use of a good, separate ventilation system. This is of particular importance when using a nitrous oxide flame and when aspirating a sample containing a toxic element or a material with toxic combustion products.

8.1.2 *Dilute Standard (0.10 mg lead per ml).* Pipet 10.0 ml of the master standard into a 100-ml volumetric flask and dilute to volume with 1:1 HCl.

8.1.3 *Working Lead Standards.* Prepare a 1.00 μg/ml working standard solution by pipetting 1.00 ml of the dilute standard into a 100-ml volumetric flask, adding 4.00 ml of 1:1 HCl and diluting to the mark with 1:10 HCl. Repeat the procedure with the appropriate amounts of the dilute standard and 1:1 HCl (total volume of 5.00 ml) to prepare standards containing 2.0 μg per ml, 3.0 μg per ml, 4.0 μg per ml and 5.0 μg per ml of lead. Working standards must be prepared daily.

8.1.4 *Calibrating Sample Preparation Procedure.* Five known amounts of lead are added to one inch strips of glass, or an equivalent portion of a membrane filter, and carried through the procedure to establish a calibration graph for the entire sample preparation procedure. If this calibration curve is in agreement with that obtained using the working standards, one need only perform this task periodically to insure that the sample preparation procedure is being carried out properly. If the calibration graph for the sample preparation differs significantly from that obtained with the working standards, then the sample preparation calibration graph should be used to determine the lead content of the samples and this calibration should be prepared with each set of samples analyzed.

8.2 BLANK. A blank filter must be carried through the entire procedure.

9. Calculations

9.1 GLASS FIBER FILTER. After a correction for the blank has been made, concentrations are calculated by multiplying the micrograms of lead per ml in the same aliquot by ten and by the total number of ml in the urban or non-urban sample (See **7.1.1**). The result is divided by the number of cubic meters of air represented by the portion of the filter taken for analysis.

9.2

$$\mu g\ Pb/m^3 = \frac{\mu g\ Pb/ml \text{ in aliquot} \times 10 \times (\text{total ml in urban or non-urban sample})}{(\text{total corrected } m^3 \text{ of air})\ddagger}$$

9.3 MILLIPORE FILTER. Make correction for blank. Calculate the concentration as follows (see **7.1.2**).

$$\mu g\ Pb/m^3 = \frac{\mu g\ Pb/ml \text{ in aliquot} \times 25}{(\text{corrected } m^3 \text{ of air for sample})}$$

10. Storage

10.1 Untreated filter samples may be stored indefinitely. Acid extracts may be stored for a month in polypropylene containers.

11. References

1. THOMPSON, R.J., G.B. MORGAN and L.J. PURDUE. 1970. Analysis of Selected Elements in Atmospheric Particulate Matter by Atomic Absorption. Atomic Absorption Newsletter, 9:53–57.
2. SLAVIN, W. 1968. Atomic Absorption Spectroscopy, Inter-Science Publishers, New York, New York.
3. HWANG, J.Y. 1971. Lead Analysis in Air Particulate Samples by Atomic Absorption Spectrometry. Canadian Spectroscopy, 16:1–4, March.
4. KNEIP, T.J., M. EISENBUD, C.D. STREHLOW and P.C. FREUDENTHAL. 1970. Airborne Particulates in New York City, APCA Journal, 20:144–149, March.
5. GLEIT, C,E. 1963. Am. J. of Med. Electronics, 2:112.
6. LOCKHART, L.B., JR., R.L. PATTERSON, JR. and W.L. ANDERSON. 1963. Characteristics of Air Filter Media Used for Monitoring Particulates, NRL Report 6054 Dec.

Subcommittee 6
T. J. KNEIP, *Chairman*
R. S. AJEMIAN
J. R. CARLBERG
J. DRISCOLL
L. KORNREICH
J. W. LOVELAND
J. L. MOYERS
R. J. THOMPSON

‡This is that volume of air in m³ represented by the fraction of filter taken for analysis. This volume is corrected to 25 C and 760 torr.

316.

Tentative Method of Analysis for Manganese Content of Atmospheric Particulate Matter by Atomic Absorption Spectroscopy

12132-02-73T

1. Principle of the Method

1.1 Samples can be collected by drawing a known volume of air through membrane or glass fiber filters. The samples are ashed and extracted with a mixture of nitric and hydrochloric acids. The analysis is subsequently made by atomic absorption spectroscopy using the 279.5 nm manganese line (**1**).

2. Range and Sensitivity

2.1 This method is applicable to the determination of manganese in an optimum range of 0.1 to 20.0 μg of manganese/ml of solution. An atmospheric concentration of 0.005 μg of manganese/m^3 can be detected using the method as described. For this concentration, a minimum air sample volume of 200 m^3 is recommended for samples taken on 8″ × 10″ filters with a Hi-Vol sampler (see Section **7:1**).

3. Interferences

3.1 Silica extracted from the glass fiber filter and from the collected particulate matter may cause a significant interference with the measurement of manganese. This interference can be overcome by allowing the acid extracts to stand overnight and centrifuging at about 2000 RPM for 30 min (**1**).

3.2 The absorption of light by manganese may be suppressed by the presence of excessive amounts of iron and nickel (**2**).

4. Precision and Accuracy

4.1 The precision of the method has not been reported using air samples; however, in the determination of manganese by a nearly identical method, a relative standard deviation of 22% was obtained for urban concentrations (**3**).

4.2 The recovery by this method has not been reported.

5. Apparatus

5.1 SAMPLE COLLECTION.

5.1.1 *Membrane filters.* Filters such as Millipore, appropriately sized for sample volume and rate, with a pore size of 0.8 μ, are satisfactory for air sampling.*

5.1.2 *Glass fiber filters.* These filters may also be used with a high volume air sampler. The filter blank should be investigated to establish the extent of manganese contamination and the potential matrix effect.

5.2 GLASSWARE. Borosilicate glassware should be used throughout the analysis. The glassware must be cleaned with 1:1 HCl and rinsed with distilled water. Beakers, pipets and auxiliary glassware will be required.

5.3 EQUIPMENT.

5.3.1 *Acetylene Gas (Cylinder).*

5.3.2 *Air Supply.*

5.3.3 *Atomic Absorption Spectrophotometer.*

5.3.4 *Centrifuge.*

5.3.5 *Manganese Hollow Cathode Lamp.*

5.3.6 *Low Temperature Asher-Tracerlab LTA-600 or equivalent.*

---

*Collection efficiencies greater than 99% are obtained for a 0.3 μ DOP smoke at face velocities from 7.2–53 cm/sec. (**4**).

6. Reagents

6.1 PURITY. ACS reagent grade chemicals or equivalent shall be used in all tests. References to water shall be understood to mean double distilled water or equivalent. Care in selection of reagents and in following the above precautions is essential if low blank values are to be obtained.

6.2 CONCD HYDROCHLORIC ACID (36.5–38.0%). sp gr 1.18.

6.3 CONCD NITRIC ACID (69.0–71.0%). sp gr 1.42.

6.4 MANGANESE METAL.

6.5 PERCHLORIC ACID. 72%.

6.6 HYDROCHLORIC ACID 1:10. Dilute 100 ml of concd acid to one liter with water.

6.7 PERCHLORIC-NITRIC ACID MIXTURE. Add 10 ml of concd perchloric to 90 ml of concd nitric.

7. Procedure

7.1 SAMPLE PREPARATION.

7.1.1 An 8″ × 10″ glass fiber filter, of which a 7″ × 9″ (63 square inches) section is exposed on the high-volume sampler, is divided into sections. The amount of a filter used for the analysis of metals depends upon the type of sample being prepared, urban or non-urban, individual or composite. Table 311:I lists the amount of filter used for each type of sample.

The strips for metal analysis are ashed in a low-temperature asher. In this procedure, an oxygen plasma is formed by passing oxygen through a high-frequency electromagnetic field and the activated oxygen reacts with the sample **(5)**. Slow burning takes place at temperatures ranging from 50 to 250 C. A pyrex boat is placed into each of five compartments of the low-temperature asher and the strips are ashed at approximately 150 C for 1 hr at 250 W at 1 mm chamber pressure with an oxygen flow of 50 cc per min.

The ashed filter is placed in a glass thimble (Figure 311:1), which is then placed in an extraction tube (Figure 311:2). A 125-ml Erlenmeyer flask with a $^{24}/_{40}$ female $\mathfrak{F}$ joint is charged with 8 ml of constant boiling (about 19%) hydrochloric acid and 32 ml of 40% nitric acid. The flask is attached to the extraction tube, and the extraction tube is fitted with an Allihn condenser. The acid is refluxed over the sample for 3 hr. The sample and extraction thimble remain at the temperature of the boiling acid throughout the extraction.

The extraction tube and condenser are removed from the Erlenmeyer flask and the flask is fitted with a thermometer adapter, which serves as a spray retainer. The extracted liquid is concentrated to 1 to 2 ml on a hot plate, and allowed to cool and stand overnight. The concentrated material is quantitatively transferred to a graduated 15-ml centrifuge tube with three washings of 5 to 10 drops 1:10 HCl. Urban samples are diluted to 4.4 ml/1″ strip (40 ml/63 sq. in of filter). Non-urban samples are diluted to 3.0 ml/2″ strip (13.3 ml per 63 sq. in of filter). Following dilution, the samples are centrifuged at 2000 RPM for about 30 min and the supernatant liquid decanted into polypropylene tubes that are then capped and stored pending analysis. One ml from each solution is diluted with 1:10 HCl to 10 ml for atomic absorption analysis.

7.1.2 The membrane filters are wet ashed with 10 to 20 ml of the mixture of nitric and perchloric acid in a suitably sized beaker. In the case of difficulty in oxidizing the organic matter, a first oxidation may be performed with a mixture of 20 ml of nitric acid and 5 ml of perchloric acid, followed by addition of a second 20 ml volume of nitric acid when the volume is first reduced to nearly 5 ml. (**Note:** All safety precautions necessary to the use of perchloric acid must be followed.) After the filter has been completely ashed, the acid is boiled down to the appearance of fumes of perchloric acid. All blanks and standards must be carried through with the same volumes of acid. Transfer to a graduated polypropylene tube with 1:10 HCl, add 1.2 ml of 1:1 HCl, and dilute to 25 ml with 1:10 HCl.

7.2 ANALYSIS.

7.2.1 Set the instrument operating

conditions as recommended by the manufacturer. The instrument should be set to the wavelength of maximum intensity for the 279.5 nm line from the hollow cathode lamp.

7.2.2 Aspirate the standards prepared fresh daily. Standards must be in the same matrix as the samples. Record the absorbance and prepare a calibration curve as described in section **8.1**.†

7.2.3 Aspirate the samples directly into the instrument and record the absorbance for comparison with the standards. Should the absorbance be off scale, dilute an appropriate aliquot to 10 ml. Be certain to aspirate distilled water between each sample. A blank and standard must be aspirated with sufficient frequency to assure the accuracy of the sample determinations.

8. Standards and Calibration

8.1 STANDARD MANGANESE SOLUTION.

8.1.1 *Master Standard (1.0000 mg manganese per ml).* Dissolve 1.0000 g of reagent grade manganese in a minimum volume of a 1:1 nitric-hydrochloric mixture (10 to 25 ml). Upon complete dissolution, evaporate to 5 ml and dilute to 1 liter with 1:1 HCl in a stoppered volumetric flask. This master standard is stable for at least one year stored in a polyethylene bottle.

8.1.2 *Dilute Standard (0.100 mg manganese per ml).* Pipet 10.0 ml of the master standard into a 100-ml volumetric flask and dilute to volume with 1:1 HCl.

8.1.3 *Working Manganese Standards.* Prepare a 1.00 μg/ml working standard solution by pipetting 1.00 ml of the dilute standard into a 100-ml volumetric flask, adding 4.00 ml of 1:1 HCl and diluting to the mark with 1:10 HCl. Repeat the procedure with the appropriate amounts of the dilute standard and 1:1 HCl (total volume of 5.00 ml) to prepare standards containing 2.0 μg/ml, 3.0 μg/ml, 4.0 μg/ml and 5.0 μg/ml of manganese. Working standards must be prepared daily.

8.1.4 *Calibrating Sample Preparation Procedure.* Five known amounts of manganese are added to 1″ strips of glass, or an equivalent portion of a membrane filter, and carried through the procedure to establish a calibration graph for the entire sample preparation procedure. If this calibration curve is in agreement with that obtained using the working standards, one need only perform this task periodically to insure that the sample preparation procedure is being carried out properly. If the calibration graph for the sample preparation differs significantly from that obtained with the working standards, then the sample preparation calibration graph should be used to determine the manganese content of the samples and this calibration should be prepared with each set of samples analyzed.

8.2 BLANK. A blank filter must be carried through the entire procedure.

9. Calculation

9.1 GLASS FIBER FILTER. After a correction for the blank has been made, concentrations are calculated by multiplying the μg of manganese per ml in the sample aliquot by ten and by the total number of ml in the urban or non-urban sample (See **7.1.1**). The result is divided by the number of m³ of air represented by the portion of the filter taken for analysis.

9.2

$$\mu g\ Mn/m^3 = \frac{\mu g\ Mn/ml \text{ in aliquot} \times 10 \times (\text{total ml in urban or non-urban sample})}{(\text{total corrected } m^3 \text{ of air})‡}$$

9.3 MEMBRANE FILTER. Make correc-

†All combustion products from the flame must be removed by direct exhaustion through the use of a good, separate ventilation system. This is of particular importance when using a nitrous oxide flame and when aspirating a sample containing a toxic element or a material with toxic combustion products.

‡This is that volume of air in m³ represented by the fraction of filter taken for analysis. This volume is corrected to 25 C and 760 torr.

tion for blank. Calculate the concentration as follows (see **7.1.2**).

$$\mu g\ Mn/m^3 = \frac{\mu g\ Mn/ml \text{ in aliquot} \times 25}{(\text{corrected } m^3 \text{ of air for sample})}$$

10. Storage

10.1 Untreated filter samples may be stored indefinitely. Acid extracts may be stored for a month in polypropylene containers.

11. References

1. THOMPSON, R.J., G.B. MORGAN and L.J. PURDUE. 1970. Analysis of Selected Elements in Atmospheric Particulate Matter by Atomic Absorption. Atomic Absorption Newsletter, 9:53–57.
2. SLAVIN, W. 1968. Atomic Absorption Spectroscopy, Inter-Science Publishers, New York, New York.
3. KNEIP, T.J., M. EISENBUD, C.D. STREHLOW and P.C. FREUDENTHAL. 1970. Airborne Particulates in New York City, *APCA Journal*, 20:144–149.
4. LOCKHART, L.B., JR., R.L. PATTERSON, JR. and W.L. ANDERSON. 1963. Characteristics of Air Filter Media Used for Monitoring Airborne Particulates. NRL Report 6054.
5. GLEIT, C.E. 1963. *Am. J. of Med. Electronics*, 2:112.

Subcommittee 6
T. J. KNEIP, *Chairman*
R. S. AJEMIAN
J. R. CARLBERG
J. DRISCOLL
L. KORNREICH
J. W. LOVELAND
J. L. MOYERS
R. J. THOMPSON

317.

Tentative Method of Analysis for Elemental Mercury in Ambient Air by Collection on Silver Wool and Atomic Absorption Spectroscopy

42242-01-74T

1. Principle of the Method

1.1 Elemental mercury vapor is collected on silver wool and released by heating the wool to 400 C.

1.2 The mercury released is swept by a carrier gas through an absorption cell of an atomic absorption instrument, and the response at 253.7 nm is measured either as a peak height or peak area (**1**).

2. Range and Sensitivity

2.1 The procedure is applicable to the determination of elemental mercury in concentrations of 0.02 to 10 $\mu g/m^3$ at a collection flow rate of 100 to 200 cm^3/min over a 24-hr period. The lower limit of detection is about 0.001 μg of mercury (**1**).

3. Interferences

3.1 High levels (μg quantities) of chlorine (**2**) and SO_2 (**3**) will poison the silver wool collectors. A tube packed with ascarite, preceding the collector will remove the chlorine (**2**).

3.2 Ambient levels of dimethyl mercury, sulfur dioxide, hydrogen sulfide and nitrogen dioxide do not seriously interfere. However, hydrogen sulfide in concentrations above 13 $\mu g/m^3$ may show an interference (**1**). Acetone, benzene, and ethanol do not interfere (**4**).

4. Precision and Accuracy

4.1 Recovery of mercury from the silver wool collectors by this method is at least 98% over the range of 0.006 to 0.6 μg with a relative standard deviation of 4% (**1**).

4.2 The precision of the method has been reported to be about 10% relative standard deviation for ambient mercury vapor concentrations of about 0.03 $\mu g/m^3$ (**1, 5, 6**).

5. Apparatus

5.1 Sample Collection (See Figure 317:1).

5.1.1 *Silver wool collector.* Pyrex tube 10-cm long × 5-mm ID equipped with ball joints on the ends and packed with 1 to 2 g of clean silver wool (Fisher micro-analysis grade) and permanently wound with 100 cm of 22-gauge Nichrome heating wire. Clean the silver wool initially by placing it in a furnace at 800 C for 2 hr prior to packing it in the pyrex tube.

5.1.2 *Pump.* Capacity to pull from 50 to $500/cm^3$ of air per min.

5.1.3 *Flowmeter.* (50 to 500 $cm^3/$ min).

5.1.4 *Flow controller.*

5.1.5 *Kel-F coated 3-way valve.*

5.1.6 *¼" glass tubing.*

5.1.7 *Caps for sealing ends of Pyrex tube.*

5.2 Equipment Train. (See Figure 317:2).

5.2.1 *Atomic Absorption Spectrophotometer.*

5.2.2 *Absorption Cell.* 20-cm long by about 3-cm ID, with fused silica windows. Lag with heating tape to maintain temperature of the surface near 90 C.

5.2.3 *Mercury hollow cathode lamp.*

5.2.4 *Recorder or meter read out.*

5.2.5 *Integrator. Digital (optional).*

5.2.6 *Variable transformer.*

5.2.7 Two charcoal filters in drying tubes and one silver wool filter (as in **5.1.1**) without Nichrome heating wire.

5.2.8 *Teflon tubing.*

5.2.9 *Carrier gas.* (Oil free air or N_2).

5.2.10 *Oven.* Capable of maintaining 800 C.

5.2.11 *Kel-F coated 3-way valves.*

5.2.12 *Flowmeter.* (50 to 500 $cm^3/$ min).

5.3 Calibration Equipment.

5.3.1 *Pyrex bottles.* One liter.

5.3.2 *Serum caps. For Pyrex bottles.*

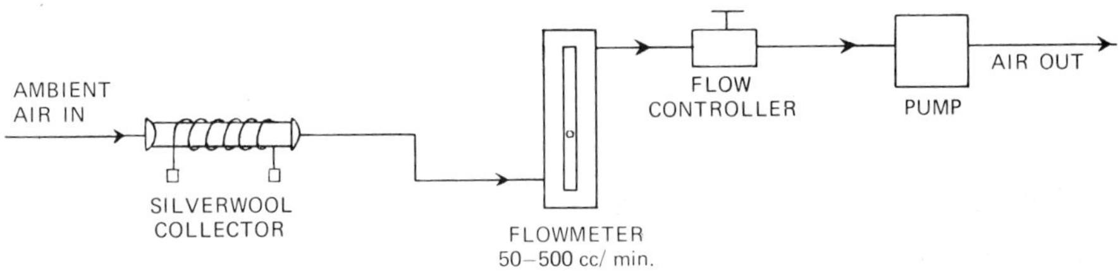

Figure 317:1—Sample Collection Train

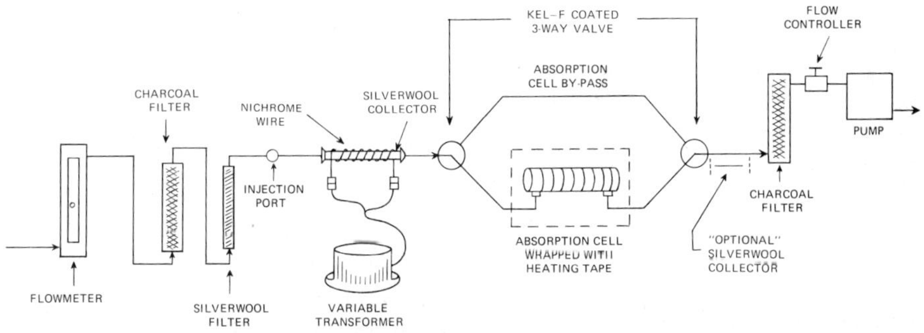

Figure 317:2—Schematic Diagram of Analytical Gas Train

5.3.3 *Constant temperature bath* (20.0 ± 0.1).

5.3.4 *Certified NBS thermometer.*

5.3.5 *Syringes*. Gas tight: 0.02 to 100 cm^3.

6. Reagents

6.1 MERCURY. ACS Reagent Grade.

6.2 SILVER WOOL. (Fisher micro-analysis grade).

6.3 CHARCOAL. 6 to 20 mesh.

7. Procedure

7.1 SAMPLING.

7.1.1 Connect the silver wool collector prepared in **5.1.1** into the Analytical Gas Train of Figure 317:2. Sweep with gas at about 200 cm^3 per min while using bypass mode. Release any residual mercury from the silver wool by applying 24 V from the variable transformer across the Nichrome wire for 30 sec, which yields a temperature of 400 C. After one 30-sec heating period, the collector is mercury free and ready for future use. Cap both ends until ready for sampling, making sure that the caps do not touch the silver wool.

7.1.2 Set up the collection train as shown in Figure 317:1 which is manifolded with ¼" ID glass, Teflon tubing and Kel-F coated valves.* Flow rates through the manifold are produced by a pump behind the absorption cell and monitored by the flowmeter.

7.1.3 Collect the mercury vapor in ambient air by pulling air through a cleaned collector, containing one or two grams of silver wool at a known rate for a fixed length of time. For field sampling purposes, a flow rate of between 100 and 200 cm^3/min. for 24 hr is recommended.† After use, disengage the collector from the train and cap it to prevent contamination.

*Teflon tubing should be as short as possible and no rubber tubing or grease should be used since mercury will be absorbed by these materials.

†The flow rate, collection time, and number of collectors can be varied to suit the expected mercury level.

7.2 ANALYSIS.

7.2.1 Set the atomic absorption spectrophotometer operating conditions as recommended by the manufacturer. The instrument should be set to the wavelength of maximum intensity for the 253.7 nm mercury line.

7.2.2 Insert a silver wool collector in the Analytical Gas Train and sweep the system for a few minutes with carrier gas until no mercury is detected. This should be done prior to running calibrations and samples.‡

7.2.3 Prepare and run standards at the beginning of each analysis period. Standards should be run to check the procedure after every tenth sample is run. Record the absorbance value or, if an integrator is used, record the peak area and prepare a calibration curve as described in Section **8.1**.

7.2.4 Clamp the collector to be analyzed into the analytical system shown in Figure 317:2 by ball-joint clamps, adjust the carrier gas flow (which is critical to the repeatability) to exactly 200 cm^3/min, and heat the collector for 30 sec at 24 V. Silver wool and charcoal filters are inserted between the flowmeter and collector to prevent sample contamination from the laboratory air. Record the absorbance peak value or peak area obtained from the spectrophotometer and determine the amount of mercury from the calibration curve. Allow the collector to cool, remove it from the analytical system, and cap it for future use.§ ||

‡Residual mercury can be attributed to contamination, possibly from slow seepage of air through the joints of the apparatus during nonuse.

§If there is some uncertainty about the concentration range, place a clean (mercury free) silver wool collector between the spectrophotometer and the charcoal filter to recapture the sample in order to rerun the sample at the proper calibration range.

||The entire analysis process takes less than five min per sample at 200 cm^3/min carrier gas flow rate. The absorption cell bypass, shown in Figure 317:2, is convenient when collectors are being cleaned and the amount of mercury released does not need to be monitored.

8. Standards and Calibration

8.1 STANDARD MERCURY SAMPLES. Add enough Reagent Grade mercury to cover the bottoms of 3 one-liter pyrex flasks. The flasks should be equipped with serum caps.

8.1.1 Place the flasks in a constant temperature bath equipped with a certified NBS thermometer and maintain the entire flask at 20.0 ± 0.1 C. One cc of air at 20 C and one atmosphere pressure at equilibrium over liquid mercury contains 13.19 ng of mercury. Concentrations of mercury at temperatures other than 20 C are given in Table 317:I.

8.1.2 *Calibrating Procedure.* Use several gas-tight syringes with volumes of from 0.02 to 100 cm^3 and withdraw from the flasks at least five different volumes of mercury saturated air corresponding to the expected field values. The sample should be withdrawn slowly, expelled back, and withdrawn slowly again. The temperature of the syringe must be equal to or slightly higher than 20 C to prevent possible condensation of mercury in the syringe.

Inject a given volume slowly into the analytical gas train previously cleaned as in **7.2.2** (See Figure 317:2) at the injection port while sweeping the train at exactly 200 cm^3/min with carrier gas. The known amount of mercury is collected on the silver wool and then released by applying 24 V to the Nichrome heater for 30 sec. The system is now ready for the next injection of standard sample when the tube is slightly hot to the touch of the finger. Record the peak values or the area-time counts from all five standards and plot the calibration curve.¶

Relieve the partial vacuum in the flask reservoir by injecting the volume withdrawn with air back into the reservoir. Withdrawals should be alternated among the three reservoirs. If more than 10% of the total volume be withdrawn, equilibrium will be established between the liquid and gaseous mercury in about 15 min after replenishing the withdrawn volume with air, provided that a clean liquid mercury surface is present.

8.2 BLANK. If proper precautions to eliminate mercury contamination have been taken before the analysis step in **7.2.2** and the calibration step in **8.1.2**, the blank value is zero.

9. Calculations

9.1 Calculate the concentration of mercury vapor in air by dividing the micrograms found by the total number of cubic meters of air taken for analysis.

9.2 $\mu g\ Hg/m^3 = \mu g\ Hg\ found/cm^3/min \times 60\ min/hr \times hrs. \times 10^{-6}\ m^3/cm^3$.

¶It has been found that better precision can be obtained by using the area measurement and by calibrating with one collector. It is advisable to calibrate and field test with the same collector if better than 4% relative standard deviation is necessary.

Table 317:I. Elemental mercury vapor density at 10 to 30 C and one atmosphere pressure

t °C	Vapor Density ng/cm³	t °C	Vapor Density ng/cm³	t °C	Vapor Density ng/cm³	t °C	Vapor Density ng/cm³
10.0	5.57	15.5	9.01	21.0	14.33	26.5	22.39
10.5	5.82	16.0	9.41	21.5	14.93	27.0	23.30
11.0	6.09	16.5	9.82	22.0	15.56	27.5	24.24
11.5	6.36	17.0	10.25	22.5	16.21	28.0	25.22
12.0	6.65	17.5	10.69	23.0	16.89	28.5	26.23
12.5	6.95	18.0	11.15	23.5	17.59	29.0	27.28
13.0	7.26	18.5	11.63	24.0	18.32	29.5	28.36
13.5	7.58	19.0	12.13	24.5	19.07	30.0	29.49
14.0	7.92	19.5	12.65	25.0	19.86		
14.5	8.27	20.0	13.19	25.5	20.67		
15.0	8.63	20.5	13.75	26.0	21.52		

10. Storage

10.1 Collected field samples, properly capped (air tight) may be stored 6 months without gain or loss of mercury.

11. References

1. LONG, S.J., D.R. SCOTT and R.J. THOMPSON. 1973. Atomic Absorption Determination of Elemental Mercury Collected from Ambient Air on Silver Wool. *Anal. Chem.*, 45:2227.
2. SPITTLER, T.M., R.J. THOMPSON, and D.R. SCOTT. 1973. Division of Water, Air and Waste, 165th National Meeting, ACS, Dallas, Texas.
3. CHASE, D.L., D.L. SGONTZ, E.R. BLOSSNER, and W.M. HENRY. 1972. Development and Evaluation of an Analytical Method for the Determination of Total Atmospheric Mercury. Battelle Columbus Laboratories, EPA Contract No. EHSD 71-32, Environmental Protection Agency.
4. SCARINGELLI, F.P., J.C. PUZAK, B.I. BENNETT, and R.L. DENNEY. 1974. Determination of Total Mercury in Air by Charcoal Adsorption and Ultraviolet Spectrophotometry. *Anal. Chem.*, 46:278.
5. WILLISTON, S.H., 1968. *J. Geophys. Res.*, 73:7051.
6. FOOTE, R. S. 1972. *Science*, 177:513.

Subcommittee 6
T. J. KNEIP, *Chairman*
R. S. AJEMIAN
J. R. CARLBERG
J. DRISCOLL
L. KORNREICH
J. W. LOVELAND
J. L. MOYERS
R. J. THOMPSON

318.

Tentative Method of Analysis for Elemental Mercury in the Working Environment by Collection on Silver Wool and Atomic Absorption Spectroscopy

42242-02-74T

1. Principle of the Method

1.1 Elemental mercury vapor is collected on silver wool and released by heating the wool to 400 C.

1.2 The mercury released is swept by a carrier gas through an absorption cell of an atomic absorption instrument, and the response at 253.7 nm is measured either as a peak height or peak area **(1)**.

2. Range and Sensitivity

2.1 The procedure is applicable to the determination of elemental mercury in concentrations of 5 to 500 $\mu g/m^3$ at a collection flow rate of 50 to 200 cc/min over a 10 to 60 min period. The lower limit of detection is about 0.001 μg of mercury **(1)**. The flow rates and collection time have been selected to cover concentration ranges from 0.1 to 10 times the TLV of 50 $\mu g/m^3$ for elemental mercury in air.

3. Interferences

3.1 High levels of hydrogen sulfide (13 to 650 $\mu g/m^3$) and to a lesser extent nitrogen dioxide (25 to 100 $\mu g/m^3$) **(1)** will poison the silver wool collectors. Chlorine at 20 $\mu g/m^3$ also interferes **(2)**. A tube packed with Ascarite, preceding the collector will remove the chlorine **(2)** and other acid gases.

3.2 Ambient levels of dimethyl mercury, sulfur dioxide, and nitrogen dioxide do not seriously interfere **(1)**. Acetone, benzene, and ethanol at normal levels do not interfere **(3)**.

4. Precision and Accuracy

4.1 Recovery of injected mercury vapor collected on 1 g silver wool collectors by this method is at least 98% over the range of 0.032 to 0.160 μg. Precision of the analysis step over the range of 0.013 to 0.594 μg

is better than 3% relative standard deviation (RSD).

4.2 The precision of the total method (Sampling and Analysis) at sampling flow rates of 100 cc/min, using 2 collectors in series (1 g and 2 g) at levels of 0.3 $\mu g/m^3$ to 1.96 $\mu g/m^3$ was better than 5% RSD (**4**). At these levels the second collector never contained more than 4% of the mercury. At mercury levels of 99 $\mu g/m^3$ 9.5 μg of mercury was collected on three 1-g collectors in series with 83% collected on the first tube and 17% on the second tube. Insignificant amounts (~2ng) were detected on the third tube (**4**).

5. Apparatus

5.1 SAMPLE COLLECTION (See Figure 318:1).

5.1.1 Clean the silver wool (Fisher micro-analysis grade) initially by placing it in a furnace at 800 C for 2 hr prior to packing it in the pyrex tube.

5.1.2 *Silver wool collector.* Pyrex tube 10 cm long × 5 mm ID (8 mm OD) equipped with ball and socket joints on the end to match those in the analytical train and packed with 1 to 2 g of clean silver wool (Figure 318:2). Permanently wind the tube with 86 cm of 22 gauge Nichrome heating wire. Keep ends capped until used in a personal sampler or sampling train of Figure 318:1.

5.1.3 *Pump.* Capacity to pull from 50 to 200 cc of air per min, for personal sampler (Model SP-2 by Sipin Co. or equivalent)

5.1.4 *Flowmeter.* (50 to 500 cc/min).

5.1.5 *Flow controller.*

5.1.6 *Kel-F coated 3-way valve.*

5.1.7 *¼" glass tubing.*

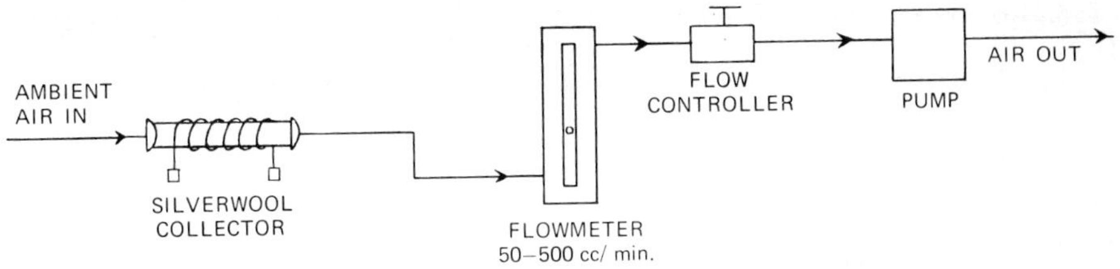

Figure 318:1—Sample Collection Train

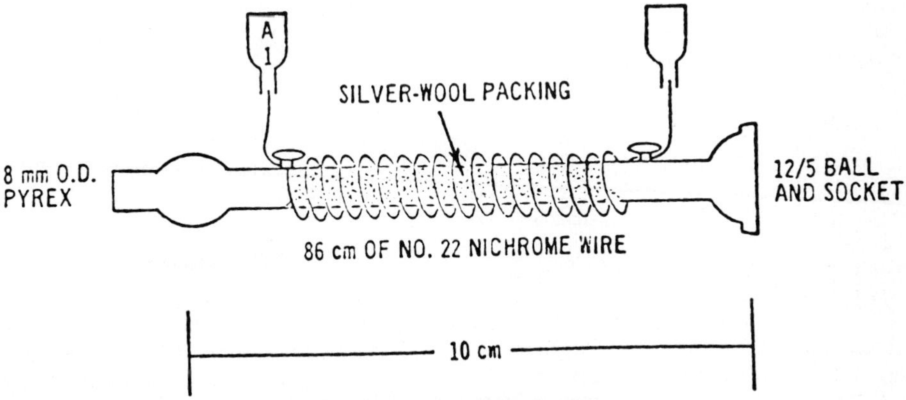

Figure 318:2—Field Collection Tube

5.1.8 *Caps for sealing ends of pyrex tube.*

5.1.9 *Pinch clamps for ball and socket fittings.*

5.2 ANALYSIS TRAIN. (See Figure 318:3).

5.2.1 *Atomic Absorption Spectrophotometer (AAS).*

5.2.2 *Absorption Cell.* Should be appropriate for the available AAS: *i.e.*, 20 cm long by about 3 cm ID with fused silica windows will fit a P.E. 403 or 303. Wrap with heating tape to maintain temperature of the surface near 90 C.

5.2.3 *Mercury hollow cathode lamp.*

5.2.4 *Recorder or meter read out.*

5.2.5 *Integrator-digital (optional).*

5.2.6 *Variable transformer.*

5.2.7 *Two charcoal filters in drying tubes and one silver wool filter.* (As in **5.1.2**)

5.2.8 *Teflon tubing.*

5.2.9 *Carrier gas.* (Oil free air or N_2).

5.2.10 *Oven capable of maintaining 800 C.*

5.2.11 *Kel-F coated 3-way valves.*

5.2.12 *Flowmeter.* (50 to 500 cc/min).

5.2.13 *Injection Port.* Glass T with septum attached.

5.3 CALIBRATION EQUIPMENT.

5.3.1 *One-liter pyrex bottles.*

5.3.2 *Serum caps for pyrex bottles.*

5.3.3 *Constant temperature bath.* (20.0 ± 0.1 C).

5.3.4 *Certified Oven capable of maintaining 800 C.*

5.3.5 *Syringes, gas tight.* 1.0 to 100 cc.

6. Reagents

6.1 MERCURY. ACS Reagent Grade.

6.2 SILVER WOOL. (Fisher micro-analysis grade).

6.3 ACTIVATED CHARCOAL. 6 to 20 mesh.

7. Procedure

7.1 SAMPLING.

7.1.1 *Preparation of Collector Prior to Sampling.* Connect the silver wool collector prepared in **5.1.2** into the Analytical Gas Train of Figure 318:3. Sweep with gas at about 200 cc per min while using bypass mode. Release any residual mercury from the silver wool by applying 24 V from the variable transformer across the Nichrome wire for 30 sec, which yields a temperature of 400 C. After one 30 sec heating period, the collector is mercury free and ready for future use. Cap both ends until ready for sampling, making sure that the caps do not touch the silver wool.

7.1.2 Set up the collection train manifolded with ¼" ID glass, Teflon tubing and Kel-F coated valves as shown in Figure 318:1. Flow rates through the manifold are

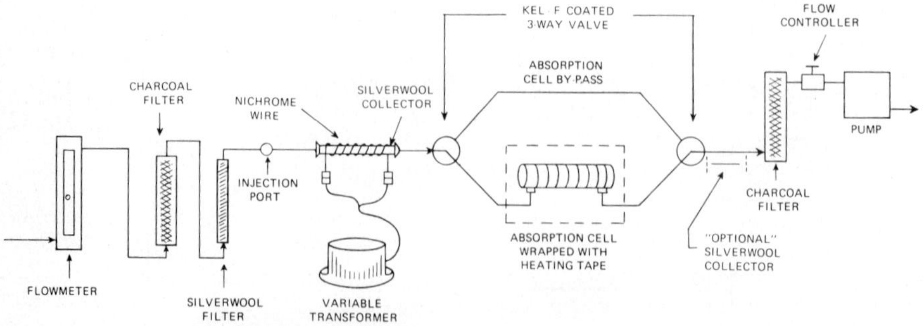

Figure 318:3—Schematic Diagram of Analytical Gas Train.

produced by a pump behind the collector and monitored by the flowmeter. A suitable personal portable battery pack pump and sampler having all the essential parts of the sampling train can also be used.

Note 1: Teflon tubing should be as short as possible and *no rubber tubing or grease* should be used since mercury will be absorbed by these materials.

7.1.3 Collect the mercury vapor in the working place by pulling air at a known rate for a fixed length of time through two cleaned collectors in series, each containing one gram of silver wool. For sampling purposes, a flow rate up to 200 cc per min. is recommended (see Note 2). After use, disengage the collectors from the train and cap to prevent contamination.

Note 2: The flow rate (50 to 200 cc/min) collection time (1 hr or less) can be varied to suit the expected mercury level.

7.1.4 Take at least 3 blank silver wool tubes to the field and expose under sealed conditions for the sampling period and return them unused for determining the blank.

7.2 ANALYSIS.

7.2.1 Set the atomic absorption spectrophotometer operating conditions as recommended by the manufacturer. The instrument should be set to the wavelength of maximum intensity for the 253.7 nm mercury line.

7.2.2 Insert a clean silver wool collector in the Analytical Gas Train and with no heat applied to the collector sweep the system for a few min with carrier gas until no mercury is detected. This should be done prior to running calibrations and samples (See Note 3).

Note 3: Residual mercury contamination, may be introduced possibly from slow seepage of air through the joints of the apparatus during non-use.

7.2.3 Prepare and run standards (See Section **8.0**) at the beginning of each analysis period. Standards should be run to check the procedure after every tenth sample is run. Record the absorbance value or, if an integrator is used, record the peak area and prepare a calibration curve as described in section 8.1.

7.2.4 Clamp the collector to be analyzed into the analytical system shown in Figure 318:3 by ball-joint clamps, adjust the carrier gas flow (which is critical to the repeatability) to an optimum flow rate, which must be maintained throughout the analytical runs, (See Note 4) and heat the collector for 30 sec at 24 V. Silver wool and charcoal filters are inserted between the flowmeter and collector in order to prevent sample contamination from the laboratory air. Record the absorbance peak value or peak area obtained from the spectrophotometer and determine the amount of mercury from the calibration curve. Allow the collector to cool, remove it from the analytical system, and cap it for future use. (See Note 5) Run the second silver wool collector as well as the three blanks under identical conditions.

Note 4: The entire system should be optimized with regard to absorption cell size, flow rate and temperature at each installation. The entire analysis process takes less than 5 min per sample at 200 cc/min carrier gas flow rate and 400 C on the silver wool collector for 30 sec (**1**). The absorption cell bypass, shown in Figure 318:3, is convenient when collectors are being cleaned, and the amount of mercury released does not need to be monitored.

Note 5: If there is some uncertainty about the concentration range, place a clean (mercury free) silver wool collector between the spectrophotometer and the final charcoal filter to recapture the sample in order to rerun the sample at the proper calibration range.

8. Standards and Calibration

8.1 STANDARD MERCURY SAMPLES. Add enough Reagent Grade mercury to cover the bottoms of 3 one-liter glass or pyrex flasks. The flasks should be equipped with serum caps.

8.1.1 Place the flasks in a constant temperature bath equipped with a certified NBS thermometer and maintain the entire flask at 20.0 ± 0.1 C. One cc of air at 20 C and one atmosphere pressure at equilibrium over liquid mercury contains 13.19 ng of mercury. Conc of mercury at temperatures other than 20 C are given in Table 318:I.

8.1.2 *Calibrating Procedure*. Use several gas-tight syringes with volumes of from 1.0 to 100 cc and withdraw from the flasks at least 5 different volumes (1, 2, 5, 10, 50 and 100 cc) of mercury saturated air to correspond to the expected work place values. The sample should be withdrawn slowly, expelled back, and withdrawn slowly again. The temperature of the syringe must be equal to or slightly higher than 20 C to prevent condensation of mercury in the syringe. Inject a given volume slowly into the injection port of the analytical gas train, previously cleaned as in **7.2.2** (See Figure 318:3) while sweeping the train at exactly the flow rate used for collectors in the field with carrier gas. The known amount of mercury is collected on the silver wool, the flow rate adjusted to the optimum value (Section **7.2.4**) and the mercury released by applying 24 volts to the Nichrome heater for 30 sec. The system is ready for the next injection of standard sample when the tube is slightly hot to the touch of the finger. Record the peak values or the integrated area-time counts from all 5 standards and plot the calibration curve according to the least squares fit. (See Note 6) Relieve the partial vacuum in the flask reservoir by reinjecting the volume withdrawn with clean uncontaminated air. Withdrawals should be alternated among the three reservoirs. If more than 10% of the total volume is withdrawn, equilibrium is established between the liquid and gaseous mercury in about 15 min after replenishing the withdrawn volume with air, provided that a clean liquid mercury surface is present.

Note 6: It has been found that better precision can be obtained by using the area measurement and by calibrating with one collector. It is advisable to calibrate and field test with the same collector if better than 4% relative standard deviation is necessary.

8.2 Blank. If proper precautions to eliminate mercury contamination have been taken before the analysis step in **7.2.2** and the calibration step in **8.1.2**, the analysis blank value is negligible. In the event that the field blanks are significant, the average value should be subtracted from the measured results on the samples.

9. Calculations

9.1 Calculate the concentration of mercury vapor in air by dividing the micrograms found by the total number of cubic meters of air taken for analysis.

9.2

$$\mu g\ Hg/m^3 = \frac{(\mu g\ Hg\ found - \mu g\ Hg\ blank)}{\dfrac{cc/min \times min\ sampled}{m^3/cc \times 10^{-6}}}$$

10. Storage

10.1 The collectors are stable for several weeks to several months depending on

TABLE 318:I. Elemental Mercury Vapor Density at 10 to 30 C and One Atmosphere Pressure

t °C	Vapor Density ng/cc	t °C	Vapor Density ng/cc	t °C	Vapor Density ng/cc	t °C	Vapor Density ng/cc
10.0	5.57	15.5	9.01	21.0	14.33	26.5	22.39
10.5	5.82	16.0	9.41	21.5	14.93	27.0	23.30
11.0	6.09	16.5	9.82	22.0	15.56	27.5	24.24
11.5	6.36	17.0	10.25	22.5	16.21	28.0	25.22
12.0	6.65	17.5	10.69	23.0	16.89	28.5	26.23
12.5	6.95	18.0	11.15	23.5	17.59	29.0	27.28
13.0	7.26	18.5	11.63	24.0	18.32	29.5	28.36
13.5	7.58	19.0	12.13	24.5	19.07	30.0	29.49
14.0	7.92	19.5	12.65	25.0	19.86		
14.5	8.27	20.0	13.19	25.5	20.67		
15.0	8.63	20.5	13.75	26.0	21.52		

storage environment. Collectors in an atmosphere of 0.050 μg/m³ picked up 0.0005 μg or less of mercury in 3 months. Others in rooms with 1 μg/m³ levels showed a pick up of 0.017 μg after 14 days and 0.0025 μg after 21 days (**4**).

11. References

1. LONG, S.J., D.R. SCOTT and R.J. THOMPSON. 1973. Atomic Absorption Determination of Elemental Mercury Collected from Ambient Air on Silver Wool. *Anal. Chem.*, 45:2227.
2. SPITTLER, T.M., R.J. THOMPSON and D.R. SCOTT. Division of Water, Air and Waste, 165th National Meeting, ACS, Dallas, Texas, April 1973.
3. SCARINGELLI, F.P., J.C. PUZAK, B.I. BENNETT, and R.L. DENNY. 1974. Determination of Total Mercury in Air by Charcoal Adsorption and Ultraviolet Spectrophotometry. *Anal. Chem.* 46:278.
4. LONG, S.M. and SCOTT, D.R., Personal Communication.

Subcommittee 6
T. J. KNEIP, *Chairman*
R. S. AJEMIAN
J. DRISCOLL
F. I. GRUNDER
L. KORNREICH
J. W. LOVELAND
J. L. MOYERS
R. J. THOMPSON

319.

Tentative Method of Analysis for Molybdenum Content of Atmospheric Particulate Matter by Atomic Absorption Spectrophotometry

12134-02-73T

1. Principle of the Method

1.1 Molybdenum may exist in the atmosphere as the element, the oxide or as molybdates. Concentrations of molybdenum in ambient air range from below detectable amounts to maximum values in the order of 0.34 μg/m³ of air (**1**). The satisfactory determination of molybdenum depends largely on the formation of an extractable green complex by the reaction of molybdenum with 4-methyl-1, 2-dimercaptobenzene, also called toluene-3, 4-dithiol or dithiol (**2**). Iron interference can be prevented by reducing ferric iron to the ferrous state with potassium iodide and decolorizing the liberated iodine with sodium thiosulfate. The dithiol complex is extracted with methyl isobutyl ketone (MIBK). The MIBK extract is analyzed with an atomic absorption spectrophotometer using the operating conditions as recommended by the manufacturer (**3**).

2. Range and Sensitivity

2.1 The detection limit for molybdenum in aqueous solutions is about 0.03 μg/ml.

2.2 The detection limit for molybdenum in organic phase is lower than that in aqueous solutions.

2.3 The sensitivity is 0.6 μg of molybdenum/ml for 1% absorption. A concentration of 0.15 μg of molybdenum/m³ of air can be measured if a 20 m³ sample is taken for analysis and is extracted into 5 ml of MIBK.

2.4 To increase the sensitivity a larger air sample can be taken or a smaller amount of MIBK can be used.

3. Interferences

3.1 No optical interferences will be encountered with this method.

3.2 Interferences may be encountered

when using an air-acetylene flame to determine molybdenum.

4. Precision and Accuracy

4.1 No data available in the literature for the range required.

5. Apparatus

5.1 An atomic absorption spectrophotometer for use at 313.3 nm.
5.2 Hollow cathode lamp for molybdenum.
5.3 Acetylene gas as fuel, nitrous oxide gas as oxidant.
5.4 125-ml separatory funnels, volumetric and general laboratory glassware.
5.5 Automatic shaker, optional.

6. Reagents

6.1 PURITY. All reagents must conform to ACS analytical purity specifications. Distilled water should conform to ASTM standards for referee reagent water.

6.2 STANDARD MOLYBDENUM SOLUTION (1.0 MG Mo/ML). Dissolve 3.000 g of molybdenum trioxide in 200 ml of 10% sodium hydroxide and dilute to 2 l with distilled water. Store in a polyethylene container. (This solution is stable indefinitely.) Standardize this solution by titrating with standard ceric sulfate using O-phenanthroline indicator (2 drops of 0.025 M indicator are used) **(4, 5)**. Dilute an aliquot with distilled water to obtain a solution containing approximately 25 μg of molybdenum/ml. Molybdenum trioxide of over 99.5% purity is available. Consequently sufficiently accurate standards may be prepared without standardization.

6.3 HYDROCHLORIC ACID, sp gr 1.19.

6.4 NITRIC ACID, sp gr 1.42.

6.5 POTASSIUM IODIDE. 50% (W/V) in distilled water.

6.6 SODIUM THIOSULFATE. 10% (W/V) in distilled water.

6.7 DITHIOL REAGENT. Dissolve 1 g of dithiol in 500 ml of 1% sodium hydroxide solution. Stir occasionally over 1 hr. Add about 8 ml of thioglycolic acid dropwise until a faint permanent opalescent turbidity forms. Store at 5 C. (**Caution:** Dithiol should not be allowed to contact the skin. Gloves should be worn when the reagent is used **(2)**.) Discard this reagent after one week.

7. Procedure

7.1 Wet ash samples collected on membrane or cellulose filters or extract glass fiber filter samples in a Phillips beaker using enough concentrated nitric acid to cover the filter (take whole sample or suitable aliquot to give about 1.0 μg of molybdenum/ml). In the case of glass fiber filters heat the beaker below the boiling point of the acid to extract the molybdenum, decant the liquid into a second beaker and wash the Phillips beaker and filter with nitric acid and distilled water. Repeat washing 3 times. (Add all washings to the second beaker.)

7.2 Place the second beaker on a low temperature hot plate (125 C) and heat until the brown fumes of oxides of nitrogen appear then transfer the beaker to a high temperature hot plate (400 C) and heat until the fumes of nitrogen oxides disappear. Repeat the ashing treatment the second time with further additions of 2 ml of concentrated nitric acid.

7.3 Take up the residue with 3 ml of reagent grade HCl and dilute to 50 ml with distilled water. Add 1 ml of 50% KI solution to reduce the ferric iron, swirl, and let stand for 10 min. The appearance of a reddish brown color indicates the liberation of iodine. Decolorize the iodine by dropwise addition of 10% sodium thiosulfate solution and 2 drops are added in excess.

7.4 Transfer to a 125-ml separatory funnel and add 2 ml of the dithiol reagent, shake for 30 sec and let stand for 10 min; 5 ml of MIBK are pipetted into each funnel. The funnels are shaken vigorously and continuously for 3 min. Allow 10 min for settling after which the aqueous layer is drained off and discarded. Water is removed from the drain cock bore by inserting filter paper into the bore. The

MIBK layer is drawn off and analyzed with an atomic absorption spectrophotometer.

7.5 A blank should be carried through entire procedure.

8. Calibration

8.1 Prepare a calibration curve using standards containing 0 to 10 μg of molybdenum and carry them through the complete procedure, Section **7**.

8.2 Plot absorbance *vs* concentration or use calibration procedure as recommended by the manufacturer.

9. Calculations

9.1 Use the absorbance of the samples to obtain the molybdenum concentration from the calibration curve.

9.2 Apply the aliquot factor.

9.3 Concentration of molybdenum in

$$\mu g/m^3 = \frac{\mu g \text{ Mo in total sample}}{\text{m}^3 \text{ of air represented by sample at 760 Torr and 25 C.}}$$

10. Effect of Storage

10.1 Molybdenum collected on solid filter media can be stored indefinitely.

11. References

1. AIR QUALITY DATA FROM THE NATIONAL AIR SAMPLING NETWORK AND CONTRIBUTING STATE AND LOCAL NETWORKS. 1966. U.S. Department of Health, Education and Welfare, Public Health Service, National Air Pollution Control Administration.
2. DELAUGHTER, BUFORD. 1965. The Determination of Sub-PPM Concentrations of Chromium and Molybdenum in Brines. *Atomic Absorption Newsletter*, Vol. 4, No. 5 May.
3. *Analytical Methods for Atomic Absorption Spectrophotometry*. 1971. Perkin-Elmer Corp., Norwalk, Conn.
4. FURMAN, N.H., and W.M. MURRY, JR. 1936. *J. Am. Chem. Soc.* 58, 1689.
5. *Scotts Standard Methods of Chemical Analyses*. 1962. Sixth Edition, 1, 1205, D. Van Nostrand, N.Y.

Subcommittee 7
R. E. KUPEL, *Chairman*
M. M. BRAVERMAN
J. M. BRYANT
H. E. BUMSTED
A. CAROTTI
H. M. DONALDSON
L. DUBOIS
W. W. WELBON

320.

Tentative Method of Analysis for Nickel Content of Atmospheric Particulate Matter by Atomic Absorption Spectroscopy

12136-01-73T

1. Principle of the Method

1.1 Samples are collected by drawing a known volume of air through a membrane or glass fiber filter. The filter samples are ashed, extracted with acid, and the analysis is subsequently made by atomic absorption spectroscopy using the 232.0 nm nickel line (**1**).

2. Range and Sensitivity

2.1 This method is applicable to the determination of nickel in quantities of 0.1 to

20.0 μg of nickel/ml of solution. An atmospheric concentration of 0.005 μg of nickel/m^3 can be detected using the method as described. For this concentration a minimum air sample volume of 2000 m^3 is recommended for H-Vol samples (See **7.1**).

3. Interferences

3.1 Silica extracted from the glass fiber filter and from the collected particulate matter can cause a significant interference with the measurement of nickel. This interference can be overcome by allowing the acid extracts to stand overnight and centrifuging at about 2000 RPM for 30 min.

3.2 Should large amounts of antimony and/or beryllium be suspected, their possible spectroscopic interference should be investigated. Usually ambient levels of these elements are not appreciable and the effects on the analyses can be considered negligible.

4. Precision and Accuracy

4.1 The precision of the method has not been reported using air samples; however, in the determination of nickel by a nearly identical method, an average standard deviation of 13% was obtained at normal urban concentrations **(2)**.

4.2 The recovery by this method is 88% provided the matrix of the sample is not appreciably different from that of the standards **(1)**. Interferences due to the presence of other metals can reduce the accuracy of the method **(3, 4, 5, 6)**.

5. Apparatus

5.1 SAMPLE COLLECTION. The fraction of the filter used in the analysis should be equivalent to the minimum air volume indicated as required in section **2.1**.

5.1.1 *Membrane filters,* such as Millipore, appropriately sized for sample volume and rate, with a pore size of 0.8 μ, are satisfactory for air sampling.

5.1.2 *Glass fiber filters* can also be used with a high volume air sampler. The filter blank should be investigated to establish the extent of nickel contamination and the potential matrix effect.

5.2 GLASSWARE. Borosilicate glassware should be used throughout the analysis. The glassware must be cleaned with 1:1 HCl and rinsed with distilled water. Beakers, pipets and auxiliary glassware will be required.

5.3 EQUIPMENT.

5.3.1 *Acetylene Gas (Cylinder).*

5.3.2 *Air Supply.*

5.3.3 *Atomic Absorption Spectrophotometer.*

5.3.4 *Centrifuge.*

5.3.5 *Low Temperature Asher-Tracerlab LTA-600 or equivalent.*

5.3.6 *Nickel Hollow Cathode Lamp.*

6. Reagents

6.1 PURITY. ACS reagent grade chemicals or equivalent shall be used in all tests. References to water shall be understood to mean double distilled water or equivalent. Care in selection of reagents and in following the above precautions is essential if low blank values are to be obtained.

6.2 CONC HYDROCHLORIC ACID (36.5–38.0%).

6.3 CONC NITRIC ACID (69.0–71.0%).

6.4 NICKEL SHOT.

6.5 PERCHLORIC ACID 72%.

6.6 HYDROCHLORIC ACID 1:10. Dilute 100 ml of conc acid to one liter with water.

6.7 PERCHLORIC-NITRIC ACID MIXTURE. Add 10 ml of conc perchloric to 90 ml of conc nitric.

7. Procedure

7.1 SAMPLE PREPARATION.

7.1.1 An 8″ × 10″ glass fiber filter, of which a 7″ × 9″ (63 sq. in.) section is exposed on the high-volume sampler, is divided into sections. The amount of a filter used for the analysis of metals depends upon the type of sample being prepared, urban or non urban, individual or composite. Table 320:I lists the amount of filter used for each type of sample.

Table 320:I. Amount of Filter Used for Each Type of Sample

Sample	Amount of Filter
Urban individual	1 × 7-inch strip
Urban composite	½ × 7-inch strip
Non-urban individual	2 × 7-inch strip
Non-urban composite	1 × 7-inch strip

The strips for metal analysis are ashed in a low-temperature asher. In this procedure, an oxygen plasma is formed by passing oxygen through a high-frequency electro-magnetic field and the activated oxygen reacts with the sample (**7**). Slow burning takes place at temperatures ranging from 50 to 250 C. A pyrex boat holding the filter section is placed into each of five compartments of the low-temperature asher and the strips are ashed at approximately 150 C for 1 hr at 250 W at 1 mm chamber pressure with an oxygen flow of 50 cc/min.

The ashed filter is placed in a glass thimble (Figure 311:1), which is then placed in an extraction tube (Figure 311:2). A 125-ml Erlenmeyer flask with a 24/40 female T joint is charged with 8 ml of constant boiling (about 19%) hydrochloric acid and 32 ml of 40% nitric acid. The flask is attached to the extraction tube, and the extraction tube is fitted with an Allihn condenser. The acid is refluxed over the sample for 3 hr. The sample and extraction thimble remain at the temperature of the boiling acid throughout the extraction.

The extraction tube and condenser are removed from the Erlenmeyer flask, and the flask is fitted with a thermometer adapter, which serves as a spray retainer. The extracted liquid is concentrated to 1 to 2 ml on a hot plate, and allowed to cool and stand overnight. The concentrated material is quantitatively transferred to a graduated 15-ml centrifuge tube with three washings of 5 to 10 drops of 1:10 HCl. Urban samples are diluted to 4.4 ml/1″ strip (40 ml/63 sq. in. of filter). Nonurban samples are diluted to 3.0 ml/2″ strip (13.3 ml/63 sq. in. of filter). Following dilution, the samples are centrifuged at 2000 rpm for 30 min and the supernatant liquid decanted into polypropylene tubes that are then capped and stored pending analysis. One ml from each solution is diluted with 1:10 HCl to 10 ml for atomic absorption analysis.

7.1.2 The Millipore filters are wet ashed with 10 to 20 ml of the mixture of nitric and perchloric acid in a suitably sized beaker. In the case of difficulty in oxidizing the organic matter, a first oxidation may be performed with a mixture of 20 ml of nitric acid and 5 ml of perchloric acid, followed by addition of a second 20 ml volume of nitric acid when the volume is first reduced to nearly 5 ml. (**Note:** All safety precautions necessary to the use of perchloric acid must be followed.) After the filter has been completely ashed, the acid is boiled down to the appearance of fumes of perchloric acid. All blanks and standards must be carried through with the same volumes of acid. Transfer to a graduated polypropylene tube with 1:10 HCl, add 1.2 of 1.1 HCl and dilute to 25 ml with 1:10 HCl.

7.2 Analysis

7.2.1 Set the instrument operating conditions as recommended by the manufacturer. The instrument should be set to the wavelength of maximum intensity for the 232.0 nm line from the hollow cathode lamp.

7.2.2 Aspirate the standards prepared fresh daily. Standards must be in the same matrix as the samples. Record the absorbance and prepare a calibration curve as described in section **8.1**.

(**Note:** All combustion products from the flame must be removed by direct exhaustion through the use of a good, separate ventilation system. This is of particular importance when using a nitrous oxide flame and when aspirating a sample containing a toxic element or a material with toxic combustion products.)

7.2.3 Aspirate the samples directly into the instrument and record the absorbance for comparison with the standards. Should the absorbance be off scale, dilute an appropriate aliquot to 10 ml. Be certain to aspirate distilled water between each sample. A blank and standard must be aspirated with sufficient frequency to assure the accuracy of the sample determinations.

8. Standards and Calibration

8.1 STANDARD NICKEL SOLUTION.

8.1.1 *Master Standard (1.000 mg nickel/ml).* Dissolve 1.000 g of reagent grade nickel shot in a 1:1 nitrichydrochloric mixture (10 to 25 ml). Upon complete dissolution evaporate to 5 ml and dilute to 1 liter with 1:1 HCl in a stoppered volumetric flask. This master standard is stable for one year and can be stored in a polyethylene bottle.

8.1.2 *Dilute Standard (0.010 mg nickel/ml).* Pipet 10 ml of the master standard into a 1-liter volumetric flask and dilute to volume with 1:10 HCl.

8.1.3 *Working Nickel Standards.* Prepare a 1.00 μg/ml working standard solution by pipetting 10.0 ml of the dilute standard into a 100-ml volumetric flask and diluting to the mark with 1:10 hydrochloric acid. Five known amounts of nickel ranging from 1.0 to 5.0 micrograms are added to the filters (membrane or glass) and carried through the procedure to establish a calibration graph. The calibration should be performed with each group of samples analyzed. The calibration graph must be a straight line. The working standard should be prepared fresh daily.

8.2 BLANK. A blank filter must be carried through the entire procedure.

9. Calculation

9.1 GLASS FIBER FILTERS. After a correction for the blank has been made, concentrations are calculated by multiplying the micrograms of nickel per ml in the sample aliquot by ten and by the total number of ml in the urban or non-urban sample (see **7.1.1**). The result is divided by the number of cubic meters of air represented by the portion of the filter taken for analysis.

9.2 $$\mu g\ Ni/m^3 = \frac{\mu g\ Ni/ml\ in\ aliquot \times 10 \times (total\ ml\ in\ urban\ or\ non\text{-}urban\ sample)}{(total\ corrected\ m^3\ of\ air)^*}$$

10. Storage

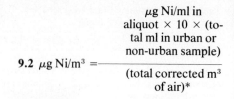

10.1 Untreated filter samples may be stored indefinitely in polyethylene bottles. Acid extracts may be stored for a month in polypropylene containers.

11. References

1. THOMPSON, R.J., G.B. MORGAN, L.J. PURDUE. 1970. Analysis of Selected Elements in Atmospheric Particulate Matter by Atomic Absorption. *Atomic Absorption Newsletter*, 9:53–57.
2. KNEIP, T.J., M. EISENBUD, C.D. STREHLOW and PETER C. FREUDENTHAL. 1970. Airborne Particulates in New York City. *APCA Journal*, 20:144–149.
3. SLAVIN, W. 1968. *Atomic Absorption Spectroscopy*. Inter-Science Publishers, New York, New York.
5. BURELL, D.C. 1965. Determination of Nickel & Cobalt in Natural Waters by Atomic Absorption Spectrometry. *Atomic Absorption Newsletter*, 4:7.
5. BEYER, M. 1965. Determination of Manganese, Copper, Chromium, Nickel & Magnesium in Cast Iron & Steel. *Atomic Absorption Newsletter*, 4:3.
6. SACHDEV, S.L., J.W. ROBINSON, P.W. EST. 1967. Determination of Manganese, Iron, Cobalt & Nickel in Air & Water by Atomic Absorption Spectroscopy. *Anal. Chem. Acta.*, 38:499–506.
7. GLEIT, C.E. *Am. J. of Med. Electronics*, 2:112.

Subcommittee 6
T. J. KNEIP, *Chairman*
R. S. ADEMIAN
J. R. CARLBERG
J. DRISCOLL
J. L. MOYERS
L. KORNREICH
J. W. LOVELAND
R. J. THOMPSON

*25 C and 760 Torr. The total volume of air in cubic meters is for that fraction of the filter taken.

321.

Tentative Method of Analysis for Vanadium Content of Atmospheric Particulate Matter by Atomic Absorption Spectroscopy

12164-01-73T

1. Principle of the Method

1.1 Samples can be collected by drawing a known volume of air through membrane or glass fiber filters. The samples are ashed and extracted with a mixture of nitric and hydrochloric acids. The analysis is subsequently made by atomic absorption spectroscopy using the 318.4 nm vanadium line (**1**). The method of additions is used to obtain greater accuracy. A nitrous oxide-acetylene flame is required.

2. Range and Sensitivity

2.1 This method is applicable to the determination of vanadium in an optimum range of 1.0 to 20.0 μg of vanadium per ml of solution. An atmospheric concentration of 0.2 μg of vanadium/m^3 can be detected using the method as described. For this concentration, a minimum air sample volume of 2000 m^3 is recommended for samples taken on 8″ × 10″ filters with a Hi-Vol sampler (see **7.1**).

3. Interferences

3.1 Silica extracted from the glass fiber filter and from the collected particulate matter may cause a significant interference with the measurement of vanadium. This interference can be overcome by allowing the acid extracts to stand overnight and centrifuging at about 2000 RPM for 30 min.

3.2 The nitrous oxide-acetylene flame must be used to reduce oxide formation and resulting loss of sensitivity. Aluminum, silicon, zinc, potassium and sodium are reported to enhance the vanadium signal (**2, 3**). As these elements are present in substantial and varying amounts in the sample, the method of standard additions must be used.

4. Precision and Accuracy

4.1 The precision of the method has not been reported using air samples; however, in the determination of vanadium by a nearly identical method, an average standard deviation of 21% was obtained at normal urban concentrations (**4**).

4.2 The recovery by this method has not been reported.

5. Apparatus

5.1 Sample Collection.

5.1.1 *Membrane filters.* Filters such as Millipore, appropriately sized for sample vollipore, appropriately sized for sample volume and rate, with a pore size of 0.8 μ, are satisfactory for air sampling.*

5.1.2 *Glass fiber filters.* These may also be used with a high volume air sampler. The filter blank should be investigated to establish the extent of vanadium contamination and the potential matrix effect.

5.2 Glassware. Borosilicate glassware should be used throughout the analysis. The glassware must be cleaned with 1:1 HCl and rinsed with distilled water. Beakers, pipets and auxiliary glassware will be required.

5.3 Equipment.

*Note: Collection efficiencies greater than 99% are obtained for a 0.3 μ DOP smoke at face velocities from 7.2–53 cm/sec (**5**).

5.3.1 *Acetylene Gas.* (Cylinder)

5.3.2 *Air Supply.*

5.3.3 *Atomic Absorption Spectrophotometer.*

5.3.4 *Centrifuge.*

5.3.5 *Vanadium Hollow Cathode Lamp.*

5.3.6 *Low Temperature Asher-Tracerlab LTA-600* or equivalent.

5.3.7 *Nitrous Oxide Gas.* (Cylinder)

5.3.8 *Nitrous Oxide Burner Acessory.*

6. Reagents

6.1 PURITY ACS reagent grade chemicals or equivalent shall be used in all tests. References to water shall be understood to mean double distilled water or equivalent. Care in selection of reagents and in following the above precautions is essential if low blank values are to be obtained.

6.2 CONC HYDROCHLORIC ACID (36.5–38.0%). sp. gr.—1.18.

6.3 CONC NITRIC ACID (69.0–71.0%). sp. gr.—1.42.

6.4 AMMONIUM VANADATE.

6.5 PERCHLORIC ACID. 72%.

6.6 HYDROCHLORIC ACID 1–10. Dilute 100 ml of conc acid to one liter with water.

6.7 PERCHLORIC-NITRIC ACID MIXTURE. Add 10 ml of conc perchloric acid to 90 ml of conc nitric acid.

7. Procedure

7.1 SAMPLE PREPARATION.

7.1.1 An 8″ × 10″ glass fiber filter, of which a 7″ × 9″ (63 square inches) section is exposed on the high-volume sampler is divided into sections. The amount of a filter used for the analysis of metals depends upon the type of sample being prepared, urban or non-urban, individual or composite. Table 311:I lists the amount of filter used for each type of sample. The strips for metal analysis are ashed in a low-temperature asher. In this procedure, an oxygen plasma is formed by passing oxygen through a high-frequency electromagnetic field and the activated oxygen reacts with the sample (**6**). Slow burning takes place at temperatures ranging from 50 to 250 C. A pyrex boat is placed into each of five compartments of the low-temperature asher and the strips are ashed at approximately 50 C for 1 hr at 250 watts at 1 mm chamber pressure with an oxygen flow of 50 cc per min. The ashed filter is placed in a glass thimble (Figure 311:1), which is then placed in an extraction tube (Figure 311:2). A 125-ml Erlenmeyer flask with a 24/40 female $ joint is charged with 8 ml of constant boiling (about 19%) hydrochloric acid and 32 ml of 40% nitric acid. The flask is attached to the extraction tube, and the extraction tube is fitted with an Allihn condenser. The acid is refluxed over the sample for 3 hr. The sample and extraction thimble remain at the temperature of the boiling acid throughout the extraction. The extraction tube and condenser are removed from the Erlenmeyer flask and the flask is fitted with a thermometer adapter, which serves as a spray retainer. The extracted liquid is concentrated to 1 to 2 ml on a hot plate, and allowed to cool and stand overnight. The concentrated material is quantitatively transferred to a graduated 15-ml centrifuge tube with three washings of 5 to 10 drops 1:10 HCl.† Urban samples are diluted to 4.4 ml/1″ strip (40 ml/63 sq in of filter). Non-urban samples are diluted to 3.0 ml/2″ strip (13.3 ml/63 sq in of filter). Following dilution, the samples are centrifuged at 2000 RPM for about 30 min and the supernatant liquid decanted into polypropylene tubes that are then capped and stored pending analysis. One ml from each solution is diluted with 1:10 HCl† to 10 ml for atomic absorption analysis.

7.1.2 The membrane filters are wet ashed with 10 to 20 ml of the mixture of nitric and perchloric acid in a suitably sized beaker. In the case of difficulty in oxidizing the organic matter, a first oxidation may be performed with a mixture of 20 ml of nitric acid and 5 ml of perchloric acid,

†Nitric acid may be used if HCl results in corrosion of parts in the burner system.

followed by addition of a second 20 ml volume of nitric acid when the volume is first reduced to nearly 5 ml. (**Note:** All safety precautions necessary to the use of perchloric acid must be followed.) After the filter has been completely ashed, the acid is boiled down to the appearance of fumes of perchloric acid. Transfer to a graduated polypropylene tube with 1:10 HCl, add 1.2 ml of 1:1 HCl, and dilute to 25 ml with 1:10 HCl or nitric acid as appropriate.

7.2 ANALYSIS.

7.2.1 Set the instrument operating conditions as recommended by the manufacturer. The instrument should be set to the wavelength of maximum intensity for the 318.4 nm line from the hollow cathode lamp.

7.2.2 Aspirate the samples directly into the instrument and record the absorbance.‡ Should the absorbance be off scale, dilute an appropriate aliquot to 10 ml with 1:10 HCl (or HNO_3 as appropriate). Be certain to aspirate distilled water after each sample. A blank and standard must be aspirated with sufficient frequency to assure the accuracy of the sample determinations.

7.2.3 *Method of Standard Additions.* Because of the possibility that the vanadium absorption may be strongly affected by matrix variations, it is recommended that the method of standard additions be used. This method has been discussed by Slavin **(2)** Dean and Rains **(7)** and by Elwell and Gidley **(8)**. Care must be exercised to avoid variations in matrix due to the additions used and to identify the true background signal. Absorption due to scattering or broad non-specific molecular absorption must be evaluated in selecting the true background signal. Use of an instrument is recommended which provides background correction.

The master standard is diluted with 1:1 HCl and 1:10 HCl as in **8.1.2** and **8.1.3** to prepare a solution to be used for the additions. The concentration to be prepared is determined by estimating the vanadium concentration in the sample by comparison to the working standards, and diluting the master standard to give a value in μg/l such that 10 μl of standard will contain a quantity of vanadium equal to that in each of the sample aliquots taken. The sample is divided into a minimum of three aliquots (four are to be preferred). Calibrated micropipets are used to add sufficient vanadium to give added concentrations in the final solution 0.5, 2.0 and 3 times the initial concentration without significant sample volume changes. This maintains the matrix unchanged.

The samples are aspirated and the absorbance readings obtained. The values obtained are plotted as shown in Figure 321:1 and a graphical extrapolation is done to the previously determined background absorbance **(2)**. The x-intercept at the background absorbance is taken as the sample concentration. If the relationship between the instrument response and concentration is non-linear, the samples should be diluted with 1:10 HCl (or HNO_3) sufficiently to bring them into the linear range.

It is recommended that a least square line be computed for the calibration, and the sample value be calculated using the equation for this line and the correct background absorbance as the zero signal.

Alternatively several sections of the filter may be taken and known amounts of the vanadium standard solution may be added to the acid mixture prior to dissolution of all but one of these filter sections. The results when plotted or calculated as described above will provide the necessary calibration line.

7.2.4 *Blank.* A blank filter must be carried through the entire procedure.

8. Standards and Calibration

8.1 MASTER STANDARD. (1.0000 MG VANADIUM PER ML). Dissolve 2.2963 g of ammonium vanadate (NH_4VO_3) in 100 ml of water plus 10 ml of HNO_3 and dilute to

‡All combustion products from the flame must be removed by direct exhaustion through the use of a good, separate ventilation system. This is of particular importance when using a nitrous oxide flame and when aspirating a sample containing a toxic element or a material with toxic combustion products.

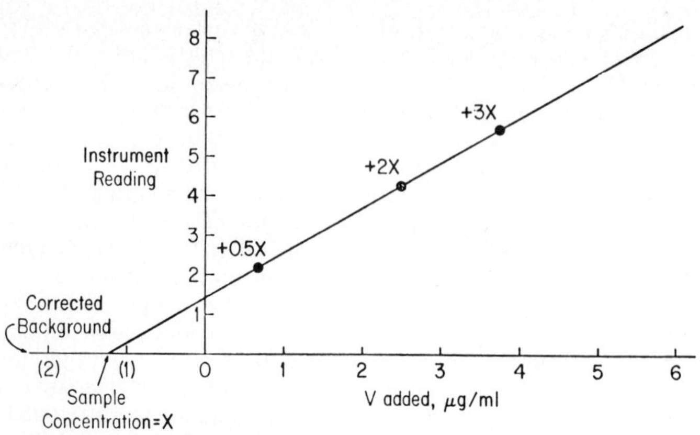

Figure 321:1—Plot of instrument reading vs. amounts of Vanadium added.

1 liter in a stoppered volumetric flask. This master standard is stable for at least one year stored in a polyethylene bottle.

8.2 DILUTE STANDARD. (0.100 mg vanadium per ml). Pipet 10.0 ml of the master standard into a 100-ml volumetric flask and dilute to volume with 1:1 HCl (or HNO_3 as appropriate).

8.3 WORKING VANADIUM STANDARDS. Prepare a 1.00 μg/ml working standard solution by pipetting 1.00 ml of the dilute standard into a 100-ml volumetric flask, adding 4.00 ml of 1:1 HCl and diluting to the mark with 1:10 HCl (or HNO_3 as appropriate). Repeat the procedure with the appropriate amounts of the dilute standard and 1:1 HCl (or HNO_3 as appropriate) (total volume of 5.00 ml) to prepare standards containing 2.0 μg/ml, 3.0 μg/ml, 4.0 μg/ml and 5.0 μg/ml of vanadium. Working standards must be prepared daily. These standards are used for establishing the instrument response and for comparison to sample absorbances as specified in section **7.2.2**.

9. Calculations

9.1 Concentrations are calculated by multiplying the μg of vanadium per ml in the sample aliquot by 10 and by the total number of ml in the urban or non-urban sample (Sec **7.1.1**) or by the appropriate dilution factors for the volumes used with membrane filters. The result is divided by the number of m^3 of air represented by the portion of the filter taken for analysis.

9.2 μg V/m^3 =

$$\frac{\mu\text{g V/ml in aliquot} \times 10 \times (\text{total ml in urban or non-urban sample})}{(\text{total corrected m}^3 \text{ of air})\S}$$

10. Storage

10.1 Untreated filter samples may be stored indefinitely. Acid extracts may be stored for a month in polypropylene containers.

11. References

1. THOMPSON, R.J., G.B. MORGAN, and L.J. PURDUE. 1970. Analysis of Selected Elements in Atmospheric Particulate Matter by Atomic Absorption. *Atomic Absorption Newsletter*, 9:53–57.
2. SLAVIN, W. 1968. *Atomic Absorption Spectroscopy*, Inter-Science Publishers, New York, New York.
3. QUICKERT, N., A. ZDROJEWSKI and L. DUBOIS. 1973. The Accurate Measurement of Vanadium in Airborne Particulates, *International Journal of Environmental Analytical Chemistry*. (In press).
4. KNEIP, T.J., M. EISENBUD, C.D. STREHLOW and P.C. FREUDENTHAL. 1970. Airborne Particulates in New York City. *APCA J*, 20:144–149.

§This is that volume of air in m^3 represented by the fraction of filter taken for analysis. This volume is corrected to 25 C and 760 torr.

5. Lockhart, L.B., Jr., R.L. Patterson, Jr. and W.L. Anderson. 1963. Characteristics of Air Filter Media Used for Monitoring Airborne Particulates. *NRL Report 6054*.

6. Gleit, C.E. 1963. *Am. J. of Med. Electronics*, 2:112.

7. Dean, J.A. and T.C. Rains. 1969. *Flame Emission and Atomic Absorption Spectrometry*, Vol. 1, p. 377, Marcel Dekker, New York.

8. Elwell, W.T., and J.A.F. Gidley. 1966. *Atomic-Absorption Spectrophotometry*, 2nd Revised Edition, p. 56, Pergamon Press, New York.

Subcommittee 6
T. J. Kneip, *Chairman*
R. S. Ajemian
J. R. Carlberg
J. Driscoll
L. Kornreich
J. W. Loveland
J. L. Moyers
R. J. Thompson

322.

Tentative Method of Analysis for Zinc Content of Atmospheric Particulate Matter by Atomic Absorption Spectroscopy

12167-01-73T

1. Principle of the Method

1.1 Samples can be collected by drawing a known volume of air through membrane or glass fiber filters. The samples are ashed and extracted with a mixture of nitric and hydrochloric acids. The analysis is subsequently made by atomic absorption spectroscopy using the 213.8 nm zinc line (**1**).

2. Range and Sensitivity

2.1 This method is applicable to the determination of zinc in an optimum range of 0.01 to 5 μg of zinc/ml of solution. An atmospheric concentration of 0.005 μg of zinc/m^3 can be detected using the method as described. For this concentraton, a minimum air sample volume of 2000 m^3 is recommended for samples taken on 8″ × 10″ filters with a Hi-Vol sampler. (see Section **7.1**, and Ref. **2, 3**).

3. Interferences

3.1 Silica extracted from the glass fiber filter and from the collected particulate matter may cause a significant interference with the measurement of zinc. This interference can be overcome by allowing the acid extracts to stand overnight and centrifuging at about 2000 RPM for 30 min.

3.2 High zinc blanks have been reported for glass fiber filters (**2**).

4. Precision and Accuracy

4.1 The precision of the method has not been reported using air samples; however, in the determination of zinc by a nearly identical method, a relative standard deviation of 10% was obtained for urban concentrations (**2, 4**).

4.2 The recovery by this method has not been reported.

5. Apparatus

5.1 SAMPLE COLLECTION.

5.1.1 *Membrane filters.* Filters such as Millipore, appropriately sized for sample volume and rate, with a pore size of 0.8 μ, are satisfactory for air sampling.*

5.1.2 *Glass fiber filters.* These may also be used with a high volume air sampler. The filter blank should be investigated to establish the extent of zinc contamination and the potential matrix effect.

5.2 GLASSWARE. Borosilicate glassware should be used throughout the analysis. The glassware must be cleaned with 1:1 HCl and rinsed with distilled water. Beakers, pipet and auxiliary glassware will be required.

5.3 EQUIPMENT.

5.3.1 *Acetylene Gas.* (Cylinder)

5.3.2 *Air Supply.*

5.3.3 *Atomic Absorption Spectrophotometer.*

5.3.4 *Centrifuge.*

5.3.5 *Zinc Hollow Cathode Lamp.*

5.3.6 *Low Temperature Asher-Tracerlab LTA-600 or equivalent.*

6. Reagents

6.1 PURITY. ACS reagent grade chemicals or equivalent shall be used in all tests. References to water shall be understood to mean double distilled water or equivalent. Care in selection of reagents and in following the above precautions is essential if low blank values are to be obtained.

6.2 CONC HYDROCHLORIC ACID (36.5–38.0%). sp gr—1.18.

6.3 CONC NITRIC ACID (69.0–71.0%). sp gr—1.42.

6.4 ZINC METAL.

6.5 PERCHLORIC ACID. 72%.

6.6 HYDROCHLORIC ACID 1–10. Dilute 100 ml of conc acid to one liter with water.

6.7 PERCHLORIC-NITRIC ACID MIXTURE. Add 10 ml of conc perchloric to 90 ml of conc nitric.

7. Procedure

7.1 SAMPLE PREPARATION.

7.1.1 An 8″ × 10″ glass fiber filter, of which a 7″ × 9″ (63 square inches) section is exposed on the high-volume sampler, is divided into sections. The amount of a filter used for the analysis of metals depends upon the type of sampling being prepared, urban or non-urban, individual or composite. Table 311:I lists the amount of filter used for each type of sample.

The strips for metal analysis are ashed in a low-temperature asher. In this procedure, an oxygen plasma is formed by passing oxygen through a high-frequency electromagnetic field and the activated oxygen reacts with the sample (**5**). Slow burning takes place at temperatures ranging from 50 to 250 C. A pyrex boat is placed into each of five compartments of the low-temperature asher and the strips are ashed at approximately 150 C for 1 hr at 250 W at 1 mm chamber pressure with an oxygen flow of 50 cc per min.

The ashed filter is placed in a glass thimble (Figure 311:1), which is then placed in an extraction tube (Figure 311:2). A 125-ml Erlenmeyer flask with a 24/40 female ₮ joint is charged with 8 ml of constant boiling (about 19%) hydrochloric acid and 32 ml of 40% nitric acid. The flask is attached to the extraction tube, and the extraction tube is fitted with an Allihn condenser. The acid is refluxed over the sample for 3 hr. The sample and extraction thimble remain at the temperature of the boiling acid throughout the extraction.

The extraction tube and condenser are removed from the Erlenmeyer flask and the flask is fitted with a thermometer adapter, which serves as a spray retainer. The extracted liquid is concentrated to 1 to 2 ml on a hot plate, and allowed to cool and stand overnight. The concentrated material is quantitatively transferred to a graduated 15-ml centrifuge tube with three washings of 5 to 10 drops 1:10 HCl. Urban

*Collection efficiencies greater than 99% are obtained for a 0.3 μ DOP smoke at face velocities from 7.2 to 53 cm/sec (**6**).

samples are diluted to 4.4 ml per 1″ strip (40 ml per 63 sq. in. of filter). Non-urban samples are diluted to 3.0 ml per 2″ strip (13.3 ml/63 sq. in. of filter). Following dilution, the samples are centrifuged at 2000 RPM for about 30 min and the supernatant liquid decanted into polypropylene tubes that are then capped and stored pending analysis. One ml from each solution is diluted with 1:10 HCl to 10 ml for atomic absorption analysis.

7.1.2 The Millipore filters are wet ashed with 10 to 20 ml of the mixture of nitric and perchloric acid in a suitably sized beaker. In the case of difficulty in oxidizing the organic matter, a first oxidation may be performed with a mixture of 20 ml of nitric acid and 5 ml of perchloric acid, followed by addition of a second 20 ml volume of nitric acid when the volume is first reduced to nearly 5 ml. (**Note:** All safety precautions necessary to the use of perchloric acid must be followed.) After the filter has been completely ashed, the acid is boiled down to the appearance of fumes of perchloric acid. All blanks and standards must be carried through with the same volumes of acids. Transfer to a graduated polypropylene tube with 1:10 HCl, add 1.2 ml of 1:1 HCl, and dilute to 25 ml with 1:10 HCl.

7.2 ANALYSIS.

7.2.1 Set the instrument operating conditions as recommended by the manufacturer. The instrument should be set to the wavelength of maximum intensity for the 213.8 nm line from the hollow cathode lamp.

7.2.2 Aspirate the standards prepared fresh daily. Standards must be in the same matrix as the samples. Record the absorbance and prepare a calibration curve as described in section **8.1**.†

7.2.3 Aspirate the samples directly into the instrument and record the absorbance for comparison with the standards. Should the absorbance be off scale, dilute an appropriate aliquot to 10 ml. Be certain to aspirate distilled water between each sample. A blank and standard must be aspirated with sufficient frequency to assure the accuracy of the sample determinations.

†All combustion products from the flame must be removed by direct exhaustion through the use of a good, separate ventilation system. This is of particular importance when using a nitrous oxide flame and when aspirating a sample containing a toxic element or a material with toxic combustion products.

8. Standards and Calibration

8.1 STANDARD ZINC SOLUTION.

8.1.1 *Master Standard (1.000 mg zinc per ml.)* Dissolve 1.000 g of reagent grade zinc in a minimum volume of a 1:1 nitric-hydrochloric mixture (10 to 25 ml). Upon complete dissolution, evaporate to 5 ml and dilute to 1 l with 1:1 HCl in a stoppered volumetric flask. This master standard is stable for at least one year stored in a polyethylene bottle.

8.1.2 *Dilute Standard (0.100 mg zinc per ml).* Pipet 10.0 ml of the master standard into a 100-ml volumetric flask and dilute to volume with 1:1 HCl.

8.1.3 *Working Zinc Standards.* Prepare a 1.00 μg/ml working standard solution by pipetting 1.00 ml of the dilute standard into a 100-ml volumetric flask, adding 4.00 ml of 1:1 HCl and diluting to the mark with 1:10 HCl. Repeat the procedure with the appropriate amounts of the dilute standard and 1:1 HCl (total volume of 5.00 ml) to prepare standards containing 2.0 μg/ml, 3.0 μg/ml, 4.0 μg/ml and 5.0 μg/ml of zinc. Working standards must be prepared daily.

8.1.4 *Calibrating Sample Preparation Procedure.* Five known amounts of zinc are added to 1″ strips of glass, or an equivalent portion of a membrane filter, and carried through the procedure to establish a calibration graph for the entire sample preparation procedure. If this calibration curve is in agreement with that obtained using the working standards, one need only perform this task periodically to insure that the sample preparation procedure is being carried out properly. If the calibration graph for the sample preparation differs significantly from that obtained with the working standards, then the sample preparation calibration graph should be used to determine the zinc content of the samples and this calibration should be prepared with each set of samples analyzed.

8.2 Blank. A blank filter must be carried through the entire procedure.

9. Calculations

9.1 Glass fiber filter. After a correction for the blank has been made, concentrations are calculated by multiplying the micrograms of zinc/ml in the sample aliquot by ten and by the total number of ml in the urban or non-urban sample (See **7.1.1**). The result is divided by the number of cubic meters of air represented by the portion of the filter taken for analysis.

9.2

$$\mu g\ Zn/m^3 = \frac{\mu g\ Zn/ml \text{ in aliquot} \times 10 \times (\text{total ml in urban or non-urban sample})}{(\text{total corrected } m^3 \text{ of air})\ddagger}$$

9.3 Millipore filter. Make correction for blank. Calculate the concentration as follows (see **7.1.2**).

$$\mu g\ Zn/m^3 = \frac{\mu g\ Zn/ml \text{ in aliquot} \times 25}{(\text{corrected } m^3 \text{ of air for sample})}$$

‡This is that volume of air in m^3 represented by the fraction of filter taken for analysis. This volume is corrected to 20 C and 760 torr.

10. Storage

10.1 Untreated filter samples may be stored indefinitely. Acid extracts may be stored for a month in polypropylene containers.

11. References

1. Thompson, R.J., G.B. Morgan, L.J. Purdue. 1970. Analysis of Selected Elements in Atmospheric Particulate Matter by Atomic Absorption. *Atomic Absorption Newsletter* 9:53–57.
2. Kneip, T.J., M. Eisenbud, C.C. Strehlow and Peter C. Freudenthal. 1970. Airborne Particulates in New York City. *APCA Journal*, 20:144–149.
3. Jackson, G.B., and H.N. Myrick. 1970. Analytical Methods Utilizing AAS for Iron, Lead and Zinc Metallic Ions in Airborne Particulates, *American Laboratory*, 19–27.
4. Kometani, T.Y., J.L. Bove, B. Nathanson, S. Siebenbert, and M. Magyar. 1972. Dry Ashing of Airborne Particulate Matter on Paper and Glass Fiber Filters for Trace Metal Analysis by Atomic Absorption Spectrometry. *Environ. Science Tech.* 6:617–620.
5. Gleit, C.E. 1963. *Am. J. of Med. Electronics*, 2:112.
6. Lockhart, L.B., Jr., R.L. Patterson, Jr. and W.L. Anderson. 1963. Characteristics of Air Filter Media Used for Monitoring Airborne Particulates. *NRL Report* 6054.

Subcommittee 6
T. J. Kneip, *Chairman*
R. S. Ajemian
J. R. Carlberg
J. Driscoll
L. Kornreich
J. W. Loveland
J. L. Moyers
R. J. Thompson

401.

Tentative Method of Analysis for Ammonia in the Atmosphere (Indophenol Method)

42604-01-72T

1. Principle of the Method

1.1 Ammonia in the atmosphere is collected by bubbling a measured volume of air through a dilute solution of sulfuric acid to form ammonium sulfate.

1.2 The ammonium sulfate formed in the sample is analyzed colorimetrically by reaction with phenol and alkaline sodium hypochlorite to produce indophenol **(1)**, a blue dye. The reaction is accelerated by the addition of sodium nitroprusside as catalyst. The reaction has been postulated to be as follows **(2)**.

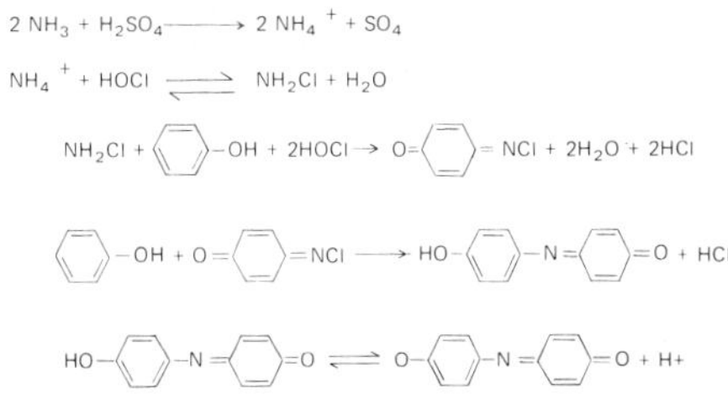

2. Range and Sensitivity

2.1 With a sampling rate of 1 to 2 liters/min a conc range of 20 to 700 $\mu g/m^3$ (0.025 to 1 ppm) of air may be determined with a sampling time of one hr.

2.2 The limit of detection of the analysis is 0.02 μ NH_3/ml.

3. Interferences

3.1 Ammonium compounds in suspended particulate matter will be determined if they are not removed by prefiltration.

3.2 Prefilters may remove some gaseous ammonia. See section **7.1.1**.

3.3 Ferrous, chromous, and manganous ions if present in mg amounts cause positive interference in the analytical procedure because of precipitation. Copper ions inhibit color development strongly and therefore cause negative interference. Addition of ethylenediamine tetraacetic acid (EDTA) prevents these effects **(1)**. Nitrite and sulfite interfere if present in 100-fold excess. Based on tests with solutions, formaldehyde causes a negative interference of 10 to 15%. Interfering particulate matter in the air can be removed by filtration of the air sample **(3)**.

4. Precision and Accuracy

4.1 Replicate samples collected with glass prefilters and analyzed manually, showed a relative coefficient of variation of 30% in the 0.7 to 21 $\mu g/m^3$ (1 to 30 ppb) range **(3)**. This coefficient may vary with conc of atmospheric ammonia and goes as low as 5% in the 700 $\mu g/m^3$ (1 ppm) range.

4.2 No accuracy data are available.

5. Apparatus

5.1 Vacuum Pump. Any vacuum pump which will maintain a vacuum of 450 Torr with an air flow of 5 liters per min.

5.2 Flow Measuring Device. A calibrated flow meter or a critical orifice to measure or control the air flow from 1 to 2 l/min. The flow meter should be calibrated under conditions of use.

5.3 Absorber. A fritted bubbler, **(4)** midget impinger, **(5)** or other gas scrubber designed for a flow rate of 1 to 2 l/min with 10 ml or more absorbing solution.

5.4 Prefilter Holder. An open face

Teflon, Lexan, or similar filter holder which can be connected in line before the absorber. Metal or other plastic holder may be used if it is known that it does not absorb ammonia.

5.5 PREFILTERS. Organic-free glass fiber filters used for air sampling for suspended particulates are satisfactory. The filters are washed with distilled water and dried prior to use.

5.6 SPECTROPHOTOMETER. Capable of measuring the absorbance at 630 nm.

5.7 GLASSWARE. Low actinic glassware or vessels must be used for analysis. The glassware must be rinsed with 1.2 N HCl, washed at least 6 times with reagent water before immediate use.

6. Reagents

6.1 PURITY. All chemicals should be of analytical reagent grade **(6)**.

6.2 WATER. Must be double distilled and ammonia free **(5)**.

6.3 ABSORBING SOLUTION. Dilute 2.3 ml of conc H_2SO_4 (18 M) to 1 l with distilled water to obtain 0.1 N H_2SO_4.

6.4 ANALYTICAL REAGENTS.

6.4.1 *Sodium nitroprusside* **(7)**. Dissolve 2 g sodium nitroprusside in 100 ml of distilled water. The solution keeps well in the refrigerator for 2 months.

6.4.2 *6.75 M Sodium hydroxide*. Dissolve 270 g sodium hydroxide in about 1 l of distilled water. Boil down to 600 ml in order to volatilize the ammonia contained in the reagent. Cool and fill to one liter. Store in polyethylene bottle. **Caution:** *This solution is extremely caustic. Prevent contact with skin or eyes.*

6.4.3 *Sodium hypochlorite solution.* Dilute 5 to 6% analytical reagent sodium hypochlorite with distilled water to give a 0.1 N solution (3.7%). Strength is determined before dilution by iodimetric titration or colorimetry after appropriate dilution. The solution keeps well for 2 months in a refrigerator.

6.4.4 *Phenol solution 45% v/v*. Melt phenol by immersing a bottle containing the material in a water bath at 60 C. Pour 45 ml (50 g) into a 100 ml warmed cylinder and fill to mark with methanol. This solution keeps well for 2 months in a refrigerator.

6.4.5 *Buffer*. Dissolve 50 g of $Na_3PO_4 \cdot 12H_2O$ and 74 ml of 6.75 M NaOH in a liter of distilled water.

6.4.6 *Working hypochlorite solution.* Mix 30 ml of 0.1 N sodium hypochlorite and 30 ml of 6.75 M sodium hydroxide and dilute to 100 ml with distilled water. Prepare fresh daily.

6.4.7 *Working phenol solution.* Mix 20 ml of the 45% phenol solution with 1 ml of 2% sodium nitroprusside and dilute to 100 ml with distilled water. Prepare fresh every 4 hr.

6.5 AMMONIA STANDARD SOLUTION.

6.5.1 *Ammonia stock solution.* Dissolve 3.18 g of NH_4Cl or 3.88 g of $(NH_4)_2SO_4$ in 1 l of distilled water (1 ml equal to 1 mg NH_3). Add a drop of $CHCl_3$ for better preservation. The solution is stable for 2 months.

6.5.2 *Ammonia working solution.* Dilute 10 ml of the stock solution to 1 l with absorbing solution in a volumetric flask (1 ml equal to 10 μg NH_3). Prepare daily.

6.6 GLASS CLEANING SOLUTION. Dilute 10 ml of conc HCl (12 M) to 100 ml with distilled water. Molarity approximately 1.2 M.

7. Procedure

7.1 SAMPLE COLLECTION. **(3)** 10 ml absorbing solution are placed in each bubbler for samples and field blanks. Cap bubblers for transport. Assemble in order, prefilter, and holder, flowmeter, bubbler, and pump. Sample at the rate of 1 to 21/min for a sufficient time to obtain an adequate sample, usually one hr. Record sampling-time and flow rate. After sample collection, recap the bubblers.

7.1.1 *Prefilters.* If prefilters are not used the method will determine both gaseous ammonia and ammonium contained in particulates. At high humidity, acid gas will promote reaction on the filter causing loss of ammonia gas from the sample. In the absence of acid gases, ammonia collected momentarily on the filter during high humidity will be stripped off during sampling with little loss. The filter must be prevented from being wetted by rain **(3)**.

7.2 Analysis. **(6)** If bubbler is marked at 25.0 ml level, the color may be developed in the flasks. If not, transfer contents to a 25.0 ml glass stoppered graduate cylinder being sure to blow out residual sample from the fritted bubblers if they are used. Maintain all solutions and sample at 25 C. Add 2 ml buffer. Add 5 ml of the working phenol solution, mix, fill to about 22 ml, then add 2.5 ml of the working hypochlorite solution and rapidly mix. Dilute to 25 ml, mix and store in the dark at 25 C for 30 min to develop color. Measure the absorbance of the solution against a reagent blank at 630 nm, using 1-cm cells.

7.3 Field Blanks. At least one bubbler of collecting solution is carried into the field and treated in the same fashion as the actual samples except that no air is drawn into the bubbler. It is treated in analysis as if it was a sample. The value of the field blank(s) is compared with the reagent blank to determine whether sampling glassware is introducing appreciable contamination.

8. Calibration and Standardization

8.1 Preparation of Standards. Pipet 0.5, 1.0 and 1.5 ml of the working standard solution into 25 ml glass stoppered graduated cylinders. These correspond to 5, 10 and 15 μg of ammonia/25 ml of solution. Fill to the 10-ml mark with absorbing solution. A reagent blank with 10 ml of absorbing solution is also prepared. Add reagents to each cylinder as in the procedure for analysis. Read the absorbance of each standard against the reagent blank.

8.2 Standard curve. Plot the absorbence as the ordinate versus the concentration as the abscissa on linear graph paper. Alternatively, determine the slope by the method of least squares.

9. Calculations

$$\mu g/m^3\ NH_3 = \frac{W}{Vo}$$

W = μg NH_3 in 25 ml from standard curve.

Vo = Volume of air sampled in m^3 at 25 C and 760 Torr.

$$Vo = \frac{(F)}{(1000)} \times (t) \times \frac{(Ps)}{(760)} \times \frac{298}{273 + Ts}$$

where

F = flow rate (liters/minute).
t = elapsed sampling time in minutes.
Ps = Atmospheric pressure in Torr at sampling point.
Ts = Temperature C at sampling point.

10. Effects of Storage

10.1 Samples of particulate matter may be stored indefinitely if protected from contamination.

10.2 Changes within the precision of the method occur when storing the collected liquid samples during 2 days. Significantly lower values have been found on replicates stored for several days.

11. References

1. Tetlow, J.A., and Wilson, A.L. 1964. An Absorptionmetric Method for Determining Ammonia in Boiler Feed-Water. *Analyst*, Vol. 89:453–465.
2. Rommers, P.J., and Visser, J. 1969. Spectrophotometric Determination of Micro Amounts of Nitrogen as Indophenol. *Analyst*, Vol. 94:653–658.
3. Tentative Method for Test for Ammonia in the Atmosphere. *ASTM* D-XXX-71 T.
4. Axelrod, H.D., A.F. Wertburg, R.J. Teck, and J.P. Lodge, Jr. 1971. A new bubbler design for atmospheric sampling.
5. American Public Health Association. 1977. Method of Air Sampling and Analysis. 2nd ed. p. 556. Washington, D.C.
6. American Chemical Society. 1950. Reagent Chemicals, ACS Specifications. Washington, D.C.
7. Harwood, J.E., and Kuhn, A.L. 1970. A colorimetric method for ammonia in natural waters. Water Res. 4:305.

Subcommittee 3

E. L. Kothny, *Chairman*
W. A. Cook
J. F. Cuddeback
B. Dimitriades
E. F. Ferrand
P. W. McDaniel
G. D. Nifong
B. E. Saltzman
F. T. Weiss

402.

Tentative Method of Analysis for Ammonia in the Atmosphere (Nitrite Method)

42604-02-73T

1. Principle of the Method

1.1 Particulate ammonium compounds are collected from the atmosphere by passing a measured volume of air through a neutral glass fiber filter. Gaseous ammonia is collected downstream from the filter by bubbling the air through a dilute solution of sulfuric acid to form ammonium sulfate. The ammonium sulfate solution obtained by extraction of the filter or by absorption into the dilute acid is reacted with hypochlorite in the presence of bromide as catalyst in an alkaline medium to form nitrite (**1**). After destruction of excess hypochlorite with sodium arsenite, the nitrite formed in the reaction is transformed into an azo dye and analyzed by spectrophotometry. A constant 65% of the absorbed ammonia is converted to nitrite (**2**).

1.2 Ammonium salts contained on hi-vol filter samples, membrane filters, and from electrostatic precipitators may also be analyzed by this method.

1.3 This method is preferred over the indophenol method (**3, 4**) because it is less subject to interferences present in urban atmospheres, such as formaldehyde.

2. Range and Sensitivity

2.1 The optimum concentration for ambient air sampled at the rate of 1 to 2 l/min for 1 hr, is 14 to 220 $\mu g\ NH_3/m^3$ (0.02 to 0.3 ppm). When higher concentrations are encountered, they may be determined by taking aliquots of the collected sample or by reducing the sampling time.

2.2 The sensitivity in molar absorptivity of ammonia is $e = 38{,}000$. The Sandell sensitivity (for $A = 0.001$ in a 1-cm cell) is 0.00045 $\mu g\ NH_3$/ml. These values correspond to approximately 0.01 $\mu g\ NH_3$/25 ml of solution.

3. Interferences

3.1 Sea salt components, such as sodium, magnesium, calcium, potassium, chloride, bromide, sulfate, as well as trace amounts of other inorganic components, do not interfere.

3.2 Formaldehyde up to 100 ppm (2.5 mg in 25 ml of final solution) does not interfere.

3.3 Nitrite, hydrolyzable amino compounds such as certain amino acids and other N-compounds may interfere (*e.g.*, 5% of the urea, 9% of methylamine, 32% of ethylamine, 71% of d,l-a-alanine, 76% of hydroxylamine (**2**) are converted to nitrite).

3.4 Particulate matter containing ammonium salts will interfere if not removed by filtration. Filters used for that purpose should have a neutral to slightly alkaline (pH 7 to 9) reaction when moistened to prevent reaction with gaseous ammonia.

4. Precision and Accuracy

4.1 Relative standard deviations of the analytical method of 1.8%, 1.6%, and 1.5% were obtained for the determination of the absorbances of the blank, sample and sample spiked with 0.82 $\mu g\ NH_3$/25 ml (**2**), respectively.

4.2 The sampling efficiency of glass impingers for gaseous ammonia is better than 95% at levels above 400 $\mu g\ NH_3/m^3$. At lower levels the sampling efficiency depends on the collecting device design, amount and concentration of the collecting solution. For midget glass impingers with 10 ml of absorbing solution the efficiency at levels of 10 to 50 $\mu g\ NH_3/m^3$ is about 50 to 70%.

4.3 Accuracy data are not available.

5. Apparatus

5.1 VACUUM PUMP. Any vacuum pump which will maintain a vacuum of 0.6 atmosphere (450 Torr) with an airflow of up to 10 l/min when using a critical orifice.

5.1.1 Alternatively, any vacuum pump capable of maintaining an airflow of up to 10 l/min is satisfactory when using a calibrated flowmeter.

5.2 PREFILTER HOLDER. An open face filter holder which can be connected by butt joints to the absorber.

5.3 ABSORBER. A fritted bubbler, midget impinger **(5)** or other gas scrubber designed for a flow rate of 1 to 2 l/min with 10 ml or more absorbing solution.

5.4 SPECTROPHOTOMETER. An instrument capable of measuring the color at 540/550 nm.

5.5 GLASSWARE. Use only actinic (dark colored) glassware and keep all vessels protected from daylight during the analysis. Rinse all glassware with 1.2 N HCl, wash at least six times with reagent water and cover with a plastic wrapper (such as Saran wrap or equivalent) or stopper with ground glass stoppers.

6. Reagents

6.1 PURITY. All chemicals should be analytical reagent grade **(6)**.

6.2 WATER. Double distilled water free of ammonia and other amino compounds which may interfere, should be used for all operations **(7)**.

6.3 FILTER PAPER. Glass fiber filter papers (such as Gelman A, Whatman, MSA, etc.) punched-out to the appropriate size, *e.g.*, 37 or 47 mm diam. Rinse with distilled water until no filter fragments can be detected visually in the water. Check the filtrate for ammonium ions. The minimum volume of water necessary for a thorough elimination of the ammonium ion is usually 400 ml. Establish this value by prior trials.

6.3.1 To insure no ammonia was removed from the sample or generated from the particulates, the pH of the filter must be tested after passing air. To test, pass ambient air through the filter for 24 hr. Moisten the filter with a drop of water and then test with a drop of universal indicator solution. It should react neutral to slightly alkaline (pH 7 to 9).

6.4 ABSORBING SOLUTION. Dilute 2.8 ml of conc sulfuric acid (18 M) to 1 l with distilled water to obtain 0.1 N H_2SO_4.

6.5 ALKALINE CATALYST MIXTURE. Dissolve 160 g NaOH and 8 g NaBr in 1 l of distilled water, boil down to 600 ml in order to drive out ammonia, cool and fill to the mark in a 1-l volumetric flask. Store the solution in plastic containers to avoid contamination by glass. **Caution!** Strongly caustic solution. Avoid contact with skin or eyes.

6.6 SODIUM HYPOCHLORITE, 1 N. Dilute 5 to 6% (approximately 1.5 N) reagent grade hypochlorite solution with distilled water to give a 1 N (3.7% w/w) solution. The amount of distilled water (v) and reagent (r) is calculated according to the available strength (c) in % w/w:

$$v = 100 - 370/c$$
$$r = 100 - v$$

Filter through a washed glass fiber filter or a frit. The strength of the solution is determined by iodimetric titrations or colorimetry before use or dilution (see sections **6.6.1** and **6.6.2** below). The solutions keep well for 2 weeks at room temperature without a significant change in concentration. Discard the solutions if the conc in either one drops below 3.5% w/w. These reagents should be kept in a refrigerator for maximum stability.

6.6.1 *Determination of hypochlorite concentration by titration*. Put about 50 ml of 1% KI solution in an Erlenmeyer flask and add 1 ml of the hypochlorite solution to be analyzed. Add 10 drops of 1:1 HCl and titrate the liberated iodine with 0.1 N $Na_2S_2O_3$ (dissolve 12.5 g $Na_2S_2O_3 \cdot 5\,H_2O$ in 500 ml of distilled water in a volumetric flask). Calculate the concentration of hypochlorite as follows:

$$C = \frac{MW}{2} \times \frac{g}{1000\ ml} \times 100\ ml \times N_h \times$$
$$N_t \times m = 0.37\ m$$

Where:

C = concentration of hypochlorite in % w/w (or g/100 ml):
m = ml of 0.1 N $Na_2S_2O_3$ used for the titration.
MW = molecular weight of hypochlorite (= 74).
g/1000 ml = conversion factor for obtaining atomic weight in g/ml.
N_h = final normality of the hypochlorite solution (= 1).
N_t = normality of the thiosulfate solution (= 0.1)

6.6.2 *Determination of hypochlorite concentration by colorimetry*. Dilute 1 ml of standard 0.1 N iodate solution to 100 ml in a volumetric flask. With this solution construct a calibration curve by adding from 0 to 20 ml of this 0.001 N dilution to 100-ml volumetric flasks containing about 50 ml of 1% KI. Add 10 drops 1:1 HCl, fill to the mark with distilled water and mix well. Dilute 1 ml of the iodine solution so obtained with 1% KI solution to 100 ml and measure the absorbance at 352 nm in a 1-cm cell. Dilute 1 ml of the hypochlorite solution to be analyzed with distilled water in a 100-ml volumetric flask. Add 1 ml of this approximately 0.01 N solution to a 100-ml volumetric flask containing about 50 ml of 1% KI. Then follow the same procedure as outlined above, starting from the point of HCl addition. Read the concentration from the calibration curve.

6.7 ALKALINE OXIDIZING REAGENT. Mix 5 ml of 1 N sodium hypochlorite with 95 ml of alkaline catalyst mixture. This solution should be used from 15 min to 4 hr after preparation. Discard after use.

6.8 SODIUM ARSENITE SOLUTION, 0.2 M. Dissolve 20 g As_2O_3 and 6 g NaOH in about 75 ml of hot distilled water. Dilute the solution to 500 ml with distilled water and filter through washed glass fiber filter paper.

6.9 BUFFERED COLOR FORMING REAGENT. Dissolve 10 g sulfanilamide, 126 g citric acid and 0.5 g naphthyl-1-ethylenediamine hydrochloride in 740 ml 5 N H_2SO_4. Fill to 1 l with distilled water. Five ml of this mixture should exactly neutralize the 5 ml of the alkaline oxidizing reagent plus 1 ml of the arsenite solution and provide enough residual acidity for allowing a smooth azo dye formation. Citric acid is added as buffer and for compensating small measuring errors. Store the solution in low actinic glass preferably in a refrigerator. The solution keeps well, but should be discarded when a color develops which would cause an absorbance increase in the blank of 0.05 units when measured against water in a 1-cm cell.

6.10 AMMONIUM CALIBRATING SOLUTION (100 μG NH_3/ML). Dry ammonium sulfate in an oven at 80 C for 2 hr, weigh and dissolve 0.39 g in 1 liter distilled water. Add 1 ml chloroform for better preservation and keep out of daylight. Discard after 1 year. Prepare dilutions as needed, just before use.

6.11 GLASS CLEANING SOLUTION (1.2 N HCL). Dilute 10 ml of conc HCl to 100 ml with distilled water in a stoppered graduated cylinder.

6.12 SULFURIC ACID, 5 N. Carefully add 136 ml of conc sulfuric (98%, density = 1.84) to 800 ml of distilled water, cool and dilute to 1 liter. Check the sulfuric acid before use for ammonium content and switch to another batch or brand if blanks obtained are too high.

7. Procedure

7.1 Assemble in order the prefilter in its holder, the absorber with 20 ml absorbing solution, filter, flow measurement or control device, and pump.

7.2 Sample at the rate of 1 to 10 l/min for a sufficient time to obtain an adequate sample, usually from 10 min to 2 or more hr. Do not expose the prefilter to high humidities (*e.g.* RH in excess of 90%, steam, rain). For reducing high humidities, the sampling train can be placed inside a warmed box, *e.g.*, at 10 to 15 C above the intake temperature. After collection, replace the prefilter and place the exposed filter in a plastic petri dish for return to the laboratory. Compensate for any evaporation loss by filling the absorber to the 20 ml mark with distilled water.

7.3 SAMPLE PREPARATION. The analysis can be performed on the prefilter **(8)** for assessing particulate ammonia compounds and on the absorbing solution for assessing gaseous ammonia. Soak the filter in 10 ml of distilled water for 30 min. Insert a 7-cm Whatman 42 filter paper in a funnel and wash free from ammonia first with 1.2 N HCl, then with distilled water. Discard the washings. An adequate washing procedure is indicated by low blank values determined on previous trials. Filter the extract through this prepared filter, rinse with distilled water and fill to 25 ml.

7.4 ANALYSIS OF AMMONIA. Add 5.0 ml of the alkaline oxidizing reagent to 10 ml of the filtrate contained in a 25-ml graduated stoppered cylinder. Reserve the remainder of the filtrate (**7.3**) for the nitrate blank. Fill the graduated cylinder to 19 ml with distilled water, and let the mixture stand for at least 17 min. Then add 1 ml of 0.2 M sodium arsenite and wait at least 3 min. Develop the azo dye by adding 5.9 ml of the buffered color forming reagent. After 10 min, measure the color at 545 nm using the appropriate cell length. Compare against the corresponding nitrite blank as prepared in (**7.4.1**). Deduct the ammonia blank as prepared in (**7.4.2**). Absorbance should read between 0.2 to 0.8 for optimum accuracy.

7.4.1 *Nitrite blanks.* To compensate for the amount of nitrite present in sample and reagent, add 1 ml sodium arsenite solution *before* addition of the 5.0 ml of alkaline oxidizing reagent to 10 ml of sample solution. Dilute to 20 ml and let stand for 3 min. Add 5.0 ml of the buffered color forming reagent. Allow 10 min developing time and use as the blank for samples and ammonia blank, respectively.

7.4.2 *Ammonia blanks.* Unexposed absorbing solution and filters from the same filter batch are treated as in **7.4** and **7.4.1** except that no air is drawn through them. Compare against the corresponding nitrite blank to provide a correction for ammonia from the reagents and filter material.

8. Calibration and Standardization

8.1 WORKING STANDARD AMMONIA SOLUTION. Dilute 5 ml of ammonia calibrating solution to 500 ml with absorbing solution (= 1 μg NH_3/ml).

8.2 CALIBRATION POINTS. Pipet 1, 2, 5, and 10 ml of the working standard and correspondingly 13, 12, 9, and 4 ml absorbing solution into 25-ml volumetric flasks or stoppered cylinders. Follow the procedure explained for sample analysis.

8.3 CALIBRATION CURVE. Plot the net absorbances as the ordinates versus the standard concentrations as the abscissa on linear graph paper and draw the curve. Alternatively, determine the slope F by the method of least squares and calculate the expression relating optical density A_{545} and concentration (c), which in its simplest form is:

$$F = A_{545}/c.$$

9. Calculation

9.1 AMMONIA IN SOLUTION. If A is the absorbance corresponding to the solutions derived from the sample compared against the nitrite blank of same sample, B the ammonia blank of either filter or absorbing solution, and F the slope as determined in (**8.3**), then the concentration of ammonia c is:

$$c = \frac{A - B}{F} \mu g\ NH_3$$

9.2 AMMONIA CONTENT OF THE AIR SAMPLE. The concentration of ammonia C_a in μg per m^3 can be calculated from the ammonia concentration c from the extract, knowing the volume V_{25} in m^3 of air sampled at 25 C and 760 Torr by:

$$C_a = \frac{2.5 \times c}{V_{25}} \mu g\ NH_3/m^3$$

The corrected value of V_{25} can be calculated knowing the flow rate **f** in 1 per min, the elapsed time **t** in min, the pressure **p** in Torr and the absolute temperature **T** in °K, these two latter values measured at the sampling point:

$$V_{25} = \frac{f \times t \times p \times 298}{1000 \times 760 \times T} =$$

$$392 \frac{f \times t \times p}{T} \times 10^{-6} m^3$$

10. Effect of Storage

10.1 The absorbing solution should be analyzed within 24 hr of collection.

10.2 Sample filters can be stored indefinitely.

10.3 The extractions should be performed within 4 hr before the analysis.

11. References

1. STRICKLAND and PARSONS. 1955, Manual of sea water analysis. Department of Fisheries, Canada.
2. TRUESDALE, V.W. 1971. *Analyst* 96:584.
3. AMERICAN PUBLIC HEALTH ASSOCIATION, INC. 1977. Methods of Air Sampling and Analysis. 2nd ed. p. 511. Washington, D.C.
4. HARWOOD, J.E. and A.L. KUHN. 1970. Water Research, 4:805.
5. AMERICAN PUBLIC HEALTH ASSOCIATION, INC. 1976. Methods of Air Sampling and Analyses. 2nd ed. p. 557. Section 5.3, Figure 411:1 Washington, D.C.
6. ACS REAGENT CHEMICALS. American Chemical Society Specifications, Washington, D.C., 1966
7. AMERICAN PUBLIC HEALTH ASSOCIATION, INC. 1977. Methods of Air Sampling and Analysis. 2nd ed. p. 53. Washington, D.C.
8. *Ibid*. p. 557. Section 5.1; p. 558, Sections 7.1 and 7.2.

Subcommittee 3

E. L. KOTHNY, *Chairman*
W. A. COOK
J. E. CUDDEBACK
B. DIMITRIADES
E. F. FERRAND
R. G. KLING
P. W. MCDANIEL
G. D. NIFONG
F. T. WEISS

403.

Tentative Method of Analysis for Nitrate in Atmospheric Particulate Matter (2,4-Xylenol Method (1))

12306-01-70T

1. Principle of the Method

1.1 Particulate matter is collected from the air on an 20 × 25 cm (8″ × 10″) glass fiber filter. Nitrates are extracted with water. 2,4-xylenol nitrated by the acidified extract is then extracted with toluene and finally reextracted from the toluene with sodium hydroxide solution. The determination is made colorimetrically at 435 nm (**2**).

2. Range and Sensitivity

2.1 Samples of outside atmospheres usually contain 0.1 to 10 $\mu g/m^3$ of nitrate. Usually a 24 hr sample is collected with a high-volume sampler, although much smaller samples are adequate; results are calculated to represent 24 hr averages. Two aliquoting steps are provided in the procedure.

2.2 This method has a working range of 10 to 100 μg of nitrate ion and a sensitivity of 5 μg when the specified aliquot is taken of the sample (Section **7.2**) and of the sample filtrate, (Section **7.3**).

2.3 If greater sensitivity is desired, the aliquot of sample plus sulfuric acid may be increased proportionally up to the entire sample, or the amount of toluene and sodium hydroxide reduced.

3. Interferences

3.1 When sample extract is colored a sample blank consisting of an equal volume of the sample extract without 2,4-xylenol is carried through the procedure and the absorbance of the blank is subtracted from the sample.

3.2 Chlorides at concentrations above 100 μg/ml in the aqueous extract cause a negative interference.

3.3 Equivalent amounts of nitrites with nitrates may cause results to be as much as 80% high. However, significant amounts of nitrites are not usually found in the atmospheric particulate matter.

3.4 Oxidizing agents may cause positive interference by forming colored complexes with xylenol or negative interference by oxidation of the xylenol leaving insufficient xylenol for complete reaction with all the nitrate.

4. Precision and Accuracy

4.1 The method has a coefficient of variation of 8%.

5. Apparatus

5.1 Any device capable of collecting particulate matter from a volume of at least 20 m^3 of air within the desired period of time may be employed.

5.1.1 *High-volume sampler*. A motor operated blower with a sampling head for an 8″ × 10″ (20 × 25 cm) glass fiber filter and capable of an initial air flow of 60 cfm (1.7 m^3/min) is suitable for particulate matter sampling. A flow-measuring device, usually a rotameter, is required to indicate flow rate.

5.1.2 Membrane filters and electrostatic precipitators may also be used for sampling.

5.2 REFLUXING APPARATUS. A 125-ml flask fitted with reflux condenser and heating mantle or hot plate.

5.3 WATER BATH. (Thermostated at 60 C).

5.4 TEST TUBES. 25-mm with polyethylene caps.

5.5 SPECTROPHOTOMETER. Suitable for measurement at 435 nm.

5.6 CUVETTES. 1.90 cm (¾″) light path. A 1 cm. light path cuvette may be used with decreased sensitivity.

5.7 GLASSWARE.

6. Reagents

6.1 PURITY. All chemicals should be analytical reagent grade **(3)**.

6.2 SULFURIC ACID SOLUTION, 85%. Cautiously and slowly, add 480 ml sulfuric acid to 117 ml distilled water, mixing well and cooling during addition.

6.3 XYLENOL SOLUTION, 1%. 1 ml of 2,4-xylenol in 100 ml glacial acetic acid. Keep in a bottle of low actinic glass.

6.4 SODIUM HYDROXIDE SOLUTION, 0.4N. Dissolve 16 g sodium hydroxide in distilled water and dilute to 1 l. Store in plastic bottle.

6.5 STOCK STANDARD SOLUTION, 100 μg NO_3/ml. Dissolve 0.163 g potassium nitrate in distilled water and dilute to 1 l.

6.6 DILUTE STANDARD SOLUTION, 20 μg NO_3/ml. Pipet 5 ml of stock solution into a 25-ml volumetric flask and dilute to mark.

6.7 COTTON. Surgical grade.

7. Procedure

7.1 SAMPLE COLLECTION. A common procedure is to employ the high volume sampler with an 8″ × 10″ (20 × 25-cm) glass filter that is operated over a 24 hr period. The average of the flow rates at the beginning and end of the sampling period is taken as the rate for the entire sampling period. This permits collection of nitrate from some 2,000 m^3 of air. Usually only 1% of the sample is taken for analysis. From 10 to 100 μg NO_3^- are required for the analysis.

7.2 SAMPLE PREPARATION. The sample filter may be folded along the 10″ axis for transportation and storage. When this is done, aliquots of the filter should be taken across the fold. A convenient aliquot is a 1.90-cm (¾″) × 19.3-cm (8″) strip. This is refluxed with 25 ml of distilled water for 90 min. Filter through Whatman No. 1 paper previously washed free from nitrate and rinsed with distilled water to obtain 50 ml of filtrate. An exposed sample filter is similarly refluxed and treated to provide a blank. (Unwashed filter paper may add up to 60 μg of nitrate ion to the filtrate.)

7.3 ANALYTICAL PROCEDURE **(2)**. Take

a 5-ml aliquot of sample filtrate in a 25-mm test tube, add slowly 15 ml of 85% sulfuric acid; mix cautiously and allow to cool. Add 1 ml of xylenol reagent, mix, and place in a 60 C water bath or oven for 30 min. Cool and transfer with distilled water to a 250-ml separatory funnel. Dilute to about 80 ml with distilled water and then add 10 ml of toluene. Shake gently for about 2 min. Avoid vigorous shaking to prevent formation of emulsions. Allow to stand until layers separate (about 10 min), then discard the lower aqueous layer. Rinse the toluene with about 20 ml of distilled water. Discard the aqueous layer. Add exactly 10 ml of 0.4N sodium hydroxide solution and shake gently for 5 min. Allow to stand until clear. Place cotton pledgets in funnel stems. Draw lower aqueous layer through the cotton pledget and into the cuvette. Measure the absorbance against a reagent blank in a suitable photometer at 435 nm. Determine the nitrate concentration by reference to the standard curve.

8. Calibration and Standardization

8.1 Prepare a standard curve by using 5 ml portions of a series of standards containing 0 to 100 μg of nitrate. A straight line is obtained by plotting the absorbance at 435 nm against the number of μg of nitrate ion.

8.1.1 One calibration standard should be included with each batch of samples. Deviations up to 10% from the standard curve can be expected.

9. Calculations

9.1 Micrograms of nitrate per $m^3 = C/FV$

where

F = sample aliquot fraction (= fraction taken of exposed filter area times fraction of aqueous extract). The exposed filter area is calculated from the actual dimensions of the frame opening.

C = number of μg NO_3 found after subtraction of blank, and

V = sample air volume in m^3 at 760 Torr and 25 C.

10. Effect of Storage

10.1 The 1% xylenol solution is stable for months when kept in a brown bottle.

11. References

1. Huey, N.A. 1965. Determination of Nitrate in Atmospheric Suspended Particulates: 2,4-Xylenol Method. Selected Methods for the Measurement of Air Pollutants, Division of Air Pollution, Public Health Service, PHS Publication No. 999-AP-11, pages J-1–J-4, May.
2. Barnes, H. 1950. A Modified 2:4 Xylenol Method for Nitrate Estimation. Analyst 75:388.
3. American Chemical Society. 1966. *Reagent Chemicals, American Chemical Society Specifications*. Washington, D.C., 1966.

Subcommittee 3
B. E. Saltzman, *Chairman*
W. A. Cook
B. Dimitriades
E. F. Ferrand
E. L. Kothny
L. Levin
P. W. McDaniel

404.

Tentative Method of Analysis for Nitrate in Atmospheric Particulate Matter (Brucine Method)

12306-02-72T

1. Principle of the Method

1.1 Particulate matter is collected from the air on a 20 × 25 cm (8″ × 10″) glass fiber filter. Nitrates are extracted with water. The nitrate in the acidified extract is reacted with brucine to form a color which is determined spectrophotometrically at 410 nm (**1, 2, 3, 4**).

1.2 This method uses the same extraction and sampling procedure as the m-xylenol method of the Intersociety Committee (**5**). The brucine procedure is faster and more reproducible, but the xylenol method has an extraction step which permits separation of some interferences. The phenoldisulfonic acid procedure (**6**) is subject to interference by chlorides and has a neutralization step which lengthens this method.

2. Range and Sensitivity

2.1 Samples of outside atmospheres usually contain 0.1 to 10 μg nitrate/m^3. Usually a 24 hr sample is collected with a high-volume sampler, although much smaller samples are adequate. Results are calculated to represent 24 hr averages, if a number of samples are collected within a 24 hr period.

2.2 This method has a working range of 20 to 100 μg of nitrate ion and a sensitivity of 4 μg in the sample solution aliquot used for the test.

2.3 For greater sensitivity, the aliquot of sample to be extracted may be increased proportionately up to the entire sample, or the amount of diluting acid may be reduced.

3. Interferences

3.1 When the sample extract is colored, a sample blank consisting of an equal volume of the sample extract without brucine is carried through the procedure and the absorbance of the blank is subtracted from the sample.

3.2 Nitrites may cause results to be high (**2**). However, significant amounts of nitrites (such as 4 μg nitrite ion in the solution aliquot) are not usually found in the atmospheric particulate matter.

3.2.1 The interference of nitrites is eliminated (**7**) using a modified reagent (see Section **6.4**).

3.3 Colored substances may form by interaction of organic substances with the concd sulfuric acid used for the reaction.

3.4 Chlorides cause erratic low results in high concentrations (1 to 60 g/l). However, significant amounts of chlorides (above 10 mg in 10 ml) are not usually found in atmospheric particulates.

3.4.1 With suitable modifications, (see Section **6.4**) the interference of chlorides is eliminated (**7, 8**).

4. Precision and Accuracy

4.1 The method has a precision of 2 μg NO_3^- over the working range of 20 to 100 μg NO_3^-. The blank is usually negligible.

4.2 The results for nitrate by this method show excellent agreement with the phenol disulfonic acid method (**3**).

5. Apparatus

5.1 Any device capable of collecting particulate matter from a volume of at least 20 m^3 of air within the desired period of time may be employed.

5.1.1 *High-volume sampler*. A motor operated blower with a sampling head for a 20 × 25 cm (8″ × 10″) glass fiber filter and capable of an initial air flow of 1.5 to 1.7 m^3/min (53 to 60 cfm) is suitable for particulate matter sampling. A flow-measuring device, usually a rotameter, is required to indicate flow rate (**9**).

5.1.2 Membrane filters at a sampling rate of 10 liter/min and electrostatic precipitators at a sampling rate of about 100 liter/min may also be used. However, the high-volume sampler is recommended.

5.2 REFLUXING APPARATUS. A 125 ml flask fitted with reflux condenser and heating mantle or hot plate.

5.3 BOILING WATER BATH.

5.4 GRADUATED CYLINDERS, 50 ML.

5.5 SPECTROPHOTOMETER SUITABLE FOR MEASUREMENT AT 410 NM.

5.6 CUVETTES. 2.54 cm light path—A 1-cm light path cuvette may be used with decreased sensitivity.

6. Reagents

6.1 PURITY. All chemicals should be analytical reagent grade (**10**).

6.2 CONCD SULFURIC ACID, 98%, AND DILUTED SULFURIC ACID, 88%. Mix cautiously 1000 ml 98% H_2SO_4 into 250 ml of distilled water.

6.3 BRUCINE (FREE BASE), 2.5% IN CHLOROFORM SOLUTION. Filter before use. If the free base is not available, brucine hydrochloride, 1% in 0.05 N HCl may be substituted. This latter reagent requires the alternate procedure. Section **7.3.1**. Both reagents are stable for months when kept in a brown bottle (**1**) or in a refrigerator.

6.4 If nitrites and chlorides are present in higher than traces, the following reagent is recommended:

6.4.1 *Stock solution*. 1 g brucine sulfate and 0.1 g sulfanilic acid dissolved in 70 ml of hot water and 3 ml conc HCl. Cool and dilute to 100 ml. This reagent is stable for 3 to 6 months if kept in a refrigerator.

6.4.2 MIXED REAGENT. Just prior to use mix 400 ml of a 30% NaCl solution with 100 ml of stock solution, Section **6.4.1**. Prepare fresh daily.

6.5 STOCK SOLUTION, 100 *mg* NO_3^-/ML. Dissolve 0.163 g potassium nitrate in distilled water and dilute to 1 l. Sterile solutions should stay stable.

6.6 DILUTE STANDARD, 10 μg NO_3^-/ML. Pipet 10 ml of stock solution into a 100-ml volumetric flask and dilute to mark. Prepare fresh.

7. Procedure

7.1 SAMPLE COLLECTION. A common procedure is to employ the high volume sampler with a 20 × 25 cm (8″ × 10″) glass filter that is operated over a 24 hr period. The average of the flow rates at the beginning and end of the sampling period is taken as the rate for the entire sampling period. This permits collection of nitrate from some 2,000 m^3 of air. Usually only 1% of the sample is taken for analysis. From 20 to 100 μg NO_3^- are required for the analysis.

7.2 SAMPLE PREPARATION. The sample filter may be folded along the 10″ axis for transportation and storage. When this is done, aliquots of the filter should be taken across the fold. A convenient aliquot is a 1.90 cm (¾″) × 19.3 cm (8″) strip. This is refluxed with 50 ml of distilled water for 30 min. Insert a Whatman #42 filter paper in a funnel and wash free from nitrate with hot distilled water from a glass wash bottle. Discard the washings. Unwashed filter paper may add up to 60 μg NO_3^- to the filtrate. An adequate washing procedure is indicated by low blank values determined on previous trials. Filter the extract through this prepared paper, rinse with distilled water to obtain 100 ml of filtrate. An unexposed sample filter from the same filter batch is similarly refluxed and treated to provide a blank. This larger volume allows other analyses to be made on the same extract.

7.3 ANALYTICAL PROCEDURE (**3, 7**). Three methods are given depending on which brucine reagent has been prepared.

7.3.1 Take a 10 ml aliquot of sample filtrate into a 50 ml graduated cylinder. Add 0.2 ml of 2.5% solution of brucine in chloroform and with caution 20 ml of conc sulfuric acid. After 20 min, cool quickly in an ice bath and dilute to 50 ml with sulfuric acid.

7.3.2 Alternatively 0.5 ml of 1% solution of brucine hydrochloride is added to a cooled mixture of a 10 ml aliquot of sample filtrate with 20 ml of sulfuric acid. Place in a boiling water bath for 20 min, then cool quickly.

7.3.3 For higher than trace amounts of nitrites and chlorides proceed as follows: Take a 10 ml aliquot of sample filtrate in a 50-ml cylinder. Add 2.5 ml of mixed reagent, Section 6.4.2, mix thoroughly, add 10 ml of 88% H_2SO_4 and place in a boiling water bath for 20 minutes. Then cool quickly.

7.3.4 Read on a suitable spectrophotometer at 410 nm any time within 24 hr against the blank.

8. Calibration and Standardization

8.1 Prepare a standard curve by diluting 2 to 10 ml aliquots of dilute standard solution to 10 ml and proceeding as in 7.3. A straight line is generally obtained by plotting the absorbance at 410 nm against the number of micrograms of nitrate ion.

9. Calculations

9.1 Micrograms of nitrate per m^3 = C/FV.

Where

F = sample aliquot fraction (fraction taken of exposed filter area times fraction of aqueous extract). The exposed filter area is calculated from the actual dimensions of the frame opening.

C = number of μg NO_3^- found after subtraction of blank, and

V = sample air volume in m^3 at 760 Torr and 25 C.

10. Effect of Storage

10.1 Sample filters can be stored indefinitely.

10.2 The extractions should be performed within 4 hr before the analysis.

11. References

1. Haase, L.W. 1926. Chem. Ztg. 50:372.
2. Lunge, G., and Lwoff, A.Z. Agnew. 1894. Chem. 12:345.
3. Fisher, F.L., Ibert, E.R., and Beckman, H.F. 1958. *Anal. Chem.* 30:1972.
4. Jenkins, D., and Medsker, L. 1964. *Anal. Chem.* 36:610.
5. American Public Health Association. 1977. Methods of Air Sampling and Analysis, 2nd ed. p. 518. Washington, D.C.
6. Ibid p. 534.
7. Kahn, L., and Brezenski, F.T. 1967. *Env. Sci. Tech.* 1:488.
8. McFarren, E.F., and Lishka, R.J. 1968. Trace inorganics in water. American Chemical Society, p. 253.
9. American Public Health Association. 1976. Methods of Air Sampling and Analysis, 2nd ed. p. 578. Washington, D.C.
10. American Chemical Society Reagent Chemicals. 1966. ACS Specifications, Washington, D.C.

Subcommittee 3

E. L. Kothny, *Chairman*
W. A. Cook
B. Dimitriades
E. F. Ferrand
G. D. Nifong
P. W. McDaniel
B. E. Saltzman
F. T. Weiss

405.

Tentative Method of Analysis for Nitric Oxide Content of the Atmosphere

42601-01-71T

1. Principle of the Method

1.1 After removing chemically nitrogen dioxide normally present in the atmosphere, the nitric oxide is converted to an equimolar amount of nitrogen dioxide by oxidation.

1.2 A total conversion in the gas phase can be obtained with chromic oxide supported on inert inorganic materials (**1, 2**).

1.3 The nitrogen dioxide so generated is determined according to the Tentative Method of Analysis for Nitrogen Dioxide Content of the Atmosphere (**3**), by absorption in Griess-Saltzman reagent to produce a pink color.

2. Range and Sensitivity

2.1 This method is intended for the manual determination of nitric oxide in the atmosphere in the range of 0.005 to about 5 parts per million (ppm) by volume or 6 to 6000 $\mu g/m^3$.

2.2 The sensitivity is 0.01 μg NO/10 ml of absorbing solution.

3. Interferences

3.1 Interfering concentrations of ozone ordinarily cannot coexist with nitric oxide because of their rapid gas phase reaction (**4**).

3.2 Nitrogen dioxide is removed in the first absorber of the train before the oxidizer and produces a slight interference at higher concentrations (**5**). (3 to 4% of the incoming nitrogen dioxide is converted to nitric oxide.)

3.2.1 The use of a fritted bubbler with absorbing reagent for removal of nitrogen dioxide is not recommended (**6**).

3.2.2 Soda lime not only absorbs nitrogen dioxide but also partially scrubs nitric oxide out of the gas stream and is not recommended (**7**).

3.3 Sulfur dioxide is removed by the oxidizer (**8**) and ordinarily produces no interference. If it is present in very high amounts, *e.g.*, stack gas levels, the oxidizer will deplete rapidly and must be changed more frequently. The oxidizer will indicate depletion by a change from orange to a brownish color.

3.4 Strongly oxidizing gases (*e.g.*, PAN, halogens, etc.) may affect the stability of the color of the Griess-Saltzman reagent. In such cases the absorbance of the solutions should be determined within one hour to minimize any loss (**9**). Interferences from oxidizing gases generally found in polluted atmosphere are negligible.

4. Precision and Accuracy

4.1 The precision of the method depends on the conversion efficiency of the oxidizer and other variables such as volume measurement of the sample, sampling efficiency of the fritted bubbler and absorbance measurement of the color. It may vary within ± 0.01 μg or ± 3%, whichever is greater.

4.2 Under controlled conditions, the conversion efficiency of the oxidizer varies within 98 to 100% (**1, 8**).

4.3 At present, accuracy data are not available.

5. Apparatus

5.1 NITROGEN DIOXIDE ABSORBER. A 20-mm ID × 50-mm long polyethylene tube with connecting caps at both ends, is filled with pellets of nitrogen dioxide absorbent. The pellets are held in place with glass wool plugs.

5.2 HUMIDITY REGULATOR. In order to provide steady 40 to 70% relative humidity control for the efficient working range of the chromic oxide mass, a 20-mm ID × 50-mm long polyethylene tube is filled with constant humidity buffer mixture in the same manner as above, **5.1 (6)**.

5.3 OXIDIZER **(1, 2, 7)**. A 15-mm ID glass tube with connecting ends is filled with 10 ml in depth of oxidizer pellets between two glass wool plugs (**Caution:** *Protect eyes and skin when handling this reagent.*)

5.3.1 The combined train of **5.1**, **5.2**, **5.3** can be used at least $\frac{100}{\text{ppm } NO_2}$ hrs and should be changed whenever it becomes visibly wet from sampling damp air or discolored from sampling strong reducing gases.

5.4 PLASTIC BAGS. Mylar or Tedlar plastic film bags may be used for collecting and storage of gaseous samples. However, losses will occur in presence of oxygen due to the oxidation of NO to NO_2.

5.5 SYRINGES. 5 ml glass syringes with Teflon plunger are convenient for sampling NO concentrations above 100 ppm.

5.6 Evacuated bottles, syringes for sampling NO_2 and other equipment as stated in the Tentative Method of Analysis for Nitrogen Dioxide Content of the Atmosphere **(3)**.

6. Reagents

6.1 PURITY. All chemicals should be analytical reagent grade **(10)**.

6.2 NITROGEN DIOXIDE ABSORBENT **(1, 5, 7)**. 1.5 mm (1/16″) pellets or 10 to 20 mesh porous inert material, such as firebrick, alumina, zeolites, etc., is soaked in 20% aqueous triethanolamine, drained, extended on a wide petri dish, and dried for 30 to 60 min at 95 C in an oven. The pellets should be free flowing.

6.3 CONSTANT HUMIDITY GRAINS (CONSISTING OF A 50 + 50% ANHYDROUS AND HYDRATED SODIUM ACETATE MIXTURE). Stir slowly and add drop by drop 13 ml of water into a beaker containing 40 g of anhydrous sodium acetate in order to obtain coarse grained crystal pellets.

6.4 OXIDIZER **(1, 2, 7)**. Glass, firebrick, or alumina, mesh size 15 to 40 is soaked in a solution containing 17 g of chromium trioxide in 100 ml of water. Then it is drained, dried in an oven at 105 to 115 C and exposed to 70% relative humidity. This can be done best by exposing a thin layer of pellets contained in a petri dish to a saturated solution of sodium acetate contained in a dessicator. The reddish color changes to a golden orange when equilibrated.

7. Procedure

7.1 Assemble a sampling train comprised, in order, of rotameter, nitrogen dioxide absorber, humidity regulator, oxidizer, fritted absorber and pump. Pipet 10 ml of Griess-Saltzman reagent into the fritted absorber. Draw an air sample through at a rate of 0.4 l/min or less to develop sufficient final color (about 10 min). Record the total air volume sampled. If the sample temperature and pressure deviate greatly from 25 C and 760 Torr, measure and record these values. Read the absorbance of the solution at 550 nm after 15 min color developing time.

8. Calibration and Standardization

8.1 Either of two methods of calibration may be employed. The most convenient method is standardizing with nitrite solution **(11)**, the other method is by standardizing with known gas mixtures. With this latter method, stoichiometric and efficiency factors are eliminated from the calculations. Concentration of the standards should cover the expected range of sample concentrations.

8.1 STANDARDIZATION BY NITRITE SOLUTION **(3)**. After conversion of NO to NO_2, standardization can be accomplished as indicated in the Tentative Method of Analysis for Nitrogen Dioxide Content of the Atmosphere **(3)**. Construct the calibration curve by adding graduated amounts of dilute nitrate solution, equivalent to concentrations expected to be sampled, to a series of 25 ml volumetric flasks. Plot the absorbances against microliters of nitrogen dioxide per 10 ml of absorbing solution. If preferred, transmittance in per cent may be plotted on the logarithmic scale versus nitric oxide concentration on the linear coordinate on a semilog graph paper. The plot follows Beer's Law up to absorbance one (or 10% transmittance).

8.2 GASEOUS STANDARDIZATION METHOD. Because of the reactivity of nitric oxide with oxygen and more so at high pressure when stored in a cylinder, only dynamic mixing with nitrogen in a diluting stream or static mixtures in bags prior to calibration is recommended.

8.2.1 *Dynamic dilution.* An asbestos plug dilutor (**12**) is used to feed a slow stream of 1% NO in N_2 into the air by means of a three-way stopcock placed upstream from the rotameter. One bore of the stopcock is plugged with asbestos and the gas pressure is maintained by a constant-head overflow. The 1% NO is obtained by pressurizing with N_2 a precisely measured volume of NO admitted into an evacuated stainless steel tank by means of a precision gauge. Trace amounts of oxygen present in N_2 will not cause errors when using this concentration level.

8.2.2 *Bag dilution.* A sample of pure NO is drawn with a syringe of a convenient size and then transferred to a Mylar or Tedlar bag while filling with nitrogen measured through a wet test meter. After a convenient mixing period, the procedure is repeated with a second bag to cover a lower range of concentrations. Then fill into the bag contents an amount of pure oxygen corresponding to ¼ of the amount of nitrogen and analyze within 20 min.

9. Calculations

9.1 For convenience, standard conditions are taken as 760 Torr and 25 C, where the molar volume is 24.47. Ordinarily the correction of the sample volume to these standard conditions is slight and may be omitted. However, if conditions deviate significantly, corrections might be made by means of the perfect gas equation.

9.2 The value of the intercept of the microliters nitric oxide with absorbance equal to one from the calibration graph can be used as factor K for calculating the concentration in parts per million, rather than reading concentrations directly from graphs:

$$\text{nitric oxide, ppm} = \frac{\text{absorbance of 10 ml} \times K}{\text{volume of air sample in liters}}$$

9.3 The absorbance must be corrected for fading of color when there is a prolonged interval between sampling and measurement of absorbance (Section **10.2**).

10. Effects of Storage

10.1 Atmospheric air samples can be stored in Mylar and Tedlar bags for 1 to 2 hr without significant loss of their nitric oxide content.

10.2 The colors obtained by the sampling procedure must be protected from light. The well stoppered tubes lose 3 to 4% of the absorbance daily, provided oxidizing gases have not been sampled (**8**). In such cases the colors should be determined within one hour after sampling to minimize any loss.

11. References

1. LEVAGGI, D.A., KOTHNY, E.L., BELSKY, T., DE VERA, E.R., and MUELLER, P.K. 1971. A precise method for analyzing accurately the content of nitrogen oxides in the atmosphere. Presented 17th Annual Meeting Institute of Environmental Sciences, Los Angeles, California.
2. LEVAGGI, D.A., KOTHNY, E.L., BELSKY, T., DE VERA, E.R., and MUELLER, P.K. A direct procedure to oxidize nitric oxide quantitatively at atmospheric concentrations in the presence of nitrogen dioxide. To be published in Env. Sci. and Tech.
3. AMERICAN PUBLIC HEALTH ASSOCIATION, INC. 1977. Methods of Air Sampling and Analysis. 2nd ed. p. 406. Washington, D.C.
4. JOHNSTON, H.S., and CROSBY, H.J. 1954. Kinetics of the fast gas phase reaction between ozone and nitric oxide. J. Chem. Phys. 19:799, 1951; 22:689.
5. LEVAGGI, D.A., SIU, W., KOTHNY, E., and FELDSTEIN, M. The quantitative separation of ni-

tric oxide from nitrogen dioxide at atmospheric concentration ranges. To be published in Env. Sci. and Tech.

6. Huygen, I.C. 1970. Reaction of nitrogen dioxide with Greiss type reagents. *Anal. Chem.* 42:407.
7. Belsky, T. 1970. Experimental evaluation of triethanolamine and chromium trioxide in the continuous analysis of NO in the air. AIHL Report No. 85.
8. Saltzman, B.E., and Wartburg, A.F. 1965. Absorption tube for removal of interfering sulfur dioxide in analysis of atmospheric oxidant. *Anal. Chem.* 37:779.
9. Saltzman, B.E. 1954. Colorimetric microdetermination of nitrogen dioxide in the atmosphere. *Anal. Chem.* 12:1919.
10. American Chemical Society. 1966. Reagent chemicals. American Chemical Society Specifications, Washington, D.C.
11. Scaringelli, F.P., Rosenburg, E., and Rehme, K.A. 1970. Comparison of permeation tubes and nitrite ion as standards for the colorimetric determination of nitrogen dioxide. Env. Sci. and Tech. 4:924.
12. Saltzman, B.E. 1961. Preparation and analysis of calibrated low concentrations of 16 toxic gases. *Anal. Chem.* 33:1100.

Subcommittee 3
E. L. Kothny, *Chairman*
W. A. Cook
B. Dimitriades
E. F. Ferrand
C. A. Johnson
L. Levin
P. W. McDaniel
B. E. Saltzman

406.

Recommended Method of Analysis for Nitrogen Dioxide Content of the Atmosphere (Griess-Saltzman Reaction)*

42602-01-68T

1. Principle of the Method

1.1 The nitrogen dioxide is absorbed in an azo dye forming reagent (**1**). A stable pink color is produced within 15 min which may be read visually or in an appropriate instrument at 550 nm.

2. Range and Sensitivity

2.1 This method is intended for the manual determination of nitrogen dioxide in the atmosphere in the range of 0.005 to about 5 parts per million (ppm) by volume or 0.01 to 10 μg/l, when sampling is conducted in fritted bubblers. The method is preferred when high sensitivity is needed.

2.2 Concentrations of 5 to 100 ppm in industrial atmospheres and in gas burner stacks also may be sampled by employing evacuated bottles or glass syringes. For higher concentrations, for automotive exhaust, and/or for samples relatively high in sulfur dioxide content, other methods should be applied.

3. Interferences

3.1 A tenfold ratio of sulfur dioxide to nitrogen dioxide produces no effect. A 30-fold ratio slowly bleaches the color to a slight extent. The addition of 1% acetone to the reagent before use retards the fading by forming another temporary product with sulfur dioxide. This permits reading within 4 to 5 hr (instead of the 45 min re-

*This is a version of ASTM Method D 1607, adopted 1960 and revised 1969. Adapted from *Selected Methods for the Measurement of Air Pollutants*, PHS Publication No. 999-AP-11, May 1965.

quired when acetone is not added) without appreciable interferences. Interference from sulfur dioxide may be a problem in some stack gas samples (see **2.2**).

3.2 A fivefold ratio of ozone to nitrogen dioxide will cause a small interference, the maximal effect occurring in 3 hr. The reagent assumes a slightly orange tint.

3.3 Peroxyacylnitrate (PAN) can give a response of approximately 15 to 35% of an equivalent molar concn of nitrogen dioxide (**2**). In ordinary ambient air the concentrations of PAN are too low to cause any significant error.

3.4 The interferences from other nitrogen oxides and other gases which might be found in polluted air are negligible. However, if the evacuated bottle or syringe method is used to sample concentrations above 5 ppm, interference from NO (due to oxidation to NO_2) is possible (see **7.1.3**).

3.5 If strong oxidizing or reducing agents are present, the colors should be determined within 1 hr, if possible, to minimize any loss.

4. Precision and Accuracy

4.1 A precision of 1% of the mean can be achieved with careful work (**3**); the limiting factors are the measurements of the volume of the air sample and of the absorbance of the color.

4.2 At present, accuracy data are not available.

5. Apparatus

5.1 ABSORBER. The sample is absorbed in an all-glass bubbler with a 60μ max pore diameter frit similar to that illustrated in Figure 406:1.†

5.1.1 The porosity of the fritted bubbler, as well as the sampling flow rate, affect absorption efficiency. An efficiency of over 95% may be expected with a flow rate of 0.4 l/min or less and a max pore diam of 60μ. Frits having a max pore diameter less than 60μ will have a higher efficiency but will require an inconvenient pressure drop for sampling (see formula in **5.1.2**). Considerably lower efficiencies are obtained with coarser frits, but these may be utilized if the flow rate is reduced.

5.1.2 Since the quality control by some manufacturers is rather poor, it is desirable to measure periodically the porosity of an absorber as follows: Carefully clean the apparatus with dichromate-conc sulfuric acid solution and then rinse it thoroughly with distilled water. Assemble the bubbler, add sufficient distilled water to barely cover the fritted portion, and measure the vacuum required to draw the first perceptible stream of air bubbles through the frit. Then calculate the max pore diam as follows:

$$\text{Max pore diam } (\mu) = \frac{30s}{P}$$

Where:

s = surface tension of water at the test temp in dynes/cm (73 at 18 C, 72 at 25 C, and 71 at 31 C), and
P = measured vacuum, Torr

5.1.3 Rinse the bubbler thoroughly with water and allow to dry before using. A rinsed and reproducibly drained bubbler may be used if the volume (r) of retained water is added to that of the absorbing reagent for the calculation of results. This correction may be determined as follows: Pipet into a drained bubbler exactly 10 ml of a colored solution (such as previously exposed absorbing reagent) of absorbance (A_1). Assemble the bubbler and rotate to rinse the inside with the solution. Rinse the fritted portion by pumping gently with a rubber bulb. Read the new absorbance (A_2) of the solution. Then:

$$10A_1 = (10 + r)\,A_2$$

$$\text{or: } r = 10\left(\frac{A_1}{A_2} - 1\right)$$

†Corning Glass Works Drawing XA-8370 specifies this item with 12/5 $ ball and socket joints. Ace Glass, Inc. specifies this item as No. 7530.

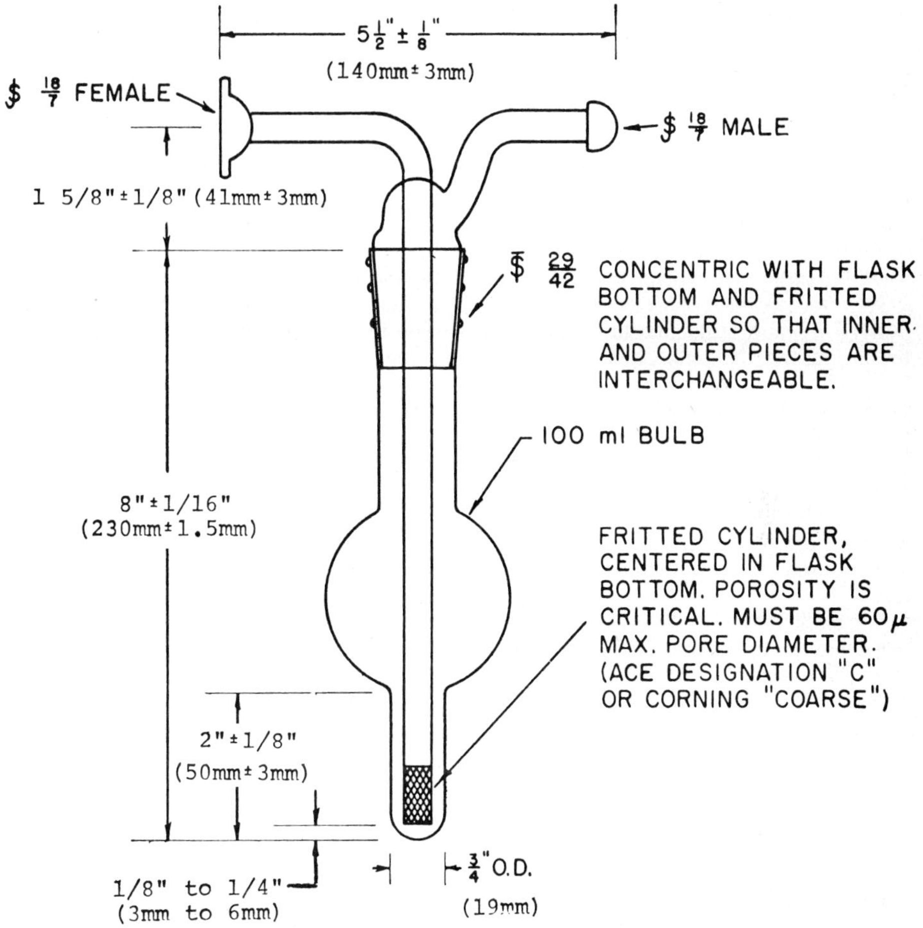

Figure 406:1—Fritted bubbler for sampling nitrogen dioxide.

5.2 AIR METERING DEVICE. A glass rotameter capable of accurately measuring a flow of 0.4 l/min is suitable. A wet test meter is convenient to check the calibration.

5.3 SAMPLING PROBE. A glass or stainless steel tube 6 to 10 mm in diam provided with a downward-facing intake (funnel or tip) is suitable. A small loosely fitting plug of glass wool may be inserted, when desirable, in the probe to exclude water droplets and particulate matter. The dead volume of the system should be kept minimal to permit rapid flushing during sampling to avoid losses of nitrogen dioxide on the surfaces.

5.4 GRAB-SAMPLE BOTTLES. Ordinary glass-stoppered borosilicate glass bottles of 30- to 250-ml sizes are suitable if provided with a mating ground joint attached to a stopcock for evacuation. Calibrate the volume by weighing with connecting piece, first empty, then filled to the stopcock with distilled water.

5.5 GLASS SYRINGES. Fifty or 100 ml syringes are convenient (although less accurate than bottles) for sampling.

5.6 AIR PUMP. A vacuum pump capable of drawing the required sample flow for intervals of up to 30 min is suitable. A tee connection at the intake is desirable. The inlet connected to the sampling train should have an appropriate trap and needle valve, preferably of stainless steel. The second inlet should have a valve for bleeding in a large excess flow of clean air to prevent condensation of acetic acid vapors from the absorbing reagent, with consequent corrosion of the pump. Alternatively, soda lime may be used in the trap. A filter and critical orifice may be substituted for the needle valve (**4**).

5.7 SPECTROPHOTOMETER OR COLORIMETER. A laboratory instrument suitable for measuring the pink color at 550 nm, with stoppered tubes or cuvettes. The wavelength band width is not critical for this determination.

6. Reagents

6.1 PURITY OF CHEMICALS. All chemicals should be analytical reagent grade (**5**).

6.2 NITRITE-FREE WATER. All solutions are made in nitrite-free water. If available distilled or deionized water contains nitrite impurities (produces a pink color when added to absorbing reagent), redistill it in an all-glass still after adding a crystal each of potassium permanganate and of barium hydroxide.

6.3 N-(1-NAPHTHYL)-ETHYLENEDIAMINE DIHYDROCHLORIDE, STOCK SOLUTION (0.1%). Dissolve 0.1 g of the reagent in 100 ml of water. Solution will be stable for several months if kept well-stoppered in a brown bottle in the refrigerator. (Alternatively, weighed small amts of the solid reagent may be stored.)

6.4 ABSORBING REAGENT. Dissolve 5 g of anhydrous sulfanilic acid (or 5.5 g of $NH_2 \cdot C_6H_4SO_3H \cdot H_2O$) in almost a liter of water containing 140 ml of glacial acetic acid. Gentle heating is permissible to speed up the process. To the cooled mixture, add 20 ml of the 0.1% stock solution of N-(1-naphthyl)-ethylenediamine dihydrochloride, and dilute to 1 l. Avoid lengthy contact with air during both preparation and use, since discoloration of reagent will result because of absorption of nitrogen dioxide. The solution will be stable for several months if kept well-stoppered in a brown bottle in the refrigerator. The absorbing reagent should be allowed to warm to room temp before use.

6.5 STANDARD SODIUM NITRITE SOLUTION (0.0203 G/LITER). One ml of this working solution of sodium nitrite ($NaNO_2$) produces a color equivalent to that of 10 μ l of nitrogen dioxide (10 ppm in 1 l of air at 760 Torr and 25 C: See **9.2.1**). Prepare fresh just before use by dilution from a stronger stock solution containing 2.03 g of the reagent grade granular solid (calculated as 100%) per l. It is desirable to assay (**5**) the solid reagent, especially if it is old. The stock solution is stable for 90 days at room temps, and for a year in a brown bottle under refrigeration.

7. Procedure

7.1 SAMPLING. Three methods are described below. Conc below 5 ppm are sampled by the bubbler method. Higher concentrations may be sampled by the evacuated bottle method, or more conveniently (but less accurately) by the glass syringe method. The latter method is more useful when appreciable concentrations (*e.g.*, 20 ppm) of nitric oxide are expected.

7.1.1 *Bubbler Method.* Assemble, in order, a sampling probe (optional), a glass rotameter, fritted absorber, and pump. Use ground-glass connections upstream from the absorber. Butt-to-butt glass connections with slightly greased vinyl or pure gum rubber tubing also may be used for connections without losses if lengths are kept minimal. The sampling rotameter may be used upstream from the bubbler provided occasional checks are made to show that no nitrogen dioxide is lost. The rotameter must be kept free from spray or dust. Pipet 10.0 ml of absorbing reagent into a dry fritted bubbler (see **5.1.3**). Draw an air sample through it at the rate of 0.4 liter/min (or less) long enough to develop sufficient final color (about 10 to 30 min). Note the total air volume sampled. Measure and record the sample air temp and pressure.

7.1.2 *Evacuated Bottle Method.* Sample in bottles of appropriate size containing 10.0 ml (or other convenient volume) or absorbing reagent. For 1 cm spectrophotometer cells, a 5:1 ratio of air sample volume to reagent volume will cover a concn range up to 100 ppm; a 25:1 ratio suffices to measure down to 2 ppm. Wrap a wire screen or glass-fiber-reinforced tape around the bottle for safety purposes. Grease the joint lightly with silicone or fluorocarbon grease. If a source of vacuum is available at the place of sampling, it is best to evacuate just before sampling to eliminate any uncertainty about loss of vacuum. A three-way Y stopcock connection is convenient. Connect one leg to the sample source, one to the vacuum pump, and the third to a tee attached to the bottle and to a mercury manometer or accurate gauge. In the first position of the Y stopcock, the bottle is evacuated to the vapor pressure of the absorbing reagent. In the second position of the Y stopcock the vacuum pump draws air through the sampling line to flush it thoroughly. The actual vacuum in the sample bottle is read on the manometer. In the third position of the Y stopcock the sampling line is connected to the evacuated bottle and the sample is collected. The stopcock on the bottle is then closed. Allow 15 min with occasional shaking for complete absorption and color development. For calculation of the standard volume of the sample, record the temp and the pressure. The latter is the difference between the filled and evacuated conditions, and the uncorrected volume is that of the bottle plus that of the connection up to the stopcock minus the volume of absorbing reagent.

7.1.3 *Glass Syringe Method.* Ten ml of absorbing reagent is kept in a capped 50 (or 100) ml glass syringe, and 40 (or 90) ml of air is drawn in at the time of sampling. The absorption of nitrogen dioxide is completed by capping and shaking vigorously for 1 min, after which the air is expelled. (When appreciable concentrations, *e.g.*, 20 ppm of nitric oxide are suspected, interference caused by the oxidation of nitric oxide to nitrogen dioxide is minimized by expelling the air sample immediately after the absorption period.) Additional air may be drawn in and the process repeated several times, if necessary, to develop sufficient final color.

7.2 MEASUREMENT OF COLOR. After collection or absorption of the sample, a red-violet color appears. Color development is complete within 15 min at room temps. Compare with standards visually or transfer to stoppered cuvettes and read in a spectrophotometer at 550 nm, using unexposed reagent as a reference. Alternatively, distilled water may be used as a reference, and the absorbance of the reagent blank deducted from that of the sample.

7.2.1 Colors too dark to read may be quantitatively diluted with unexposed absorbing reagent. The measured absorbance is then multiplied by the dilution factor.

8. Calibration and Standardization

8.1 Either of two methods of calibration may be employed. The most convenient is standardizing with nitrite solution. Greater accuracy is achieved by standardizing with accurately known gas samples in a precision flow dilution system **(3, 6, 7)**. The recently developed permeation tube technique **(8)** appears promising. If the gaseous method is used, the stoichiometric factor is eliminated from the calculations. Concentrations of the standards should cover the expected range of sample concentrations.

8.2 NITRITE SOLUTION METHOD. Add graduated amts of the $NaNO_2$ solution up to 1 ml (measured accurately in a graduated pipet or small buret) to a series of 25-ml volumetric flasks, and dilute to the marks with absorbing reagent. Mix, allow 15 min for complete color development and read the colors (see **7.2**).

8.2.1 Good results can be obtained with these small volumes of standard solution if they are carefully measured. Making the calibration solutions up to 25 ml total volume, rather than the 10 ml volume used for samples, facilitates accuracy. If preferred, even larger volumes may be used with correspondingly larger volumetric flasks.

8.2.2 Plot the absorbances of the standard colors against the μl of nitrogen dioxide per ml of absorbing reagent. The latter values are equal to the corresponding ml of standard nitrite solution times 0.4 (see **9.2.2**). If preferred, transmittance may be plotted instead of absorbance, using semilogarithmic graph paper. The plot follows Beer's law. Draw the straight line through the origin giving the best fit, and determine the slope, K, which is μliter of NO_2 intercepted at absorbance of exactly 1.0 or at 10% transmittance. For 1 cm cells, the value of K is about 0.73.

8.3 GASEOUS STANDARD METHOD. Two techniques are outlined below. Consult the original references for complete details, and section **9.3** for calculations.

8.3.1 One system **(3)** for gaseous standardization is as follows: About 5 ml of pure liquid nitrogen dioxide is placed in a small glass bubbler (10 mm in diam and 100 mm in length) provided with ground glass stopcocks and spherical joints on both intake and outlet tubes. The bubbler is immersed in a thermos bottle ice bath and connected to an air line. A small pump with flowmeter delivers a steady stream of a few ml/min of air to the bubbler, thence through two flowmeters which permit the discarding of up to 90% of the NO_2 and finally to a large stream of carbon filtered air (1000 to 1500 l/min) from a small blower. All this air passes through a 10 cm Biram anemometer mounted on the end of a pipe 10 cm in diam. It has been found empirically with this arrangement that the anemometer reading in ft/min times 1.64 is equivalent to liters/min. The bubbler is weighed to 0.1 mg at the beginning and end of an accurately measured time period. The stopcocks are closed each time before the bubbler is removed from the ice bath and wiped dry for weighing.

8.3.2 Another system of preparing known dilutions of nitrogen dioxide **(6, 7)** consists in making a preliminary dilution (about 0.4%) of nitrogen dioxide in air in a stainless steel tank at 1000 lb. pressure. Subsequent dilution, by air in a flow system at atmospheric pressure, of the analyzed tank mixture controlled by an asbestos plug and manometer yields concentrations of 0.1 to 10 ppm NO_2.

8.3.3 Sample the gas mixtures by the bubbler method, **7.1.1**, and read the colors, **7.2**. Select concentrations and sample volumes to produce colors covering the accurate absorbance range of the spectrophotometer.

8.3.4 Standardization by gaseous samples can be based either upon a weight-volume relationship if the source of nitrogen dioxide is weighed, or a volume-volume relationship if the source is an analyzed tank mixture.

8.3.5 Calculate the concn of the sample air stream, C, in ppm by volume:

$$C = \frac{10^6 (W_1 - W_2)}{t} \times \frac{0.532}{F_1} \times \frac{F_2}{F_3}$$

$$\text{or } C = C_t \times \frac{F_2}{F_3}$$

Where:

W_2 and W_1 = final and initial weights, in grams, of nitrogen dioxide bubbler (or of permeation tube); weight loss usually is 0.01 to 0.05 g.

t = time interval in min, between the weighings.

0.532 = 24.47/46.0 = ideal volume at standard conditions, in liters, of 1 g of nitrogen dioxide.

F_1 = flow rate, l/min, of air passed through nitrogen dioxide bubbler (or over permeation tube), corrected to standard conditions.

F_2 = flow rate, l/min, of concentrated gas mixture injected into sample air stream.

F_3 = total flow rate, l/min, of sample air stream.

C_t = analyzed concentration, ppm by volume of an ideal gas basis, of the tank mixture.

8.3.6 For each standard color, calculate the μl of nitrogen dioxide/ml of absorbing reagent:

$$\mu l\ NO_2/ml = Cv$$

Where:

v = volume of air sample, at standard conditions, in liters/ml of absorbing reagent.

8.3.7. Plot the absorbances of the colors against the μl of gaseous nitrogen dioxide/ml of absorbing reagent. Draw the straight line through the origin giving the best fit, and determine the slope, K (the value of $\mu l/ml$ intercepted at absorbance of exactly 1.0).

9. Calculations

9.1 For convenience, standard conditions are taken as 760 Torr and 25 C, at which the molar gas volume is 24.47 liters. (This is identical with standard conditions for Threshold Limit Values of the American Conference of Governmental Industrial Hygienists; it is very close to the standard conditions used (**9**) for air-handling equipment, of 29.92 in Hg, 70 F, and 50% relative humidity, at which the molar gas volume is 24.76 l, or 1.2% greater.)

9.1.1 Ordinarily the correction of the sample volume to these standard conditions is slight and may be omitted; however, for greater accuracy, it may be made by means of the perfect gas equation.

9.2 Standardization by nitrite solution is based upon the empirical observation (**1, 6**) that 0.72 mole of sodium nitrite produces the same color as 1 mole of nitrogen dioxide.‡

9.2.1 This factor is applied to calculate the equivalence of the nitrite solution to the volume of NO_2 absorbed as follows: One ml of the working standard solution contains 2.03×10^{-5} g $NaNO_2$. Since the molecular weight of $NaNO_2$ is 69.00 g, this is equivalent to:

$$\frac{2.03 \times 10^{-5}}{69.00} \times \frac{24.47}{0.72} =$$

1.00×10^{-5} l, or 10 μl of NO_2.

9.2.2 In the calibration section **8.2**, the standard containing 1 ml of nitrite solution (10 μl NO_2) per 25 ml total volume is equivalent to 10/25 or 0.4 μl of NO_2 per ml.

9.3 Compute the concn of nitrogen dioxide in the sample as follows:

$$\text{Nitrogen dioxide, ppm} = \text{absorbance} \times K/v$$

Where:

K = standardization factor from **8.2.2** or **8.3.7**.

v = volume of air sample, at standard conditions, in liters/ml of absorbing reagent.

If preferred, the graph from **8.2.2** or **8.3.7** may be used instead as follows:

$$\text{Nitrogen dioxide (ppm)} = \mu\text{liter } NO_2 \text{ per ml of absorbing reagent}/v$$

9.3.1 If v is a simple multiple of K, calculations are simplified. Thus, for the K value of 0.73 previously cited, if exactly 7.3 l of air are sampled through a bubbler containing 10 ml of absorbing reagent, $K/v = 1$, and the absorbance is also ppm directly.

9.3.2 For exact work, an allowance may be made in the calculations for sampling efficiency and for fading of the color using the following equation:

$$\text{Nitrogen dioxide, ppm} = \text{corrected absorbance} \times K/vE$$

Where:

E = sampling efficiency. For a bubbler, E is estimated from prior tests using two absorbers in series (**7**) (sec **5.1.1**). For a bottle or syringe, E = 1.0.

The absorbance is corrected for fading of the color (sec **10**), when there is a prolonged interval between sampling and measurement of the absorbance.

‡Recently Stratmann and Buck (**10**) reported a stoichiometric relationship of 1.0. Subsequently they found (**11**) decreasing values at concentrations above 0.3 ppm, approaching approximately the 0.7 figure at a few ppm. Shaw (**12**) confirmed the 0.72 value and suggested that higher values could be obtained erroneously if inadequate corrections for blanks were made. It is recommended that no change be made in the widely used 0.72 value at present if no change is made in the construction and operation of the fritted absorber (Section 5 and Figure 406:1). Other types of absorbers may yield different empirical factors.

10. Effects of Storage

10.1 Colors may be preserved, if well stoppered, with only 3 to 4% loss in absorbance per day; however, if strong oxidizing or reducing gases are present in the sample in concentrations considerably exceeding that of the nitrogen dioxide, the colors should be determined as soon as possible to minimize any loss. (See Section 3 for effects of interfering gases.)

11. References

1. SALTZMAN, B.E. 1954. Colorimetric microdetermination of nitrogen dioxide in the atmosphere. *Analytical Chemistry*. 26:1949–1955.
2. MUELLER, P.K., F.P. TERRAGLIO, and Y. TOKIWA. 1965. Chemical Interferences in Continuous Air Analysers. Presented 7th Conference on Methods in Air Pollution Studies, Los Angeles, Calif., Jan.
3. THOMAS, M.D. and R.E. AMTOWER. 1966. Gas dilution apparatus for preparing reproducible dynamic gas mixtures in any desired concentration and complexity. *J. Air Pollut. Contr. Assoc.* 16:618–623.
4. LODGE, J.P., JR., J.B. PATE, B.E. AMMONS, and G.A. SWANSON. 1966. The use of hypodermic needles as critical orifices in air sampling. *J. Air Pollut. Contr. Assoc.* 16:197–200.
5. AMERICAN CHEMICAL SOCIETY. 1966. Reagent Chemicals, American Chemical Society Specifications. Washington, D.C.
6. SALTZMAN, B.E., and A.F. WARTBURG, JR. 1965. Precision flow dilution system for standard low concentrations of nitrogen dioxide. *Analytical Chemistry*. 37:1261–1264.
7. SALTZMAN, B.E. 1961. Preparation and analysis of calibrated low concentrations of sixteen toxic gases. *Analytical Chemistry*. 33:1103–1104.
8. O'Keefe, A.E., and G.C. ORTMAN. 1966. Primary standards for trace gas analysis. *Analytical Chemistry*. 38:760–763.
9. ASTM Committee D-22. 1962. Terms Relating to Atmospheric Sampling and Analysis, D 1356–60. ASTM Standards on Methods of Atmospheric Sampling and Analysis, 2nd ed. Phila., Pa.
10. STRATMANN, H., and M. BUCK. 1966. Messung von Stickstoffdioxid in der Atmosphere. *Air & Water Pollut. Int. J.* 10:313–326.
11. STRATMANN, H. 1966. Personal Communication, September.
12. SHAW, J.T. 1967. The measurement of nitrogen dioxide in the air. *Atmospheric Environment*. 1:81–85.

Subcommittee 3
B. E. SALTZMAN, *Chairman*
W. A. COOK
B. DIMITRIADES
E. L. KOTHNY
L. LEVIN
P. W. MCDANIEL
J. H. SMITH

407.

Tentative Method of Analysis for Total Nitrogen Oxides as Nitrate (Phenoldisulphonic Acid Method)*

42603-01-70T

1. Principle of the Method

1.1 Nitric oxide (NO), nitrous anhydride (N_2O_3), nitrogen dioxide (NO_2), nitrogen tetroxide (N_2O_4), also vapor or mist of nitric acid (HNO_3) and nitrous acid (HNO_2), but not nitrous oxide (N_2O), may be collected and oxidized to the nitrate ion in an evacuated flask containing sulfuric acid and hydrogen peroxide. The yellow compound resulting from reaction of the nitrate ion with phenoldisulfonic acid is measured colorimetrically at 400 nm.

1.2 This is a long established method. The first extensive investigation of the method for nitrate determination in water was conducted by Chamot *et al* (**2**). The method was first adapted to determination of oxides of nitrogen in air by Cook (**3**), al-

*This procedure is comparable to ASTM Method D1608-60 (**1**) with inclusion of certain details from the APHA Standard Methods for the Examination of Water and Waste Water (**4**).

though the sampling technique then suggested has since been superseded by the method here described.

2. Range and Sensitivity

2.1 The phenoldisulfonic acid method is sensitive to 1 μg of nitrate in a water sample (**4**).

2.2 A 1000-ml air sample as collected in an evacuated flask of this volume will permit accurate determination of 50 or more ppm (by volume) of oxides of nitrogen as nitrogen dioxide (NO_2). Accordingly, the application of this method to ambient air is appropriate only to industrial locations where the concentration may exceed 50 ppm.

3. Interferences

3.1 Inorganic nitrates and other compounds easily oxidized to nitrates, such as nitrites and organic nitrogen compounds, give high results.

3.2 Certain reducing compounds, such as sulfur dioxide (SO_2), may consume a sufficient amount of the hydrogen peroxide in the absorbing solution that an inadequate amount is available to oxidize all of the oxides of nitrogen to the nitrate.

3.3 Chlorides and other halides lower the results, as does lead, but ordinarily insufficient amounts are present in air samples to require special treatment. If the chloride content exceeds 3 mg, silver nitrate should be used to reduce the content to about 0.1 mg (**5**).

3.4 Any substance that may increase the absorbance at 400 nm should be absent.

4. Precision and Accuracy

4.1 The precision of the method for ambient air can be estimated from the repeatability of 5% of the mean found on its application.

4.2 Below 50 ppm, accuracy is reduced due to incomplete absorption and relatively high blanks.

5. Apparatus

5.1 GAS SAMPLING FLASK OR BOTTLE. Standard 1000-ml round-bottom flask or bottle of borosilicate glass with a male standard taper 24/40 neck and a female cap with a sealed-on tube. Glass-fiber-reinforced tape should be wrapped around container for safety purposes. Determine the volume of the container by measuring the volume of water required to fill it.

5.2 SAMPLING ASSEMBLY. A glass or stainless steel tube 6 to 10 mm in diam provided with a downward-facing intake (funnel or tip) is suitable. A small loosely fitting plug of glass wool may be inserted, when desirable, in the probe to exclude water droplets and particulate matter. A T-tube and three-way Y stopcock are required. One leg of the stopcock may be of small bore capillary tubing for prolonging the sampling period.

5.3 SUCTION PUMP, capable of producing vacuum to the vapor pressure of water.

5.4 PHOTOMETER. A commercial photoelectric filter photometer or, preferably, a spectrophotometer suitable for measurement at 400 nm.

5.5 LABORATORY GLASSWARE. Microburet, 10-ml capacity; buret, 50-ml capacity; pipets, 25-ml capacity; volumetric flasks, 50-ml capacity; evaporating dishes or casseroles, heat resistant glass, 200-ml capacity.

5.6 STEAM BATH.

5.7 MERCURY MANOMETER, OPEN END.

6. Reagents

6.1 PURITY. All chemicals should be analytical reagent grade (**6**). Select hydrogen peroxide with a low nitrate content.

6.2 NITRATE AND NITRITE-FREE WATER. All water is to be free of nitrate and nitrite ions. If available distilled or deionized water contains nitrate or nitrite impurities, redistill it in an all-glass still after adding a crystal each of potassium permanganate and of barium hydroxide.

6.3 HYDROGEN PEROXIDE (3%). Dilute 10 ml of 30% hydrogen peroxide (H_2O_2) to 100 ml.

6.4 SULFURIC ACID (SP GR 1.84). Conc sulfuric acid (H_2SO_4).

6.5 SULFURIC ACID (3:997). Mix 3 ml of H_2SO_4 (sp gr 1.84) with water and dilute to 1 l.

6.6 ABSORBENT SOLUTION. 1.0 ml of H_2O_2 (3%) in 100 ml of H_2SO_4 (3:997). For high concentrations of oxides of nitrogen, the amount of H_2O_2 should be increased to 3 ml. Since dilute H_2O_2 may be unstable, these solutions should not be kept for a prolonged period.

6.7 AMMONIUM HYDROXIDE (SP GR 0.90). Conc ammonium hydroxide (NH_4OH). Keep stoppered to avoid loss of strength.

6.8 PHENOLDISULFONIC ACID REAGENT. Dissolve 25 g of phenol in 150 ml of conc sulfuric acid (H_2SO_4, sp gr 1.84) by heating on a steam bath (100 C). Cool, add 75 ml of fuming sulfuric acid (15% SO_3) and heat on the water bath for 2 hr. Cool and store in a brown glass bottle. The solution should be colorless; it deteriorates on long standing.

6.9 POTASSIUM NITRATE, STANDARD SOLUTION (1 ML = 0.1880 MG NO_2). Dry potassium nitrate (KNO_3) in an oven at 105 + 1 C for 2 hr. Dissolve 0.4131 g of the salt in water and dilute to 1 l in a volumetric flask.

6.10 POTASSIUM NITRATE, STANDARD SOLUTION (1 ML = 0.0188 MG NO_2). Dilute 10 ml of KNO_3 solution (1 ml = 0.1880 *mg* NO_2) to 100 ml with water in a volumetric flask and mix well. (0.0188 mg NO_2 in an air sample of 1000 ml = 10 ppm.)

6.11 SODIUM HYDROXIDE SOLUTION (42 G/L). Dissolve 42 g of sodium hydroxide (NaOH) in water and dilute to 1 l. Store in a plastic bottle and keep well stoppered.

6.12 EDTA REAGENT. Rub 50 g disodium ethylenediamine tetraacetate dihydrate with 20 ml distilled water to form a thoroughly wetted paste. Add 60 ml conc NH_4OH and mix well to dissolve the paste.

7. Procedure

7.1 SAMPLING. Sampling may be conducted by use of an evacuated flask or bottle. Pipet 25.0 ml of absorbent solution into the sampling container. If a source of vacuum is available at the place of sampling, it is best to evacuate just before sampling to eliminate any uncertainty about loss of vacuum. The T-tube is connected to the tube of the container cap. The mercury manometer or accurate vacuum gauge is attached to one branch of the T-tube and the three-way Y stopcock to the other. The vacuum pump is attached to a second branch of the stopcock and the sampling probe to the third. By successive positionings of the stopcock plug, the container is first evacuated to the incipient boiling point of the absorbing solution, and the manometer reading recorded, then air is drawn through the sampling line or probe to flush it, and finally the sample is collected in the evacuated container. The cap of the container is turned to seal the sample and the T-tube disconnected. The temperature is recorded.

7.2 LABORATORY ANALYSIS.

7.2.1 The glass flask or bottle in which the sample has been collected should remain in contact with the absorbent overnight to complete oxidation to the nitrate. Retaining the sample in the closed sample container longer than overnight and even up to a week is desirable to increase extent of absorption (**7**). Transfer the absorbent solution quantitatively from the container into a 200-ml evaporating dish.

7.2.2 A blank should be treated in the same manner as the sample. If a 1000-ml container was used for sampling, 25 ml of the unused absorbent solution should be pipetted into a 200-ml evaporating dish and the same amount of water added as was used in transferring the sample. The blank and samples should then be treated as described in Section **7.2.3**.

7.2.3 Add NaOH solution to the sample solution and to the blank in the evaporating dish until just basic to litmus paper. Avoid adding excess NaOH as this may dissolve some silicate from the dish and later cause turbidity. Evaporate the contents of the evaporating dish to dryness on a hot water or steam bath and allow to cool. Using a glass rod, rub the residue thoroughly with 2.0 ml phenoldisulfonic acid reagent to insure solution of all solids. If necessary, heat mildly on the hot water bath a short time to dissolve the entire residue. Cool and add 20 ml distilled water, stir, then add sufficient fresh, cool NH_4OH dropwise (about 6 or 7 ml) with constant stirring to give a basic reaction with litmus. If turbidity should occur, filter the solution through 7-cm, rapid, medium-texture filter paper into a 50-ml volumetric flask. Wash the evaporating dish three times with 4 to 5 ml of water and pass the washings through the filter. Since some yellow color may be left on the filter paper, this step should be done in a reproducible manner both for the samples and the calibration curves. Instead of filtering to remove the turbidity, the EDTA reagent may be added dropwise with stirring until the turbidity redissolves. Make up the volume to 50 ml in a volumetric flask with water and mix thoroughly.

7.2.4 Read the absorbance of the sample solution against the blank in the photometer at 400 nm. If a greater dilution is required, dilute the blank to the same volume.

7.2.5 Convert the absorbance found by means of the calibration curve to mg of NO_2.

8. Calibration and Standardization

8.1 Prepare a calibration curve of mg of NO_2 plotted against absorbance for the range of 50 to 500 ppm NO_2 based on 1000 ml samples of dry gas under standard conditions of 760 Torr and 25 C.

8.1.1 Using the microburet for the first 2 volumes and the 50-ml buret for the last 3, transfer 0.0, 5.0, 10.0, 35.0 and 50.0 ml of KNO_3 solution (1 ml = 0.0188 mg NO_2) into 200-ml evaporating dishes. Pipet 25 ml of the acid absorbent solution into each evaporating dish. Add NaOH solution until just basic to litmus paper. Then proceed as directed in Sections **7.2.3** and **7.2.4**.

9. Calculations

9.1 For convenience, standard conditions are taken as 760 Torr and 25 C, at which the molar gas volume is 24.47 liters. (This is identical with standard conditions for Threshold Limit Values of the American Conference of Governmental Industrial Hygienists; it is very close to the standard conditions used for air-handling equipment, of 29.92 in Hg, 70 F, and 50% relative humidity, at which the molar gas volume is 24.76 l, or 1.2% greater.)

9.1.1 The volume of the gas sample may be corrected to standard conditions by the following calculations:

$$V_s = \frac{(V_f - V_r)\,P \cdot 298.2}{760\,(t + 273.2)}$$

where:

V_s = volume of gas sample corrected to standard conditions of 760 Torr and 25 C, in milliliters,
V_f = volume of sampling flask up to stopcocks in milliliters,
V_r = volume of absorbent reagent,
P = vacuum in sampling container as measured by the manometer, in mm, and
t = sample temperature, C.

9.2 Calculation of Concentration of NO_2 in Parts per Million by Volume

Oxides of nitrogen as NO_2

$$= \frac{24.47W \times 10^6}{46.0\ V_s}\ \text{ppm}$$

$$= \frac{532W \times 10^3}{V_s}\ \text{ppm}$$

where:

V_s = volume of gas sample corrected to standard conditions, in milliliters,
W = milligrams of oxides of nitrogen found (as NO_2),
24.47×10^3 = standard molar volume (760 Torr at 25 C), in milliliters,
and

46.0 = formula weight of NO_2.

10. Effect of Storage

10.1 The sample as collected either in the acid or alkaline absorbing solution with H_2O_2 can be held for analysis at a later time.

10.2 The yellow color formed with the phenoldisulfonic acid is stable.

11. References

1. ASTM COMMITTEE D-22. 1968. Method of Test for Oxides of Nitrogen in Gaseous Combustion Products (Phenoldisulfonic Acid Procedure) ASTM Designation D 1608-60 (Reapproved 1967) Book of ASTM Standards, Part 23, pp. 461–465.
2. CHAMOT, E.M., D.S. PRATT, and H.W. REDFIELD. 1909, 1910, 1911. A Study on the Phenoldisulfonic Acid Method for the Determination of Nitrates in Water. *J. Amer. Chem. Soc.* 31:922, 32:630, 33:336.
3. COOK, W.A. 1936. Chemical Procedures in Air Analysis. Methods for Determination of Poisonous Atmospheric Contaminants. Inorganic Substances. 1935–36 Year Book, Supplement to Amer. J. Public Health, 26:80.
4. AMERICAN PUBLIC HEALTH ASSOCIATION. 1976. Standard Methods for the Examination of Water and Waste Water. 14th ed. Washington, D.C.
5. JACOBS, M.B. 1967. The Analytical Toxicology of Industrial Inorganic Poisons. Interscience Publishers. John Wiley & Sons, New York, N.Y.
6. AMERICAN CHEMICAL SOCIETY. 1966. Reagent Chemicals, American Chemical Society Specifications. Washington, D.C.
7. SALTZMAN, B.E. 1954. Colorimetric Microdetermination of Nitrogen Dioxide in the Atmosphere. (Note on phenoldisulfonic acid procedure on p. 1953) *Anal. Chem.* 26:1949.

Subcommittee 3

B. E. SALTZMAN, *Chairman*
W. A. COOK
B. DIMITRIADES
E. F. FERRAND
E. L. KOTHNY
L. LEVIN
P. W. MCDANIEL
C. A. JOHNSON, Air Conditioning & Refrigeration Institute Liaison

408.

Tentative Method of Analysis for Atmospheric Nitrogen Dioxide (24-hour-average)

42602-03-73T

1. Principle of the Method

1.1 Nitrogen dioxide is absorbed from the air by aqueous triethanolamine solution, subsequent analysis is performed using an azo-dye forming reagent (**1, 2, 3**). The color produced by the reagent is measured in a spectrophotometer at 540 nm.

2. Range and Sensitivity

2.1 The atmospheric conc range for which this method may be used with confidence is 10 to 1000 $\mu g/m^3$ (0.005 to 0.50 ppm), based on a 24 hr sampling period. Method performance at NO_2 levels above 1000 $\mu g/m^3$ (0.5 ppm) has not yet been established.

2.2 The sensitivity of the method is dependent on that of the Griess-Saltzman reagent (**3**). For a 1 cm sample cell path and 0.1 absorbance units this is equivalent to 0.14 $\mu g/ml$ of NO_2 in the absorbing solution.

3. Interferences

3.1 Sulfur dioxide in concentrations up to 2000 $\mu g/m^3$ (0.7 ppm) does not interfere, if hydrogen peroxide is added after sampling and before color development (as described in Section **7.2**). Ozone causes no interference at atmospheric concentration ranges (up to 1000 $\mu g/m^3$). Nitric oxide at concentrations up to 800 $\mu g/m^3$ (0.6 ppm) as a 24 hr average causes no interference (**2**).

3.2 Organic nitrites and peroxyacyl nitrate (PAN) which might be present in the air would produce a positive interference. There is no data on the magnitude of the interference; however, in view of the 24 hr average conc of organic nitrites and PAN that have been reported in the literature, it would appear that the interference would be negligible (**2**). PAN occurs in the Los Angeles atmosphere in concentrations up to 0.080 ppm and weekly average concentrations of 0.014 ppm have been recorded.

4. Precision and Accuracy

4.1 The precision of the method in the field, expressed as standard deviation, was ± 12 $\mu g/m^3$ (± 0.006 ppm) on a 24 hr sampling for a mean of 81 $\mu g/m^3$ (0.043 ppm) as compared with a continuous colorimetric analyzer (**2**).

4.2 At present, accuracy data are not available. However, the accuracy can be considered comparable to that obtained in short term sampling with the Griess-Saltzman reagent (**4**) which is less than 9% deviation when using the factor 0.72 (**1**) or less than 2% when using the factor 0.764 (**2, 4**).

4.3 Absorption efficiency is 95 to 99% (**2**) in the plastic EPA bubbler (**5**) with 50 ml of absorbing solution.

5. Apparatus

5.1 ABSORBER. The sample is absorbed in a polypropylene tube 163 × 32 mm ID fitted with polypropylene two part closures. A gas dispersing tube with a 60 to 100 μm porosity is used in conjunction with the tube (**5**). Alternatively, an all glass system may be used (**3**).

5.2 AIR METERING DEVICE. A rotameter capable of measuring flow rates up to 0.4 liters/min, calibrated against a wet test meter, is suitable.

5.2.1 Calibrated orifices, such as hypodermic needles can also be used for controlling the flow. In such case, a minimum pressure differential of 500 Torr across the needle is required. A membrane filter should be placed in front of the hypodermic needle to prevent water droplets from adhering to the needle, causing alteration of the gas flow.

5.3 SAMPLING LINES. Sampling lines should be constructed of Teflon, glass or stainless steel.

5.4 AIR PUMP. A vacuum pump capable of sampling air through the absorber at the required rate of 0.4 l/min for 24 hr, is suitable. The pump should be equipped with a needle valve for accurate control of sampling rate.

5.4.1 If a critical orifice is used as a flow control device, the pump should be also capable of maintaining a minimum vacuum of 500 Torr across the orifice at the required rate of 0.4 l/min.

5.5 SPECTROPHOTOMETER. The instrument used must be capable of measuring solution absorbance at 540 nm. A cell path length of one cm is suitable.

6. Reagents

6.1 Analytical reagent grade chemicals should be used (**6**). The water should be free of nitrite and conform to ISC standards (**7**).

6.2 LIQUID ABSORBER. Add 15 g of triethanolamine to approximately 500 ml of distilled water; then add 3 ml of n-butanol. Mix well and dilute to 1 l with distilled water. The n-butanol acts as a surfactant. If excessive foaming occurs during sampling, the amount added of n-butanol should be decreased. The reagent is stable for two months if kept in a brown bottle, preferably in the refrigerator.

6.3 HYDROGEN PEROXIDE. Dilute 0.2 ml of 30% hydrogen peroxide to 250 ml with distilled water. **Caution:** Hydrogen peroxide is a strong oxidant and will damage skin and clothing.

6.4 SULFANILAMIDE SOLUTION. Dissolve 10 g of sulfanilamide in 400 ml of distilled water, then add 25 ml of conc (85%) phosphoric acid. Make up to 500 ml. This solution is stable for several months if stored in a brown bottle and up to one year in the refrigerator.

6.5 N-(1-NAPHTHYL)-ETHYLENEDIAMINE DIHYDROCHLORIDE (NEDA). Dissolve 0.1 g of NEDA in 100 ml of water. The solution is stable for a month, if kept in a brown bottle in the refrigerator.

6.6 NITRITE STOCK SOLUTION. Dissolve

0.135 g of sodium nitrite in distilled water and make up to 1 l. The sodium nitrite should be of known purity or analyzed before use. The solution contains 90 μg of nitrite per ml which corresponds to 100 μg NO_2 (gas)/ml (See **9.2**). The solution should be made up once a month.

7. Procedure

7.1 SAMPLING. Add 50 ml of the absorbing solution to the bubbler. Connect the bubbler to the sampling train and turn the pump on. Adjust the flow rate quickly to between 150 and 200 ml/min if a needle valve is used. If a critical orifice is used, check the flow rate to determine conformance with the calibration. Sample for 24 hr, recheck the flow rate at the end of that period. The sample, if capped may be kept for as long as 3 weeks before analysis without losses **(2)**.

7.2 COLOR DEVELOPMENT. Make up any loss in the absorbing solution to 50 ml with distilled water. Transfer a 10 ml aliquot of the solution to a 25-ml graduate cylinder. Similarly, transfer a 10 ml aliquot of unexposed absorbing solution to a second 25-ml graduated cylinder. To each cylinder add 1.0 ml of dilute hydrogen peroxide solution and mix well. Then add 10 ml of the sulfanilamide solution and 1.0 ml of the NEDA solution to sample and blank, mix well and allow color to develop for 10 min. Measure the absorbance of the sample solution against the blank at 540 nm. The NO_2 is then determined from the calibration curve.

8. Calibration and Standardization

8.1 A calibration curve may be obtained by one of two different methods. The simplest and most convenient method is standardization with nitrite solutions. However, accurately known gas mixtures would provide more realistic standards for calibration of the entire analytical system. The known gas mixtures may be prepared by use of a flow dilution system **(3, 8, 9)**, or permeation tubes **(10)**.

8.2 STANDARDIZATION WITH NITRITE SOLUTION. Transfer 1.0 ml of the nitrite stock solution into a 50-ml volumetric flask and dilute to the mark with absorbing solution. The working solution so obtained contains 1.8 μg NO_2^-/ml (corresponding to 2 μg NO_2 (gas)/ml). To a series of 25-ml graduated cylinders introduce 0, 1, 3, 5, and 7 ml of the nitrite working solution. Dilute to 10 ml with absorbing solution. From this point treat the standards as described in **7.2**. A graph of absorbance versus NO_2 concentration is constructed. Weekly checks should be made to determine if the response of the reagents has changed.

9. Calculations

9.1 The air sampling volume is corrected to 25 C and 760 Torr. The normal deviations from these conditions add only small corrections.

9.2 Standardization by the nitrite solution requires use of an empirical factor that relates μg of NO_2 (gas) to μg of nitrite ion. The conversion factor in this method is 0.90, *i.e.*, 90% of the NO_2 in the air sample is eventually converted to nitrite ion which reacts with the color producing azo-dye reagent **(2, 4)**. This factor has been incorporated in the nitrite stock solution (Section **6.6**).

9.3 Computation of the average concentration of NO_2 in air for the 24 hr sample period may be made as follows:
Read the μg NO_2 directly from the graph.

$$c = \frac{\mu g\ NO_2 \times 10^3}{r \times t \times k}\ \mu g\ NO_2/m^3$$

$$p = \frac{\mu g\ NO_2 \times 0.532}{r \times t \times k} = \frac{c \times 0.532}{10^3}\ ppm$$

where:

c = concentration in μg NO_2/m^3
p = concentration in ppm NO_2 (μl/1)
r = sampling rate in liters/min
t = sampling time in min
k = Dilution factor (*i.e.*, 0.2 if one fifth of the sample is analyzed)
0.532 = μl NO_2/μg NO_2 at 25 C and 760 Torr
10^3 = conversion factor liters/m^3

10. Effects of Storage

10.1 After sample collection the absorbing solution may be stored for up to three weeks without loss if kept capped in the dark. However, after the color producing reagents are added, the sample must be analyzed within a few hours. The absorbance of the colored solution decreases by about 4% per day.

11. References

1. SALTZMAN, B.E. 1954. Colorimetric microdetermination of nitrogen dioxide in the atmosphere. *Anal. Chem.* 26:1949.
2. LEVAGGI, D.A., W. SIU, and M. FELDSTEIN. 1973. A new method for measuring average 24-hour nitrogen dioxide concentrations in the atmosphere. *J. Air Poll. Cont. Assoc.* 23:30.
3. AMERICAN PUBLIC HEALTH ASSOCIATION INC., INTERSOCIETY COMMITTEE. 1977. Methods of Air Sampling and Analysis. p. 527. Washington, D.C.
4. SCARINGELLI, F.P., E. ROSENBURG, and K.A. REHME. 1970. Comparison of permeation tubes and nitrite ions as standards for the colorimetric determination of nitrogen dioxide. *Env. Sci. Tech.* 4:924.
5. FEDERAL REGISTER. 1971. 36(84), part II, 8201.
6. AMERICAN CHEMICAL SOCIETY. 1966. Reagent Chemicals, American Chemical Society specifications. American Chemical Society, Washington, D.C.
7. AMERICAN PUBLIC HEALTH ASSOCIATION INC., INTERSOCIETY COMMITTEE. 1977. Methods of Air Sampling and Analysis, p. 52. Washington, D.C.
8. SALTZMAN, B.E. 1961. Preparation and analysis of calibrated low concentrations of 16 toxic gases. *Anal. Chem.* 33:1100.
9. AMERICAN PUBLIC HEALTH ASSOCIATION. 1977. Methods of Air Sampling and Analysis, 2nd ed. Washington, D.C.
10. SALTZMAN, B.E., W.R. BURG, and G. RAMASWAMY. 1971. Performance of permeation tubes as standard gas sources. *Env. Sci. Tech.* 5:1121.

Subcommittee 3

E. L. KOTHNY, *Chairman*
W. A. COOK
J. E. CUDDEBACK
B. DIMITRIADES
E. F. FERRAND
R. G. KLING
P. W. MCDANIEL
G. D. NIFONG
F. T. WEISS

409.

Tentative Method for Calibration of Continuous Colorimetric Analyzers for Atmospheric Nitrogen Dioxide and Nitric Oxide

42602-02-72T; 42601-02-72T

1. Principle of the Method

1.1 In instruments for the continuous colorimetric analysis of nitrogen dioxide and nitric oxide a metered sample flow is contacted with a metered liquid reagent flow to produce a color corresponding to the NO_2 content. NO is converted to NO_2 by a solid oxidant in the train upstream from the absorber. A typical arrangement is shown in Figure 409:1.

1.2 The calibration procedure is divided into 2 phases: static and dynamic. Static calibration is a test of the optical and measuring system. Reagent solutions are used to simulate actual flow and reagent conditions encountered during normal operation. A static calibration curve is a plot of analyzer responses versus reagent solutions equivalent to specific NO_2 concentrations. Adjustment of the instrument variables, *e.g.*, air and liquid flow rates, is performed after static calibration to make output response conform to predetermined concentrations in ppm or μg per m^3 of pollutant.

1.3 Reagent solutions equivalent to spe-

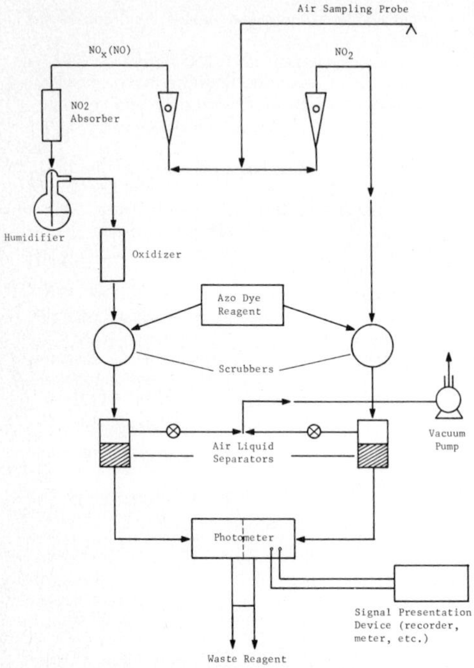

Figure 409:1—Typical Flow Diagram of NO_x Continuous Ambient Air Analyzers

cific NO_2 concentrations are prepared by reacting known quantities of sodium nitrite solutions with the azo-dye reagent and introducing these calibrating solutions into the measuring cell of the photometer at a rate that is commensurate to that of the analyzer.

1.4 Dynamic calibration is a performance test of the complete instrument when sampling a gas of known concentration. The efficiency of the absorbing system and the correctness of air and liquid flow rates are specifically tested.

1.5 A flow dilution panel is used to generate calibrated concentrations of gases which are fed into the instrument. The mixture is analyzed for NO_2 by a manual method **(2)**. The nitric oxide content is determined similarly after first removing any NO_2 with a highly selective absorber and then oxidizing the NO to NO_2 with a high-efficiency solid oxidant **(3, 4)**.

2. Range and Sensitivity

2.1 Continuous colorimetric analyzers for nitrogen dioxide and nitric oxide usually measure these gases in the range of 0.01 to 2 ppm by volume (equivalent to approximately 12 to 2450 μg NO per m^3 or 19 to 3760 μg NO_2/m^3 **(1)**. The range and sensitivity of the dilution procedure are determined by that of the manual method (0.005 to about 2 ppm) and by the volumes or flow rates that can be reasonably handled. Modifications of the flow rates and sampling times may reduce or extend the range. The absorbance of the dye solution in the spectrophotometer must be kept in the linear range of the calibration curve.

3. Interferences

3.1 The calibrating gases are essentially free of interfering compounds. However, high levels of NO_2 in the NO calibrating gas may generate some additional NO in the NO_2 scrubber **(4)**.

3.2 Relative humidities in excess of 90% for extended periods inactivate the NO oxidizer. Operating the oxidizer 5 C above ambient temperatures (*e.g.*, by heating with a 7 W lamp) eliminates this interference.

4. Precision and Accuracy

4.1 Precision of the dilution panel. A precision of 1% of the mean can be achieved in the manual sampling with careful work **(5)**; the limiting factors are the measurements of the volume of the air sample and the absorbance of the colored reagent. During instrument operation the precision will usually drop 2 to 5% because of airflow variations. These are produced by pressure drops and unstable or pulsating flows.

4.2 Statistical precision of dynamic calibration. The dynamic response curve (Section **8**) will be within ± 10% of the static calibration curve 95% of the time **(6)**.

4.3 At present, accuracy data are not available.

5. Apparatus

5.1 A dilution panel (Figure 409:2) capable of delivering test gases in the range of zero to 2 ppm with steady flow rates is suitable. Dilution panels should be designed to have low pressure drops in the various flow components to prevent errors in flow rate measurement.

5.1.1 Portable instrument carrying case (dimensions approximately 25 × 50 × 60 cm fiberglass or aluminum).*

5.1.2 Perforated panel, 0.64-cm (¼″) thick, for mounting the dilution components.

5.1.3 Air-Vacuum pump capable of stable maximum flow rate of at least 5 l/min, bolted through a vibration-damping pad to the case. Control switch mounted on panel.

5.1.4 Safety leg, removable and bent, to offer support during operation.

5.1.5 *Flowmeters with sapphire balls.* F_1 with maximum flow rate of 0.25 liter/min; F_2 with maximum flow rate of 1.5 liter/min; F_3 with maximum flow rate of 9 liter/min.

5.1.6 Reducing adapter (RA) of Teflon to connect F_3 flowmeter to flow lines.

5.1.7 Gas flow lines of 8 mm outer diam glass tubing or 10 mm outer diam Teflon. Flexible PVC tubing (10 mm outlet diam) may be used for butt-joint connection of inert tubing and for air lines. Ball and socket joints (*e.g.*, 12/5 mm) with 8 mm OD stems are preferred for connecting the calibration gas flow lines to the dilution panel, and for sampling lines.

5.1.8 Mixing Chamber (200 to 300 ml) fabricated of glass with two deep-set inlets and a shallow outlet, all with 12/5 mm ball ends.

5.1.9 *Stopcock and valves.* V_1 is a 3-way glass stopcock† with Teflon T-bore plug;§ V_2, a 2-way glass stopcock with Teflon 90° bore plug; V_3, a fine control stainless steel needle valve.

5.1.10 Timer for manual sampling, 0 to 99.9 min.||

5.1.11 *NO_2 absorber.* A 20 mm interior diam 100 to 150 mm long polyethylene tube with connecting caps at both ends, is filled 50 mm in depth with NO_2 absorbent granules (Section **6.2**). The granules are held in place with glass wool plugs. The granular size of the solid support and the packing of the glass wool should be chosen so that the pressure drop across the tubes is negligible.

5.1.12 *NO oxidizer.* An 8 to 15 mm internal dia glass tube with connecting ends is filled 20 mm to maximally 80 mm in depth with oxidizer granules (Section **6.3**) between two glass wool plugs. (**Caution:** Protect eyes and skin when handling this reagent.)

5.1.13 *Gas absorbers, or bubblers.* A pair with matched flow rate characteristics is used to minimize flow rate corrections

*H. Koch and Sons, Corte Madera, Calif. or equivalent.

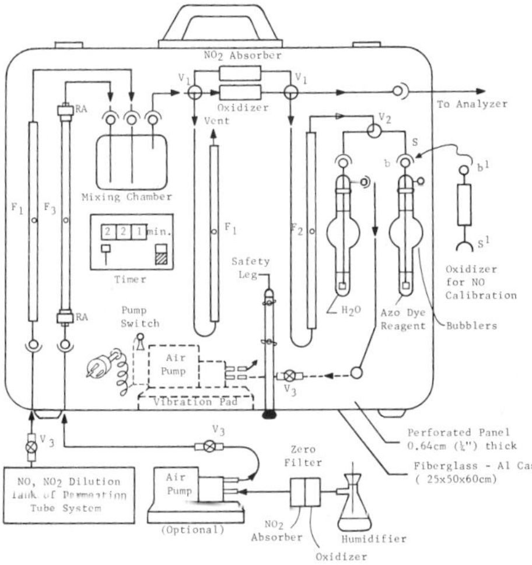

Figure 409:2—Diagram of Typical Portable Dilution Apparatus

†Kontes Glass Co., Vineland, N.M. cat# K-833500, plug ₸ 20/44 mm or equivalent.

§Ace Glass Inc., Vineland, N.M. cat# 8145-T, 3 mm bore ₸ 20/44 or equivalent.

||Scientific Products, Evanston, Ill. Tek-Timer C#C or equivalent.

on switching. Another absorber is used with water to maintain constant flow conditions to the continuous analyzer inlet.

5.1.14 *See Section* **5.5**

5.2 DILUENT AIR.

5.2.1 An oil-less (carbon vane or diaphragm) vacuum pump capable of stable maximum flow rate of 5 to 10 l/min fitted with vibration pad, PVC flexible tubing, fine control needle valve V_3 and a 12/5 mm socket, is needed.

5.2.2 Cylinder air (if air pump is not used), fitted with two-stage pressure regulator, PVC flexible tubing, and fine control needle valve V_3, is suitable. The zero filter (**5.2.4**) is not necessary, but the humidifier should be used with an appropriate safety device to prevent pressure build-up in the glass vessel.

5.2.3 A humidifier, made with a 250 ml Erlenmeyer flask, with inlet tube placed 2 cm above water level to give a relative humidity of 70 to 80% is placed between diluent air source and zero filter (**5.2.4**), where applicable.

5.2.4 The zero filter is made assembling a NO oxidizer (**5.1.12**) and a NO_2 absorber (**5.1.11**) in series.

5.3 DILUTED NO AND NO_2 CALIBRATING GASES. For routine calibration of analyzers 10 to 20 ppm NO in nitrogen (< 10 ppm water and < 3 ppm oxygen), and NO_2 in nitrogen or air (< 10 ppm water) are commercially available.

5.4 PERMEATION TUBES.

5.4.1 Nitric oxide is not available in permeation tubes because of its low boiling point, −152 C.

5.4.2 Permeation tubes of NO_2 are available;†† ** although the permeation rates may change with time due to reactions with the tube material. They require the use of thermostats. Permeation systems described for SO_2 (**7**) are also applicable for NO_2 generation (**8**). A portable permeation apparatus for SO_2 calibration was developed by APCO personnel‡‡. Another compact, lightweight gas mixing thermostat for field and laboratory use is available commercially.**

5.5 Additional apparatus needed to perform the reference manual sampling described in the Tentative Method of Analysis for Nitrogen Dioxide Content of the Atmosphere (Griess-Saltzman Reaction) (**2**).

6. Reagents

6.1 PURITY OF CHEMICALS. All reagents should be of analytical grade (**9**).

6.2 NITROGEN DIOXIDE ABSORBENT (**4, 10**). 1.5 mm (1/16″) granules or 10 to 20 mesh porous inert material such as firebrick, alumina, zeolites, etc., is soaked in 20% aqueous triethanolamine, drained, extended on a wide petri dish, and dried for 30 to 60 min at 90 to 95 C in an oven. The granules should be free flowing.

6.3 OXIDIZER (**3, 4, 10**). Glass, firebrick, or alumina, mesh size 15 to 40 is soaked in a 17% w/w aqueous chromium trioxide solution, then drained and dried in an oven at 105 to 115 C and subsequently exposed to 70% relative humidity. This can be done best by exposing a thin layer of pellets contained in a petri dish to a saturated solution of sodium acetate contained in a desiccator. The reddish color of the oxidizer changes to a golden orange when properly equilibrated.

6.4 $NaNO_2$ STOCK SOLUTION. Dissolve 2.03 g $NaNO_2$ (a correction is applied if purity is less than 100%) in distilled water in a 1000-ml volumetric flask.

6.5 OTHER. Additional reagents needed to perform the reference manual sampling, described in the Tentative Method of Analysis for Nitrogen Dioxide Content of the Atmosphere (Griess-Saltzman Reaction) (**2**).

††Metronics Associates, Inc., Palo Alto, Calif., Bureau of Standards, Washington, D.C.

**Analytical Inst. Development, Inc., Westchester, Pa. Model 303.

‡‡Rodes, CE, Bowen, JA, and Burmann F: Automated Air Analysis Section, Air Pollution Control Office, 225 S. Magnum St., Durham, NC 27701.

7. Static Calibration Procedure (11, 12)

7.1 From a stock solution containing 2.03 g of $NaNO_2$/l, take 5 ml and dilute to 500 ml with distilled water (nitrite-free) in a low-actinic volumetric flask. This is the working solution of $NaNO_2$ equivalent to 10 μl of NO_2/ml (or 20 μg of NO_2/ml). Standardization by nitrite solution is based upon the empirical observation that 0.72 mole of sodium nitrite produces the same color as 1 mole of nitrogen dioxide **(2, 13)**. The conversion factor may be affected by design variables and may increase up to a value of unity **(14, 15)**.

7.2 The reagent used in the following procedure should be formulated in accordance with the manufacturer's recommendations.

7.2.1 Turn off the air and liquid pumps from the analyzer. Disconnect the reference cell solution lines at the air-reagent junctions and drain the solution back into the reagent reservoir. Transfer the spent-reagent lines from the used-reagent bottle and drain into a waste bucket. If the analyzer being calibrated uses a carbon column, then the inlet to the column is bypassed and the line placed into a waste bucket. Disconnect the inlet lines of the air-liquid separators and attach the delivery ends of a hypodermoclysis set.†† The upper end of the set is attached to a 125-ml low-actinic separatory funnel which is supported near the instrument by a ring stand. The liquid flow rates to each measuring cell can be controlled by the small clamps which come with the administration set. Turn on the solution pumps. At this point, reagent flows only through the reference cells. Siphon reagent almost to the mark into 5 250-ml low-actinic volumetric flasks. Use one flask as "blank." Pour about 25 ml of blank reagent into the separatory funnel. Squeeze out any air bubbles trapped between the funnel and the screw clamps. Flush the connecting lines and cells with about 25 ml of blank reagent. Repeat with a second 25 ml portion. Fill the separatory funnel with blank reagent and adjust the discharge rates into the cells at about the same flow rate the analyzer uses (1 ml/min or about 20 drops/min).

7.2.2 While wating for the instrument to stabilize, pipet from the working solution of $NaNO_2$ (10 μl NO_2/ml) 1.0, 2.0, 3.0, and 5.0 ml portions into the 4 remaining 250 ml volumetric flasks containing reagent. Dilute each to mark with reagent and mix well. These solutions are equivalent to 0.04, 0.08, 0.12, and 0.20 μl of NO_2/ml, respectively. Maximum color develops in about 15 min. After the recorder trace has stabilized, readjust to zero or slightly above zero to allow for negative zero shifts. Repeat the procedure successively with the 4 nitrite dilutions prepared previously. Check the conc of each dilution by measuring its absorbance with a calibrated spectrophotometer before using. In this way errors in dilution are found before introduction in the measuring cells. Note the recorder readings for each concentration when the trace reaches a stable level. Reintroduce blank solution to check for zero drift.

7.2.3 Plot the net recorder readings on linear graph paper against the concentration in μl NO_2/ml of the calibrating solutions. Convert the readings to absorbance if they are in other units. Determine the slope, *b*, of the line. If the instrument has a variable resistor to control the upper-limit of the span an additional step is often desirable. Determine, from the recently developed static response curve, the concentration (μl NO_2/ml) which corresponds to an absorbance of 1.00. Prepare and introduce this concentration in the same manner as the others and obtain a stable trace. Adjust the upper-limit span control of the recorder to give a trace corresponding to absorbance 1.00 and check the linearity by reintroducing a dilution with a lower absorbance. Re-zero with blank solution. Replumb the analyzer to original operating

††These sets are used to administer fluids under the skin, are inexpensive and very convenient in aiding the delivery of solutions under controlled flow rates. They are readily available from medical supply houses.

condition, but bypass the NO_2 filter, solid oxidant, and humidifier on the NO channel.

7.2.4 The recommended reagent flow rate is kept constant and the corresponding air flow rate is calculated as described in **8.3.2**. The calculation of air flow rate from the slope of the response curve is meaningless for instruments whose photometer output is not proportional to absorbance. For these cases, a dynamic calibration at several concentrations, after choosing some convenient air flow rate, is performed. From the resulting dynamic response curve, a non-linear template is prepared and subsequently used to transform recorded readings directly to concentrations. An alternate method is to attempt to linearize the instrument output by adjusting the electronics of the photometer by some combination of photocell voltage and span upper-limit setting until a linear output is found over the range of pollutant concentration of interest.

8. Dynamic Calibration Procedure

8.1 Adjust previously determined air and liquid flow rates to their proper settings. The analyzer is now ready for dynamic calibration.

8.1.1 *Nitrogen Dioxide*. Place the dilution panel as close to the analyzer as possible to minimize pressure drop in the tubing used to connect the 2 systems. Calculate the total air flow of the analyzer. To this add the air flow of the manual sampler (400 ml/min) and a slight excess, *e.g.*, 30 to 50 ml/min. Turn on the dilution panel air pumps and set the dilution air flow to provide the amount just calculated. Connect the sampling probe of the analyzer to the dilution panel outlet.

8.1.2 Adjust the dilution air flow so that a slight excess is indicated on the vent flowmeter. In the ideal situation, the position of the flowmeter floats should be the same with or without the connection of the sample probe. After the recorder trace is stable, adjust the chart reading to zero or to a desired baseline. Clean diluent air (NO and NO_2 free) is produced by attaching an oxidizer **5.1.12** and a NO_2 absorber **5.1.11** in series on the inlet end of the dilution air pump or by using synthetic air. Switch to the oxidizing system on the dilution panel to oxidize residual NO to NO_2 in the NO_2 source. Before use, the NO_2 source either from the pressure tank equipped with a two-stage regulator or from the permeation panel should be bled into a soda lime tube for at least an hr at a flow rate of about 50 ml/min, in order to condition the delivery tubing. Adjust the NO_2 feed rate to obtain a recorder response above 50% of the chart. When the recorder trace is stable, assay the diluted NO_2 stream manually **(2)** with bubblers (Figure 409:2) containing 10 ml of azo-dye reagent. Sample at 400 ml/min and collect sufficient sample to give an absorbance within the range of 0.1 to 0.7. Five min is usually sufficient. Collect a duplicate sample and allow 15 min for full color development of each one. Read the color at 550 nm and determine the equivalent NO_2 concentration from the spectrophotometer calibration curve. Reduce the NO_2 gas input to achieve a lower reading. After the recorder has stabilized, assay the NO_2 stream in duplicate as before.

8.1.3 For each channel, plot the net recorder chart readings versus the μl of NO_2/ml obtained by manual sampling. Determine the slope of each line. The ratio of the slopes of the dynamic to static calibration curves represents the absorption efficiency or stoichiometric response of each contact column. The ratio is usually close to 1.0 and large deviations are indicative of errors in air or liquid flows, or in dilutions made during the static calibration. The cause of the deviation is usually found when the fluid flows are rechecked or the static calibration is repeated. Occasionally, the error may lie in some aspect of the dynamic calibration itself.

8.1.4 *Nitric Oxide*. Place the NO_2 absorber, the humidifier, and the oxidant back in line on the analyzer. Readjust the air flow to the required values. The analyzer is now ready to monitor NO from the dilution panel. The oxidizer **5.1.12** is placed between the manual sampling exit and the bubbler containing the azo-dye reagent. The NO_2 absorber **5.1.11** is switched in-line on the dilution panel to remove any traces of NO_2 which may be present in the NO stream. The rest of the calibration follows the procedure used for NO_2. This method of calibrating for NO does not determine the oxidation efficiency of the oxidizer. Because the same oxidizer is used in the manual sampling as well as in the ambient analyzer, incomplete conversion would not be noticed. However, conversion is complete (**10**). At least once a month, replace the oxidizer in use by a fresh one to check the one in use. When the dynamic calibrations are completed, disconnect the transfer line connecting the dilution panel to the analyzer and reconnect the external sampling line. Readjust the air flow to the required values. The analyzer is now ready to monitor ambient NO and NO_2.

9. Calculations

9.1 See manual method (**2**) for calculating on-stream concentrations.

9.2 For the static calibration it was convenient to use a solution of $NaNO_2$ which would yield theoretically with conversion factor 0.72 (**2**) (see **7.1**) the equivalent of 10 μl gaseous NO_2/ml of working solution at 25 C. The concentration in moles NO_2 per ml of solution corresponding to 10 μl gaseous NO_2 per ml is calculated as follows:

$$\frac{10 \times 10^{-6} \text{ liters } NO_2}{\text{ml of solution}} \times \frac{\text{mole}}{22.4 \text{ liter}} \times \frac{273°K}{298°K} = 408 \times 10^{-7} \text{ mole } NO_2/\text{ml of solution.}$$

The corresponding concentration in g $NaNO_2$/liter is given by:

$$\text{moles } NO_2/\text{m]} \frac{\text{moles } NO_2^-}{\text{mole } NO_2} \text{(conversion factor)} \times 10^3 \text{ ml/liter} \times \text{MW } (NaNO_2) =$$

$$4.08 \times 10^{-7} \times 0.72 \times 10^3 \times 69.00 = 2.03 \times 10^{-2} \text{ g } NaNO_2 \text{ per liter working solution.}$$

The working solution is made by dilution of a 100 times concentrated stock solution (containing 2.03 g $NaNO_2$/liter, section **6.4**).

9.3 DYNAMIC CALIBRATION.

9.3.1 By definition $\text{ppm (v/v)} = \dfrac{\text{microliters of pollutant}}{\text{liters of sample}}$

In a dynamic system, such as in an oxides of nitrogen analyzer, a material balance gives:

$$\text{ppm } NO_2 \text{ (v/v)} = \frac{\text{(microliters } NO_2\text{/ml of solution) (ml of solution per minute)}}{\text{liters of air per minute}}$$

9.3.2 If a direct readout on the chart is desired, that is, the chart reading in absorbance units is equivalent to ppm NO_2, a suitable reagent flow rate is chosen and the air flow is calculated using the static calibration curve. Often a scale factor, f_s is used so that the output reading is a simple fraction of input concentration.

$$\text{Chart reading A} = f_s \text{ ppm } NO_2 = f_s \text{ (microliters } NO_2\text{/ml)} = \frac{Q_r}{Q_a} \text{ or}$$

$$Q_a = f_s \text{ (microliters } NO_2\text{/ml)} \times \frac{Q_r}{A} \text{ but}$$

$$\text{(microliters } NO_2\text{/ml)/A} = \frac{1}{b}$$

therefore, $Q_a = Q_r \times \frac{f_s}{b}$

where:

A = Absorbance units
Q_r = reagent flow rate (ml/min)
Q_a = air flow rate (liter/min)
f_s = a fractional number
b = slope of static calibration curve.

10. Effect of Storage

10.1 Mixtures appear unstable in ordinary steel cylinders. Nitrogen dioxide mixtures in dry air can be kept for periods of months without noticeable change, if prepared in stainless steel cylinders. Nitric oxide mixtures in dry nitrogen can be stored for several months.

11. References

1. ASTM. 1970. Continuous measurement of nitric oxide, nitrogen dioxide, and ozone in the atmosphere. Method D 2012-69, Annual Standards, Part 23, 620–27, November.
2. American Public Health Association. 1977. Methods of Air Analysis and Sampling. 2nd ed. p. 527. Washington, D.C.
3. Ibid. p. 527.
4. Levaggi, D.A. *et al.* A precise method for accurately analyzing the content of nitrogen oxides in the atmosphere. Presented at the 17th Annual Meeting of the Institute of Environmental Sciences, Los Angeles, April.
5. Thomas, M.A., and Amtower, R.E. 1966. Gas dilution apparatus for preparing reproducible dynamic gas mixtures in any desired concentration and complexity. *J. Air Pollut. Contr. Ass.* 16:618.
6. Air and Industrial Hygiene Laboratory. 1970. Continuous air analysis section, California State Department of Public Health, Berkeley, California. Routine field calibration of continuous analyzers for the California Air Resources Board.
7. American Public Health Association. 1977. Methods of Air Analysis and Sampling. 2nd ed. p. 696. Washington, D.C.
8. Saltzman, B.E., Burg, W.R., and Ramaswamy, G. 1971. Performance of permeation tubes as standard gas sources. *Env. Sci. Tech.* 5:1121.
9. American Chemical Society Specifications. 1966. ACS Reagent Chemicals. Washington, D.C.
10. Belsky, T. 1971. Unpublished results. Air and Industrial Hygiene Laboratory, California State Department of Public Health, Report No. 85, January.
11. Pierce, L.B., Nishikawa, K., and Fansah, N.O. 1967. Validation of calibration techniques. 8th Conf. on Methods in Air Poll. Ind. Hyg. Studies, Oakland, February.
12. Air and Industrial Hygiene Laboratory. 1966. California State Department of Public Health. Guide to operation of atmospheric analyzers, Recommended Method No. 7.
13. Scaringelli, F.P., Rosenberg, E., and Rehme, K.A. 1970. Comparison of permeation devices and nitrite ion as standards for the colorimetric determination of nitrogen dioxide. *Env. Sci. Tech.* 4:924.
14. Thomas, M.D., *et al.* 1956. Automatic apparatus for the determination of nitric oxide and nitrogen dioxide in the atmosphere. *Anal. Chem.* 28:1810.
15. Huygen, I.C. 1970. Reaction of nitrogen dioxide with Griess type reagents. *Anal. Chem.* 42:407.

Subcommittee 3
E. L. Kothny, *Chairman*
W. A. Cook
B. Dimitriades
E. F. Ferrand
G. D. Nifong
P. W. McDaniel
B. E. Saltzman
F. T. Weiss

410.

Tentative Method for Continuous Monitoring of Atmospheric Oxidant* with Amperometric† Instruments

44101-01-69T

1. Principle of the Method

1.1 Ozone and other atmospheric oxidants present in the atmospheric air sample react with iodide ions present in the sensing reagent, to produce iodine which forms triiodide ion. Triiodide ion in turn is electrically converted back into iodide ions. The amperometric conversion of the triiodide to iodide ion is accomplished through cathodic reduction utilizing either an electrolytic **(7, 13)** or a galvanic cell **(3, 5)**.

1.2 In the electrolytic procedure an iodide-containing electrolyte is used. A small potential (0.025 to 0.3 V) is applied between electrodes. The initial current polarizes the cathode with a layer of hydrogen which prevents further current flow. A fraction **(9)** of the triiodide ion in solution reacts with some of the hydrogen. Consumption of hydrogen causes flow of polarization current to replace the reacted hydrogen and reestablish polarization. The polarization current is proportional to the triiodide ion concentration according to the configuration of the electrodes. Details are discussed in sections **5.2** and **7.1**.

1.3 In the galvanic cell cathodic reduction of iodine occurs spontaneously along with parallel oxidation of the carbon anode. The galvanic oxidant detector operates as an electro-chemical fuel cell in which the fuels are oxidant (in the sample) and carbon (in the anode). Details are discussed in sections **5.3** and **7.2**.

2. Range and Sensitivity

2.1 Amperometric oxidant detectors operate most reliably at atmospheric oxidant concentration ranges equivalent to 0.01 to 0.2 ppm of ozone (0.02 to 0.4 μg/l).

2.2 Usable response may be obtained from oxidant levels equivalent to as much as 1 ppm ozone. In the range 0.2 to 1 ppm results in comparison to colorimetric procedures show some scatter and differences.

3. Interferences

3.1 Amperometric oxidant detectors respond to substances that are capable of either oxidizing iodide ion to iodine (positive response) or reducing iodine to iodide ion (negative response) in the sensing solution. The most common oxidizing interference is nitrogen dioxide, whereas sulfur dioxide, and hydrogen sulfide are among the most common negative interferences.

3.2 Presence of nitrogen dioxide in the sample causes positive interference equivalent to 0.03 to 0.20 moles of ozone/mole of nitrogen dioxide **(4, 5, 14)**. The exact level of the detector's specific response to nitrogen dioxide depends on design of detector, composition of sensing solution, operating conditions, and other unknown factors. Because of these unknown factors, the specific response to nitrogen dioxide for each detector must be determined experimentally rather than taken from literature data. The oxidant measurements may be corrected by subtracting

*The term "oxidant" is used to describe substances, other than nitrogen dioxide, that oxidize iodide ions to iodine under neutral pH conditions.

†The term amperometric more accurately describes this process which previously was called coulometric.

from the total detector response the amount of response caused by nitrogen dioxide.

3.3 Presence of sulfur dioxide in the sample causes negative interference equivalent to approximately 1.0 mole of ozone/mole of sulfur dioxide. Two methods can be used for eliminating the sulfur dioxide interference.

3.3.1 *First method.* Add to the total detector response an increment equal in value to the reduction caused by the sulfur dioxide. To determine the latter, the sulfur dioxide level in the sample and the specific response of the detector to sulfur dioxide must be determined previously. This correction method is recommended only when sulfur dioxide levels do not exceed those of oxidants.

3.3.2 *Second method.* Remove sulfur dioxide from the sample stream before it enters the oxidant sensor cell. Such removal without affecting the oxidant, can be accomplished using solid **(11, 13)** or liquid **(3, 5)** oxidants. Due to lack of adequate data regarding performance of these oxidants, performance for each oxidant system must be established experimentally rather than from literature data. This correction method is recommended when sulfur dioxide levels exceed those of oxidant.

3.4 Presence of hydrogen sulfide and/or other reducing compounds in the sample causes negative interference of varying magnitude. To correct oxidant-measurement results for these interferences add to the total detector response an increment equal in value to the reduction caused by the reducing compounds present. The latter is determined from both the total concentration of the reducing compounds in the sample and the specific response of the detector to these compounds. If a sulfur dioxide scrubber is used, **3.3.2**, correction should be applied for the presence of reducing compounds in the sample stream that emerges from the sulfur dioxide scrubber.

4. Precision and Accuracy

4.1 Because of the nonspecific nature of atmospheric oxidant, precision of the analytical method is defined in terms of stability and reproducibility of the detector's response to ozone. In continuous monitoring, the hourly variation in response to ozone present at a constant level between 0 and 0.50 ppm should be equivalent to ± 0.01 ppm **(9)** or less **(5)**. Day-to-day variability is considerably larger and its magnitude varies with individual instruments **(9, 12)**.

4.2 Accuracy of the method has not been established.

5. Apparatus

5.1 MATERIALS. The sampling lines and all parts of the instrument that come in contact with the sample stream should be made of glass, Teflon **(1)** or stainless steel. Polyvinyl chloride tubing can be used for butt-to-butt connections of more inert tubing. Fluorocarbon or fluorosilicon grease may be used sparingly, if necessary. Sample bags **(1)** should be made of Teflon (fluorinated ethylene-propylene copolymer), Mylar (polyethylene terephthalate), or Tedlar (polyvinyl fluoride) and should be constructed and used so as to maintain a surface-to-volume ratio below 20 m^2/m^3.

5.2 ELECTROLYTIC CELL DETECTOR. Figure 410:1 shows the construction of a

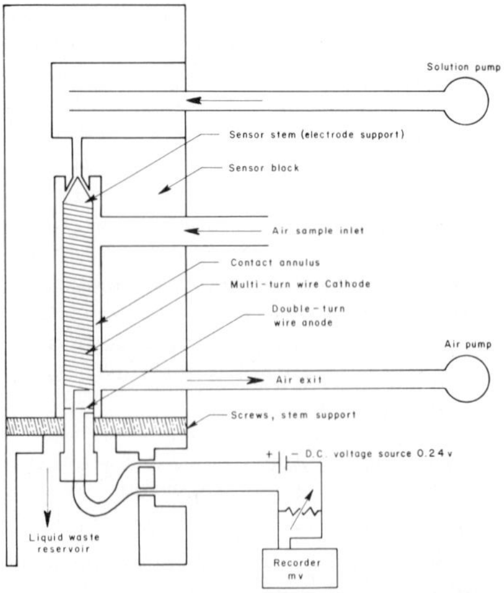

Figure 410:1—Electrolytic oxidant detector assembly.

commercially available electrolytic cell assembly **(7)**. Description of other commercially available instruments more recently introduced by different manufacturers is not included here but can be found in literature **(13)**.

5.2.1 *Sensing cell.* The sensing cell consists of a 10-cm long plastic (clear) block with a hollow cylinder into which a glass rod is placed coaxially. The glass rod serves as a support stem upon which are wound a multi-turn platinum wire cathode and double-turn anode. The annular space around the electrode support stem serves to bring the reagent solution and sample gas into intimate contact. The two electrodes are connected to a source of DC voltage (0.24 V) and the electrolytic current is converted into a potential drop across a resistor and is recorded on a potentiometric recorder.

5.2.2 *Air pump.* A constant-volume suction pump capable of operating continuously at sample-flow rates between 100 and 200 ml/min is required. The gas flow should be measured accurately with a soap-film flow meter at least once a week.

5.2.3 *Reagent solution pump.* A constant delivery pump capable of operating continuously at liquid flow rates between 1 and 5 ml/hr is required. Reagent flow rate should be measured accurately at least once a week by the following procedure: The liquid waste reservoir flask is replaced by a graduated cylinder and the volume of reagent collected over a precisely known period of time is measured.

5.2.4 *Air metering device.* The electrolytic detectors utilize constant-volume pumps; therefore, air metering device is not necessary.

5.3 GALVANIC CELL DETECTOR. Figure 410:2 shows the arrangement of a commercially available galvanic cell assembly **(3, 5)** which is an alternative system to the electrolytic cell detector.

5.3.1 *Sensing cell.* The sensing cell, Figure 410:3, consists of a capped glass tube containing two electrodes in contact with the electrolytic reagent solution. The anode electrode is made of carbon-electrolyte paste and is located at the bottom of the cell. The cathode electrode consists of a platinum gauze jacket that surrounds a short piece of open glass tubing. The short glass tube and platinum jacket are held in place, as shown in Figure 410:3, inside a longer glass tube that is attached to the cell's cap. The air sample stream is delivered through a capillary Teflon tube into the solution of electrolyte. The two electrodes are connected externally through a resistor; the galvanic current is measured as potential drop across the resistor and is registered on a potentiometric recorder.

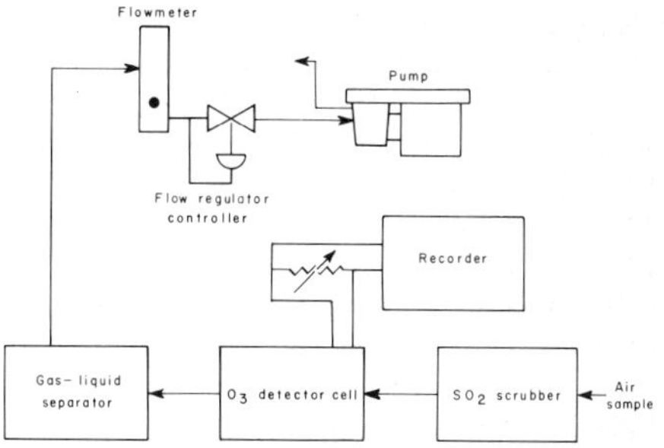

Figure 410:2—Galvanic oxidant detector assembly.

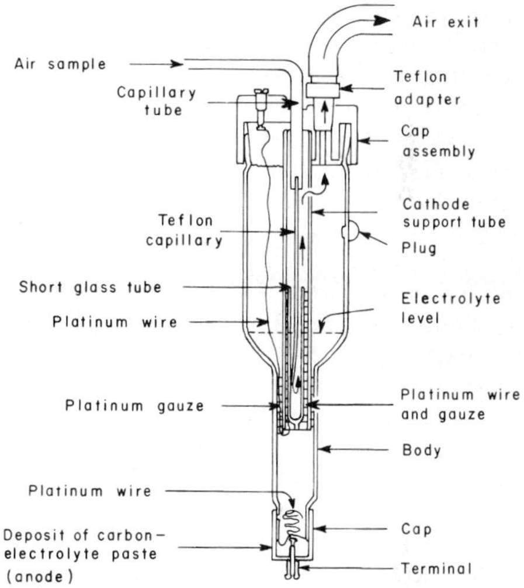

Figure 410:3—Galvanic oxidant sensing cell.

5.3.2. *Air pump.* See **5.2.2**.

5.3.3. *Air metering device.* When a constant-volume pump is not used, a needle valve and a glass rotameter capable of measuring gas flows of 100 to 200 ml/min are required. The rotameter preferably should be placed upstream from the sensor cell and should be calibrated with a wet test meter or with a soap-film flow meter. It is permissible to place the rotameter downstream from the sensor cell if a small loosely fitting plug of glass wool is placed in the sample line, ahead of the air metering device to prevent electrolyte droplets from being carried over and deposited in the air metering device. In either case the rotameter should also be periodically dismantled, cleaned, and calibrated.

5.4 Ozonizers. Ozonization of an air or oxygen stream can be accomplished by exposure to UV radiation. A convenient ozonizer can be made using a 4 W UV lamp‡ and appropriate circuitry for controlling lamp current between 0 and 500mA **(8)**. Varying levels of ozone can be obtained reproducibly by adjusting and stabilizing lamp current and flow rate of gaseous stream.

5.5 Sulfur dioxide scrubber using solid oxidant. A 100-mm Schwartz drying tube packed with 40 cm^2 of solid oxidant paper (Sec **6.6.1**) can be used **(11)**. Alternatively, a 15 cm tube × 1.5 cm (ID) packed with oxidant-coated quartz chips (Sec **6.6.2**) can also be used **(13)**.

5.6 Sulfur dioxide scrubber using liquid oxidant Sec **6.7**). A design for the galvanic detector is shown in Figure 410:4 **(3)**.

5.7 Charcoal column. A 13-cm tube × 1 cm (ID) or a cartridge packed with granular activated charcoal can be used to remove traces of oxidants from the air sample to provide zeroing gas.

6. Reagents

6.1 Purity. All chemicals should be of analytical reagent grade **(2)**.

6.2 Double-distilled water. Double-distilled water free of reducing compounds, may be prepared in an all-glass still from distilled water to which crystals of potassium permanganate and barium hydroxide have been added. Before use, double-distilled water should be deaerated by boiling.

‡such as Westinghouse 794 H, or G.E. OZ4-S-11.

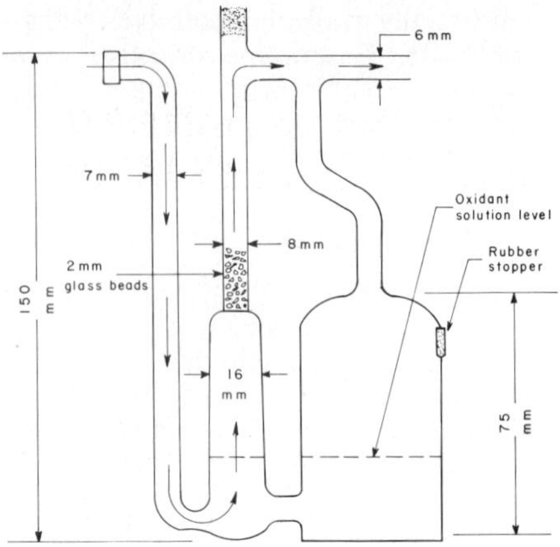

Figure 410:4—Sulfur dioxide scrubber using liquid oxidant.

6.3 Reagent solution for electrolytic detector. A 500-ml volumetric flask is approximately half-filled with double-distilled deaerated water to which the following chemicals are added:

10 g potassium iodide (KI)
25 g potassium bromide (KBr)
1.25 g monobasic sodium phosphate, monohydrate ($NaH_2PO_2 \cdot H_2O$)
3.5 g dibasic sodium phosphate, heptahydrate ($Na_2HPO_2 \cdot 7H_2O$)

The chemicals and water are mixed thoroughly and water is added to the 500-ml mark. Minor variations in the reagent composition will not affect the detector's performance. The reagent solution should be filtered before use if solid particles are visibly present. The solution can be stored in a stoppered bottle without deterioration for up to 4 months.

6.4 Reagent solution for galvanic cell. A 1-l volumetric flask is filled with approximately 800 ml double-distilled water to which the following chemicals are added:

14 g dibasic sodium phosphate (Na_2HPO_4)
14 g monobasic sodium phosphate (NaH_2PO_4)

360 g potassium bromide (KBr)
0.15 g potassium iodide (KI)

The chemicals are dissolved by mixing thoroughly and then water is added to the 1-l mark. Minor variations in reagent composition will not affect the detector's performance. The reagent solution can be stored in a stoppered bottle without deterioration for up to 4 months.

6.5 ANODE PASTE FOR GALVANIC DETECTOR. To prepare paste for one anode electrode, electrolyte solution is added to 5 g of Darco G-60 activated carbon with continuous mixing until a paste of thick but plastic consistency is obtained.

6.5.1 *Solid state sulfur dioxide scrubber using oxidant-coated paper.* Glass fiber filter paper is impregnated with chromium trioxide as follows **(11)**: spread 7.5 ml of aqueous solution containing 1.25 g of chromium trioxide and 0.35 ml of conc sulfuric acid uniformly over 200 cm^2 of paper, and dry in an oven at 80 to 90 C for 1 hr. Cut the paper into 6 × 12 mm strips, each folded once lengthwise into a V-shape, pack into a 70-ml U-tube and condition by drawing dry air through the tube overnight. The scrubber is effective for at least 1 month. If the packing becomes visibly wet from sampling humid air, it must be dried with dry air before further use.

6.5.2 *Solid state sulfur dioxide scrubber using oxidant-coated quartz chips.* The following preparation instructions have been recommended **(13)**; however, no supporting data are available. Porous quartz chips (Hengar granules), 4 to 10-mesh are soaked in a solution containing equal weights of chromium trioxide, water and 85% phosphoric acid. Following soaking, excess solution is drained off before loading the scrubber column.

6.6 SULFUR DIOXIDE SCRUBBER SOLUTION. The following instructions for preparation and use of a liquid sulfur dioxide scrubber have been recommended **(3)**; however, no supporting data are available.

A 1-l measuring flask is filled with approximately 800 ml distilled water to which the following chemicals are added:

100 g of chromium trioxide (CrO_3)
3 ml conc sulfuric acid (H_2SO_4)

After solution fill with water to the 1-l mark. Minor variations in reagent composition will not affect scrubbing efficiency. The scrubber solution can be stored in a stoppered flask for up to one week.

7. Procedure

7.1 ELECTROLYTIC DETECTOR **(7)**. The sensor cell is connected to the flask containing the reagent solution and the air and reagent pumps are turned on. If constant-volume pumps are not used, both the liquid reagent and air sample flow rates should be adjusted to approximately 1.25 ml/hr and 150 ml/min, respectively. The liquid reagent flow should be checked for constancy (Sec **5.2.3**) and the gas sample flow should be measured accurately (Sec **5.2.2**). Ambient temperature and pressure should be measured and recorded to allow correction of results for effects of temperature and pressure deviations on gas sample flow. Under recommended flow conditions the detector should yield 63% of its final response level within 30 sec **(9)** from the moment of first indication of response change. Longer response times may be caused by oxidant losses on "unconditioned" surfaces of sample lines. These surfaces can be conditioned by operating the instrument on an air sample containing about 1 ppm ozone until response time is minimized. Following each startup, the instrument should be allowed a stabilization period of at least 24 hr before it is used on a sample.

7.2 GALVANIC DETECTOR. The detector is assembled as shown in Figure 410:2 and the air pump is turned on. The system is checked for leaks and after air tightness is secured, the gas flow rate is adjusted to approximately 150 ml/min. The gas flow rate is then measured accurately as described in section **5.2.2**. Ambient temperature and pressure should be measured and recorded to allow correction of results for effects of temperature and pressure deviations on gas sample flow. Under recommended flow conditions the detector should yield 90% of its final response level within 45 sec from the moment of first indication of response change **(5)**. Longer response time may be caused by oxidant losses on "unconditioned" surfaces of sample lines. These surfaces can be conditioned as described in section **7.1**. Following each startup, the instrument should be allowed a stabilization period of at least 8 hr before it is used on a sample.

8. Calibration and Standardization

8.1 Because of the nonspecific nature of atmospheric oxidant, the response of an amperometric detector to atmospheric oxidant is expressed in terms of an ozone concentration level (ppm) that causes equal response. Therefore, the amperometric detectors must be calibrated to determine their specific response to ozone. Theoretically, the specific response to ozone is fixed and is equivalent to 2 electrons/ozone molecule. In practice, this response has values that are below the theoretical level and vary from instrument to instrument for unknown reasons **(4, 12, 14)**. Therefore, each detector must be calibrated experimentally.

8.2 Amperometric detectors are calibrated using standard gaseous mixtures of ozone. Such mixtures arc prepared by passing pure, filtered air or oxygen through an ozonizer (Sec **5.4**) and receiving the ozonized mixture into a properly conditioned plastic bag **(1)**. Ozone concentration of varying levels can be obtained either by diluting the ozonizer output with pure, filtered air to varying degrees or by varying the air-flow rate through the ozonizer **(8)**. In this manner a series of mixtures can be obtained that covers the ozone concentration range of 0 to 2 ppm. Following preparation, the concentrations of the ozone mixtures in the bags are determined by the Tentative Method for the Determination of Oxidizing Substances in the Atmosphere **(6)**. For calibration the detector is first zeroed using a stream of air that has been freed from oxidizing and/or reducing contaminants by passing through a filter and an activated charcoal column (Sec **5.7**). The detector is then connected to the bag that contains the standard ozone-air mixture, using a piece of conditioned Teflon (or other inactive material) tubing of minimum length. The resultant deflection data are treated mathematically as discussed in section **9** to define the detector's response as a function of ozone concentration.

9. Calculations

9.1 The instruments are calibrated by the manufacturers in accordance with the theoretical coulometric yield of 2 electrons/molecule of reacted ozone. The theoretical calibration relationships are given in equations 1 and 2.

$$V = rI = r \frac{P \times F \times C \times 2 \times f}{R(t + 273)} \quad (1)$$

$$V = 5.16 \times 10^{-5} \times \frac{r \times P \times F \times C}{t + 273} \quad (2)$$

where:

- I = detector response, milliamperes;
- V = detector response, millivolts;
- F = sample flow rate, ml/minute;
- r = resistance of resistor used to convert signal current to voltage drop, ohms;
- P = barometric pressure, mm/Hg;
- t = ambient temperature of the constant volume pump or the rotameter, °C;
- C = concentration of ozone in sample, ppm;
- f = the amount of electricity produced from coulometric conversion of one electrochemical equivalent of ozone (96,494 coulombs), and
- R = the universal gas constant in appropriate units.

9.1.1 In the electrolytic cell, response is affected by liquid flow rate although the later is not included in the theoretical calibration equation. Also, response seems to be affected by temperature to a greater extent than expected from the perfect gas law.

9.2 In practice, the response is a linear function of ozone concentration at ozone levels only below 1 ppm. This linear function is different from that described by equations 1 and 2 and is determined from the calibration data in the following manner: Plot response values against ozone concentrations in the standardized ozone mixtures. Apply the method of least squares to obtain the slope and intercept of the straight line that fits the data best. If the intercept value is significantly different from zero, apply the defined linear function only within the range of ozone concentrations used in calibration and obtain additional calibration points using ozone concentrations closer to zero.

9.2.1 To convert response data into ozone concentration, equation 3 is used. If response is proportional to ozone concentration, the value of k_1 is zero.

$$C = \frac{(t + 273)}{FP} [k_1 + k_2 V] \qquad (3)$$

The values of k_1 and k_2 are calculated from the intercept and slope, respectively, of the calibration line (section **9.2**).

9.3 If the recorder has appropriate zero and span controls, it can be adjusted to give corrected readings under calibration conditions. If sample temperature and pressure deviate significantly, correct the sample flow rate to calibration conditions according to the perfect gas law.

10. Effects of Storage

10.1 With the exception of the sulfur dioxide scrubber solution, all reagents used in amperometric oxidant detectors can be stored in stoppered flasks without deterioration for up to 4 months. Storage of sulfur dioxide scrubber solution should not exceed 1 week (section **6.7**).

10.2 Fungus growth may occur occasionally, especially in instruments that have been shut down for extended periods of time without having first removed the reagent and washed the instrument with distilled water. When such growth appears, the instrument should be taken apart, cleaned, rinsed with distilled water and charged with fresh reagent.

11. References

1. ALTSHULLER, A.P., and A.F. WARTBURG. 1961. The Interaction of Ozone with Plastic and Metallic Materials in Dynamic Flow System. *Int. J. Air Water Poll.* 4:70–78.
2. AMERICAN CHEMICAL SOCIETY, Washington, D.C. Reagent Chemicals, American Chemical Society Specifications.
3. BECKMAN INSTRUMENTS, INC., Fullerton, California.
4. CHERNIAK, I., and R.J. BRYAN. 1965. A Comparison Study of Various types of Ozone and Oxidant Detectors which are Used for Atmospheric Air Sampling. *J. Air Poll. Control Assoc.* 15:351–354.
5. HERSCH, P., and R. DEURINGER. 1963. Galvanic Monitoring of Ozone in Air. *Anal. Chem.* 35:897–899.
6. AMERICAN PUBLIC HEALTH ASSOCIATION, INC. 1977. Methods of Air Sampling and Analysis. 2nd ed. pp. 527. Washington, D.C.
7. MAST, G.M., and H.E. SAUNDERS. 1962. Research and Development of the Instrumentation of Ozone Sensing. ISA Transactions, 1:325–328.
8. PIERCE, L. B., K. NISHIKAWA, and N.O. FANSAH. 1967. Validation of Calibration Techniques. Eighth Conference on Methods in Air Pollution and Industrial Hygiene Studies. Oakland, California.
9. POTTER, L., and S. DUCKWORTH. 1965. Field Experience with the Mast Ozone Recorder. *Air Pollution Control Assoc. J.* 15:207–209.
10. SALTZMAN, B.E., and N. GILBERT. 1959. Iodometric Microdetermination of Organic Oxidants and Ozone. *Anal. Chem.* 31:1914–1920.
11. SALTZMAN, B.E., and A.F. WARTBURG. 1965. Absorption Tube for Removal of Interfering Sulfur Dioxide in Analysis of Atmospheric Oxidant. *Anal Chem.* 37:779–782.
12. STEPHENS, E.R. 1965. Infrared Calibration of Coulometric Analyzers. Seventh Conference on Methods in Air Pollution Studies. California State Department of Public Health, Los Angeles, California.
13. SCHULZE, F. 1966. Versatile Combination Ozone and Sulfur Dioxide Analyzer. *Anal. Chem.* 38:748–752.
14. WARTBURG, A. F., A.W. BREWER, and J. P. LODGE, JR. 1964. Evaluation of a Coulometric Oxidant Sensor. *Int. J. Air Wat. Poll.* 8:21–28.

Subcommittee 3

B. E. SALTZMAN, *Chairman*
W. A. COOK
B. DIMITRIADES
E. L. KOTHNY
L. LEVIN
P. W. MCDANIEL
J. H. SMITH

411.

Tentative Method for the Manual Analysis of Oxidizing Substances in the Atmosphere

44101-02-70T

1. Principle of the Method

1.1 Micro amounts of ozone and other oxidants liberate iodine when absorbed in a 1% solution of potassium iodide buffered at pH 6.8 ± 0.2. The iodine is determined spectrophotometrically by measuring the absorption of triiodide ion at 352 nm.

1.2 The stoichiometry is approximated by the following reaction:

$$O_3 + 3\,KI + H_2O \rightarrow KI_3 + 2\,KOH + O_2$$

2. Range and Sensitivity

2.1 This method covers the manual determination of oxidant concentrations between 0.01 to 10 ppm (0.01 to 10 μl/l) as ozone (**1**).

2.2 When 10 ml of absorbing solution is used, between 1 and 10 μl of ozone, corresponding to absorbances between 0.1 and 1 in a 1 cm cell, are collected.

3. Interferences

3.1 Sulfur dioxide produces a negative interference equivalent to 100% of that of an equimolar concentration of oxidant.

3.1.1 Up to 100-fold ratio of sulfur dioxide to oxidant may be eliminated without loss of oxidant by incorporating a chromic acid paper absorber in the sampling train upstream from the impinger (**2**).

3.1.2 The absorber removes sulfur dioxide without loss of oxidant but will also oxidize nitric oxide to nitrogen dioxide.

3.1.3 When sulfur dioxide is less than 10% of the nitric oxide concentration, the use of chromic acid paper is not recommended. In this case the effect of sulfur dioxide on the oxidant reading can be corrected for by concurrently analyzing for sulfur dioxide (**3**) and adding this concentration to the total oxidant value.

3.2 Nitrogen dioxide is known to give a response in 1% KI (**1**), equivalent to 10% of that of an equimolar concentration of ozone. The contribution of the nitrogen dioxide to the oxidant reading can be eliminated by concurrently analyzing for nitrogen dioxide (**4**) and subtracting one-tenth of the nitrogen dioxide concentration from the total oxidant value.

3.3 Peroxyacetyl nitrate gives approximately a response equivalent to 50% of that of an equimolar concentration of ozone (**5**). Concentrations in the atmosphere may range up to 0.1 ppm.

3.4 Other oxidizing substances besides ozone will liberate iodine with this method: *e.g.*, halogens, peroxy compounds, hydroperoxides, organic nitrites and hydrogen peroxide (**6, 7**).

3.5 Hydrogen sulfide, reducing dusts or droplets can act as negative interferences.

4. Precision and Accuracy

4.1 The precision of the method within the recommended range (Section **2.2**) is about ± 5% deviation from the mean. The major error is from loss of iodine during sampling periods; this can be reduced by using a second impinger.

4.2 A check of reproducibility under various operating conditions may be made in a flow system. Ozone may be supplied directly from an ultraviolet ozonizer (**8**), or from a Mylar bag reservoir (**9**) previously filled with an ozone-air mixture.

4.3 The accuracy of this method has not been established because of the difficulty in preparing known micro concentrations of ozone or oxidant.

4.4 The calibration is based on the assumed stoichiometry of the reaction with the absorbing solution (Section **1.2**).

5. Apparatus

5.1 SAMPLING PROBE. Sampling probes should be of Teflon, glass or stainless steel. Ozone is destroyed by contact with polyvinyl chloride tubing and rubber even after a conditioning period. Short sections of polyvinyl chloride tubing can be used to secure butt-to-butt connections of more inert tubing.

5.2 AIR METERING DEVICE. A glass rotameter capable of measuring gas flows of 0.5 to 3 l/min calibrated with a wet test meter to assure an accuracy of ± 2%.

5.3 ABSORBER. All-glass midget impingers graduated with 5 ml graduations should be used (see Figure 411:1). Impingers should be kept clean and dust free. Cleaning should be done with laboratory detergent followed by rinses with tap and distilled water.

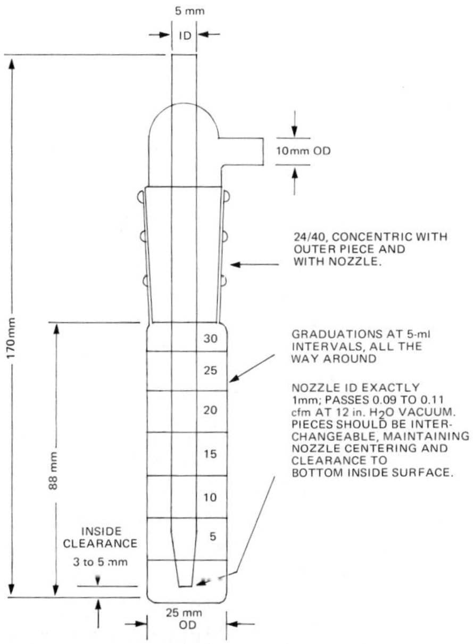

Figure 411:1—All-glass midget impinger (this is a commercially stocked item).

5.3.1 Do not use fritted glass bubblers as these produce less iodine **(10)**.

5.4 AIR PUMP. Any suction pump capable of drawing the required sample flow for intervals up to 30 min. It is desirable to have a needle valve or critical orifice **(11)** for flow control. A trap should be installed upstream of the pump to protect against accidental flooding with absorbing solution and consequent corrosion.

5.5 SPECTROPHOTOMETER. Any laboratory instrument capable of accurately measuring the absorbance of the triiodide ion at 352 nm. Stoppered cuvettes or tubes transparent in the near ultraviolet region should be used to hold the solutions.

6. Reagents

6.1 PURITY. Reagent grade chemicals shall be used in all tests. Unless otherwise indicated, all reagents shall conform to the specifications of the Committee on Analytical Reagents of the American Chemical Society, when such specifications are available **(12)**. Other grades may be used provided it is first ascertained that the reagent is of sufficiently high purity to permit its use without lessening the accuracy of the determination.

6.2 DOUBLE DISTILLED WATER. Double distilled water should be used for all reagents. The double distilled water can be prepared in an all-glass still by adding potassium permanganate to produce a faint pink color and barium hydroxide to alkalize the distilled water before the second distillation.

6.3 ABSORBING SOLUTION (1% KI IN 0.1 M PHOSPHATE BUFFER). Dissolve 13.6 g of potassium dihydrogen phosphate (KH_2PO_4), 14.2 g of disodium hydrogen phosphate (Na_2HPO_4) or 35.8 g of the dodecahydrate salt ($Na_2HPO_4 \cdot 12\ H_2O$), and 10.0 g of potassium iodide in sequence and dilute the mixture to 1 l with double distilled water. Keep at room temperature for at least 1 day before use. Measure pH and adjust to 6.8 ± 0.2 with NaOH or KH_2PO_4. This solution can be stored for several months in a glass stoppered brown bottle at room temperature without deterioration. It should not be exposed to direct sunlight.

6.4 STOCK SOLUTION 0.025M I_2 (0.05N). Dissolve 16 g of potassium iodide and 3.173 g of resublimed iodine successively and dilute the mixture to exactly 500 ml with double distilled water. Keep at room temperature at least 1 day before use. Standardize shortly before use against 0.025 M $Na_2S_2O_3$. The sodium thiosulfate is standardized against primary standard biiodate ($KH(IO_3)_2$) or potassium dichromate ($K_2Cr_2O_7$).

6.4.1 *0.001M I_2 solution.* Pipet exactly 4.00 ml of the 0.025 M stock solution into a 100-ml low actinic volumetric flask and dilute to the mark with absorbing solution. Protect from strong light. Discard after use.

6.4.2 *Calibrating iodine solution.* For calibration purposes exactly 4.09 ml of the 0.001 M I_2 solution (or equivalent volume for other molarity) is diluted with absorbing solution just before use to 100 ml (final volume) to make the final concentration equivalent to 1 μl of O_3/ml (see Section **9**). Discard this solution after use.

6.5 SULFUR DIOXIDE ABSORBER. Flash-fired glass fiber paper is impregnated with chromium trioxide, as follows (**2**): Drop 15 ml of aqueous solution containing 2.5 g chromium trioxide and 0.7 ml conc sulfuric acid uniformly over 400 cm^2 of paper, and dry in an oven at 80 to 90 C for 1 hr, store in a tightly capped jar. Half of this paper suffices to pack one absorber. Cut the paper in 6 × 12 mm strips, each folded once into a V-shape, pack into an 85 ml U-tube or drying tube and condition by drawing air, that has been dried over silica gel, through the tube overnight. The absorber is active for at least one month. When it becomes visibly wet from sampling humid air, it must be dried with dry air before further use.

7. Procedure

7.1 Assemble a train consisting of a rotameter, U-tube with chromium trioxide paper (optional), midget impinger, needle valve or critical orifice (**11**) and pump. Connections upstream from the impinger should be ground glass or inert tubing butt jointed with polyvinyl tubing. Fluorosilicon or fluorocarbon grease should be used sparingly. Pipet exactly 10 ml of the absorbing solution into the midget impinger. Sample at a rate of 0.5 to 3 l/min for up to 30 min. The flow rate and the time of sampling should be adjusted to obtain a sufficiently large conc of oxidant in the absorbing solution. Approximately 1 μl of ozone can be obtained in the absorbing solution at an atmospheric concentration of 0.01 ppm by sampling for 30 min at 3 l/min. Calculate the total volume of the air sample. Also measure the air temperature and pressure. Do not expose the absorbing reagent to direct sunlight.

7.2 MEASUREMENT OF COLOR. If appreciable evaporation of the absorbing solution occurs during sampling, add double-distilled water to bring the liquid volume to 10 ml.

7.3 Within 30 to 60 min after sample collections, read the absorbance in a cuvette or tube at 352 nm against a reference cuvette or tube containing double-distilled water.

7.4 BLANK CORRECTION. Measure the absorbance of the unexposed reagent and subtract the value from the absorbance of the sample.

8. Calibration and Standardization

8.1 The sample is absorbed and read in 10 ml of absorbing reagent. Calibrating solutions are made up to 10 ml to facilitate the calculations. For greater precision these can be made up to 25 ml or more.

8.1.1 Using iodine solutions is more convenient than preparing accurately known gas samples for standardizing. See Section **9** for stoichiometric relationships.

8.1.2 To obtain a range of conc values, add graduated amounts of calibrating solution up to 10 ml to a series of 10-ml volumetric flasks. Dilute to volume with absorbing reagent.

8.2 Read the absorbance (Section **7.3**).

8.3 Plot the absorbances of the obtained colors against the concentration of O_3 in μl/10 ml absorbing reagent. The plot follows Beer's law. Draw the straight line through the origin giving the best fit. Do not extrapolate beyond the highest concentration.

9. Calculations

9.1 Standard conditions are taken as 760 Torr and 25 C, at which the molar gas volume is 24.47 liters.

9.2 Record the volume of sample collected in liters. Generally the correction of the sample volume to standard conditions is slight and may be omitted. However, for greater accuracy corrections may be calculated by means of the perfect gas laws.

9.3 The calibrating iodine solutions are calculated on the basis of equivalence of O_3 and I_2, as indicated in **1.2**. Hence, a 100 ml portion of final solution equivalent to 1 μl O_3/ml contains the following amount of iodine:

$$I_2 = \frac{100 \times 10^{-6}}{24.47} = 4.087 \times 10^{-6} \text{ moles}$$

This is equivalent to 4.09 ml of 0.001 M (or 0.002N) I_2 solution.

9.4 The total μl of O_3/10 ml of reagent are read from the calibration curve.

9.5 The conc of O_3 in the gas phase in μl/l or ppm is given by:

$$O_3 \text{ ppm} = \frac{\text{total } \mu\text{l ozone per 10 ml}}{\text{volume of air sample in liters}}$$

9.6 The conc of O_3 in terms of μg/l at 760 Torr and 25 C is obtained when desired from the value of μl/l (Section **9.5**) by:

$$\mu\text{g } O_3/\text{liter} = \frac{\text{ppm} \times 48.00}{24.47} = 1.962 \times \text{ppm}$$

10. Effects of Storage

10.1 Ozone liberates iodine through both a fast and a slow set of reactions. Some of the organic oxidants also have been shown to cause slow formation of iodine (**6, 7**). Some indication of the presence of such oxidants and of gradual fading due to reductants can be obtained by making several readings during an extended period of time, *e.g.*, every 20 min.

10.2 Occasionally mold may grow in the absorbing reagent. When this occurs discard the reagent because reducing substances and a change in pH make it useless.

11. References

1. BYERS, D.H., and B.E. SALTZMAN. 1958. Determination of Ozone in Air by Neutral and Alkaline Iodide Procedures. *J. Am. Indust. Hyg. Assoc.* 19:251–7.
2. SALTZMAN, B.E., and A.F. WARTBURG. 1965. Absorption Tube for Removal of Interfering Sulfur Dioxide in Analysis of Atmospheric Oxidant. *Anal. Chem.* 37:779.
3. AMERICAN PUBLIC HEALTH ASSOCIATION. 1977. Methods for Air Sampling and Analysis. 2nd ed. p. 696. Washington, D.C.
4. Ibid. p. 527.
5. MUELLER, P.K., F.P. TERRAGLIO, and Y. TOKIWA. 1965. Chemical Interferences in Continuous Air Analysis. Proc., 7th Conference on Methods in Air Pollution Studies, State of California Department of Public Health, Berkeley, California.
6. SALTZMAN, B.E., and N. GILBERT. 1959. Iodometric Microdetermination of Organic Oxidants and Ozone. *Anal. Chem.* 31:1914–20.
7. ALTSCHULLER, A.P., C.M. SCHWAB, and M. BARE. 1959. Reactivity of Oxidizing Agents with Potassium Iodide Reagent. *Anal. Chem.* 31:1987–90.
8. NISHIKAWA, K. 1963. Portable Gas Dilution Apparatus for the Dynamic Calibration of Atmospheric Analyzers. Proc., 5th Conference on Methods in Air Pollution Studies, State of California Department of Public Health, Berkeley, California.
9. ALTSHULLER, A.P., and A.F. WARTBURG. 1961. The Interaction of Ozone with Plastic and Metallic Materials in a Dynamic Flow System. *Intern. J. Air Water Poll.* 4:70–78.
10. COHEN, I.C., A.F. SMITH, and R. WOOD. 1968. A Field Method for the Determination of Ozone in the Presence of Nitrogen Dioxide. *Analyst* 93:509.
11. LODGE, JR., J.P., J.B. PATE, B.F. AMMONS, and G.A. SWANSON. 1966. The Use of Hypodermic Needles as Critical Orifices in Air Sampling. *Air Poll. Cont. Assoc. J.* 16:197–200.
12. AMERICAN CHEMICAL SOCIETY SPECIFICATIONS. Reagent Chemicals, Am. Chem. Soc., Washington, D.C. For suggestions on the testing of reagents not listed by the American Chemical Society, see "Reagent Chemicals and Standards," by Joseph Rosin, D. Van Nostrand Co., Inc., New York, N.Y., and the "United States Pharmacopoeia".

Subcommittee 3
B. E. SALTZMAN, *Chairman*
W. A. COOK
B. DIMITRIADES
E. F. FERRAND
E. L. KOTHNY
L. LEVIN
P. W. MCDANIEL
C. A. JOHNSON, Air Conditioning & Refrigeration Institute Liaison

412.

Tentative Method for the Continuous Analysis of Atmospheric Oxidants (Colorimetric)

44101-03-71T

1. Principle of the Method

1.1 Oxidants are defined as those substances, with the exception of NO_2, which liberate iodine from the neutral buffered KI solution (**1**). In the atmosphere, the larger part of the oxidants is made up of ozone. Smaller amounts are contributed by PAN, **1.1.1**, and other oxidizing substances, **1.1.2**. The contribution of NO_2 to the readings should be subtracted, **3.2**.

1.1.1 Peroxyacyl nitrates give a response approximately equivalent to that of an equimolar conc of ozone (**2**). Conc in the atmosphere may range up to 0.1 ppm.

1.1.2 Other oxidizing substances besides ozone will liberate iodine with this method: *e.g.*, halogens, peroxy compounds, hydroperoxides, organic nitrates, hydrogen peroxide (**1, 3**), and possibly singlet oxygen.

1.2 The triiodide ion which forms is determined spectrophotometrically at 352 nm.

1.3 The reaction with ozone at pH 6.8 ± 0.2 is approximated by:

$$0_3 + 3\ KI + H_2O \rightarrow KI_3 + O_2 + 2KOH$$ (**4,* 5, 6**).

1.4 Air samples are scrubbed in a wetted wall absorber by the reagent solution. The resulting color is measured photometrically and recorded continuously on a strip chart recorder. As long as constant air-to-liquid ratio is maintained, the instrument can be calibrated to read directly in concentration of ozone as a reference oxidant. Calibration is based on the "Tentative Method for the Manual Determination of Oxidizing Substances in the Atmosphere" (**7**).

*These authors reported that 1 mole O_3 liberates 1.5 mole I_2. However, this contradicts older and newer findings. See reference **5**.

2. Range and Sensitivity

2.1 Colorimetric oxidant detectors operate most reliably at atmospheric oxidant levels ranging from 0.01 to 1.0 ppm of ozone (20 to 2000 $\mu m/m^3$) or other equivalent oxidant (**8**).

2.2 Usable response may be obtained from oxidant levels equivalent to as much as 100 ppm ozone. At such high oxidant levels the stability and speed of detector response are markedly inferior to those attained in the 0.01 to 1.0 ppm range.

3. Interferences

3.1 Sulfur dioxide produces a negative interference equivalent to 100% of an equimolar concentration of oxidant (**9**). To eliminate interferences either of the following two procedures may be used.

3.1.1 Up to hundred-fold ratio of sulfur dioxide (and probably H_2S) to oxidant may be eliminated without loss of oxidant by incorporating a conditioned chromic acid paper absorber in the sampling train upstream from the wetted wall absorber **(10, 11)**.

3.1.2 Alternatively, the effect of sulfur dioxide (or H_2S) on the oxidant reading can be corrected for by concurrently analyzing for sulfur dioxide **(12)** (or H_2S) and adding this concentration to the total oxidant value. This procedure is not applicable when sulfur dioxide (or H_2S) levels exceed those of the oxidant.

3.2 Nitrogen dioxide is known to give a response in 10% KI equivalent to about 20% **(5, 13, 14)** of that of an equimolar concentration of ozone. The contribution of the nitrogen dioxide to the oxidant reading can be eliminated by measuring the instrument response to a known stream of NO_2 and by concurrently analyzing for nitrogen dioxide **(14, 15)** subtracting the contribution of this concentration from the total oxidant value. The exact level of the specific response to nitrogen dioxide depends on design of the absorbing column, composition of sensing solution, operating conditions, and other unknown factors.

3.3 Nitric oxide does not interfere unless an SO_2 absorber as described in **3.1.1** is used. The SO_2 absorber will oxidize nitric oxide to nitrogen dioxide. Therefore, when sulfur dioxide is less than 20% of the nitric oxide concentration, the use of the chromic acid paper is not recommended.

3.4 Hydrogen sulfide, reducing dusts or droplets can act as negative interferences (see correction for their effect, **3.1.1** or **3.1.2**).

4. Precision and Accuracy

4.1 Precision of calibration with O_3 stream: in continuous monitoring, the hourly variation of the instrument response to a constant level of ozone set between 0 and 0.5 ppm should be equivalent to ± 0.01 ppm or less. Day-to-day variability of instrument response to a constant level of ozone should be less than ± 0.03 ppm. This performance is achieved by available instruments.

4.2 The accuracy of this method has not been established because of the difficulty in preparing known concentrations of ozone within the range of this method.

5. Apparatus (8)

5.1 AIR SAMPLING LINES. The air sampling lines and all parts of the instrument that come in contact with the airstream should be made of glass, Teflon or stainless steel. Polyvinyl chloride tubing can be used only for butt connections of more inert tubing. Fluorocarbon or fluorosilicon grease may be used sparingly, if necessary. The ozone may be stored up to one hr in bags **(16)** made of Teflon or Tedlar. The surface to volume ratio of the bags should be less than 20 m^2/m^3.

5.2 AIR PUMP. A constant-volume or a self-lubricating carbon-vane pressure-suction pump with vacuum gauge and a vacuum relief valve operating continuously at sample flow rates between 2 to 5 liter/min is suitable. It should be equipped with a bleeder inlet to the suction line, having a needle valve for flow adjustment and to protect the pump from corrosion and overheating.

5.3 LIQUID LINES. The reagent lines and all parts of the instrument that come in contact with the unexposed reagent should be made of glass, Teflon, polyethylene, polyvinyl chloride or other plastic polymer, but not stainless steel or other metallic alloy, because of the corrosive nature of free iodine. The exposed reagent lines up to the photometric cell should be made of glass.

5.4 REAGENTS SOLUTION PUMP. Use a self-priming constant delivery pump capable of operating continuously at the required reagent flow rates with not more than 1% variation in any 24-hr period. Reagent flow rate should be measured accurately at least once a week by replacing the liquid waste reservoir bottle with a graduated cylinder. The time to pump a precisely known volume is measured.

5.5 AIR METERING DEVICE. A needle valve and a glass rotameter is used to control and measure the flow. The calibrated rotameter should be placed preferably in a position upstream from the contact column. It is permissible to place the rotameter downstream from the contact column if a small, loosely fitting plug of glass wool is placed in the sample line ahead of the needle valve to prevent droplets from being carried over and deposited in the seat of the valve or in the air metering device. In either case, the rotameter should also be periodically dismantled, cleaned, and calibrated. Rotameters can be calibrated with wet test meters, spirometers, or other methods of similar accuracy.

5.6 CONTACT COLUMN. A 500 to 600 mm straight glass tube with a 6 to 10 mm ID. A 4 mm pitch helix of 1-mm glass rod which fits snugly inside, increasing the area of the absorbing surface and assuring that this surface is completely and continuously wetted, is commonly used. Alternatively, a 3-pitch, 120 mm diam spiral of 6 to 8 mm ID glass tubing, a midget impinger adapted with a reagent through-flow tubing, or a spray-jet bulb could be used **(17)**. The interferences of substances, such as NO_2, may be lessened when an impinger or spray bulb is used. The hold-up of liquid in the absorber should be minimized for rapid response.

5.7 PHOTOELECTRIC SENSOR. A dual beam type photometer system with an air path or a flow-through reference optical cell plus a flow-through measuring cell is suitable.

5.7.1 A thermostatically controlled heating system is optional for the photocell compartment to stabilize the response. The light filters and the two photocell detectors must be capable of providing a measurement at 325 nm ± 20 nm with a maximum deviation during the calibration procedure of 1% of full span. The baseline drift over any 24-hr period should be less than 2% of full span.

5.7.2 *Optical cells*. They should be transparent in the UV at 352 nm and an optical pathlength of 10 to 30 mm is recommended. The cells should be designed so that bubbles pass freely through without being trapped.

5.8 RECORDER. A strip chart recorder of the servo-balance type capable of sensing the potential created by the balanced photocell detector output is commonly used. Provision is made for electrical zero and infinity adjustment. Less than 0.25% baseline drift and noise over any 72-hr period is desirable. The electronic response time should be less than a min.

5.9 IODINE ABSORBING COLUMN. A cylinder filled either with 8 to 20 mesh activated charcoal (free of soluble salts, iron and sulfides) **(18)**, or with nylon wool **(17)**. The column should be capable of removing the free iodine formed in the reagent solution for up to a period of two months.

5.10 SULFUR DIOXIDE SCRUBBER. A drying tube of the approximate size for the airflow, packed with solid oxidant paper (described in section **6.5**) can be used to remove sulfur dioxide **(7, 10, 17)**.

5.11 CHARCOAL FILTER. A 13-cm tube, 1 cm ID or a cartridge **(19)** packed with granular activated charcoal and soda lime can be used to remove traces of oxidants and NO_2 from the air sample, for obtaining a zero reading.

5.12 ULTRAVIOLET OZONATOR **(23)**. The ozonator consists of a 4 W ozone lamp (such as GE G4511 or similar), a voltage stabilizer, a ballast transformer, a rheostat and an 0 to 1000 mA AC-meter in series. It is powered by a 117 AC source. The ozone chamber is fabricated from a 45/50 standard taper joint with the lamp socket and the inlet and outlet gas stream tubes sealed in. Varying levels of ozone can be obtained by adjusting lamp current and flow rate of gaseous streams.

6. Reagents

6.1 PURITY. All chemicals should be of analytical reagent grade **(20)**.

6.1.1 *Potassium iodide*. Some brands of KI are not suitable because of the addition of iodine-reducing substances, such as hypophosphite.

6.2 Double distilled water is recommended for all reagents. If only single distilled water is available the conductivity of the water should be 2.0×10^{-6} mhos/cm or less. Deionized water is not recommended.

6.3 ABSORBING SOLUTION.

6.3.1 *0.5 M buffer solution*. Dissolve 284 g of disodium phosphate, anhydrous (or 716 g disodium phosphate dodecahydrate), and 272 g potassium dihydrogen phosphate, anhydrous, in approximately 3 l distilled water. Make up to almost 4 l. Measure the pH. If different from pH 6.6 to 7.0, adjust if necessary with NaOH or KOH pellets, granular KH_2PO_4 or 10% phosphoric acid and make up to volume. Store in an amber bottle out of direct sunlight. This reagent has a shelf life in excess of 1 year. Optionally, 0.1 g of a non-reducing preserving agent may be added, *e.g.*, Na o-phenylphenate, Na pentachlorophenate.

6.3.2 *50% w/v alkaline KI solution.* Dissolve 2 kg of potassium iodide (free of iodine color-reducing substances, as described in **6.1.1**) in approximately 3 l of water. Add a sufficient amount of sodium hydroxide pellets to make the reagent slightly alkaline (pH 9 to 11) and make up to 4 l. Mix well and store in an amber bottle out of direct sunlight. This reagent has a shelf life in excess of one year.

6.3.3 *10% KI absorbing stock solution.* Place 2 l of 0.5 M buffer and 2 l of the 50% alkaline KI solution in a large jug or carboy and dilute to 10 l with water. Mix well. The absorbing solution should be stored in light-shielded bottles out of direct sunlight and may be kept at room temperature for several months.

6.4 CALIBRATING IODINE SOLUTION. For calibration purposes, exactly 10.2 ml of a 0.001 M I_2 (≡ 0.002 N) solution **(7)** (or an equivalent volume of another molarity) is diluted just before use with 10% KI solution to 250 ml (final volume) to make the final concentration equivalent to 1 μl 0_3/ml. Discard this solution after use. The following steps should be performed rapidly, just prior to the calibration of the instrument.

6.4.1 Prepare the static calibrating solutions equivalent to 0.1, 0.2, 0.3, 0.4 μl O_3/ml in 10% KI absorbing solution as follows: Dilute 10, 20, 30, 40 ml of the calibrating iodine solution (which is equivalent to 1 μl O_3/ml) with buffered 10% KI absorbing solution in a series of 100 ml flasks, to volume. Other equivalent solutions are prepared accordingly. Discard these solutions after use.

6.5 SULFUR DIOXIDE ABSORBER. Flash-fired glass fiber paper is impregnated with chromium trioxide, as follows **(10)**: Drop 15 ml of aqueous solution containing 2.5 g chromium trioxide and 0.7 ml concentrated sulfuric acid uniformly over 400 cm^2 of paper, and dry in an oven at 80 to 90 C for 1 hr, store in a tightly capped jar. Half of this paper suffices to pack one absorber. Cut the paper into 6 × 12 mm strips, each folded once into a V-shape, pack into an 85 ml U-tube or drying tube and condition by drawing air, that has been dried over silica gel, through the tube overnight. At least once a month substitute a fresh absorber to check the one in use. When it becomes visibly wet from sampling humid air, it must be dried with dry air before further use.

7. Procedure (21)

7.1 A commonly used system is described below but similar systems are acceptable.

7.1.1 LIQUID FLOW. The system (Figure 412:1) consists of a 4-l bottle (A) from which the reagent solution is forced by the constant-delivery pump (B), (C), (D), through the iodine absorbing column (E) to the top of the contact column (H). It is optional whether or not the solution passes first through a reference colorimetric cell (F) attached to the light source (G). The pump should be adjusted to deliver about 4 ml/min and checked for constancy. From the contact column the solution flows by gravity through the optical cell (F′) of the photoelectric sensor, which is connected electrically to the amplifier and recorder. The waste reagent returns to the reagent bottle (A).

7.1.2 An alternative method of dispensing the reagent to the contact column (H) is illustrated in Figure 412:2 (Alternative Liquid Feed System), where an excess of liquid is delivered to a constant-level reservoir (R), the excess returning to the reagent bottle through an overflow (S). Constant delivery to the contact column (H) is accomplished by gravity through a glass stopcock (K) and a rotameter (I^1). The flow is adjusted to approximately 4

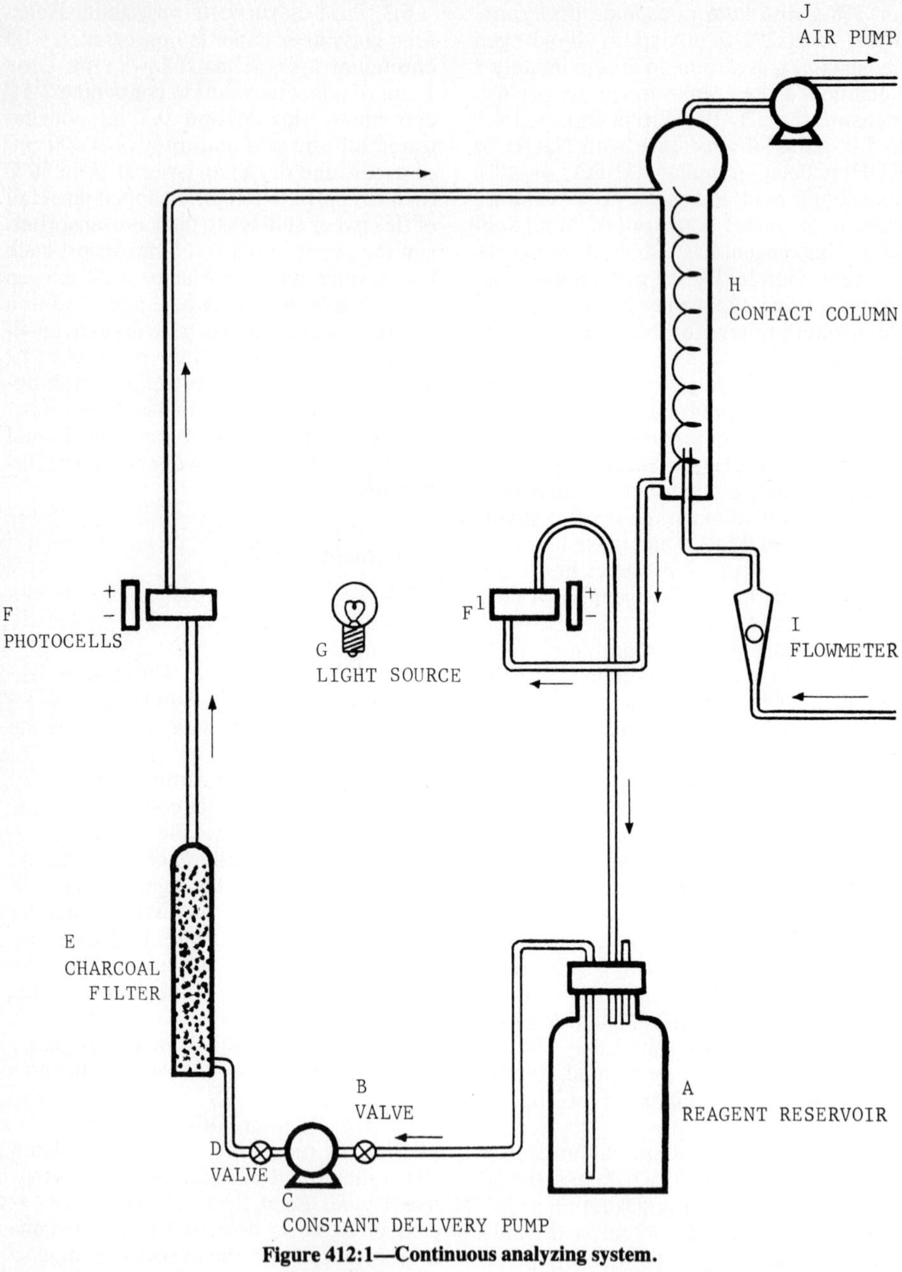

Figure 412:1—Continuous analyzing system.

ml/min and checked for constancy. This system is sensitive to variation in temperature. At a constant rotameter setting the liquid flow will increase about 0.1 ml/min/degree F rise (22).

7.2 AIRFLOW. An air pump draws sample air through the sulfur dioxide scrubber (optional) and the air metering device (I), and flows counter-currently to the reagent flow from the bottom of the

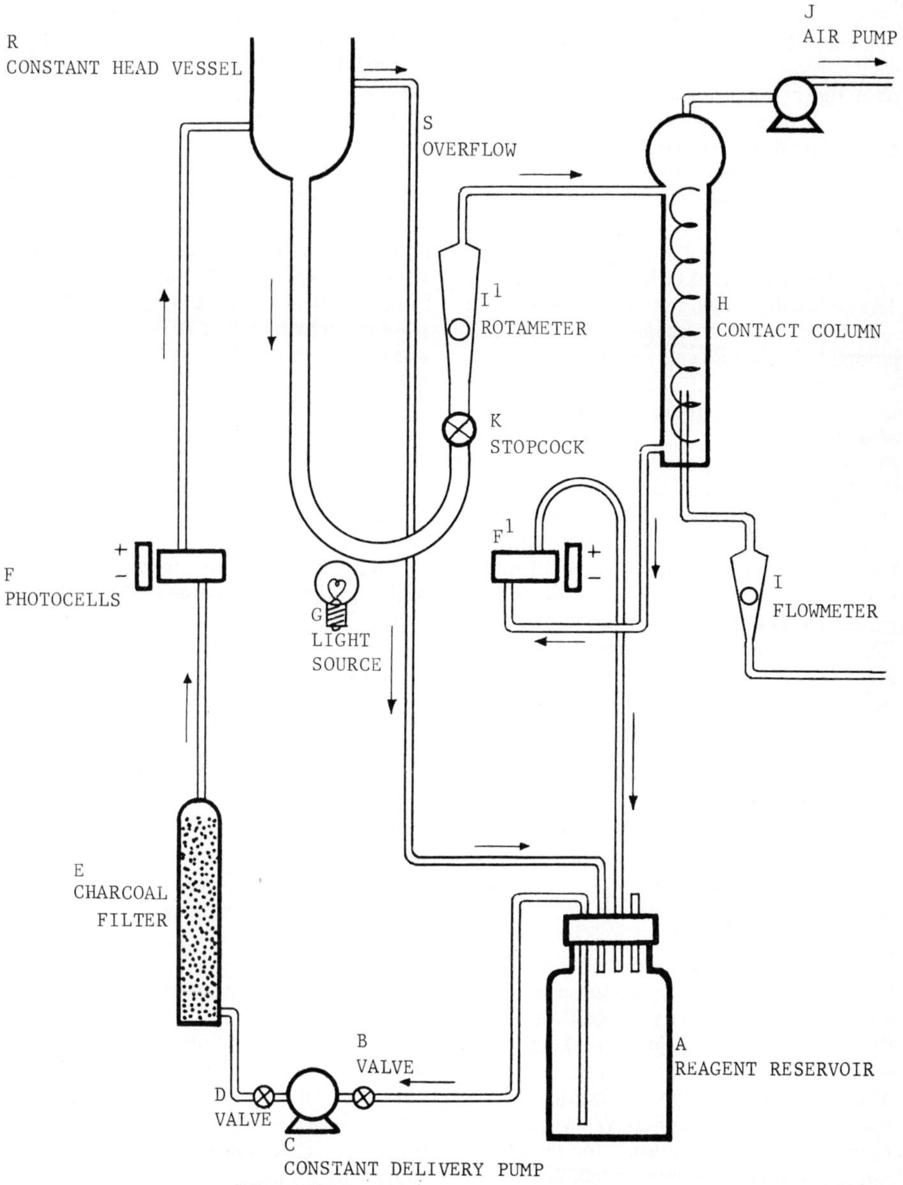

Figure 412:2—Alternative liquid feed system.

contact column (H). A surge jar prevents accidental entry of the scrubbing solution into the air pump. The airflow is regulated to about 4 l/min, and should be measured accurately. Ambient temperature and pressure should be recorded to allow fluctuation corrections for the results of gas flow measurements.

8. Calibration and Standardization

8.1 Dynamic calibration with known concentrations of ozone is required to check the overall response of the instrument. Static calibration is optional for establishing the overall response of the photometer-recorder system.

8.1.1 *Static calibration.* This test is performed by introducing varying amounts of iodine solutions equivalent to atmospheric ozone concentrations into the photometer cell. Neither the accuracy of the reagent or airflow rates or the efficiency of the absorption column are checked. The following steps can be followed:

a. Shut off the reagent and airflow pumps. Then connect a separatory funnel, mounted on ringstand, to the sample cell inlet. Flush the system with blank 10% KI absorbing solution and allow it to flow through the sample cell at approximately the flow rate at which the instrument is to operate. Adjust the recorder baseline.

b. Drain the contents of the separatory funnel and replace by the static calibrating solutions, **6.4**. Allow the solution to flow through the sample cell at approximately the rate at which the instrument is to operate. Record analyzer reading. Plot the points on coordinate paper (Section **9**).

c. To establish the linearity of response of the instrument, determine at which concentration in μl O_3/ml, the curve intersects absorbance = 1, or 10% transmittance (90% of chart). Prepare that concentration of static calibrating solution, **6.4.1**, and determine the instrument response according to step 8.1.1.b. Then adjust the recorder to that reading.

8.2 DYNAMIC CALIBRATION. The ozone for calibration is generated in an ultraviolet ozonator, **5.12**. The concentrations of the ozone mixtures so generated are determined by the Tentative Method for the Manual Determination of Oxidizing Substances in the Atmosphere **(7)**. Under recommended flow conditions, the detector's initial response (lag time) is reached within 60 sec. Longer response time may be caused by oxidant destruction on unconditioned surfaces of sample lines. Conditioning can be accomplished by operating the instrument on air samples containing about 1 ppm ozone until response time is minimized. Following each start-up, the instrument can be stabilized by drawing ambient air for at least 24 hr before it is calibrated or used for recording purposes.

8.2.1 Adjust the analyzer operating parameters approximately, so that the chart reading will give the concentrations directly in ppm O_3. The liquid flow rate (f_1) in ml/min is set according to the manufacturer's specifications. Then, the airflow rate (f_a) in liters/min is determined by multiplying the liquid flow rate f_1 by the inverse average slope value $1/b = c_s$. (c_s is the concentration in μl O_3/ml of absorbing solution which will give an analyzer response of absorbance one, according to **8.1.3**).

$$f_a = c_s \times f_1 = f_1/b$$

8.2.2 Zero the instrument by switching a charcoal filter, **5.12**, into the sampling line.

8.2.3 Connect the analyzer to one leg of the ozone dilution panel. (The other leg is drawn by the impinger for determining the ozone concentration). When the trace on the recorder is stabilized, the measuring impinger with absorbing solution is switched into place (sec **5.10**), and the ozone concentration is then determined colorimetrically **(20)**.

8.2.4 If the response of the dynamically determined points deviate by more than 10% from the static calibration curve, either the efficiency of the contact column is low or other operating parameters are off. Check the following parameters until the cause is discovered: liquid flow rate, airflow rate, change in static response, recorder malfunction.

8.2.5 The maximum deviation of any individual sampling point from the best fit curve drawn through the dynamic calibration points is generally not more than 0.025 ppm **(24)**. Those points that deviate more than the stated amount, are questionable and should be repeated.

8.2.6 Determine the lag and time to 95% of ultimate response of the analyzer under simulated operating conditions as follows: Disconnect the ozone sample line. Allow the analyzer to sample filtered air until the analyzer trace is on baseline. Then reconnect to the ozone generator.

Record the lag time by determining the interval between the time the ozone generator was reconnected and the observed start of the analyzer response on the chart. Then record the time to 95% of ultimate response by measuring the interval from the input concentration change to 95% of the ultimate steady state concentration. Several other concentrations can be checked by decreasing the current or turning off one ozone lamp.

9. Calculations

9.1 In practice, the response with most instruments is a linear function of absorbance versus ozone cocentration only at ozone levels below 1 ppm. The linear function is determined from the calibration data in the following manner:

9.2 For recorders utilizing exponential (absorbance) chart paper, plot the net response in absorbance units (sample less baseline readings on recorder chart) versus ppm O_3 (or μl O_3/ml) on linear coordinate paper.

9.3 For recorders utilizing linear chart paper, the net recorder response in per cent (sample less baseline readings divided by chart divisions less baseline readings) can be plotted on the logarithmic scale versus ppm O_3 (or μl O_3/ml) on the linear coordinate of a semi-log paper.

9.4 The method of least squares **(25)** might be used to obtain the slope and intercept of the straight line that fits the data best. If the intercept value is significantly different from zero, apply the define linear function only within the range of ozone concentration used in calibration and obtain additional calibration points using ozone concentrations closer to zero.

9.5 Alternatively, a simpler method for calculating the relationship between response and concentration, or response slope (b) can be done by dividing the sum of the absorbance values (Σ A), from all the points, by the sum of all the corresponding concentrations (Σ C). Then draw a straight line from the zero intercept to the point a 1 ppm O_3 (or at 1 μl O_3/1) and (b).

$$(b) = \frac{(\Sigma\ A)}{(\Sigma\ C)}$$

10. Effects of Storage

10.1 The 10% buffered KI solution must be free of fungus growth. It should be filtered through activated carbon for removing free iodine before use for calibration. When fungus growth occurs, discard the reagent. The addition of fungicide to the reagent is not recommended.

10.2 Fungus growth in instruments that have been shut down for extended periods of time may occur occasionally. When such growth appears, the lines and parts in contact with the absorbing solution should be taken apart, cleaned with detergent, rinsed with distilled water and the instrument charged with fresh reagent.

10.3 Ozone and many of the other oxidants from the atmosphere are unstable and decay within short periods of time. Whenever possible these samples should be analyzed *in situ*. If bag samples are taken, **5.1**, a measurable decay occurs within 30 min.

11. References

1. SALTZMAN, B.E. and N. GILBERT. 1959. Iodometric microdetermination of organic oxidants and ozone. *Anal. Chem.* 31:1914.
2. MUELLER, P.K., F.P. TERRAGLIO, Y. TOKIWA. 1965. Chemical interferences in continuous air analysis. Proceedings 7th Conference on Methods in Air Pollution Studies, State of California Department of Public Health, Berkeley, California.
3. ALTSHULLER, A.P., C.M. SCHWAB, M. BARE. 1959. Reactivity of oxidizing agents with potassium iodide reagent. *Anal. Chem.* 31:1987.
4. BOYD, A.W., C. WILLIS, R. CRY. 1970. *Anal. Chem.* 42:670.
5. TOKIWA, Y., M. IMADA, P.K. MUELLER. 1971. Stoichiometry of O_3 analysis. Report No. 101, Air and Industrial Hygiene Laboratory, State of California Department of Public Health, Berkeley, California.
6. BERGSHOEFF, G. 1970. Ozone determination in air. Presented at Pittsburgh Conference on Analytical Chemistry and Applied Spectroscopy, Cleveland, Ohio.
7. AMERICAN PUBLIC HEALTH ASSOCIATION, INC. 1977. Methods of Air Sampling and Analysis. 2nd ed. p. 556. Washington, D.C.
8. LITTMAN, F.E., and R. BENOLIEL. 1953. Continuous oxidant recorder. *Anal. Chem.* 25:1480.

9. CHERNIACK, I. and R.J. BRYAN. 1964. A comparison study of various types of ozone and oxidant detectors which are used for atmospheric air sampling. Paper No. 64–70, 57th Annual Meeting of the APCA, Houston, Texas.
10. SALTZMAN, B.E. and A.F. WARTBURG. 1965. Absorption tube for removal of interfering sulfur dioxide in analysis of atmospheric oxidant. *Anal. Chem.* 37:779.
11. HODGESON, J.A. 1970. Review of analytical methods for atmospheric oxidants measurements. Presented at 11th Conference on Methods in Air Pollution and Industrial Hygiene Studies, State of California Department of Public Health, Berkeley, California.
12. AMERICAN PUBLIC HEALTH ASSOCIATION, INC. 1977. Methods of Air Sampling and Analysis. 2nd ed. p. 696. Washington, D.C.
13. BOVEE, H.H. and R.J. ROBINSON. 1961. Sodium diphenylamine sulfonate as an analytical reagent for ozone. *Anal. Chem.* 33:1115.
14. SIU, W., M.R. AHLSTROM, and M. FELDSTEIN. 1969. Comparison of coulometric and colorimetric oxidant analyzer data. Presented at 10th Conference on Methods in Air Pollution and Industrial Hygiene Studies, State of California Department of Public Health, San Francisco, California.
15. AMERICAN PUBLIC HEALTH ASSOCIATION, INC. 1977. Methods of Air Sampling and Analysis. 2nd ed. p. 527. Washington, D.C.
16. ALTSHULLER, A.P. and A.F. WARTBURG. 1961. The interaction of ozone with plastic and metallic materials in dynamic flow system. *Int. J. Air Water Poll.* 4:70.
17. SCHULTZ, F. 1966. Versatile combination ozone and sulfur dioxide analyzer. *Anal. Chem.* 38:748.
18. FANSAH, N.O. and Y. TOKIWA. 1969. Removing impurities from activated charcoal. Report No. 66, Air and Industrial Hygiene Laboratory, State of California Department of Public Health, Berkeley, California.
19. MINE SAFETY APPLIANCES. Type DZ 78006 filter and Cr 10340 cartridge.
20. AMERICAN CHEMICAL SOCIETY, WASHINGTON, D.C. 1950. Reagent Chemicals, American Chemical Society Specifications.
21. SPECIFICATIONS FOR INSTRUMENTS FOR CONTINUOUS MEASUREMENT OF ATMOSPHERIC CONTAMINANTS. 1968. Air Pollution Control District, Los Angeles, California.
22. GORDON, C. 1963. Effect of temperature on solution flow. Proceedings 5th Conference on Methods in Air Pollution and Industrial Hygiene Studies, Berkeley, California.
23. NISHIKAWA, K. 1963. Portable gas dilution apparatus for the dynamic calibration of atmospheric analyzers. Proc. 5th Conference on methods in air pollution studies. State of Calif., Dept. of Public Health, Berkeley, California.
24. GUIDE TO OPERATION OF ATMOSPHERIC ANALYZERS. 1960. Recommended Method No. 7, Air and Industrial Hygiene Laboratory, State of California Department of Public Health, Berkeley, California.
25. BROCKMAN, W.E. 1968. Least squares, an easy method. *Inst. and Contr. Methods* 41:105.

Subcommittee 3
E. L. KOTHNY, *Chairman*
W. A. COOK
D. DIMITRIADES
E. F. FERRAND
C. A. JOHNSON
L. LEVIN
P. W. MCDANIEL
B. E. SALTZMAN

413.

Tentative Method of Analysis for Ozone in the Atmosphere by Gas-Phase Chemiluminescence Instruments

44201-01-73T

1. Principle of the Method

1.1 The method is based on the gas-phase chemiluminescent reaction of ozone with ethylene to produce an excited species which emits light in the visible range (**1, 2, 3, 4**). The reaction is rapid and specific for ozone. The chemiluminescent reaction takes place in a chamber which has a light transparent end-face coupled to a photomultiplier tube. The resulting signal produced by the photomultiplier tube is proportional to ozone concentration. The signal is further amplified and either read directly on a recorder or converted into a digital display, depending upon the construction of the specific instrument.

1.2 The chemiluminescent instrument is calibrated using an ozone generator which is, in turn, standardized against the buffered neutral 1% potassium iodide method **(5)**.

2. Range and Sensitivity

2.1 Chemiluminescent instruments are available with several measurement ranges in the same instrument. Generally these are 0 to 40, 0 to 400, and 0 to 4000 μg ozone/m^3 (0 to 0.02, 0 to 0.2 and 0 to 2 ppm, respectively, by volume).

2.2 The lower limit of detection has been reported at 6 or 8 μg ozone/m^3 (0.003 or 0.004 ppm, respectively) in air **(3, 4)**.

3. Interferences

3.1 Exhaust gases from the instrument should be vented at a safe distance from the intake to prevent interference from the excess ethylene used when operating the instrument.

3.2 A slight humidity dependence is observed when operating the instrument at the lowest measurement range.

3.3 Other components normally found in ambient air do not interfere **(3, 4)**.

4. Precision and Accuracy

4.1 Precision is defined in terms of stability and repeatability of response to ozone. Response instability due to baseline drift should be equivalent to no more than ± 20 μg ozone per m^3 (± 0.01 ppm by volume) **(4)**. Noise level should be equivalent to no more than ± 16 μg ozone per m^3 (± 0.008 ppm by volume) **(4)**. Repeatability of response to a standardized ozone-air mixture should have a coefficient of variation not greater than 1% **(4)**. This latter precision figure represents the composite error that is associated with both the ozone measurement and the standardization operations.

4.2 Accuracy data are not available.

5. Apparatus

5.1 DETECTOR CELL, Figure 413:1 **(6)**, is a drawing of a typical detector cell showing flow paths of gases, the mixing zone, and placement of the photomultiplier tube. Other flow paths in which the air and ethylene streams meet at a point near the photomultiplier tube are also usable.

5.2 VACUUM PUMP. A vacuum pump that will maintain a vacuum of 500 Torr with an air flow of 5 l/min is suitable.

5.3 BAROMETER. Any device capable of measuring atmospheric pressures in Torr with an error of ± 1 Torr.

5.4 AIR FLOWMETER. A calibrated device for measuring air flows between 0 and 1.5 l/min.

5.5 ETHYLENE FLOWMETER AND CONTROL. A device capable of metering and controlling ethylene flows between 0 and 80 ml/min. The actual flow should follow the manufacturer's recommendation. At any flow in this range, the device should be capable of maintaining constant flow rate within ± 3 ml/min.

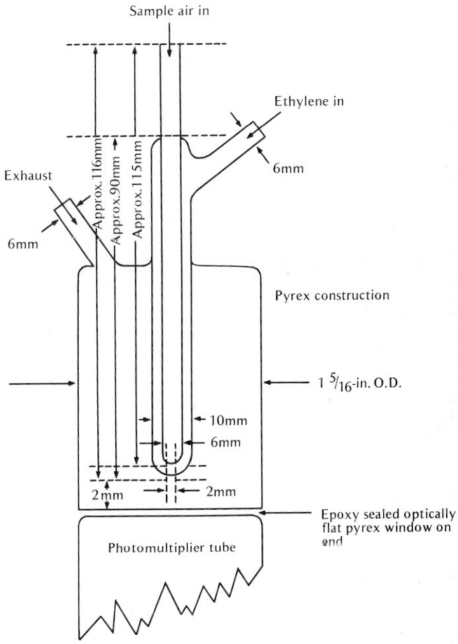

5.6 Air inlet filter (Figure 413:3). A Teflon filter capable of removing all particles greater than 5μ in diameter.*

5.7 Photomultiplier tube. A high-grain low dark current (not more than 10^{-9} ampere) photomultiplier tube having its maximum gain at about 430 nm. Follow manufacturer's instructions for tube replacement.

5.8 High voltage power supply, capable of delivering up to 2,000 V.

5.9 Direct current amplifier, capable of full scale amplification of currents from 10^{-10} to 10^{-7} ampere; an electrometer is commonly used with an output of 0 to 10 mV or greater.

5.10 Recorder, capable of full scale display of voltages from the DC amplifier. These voltages commonly are in the 1 mV to 1 V range.

5.11 Ozone source and dilution system. The ozone source **(6, 7)** consists of a quartz tube into which purified air is introduced (Figure 413:2, 3) and then irradiated with a very stable low pressure mercury lamp.† The level of irradiation is controlled by an adjustable metal sleeve which fits around the lamp **(7)**. Ozone concentrations are varied by adjustment of the sleeve. At a fixed level of irradiation and at constant temperature and humidity, ozone is produced at a constant rate. By careful control of the flow of air through the quartz tube, atmospheres can be generated which contain constant concentrations of ozone. The concentration of ozone in the test atmosphere is determined by the neutral buffered potassium iodine method **(5)**.

*Mace Corporation, 1810 Floradale Ave., South El Monte, California 91733 or equivalent.

†Pen-Ray model KCQ9G-1, Ultraviolet Products Inc., San Gabriel, California 91771 manufactured in lengths from 62 to 200 mm (2½" to 8"), or equivalent.

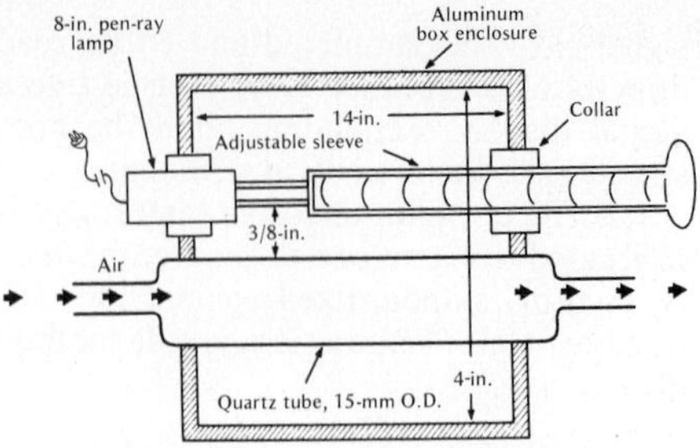

Figure 413:2—Ozone generator.

5.12 Apparatus for calibration. The calibration equipment, described in ISC Method 44101-02-70T, Section 5 **(5)**, consists of the following:

5.12.1 *Air metering device.*

5.12.2 *Absorber.*

5.12.3 *Air pump.*

5.12.4 *Spectrophotometer.*

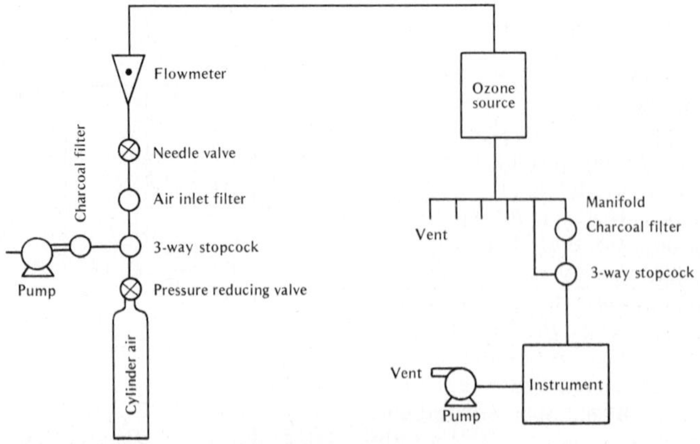

Figure 413:3—Air inlet filter.

6. Reagents

6.1 PURITY. Reagent grade chemicals shall be used in all tests. Unless otherwise indicated, all reagents shall conform to the specifications of the American Chemical Society (**8**).

6.2 ETHYLENE, C.P.

6.3 FILTERED AIR. Purified by passing through an activated charcoal filter.

6.4 ACTIVATED CHARCOAL FILTER. A 13 cm tube × 1 cm ID or a cartridge filled with granular activated charcoal can be used to remove traces of oxidants and other interfering gases from the air.‡

6.5 CALIBRATING REAGENTS. Prepare as described in method 44101-02-70T (**5**). These include: reagent water; 1% neutral potassium iodide solution; and calibrating iodine solution.

7. Procedure

7.1 Operate any commercial instrument following procedures given in the manufacturer's manual. Perform calibration as directed in Section **8**. By proper adjustments of zero and span controls, direct reading of ozone concentration may be obtained.

8. Calibration

The calibration procedure is applicable to all types of chemiluminescent instruments.

8.1 GENERATION OF TEST ATMOSPHERES. Assemble the apparatus as shown in Figure 413:3. The concentration of ozone produced by the generator (Figure 413:2), can be varied by changing the position of the adjustable sleeve, which covers more or less the full length of the quartz-mercury lamp. The ultraviolet radiation of the lamp traverses the quartz wall and ozonises the air or oxygen flowing through the tube. The quartz-mercury lamp must be fed through a power stabilizer for maintaining a constant output. For calibration of ambient air analyzers, the ozone source should be capable of producing ozone concentrations in the range of 100 to 2000 μg ozone per m^3 (0.05 to 1 ppm, by volume) at an air flow rate of at least 5 l/min. At all times the air flow through the generator must be greater than the total flow required by the sampling systems. Zero air can be obtained by charcoal filtration of ambient air or zero cylinder air.

8.2 SAMPLING AND ANALYSIS OF TEST ATMOSPHERES. The chemiluminescent instrument should be conditioned before use (when using commercial instruments, follow manufacturer's instructions; conditioning time may vary from 5 min at the maximum output from the ozoniser to 30 min at atmospheric levels). The air flow rate of the analyzer should be the same as used for sampling. Sampling lines should be Teflon or glass only (**9**). Ozone is destroyed by contact with polyvinyl chloride and rubber tubing even after a conditioning period, and some metals show a constant small demand (**9**) which affects the measurement of atmospheric levels of ozone. The manifold that distributes the test atmospheres must be sampled simultaneously by the KI sampling train and the instrument to be calibrated. Check the assembled system for leaks. Record the chemiluminescent instrument response at five or six equally spaced ozone concentrations starting with zero. Establish the ozone concentrations by analysis, using the 1% neutral buffered potassium iodide method (**5**).

8.3 BLANK. With the ozone lamp off, flush the system for several min to remove residual ozone. Pipet 10 ml of absorbing reagent into each absorber. Sample the manifold with the sampling train at 1 l/min for 10 min. Immediately transfer the exposed iodide solution to a clean (previously flushed with absorbing solution and drained for a min) 1-cm cell. Determine the absorbance at 352 nm against unexposed absorbing reagent as the reference. When consistent low blanks are obtained then proceed to calibrate the chemiluminescent instrument by the following section.

8.4 TEST ATMOSPHERES. With the ozone lamp well equilibrated, adjust the ozone concentration to the range desired

‡Filter type DZ 78006, Mine Safety Appliances, Allison Park, Pennsylvania 15101 or equivalent.

for calibration. Wait 10 min for equilibration at this setting and measure according to the preceding section **8.3**.

8.5 INSTRUMENT CALIBRATION CURVE. Instrument response from the photomultiplier tube is ordinarily in current or voltage. Plot the response as the y-axis against the corresponding ozone concentrations as determined by the 1% neutral buffered potassium iodide method, in $\mu g/m^3$ (or ppm) as the x-axis.

9. Calculations

9.1 Calculate the total volume, in liters, of air sampled into the absorbers corrected to standard conditions as follows:

$$V_R = F \times T \times \frac{P}{760} \times \frac{298}{t + 273}$$

where:

V_R = volume of air at standard conditions, in liters.
F = flowmeter air rate, liters per minute.
T = time, in minutes.
P = barometric pressure in Torr.
t = sampling temperature in °C.

9.2 The total micrograms of ozone are read from the spectrophotometric calibration curve (Section **8.1**).

9.3 The concentration of ozone in the gas phase is given by:

Ozone, $\mu g/m^3$ =

$$\frac{\text{Total } \mu g\ O_3 \text{ per 10 ml}}{V_R} \times 10^3$$

9.4 Conversion between $\mu g/m^3$ values for ozone and ppm can be made as follows:

$$\text{ppm ozone} = \frac{\mu g\ O_3/m^3 \times 24.45}{48.00} \times 10^{-3} = \mu g\ O_3/m^3 \times 5.1 \times 10^{-4}$$

where:

24.45 = molar volume of a gas at 25 C and 760 Torr
48.00 = molecular weight of ozone in grams

10. References

1. NEDERBRAGT, G.W., A. VAN DER HORST, and J. VAN DUIJN, 1965. Rapid Ozone Determination near an Accelerator. *Nature* 206:87.
2. WARREN, G.J., and G. BABCOCK. 1970. *Rev. Sc. Instru.* 41:280.
3. HODGESON, J.A., B.E. MARTIN, and R.E. BAUMGARDNER. 1970. Comparison of Chemiluminescence Methods for Measurement of Atmospheric Ozone. Eastern Anal. Symposium. New York City, Preprint No. 77. (National Environmental Research Center, EPA, Research Triangle Park, NC 27711).
4. CARROLL, H., B. DIMITRIADES, and M. RAY. 1972. A Chemiluminescence Detector for Ozone Measurement. Bur. of Mines, Dept. of Interior, R.I. 7650.
5. AMERICAN PUBLIC HEALTH ASSOCIATION, INC. 1977. Methods of Air Sampling and Analysis. 2nd ed. p. 556. Washington, D.C.
6. FEDERAL REGISTER, Vol. 36, No. 228, Thursday, Nov. 25, 1971, pp. 22391–22395.
7. HODGESON, J.A., R.K. STEVENS, and B.E. MARTIN. 1972. A Stable Ozone Source Applicable as a Secondary Standard for Calibration of Atmospheric Monitors. pp. 149–158. In *Air Quality Instrumentation*, Vol. I, John W. Scales, Ed., Pittsburgh, Instrument Society of America.
8. ACS REAGENT CHEMICALS. 1966. American Chemical Society Specifications, Washington, D.C.
9. ALTSHULLER, A.P., and A.F. WARTBURG. 1961. The Interaction of ozone with plastic and metallic materials in a dynamic flow system. *Int. J. Air and Water Poll.* 4:70.

Subcommittee 3
E. L. KOTHNY, *Chairman*
W. A. COOK
J. E. CUDDEBACK
B. DIMITRIADES
E. F. FERRAND
R. G. KLING
P. W. MCDANIEL
G. D. NIFONG
F. T. WEISS

414.

Tentative Method of Analysis for Peroxyacetyl Nitrate (Pan) in the Atmosphere (Gas Chromatographic Method).

44301-01-70T

1. Principle of the Method

1.1 Gas chromatography, which is used in this method for measuring peroxyacetyl nitrate (PAN) (**4, 6, 7, 8, 9**) in ambient air, is a very powerful procedure for separating mixtures into their chemical components. It can be adapted for quantitative measurement of individual components after separation (**3, 13, 14**). The present method exploits the exceptional ability of PAN to capture free electrons to provide a method for measuring ppb concentrations of PAN in a 2 or 3 ml sample of ambient air. Higher homologues in the series such as peroxypropionyl nitrate can be detected but procedures for calibration have not been worked out since the amounts present are much smaller than the amounts of PAN found. The recently reported peroxybenzoyl nitrate (**2**) might be measured by an adaptation of the present technique.

1.2 Three major components are required: (a) a chromatographic column to separate PAN from the many other compounds present; (b) a sample introduction system, in this case a gas sample valve activated by a timer and solenoid; (c) an electron capture detector with associated amplifier and recorder. The instrument injects a sample every 15 min and about 1.5 min are required to develop the chromatogram. The minimum detectable quantity is about one ppb although this could be lowered if there were need.

2. Range and Sensitivity

2.1 Since the detector used in this method is based on the capture of electrons, large amounts of substances which have a high affinity for electrons can saturate the detector by capturing all the electrons. As saturation is approached, the response becomes very non-linear. To maintain linear response it is necessary to restrict the sample so that the steady current through the detector is reduced by no more than about 25%. This limits the upper end of the concentration range. The maximum depends on sample size, column length, temperature, and carrier gas flow rate, and is usually less than 1 ppm by volume.

2.2 On the other hand maximum sensitivity permits detection of concentrations of less than one ppb (10^{-9} by volume or 4.95 $\mu g/M^3$ at 25 C and 760 Torr).

3. Interferences

3.1 There are no known interferences with this PAN method. An interfering substance would have to meet three conditions; (a) it must have high electron affinity; (b) it must have an elution time very close to that of PAN; and (c) it must be present in the sample at a concentration detectable by this procedure. These stringent conditions eliminate virtually everything.

4. Precision and Accuracy

4.1 Repetitive sampling of standard samples containing about 60 ppb of PAN gave a relative standard deviation of less than 2% for the amount present. Accuracy of course is dependent on the validity of the calibration which in turn depends on the accuracy of the infrared method (section **8**). This has been compared with a colorimetric method based on hydrolysis to nitrite ion with good agreement (**12**). It is probable that the overall accuracy of the method is within 5%.

5. Apparatus

5.1 The procedures for measuring PAN described by Darley *et al.* (**1**) have been modified and used in the automated system. An Aerograph Model 681 gas chromatograph* with an electron capture detector or any similar instrument is equipped with an automatic system which injects a sample of ambient air into the following column every 15 min. A 9″ long column of ⅛″ I.D. Teflon tubing packed with 5% carbowax E400 on 100 to 120 mesh HMDS treated chromosorb-W is used. It is operated at 25 C oven temperature with a nitrogen carrier gas flow of 40 ml/min. Under these conditions the retention time for PAN is 60 secs.

5.2 A stainless steel hexaport valve (similar to the valve supplied by Varian Aerograph, Walnut Creek, California) with an external 2-ml stainless steel sample loop is used instead of the glass sample valve described by Darley *et al.* (**1**). With this system there is no detectable decomposition of PAN.

5.3 The automatic sampling mechanism attached to the outside of the chromatograph case, Figure 414:1, consists of: (a) a timing mechanism; (b) sample valve; (c) solenoid activator; and (d) time delay relay. The timing mechanism consists of a micro switch, a cam to operate the microswitch, and a standard Synchron 4 RPH motor. The cam is designed to energize the solenoid for 90 sec during each 15 min period. The shaft of the hexaport valve is attached rigidly to the solenoid shaft and positioned so that the air sample flows through the 2 ml sample loop during the time the solenoid is energized. A coil spring of proper tension installed between the solenoid and sample valve returns the valve to injection position when the solenoid is de-energized.

5.4 Calibration requires reference to a known PAN mixture which is most easily measured using a 10 cm path length gas cell and a spectrometer capable of scanning the 1161 cm^{-1} band. Since this system cannot measure concentrations less than a few hundred ppm, provision must be made for quantitative dilution by a factor of about 1,000.

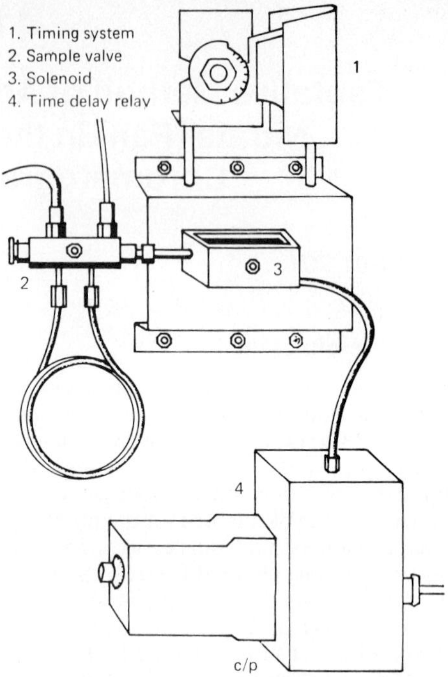

Figure 414:1—Automatic sampling system consisting of: (a) standard timing system; (b) hexaport valve and 2 ml sample loop; (c) solenoid to activate valve; (d) time delay relay to control recorder chart drive motor.

6. Reagents

6.1 The only reagents required for this determination are the column packing, nitrogen carrier gas, and PAN for calibration. The column packing is made by coat-

*Varian Aerograph, 2700 Mitchell Drive, Walnut Creek, California. This model is no longer available but other chromatographs can be adapted to this purpose.

ing 100 to 120 mesh HMDS treated chromosorb-W with 5% by weight of carbowax E400. Both these materials can be obtained from Varian Aerograph or other supply sources. Dry "prepurified" nitrogen can be obtained in 220 ft^3 cylinders from various suppliers.

6.2 PAN is synthesized, purified as described by Stephens *et al.*, **(11)** and stored in 34-l stainless steel cylinders for use in the dynamic calibration system for the chromatograph **(5)**. The cylinders, pressurized with nitrogen to 100 psig, contain 500 to 1000 ppm PAN and are stored at 60 F (16 C). An infrared spectrophotometer with a 10 cm cell is employed to determine the concentration of PAN in the cylinders, using the absorptivities of 13.9×10^{-4} ppm^{-1}m^{-1} at 8.6 μ (1161 cm^{-1}) **(9)**.

6.3 A simplified alternate method which requires much less apparatus may also be used. Small quantities of PAN are prepared directly in a 10-cm infrared cell with a pyrex glass body by photolyzing traces of ethyl nitrite, CH_3CH_2ONO, in pure oxygen. The PAN formed is then measured in the cell and is not separated from unreacted nitrite or from by-products. But it must be diluted by a factor of several thousand to reach the electron capture concentration range. Blacklight fluorescent lamps provide a good energy source for the necessary 300 to 400 nm radiation but it might well be possible to use the sun for this photolysis. The cell is first flushed with pure oxygen and then small amounts (50 μl) of vapor from the space above the liquid in the ethyl nitrite bottle are added at 15 min intervals while irradiating. The ethyl nitrite concentration must be kept low to maximize yield.

7. Procedure

7.1 Dry nitrogen carrier gas is flushed continuously through the 2-ml sample loop of the hexaport valve for 13.5 min of each period when the solenoid is de-energized. When the solenoid is energized the sample loop is open to the atmosphere to be sampled and a Neptune pump Model 4-K or similar pump draws a continuous flow of air through the loop for 90 sec. When the solenoid is again de-energized, the carrier gas flushes the 2-ml sample into the column.

7.2 The time delay relay, Figure 414:1, energizes the chart drive motor on the strip chart recorder at the end of the 90 sec sample period when the solenoid is de-energized. The chart drive motor is allowed to run for 90 sec which is ample time to record the chromatogram. The column is too short to separate peroxypropionyl nitrate (PPN) and peroxybutyryl nitrate (PBN), although a shoulder on the PAN peak, assumed to be PPN, appears occasionally.

7.3 A typical chromatogram for the period between 8:00 and 9:00 p.m. on March 27, 1968, Figure 414:2, shows that significant concentrations may be measured long after sundown.

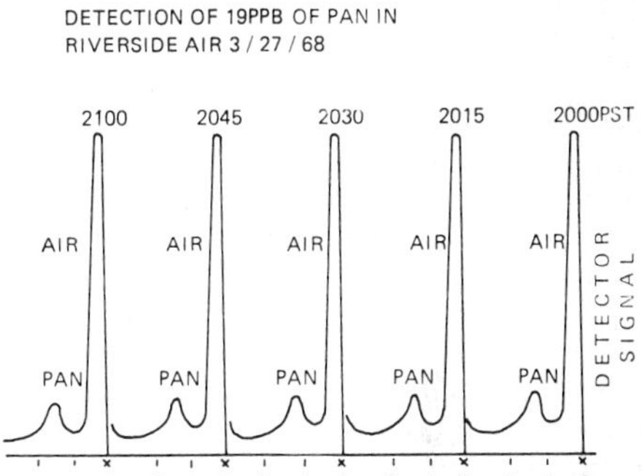

Figure 414:2—A chromatogram produced by the automated PAN analyzer from ambient air samples during the period 8.00 to 9.00 p.m. on March 27, 1968. The large air peak is followed by a PAN peak in each of the 15 minute sample periods.

8. Calibration and Standards

8.1 The chromatograph is calibrated with a flow dilution panel (Figure 414:3) from a supply containing PAN at a concentration high enough to be measured in a 10-cm infrared cell (~ 1000 ppm).

8.2 In the first dilution of the dynamic system **(5)** (Figure 414:3), 1 part PAN from the storage cylinder is diluted with 100 parts activated-charcoal-filtered air. A similar 1 to 100 dilution with filtered air in the second step of the dilution system reduces the concentration of PAN in a constant flow of air to the ppb range. Calibration curves are plotted from calculated concentrations of PAN and peak height. The oxidant recorder shown in Figure 414:3 is not essential.

9. Calculation

9.1 The tank concentration is calculated from the infrared spectrum **(10)** by applying the Beer-Lambert equation to the absorbance at 8.6μ (8.6×10^{-6} m).

$$C\ (\text{ppm}) = A/(0.1\ \text{m}) \times (13.9 \times 10^{-4}\ \text{ppm}^{-1}\ \text{m}^{-1})$$

where C is conc. in ppm

A is absorbance = $\log_{10}(^{I_0}/_{I})$

For accurate measurement this concentra-

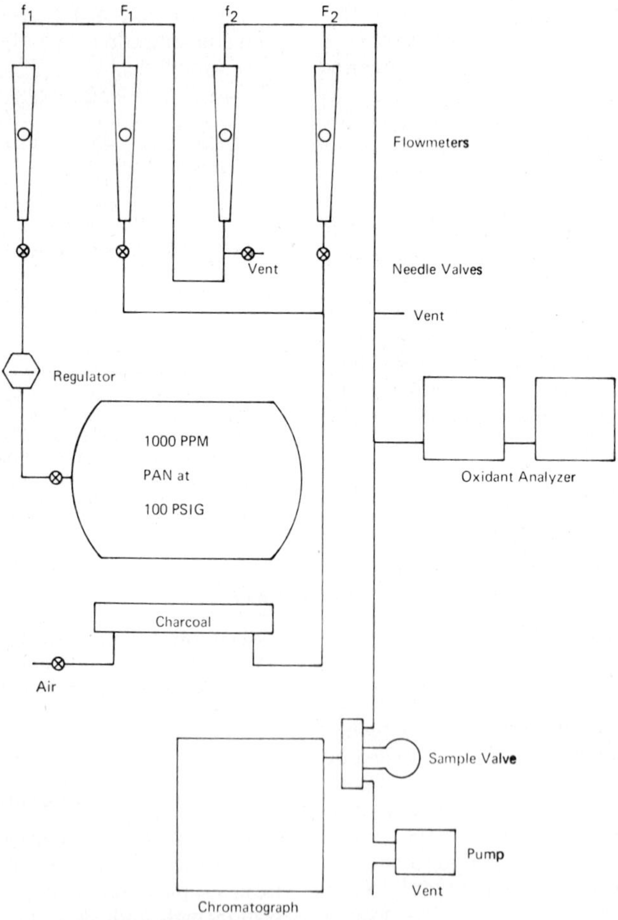

Figure 414:3—Dynamic calibration system.

tion must be of the order of a 1000 parts/million. This is about 10,000 times higher than the concentration required for calibration so dilution is carried out in two steps of about one hundredfold using the dilution panel shown in Figure 414:3. The final concentration is then calculated by the formula:

$$C_e = C\,[f_1/(F_1 + f_1)]\,[f/_2(F_2 + f_2)]$$

In which f_1 and f_2 are rates in the small streams and F_1 and F_2 are flow rates in the large streams.

9.2 Calculated concentration of PAN and peak height show a linear relationship in the range from 1 to about 50 ppb; therefore, concentration within this range can be calculated by multiplying the peak height by a constant. The constant varies from one instrument to another and varies inversely with changes in standing current of a particular instrument.

9.3 The automatic system has been operated continuously 24 hr a day for 11 months with only brief interruptions to clean the electron capture detector. Contamination of the detector causes a slow decline in standing current and a concomitant reduction in sensitivity. It is desirable to calibrate the instrument about once a week to compensate for the gradual reduction in sensitivity. When the detector is thoroughly cleaned the original sensitivity is regained.

10. Effect of Storage

10.1 No reliable means of retaining PAN-containing air samples is known at this time.

11. References

1. DARLEY, E.F., K.A. KETTNER, and E.R. STEPHENS. 1963. Analysis of peroxyacetyl nitrates by gas chromatography with electron capture detection. *Anal. Chem.* 35:589–591.
2. HEUSS, J.M., and W.A. GLASSON. 1968. Hydrocarbon reactivity and eye irritation. *Env. Sci. & Tech.* 2:1109.
3. MAYRSOHN, H., and C. BROOKS. 1965. The analysis of PAN by electron capture gas chromatography. Western Regional meeting of the Amer. Chem. Soc., Nov. 18.
4. MIDDLETON, J.T. 1961. Photochemical air pollution damage to plants. *Ann. Rev. Plant Physiol.* 12:431.
5. PLATA, R.L. 1968. Calibration and comparison of coulometric and flame ionization for monitoring PAN in experimental atmospheres. Ninth Conference on Methods in Air Pollution and Industrial Hygiene Studies, Huntington-Sheraton Hotel, Pasadena, California.
6. STEPHENS, E.R., P.L. HANST, R.C. DOERR, and W.E. SCOTT. 1956. Reactions of nitrogen dioxide and organic compounds in air. *Ind. Eng. Chem.* 48:1498–1504.
7. STEPHENS, E.R., W.E. SCOTT, P.L. HANST, and R.C. DOERR. 1956. Recent developments in the study of the organic chemistry of the atmosphere. *J. Air Poll. Cont. Assoc.* 6:159–165.
8. STEPHENS, E.R., E.F. DARLEY, O.C. TAYLOR, and W.E. SCOTT. 1960. 1961. Photochemical reaction products in air pollution, Proc. API 40 III: 325–338; *Int. J. Air and Water Poll.* 4:79–100.
9. STEPHENS, E.R. 1961. The photochemical olefin-nitrogen oxides reaction in the lower and upper atmosphere. *Interscience*, New York.
10. STEPHENS, E.R. 1964. Absorptivities for infrared determination of peroxyacyl nitrates. *Anal. Chem.* 36:928–929.
11. STEPHENS, E.R., F.R. BURLESON, and E.A. CARDIFF. 1965. The production of pure peroxyacyl nitrates. *J. Air Poll. Control Assoc.* 15:87–89.
12. STEPHENS, E.R. and M.A. PRICE. 1967. Colorimetric analysis of PAN. *Eighth Conference on Methods in Air Pollution and Industrial Hygiene Studies*. Oakland, California.
13. TAYLOR, O.C., E.R. STEPHENS, and E.A. CARDIFF. 1968. Automatic chromatographic measurement of PAN. *61st Annual Meeting of the Air. Poll. Cont. Assoc.* St. Paul, Minnesota.
14. TINGEY, D.T., and A.C. HILL. 1968. The occurrence of photochemical phytotoxicants in the Salt Lake Valley, *Utah Academy Proceedings* 44:387–395.

Subcommittee 4

R. G. SMITH, *Chairman*
R. J. BRYAN
M. FELDSTEIN
B. LEVADIE
F. A. MILLER
E. R. STEPHENS

501. Recommended Method of Analysis for Suspended Particulate Matter in the Atmosphere: (High-Volume Method)

11101-01-70T

1. Principle of the Method

1.1 Air is drawn into a covered housing and through a filter by means of a high-flow-rate blower at a flow rate (1.13 to 1.70 m^3/min or 40 to 60 ft^3/min) that allows suspended particles having diameters of less than 100μ (Aerodynamic diameter) to pass to the filter surface (**1**).

1.2 The mass concentration of suspended particulate in the ambient air ($\mu g/m^3$) is computed by measuring the mass of collected particulate and the volume of air sampled.

1.3 This method is applicable to measurement of the mass concentration of suspended particulate in ambient air. This method does not control the flow of air during sampling and for this reason is most applicable to trend measurement. The size of the sample collected is usually adequate for other analyses.

1.4 Other Analyses. In practice, the primary use of the high-volume sampler has been that of determining the mass concentration of particulates in ambient air. Much additional information may be obtained from these samples with the additional expenditure of time and effort. Other recognized analyses are enumerated as follows: (a) gross β activity, (b) organic materials, (c) nitrates, (d) sulfates, (e) metals, (f) ammonium ion and (g) fluoride ion.

2. Range and Sensitivity

2.1 When the sampler is operated at an average flow rate of 1.70 m^3/min (60 ft^3/min) for 24 hr, an adequate sample will be obtained even in an atmosphere having concentrations of suspended particulate as low as 1 $\mu g/m^3$. If particulate levels are unusually high, a satisfactory sample may be obtained in 6 to 8 hr or less. For determination of average concentrations of suspended particulate in ambient air, a standard sampling period of 24 hr is recommended.

2.2 Weights are determined to the nearest milligram; air flow rates are determined to the nearest 0.03 m^3/min (1.0 ft^3/min), times are determined to the nearest 2 min and mass concentrations are reported to the nearest microgram per cubic meter.

3. Interferences

3.1 Particulate that is oily, such as photochemical smog or wood smoke, may block the filter and cause a rapid drop in air flow at a non-uniform rate. Dense fog or high humidity can cause the filter to become too wet and severely reduce the air flow through the filter.

3.2 Glass-fiber filters are comparatively insensitive to changes in relative humidity, but collected particulate can be hygroscopic (**2**).

4. Precision and Accuracy

4.1 At an average mass concentration of 112 $\mu g/m^3$ of particulate matter in ambient air the standard deviation is 10 $\mu g/m^3$ (corresponding to a relative standard deviation of 9%); at an average of 39 $\mu g/m^3$ the standard deviation is 6 $\mu g/m^3$ (corresponding to a relative standard deviation of 15%) (**3**).

4.2 The accuracy with which the sampler measures the true average concentration depends upon the degree of constant

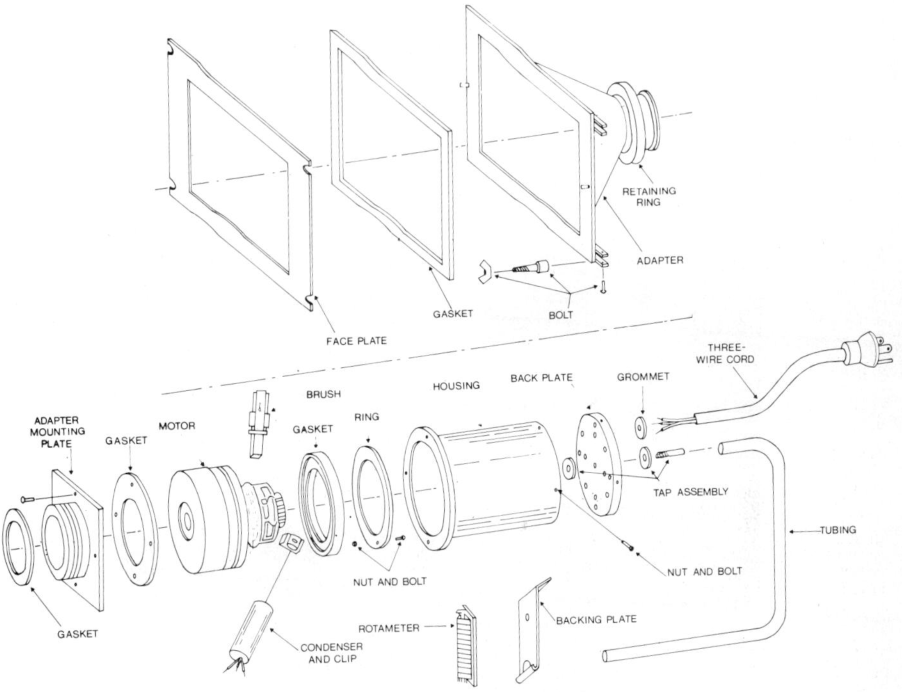

Figure 501:1—Exploded view of typical high-volume air sampler parts.

air flow rate maintained in the sampler. The air flow rate is affected by the concentration and the nature of the dust in the atmosphere, which may clog the filter and significantly reduce the air flow rate. Under these conditions, the error in the measured average concentration may be as much as ± 50% or more of the true average concentration, depending on the amount of reduction of air flow rate and on the variation of the mass concentration of dust with time during the 24 hr sampling period (**4**).

5. Apparatus

5.1 SAMPLING.

5.1.1 *Sampler*. The sampler consists of three units: (a) the face plate and gasket, (b) the filter adapter assembly, and (c) the motor unit. Figure 501:1 shows an exploded view of these parts, their relationship to each other, and how they are assembled. The sampler must be capable of passing environmental air through an approximate 400 cm^2 area of a clean 20.3 × 25.4 cm (8″ × 10″) glass-fiber filter at a rate of at least 1.13 m^3/min (40 ft^3/min). The motor must be capable of continuous operation for 24 hr periods with input voltages ranging from 110 to 120 V, 50 to 60 cycles alternating current and must have third-wire safety ground. The housing for the motor unit may be of any convenient construction so long as the unit remains airtight and leak-free. The life of the sampler motor can be extended by lowering the voltage by about 10% with a small "buck or boost" transformer between the sampler and power outlet.

5.1.2 *Sampler Shelter*. It is important that the sampler be properly installed in a suitable shelter. The shelter is subjected to extremes of environmental conditions such as high and low temperatures, extremes of humidity, and all types of air pollutants. For these reasons the materials of the shelter must be chosen carefully. Prop-

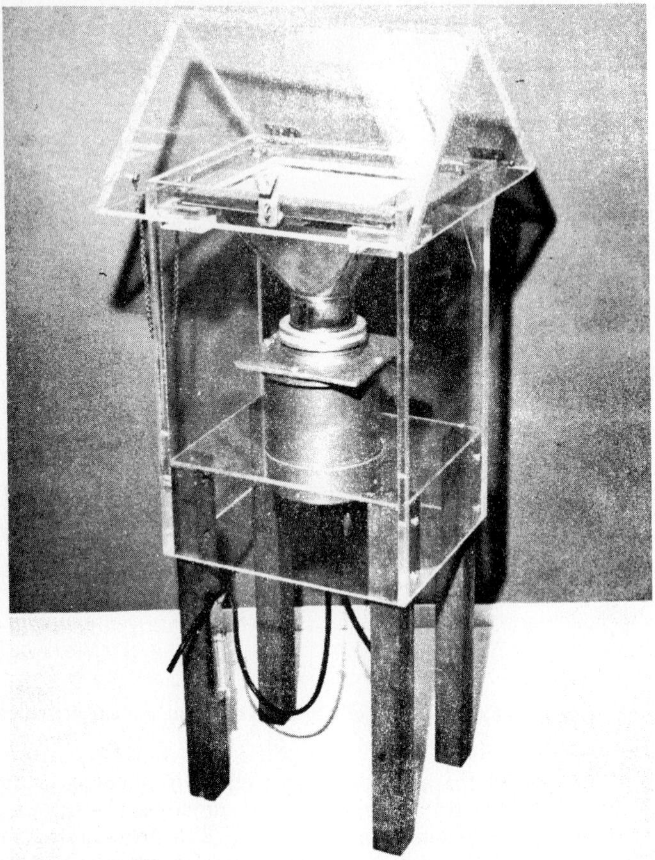

Figure 501:2—Assembled sampler and shelter (clear plastic is for display purposes and is not recommended for shelter construction)

erly painted exterior plywood or heavy gauge aluminum serve well. The sampler must be mounted vertically in the shelter so that the filter is parallel with the ground. The shelter must be provided with a roof so that the filter is protected from precipitation and debris. The internal arrangement and configuration of a suitable shelter with a gable roof are shown in Figure 501:2. The clearance area between the edge of the roof and the main housing should be 645 ± 65 cm^2 (100 ± 10 in^2). The main housing should be rectangular and have dimensions of 29.2 × 35.6 cm (11½″ × 14″).

5.1.3 *Folders.* 21.6 × 27.9 cm (8½″ ×11″) manila cardboard folders, creased.

5.1.4 *Rotameter.* Marked in arbitrary units, frequently 0 to 70, and capable of being calibrated; or

5.1.5 *Draft Gauge.* Marked in any suitable set of units, but having range equivalent to 12.5 cm water; or

5.1.6 *Recording Flowmeter.* Calibrated in flow units.

5.2 Calibration

5.2.1 *Orifice Calibration Unit.* Consisting of a metal tube 7.62 cm (3″) ID and 15.9 cm (6¼″) long with a static pressure tap 5.08 cm (2″) from one end. The tube end nearest the pressure tap is flanged to about 10.8 cm (4¼″) OD with a male thread of the same size as the inlet end of the high-volume air sampler. A single metal plate 9.2 cm (3⅝″) in diam. and 0.24 cm (3/32″) thick with a central orifice 2.86 cm (1⅛″) in diam is held in place at the air inlet end with a female threading ring. The other end of the tube is flanged to hold a loose female threaded coupling, which screws on to the inlet of the sampler. An 18-hole metal plate, an integral part of the unit, is positioned between the orifice and sampler to simulate the resistance of a clean glass-fiber filter.

5.2.2 *Differential Manometer.* Capable of measuring to at least 41 cm (16″) of water.

5.2.3 *Flow Measuring Device.* Positive displacement type or rate meter type, calibrated in m^3 or ft^3, to be used as a primary standard.

5.2.4 *Barometer.* Capable of measuring atmospheric pressure to the nearest 0.1 cm or access to such in local vicinity.

5.3 ANALYSIS REQUIREMENTS.

5.3.1 *Balance Room Environment.* Maintained at a particular temperature between 15 and 35 C to within ± 1 C and at a particular relative humidity below 50% and to within ± 5%

5.3.2 *Analytical Balance.* Equipped with a weighing chamber designed to handle unfolded 20.3 × 25.4 cm (8″ × 10″) filters and having a sensitivity of 0.1 mg.

5.3.3 *Light Table.* Of the type used to view X-ray films.

5.3.4 *Numbering Machine.* Capable of printing identification numbers of 4 to 8 digits on the filter.

6. Reagents

6.1 FILTER MEDIA. Glass-fiber filters having a collection efficiency of at least 99% for particles of 0.3 diam, as measured by the DOP test, are suitable for the quantitative measurement of conc of suspended particulate (**5**), although some other medium, such as paper, may be desirable for some analyses. Care must be exercised to prevent use of filters that contain high background conc of the pollutant being investigated. Careful quality control is required to determine background values of these pollutants.

7. Procedure

7.1 SAMPLING.

7.1.1 *Filter Preparation.* Place each filter on a light table and inspect for pinholes, dark particles, or other imperfections. Filters with visible imperfections should not be used. A small brush is useful for removing particles. Equilibrate the filters to the conditions of the balance room for at least 24 hr after numbering. Weigh the filters to the nearest mg; record tare weight and filter identification number. Do not bend or fold the filter before collection of the sample.

7.1.2 *Sample Collection.* Open the shelter, loosen the wing nuts, and remove the face plate from the filter holder. Install a numbered, pre-weighed, glass-fiber filter in position (rough side up), replace the face plate without disturbing the filter, and fasten securely. Undertightening will allow air leakage, overtightening will damage the sponge-rubber face plate gasket. A very light application of talcum powder may be used on the sponge-rubber face plate gasket to prevent the filter from sticking. During inclement weather the sampler may be removed to a protected area for filter change. Close the roof of the shelter, run the sampler for about 5 min, connect the rotameter or draft gauge to the nipple on the back of the sampler. Read the rotameter ball with rotameter in a vertical position. Estimate to the nearest whole number. If the ball is fluctuating rapidly, tip the rotameter and slowly straighten it until the ball gives a constant reading. If a draft gauge is used, record an average reading. Remove the rotameter or gauge from the tube; record the initial reading and the starting time and date on the filter folder. The rotameter or gauge should never be connected to the sampler except when the flow is being measured. Let the sampler run for the desired time (usually 24 hr) and take a final reading in similar fashion. Record the final reading and ending time and date on the filter folder. Remove the face plate as described above and carefully remove the filter from the holder, touching only the outer edges. Fold the filter lengthwise so that only surfaces with collected particulate are in contact and place in a filter folder. Record on the folder the filter number, location, and any other factors, such as meteorological conditions or razing of nearby buildings that might affect the results. If the sample is defective, void it at this time. In order to obtain a valid sample, the high-volume sampler must be operated with the same rotameter and tubing that were used during its calibration. This is not necessary with the draft gauge if its accuracy is assured.

7.1.3 *Maintenance.*

a. Sampler Motor. Replace brushes before they are worn to the point motor damage can occur. Motors in good condition may be expected to operate approximately 600 hr at 100 volts. Extreme conditions of operations may require more request changes.

b. Face Plate Gasket. Replace when the margins of samples are no longer sharp. Seal the gasket to the face plate with rubber cement or double-sided adhesive tape.

c. Rotameter. Clean as required, using alcohol.

d. Draft Gauge. Maintenance usually not required unless damaged.

e. Recording Flowmeter. Maintenance usually not required unless damaged.

7.2 ANALYSIS.

7.2.1 Equilibrate the exposed filters for at least 24 hr in the environment of the balance room, then reweigh. After they are weighed, the filters may be saved for detailed chemical analysis.

8. Calibration

8.1 GENERAL. To be of any use, the high-volume sampler must provide a means for reliably determining air-flow rate. In practice this is accomplished by calibrating a multi-holed orifice plate which is an integral part of the sampler, located at its discharge. This plate as stated, forms an orifice the downstream pressure of which is always ambient pressure. In all calibration methods, therefore, it is desired to relate the differential of upstream pressure to ambient air flow rate. Most commonly this is done with a rotameter provided with the unit. A small fraction of the discharge flow is routed through this meter due to the differential pressure mentioned above, this flow being measured and related to total ambient flow. An equally valid method is to directly measure the differential pressure in some convenient manner, such as a draft gauge, and in like fashion relate this reading to ambient flow.

8.2 SAMPLING. Whether one chooses to use the rotameter or draft gauge method of routine air-flow determination, each sampler must be calibrated with its associated rotameter or draft gauge. This calibration is performed with one of several calibration units **(3)**. Before one of these units may be used it in turn must have been calibrated against some reliable air flow meter. There are a number of devices commercially available which conform to the ASME Power Test Code **(7)**, or the user may take advantage of available commercial services.

8.2.1 *Orifice Calibration Unit.* In calibrating this unit, of whatever configuration, it is important that it be situated in the air flow stream in essentially the same sequence as will be the case when calibrating a sampler at a later date. A typical procedure is described for a typical inlet type orifice calibrated with a rotating lobe-type positive displacement flow meter. Attach the orifice calibration unit to the intake end of the positive displacement primary standard and attach a high-volume motor blower unit to the exhaust end of the primary standard. Connect one leg of a differential manometer to the single pressure tap of the orifice calibration unit and leave the other leg open to the atmosphere. Operate the high-volume motor blower unit so that a series of different, but constant, air flows (usually 6) are obtained for definite time periods. Record the reading on the differential manometer at each air flow. The different constant air flows can be obtained by placing a series of load plates, one at a time, between the calibration unit and the primary standard.* Placing the orifice before the inlet reduces the pressure at the inlet of the primary standard below atmospheric therefore a correction must be made for the increase in air volume caused by this decreased inlet pressure. Attach one leg of a second differential manometer to an inlet pressure tap of the primary standard and leave the other leg open to the atmosphere. During each of the constant air flow measurements made above, measure the true inlet pressure of the primary standard with this second differential manometer. Measure atmospheric pressure with a barometer. Correct the measured air volume to true air volume as directed in **9.1.1** then obtain true air flow rate, Q, as directed in **9.1.2**. Plot the differential manometer readings of the orifice calibration unit versus Q.

8.2.2 *High-Volume Sampler.*

a. Rotameter Method. Assemble a high-volume sampler with a clean filter in place and run for at least 5 min. Attach a rotameter, read the ball, adjust so that the ball reads 65, and seal the adjusting mechanism so that it cannot be changed easily. Shut off motor, remove the filter, and attach the orifice calibration unit in its place. Operate the high-volume sampler at a series of different, but constant, air flows (usually 6). Record the reading of the differential manometer on the orifice calibration unit and record the reading of the rotameter at each flow. Convert the differential manometer reading to m^3/min, Q, then plot rotameter reading versus Q. Identify this plot with the particular sampler and rotameter combination used in its preparation.

b. Draft Gauge Method. Assemble a high-volume sampler with the orifice calibration unit installed. Run for at least 5 min. Attach the draft gauge to the fitting provided in the multi-holed orifice plate built into the discharge of the sampler (this is the fitting normally used for the rotameter). Operate the high-volume sampler at a series of different, but constant, air flows (usually 6). Record the readings of the differential manometer on the orifice calibration unit and that of the draft gauge at each flow. Convert the differential manometer reading to m^3/min, Q, then plot draft gauge reading versus Q. Identify this plot with the particular sampler and gauge combination used in its preparation.

c. Recorder Method. Assemble a high-volume sampler with the orifice calibration unit installed. Run for at least 5 min. Attach the recorder to the fitting provided in the multi-holed orifice plate built into the discharge of the sampler (this is the fitting normally used for the rotameter). Operate the high-volume sampler at a series of different, but constant, air flows (usually 6). Record the readings of the differential manometer on the orifice calibration unit and that of the recorder at each flow. Convert differential manometer reading to m^3/min, Q, then plot recorder reading versus Q. Identify this plot with the particular sampler and recorder used in its preparation.

d. If the pressure or temperature during high-volume sampler calibration is substantially different from the pressure or temperature during orifice calibration, a correction of the flow rate, Q, may be required. If necessary, obtain the corrected flow rate as directed in **9.2.1**. This correction applies only to orifice meters having a constant orifice coefficient. The coefficient for the calibrating orifice described in **5.2.1** has been shown experimentally to be constant over the normal operating range of the high-volume sampler (0.6 to 2.2 m^3/min–20 to 78 ft^3/min).

9. Calculations

9.1 Calibration of Orifice.

9.1.1 Calculate the true air volume measured by the positive displacement pri-

*The use of a variable transformer for regulating the speed of the high volume sampler motor is not recommended, according to the results of a study of calibration methods by Lynam, Pierce and Cholak **(8)**. Resistances in the form of resistance plates or resistance filters should be employed as the standard procedure for varying the flow rate during calibration of high volume samplers.

mary standard, making corrections for pressure as follows:

$$P_aV_a = P_MV_M$$
$$P_M = P_a - P_m$$

Combining these relationships:

$$V_a = \frac{(P_a - P_m)\,V_M}{P_a}$$

Where

V_a = volume of air at atmospheric pressure, m^3
P_a = barometric pressure, cm Hg
P_M = pressure at inlet of the primary standard, cm Hg
P_m = pressure difference between inlet of primary standard and atmospheric, cm Hg
V_M = volume measured by primary standard, m^3

Convert inches of water to inches of mercury: inches water $\times$ 73.48 $\times$ 10^{-3} = inches Hg
Convert inches mercury to centimeters mercury: inch Hg $\times$ 2.54 = cm Hg

9.1.2 Calculate flow rate

$$Q = \frac{V_a}{T}$$

Where:

Q = flow rate, m^3/min.
T = time of flow, min.

9.2 CALIBRATION OF SAMPLER.

9.2.1 Calculate corrected flow rate, if required.

$$Q_2 = Q_1\left[\frac{T_2P_1}{T_1P_2}\right]^{\frac{1}{2}}$$

where:

Q_2 = corrected flow rate, m^3/min
Q_1 = flow rate during high-volume sampler calibration (Section **8.2.2**), m^3/min
T_1 = absolute temperature during orifice unit calibration (Section **8.2.1**), °K or °R
P_1 = barometric pressure during orifice unit calibration (Section **8.2.1**), cm Hg
T_2 = absolute temperature during high-volume sampler calibration (Section **8.2.2**), °K or °R
P_2 = barometric pressure during high-volume sampler calibration (Section **8.2.2**), cm Hg

9.3 SAMPLE VOLUME.

9.3.1 Convert the initial and final rotameter or gauge readings to m^3/min using the appropriate calibration curve.

9.3.2 Calculate volume of air sampled.

$$V = \frac{(Q_i + Q_f) \times T}{2}$$

Where:

V = air volume sampled, m^3
Q_i = initial air flow rate, m^3/min
Q_f = final air flow rate, m^3/min
T = sampling time, minutes

9.4 Calculate mass concentration of suspended particulates:

$$S.P. = \frac{(W_f - W_i) \times 10^6}{V}$$

Where:

S.P. = mass concentration of suspended particulate, $\mu g/m^3$
W_i = initial weight of filter, g
W_f = final weight of filter, g
V = air volume sampled, m^3
10^6 = conversion of g to μg

10. Effect of Storage

10.1 Samples of particulate matter may be stored indefinitely if protected from contamination.

11. References

1. ROBSON, C.D., and K.E. FOSTER. 1962. Evaluation of Air Particulate Sampling Equipment. *Am. Ind. Hyg. Assoc. J.* 23:404.
2. TIERNEY, G.P., and W.D. CONNER. 1967. Hygroscopic Effects on Weight Determinations of Particulates Collected on Glass-Fiber Filters. *Am. Ind. Hyg. Assoc. J.* 28:363.
3. FORTUN, S.F.G. 1964. Factors Affecting the Precision of High-Volume Air Sampling, M.S. Thesis, University of Florida, Gainesville, Florida, April.

4. Harrison, W.K., J.S. Nader, and F.S. Fugman. 1960. Constant Flow Regulators for High-Volume Air Sampler. *Am. Ind. Hyg. Assoc. J.* 21:115–120.
5. Pate, J.B., and E.C. Tabor. 1962. Analytical Aspects of the Use of Glass-Fiber Filters for the Collection and Analysis of Atmospheric Particulate Matter. *Am. Ind. Hyg. Assoc. J.* 23:144–150.
6. Henderson, J.S. 1967. Eighth Conference on Methods in Air Pollution and Industrial Hygiene Studies, Oakland, California.
7. ASME Power Test Code. 1959. Part 5, Chapter 4. "Flow Measurement by Means of Standardized Nozzles and Orifice Plates." (PTC 19.5.4).
8. Lynam, D.R., J.O. Pierce, and J. Cholak. 1969. Calibration of the High Volume Sampler. *Am. Ind. Hyg. Assoc. J.* 30:83–88.

Subcommittee 10
R. S. Sholtes, *Chairman*
R. B. Engdahl
R. A. Herrick
C. Phillips
E. Stein
J. Wagman
P. F. Woolrich

502.

Tentative Method of Analysis for Dustfall from the Atmosphere

21101-01-70T

1. Principle of the Method

1.1 Large particulate matter, which becomes suspended in the atmosphere by wind forces or mechanical means, and other particulate matter cleansed from the atmosphere by rain or agglomeration, is measured using an open-mouth container exposed for a period of approximately one month. This static measurement technique is simple but the results are strongly dependent on the design and placement (especially the height) of the container. In cases where settleable dust is generated by local sources, the inherent dustfall rate is widely variable. Reported data indicate that dustfall measurement is not precise. In most situations this may be less important than the inherent variability in dustfall rate at any give location **(1–5)**.

1.2 Following collection, samples can be analyzed in many ways to suit the needs of specific situations. The gross measurement of soluble and insoluble fractions is normally undertaken. Further physical analyses can be performed on the insoluble dustfall, *e.g.*, particle size distribution, specific gravity. Chemical analysis can be performed on either fraction to determine the elemental composition, or to determine compounds of specific interest in local situations.

2. Range and Sensitivity

2.1 The lower limit of measurement is approximately 0.2 g/m^2/month. The upper limit is dependent upon the ratio of the depth to the diameter of the collector and is not likely to be reached.

3. Interferences

3.1 The major interferences in the method are algae, fungi, bird droppings, and insects. Any of these four can be present in the container after the period of exposure to such a degree that they overwhelm the analysis. Because of the unattended nature of the sampling device, tampering has frequently been observed. The container is frequently a target for rocks, snowballs, and fire arms.

4. Precision and Accuracy

4.1 Replicate samples by various investigators indicate that a precision of ± 15% is attainable for any given combination of collecting element and retention fluid. Results of greater than 2 to 1 variation have been found with replicate samples taken by different methods.

4.2 No information on accuracy is available.

5. Apparatus

5.1 Collector. The collector of choice is a polyethylene container with a tapered cylinder and a sealable lid. The dimensions are 7⅜″ diam × 8¼″ high. A suitable holder for the collector should allow convenient changing of the samples, adequate support, and a bird ring approximately 1.5 × the diam of the collector and positioned so it is approximately 3″ higher than the top of the collecting jar.

5.2 Sieves. A chemically inert 20-mesh screen shall be used to remove extraneous material, *e.g.*, leaves, insects, etc., prior to analysis.

5.3 Filtering Apparatus. The use of gooch crucibles, alundum crucibles, filter paper, or membrane filters is acceptable to separate the insoluble fraction from the soluble fraction. The choice of method is frequently based on the further analysis which is planned.

6. Reagents

6.1 Distilled Water. Water conforming to specifications for reagent water (ASTM designation D 1193).

6.2 Algacide. In those areas where algae or fungus growth is found not to occur, it is recommended that no chemical be added to the collecting liquid. When secure control of the dustfall collectors is assured, the addition of 0.10 g of reagent grade mercuric chloride is recommended. This will act as both an algacide and fungicide. In instances where there is the possibility that the collecting fluid may be consumed by humans or animals, 0.10 g of copper sulfate pentahydrate shall be used as an algae inhibitor. This has no fungicidal properties. The addition of any foreign material will complicate the analytical procedures.

7. Procedure

7.1 Sampling Site. Freedom from tampering and accessibility to the operator changing the samples are prime considerations in site selection. The most common locations for dustfall sampling sites are on the roof of low buildings or on utility poles. When a number of dustfall samples are collected concurrently for the purpose of a survey, it is important that all collectors should be located consistently. The following general recommendations should be observed:

a. the top of the dustfall container shall be a minimum of 8 ft and a maximum of 50 ft above the ground or 4 ft minimum above any other surface such as a roof. Higher objects such as parapets, signs, penthouses, and the like must not be more than 30° from the horizontal, i.e., a line drawn from the sampling jar to the nearest edge of the highest point on any building should form not more than a 30° angle with the horizontal;

b. public buildings such as schools, fire stations, libraries, etc., are most favorable to public agencies because of their accessibility and security;

c. the influence of a source close to the collecting jar can dominate the results. When selecting the site, detailed notes should be made on the direction and distance of stacks, parking lots, material storage piles, roads, or other sources likely to make a significant contribution to dustfall as measured at the location;

d. the support for the collecting jar should be mechanically stable and firm enough to prevent tipping or swaying from the wind.

7.2 Collection of Sample. Prior to exposure, 200 to 1000 ml of liquid should be added to the collection jar as dictated by expected evaporation rate over the exposure period. This liquid may be distilled water, algacide or fungicide as deemed necessary. The cover should be placed on the jar, and it should not be exposed until it is placed in its holder at the sampling location. Following the collection period, which is normally 30 days, ± 3 days, the cover shall be placed on the collecting jar prior to removal from its support stand and transported to the laboratory with the cover.

7.3 ANALYSIS. The first step in analysis is the transfer of the sample from the collecting jar to laboratory glassware. The sample should be screened through a 20-mesh screen to remove extraneous material. The inside of the collecting jar shall be scrubbed or policed to remove all adhering materials. In the event that the collection jar is dry or that only a small amount of water remains, make up the volume to at least 200 ml before the next step in the analysis. If the collection jar contained more than 200 ml of liquid at the time of its collection, proceed to the filtration step. If the volume must be made up to 200 ml, the sample should stand for a period of 24 hr at room temperature to allow the soluble material to dissolve. The filtration step shall be carried out according to routine laboratory procedures and the weight of insoluble material determined. For determination of the soluble material, the entire filtrate portion or suitable aliquot shall be evaporated to dryness and the total weight of soluble material in the sample shall be reported.

8. Calibration and Standards

8.1 There is no calibration procedure available for dustfall, and standards do not exist.

9. Calculations

9.1 Calculate the dustfall rate in terms of grams per square meter per month (30-day basis) as follows:

$$\text{Dustfall} = \frac{W}{a} \times \frac{30}{t}$$

Where dustfall equals grams per square meter per month, W = weight analyzed, grams; a = open area of sampling container at top, 0.0275 square meters; t = time of exposure days. ($1 g/m^2 = 2.86$ tons/sq. mile).

10. Effect of Storage

10.1 The major effect of storage of samples following collection and prior to analysis is the growth of algae or the chemical breakdown of organic material. It is recommended that the samples be analyzed within one week following their collection.

11. References

1. ASTM STANDARDS. Part 23, Standard Method for Collection and Analysis of Dustfall, ASTM D-1739-62.
2. RECOMMENDED STANDARD METHOD FOR CONTINUING DUSTFALL SURVEY (APM-1, Revision 1). 1966. TR-2 Air Pollution Measurements Committee, Robert A. Herrick, principal author, *J. Air Poll. Control Assoc.*, 16. No. 7.
3. STOCKHAM, J., S. RADNER, and E. GROVE. 1966. The Variability of Dustfall Analysis Due to the Container and the Collecting Fluid. *J. Air Poll. Control Assoc.*, 16, No. 5.
4. NADER, J.S. 1958. Dust Retention Efficiencies of Dustfall Collectors, *J. Air Poll. Control Assoc.*, 8, 35.
5. SANDERSON, H.P., P. BRADT, and M. KATZ. 1963. A Study of Dustfall on the Basis Replicated Latin Square Arrangements of Various Types of Collectors. Presented at the 56th Annual Meeting of Air Poll. Control Assoc., June 9–13, Detroit, Michigan.

Subcommittee 10
R. S. SHOLTES, *Chairman*
R. A. HERRICK
D. LUNDGREN
C. PHILLIPS
E. STEIN
J. WAGMAN
P. F. WOOLRICH

503.

Tentative Method of Analysis for Atmospheric Soiling Index by Transmittance (Paper Tape Sampler Method)

11201-01-72T

1. Principle of the Method (1, 2)

1.1 This method is based on measurement by light transmission of the extent of soiling or darkening of clean filter paper caused by suspended particulate matter removed by filtration from the atmosphere. The optical density values of the deposit are indicative of the soiling characteristics of particles of less than about 10 μm to below 0.1 μm in size and, therefore, describe the smoke or haze content of the atmosphere.

1.2 The atmosphere under test is screened to remove large particles. The screened air is then passed through a known circulate area of filter paper at a measured sampling rate for a 1-hr period or longer, as required. This operation is repeated automatically on fresh areas of paper at equal time periods and constant sampling rate.

1.3 The optical density of the circular spots is measured by transmittance in comparison with that of the clean paper in a spectrophotometer or colorimeter using light of wave length of approximately 400 nm. In an aerosol, the optical density is proportional to the mass concentration only when the particle size, shape, refractive index, color, etc., remain constant.

1.4 The reporting unit is COHs (coefficient of haze) per 1000 linear feet. This unit is indirectly related to the mass of deposit. To correlate optical density measurements with suspended particulate conc, a sample should be collected concurrently using a High Volume Sampler, ISC Method 11101-01-70T, (**3**). For conversion to the metric system, multiply COHs per 1000 linear feet by the factor 3.28 to obtain COHs per 1000 linear meters.

2. Range and Sensitivity

2.1 The amount of deposited matter evaluated by this method should be limited to 50% attenuation of incident light or a maximum optical density of about 0.30. For testing under heavily contaminated atmospheric conditions, the sampling cycle is set at 1 hr by suitable adjustment of the instrument. Conversely, in relatively clean atmospheres, the sampling cycle may be increased to 2-, 3-, 4-, or 6-hr periods. The most useful range for measurement of the light attenuation of the circular deposits is between optical densities of 0.05 and 0.30.

3. Interferences

3.1 In the evaluation of the spot sample the magnitude of the result is dependent on the characteristics of the filter paper and on the wavelength of the light, whether monochromatic or polychromatic. For the filtered layer, Beer's Law does not apply because of the multiple scattering from closely packed particles. However, a unit of measurement may be employed to characterize such deposits if the volume of air filtered, the area of the deposit, and its optical density are accurately known. Variations in transmittance of filter paper may constitute a source of error. Normally, filter paper deposits are essentially various shades of black due to the presence, primarily, of carbon. They may be tinted brown, reddish brown or light grey but such colors are not the rule and are due to local sources of contamination.

4. Precision and Accuracy

4.1 The precision and accuracy of this method depend upon the accuracy of the air flow rate measurement extent of variation of filtration velocity, filtration efficiency, and the optical system. All these factors should be controlled and the extent of variation kept within reasonably narrow limits. The optical measurements can be made with a high degree of precision if a steady source of monochromatic light is used, or if light from a stable source is passed through narrow band-pass filters. Periodic checks of the sampling instrument and colorimeter or spectrophotometer should be carried out for satisfactory operation.

5. Apparatus

5.1 General Description. The instrument consists of an assembly containing a sampling probe with filter screen* for removing coarse material, clamping device or head to hold the inlet plug securely to the paper, paper holder to accommodate a roll of filter paper, interval timer actuated by a clock, calibrated flowmeter, electrically driven air pump, and timer control device for adjustment of the sampling cycle (see Figure 503:1). The optical density of the deposit on the filter paper is determined by measurement of light transmittance in an optical instrument of conventional design, such as a

*A suitable filter screen is No. 20 Wire Cloth Sieve.

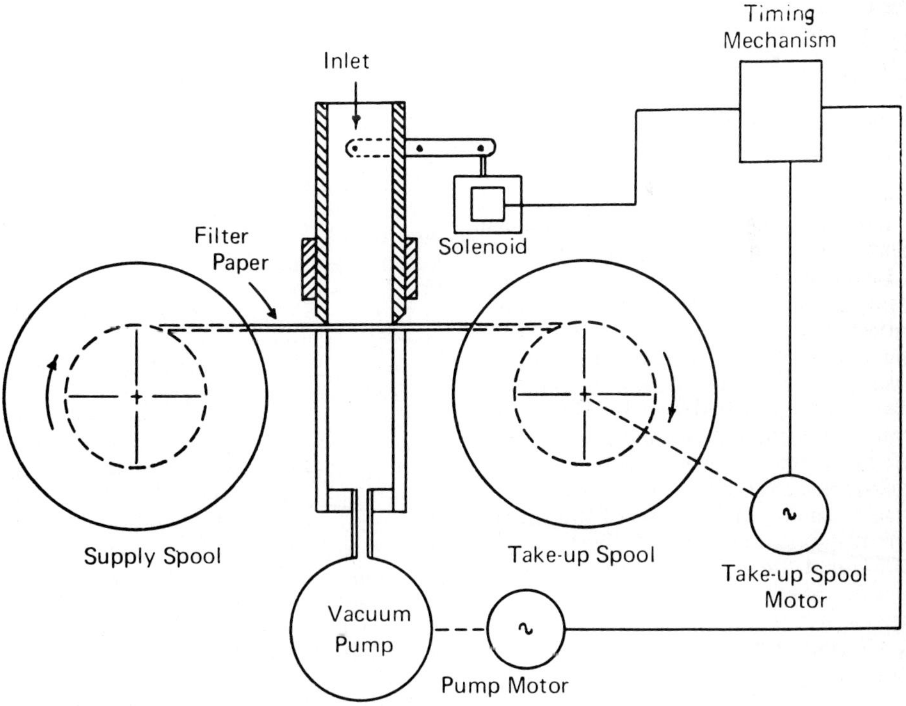

Figure 503:1—Diagrammatic representation of the automatic paper tape sampler.

densitometer, spectrophotometer or colorimeter.

5.2 Flow Meters. A suitable meter capable of measuring air flow in the range of 0.10 to 0.50 ft^3/min shall be used downstream of the filter (2.8 to 14.1 l/min).

5.3 Pumps. A suitable flow rate positive displacement pump which will maintain a filtration velocity of 45 fpm with a variation not to exceed ± 10% shall be used (*e.g.*, 15 CFH through a 1″/diam filter area) (425 l/hr through a 2.54 cm diam area).

5.4 Clamping Device and Interval Timer. The inlet head shall be equipped with a clamping device to fasten securely over a constant exposed area of filter paper. At the end of each sampling cycle a timing device releases the paper, the tape is advanced to expose a fresh area, and the cycle of operation is repeated. The timing cycle must be capable of adjustment to sampling intervals of 1, 2, 3, 4 and 6 hr. Parts that come in contact with the filter paper shall be of stainless steel, plastic, or other corrosion-resistant noncontaminating materials.

5.5 Sampling Inlet Tube. If used, the length shall not exceed 10 ft without sharp bends and the tube diam shall be such that the transport velocity is about 700 fpm. (213 m/min.)

5.6 Densitometer. Any conventional instrument is suitable if equipped with a stable light source and photoelectric optical system suitable for determining transmitted light from clean paper tape in comparison with spot samples. For convenience, the instrument may be provided with an optical density scale. The monochromatic light source shall provide light of about 400 nm wavelength. If monochromatic light is not available, white light may be employed with a set of suitable filters that pass a band of light in the region of 375 to 450 nm.

6. Reagents

6.1 Filter Media. Whatman No. 4 or equivalent filter paper shall be used.

Warning: Sample deposits on this paper are not suitable for chemical analysis.

7. Procedure

7.1 Procedure for Spot Samples.

7.1.2 After assembling the apparatus described in Section **5**, place a roll of clean filter paper tape in the holder. Set the air sampling rate as specified in Section **5.3**. The time cycle is set to avoid spots of optical density exceeding 0.30. This specification is usually met by setting the timer to yield a spot sample very 2 hr.

7.1.3 The air flow rate shall be checked daily and adjusted as necessary. The gas flow meter shall be periodically calibrated with a standard wet test gas meter.

7.1.4 For nonautomatic reading units, the exposed paper tape is removed at convenient intervals from the rewind roller. The optical density of each sample spot is determined using a densitometer or colorimeter as specified in Section **5.6**.

8. Calibration

8.1 The values based on optical density of the deposit are indicative of the soiling characteristics of particles of less than about 10 μm to below 0.1 μm in size and, therefore, describe the smoke or haze content of the atmosphere. There is no direct relation between such values and mass concentration of particulates per unit volume of air, although empirical relations may be calculated for specific cases. If particle size, shape, refractive index, and other factors are reasonably constant, these empirical relations should be useful in comparing deposits from different types of sources and in different locations.

8.2 A paper tape sampler is not calibrated in the absolute sense because a COH reading cannot be explicitly related to any other atmospheric particulate or gas measurement. The instrument components such as the timer should be calibrated in the normal manner. The air flow meter should be calibrated by placing a calibrated laboratory test meter at the inlet to the sampling unit.

9. Calculation

9.1 Calculate the optical density as follows:

$$\text{Optical Density} = \frac{\log I_0}{I}$$

where:

I_0 = the intensity of transmitted light through the clean paper, and
I = the intensity of transmitted light through the sample.
Log refers to the logarithm to base 10.

9.2 The optical density values are converted to COH units defined as 100 times the optical density of the deposit. For example, an optical density of 0.301 is equal to 30.1 COH units. Since the deposition of particulate matter on the paper tape is a function of the total volume of air aspirated through the filter and the area of the filter, the size of the air sample is properly expressed in lineal units of air; that is, the volume of air divided by the area of the filter. For convenience, the results are reduced to multiples of 1000 linear ft (LF) of air sampled, so that the soiling for any time period is described in terms of COH units per 1000 linear feet of air, COHs/1000 LF.

9.3 If the quantity of air sampled during each spot sample is equal to L, in thousands of linear ft, then,

$$L = \frac{\text{flow (ft}^3\text{/min)} \times \text{sampling time in min}}{1000 \times \text{circular area of spot (ft}^2\text{)}}$$

COHs per 1000 LM =

$$\frac{\text{optical density} \times 100}{L}$$

(Metric system) COHs per 1000 LM =

$$\frac{\text{optical density} \times 100}{L \times 0.305}$$

9.4 Example. Suppose measurements with the gas flow meter have established the average flow rate through the spot area of paper as 0.24 ft³/min, the area exposed is a sampling circle 1″ in diam or 5.45×10^{-3} ft², and the timing spindles of the instrument are set for 2-hr samples, then,

$$L = \frac{0.24 \times 120}{1000 \times 5.45 \times 10^{-3}}$$

$$= 5.28 \text{ (thousands of ft)}$$

If a given spot shows a light transmission value of 75%, then,

Optical density =

$$\log \frac{100}{75} = \log 1.33 = 0.125$$

COHs/1000 LF =

$$\frac{0.125 \times 100}{5.28} = 2.4 \text{ or in metric units}$$

COHs/1000 LM =

$$\frac{0.125 \times 100}{5.28 \times 0.305} = 7.75$$

Experience has shown that paper tapes will undergo color changes with time when exposed to light.

11. References

1. Standard Method of Test for Particulate Matter in the Atmosphere: Optical Density of Filtered Deposit. ASTM Designation D 1704-61 (Reapproved 1969).
2. Recommended Method for Continuing Air Monitoring for Fine Particulate Matter by Filter Media Using the Low Volume Sampler. Standard Method APM 2.2.
3. American Public Health Association. 1977. Methods of Air Sampling and Analysis. 2nd. ed. p. 578. Washington, D.C.

Subcommittee 10
R. A. Herrick, *Chairman*
S. Kinsman
J. Lodge
D. Lundgren
C. Phillips
R. S. Sholtes
E. Stein
J. Wagman

504.

Tentative Method for Determination of the Size Distribution of Atmospheric Particulate Matter by Weight (Cascade Impactor Method)

11101-01-74T

1. Principle of the Method

1.1 The size distribution of the mass of particulate matter suspended in the atmosphere can be determined—at least in part—using a multi-stage inertial impactor **(1–14)**. Impactors utilize the fact that particles of different mass possess different inertia at a given velocity. Particles having inertia greater than a critical value can be selectively removed from an air stream at each of the stages of a multi-stage inertial impactor, most commonly called a cascade impactor. These impactors may have any number of stages but four to six stages are most common. Particle size data obtained with an impactor are best interpreted and expressed in terms of equivalent or aerodynamic size. The equivalent (or aerodynamic) size of any particle regardless of its shape or density, is defined as the diam of a unit density (1 g/cm^3) sphere having the same settling velocity as the actual particle when subjected to a gravitational or centrifugal force field. Aerodynamic particle size is an important parameter for estimation of health effects, for selection of air pollution control equipment, and for atmospheric modeling of particulate pollutants.

2. Range and Sensitivity

2.1 Cascade impactors are best suited to determination of particle size distribution for aerosols having mass median diam (MMD) between 0.5 μm and 5 μm. The MMD of atmospheric particles will normally lie within the above size range limits.

2.2 By sampling for appropriate periods of time particle size distributions can be determined for atmospheric particle concentration in the 10 μg/m^3 to 1000 μg/m^3 range. This range includes all but the very cleanest ambient air. Sampling periods of 1 day or more for very clean air or 1 hr for very dirty air may be required for this concentration range.

2.3 The sensitivity of a particle size distribution determination is directly related to the weight of particles collected on the various collection surfaces (or stages) and to the sensitivity of weighing these particles. For example, if the ambient concentration is 30 μg/m^3, and if an impactor with a flow rate of 2 m^3/hr is operated for 24 hr ($30 \times 2 \times 24 = 1440$ μg = 1.44 mg), and if a 10% point on the distribution curve is to be determined, then a 0.14 mg deposit (0.1×1.44 mg) must be weighed. This will require a weighing sensitivity of 0.01 or 0.02 mg.

3. Interferences

3.1 Interferences are mainly operational problems caused by insects, lint fibers, leaves, etc. which block part of the sampler inlet or part of an impactor nozzle. Condensation of water or other vapors or the evaporation of volatile matter from the collection surface can also be an interference. Particle bounce off from the collection surface or any type of particle losses can be considered interferences.

4. Precision and Accuracy

4.1 Precision and accuracy of the cascade impactor method are both closely related to the accuracy with which the particle collection deposits can be determined. Under ideal operating conditions the overall method accuracy is limited to about ± 10%. This accuracy limit applies to the total particulate concentration determination and to the particulate MMD when the median lies within about a 0.5 μm to 5 μm diam range. Inability to obtain a representative particle sample and increasing particle losses cause the method accuracy to rapidly decrease for an aerosol MMD much above 5 μm. Particle losses are defined here as particle collection on any internal impactor surfaces other than the surfaces specifically intended for collection of these particles. Method accuracy decreases below about 0.5 μm MMD because of difficulties associated with construction, calibration and operation of the very fine nozzles required to fractionate particles in this size range. A simple cascade impactor is not suitable for aerosol size distribution determination if the aerosol MMD is less than 0.2 μm or greater than 20 μm.

4.2 The above discussion of accuracy assumes a known, constant air sampling rate, no air leakage and, most important, no particle bounce or re-entrainment of deposited particulate matter. This last assumption is often not satisfied and can lead to large errors.

4.3 The precision of the method can be very good under ideal sampling conditions. Lack of precision is mainly due to operator variability, non ideal operating conditions and poor manufacture of the cascade impactor. Precision for a given instrument is very dependent upon maintaining a constant flow rate through the impactor. Precision between instruments depends upon manufacturing precision, with the impactor nozzle dimensions being most critical.

4.4 Overall precision or accuracy of this method cannot be simply defined or completely discussed because of the extremely variable nature of the particulate matter to be collected. These particle variables include concentration, median size, size range, particle shape, particle volatility and reactivity, particle chemical composition and particle physical properties.

5. Apparatus

5.1 SAMPLING.

5.1.1 *Cascade Impactor.* Several types of cascade impactors are commercially available (**15–20**). Impactors with either circular nozzles or rectangular nozzles are capable of providing the desired information and neither has a sound theoretical advantage over the other. For most ambient air sampling the following cascade impactor specifications are recommended:

a. air flow rate—1 m^3/hr minimum (about 0.6 ft^3/min).

b. number of impactor stages—3 minimum.

c. finest stage particle cut-size*—0.3 to 0.6 μm diam (based on unit density sphere).

d. coarsest stage particle cut-size*—8 to 20 μm diam.

e. ratio between consecutive stage particle cut-sizes*—2/1 to 4/1.

f. manufacturing tolerance on impactor stage nozzle cross sectional areas—± 5%.

g. total internal impactor wall losses—less than 10% for 3 μm diam particles.

h. nozzle calibration precision—± 10%.

i. impactor leak rate—less than 1% at design flow rate and pressure drop.

j. materials of construction—aluminum, stainless steel, or any other non corrosive material that can be cleaned with solvents such as acetone.

k. impactor nozzle-to-impaction surface distance at least 1 jet width for rectangular jets and at least 0.5 throat diameter for round impactor jets.

l. impactor jet Reynolds number should be about 3000 and certainly within the range from 500 to 20,000.

m. impactor nozzles should have a relatively straight throat section of about 1 or

*Cut-size is used to mean the particle size for which the stage collection efficiency is 50% (also noted as stage 50% cut-size)—see Figure 504:1.

more jet widths in length. That is, it should not have a continuous straight taper up to the nozzle discharge point.

5.1.2 *Particle Collection Surfaces.* Collection surfaces should be inert or be coated with a relatively inert material, capable of being cleaned with acetone or a similar solvent, capable of being coated with a sticky material or other substances which will insure good particle adhesion, and light enough to be tared on a semi-micro analytical balance. It is desirable that the surfaces to be weighed be as light as possible and be easily handled, cleaned, stored, and transported.

5.1.3 *Filter Unit.* A cascade impactor must be followed by a filter unit. This unit should accommodate a high efficiency glassfiber media or other appropriate type filter media and be of sufficiently large size so as to avoid high filter pressure drop or rapid filter loading. A filter size of 47 mm diameter would be appropriate for a 2 m^3/hr flow rate.

5.1.4 *Flow Measuring Device.* Gas meter, orifice meter or rotameter with an accuracy of ± 5% at the design flow rate.

5.1.5 *Timer.* Any standard type.

5.1.6 *Vacuum Pump.* Rotating vane or positive pressure type, capable of maintaining a constant flow of air through the impactor unit and after-filter over the entire pressure drop range normally encountered.

5.1.7 *Analytical Balance.* A semi-micro analytical balance with a sensitivity of 0.01 mg is required. The balance must be located on a stone type table in an enclosure or in an area free of drafts. The temperature and humidity of the balance room should be maintained at about 20 ± 5 C and 40% ± 10% R.H.

6. Reagents

6.1 Distilled Water, used for cleaning of particle collection surfaces. Water should conform to specifications for reagent water.

6.2 Acetone, used for cleaning of particle collection surfaces. Acetone should be reagent grade.

6.3 Filter Media. Glass-fiber filter media having a weight collection efficiency of at least 99% for particles of 0.3 μ diam are suitable. Other media may be desirable or needed for certain analyses. If the filter efficiency is unknown it should be determined.

6.4 Impactor Collection Surface Adhesive Coatings. When an impactor coating is required, a nonvolatile, nonhygroscopic high viscosity fluid or grease should be used. An important requirement of any coating is the stability required for good weight reproducibility. Examples of suitable coatings are Dow Corning series 200 Fluid (about 100,000 centistoke or greater viscosity) and High Vacuum Silicone Grease.

7. Procedure

7.1 Preparation.

7.1.1 Before each sampling endeavor the cascade impactor should be inspected and cleaned together with the after filter and collection surfaces. Special care should be taken to clean the impactor nozzles. If the impactor unit is fairly clean a wiper such as a lint-free lens tissue dipped in acetone can be used to wipe off all internal surfaces. Tweezers can be used to reach into corners.

7.1.2 Sampling at a known and constant air flow rate is very critical when using an impactor and the unit flow should be frequently checked and the flow rate indicator calibrated. Leak testing and flow rate indicator calibration can be accomplished by placing a calibrated dry gas meter before the impactor inlet and a second similar meter after the unit filter. Corrected gas meter readings should agree within 1%.

7.1.3 Impaction surfaces should be cleaned and prepared for use. The impaction surfaces and the filters should be marked and placed in the balance-room environment for at least 24 hr to equilibrate. The balance room should be maintained at a temperature of about 20 ± 5 C and a relative humidity of 40% ± 10%.

7.1.4 Each filter and impactor collection surface should be weighed twice on a semi-micro analytical balance to the nearest 0.01 mg. An average of the two readings should be used for the tare weight. Record tare weight and identification number.

7.1.5 In a clean area assemble the sampling train consisting of the cascade impactor, after filter, flow meter, flow regulator and vacuum pump. With all impaction surfaces and after-filter media in place, briefly operate the unit in order to set the desired flow rate and check the unit's operation. Disassemble the sampling train and seal off the impactor and after filter inlet and outlet.

7.2 SAMPLING.

7.2.1 Set up the unit in the desired sampling location. If air is ducted to the impactor the ducting should be as short as possible and free of bends or area changes. Where used, this ducting should be metallic, have smooth interior surfaces and be amenable to thorough cleaning. If insects, leaves or large lint is a problem, a medium mesh screen should be used at the sampling line inlet. The screen area should be large in comparison to the sampling tube or impactor inlet area.

7.2.2 Start the vacuum pump at the desired time. Record starting time, air temperature, unit flow rate, and any other needed information. Set and operate the unit at the desired flow rate. The flow rate should be checked and adjusted as often as necessary to keep it within ± 5% of the desired rate.

7.2.3 At the end of the desired sampling period stop the unit. Record all necessary information such as sampling time, flow rate, air temperature, etc.

7.2.4 Disassemble sampling train and block off the impactor and filter inlets and outlets to prevent extraneous dust from entering. Return unit to laboratory area for removal of filter and impaction collection surfaces.

7.3 ANALYSIS.

7.3.1 Wipe extraneous dust and dirt off exteriors of impactor and filter holder. Disassemble or otherwise remove filter and impactor collection surfaces. Note identification number and positions of various impactor surfaces. Note appearance of collection surfaces and interior of cascade impactor.

7.3.2 Transport impactor surfaces and filters to balance room to equilibrate for at least 24 hr before reweighing. Weigh impactor surfaces and filters to the nearest 0.01 mg. Record weights and identification numbers.

7.3.3 Collection deposits may be analyzed chemically or by other means after weights have been obtained. If major modifications in the described method are required in order to perform a certain chemical analysis, two separate impactor trains should be prepared and run together—one unit for weight determinations and the second for chemical analysis.

8. Calibration

8.1 GENERAL.

8.1.1 In order to be useful, a cascade impactor must have known particle collection characteristics. Accurate calibration of a cascade impactor is both difficult and time consuming, in addition to requiring specialized equipment and knowledge. However, all identical impactor units will have the same calibration if they are all operated in the identically same manner. It is advantageous to purchase a cascade impactor from a manufacturer who has carried out a thorough and accurate calibration of a unit that is identical to those being sold. This will save the user considerable time and expense.

8.1.2 Particle collection characteristics of impactors are critically dependent upon air flow rate, nozzle dimensions and the position and type of particle collection (impaction) surfaces. Of general importance is the cleanliness of the impactor interior, absence of air leaks, and low internal particle loss (loss meaning particle collection on other than the specifically intended collection surface). A basic calibration and operation requirement is that all particles striking a collection surface must stick to it and not become re-entrained in the air stream or otherwise lost.

8.1.3 For many types of particles a viscous coating on the collection surface or a special type of collection surface is required to assure particle adhesion. For coating a smooth surface any viscous liquid may be suitable if it meets the important requirement of being weight stable (coated surface weight stability of ± 0.01 mg). Additional surface coating requirements are: (a) must uniformly wet the surface, (b) be non hygroscopic, (c) be non volatile after the initial weight equilibration time, (d) be non reactive with deposited particles or atmospheric gases, and (e) be generally safe and convenient to use.

8.1.4 As the collection surface becomes covered with collected particles, subsequent particles may bounce off or previously collected particles may become re-entrained. The degree to which this occurs varies greatly with the type and size of particle and velocity of the air stream. A minimal problem exists with nonvolatile oil droplets while a serious problem exists with mineral particles. An important method of overcoming the problem of collection surface overloading is by providing new collection surface area. Other factors being equal, the greater the exposed collection surface the greater the amount of particulate matter that may be collected without re-entrainment and, therefore, the more accurate the impactor test results.

8.1.5 Definite rules cannot be given as to if, when, or how much particle bounce-off or re-entrainment has occurred or will occur. Therefore, precautions must be taken to prevent these problems as a correction is not possible.

8.2 METHOD OF CALIBRATION

8.2.1 Uniform size (homogeneous) aerosols should be used to calibrate each stage of a cascade impactor (**11, 21**). The calibration should be conducted in a reputable laboratory and directly supervised by a competent individual using known and accepted techniques. Calibration data should be certified by the individual or firm conducting the calibration and must include: (a) accurate dimensions of each impactor nozzle, (b) description of the impaction surface size, type, coating and position with respect to the impactor nozzle, (c) flow rate for each impactor test, (d) method of aerosol generation, (e) particle collection efficiency (for each stage and for each test), (f) test particle size, shape and density and method of measurement (for each test particle used), (g) the per cent loss of particle within the impactor, and (h) estimated accuracy of each measured quantity, size and dimension.

8.2.2 The cascade impactor should be calibrated over the particle size range from about 0.3 μ to about 20 μ diameter (equivalent unit density sphere). Each impactor stage should be tested at a minimum of three test conditions. For each test condition the impactor flow rate and the test particle size, density and shape must be fixed. A minimum of two tests should be run at each test condition and the average collection efficiency used for plotting of the test data. For at least one test condition the resulting stage collection efficiency must fall between 50% and 80% and for a second test condition between 20 and 50%.

8.2.3 A best fitting straight line should be drawn through the three or more test condition points on a log-probability plot of collection efficiency versus the inertial impaction parameter. This line should be used to determine the impaction parameter value for the stage 50% collection efficiency (see Figure 504:1). Use of the impaction parameter allows calibration tests to be run at flow rates other than the design flow rate provided that the rate is not less than one half or more than twice the flow rate for which the impactor unit is being calibrated. The inertial impaction parameter (K) is defined by the following equation:

$$K = [C\rho_p V_j D_p^2/(18\mu_g D_j]^{\frac{1}{2}}$$

where:

C = Cunningham correction factor, dimensionless

$$C = 1 + \frac{(1.63 \times 10^{-5})}{D_p} + \frac{(0.55 \times 10^{-5})}{D_p}$$

$= \exp(-0.67 \times 10^5 D_p)$ for air at standard conditions

ρ_p = particle density, gm/cm^3

V_j = average velocity at discharge of impactor jet, cm/sec

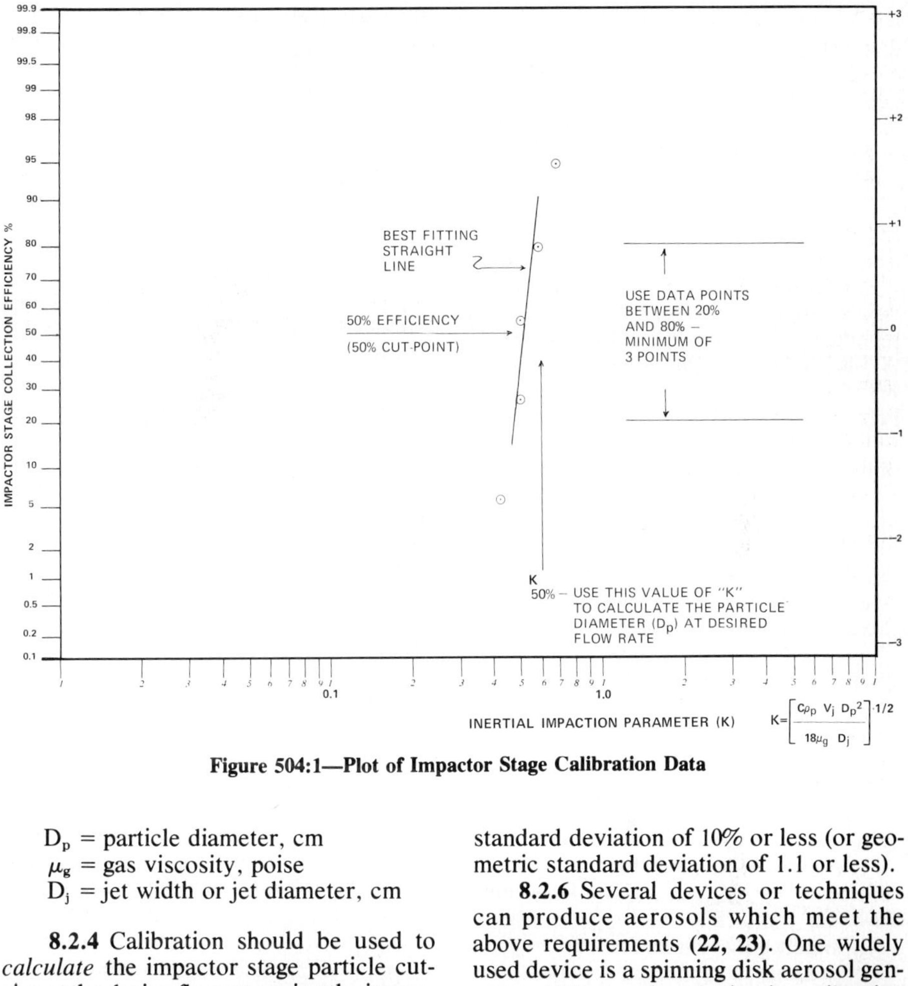

Figure 504:1—Plot of Impactor Stage Calibration Data

D_p = particle diameter, cm
μ_g = gas viscosity, poise
D_j = jet width or jet diameter, cm

8.2.4 Calibration should be used to *calculate* the impactor stage particle cut-size at the design flow rate using the impaction parameter value for the 50% efficiency value on the calibration data plot (see Figure 504:1). Use of the impaction parameter is necessary to eliminate the requirement of using unit density particles and also to correct for variations in air flow rate and gas viscosity, both in the impactor calibration and in its use.

8.2.5 Calibration aerosols may be produced by any suitable technique if the *aerosol* meets the following specifications:

a. particle shape: spherical.

b. particle density: determined to have a standard deviation of 10% or less.

c. particle diam: determined to have a standard deviation of 10% or less (or geometric standard deviation of 1.1 or less).

8.2.6 Several devices or techniques can produce aerosols which meet the above requirements (**22, 23**). One widely used device is a spinning disk aerosol generator (**22**). Another device is a vibrating orifice aerosol generator (**23**). Solid spherical particles made of a fluorescent dye such as uranine can be produced by these devices. Another aerosol generation technique involves the use of very uniform size polystyrene spheres suspended in a volatile liquid. This same technique is also used with larger size plant pollen and spores. In all cases, the *aerosol* produced must meet the specifications listed above.

8.3 MEASUREMENT ACCURACY REQUIREMENTS.

8.3.1 *Air Flow Rate.* For each test the impactor flow rate should be determined within ± 5%. The air flow rate during a test should vary from the mean by less than 5%.

8.3.2 *Collection Efficiency*. The average collection efficiency for two or more identical tests should have a precision of ± 5% (*i.e.*, 40 ± 5% or 75 ± 5%). If the collection efficiencies for two identical tests differ by more than 10% (*i.e.*, 30% and 41% or 61% and 72%) then two additional tests should be run.

8.3.3 *Particle Density*. Particle density should be determined by a suitable technique with an error of 10% or less.

8.3.4 *Particle Diameter*. Median particle diam should be determined with a light microscope, electron microscope or sedimentation rate apparatus with a maximum error of 10% for any given measurement and with an average error of 5% or less based on all diam determinations.

8.3.5 *Impactor Dimensions*. All critical impactor dimensions such as nozzle sizes should be determined with 1% or 0.0005″ whichever is larger.

8.3.6 *Impactor Nozzle Cut-Point*. Each impactor nozzle 50% cut-point size should be determined with a maximum error of 20% for any given nozzle and with an average nozzle cut-size error of 10% or less.

8.3.7 *Nozzle Dimensional Uniformity*. For any cascade impactor which utilizes two or more openings for a given stage, all nozzle openings in a given stage must be of the same shape and type and be dimensionally equal within the following limits: discharge area of each separate stage nozzle should differ from the mean stage nozzle area by 5% or less.

9. Calculations

9.1 Air Flow. Calculations of air flow rate and sampled volume are described in detail in Method 501—Tentative Method of Analysis for Suspended Particulate Matter in the Atmosphere **(24)**.

9.2 Impactor Stage Weight Fraction.

9.2.1 Weight gained by each impactor stage collection surface and by the filter are divided by the corrected total sampled volume of air and the results expressed as $\mu g/m^3$.

9.2.2 For the impactor 1st stage, the interpretation of particle weight collected is amount ($\mu g/m^3$) of particulate matter of equivalent size *greater than* the 1st stage nozzle 50% cut-point. A mean size should not be associated with this size fraction.

9.2.3 For the impactor 2nd stage, the interpretation of particle weight collected is amount in μg per m^3 of particle matter greater than the 2nd stage 50% cut-point but less than the 1st stage 50% cut-point. It is unnecessary to assign any mean size to this fraction as the results are best plotted on a cumulative basis as described below. The 3rd and all later stages are calculated in the same manner as described for the 2nd stage.

9.2.4 Filter mass is calculated as μg per m^3 and interpreted as particulate matter of the size *less than* the last stage 50% cut-point.

9.3 Total Particle Concentration. Sum up all particle weight fractions to get the total particle concentration.

9.4 Cumulative Data Pre-

Table 504:I. Cascade Impactor Data Table

Stage #	Stage cut-size (microns)	Particle Weight Deposited (mg)	Fraction Deposited (%)	Cumulative %
1	15	0.90	10	10
2	5	1.26	14	24
3	1.5	1.26	14	38
4	0.5	1.98	22	60
After Filter		3.60	40	100
		total = 9.00 mg		

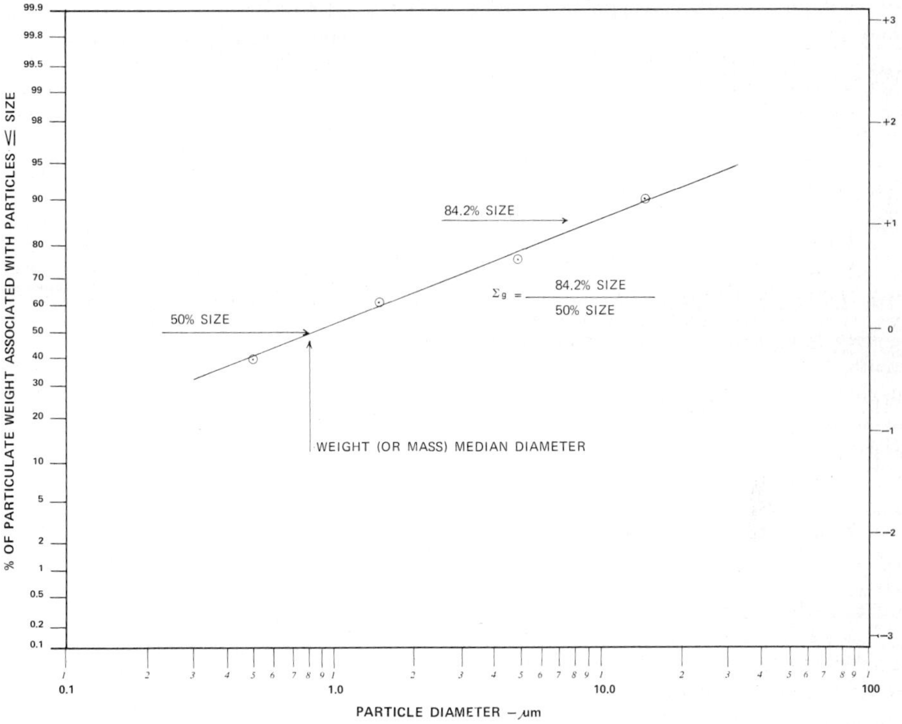

Figure 504:2—Cumulative Distribution Curve of Cascade Impactor Data

SENTATION. In preparing impactor data for plotting it is desirable first to tabulate the data, as shown in Table 504:I. Data are then plotted as a cumulative distribution curve on a normal probability plot or on a logarithmic probability plot. One method of data plotting is to start with the 1st stage collection and plot the 1st stage weight as per cent of the total weight having a size greater than the 1st stage 50% cut-point size (equivalent diameter). The cumulated 1st and 2nd stage weights are plotted as per cent greater than the 2nd stage 50% cut-size. This procedure is carried on to the last stage. Particle weight on the filter represents the per cent less than the last stage cut size. An example plot is shown in Figure 504:2. A smooth curve or straight line is next drawn through the actual data points. In nature aerosols are often approximately log-normally distributed, therefore, the cumulative-probability plot is often used.

9.5 DATA INTERPRETATION.

9.5.1 *Mass (Weight) Median Diameter.* The particle of distribution MMD is read directly off the distribution curve at the point where the curve crosses the 50% line. It is easier and more accurate to read the particle MMD directly off the distribution curve than to calculate it from the data.

9.5.2 *Measure of Particle Distribution Spread.* If a distribution is log-normal the spread of the distribution is described by the geometric standard deviation (σ_g) which can be read directly from a cumulative plot on log-probability paper **(25)** (See Figure 504:2). If a distribution is normal the distribution spread is described by the standard deviation (σ) which can be read from a cumulative plot on normal probability paper. If the distribution does not appear to be normally or log-normally distributed the entire distribution data should be presented and used.

9.5.3 *Particulate Matter in a Given Size Range*. The per cent and quantity ($\mu g/m^3$) of particles by weight in any particle size range can be estimated by using a best fitting smooth curve or straight line through the cumulative data plot points. The distributions curve should not be extrapolated much beyond the data end points.

10. Effect of Storage

10.1 Samples of non volatile particulate matter may be stored indefinitely if protected from contamination.

11. References

1. MAY, K.R. 1945. The Cascade Impactor: An Instrument for Sampling Coarse Aerosols, *J. Sci. Instr.* 22:187.
2. SONKIN, L.S. 1946. A Modified Cascade Impactor, *J. of Ind. Hyg. and Tox.* 28:269.
3. RANZ, W.E. and J.B. WONG. 1952. Jet Impactors for Determining the Particle-Size Distributions of Aerosols. *A. M. A. Arch. Ind. Hyg. Occup. Med.* 5:464.
4. WILCOX, J.D. 1953. Design of a New Five-Stage Cascade Impactor. *A. M. A. Arch. of Ind. Hyg. and Occup. Med.* 7:376.
5. MAY, K.R. 1956. A Cascade Impactor with Moving Slides. *A. M. A. Arch of Ind. Health.* 13:481.
6. RANZ, W.E. 1956. Principles of Inertial Impaction. Engineering Research Bulletin No. 66, The Pennsylvania State University.
7. ANDERSEN, A.A. 1958. New Sampler for the Collection, Sizing, and Enumeration of Viable Airborne Particles. *J. Bact.* 76:471.
8. BRINK, JR., J.A. 1958. Cascade Impactor for Adiabatic Measurements. *Ind. and Eng. Chem.* 50:645.
9. MITCHELL, R.I. and J.M. PILCHER. 1959. Improved Cascade Impactor for Measuring Aerosol Particle-Sizes in Air Pollutants, Commercial Aerosols and Cigarette Smoke. *Ind. and Eng. Chem.* 51:1039.
10. LIPPMANN, MORTON. 1961. A Compact Cascade Impactor for Field Survey Sampling. *Amer. Ind. Hyg. Assoc. J.* 22:348.
11. LUNDGREN, D.A. 1967. An Aerosol Sampler for Determination of Particle Concentration as a Function of Size and Time. *J. Air Poll. Cont. Assoc.* 17:225.
12. MERCER, T.T., M.I. TILLERY, and G.J. NEWTON. 1970. A Multistage, Low Flow Rate Cascade Impactor. *Aerosol Science* 1:9.
13. MARPLE, V.A. 1970. A Fundamental Study of Inertial Impactors. Ph.D. Thesis University of Minnesota.
14. JAENICKE, RUPRECHT. 1971. The Double-Stage Impactor, A Further Application of the Impactor Principle, Staub, 31:6.
15. UNICO—Union Industrial Equipment Company.
16. CASELLA—Wilson Company.
17. ANDERSEN—Inc., Atlanta, Georgia.
18. BATELLE—Scientific Advances.
19. LUNDGREN—Environmental Research, Corp., St. Paul, Minnesota.
20. BGI INCORPORATED, Waltham, Massachusetts.
21. FLESCH, J.P., C.H. NORRIS, and A.E. NUGENE, JR. 1967. Calibrating Particulate Air Samplers with Monodisperse Aerosols: Application to the Andersen Cascade Impactor. *Am. Ind. Hyg. Assoc. J.* 28:507.
22. WHITBY, K.T., D.A. LUNDGREN, and C.M. PETERSON. 1965. Homogeneous Aerosol Generators. *Intern. J. Air Water Poll.* 9:263.
23. BERGLUND, R.N. and B.Y.H. LIU. 1973. Generation of Monodisperse Aerosol Standards. *Env. Sci. & Tech.* 7:147.
24. AMERICAN PUBLIC HEALTH ASSOCIATION, INC. 1977. Methods of Air Sampling and Analysis. 2nd Edition. Washington, D.C. 20036.
25. HERDAN, G. 1960. Small Particle Statistics. 2nd Edition, Butterworth & Co. Ltd., 88 Kingsway, London.

Subcommittee 10

R. A. HERRICK, *Chairman*
S. KINSMAN
J. LODGE
D. LUNDGREN
C. PHILLIPS
W. C. RUTLEDGE
R. S. SHOLTES
E. STEIN
J. J. STUKEL
J. WAGMAN
W. D. CONNER *(Consultant)*

505.

Tentative Method of Analysis for Suspended Particulate Matter in the Atmosphere (The Wind Direction Controlled Air Sampler)

11101-03-76T

1. Principle of the Method

1.1 A wind direction controlled air sampler samples air when the wind is coming from a specific direction. The vane assembly sends a signal to the control unit when the air is coming from the selected direction sector, and the control unit then activates a power output used to run the sampling device.

1.2 APPLICABILITY. The vane and control unit assembly can be used in conjunction with any sampler requiring electrical power to operate. The unit can be used to locate the source of particulates or gases in monitoring applications requiring specific directional air sampling.

2. Range and Sensitivity

2.1 The wind direction controlled air sampler should respond to low wind speeds (approximately 0.5 mph) and should still be operational in winds of about 60 mph.

2.2 The range and sensitivity of the sampling unit depends upon the type of sampler used.

3. Interferences

3.1 The only interference occurs when certain types of directional components are used, and the wind fluctuates at the edge of the turn-on sector zone. A directional hysteresis and an actuation time delay may be used to minimize the on-off chatter of this type of equipment.

4. Precision and Accuracy

4.1 The hysteresis mentioned in Section 3 which minimizes on-off chatter typically accounts for ± 2 degrees in the sector zone angle.

4.2 The precision and accuracy of the sampling unit depends upon the type of sampler used.

5. Apparatus

5.1 A typical unit consists of the three components (shown in Figure 505:1) 1) the vane assembly and support, 2) the control unit, and 3) the sampler.

5.2 THE VANE ASSEMBLY AND SUPPORT.

5.2.1 A typical vane assembly is shown in Figure 505:2. The vane assembly must have low inertia, and must be able to respond to low wind movements.

5.2.2 The assembly should have a component which measures the wind direction within a specified controlled angle, and which can be easily changed to a different control angle. A typical unit (1) uses sensor rings which are commercially available in many control angles from 15 to 180 degrees. The senior ring angle is the total of the angles on both sides of the control direction.

5.2.3 Any mast and tripod system can be used, as long as it does not interfere with the operation of the unit. Support systems are commercially available, however, with the sizes of the mast and tripod dependent upon the operating height of the instrument.

5.3 THE CONTROL UNIT.

5.3.1 The control unit receives a signal from the vane assembly when the wind direction lies within the sector zone angle set by the direction component, and it activates a power output.

5.3.2 The signal from the vane is car-

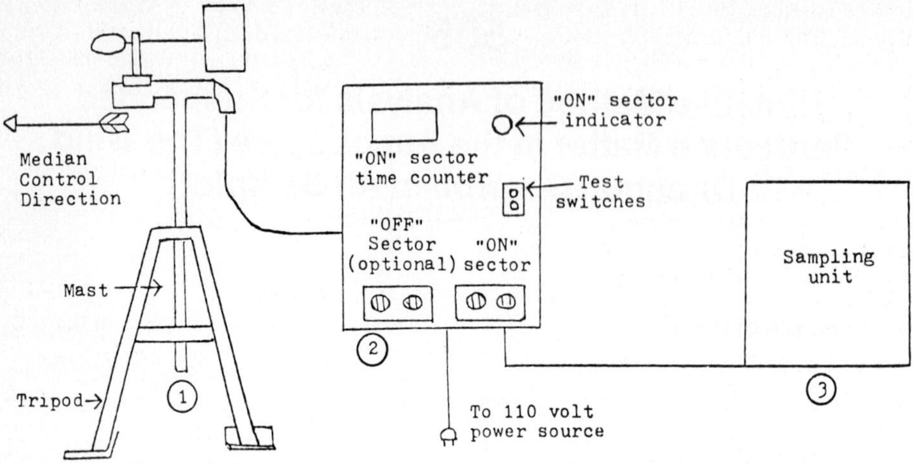

Figure 505:1—The components of the wind direction controlled air sampler.

ried by a cable, usually 50 to 500 feet in length.

5.3.3 The control unit must be sheltered, although commercially available weatherproof units eliminate this requirement.

5.3.4 To minimize the harmful chattering effects of wind fluctuations, in and out of the control sector, a thermal time delay relay having a nominal delay value is incorporated into the control circuit. The delay value is replacable, and certain values from 2 to 20 sec are available.

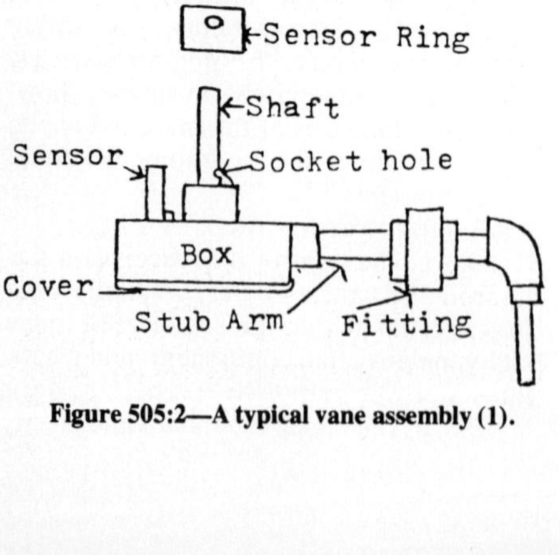

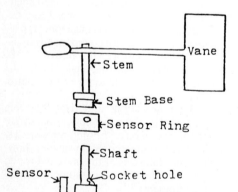

Figure 505:2—A typical vane assembly (1).

5.3.5 The control unit usually contains a timer which counts the amount of time the "ON" sector is activated, an indicator which shows when the "ON" sector is activated, and test switches are provided for checking circuit operation.

5.3.6 An "OFF" sector receptacle is an optional feature, and can be used to shut down sampling equipment when the wind is from a specific direction.

5.4 THE SAMPLING UNIT. Any sampling device which requires electrical power to operate may be used.

6. Reagents

None.

7. Procedure

7.1 SITE SELECTION. The fact that the vane assembly can be separated from the sampler by relatively large distances gives this unit much flexibility.

7.1.1 The vane assembly should be placed at a site where there are no obstacles or localized wind effects, unless the survey is attempting to measure these effects. If obstacles cannot be avoided, make sure that a line drawn from the vane assembly to the highest point of any nearby building does not form an angle greater than 15° with the horizontal. The roof of an easily accessible building is often a favorable site. If the vane assembly is placed on a building roof, the sampling unit may be placed at the same location or at ground level.

7.1.2 When sampling for the effects of a specific source, place the sampling unit at ground level 8 to 10 stack heights downwind in the prevailing wind direction. Place the vane assembly in an open area, as discussed above.

7.2 Assemble the vane assembly and support according to manufacturer's instructions, and place it in the desired position. The working height of the vane assembly should be between 6 and 12 feet above the supporting surface.

7.3 Level the tripod by shimming the feet until the vane stem is vertical, and fasten down securely.

7.4 Rotate the mast until the directional component is in the desired position, and secure the unit in that position.

7.5 Connect the cable from the vane assembly to the control unit, and plug the power cord into the 110 volt AC receptacle.

7.6 Use the test switches according to manufacturer's instructions to check circuit operation.

7.7 Place the sampling unit in the desired location, and connect it to the "ON" sector receptacle on the control unit. The system should now be operational.

8. Calibration

8.1 The calibration will be that associated with the sampling unit.

9. Calculations

9.1 The calculations required will be those associated with the sampling unit.

10. Effect of Storage

Not applicable.

11. References

1. Operating Instructions, Model MDA-1 Wind Direction Control System. 1964. Wong Laboratories, 3357 Madison Rd., Cincinnati, Ohio 45209.

Subcommittee 9
A. L. Linch, *Chairman*
J. V. Harman
E. R. Hendrickson
Morris Katz
G. O. Nelson
J. N. Pattison
A. L. Vander Kolk
R. B. Weidner

506.

Tentative Method of Analysis for Atmospheric Visibility (Integrating Nephelometer Method)

11203-01-73T

1. Principle of the Method

1.1 The scatter of light by suspended particulate matter in the atmosphere decreases the luminance of bright objects and increases the luminance of dark objects when viewed through a polluted atmosphere (**2, 4, 6**). This loss in apparent contrast degrades the atmospheric visibility. The integrating nephelometer determines atmospheric visibility by continuously monitoring the amount of light scattered by the suspended particulates in the atmosphere (**1, 3, 5**).

1.2 The air sample within the viewing volume of the nephelometer is illuminated by a high intensity lamp and the amount of light scattered by the aerosols is measured by a sensitive photon detector. The geometric configuration of the integrating nephelometer is such that the instrument integrates the scatter over a wide range of angles with a specific weighted integral. Proper weighting of the integral requires a light diffusing material mounted in front of the lamp to provide a cosine characteristic to the light source.

2. Range and Sensitivity

2.1 The integrating nephelometer used for measuring atmospheric visibility should be able to measure the scattering coefficient over a range of $0.1 \times 10^{-3}m^{-1}$ to $10 \times 10^{-3}m^{-1}$ corresponding to a visual range of 40 km to 0.4 km. The upper visibility range (low scatter range) of the nephelometer is determined by the adequacy of the instrument's optical and electronic design for eliminating stray light and electronic noise. The lower visibility range (upper scatter range) is defined by the instrument electronics. Nephelometers are available with scatter coefficient sensitivities down to $0.01 \times 10^{-3}m^{-1}$. These nephelometers will measure the Raleigh scatter from particle-free air.

3. Interferences

3.1 The measurement of atmospheric visibility by integrating nephelometer assumes that light attenuation by suspended particulates in the atmosphere is predominately due to light scatter and absorption is negligible. Although this is generally accepted there is no good data for verification.

3.2 Since the integrating nephelometer measures the scatter by the particulates at a single point, a spatially homogeneous aerosol must exist for the measurement to represent the observed visual range.

4. Precision and Accuracy

4.1 Integrating nephelometers are available with r.m.s. precision in actual use of between 3 and 4% of the reading of the scattering coefficient for the normal ambient atmospheric conditions. A slight degradation in precision may occur for readings in extremely clean air, with the exact degree of degradation depending on the care exercised in calibration of the instrument.

4.2 The accuracy of the integrating nephelometer is estimated to be ± 10% of reading of scattering coefficient for normal ambient atmospheric aerosols. The accuracy is expected to be somewhat poorer for aerosols composed primarily of large particles or droplets (*e.g.*, dust storms, fogs) with a maximum expected error of 35%.

4.3 On a time scale of several days the dominant instability is due to electronic drift of the background (zero) adjustment and of the instrument gain. The maximum zero drift to be expected over a period of several days corresponds to a change in scattering coefficient of about $0.1 \times 10^{-4}m^{-1}$, *i.e.*, less than a few per cent of a typical urban aerosol reading and less than 10 to 20% of the reading encountered in clean background locations (desert or arctic air). The gain drift over the same period of time is less than 3% of reading.

Longer term stability is influenced by aging of the lamp and slow accumulation of dirt. These effects are imperceptible over periods of less than 1 to 2 weeks of normal operation in urban air and periodic recalibration against the instrument calibrate reference corrects for them.

5. Apparatus

5.1 The integrating nephelometer may be of the enclosed type that draws an air sample through the instrument or the open type that measures the scattering coefficient of the atmospheric aerosols *in situ*. The enclosed system generally has greater sensitivity and is more suitable for measurement of visibility at low atmospheric aerosol levels. It is not suitable for aerosols composed primarily of large particles, *e.g.*, very wet fogs, for which there is impaction of large drops at points where the flow is required to turn within the instrument. The open integrating nephelometer is more suitable for measurement of visibility at high atmospheric aerosol levels and for fog. Both types are suitable for measurement of visibility at normal aerosol levels.

5.2 The instrument should be capable of integrating the scatter over a range of angles of 10 to 170 degrees. The lamp should illuminate the aerosol with diffuse white light such that the illumination varies as the cosine of the illumination angle. The spectral response of the photodetector should be restricted to the visual range (400 to 700 nm). The range and sensitivity of the instrument should be as described in Section **2** above.

5.3 Auxiliary equipment required for primary calibration are a container of Refrigerant-12 (CCl_2F_2)* with the necessary valve, a glass fiber filter and holder, and tubing.

6. Reagents

6.1 No reagents, as such, are required for the operation of the integrating nephelometer. The only chemical used at all is Refrigerant-12 which is required for the primary calibration of the instrument. Ordinary refrigerant quality cans of the liquefied gas are adequate if the gas is filtered through a glass fiber filter (Gelman, type E or equivalent) to remove any particulate matter that may be present in the gas emanating from the pressurized container.

7. Procedure

7.1 Calibrate the nephelometer as described in Section **8**. For specific operating instructions, refer to the manufacturer's manual.

*Freon-12, Genetron-12, Neon-12, etc., all are dichlorodifluoromethane, CCl_2F_2, referred to hereafter as Refrigerant-12 or CCl_2F_2.

8. Calibration and Standards

8.1 Primary calibration of the integrating nephelometer is obtained by using filtered air and Refrigerant-12 as standards. As a result of such a calibration, the instrument calibrator may be standardized for use as a secondary standard between the infrequent primary calibrations. To perform the absolute or primary calibration the instrument view volume is purged with particle-free air to obtain 0 ($0.23 \times 10^{-4}m^{-1}$), and then purged with particle-free Refrigerant-12 to obtain a span point. The scatter coefficient of CCl_2F_2 is $0.36 \times 10^{-3}m^{-1}$ at sea level, $0.31 \times 10^{-3}m^{-1}$ at 5000 feet and $0.26 \times 10^{-3}m^{-1}$ at 10,000 feet at a wavelength of 470 nm. The effective wavelength of an integrating nephelometer with Xenon flash lamp, S-11 response phototube and UV cutoff filter is 470 nm for gases. Adjust the 0 and span of the instrument to obtain the required readings, and record the instrument reading on the instrument calibrator for use in determining instrument calibrate in the field.

8.2 To calibrate an open view volume integrating nephelometer with gases the instrument must be placed in a container suitable for purging with the gases.

9. Calculations

9.1 None required. Instrument reads directly in scattering coefficient (visual range).

10. Effect of Storage

10.1 Not applicable.

11. References

1. AHLQUIST, N.C., and R.J. CHARLSON. 1967. A new instrument for evaluating the visual quality of the air, *J. Air Pollution Control Association*, 17:467–469.
2. BEUTTEL, R.G., and A.W. BREWER. 1949. Instruments for the measurement of the visual range, *J. Sci. Inst.*, 26:357–359.
3. CHARLSON, R.J., ET AL. 1967. The direct measurement of atmospheric light scattering coefficient for studies of visibility and air pollution, *Atmos. Environ*, 1:469–678.

4. CLAYTON, G.D., and P.M. GIEVER. 1955. Instrumental measurement of visibility in air pollution studies, *Anal. Chem.*, 27:708–713.
5. HORVATH, H., and K.E. NOLL. 1969. The relationship between atmospheric light scattering coefficient and visibility, *Atmos. Environ.*, 3:543–552.
6. MIDDLETON, W.E.K. 1952. *Vision through the atmosphere*. Univ. of Toronto Press.

Subcommittee 10

R. A. HERRICK *Chairman*
S. KINSMAN
J. LODGE
D. LUNDGREN
C. PHILLIPS
W. C. RUTLEDGE
R. S. SHOLTES
E. STEIN
J. J. STUKEL
J. WAGMAN
W. D. CONNER (*Consultant*)

507.

Tentative Method of Analysis for Suspended Particulate Matter in the Atmosphere (The Integrating Nephelometer)

11203-02-76T

1. Principle of the Method

1.1 The integrating nephelometer measures the light scattering of suspended particulate matter in the ambient atmosphere. The air sample is drawn continuously through a typical instrument at a rate of about 5 cfm and the moving airstream within the instrument is observed by an optical detector.

1.2 The air sample within the nephelometer is periodically illuminated by a flashlamp (usually xenon), and the amount of light scattered by the aerosol is measured by a photomultiplier detector. The instrument integrates the scattering over a range of angles (typically 9° to 170°), and thus it effectively measures the extinction coefficient, b, due to scattering. The measurements, being instantaneous, are not influenced by the flow rate of the sample.

1.3 APPLICABILITY. The integrating nephelometer will measure the scattering coefficient for nearly all normal ambient atmospheric aerosols. The only exceptions are those aerosols composed primarily of large particles (about 100 μ, such as very wet fogs) for which there is deposition or impaction in the air flow system.

1.4 If only the solid particulate content is of interest, the inlet air can be heated slightly to vaporize the moisture.

1.5 The scattering coefficient can be related to the visibility (or more precisely the meteorological range) if the particulates do not absorb light to any significant degree. Since this is the usual case in the ambient atmosphere, the integrating nephelometer can be used to determine the local visibility (**1, 2**).

2. Range and Sensitivity

2.1 The integrating nephelometer can be used to measure the scattering coefficient, b, from approximately $0.1 \times 10^{-4}m^{-1}$ to about $100 \times 10^{-4}m^{-1}$. This upper limit is defined by the instrument electronics. It may be increased by an electronic change, but the instrument linearity at the higher levels is not usually guaranteed. The scattering coefficient for particle-free air depends upon the effective wavelength of the instrument, and it decreases with altitude at a rate proportional to the atmospheric density. A typical value for sea level particle free air is $0.23 \times 10^{-4}m^{-1}$. The range above corresponds approximately to the scattering produced by ambient aerosol concentrations of 0 to about 3800 $\mu g/m^3$.

2.2 The integrating nephelometer is sensitive to the Rayleigh scattering from particle-free air, and thus particulate concentrations are limited to 5 $\mu g/m^3$ or greater.

2.3 The sensitivity for visibility is nonlinear, and varies from about 125 miles visibility at the clean air level to about 0.3 miles visibility.

3. Interferences

3.1 The types of interferences present depend on the use to which the integrating nephelometer is applied. For the measurement of the scattering coefficient of the atmosphere, or for the measurement of visibility, there are no interferences.

3.2 If the scattering due solely to the atmospheric aerosol is desired, then the concentration due to gases must be removed. The scattering due to particle free air at sea level is discussed in Section **2**, and this may be subtracted from the overall scattering coefficient to arrive at the scattering coefficient due to the aerosol. The effect of other gases in the ambient atmosphere is negligible.

3.3 Moisture on hygroscopic particles may be an interference at relative humidities about 65 to 70% (**4**), if only the effect of the dry aerosol is of interest. The effect of the water contained in and on the particles can be sufficient to swamp the dry aerosol signal. Slight heating of the inlet air and of the instrument to decrease the relative humidity to less than 65% cures this problem.

4. Precision and Accuracy

4.1 The r.m.s. precision of the integrating nephelometer in actual use is between 3 and 4% of the reading of scattering coefficient for the normal ambient atmospheric conditions (**3**). A slight degradation in precision may occur for readings in extremely clean air, with the exact degree of degradation depending on the care exercised in calibration of the instrument. The precision in visibility and mass concentration is the same as that for scattering coefficient.

4.2 The accuracy of the integrating nephelometer is better than ± 10% of the scattering coefficient reading for the normal ambient atmospheric aerosol (**1**), based on theoretical computation of the scattering and actual experience with the linearity of the instrument. The accuracy is somewhat poorer for aerosols composed primarily of large droplets (dust storms, fogs) with a maximum error of approximately 35%.

5. Apparatus

5.1 The integrating nephelometer consists of three major assemblies, (shown in Figure 507:1) 1) an optical assembly, 2) an electronics and control package, and 3) an air blower and filter unit. Assemblies 1 and 3 are connected to 2 by electrical cables and a flexible hose. Flexible tubing connects 1 and 3. In some units, assemblies 2 and 3 are installed in a single cabinet along with a recorder, and are interconnected at the factory.

5.2 A can of pressurized Freon 12 gas with the necessary valve, a glass fiber filter and holder, and sufficient tubing to connect these to the optical assembly are required for the primary calibration described in Section **8**.

5.3 The heated version of the integrating nephelometer contains an additional heater assembly which is installed on the air inlet to the optical assembly and an insulating jacket which is permanently installed around the optical assembly.

6. Reagents

6.1 No reagents are required for the operation of the integrating nephelometer.

6.2 Freon-12 gas (CCl_2F_2) is required for the primary calibration of the instrument. Ordinary refrigerant quality cans of pressurized gas are adequate if the gas is filtered through a glass fiber filter (Gelman, Type E or equivalent) to remove any particulate matter which may be present in the gas.

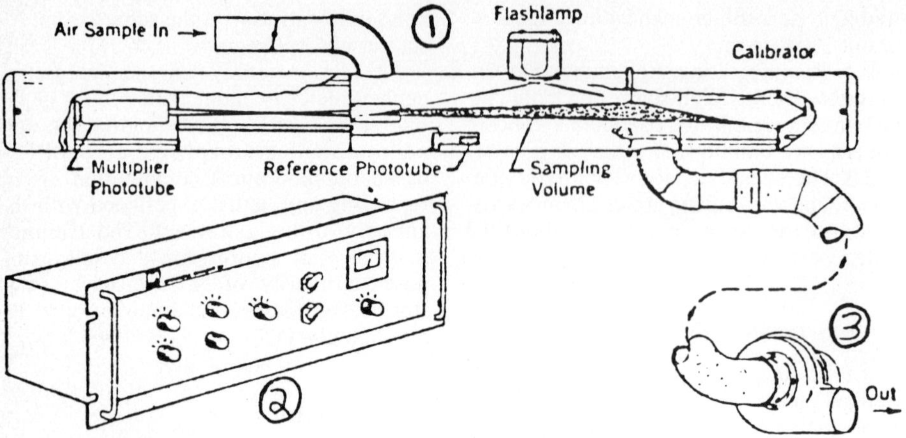

Figure 507:1—The assemblies of an MRI Integrating Nephelometer (1).

7. Procedure

7.1 The integrating nephelometer should be installed according to the manufacturer's instructions. The instrument should first be calibrated as described in Section **8**.

7.2 OPERATIONAL CHECK.

7.2.1 Turn on the electronics and the purge pump, and close the solenoid (or similar device) on the exhaust line.

7.2.2 If a sample heater is installed, verify that it is turned off or unplugged. (It is important that the heater be on only when air is flowing through it. Operation of the heater with no air flow may burn it out.) The flashlamp on the optical assembly should now flash periodically.

7.2.3 Seal the inlet to drafts by disconnecting the inlet hose and stuffing a couple of sheets of tissue (*e.g.* Kimwipes) into the tube. This lets the optical assembly fill with clean air from the purge pump.

7.2.4 Allow the instrument approximately 20 min to warm up, or longer if the manufacturer recommends.

7.2.5 Follow the recommended procedure for removing any background scatter from a dirty instrument to make it read correctly in particle free air.

7.2.6 Check that the internal calibrator is properly adjusted with the reference value. It may be necessary to readjust this and the background setting several times until they both read the proper values. Allow at least one minute for a complete response following each change. If these adjustments cannot be completed satisfactorily, refer to the manual.

7.2.7 Remove the tissue plug and reconnect the inlet hose and clamp it to the inlet of the optical assembly.

7.3 THE INSTRUMENT IS READY FOR OPERATION.

7.3.1 Turn on the blower to bring in the sample air.

7.3.2 If an inlet air sample heater is used, turn it on and let it heat up for at least 15 min.

7.3.3 Set the flashes per sec and time constant settings to recommended settings for stationary monitoring. Mobile operations and special applications may require other settings, which should be listed in the instruction manual.

7.3.4 Set the appropriate scale on the meter, depending upon the estimated amount of pollution in the air.

7.3.5 For measurement of the mass concentration of particulates, the instrument should first be calibrated for the aerosol size distribution occurring at a specific location by correlation with other measuring techniques.

7.3.6 The instrument can now be left on for recording with no further adjustment necessary. Periodically (every 2 to 4 weeks) repeat the above check of the background and gain settings. Every 2 to 3 months, or if the instrument is disassembled, a complete calibration should be performed according to the procedures given in Section **8**.

8. Calibration

8.1 SAMPLING CALIBRATION. This consists of a determination of the degree of particle loss due to deposition on the walls of the sample air hose leading from the sampling point to the instrument. Proper sampling line installation should minimize this error. The loss of particles inside the instrument is negligible for normal aerosols. To calibrate the system for sampling line losses, the readings of the installed integrating nephelometer are compared with those of another integrating nephelometer which is temporarily mounted at the location of the sampling line inlet opening, but which has no sampling line attached to it, and hence has no sampling losses. As a result of such a test, an efficiency factor can be determined for the installed instrument. The readings obtained on the installed instrument can then be corrected by the efficiency factor to obtain the correct values.

8.2 There are two other calibrations which apply to the integrating nephelometer. The first is a primary calibration using filtered air and Freon-12 as standards. As a result of this calibration, the internal optical calibrator can then be calibrated, and this can then be used as a secondary standard between the infrequent primary calibrations.

8.2.1 *Primary Calibration.* Disconnect the air intake and exhaust hoses from the optical assembly, and close the air outlet tube (with a rubber stopper or similar plug). The inlet tube should be closed by stuffing it with a soft cloth or with tissues.

a. Set the flashes per second switch inside the electronics unit to the recommended setting, and follow instructions in the manual for other switches.

b. Disconnect the inlet heater power, if a heater is attached.

c. Allow the instrument to warm up and purge for at least one hour.

d. Calibrate with clean air by setting the flashes per second switch and the time constant to recommended values.

e. After at least a one minute wait, follow instructions for adjustments made in the background scattering settings. The fast time constant recommended will cause a variation in reading with each flash. Strike an average for use in setting the background, keeping the needle excursions between recommended values on the scale.

f. Change the switches to the required calibration settings, and after the meter has stabilized at the new reading, note and record this reference calibration reading for later use.

g. Make other necessary setting changes, and then disconnect the clean-air line at the filter on the blower/filter unit and connect it to the extra filter unit provided. Connect the other end of this filter to the Freon-12 container with suitable tubing.

h. Open the Freon valve and purge the optical chamber for at least 10 to 12 minutes, or until the meter reading becomes constant. Record the meter reading for Freon for future reference.

i. Turn off the Freon flow and reconnect the clean air line to the filter on the blower/filter unit. Allow at least 20 min for purging, and then recheck the clean air point. Change the switches back to the required calibration settings, and recheck the reading.

j. Determine the correct calibration value from calculations listed in the instruction manual.

8.2.2 *Secondary calibration.* Use the correct calibrator value determined above as instructed to make the desired setting adjustments.

a. Recheck the background scattering setting and reset for the proper clean air value (at that altitude) if necessary. Repeat these last two adjustments until the instrument reads correctly on both settings. The instrument is now calibrated.

9. Calculations

9.1 The integrating nephelometer measures the scattering coefficient, b, directly. The effective wavelength, λ, of measurement is about 500 nm for aerosols. The variation of b with wavelength is proportional to $\lambda^{-\alpha}$ for aerosols, with α having a value near 2. This means that the aerosol scattering coefficient at the effective wavelength of the eye (about 550 nm) is $b_{550} \approx b_{500}/1.21$ where b_{500} is the nephelometer measurement. It can be shown that the visibility, L_v, can be expressed in terms of b_{550} by $L_v = \text{Constant}/b_{550}$ (**1**). The constant depends on the contrast definition of the eye, and is 3.9 for optimal viewing of ideal black targets against a white background (**2**). Considering the ideal case as an upper visibility limit and correcting for the effective wavelength of the nephelometer gives

$$L_v\ (\text{m}) = \frac{4.7}{b_{500}\ (\text{m}^{-1})}$$

This visibility, converted to miles, may be displayed on the scales of the instrument.

10. Effect of Storage.

10.1 Not applicable.

11. References

1. Charlson, R.J., N.C. Ahlquist, H. Selvidge, and P.B. MacCready. 1969. Monitoring of Atmospheric Aerosol Parameters with the Integrating Nephelometer, Air Poll. Cont. Assoc. J. 19:937–942.
2. Charlson, R.J. 1969. Atmospheric Visibility Related to Aerosol Mass Concentration. A Review. Envir. Sci. and Tech. 3:913–918.
3. Tombach, I.H. Procedure for Particulate Measurement with the Integrating Nephelometer, Meteorology Research Inc., 464 W. Woodbury Rd., Altadena, Calif. 91001, 1971.
4. Tombach, I.H. Measurement of Some Optical Properties of Air Pollution with the Integrating Nephelometer, Joint Conf. on Sensing of Envir. Poll., Palo Alto, Calif., 1971.

Subcommittee 9
A. L. Linch, *Chairman*
J. V. Harman
E. R. Hendrickson
Morris Katz
G. O. Nelson
J. N. Pattison
A. L. Vander Kolk
R. B. Weidner

508.

Tentative Method for Measuring In-Stack Opacity of Visible Emissions by the Transmissometer Technique

112-01-76T

1. Principle and Applicability

1.1 A transmissometer provides a direct measure of the relative amount of light transmitted through or attenuated by visible emissions in a stack or duct. Visible emissions would include steam. Light from a lamp is projected across the stack of a pollution source to a light sensor. The light is attenuated due to absorption and scatter by the visible matter in the effluent. The percentage attenuated is the opacity of the in-stack pollutant plume. Emissions which do not attenuate any light will be invisible and will have a transmittance of 100% and an opacity of 0. Emissions which attenuate all of the light will have a transmittance of 0 and an opacity of 100%.

1.2 A transmissometer can also provide

a measure of the mass concentration of the particulates in a plume if the size, shape and composition of the particulates are sufficiently invariant. Empirical relations have been observed between the mass concentration and transmittance of particulates in many sources (**1, 2, 3, 4**).

1.3 A transmissometer is also capable of monitoring visible plume opacities of sources where in-stack and out-of-stack plume particulate characteristics are similar by meeting specifications of installation and output readout (**5, 6**).

1.4 Two important optical characteristics of transmissometers must be specified to obtain similar performance from instruments. These have been identified as the operating wavelength and light collimation of the instruments. The operating wavelength is important because fine particulates attenuate shorter wavelength radiation more than longer wavelengths (**6, 7**). Collimation requires an optical design that limits the light viewing and projection angles of the transmissometer. No restriction on the viewing and projection angles results in instruments with poor sensitivity and nonstandard performance.

1.5 Specifications for continuous measurement of visible emissions are given in terms of optical instrumentation design, (sec **5.2** & **5.3**) measurement of visible performance, (sec **5.4** & **5.5**) and installation parameters (sec **5.6**).

2. Range and Sensitivity

2.1 The transmissometer output should be linear with respect to opacity and provide a display of the in-stack opacity on a standard of 0 to 100% linear scale.

2.2 Individual measurements should be distinguishable to at least a difference of 2% transmittance.

3. Interferences

3.1 Interference due to the absorption and emission of radiation by hot gases can occur but is at a minimum in the visible region of the spectrum. Proper selection of the operating wavelength of the transmissometer will minimize such interference due to absorption and radiation from gases (CO_2, H_2O, SO_2, etc.) in the emissions.

3.2 Light scattered by particulate emissions or stray light which reaches the viewing end of the transmissometer is also considered an interference. Limitation of the viewing and projection angles will minimize this interference. The light collimation is very important because measurement of true transmittance and opacity of aerosols requires the exclusion from the measurement of light scattered by the aerosols (**5, 8**).

4. Precision and Accuracy

4.1 Accuracy of the method with prescribed specifications is ±5% of span maximum.

4.2 Repeatability of readings should be obtainable at 2%.

5. Apparatus

5.1 A transmissometer schematic is depicted in Figure 508:1.

5.2 A definition of terms with respect to optical instrumentation includes the following:

5.2.1 *Spectral Response*. The relative response of a transmissometer to radiation of different wavelengths.

5.2.2 *Mean Spectral Response*. The wavelength which bisects the total area under the curve obtained pursuant to paragraph **8.1.1**.

5.2.3 *Angle of View*. The maximum (total) angle of radiation that is seen by the photodetector assembly of an optical transmissometer.

5.2.4 *Angle of Projection*. The maximum (total) angle of radiation that is projected by the lamp assembly of an optical transmissometer.

5.2.5 *Pathlength*. The depth of effluent in the light beam between the receiver and the transmitter of the single-pass transmissometer, or the depth of effluent between the transceiver and reflector of a double-pass transmissometer.

5.3 The opacity measurement system

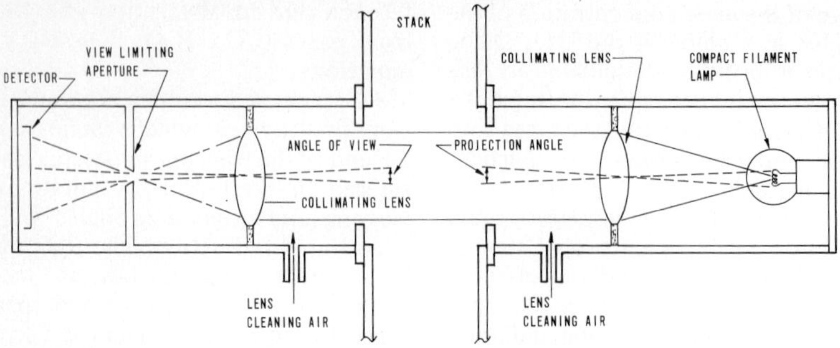

Figure 508:1—Schematic of a transmissometer with collimating optics.

shall conform to the design specifications set forth as follows:

5.3.1 *Peak spectral response.* Response at any wavelength below 400 nm or above 700 nm shall be less than 10% of the peak response to the measurement system at all wavelengths between 400 nm and 700 nm.

5.3.2 *Mean spectral response.* The mean spectral response of the measurement system shall occur between 500 nm and 600 nm.

5.3.3 *Angle of view.* The total angle of view shall be no greater than 5 degrees.

5.3.4 *Angle of projection.* The total angle of projection shall be no greater than 5 degrees.

5.4 *System Performance Definitions.* Definitions of terms with respect to system performance specifications includes the following:

5.4.1 *Measurement System.* The total equipment required for the continuous determination of a pollutant concentration in a source effluent. The system consists of three major subsystems:

a. Sampling Interface. That portion of the measurement system that performs one or more of the following operations: delineation, acquisition, transportation, and conditioning of a sample of the source effluent, or protection of the analyzer from the effluent.

b. Analyzer. That portion of the system which senses the pollutant and generates a signal output that is a function of the pollutant concentration.

c. Data Recorder. That portion of the measurement system that processes the analyzer output and provides a permanent record of the output signal in terms of pollutant concentration.

5.4.2 *Span.* The value of opacity at which the measurement system is set to produce the maximum data display output. For the purpose of this method, the span shall be set at an opacity of 50% for a pathlength equal to the stack exit diameter of the source.

5.4.3 *Calibration Error.* The difference between the opacity reading indicated by the measurement system and the known values of a series of test standards. For this method the test standards are a series of calibrated neutral density filters.

5.4.4 *Zero Drift.* The change in measurement system output over a stated period of time of normal continuous operation when the pollutant concentration at the time of the measurements is zero.

5.4.5 *Calibration Drift.* The change in measurement system output over a stated period of time of normal continuous operation when the pollutant concentration at the time of the measurements is the same known upscale value.

5.4.6 *System Response.* The time interval from a step change in opacity in the stack at the input to the measurement system to the time at which 95% of the corresponding final value is reached as displayed on the measurement system data presentation device.

5.4.7 *Operational Test Period.* A minimum period of time over which a measurement system is expected to operate within certain performance specifications without unscheduled maintenance, repair, or adjustment. This period will be 168 hr unless specified otherwise.

5.5 SYSTEM PERFORMANCE SPECIFICATIONS.

5.5.1 *Accuracy (calibrate filters): ±2% of span* maximum.*

5.5.2 *Zero drift (24 hr): ±2% of span maximum.*

5.5.3 *Span drift (24 hr): ±2% of span maximum.*

5.5.4 *Response time: 1 to 10 sec (variable).*

5.5.5 *Alignment drift: ±2% of span maximum.*

5.5.6 *Soiling drift: ±2% of span maximum.*

5.5.7 *Temperature range: −20 to +120 F minimum.*

5.5.8 *Voltage range: 115V ±2% minimum.*

5.5.9 *Operational period: 168 hr.*

5.6 INSTALLATION SPECIFICATIONS.

5.6.1 *Location.* The transmissometer must be located across a section of duct or stack that will provide a visible emission flow through the optical volume of the transmissometer that is representative of the particulate matter flow through the duct or stack. It is recommended that the pathlength or depth of effluent for the transmissometer include the entire width or diameter of the duct or stack. Installations using a shorter pathlength must use extra caution in determining the measurement location representative of the particulate matter flow through the duct or stack.

5.6.2 The transmissometer location shall be downstream from all particulate control equipment.

5.6.3 The transmissioner shall be located as far from bends and obstructions as practical.

5.6.4 A transmissometer that is located in the duct or stack following a bend shall be installed in the plane defined by the bend where possible.

*100% opacity

5.6.5 The transmissometer should be installed in an accessible location.

5.6.6 *Slotted Tube.* Installations that require the use of a slotted tube shall use a slotted tube of sufficient size and blackness so as not to interfere with the free flow of effluent through the entire optical volume of the transmissometer or reflect light into the light detector of the transmissometer.

6. Calibration Equipment.

6.1 NEUTRAL DENSITY FILTERS. Filters with neutral spectral characteristics and known optical densities to visible light. One each low, mid, and high range filters, 5- or 6″ sq or 6″ diam are required. Calibrated filters with accuracies certified by the manufacturer to within 3% shall be used. It is recommended that filter calibrations be checked with a well-collimated photopic transmissometer of known linearity prior to use.

6.2 CHART RECORDER. Analog chart recorder with input voltage range and performance characteristics compatible with the measurement system output.

6.3 OPACITY MEASUREMENT SYSTEM. An in-stack transmissometer (folded or single path) with the optical design specifications designated below, associated control units and apparatus to keep optical surfaces clean.

7. Procedure

7.1 For specific operating instructions, refer to the manufacturer's manual.

7.2 Direct readout of % transmittance by in-stack measurement technique should be normalized to obtain measurement for unit pathlength of 1 m using equations in Section **9**.

8. Specification Test Procedure

8.1 OPTICAL DESIGN. The general test procedures to be followed to demonstrate conformance with requirements are as follows:

8.1.1 *Spectral Response.* Obtain spectral data for detector, lamp, and filter components used in the measurement system from their respective manufacturers.

8.1.2 *Angle of View.* Set the receiver up as specified by the manufacturer. Draw an arc with radius of 3 m. Measure the receiver response to a small (less than 3 cm) non-directional light source at 5-cm intervals on the arc for 26 cm on either side of the detector centerline. Repeat the test in the vertical direction.

8.1.3 *Angle of Projection.* Set the projector up as specified by the manufacturer. Draw an arc with radius of 3 m. Using a small photoelectric light detector (less than 3 cm), measure the light intensity at 5-cm intervals on the arc for 26 cm on either side of the light source centerline of projection. Repeat the test in the vertical direction.

8.2 SYSTEM CALIBRATION.

8.2.1 *Calibration Error and Response Time Test.* Set up and calibrate the measurement system as specified by the manufacturer's written instructions prior to installation on the stack using the instrument's maximum pathlength or a 3-m pathlength, whichever is less. Span the instrument for 0 to 100% opacity.

8.2.2 *Calibration Error Test.* Insert a series of neutral density filter standards in the transmissometer path at the midpoint. A minimum of 3 neutral-density filters with opacities (low, mid, and high range) calibrated within 3% must be used. Make a total of 5 non-consecutive readings for each filter. Record the measurement system output readings in % opacity. See Figure 508:2.

8.2.3 *System Response Test.* Insert the filter in the transmissometer path 5 times and record the time required for the system to respond to 95% of final zero and span values. See Figure 508:3.

8.3 SYSTEM FIELD TEST.

8.3.1 *Field Test for Zero Drift and Calibration Drift.* Install on the stack and operate the measurement system in accordance with the manufacturer's written instructions and drawings as follows:

a. Conditioning Period. Offset the zero setting at least 10% of span so that negative zero drift can be quantified. Operate the system for an initial 168 hr conditioning period in a normal operational manner.

b. Operational Test Period. Operate the system for an additional 168-hr period. The system shall monitor the source effluent at all times except when being zeroed or calibrated. At 24-hr intervals the zero and span shall be checked according to the

PROPOSED RULES

Calibrated Neutral Density Filter Data
(See paragraph 8.1.1)

Low Range ___% opacity Mid-Range ___% opacity High Range ___% opacity

Run #	Calibrated Filter[1]	Analyzer Reading % Opacity	Differences[2] % Opacity
1			
2			
3			
4			
5			
6			
7			
8			
9			
10			
11			
12			
13			
14			
15			

	Low	Mid	High
Mean difference	___	___	___
Confidence interval	± ___	± ___	± ___
Calibrated error = $\frac{\text{Mean Difference}^3 + \text{C.I.}}{\text{Calibrated Filter Value}} \times 100$	___	___	___

[1]Low, mid or high
[2]Calibration filter opacity - analyzer reading
[3]Absolute value

Calibration Error Test

Figure 508:2—Calibrated Error Test Data

Date of Test ______________

Span Filter ______________ % Opacity

Analyzer Span Setting ______________ % Opacity

1 ______________ seconds
2 ______________ seconds
3 ______________ seconds

1 ______________ seconds
Downscale 2 ______________ seconds
3 ______________ seconds

Average response ______________ seconds

Figure 508:3—System Response Test Data

applicable manufacturer's written procedures. These procedures may vary from manufacturer to manufacturer, but shall include a procedure for producing a simulated zero opacity condition, and a simulated upscale (span) opacity condition as viewed by the receiver. After the zero and span check, clean all optical surfaces open to the effluent, realign optics, and make any necessary adjustments to the calibration of the system. These zero and calibration corrections and adjustments are allowed only at 24-hr intervals or at such shorter intervals as manufacturer's written instructions specify. Automatic corrections made by the measurement system without operator intervention are allowable at any time. During this 168-hr operational test period, record the following at 24-hr intervals: (a) the zero reading and span readings after the system is calibrated (these readings should be set at the same value at the beginning of each 24-hr period); (b) the zero reading after each 24 hr of operation, but before cleaning and adjustment; and (c) the span reading after cleaning and zero adjustment, but before span adjustment. (See Figure 508:4.)

Zero Setting _______________ (See paragraph 8.2.1)

Span Setting _______________

Date and Time	Zero Reading (Before cleaning and Adjustment)	Zero Drift (ΔZero)	Span Reading (After cleaning and zero adjustment but before span adjustment)	Calibration Drift (ΔSpan)

Zero Drift = [Mean Zero Drift * _______ + CI (Zero) _______

Emission Standard] x 100 + _________

Calibration Drift + [Mean Span Drift * _______ + CI(Span) _______

Emission Standard] x 100 + _________.

*Absolute Value

9. Calculations

9.1 For calculation purposes, the following definitions are included:

9.1.1 *Transmittance.* The fraction of incident light that is transmitted through an optical medium of interest.

9.1.2 *Opacity.* The fraction of incident light that is attenuated by an optical medium of interest. Opacity (O) and transmittance (T) are related as follows:

$$O = 1 - T$$

9.1.3 *Optical Density.* A logarithmic measure of the amount of light that is attenuated by an optical medium of interest. Optical density (D) is related to the transmittance and opacity as follows:

$$D = \log_{10} T$$

$$D = -\log_{10} (1 - O)$$

9.1.4 *Plume opacity.* Measured opacity readings can be normalized to unit pathlength in-stack opacity readings by use of the equation:

$$\log_{10}(1 - O_1) = (\ell_1/\ell_2) \log_{10}(1 - O_2)$$

where

O_1 = plume opacity (calculated)
O_2 = in-stack opacity (measured/fixed)
ℓ_1 = pathlength of plume diameter (chosen)
ℓ_2 = in-stack transmissometer pathlength (measured/Fixed)

This same equation can be utilized to relate in-stack opacity to out-of-stack opacity which would normally be observed visually by EPA Method 9 entitled "Visual Determination of the Opacity of Emissions or Stationary Sources (**9**)". This same reference contains the proposed rulemaking for "Performance Specifications and Specification Test Procedures for Transmissometer Systems for Continuous Measurement of the Opacity of Stack Effluents".

10. Effect of Storage

10.1 Not applicable

11. References

1. BUHNE, K. 1971. Investigations into the Directional Dependence Photoelectric Smoke Density Measuring Instruments, Staub-Reinhalt, Luft Vol. 31, No. 7, July.
2. DUWELL, L. 1968. Latest State of Development of Control Instruments, Instruments for the Continuous Monitoring of Dust Emissions, Staub-Reinhalt, Luft Vol. 28, No. 3, March.
3. HURLEY, T.F. and P.L. BAILEY. 1958. The Correlation of Optical Density with the Concentration and Composition of the Smoke Emitted from a Lancashire Boiler, *J. Inst. Fuel*, 31, 534–540.
4. ENSOR, D.S. and M.J. PILAT. 1972. Relationship of Plume Opacity to the Properties of Particulates Emitted from Kraft Recovery Furnaces, *Tappi*, 55, 88–92.
5. PETERSON, C.M. and M. TOMAIDES. 1972. In-Stack Transmittance Techniques for Measuring Opacities of Particulate Emissions from Stationary Sources, Final Report, Contract 68-02-0309. Environmental Protection Agency.
6. CONNER, W.D. and J.R. HODKINSON. 1972. Optical Properties and Visual Effects of Smoke-Stack Plumes, U.S. Environmental Protection Agency, Publication No. AP-30.
7. HODKINSON, J.R. 1966. The Optical Measurement of Aerosols, Chapter 10, Aerosol Science, C.N. Davies (editor), Academic Press, New York.
8. ENSOR, D.S. and M.J. PILAT. 1971. The Effect of Particle Size Distribution on Light Transmittance Measurement, *Am. Ind. Hyg. Association J.*, 32, 287–292.
9. FEDERAL REGISTER. 1974. Vol. 39, No. 177, General Services Administration, Washington, D.C., September 11.

Subcommittee 11
M. DEAN HIGH, *Chairman*
JEROME FLESCH
RICHARD GERSTLE
WILLIAM T. INGRAM
J. B. KOOGLER
L. R. PERKINS
C. J. THEOPHIL

509.

Tentative Method of Analysis for Carbonate and Non-Carbonate Carbon in Atmospheric Particulate Matter*

12116-01-72T

1. Principle of the Method

1.1 Carbonate particulate matter is determined by measuring the CO_2 released by acidification in a system closed to the atmosphere. Non-carbonate carbon is determined by subtracting the carbonate CO_2 from the total CO_2 produced by complete combustion in an oxygen atmosphere. Although it is possible to determine the two forms of carbon sequentially on the same sample, it is many times more efficient to divide each sample and perform the two analyses separately in batches.

1.2 Air particulate samples are collected on glass fiber filters in a high-volume air

Warning. The reaction of strong acids on reactive metals with the release of hydrogen gas should never be done in a closed glass system. Each g of aluminum can release 1.33 liters of H_2 at 20 C. Venting of the H_2 should also be done to a fume hood to prevent dangerous concentrations of H_2 near laboratory ceilings.

*Adapted from "Carbonate and Non-Carbonate Carbon in Atmospheric Particles" (7). Draft of method submitted by T. Belsky, Air and Industrial Hygiene Lab., California Dept. of Public Health, 2151 Berkeley Way, Berkeley, California.

sampler or on aluminum or other metallic foil in an impactor type (**5**) air sampler. Particulate matter in the latter samples are fractionated into several aerodynamic size ranges depending on the number of impactor stages. This method was designed (**7**) to handle the very small amounts of material on the foils from each stage. Portions of a foil or filter are cut into thin strips and placed in a porcelain boat. Acid is added to the boat inside a tube under a flow of helium and the CO_2 produced is collected in a freeze trap at liquid nitrogen (−196 C) or liquid argon (−186 C) temperature. Water is trapped out in a dry ice trap (−78 C) upstream from the colder trap. Hydrogen gas produced from the reaction of acid on aluminum is not captured and is vented along with the helium flow. Copper foil may be used instead of aluminum to eliminate the generation of hydrogen gas. The CO_2 is subsequently measured by injection (through a gas valve) into a gas chromatograph previously calibrated for CO_2. The remaining portion of the sample in another porcelain boat is oxidized in a combustion tube to yield the total carbon as CO_2 that is also measured by the gas chromatograph.

1.3 Total carbon and carbonate analysis is very old and is much used, as the extensive literature shows. Every conceivable carbon-containing material has probably been examined and many modifications of the basic techniques are used to analyze specific samples. Carbonate carbon is always released by acid and the non-carbonate fraction is released by oxidation, either by wet or dry reagents or by combustion. Many variations are also found in the method of measuring the CO_2 produced. Thomas and Hieftje (**9**) list several methods with references in their paper on carbonate-containing samples. The methods described here have features in common with those of Miles (**6**), Lee and Lewis (**4**), Van Hall and Stenger (**10**), Dugan and Aluise (**1**), and others.

1.4 The carbon content of suspended particulate matter varies widely depending upon sources, seasons, location, etc. The chemical form of this carbon also varies widely. This method provides information on the distribution of the above two forms of carbon.

2. Range and Sensitivity

2.1 The lower limit of the range of this method is determined by the amount of carbon in the blank runs, the smallest amount of CO_2 that can be measured, and the maximum amount of foil or filter that can be handled in the analysis. Purification of gases and reagents, and careful handling of samples and apparatus will yield low blanks. Miles (**6**) lists many exhaustive steps to achieve low background levels. Mueller *et al* (**7**) had acidification and oxidation blanks of about 0.16 μg C/cm^2 and 0.8 μg C/cm^2, respectively for aluminum foil of about 60 cm^2. The glass fiber filters gave blanks about three times larger than the aluminum foils and probably reflects some residual resin used in its manufacture even though it had been flash-fired.

2.2 One μg of CO_2 (0.27 μg C) can routinely be measured on a gas chromatograph with a thermal conductivity detector. A microthermistor or helium ionization detector can do much better, but is not useful unless the blank runs are very low. Mueller *et al* (**7**) report a lower limit of 0.056 μg C/m^3 for a 17.8 m^3 sample and 0.021 μg C/m^3 for a 4.75 m^3 sample from a Lundgren impactor.

3. Interferences

3.1 One of the advantages of using a gas chromatograph as part of the detection apparatus is that interfering gases can be separated from the CO_2 peak. The use of cryogenic trapping also removes many possible interferents such as H_2O and NO_2 at −78 C and H_2, O_2, N_2, CO, CH_4, C_2H_6, and C_2H_4 by nontrapping at −172 C with He or O_2 flushing. Acid gases such as HCl and HBr, and H_2S (from sulfides) are the most likely interferents produced by acidification of the samples. NO, NO_2, and SO_2 may be produced in the oxidation step and will be trapped along with CO_2. All can be separated from CO_2 on the GC column. Alternatively, these interferents can

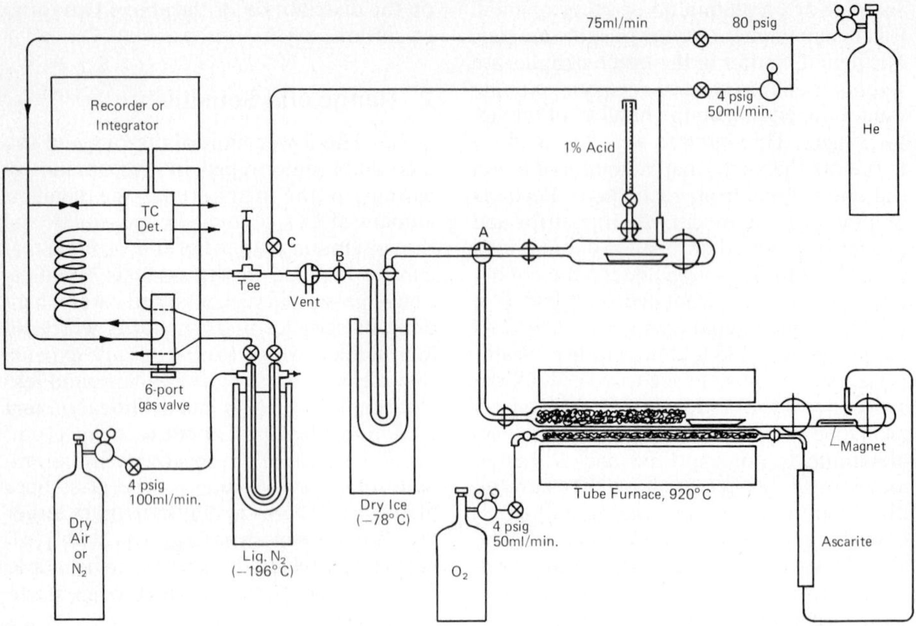

Figure 509:1—Apparatus for analysis of carbonate and non-carbonate carbon.

be removed by additions to the apparatus (see Thomas and Hieftje **(9)** and Kaplan *et al* **(3)**).

4. Precision and Accuracy

4.1 The relative error for CO_2 recovery in the freeze-trap is about 1.3%; for CO_2 release from carbonate, about 3.3%; for the combustion efficiency of carbon-containing materials, 2.1 to 5.6% with an average of 3.3%; and for the manual measurement of peak area on the chromatogram, about 3%. The relative error of the last manipulation can be decreased to about 1.5% with a Disc integrator, and to about 0.5% with an electronic integrator. The overall relative error for the determination of the two main forms of carbon is about ± 8%. The final calculation presented as $\mu g\ C/m^3$ of air will have a relative error of about ± 10% because of the added uncertainties of measuring the total volume of air sampled.

4.2 The efficiency of CO_2 recovery is about 95%. The efficiency of CO_2 release from calcium carbonate is about 82%. The efficiency of combustion is about 99.7 % for graphite; about 106% for potassium acid phthalate; and about 101% for lactose. The latter two figures may represent incomplete drying of the test samples.

5. Apparatus

5.1 The arrangement of the following apparatus is shown in Figure 509:1.

5.1.1 *Gas Supply and Purification.* Flow regulation for the gases is provided by two-stage pressure regulators in series with Nupro needle valves. The purification train consists of an 8-mm ID × 210-mm combustion tube maintained at 920 ± 5 C in a tube furnace and contains copper oxide catalyst for combusting organic vapors in the oxygen. This combustion tube may be placed alongside the sample combustion tube in one furnace similar to the arrangement of Van Hall and Stenger **(10)** so that only one tube furnace is needed. A 120-ml tube filled with Ascarite is connected to the downstream end of the oxygen purification furnace tube to absorb carbon dioxide. The helium line usually can be connected to the acidification tube without purification unless the blank runs show unacceptably high levels of CO_2 or interferents.

5.1.2 *Sample Acidification.* A 10.0 ml buret with a Teflon stopcock and dry glass ball joint near the delivery tip is clamped to a 13-mm ID × 130-mm pyrex tube constructed with ball joints as shown in Figure 509:1. Helium flows into this tube through a smaller side tube. This side tube should protrude through the wall of the larger tube and should be bent toward the open end for efficient flushing of air entrained during sample introduction. One end of the large tube is fitted with a glass ball joint (sample port) and a removable pyrex cap for inserting and removing the sample combustion boat. The other end is connected with a ball joint to the 3-way stopcock that selects the gas stream from either the acidification or the combustion tube.

5.1.3 *Sample Combustion.* A 120-mm long section of the 13-mm ID × 210-mm quartz combustion tube is packed with the copper oxide catalyst. The 90-mm section of the tube adjacent to and upstream of the copper oxide section contains the sample boat during combustion. Oxygen flows through this combustion tube which is maintained at 920 ± 5 C by a tube furnace. A similar 13-mm ID × 130-mm pyrex tube as described in **5.2** is attached to the sample end of the quartz tube. The purified oxygen stream enters the combustion system through this adjacent tube. According to Miles (**6**) the quartz tube should be specified as free-drawn rather than formed on carbon rods to minimize the background carbon contamination. The downstream end of the quartz tube is joined to the 3-way stopcock with a ball joint. All connections are dry; no grease or organic material is used for sealing.

5.1.4 *Water Trap.* A U-tube freeze-trap of about 13 mm OD × 300 mm extended length is placed between—and joined to—the 3-way stopcocks with ball joints as shown in Figure 509:1. The type and size of the freeze-trap was chosen to minimize both the flushing time and the possibility of occluding the gas flow due to ice formation in the trap. The ball joints allow for easy removal of accumulated water which is on the order of one ml/8 hr day. The freeze-trap is cooled by a dry ice-methanol bath at −78 C in a Dewar flask.

5.1.5 *CO_2 Trap.* The freeze-trap for CO_2 is a U-tube made from 3.18 mm OD × 1.65 mm ID type 304 stainless steel tubing about 55 cm long. The volume of this tube is about 1 ml. The lower 80 mm is packed with Chromosorb W of about 45/60 mesh size. A length of 6.35 mm ID polyvinyl chloride tubing is placed around the steel tube to maintain the trap temperature above the boiling point of oxygen (−183 C). Dry nitrogen is passed through the plastic tubing at 100 ml/min. Most of the freeze-trap is submerged in liquid nitrogen (−196 C) during the trapping stage. Two quick-acting toggle valves are used to isolate the CO_2 which is volatilized with a hot water bath before injection into the chromatograph. The CO_2 trap is connected to the chromatographic column through a 6-port gas sampling valve.

5.6 CO_2 MEASUREMENT. Carbon dioxide is separated from other compounds in a gas chromatograph on a 2.4 m × 1.65 mm ID stainless steel, Poropak Q column at 40 C and 75 ml/min helium flow rate. Detection is accomplished by a conventional hot wire, thermal conductivity cell held at 55 C and at a current of 150 ma, the resulting signal being applied to a 1 mV potentiometric recorder or other signal presentation device.

6. Reagents

6.1 HELIUM 99.995% MINIMUM.

6.2 AIR (OR NITROGEN) DRY. Less than 100 ppm H_2O, to prevent obstruction of the gas flow around the CO_2 trap.

6.3 OXYGEN 99.6% MINIMUM. Trace CO_2 and organic content is removed by passage through a combustion tube and absorber.

6.4 CARBON DIOXIDE 99.9%.

6.5 ASCARITE 20/30 MESH.

6.6 Chromosorb W 45/60 mesh, acid washed.

6.7 Poropak Q 100/120 mesh.

6.8 Quartz Wool.

6.9 Phosphoric Acid 1%. Diluted from 85% acid, ACS reagent grade. This solution should be degassed by boiling or flushing with N_2 to remove dissolved CO_2.

6.10 Nitrogen liquid.

6.11 Copper Turnings light, purified, ACS reagent grade. The copper oxide catalyst is prepared by packing the quartz combustion tube with the copper turnings held in place with a quartz wool plug. A 50% oxygen—50% helium gas mixture is passed through the packed tube at 50 ml/min for 15 hr at 600 ± 5 C followed by 5 hr of 100% oxygen at the same flow rate and temperature.

6.12 Calcium Carbonate ACS reagent grade.

6.13 Graphite spectographic grade.

6.14 Potassium Acid Phthalate 99.92%.

6.15 Lactose ACS reagent grade. The potassium acid phthalate and the lactose are used solely to test the efficiency of combustion and the trapping of CO_2. Other organic compounds of high purity may be substituted. Graphite is unique in its resistance to oxidation at elevated temperatures.

6.16 Distilled or de-ionized water. Free of organic matter.

7. Procedure

7.1 Sampling. Atmospheric particulate matter is usually collected on flash-fired glass fiber filters in a high-volume air sampler (**2**), or on metallic film (aluminum, for economy) in an impactor air sampler (**5**). The sample substrate material should be washed in a reagent grade solvent mixture such as benzene-methanol (3:1) to remove polar and non-polar organic contamination and carefully dried. Subsequent handling should be done with clean forceps or non-contaminating gloves since appreciable amounts of organic matter can be obtained from bare hands. Exposure to dust fall should be kept to a minimum and the samples should be kept in glass containers or in smooth, non-fragmenting plastic containers such as polystyrene. The cutting of the samples into thin strips for placement in the boats may be done with a clean pair of scissors or a clean razor blade on a sheet of Teflon or aluminum. A "clean bench" with filtered, laminar airflow is a useful adjunct in the handling of the samples since sub-millimeter sized organic particles may contain 50 to 100 μg of carbon.

7.1.1 A series of samples sufficient for a day's run is prepared by cutting half of each sample into 5 mm × 25 mm strips and placing them into porcelain boats to be used for either the carbonate or non-carbonate analysis. The boats are kept in their respective containers until analyzed. If the amount of material on the foils or filters is too small, the whole sample may be cut into strips and analyzed first for the carbonate fraction and then saved for the non-carbonate fraction. In this latter case, it is necessary to place the boats containing phosphoric acid into a closed container at about 120 C in an inert atmosphere to evaporate off the water.

7.2 Carbonate Carbon. The gas chromatograph (GC) is placed into normal operation, the dry ice bath is added to the water freeze-trap, and the helium pressure in the acidification system is set at 4 psig and the flow rate at 50 ml/min. The buret is filled with 1% phosphoric acid. The 3-way stopcock, **A**, is turned to the carbonate mode and stopcock **B** is turned to the vent position as shown in Figure 509:1. Toggle valve **C** is left open to allow helium to flow through the U-tube. The sample boat is added to the acidification tube and helium is allowed to flush out any residual CO_2 for 10 min. If the U-tube CO_2 trap has been flushed out with helium for about 5 min during the setup of the GC, then liquid nitrogen is used to immerse most of the U-tube during the 10 min flushing of the reaction tube. No dry air or nitrogen flow is needed in the plastic jacket in this phase of analysis, and the open end should be plugged to prevent freeze-out of ambient water vapor.

7.2.1 Stopcock **B** is turned to allow the effluent of the reaction tube to enter the CO_2 trap; valve **C** is closed. About 1 ml of acid is added to the boat and 15 to 30 min is allowed for the release of CO_2 and its subsequent capture in the freeze-trap. The reaction time is short for the very small-sized airborne particles if the acid wets the foil or filter surface. Some of the surfaces may be occluded by the layering of the thin strips in the boat, and the longer time may be required.

7.2.2 After it is decided that the reaction is complete, open valve **C** and turn stopcock **B** to the vent position (hydrogen is still evolving if aluminum foil is being used). Close both toggle valves on the CO_2 U-tube, remove the liquid nitrogen Dewar flask, and replace it with another Dewar flask containing hot water to volatilize the trapped CO_2. After 3 min, push in the 6-port gas sampling valve and flip open the two toggle valves on the U-tube. This places the tube contents in-stream onto the GC column where the CO_2 is separated from residual H_2O and acid gases. The output of the GC is displayed on the 1 mV recorder. If an electronic integrator is not used to receive the GC signal, the attenuation is determined empirically. Usually, the range of CO_2 produced is similar in related samples.

7.2.3 The boat with the reacted sample is removed and its contents are either discarded or saved for further analyses. The 6-port gas valve is pulled out to its original position which allows helium to flow in the GC column independent of the sample U-tube. The carbonate system is now ready for the next sample boat and the procedure is repeated from **7.2**.

7.3 Non-Carbonate Carbon. The time required for most commercial tube furnaces to reach a stable temperature of 920 C is about 2 hr. If time is to be saved it may be preferable to leave the furnace at 920 C continuously, or at an intermediate temperature overnight. As an alternative, there are tube furnaces available commercially that use tungsten lamps and parabolic reflectors to reach high temperatures in a few min. These furnaces also cool down very rapidly since they are of low mass and also have air or water cooling.

7.3.1 After the GC is operative, the furnace is set at 920 C, the helium and oxygen pressure regulators are set at 4 psig and each flow rate is set at 50 ml/min, and the dry ice bath is installed on the water trap. Stopcock **A** is set in the combustion mode and stopcock **B** is turned to the vent position. Toggle valve **C** is left open to allow helium to flow through the U-tube. The sample boat is placed in the pyrex tube adjacent to the quartz combustion tube. A glass-encased bar magnet is also included behind the boat in order to push it into the high-temperature zone after flushing is completed. The combustion system is flushed with oxygen for 10 min. During this time liquid nitrogen in a Dewar flask is placed around the U-tube and 100 ml/min of dry air or nitrogen is passed through the polyvinyl chloride jacket surrounding the U-tube to raise its temperature so that oxygen is not trapped-out with CO_2. If oxygen is still being held in the trap it will be necessary to increase the flow rate or heat the gas slightly.

7.3.2 Stopcock **B** is turned to allow the effluent of the combustion tube to enter the CO_2 trap; valve **C** is closed. A magnet outside the sample compartment is used to transfer the sample boat into the high-temperature region of the tube. Allow 15 to 30 min for the production of CO_2 and its trapping in the freeze-trap. Turn stopcock **B** to vent position and open valve **C**. Allow the helium flow to purge the trap for about 5 min to remove the excess O_2. From this point on the procedure is identical to **7.2.2** and **7.2.3**.

7.4 Quantitation of CO_2. The conversion of the GC output into μg of carbon is done by one of several ways. The peak areas of the recorder trace (multiplied by attenuation factors) can be measured in several manual ways—planimetry, triangulation, or cut-and-weigh—with a probable error of 3 to 4% depending on the skill of the operator. They can be measured with a semi-automatic method—Disc integration—that gives an error of about 1.5%. Automatic electronic integrators have a probable error of about 0.5% and can accept a wide range in signal size from the GC. The data is presented as a number of counts and during calibration the number of counts/μg carbon is determined by injecting known amounts of standard material. Peak area/μg carbon is determined in like fashion for the manual methods.

8. Calibration and Standards

8.1 The gas chromatograph is calibrated by injecting CO_2 through a septum held in a tee that is located just upstream of the CO_2 freeze-trap.

8.1.1 A 0.25-ml Pressure-Lok gas-tight syringe is calibrated by filling with 200 μl of distilled water at room temperature and weighing to the nearest 0.1 mg. 100 μl of water is discharged and the syringe reweighed. The remaining 100 μl of water is discharged and the syringe weighed again. The mean error in measuring volume with this syringe is calculated to be less than 2%. The syringe should be washed with reagent grade acetone and thoroughly dried before being used for calibration. The syringe is carefully filled with the desired volume of pure CO_2 at room temperature.

8.1.2 The trap is pre-cooled with liquid nitrogen and the helium flow rate is set at 50 ml/min. The efficiency of trapping should be tested both with and without the 100 ml/min of dry air flowing through the jacket around the trap in order to detect any differences. At the end of precooling, the CO_2 is injected through the septum into the helium gas stream. An interval of 5 min is allowed for collection. The trap is closed by the two toggle valves, the liquid nitrogen removed, and a hot water bath added for 3 min. The contents of the trap are introduced to the GC by way of the 6-port gas valve as described in **7.2.2**. Injections of varying amounts of CO_2 and at various GC attenuations are made for the range expected during analysis. The GC output (peak area or counts) is plotted against the weight of carbon in the CO_2 injected.

8.2 The efficiency of trapping CO_2 through the combustion tube is determined by maintaining the oxygen flow rate at 50 ml/min and the freeze-trap at slightly above the boiling point of oxygen with the 100 ml/min flow rate of nitrogen through the jacket. Known amounts of CO_2 are added upstream of the combustion tube and the times of collection in the trap are varied to find the maximum or optimum efficiency levels. The absolute value of the collection efficiency is not critical since the final calculation of μg C/m^3 is normalized for the overall efficiency factors in the analytical procedure. It is important, however, that the individual efficiency factors remain constant during the course of analysis.

8.3 The efficiency of CO_2 release from carbonates by acidification is determined by weighing a 100 to 200 μg of dry calcium carbonate in a small aluminum pan in a microbalance (Cahn, Model M-10, or equivalent) and placing the pan in a Coors No. 2 porcelain combustion boat. Clean, dry stainless steel spatula and forceps are used in handling these materials. The boat is placed in the acidification tube and reacted and analyzed as described in section **7.2**. This, and the other calibrations, should be repeated a sufficient number of times to determine standard deviations or probable errors for each analytical step. The mean value of the blank runs is, of course, subtracted from the values of the sample runs.

8.4 The efficiency of CO_2 production from combustion is determined in similar fashion by placing known amounts of dry, high-purity organic compounds or graphite in the combustion furnace and running through the oxidation procedure. Other compounds may be used besides the lactose or potassium acid phthalate as representative organic material as long as adsorbed water is removed by drying.

8.5 Background levels of CO_2 are determined by performing the analyses without sample particulate matter. If a detailed knowledge of the sources of background CO_2 is desired, a series of analyses may be run in such manner to increasingly include the various steps of the sample-handling procedure. The first step would be to use just the clean porcelain boat. The second step would include a clean, unused foil or filter in the boat. And, the third would include a foil or filter that had been taken to the sampling site, placed on the sampling apparatus and then removed. This procedure is recommended if different persons will be handling the samples from the time of initial preparation and cleaning to the time of analysis. The source of high blanks can then be determined and corrected.

8.6 There is a strong possibility that some acidic material may be present on glass fiber filters from manufacturing processes. Basic washing, followed by rinsing and drying, is recommended to remove this material which may cause loss of carbonate particles on collection. Similarly, it may be desirable to find a source of glass fiber filters that contain no residual resin so that background organic levels can be lower. Flash-firing of the filters is apparently insufficient to remove the resin.

9. Calculations

9.1 The net GC output which is the sample output minus the output of the blank run, must be transformed into units of μg of carbon/m^3 of air sampled. Calculate some efficiency factors to achieve this.

Let

f_t = efficiency factor of trapping CO_2 in the freeze-trap;

f_a = efficiency factor of releasing CO_2 from carbonates;

f_o = efficiency of producing CO_2 from oxidation of carbonaceous matter.

Then,

$$f_t = \frac{\text{(Net GC output)}}{(\text{GC output}/\mu\text{g C})\ (\mu\text{g C added})}$$

$$f_{a,o} = \frac{\text{(Net GC output)}}{(\text{GC output}/\mu\text{g C})\ (\mu\text{g C added})\ (f_t)}$$

with the efficiency factors varying between 0 and 1.0. Occasionally, they will be larger than 1.0 which may indicate that some residual water was still present in the calibration standards.

If V = the total volume of air (in m^3) that was sampled, then

$$\mu\text{g C/m}^3 = \frac{\text{(Net GC output)}}{(\text{GC output}/\mu\text{g C}) f_t f_{a,o}\, V}$$

9.2 Example calculation for an oxidative step.

If f_t = 0.98
f_o = 0.92
V = 100 m^3 at 25 C and 760 Torr.
Net GC output = 425 mv-sec
GC output/μg C = 1.0 mv-sec/μg C

then, C_a, the ambient concentration of carbon/m^3 is

$$C_a = \frac{425 \text{ mv-sec}}{(1.0 \text{ mv-sec}/\mu\text{g C})\ (0.98)\ (0.92)\ 100 \text{ m}^3}$$

$$C_a = 4.7\ \mu\text{g C/m}^3$$

10. Effects of Storage

10.1 All of the reagents are stable with time if they are kept in sealed containers to prevent adsorption of atmospheric water, CO_2, or contaminants. The long-term effects of storing the foils or filters is not yet known, but, it is supposed that airborne acids—either from the laboratory or ambient pollution—will degrade the carbonate matter trapped on them. Some of the non-carbonate organic compounds are likely to be unstable and the possibility exists that some of the degradation products will be volatile. It is recommended that the samples be kept in closed containers in the dark, and cooled in a refrigerator, if the time between collection and analysis is long.

11. References

1. DUGAN, G., and V.A. ALUISE. 1969. An Analyzer for the Dynamic Microdetermination of C, H, N, and O. *Anal. Chem.* 41:495.
2. AMERICAN PUBLIC HEALTH ASSOCIATION, INC.

1977. Methods of Air Sampling and Analysis. 2nd Edition. p 00. Washington, D.C. 20036.
3. KAPLAN, I.R., J.W. SMITH, and E. RUTH. 1970. Carbon and Sulfur Concentration and Isotopic Composition in Apollo 11 Lunar Samples. Proc. of the Apollo 11 Lunar Sci. Conf. 2:1317.
4. LEE, W.R., and L.L. LEWIS. 1969. The Determination of Carbon in Thin Films on Steel Surfaces. Gen. Motors Corp., Warren, Mich. Res. Publ. GMR-890, June.
5. LUNDGREN, D.A. 1967. An Aerosol Sampler for Determination of Particle Concentration as a Function of Size and Time. APCA Jour. 17:225.
6. MILES, C.C. 1969. An Oxyacidic Flux Method for Determination of Carbon in Sodium. *Anal. Chem.* 41:1041.
7. MUELLER, P.K., R.W. MOSLEY, and L.B. PIERCE. 1971. Carbonate and Non-Carbonate Carbon in Atmospheric Particles. Proc. 2nd International Clean Air Congress, pp. 532–539, Academic Press, N.Y.
8. MUELLER, P.K., R.W. MOSLEY, and L.B. PIERCE. 1972. Chemical Composition of Pasadena Aerosol by Particle Size and Time of Day. IV Carbonate and Noncarbonate Carbon Content. *J. Coll. and Interface Sci.*, 39, No. 1, April.
9. THOMAS, J., JR., and G.M. HIEFTJE. 1967. Rapid and Precise Determination of CO_2 from Carbonate-Containing Samples Using Modified Dynamic Sorption Apparatus. *Anal. Chem.* 38:500.
10. VAN HALL, C.E., and V.A. STENGER. 1967. An Instrumental Method for Rapid Determination of Carbonate and Total Carbon in Solutions. *Anal. Chem.* 39:503.

Subcommittee 5
E. SAWICKI, *Chairman*
P. R. ATKINS
T. A. BELSKY
R. A. FRIEDEL
D. L. HYDE
J. L. MONKMAN
R. A. RASMUSSEN
L. A. RIPPERTON
J. E. SIGSBY
L. D. WHITE

510.

Tentative Method of Analysis for Wind-Blown Nuisance Particles in the Atmosphere (The Atmospheric Adhesive Impactor)

21200-01-76T

1. Principle of the Method

1.1 The atmospheric adhesive impactor (AAI) is a sampling device used to collect wind-blown particles greater than about 20 μ from the ambient atmosphere. The particles are impacted by the wind, and retained on an adhesive collection media wrapped around a standard diameter cylindrical holder. The efficiency of collection of smaller particles varies greatly with the wind speed.

1.2 APPLICABILITY. The device can be used in a continuing survey or monitoring program. At the conclusion of the sampling period, the AAI is visually compared with precounted visual particle standards. The directional properties of the device enable its use in locating the source(s) of large particles by multiple sampling in appropriate locations surrounding the suspected source or sources.

2. Range and Sensitivity

2.1 The sampling results will vary from less than 1000 to more than 60,000 particles per square inch per week. Values over 60,000 particles per square inch are unreliable, and for very heavily contaminated areas, shorter exposure times should be used to prevent oversaturation.

2.2 For monitoring purposes, the standard sampling period is 7 days. Longer or shorter periods of sampling are normally adjusted to the nearest whole day.

2.3 When attempting to locate the

source of particulates, shorter observation periods (such as a day or less) are useful.

3. Interferences

3.1 The adhesive and its face stock are subjected to age, high and low temperatures, high and low humidity, precipitation and other adverse environmental factors. Variation due to these factors can be minimized by using the material suggested in Section **6**.

3.2 Any dust to which the AAI is subjected between the end of the sampling period and the analysis is an interference. This can be minimized by carrying the AAI's in a dust free carrying case, and by performing the analysis as soon as possible after the end of the sampling period.

3.3 Buildings, signs, or other objects can have a screening effect or cause turbulence. Hence, proper care must be taken in site selection to ensure reliable readings.

3.4 When the bracket mounted AAI is used with a utility pole as a support, a shading influence would be expected. A survey **(1)** shows that this influence does not significantly decrease the normal accuracy discussed in Section **4**, if the cylinder is at least 16″ from the pole.

4. Precision and Accuracy

4.1 The total concentration of wind-blown particles at two or more separate locations are considered to be significantly different when the total concentration of particles varies by more than 20%.

4.2 The collection efficiency of a 2 ¾″ diameter adhesive sampler was determined under controlled conditions by constructing a laboratory wind tunnel and a special apparatus to determine the actual concentration of particles which passed by the sampler **(4)**. Results are shown in Figure 510:1. The collection efficiencies relate closely to the theoretical collection efficiencies of a 35 to 45 μ particle, although the mean particle diameters of the samples were between 31 and 36 μ. A mean diameter size of 39 μ has been reported for particles captured by the AAI sampler in the atmosphere **(2, 3)**. Hence, the above data and Figure 510:1 show that the collection efficiencies are approximately 25% at 5 mph wind speed, 40% at 10 mph wind speed, and 55% at 15 mph wind speed for 40 μ particulates.

5. Apparatus

5.1 The Adhesive Impactor. This unit consists of two principal parts.

5.1.1 *An extra-wide mouthed glass jar,* 2 ¾″ diameter and 3 ⅜″ height. The manufacturer's specifications should suffice in meeting the tolerances in diameter required for this method. A more exact discussion of the effect of cylinder size upon sampling is available **(3)**.

5.1.2 *The adhesive collection media.* The material is discussed in Section **6**. A 1 ¾″ wide by 10″ long strip is mounted (sticky side out) around the side of the inverted glass holder.

5.2 Supports. The support to be used depends upon the sampling site. There are two supports commonly used.

5.2.1 *A wooden bracket,* which is easily attached to another suitable support, usually a utility pole. The bracket consists of a wood board 18″ long (by about 2 ½″ × 2 ½″). A metal "L" shaped 8″ × 10″ shelf bracket is attached to the bottom of the board. The 8″ perpendicular section is attached to the utility pole. The screw top of the jar is attached to the top of the board, centered 16 ½″ from the utility pole.

5.2.2 *A wood or metal pole support* is useful for movable sampling. A length of ½″ iron pipe or a 1″ wooden dowel is cemented into a discarded auto battery box or concrete block. The height of this pole support depends upon the working height of the sampler, which is discussed in Section **7**. The screw cap is then soldered to a ½″ iron pipe flange at the top of the pole, or screwed to the top of the wooden dowel.

5.3 Visual Comparator. The analysis of the samples may be performed on a particle comparator, such as the Gruber Comparator, or a similar arrangement. The Gruber Comparator consists of a cylinder, 4 ½″ in diameter, attached to a slotted

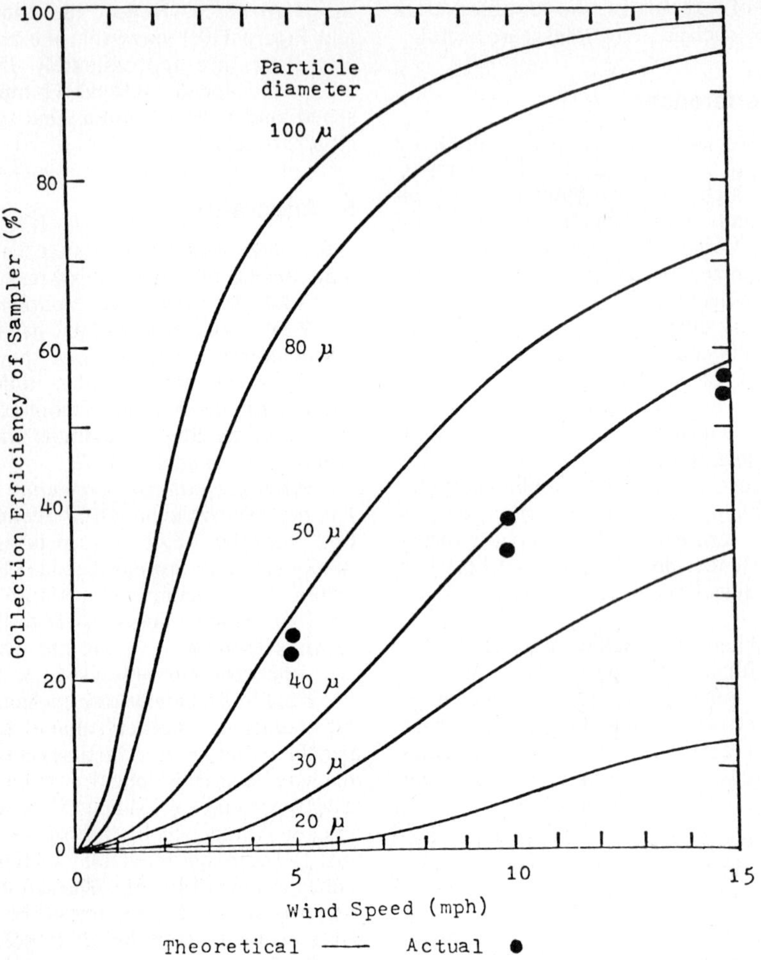

Figure 510:1—Theoretical and Actual Collection Efficiencies of the Atmospheric Adhesive Impactor (adapted from (4)).

gear indicator by means of a drive shaft. The gear contains eight slots, 45 degrees apart, as are the eight points of the compass. The pictorial standards are mounted on the cylinder, directly under an observation slot, by means of black plastic tape. The sample is attached to the cylinder, and the compass directions are properly aligned. The standards and sample are viewed through a fluorescent, illuminated magnifier of 4 power magnification with rigid 6″ arms and a portable base. The viewer is a standard illuminated inspection glass. The cylinder is rotated manually as each point of the compass is read.

6. Reagents

6.1 Research (**6**) has shown that a strip of Fasson Products R-135 Pli-A-Print adhesive is the most suitable adhesive material for this method. The material is available from the Fasson Products Division of Avery Paper Company, Painesville, Ohio.

7. Procedure

7.1 Site Selection. A sampling station should be selected so that the AAI is not unduly affected by interference from buildings or other high objects or structures. Accessibility and security are also of importance. The following specific recommendations are to be used as a guide.

7.1.1 The impactor shall be a minimum of 5′ and a maximum of 15′ above the supporting surface, with 10′ the usual working height.

7.1.2 Where a utility pole is used, the impactor is 16 ½″ from the pole and 10′ above grade.

7.1.3 The roof of low buildings easily accessible is a favorable location, but it is not necessarily representative of lower or ground levels.

7.1.4 When higher buildings in the immediate vicinity cannot be avoided, the top of any building should not be more than 15° above the horizontal from the impactor. That is, a line drawn from the AAI to the nearest edge of the highest point on any building should form not more than a 15° angle with the horizontal. A similar 30° angle with objects below the AAI is also desirable.

7.2 Number. The number of AAI's in a given area and their spacing throughout the area will be determined by the purpose of the survey.

7.3 Sample Exposure and Handling.

7.3.1 The impactor is mounted on its appropriate jar lid, which has been attached to one of the supports described in Section **5**.

7.3.2 The impactor may be prepared in advance and the release liner removed from the adhesive at the start of the test period.

7.3.3 A compass is used to determine the north direction, and this is marked on the adhesive paper with a permanent type ink marker.

7.3.4 The impactor should be adequately protected, both before and after the sampling period, from extraneous materials by placing it in a suitable container during transportation.

7.4 Analysis. The concentration of particles per square inch is determined for the eight compass directions by visual comparison with photographic standards of known concentration. The apparatus used for this comparison is described in Section **5**. The following procedure is used for the Gruber Comparator.

7.4.1 The AAI is attached to the cylinder of the comparator, and the compass directions are properly aligned.

7.4.2 A known compass direction (such as north) is centered under the inspection glass.

7.4.3 The photographic standard is rotated until the closest match between the photographic standard and the sample is obtained.

7.4.4 The concentration in particles per square inch is recorded for that compass direction.

7.4.5 Steps **7.4.2, 7.4.3**, and **7.4.4** are repeated until all eight compass directions have been read and recorded.

7.5 Recording.

7.5.1 The recording format will depend upon the purpose of the survey and the length of the sampling period. For a sampling period of seven days, when the purpose of the monitoring is a continuing ambient survey, the data are recorded for each direction, and the average of the eight directional values is calculated and recorded. All data are recorded as particles per square inch per week. The monthly average for the sampling station may also be calculated.

7.5.2 Weather data is included as an aid to comparison since the wind velocity and direction, and precipitation affect the concentration of particulates in the atmosphere. Table 510:I is an example of a suitable data recording format.

8. Calibration

8.1 The photographic standards used on the Gruber Comparator are available from the manufacturer.

8.2 The standards may be prepared by random scattering of typical particulate matter on sticky paper. This may be accomplished by use of an air spray or by

Table 510:I. Recording format for AAI results.

Summary of wind-blown nuisance particles for week

From:________________ To:________________

Site:________________

Sample Data	N	NE	E	SE	S	SW	W	NW
Particles per sq. in. per week								
Weekly Average:________ particles per sq. in per week								
Wind Data	Hours of Wind							
0–3.5 mph								
4–7.5 mph								
8–12.5 mph								
13–18.5 mph								
19 mph								

Remarks and additional Weather information (e.g., precipitation):

physical scattering. These catches are then sprayed with clear plastic preservative, and, after drying, portions are selected which visually approximate the desired particle count. The portions are cut into 1″ squares and then counted under 14× magnification. By necessity, a trial and error method of selection must be used until the desired standards are found. The same procedure can be followed on previously collected samples which are accurately counted.

9. Calculations

9.1 The only calculations required are the weekly and monthly averages of recorded data.

10. Effect of Storage.

10.1 Not applicable.

11. References

1. GRUBER, C.W., and G.A. JUTZE. 1957. The Use of Sticky Paper in an Air Pollution Monitoring Program, *Air Poll. Cont. Assoc. J.* 7:115–117.
2. PRITCHARD, W.L., C.E. SCHUMANN, and C.W. GRUBER. 1965. Particulate Sampling by Adhesive-Coated Materials, First Annual Progress Report, USPHS Research Grant AP00370-01, Division of Air Pollution Control, City of Cincinnati.
3. PRITCHARD, W.L., C.E. SCHUMAN, and C.W. GRUBER. 1966. Particulate Sampling by Adhesive-Coated Materials, Second Annual Progress Report, USPHS Research Grant AP00370-02, Division of Air Pollution Control, City of Cincinnati.
4. PRITCHARD, W.L., C.E. SCHUMANN, and C.W. GRUBER. 1967. Particulate Sampling by Adhesive-Coated Materials, Third Annual Progress Report, USPHS Research Grant AP00370-03, Division of Air Pollution Control, City of Cincinnati.
5. GRUBER, C.W., W.L. PRITCHARD, and C.E. SCHUMANN. 1970 Recommended Standard Method for Measuring Wind-Blown Nuisance Particles, *Air Poll. Cont. Assoc. J.* 20:161–163.
6. PRITCHARD, W.L., C.E. SCHUMANN, and C.W. GRUBER. 1967. Selection of Adhesive-Coated Materials for Particulate Sampling, *Air Poll. Cont. Assoc. J.* 17:305–307.

Subcommittee 9
A. L. LINCH, *Chairman*
J. V. HARMAN
E. R. HENDRICKSON
MORRIS KATZ
G. O. NELSON
J. N. PATTISON
A. L. VANDER KOLK
R. B. WEIDNER

601.

Tentative Method of Analysis for Gross Alpha Radioactivity Content of the Atmosphere

11301-01-69T

1. Principle of the Method

1.1 Air particulates collected on a filter paper possessing a high surface retention are counted with an alpha sensitive detector to establish the gross concentration of alpha emitters present in the sampled ambient air. Since a given alpha emitter cannot be present in a greater concentration than the gross alpha concentration of a mixture of unidentified radionuclides, a gross analysis may eliminate the need for a more time-consuming and expensive analysis for specific radionuclides. While a gross alpha analysis does not yield specific information on the radionuclide composition of a sample, it is simple, rapid, and inexpensive.

2. Range and Sensitivity

2.1 The method has the advantage of being suitable over the entire range of conc of airborne alpha emitters ordinarily encountered in the environment. Since most alpha counters have inherently low backgrounds, sensitivity is primarily a function of the volume of air sampled. For example, a sample of 100 m^3 of air with an alpha particulate concentration of 0.006 pCi/m^3 (6.0×10^{-15} μCi/cc) will contain 1.32 dpm, sufficient to produce acceptable precision in less than an hour's counting time in most alpha counters.

3. Interferences

3.1 The principal interference is from the progeny of naturally occurring radon and thoron usually present in the atmosphere. Since the effective radioactive half-lives of the progeny of these naturally occurring radionuclides are controlled by relatively short half-life radionuclides, namely 26.8 min ^{214}Pb and 19.7 min ^{214}Bi for radon, and 10.6 hr ^{212}Pb for thoron, their interferences may be nullified by waiting until these daughter activities are negligible before counting. This would require about 3 days for the long-lived thoron progeny. Less accurate, but more rapid, evaluations of long-lived gross alpha concentrations can be secured by a dual counting procedure (**1, 2**).

3.2 Another way to differentiate between these natural progeny and other alpha emitters is to employ a solid-state semiconductor detector, the signal from which is proportional to the energy of the detected alpha particles. Most long-lived alpha emitters have energies below 6 MeV, whereas the short-lived daughters of radon and thoron have greater energies. It is therefore possible to gate the associated electronics to count only the pulses from lower energy alpha emissions (**3**).

3.3 Inert dust loadings and the use of improper filter media are also potential sources of error. Alpha absorption and energy degradation (in the case of solid-state counting) become pronounced in samples with total dust loading larger than 1 mg/cm^2. The total volume of air which can be sampled in a dust laden atmosphere is thereby limited. Absorption of alpha activity takes place if particles have penetrated into the filter matrix. The use of a membrane type filter is recommended to avoid this problem.

4. Precision and Accuracy

4.1 The precision of the method is essentially a function of the volume of air sampled, the inert dust loading of the air, the length of time for which the sample is counted, and the background of the counter. For example, the precision of count of the very low activity sample described above (with an activity of only 1.32 dpm), counted for an hr at 50% geometry in a counter with a background of 0.20 cpm, would be 1.32 ± 0.52 dpm (at the 95% Confidence Level).

4.2 The accuracy of a gross alpha analysis is dependent on the accuracy of measurement of the volume of air sampled, and the appropriateness of the self-absorption correction.

5. Apparatus

5.1 The basic requirement for gross alpha analysis is an air mover, a calibrated and properly situated device for measuring air flow rate or sample volume, a particulate filter, an alpha detector, and a counter. The size of the sample should not be any larger than the sensitive area of the alpha detector. Due to the limited range of alpha radiation, a high surface retention filter (*i.e.*, low penetration of particles into the filter matrix) such as a membrane type filter on which collection is essentially on the surface is recommended.

5.2 A windowless or very thin window alpha sensitive detector should be used to count the filter sample. Spurious counts may be produced in windowless counters by the buildup of static charge on filters. This may be eliminated by the use of antistatic fluid. ZnS coated photomultiplier tubes are widely used for alpha counting. Although presently limited to small diameter samples, solid-state detectors are extremely attractive for gross alpha counting because of their stability and low backgrounds (in the order of 0.01 cpm).

6. Reagents

6.1 None.

7. Procedure

7.1 Operate the sampler in a location representative of the atmosphere for which the conc is to be established and for a period long enough to collect measurable alpha activity considering the minimum detectable concentration.

7.2 Place the filter in a protective cover or container upon removal from its holder to minimize the possibility of dislodging the collected activity.

7.3 Prior to counting, place the filter sample in a planchet or holder for reproducible positioning under the detector.

7.4 Count the sample for an interval (or total count) sufficient for a statistically significant result.

8. Calibration and Standards

8.1 The efficiency of an alpha counter should be determined with an alpha standard as close as possible to the sample in size and composition. If the latter is unknown, a standard may be prepared from uranium solution.

8.2 The counter's background should be determined over an interval similar to or longer than the sample counting time. Alpha counters are usually quite stable and it is satisfactory to make only daily determinations of background to ascertain that major shifts due to malfunction, drift, or contamination have not occurred unless anomalous counting results are encountered.

9. Calculation of Concentration

9.1 CALCULATION.

$$\text{Counter Efficiency} = \frac{\text{net counting rate of standard (cpm)}}{\text{disintegration rate of standard (dpm)}} \tag{1}$$

$$\text{Net Sample Disintegration Rate (dpm)} = \frac{\dfrac{\text{gross counts}}{\text{sample count time (min.)}} - \dfrac{\text{background counts}}{\text{background count time (min.)}}}{\text{counter efficiency}} \tag{2}$$

$$\text{Activity (pCi/m}^3\text{)} = \frac{\text{sample disintegration rate}}{\text{volume sampled (m}^3\text{)} \times 2.2 \text{ (dpm/pCi)}} \quad (3)$$

9.2 The statistical significance of this result should also be indicated. It can be determined from the standard deviation (σ) of the count as follows:

$$\sigma \text{ (cpm)} = \pm \sqrt{\frac{\text{cpm (gross)}}{\text{sample count time (min.)}} + \frac{\text{cpm (background)}}{\text{background count time (min.)}}} \quad (4)$$

$$\sigma \text{ (dpm)} = \frac{\pm\, \sigma \text{ (cpm)}}{\text{counter efficiency}} \quad (5)$$

9.3 The statistical significance of a count should be expressed as the uncertainty (or error) of the acount at a specified confidence level, such as 95% (1.96 σ).

10. Effects of Storage

10.1 Most properly protected samples should experience no effect other than radioactive decay during storage. Some alpha emitters, such as ^{210}Po, are found to "creep" by recoil action and thereby lose activity from the sample to the immediate surroundings.

11. References

1. SETTER, L.R. and G.I. COATS. 1961. The Determination of Airborne Radioactivity. *AIHA J.* 22:64.
2. SCHULTE, H.F. 1968. Chapter on Monitoring Airborne Radioactivity in *Air Pollution*, 2nd Edition, Vol. 11, Academic Press, New York.
3. LINDEKEN, C.L. and K.F. PETROEK. 1966. Solid-State Pulse Spectroscopy of Airborne Alpha Radioactivity Samples. *Health Physics* 12:683.

Subcommittee 8
B. SHLEIEN, *Chairman*
R. G. HEATHERTON
P. F. HILDEBRANDT
L. B. LOCKHART, JR.
R. S. MORSE
W. STEIN
D. F. VAN FAROWE
A. HULL, *Health Physics Society Liaison*

602.

Tentative Method of Analysis for Gross Beta Radioactivity Content of the Atmosphere

11302-01-69T

1. Principle of Method

1.1 Air particulates collected on a filter paper possessing a high retention are counted with a beta sensitive detector to establish the gross concentration of the beta emitters present in the sampled ambient air. Since a given beta emitter cannot be present in a greater concentration than the gross beta concentration of a mixture of unidentified radionuclides, a gross analysis may eliminate the need for a more time consuming and expensive analysis for specific radionuclides. While a gross beta analysis does not yield specific information on the radionuclide composition of the sample, it is simple, rapid, and inexpensive.

2. Range and Sensitivity

2.1 The method has the advantage of being suitable over the entire range of concentrations of airborne beta emitters ordinarily encountered in the environment. With proper control of counter background, sensitivity is primarily a function of the volume of air sampled. A sample of 10 m^3 of air with a beta particulate concentration of 3 pCi/m^3 (3.0×10^{-12} μCi/cc) will contain 66.6 dpm, sufficient to produce acceptable precision in ten min counting time.

3. Interferences

3.1 The principal interference is from the progeny of naturally occurring radon and thoron usually present in the atmosphere. Since the effective radioactive half-lives of the progeny of these naturally occurring radionuclides are controlled by relatively short half-life radionuclides, namely 26.8 min ^{214}Pb and 19.7 min ^{214}Bi for radon, and 10.6 hr ^{212}Pb for thoron, their interferences may be nullified by waiting until these daughter activities are negligible before counting. This would require about 3 days for the long-lived thoron progeny. Less accurate, but more rapid, evaluations of long-lived gross beta concentrations can be secured by a dual counting procedure (**1, 2**).

3.2 Inert dust loadings may cause serious interference, particularly for measuring low energy beta emitters. Reasonably accurate self-absorption corrections may be made providing these loadings do not exceed a few mg/cm^2.

4. Precision and Accuracy

4.1 The precision of the method is essentially a function of the volume of air sampled, the inert dust loading of the air, the length of time for which the sample is counted, the background of the counter. For example, the precision of count of the sample described above (with an activity of 66.6 dpm), counted for 10 min at 30% overall efficiency in a counter with a background of 10 cpm, would be 66.6 ± 13.2 dpm (at a 95% Confidence Level).

4.2 The accuracy of a gross beta analysis is dependent on the accuracy of measurement of the volume of air sampled, the appropriateness of the self-absorption corrections, and the similarity of energy of the calibration standard to the energy of the unknown sample radionuclides. Since beta energies vary over an order of magnitude, the gross beta activities (and concentrations) must be reported in terms of the calibration standard employed.

5. Apparatus

5.1 The basic requirements for gross beta analysis are an air mover, a calibrated and properly situated device for measuring flow rate or air volume sampled, a particulate filter, a beta detector and a counter. A suitable high efficiency filter should be employed.

5.2 If filters of appreciably less than 100% collection efficiency are employed, it is necessary that their collection efficiency as a function of filter face velocity and the physical properties of the material being sampled be known.

5.3 A beta-sensitive detector such as an end window Geiger-Muller tube, thin window proportional flow counter, plastic scintillator, or solid state detector can be used for counting the sample.

6. Reagents

6.1 None.

7. Procedures

7.1 Operate the air sampler in a location representative of the atmosphere for which the gross beta concentration is to be determined, for a period long enough to collect measurable beta activity considering the minimum concentration.

7.2 To minimize the possibility of dislodging collected activity, store the sample in a protective cover or container upon removal from its holder.

7.3 For counting, place the filter sample in a planchet or holder at a reproducible distance under the window or sensitive surface of a beta detector. Care should be taken to avoid contaminating the detector. This can be done by covering the sample with thin mylar sheet.

7.4 Count the filter sample for an interval (or total count) sufficient for a statistically significant result.

8. Calibration and Standards

8.1 The efficiency of a beta counter should be determined with a beta standard resembling the sample in size and matrix, and containing the beta radionuclide of principal concern or one of similar energy. If the latter is unknown, ^{137}Cs is a desirable standard, since its average energy corresponds to that of aged mixed fission products in fallout. It is desirable that a number of other beta radionuclides covering a range of energies be counted in order to determine the energy dependence of the counter. If counting is done through a thin mylar cover, **7.3**, calibration must be done through an equivalent absorber.

8.2 Determine the background of the counter over an interval similar to or longer than the sample counting time. Most beta counters are relatively stable, and it is sufficient to make one daily determination of background to ascertain that major shifts due to malfunction, drift or contamination have not occurred.

9. Calculation of Concentration

9.1 CALCULATION.

$$\text{Counter Efficiency} = \frac{\text{net counting rate of standard (cpm)}}{\text{disintegration rate or standard (dpm)}}$$

$$\text{Net Sample Disintegration Rate (dpm)} = \frac{\dfrac{\text{gross counts}}{\text{sample count time (min)}} - \dfrac{\text{background counts}}{\text{background count time (min)}}}{\text{counter efficiency}}$$

$$\text{Activity (pCi/m}^3) = \frac{\text{sample disintegration rate (dpm)}}{\text{volume sampled (m}^3) \times 2.2 \text{ (dpm/pCi)}}$$

9.2 The statistical significance of this result should also be indicated. It can be determined from the standard deviation (σ) of the count as follows:

$$\sigma\text{(cpm)} = \pm \sqrt{\frac{\text{cpm (gross)}}{\text{sample count time (min.)}} + \frac{\text{cpm (background)}}{\text{background count time (min.)}}}$$

$$\sigma\text{(dpm)} = \frac{\pm\,\sigma\text{(cpm)}}{\text{counter efficiency}}$$

9.3 The statistical significance of a count should be expressed as the uncertainty (or error) of the count at a specified confidence level, such as 95% (1.96σ).

10. Effects of Storage

10.1 Samples should be stored so they are protected from the loss of loosely embedded particles. A properly protected sample should experience no effects other than radioactive decay during storage.

11. References

1. SETTER, L.R. and G.I. COATS. 1961. The Determination of Airborne Radioactivity, *AIHA J.* 22:64.
2. SCHULTE, H.F. 1968. Chapter on Monitoring Airborne Radioactivity in *Air Pollution*, 2nd Ed. 11, Academic Press, New York.

Subcommittee 8
B. SHLEIEN, *Chairman*
R. G. HEATHERTON
P. F. HILDEBRANDT
L. B. LOCKHART, JR.
R. S. MORSE
W. STEIN
D. F. VAN FAROWE
A. HULL, *Health Physics Society Liaison*

603.

Tentative Method of Analysis for Iodine-131 Content of the Atmosphere (Particulate Filter—Charcoal Gamma)

11316-01-69T

1. Principle of the Method

1.1 Radioactive iodine may be introduced into the atmosphere as an elemental or chemically combined vapor, as a vapor adsorbed on particulate matter, or as a particulate iodine compound. It thus presents a unique sampling problem. The proposed method involves sampling iodine in its solid and gaseous slates with an "absolute" particulate filter in series with an activated charcoal cartridge. The ^{131}I activity is then quantitatively determined by gamma spectroscopy. Iodine adsorbed on solid particles and possibly some vaporous iodine will be deposited on the "absolute" filter while the remaining vaporous iodine will be adsorbed on the activated charcoal.

1.2 Samples may be collected continuously for a week or more at low flow rates for routine environmental surveillance or for 24 hr or less at high flow rate in order to evaluate a suspected or known incident.

2. Sensitivity and Range

2.1 LOWER RANGE. Minimum detectable quantity of ^{131}I by gamma spectroscopy is approximately 10 pCi. At a sampling rate of 0.01 m^3/min the practical minimum detectable quantity for a 7 day sample is 0.1 pCi/m^3 (1.0×10^{-13} μCi/cc).

2.2 UPPER RANGE. This is dependent on the state of iodine and the adsorptive capacity of the charcoal trap.

3. Interferences

3.1 The major interference may be the presence of other gamma emitting radionuclides collected on the filter. The effect of short-lived radionuclides can be diminished or eliminated by permitting a sufficiently long decay time (3 to 4 days) prior to counting. Decay measurements of the ^{131}I photopeak can be used to determine ^{131}I in the presence of long-lived gamma emitters. If decay measurement is not employed, gamma spectral stripping techniques can be used to correct interferences due to long-lived gamma emitters.

3.2 Another interference may be due to adsorption of radioactive gases or vapors on the charcoal cartridge. The same techniques used with interferences on the filter are applicable.

4. Precision and Accuracy

4.1 A flow measurement accuracy of about 3% is obtainable with most commonly employed devices when properly calibrated.

4.2 The minimum detectable quantity (10 pCi of ^{131}I) can be determined with a precision of 10 at the 95% confidence level by employment of a well shielded counting system and a sufficiently long counting time.

4.3 Although collection of elemental iodine vapor on charcoal under the described conditions will approximate 100%, other vaporous iodine compounds such as organic iodides or hydrogen iodide may have much lower collection efficiences.

4.4 A calibration accuracy of ± 5 to 20% is achievable. In the case of calibration of the charcoal cartridges, this depends on how well the distribution of ^{131}I in the calibration standard conforms to the actual distribution of ^{131}I on the sample cartridge.

5. Apparatus

5.1 The sampling apparatus consists of a particulate filter, charcoal cartridge and a vacuum source in series, associated holders and connections, and a flow measuring device.

5.2 Typical charcoal cartridges **(1)** useful for long-term sampling can be made from plastic tubing ⅝″ in diameter × 1½″ deep. These contain 3 g of 12 to 30 mesh activated charcoal per cartridge. The charcoal is held in place by wire mesh screen at each end of the cartridge with a low resistance paper filter at the effluent end to prevent charcoal particles from being drawn into the pump.

These cartridges are designed for ease of insertion, inlet side down, into a standard 3-1/32″ diameter × 1½″ deep scintillation counting well-type detector. A sampling rate of 0.01 m^3/min is useful for long-term sampling with the above cartridges.

5.3 For short term sampling a greater flow rate (40 ft^3/min) and filters of larger area are desirable. An organic solvent type of gas-fume respirator charcoal canister has been successfully used for this purpose **(2)**.

5.4 Counting equipment includes a shielded sodium iodide, thallium activated scintillation crystal coupled to a multichannel analyzer. For small charcoal cartridges a "well" type detector provides an excellent counting geometry. Alternate counting equipment might include a single channel analyzer or a solid state detector.

6. Reagents

6.1 None.

7. Procedure

7.1 SAMPLING. Air is drawn by the pumping system into the filter holder air inlet. A bypass air inlet may be installed to prevent failure of the pumping system in the event of excessive loading of the filter. Airflow may be determined by use of a rotometer at the initiation and termination of the sampling period with a simple average being taken as the characteristic flow rate. The rotometer should not be left in the system while sampling. Flow can be switched through the rotometer for the measurement period. An alternate method employs a dry gas meter inserted in the line. If the airflow measurement is made between the vacuum device and the filtering media, correction must be made for the reduced and variable density of the gas passing through the meter. Vacuum readings may be taken with a gage at the initiation and termination of the sampling period with a simple average being taken as the characteristic pressure drop.

7.2 COUNTING.

7.2.1 For equipment see paragraph 5.4.

7.2.2 Counting of the sample must be performed in a counting geometry approximating that used for calibration.

7.2.3 The predominant ^{131}I photopeak at 0.364 MeV is employed for ^{131}I quantification.

7.2.4 The sample is counted for a suitable time so that a counting error of 10% at the 95% confidence level is achieved.

8. Calibration

8.1 Calibration is achieved by counting a ^{131}I standard in the same geometry in which the sample is to be counted.

8.2 For particulate filter calibration a liquid ^{131}I standard should be distributed as uniformly as possible on a filter paper.

8.3 For charcoal cartridges and canisters, distribution of ^{131}I within the container will affect the calibration and may vary, depending upon the amount and chemical state of the ^{131}I present, the flow rate, the relative humidity, and the length of the sampling period. Unless specific data are available concerning its distribution, it may be assumed that most ^{131}I adsorption is in the first ¼″ of charcoal **(1)**.

8.4 To prepare a calibration standard for the charcoal cartridge, deposit a known amount of ^{131}I on a paper filter equal in area to the cross section of the cartridge and place it within a cartridge at a depth of ⅛″. If the internal distribution of ^{131}I is known and is not concentrated in the first ¼″, the standard impregnated filter should be placed at a suitable depth.

8.5 For the charcoal canister a calibration standard is prepared in a similar manner.

9. Calculation

9.1 Decay measurements of the ^{131}I photopeak can be used to determine ^{131}I in the presence of long-lived gamma emitters without resorting to more complex techniques. The filter and/or cartridge is counted at appropriate time intervals (every 2 to 4 days is sufficient). The observed counting rate (counts/min) at each counting time under the ^{131}I photopeak (approximately from 0.32 to 0.40 MeV) is plotted on the logarithmic axis of semi-log paper, while the time elapsed since the end of collection is plotted on the arithmetic axis. Connect the points with a straight line. The line should have a slope indicating the half-life of ^{131}I (8.05 days). Extend the line to the counting rate axis. The point at which the line intercepts the logarithmic axis indicates the counting rate due to ^{131}I at the end of collection.

9.2 If immediate results are required or if there is a major interference in the ^{131}I photopeak area due to other radionuclides, spectral stripping techniques should be employed.

9.3 In order to correct for interfering radionuclides having energies greater than 0.364 MeV, a method of Compton continuum subtraction may be used. Draw the photopeak curve (which should be Gaussian in shape) and then a smooth line to estimate the base of the peak. Sum up the observed counting rate under the photopeak (approximately from 0.32 to 0.40 MeV) and subtract the sum of the base line counting rate over the same energy interval to obtain the net counting rate due to ^{131}I. The activity is then determined by dividing this net counting rate by the counting efficiency factor (net counting rate of the standard divided by the disintegration rate of the standard).

9.4 If radionuclides having higher energy gammas than 0.364 MeV are present and one wishes to concurrently determine their activities, more complex spectrum stripping may be employed (**3**). Spectrum stripping consists of identification and quantification of the highest energy radionuclide, determination of its curve from a standard, and subtraction of its curve from the sample curve, either graphically or mathematically. In a complex mixture the spectrum may be "stripped" several times before the ^{131}I peak is evident. However, each substraction results in accumulated errors, which if very large, may compel radiochemical separation for quantitative results.

9.5 The activity per unit volume of air at the time of counting (A_c) is determined as follows:

$$A_c = \frac{s \times {}^{A}STD}{STD \times T \times r} \qquad (1)$$

Where:

s (cpm) = net gamma counting rate of sample (total sample counting rate minus background counting rate) for 0.364 MeV photopeak.

STD (cpm) = net gamma counting rate of standard (total standard counting rate minus background counting rate) for 0.364 MeV photopeak.

^{A}STD = activity of standard (pCi)

T = total sampling time (min)

r = average volume sampling rate (m^3/min)

Ac = activity of sample at time of count (pCi/m^3)

9.6 Correct for activity at the end of sampling (A_1) due to decay of ^{131}I as follows:

$$A_1 = \frac{A_{\bar{c}}}{e^{-\gamma t_1}} \qquad (2)$$

Where:

e = 2.72

t_1 = time between end of sampling and counting (days)

$\lambda = .0858$ day^{-1} (^{131}I decay factor)
If ^{131}I is determined by the decay method, the activity at end of sampling may be directly discerned from the resultant graph.

9.7 If the sampling period is a significant portion of the half-life of ^{131}I and deposition was uniform throughout the sampling period, a correction can be made for the fraction of the sample activity which decayed during the sampling period as follows:

$$A = A_1 \frac{t\lambda}{1 - e^{-\lambda t}} \quad (3)$$

Where: t = sampling time (days)

9.8 The statistical significance of the result should be indicated. It can be determined from the standard deviation(σ) of the count rate as follows:

$$\sigma_s(\text{cpm}) = \pm\sqrt{\frac{\text{gross counting rate (cpm)}}{\text{sample count time (min.)}} + \frac{\text{background counting rate (cpm)}}{\text{background count time (min.)}}} \quad (4)$$

Where

σ_s = Standard deviation of the net counting rate.

The statistical significance of a count should be expressed as the uncertainty (or error) of the count at a specified confidence level, such as 95% (1.96σ).

10. Effect of Storage

10.1 A properly protected sample should experience no effects other than radioactive decay during storage.

11. References

1. SILL and FLYGARE. 1960. Iodine Monitoring at the National Reactor Testing Station. *Health Physics* 2:261.
2. ETTINGER and DUMMER. Iodine-131 Sampling with Activated Charcoal and Charcoal-Impregnated Filter Paper. LA3363, UC-41 Health and Safety TID-4500.
3. ENVIRONMENTAL HEALTH SERIES, U.S. DEPT. OF HEALTH, EDUCATION AND WELFARE, PUBLIC HEALTH SERVICE. 1967. Radioassay Procedures for Environment Samples. Publication PHSP-RH-999-2, Rockville, Maryland.

Subcommittee 8
B. SHLEIEN, *Chairman*
R. G. HEATHERTON
P. F. HILDEBRANDT
L. B. LOCKHARD, JR.
R. S. MORSE
W. STEIN
D. F. VAN FAROWE
A. HULL, *Health Physics Society Liaison*

604.

Tentative Method of Analysis for Lead-210 Content of the Atmosphere

11342-01-68T

1. Principle of the Method

1.1 Lead-210 (21 year half-life) is a trace component of the normal atmosphere and is produced through decay of radon-222 which diffuses into the air from radium sources in the soil; it is removed by washout, by precipitation and by dry deposition processes ("fallout"). Though normal radon concentrations over land are generally in the range of 10 to 1000 pCi/m^3, the concn of airborne lead-210 rarely exceeds 25 pCi/1000 m^3 (**1**). Consequently, large volumes of air must be sampled to obtain a statistically significant count rate of lead-210. (If one is only interested in levels of activity which approach or exceed the max permissible concn of 1 pCi/m^3 for lead-210, considerably smaller air volumes need be sampled.) Analysis is performed by dissolution of a filter through which a known volume of air has passed, addition of inactive bismuth as a carrier for the 5.0-"day" bismuth-210 daughter of lead-210, separation and radiochemical purification of the bismuth-210, and determination of its count rate relative to a bismuth-210 standard (**2, 3**). The filter sample should be retained sufficiently long before analysis to assure near secular equilibrium between lead-210 and bismuth-210 (*e.g.*, 30 days).

2. Sensitivity and Range

2.1 With the described procedure, lead-210 concentrations as low as 1 pCi/1000 m^3 can be determined with a standard error of less than ± 10%. Sensitivity can be improved by increasing the volume of air sampled or by increasing the counting time; a practical lower limit might be concentrations of 0.1 pCi/1000 m^3 with collections of 10^5 m^3. For higher levels of activity, the sample volume may be decreased proportionately.

3. Interferences

3.1 If no recent nuclear weapons tests have been carried out in the atmosphere, contamination of bismuth-210 by fission products will be negligible. Measured decontamination from strontium-90, cesium-137 and rare earths exceeds 99.9%.

3.2 Separation from fresh fission debris which contain radionuclides precipitated as sulfides will be poor. Additional decontamination steps include the addition of ruthenium, tin, antimony, tellurium and other carriers, the extraction of the bismuth sulfide precipitate by ammonium polysulfide, and the insertion of a bismuth oxychloride precipitation prior to the final ammonium carbonate precipitation.

3.3 If negligible amounts of fission products having half-lives of less than 25 days are present, the standard purification procedure can be used and bismuth-210 activity deduced from decay data.

4. Precision and Accuracy

4.1 The precision of the method depends primarily on the reproducibility of sample mounting, the stability of the counting system, and the statistical limitations imposed by the low count rates. The sample mounting procedure has been determined to introduce an uncertainty of ± 1% (standard deviation) in the results. With reasonable count rates and a reliable low background counter, the total uncertainty should be maintainable within ± 2% (standard deviation) relative to a radioactive standard. The primary standard solutions from which the working standards are made usually have a stated absolute accuracy no greater than ± 3%.

4.2 The measured air conc of lead-210 may be 1 to 3% high as a result of the formation of lead-210 from short-lived radon daughters (lead-214 and bismuth-214) collected by the filter. This error can be estimated from knowledge of the average air conc of radon during the collection period.

5. Apparatus

5.1 Positive displacement blower having capacity in excess of 25 cfm under expected load.

5.2 Flow meter and/or calibrated pressure gauge.

5.3 Filter holder for 6″ × 9″ or 8″ × 10″ high efficiency glass fiber filters.

5.4 Analytical laboratory with analytical balance, fume hood, muffle furnace, hot plates, pyrex and Vycor glassware, etc.

5.5 Sensitive low-background (< 2 counts/min) Beta counting equipment (anticoincidence scaler, detector, automatic count recorder).

5.6 Standard planchets compatible with above counting equipment.

6. Reagents

6.1 Sulfuric Acid (H_2SO_4). 36 N and 6N.

6.2 Nitric Acid (HNO_3). 12N and 6N.

6.3 Hydrogen Fluoride (HF). 48%.

6.4 Bismuth^{+3} carrier solution. Dissolve 22.36 g bismuth oxychloride (BiOCL) in 500 ml 2N hydrogen chloride (HCl) and dilute to 1000 ml to give a solution containing the equivalent of 20.0 mg bismuth oxide (Bi_2O_3) per ml. If desired, the solution may be standardized by treating an aliquot as described under **7.2.6** to **7.2.9**.

6.5 Holdback carrier solution. Prepare a single solution containing 5.0 g each of yttrium nitrate—$Y(NO_3)_3$, strontium nitrate—$Sr(NO_3)_2$, lead nitrate—Pb $(NO_3)_2$, and cesium nitrate—$CsNO_3$ dissolved in 1000 ml of 1% nitric acid (HNO_3).

6.6 Thioacetamide CH_3CSNH_2.

6.7 Ammonium Persulfate $(NH_4)_2$ S_2O_8.

6.8 Ammonium Hydroxide NH_4OH.

6.9 Ammonium Carbonate $(NH_4)_2$ CO_3.

7. Procedure

7.1 Sampling.

7.1.1 The particulate matter in a 10,000 m^3 or larger sample of air is collected on a high efficiency glass fiber filter paper 6″ × 9″ or 8″ × 10″ in size mounted in a suitable leak-tight holder and protected from rain and snow by a louvered shelter or canopy.

7.1.2 Air is drawn through the filter at a rate of about 1000 m^3/day through use of a properly calibrated or instrumented positive-displacement blower. Filters should be changed before the pressure increase across the filter indicates a 10% flow decrease, in order to obtain a representative sample and to minimize error in the estimation of sample volume. Under these conditions of flow more than one filter may be required to collect a 10,000 m^3 sample, dependent on the level of atmospheric dust loading.

7.1.3 Records are to be kept of the time of the beginning and end of collection, the duration of collection, and the initial and final flow rates.

7.2 Analysis.

7.2.1 Fold glass fiber filter sample and place in a 400-ml Vycor beaker. Add 10 ml concd nitric acid, 20 ml concd sulfuric acid, and 10 ml hydrogen fluoride (48%). Add by pipet 5.00 ml of standardized bismuth^{+3} carrier solution and a 20 ml portion of the solution containing the lead, cesium, strontium and yttrium holdback carriers.

7.2.2 Cover with a Vycor cover glass and heat to fumes of sulfur trioxide. Add 10 ml more nitric acid and 2 ml hydrogen fluoride and again heat to fumes of sulfur trioxide. Ignore any residue as experience has shown it to be essentially free of radionuclides of interest **(3)**. Add 0.5 g ammonium persulfate to destroy the nitrosyl complex. Fume 5 min. Cool and dilute to 200 ml with distilled water.

7.2.3 After 2 hr, filter off lead sulfate onto Whatman No. 40 paper (plus paper pulp). Do not wash. Discard precipitate. Record time of this separation.

7.2.4 Carefully add ammonium hydroxide until the solution is basic. Redissolve precipitate by addition of a small amt of 6N sulfuric acid. Add 0.2 g thioacetamide and boil 30 to 40 min to precipitate bismuth sulfide. Filter through Whatman No. 40 paper and wash with hot water.

7.2.5 Dissolve bismuth sulfide from the filter by dropwise addition of 3 to 5 ml of conc nitric acid. Wash with 50 ml water, collecting solution in a clean 400-ml beaker. Dilute to 200 ml.

7.2.6 Add 0.2 g thioacetamide and boil 30 min to precipitate bismuth sulfide. Filter.

7.2.7 Repeat **7.2.5**.

7.2.8 Slowly add ammonium hydroxide to the solution until faintly turbid, then add a 10 ml excess of a freshly prepared saturated solution of ammonium carbonate. Digest precipitate at 100 C for 10 to 15 min, then filter onto Whatman No. 40 paper plus paper pulp. Wash precipitate with 50 ml of 0.5% ammonium carbonate wash solution.

7.2.9 Transfer paper and precipitate to a Vycor or porcelain crucible, place in cold muffle furnace, and gradually raise temp to 600 C. Do not overheat. (If bismuth oxide appears dark with metallic bismuth, cool, add 1 ml of 1:1 nitric acid and evaporate to dryness on a hot plate. Again place in a cold muffle and slowly raise temp to 600 C.)

7.2.10 Cool, weigh 50.0 ± 0.1 mg bismuth oxide onto a standard planchet, mount with dilute solution of a binder, and count for beta activity.

7.3 COUNTING.

7.3.1 Counting must be done under highly reproducible conditions using, preferably, a low-background counter (< 2 counts/min). A minimum absorber thickness (or window plus absorber) of 8 mg/cm_2 is required to eliminate response to polonium-210 alphas. Record initial count during several 30 or 60 min periods; correct activity for background and for decay to time of lead-bismuth separation (Step **7.2.3** above). Recount after several days to confirm 5.0-day half-life of bismuth-210. Due to the low levels of bismuth-210 being measured, it is necessary that blank analyses on the unexposed filter material be made and appropriate corrections applied to the measured count rate.

8. Calibration

8.1 The efficiency of the counter is determined by adding a known quantity of a standardized carrier-free lead-210 bismuth-210 solution to a known amount of bismuth carrier, together with a lead hold-back carrier, and carrying through the chemical separation procedure described above (**4**). The final wt of the sample mounted (50.0 ± 0.1 mg) and its physical arrangement on the sample planchet must reproduce as closely as possible the conditions under which unknowns will be counted. Thus, the counter efficiency correction will include corrections for counter response, absorber thickness, self-absorption, back-scattering, geometry, and so forth.

$$B_c = \frac{D_s}{A_s} = \frac{D_t}{A_s \times Y_s}$$

Where:

D_s and
A_s = disintegration rate (dis/min) and measured count rate (counts/min) of standard, respectively
D_t = total disintegration rate of added activity
Y_s = yield correction =

$$\frac{\text{carrier added (mg bismuth oxide)}}{\text{carrier mounted (mg bismuth oxide)}}$$

8.2 Permanent standards can be prepared by precipitating the lead-210 as lead sulfate using a known quantity of lead carrier, and allowing the bismuth-210 to grow into secular equilibrium. For thin sample mounts (*e.g.*, 50 mg on a 5-cm diam planchet) the overall counter efficiency correction should be equal to that obtained with bismuth carrier.

9. Calculation

Lead-210 (pCi/m^3) =
Bismuth-210 (pCi/m^3) =

$$\frac{A_{Bi} \times e^{\lambda t} \times B_c \times Y_c \times k}{V_{air}}$$

Where:

A_{Bi} = measured beta activity of sample (counts/min)
$e^{\lambda t}$ = decay correction, where t = elapsed time (hours) between separation and counting and $\lambda = 5.78 \times 10^{-3} hr^{-1}$
B_c = counter efficiency correction for converting counts/min to dis/min of bismuth-210
Y_c = chemical yield correction of sample =

$$\frac{\text{carrier added (mg bismuth oxide)}}{\text{carrier mounted (mg bismuth oxide)}}$$

k = 0.45 pCi per dis/min

and:

V = volume of air sampled (m^3).

10. Effect of Storage

10.1 The storage of collected samples will present no problems. For long-term storage, corrections must be applied for decay of the 21-year lead-210. A minimum 30-day storage period is advised to assure secular equilibrium between lead-210 and its daughter bismuth-210.

11. References

1. PATTERSON, R.L., and L.B. LOCKHART, JR. 1964. Geographical Distribution of Lead-210 (RaD) in the Ground Level Air. In *The Natural Radiation Environment* (Adams, J.A.S., and W.M. Lowder, Eds.). Chicago: University of Chicago Press. pp. 383–392.
2. KING, P., L.B. LOCKHART, JR., R.A. BAUS, R.L. PATTERSON, JR., H. FRIEDMAN, and I.H. BLIFFORD, JR. 1952. The Collection of Long-Lived Natural Radioactive Products from the Atmosphere. *Naval Research Laboratory Report 4069*.
3. BAUS, R.A., P.R. GUSTAFSON, R.L. PATTERSON, JR., and A.W. SAUNDERS, JR. 1957. Procedure for the Sequential Radiochemical Analysis of Strontium, Yttrium, Cesium, Cerium and Bismuth in Air-Filter Collections. *Naval Research Laboratory Memorandum Report 758*.
4. SAUNDERS, A.W., JR., R.L. PATTERSON, JR., and L.B. LOCKHART, JR. 1967. Standardization of B-Counting Procedures Used at NRL in the Radio-Chemical Analysis of Air Filter Samples. *Naval Research Laboratory Memorandum Report 1764*.

Subcommittee 8
B. SHLEIEN, *Chairman*
O. G. HANSON
R. G. HEATHERTON
P. F. HILDEBRANDT
L. B. LOCKHART
R. S. MORSE
D. E. VANFAROWE

605.

Tentative Method of Analysis for Plutonium Content of Atmospheric Particulate Matter

11322-01-70T

1. Principle of Method (1)

1.1 This procedure may be applied to the analysis of ^{238}Pu, ^{239}Pu and ^{240}Pu* in air filters.

1.2 The plutonium is equilibrated with ^{236}Pu tracer, and isolated by coprecipitation with cerium and yttrium fluorides. The plutonium is further purified by a double anion-exchange column technique; first by absorbing from a hydrochloric acid medium, the second time from a nitric acid medium. Finally the plutonium is electrodeposited onto a platinum disc. The plutonium isotopes are resolved by a solid state alpha spectrometer. The ^{236}Pu is used to determine the chemical yield.†

2. Sensitivity and Range

2.1 The minimum detectable quantity of plutonium is 0.015 pCi if a 1000 min counting time is employed.

2.2 Under practical conditions, a concentration of 0.08×10^{-3} pCi/m^3 (0.08×10^{-15} μCi/cc) can be determined with a standard error of less than ± 10%.

3. Interferences

3.1 None.

4. Precision and Accuracy

4.1 The precision of the method is essentially a function of volume of air sampled, and the length of time for which the sample is counted.

4.2 The primary standard usually is stated to have an absolute accuracy of no greater than ± 3% (standard deviation).

*The energies of the alpha particles from ^{239}Pu and ^{240}Pu are not sufficiently different to be separated by alpha spectroscopy. Therefore, when ^{239}Pu is referred to, it is meant to represent the sum of the activities of ^{239}Pu and ^{240}Pu.

†Available on special order from Oak Ridge.

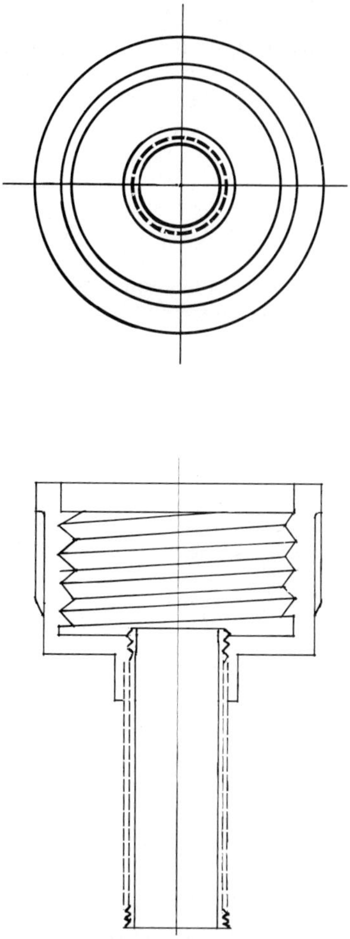

Figure 605:1—Electrolytic cell.

5. Apparatus

5.1 VIRGIN PLATINUM DISCS. 17.6 mm diam × 0.005″, mirror finish one side.

5.2 COPPER DISC, 17.6 mm diam for backing.

5.3 ELECTROLYTIC CELL‡ (Figure 605:1).

5.3.1 The major feature of the electrolytic cell is an elongated 20-mm cap for 30 ml (1 oz) and 60-ml (2 oz) polyethylene bottles. This cap has space for an 18 mm (11/16″) diam plating disc and has a threaded brass bushing for making electrical connection.

5.3.2 In use, the bottom is cut off a 1-oz or 2-oz bottle (20-mm cap size), and any flashing or other roughness on the top is smoothed by rubbing once or twice over a piece of sandpaper. The plating disc is placed on the cap and the bottle screwed in firmly.

5.3.3 Electrical connection may be made by a clip, or by threading the bushing into a metal plate which then also acts as a support for the cell.

5.4 ELECTROLYTIC ANALYZER.

5.5 SMALL ION-EXCHANGE COLUMN (Figure 605:2).

6. Reagents

6.1 PURITY. Reagents must be ACS analytical reagent quality. Distilled water should conform to ASTM Standards For Referee Reagent Water.

6.2 PU-236 TRACER SOLUTION. About 10 dpm/g in dispensing bottle.

6.3 CERIUM CARRIER, 10 mg/ml—31.0 g $Ce(NO_3)_3 \cdot 6H_2O$/l 1:99 HNO_3.

6.4 YTTRIUM CARRIER, 10 mg/ml—43.2 g $Y(NO_3)_3 \cdot 6H_2O$/l 1:99 HNO_3.

6.5 0.4N HCL—0.01N HF.

6.6 0.4N HNO_3—0.01N HF.

6.7 METHYL RED INDICATOR SOLUTION. Dissolve 100 mg of dye in 65 ml of ethyl alcohol and dilute to 100 ml with water.

6.8 FRESHLY PREPARED 5% $NaNO_2$ SOLUTION.

6.9 DOWEX 1 × 4 (100 TO 200 MESH, CHLORIDE FORM).§

6.10 DOWEX 1 × 4 (100 TO 200 MESH, NITRATE FORM). Convert to nitrate form by washing the resin with 1:1 HNO_3 until the washings show no trace of chloride when tested with $AgNO_3$.

6.11 SATURATED BORIC ACID SOLUTION.

6.12 HYDROXYLAMINE.

‡Based on design of the Laboratory of Radiation Biology, University of Washington. These cells are commercially available from Control Molding Corporation, 84 Granite Avenue, Staten Island, New York 10303.

§Analytical grade resin, commercially available from numerous suppliers.

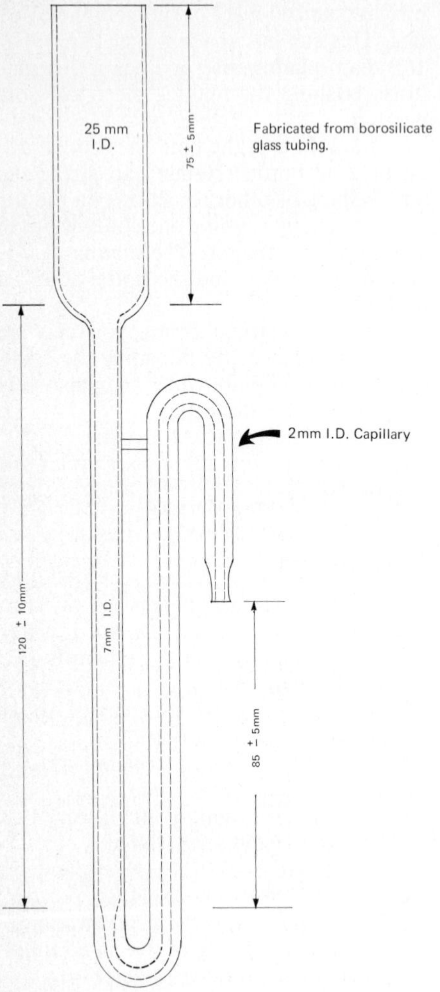

Figure 605:2—Small ion exchange column.

7. Procedure

7.1 SAMPLING.

7.1.1 The particulate matter in a 10,000 m^3 or larger sample of air is collected on a high efficiency filter 6″ × 9″ or 8″ × 10″ in size mounted in a suitable leak-tight holder and protected from rain and snow by a louvered shelter or canopy.

7.1.2 Air is drawn through the filter at a rate of about 1000 m^3/day through use of a properly calibrated or instrumented blower. Filters should be changed before the pressure increase across the filter indicates a 10% flow decrease, in order to obtain a representative sample and to minimize the error in the estimation of sample volume. Under these conditions of flow, several filters may be required to collect a 10,000 m^3 sample depending on the level of atmospheric dust loadings.

7.1.3 Records are to be kept of the time of the beginning and end of collection, during the collection, and the initial and final flow rates.

7.2 ANALYSIS.

7.2.1 Place the sample in a 600- or 800-ml beaker and add 25 ml of 1:10 HNO_3. Add an exactly weighed amount (1 to 3 dpm) ^{236}Pu tracer solution and 1 ml each of cerium and yttrium carriers.

7.2.2 Add 300 ml of HNO_3 and 25 ml of HCl and cover (HCl aids in dissolving iron oxides). Allow the mixture to react, then heat on a hot plate to decompose organic material.

7.2.3 Evaporate to a small volume, adding 30% H_2O_2 repeatedly to complete the oxidation. Cool, add 5 ml of HNO_3 and 10 ml of 60% $HClO_4$.

7.2.4 Evaporate to $HClO_4$ fumes, cool. Transfer to a 250-ml Teflon beaker with 1:10 HNO_3.

7.2.5 Add 10 ml of HNO_3 and 10 ml of HF and heat to expel silica. Evaporate to $HClO_4$ fumes, cool.

7.2.6 Dilute the perchloric acid solution in the Teflon beaker to about 20 ml, add a few drops of 30% H_2O_2 to decompose any MnO_2 that may have formed on the side of the beaker. Cool and transfer the solution to a Lusteroid‖ (or polyethylene) centrifuge tube with 2:1 HCl.

7.2.7 Add about 100 mg of hydroxylamine hydrochloride, stir, and let stand for about 5 min.

7.2.8 Add 5 to 10 ml of HF, stir, and allow the precipitate to coagulate. Centrifuge, decant and discard the supernate.

7.2.9 Wash the fluoride precipitate with about 5 ml of 2:1 HCl. Centrifuge, decant and discard the supernate.

‖Commercially available from numerous suppliers.

7.2.10 Add 5 to 10 ml of HF to the precipitate and stir. Centrifuge, decant, and discard the supernate.

7.2.11 Dissolve the fluoride precipitate in 5 ml of saturated H_3BO_3 solution and heat in a hot water bath. Add 2 to 5 ml of HCl and heat in a hot water bath with occasional stirring to complete solution. Transfer to a 40-ml glass centrifuge tube with water.

7.2.12 Add 1:1 NH_4OH dropwise to precipitate cerium and yttrium hydroxides and place in a hot water bath to coagulate the precipitate. Centrifuge and discard the supernate.

7.2.13 Dissolve the precipitate in about 5 ml of HCl. Transfer the solution to a 100-ml beaker with 2:1 HCl.

7.2.14 Add 2 drops of HNO_3 to oxidize the plutonium. Heat until the reaction stops. Cool in an ice water bath.

7.2.15 Add about 1 ml of the analytical grade Dowex 1 × 4 (100 to 200 mesh, chloride form) resin in 2:1 HCl to the beaker and stir.

7.2.16 Prepare a small ion-exchange column with a glass wool plug and about 2 ml of the chloride form resin and wash with 2:1 HCl until all the resin settles.

7.2.17 Rinse the cover and sides of the beaker with 2:1 HCl. Decant the solution into the anion-exchange column, then wash the resin in the beaker into the column with 2:1 HCl. Wash the beaker and column with at least 50 ml of 2:1 HCl. Discard the effluent and washings.

7.2.18 Elute the plutonium from the column into a 100-ml beaker with at least 50 ml of 0.4*N* HCl–0.01*N* HF solution. Discard the resin.

7.2.19 Evaporate the eluate to dryness several times with repeated additions of HNO_3.

7.2.20 Add 10 ml of 1:1 HNO_3 and 0.25 ml of freshly prepared 5% $NaNO_2$. Heat the solution to boiling, then cool in an ice water bath.

7.2.21 Add about 1 ml of analytical grade Dowex 1 × 4 (100 to 200 mesh, nitrate form) resin in 1:1 HNO_3 to the solution and stir.

7.2.22 Prepare a small ion-exchange column with a glass wool plug and about 2 ml of the nitrate form resin and wash with 1:1 HNO_3 until all the resin settles.

7.2.23 Rinse the cover and sides of the beaker with 1:1 HNO_3. Decant the solution into the anion-exchange column, then wash the resin in the beaker into the column with 1:1 HNO_3. Wash the beaker and column with at least 50 ml of 1:1 HNO_3. Discard the effluent and washings.

7.2.24 Elute the plutonium from the column into a 100-ml beaker with at least 50 ml of 0.4*N* HNO_3–0.01*N* HF. Discard the resin.

7.2.25 Evaporate the eluate to dryness several times with repeated additions of HCl.

7.2.26 Add 1 ml of HCl, heat gently and transfer the solution to an electrolytic cell. Rinse and police the beaker with 2 1-ml portions of water and add to the cell.

7.2.2 Add 1 drop of methyl red indicator solution and make the solution just basic with 1:1 NH_4OH. Adjust the solution to just acid with 1:5 HCl and add 2 drops in excess. Dilute to 5 ml with water.

7.2.28 Electroplate at a current of 1.2 amperes for 1 hr.

7.2.29 Quench the electrolyte with 1 ml of NH_4OH at the end of the electroplating period. Dismantle the cell and rinse the disc with water, then ethanol.

7.2.30 Dry and flame the disc over a burner to convert $Pu(OH)_4$ to PuO_2.

7.3 COUNTING.

7.3.1 Count the disc with a solid state silicon surface barrier alpha detector linked to a multichannel analyzer in order to resolve ^{236}Pu, ^{238}Pu, ^{239}Pu and ^{240}Pu.¶

7.3.2 With the sample electroplated on a 3.1 cm² area, the detector counting efficiency is approximately 31% and the resolution 75 KeV (full width at half peak maximum).

7.3.3 Count for a sufficient length of time. For low level environmental samples 1000 min counting time is often required.

¶The major alpha energies for the respective plutonium isotopes are: For ^{236}Pu, 5.77 and 5.72 MeV; for ^{238}Pu, 5.50 and 5.46 MeV; and for ^{239}Pu and ^{240}Pu, 5.16 and 5.14 MeV. Resolution between the energies given for any specific plutonium isotope is difficult.

8. Calibration

8.1 Since an isotopic tracer technique is used in the analysis no absolute calibration of the detector is necessary. A known amount of a calibrated standard of ^{236}Pu is added to the sample and the unknown activities calculated from the relative count rates of the ^{236}Pu, ^{239}Pu, and ^{238}Pu.

8.2 A source containing a mixture of ^{236}Pu and ^{239}Pu is used to align the multichannel analyzer.

9. Calculation

9.1 The activity (in pCi) of ^{239}Pu (or ^{238}Pu) per m³ is calculated as follows:

$$\text{pCi/m}^3 = \frac{(A - B)E}{(C - D)F} \quad (1)$$

where:

A = cpm in plutonium-239 (or plutonium-238) spectral region
B = reagent blank** cpm in plutonium-239 (or plutonium-238) spectral region
C = cpm in plutonium-236 spectral region
D = counter background (cpm) in plutonium-236 spectral region
E = pCi plutonium-236 added
F = sample volume in cubic meters

9.2 The counting error is calculated as follows:

**Consists of ^{236}Pu added to a blank sample filter and carried through the procedure.

Error at 95% Confidence Level =

$$\frac{1.96E}{F}\sqrt{\frac{\frac{A}{t_1} + \frac{B}{t_2} + \left(\frac{A - B}{C - D}\right)^2 \left(\frac{C}{t_1} + \frac{D}{t_3}\right)}{(C - D)^2}}$$

where:

t_1 = sample counting time
t_2 = reagent blank counting time
t_3 = background counting time

10. Effects of Storage

10. Because of the long half lives of the radionuclides little decay occurs during usual storage.

11. References

1. With minor adaptations the radiochemical portions of the procedure are essentially the same as used at the Health and Safety Laboratory, New York Operations Office, U.S. Atomic Energy Commission. "HASL Manual of Standard Procedures with Revisions and Modifications to August 1965." NYO-4700 (Rev.)

Subcommittee 8
B. SHLEIEN, *Chairman*
R. G. HEATHERTON
P. F. HILDEBRANDT
L. B. LOCKHART, JR.
R. S. MORSE
W. STEIN
D. F. VAN FAROWE
A. HULL, *Health Physics Society Liaison*

606.

Tentative Method of Analysis for Radon-222 Content of the Atmosphere

11327-01-68T

PART A: *Estimation of Airborne Radon-222 by Filter Paper Collection and Alpha Activity Measurement of Its Daughters.*

PART B: *Estimation of Airborne Radon-222 by Filter Paper Collection and Beta Activity Measurement of Its Daughters.*

PART C: *Determination of Airborne Radon-222 by Its Adsorption from the Atmosphere and Alpha Activity Measurement.*

PART D: *Determination of Airborne Radon-222 by Its Adsorption from the Atmosphere and Gamma Measurement.*

INTRODUCTION

A summary of the techniques presently used for measurement of airborne radon is presented in this section. These methods are useful in determining the potential hazard of radon in certain occupational environments, in evaluating the radiation hazard of ambient radon to population groups, and in atmospheric tracer studies involving radon.

Interest in naturally occurring radioactivity arises from the fact that the substances have always been part of man's radiation environment and thus may serve as a standard against which radiation hazards from other sources can be compared. Studies of gaseous radon,* together with its particulate daughter products, provide a source of natural radiation data with which to evaluate the inhalation hazards to human populations from man-made radionuclides. In certain industrial environments, where uranium is mined and processed and radium products for medical and industrial applications are produced and used, the hazards from radon and its daughters constitutes an occupational health problem. Loss and mishandling of radium in medical and industrial usage have on occasion been a matter of public health concern and have presented a major decontamination problem involving radon control and hazards evaluation. In addition to studies oriented toward health, radon may also be used as an atmospheric tracer in meteorologic, geophysic, and air pollution investigations.

Radon is the daughter product of radium-226 in the uranium-238 decay chain and is ubiquitous in the terrestrial environment. Gaseous radon atoms, released to the atmosphere from soil, decay to polonium, lead and bismuth daughter atoms. These radionuclides in turn become attached to solid matter in the environment, forming a radioactive aerosol. The decay series for radon is presented in Figure 606:1. Radon, together with its short- and long-lived daughter products, constitutes the most significant portion of the natural radioactivity inhaled by man **(21)**.

*Radon in these methods refers to the isotope radon-222 and is not to be confused with other radon isotypes, *i.e.*, radon-220 (thoron) or radon-219 (actinon).

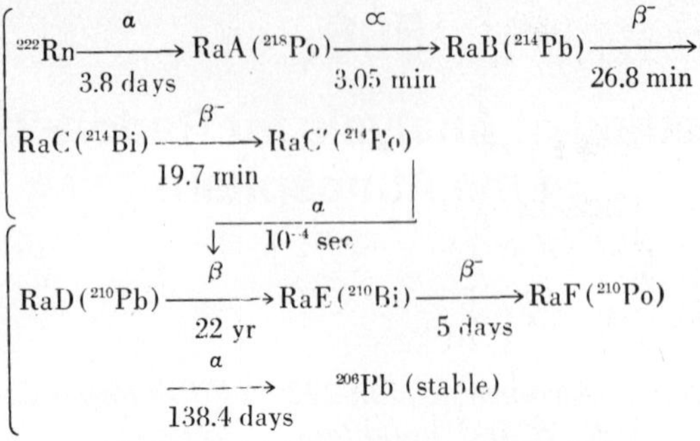

Figure 606:1—Decay Series of Radon-222.

Table 606:I presents a summary of some pertinent concentrations and dose estimates from airborne radon. The wide range of radon concentrations and inhalation doses, evident from these data, is due to differences in the natural and industrial environments investigated as well as assumptions used in making the calculations.

Considerable amounts of data on airborne radon concentrations have been gathered and several texts contain reviews of the subject (**12, 19, 30**). Due to its wide distribution and the large variations in airborne concentration, studies of population and industrial exposure to various levels of radon concn may provide a better understanding of the inhalation hazards from radioactive aerosols and of the environmental and biologic processes operative.

The purpose of this section is to provide an investigator with a broad outline of methods employed in the determination of airborne radon, as well as their applications and limitations, so there may be a basis for choosing one method for a particular field situation. At the same time, detailed descriptions of the more commonly employed procedures are presented which can serve as a guide for technicians charged with routine determination of airborne radon.

It is beyond the scope of this method to describe all analytic procedures for airborne radon determination in detail. Those methods which are completely described offer the advantages of simplicity, rapidity and easy field application and/or the advantages of high sensitivity and lack of dependence on radon and radon-daugh-

Table 606:I. Representative Concentrations and Estimated Doses from Airborne Radon

A. *Concentrations*	
1. Ground Level Air Over Continents	(**21**) 30 to 300 pCi m^3
	(**20**) 100 to 3900 pCi/m^3 surface
2. Over Oceans (between one and two orders of magnitude lower than land concentrations) (**21**)	
3. Uranium Mines in Colorado Plateau (**2**) 3 × 10^6 pCi/m^3 (average) to 58 × 10^6 pCi/m^3	
B. *Doses* To a Nonoccupationally Exposed Population*	
1. To Lung	(**22**) 100–900 mrem/yr
	(**8**) 15– 1500 mrem/yr
2. To Bronchial Epithelium	(**10**) 140 mrad/yr
3. To Respiratory Bronchioles	(**5**) 20– 30 mrad/yr†

*Includes dose from daughter products. Wide range due to different assumptions used in calculations, i.e., average concn, degree of equilibrium, retention of radon daughters in lungs, and weight of lungs.

†Based on living accommodations with adequate ventilation, 6000 hr exposure. Tissue depth 15 to 30μ from top of mucous layer.

ter equilibrium. Alternative methods, described to a lesser degree, have limitations of greater complexity or less sensitivity than those more completely detailed. They may, however, find applicability in specific environmental situations. Methods employable only in high radon concentrations are referenced.

Choice of the method to use in determining airborne radon concentrations from the various techniques available depends on: (a) the levels of radon concentration to be measured, (b) the accuracy required, (c) the equipment available, and (d) complexity of technique. Table 606:II summarizes suggested methods together with comments on each method.

PART A: *Estimation of Airborne Radon-222 by Filter Paper Collection and Alpha Activity Measurement of Its Daughters.*

1. Principle of the Method

1.1 This method is based on collection and measurement of the short-lived radon daughters whose activity is then related to the parent gaseous radon concentration. Radioactive equilibrium between radon and its daughters is assumed. In particular, the alpha activity of RaC′ (Polonium-214), and to a lesser extent that of RaA (Polonium-218), is measured and, assuming equilibrium conditions between radon and its short-lived daughters, the concn of the parent radionuclide, radon-222, can be calculated. The average alpha half-life of the combined radon daughters is approximately 36 ± 5 min **(17, 24, 27)** so that after 4 hr of sampling the collection rate of the short-lived radon daughters on the filter is essentially equivalent to their decay rate on the filter. Collection times need not be as long as 4 hr since corrections for differences in the rate of daughter buildup and decay on the filter can be made. Radon daughter activity must be corrected for per cent buildup on the filter, decay on the filter between the end of the sampling period and the time of counting, interference from radon-220 (thoron) daughters and long-lived alpha emitters, and efficiency of the counting equipment.

1.2 This method is simple, rapid and inexpensive. Errors may arise due to disequilibrium between radon and its daughters, nonuniformity of attachment of radon daughters to particulate matter, or differences in the relative activity of the various daughter products at the time of counting.

2. Range and Sensitivity

2.1 Lower Range. Assuming a 4 hr sampling period, a delay between the end of sampling and the "exponential center" of the counting time of 5 min, a sampling rate of 10 l/min, a counter efficiency of 0.40, and a counting time of 10 min, the lower detection limit for this method is approximately 30 pCi/m^3 with a counting error of 10% at the 95% confidence limit.

2.2 Upper Range. Not limited.

3. Interferences

3.1 Disequilibrium conditions between radon and its daughters, and between the daughters themselves, may interfere with accurate estimation of radon.

3.2 Alpha-emitting radionuclides, particularly the daughters of radon-220 (thoron), will interfere in estimation of radon activity. This is corrected for by recounting the sample after 4 hr. Corrections need not be made for the decay of radon-220 (thoron) daughters during this period because of their relatively long half-life and low concentration. If a correction is required (high ratio of radon-220 to radon-222) the t_2 count is divided by 0.78 to give the corrected radon-220 (thoron) interference.

3.3 Other interfering alpha emitters,

Table 606:II. Summary of Methods for Radon Determination

Mode of Collection	Theoretical Basis	Counting Method	Sensitivity*	Application	Advantages and/or Disadvantages
I. Filter Paper.	Particulate daughter products collected on filter paper and their activity related to radon concentration. Equilibrium assumed to exist between radon and its daughters.	(a) Alpha **(6, 24)**	30 pCi/m³	Field Studies	Simple, rapid and inexpensive. Assumes a radon radon-daughter equilibrium.
		(b) Beta **(13, 14)**	15 pCi/m³	Field Studies	Useful for high dust loadings due to greater β particle range.
II. Adsorption on charcoal bed.	Gaseous radon collected by adsorption from atmosphere by passage over activated charcoal at −78 C.	(a) Alpha **(18)**	~ 3 pCi/sample	In studies of low levels of radon or where high sensitivity is required.	No assumptions as to equilibrium state are required. Equipment and technique complex.
		(b) Gamma **(11, 27)**	120 pCi/m³	Survey studies where direct determination of radon is required.	Requires no transfer of radon from collection to counting.
III. Entrapment of air containing radon.	Radon and daughters not isolated from air. Counted directly.	(a) Ionization Chamber **(28)**	~ 1.0×10^4 pCi/m³ by direct reading mode.	At relatively high levels.	Direct reading. Sensitivity can be increased by pulse height counting.
		(b) Alpha **(7)**	2.5×10^4 pCi/m³	In areas of high concentration.	Least sensitive.

*Approximate concentrations for 10% counting errors at 95% confidence level using collection and analysis methods cited in reference.

usually present in such low concentrations that they offer minimal interference with the determination of radon, are polonium-210, plutonium-239, -240, and -238, uranium-238 and -235, and radium-226.

4. Precision and Accuracy

4.1 A precision of 10% at the 95% confidence level is obtainable due to counting errors. Assuming no leakage in the filtration device, flow measurements should be correct to ± 2% of the full scale reading on the flow meter.

5. Apparatus

5.1 Collection. Several studies of radon daughter collection efficiencies with a variety of filters have been made **(15, 29)**. A membrane filter with a 0.8 μ pore size† has been shown to be very successful in trapping radon daughters efficiently **(1, 4)** and offering negligible self-adsorption during counting **(25)**. A simple inexpensive device for collection of airborne radon daughters consists of a self-lubricating vacuum pump with a sampling rate of 1 to 2 cfm,‡ a flow rate meter,§ manometer (if required) and membrane filter holder.‖ This collector (Figure 606:2) is capable of sampling air at about 20 l/min through a 47 mm diam membrane filter. The sampler is trouble-free for long durations and requires little maintenance.

5.2 Alpha Counting Equipment. A Zinc Sulfide Screen Alpha Scintillation counter consists of a silver activated zinc

†Millipore Filter Corp., Bedford, Mass.; Gelman, Ann Arbor, Mich.; and Schleicher and Schuell, Keene, N.H.

‡Gast, Benton Harbor, Mich., and Leiman Bros., Inc., E. Rutherford, N.J.

§Fisher Porter, Warminister, Pa.; Brooks Inst. Co., Hatfield, Pa.; and RGI Inc., Vineland, N.J.

‖Unico, Port Chester, N.Y.

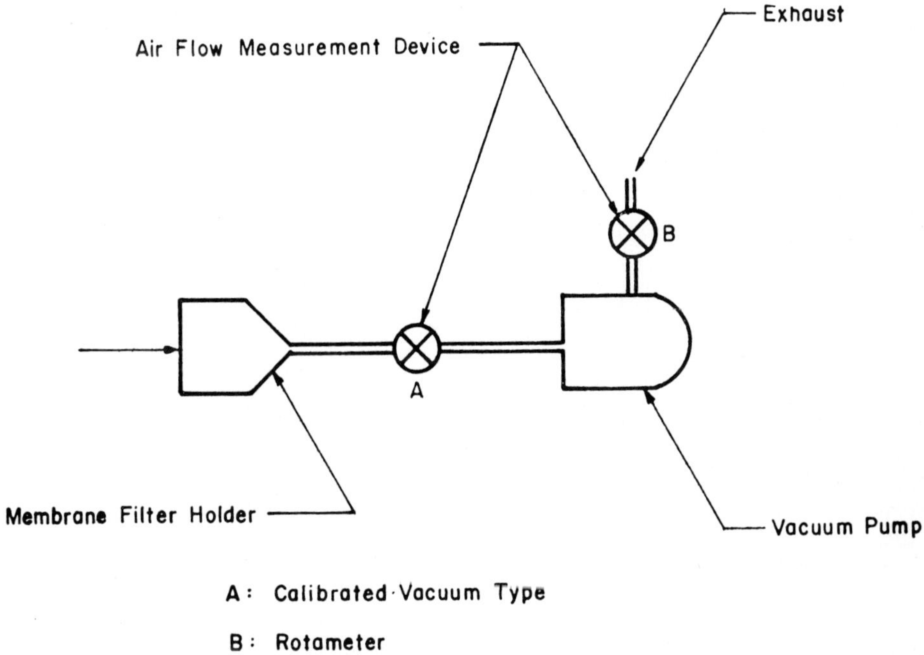

Figure 606:2—Membrane filter collector for radon daughters.

sulfide screen¶ mounted on the face of a 2″ photomultiplier tube. The photomultiplier tube and preamplifier (both housed in a light-tight enclosure) are connected to a stable high voltage supply. The interaction between the alpha particles and the zinc sulfide screen causes emission of actinic light which is converted to electrical pulses by the photomultiplier tube and power supply. In one such instrument the operating voltage plateau extends from approximately 700V to 950V. Electronic noise is less than 0.5 count/min (**3**). Counter efficiency should be determined with a suitable alpha emitter spread over a membrane filter or a plated alpha source of proportions similar to that of a collected dust sample. Polonium-210 sources give counter efficiencies of approximately 30 to 40%, depending on the source to zinc sulfide screen distance.

5.3 An alternate counting system to the one described above employs an internal proportional gas flow counter. Counting efficiency is somewhat higher (approximately 50 to 51%), but the possibility of static charge on the filter and of counter contamination exists.

6. Reagents

6.1 None.

7. Procedure

7.1 Sample for a suitable collection period: For samples collected for more than 4 hr, radon daughter decay on the filter and radon daughter buildup on the filter will be in equilibrium.

7.2 Remove sample from filter holder. Record sampling rate (l/min), duration of sampling, and the time of sample removal.

7.3 Count as soon as possible. Record the time from the end of the sampling period to the "exponential center" of the counting period, *i.e.*, to time 31.6% into the counting period. To keep the standard deviation of the alpha counting to less than 5%, the following table can be used as a counting guide (**24**). These suggested times assume that background levels of less than 0.5 cpm, usually obtainable with alpha counters, are maintained. Subtract counter background to yield the net counts for the sample. Divide by the length of the count to yield net count/min.

Table 606:III. Counting time for a standard deviation of five %.

Count/Min	Counting Time (min)
400	1
200	2
100	4
50	8
40	10

7.4 Recount the sample after the decay of the radon daughters, *i.e.*, 4 hr after end of sampling. Second count represents activity of longer-lived alpha emitters (thoron daughters and other long-lived alphas). Make subtractions as above to yield net count/min.

8. Calibration

8.1 Alpha counter should be calibrated with a standard of known disintegration rate distributed over a geometric area similar to the area of dust collection. The "counting efficiency" is equivalent to the ratio of the counting rate (in count/min) of the instrument to the known disintegration rate (in dpm of the standard). Suitable alpha standards are Uranium Oxide, Polonium-210 in radioactive equilibrium with Lead-210 or Plutonium-239.

9. Calculations

9.1 Calculate the count rate and counting error at t_1 (initial count) due to the sample as illustrated below:

Counting Rate Sample and Instrument Background in count/min (cpm), $(R_S + R_B)$ =

$$\frac{\text{Total Counts Sample with Background}}{\text{Total Time Counted}} \quad (1)$$

¶U.S. Radium Corp., Morristown, N.J.

Instrument Background Counting Rate (R_B) =

$$\frac{\text{Total Counts Instrument Without Sample}}{\text{Total Time Counted}} \quad (2)$$

Counting Rate Sample (R_S) =
$(R_S + R_B) - (R_B)$ (3)

Error Due to Counting (at 95 per cent confidence level) =

$$1.96\sqrt{\frac{(R_S + R_B)}{\text{Time Counted}} + \frac{R_B}{\text{Time Counted}}} \quad (4)$$

9.2 Calculate the count rate and counting error at t_2 (after 4 hr decay). This activity is due to long-lived radon-220 (thoron) daughters and other long-lived alpha emitters. Calculate as above. Corrections need not be made for radon-220 (thoron) daughter decay in most cases (see **Interferences**, Section **3**).

9.3 The activity due to short-lived radon daughters at t_1 is the difference between the counts t_1 and t_2.

Counting Rate Radon Daughters $(R_S') = (R_S \text{ at } t_1) - (R_S \text{ at } t_2)$ (5)

Error Due to Counting is:

$$\sqrt{(\text{Counting Error at } t_1)^2 + (\text{Counting Error at } t_2)^2} \quad (6)$$

This error term is carried throughout further calculations and the same mathematical processes are repeated on it as on the counting rate.

9.4 Correct for disequilibrium due to sampling less than 4 hr and for decay of radon daughters from end of sampling to exponential center time of initial count. These corrections can be made from the graph presented in Figure 606:3 **(24)**.

Corrected Counting Rate Radon Daughters (R_S'') =

$$\frac{\text{Counting Rate Radon Daughters } (R_S')}{\text{Correction Factor } (d = O, T = \infty)} \quad (7)$$

9.5 Calculation of Disintegration Rate Radon Daughters. Disintegration Rate Radon Daughters in Disintegrations per Min (dpm) (Ar) =

$$\frac{\text{Corrected Counting Rate Radon Daughters } (R_S'')}{\text{Counting Efficiency}} \quad (8)$$

9.6 Calculation of Radon Concentration. Radon in pico curies per cubic meter (pCi/m³) can be estimated from radon daughter alpha activity by the following formula: **(6)**

$$Rn = 5.95\ Ar/F,$$

where

Ar = radon daughter activity in disintegrations/min (dpm), and

F = flow rate in liters/min.

10. Effects of Storage

10.1 (Not Applicable).

11. References (See Part D, Section 11.)

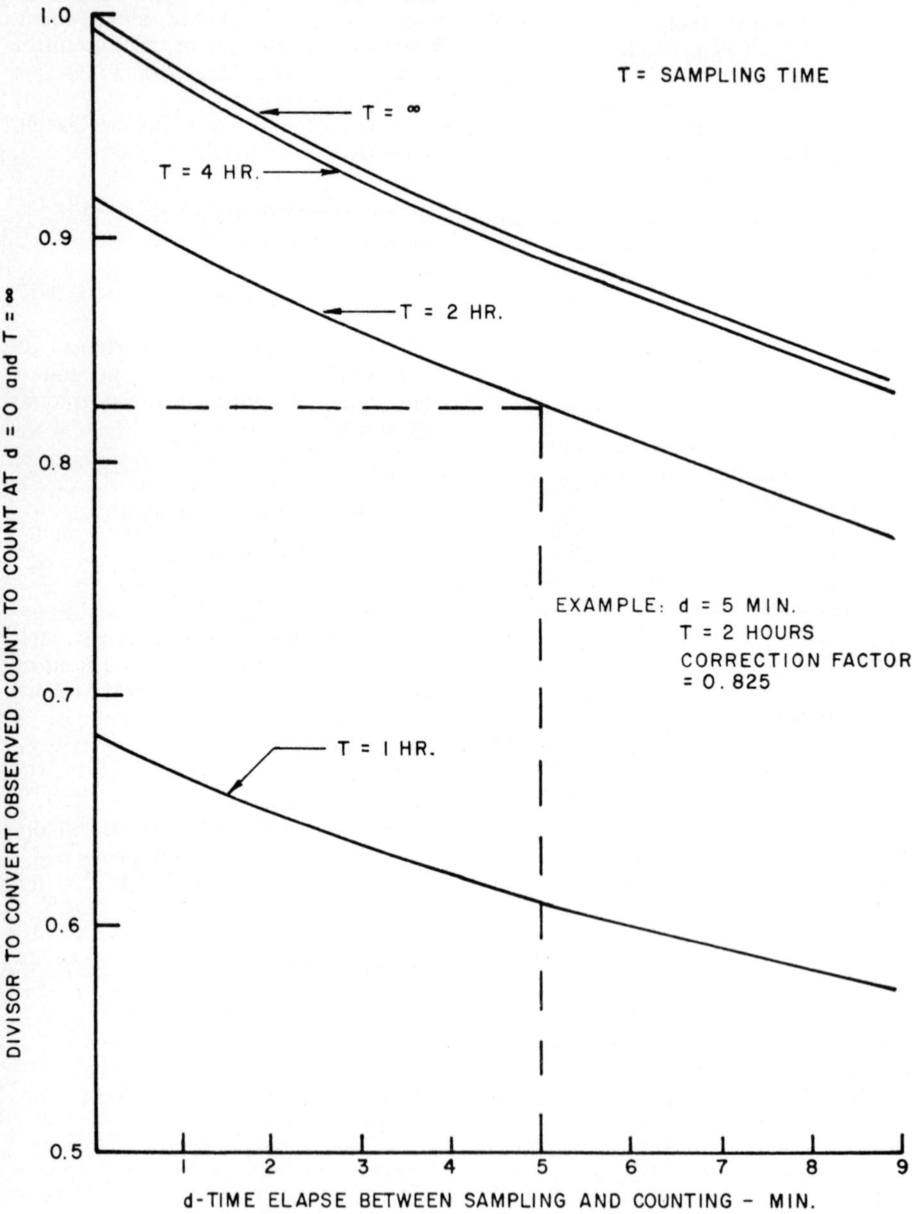

Figure 606:3—Conversion chart—radon daughter products (24).

PART B: *Estimation of Airborne Radon-222 by Filter Paper Collection and Beta Activity Measurement of Its Daughters*

1. Principle of the Method

1.1 Principle of the method is similar to that of the alpha method except the beta activity of the radon daughters RaB (lead-214) and RaC (bismuth-214) is determined. These activities are then related to the parent radon. Again radioactive equilibrium between radon and its daughters is assumed. Beta counting of radon decay products collected on filters offers several advantages over alpha-counting:

1.1.1 Larger samples can be handled before dust loading becomes a problem because of the greater range of beta particles.

1.1.2 Many laboratories are equipped to do beta-counting and are familiar with beta-counting techniques and standardization of beta counters.

1.1.3 Stable, long-lived GM tubes are available for both field and laboratory use: anticoincident counters of low background are available for laboratory use.

2. Range and Sensitivity

2.1 Radon concentrations as low as 15 pCi/m^3 can be measured with a counting error of 10% at the 95% confidence level, under the flow conditions (0.3 m^3/min) obtainable with a 2-inch diam glass fiber filter of essentially 100% efficiency **(15)** and with a counter having a 30 cpm background.

3. Interferences

3.1 Disequilibrium conditions between radon and its daughters, and between the daughters themselves, may interfere with accurate estimation of radon.

3.2 The major interference is from radon-220 (thoron) daughters in the air; they are adequately corrected for in the procedure described below. If quantities of other beta emitters are present, *e.g.*, after a nuclear test, they can be corrected for separately by making a third beta measurement after 12 to 16 hr decay and solving a simple equation to obtain the relative contributions of thoron daughters and long-lived beta emitters to the total. For 20-min collections such corrections for long-lived beta emitters are generally negligible.

4. Precision and Accuracy (See 2 above.)

5. Apparatus (See Part A.)

6. Reagents

6.1 None.

7. Procedure

7.1 Filter air at a known, constant rate near 0.5 m^3/min through a 2″ diam glass fiber filter for exactly 20.0 min, using a positive displacement blower and flowmeter or calibrated pressure gauge. Remove filter from blower and beta count through a 75 mg/cm^2 absorber (total Al plus counter window) for 10.0 min starting 60 sec following end of collection. For a 20-min collection, an air concentration of 1 pCi/m^3 of radon in secular equilibrium with its short-lived daughters when sampled at 1 m^3/min will give a total of 332.2 disintegrations of RaB (lead-214) and 416.9 disintegrations of RaC (bismuth-214) during the 10-min counting period (t_{1-11}). From the calibration procedure described in References **13** and **14**, and data shown in Tables 606:IV, V, an overall beta-counting efficiency for the RaB/RaC complex can be derived, based on standardization with Cesium-137, Lead-210, and Protactinium-234 (UX_2). An approximate calibration can be obtained by assuming the RaB/RaC complex (including conversion electrons) is counted with 53% of the efficiency of a protactinium-234 standard of the same geometry (assuming 620 dpm of protactinium-234/mg of U_3O_8). A second count of the sample after 5 hr permits determination of longer-lived betas from the decay of atmospheric radon-220 (thoron).

Table 606:IV. Calculated efficiency of beta counting of RaB (lead-214) and RaC (bismuth-214) deposited on a glass fiber filter*

Activity	Beta Energy Max MeV	Counting Efficiency (%)		Corrected for Conversion Electrons (%)
RaB (lead-214)	0.65 (100%)	1.00 × 1.82	1.82	1.87
	25% low energy conversion electrons			
RaC (bismuth-214)	3.17 (23%) 1.65 (77%)	0.23 × 13.0 0.77 × 6.2	7.76	8.15

*Counted through 75 mg/cm^2 total absorber (Al plus mica window).

8. Calibration and Standards

8.1 A more complex extension of the above technique allows determination of atmospheric radon during periods of radon-radon daughter disequilibrium. Beta activity is also determined after 61 to 71 min have elapsed in addition to the times mentioned above. The ratio of RaC to RaB can then be characterized by the ratio of the 61 to 71 min count to the 1 to 11 min count. The advantage of this technique is that the activity of RaA can be estimated from the atom ratio of RaC and RaB. Because of its short half-life, RaA is very likely to be in equilibrium with gaseous radon. Therefore, the investigator gains an indication of the equilibrium state between gaseous radon and its short-lived daughters which is particularly useful in meteorologic studies. However, the complexity of this approach makes it most applicable to the specific situations (**13**).

9. Calculation

9.1 The following equation takes into account the disequilibrium of RaB and RaC on the filter and corrects for the small background of ThB/ThC activity (10.6 hr half-life) and the traces of fission products in the air.

$$\text{Radon-222 in pCi/m}^3 = \frac{\text{Counts } t_{1-11} - 1.4 \text{ counts } t_{301-311}}{\text{calc. counts } t_{1-11} \times F}$$

Where:

Counts t_{1-11} and $t_{301-311}$ = total measured beta counts above background during the 1 through 11 min and 301 through 311 min periods following collection.

Calculation counts t_{1-11} are the total counts calculated for the RaB/RaC complex from a radon-222 concentration of 1 pCi/m^3 in the air when counted under the standard conditions of absorber and geometry.

F is the average air flow in m^3/min.

10. Effect of Storage

10.1 Not applicable.

11. References (See Part D, Section 11.)

Table 606:V. Determination of β counting rates for 20-min collections of RaB/RaC complex in equilibrium with 1 pCi/m^3 radon and filtered at rate of 1 m^3/min*

	Dis/10 min (t_{1-11})	Counting Efficiency (%)	Counts/10 min (t_{1-11})
RaB	332.2	1.87	6.2
RaC	416.9	8.15	34.0
Total	749.1	—	40.2

*Counted on a counter having counting efficiency of 10.15% for ^{234}Pa (UX_2) under same conditions of absorber thickness and geometry.

Part C: *Determination of Airborne Radon-222 By Its Adsorption From The Atmosphere and Alpha Activity Measurement*

1. Principle of the Method

1.1 The adsorption of gaseous radon on activated charcoal (at low temps) has been investigated for some time **(23, 26)** and forms the basis of sampling techniques used in this method. The adsorbed radon is then de-emanated (at elevated temps) from the collection trap, transferred to a suitable counting cell, allowed to reach equilibrium with its daughters and alpha counted **(18)**.

2. Range and Sensitivity

2.1 Counting rate of 16 cpm has been observed with errors of 6% or less at the 95% confidence level. This counting rate is equivalent to approximately 3 pCi of radon **(18)**.

3. Interference

3.1 Radon from radium present in apparatus and cylinders of helium and aged air.

4. Precision and Accuracy

4.1 Six% at 95% confidence level for counting. Flow measurement ± is 4% with rotameters, and 0.3% with wet test meters. Transfer procedure ± 4%.

5. Apparatus

5.1 COLLECTION. A schematic of the sampling apparatus is presented in Figure 606:4.

5.2 The operation and efficiency of this sampler have been described in detail **(8)**. The air filter** is employed to remove radon daughters prior to entrance of the air into the adsorption portion of the apparatus. The drying agent†† and water traps enhance the adsorptive capacity of the charcoal and prevent clogging of the apparatus. Absorption of radon by these components is minimal. Design of the radon traps themselves depends on preference and on the method by which the activity of the adsorbed radon is to be determined. A backup trap is used initially to determine sampler efficiency; the trap is optional afterwards. When cooled to −78 C (dry ice in 1:1 chloroform: carbon tetrachloride mixture) quantitative recoveries of atmospheric radon have been achieved on 30 g of 6 to 14 mesh activated charcoal‡‡ at approximately 8 l/min **(27)**.

5.3 DE-EMANATION AND TRANSFER. The collection trap is then placed in the apparatus shown in Figure 606:5 **(3)** for de-emanation and transferred to the radon counting cell.

5.4 MEASUREMENT. The radon counting cell has been described in detail **(16)**. Basically it consists of a Kovar§§ metal cylinder enclosed on one end and coated on the inner surface with silver activated zinc sulfide. A quartz window with a transparent tin oxide coating on its inner surface is sealed to the counter shell.

Alpha activity within the cell is measured with equipment similar to that described in Part A, Section **5.2**, except that the photomultiplier tube is not covered with a scintillating screen.

6. Reagents

6.1 None

7. Procedure

7.1 COLLECTION.

7.1.1 Sample at rate of approximately 10 l/min.

7.1.2 After sampling is completed, seal collection trap, removed from sampling apparatus, and allow to warm to room temp.

**See Part A, Section 5.1 and footnote†.

††Hammond Drierite Co., Xenia, Ohio.

‡‡Matheson, Colman and Bell, Norwood, Ohio; Barneby-Cheney Columbus, Ohio.

§§Stupakoff Ceramic and Mfg. Co., Latrobe, Pa.

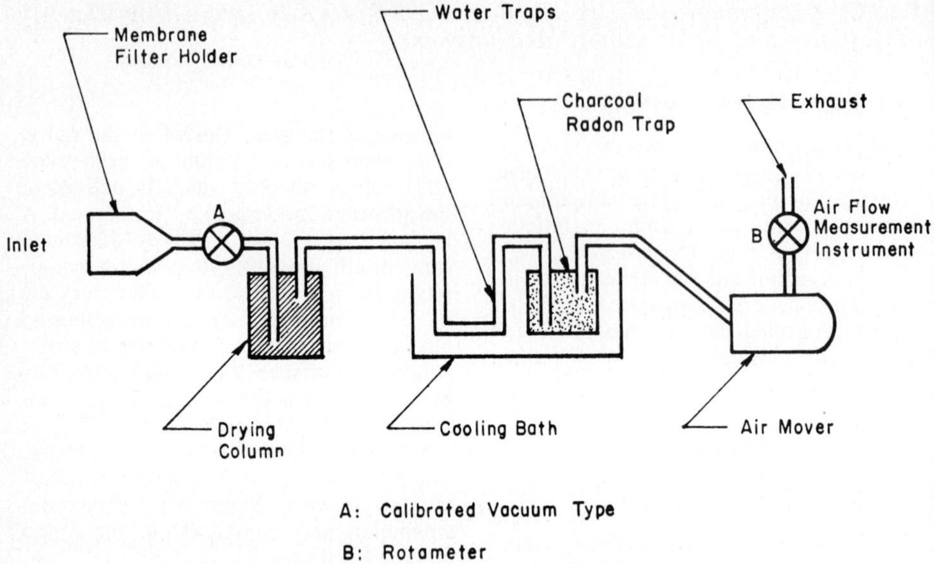

Figure 606:4—Sampling apparatus for collection of gaseous airborne radon.

7.2 TRANSFER.

7.2.1 De-emanate adsorbed radon from collection trap by placing the trap in a preheated tube furnace and heating for 5 min at 500 C.

7.2.2 Flush the transfer line and radon counting cell with helium and then evacuate this portion of the apparatus.

7.2.3 Add a 15 ml portion of helium from the reservoir to the collection trap. Allow to mix for 3 min.

7.2.4 Transfer helium-radon mixture

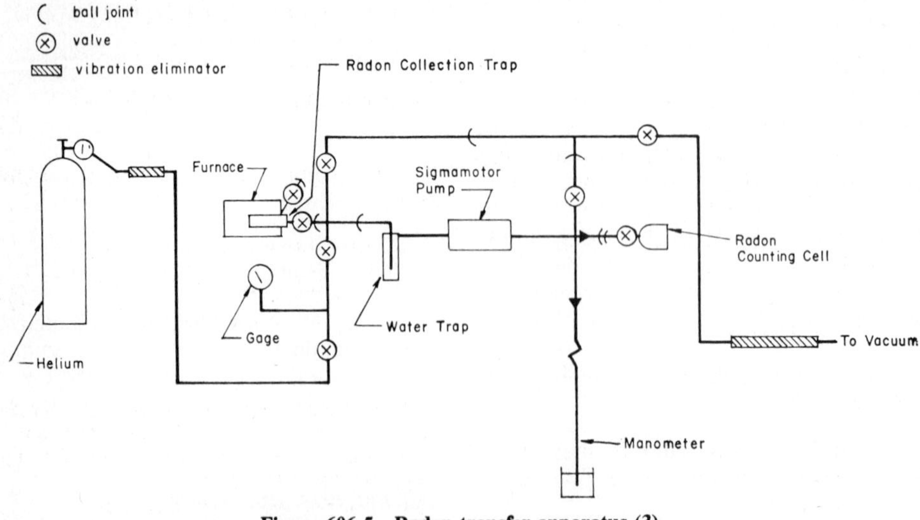

Figure 606:5—Radon transfer apparatus (3).

to radon counting cell by means of transfer pump.

7.2.5 Repeat process until a pressure of 1 atm is reached (about six 15 ml transfers).

7.3 Counting.

7.3.1 After transfer is completed, place the window of the radon cell in contact with the photomultiplier window of the counting apparatus. Complete apparatus to stand in darkness for 4 hr to allow equilibrium to be reached between radon and its daughters. Care should be taken not to expose the radon counting cell to light during the ingrowth period or counting.

8. Calibration

8.1 Efficiency of the collection apparatus, de-emanation and transfer apparatus, and radon counting cell can be determined by de-emanation of standard radium-226 solutions of known activity contained in a suitable bubbler (**9**). The collection apparatus can be checked by passing a small portion of aged air (radon free) through the radon bubbler and mixing it with more aged air so that it passes through the collection apparatus at flow rates employed in field sampling. The de-emanation, transfer and counting procedure are then performed, and the radon recovery calculated. These procedures are quantitative.

8.2 To calculate the radon counting efficiency a similar process is employed, but transfer of ingrown radon from the standard radium-226 solution to the counting cell is achieved with helium. Counting efficiency for alpha particles is approximately 80% (**3**).

9. Calculation

9.1 Calculate the count rate due to the sample:

Counting Rate Sample and Instrument Background in Counts/Min (cpm), $(R_S + R_B)$ =

$$\frac{\text{Total Counts Sample with Background}}{\text{Total Time Counted}} \quad (1)$$

Instrument Background Counting Rate (R_B) =

$$\frac{\text{Total Counts Instrument Without Sample}}{\text{Total Time Counted}} \quad (2)$$

Counting Rate Sample (R_S) = $(R_S + R_B) - (R_B)$ (3)

Error Due to Counting (at 95 per cent confidence level) =

$$1.96\sqrt{\frac{(R_S + R_B)}{\text{Time Counted}} + \frac{(R_B)}{\text{Time Counted}}}$$

9.2 Calculate the count rate due to radon:

Counting Rate Due to Radon (R_{Rn}) =

$$\frac{\text{Count Rate Due to Sample } (R_S)}{3\| \|} \quad (5)$$

9.3 Calculate radon activity:

Radon Activity (A_{Rn}) = in pCi

$$\frac{\text{Counting Rate Due to Radon (cpm)}}{\text{Counting Efficiency\#\#} \times 2.22^{***}} \quad (6)$$

9.4 Calculate radon concn unit volume of air:

Radon concn in pCi/M^3 =

$$\frac{\text{Radon Activity (in pCi)}}{\text{Total m}^3\text{ Sampled}} \quad (7)$$

‖ ‖To correct for the two alpha-emitting daughter products of radon (RaA and RaC′), which are in equilibrium with radon at time of counting, *i.e.*, 4 hr after transfer.

##To convert count per min to dpm.

***To convert dpm to pCi (pico curies).

10. Effect of Storage

10.1 Not applicable.

11. References (See Part D, Section 11.)

PART D: *Determination of Airborne Radon-222 by its Absorption from the Atmosphere and Gamma Measurement*

1. Principle of the Method

1.1 Gaseous radon, quantitatively absorbed on activated charcoal, can be determined without de-emanation and transfer to a counting cell from the gamma activity of its RaC (bismuth-214) daughter (**11, 27**).

The count rate of the 0.61 MeV and 1.76 MeV peaks of RaC are determined. The efficiency used to count these peaks is sufficient to detect RaC from quantities which have reached equilibrium from environmentally collected radon.

2. Range and Sensitivity

2.1 This method is useful in field analyses of radon. It is not dependent on radon-radon daughter equilibrium in the atmosphere. Concentrations as low as 120 pCi/m^3 of radon can be determined.

3. Interference

3.1 None.

4. Precision and Accuracy

4.1 120 pCi/m^3 of radon can be determined with an error of 10% at the 95% confidence limit.

5. Apparatus

5.1 Sampling is carried out in the same manner as described for alpha measurement (Part C, Section **5**.)

6. Reagents

6.1 None.

7. Procedure

7.1 The collection trap is counted with a sodium iodide (thallium activated) crystal, suitably shielded, and linked to a multichannel pulse height analyser. Four hours should elapse prior to counting to allow RaC to reach equilibrium with radon.

8. Calibration and Standards

8.1 Ease of counter calibration is facilitated by the calculation means described below since corrections due to geometry differences between the standard and absorbed radon were not made. Radon loss due to transfer of the charcoal from one container to another was not fully investigated.

8.2 In the first study cited above, the charcoal and absorbed radon were removed from the collection trap and transferred to another container prior to counting.

9. Calculations

9.1 Counting efficiency can be determined by de-emanating a known amt of radon from a standard radon-226 solution and passing it through the sampling train at the flow rate and for the sampling times used in the field. This was previously described in Part C, Section **5.3**.

9.2 The activity of radon in the sample, counted between 3 to 5 hr after the end of sampling so that corrections for disequilibrium need not be made, is calculated as follows:

Rn in pCi =

$$\frac{\text{net cpm under 1.76 MeV peak}}{\text{counting efficiency (1.76 MeV)} = 2.22}$$

or

$$\frac{\text{net cpm under 0.61 MeV peak}}{\text{counting efficiency (0.61 MeV)} = 2.22}$$

10. Effect of Storage

10.1 For storage, corrections must be applied for decay of 3.8-day radon-222.

11. References

1. ANDERSON, D.E. 1960. Efficiencies of filter papers for collecting radon daughters. *Am. Indust. Hyg. A. J.* 21:428.
2. AYERS, H.E. 1954. Control of Radon and Its Daughters in Mines by Ventilation. U.S. Atomic Energy Commission, Div. of Technical Information. Report #AECU-2858. Oak Ridge, Tenn.
3. BLANCHARD, R.L. 1964. An Emanation System for Determining Small Quantities of Radium-226. Public Health Service Publ. #999-RH-9.
4. FITZGERALD, J.D., and C.C. DETWILER. 1956. Optimum Particle Size Penetration through the Millipore Filter. U.S. Atomic Energy Commission, Div. of Technical Information, Report #KAPL-1592. Oak Ridge, Tenn.
5. HAQUE, A.K.M.M., and A.J.L. COLLINSON. 1967. Radiation dose to the respiratory system due to radon and its daughter products. *Health Physics.* 13:431.
6. HARLEY, J.H. 1953. Sampling and measurement of airborne daughter products of radon. *Nucleonics.* 11:12.
7. HARRIS, W.B., H.D. LAVINE, and S.I. WATNICK. 1957. Portable radon detector for continuous air monitoring. *A.M.A. Arch. Ind. Health.* 16:493.
8. HAVLOVIC, V. 1965. Natural radioactive aerosols in the ground-level air of a Czechoslovak locality with respect to the radiological exposure of its population. *Health Physics.* 11:553.
9. HOLADAY, D.A., D.E. RUSHING, P.F. WOOLRICH, H.L. KUSNETZ, and W.F. BALE. 1957. Control of Radon and Daughters in Uranium Mines and Calculation of Biological Effects. Public Health Service Publ. #494.
10. JACOBI, W. 1964. The dose to the human respiratory tract by inhalation of short-lived ^{222}Rn and ^{220}Rn-Decay Products. *Health Physics.* 10:1163.
11. JONES, G.E., and L.M. KLEPPE. 1966. A Simple and Inexpensive System for Measuring Concentrations of Atmospheric Radon-222. U.S. Atomic Energy Commission, Div. of Technical Information. Report #UCRL-16952. Oak Ridge, Tenn., 1966.
12. JUNGE, C.E. 1963. *Air Chemistry and Radioactivity.* New York: Academic Press.
13. LOCKHART, L.B., JR., and R.L. PATTERSON, JR. 1965. Determination of Radon Concentration in the Air through Measurement of Its Solid Decay Products. U.S.A.E.C., Div. of Technical Information. Report #NRL 6229. Oak Ridge, Tenn.
14. LOCKHART, L.B., JR., and R.L. PATTERSON, JR. 1966. The Extent of Radioactive Equilibrium Between Radon and Its Short-Lived Daughter Products in the Atmosphere. U.S.A.E.C., Div. of Technical Information. Report #NRL 6374. Oak Ridge, Tenn.
15. LOCKHART, L.B., JR., and R.L. PATTERSON, JR. 1964. Characteristics of Air Filtering Media Used for Monitoring Airborne Radioactivity. U.S.A.E.C., Div. of Technical Information. Report #NRL-6054. Oak Ridge, Tenn.
16. LUCAS, H.F. 1957. Improved low-level scintillation counter for radon. *Rev. of Sci. Inst.* 28:680.
17. MERCER, T.T. 1954. Atmospheric Monitoring of Alpha Emitters Using Molecular Membrane Filters. U.S.A.E.C., Div. of Technical Information. Report #UR-294. Oak Ridge, Tenn.
18. MOSES, H., A.F. STEHNEY, and H.F. LUCAS, JR. 1960. The effect of meteorological variables upon the vertical and temporal distributions of atmospheric radon. *J. of Geophy. Res.* 65:1223.
19. ISRAEL, H., and A. KREBS. 1962. Editors. *Nuclear Radiation in Geophysics.* Berlin: Springer-Verlag.
20. PEARSON, J.E., and H. MOSES. 1966. Atmospheric radon-222 concentration variation with height and time. *J. Applied Meteorology.* 5:175.
21. REPORT OF THE UNITED NATIONS SCIENTIFIC COMMITTEE ON THE EFFECTS OF ATOMIC RADIATION. 1966. Supplement No. 14 (A/6314), p. 23. New York.
22. REPORT OF THE UNITED NATIONS SCIENTIFIC COMMITTEE ON THE EFFECTS OF ATOMIC RADIATION. 1962. Supplement No. 16 (A/5216), Part E. New York.
23. RUTHERFORD, E. 1906. Absorption of radioactive emanations from charcoal. *Nature.* 74:634.
24. SETTER, L.R., and G.I. COATS. 1964. The determination of airborne radioactivity. *Am. Indust. Hyg. A. J.* 22:64.
25. SHAPIRO, J. 1956. Studies of the Radioactive Aerosol Produced by Radon in Air. U.S.A.E.C., Div. of Technical Information. Report #UR-461. Oak Ridge, Tenn.
26. SHATTERLY, J. 1906. Some experiments on adsorption of radium emanations by coconut charcoal. *Phil Mag.* 20:778.
27. SHLEIEN, B. 1963. The simultaneous determination of atmospheric radon by filter paper and charcoal adsorptive techniques. *Am. Indust. Hyg. A. J.* 24:180.
28. SHLEIEN, B. 1967. Based on laboratory calibration of 4.3 liter ionization chamber. Northeastern Radiological Health Laboratory. Winchester, Mass.
29. SMITH, W.J., and N.F. SURPENANT. Properties of various of filtering media for atmospheric dust sampling. *A.S.T.M. Proc.* 53:1122.
30. ADAMS, J.A.S., and M.W. LAWDER. 1964. Editors. *The Natural Radiation Environment.* University of Chicago Press. Chicago, Ill.

Subcommittee 8
B. SHLEIEN, *Chairman*
O. G. HANSON
R. G. HEATHERTON
P. F. HILDEBRANDT
L. B. LOCKHART
R. S. MORSE
D. E. VAN FAROWE

607.

Tentative Method of Analysis for Strontium-89 Content of Atmospheric Particulate Matter

11332-01-70T

1. Principle of Method (1)

1.1 ^{89}Sr is a short-lived radionuclide (50.5 day half-life) produced by nuclear fission.

1.2 ^{89}Sr is determined at the same time as ^{90}Sr, using the separated strontium fraction from the ^{90}Sr procedure. Thus the radiochemical procedures are those described in the analysis of ^{90}Sr. The ^{89}Sr content of the sample is calculated by correcting the total counting rate of the $SrCO_3$ fraction for the ^{90}Sr present and for the ^{90}Y which grows in between separation and counting. The procedure is simple chemically but requires rather complex calculations, particularly in obtaining the estimate of counting error.

2. Sensitivity and Range

2.1 A minimum of 5 pCi per sample of ^{89}Sr can be determined with an error of less than ± 10% @ the 95% confidence level under usual conditions (counter background of less than 1 cpm and a 60 min counting time).

3. Interferences

3.1 Same as for ^{90}Sr.

4. Precision and Accuracy

4.1 Same as for ^{90}Sr.

4.2 Because the determination of ^{89}Sr is based on the difference in total count rate in the strontium fraction and that due to ^{90}Sr and ^{90}Y, the precision and accuracy is limited by a greater error than for ^{90}Sr analysis.

5. Apparatus

5.1 Same as for ^{90}Sr.

6. Reagents

6.1 Same as for ^{90}Sr.

7. Procedure

7.1 ANALYSIS.

7.1.1 As soon as possible transfer the supernatant liquid obtained in steps **7.4.2** and **7.4.4** under ^{90}Sr Procedure, Second Milking, p. 00 to a 150-ml beaker, heat to boiling and add 5 ml of a saturated sodium carbonate solution slowly with stirring.

7.1.2 Cool and filter through a 2.8 cm glass fiber filter paper using suction. Wash thoroughly with distilled water. Record the time of separation.

7.1.3 Dry and weigh the paper and precipitate.

NOTE: A 10-ml aliquot of the original strontium carrier solution (20 mg/ml) is standardized as the carbonate by preparing it for weighing in the same way as described above. Calculate the strontium recovery.

7.1.4 Mount the precipitate on a plastic disc, cover with Mylar film and fasten with a plastic ring.

7.2 COUNTING.

7.2.1 Beta count, recording the hour and date.

7.2.2 Beta count on the same instrument as the yttrium oxalate precipitate.

8. Calibration

8.1 The counting efficiencies ^{90}Y, ^{90}Sr and ^{89}Sr are required for calculation of results. Counting efficiency for ^{90}Y is described under Section **8** of the ^{90}Sr Sampling, p. 00.

Counting efficiencies versus weight of strontium carbonate* precipitate must be determined for ^{90}Sr and ^{89}Sr.

8.2 The ^{90}Sr counting efficiency is determined as follows: Using the combined supernates from step **7.4.4** of the ^{90}Y calibration procedure, carry it through steps **7.1.1** to **7.2.1** of the ^{89}Sr procedure and calculate the efficiency as follows:

$$^{90}Sr = \frac{A - BCDE}{BE}$$

A = net counts per minute beta.
B = dpm ^{90}Sr.
C = ^{90}Y ingrowth factor $(1 - e^{-\lambda t})$ where t is the time of separation of yttrium (second milking) to the time of counting.
D = ^{90}Y counting efficiency.
E = chemical yield of strontium carbonate.

8.3 The ^{89}Sr counting efficiency is determined by carrying a ^{89}Sr standard solution through steps **7.1.1** to **7.2.1** of the ^{89}Sr procedure and calculating the efficiency as follows:

$$^{89}Sr = \frac{A}{B \times C}$$

A = net counts per minute ^{89}Sr.
B = dpm ^{89}Sr.
C = chemical yield of strontium carbonate.

9. Calculations

9.1 ^{89}Sr activity (pCi/m^3) =

$$\frac{1}{B \times C}\left[\frac{A}{D \times E \times F} - G(H + I \times J)\right]$$

*Covering a range of weights which could be present in samples to be counted.

where:

A = net beta count rate of total radiostrontium (cpm).
B = correction factor $e^{-\lambda t}$ for strontium-89 decay, where t is the time from sample collection to the time of counting.
C = counting efficiency for counting ^{89}Sr as strontium carbonate mounted on a glass fiber filter (cpm/pCi).
D = strontium carrier recovery.
E = total air sample volume (m^3).
F = fraction of total volume analyzed.
G = ^{90}Sr activity as computed from the ^{90}Y activity (pCi/m^3).
H = counter efficiency for counting ^{90}Sr as strontium carbonate mounted on a glass fiber filter (cpm/pCi).
I = counter efficiency for counting ^{90}Y as yttrium oxalate mounted on a glass fiber filter (cpm/pCi).
J = correction factor $1 - e^{-\lambda t}$ for yttrium-90 ingrowth, where t is the time from separation of the strontium (first milking) to the time of counting.

9.2 ^{89}Sr error. (pCi/m^3 @ 95% confidence level) =

$$\left\{\frac{\left[\frac{1.96}{B \times C \times D \times E \times F}\right]^2\left[\frac{A_1 + A_2}{A_3}\right] +}{\left[G\,G_1\right]^2}\right\}^{1/2}$$

where:

A_1 = gross beta count rate of total radiostrontium (cpm).
A_2 = beta background count rate of total radiostrontium (cpm).
A = minutes total radiostrontium counted.
B = correction factor $e^{-\lambda t}$ for ^{89}Sr decay, where t is the time from sample collection to the time of counting.

C = counting efficiency for counting ^{89}Sr as strontium carbonate mounted on a glass fiber filter (cpm/pCi).
D = strontium carrier recovery.
E = total air sample volume (m^3).
F = fraction of total volume analyzed.
G = ^{90}Sr activity as computed for the ^{90}Y activity (pCi/m^3).
G_1 = ^{90}Sr error associated with G(pCi/m^3).

10. Effect of Storage

10.1 Because of its short half-life strontium-89 determinations must be corrected for the period spent in storage.

11. References

1. With minor adaptations the radiochemical portions of the procedure is essentially the same as used at the Health and Safety Laboratory, New York Operations Office, U.S. Atomic Energy Commission. "HASL Manual on Standard Procedures with Revisions and Modifications to August 1965." NYO-4700 (Rev.)

Subcommittee 8
B. SHLEIEN, *Chairman*
R. G. HEATHERTON
P. F. HILDEBRANDT
L. B. LOCKHART, JR.
R. S. MORSE
W. STEIN
D. F. VAN FAROWE
A. HULL, *Health Physics Society Liaison*

608.

Tentative Method of Analysis for Strontium-90 Content of Atmospheric Particulate Matter

11333-01-70T

1. Principle of Method (1)

1.1 ^{90}Sr (27.8 year half-life) is a long-lived radionuclide produced by nuclear fission which together with its daughter, yttrium-90 constitutes one of the hazardous artificially produced beta emitting radionuclides which may be present in ground level air.

1.2 Analysis is performed by dissolution of a filter through which a known volume of air has been passed. The required volume of air sampled is dependent on the ^{90}Sr concentration and should be large enough to obtain a statistically significant result.

1.3 Strontium is separated from calcium, other fission products, and natural radioactive elements. Nitric acid separations remove the calcium and most of the other interfering ions. Radium, lead, and barium are removed with barium chromate. Traces of other fission products are scavenged with iron hydroxide. After the ^{90}Sr + ^{90}Y equilibrium has been attained, the ^{90}Y is precipitated as the hydroxide and converted to the oxalate for counting. Strontium chemical yield is determined gravimetrically as strontium carbonate.

2. Sensitivity and Range

2.1 0.5 pCi ^{90}Sr can be determined by this method with an error of less than ± 10% at the 95% confidence level under usual conditions (counter background of less than 1 cpm and a 60 min count).

3. Interferences

3.1 Separation from other fission products and naturally occurring radionuclides is required.

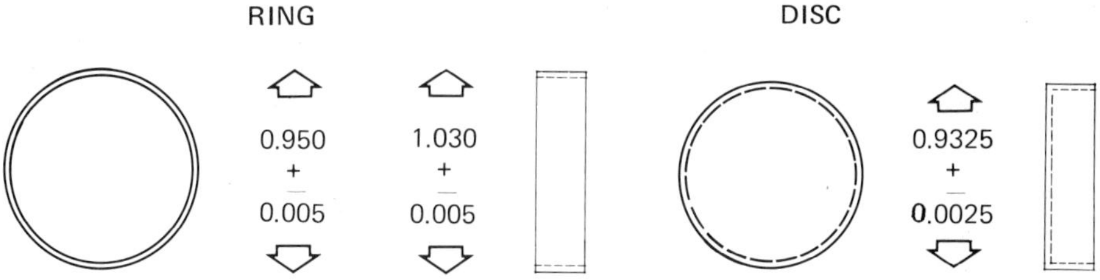

Figure 608:1—Ring and Disc

3.2 Calcium and barium removal is required when gravimetric technique is to be used for yield determination.

4. Precision and Accuracy

4.1 The precision of the method depends primarily on the reproducibility of sample mounting, yield determination, stability of the counting system, and the statistical limitations imposed by low counting rates. Standardization of carriers and counters will also affect the accuracy of the measurements.

5. Apparatus

5.1 TEFLON FILTER STICKS AND FUNNELS.

5.2 RINGS AND DISCS (NYLON).* (Figure 608:1).

5.2.1 Plastic rings and discs in nominal 1″ to 2″ diam are molded of nylon. The 1″ units are made according to dimensions given in the figure, and the 2″ units follow the same form with an overall diameter of 2.03″.

5.2.2 The discs are molded as cups to allow use of the discs either for mounting solid samples using the ring or to evaporate liquid samples in the cup.

5.3 MAGNETIC STIRRERS WITH TEFLON-COATED MAGNET BARS.

5.4 MYLAR FILM.

5.5 FILTERS, (for use in radiochemical technique).

5.6 FISHER FILTRATOR.

*These rings and discs are commercially available from Control Molding Corporation, 84 Granite Avenue, Staten Island, New York, 10303.

6. Reagents

6.1 PURITY. Reagents must be ACS analytical reagent quality. Distilled water should conform to ASTM Standards For Referee Reagent Water.

6.2 STRONTIUM CARRIER. 20 mg Sr/ml 48.4 g Sr $(NO_3)_2$/l 1:99 HNO_3.

6.3 YTTRIUM CARRIER. 20 mg Y/ml. See APPENDIX to this procedure for preparation.

6.4 IRON CARRIER. 5 mg Fe/ml—Dissolve 5 g iron wire in 1:1 HCl and dilute to 1 l with 1:99 HCl.

6.5 BARIUM CARRIER. 20 mg Ba/ml—30.4 $BaCl_2$/l 1:99 HCl.

6.6 CALCIUM CARRIER. 200 mg Ca/ml—Dissolve 500 g $CaCO_3$ in a minimum of 1:1 HCl and dilute to 1 l with 1:99 HCl.

6.7 SATURATED SODIUM CARBONATE SOLUTION.

6.8 6M AMMONIUM ACETATE SOLUTION. 463 g NH_4OAc/l water.

6.9 6M ACETIC ACID SOLUTION. 345 ml glacial HOAc/l water.

6.10 0.3M SODIUM CHROMATE SOLUTION. 50 g Na_2CrO_4/l water.

6.11 SODIUM HYDROXIDE SOLUTION. 240 g NaOH/l water.

7. Procedure

7.1 SAMPLING.

7.1.1 The particulate matter in a 10,000 m^3 or larger sample of air is collected on a high efficiency filter 6″ × 9″ or 8″ × 10″ in size mounted in a suitable leak-tight holder and protected from rain and snow by a louvered shelter or canopy.

7.1.2 Air is drawn through the filter at a rate of about 1000 m^3/day through use of a properly calibrated or instrumented blower. Filters should be changed before the pressure increase across the filter indicates a 10% flow decrease in order to obtain a representative sample and to minimize the error in the estimation of sample volume. Under these conditions of flow several filters may be required to collect a 10,000 m^3 sample depending on the level of atomspheric dust loadings.

7.1.3 Records are to be kept of the time of the beginning and end of collection, during the collection, and the initial and final flow rates.

7.2 Sample Preparation.

7.2.1 *Ashing (Wet).*

a. Place the filter in a 600 to 800-ml beaker and add 25 ml of 1:10 HNO_3. Add exactly 1 ml of strontium carrier and 1 ml of barium carrier.

b. Add 300 ml of HNO_3 and 25 ml of HCl and cover. Allow the mixture to react, then heat on a hot plate to decompose organic material.

c. Evaporate to a small volume, adding 30% H_2O_2 repeatedly to complete the oxidation.

d. Transfer to a 250-ml Teflon beaker using dilute nitric acid.

e. Add 10 ml of HF and heat to expel silica. Heat to dryness.

f. Dissolve the salts in dilute HNO_3 and transfer to a 250-ml beaker. Evaporate to dryness and continue with step **7.2.2.g** under dry ashing.

7.2.2 *Ashing (Dry).*†
Muffle furnaces can be used for ashing the sample. A muffle may be operated at 900 F (500 C) continuously. Stainless steel and monel crucibles have proven satisfactory for all ashing. They are relatively inexpensive and sturdy. Cleaning can be accomplished readily with detergents or with dilute mineral acids (usually HCl).

a. Transfer the ash to a 50-ml platinum crucible.

b. Add approximately 4 times the ash volume (visually determined) of Na_2CO_3 and mix thoroughly. Fuse to a clear melt in a muffle at 900 F. Cool.

c. Grind to a powder using a mortar and pestle. Transfer the ground fused material to a 600-ml beaker and add 300 ml of water. Add 1 ml of strontium carrier solution, and 1 ml of barium carrier.

Heat gently just below boiling, while stirring, for about an hour.

d. Cool and filter with suction on a 5.5 cm glass fiber paper backed by a #42 Whatman paper. Wash the residual fused material with water. Discard the filtrate.

e. Dissolve the carbonates on the paper in 1:1 HNO_3 with the suction off, collecting the solution in the filter flask. Apply suction and wash with 1:1 HNO_3. Discard the filter and residue.

f. Transfer the solution to the original 250-ml beaker and evaporate to dryness.

g. Add 20 ml of water and 13 ml of 90% fuming HNO_3 to dissolve solid matter, then add 57 ml additional of 90% fuming HNO_3 with magnetic stirring using a Teflon-coated bar. (Do this in a well-ventilated hood.) Stir for 30 min.

h. Cool and allow the nitrates to settle. Filter by suction on a 5.5-cm glass fiber filter using a Buchner funnel. Drain the filter thoroughly. Discard the filtrate.

i. With suction off, wash the precipitate remaining in the beaker into the funnel with water and collect the solution. Apply suction and thoroughly drain the filter.

j. Transfer the solution to a 150 ml beaker. Evaporate slowly to dryness, then dissolve the residue in 23 ml of water.

k. Add 77 ml of 90% fuming HNO_3 with magnetic stirring using a Teflon-coated bar. (Do this in a well-ventilated hood.) Stir for 30 min.

l. Cool and allow the strontium nitrate precipitate to settle. Filter by suction on a 2.8-cm glass fiber filter with a Fisher filtrator and a Teflon funnel. Remove as much nitric acid as possible.

m. Dissolve the remaining precipitate in the beaker with water. Pour through the Teflon funnel and collect the solution in a

†When gravimetric technique is to be used for yield determination, all of the calcium must be separated from strontium. It is important, therefore, that a sufficient number (usually 2 to 3) of nitric acid separations be performed to insure removal of calcium. Lead and barium also precipitate under these conditions but are removed in the subsequent hydroxide and chromate scavenging steps.

40-ml short cone, heavy wall centrifuge tube. Wash the filter with water, keeping the volume below 20 ml.

7.3 FIRST MILKING.

7.3.1 Add 1 ml of iron carrier solution. Heat in a water bath regulated at approximately 90 C.

7.3.2 Adjust the pH to 8 with 1:1 NH_4OH. Cool to room temperature and allow the precipitate to settle.

7.3.3 Centrifuge for 5 min. Decant, pouring the supernatant liquid into a second 40-ml centrifuge tube.

7.3.4 Dissolve the precipitate with a few drops of HCl, dilute to about 5 ml and heat in a water bath regulated at approximately 90 C.

7.3.5 Adjust the pH to 8 with 1:1 NH_4OH. Cool to room temperature and allow the precipitate to settle.

7.3.6 Centrifuge for 5 min. Decant, combining the supernate with that from step 3. Discard the residue.

7.3.7 Add 1 ml barium carrier solution to the solution, if not added previously. Add 1 ml of 6*M* acetic acid and 2 ml of 6*M* ammonium acetate solution. Adjust the pH to 5.5 with 1:1 HCl or NH_4OH. **(See Note A)**

7.3.8 Heat in a water bath to about 90 C. Add dropwise with stirring, 1 ml of 0.3*M* Na_2CrO_4 solution. Allow the precipitate to settle. If necessary, add with stirring additional 0.3*M* Na_2CrO_4 solution to give the supernate a yellow chromate color. **(See Note B)**

7.3.9 Allow the sample to cool. Centrifuge and decant the supernate into a 2-oz. polyethylene bottle. Discard the residue.

7.3.10 Add 3 drops of HCl and exactly 1 ml of yttrium carrier solution to the supernate.

7.3.11 Store for 2 weeks to allow equilibration of ^{90}Y **(See Note C)**.

NOTES:

A. The pH of the solution at this point is critical. Barium chromate will not precipitate completely in more acid solutions and strontium will partially precipitate in more basic solutions.

B. If large quantities of barium are present in the sample, handling may be facilitated by centrifuging before the $BaCrO_4$ precipitation is complete. The supernate is then decanted into another 40-ml centrifuge tube. The dropwise addition of 0.3*M* Na_2CrO_4 is continued until precipitation is complete, and the analysis is continued with step **7.3.9**.

C. If less than full ^{90}Y build-up is satisfactory, the time from the iron hydroxide precipitation (step **7.3.1**) should be noted as the start of build-up.

7.4 SECOND MILKING.

7.4.1 Transfer the equilibrated solution to a 40-ml short cone, heavy wall centrifuge tube. Heat in a water bath regulated at approximately 90 C. Adjust the pH to 8 with 1:1 NH_4OH, stirring continuously. Add 6 drops of H_2O_2 (30%). Continue heating to remove excess H_2O_2 (about an hr).

7.4.2 Cool to room temperature and centrifuge for 5 min. Decant, returning the supernatant liquid to the original polyethylene bottle. Record the hour and date.

7.4.3 Add 15 to 20 ml of water to the precipitate in the centifuge tube. Dissolve by adding HCl dropwise with stirring. Heat in a water bath and adjust the pH to 8 with 1:1 NH_4OH, stirring continuously.

7.4.4 Cool to room temperature and centrifuge for 5 min. Combine the supernate in the polyethylene bottle and save for yield determination.

7.4.5 Add 2 drops of HCl to dissolve the precipitate (from step **7.4.4**) then add 25 to 30 ml of water. Heat in a water bath regulated at approximately 90 C. Add 1 ml of saturated oxalic acid solution dropwise with vigorous stirring. Add 2 to 3 drops of NH_4OH. The pH should be 2 to 3. Allow the precipitate to digest for about an hr.

7.4.6 Cool to room temperature. Filter by suction on a weighed 2.8-cm glass fiber filter (Whatman #41 paper is preferable for low activity samples) using a Teflon funnel.

7.4.7 Dry the precipitate in an oven at 110 C. Weigh. Mount on a plastic disc. See specifications. Cover with Mylar and fasten with a plastic ring.

7.4.8 Beta count, recording the hour and date. Standardize with an ^{90}Y standard which has been precipitated and mounted in an identical manner.

7.5 Gravimetric Strontium Yield Measurement.‡

7.5.1 Transfer the solution from steps **7.4.2** and **7.4.4** under *Second Milking* to a 150-ml beaker and heat to a gentle boil.

7.5.2 With continuous stirring, add approximately 5 ml of saturated Na_2CO_3 solution. Heat gently for 10 min.

7.5.3 Cool and filter on a weighed 5.5-cm glass fiber filter. (When using the Buchner funnel to filter, care must be taken to prevent the precipitate from creeping under the filter paper. This can be done by filtering only through the center portion of the paper.)

7.5.4 Wash the paper and precipitate thoroughly with water.

7.5.5 Dry at 105 C and weigh.

7.5.6 Standardize a 10 ml aliquot of the original strontium carrier solution by precipitating it as the carbonate in the same way as described above. Calculate the strontium recovery.

7.6 Gravimetric Yttrium Yield Measurement.‡

7.6.1 Determine the weight of yttrium oxalate from the measurements in steps **7.4.6** and **7.4.7** of *Second Milking*.

7.6.2 Standardize triplicate 10 ml aliquots of the original yttrium carrier solution each time a fresh batch is made, by precipitating the oxalate as described, and filtering on a weighed 5.5-cm glass fiber filter. Calculate the yttrium recovery.

7.7 Counting.

7.7.1 The sample may be counted in a low-background beta counter. A low-background beta counter is shielded by approximately 2″ of lead, or equivalent, to stop external gamma rays, and by coincidence circuitry to reduce the contribution from cosmic interactions in the shielding material. Typical background for such systems is less that 1.0 cpm and the counting efficiency for strontium-90 and yttrium-90 is about 50%.

7.7.2 Scintillation counting may be carried out with beta phosphors **(2)**. The phosphor discs are 0.010″ thick and available in various diam.§ The beta phosphor is mounted directly on the sample and is discarded after counting.

‡If ^{89}Sr is to be determined, proceed to Step **7.1** of procedure for ^{89}Sr.

8. Calibration

8.1 The efficiency is determined by counting the ^{90}Y separated from a ^{90}Sr standard solution in equilibrium. Follow the procedure outlined from step **7.3.10** to **7.4.7** with the elimination of step **7.3.11**. The counter background is determined daily. The value used for background is the mean daily background for a period of 8 weeks.

8.2 Calculate the efficiency for ^{90}Y as follows:

$$^{90}Y = \frac{A}{B \times C \times D}$$

A = net counts per minute ^{90}Y.
B = dpm ^{90}Sr.
C = correction factor for ^{90}Y decay ($e^{-\lambda t}$) where t is time of separation of yttrium (first milking) to the time of counting.
D = chemical yield of yttrium oxalate.

9. Calculations

9.1 The gross counting data obtained on the ^{90}Y precipitation must be corrected to give the proper disintegration rate representing the ^{90}Sr in the sample. The corrections include those for build-up of ^{90}Y, counter background, efficiency, strontium yield and yttrium-90 decay. Since the ^{90}Y beta is very energetic and is always counted with the same weight of precipitate, no corrections for self absorption are necessary.

9.2 Ordinarily no correction is made for the degree of build-up of ^{90}Y during the

§Commercially available from Pilot Chemicals, Inc., 36 Pleasant Street, Watertown, Mass. 02172; William B. Johnson and Associates, Inc., P.O. Box 415, Mountain Lakes, N.J. 07046; Nuclear Enterprises, Ltd., 550 Berry Street, Winnipeg 21, Canada.

equilibration. A 2 weeks build-up period gives over 97% of the expected equilibrium value. When shorter build-up periods are used to hasten analysis, however, appropriate correction must be made. Build-up corrections are shown in Figure 608:2.

9.3 The decay of yttrium-90 from the time of second milking to the time of counting must be corrected for. This is done graphically by reading off the decay factor from the graph shown in Figure 608:3.

9.4 Strontium-90 activity (pCi/m^3) =

$$\frac{A}{B \times C \times D \times E \times F \times G \times H}$$

where:

A = net count rate of yttrium-90 (cpm).
B = total air sample volume (m^3).
C = fraction of total volume analyzed.
D = counter efficiency for counting yttrium-90 as yttrium oxalate mounted in the same manner as the sample (cpm/pCi).
E = chemical recovery of yttrium.
F = correction factor $e^{-\lambda t}$ for yttrium-90 decay, where t is the time for separation of yttrium (second milking) to the time of counting.
G = strontium carrier recovery.
H = correction factor $1-e^{-\lambda t}$ for yttrium ingrowth where t is the time from the first separation of yttrium from strontium (first milking) to the separation following the ingrowth period (second milking).

9.5 The standard deviation for the net

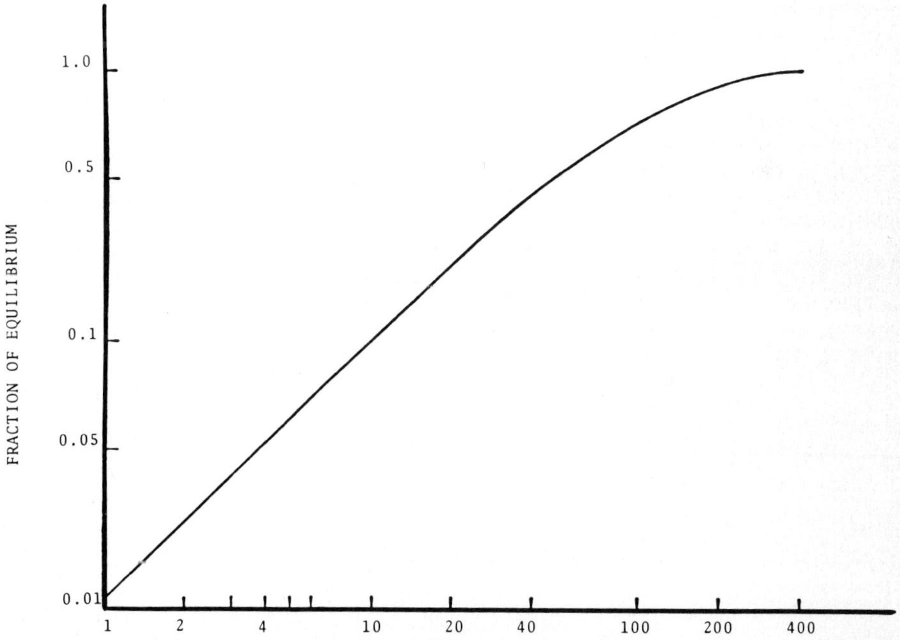

Figure 608:2—Buildup of Yttrium-90

counting rate is strontium-90 error @ 95% confidence level (pCi/m³) =

$$\frac{1.96}{B \times C \times D \times E \times F \times G \times H}\left[\frac{RB}{tB}+\frac{RG}{tG}\right]^{1/2}$$

where:

RB = counting rate of background
tB = time the background was counted
RG = gross counting rate of sample
tG = time the sample was counted.

10. Effects of Storage

10.1 Because of the radioactive decay of strontium-90, corrections are required if the sample is stored for a long period of time.

11. References

1. With minor adaptations the radiochemical portions of the procedure is essentially the same as used at the Health and Safety Laboratory, New York Operations Office, U.S. Atomic Energy Commission. "HASL Manual on Standard Procedures with Revisions and Modifications to August 1965." NYO-4700 (Rev.)

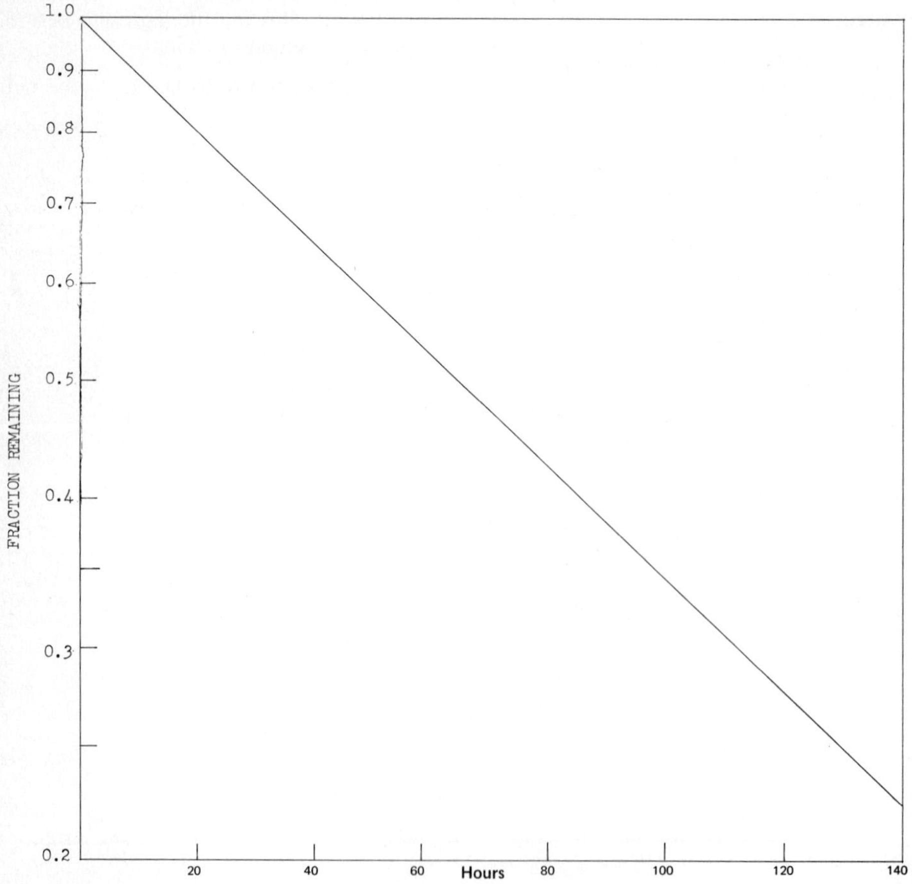

2. HARLEY, J.H., N.A. HALLDEN, and I.M. FISENNE. 1962. Beta Scintillation Counting with Thin Plastic Phosphors, *Nucleonics*, 20:59.

APPENDIX

Compounds of yttrium tend to contain some natural radioactivity which can be troublesome to deal with when determining low levels of radioactivity. This material should be tested for radioactive contamination prior to use. Commercial material of an acceptable purity is available, however in some instances further purification may be necessary.

PURIFICATION OF YTTRIUM CARRIER

1. Dissolve 86.2 g of yttrium nitrate, $Y(NO_3)_3 \cdot 6H_2O$ in 80 ml of water. Add a few drops of HNO_3. Transfer to a 1000-ml separatory funnel using 2 to 20 ml portions of water.
2. Add 120 ml of saturated NH_4NO_3 to the separatory funnel. Add 240 ml of tributyl phosphate (TBP) to the separatory funnel and shake for 5 min. Allow the phases to separate for 10 min.
3. Draw off the aqueous lower layer into a second separatory funnel. Add 240 ml of fresh TBP and shake for 5 min. Allow phases to separate and discard the lower aqueous layer.
4. Combine both TBP phases in one separatory funnel, add 200 ml of water and shake for 5 min. Allow the phases to separate and transfer the lower aqueous layer to a clean separatory funnel.
5. Repeat the water wash and combine the aqueous fractions. Discard the TBP.
6. Add 50 ml of CCl_4 to the water solution, shake for 1 min and allow to separate. Discard the CCl_4.
7. Dilute to 1 l with water and store in polyethylene bottle.

YTTRIUM CARRIER COUNTING CHECK

1. Pipet 1 ml of yttrium carrier into each of 3 40-ml centrifuge tubes. Dilute to 20 ml with water.
2. Heat in a water bath to about 80 C. With stirring, add 1:1 NH_4OH to a pH of 8. Digest for 10 min and cool in a cold water bath.
3. Centrifuge for 5 min. Decant and discard the supernate.
4. Break up the precipitate with a few ml of water. Dilute to 20 ml with water. Add a few drops of conc HCl to just dissolve the precipitate. Heat the solution in a water bath to about 80° C and add 1 ml of saturated oxalic acid while stirring.
5. Cool and allow the precipitate to digest. Filter on a 2.8 cm #42 Whatman paper. Discard the filtrate.
6. Dry on a hot plate. Mount with a ring and disc and β count.

Subcommittee 8
B. SHLEIEN, *Chairman*
R. G. HEATHERTON
P. F. HILDEBRANDT
L. B. LOCKHART, JR.
R. S. MORSE
W. STEIN
D. F. VAN FAROWE
A. HULL, *Health Physics Society Liaison*

609.

Tentative Method of Analysis for Tritium Content of the Atmosphere*

11314-01-70T

1. Principle of the Method

1.1 Air is pumped at a sampling rate of approximately 100 to 150 cm^3/min through a silica gel column which collects moisture from the air. This flow rate is sufficient for continuous sampling. Other methods of moisture collection such as cold traps or dehumidifiers are also suitable for short term sampling. The collected moisture sample is then removed from the column or collection device and is analyzed for tritium. A suitable volume of collected moisture is mixed with scintillation solution, and the mixture counted in a liquid scintillation beta counter. The air flow rate and the total volume of moisture collected during the sampling period, or the absolute humidity, must be determined to convert the concentration of tritium in the collected moisture sample to its concentration in air. For long sampling periods the determination of air flow rate is more convenient than humidity measurements.

2. Range and Sensitivity

2.1 The method is suitable for sampling periods of several days, over the entire significant range of tritium concentration in air. Sensitivity is a direct function of the volume of air sampled and an inverse function of the prevailing temperature and relative humidity (and therefore moisture content) of the air. A sample of moisture from air at an average temperature of 30 C (86 F) and at 100 per cent relative humidity, and containing tritium in a concentration of 2×10^{-8} μCi/cm^3 (10% of the Maximum Permissible Concentration) would contain 666 pCi/ml of moisture; sufficient to produce acceptable precision in less than 2 min of counting in most liquid scintillation counting units.

3. Interferences

3.1 None.

4. Precision and Accuracy

4.1 The precision of the method is, over the range of concentrations of interest, a function of counter efficiency and counting statistics. For example, the precision of the count of the sample described above (2×10^{-8} μCi/cm^3, 30 C, 100% RH) would be ± 10%, counted for 2 min counting time at 15% counting efficiency in a dual-channel liquid scintillation counter with a background of 25 cpm in the tritium channel (at the 95% confidence level).

5. Apparatus

5.1 LIQUID SCINTILLATION BETA COUNTER.

5.2† ALUMINUM CYLINDER, 12″ × 1¼″ diam. for 6 to 16 mesh non-indicating silica gel.

5.3‡ LOW VOLUME AIR PUMP AND FLOW RATE INDICATOR, (or temperature and relative humidity device).

---

*This procedure assumes that only tritiated water vapor is collected.

†If cold finger or de-humidifier used, Items 5.2 to 5.5 are unnecessary. Device for establishing temperature and relative humidity would be required.

‡If routine particulate and/or ^{131}I sampling is also to be conducted at same location, it may be convenient to utilize a side-stream from a larger air pump for tritium sampling.

5.4 INDICATING SILICA GEL CYLINDER, (3″ × 2″ diam).

5.5 DISTILLING FLASK AND CONDENSER.

5.6 PIPETS.

5.7 VIALS, POLYETHYLENE, (25 ml for liquid scintillation counters).

5.8 RUBBER STOPPERS, TUBING.

6. Reagents

6.1 SCINTILLATION SOLUTION§(**1**):

Naphthalene recrystallized from alcohol	120 g
2, 5 diphenyloxazol (PPO) scintillation grade	4 g
p-Bis-2-(5-phenyloxazolyl)-benzene (POPOP) scintillation grade	0.05 g
Spectroscopic grade p-dioxane	1,000 ml

7. Procedure

7.1‖ Dry 6 to 16 mesh, non-indicating silica gel for 24 hr at 150 C prior to moisture collection use. Regeneration of used silica gel may be accomplished by the same technique with a "memory" of $< 5\%$ of previously sampled moisture. Although this residual moisture may be lowered by drying at a higher temperature, the absorption capacity of the gel may also be impaired by such a drying procedure.

7.2 Fill 12″ aluminum cylinder with the non-indicating silica gel (180 g). Since silica gel can absorb moisture up to 40% of its own weight, at 100% relative humidity, (**3**) 180 g should not saturate during a sampling period as long as 2 weeks except in extremely warm and humid conditions. Seal ends with small wad of glass wool and #7 rubber stoppers.

7.3 Remove stoppers at field sampling location. Insert one-hole stoppers in cylinder ends and connect tubing to air pump.

7.4 Insert indicating silica gel cylinder in-line between the collection cylinder and air pump. (A color change during sample collection indicates saturation of the non-indicating silica gel.)

7.5 Following sample collection, remove non-indicating silica gel cylinder and seal ends with #7 rubber stoppers for transport to laboratory.

7.6 Pour silica gel into a 500-ml distilling flask.

7.7 Heat until *all* moisture is removed. Weigh or measure total moisture collected.

7.8 Add 20 ml of scintillation solution to a polyethylene vial containing 4 ml of distillate.

7.9 Count for not less than 2 min in a liquid scintillation beta counter.

8. Calibration and Standards

8.1 Using distilled tritium-free water, dilute an NBS¶ tritiated water primary standard to obtain known calibration standards for the liquid scintillation counter. A concentration between 10^3 to 10^4 dpm/ml is suggested to give a practicable calibration standard.

8.2 To determine counting efficiency mix 4 ml of the tritiated water standard with 20 ml of scintillation solution, and count under the same conditions as the sample. Counting Efficiency should be redetermined if scintillation solution of a different manufacture or lot number is used.

8.3 To determine background, add 4 ml of distilled tritium-free water to 20 ml of scintillation mix and count. Background samples should be included at the start of every counting run and between groups of not more than ten samples.

9. Calculations

9.1 Efficiency:

$$E = \frac{S - B}{D} \quad (1)$$

§Performance of liquid scintillation unit with bi-alkali photo tubes may be improved with use of Bis-MSB secondary scintillator (**2**).

‖If cold finger or de-humidifier used to collect sample, start procedure at 7.8 (substituting collected moisture for distillate).

¶National Bureau of Standards, SRM No. 4927, or standard of equivalent calibration accuracy.

where

E = counting efficiency.
S = counts per minute of standard (in scintillation solution).
B = background counts per minute of distilled water (in scintillation solution).
D = disintegrations per minute of standard.

9.2 To determine tritium concentration of moisture:

$$C_m(\mu Ci/ml) = \frac{C_S - B}{E \times V \times K} \quad (2)$$

where

C_m = concentration of tritium in collected moisture in μCi/ml.
C_s = gross sample counts per minute; i.e., sample count plus background.
V = volume (ml) of sample counted.
K = 2.22×10^6 to convert disintegrations per minute to μCi).

9.3 Error due to counting (at the 95% confidence level) is calculated as follows:

$$E_{95} = 1.96 \sqrt{\frac{C_S}{\text{Time counted}} + \frac{B}{\text{Time counted}}} \quad (3)$$

9.4 The concentration of tritium oxide in air is calculated as follows if silica gel

Table 609:I. Weight of a cubic meter of aqueous vapor at different temperatures and percentages of saturation,* g.

	Percentage of Saturation									
(°F)	10	20	30	40	50	60	70	80	90	100
	←				grams					→
−20	0.039	0.076	0.114	0.151	0.190	0.229	0.266	0.304	0.341	0.380
−15	0.050	0.101	0.149	0.199	0.250	0.300	0.350	0.398	0.449	0.499
−10	0.064	0.130	0.197	0.261	0.325	0.392	0.458	0.522	0.586	0.653
−5	0.085	0.169	0.254	0.339	0.424	0.508	0.593	0.678	0.762	0.847
0	0.110	0.220	0.330	0.440	0.550	0.662	0.772	0.882	0.992	1.101
5	0.140	0.279	0.419	0.559	0.698	0.838	0.978	1.118	1.257	1.397
10	0.179	0.355	0.534	0.710	0.888	1.067	1.243	1.422	1.598	1.777
15	0.227	0.451	0.678	0.902	1.129	1.356	1.580	1.807	2.031	2.258
20	0.284	0.566	0.847	1.131	1.415	1.697	1.978	2.262	2.546	2.828
25	0.355	0.710	1.065	1.420	1.777	2.132	2.487	2.842	3.197	3.552
30	0.444	0.886	1.328	1.772	2.217	2.659	3.101	3.545	3.989	4.431
35	0.543	1.083	1.626	2.166	2.709	3.252	3.792	4.335	4.875	5.418
40	0.653	1.305	1.958	2.611	3.261	3.914	4.566	5.219	5.872	6.524
45	0.781	1.564	2.345	3.128	3.909	4.690	5.473	6.254	7.037	7.818
50	0.934	1.866	2.801	3.733	4.667	5.601	6.533	7.468	8.400	9.334
55	1.111	2.221	3.332	4.443	5.551	6.662	7.772	8.883	9.994	11.104
60	1.314	2.631	3.948	5.262	6.577	7.894	9.210	10.525	11.839	13.156
65	1.553	3.105	4.660	6.213	7.765	9.318	10.871	12.426	13.978	15.531
70	1.827	3.655	5.482	7.310	9.137	10.964	12.792	14.619	16.447	18.274
75	2.143	4.284	6.428	8.569	10.713	12.856	14.997	17.141	19.282	21.425
80	2.503	5.008	7.511	10.016	12.519	15.022	17.528	20.031	22.536	25.039
85	2.917	5.833	8.750	11.665	14.583	17.500	20.415	23.333	26.248	29.165
90	3.387	6.774	10.161	13.548	16.934	20.321	23.708	27.095	30.482	33.869
95	3.920	7.843	11.764	15.686	19.607	23.527	27.450	31.371	35.293	39.214
100	4.527	9.052	13.580	18.105	22.632	27.159	31.684	36.212	40.737	45.264

*Adapted from Reference (**4**).

(having 100% collection efficiency) is used and if the flow rate is known.

$$C_a(\mu Ci/cm^3) = \frac{C_m \times W}{F \times t} \quad (4)$$

where

C_a = concentration of tritium oxide in air in μCi/cm^3.
F = flow rate (cm^3/min).
t = sampling time (min).
W = total volume of water collected (in ml).

9.5 If a less than 100% efficient collector, such as a cold trap or dehumidifier is used, or the flow rate is unknown, the following equation is used to determine tritium oxide concentration in air:

$$C_a(\mu Ci/cm^3) = C_m A \times 10^{-6} \quad (5)$$

where

A = absolute humidity during sample collection (g/m^3)**

This is the mass of water vapor present in unit volume of the atmosphere, usually expressed as grams per cubic meter. Representative values applicable at sea level are indicated in Table 607:I (4**).

10. Effects of Storage

10.1 No effects are encountered by storage if silica gel collection cylinders are sealed with rubber stoppers and stored in a non-tritiated atmosphere.

11. References

1. BUTLER, F.F. 1961. Determination of Tritium in Water and Urine. *Anal. Chem.* 33:3, 409–414.
2. MOGHISSI, A.A., H.L. KELLEY, J.I. REGNIER, and M.W. CARTER. 1969. Low Level Counting by Liquid Scintillation—I. Tritium Measurement in Homogeneous Systems. *Int'l. Jrl. Appl. Rad. and Isotopes* 20:145–56.
3. PERRY, R.H., C.H. CHILTON, and S.D. KIRKPATRICK. 1963. Chemical Engineers Handbook Section 16.
4. MARVIN, C.F. 1941. Psychrometric Tables. Weather Bureau Publication #735.

Subcommittee #8
B. SHLEIEN, *Chairman*
R. G. HEATHERTON
P. F. HILDEBRANDT
L. B. LOCKHART, JR.
R. S. MORSE
W. STEIN
D. E. VAN FAROWE
A. HULL, *Health Physics Society Liaison*

701.

Tentative Method of Analysis for Hydrogen Sulfide Content of the Atmosphere

42402-01-70T

1. Principle of the Method

1.1 Hydrogen sulfide is collected by aspirating a measured volume of air through an alkaline suspension of cadmium hydroxide (**1**). The sulfide is precipitated as cadmium sulfide to prevent air oxidation of the sulfide which occurs rapidly in an aqueous alkaline solution. STRactan 10 is added to the cadmium hydroxide slurry prior to sampling to minimize photo-decomposition of the precipitated cadmium sulfide (**2**). The collected sulfide is subsequently determined by spectrophotometric measurement of the methylene blue produced by the reaction of the sulfide with a strongly acid solution of N,N-dimethyl-p-phenylenediamine and ferric chloride (**3, 4, 5**). The analysis should be completed with 24 to 26 hr following collection of the sample.

1.2 Hydrogen sulfide may be present in the open atmosphere at concentrations of a few ppb or less. The reported odor detection threshold is in the 0.7 to 8.4 $\mu g/m^3$ (0.5 to 6.0 ppb) range (**6, 7**). Concentrations in excess of 139 $\mu g/m^3$ (100 ppb) are seldom encountered except as a result of an accident such as the Columbus, Ohio, theater fire or the Posa Rica disaster in Mexico.

1.3 Collection efficiency is variable below 10 $\mu g/m^3$ and is affected by the type of scrubber, the size of the gas bubbles, and the contact time with the absoring solution and the concentration of H_2S (**8, 9, 10**).

2. Sensitivity and Range

2.1 This method is intended to provide a measure of hydrogen sulfide in the range of 1.1 to 100 $\mu g/m^3$. For concentrations above 70 $\mu g/m^3$ the sampling period can be reduced or the liquid volume increased either before or after aspirating. (This method is also useful for the mg/m^3 range of source emissions. For example, 100 ml cadmium sulfide-STRactan 10 media in Greenberg-Smith impingers and 5 min sampling periods have been used successfully for source sampling.) The minimum detectable amount of sulfide is 0.008 $\mu g/ml$, which is equivalent to 1.1 $\mu g/m^3$ in an air sample of 1 m^3 and using a final liquid volume of 25 ml. When sampling air at the maximum recommended rate of 1.5 l/min for 2 hr, the minimum detectable sulfide concentration is 1.1 $\mu g/m^3$ at 760 Torr and 25 C.

3. Interferences

3.1 The methylene blue reaction is highly specific for sulfide at the low concentrations usually encountered in ambient air. Strong reducing agents (*e.g.*, SO_2) inhibit color development. Even sulfide solutions containing several μg $S^=$/ml show this effect and must be diluted to eliminate color inhibition. If sulfur dioxide is absorbed to give a sulfite concentration in excess of 10 $\mu g/ml$, color formation is retarded. Up to 40 $\mu g/ml$ of this interference, however, can be overcome by adding 2 to 6 drops (0.5 ml/drop) of ferric chloride instead of a single drop for color development, and extending the reaction time to 50 min.

3.2 Nitrogen dioxide gives a pale yellow color with the sulfide reagents at 0.5 $\mu g/ml$ or more. No interference is encountered when 0.3 ppm NO_2 is aspirated through a midget impinger containing a slurry of cadmium hydroxide-cadmium sulfide-STRactan 10. If H_2S and NO_2 are simultaneously aspirated through cadmium hydroxide-STRactan 10 slurry, lower H_2S results are obtained, probably because of gas phase oxidation of the H_2S prior to precipitation as CdS **(10)**.

3.3 Ozone at 57 ppb reduced the recovery of sulfide previously precipitated as CdS by 15% **(10)**.

3.4 Sulfides in solution are oxidized by oxygen from the atmosphere unless inhibitors such as cadmium and STRactan 10 are present.

3.5 Substitution of other cation precipitants for the cadmium in the absorbent (*i.e.*, zinc, mercury, etc.) will shift or eliminate the absorbance maximum of the solution upon addition of the acid-amine reagent.

3.6 Cadmium sulfide decomposes significantly when exposed to light unless protected by the addition of 1% STRactan to the absorbing solution prior to sampling **(2)**.

4. Precision and Accuracy

4.1 A relative standard deviation of 3.5% and a recovery of 80% has been established with hydrogen sulfide permeation tubes **(2)**.

5. Apparatus

5.1 ABSORBER, midget impinger.

5.2 AIR PUMP, with flowmeter and/or gasmeter having a minimum capacity of 2 l/min through a midget impinger.

5.3 COLORIMETER, with red filter or spectrophotometer at 670 nm.

5.4 AIR VOLUME MEASUREMENT. The air meter must be capable of measuring the air flow within ± 2%. A wet or dry gas meter, with contacts on the 1-ft^3 or 10-l dial to record air volume, or a specially calibrated rotameter, is satisfactory. Instead of these, calibrated hypodermic needles may be used as critical orifices if the pump is capable of maintaining greater than 0.7 atmosphere pressure differential across the needle **(11)**.

6. Reagents

6.1 PURITY. Reagents must be ACS analytical reagent quality. Distilled water should conform to the ASTM Standards for Referee Reagent Water.

6.2 Reagents should be refrigerated when not in use.

6.3 AMINE-SULFURIC ACID STOCK SOLUTION. Add 50 ml conc sulfuric acid to 30 ml water and cool. Dissolve 12 g of N,N-dimenthyl-p-phenylenediamine dehydrochloride (para-aminodimethylaniline) (redistilled if necessary) in the acid. Do not dilute. The stock solution may be stored indefinitely under refrigeration.

6.4 AMINE TEST SOLUTION. Dilute 25 ml of the Stock Solution to 1 l with 1:1 sulfuric acid.

6.5 FERRIC CHLORIDE SOLUTION. Dissolve 100 g of ferric chloride, $FeCl_3$ $6H_2O$, in water and dilute to 100 ml.

6.6 AMMONIUM PHOSPHATE SOLUTION. Dissolve 400 g of diammonium phosphate, $(NH_4)_2HPO_4$, in water and dilute to 1 l.

6.7 STRACTAN 10, (ARABINOGALACTAN). Available from Stein-Hall and Co., Inc., 385 Madison Avenue, New York, New York.

6.8 ABSORBING SOLUTION. Dissolve 4.3 g of $3CdSO_4 \cdot 8\ H_2O$, and 0.3 g sodium hydroxide in separate portions of water, mix, add 10 g STRactan 10 and dilute to 1 l. Shake the resultant suspension vigorously before removing *each* aliquot. The STRactan-cadmium hydroxide mixture should be freshly prepared. The solution is only stable for 3 to 5 days.

6.9 H_2S PERMEATION TUBE. Prepare or purchase* a triple-walled or thick walled Teflon permeation tube **(10, 12, 13, 14, 15)** which delivers hydrogen sulfide at a maximum rate of approximately 0.1 μg/min at 25 C. This loss rate will produce a standard atmosphere containing 50 $\mu g/m^3$ (36 ppb H_2S when the tube is swept with a 2 l/min air flow. Tubes having H_2S permeation rates in the range of 0.004 to 0.33 μg/min will produce standard air concentrations in

*Available from Metronics, Inc., 3201 Porter Drive, Palo Alto, California 94304, or PolyScience Corp., 909 Pitner Ave., Evanston, Ill. 60202.

the realistic range of 1 to 90 $\mu m/m^3$ H_2S with an air flow of 1.5 l/min.

6.9.1 *Concentrated, Standard Sulfide Solution.* Transfer freshly boiled and cooled 0.1M NaOH to a liter volumetric flask. Flush with nitrogen to remove oxygen and adjust to volume. (Commercially available, compressed nitrogen contains trace quantities of oxygen in sufficient concentration to oxidize the small concentrations of sulfide contained in the standard and dilute standard sulfide standards. Trace quantities of oxygen should be removed by passing the stream of tank nitrogen through a pyrex or quartz tube containing copper turnings heated to 400 to 450 C). Immediately stopper the flask with a serum cap. Inject 300 ml of H_2S gas through the septum. Shake the flask. Withdraw measured volumes of standard solution with a 10-ml hypodermic syringe and fill the resulting void with an equal volume of nitrogen. Standardize with standard iodine and thiosulfate solution in an iodine flask under a nitrogen atmosphere to minimize air oxidation. The approximate concentration of the sulfide solution will be 440 μg $S^=$/ml of solution. The exact concentration must be determined by iodine-thiosulfate standardization immediately prior to dilution.

For the most accurate results in the iodometric determination of sulfide in aqueous solution, the following general procedure is recommended:

a. Replacement of oxygen from the flask with an inert gas such as carbon dioxide or nitrogen.

b. Addition of an excess of standard iodine, acidification and back titration with standard thiosulfate and starch indicator (**16**).

6.9.2 *Diluted Standard Sulfide Solution.* Dilute 10 ml of the concentrated sulfide solution to 1 liter with freshly boiled, distilled water. Protect the boiled water under a nitrogen atmosphere while cooling. Transfer the deoxygenated water to a flask previously purged with nitrogen and immediately stopper the flask. This sulfide solution is unstable. Therefore, prepare this solution immediately prior to use. This test solution will contain approximately 4 μg $S^=$/ml.

7. Standardization Procedure

7.1 COLLECTION OF SAMPLE. Aspirate the air sample through 10 ml of the absorbing solution in a midget impinger at 1.5 l/min for a selected period up to 2 hr. The addition of 5 ml of 95% ethanol to the absorbing solution just prior to aspiration controls foaming for 2 hr (induced by the presence of STRactan 10). In addition, 1 or 2 Teflon demister discs may be slipped up over the impinger air inlet tube to a height approximately 1 to 2″ from the top of the tube.

7.2 ANALYSIS. Add 1.5 ml of the amine-test solution to the midget impinger through the air inlet tube and mix. Add 1 drop of ferric chloride solution and mix. (**Note:** See section **3.1** if SO_2 exceeds 10 μg/ml in the absorbing media.) Transfer the solution to a 25-ml volumetric flask. Discharge the color due to the ferric ion by adding 1 drop ammonium phosphate solution. If the yellow color is not destroyed by 1 drop ammonium phosphate solution, continue dropwise addition until solution is decolorized. Make up to volume with distilled water and allow to stand for 30 min. Prepare a zero reference solution in the same manner using a 10 ml volume of unaspirated absorbing solution. Measure the absorbance of the color at 670 nm in a spectrophotometer or colorimeter set at 100% transmission against the zero reference.

8. Calibration

8.1 AQUEOUS SULFIDE. Place 10 ml of the absorbing solution in each of a series of 25-ml volumetric flasks and add the diluted standard sulfide solution, equivalent to 1, 2, 3, 4, and 5 μg of hydrogen sulfide to the different flasks. Add 1.5 ml of amine-acid test solution to each flask, mix and add 1 drop of ferric chloride solution to each flask. Mix, make up to volume and allow to stand for 30 min. Determine the absorbance in a spectrophotometer at 670 nm, against the sulfide-free reference solution. Prepare a standard curve of absorbance vs. μg H_2S/ml.

8.2 GASEOUS SULFIDE. Commercially available permeation tubes containing liquefied hydrogen sulfide may be used to prepare calibration curves for use at the upper range of atmospheric concentration. Triple-walled tubes, drilled rod and micro bottles which deliver hydrogen sulfide within a minimum range of 0.1 to 1.2 μg/min at 25 C have been prepared by Thomas **(10)**; O'Keeffe **(12, 13)**; Scaringelli **(14, 15)**. Preferably the tubes should deliver hydrogen sulfide within a loss rate range of 0.003 to 0.28 μg/min to provide realistic concentrations of H_2S (1.5 to 139 $\mu g/m^3$, 1.1 to 100 ppb) without having to resort to a dilution system to prepare the concentration range required for determining the collection efficiency of midget impingers. Analyses of these known concentrations give calibration curves which simulate all of the operational conditions performed during the sampling and chemical procedure. This calibration curve includes the important correction for collection efficiency at various concentrations of hydrogen sulfide.

8.2.1 Prepare or obtain a Teflon permeation tube that emits hydrogen sulfide at a rate of 0.1 to 0.2 μg/min (0.07 to 0.14 μl/min at standard conditions of 25 C and 1 atmosphere). A permeation tube with an effective length of 2 to 3 cm and a wall thickness of 0.318 cm will yield the desired permeation rate if held at a constant temperature of 25 C ± 0.1 C. Permeation tubes containing hydrogen sulfide are calibrated under a stream of dry nitrogen to prevent the precipitation of sulfur in the walls of the tube.

8.2.2 To prepare standard concentrations of hydrogen sulfide, assemble the apparatus consisting of a water-cooled condenser, constant temperature bath maintained at 25 C ± 0.1 C cylinders containing pure dry nitrogen and pure dry air with appropriate pressure regulators, needle valves and flow meters for the nitrogen and dry air, diluent-streams. The diluent gases are brought to temperature by passage through a 2-meter-long copper coil immersed in the water bath. Insert a calibrated permeation tube into the central tube of the condenser maintained at the selected constant temperature by circulating water from the constant-temperature bath and pass a stream of nitrogen over the tube at a fixed rate of approximately 50 ml/min. Dilute this gas stream to obtain the desired concentration by varying the flow rate of the clean, dry air. This flow rate can normally be varied from 0.2 to 15 l/min. The flow rate of the sampling system determines the lower limit for the flow rate of the diluent gases. The flow rates of the nitrogen and the diluent air must be measured to an accuracy of 1 to 2%. With a tube permeating hydrogen sulfide at a rate of 0.1 μl/min, the range of concentration of hydrogen sulfide will be between 6 to 400 $\mu g/m^3$ (4 to 290 ppb), a generally satisfactory range for ambient air conditions. When higher concentrations are desired, calibrate and use longer permeation tubes.

8.2.3 *Procedure for preparing simulated calibration curves.* Obviously one can prepare a multitude of curves by selecting different combinations of sampling rate and sampling time. The following description represents a typical procedure for ambient air sampling of short duration, with a brief mention of a modification for 24 hr. sampling.

The system is designed to provide an accurate measure of hydrogen sulfide in the 1.4 to 84 $\mu g/m^3$ (1 to 60 ppb) range. It can be easily modified to meet special needs.

The dynamic range of the colorimetric procedure fixes the total volume of the sample at 186 liters; then, to obtain linearity between the absorbance of the solution and the concentration of hydrogen sulfide in ppm, select a constant sampling time. This fixing of the sampling time is desirable also from a practical standpoint: in this case, select a sampling time of 120 min. Then to obtain a 186-l sample of air requires a flow rate of 1.55 l/min. The concentration of standard H_2S in air is computed as follows:

$$C = \frac{Pr \times M}{R + r}$$

Where

C = Concentration of H_2S in ppm,
Pr = Permeation rate in μg/min,
M = Reciprocal of vapor density, 0.719 μl/μg,
R = Flow rate of diluent air, liter/min,
r = Flow rate of diluent nitrogen, liter/min

Data for a typical calibration curve are listed in Table 701:I.

A plot of the concentration of hydrogen sulfide in ppm (x—axis) against absorbance of the final solution (y—axis) will yield a straight line, the reciprocal of the slope of which is the factor for conversion of absorbance to ppm. This factor includes the correction for collection efficiency. Any deviation from the linearity at the lower concentration range indicates a change in collection efficiency of the sampling system. If the range of interest is below the dynamic range of the method the total volume of air collected should be increased to obtain sufficient color within the dynamic range of the colorimetric procedure. Also, once the calibration factor has been established under simulated conditions the conditions can be modified so that the concentration of H_2S is a simple multiple of the absorbance of the colored solution.

For long-term sampling of 24-hr duration, the conditions can be fixed to collect 1200 liters of sample in a larger volume of STRactan 10-cadmium hydroxide. For example, for 24 hrs at 0.83 l/min, approximately 1200 liters of air are scrubbed. An aliquot representing 0.1 of the entire amount of sample is taken for the analysis.

Table 701:I. Typical Calibration Data

Concentrations H_2S, ppb	Amount of H_2S in μl/186 liters	Absorbance of Sample
1	.144	.010
5	.795	.056
10	1.44	.102
20	2.88	.205
30	4.32	.307
40	5.76	.410
50	7.95	.512
60	8.64	.615

The remainder of the analytical procedure is the same as described in the previous paragraph.

8.2.4 The permeation tubes must be stored in a wide-mouth glass bottle containing silica gel and solid sodium hydroxide to remove moisture and hydrogen sulfide. The storage bottle is immersed to two-thirds its depth in a constant temperature water bath in which the water is controlled at 25 C ± 0.1 C.

Periodically, (every 2 weeks or less) the permeation tubes are removed and rapidly weighed on a semimicro balance (sensitivity ± 0.01 mg) and then returned to the storage bottle. The weight loss is recorded. The tubes are ready for use when the rate of weight loss becomes constant (within ± 2%).

9. Calculation

9.1 Determine the sample volume in liters from the gas meter or flow meter readings and time of sampling. Adjust volume to 76 cm mercury and 25 C (V_s).

$$H_2S = \frac{\mu g \times 10^3}{V_{s,1}} = \mu g/m^3$$

10. Effect of Light and Storage

10.1 Hydrogen sulfide is readily volatilized from aqueous solution when the pH is below 7.0. Alkaline, aqueous sulfide solutions are very unstable because $S^=$ is rapidly oxidized by exposure to the air. Therefore, the dilute, alkaline sulfide standard must be carefully prepared under a nitrogen atmosphere. The preparation of a standard curve should be completed immediately upon dilution of the concd standard sulfide solution. Aqueous sulfide standards may be protected from air oxidation by the addition of 0.1M ascorbic acid (**17**). Ascorbic acid is to be used only in standards which are to be analyzed by titration. Ascorbic acid interferes with the development of the methylene blue color.

10.2 Cadmium sulfide is not appreciably oxidized even when aspirated with pure oxygen in the dark. However, exposure of an impinger containing cadmium sulfide to laboratory or to more intense light sources produces an immediate and variable photodecomposition. Losses of 50 to 90% of added sulfide have been routinely reported by a number of laboratories. Even though the addition of STRactan 10 to the absorbing solution controls the photodecomposition (**2**), it is necessary to protect the impinger from light at all times through use of low actinic glass impingers, paint on the exterior impingers or an aluminum foil wrapping.

11. References

1. Jacobs, M.B., M.M. Braverman, and S. Hochheiser. 1957. Ultramicrodetermination of sulfides in air. *Anal. Chem.* 29:1349.
2. Bamesberger, W.L. and D.F. Adams. 1969. Improvements in the collection of hydrogen sulfides in cadmium hydroxide suspension. *Environ. Sci. & Tech.* 3:258.
3. Mecklenburg, W. and R. Rozenkranzer. 1914. Colorimetric determination of hydrogen sulfide. *A. Anorg. Chem.* 86:143.
4. Almy, L.H. 1925. Estimation of hydrogen sulfide in proteinaceous food products. *J. Am. Chem. Soc.* 47:1381.
5. Sheppard, S.E. and J.H. Hudson. 1930. Determination of labile sulfide in gelatin and proteins. *Ind. Eng. Chem., Anal, Ed.* 2:73.
6. Adams, D.F., F.A. Young, and R.A. Luhr. 1968. Evaluation of an odor perception threshold test facility. *Tappi.* 51:62A.
7. Leonardos, G., D. Kendall, and N. Barnard. 1969. Odor threshold determinations of 53 odorant chemicals. *J. Air Pollut. Contr. Assoc.* 19:91.
8. Boström, C.E. 1965. The absorption of sulfur dioxide at low concentrations (ppnm) studied by an isotopic tracer method. *Air & Water Pollut. Int. J.* 9:333.
9. Boström, C.E. 1966. The absorption of low concentrations (ppnm) of hydrogen sulfide in a $Cd(OH)_2$ suspension as studied by an isotopic tracer method. *Air & Water Pollut. Int. J.* 10:435.
10. Thomas, B.L. and D.F. Adams. Unpublished information.
11. Lodge, J.P., J.B. Pate, B.E. Ammons, G.A. Swanson. 1966. The use of hypodermic needles as critical orifices. *J. Air Poll. Control Assoc.* 16:197.
12. O'Keeffe, A.E. and G.C. Ortman. 1966. Primary standards for trace gas analysis. *Anal. Chem.*, 38:760.
13. O'Keeffe, A.E. and G.C. Ortman. 1967. Precision picogram dispenser for volatile substances. *Anal. Chem.* 39:1047.
14. Scaringelli, F.P., S.A. Frey, and B.E. Saltzman. 1967. Evaluation of teflon permeation tubes for use with sulfur dioxide. *Am. Ind. Hyg. Assoc. J.* 28:260.
15. Scaringelli, F.P., E. Rosenberg, and K. Rehme. 1969. Stoichiometric comparison between permeation tubes and nitrite ion as primary standards for the colorimetric determination of nitrogen dioxide. Presented before the Division of Water, Air and Waste Chemistry of the American Chemical Society, 157th National Meeting, Minneapolis, Minn., April.
16. Kolthoff, I.M. and P.J. Elving. 1961. Eds. Treatise on Analytical Chemistry, Part II, Analytical Chemistry of the Elements, V. 7, Interscience Publishers, New York.
17. Bock, R. and H.-J. Puff. 1968. Bestimmung von sulfid mit einer sulfidionenempfindlichen elektrode. *Z. Anal. Chem.* 240:381.

Subcommittee 1

D. F. Adams, *Chairman*
D. Falgout
J. O. Frohliger
J. B. Pate
A. L. Plumley
F. P. Scaringelli
P. F. Urone

702.

Tentative Method of Analysis of the Sulfation Rate of the Atmosphere (Lead Dioxide Cylinder Method)

42410-01-70T

1. Principle of the Method

1.1 The lead dioxide candle method for determination of total atmospheric sulfation has been used extensively in Great Britain since it was first proposed by Wilsdon and McConnell (**1**) in 1934. The method depends on the reaction between sulfur dioxide and lead dioxide which produces lead sulfate. The reaction may be represented stoichiometrically by the following chemical equation:

$$SO_2 + PbO_2 \rightarrow PbSo_4.$$

1.2 Hydrogen sulfide reacts with lead dioxide according to Russell (**2**).

$$PbO_2 + H_2S \rightarrow PbO + H_2O + S$$

The sulfur thus formed may then further react to form lead sulfate.

$$PbO_2 + O_2 + S \rightarrow PbSO_4$$

Both of these reactions are thermodynamically favorable and Hickey (**3**) did in fact find sulfate after lead candles were exposed to hydrogen sulfide.

1.3 In addition, sulfate aerosols such as ammonium sulfate and sulfuric acid mist which may be suspended in the atmosphere can contribute to the total sulfation if they impinge on the surface of the candle.

1.4 The lead sulfate formed is insoluble in water but it can be dissolved in sodium carbonate solution. Sulfate can then be determined without interference from lead dioxide.

1.5 The method gives relative exposure values for different parts of a community for the period of sampling. If proper care is taken to follow a standard procedure and to use uniform reagents and equipment, the procedure will give data which can be compared with that which is obtained in different communities and over long periods of time.

2. Range and Sensitivity

2.1 The reaction of sulfur dioxide with lead dioxide has been reported to be first order with respect to sulfur dioxide concentrations up to 700 ppm (**1**). If the concentration of sulfur dioxide, the wind speed and the temperature are held constant the rate of formation of lead sulfate is a constant until at least 15% of the total available lead peroxide is reacted. Hickey (**3**) reported that the reaction of sulfur dioxide with lead dioxide can be classed with the metal-oxygen reactions which produce a protective film, such as the formation of aluminum oxide. Therefore, the quantity of sulfur dioxide that a lead candle will absorb before the remaining available surface area becomes rate controlling is inversely proportional to the particle size of the lead dioxide reagent.

2.2 Using the data of Stalker (**4**) Hickey calculated coefficients for sulfur dioxide on lead dioxide. Taking these coefficients and his own reaction saturation data, Hickey plotted curves of allowable exposure time vs. sulfur dioxide concentration. These curves are reproduced here (Figures 702:1 and 702:2) and may be used to estimate the allowable exposure given the approximate sulfur dioxide concentration. The manufacturer and average particle size of the lead dioxide is identified for each curve. Figure 702:1 may be used to find the maximum allowable exposure time for the case where prevailing conditions are expected to promote a rapid rate of reac-

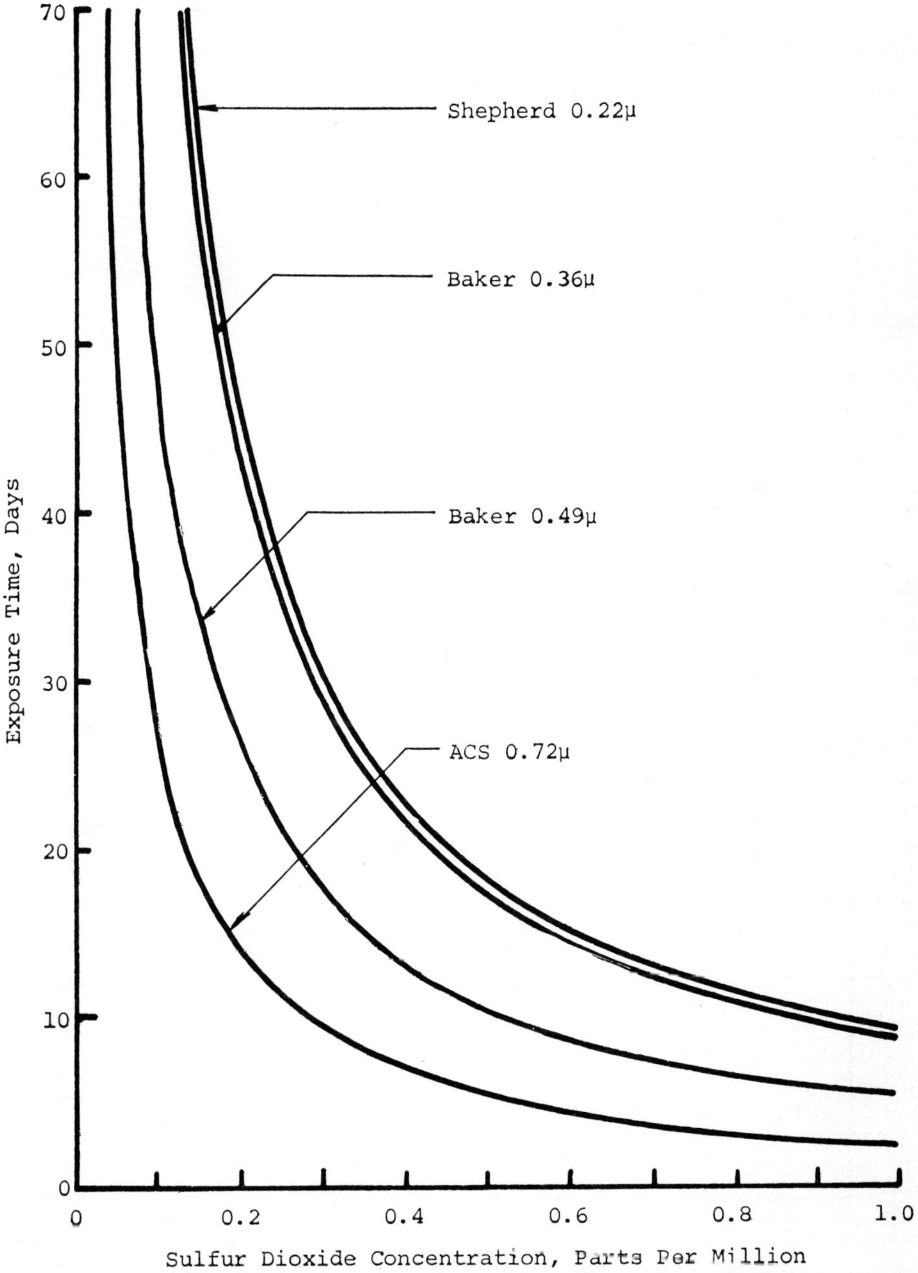

Figure 702:1—Allowable exposure time vs. sulfur dioxide concentration with an absorption coefficient of 32.0 mg SO_2/ppm/day/100 cm^2 (2).

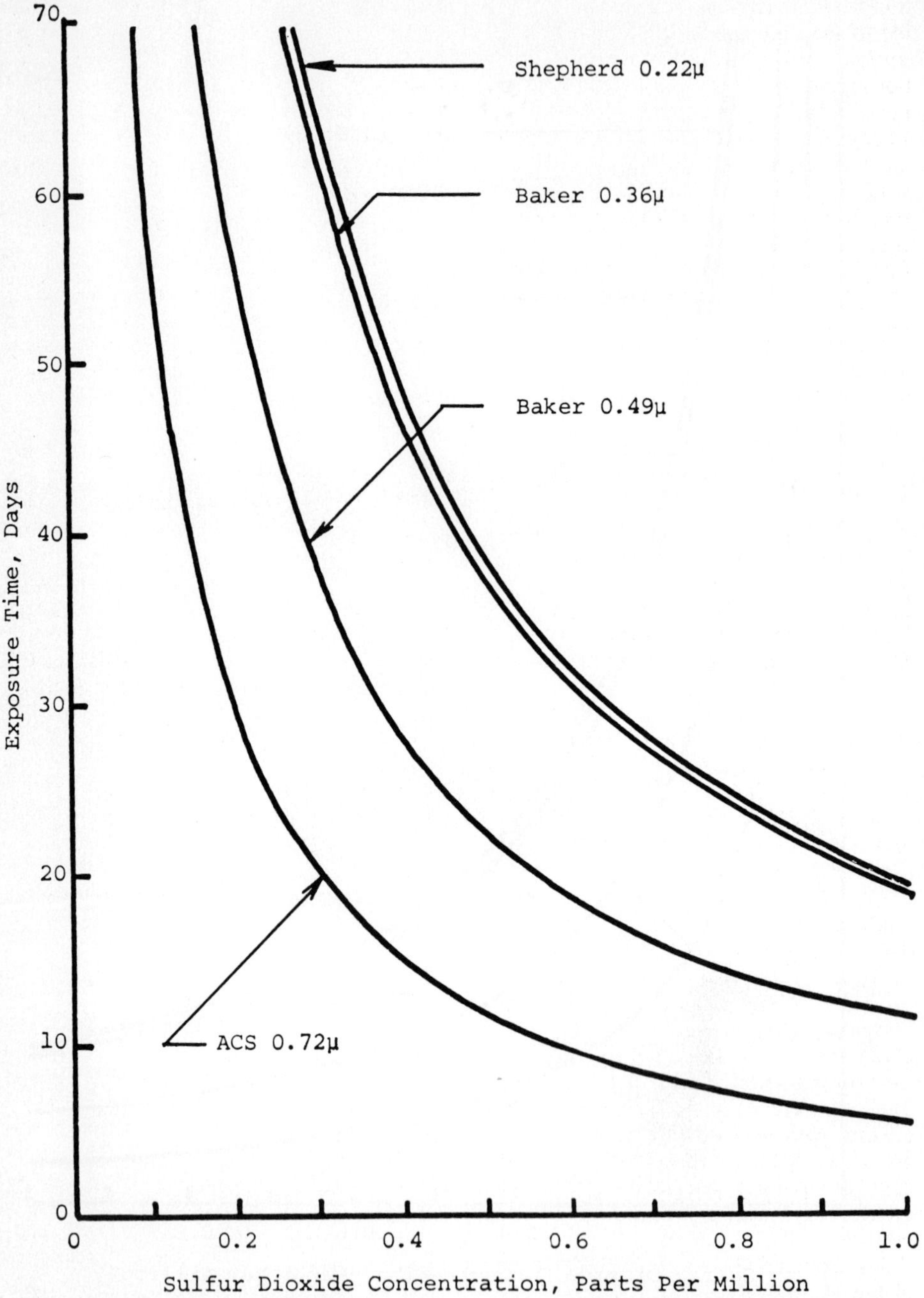

Figure 702:2—Allowable exposure time vs. sulfur dioxide concentration with an absorption coefficient of 15.1 mg SO_2/ppm/day/100 cm^2 (2).

tion between lead dioxide and sulfur dioxide. If the expected sulfur dioxide concentration is 0.2 ppm and the reaction conditions are favorable then the candles should not be exposed longer than 14 days. After 14 days the amount of surface area remaining becomes rate controlling. Figure 702:2 is used when the reaction rate is slow. It should be emphasized that the sorption coefficients are not universally applicable. For this reason, calculation of the average concentration of sulfur dioxide from sulfation rate data is not recommended.

2.3 The lower limit of sensitivity of the method can be calculated given the lower limit of sensitivity of the procedure for determination of sulfate.

3. Interferences

3.1 In their original work Wilsdon and McConnell (**1**) reported that the rate of sulfation is inversely proportional to the fourth root of the incident wind velocity. Thus the sorption coefficients given by Stalker should be reduced for high wind speeds.

3.2 Wilsdon and McConnell reported that the rate of sulfation is reduced by 0.4% for each C degree rise of temperature. Hickey states that within the limitations of his work the effect of temperature on the reaction rate is indiscernible in the range 25 to 40 C.

3.3 Methyl mercaptan is fairly easily oxidized and lead dioxide is a strong oxidant. The reaction:

$$PbO_2 + 2CH_3SH \rightarrow CH_3SSCH_3 + PbO + H_2O$$

could occur but it would not be expected to produce sulfate. Methyl mercaptan does interfere with the reaction of sulfur dioxide with lead dioxide however. Data are not available for the effects of other organic compounds, nor have other specific interferences with sulfate formation been reported.

3.4 Bowden (**5**) reported that the shape of the shelter affects the response of the method. Round plastic shelters gave 20% more sulfation than square shelters exposed at the same site at the same time. In addition, the reactivity of the lead dioxide reagent varies from batch to batch. It is recommended that enough lead dioxide of one batch be stocked for the entire survey. If this is not possible a supply of standard lead dioxide should be stored so that the relative reactivity of each new batch can be determined.

4. Precision and Accuracy

4.1 Comparisons of a number of cylinders exposed simultaneously in the same location have generally shown good concordance in the amount of sulfur taken up. Thomas and Davidson (**6**) reported that sulfation values for triplicate cylinders exposed at four sites did not vary more than 10%.

4.2 Keagy *et al* (**7**) reported that the average annual sulfation rate could be estimated within 10% at the 95% confidence level. In this study 2 cylinders/sq mile were exposed for 30 day periods. In this same study the exposure of 1 cylinder/9 sq miles in a uniformly spaced pattern gave about the same annual mean but with a much smaller degree of confidence.

5. Apparatus

5.1 ONE INCH WIDE SHORT BRISTLED PAINT BRUSH.

5.2 TUBULAR GAUZE.

5.3 MEDIUM POROSITY FILTERING CRUCIBLES.

5.4 MECHANICAL STIRRER.

5.5 PYREX BEAKERS, 400-ml.

5.6 FILTERING FLASKS.

5.7 DESICCATOR.

5.8 MUFFLE FURNACE.

5.9 DRYING OVEN (105°).

5.10 WHATMAN NO. 12 FOLDER FILTER PAPER.

5.11 WHATMAN FILTER ACCELERATOR.

5.12 LONG STEM CONICAL FUNNELS.

5.13 VACUUM SOURCE.

5.14 EXPOSURE STATION AND CYLINDERS. A complete exposure station may be purchased (RAC Corporation, Allison Park, Penn. 15101). Instructions for assembly of a square wooden shelter are included in the appendix. (Figure 702:3)

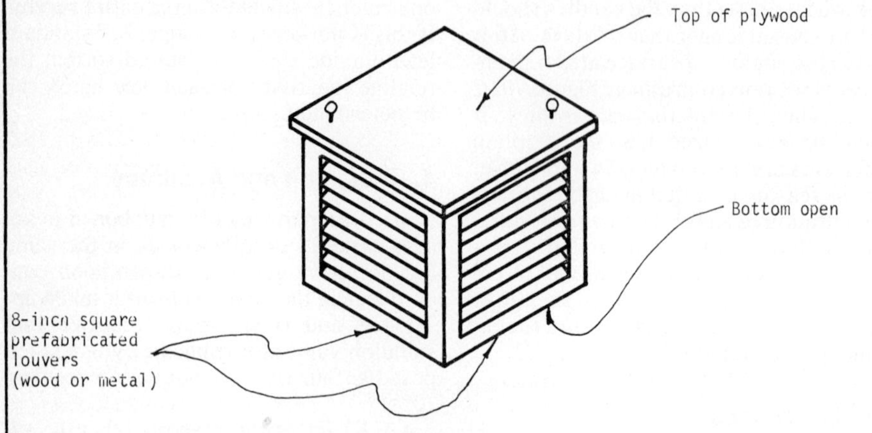

SKETCH OF SUGGESTED SHELTER

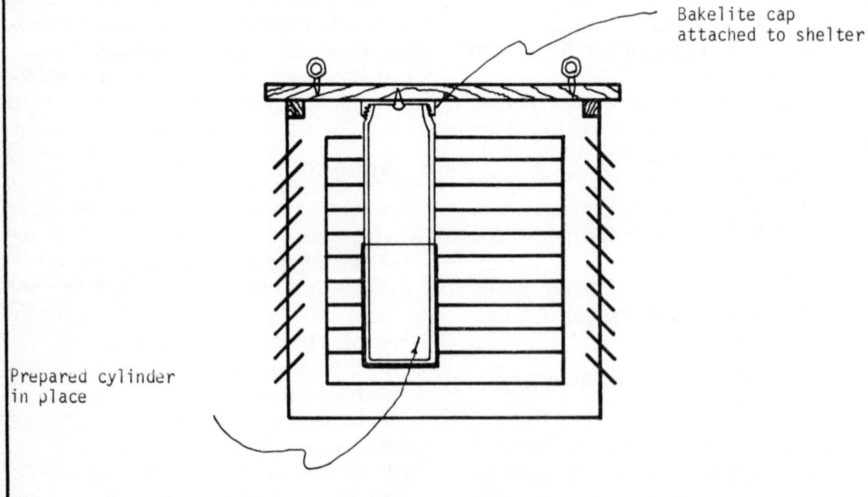

SECTION THROUGH SHELTER

FIGURE 3

6. Reagents

6.1 Purity. All reagents should be ACS analytical reagents unless otherwise noted.

6.2 5% sodium carbonate solution. Dissolve 50 g of anhydrous sodium carbonate in 1 l of distilled water.

6.3 10% barium chloride solution.

Dissolve 200 g of barium chloride in distilled water. Filter the solution through quantitative grade filter paper.

6.4 METHYL ORANGE INDICATOR SOLUTION. Dissolve 0.5 g of methyl orange in 1 l of distilled water.

6.5 1:1 HYDROCHLORIC ACID. Mix equal volumes of hydrochloric acid and distilled water.

6.6 1% SILVER NITRATE SOLUTION. Dissolve 10 g of silver nitrate in 1 l of distilled water. Store in a dark bottle.

6.7 ETHYL ALCOHOL (95%).

6.8 GUM TRAGACANTH. Use powdered form of best grade available.

6.9 LEAD DIOXIDE PASTE. Dissolve 12 g of gum tragacanth in 30 ml of ethyl alcohol. Make up to 600 ml with distilled water. Slowly add 950 g of lead dioxide with constant stirring. Continue stirring until the mixture takes the form of a smooth paste. **CAUTION:** *Lead dioxide is a powerful oxidant and a toxic material and should be handled with the precautions appropriate to such materials. In laboratories where determinations of lead are being carried on, lead dioxide dust presents a serious contamination hazard.*

7. Procedure

7.1 PREPARATION OF CYLINDERS FOR EXPOSURE. Put the gauze strips on all cylinders so that the coating of the cylinders can be begun immediately after the paste is prepared and finished directly without interruption. The lead dioxide paste is painted onto the gauze wrapped cylinders with the small paint brush (or a spatula). The candles are allowed to dry in the air or they may be carefully dried in the drying oven at 90 C. If the layer of lead dioxide paste is too thick or if the drying is done too rapidly cracks will develop in the surface and some lead dioxide may be lost.

After drying, the cylinders are kept in air tight jars until their exposure. The completed cylinder should hold 7 to 8 g of dry lead dioxide which is just enough to hide the texture of the gauze but thin enough to dry without cracking. The consistency of the paste and the method of application may be varied by experience to produce the desired result. At least 2 prepared cylinders should be stored in the airtight jars for every 40 to be exposed. These cylinders serve as a control when the analysis is run. All cylinders to be exposed simultaneously should be prepared at the same time.

7.2 EXPOSURE OF CYLINDERS. After preparation the cylinders are carried into the field in their airtight containers and placed in shelters. It is recommended that 2 cylinders be exposed at each site. The length of the exposure period may be varied between 27 and 33 days to avoid collection on weekends or holidays but the result is calculated on the basis of a 30-day month. The exposure period may be varied to any desirable time for the purposes of special studies.

7.2.1 The shelter should be located at a site which is unaffected by local sources of sulfur gases, and which is not protected from air flow by a building or trees. The elevation of the shelter should be between 6 and 15 ft above ground level. The shelters may be suspended from power or telephone poles or suported by stanchions.

7.3 ANALYSIS OF SULFATE. After exposure, return the cylinders to their air-tight containers and bring them to the laboratory. Pour 100 ml of the sodium carbonate solution into the container with the candle and the gauze. Add enough distilled water to cover the gauze entirely and stir. After a few min of soaking, remove the gauze from the glass cylinder and transfer the gauze, the solution and the lead dioxide slurry to a 400-ml beaker. Take care to get all of the lead dioxide into the beaker. Allow the gauze strip and lead dioxide to stand in the carbonate solution for about 4 hr, stir occasionally. Next, simmer the contents of the beaker on a hot plate for 30 min then add the filter accelerator and simmer with stirring until it is well dispersed.

7.3.1 Remove the beaker from the hot plate and allow the lead dioxide and gauze to settle to the bottom but do not let the mixture cool. Decant the relatively clear liquid from the mixture through the Whatman #12 folded filter paper. Collect the filtrate in a 400-ml beaker. Wash the residue in the 250-ml beaker with 2 to 25 ml portions of boiling distilled water, pouring the supernatant through the filter as before. Then wash the entire contents of the 250-ml beaker into the filter with additional boiling distilled water. The volume in the 400-ml beaker should not exceed 250 ml. After the last liquid has drained out of the filter discard the filter with appropriate safety precautions. The filter contains the lead dioxide and lead carbonate, the filtrate contains the sulfate.

7.3.2 Add 4 to 6 drops of methyl orange solution to the filtrate and neutralize it to the red endpoint with 1:1 HCl solution. Carbon dioxide is released during the neutralization and the solution has a tendency to foam, especially when the endpoint is approached.

7.3.3 Heat the acidified solution to boiling and slowly add to it 5 ml of hot $BaCl_2$ solution. Stir the sample vigorously while adding the $BaCl_2$ solution. Keep the temperature just below boiling until the liquid has become clear and the precipitate has settled out completely. In no case shall this settling period be less than 2 hr (**8**).

7.3.4 Filter the precipitate from the solution with the tared medium porosity filtering crucible. Be sure to remove all traces of the barium sulfate precipitate from the sides and bottom of the beaker. Wash the precipitate with small portions of hot distilled water until the filtrate no longer contains traces of chloride ion. This step is speeded up by use of vacuum filtration and by decanting the bulk of the clear supernatant through the filtering crucible before removing the precipitate with the last 25 to 50 ml of liquid.

7.3.5 Dry the filtering crucibles in the oven then fire them at 800 C for 30 min. Return the crucibles to the oven and allow them to cool to 105 C then place them in the desiccator. Do not put wet crucibles into a muffle furnace or red hot crucibles into a dessicator. Reweigh the crucibles to the nearest 0.1 mg.

8. Calibration and Standards

8.1 There is no calibration to be done for this procedure since it yields a relative index of exposure to sulfur compounds rather than a quantity or concentration. Several attempts have been made to relate the sulfation rate to the concentration of sulfur dioxide as determined by various quantitative procedures. In some cases the correlation is good, in others it is poor. The factors which affect the rate of sulfation of lead dioxide are:

- **a.** Depth of lead dioxide beneath the surface.
- **b.** Wind velocity.
- **c.** Conc and type of sulfur compounds.
- **d.** Wetness of lead dioxide surface.
- **e.** Particle size of lead dioxide.
- **f.** Purity of lead dioxide.

The length of the list of factors which can affect the rate of sulfation makes it impossible to correlate sulfation rate with just one of them under field conditions.

The variations which are dependent upon the lead dioxide itself can be eliminated by purchasing enough reagent from one manufacturing batch to complete the entire study. If this is not possible, the variation can be normalized by comparing the reactivity of each batch to that of a standard batch.

9. Calculation

9.1 The results are reported as mg So_3 per 100 cm^2 per month.
If

W_1 = weight of $BaSO_4$ (mg) on the exposed cylinder
W_2 = weight of $BaSO_4$ (mg) on the unexposed cylinder
a = area of exposed PbO surface (cm^2)
b = number of days of exposure

mg SO_3/100 cm^2/month

$$= 30\,(W_1 - W_2) \times \frac{(80)\,(100)}{233\ a.b}$$

$$= \frac{1029\,(W_1 - W_2)}{a.b}$$

10. Effects of Storage

10.1 The lead dioxide cylinders, both exposed and unexposed, can be stored indefinitely in airtight containers.

11. References

1. WILSDON, B.H. and F.J. MCCONNELL. 1934. The Measurement of Atmospheric Sulfur Pollution by Means of Lead Peroxide. *J. Soc. of Chem. Ind.* Dec. 21.
2. RUSSELL, E.J. 1900. Notes on the Estimation of Gaseous Compounds of Sulfur. *J. Chem. Soc.* 77:352.
3. HICKEY, H.R. and E.R. HENDRICKSON. 1965. A Design Basis for Lead Dioxide Cylinders. *J. Air Poll. Control Assoc.* 15:409–414.
4. STALKER, W.W., R.C. DICKERSON, and G.D. KRAMER. 1963. Atmospheric Sulfur Dioxide and Particulate matter. *Amer. Ind. Hyg. Assoc. J.* 24:68.
5. BOWDEN, S.R. 1964. Improved Lead Dioxide Method of Assessing Sulfurous Pollution of the Atmosphere. *Int. J. Air Water Poll.* 8:101–106.
6. THOMAS, F. and C.M. DAVIDSON. 1961. Monitoring Sulfur Dioxide with Lead Peroxide Cylinders. *J. Air Poll. Control Assoc.* 11:24–27.
7. KEAGY, D.M. *et al.* 1961. Sampling Station and Time Requirements for urban Air Pollution Surveys: Part 1: Lead Peroxide Candles and Dustfall Collectors. *J. Air Poll. Control Assoc.* 11:270–280.
8. AMERICAN SOCIETY OF TESTING MATERIALS. 1968 Book of Standards Part 23 Water; Atmospheric Analysis p 55 ASTM Phil. Pa.

APPENDIX I

Alternate Exposure Station

If it is desirable that the exposure stations be constructed rather than purchased, the equipment described below may be used. It should be noted that Bowden (7) reported that cylinders exposed in square shelters gave 80% of the response of those exposed in round shelters at the same site. The initial cost of this exposure station is less than that of the R.A.C. exposure station, but over a long survey period (several months) the extra time and expense of preparing the cylinders for exposure will negate the initial cost advantage.

APPARATUS

Olive jars with Bakelite screw caps, 5 oz., 4.4 cm diam × 15 cm length
Extra cap for above
One quart Mason Jars
Surgitube, No. 2, tubular gauze
Housing for prepared cylinders to prevent exposure to rain but permit adequate air flow (see Figure 702:3).

ASSEMBLY

Attach 1 olive jar cap to the ceiling of the shelter as shown (use 2 jars per shelter). Attach the extra cap to the inside of the Mason Jar lids. The Mason Jars serve as airtight containers for storage and transporting of the cylinders.

Number each jar near the cap end with a scriber. Each jar should be marked with a scriber at a distance from the closed end such that the total area below the mark, including the bottom surface, is equal to 100 cm^2. For the jars recommended, this distance is 6.1 cm. Cut pieces of tubular gauze 21 cm in length and sew through the gauze perpendicular to the long axis about 0.5 cm from one end. Turn the gauze strip inside out and slip it smoothly over the closed end of the jar. The top edge should extend above the scribed mark. One layer of masking tape should be wrapped around the jar with the lower edge of the tape on the scribe mark. This fixes the gauze firmly to the jar. A second layer of tape is then placed around the jar directly on top of the first. The purpose of the second strip is to limit the exposed area to exactly 100 cm^2. Any lead dioxide scale deposited on this tape during preparation of the cylinder can be removed before analysis by removing the outer layer of tape. Sew the gauze, do not staple it, because iron interferes with the sulfate analysis.

APPENDIX II

Colorimetric Sulfate Determination

To be used for most low-level exposures where the equivalent sulfur dioxide collected does not exceed 40 mg/100 cm^2 of active surface. In uncontrolled field tests, with exposure of 2 cylinders at each location, the results by gravimetric and colorimetric analysis agreed within 20% at 95% confidence level.

The following additional materials are required:

Barium chloranilate
Buffer solution (pH 4.0)
  0.05M solution of AR grade potassium acid phthalate (10.2 g/l)
Hydrochloric acid (25% solution)
Ammonium hydroxide (5% solution)
Ethyl alcohol, 95%
Filter Paper (Whatman No. 12 folded or equivalent)

Standard solution (5 mg.SO_3/m)
545 g AR grade potassium sulfate/500 ml distilled water
Spectrophotometer or filter photometer with 530 nm filter
Using a spectrophotometer, transmittance or absorbance is measured at 530 nm.

PROCEDURE

1. Transfer all material from the candle to a 250-ml beaker. The glass cylinder cannot be soaked prior to removal of the gauze and lead dioxide because the total volume of the slurry must be kept low. The lead dioxide crust should be dampened before the gauze is removed.

2. Add 25 ml of 5% sodium carbonate and 25 ml of distilled water and stir. Let stand 3 hr, stirring occasionally. Boil for 30 min, adding distilled water to maintain approximate constant volume.

3. Filter contents of the beaker through the filter paper. Transfer traces from the beaker with a policeman and wash into filter several times with small quantities of distilled water and save filtrate.

4. Adjust the pH of the filtrate to 4.0 with 25% HCl or 5% NH_4OH using a pH meter or short-range indicator paper. Dilute to exactly 100 ml with distilled water.

5. To a 10 ml aliquot containing up to 5.0 mg SO_3 in a 20 × 150 mm test tube are added 2 ml of buffer solution and 10 ml of 95% ethyl alcohol. Mix by inverting tube. Add approximately 0.1 g barium chloranilate and shake for 10 min. The excess barium chloranilate and precipitate barium sulfate are removed by filtration through filter paper. (95% ethyl alcohol is recommended. Alcohol conc and pH are critical. If any other strength alcohol is used, the standard curve must be prepared using the same strength used in the analysis.)

6. The per cent transmittance or absorbance is measured on a spectrophotometer at 530 nm versus a blank prepared in the same manner starting with Step **2**. The sulfur dioxide concentration in mg SO_3/100 cm^2 is then obtained from a calibration curve previously prepared.

A standard curve is prepared by starting at Step **2** of the procedure. To 5 beakers containing the 50 ml of liquid are added 0, 2, 4, 6 and 8 ml of standard solution and the complete procedure followed. The curve is plotted with transmittance or absorbance against mg of SO_3/100 cm^2. The standards provide a range of 0 to 32 and the curve is a straight line over that range. When new reagents are prepared, a new standard curve must be made.

7. Run identical determination on the control cylinder and subtract the results from each exposed cylinder from that batch.

8. From the corrected values, the results should be expressed in milligrams of SO_3/100 cm^2 of surface/month.

Subcommittee 1
D. F. ADAMS, *Chairman*
D. FALGOUT
J. O. FROHLIGER
J. B. PATE
A. L. PLUMLEY
F. P. SCARINGELLI
P. F. URONE

703.

Tentative Method of Analysis of the Sulfation Rate of the Atmosphere (Lead Dioxide Plate Method-Turbidimetric Analysis)

42410-02-71T

1. Principle of the Method

1.1 The lead dioxide candle method for determination of total atmospheric sulfation has been used extensively in Great Britain since it was first proposed by Wilsdon and McConnell (**1**) in 1934. The method (**2**) depends on the reaction between sulfur dioxide and lead dioxide which produces lead sulfate. The reaction may be represented stoichiometrically by the following chemical equation:

$$SO_2 + PbO_2 \rightarrow PbSO_4$$

1.2 Hydrogen sulfide reacts with lead dioxide according to Russell (**3**). Hickey and Hendrickson (**4**) did in fact find sulfate after lead candles were exposed to hydrogen sulfide.

1.3 In addition, sulfate aerosols such as ammonium sulfate and sulfuric acid mist which may be suspended in the atmosphere can contribute to the total sulfation if they impinge on the surface of the lead dioxide.

1.4 The lead sulfate formed is insoluble in water but it can be dissolved in sodium carbonate solution. The lead dioxide is separated from the lead sulfate solution by filtration prior to analysis of sulfate.

1.5 Huey *et al.* (**5**) devised a "lead plate" which is simpler and less expensive than the lead candle which has been used previously. The exposure procedure differs from the lead candles procedure in that an inverted petri dish is used to support the lead dioxide surface rather than a cylinder.

1.6 The method gives relative exposure values for different parts of a community for the period of sampling. Useful data are obtained if proper care is taken to follow this procedure and to use uniform reagents and equipment. With appropriate statistical treatment, the data can be compared with that which is obtained in different communities and over long periods of time. (See Appendix I). However, meteorological conditions may exert a major influence on the sulfation rate in communities having differing meteorological regimes.

2. Range and Sensitivity

2.1 The reaction of sulfur dioxide with lead dioxide has been reported to be first order with respect to sulfur dioxide conc up to approximately 700 ppm (**1**). Hickey and Hendrickson (**4**) reported that the reaction of sulfur dioxide with lead dioxide can be classed with the metal-oxygen reactions which produce a protective film (*e.g.*, Al_2O_3 over Al). If the concentration of sulfur dioxide, the wind speed and the temperature are held constant, the rate of formation of lead sulfate is a constant until at least 15% of the total available lead peroxide has reacted. Therefore, the quantity of sulfur dioxide that lead dioxide will absorb is inversely proportional to the particle size of the lead dixoide reagent (smaller particles have more surface area per unit weight).

2.2 Using the data of Stalker (**6**), Hickey calculated absorption coefficients for the reaction of sulfur dioxide with lead dioxide. Taking these coefficients and his own reaction saturation data he plotted curves which relate the average sulfur dioxide concentration to the maximum time during which a constant reaction rate may be expected, (see Figures 702:1 and 702:2, (**2**)). It must be emphasized that the absorption coefficients are not universally applicable. For this reason, calculation of the average conc of sulfur dioxide from sulfation rate data is not recommended. This point is discussed further in Section **8**.

2.3 The lower limit of detectability of the method is 20 μg of sulfate.

3. Interferences

3.1 In their original work Wilsdon and McConnell (**1**) reported that the rate of sulfation is inversely proportional to the fourth root of the incident wind velocity.

3.2 Wilsdon and McConnell reported that the rate of sulfation is reduced by 0.4% for each Celsius degree rise of temperature. Hickey states that within the limitation of his work the effect of temperature on the reaction rate is indiscernible in the range 25 to 40 C.

3.3 Methyl mercaptan is easily oxidized by lead dioxide. The reaction:

$$PbO_2 + 2CH_3\,SH \rightarrow CH_3SSCH_3 + PbO + H_2O$$

could occur but it would not be expected to produce sulfate. Methyl mercaptan does interfere with the reaction of sulfur dioxide with lead dioxide however. Data are not available for the effects of other organic compounds, nor have other specific interferences with sulfate formation been reported.

3.4 The reactivity of the lead dioxide reagent varies from batch to batch.

4. Precision and Accuracy

4.1 Comparisons of a number of lead plates exposed simultaneously in the same location have generally shown good concordance in the amount of sulfur taken up. Huey *et al.* (**5**) found that the relative standard deviation of 11 plates exposed simultaneously was 8%. Results with the lead plate were found to be 10% higher than those obtained with lead candles (Appendix I).

5. Apparatus

5.1 BEAKERS. (100-ml and 50-ml).

5.2 GLASS STIRRING RODS.

5.3 WATER BATH.

5.4 WHATMAN NO. 40 FILTER PAPER, 11-cm (or equivalent).

5.5 LONG STEM CONICAL FUNNEL (59°).

5.6 MEASURING SCOOP. To dispense 0.1 g of precipitant. (May be made or purchased).

5.7 PIPETS VOLUMETRIC. (1- to 20-ml).

5.8 GLASS FIBER FILTERS, FILTER PAPER.

5.9 PHOTOELECTRIC COLORIMETER.

5.10 HUEY SULFATION PLATE AND BRACKET. (SEE FIGURE 703:1). Suppliers are listed in Appendix II.

5.11 BLENDER. ONE QUART CAPACITY.

5.12 VOLUMETRIC FLASK. (50-ML).

5.13 STAINLESS STEEL SPATULA.

5.14 DISPOSABLE PETRI DISHES WITH TIGHT FITTING LIDS.

6. Reagents

6.1 PURITY OF CHEMICALS. All reagents should be ACS analytical grade unless otherwise noted.

6.2 HYDROCHLORIC ACID, 0.4 N. Dilute 33 ml of conc HCl to 1l with distilled water.

6.3 SODIUM CARBONATE.

6.4 PRECIPITANT SULFASPEND REAGENT. (Harleco, 60th & Woodland Avenue, Philadelphia) or Sulfaver powder (Hach Co., Ames, Iowa).

6.5 SULFATE STANDARD—500 μG SO_4 PER ML. Dissolve 0.739 g anhydrous Na_2SO_4 in distilled water and make up to 1 l in a volumetric flask.

6.6 LEAD DIOXIDE. It is recommended that enough lead dioxide of one batch be stocked for the entire survey. If this is not possible, a supply of one batch should be stored so that the relative reactivity of each new batch can be determined. This is done by exposing plates prepared from the "standard" batch of lead dioxide side by side with plates prepared from the new batch.

6.7 GUM TRAGACANTH. Use powdered form of best grade available.

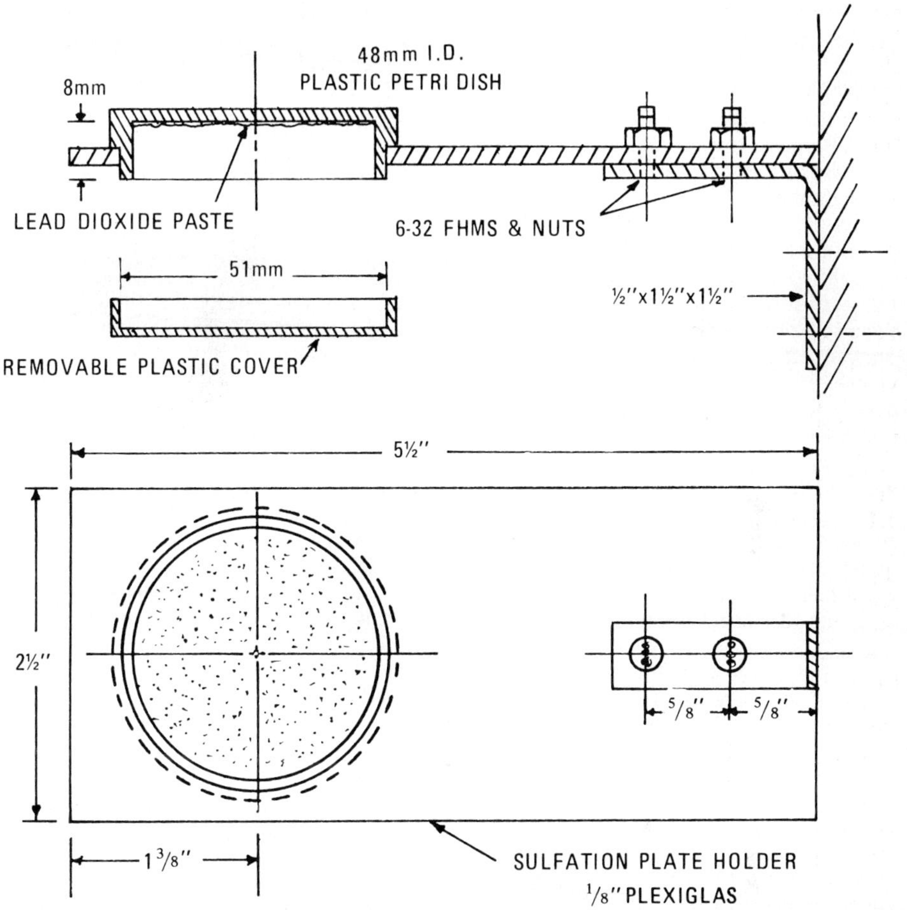

Figure 703:1—Sulfation plate and holder.

6.8 Lead Dioxide Paste. Mix 112 g of lead dioxide, 700 ml of distilled water, 0.7 g of gum tragacanth and 7 g of glass fiber filters in the blender. **CAUTION:** Lead dioxide is a powerful oxidant and a toxic material and should be handled with the precautions appropriate to such materials. In laboratories where determinations of lead are being carried on, lead dioxide dust presents a serious contamination hazard.

6.9 Sodium Carbonate Solution (5%). Dissolve 50 g of anhydrous Na_2CO_3 in a liter of distilled water.

6.10 Sodium Carbonate Solution (2%). Dissolve 20 g of anhydrous Na_2CO_3 in a liter of distilled water.

7. Procedure

7.1 Preparation of Plates for Exposure. Ten ml of the lead dioxide paste is poured into the bottom half of the petri dish. The coated dish is dried at 60 C in an atmosphere that is relatively free of SO_2. When the paste is dry (this may take 12 to 24 hr), the petri dish is sealed with its lid. At least 2 plates should be stored in a desiccator for every 40 to be exposed. These plates serve as a control when the analysis is run. All plates to be exposed simultaneously should be prepared at the same time.

7.2 EXPOSURE OF PLATES. After preparation the sealed plates are carried into the field. Remove the lid and place the plate (lead dioxide side down) into the holder. It is recommended that two plates be exposed at each site. After exposure the plates are resealed with the airtight lid and brought to the laboratory. The length of the exposure period may be varied between 27 and 33 days to avoid collection on weekends or holidays but the result is calculated on the basis of a 30-day month. The exposure period may be varied for the purposes of special studies.

7.2.1 The plate should be located at a site which is not unduly affected by local sources of sulfur gases. The elevation of the plate should be between 6 and 15 feet above ground level. The brackets should be attached to exposed utility poles. The brackets should not be mounted on buildings or trees.

7.3 ANALYSIS OF SULFATE. Carefully transfer the contents of the exposed petri dish to a 100-ml beaker. If traces of lead dioxide remain in the dish transfer them to the beaker with the aid of a rubber policeman and 1 to 3 ml of water. Add 20 ml of the 5% sodium carbonate solution. Let this mixture stand 3 hr at room temperature, stir occasionally. Heat the mixture at 100 C for 30 min (on a water bath or carefully controlled hot plate; avoid splattering), and then cool to room temperature. Transfer the mixture quantitatively to a 50-ml volumetric flask and bring to volume. Filter about 15 ml through a Whatman #40 filter paper. Depending upon the amount of sulfate expected pipet either 1.00 or 5.00 ml of the filtered solution into a clean 50-ml beaker. Add enough of the 2% sodium carbonate solution to make a total of 5.00 ml. Add 5.00 ml of the HCl solution and 10.00 ml of distilled water to make a total of 20.00 ml. Mix thoroughly. The pH of this solution should now be between 2.5 and 4.0.

Add one scoop (0.1 g) of the precipitant and stir until all the powder dissolves. Allow the beaker to stand 20 min, stir again to resuspend the precipitate. Transfer a portion of the solution to the spectrophotometer cuvette and read the absorbance at 450 nm. The spectrophotometer should be zeroed against a distilled water blank. The plates which were not exposed are treated in exactly the same matter as the sample.

8. Calibration and Standards

8.1 The analysis of the sulfate is calibrated by preparing a series of dilutions from the 500 μg/ml sulfate standard solution and plotting absorbance *vs*. concentration for these standards. Dilute 10 ml of the 500 μg/ml sulfate standard to 100 ml and then 0.4, 1, 2, 3, 4, 5, 10, and 15 ml of the diluted standard to 20 ml. These standard samples contain 20, 50, 100, 150, 200, 250, 500, and 750 μg of sulfate respectively. Add 1 scoop (0.1 g) of the precipitant to each standard and proceed as was done with the sample solutions. Plot the sulfate content of the standards against absorbance. The sulfate content of the samples is read from this plot.

8.2 Several attempts have been made to relate the sulfation rate to the concentration of sulfur dioxide as determined by various quantitative procedures. In some cases the correlation is good; in others it is poor **(6)**.

The factors which affect the rate of sulfation of lead dioxide are:

a. Depth of lead dioxide beneath the surface.

b. Wind velocity.

c. Concentration and type of sulfur compounds.

d. Wetness of lead dioxide surface.

e. Particle size of lead dixoide.

f. Purity of lead dioxide.

The length of the list of factors which can affect the rate of sulfation makes it impossible to correlate sulfation rate with just one of them under field conditions.

The variations which are dependent upon the lead dixoide itself can be eliminated by purchasing enough reagent from one manufacturing batch to complete the entire study. If this is not possible, the variation can be normalized by comparing the reactivity of each batch to that of a standard batch.

9. Calculation

9.1 The results are reported as mg SO_4 = 100 cm^2 per month

Sulfate rate (mg $SO_4/100cm^2$/month)

$$= \frac{AB}{10CDE}$$

A = μg SO_4 from standard curve
B = 30 days per standard month
C = fraction of filtrate used to produce turbidity

$$\frac{5}{50} = 0.10 \text{ (if 5 ml of filtrate used)}$$

$$\frac{1}{50} = 0.02 \text{ (if 1 ml filtrate used)}$$

D = exposure time in days
E = area of plate in cm^2

10. Effects of Storage

10.1 The lead dioxide plates, both exposed and unexposed, can be stored indefinitely in air tight containers.

11. References

1. WILSDON, B.H. and F.J. MCCONNELL. 1934. The Measurement of Atmospheric Sulfur Pollution by Means of Lead Peroxide. *J. Soc. Chem. Ind.* December 21.
2. AMERICAN PUBLIC HEALTH ASSOCIATION, INC. 1977. Methods of Air Analysis and Sampling. 2nd ed. p. 682. Washington, D.C.
3. RUSSELL, E.J. 1900. Notes on the Estimation of Gaseous Compounds of Sulfur. *J. Chem. Soc.* 77:352.
4. HICKEY, H.R. and E.R. HENDRICKSON. 1965. A Design Basis for Lead Dioxide Cylinders. *Jour. Air Poll. Control Assoc.* 15:409–414.
5. HUEY, N.A., M.A. WALLAR and C.D. ROBSON. 1969. Field Evaluation of an Improved Sulfation Measurement System. Paper No. 69-133 presented at the Air Pollution Control Association Annual Meeting, June.
6. STALKER, W.W., R.C. DICKERSON, and G.D. KRAMER. 1963. Atmospheric Sulfur Dioxide and Particulate Matter, *Amer. Ind. Hyg. Assoc. J.* 24:68.
7. BOWDEN, S.R. 1964. Improved Lead Dioxide Method of Assessing Sulfurous Pollution of the Atmosphere. *Int. J. Air Poll.* 8:101–106.

Subcommittee 1
D. F. ADAMS, *Chairman*
D. FALGOUT
J.O. FROHLIGER
J. B. PATE
A. L. PLUMLEY
R. P. SCARINGELLI
P. URONE

APPENDIX I

COMPARISON OF LEAD CYLINDER AND PLATE METHODS

Bowden **(7)** reported that the shape of the shelter affects the response of the lead candle method. Round plastic shelters gave 20% more sulfation than square shelters exposed at the same site at the same time. Huey, Wallar and Robson **(5)** compared the results of 2500 candle-plate pairs which were exposed at stations operated by the Interstate Surveillance Network of the Division of Abatement of the National Air Pollution Control Administration. The support media and shelters used with the lead candles were of the type available from Research Appliance Company. The results obtained with the lead plate were found to be 10% higher than those obtained with the lead candles. A linear regression analysis gave a correlation coefficient of 0.95.

APPENDIX II

VENDORS OF SULFATION PLATES AND BRACKETS

Doerfer Corporation
201 Washington
Cedar Falls, Iowa 50613

Research Appliance Company
Route 8 and Craighead Road
Allison Park, Pennsylvania 15101

Harleco Division
60th and Woodland Avenue
Philadelphia, Pennsylvania 19143

Precision Scientific
3737 West Coulland Street
Chicago, Illinois 60647

704.

Recommended Method of Analysis for Sulfur Dioxide Content of the Atmosphere (Colorimetric)

42401-01-69T

1. Principle of the Method

1.1 Sulfur dioxide is absorbed by aspirating a measured air sample through a solution of potassium or sodium tetrachloromercurate, TCM. This procedure results in the formation of a dichlorosulfitomercurate complex, which resists oxidation by the oxygen in the air (**1, 2**). Ethylenediaminetetraacetic acid disodium salt, EDTA, is added to this solution to complex heavy metals that can interfere by oxidation of the sulfur dioxide before formation of the dichlorosulfitomercurate (**3, 4**). This compound, once formed, is stable to strong oxidents, (*e.g.*, ozone and oxides of nitrogen). After the absorption is completed, any ozone in the solution is allowed to decay (**4**). The liquid is treated first with a solution of sulfamic acid to destroy the nitrite anion formed from the absorption of oxides of nitrogen present in the atmosphere (**5**). It is treated next with solutions of formaldehyde and specially purified acid-bleached pararosaniline containing phosphoric acid to control pH. Pararosaniline, formaldehyde, and the bisulfite anion react to form the intensely colored pararosaniline methyl sulfonic acid, which behaves as a two-color pH indicator (λ max. 548 nm at pH 1.6 $\pm$ 0.1, $\epsilon = 47.7 \times 10^3$). The pH of the final solution is adjusted to 1.6 $\pm$ 0.1 by the addition of prescribed amount of 3*M* phosphoric acid to the pararosaniline reagent (**4**).

1.1.1 Two variations are given, they differ only in the pH of the final solution. The variation described above is designated Variation A and is the method of choice. It gives the highest sensitivity. In Variation B, a larger quantity of phosphoric acid is added to yield a pH in the final solution of 1.2 $\pm$ 0.1. The wavelength of maximum absorbance under these conditions is 575 nm and the compound has a molar extinction of 37.0×10^3. Variation B is less sensitive, but has the advantage of a lower blank. It is pH dependent, but may be more suitable with less expensive spectrophotometers.

1.2 Atmospheric sulfur dioxide concentrations of interest usually range from a few pphm to a few ppm. Higher concentrations (5 to 500 ppm), employed in special studies, must be analyzed by using smaller gas samples. A rapid redox reaction occurs between Hg (II) and the sulfito ion, if concentrations of the latter exceed a certain limit, 500 μg/ml (**6**).

1.3 Collection efficiency falls off rapidly below 0.01 ppm, and varies with the geometry of the absorber, the size of the gas bubbles, and the contact time with the solution (**7, 8, 9**).

2. Range and Sensitivity.

2.1 The lower limit of detection of sulfur dioxide in 10 ml of TCM is 0.3 μl (based on twice the standard deviation) representing a conc of 0.01 ppm (26 $\mu g/m^3$) SO_2 in an air sample of 30 liters. One cannot extrapolate to lower values by taking larger volumes of air (*e.g.*, 100 liters at 0.003 ppm). The method is applicable to concentrations below 0.01 ppm if the absorption efficiency of the particular system is determined.

2.2 Beer's Law is followed through the working range from 0.1 to 1.0 absorbance units (0 to 35 μg in 25 ml final solution).

3. Interferences

3.1 The effects of the principal known interferences [oxides of nitrogen, ozone, and heavy metals (*e.g.*, iron, manganese, and chromium)] have been minimized or eliminated. The interferences by oxides of nitrogen are eliminated by sulfamic acid (**4, 5**), the ozone by time-delay (**4**), and the heavy metals by EDTA and phosphoric acid (**3, 4**). At least 60 μg of Fe III, 10μg of Mn II, and 10 μg of Cr III in 10 ml of absorbing reagent can be tolerated in the procedure. No significant interference was found with 10 μg of Cu (II) and 22 μg V(V).

4. Precision and Accuracy

4.1 The precision at the 95% confidence level is 4.6% (**4**).

5. Apparatus

5.1 ABSORBER. Satisfactory absorbers are (a) the midget or standard fritted bubbler; (b) the midget impinger; (c) the Greenburg-Smith impinger; and (d) the multiple jet bubbler (**10**).

5.2 AIR VOLUME MEASUREMENT. The air meter must be capable of measuring the air flow within ± 2%. A wet or dry gasmeter, with contacts on the 1 ft^3 or 10-l dial to record air volume, or a specially calibrated rotameter, is satisfactory. Instead of these, calibrated hypodermic needles may be used as critical orifices if the pump is capable of maintaining greater than 0.5 atmosphere pressure differential across the needle (**11**).

5.3 MANOMETER. Mercury manometer accurate to 5 mm.

5.4 SPECTROPHOTOMETER OR COLORIMETER. The instrument must be suitable for measurement of color at 548 nm or 575 nm. With Variation A, reagent blank problems may result with spectrophotometers or colorimeters having greater spectral band width than 16 nm. The wavelength calibration of the spectrophotometer should be verified.

6. Reagents

6.1 PURITY OF CHEMICALS. All chemicals must be ACS analytical reagent grade. The pararosaniline dye should meet the specifications as described in **6.8**.

6.1.1 *Distilled water*. Distilled water must conform to the ASTM Standard for Reference Reagent Water. Water must be free from oxidants. It should preferably be double distilled from all glass apparatus.

6.2 ABSORBING REAGENT, 0.04M POTASSIUM TETRACHLOROMERCURATE (TCM), K_2HgCl_4. Dissolve 10.86 g mercuric chloride (**CAUTION:** Highly poisonous. If spilled on skin, flush off with water immediately), 5.96 g of potassium chloride, 0.066 g of EDTA in water and bring to mark in a 1-liter volumetric flask. Sodium chloride (4.68 g) may be substituted for the potassium chloride, but potassium chloride is usually obtained in purer form. The pH of this reagent should not be less than 5.2. Low pH values of the absorbing reagent reduce the collection efficiency of this reagent for sulfur dixoide. There are two reasons for obtaining low pH values. One is the incorrect ratio or concentrations of $HgCl_2$ and KCl. This can be adjusted by addition of a dilute solution of KCl if the ratio is not correct. The other occurs when the EDTA is not the disodium salt. If the latter is the cause of low pH value, adjust to the correct value by the drop-wise addition of dilute alkali. The absorbing reagent is normally stable for 6 months, but if a precipitate forms, discard the solution.

6.3 SULFAMIC ACID, 0.6%. Dissolve 0.6 g of sulfamic acid in 100 ml of distilled water. This reagent can be kept for a few days if protected from air.

6.4 BUTANOL. Certain batches of 1-butanol contain oxidants that create an SO_2 demand. Check by shaking 20 ml of 1-butanol with 5 ml of 20% KI. If a yellow color appears in the alcohol phase, redistill the 1-butanol from silver oxide.

6.5 BUFFER STOCK SOLUTION (pH 4.69). In a 100-ml volumetric flask, dissolve 13.61 g of sodium acetate trihydrate in water. Add 5.7 ml of glacial acetic acid and dilute to volume with water.

6.6 1.0 N HYDROCHLORIC ACID. Dilute 86 ml of 11.6*M* acid (36% HCl) to a liter.

6.7 3M PHOSPHORIC ACID H_3PO_4. Dilute 205 ml of 14.6*M* acid (85% H_3PO_4) to a liter.

6.8 PURIFIED PARAROSANILINE, 0.2%

(NOMINAL) STOCK SOLUTION. The pararosaniline dye needed to prepare this reagent must meet the following performance specifications: the dye must have a wavelength of maximum absorbance of 540 nm, when assayed in a buffered solution of 0.1 M sodium acetate-acetic acid; the absorbance of the reagent blank, (Section **7.3**) which is temperature sensitive (0.015 A. U./C), should not exceed 0.170 absorbance unit at 22 C, when prepared according to the prescribed analytical procedure and to the specified concentration of the dye; and the reagents should give a calibration curve with a slope of 0.746 ± 0.04 absorbance units/μg/ml for 1-cm cell, when the dye is pure and the sulfite solution is properly standardized. If specially purified pararosaniline dye,* 99.0%, is available, weigh 0.200 g and completely dissolve the dye by shaking with 100 ml of 1 *N* HCl in a 100-ml graduated cylinder that is glass stoppered. If the pararosaniline dye is obtained in solution, assay the concentration according to **6.8.2**, and proceed to **6.8.3**. When the dye does not meet these specifications, it normally can be purified satisfactorily by following the procedure in **6.8.1**.

6.8.1 *Purification Procedure.* In a large separatory funnel (250-ml), equilibrate 100 ml each of 1-butanol and 1*N* HCl. Weigh 0.1 g of pararosaniline hydrochloride (PRA) in a beaker. Add 50 ml of the equilibrated acid and let stand for several minutes. To a 125-ml separatory funnel add 50 ml of the equilibrated 1-butanol. Transfer the acid solution containing the dye to the funnel and extract. The violet impurity will transfer to the organic phase. Transfer the lower (aqueous) phase into another separatory funnel and add 20 ml of 1-butanol; extract again. Repeat the extraction with three more 10-ml portions of 1-butanol. This is usually sufficient to remove almost all of the violet impurity which contributes to the reagent blank. If violet impurity still appears in 1-butanol phase after 5 extractions; discard this lot of dye. After the final extraction, filter the aqueous phase through a cotton plug into a 50-ml volumetric flask and bring to volume with 1*N* HCl. This stock reagent will be yellowish red.

6.8.2 *Assay Procedure.* The actual concentration of PRA need be assayed only once for each lot of dye in the following manner: Dilute 1 ml of the stock reagent to the mark in a 100-ml volumetric flask with distilled water. Transfer a 5 ml aliquot to a 50-ml volumetric flask. Add 5 ml of 1*M* sodium acetate-acetic acid buffer, and dilute the mixture to 50 ml volume with distilled water. After 1 hr, determine the absorbance at 540 nm with a spectrophotometer. Determine the per cent of the nominal concentration of PRA by the formula:

$$\% \text{ PRA} = \frac{\text{Absorbance} \times \text{K}}{\text{grams taken}}$$

For 1-cm cells and 0.04 mm slit width in a Beckman DU Spectrophotometer, K = 21.3 (Mean value after extensive purification of dye).

6.8.3 *Pararosaniline Reagent.* To a 250-ml volumetric flask add 20 ml of stock pararosaniline reagent. Add an additional 0.2 ml of stock for each percent the stock assays below 100%. For Variation A, add 25 ml of 3*M* H_3PO_4 and dilute to volume with distilled water. These reagents are stable for at least 9 months. For Variation B, add 200 ml of 3*M* H_3PO_4 and dilute to volume.

6.9 FORMALDEHYDE, 0.2%. Dilute 5 ml of 40% formaldehyde to a liter with distilled water. Prepare daily.

6.10 REAGENTS FOR STANDARDIZATION.

6.10.1 *Stock Iodine Solution,* 0.1*N*. Place 12.7 g of iodine in a 250-ml beaker; add 40 g of potassium iodide and 25 ml of water. Stir until all is dissolved, then dilute to a liter with water.

a *Working Iodine Solution, 0.01N:* Prepare approximately 0.01*N* iodine solution by diluting 50 ml of the stock solution to 500 ml with distilled water.

6.10.2 *Starch Indicator Solution.* Triturate 0.4 g of soluble starch and 0.002

*Specially purified dye is available from Harleco, 60th & Woodland Ave., Philadelphia, Pa. 19143 g of mercuric iodide (preservative) with a little water, and add the paste slowly to 200 ml of boiling water. Continue boiling until clear; cool, and transfer to a glass-stoppered bottle.

6.10.3 *Standard 0.1N Sodium Thiosulfate Solution.* Dissolve 25 g of sodium thiosulfate ($Na_2S_2O_3 \cdot 5\ H_2O$) in a liter of freshly boiled, cooled distilled water and add 0.1 g of sodium carbonate to the solution. Allow the solution to stand for 1 day before standardizing. To standardize, accurately weigh 1.5 g of potassium iodate, primary standard grade, that was dried at 180 C, and dilute to volume in a 500-ml volumetric flask. To a 500-ml iodine flask, pipet 50 ml of the iodate solution. Add 2 g of potassium iodide and 10 ml of a 1:10 dilution of concentrated hydrochloric acid. Stopper the flask. After 5 min, titrate with thiosulfate to a pale yellow color. Add 5 ml of starch indicator solution and complete the titration.

$$\text{Normality of thiosulfate} = \frac{\text{Wt. (grams } KIO_3) \times 10^3 \times 0.1}{\text{ml of titer} \times 35.67}$$

6.10.4 *Standard Sulfite Solution.* Dissolve 0.400 g of sodium sulfite (Na_2SO_3) or 0.300 g of sodium metabisulfite ($Na_2S_2O_5$) in 500 ml of recently boiled and cooled high-quality distilled water. (Double-distilled water that has been deaerated is preferred). This solution contains from 320 to 400 μg/ml as SO_2. The actual concentration in the standard solution is determined by adding excess iodine and back-titrating with sodium thiosulfate that has been standardized against potassium iodate or dichromate (primary standard). Sulfite solution is unstable.

a *Back-titration* is performed in the following manner: into each of two 500-ml iodine flasks, pipet accurately 50 ml of the 0.01 *N* iodine. To flask A (blank) add 25 ml of distilled water, and to flask B (sample) pipet 25 ml of the standard sulfite solution. Stopper the flasks and allow to react for 5 minutes. By means of a buret containing standard 0.01*N* thiosulfate, titrate each flask in turn to a pale yellow color. Then add 5 ml of starch solution and continue the titration to the disappearance of the blue color. Calculate the concentration of sulfur dioxide in the standard solution as follows:

$$SO_2\ \mu g/ml = \frac{(A - B)\ N\ K}{V}$$

Where:

A = number of ml for blank,
B = number of ml for sample,
N = normality of thiosulfate solution.
K = micro equivalent weight for SO_2 is 32,000
V = sample volume taken

6.10.5 *Dilute Sulfite Solution.* Immediately after standardization of sulfite solution, pipet accurately 2 ml of the freshly standardized solution into a 100-ml volumetric flask and bring to mark with 0.04*M* TCM. This solution is stable for 30 days if stored at 5 C.

7. Procedure

7.1 Collection of Sample. Place 10 ml of 0.04*M* TCM (20 ml for sampling of long duration) absorbing solution in a midget impinger, or 75 to 100 ml in one of the larger absorbers. Connect the sampling probe upstream of the absorber with glass, stainless steel, or Teflon. Rigid tubing may be joined with butted joints under polyethylene tubing. Downstream, a trap and calibrated air flow meter, and/or a gas meter provided with thermometer and manometer lead to the pump. (Instead, a hypodermic needle in parallel with a manometer can be used as critical orifice in a thermostated box, if the pump can maintain a differential pressure of at least 0.5 atmosphere across the needle). The duration and rate of aspiration depend on the concentration of sulfur dioxide. With midget impingers, rates of 0.5 to 2.5 l/min are satisfactory; with large absorbers, the rate can be 5 to 15 l/min. Rates of sampling within the above ranges generally have an efficiency of absorption of 98% or greater. For best results, rates and sampling time should be chosen to absorb 0.5 to 3.0 μg

(0.2 to 1.3 μl at 760 Torr, 25 C) of sulfur dioxide per ml of sampling solution. Shield the absorbing reagent from direct sunlight during and after sampling by covering the absorber with a suitable wrapping, such as aluminum foil, to prevent deterioration. If the sample must be stored for more than a day before analysis, keep it at 5 C in a refrigerator, if possible.

7.2 Centrifugation. If a precipitate is observed, remove it by centrifugation.

7.3 Analysis. After collection, transfer the sample quantitatively to a 25-ml volumetric flask; use about 5 ml of distilled water for rinsing. Aliquots may be taken, at this point, if the concentration or volume of reagent is large. If the presence of ozone is suspected, delay analyses for 20 min to allow ozone to decompose. For each set of determinations prepare a reagent blank by adding 10 ml of the unexposed absorbing reagent to a 25-ml volumetric flask. To each flask add 1 ml of 0.6% sulfamic acid and allow to react for 10 min to destroy the nitrite from oxides of nitrogen. Accurately pipet in 2 ml of the 0.2% formaldehyde, then 5 ml of pararosaniline reagent prescribed for Variation A or Variation B. Start a laboratory timer that has been set for 30 min. Bring all flasks to volume with freshly boiled distilled water. After the 30 min, determine the absorbances of the sample and of the blank at the wavelength of maximum absorbance, 548 nm for Variation A, or 575 nm for Variation B. Use distilled water (**not** the reagent blank) in the reference cell. Do not allow the colored solution to stand in absorbance cell; a film of dye will be deposited thereon.

7.3.1 If the absorbance of the sample solution ranges between 1.0 and 2.0, the sample can be diluted 1:1 with a portion of the reagent blank and read within a few minutes. Solutions with higher absorbance can be diluted up to six-fold with the reagent blank in order to obtain on-scale readings within 10% of the true absorbance value.

8. Calibration and Standards

8.1 Accurately pipet graduated amounts of the diluted sulfite solution (such as: 0, 1, 2, 3, 4, and 5 ml) in to a series of 25-ml volumetric flasks. Add sufficient 0.04*M* TCM to each flask to bring the volume of its contents to approximately 10 ml. Then add the remaining reagents as described in the procedure. For greatest precision a constant temperature bath is preferred. The temperature of calibration should not differ from the temperature of analysis by more than a few degrees.

8.2 The total absorbances of the solutions are plotted (as ordinates) against the total micrograms of SO_2. A linear relationship is obtained. The intercept with the vertical axis of the line best fitting the points usually is within 0.02 absorbance units of the blank (zero standard) reading. Under these conditions the plot need be determined only once to evaluate the calibration factor (reciprocal of the slope of the line). This calibration factor can be used for calculating results provided there are no radical changes in temperature or pH. At least one control sample is recommended per series of determinations to insure the reliability of this factor.

8.3 Alternative Calibration Procedure. Permeation tubes that contain liquefied sulfur dioxide are calibrated gravimetrically and used to prepare standard concentrations of sulfur dioxide in air (**12–14**). Analyses of these known concentrations give calibration curves which simulate all of the operational conditions performed during the sampling and chemical procedure. This calibration curve includes the important correction for collection efficiency at various concentrations of sulfur dioxide.

8.3.1 Prepare or obtain† a Teflon permeation tube that emits sulfur dioxide at a rate of 0.1 to 0.2 μg/min (0.04 to 0.08 μl/min at standard conditions of 25 C and 1 atmosphere). A permeation tube with an effective length of 1- to 2-cm and a wall thickness of 0.76-mm, (0.030″), will yield the desired permeation rate if held at a constant temperature of 20 C.

Permeation tubes containing sulfur diox-

†Available from Polyscience Corp., 909 Pitner Ave., Evanston, Ill. 60202, and Metronics, Inc., 3201 Porter Drive, Palo Alto, Calif. 94304.

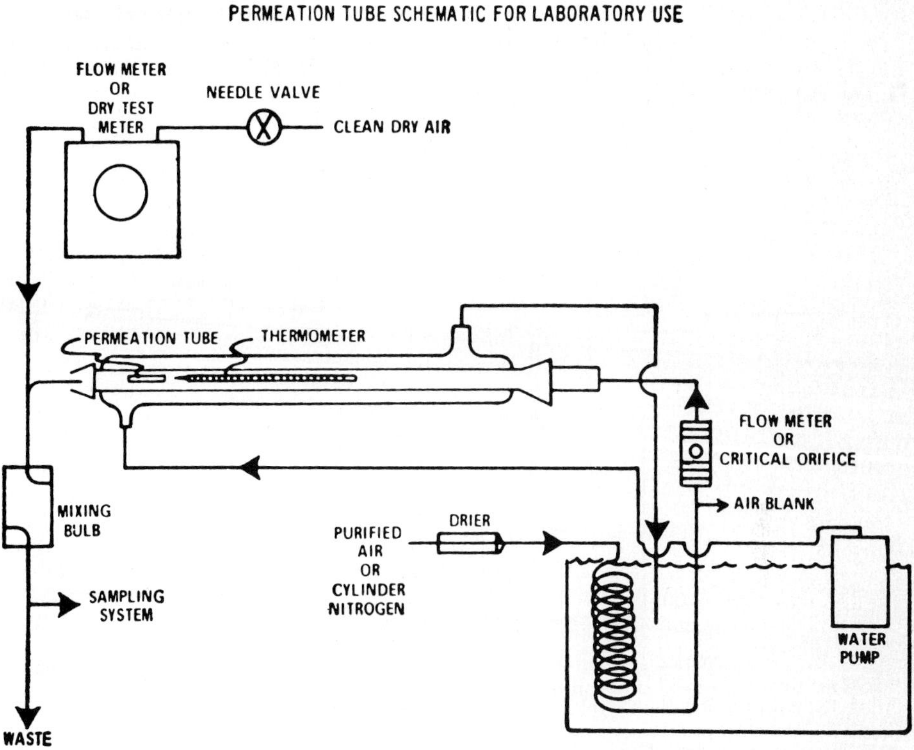

Figure 704:1—Gas dilution system for preparation of standard concentrations of sulfur dioxide for laboratory use by the permeation tube method.

ide are calibrated under a stream of dry nitrogen to prevent the formation of blisters in the walls and sulfuric acid inside the tube.

8.3.2 To prepare standard concentrations of sulfur dioxide, select either the system designed for laboratory or field use, Figure 704:1 and Figure 704:2, respectively. Assemble the apparatus as shown in one of these systems, consisting of a water-cooled condenser; constant-temperature water bath maintained at 20 C; cylinders containing pure, dry nitrogen and pure, dry air with appropriate pressure regulators; needle valves and flow meters for the nitrogen and dry air, diluent gas streams. The diluent gases are brought to temperature by passage through a 2-meter-long copper coil immersed in the water bath. Insert a calibrated permeation tube **(13)** into the central tube of the condenser maintained at 20 C by circulating water from the constant-temperature bath and pass a stream of nitrogen over the tube at a fixed rate of approximately 50 ml/minute. Dilute this gas stream to use the desired-concentration by varying the flow rate of the clean, dry air.‡ This flow rate can normally be varied from 0.2 to 15 l/min. The flow rate of the sampling system determines the lower limit for the flow rate of the diluent gases. The flow rates of the nitrogen and the diluent air must be measured to an accuracy of 1 to 2%. With a

‡Clean dry air may also be prepared by passing ambient air from a relatively uncontaminated outside source through absorption tubes packed with activated carbon and soda-lime followed by an efficient fiber glass filter, in series.

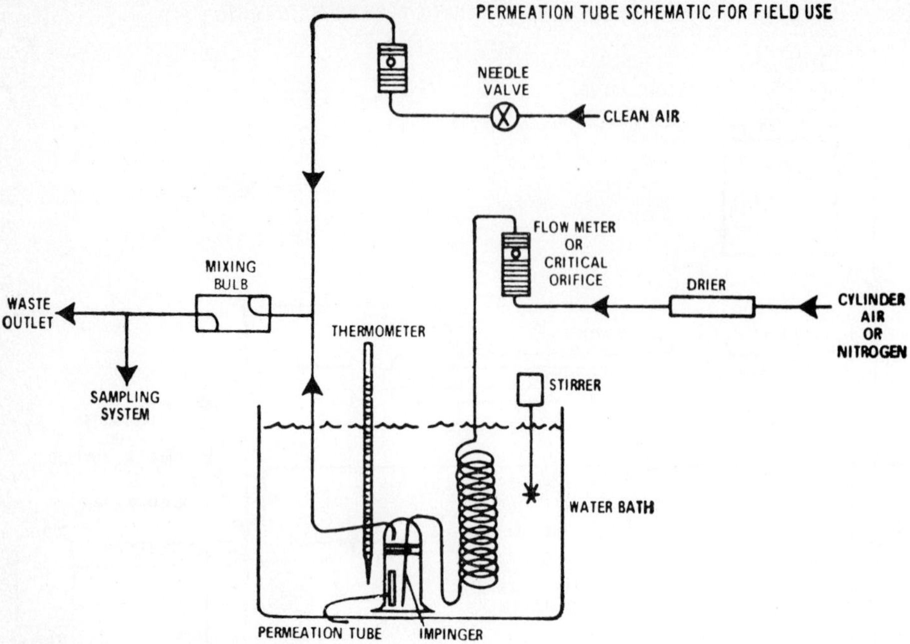

Figure 704:2—Gas dilution system for preparation of standard concentrations of sulfur dioxide for field use by the permeation tube method.

tube permeating sulfur dioxide at a rate of 0.1 μl/min (0.26 μg/min), the range of concentration of sulfur dioxide will be between 0.007 to 0.4 ppm (18 to 1047 $\mu g/M^3$), a generally satisfactory range for ambient air conditions. When higher concentrations are desired, calibrate and use longer permeation tubes.

8.3.3 *Procedure for preparing simulated calibration curves.* Obviously, one can prepare a multitude of curves by selecting different combinations of sampling rate and sampling time. The following description represents a typical procedure for ambient air sampling of short duration, with a brief mention of a modification for 24 hr sampling. The system is designed to provide an accurate measure of sulfur dioxide in the 0.01 to 0.5 ppm range. It can be easily modified to meet special needs.

The dynamic range of the colorimetric procedure fixes the total volume of the sample at 30 liters; then, to obtain linearity between the absorbance of the solution and the concentration of sulfur dioxide in parts per million, select a constant sampling time. This fixing of sampling time is desirable also from a practical standpoint. In this case, select a sampling time of 30 min. Then to obtain a 30 liter sample of air requires a flow rate of 1 l/min. A 22 gauge hypodermic needle operating as a critical orifice will control air flow at this approximate desired rate. The concentration of standard SO_2 in air is computed as follows:

$$C = \frac{Pr \times M}{R + r}$$

Where

C = Concentration of SO_2 in ppm,
Pr = Permeation rate in μg/min,
M = Reciprocal of vapor density, 0.382 μl/μg
R = Flow rate of diluent air, liter/min.
r = Flow rate of diluent nitrogen, liter/min.

Table 704:I. Typical calibration data

Concentrations of SO_2, ppm	Amount of SO_2 in μl for 30 liters	Absorbance of sample
0.005	0.15	0.01
0.01	0.30	0.02
0.05	1.50	0.117
0.10	3.00	0.234
0.20	6.00	0.468
0.30	9.00	0.703
0.40	12.00	0.937

Data for a typical calibration curve are listed in Table 704:I.

A plot of the concentration of sulfur dioxide in ppm (x—axis) against absorbance of the final solution (y—axis) will yield a straight line, the slope of which is the factor for conversion of absorbance to ppm. This factor includes the correction for collection efficiency. Any deviation from linearity at the lower concentration range indicates a change in collection efficiency of the sampling system. Actually, the standard concentration of 0.01 ppm and below of sulfur dioxide is slightly below the dynamic range of the method. If this is the range of interest, the total volume of air collected should be increased to obtain sufficient color within the dynamic range of the colorimetric procedure. Also, once the calibration factor has been established under simulated conditions the conditions can be modified so that the concentration of SO_2 is a simple multiple of the absorbance of the colored solution.

For long-term sampling of 24-hr duration, the conditions can be fixed to collect 300 liters of sample in a larger volume of tetrachloromercurate. For example, for 24 hr at 0.2 liter/min, approximately 288 liters of air are collected. An aliquot representing 0.1 of the entire amount of sample is taken for the analysis.

The remainder of the analytical procedure is the same as described in the previous paragraph.

9. Calculations

9.1 Compute the conc of sulfur dioxide in the sample by the following formula:

$$ppm = \frac{(A - A_o)0.382\ B}{V}$$

Where

A = the sample absorbance,
A_o = the reagent blank absorbance,
0.382 = the volume (μl) of 1 μg SO_2 at 25 C, 760 Torr.
B = the calibration factor, μg/absorbance unit,
V = the sample volume in liters corrected at 25 C, 760 Torr (by PV = nRT).

10. Effects of Storage

10.1 Sampling solutions of dichlorosulfitomercurate are relatively stable. When stored at 5 C for 30 days no detectable losses of sulfur dioxide occur. At 25 C losses of sulfur dioxide in solution occur at a rate of 1.5% per day. These losses of sulfur dioxide follow a first order reaction, and the reaction rate is independent of concentration. Actual field samples containing EDTA have similar decay curves and when analysis of the samples is delayed for any appreciable time, the results must be corrected for these losses.

11. References

1. West, P.W., and G.C. Gaeke. 1956. Fixation of sulfur dioxide as sulfitomercurate III and subsequent colorimetric determination. *Anal. Chem.* 28:1816.
2. Ephraims, F. 1948. *Inorganic chemistry*. Ed. P.C.L. Thorne and E.R. Roberts, 5th ed. p. 562. Interscience, N.Y.
3. Zurlo, N., and A.M. Griffini. 1962. Measurement of the SO_2 content of air in the presence of oxides of nitrogen and heavy metals. *Med. Lavoro.* 53:330.
4. Scaringelli, F.P., B.E. Saltzman, and S.A. Frey. 1967. Spectrophotometric determination of atmospheric sulfur dioxide. *Anal. Chem.* 39:1709.
5. Pate, J.B., B.E. Ammons, G.A. Swanson, and J.P. Lodge, Jr. 1965. Nitrite interference in spectrophotometric determination of atmospheric sulfur dioxide. *Anal. Chem.* 37:942.
6. Lyles, G.R., F.B. Dowling, and V.J. Blanchard. 1965. Quantitative determination of formaldehyde in parts per hundred million concentration level. *J. Air Pollut. Contr. Assoc.* 15:106.
7. Urone, P., J.B. Evans, and C.M. Noyes. 1965. Tracer techniques in sulfur dioxide colorimetric and conductimetric methods. *Anal. Chem.* 37:1104.

8. Boström, C.E. 1965. The absorption of sulfur dioxide at low concentrations (pphm) studied by an isotopic tracer method. *Air & Water Pollut. Int. J.* 9:333.
9. Boström, C.E. 1966. The absorption of low concentration (pphm) of hydrogen sulfide in a $Cd(OH)_2$ suspension as studied by an isotopic tracer method. *Air & Water Pollut. Int. J.* 10:435.
10. Stern, A.C. 1968. *Air Pollution*, Vol. II. 2nd ed. Academic Press. N.Y.
11. Lodge, J.P., Jr., J.B. Pate, B.E. Ammons, and G.A. Swanson. 1966. The use of hypodermic needles as critical orifices in air sampling. *J. Air Pollut. Contr. Assoc.* 16:197.
12. O'Keeffe, A.E., and G.C. Ortman. 1966. Primary standards for trace gas analysis. *Anal. Chem.* 38:760.
13. Scaringelli, F.P., S.A. Frey, and B.E. Saltzman. 1967. Evaluation of teflon permeation tubes for use with sulfur dioxide. *Amer. Ind. Hyg. Assoc. J.* 28:260.
14. Thomas, M.D., and R.E. Amtower. 1966. Gas dilution apparatus for preparing reproducible dynamic gas mixtures in any desired concentration and complexity. *J. Air Pollut. Contr. Assoc.* 16:618.

Subcommittee 1
D. F. Adams, *Chairman*
M. Corn
C. L. Harding
J. B. Pate
A. L. Plumley
F. P. Scaringelli
P. F. Urone

705.

Tentative Method of Analysis for Sulfur Dioxide Content of the Atmosphere *(Manual Conductimetric Method)*

42401-02-70T

1. Principle of the Method

1.1 Sulfur dioxide is collected by aspirating a measured volume of air through a dilute acidified solution of hydrogen peroxide. The sulfur dioxide is oxidized to sulfuric acid by the hydrogen peroxide, yielding two moles of hydrogen ion and one of sulfate ion for each mole of sulfur dioxide. The change of conductivity of the solution caused by the increased ionic content is determined. Because of the simplicity and flexibility of making the conductivity measurements, this method is frequently used. It is applicable to the measurement of sulfur dioxide in the range of 0.05 ppm and upward.

1.2 The electrical conductivity of a solution is dependent upon both the nature and the concentration of every ionic species present in the solution. When a voltage is applied across two electrodes immersed in a solution, the dissolved ions will migrate towards the electrodes: the cations, the positive ions, will migrate towards the negative electrode and anions, the negative ions, will move toward the positive electrode. This migration of ions through the solution constitutes a flow of electric current. The flow of current is greater in a more concentrated solution because of the larger number of ions available as charge carriers. The current flow also depends upon the capacity of the ions to carry a charge and the rate at which the ions move. The hydrogen ion moves through an aqueous solution much more rapidly than any other ion. In practice an alternating current is used to prevent polarization at the electrodes.

1.2.1 The measured conductivity is a function of the sum of the concentration times the conductances of all the ions as well as the characteristics of the conductivity cell used in the measurement.

The characteristics of the conductivity cell, the cell constant, need not be known precisely because the procedure requires that the system be calibrated. If the cell constant should change due to rough handling, or other reasons, the system must be recalibrated. It is recommended that an approximate value of the cell constant be known since the smaller the cell constant the greater the measured conductance is for a given solution. A cell with a cell constant of 0.1 is recommended.

2. Sensitivity and Range

2.1 The electrolytic conductivity of a solution in mhos, or the reciprocal of the resistance in ohms, is a linear function of the concentration of the electrolyte over a wide range. The sensitivity and the range are dependent upon the conductance equipment used in the analysis. The factors that affect the sensitivity and range include the conductance cell, volume of absorbing reagent and the volume of air sampled.

A reagent solution 10^{-5} *M* in H_2SO_4 using a conductance cell with a cell constant of 0.1 will give a conductivity of approximately 78 μmho. A 1% increase in the conductivity in 100 ml of reagent requires 0.64 μg SO_2. A 400 liter air sample containing 25 $\mu g/m^3$ (0.01 ppm) of SO_2 will give a change of conductance of 12 μ mhos under these conditions. Table 705:I shows the conductance values calculated for the reagent, and a 400 liter air sample containing 0.01 ppm SO_2 absorbed in 100 ml of reagent for cell constants of 0.1, 1.0 and 10.0 cm^{-1}.

2.2 The practical lower limit is 130 $\mu g/m^3$ (0.05 ppm) of SO_2 because of the high probability of interferences by other air pollutants (see Section **3**).

3. Interferences

3.1 Conductimetric methods respond to all ionizable substances collected in the sampling solution. In most air pollution situations, sulfur dioxide is assumed to be the principal source of the increased conductivity in the H_2O_2–H_2SO_4 sampling solution. The degree of interference by any ionizable substance can be expressed in terms of its SO_2 equivalent conductance. The SO_2 equivalent conductance of an interfering substance has been found to depend upon its concentration relative to the actual SO_2 concentration. Recent studies **(1–3)** give approximations of the SO_2 equivalents of some interfering substances. Carbon monoxide, ozone and hydrogen sulfide do not contribute to the conductance while hydrochloric acid gives an SO_2 equivalent that is about 30% of the true hydrochloric acid concentration. Chlorine (Cl_2) gives an SO_2 equivalent that is about 10% of the true chlorine concentration. The oxides of nitrogen (NO and NO_2) give an SO_2 equivalent that is about 2% of the true oxides of nitrogen concentration. Ammonia acts as a negative interference **(3)** by reacting with the hydrogen ion (equivalent conductance = 350 at 25 C) to form the ammonium ion (equivalent conductance = 74.5 at 25 C). The effect of ammonia is to give an SO_2 equivalent that is about minus 30% of the true ammonia concentration. Beyond neutrality (pH > 7), the continued absorption of ammonia increases the conductivity due to the additional formation of hydroxide ions (equivalent conductance = 192 at 25 C).

3.2 Aerosol interference is assumed to be small when the recommended scrubber (see Section **5.2**) is used **(4)**. Impingers collect aerosols (and SO_2) efficiently and should not be used. Aerosols may be removed by prefiltration with suitably pre-

Table 705:I. Change in Conductance of a 400-Liter Air Sample in 100 ml Absorbing Reagent

Cell Constant	Reagent Conductivity	0.01 ppm SO_2	Change in Conductance
cm^{-1}	micromhos	micromhos	micromhos
0.1	78	90	12
1.0	7.8	9.0	1.2

pared filters (5). Collected particulates will adsorb some SO_2 especially if they are basic in nature.

3.3 Ambient measurements giving SO_2 concentrations of 0.05 ppm or less should be qualified by recognizing that the probability of interference is high.

4. Precision and Accuracy

4.1 A collection efficiency of better than 99% with an average standard deviation of 0.3% was shown by radio-tracer techniques for seven automatic instruments using the conductivity method (2).

5. Apparatus

5.1 GENERAL. A schematic representation of a portable apparatus for the manual determination of sulfur dioxide is shown in Figure 705:1. It consists of a supply bottle, an automatic pipet, an absorber with accessories, a flowmeter, a trap, and a source of suction. A Wheatstone bridge or conductivity meter is also required.

5.2 ABSORBER. The absorber is a multiple jet bubbler* comprised of a 40-mm tube, 30-cm long, drawn down to a 7-mm drain tube at the bottom, and having a 40/50 ground joint at the top. The bubbler consists of 7 to 8 mm tubing mounted in the joint or stopper. The lower end is bent in a circle to fit loosely near the inside wall of the absorber. The lower side of the cir-

*The multijet absorber is available from Dependable Scientific Glass Works, 346 East Lambourne Street, Salt Lake City, Utah 84115.

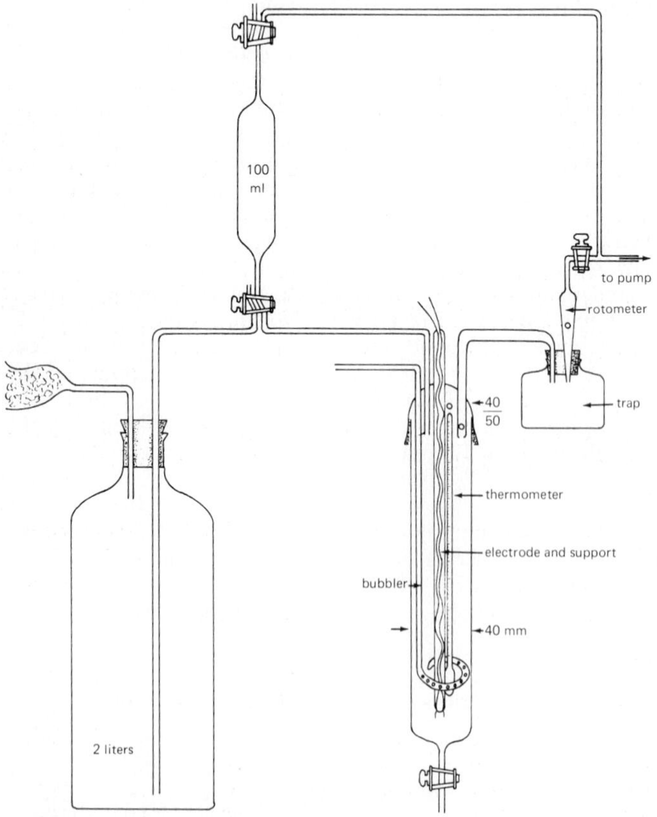

Figure 705:1—Portable Cumulative Sulfur Dioxide Analyzer for Conductivity.

cular tube is pierced with ten 1-mm holes **(1)** (fritted bubblers are not satisfactory). The platinum electrodes are mounted in the lower end of a 10-mm tube which passes through a center hole of the joint or stopper and extends about 2 cm below the circular bubbler. When the bubbler is operated at 15 l/min, the gas bubbles do not come in contact with the platinum electrodes and it is possible to read the conductivity without stopping the air stream. A mercury thermometer, supported by the stopper or joint, extends into the liquid. The absorbing efficiency of the bubbler for sulfur dioxide is approximately 99%. The conductivity electrodes and the thermometer can be omitted from the absorber if the solutions are returned to the laboratory for final measurement.

5.3 The automatic pipet is operated from the suction line, simplifying the handling of the absorbing solution.

5.4 CONDUCTIVITY BRIDGES AND CELLS.

5.4.1 *Manual Bridge*. An AC conductivity bridge provided with temperature compensation circuit. The bridge must be checked for application to dilute sulfuric acid.

a. *Temperature Compensation:* If the conductivity bridge does not have a temperature adjustment circuit, the correction must be calculated and applied. This correction must be applied whenever the temperature of the sampling solution differs significantly (> 2 C) from the calibration temperature. The temperature coefficient of the conductivity of dilute sulfuric acid solutions is 1.6 per cent/degree C.

5.4.2 *Conductivity Cells*. A cell constant of 0.1 is recommended for this procedure. However, conductivity cell constants ranging from < 0.01–> 2 are commercially available. Dip cells may also be used. When cells have closely spaced electrode plates, care must be taken that the liquid between the plates is well mixed with the main body of the absorbing solution. Cells must be kept in operating condition by storage in fresh sampling solution, and care must be taken not to transfer any contaminants to the solution to be measured.

5.5 AIR PUMP. Capable of drawing 15 to 20 l of air sample/min through the absorber.

5.6 FLOWMETER, capable of measuring the rate of flow of the air sample. The flow meter must be calibrated under conditions of use.

6. Reagents

6.1 GENERAL. Reagents must be ACS analytical grade quality. Distilled water should conform to ASTM Standard for Referee Reagent Water. Maximum total solids: < 0.1 mg/liter; maximum conductivity: 1.5 μ mhos/cm at 25 C.

6.2 SULFURIC ACID, 0.1 N for calibrating the analyzer. Standardize in the usual manner either against sodium carbonate or other base of known strength.

6.3 ABSORBING SOLUTION, 0.006% H_2O_2 and 10^{-5} N H_2SO_4. To prepare the absorbing reagent, dilute 0.25 ml of 30% stabilizer-free hydrogen peroxide and 1 ml of 0.01 N sulfuric acid to 1 l. Add 2 mg of

Table 705:II. Liters of air to be sampled in 100 ml absorbing solution

Micrograms equivalent SO_2 per ml in absorbing solution	SO_2 concentration in air, ppm 25 C 760 Torr 0.05	0.10	0.20	0.30	0.40	0.50	1.0
0.1 μg/ml	76.3 l	38.2 l	19.1 l	12.7 l	9.5 l	7.6 l	3.8 l
0.5 "	382	191	95.4	63.6	47.7	38.2	19.1
1.0 "	763	382	191	127	95.4	76.3	38.2
2.0 "	1527	763	382	254	191	153	76.3
5.0 "	3817	1908	954	636	477	382	191
10.0 "	7634	3817	1908	1272	954	763	382

any noninterfering fungicide, such as 2, 4, 5-trichlorophenate (Dowicide) **(6)**. The pH of the absorbing reagent must be between 4 and 5. If the pH is greater than 5, add 0.01 *N* H_2SO_4 dropwise until the pH is within the limits. Discard the reagent if one drop added to 10% KI on a spot plate does not produce a yellow-brown color.

7. Procedure

7.1 Assemble the sampling train using a preliminary 100 ml of the absorbing reagent. Adjust the air flowrate to a predetermined value which takes into account the estimated SO_2 concentration (see Table 705:II). Stop or bypass the air flow, drain the absorbing solution and add a fresh 100 ml of absorbing solution. Record the conductance reading and begin aspirating the sample; check the air flowrate and record the time and flowmeter reading in order to obtain the volume of air sampled.

7.2 After sampling for the desired sampling interval, adjust the temperature compensator of the conductivity meter to the temperature of the solution and record the conductivity. Total sampling periods of ½–2 hr duration are commonly employed. The solution is discharged from the absorber after the final conductance reading is recorded and the volume of the absorbing reagent is measured in a 100-ml graduated cylinder and recorded.

7.2.1 The sampling process can be repeated as often and for as long a sampling period as desired on the same 100 ml of absorbing solution as long as the conductivity readings do not go off scale. Care must be taken in adjusting the values obtained for evaporation losses which may amount to 1 to 5 ml/hr.

7.3 Titration. If desired, the analysis can be checked by titrating the spent solution with 0.002 N sodium hydroxide using a mixed methyl red-bromcresol green indicator. The endpoint is judged by comparison with the unaspirated absorbing solution having the same amount of indicator and titrated to a selected color. The blank titer is subtracted from the total titer.

8. Calibration

8.1 Liter volume solutions of H_2SO_4 containing the equivalent of 10, 5, 2, 1, 0.5 and 0.1 μg of sulfur dioxide/ml should be prepared for calibration. The volume of standardized H_2SO_4 solution (Ml_{std}) needed to prepare 1 liter of a calibrating solution containing the equivalent of 100 micrograms of sulfur dioxide/ml is calculated as follows:

$$Ml_{std} = \frac{(0.03122)\,(100)}{N\ H_2SO_4}$$

This volume of sulfuric acid must be dispensed from a buret. This stock solution containing 100 μg/ml may be used to make the more dilute solutions by quantitative dilution steps of no less than 1 to 9 ratios (*i.e.*, 100.0 ml made to 1.000 liter) and very thoroughly mixing at each step.

8.2 Measure the conductivity of the solutions in ascending or descending order by placing them successively in the absorber. Set the temperature compensator at the solution temperature and record the conductance value. Repeat the series if desired. Plot the concentration (μg SO_2/ml) vs. conductivity (μ mhos). The calibration curve should be a straight line passing through or very close to the origin. Calculate the response factor:

$$\frac{\Delta\mu g\ SO_2/ml}{\Delta\mu\ mhos}$$

The calibration curve should be checked daily during routine use with two of the solutions. New calibration solutions should be prepared when there is a 10% change in the value of the response factor.

9. Calculation

9.1 The volume of air sampled is corrected to 25 C and 760 Torr as follows (ideal gas law relationship):

$$V_0 = \frac{P_1\,V_1\,T_0}{P_0\,T_1}$$

where:

V_0 = corrected volume, liters
P_1 = observed pressure, Torr
V_1 = observed volume, liters
T_0 = 298.1 K (absolute temperature at 25 C)
T_1 = observed temperature, C + 273.1
P_0 = standard pressure, 760 Torr

The equation simplifies to:

$$V_0 = \frac{P_1 V_1}{T_1} \times 0.359$$

9.2 The number of μg of sulfur dioxide per m^3 of air is calculated from the following:

$$\mu g\ SO_2/m^3 = \frac{(C_{ob} - C_b)\ B\ V_{sol}}{V_0} \times 1000$$

where:

C_{ob} = conductance of sample, μ mho
C_b = conductance of blank, μ mho
B = response factor,

$$\frac{\Delta \mu g\ SO_2/ml}{\Delta \mu\ mhos}$$

V_{sol} = volume of absorbing solution, ml
V_0 = volume of air sampled corrected to 25 C and 760 Torr, liters

9.2.1 A convenient conversion factor frequently used is that 1 ppm SO_2 at 25 C (298 K) and 760 Torr is equal to 2.620 μg/liter, or 2,620 μg/cubic meter. One μg/l is equal to 0.3817 μl/l at 25 C and 760 Torr.

10. Effect of Storage

10.1 The absorbing solution is reasonably stable and will keep for periods up to 2 weeks in inert containers. Several workers have reported difficulties in keeping up the H_2O_2 strength. This may be caused by trace metal catalysts or trace amounts of organic substances which react slowly with the H_2O_2.

10.2 Calibration solutions should be stable indefinitely. Soft glass or basic substances might dissolve sufficiently to give reduced conductivity and calibration curves which intercept the concentration axis to the right of zero.

11. References

1. KUCZYNSKI, E.R. 1967. Effects of Gaseous Air Pollutants on the Response of the Thomas SO_2 Autometer. *Envir. Sci. and Tech.* 1:68.
2. RODES, C.E., L.A. ELFERS, H.F. PALMER, and H.F. NORRIS. 1969. Performance Characteristics of Instrumental Methods for Monitoring Sulfur Dioxide, *Air Pollut. Cont. Assoc.*, 8:575.
3. MORGAN, G.B., C. GOLDEN, and E.C. TABOR. 1967. *Air Pollut. Cont. Assoc.*, 17:300.
4. THOMAS, M.D., and R.E. AMTOWER. 1966. Gas Dilution Apparatus for Preparing Reproducible Dynamic Gas Mixtures in any Desired Concentration and Complexity, *Air Pollut. Cont. Assoc.*, 16:618.
5. DUCKWORTH, SPENCER. 1969. Particulate Interference in Sulfur Dioxide Manual Sampling by the West-Gaeke Method, 10th Conference on Methods in Air Pollution and Industrial Hygiene Studies, San Francisco, California, Feb.
6. HOCHHEISER, S. 1964. Methods of Measuring the Monitoring Atmospheric Sulfur Dioxide. U.S. Public Health Publication No. 999-AP-6, August.

Subcommittee 1

D. F. ADAMS, *Chairman*
D. FALGOUT
J. O. FROHLIGER
J. B. PATE
A. L. PLUMLEY
F. P. SCARINGELLI
P. URONE

706.

Tentative Method of Analysis for Sulfur Dioxide in the Atmosphere (Automatic Conductimetric Method)

42401-03-73T

1. Principle of the Method

1.1 Sulfur dioxide is sampled by passing a measured volume of air through, or in close proximity to, a dilute acidified solution of hydrogen peroxide in order to provide intimate contact of SO_2 with the reagent. The sulfur dioxide is oxidized to sulfuric acid by the hydrogen peroxide, yielding two moles of hydrogen ion and one mole of sulfate ion for each mole of sulfur dioxide. The change of conductivity of the solution caused by the increased ionic content is determined. The method is not dependable for sulfur dioxide concentration below 130 $\mu g/m^3$ (0.05 ppm) (**1, 2, 3**).

1.2 The electrical conductivity of a solution is dependent upon both the nature and the concentration of every ionic species present in the solution. The conductance of a solution is the reciprocal of the resistance of that solution to the flow of electric current. The current is carried by the ions present in the solution. When a voltage is applied across two electrodes immersed in a solution, the dissolved ions will migrate towards the electrodes: the cations, the positive ions, will migrate towards the negative electrode and the anions, the negative ions, will migrate towards the positive electrode. This migration of ions through the solution constitutes the flow of electrical current. The flow of current is greater in a more concentrated solution and thus the conductance is greater because of the larger number of ions available as charge carriers. The current flow, and thus the conductance, also depends upon the capacity of the ions to carry a charge and the velocity at which the ions move. The hydrogen ion moves through an aqueous solution much more rapidly than any other ion. To minimize polarization of the electrodes, which can give erroneous reading, the applied voltage is alternated.

1.2.1 The measured conductivity is a function of the sum of the products of the concentration times the conductance of all the ions present as well as the geometry and physical characteristics of the electrode system used in the measurement. The specific response of the electrodes need not be known precisely because the procedure requires that the measurement be calibrated with known concentrations of sulfur dioxide in clean air. (See Section **8**.)

1.2.2 The readout of the conductance measurement of the solutions will depend on a large number of factors including the intervals of measurement, the mode of readout, the number of sampling stations and how the data are to be compiled and evaluated. The read-out system may range from a single strip chart recorder to an automatic electronic display or involve an intricate telemetering system with multiple instrumented sampling stations.

2. Sensitivity and Range

2.1 The conductivity of a solution in mhos, or the reciprocal of the resistance in ohms, is a linear function of the concentration of a strong electrolyte over a wide range. The factors that affect the sensitivity and range of this method include the size and positions of the electrodes, the volume of absorbing reagent, the volume of air samples, the method of measurement (such as continuous or batch), and the electronic circuitry used. To insure a common basis for understanding the performance of a specific instrument and interpreting the results obtained from it, the terms used must be carefully defined and used precisely.

2.1.1 The *sensitivity* of an instrument is defined as the smallest change in concentration that produces a change in the signal that is twice the noise level or 0.5% full scale, whichever is greater. The instrument must be capable of resolving 52 $\mu g/m^3$ (0.02 ppm) above 130 $\mu g/m^3$ (0.05 ppm).

2.1.2 The *span* of an attenuation setting is defined as the difference in concentration from the limit of detection (not zero) to the upper limit of the attenuation step in units of concentration. The span should be stated for each attenuation step.

2.1.3 The *limit of detection* is defined as the smallest concentration of sulfur dioxide that will give a signal that is twice the noise level or 0.5% full scale whichever is greater. However, measurements made by an instrument with a limit of detection that is less than 0.05 ppm must still meet the interference requirements of Section **3**.

2.1.4 The *noise level* of an instrument is defined as the random peak to peak excursions from the baseline when clean air (0.00 ppm SO_2) is being sampled by the instrument.

3. Interference

3.1 Conductimetric methods respond to all ionizable substances collected in the sampling solution. In most air pollution situations, sulfur dioxide is assumed to be the principal source of the increased conductivity in the H_2O_2-H_2SO_4 absorbing solution. The degree of interference by any ionizable substance can be expressed in terms of its SO_2 equivalence. Approximations of the SO_2 equivalence of some interfering substances have been reported **(1, 2, 3)**. Carbon monoxide, ozone and hydrogen sulfide do not contribute to the conductance while hydrochloric acid (HCl) gives an SO_2 equivalent that is about 30% of the true hydrochloric acid concentration. Chlorine (Cl_2) gives an SO_2 equivalent that is about 10% of the true chlorine concentration. The oxides of nitrogen (NO and NO_2) give an SO_2 equivalent that is about 2% of the true oxides of nitrogen concentration. Ammonia (NH_3) acts as a negative interference **(2)** by reacting with hydrogen ion (equivalent conductance = 350 at 25 C) to form the ammonium ion (NH_4^+; equivalent conductance = 74.5 at 25 C). The effect of ammonia in the acidic sampling solution is to give an SO_2 equivalent concentration that is about minus 30% of the true ammonia concentration.

3.2 In instruments in which water is used as the absorbing reagent, CO_2 is readily absorbed along with the SO_2. The CO_2 contributes to the total conductance and this contribution must be subtracted from the total reading. The CO_2 contribution is periodically measured by scrubbing SO_2 from the air stream. If the CO_2 concentration does not vary more than the stated sensitivity of the instrument such corrections are reasonable.

3.3 Aerosol interference is assumed to be small when multiple jet bubblers or co- or countercurrent flow contact tubes are used. Impinging devices collect aerosols (and SO_2) efficiently and should not be used. Aerosols may be removed by prefiltration with suitably prepared filters; **(4)** however, collected particulates will absorb some SO_2 **(5, 6, 7)**.

3.4 Ambient measurements giving SO_2 concentrations of 0.05 ppm or less should be qualified by recognizing that the probability of interference by other air pollutants is high.

4. Precision and Accuracy

4.1 The stated precision and accuracy of an automated analytical system should reflect the variances in the total chemical and instrumental components under test conditions.

4.1.1 Repeatability, at the 95% confidence level shall be within ± 2% for successive identical samples of the calibration gas following the procedure described in Section **8.3.4**.

4.1.2 A collection efficiency of better than 98% with an average deviation of 0.5% was obtained from seven automatic instruments using the conductivity principle (**2**).

4.1.3 System drift shall not deviate by more than 5% from its initial setting more than once in a 24 hour period. The system drift should be determined with the procedure described in Section **8**.

4.1.4 Noise level of the instrument (peak to peak) shall not exceed 1% of full scale (peak to peak).

4.1.5 The mutual interaction of a manual *zero and span adjustment* shall not change the span (difference between the zero point and the span point) of the instrument by interacting with the unadjusted control. The mutual interaction of the zero and span adjustment circuits shall be such that a 5% change in one control shall not change the other control by more than 1%.

4.1.6 Attenuation from one scale to another shall not cause the recorder zero point to move by more than 1% of the recorder full scale.

5. Apparatus

5.1 At least eleven manufacturers of commercial continuous monitoring instruments using the conductance principal for sulfur dioxide measurement were reported in 1972 (**8**). Because of the variations in the individual instruments, this section will describe the basic components of a sulfur dioxide conductivity instrument and the specifications necessary to meet the objective of ambient air analysis. Not all instruments will need to meet all these specifications and each individual instrument should be carefully evaluated for performance. The basic components of the conductivity instrument are shown in Figure 706:1. The instrument is separated into six sections and each section will be discussed below.

5.1.1 The sample probe and delivery train (Figure 706:1, #3) should be fabricated from Teflon or borosilicate glass. The inlet should be at least 12 mm ID and protected by a Teflon screen to remove small insects from the sample air stream. The entire system must be heated to prevent condensation and minimize adsorption. If a particulate filter is used, the filter material should be checked for inertness towards SO_2.

5.1.2 The flow measurement and regulation system (Figure 706:1, #7) includes means to control both the sample air flow and reagent flow, to measure these flow

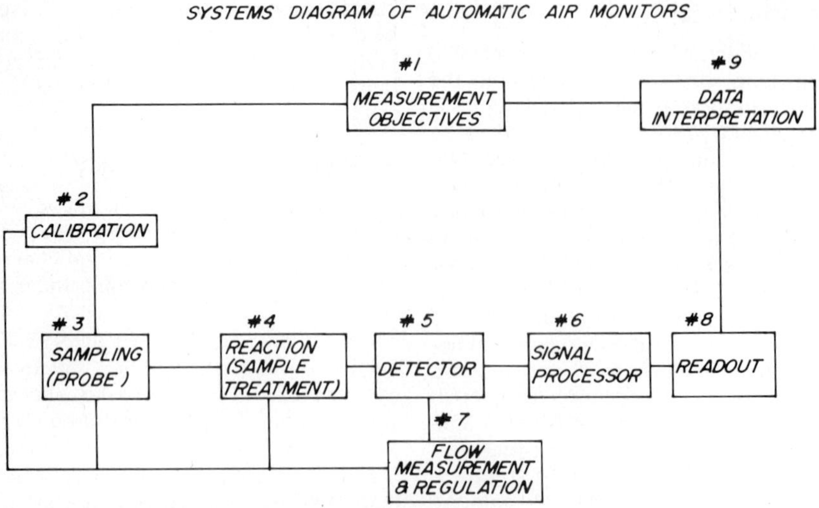

Figure 706:1—Systems Diagram of Automatic Air Monitors.

rates and to adjust the flow rates as needed.

The air flow rate should be stable and should not vary more than 3% over a 24 hr period. The residence time of the air sample in the sample probe should be no more than a few seconds.

The reagent flow should be uniform and capable of continuous operation with minimum maintenance. Reagent reservoirs should have the capacity to hold enough reagent for the time of unattended operation that is required by the operator. The reagent should be protected from contamination and deterioration during the operational period. (See Section **10.1**.)

5.1.3 The sample treatment component (Figure 706:1, #4) of all conductivity instruments scrubs the SO_2 to form an ionic species for detection by the electrode system. The scrubbing systems generally used are co-current contact tubes, multiple jet bubblers or direct impingement on the surface of the reagent. Whatever system is used in a specific instrument, the scrubber should remove 99% of the sulfur dioxide from the air stream. The volume of the scrubber should be minimal to provide the maximum sensitivity and a minimum of lag time. The lag time is defined as the time required to obtain an observable signal after a step-change of sulfur dioxide at the probe. The scrubbing device usually contributes the greatest effect on the lag time. For conductivity instruments, a lag time of 30 seconds or less is desirable.

5.1.4 The conductivity of the exposed reagent is detected by a pair of electrodes immersed in the reagent. (Figure 706:1, #5) The electrodes are usually constructed of platinum. These electrodes should be cleaned periodically according to the manufacturer's instructions. Since the measured conductance is, in part, dependent upon the geometry and spacing of the electrodes, the system should be protected from damage by jarring. Periodic calibration should isolate problems with geometry and spacing. The electrodes are located in such a way that the reagent either flows past the electrodes continuously or a small increment of reagent is added to a fixed volume of reagent and the excess allowed to overflow to the spent reagent reservoir. The continuous flowing system usually has a faster rise time, that is the time required for the signal to go from the baseline to 95% of the peak signal, than the overflow cell. A rise time of 90 sec or less is recommended.

5.1.5 The conductivity electrodes comprise one arm of an AC Wheatstone bridge circuit and as the conductivity of this arm changes, the signal at the output changes. (Figure 706:1, #6) The analyzer should incorporate an "electric function" test that can be operated manually to determine if there are malfunctions in this section of the analyzer.

The conductivity cell of the instrument must be temperature compensated or maintained at constant temperature. The temperature coefficient of the conductivity of a dilute sulfuric acid solution is 1.6%/C.

5.1.6 The readout or signal display (Figure 706:1, #8) is most commonly provided by a strip chart potentiometric recorder. Signal output modes should be sufficiently flexible to permit not only analog recording with potentiometric recorder but also permit analog to digital conversion for punched tapes and/or teleprinter input, or for telemetering to a data storage-processing center.

Performance characteristics which affect the reliability of the data must be clearly understood. These include initial response time, rise time, time to 95% of response and fall time.

In areas near large single sources where sudden fumigations may occur the response time of the instrument should be no greater than 20 sec to 95% of full response.

5.1.7 The analyzer should meet all operational specifications stated herein and those of the manufacturer at environmental temperatures ranging from 35 to 120 F and should be housed to resists damage from adverse weather conditions.

5.1.8 Unattended operation for a minimum period of one week is required. Control mechanisms for preventing the inclusion of bad data during breakdowns or malfunctions must be supplied with each instrument.

6. Reagents

6.1 Manufacturers of analyzers sometimes specify the composition of the absorbing reagent to be used with their specific instrument. In all cases the manufacturers' specifications must meet the specifications of this section (Section **6**). The reagents must be ACS reagent grade quality. Distilled water must conform to ASTM Standard D 1193 Type 1 for reagent water. If the manufacturer does not specify the composition of the absorbing reagent, the following reagent specified in Section **6.2** should be prepared and used.

6.2 Absorbing solution, 0.006% H_2O_2 and 10^{-5} N H_2SO_4. To prepare the absorbing reagent, dilute 0.20 ml of 30% stabilizer-free hydrogen peroxide and 1 ml of 0.01 N sulfuric acid to 1 l. Add 2 mg of any non-interfering fungicide, such as 2,4,5-trichlorophenate (Dowicide) **(9)**. The pH of the absorbing reagent must be between 4 and 5. If the pH is greater than 5, add 0.01 N H_2SO_4 dropwise until the pH is within the limits. If the pH is less than 4, discard the reagent and prepare a fresh batch. Discard the reagent if one drop added to acidified 10% KI on a spot plate does not produce a yellow-brown color.

6.3 Sulfuric acid, 0.1 N for static calibration of the analyzer. Dilute 2.8 ml of concentrated sulfuric acid to 1 l with distilled water. Standardize in the usual manner either against sodium carbonate or other base of known strength.

7. Procedure

7.1 Follow the instruction manual furnished with each instrument. The manual should include:

7.1.1 A clear and easily understood schematic of the air path including the principal apparatus components as outlined in Section **5**.

7.1.2 A description of the way the instrument works in a language easily understood.

7.1.3 A step by step operating procedure.

7.1.4 A trouble shooting section describing common problems and how to solve them.

7.1.5 A schematic of the electrical circuitry giving basic information for understanding the instrument.

7.1.6 A routine method of maintenance including cleaning, refilling and adjusting reagent reservoirs and flow systems is to be outlined and followed.

7.1.7 A parts list.

8. Calibration

8.1 The manufacturer should provide in the instruction manual, a detailed calibration procedure that will dynamically calibrate the instrument under operation conditions. This procedure should conform to the conditions given in Section **8.2**.

8.2 Dynamic Calibration with Permeation Devices.

8.2.1 The dynamic calibration of automated conductance instruments requires a source of known SO_2 concentration. Permeation devices are used for the calibration of automated instruments **(10)**. Because the sample flow rates of the various automated instruments vary greatly (.24 l/min to 3.8 l/min) it is necessary to design and operate the calibrating system at an adequate flow rate in order to provide the proper concentration range of the sulfur dioxide.

8.2.2 In order to maintain a constant concentration of the sulfur dioxide, the total air flow must be constant and accurately measured. The air which flows over the permeation device must be clean and dry. The flow rate must also be adequate to present the sulfur dioxide calibration gas to the probe of the instrument at atmospheric pressure, and at the proper concentration. The excess calibration gas should be vented away from the working area or scrubbed to remove the sulfur dioxide.

8.2.3 The permeation device should have an adequate output to provide the proper calibration gas concentration to cover the expected operating range of the instrument being used.

8.2.4 It is necessary that the calibration system test the entire instrument and the calibration gas should be introduced prior to any device on the instrument including the probe.

8.2.5 Although the concentration of the calibration gas can be calculated from the permeating rate of the device and the measured flow rate of the calibration gas, the accuracy of the system should be determined periodically (**11**).

8.2.6 Changes in air and liquid flows from the manufacturers recommended optimum settings indicate deteriorating instrumental conditions and will require trouble shooting and correction when the instrument no longer conforms to the specifications stated on Section **4**.

8.2.7 All adjustable control settings used in the calibration should be recorded to give permanent record. The concentration of sulfur dioxide in $\mu g/m^3$ or ppm should be plotted against the instrument response using any convenient units (i.e., chart units, inches, millivolts, mhos, etc.). The plot must be labeled to indicate the response used. This plot will be a straight line and should include four different concentrations of sulfur dioxide and clean air. The highest concentration of sulfur dioxide used in the calibration should be slightly above the highest concentrations expected in the atmosphere to be monitored. The response factor (K) can be calculated from the slope of the line of the calibration plot

$$\text{K (response factor)} = \frac{C}{R}$$

where C is the concentration of sulfur dioxide and R is the instrument response.

8.3 SELF-CONTAINED CALIBRATION. Continuous monitors which are to be operated remotely should be equipped with a self-contained calibration system capable of carrying out four tests on demand from a central control. The instrument should be able:

8.3.1 To adjust instrument zero by passing clean air (0.00 ppm SO_2) through the instrument. This mode of operation must persist for a time period equal to or greater than the sum of the lag time and twice the fall time. The fall time is defined as the time required for the signal to drop 95% of its value.

8.3.2 To adjust span calibration by passing clean air containing a known concentration of sulfur dioxide such that a response of approximately two thirds of the chart span is observed. This mode of operation is to persist for a time equal to or greater than the sum of the lag time and twice the rise time of the instrument before the adjustment is made.

8.3.3 To add a known amount of sulfur dioxide to the ambient air being measured for the purpose of detecting and quantifying the effect of interferences in the ambient air. This mode of operation should persist for a time that is equal to or greater than the lag time plus three times the rise time.

8.3.4 To measure the concentration of the SO_2 in ambient air. The observed signal obtained in this mode of operation is not valid for the time period that is equal to the lag time and twice the fall time of the instrument and should not be used in the data storage or data processing steps.

9. Calculation

9.1 Instrument response and recorder chart design or tape output should provide a minimum amount of work for data compilation and evaluation.

9.2 The instrument shall display at least one measurement per 5 min.

9.3 The instrument response (R) can be converted to the concentration of sulfur dioxide (C) using the relationship

$$C = R \cdot K$$

where K is the response factor determined from the calibration procedure in Section **8.2.7**. In most instances a response factor of 1 is more useful since the instrument response is reading directly in concentration of sulfur dioxide.

10. Effect of Storage

10.1 The absorbing solution is reasonably stable and will keep for periods up to 2 weeks in inert containers. Several workers have reported difficulties with deteriorating H_2O_2 strength. This may be caused by trace metal catalysts or trace amounts of organic substances which react slowly with the H_2O_2.

10.2 Calibration solutions should be stable indefinitely when stored in borosilicate glass or polyethylene bottles. Soft glass or basic substances might dissolve sufficiently to give reduced conductivity and calibration curves which intercept the concentration axis to the right of zero.

11. References

1. KUCZYNSKI, E.R. 1967. Effects of Gaseous Air Pollutants on the Response of the Thomas SO_2 Autometer. *Envir. Sci. and Tech.* 1:68.
2. RODES, C.E., H.F. PALMER, L.A. ELFERS and H.F. NORRIS. 1969. Performance Characteristics of Instrumental Methods for Monitoring Sulfur Dioxide. *J of the Air Pollution Control Assoc.* 19:575.
3. PALMER, H.F., C.E. RODES and C.J. NELSON. 1969. Performance Characteristics of Instrumental Methods for Monitoring Sulfur Dioxide: II. Field Evaluation, *Ibid.* 19:778.
4. THOMAS, M.D. and R.E. AMTOWER. 1966. Gas Dilution Apparatus for Preparing Reproducible Dynamic Gas Mixtures in any Desired Concentration and Complexity. *Ibid.* 16:618.
5. CHENG, R.T., J.O. FROHLIGER and M. CORN. 1971. Aerosol Stabilization for Laboratory Studies of Aerosol-Gas Interactions. *Ibid.* 21:138.
6. CORN, M. and R.T. CHENG. 1972. Interaction of Sulfur Dioxide with Insoluble Suspended Particulate Matter. *Ibid.* 22:870.
7. DUCKWORTH, SPENCER. 1969. Particulate Interference in Sulfur Dioxide Manual Sampling by the West-Gaeke Method. 10th Conference on Methods in Air Pollution and Industrial Hygiene Studies, San Francisco, California.
8. HOCHHEISER, S., F. BURMAN and G. MORGAN. 1971. Atmospheric Surveillance: The Current State of Air Monitoring Technology. *Envir. Sci. and Tech.* 5:678.
9. HOCHHEISER, S. 1964. Methods of Measuring and Monitoring Atmospheric Sulfur Dioxide. U.S. Public Health Publication No. 999 AP 6.
10. AMERICAN PUBLIC HEALTH ASSOCIATION, INC. 1977. Methods of Air Sampling and Analysis. 2nd ed. p. 18. Washington, D.C.
11. *Ibid.* p. 696.

Sub-Committee 1
D. F. ADAMS, *Chairman*
J. O. FROHLIGER
D. FALGOUT
A. M. HARTLEY
J. B. PATE
A. L. PLUMLEY
F. P. SCARINGELLI
P. URONE

707.

Tentative Method for Continuous Monitoring of Atmospheric Sulfur Dioxide with Amperometric* Instruments

42401-04-74T

1. Principle of the Method

1.1 This procedure describes an automated analyzer primarily intended for the continuous detection and determination of sulfur-containing gases in the atmosphere. It is applicable for the determination of sulfur dioxide, only if it is known that other sulfur compounds or other interferents do not exceed 5% of the sulfur dioxide concentrations. If this limit is exceeded, then appropriate sample pre-treatment procedures must be used to remove the interferences.

1.2 Air is drawn through a Teflon probe and scrubbed through a suitable liquid reactant in a detector cell. A multi-port valve is incorporated into the probe to introduce zero and calibration gases and the air sample into the detector cell.

1.3 Sulfur dioxide and other reducing gases present in the air sample react with

*The term amperometric more accurately describes this process which is commonly called coulometric.

electrically-generated, free halogen (titrant) in the detector cell. The detection cell contains two pairs of electrodes which function as anode/cathode and sensor/reference. The cell and its associated electronics operate to maintain a constant titrant concentration in the electrolyte. Any compound introduced into the cell that reacts with the titrant will change the titrant conc and produce a potential change in the solution that is detected by the sensor/reference electrode pair. The current is supplied to the anode/cathode generator pair until the original potential and titrant conc are restored. The quantity of electricity required is directly proportional to the reactant equivalents introduced into the cell.

Whenever one Faraday (96,492 absolute coulombs) passes through an electrolyte it produces one gram-equivalent of chemical change of the material in the solution. The number of coulombs (Q) expended is equal to the current in amperes, $Q/t = i$, multiplied by the titration time (t) in seconds.

In air monitoring, the electrolyte composition is selected to promote the most favorable reaction of sulfur dioxide with regard to reaction rate and specificity. Iodine or bromine is used for the titration of sulfur-containing compounds. During a titration the current passing through the cell is measured as the potential drop (iR) across a precision resistor having a low-temperature coefficient. A potentiometric strip-chart recorder may be used to record the potential-time curve. The area under this curve $[(i) \times (t)]$ is equivalent to the number of coulombs used.

2. Range and Sensitivity

2.1 The optimum operating range for amperometric sulfur dioxide detectors is 26 to 5200 $\mu g/m^3$ (0.01 to 2 ppm). The range provided by any given analyzer is affected by cell configuration, electrolyte formulation, and associated electronics.

3. Interferences

3.1 The most commonly encountered interfering pollutants include oxidizing, reducing, and olefinic compounds.

Any compund in the air sample capable of oxidizing bromide to bromine or iodide to iodine under the conditions in the cell will be sensed as excess titrant. Such compounds as ozone or oxides of nitrogen act in this manner. Similarly, compounds which reduce bromine to bromide or iodine to iodide will create a positive interference. Other compounds including unsaturated hydrocarbons, α-hydroxy acids, some amines and phenols may consume titrant by addition or substitution reactions.

3.1.1 Common oxidizing compounds such as ozone, chlorine, and nitrogen dioxide, when absorbed in the detecting cell electrolyte, create an excess of halogen titrant. The presence of such oxidizing pollutants causes a negative titration error and less titrant will be generated than corresponds to the amount of sulfur dioxide present.

3.1.2 Amperometric sulfur dioxide detectors also respond to any compounds that will reduce free halogen. Common interfering reductants include hydrogen sulfide, alkyl mercaptans, and alkyl sulfides and disulfides.

Differences in the stoichiometry of sulfur-containing compounds may also create a problem in the calculation of concentration should the analyzer be used as a total sulfur gas analyzer. Hydrogen sulfide requires four times as much titrant per sulfur atom as does sulfur dioxide. Thus sulfur dioxide-hydrogen sulfide mixtures of varying ratios all of equal sulfur atom content, will yield titration values from 4 to 1.

3.1.3 Olefins will react but with less than quantitative stoichiometry primarily because of a slow and non-stoichiometric addition reaction with the titrant (bromine or iodine) coupled with a short gas residence time in the detection cell.

3.2 Interferences may be removed from atmospheric mixtures by use of appropriate scrubbers. Several systems having varying degrees of removal efficiencies for selected compound types have either been reported in the literature **(1)** or remain proprietary information of certain instrument manufacturers **(2)**. Chemically impregnated filters or heated silver gauze **(3)** are the types most frequently used and may be supplied as a built-in component or accessory in commercially available analyzers. Some filters are more effective than others and they should be checked for effectiveness in the intended environment. For example, one available filter apears to remove more than 95% of hydrogen sulfide, mercaptans, alkyl sulfide and disulfide whereas another brand will not remove mercaptans from this mixture. Selection of the interference removing filter should be based on detailed knowledge of the range of potential interfering compounds present within the intended survey area.

4. Precision and Accuracy

4.1 Although the absolute coulomb is a primary standard, the precision and accuracy of the method will be determined by the sample delivery train, sample pretreatments, detector/cell response, interfering pollutants, and the recording system.

4.1.1 *Accuracy*. An analyzer calibrated within the range of 1 to 2 ppm full scale should have an overall accuracy of 0.02 ppm (1% relative). Short term accuracy of 0.5% of full scale should be attainable when standardized against a known dilution flow from a certified permeation tube.

4.1.2 *Precision*. Repeatibility shall be within 1% relative standard deviation for successive identical samples and within 3% over several 24-hr periods.

5. Apparatus

5.1 A general description of instrument components is given because more than one analyzer is available commercially. Not all instruments may meet the specifications described below for the measurements of sulfur dioxide. The basic components are shown schematically in Figure 707:1. Generally, an automated amperometric analyzer system will consist of a probe containing a heated filter for removal of particulate matter, a sampling valve for introducing span and calibration gases automatically or on demand, zero and calibration gases, an interference scrubber, the detector cell—an electrolysis power supply, signal processor, a means for recording the output signal, and an air sample pump.

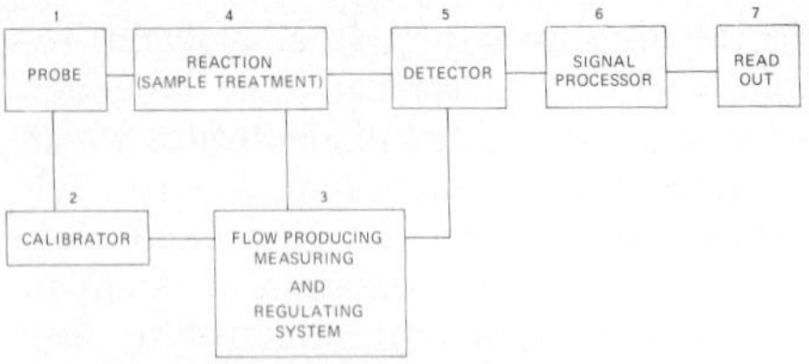

Figure 707:1—Basic components automated amperometric analyzer system.

5.1.1 The sample probe and delivery train should be fabricated from Teflon or borosilicate glass. The inlet should be protected by a Teflon screen† to remove small insects from the sample. The entire system should be heated to prevent or minimize condensation or adsorption and should include a heated, SO_2-inert filter for particulate matter removal.

5.1.2 The calibrator (Figure 707:1) includes a constant temperature gas dilution system for permeation tubes (**4**). Clean, temperature-controlled, dilution air is passed over a permeation tube containing sulfur dioxide. This primary gas stream is diluted further to obtain a response on the instrument that is at least 80% of full scale. The calibration gas is connected directly to one of the ports of the sample valve.

A permeation tube is recommended for use in a regular daily schedule of dynamic calibration. An ideal system consists of a closely thermostatted (±0.05 C) permeation tube or vessel having a constant, clean, low flow of dilution air over its surface. The output from the permeation tube calibration system should be swept continuously through a manual or automatic multiport valve having three selectable modes: (a) vented to atmosphere through an adequate cleanup scrubber, (b) swept

†Teflon monofilament screen is available from TET-Kressilk, Inc., 420 Saw Mill River Road, Elmsford, NY, 10523.

directly into the detector cell with the outside air intake either closed or protected from atmospheric contaminants with appropriate cleanup scrubbers, or (c) added to the sampled ambient air to provide a "standard addition" calibration for an assessment of the interaction between other pollutants present in the sample air and SO_2.

5.1.3 The flow measurement and regulation system (Figure 707:1) must provide a reliable and accurate flow of sample air to the detector. A visual display of the air sampling rate is desirable. The air pump must be designed for continuous duty operation.

5.1.4 Sample treatment (Figure 707:1) includes procedures for removing potential interfering substances while allowing SO_2 to pass through without loss. These devices should be evaluated in line using a dynamic source of an SO_2 reference concentration to establish per cent SO_2 loss on a real-time basis throughout the anticipated range of concentration. Similarly, the effectiveness of the pretreatment devices for removal of potential interfering gases should be tested at ten times the highest range anticipated of concentrations to determine retention efficiency and breakthrough times. Pretreatment devices should not retain more than 5% of the SO_2 in the concentration range of interest nor less than 98% (or to a lower value of 0.02 ppm interference equivalent) of the interfering compounds for which the device was designed.

5.1.5 The detection cell (Figure 707:1) contains the electrolyte, sensor and generator electrodes, a mechanism to maintain constant electrolyte level and concentration, a sample dispersion device, and a method for stirring the solution. The total system must be carefully evaluated to establish its lower limit of detection; linearity of response; zero and span drift; signal-to-noise ratio; and lag, rise, and fall times as defined in the equivalency document published in the Federal Register. **(5)**

5.1.6 The signal processor—electrolysis power supply—(Figure 707:1) should be solid state design, temperature compensated to give less than 0.5% drift in 24 hr, and should respond by less than 1% of full scale to a 10% fluctuation in line voltage. The circuit should incorporate a standard millivolt source so that an "electric function" check may be performed by actuating a "Function Test" switch. Signal output modes should be sufficiently flexible to permit analog recording within the range of 1 to 1000 mV as well as analog-to digital data conversion.

5.1.7 Signal display (Figure 707:1) is most commonly provided by a potentiometric recorder. For some systems having a wide concentration range measuring capability, an automatic signal attenuator may be used in conjunction with the recorder to permit, for example, 2.6 mg/m^3 (2 ppm) SO_2 full scale recording during periods of peak SO_2 concentrations.

5.2 It is known that variable losses of trace concentrations of sulfur gases occur on metals, glass and plastic surfaces **(6, 7)**; one must use care in selecting materials that come into contact with the air sample or reference gas. FEP Teflon is preferable to all other plastic because it quickly conditions to give repeatable results. **(7, 8)** In addition, all components of the gas flow system must be closely coupled and dead volume kept to a minimum.

5.3 SUMMARY OF MINIMUM SPECIFICATIONS.

		Units	SO_2
5.3.1	*Range*	(ppm)	10 to 1 (0 to 2)
5.3.2	*Minimum Detectable Limit*	(ppm)	0.01 (0.02)
5.3.3	*Rise Time, 95%*	(sec)	15
5.3.4	*Fall Time, 95%*	(sec)	15
5.3.5	*Lag Time, 95%*	(sec)	30
5.3.6	*Zero Drift*, 12hr‡	(ppm)	±0.02
	Zero Drift, 24hr‡	(ppm)	±0.02
5.3.7	*Span Drift*, 24hr‡	(ppm)	±0.02
5.3.8	*Precision at 20% full scale for each attenuation range*	(ppm)	±0.02

‡Determined under conditions of continuous operation while measuring zero gas and span gas over the stated period.

5.3.9	*Precision at 80% of full scale for each attenuation range*	(ppm)	±0.04
5.3.10	*Total Interference Equivalent*	(ppm)	±0.02
5.3.11	*Noise*	(ppm)	0.005
5.3.12	*Linearity§*	(ppm)	±3%

6. Reagents

6.1 CLEAN AIR. Commercial grade, dry, in a cylinder. Clean air may alternatively be supplied from ambient air through a low pressure drop, multi-component filter consisting of activated charcoal and soda lime. Proprietary filters are also available. **(2)**

6.2 PERMEATION TUBES. Prepare or obtain Teflon permeation tubes **(9–11)** that generate sulfur dioxide at rates of 3 to 100 ng/min to prepare known atmospheres within the dynamic range, 5 to 1000 ppb, of the instrument. Certified sulfur dioxide permeation tubes are available from the National Bureau of Standards. **(12)**

6.3 ELECTROLYTE. The electrolyte should either be prepared from ACS Reagent Grade chemicals and distilled water according to directions furnished by the instrument manufacturer or purchased in pre-packaged containers from the manufacturer. The electrolyte should be maintained or exchanged according to the manufacturer's instructions. Distilled water should conform to the Intersociety Committee specifications for Reagent Water.||

7. Procedure

7.1 For specific operating instructions, refer to the manufacturer's manual. Turn the power on and allow the instrument to warm up for the manufacturers recommended time. The instrument should meet minimum specifications for drift at the end of this warmup period.

7.2 Ambient air flows through the Teflon probe and the multiport valve into the detector. Adjust the air flow to the recommended rate. Energize the multiport valve on a programmed or demand time schedule to sample zero air and span gas. Adjust the instrument and recording system with the zero air and span gas.

7.3 Install and evaluate the instrument on-site to establish compliance with the minimum specifications described herein.

8. Calibration and Standards

8.1 The system of handling the sample must be tested for conformance to manufacturer's specifications. All flow measuring devices must be validated with a calibrated wet test meter. Correct or adjust all discrepancies by cleaning or replacing any defective parts.

8.2 Calibrate the instrument under the same conditions to be used in analyzing ambient concentrations of sulfur gases using a permeation tube dilution system. **(4)** Certified analyses of commercially available standard sulfur gas mixtures are not reliable.

8.3 Introduce at least four different concentrations, greater than zero, of the standard SO_2 reference gases into the probe system to give responses from 10 to 95% of full scale. Compare the response of the instrument to the same step concentration standards introduced directly into the detection cell to establish probe line loss. Adjust the recorder span to give the known calibration value. If a span control is not available on the instrument, prepare a calibration curve by plotting the several added SO_2 dilutions against the recorded instrument output.

8.4 Prepare a calibration curve, plotting standard concentrations against recorded instrument output.

8.5 To minimize error when assuming linearity and a single point quality control sample, the standard gas concentration should be selected so as to optimize the most important segment of the data. For example, if peak excursions are impor-

§Each value obtained from calibration gases must be within ± 3% of the actual concentration predicted by the line of best fit determined by the method of least squares.

||American Public Health Association, 1977. Methods of Air Sampling and Analysis. 2nd ed. p. 53. Washington, D.C.

tant, the instrument should be calibrated at full scale or at the anticipated peak value. If no particular range of values is specially important, quality control should be done at mid-range. In some amperometric instruments, reagent deterioration is initially reflected in a falloff at near full scale response. Such instruments should not be respanned to accommodate this behavior. The reagent should be replaced and the total instrument system completely recalibrated.

9. Calculation

9.1 Correct flow rate of sample to ambient air condition of 1 atmosphere and 25 C (298.1K).

9.2 Calculate concentration of standard gas mixtures in ppm by volume as follows:

$$\text{ppm } SO_2 = (0.382)\ B/A$$

A = flow rate of air + dilution air, liter/min.

B = output of permeation tube, ng/min × 0.001, where 0.001 is the factor for converting ng to μg.

0.382 = standard volume, ambient/molecular wt = 24.45/64 μl/μg = molar volume.

9.3 Calculate the concentration of sulfur dioxide in air by multiplying the net recorder response (sample less baseline reading on recorder chart) by the slope of the linear calibration curve.

9.4 Data reduction procedures used are dependent upon the parameters to be reported.

9.4.1 For determination of mean values, for any given time period, graphical, mechanical or electronic integration techniques of analog strip charts are used. Hourly mean values may also be obtained from a sufficient number of discrete values to give the confidence level desired.

9.4.2 For determination of fumigation values, valid peak values must have a duration of two times the sum of the rise and fall times of the instrument used.

9.5 Concentrations in the atmosphere should be reported to the nearest 0.002 ppm or 1% relative to the measured value whichever is smaller. A corresponding confidence level should accompany the result. The latter value is best taken from field installation evaluations of overall precision and accuracy. These values must meet minimum specifications (Section **5.3**).

10. Effects of Storage

10.1 Permeation tubes must be stored in a sealed bottle with a drier and an absorbant to remove sulfur dioxide. Storage in a refrigerator prolongs their useful life. However, the tubes must be equilibrated at operating temperature for 48 hr prior to use.

11. References

1. ADAMS, D.F., BAMESBERGER, W.L., and T.J. ROBERTSON. 1968. *J. Air Poll. Control Assoc.* 18:145–148.
2. BECKMAN INSTRUMENTS, INC., INSTRUCTION MANUAL 82043-A, Fullerton, California, 92634.
3. PHILLIPS ELECTRONIC INSTRUMENTS, Mount Vernon, New York, 10550.
4. AMERICAN PUBLIC HEALTH ASSOCIATION, INC. 1977. Methods of Air Sampling and Analysis. 2nd. ed. p. 18. Washington, D.C.
5. FEDERAL REGISTER. 1973. Proposed Ambient Air Monitoring Equivalent and Reference Methods. 38:28438, Part II, Friday, Oct. 12. Washington, D.C.
6. KOPPE, R.K., and ADAMS, D.F. 1967. Evaluation of Gas Chromatographic Columns for Analysis of Subparts Per Million Concentration of Gaseous Sulfur Compounds. *Environ. Sci. and Tech.* 1:479.
7. PESCAR, R.E., and HARTMANN, C.H. 1971. Automated Gas Chromatographic Analyses of Sulfur Compounds. Preprint, 17th Annual Analysis Instrumentation Symposium, Instrument Society of America, Houston, Texas.
8. STEVENS, R.K., MULIK, J.D., O'KEEFFE, A.E., and KROST, K.J. 1971. Gas Chromatography of Reactive Sulfur Gases in Air at the Parts Per Billion Level, *Anal. Chem.* 43:827.
9. O'KEEFFE, A.E., and ORTMAN, G.C. 1966. Primary Standards for Trace Gas Analysis, *Anal. Chem.* 38:760.
10. SCARINGELLI, F.P., FREY, S.A., and SALTZMAN, B.E. 1967. Evaluation of Teflon Permeation Tubes for Use with Sulfur Dioxide. *Am. Ind. Hyg. Assoc. J.* 28:260.
11. SCARINGELLI, F.P., O'KEEFFE, A.E., ROSENBERG, E., and BELL, J.P. 1970. Preparation of Known Concentrations of Gases and Vapors with Permeation Devices Calibrated Gravimetrically. *Anal. Chem.* 43:871.
12. NATIONAL BUREAU OF STANDARDS. 1971. New Sulfur Dioxide Permeation Tubes, *NBS Technical News Bulletin*, p. 106, April, Washington, D.C.

Subcommittee 1
D. F. ADAMS, *Chairman*
J. O. FROHLIGER
D. FALGOUT
A. M. HARTLEY
J. B. PATE
A. L. PLUMLEY
F. P. SCARINGELLI
P. URONE

708.

Tentative Method of Mercaptan Content of Atmosphere

43901-01-70T

See **Method 118. p.**

709.

Tentative Method of Gas Chromatographic Analysis for Sulfur-Containing Gases in the Atmosphere (Automatic Method with Flame Photometer Detector)

42269-01-73T

1. Principle of the Method

1.1 This procedure describes an automated, semi-continuous technique for the detection and determination of low molecular weight sulfur-containing gases in the atmosphere including hydrogen sulfide, sulfur dioxide, methyl mercaptan, and dimethyl sulfide using the specificity of the flame photometric detector.

1.2 The atmospheric sampling line is attached to an automatic, multiport switching valve and air pump. The sampled air and calibration gas are alternatively drawn through the valve sample loop. On a preset cycle the valve sample loop is switched to the carrier gas and the sample in the loop is purged into the gas chromatographic (GC) column for sample resolution. On alternate cycles the calibration gas is purged into the GC column to provide standard reference peak heights and to maintain column conditioning (priming).

1.3 The eluted sulfur compounds pass through a hydrogen flame and the sulfur is converted to S_2 which is at a higher energy state. Upon return of the S_2 molecule to the ground state, these molecules emit light of characteristic wavelengths between 300 and 425 nm (**1, 2, 3**). This light passes through a narrow-band optical filter and is detected by a photomultiplier (PM) tube. The current produced in the PM tube is amplified by an electrometer, and the magnitude of the response is displayed on a potentiometric recorder or other suitable devices. The analyzer is calibrated using H_2S, SO_2, CH_3SH, and dimethyl sulfide permeation tubes (**4, 5**) and a dual-flow gas dilution device (**6**) capable of producing reference standard atmospheres down to the limits of detection of the method.

2. Sensitivity and Range

2.1 The sensitivity, repeatability, and accuracy of the method are dependent upon many variables including the materials of construction used in the multiport valve, and the total gas chromatograph sample handling system; carrier and make-up gas flow rates; bias voltage, temperature and type of the PM tube; and the column preparation technique. The limits of detection at twice the noise level at present for hydrogen sulfide, sulfur dioxide, methyl mercaptan, and dimethyl sulfide range from 5 to 13 $\mu g/m^3$ (0.002 to 0.005 ppm) (**7, 8**) without the use of sample concentration techniques such as freeze-out loops. The sensitivity of the dectector should be reported as mass per unit time and is 1.6×10^{-4} μg/sec or 6.0×10^{-5} μl/sec, since this is a dynamic system.

2.2 The response of the system is nonlinear, but a linear relationship is obtained by plotting the response against concentration on a log scale or by using a log-linear amplifier. By using either of these techniques the linear dynamic range is approximately 130 $\mu g/m^3$ (0.05 ppm) to 500 $\mu g/m^3$ (0.2 ppm) with a 1% noise level.

3. Interferences

3.1 The FPD is based upon a spectroscopic principle. An optical filter isolates the emission wavelength for the sulfur species at 394 ± 5 nm from other extraneous light sources. Other emission bands of sulfur of almost equal intensity at 374 and 383.8 nm are also suitable for the quantification of sulfur.

3.2 Phosphorus presents a potential interference at 394 ± 5 nm. Phosphorus-containing gases are not generally found in ambient air unless the samples are obtained near known sources of phosphorus-containing insecticides. Under the conditions of the method (gas chromatographic separation) it is unlikely, however, that phosphorus-containing gases would have the same retention times as the sulfur gases and thus would not present a real source of interference. The FPD discrimination ratio is approximately 10,000:1 for hydrocarbons and at least 1,000:1 for other gases (**8**).

3.3 Air produces a spectral continuum which yields a measureable signal at 394 nm and contributes to the background noise of the method. However, it is not considered to be a significant interference under the conditions of the method.

3.4 For a substance to interfere it must meet three conditions: (a) it would have to emit light within the band pass of the filter; (b) it must have an elution time very close to those of SO_2, H_2S, CH_3SH, or dimethyl sulfide; and (c) it must be present in the sample at a concentration detectable by this procedure.

4. Precision and Accuracy

4.1 Precision or repeatability of the flame photometric detector depends upon close control of the sample flow and hydrogen flow to the detector. An airflow change of 1% changes the response approximately 2%.

4.2 Reproducible peak heights are primarily dependent upon the materials of construction used in the GC, the close control of all GC operating variables, and the technique of column preparation. It is necessary to pre-condition (prime) the total analysis system with a series of standard injections to achieve elution equilibrium. Following periods of disuse, equilibrium must again be attained by the serial injection of standard sulfur gas mixtures. In the automated, semi-continuous mode, sample air and a standard reference gas mixture should be alternately injected into the instrument on a timed cycle (typically every 5–10 min) to maintain column conditioning. The instrument should have the capability for repetitive standard gas injections through a manual override control until equilibrium is achieved as indicated by the uniformity of the resultant peak heights on the strip chart recorder.

Repetitive sampling of standard reference gases containing 78 $\mu g/m^3$ (0.03 ppm) H_2S and 104 $\mu g/m^3$ (0.04 ppm) SO_2 over several 24-hr periods gave a relative standard deviation of less than 3% of the amount present **(9)**.

4.3 Accuracy of the method will depend upon the ability to control the flow and temperature of dilution gas over certified or calibrated permeation tubes maintained in a gas dilution device **(6)**.

No data are available on precision and accuracy for atmospheric samples.

5. Apparatus

5.1 A general description of instrument components is given because more than one make of analyzer may be commercially available. Not all instruments may meet the minimum specifications required for the selected sulfur-containing compounds to be determined (the measurement objective). The basic components of a suitable automated system are shown schematically in Figure 709:1.

Generally, the automated GC-FPD system will consist of a regulated delivery system for hydrogen for the sample burner; an automatic multiport valve, for reproducible injection of the air sample and reference gas mixtures into the GC; a GC column and an isothermal or temperature-programmable column oven; an electrometer for measuring and amplifying the PM current; a means of recording the electrometer output signal, and an air sample pump.

5.1.1 The sample and delivery train (#3, Figure 709:1) should be fabricated from FEP Teflon. The inlet should be protected by a 60 mesh Teflon screen* to remove small insects from the air sample stream followed by a heated filter for removal of atmospheric particulates. The entire sample air handling system should be heated to 120 F to prevent condensation of water and minimize adsorption of sulfur gases **(11)**.

5.1.2 The calibrator (#2, Figure 709:1) includes a contant temperature, gas dilution system for permeation tubes **(6)**. Clean, temperature controlled dilution air

*Teflon monofilament screen is available from Tobler, Ernst and Traber, Inc., 420 Saw Mill River Road, Elmsford, NY, 10523.

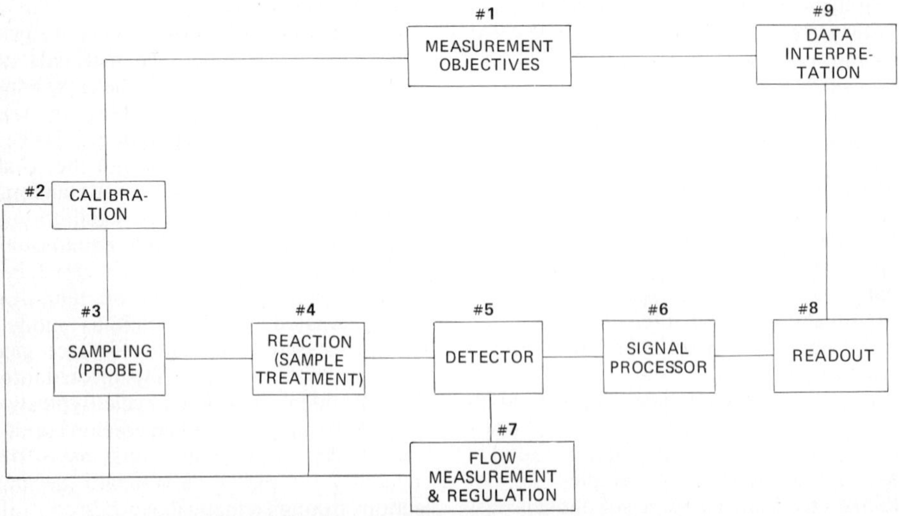

Figure 709:1—Systems Diagram of Automatic Air Monitors

is passed over permeation tubes containing the sulfur-containing compounds of interest. This primary gas flow is then further diluted with a clean, secondary air flow to provide the required 80% of full scale response for these compounds. The diluted calibration gas is plumbed into the multiport valve to maintain column condition and reference peak height or peak areas.

5.1.3 The flow producing, measuring, and regulating system (#7, Figure 709:1) controls the flow of the gas chromatographic carrier gas, and the clean air over the calibration permeation tubes. Hydrogen gas moves under its own differential pressure either from the hydrogen generator or pressure cylinder. The sample air pump must be designed for continouous operation. It is possible to pull both sample and calibration gas through their respective loops with a single sample pump. This is accomplished by utilizing a 500 ml/min flow restrictor in the air sample loop and a calibration gas flow restrictor in the calibration gas mixture loop (usually between 50 and 500 ml/min).

5.1.4 Sample treatment (#4, Figure 709:1) includes procedures for removing potentially interfering substances. For example, high concentrations of dialkyl disulfides if present in the ambient air, should be removed through use of a three-inch pre-column of Deactigel† Prolonged use of a gas chromatographic column under the prescribed conditions in the absence of broad peaks with the lighter sulfur gas peaks superimposed. If a pre-column is used, it will have to be replaced or backflushed when baseline noise or drift develops. The exact lifetime of the pre-column cannot be predicted because of the unpredictability of the concentrations of the higher molecular weight sulfur-containing compounds in varying locales.

5.1.5 *Detector*. A filter photometric detector (#5, Figure 709:1) detects the luminescence at 394 nm produced in the hydrogen flame emitted by the S_2 species when sulfur-containing compounds are burned. The air/fuel ratio is optimized at a 1:1 ratio. The carrier gas acts as an oxidant in the flame. A narrow band optical filter, with maximum transmission at 394 ± 5 nm is placed between the flame and a suitable photomultiplier (PM) tube. The sensitivity of the PM tube is primarily dependent upon its type, operating temperature, and voltage. A well regulated voltage source supplies the power required by the PM tube.

5.1.6 The amplifier (signal processor #6, Figure 709:1) may be either linear or logarithmic, depending upon the intended application and anticipated range of concentrations. The amplifier should be temperature-compensated for less than 0.5% drift in 24 hr, and respond less than one per cent of scale to a 10% fluctuation in line voltage. The amplifier should incorporate a standard, internal millivolt source so that an "electronic function" check can be performed manually by depressing a "function test" push button switch.

5.1.7 Signal readout (electrometer output #8, Figure 709:1) should be sufficiently flexible to permit analog recording within the range of 1 to 1000 mV with available strip chart recorders. Options for A-D conversion for punched tape, magnetic tape, teleprinter input, or telemetering to a data storage-processing center may be required for particular applications.

Single display is most commonly provided by a potentiometric recorder. For some detector systems having a wide conc range measuring capability, an automatic signal attenuator may be used in conjunction with the display to expand the display capability from 1 ppm to 2 ppm full-scale, for example.

5.2 Becaue of the known and variable losses of trace conc of sulfur gases on metal, glass, and plastic surfaces **(8, 12)** special care must be devoted to the material of construction used throughout the entire sample and reference gas flow paths. Heated metal surfaces must be avoided because they catalyze conversion of CH_3SH to $(CH_3)_2S_2$ **(13)**. TFE Teflon is preferable to all other materials because it quickly conditions to give reproducible results

†Applied Science Laboratories Inc., State College, PA 16801.

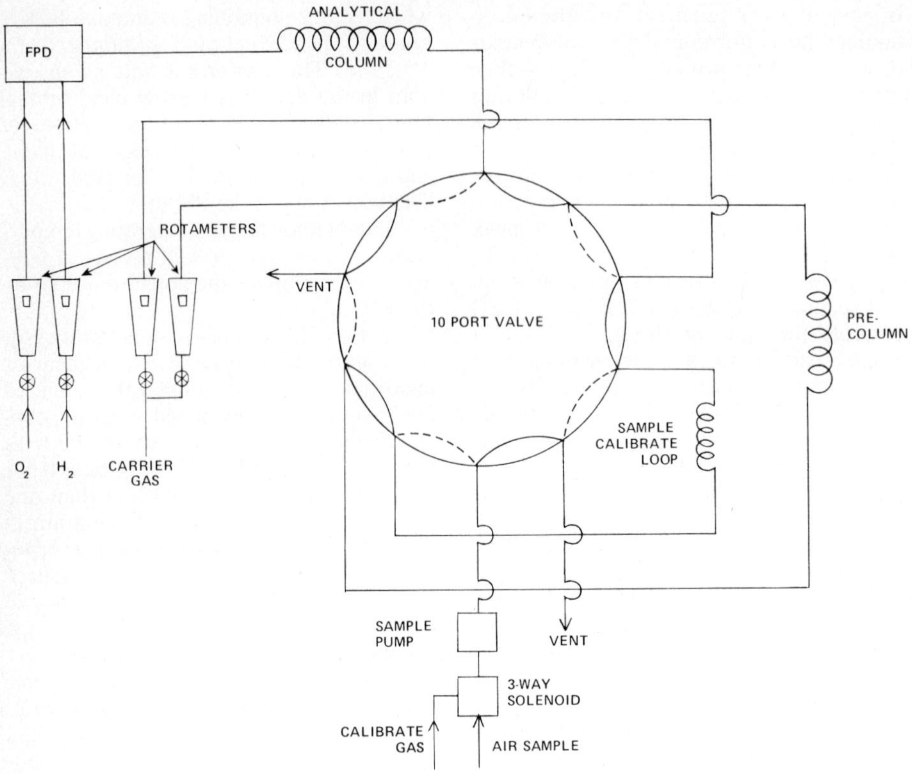

Figure 709:2—Automated 10-Port Rotary Gas Sampling Valve

(**8, 16**). In addition, all components of the gas flow system must be closely coupled and dead volume minimized to produce sharp peaks.

5.3 Sequential injection of sample and the standard calibration gas mixture into the GC system is accomplished with an automated ten-port rotary gas sampling valve constructed of Carpenters stainless steel (or Roulon-filled TFE Teflon core with a stainless steel outerbody) and incorporating two matched FEP Teflon sample loops, Figure 709:2, and associated timing mechanism (**14**).

When used with the recommended columns, oven temperature, and carrier gas flow H_2S, SO_2, CH_3SH, and dimethylsulfide are eluted in that order and a new sample or standard gas mixture can be injected automatically every 10 min.

5.4 A gas chromatographic column for separation and determination of H_2S, and SO_2. Five-inches of 1/8″-OD FEP Teflon thin wall tubing packed with Deactigel, 100 to 120 mesh (**14**). Certified columns may be obtained commercially, or they may be packed in the laboratory.

5.5 A pre-column, three-inches of 1/8″ OD FEP Teflon thin wall tubing packed with Deactigel, 100 to 120 mesh, to retain alkyl sulfides and disulfides (**14**). This pre-column is placed in the back flush circuit, Figure 709:2.

5.6 A gas chromatographic column for separation and determination of H_2S, SO_2, CH_3SH, $(CH_3)_2S$, and $(CH_3)_2S_2$. Ten feet of 1/8″-OD FEP Teflon thin wall tubing packed with 5 per cent Triton X-305 on Chromosorb G HP, 100 to 120 mesh (**15**).

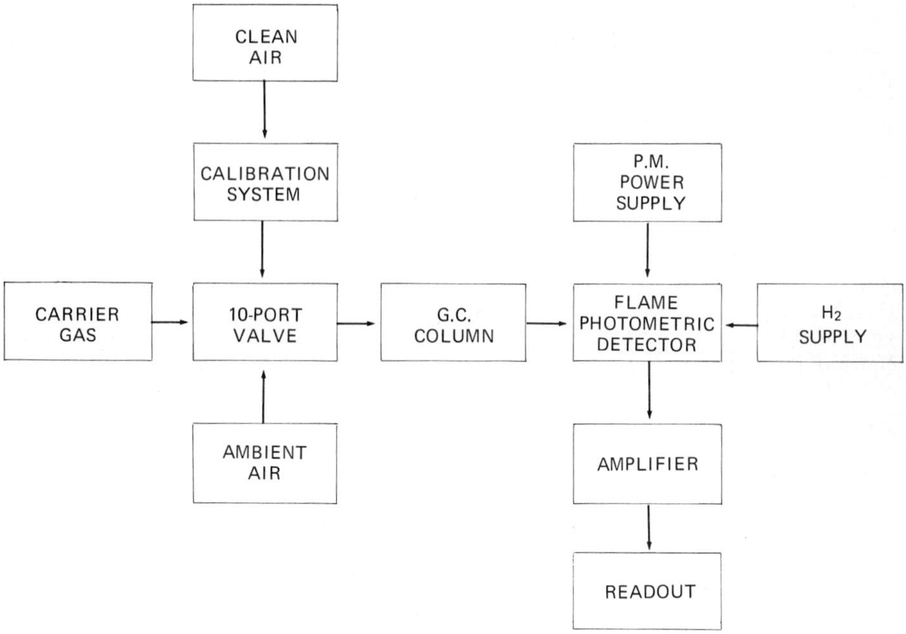

Figure 709:3—Automated FPD—GC System

Certified columns may be obtained commercially, or they may be packed in the laboratory.

5.7 A schematic diagram of the automated gas chromatographic system is shown in Figure 709:3.

5.8 SUMMARY OF MINIMUM SPECIFICATIONS.

		Units	SO_2
5.8.1	*Range.*	(ppm)	0.01–1
5.8.2	*Minimum Detectable Limit.*	(ppm)	0.01
5.8.3	*Rise Time, 95%.*	(Sec)	10
5.8.4	*Fall Time, 95%.*	(Sec)	10
5.8.5	*Lag Time.*	(Min)	5–10
5.8.6	*Zero Drift, 12-Hr‡.*	(ppm)	± 0.02
	Zero Drift, 24-Hr‡.	(ppm)	± 0.02
5.8.7	*Span Drift, 24-Hr‡.*	(ppm)	± 0.02
5.8.8	*Precision at 20% of Full Scale.*	(ppm)	± 0.02
5.8.9	*Precision at 80% of Full Scale.*	(ppm)	± 0.04
5.8.10	*Total Interference Equivalent.*	(ppm)	0.02
5.8.11	*Noise.*	(ppm)	0.005
5.8.12	*Linearity.*	(%)	3

6. Reagents

6.1 HYDROGEN. Hydrogen may be obtained from either of two sources. Extra dry grade gas, 99.9% minimum purity, in pressure cylinder equipped with a two-

‡Determined under conditions of continuous operation while measuring zero or span gas over the stated time period.

stage pressure regulator. A hydrogen generator is advised for safety reasons. A generator capable of generating a flow slightly greater than required is recommended. An automatic cut-off must be provided for the instrument as a safety measure in the event of a detector flame-out.

6.2 AIR. Water-pumped, dry grade, in pressure cylinder.

6.3 PERMEATION TUBES. Prepare or obtain Teflon permeation tubes that release H_2S, SO_2, CH_3SH, and dimethyl sulfide at rates of 0.003 to 0.1 μg/min, to prepare concentrations within the dynamic range of the FPD-GC method. Certified SO_2 permeation tubes are available from the National Bureau of Standards.

6.4 DEACTIGEL 100–120 MESH (DEACTIVATED SILICA GEL). The material as received must be further deactivated by first purging with conc hydrochloric acid (37%, 1.18 s.g.) then by flushing with distilled water and then with acetone.

6.5 TRITON X-305. Alkyl aryl polyether alcohol.

6.6 CHROMOSORB G HP. 100–120 mesh (Johns-Manville Corp.).

7. Procedure

7.1 For specific operating instructions, refer to the manufacturer's manual. The basic principles of gas chromatography and its application to the separation and analysis of air pollutants are described in Part I, Section **16**.

7.2 Ambient air and standard reference gas mixtures are continuously flushed through their respective 10-ml sample loops in the multi-port sampling valve. The multi-port valve is energized on a preset time schedule to flush, alternately, the 10-ml ambient air sample volume and the reference gas mixture into the GC column.

7.3 Commercially available instruments must be evaluated on-site to determine compliance with the specifications described in this method. The instrument and detector response are temperature-dependent and must be provided with a controlled environment for field application within a range of 60 to 80 F.

7.4 DEACTIGEL COLUMN. Operate the column isothermally at 85 C for the separation and determination of H_2S and SO_2 (**10**). Elution of these two compounds will be completed within 4 min under the specified conditions. The pre-column will be back-flushed automatically during the analysis cycle (Figure 709:2). The multi-port valve should be set to cycle once every five minutes under these conditions.

7.5 TRITON X-305 COLUMN. Use temperature-programming from 50 to 100 C at the 12 C per min for the separation and determination of H_2S, SO_2, CH_3SH, and the alkyl sulfides and disulfides. Sulfur dioxide may be partially retained by this column within the ppb conc range (**14**). Elution of these compounds will require 6 min under the specified conditions. The multiport valve should be set to cycle once every 12 min under these conditions to allow 6 min for the column oven to cool and reequilibrate at the 50 C starting temperature.

7.6 Carrier gas (air) and the FPD fuel (hydrogen) flow rates are usually controlled at 80/min each.

8. Calibration and Standards

8.1 The entire sample handling system must be tested for conformance to the manufacturer's specifications at least once a week. Validate all gas flow measurements with a wet test meter. Correct or adjust all discrepancies in flow by cleaning or replacing defective devices.

8.2 Calibrate the instrument under the same conditions to be used in the automatic determination of sulfur gases in the atmosphere using a permeation tube-dilution system (**6**).

8.3 Introduce at least four different concentrations not including zero, of the standard reference gases (preferably in mixture) to give responses from 10 to 95% of full scale. With a linear amplifier, plot the response against the corresponding concentration of sulfur compounds on log-log paper. A typical plot of sulfur dioxide is shown in Figure 709:4. With log amplifier, plot the log of the concentration

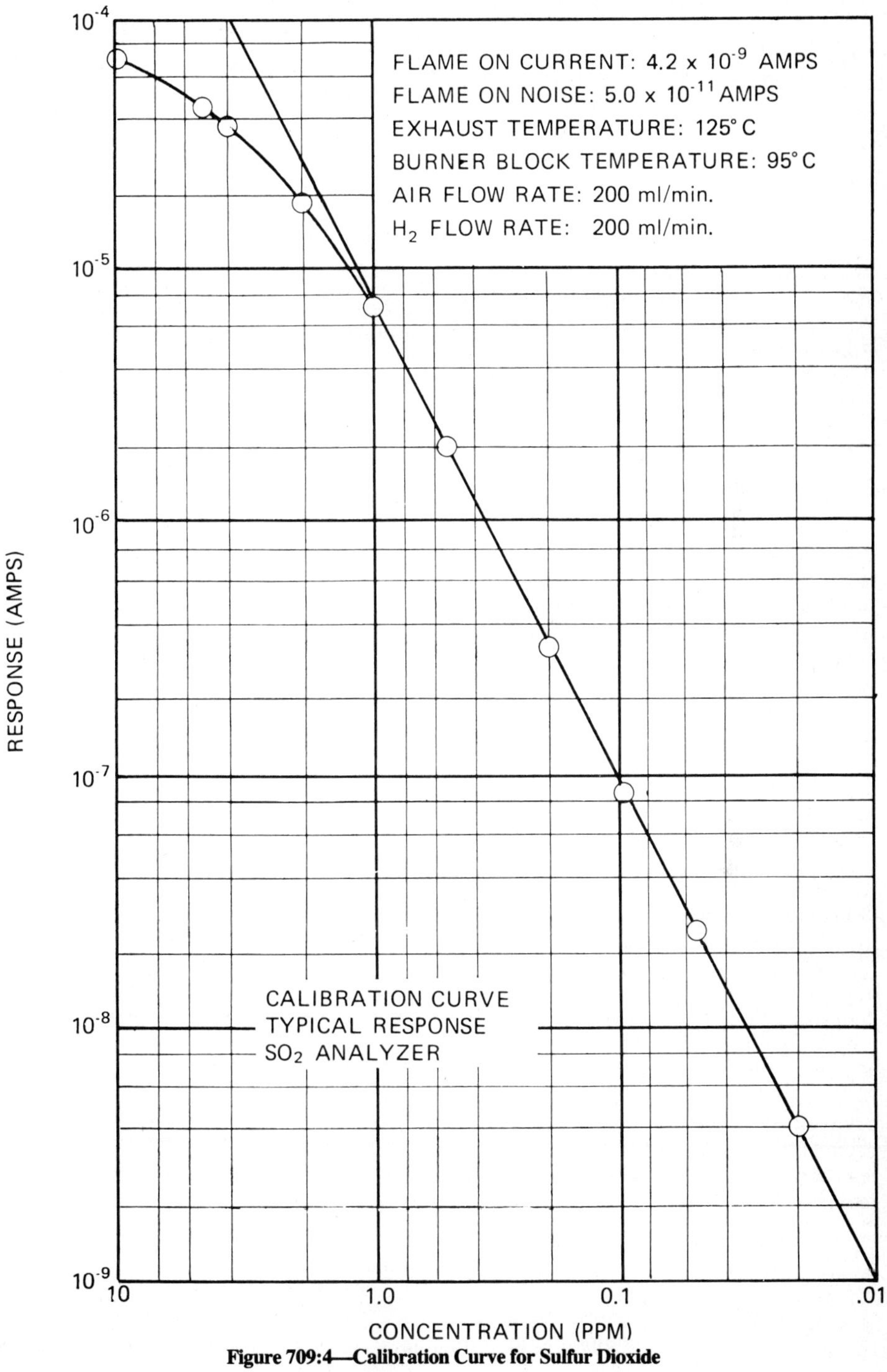

Figure 709:4—Calibration Curve for Sulfur Dioxide

against response on rectangular coordinate paper.

8.4 Check the electrometer attenuation to determine the agreement between the true attenuation and the nominal attenuation. The instrument should be calibrated using appropriate standards at each attenuator setting used.

9. Calculation

9.1 The response is recorded on a strip chart recorder (or othe suitable readout device) as each sulfur gas is eluted from the column.

9.2 The time for elution is measured for each sulfur gas.

9.3 Determine the identity of each eluted peak by comparing the measured elution times against the elution times for the standards.

9.4 PEAK HEIGHT IS USED FOR QUANTITATING H_2S. The (peak height × attenuation factor)$^{\frac{1}{2}}$ relationship is linear over the indicated range of concentration and varies for each component as well as for each column and operating condition. Peak height is most satisfactory for components such as H_2S which emerge from the column early with a tall, narrow peak.

9.5 PEAK AREA FUNCTION is used for SO_2, CH_3SH and dimethyl sulfide. Peak area function is proportional to the amount of substance present. Peak area measurements are generally more accurate and are more readily applied to components having longer retention times and broader symmetrical peaks.

Peak Area Function = W $(HA)^{\frac{1}{2}}$ where

H = peak height in mm
W = peak width in mm
A = attenuation factor from electrometer attenuation setting

9.6 Determine the concentration of each sulfur gas present in the sample, using each compound's calibration response factor:

$\mu g/m^3$ sulfur gas $= f \times$ *peak response*

where

f = calibration response factor for each sulfur compound in consistent units

f is obtained from the calibration plots as described under Section **8.2**.

Peak response for H_2S = (peak height, mm × attenuation)$^{\frac{1}{2}}$

Peak response for SO_2, CH_3SH and $(CH_3)_2S$ = peak area function (as defined above)

10. Effect of Storage

10.1 Permeation tubes should be discarded when approximately 10% of the original volume of permeant liquid remains. Replacement permeation tubes may be stored prior to use in a refrigerator in a sealed bottle containing an absorbant to remove the permeants. Re-equilibrate the tubes to operating temperature for 48 hr prior to use.

10.2 Gas chromatographic columns must be equilibrated with the sulfur-containing gases of interest just prior to use as directed in Section **4**.

11. References

1. CRIDER, W.L. 1965. Hydrogen Flame Emission Spectrophotometry in Monitoring for Sulfur Dioxide and Sulfuric Acid. *Anal. Chem.*, 37:1770.
2. BRODY, S.S., and J.E. CHANEY. 1966. Flame Photometric Detector. *J. Gas Chrom.*, 2:42.
3. CRIDER, W.L. 1967. *Analysis Instr.*, Vol. 4, L. Fowler, *et al.*, (eds.) Plenum Press, p. 67.
4. O'KEEFE, A.E., and G.C. ORTMAN. 1966. Primary Standards for Trace Gas Analysis. *Anal. Chem.*, 38:760.
5. SCARINGELLI, F.P., A.E. O'KEEFE, E. ROSENBERG, and J.P. BELL. 1970. Preparation of Known Concentrations of Gases and Vapors with Permeation Devices Calibrated Gravimetrically. *Anal. Chem.*, 42:871.
6. AMERICAN PUBLIC HEALTH ASSOCIATION. 1977. Methods of Air Sampling and Analysis. 2nd ed. p. 18. Washington, D.C.
7. Ibid., p. 88.
8. PESCAR, R.E., and C.H. HARTMANN. 1971. Automated Gas Chromatographic Analysis of Sulfur Compounds. Preprint, 17th Annual Analysis Instrumentation Symposium, Instrument Society of America, Houston, Texas, April.
9. STEVENS, R.K., A.E. O'KEEFFE. 1970. Modern Aspects of Air Pollution Monitoring. *Anal. Chem.*, 42:143A.
10. HARTMANN, C.H. 1971. Private Communications, Oct. 4.
11. CORN, M., and R.T. CHENG. 1972. Interaction of Sulfur Dioxide with Insoluble Suspended Particulate Matter, *J. Air Pol. Cont. Assoc.*, 22:870.
12. KOPPE, R.K., and D.F. ADAMS. 1967. Evaluation of Gas Chromatographic Columns for Analysis of

Subparts per Million Concentrations of Gaseous Sulfur Compounds. *Env. Sci. Technol.*, 1:479.
13. ADAMS, D.F. 1965. Unpublished information.
14. HARTMANN, C.H. 1971. Improved Chromatographic Techniques for Sulfur Pollutants. Paper presented Joint Conference on Sensing of Environmental Pollutants, Palo Alto, California, November. Thornsberry, W.L. *Anal. Chem.*, 43:452.
15. ADAMS, D.F., R.K. KOPPE. 1959. Gas Chromatographic Analysis of Hydrogen Sulfide, Sulfur Dioxide, Mercaptan, and Alkyl Sulfides and Disulfides. Tappi, 42:601.
16. STEVENS, R.K., J.D. MULIK, A.E. O'KEEFFE, and K.J. DROST. 1971. Gas Chromatography of Reactive Sulfur Gases in Air at the Parts Per Billion Level, *Anal. Chem.*, 43:827.

Subcommittee 1
D. F. ADAMS, *Chairman*
J. O. FROHLIGER
D. FALGOUT
A. M. HARTLEY
J. B. PATE
A. L. PLUMLEY
F. P. SCARINGELLI
P. URONE

710.

Tentative Method of Analysis for Sulfur-containing Gases in the Atmosphere (Automatic Method with Flame Photometer Detector)

42269-02-73T

1. Principle of the Method

1.1 This procedure describes an automated analyzer for the continuous detection and determination of sulfur compounds in the atmosphere. It is applicable for the determination of sulfur dioxide, only if it is known that other sulfur compounds do not exceed 5% of the sulfur dioxide concentrations. If this limit is exceeded, then the Intersociety Committee Gas Chromatograph Flame Photometer Detector Method (**12**) or pretreatment by suitable scrubber must be used to separate out these interferences for sulfur dioxide determination.

1.2 Air is drawn through a Teflon probe into a hydrogen-rich flame at a constant rate by means of an exhaust pump. A multiport valve is incorporated into the probe to introduce zero gases and calibration gas.

1.3 Upon entering the flame, sulfur compounds are converted to S_2, "excited state," which is at a higher energy state. Upon return of the S_2, "excited state," molecule to the ground state, these molecules emit light of a characteristic wavelength between 300 and 425 nm (**1, 2, 3**). This light passes through a narrow-band optical filter and is detected by a photomultiplier (PM) tube. The current produced in the PM tube is amplified by an electrometer, and the magnitude of the response is displayed on a potentiometric recorder or other suitable devices. The analyzer is calibrated using permeation tubes (**4, 5, 6, 7**) and gas dilution devices capable of producing standard reference atmospheres down to the limits of detection of the method.

2. Sensitivity and Range

2.1 The sensitivity or limit of detection, is dependent upon many variables that include the materials of construction, flow rates, operating voltage, temperature and special response of PM tube. The limit of detection for sulfur compounds at twice the noise levels is 5 to 13 $\mu g/m^3$ (2 to 5 ppb) calculated as SO_2. Most commercial analyzers report a limit of detectability of 10 ppb.

2.2 The response of the system is nonlinear, but a linear relationship is obtained by plotting the response against concentration on a log scale or by using a log-linear amplifier. Using either of these techniques the linear dynamic range is approximately 13 to 2600 $\mu g/m^3$ (5 to 1000 ppb).

3. Interferences

3.1 The FPD is based on a spectroscopic principle. Emission bands of sulfur of almost equal intensity at 374 and 383.8 nm are also suitable for the quantification of sulfur. An optical filter isolates the emission band for the sulfur species at 394 ± 5 nm from other extraneous light sources.

3.2 Phosphorus presents a potential source of interference at 394 nm. Phosphorus-containing compounds are not generally found in ambient air unless the samples are taken near known sources of phosphorus-containing pesticides.

3.3 The instrument responds to all sulfur compounds with a discrimination ratio of sulfur to non-sulfur compounds other than phosphorus of at least 30,000 to 1. Under atmospheric conditions found in many localities, the response from compounds other than sulfur dioxide is small, *i.e.*, less than 5%. In regions where concentrations of other sulfur compounds exceeds 5%, then selective scrubbers must be used that allow sulfur dioxide to pass quantitatively, if sulfur dioxide is to be reported.

3.4 Air produces a spectral continuum, which yields a measurable signal at these wavelengths and hence this signal must be subtracted from the total response.

3.5 Aerosols of metallic salts, if allowed to enter the flame, produce highly intense light emissions that may interfere with the determination.

4. Precision and Agency

4.1 Precision of the flame photometric detector depends critically on controlling the flow of sample and hydrogen to the detector. An air flow change of 1% changes the response approximately 2%. To insure the precision of the measurement, the instrument should be span checked at least seven times a day. Repeatability of the method over several 24 hr periods is 3%, relative standard deviation.

4.2 Accuracy will depend upon the ability to control the dilution of certified permeation tubes.

5. Apparatus

5.1 A general description of instrument components is given because more than one analyzer is commercially available. Not all instruments may meet the specification required for the measurement of sulfur compounds or sulfur dioxide. The basic components of an automated system are shown diagramatically in Figure 710:1. Generally, an automated FPD analyzer will consist of a regulated delivery system for hydrogen; a sampling valve for injecting automatically and reproducibly air samples, zero gas and reference gas; an electrometer for measuring and amplifying the PM current; an operating voltage to power the PM tube; a means of recording the electrometer output signal; and an air sample pump.

5.1.1 The sample and delivery train (Figure 710:1) should be fabricated from FEP Teflon. The inlet should be protected by 60-mesh Teflon screen to remove small insects followed by a heated filter for removal of atmospheric particulates. The air sampling handling system should be heated to 120 F to prevent condensation of water vapor and reduce losses of sulfur gases by adsorption and oxidation.

5.1.2 The calibrator (Figure 710:1) includes a constant temperature, gas dilution system for permeation tubes **(7)**. Clean temperature controlled dilution air is passed over permeation tube containing sulfur dioxide. This primary gas stream is diluted further to obtain a response on the instrument that is 80% of full scale. The dilution air is connected directly to one of the ports of the sample valve.

5.1.3 The flow producing, measuring and regulating system (Figure 710:1) moves the gases, measures and controls the flow of sample air, hydrogen fuel to the detector, pure air to the probe and over the permeation tube. A vacuum pump that is capable of producing a non-modulating

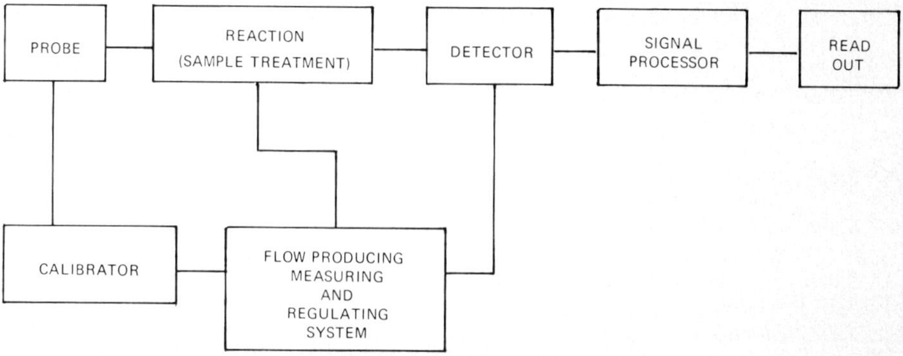

Figure 710:1—Scheme of automatic instruments.

flow creates the pressure differential for the delivery of all gases to the flame with the exception of hydrogen. Pressure fluctuation by the pump produces a noisy signal (see Section **4.1**). Hydrogen gas flows under its own differential pressure controlled by a two-stage regulator either from cylinders or from a hydrogen generator.

5.1.4 Sample treatment (Figure 710:1) includes procedures for removing potentially interfering substances and the process of converting the sulfur in the compounds to the excited state by a hydrogen flame. A one to one ratio of air to hydrogen gives the optimal sensitivity for SO_2.

5.1.5 *Detector (Figure 710:1).* A filter photometer measures the quantity of light at 394 ± 5 nm that is emitted by the S_2, "excited state," species and passes through the narrow-band filter. A well regulated voltage source supplies the power required by the PM tube. The sensitivity of the PM tube is dependent upon the type, operating temperature and voltage.

5.1.6 The amplifier, signal processor, (Figure 710:1) may be either linear or logarithmic depending upon the intended application and anticipated ranges. The amplifier should be temperature compensated for less than 0.5% drift in 24 hr, and respond less than 1% of full scale for a 10% change in line voltage. A standard mV source that is operated manually serves as an electronic function check.

5.1.7 Output of the electrometer (Figure 710:1) should be sufficiently flexible to permit analog recording within the range of 1 to 1000 mV with available strip chart recorders. AD conversion for punch tape, teleprinter input, or telemetering to a data storage-processing center may be required. For detection systems having a wide range of concentration capability, an automatic signal attenuator may be used to alter the range of a linear amplifier.

5.2 It is known that variable losses of trace conc of sulfur gases occur on metals, glass and plastic surfaces **(8, 9)**; one must be careful in selecting materials that come into contact with the air sample or reference gas. Metallic surfaces that convert CH_3SH to $(CH_3)S_2$ at elevated temperature must be avoided **(10)**. FEP Teflon is preferable to all other plastics because it quickly conditions to give repeatable results **(8, 10)**. In addition, all components of the gas flow system must be closely coupled and dead volume kept to a minimum.

5.3 SUMMARY OF MINIMUM SPECIFICATIONS.

	Units	SO_2
5.3.1 *Range*	(ppm)	0–1
5.3.2 *Minimum Detectable Limit*	(ppm)	0.01

5.3.3	*Rise Time*, 95%	(sec)	15
5.3.4	*Fall Time*, 95%	(sec)	15
5.3.5	*Lag Time*, 95%	(sec)	30
5.3.6	*Zero Drift*, 12-Hour*	(ppm)	± 0.02
	Zero Drift, 24-Hour*	(ppm)	± 0.02
5.3.7	*Span Drift*, 24-Hour*	(ppm)	± 0.02
5.3.8	*Precision* at 20% of full scale for each attenuation range	(ppm)	± 0.02
5.3.9	*Precision* at 80% of full scale for each attenuation range	(ppm)	± 0.04
5.3.10	*Total Interference Equivalent*	(ppm)	± 0.02
5.3.11	*Noise*	(ppm)	0.005
5.3.12	*Linearity*†	(ppm)	± 3%

*Determined under conditions of continuous operation while measuring zero gas and span gas over the stated period.

†Each value obtained from calibration gases must be within ± 3% of the actual concentration predicted by the line of best fit determined by the method of least squares.

6. Reagents

6.1 Hydrogen. Dry gas, with minimum purity of 99.8% in a pressurized cylinder and equipped with a two-stage regulator. An automatic cut-off must be provided as a safety measure in the event the flame is extinguished in the detector. A hydrogen generator capable of producing a flow of 80 ml/min or slightly greater than that required by the detector is strongly advised as a safety measure.

6.2 Air. Commercial grade, dry in a cylinder or a means of drying and removing trace quantities of sulfur compounds from ambient air.

6.3 Permeation Tubes. Prepare or obtain Teflon permeation tubes that generate sulfur dioxide at rates of 3 to 100 ng/min to prepare known atmospheres with the dynamic range, 5 to 1000 ppb, of the instrument. Certified sulfur dioxide permeation tubes are available from the National Bureau of Standards (**11**).

7. Procedure

7.1 For specific operating instructions, refer to the manufacturer's manual.

7.2 Ambient air flows through the multiport valve and the Teflon probe into the detector. On a programmed time schedule, the multiport valve is energized to sample zero air and span gas.

7.3 Commercially available instruments must be installed and evaluated on-site to establish compliance with the minimum specifications described herein.

8. Calibration and Standards

8.1 The system of handling the sample must be tested for conformance to manufacturer's specifications. All flow measuring devices must be validated with a wet test meter. Correct or adjust all discrepancies by cleaning or replacing defective parts.

8.2 Calibrate the instrument under the same conditions to be used in analyzing sulfur gases using a permeation tube-dilution system (**7**).

8.3 Introduce at least four different concentrations, not including zero, of the standard reference gases containing SO_2 to give responses from 10 to 95% of full scale. With a linear amplifier, plot the response against the corresponding concentration of sulfur dioxide on log-log paper. A typical calibration pilot is shown in Figure 710:2. With log amplifier, use semi-log paper and plot the response on the linear axis and concentration on the log axis.

9. Calculation

9.1 Correct flow rate of sample to ambient air conditions of 1 atmosphere and 25 C (298.1° K).

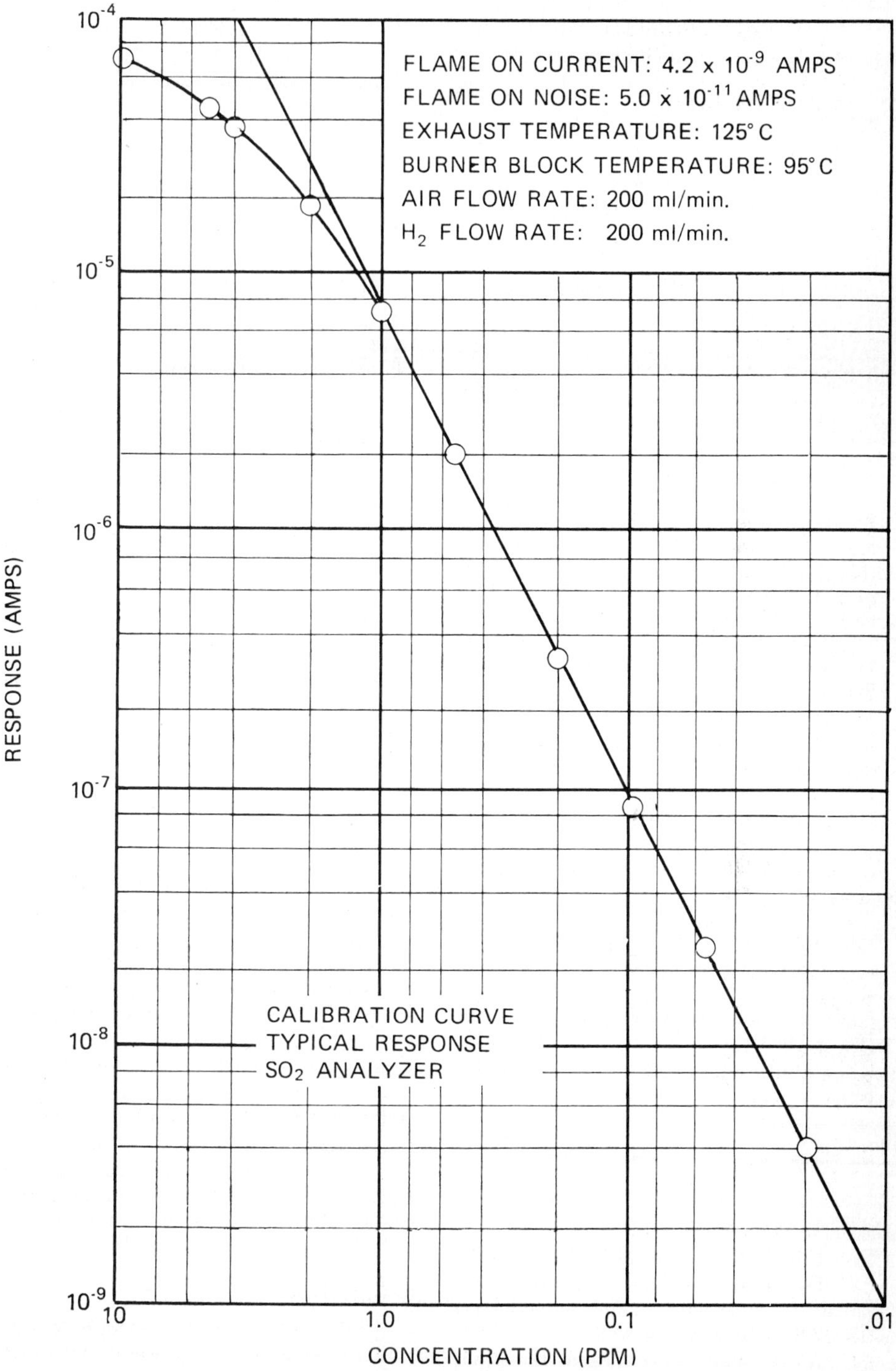

Figure 710:2—Calibration curve.

9.2 Calculate conc of standard gas mixtures in ppm by volume as follows:

$$\text{ppm } SO_2 = \frac{(0.382)\,B}{A}$$

B = output of permeation tube, ng/min × 0.001, where 0.001 is the factor for converting ng to μg.
A = flow rate of air + dilution air, liter/min.

$$0.382 = \frac{\text{molar volume}}{\text{molecular wt.}} = \frac{24.45}{64}\ \mu l/\mu g$$

9.3 Calculate the conc of sulfur compounds in air as sulfur dioxide.

The response (amp or full scale) times the calibration factor, the reciprocal of the slope of the calibration curve, is equal to the antilog of the concentration of sulfur dioxide. Or by solving the equation

$$C = a\,(R - R_0)^n:$$

where

C = the concentration ppm:
a = the coefficient (antilog of intercept)
n = slope (log-log plot) or

slope of $\frac{\Delta R}{\Delta C}$ obtained

from calibration curve
R = signal or response
R_0 = base line signal or response

9.4 Calculate the limits of detection or the sensitivity of the detector for sulfur compounds as mass per unit of time as follows:

$$L = \frac{F \cdot dc \cdot s}{1 + dR} \times 1.67 \times 10^{-5}$$

where

L = the limit of detection, μg/sec
F = flow rate, l/min
dc = incremental change in concentration, μg/m³
s = estimate of standard deviation, arbitrary unit but same as dR and equal to twice RMS noise level
dR = change in response, arbitrary units
1 = arbitrary constant

$$1.67 \times 10^{-5} = \text{conversion factor, } \frac{\text{min}}{60 \text{ sec}} \times \frac{m^3}{1000\ l}$$

10. Effects of Storage

10.1 Permeation tubes must be stored in a sealed bottle with an absorbant to remove sulfur dioxide and a drier. Storage in a refrigerator prolongs its useful life. However, the tubes must be equilibrated to operating conditions for 48 hr prior to use.

11. References

1. CRIDER, W.L. 1965. Hydrogen Flame Emission Spectrophotometry in Monitoring for Sulfur Dioxide and Sulfuric Acid. *Anal. Chem.* 37:1770.
2. BRODY, S.S., and J.E. CHANEY. 1966. Flame Photometric Detector. *J. of Gas Chrom.* 2:42.
3. CRIDER, W.L. 1967. Analysis Instr., vol. 4, L. Flower, et al., eds Plenum Press, p. 67.
4. O'KEEFFE, A.E., and G.C. ORTMAN. 1966. Primary Standards for Trace Gas Analysis. *Anal. Chem.* 38:760.
5. SCARINGELLI, F.P., S.A. FREY, and B.E. SALTZMAN. 1967. Evaluation of Teflon Permeation Tubes for Use with Sulfur Dioxide. *Am. Ind. Hyg. Assoc. J.* 28:260.
6. SCARINGELLI, F.P., A.E. O'KEEFE, E. ROSENBERG, and J.P. BELL. 1970. Preparation of Known Concentrations of Gases and Vapors with Permeation Devices Calibrated Gravimetrically. *Anal. Chem.* 42:871.
7. AMERICAN PUBLIC HEALTH ASSOCIATION, INC. 1977. Methods of Air Sampling and Analysis. 2nd ed. p. 18. Washington, D.C.
8. KOPPE, R.R., and D.F. ADAMS. 1967. Evaluation of Gas Chromatographic Columns for Analysis of Subparts Per Million Concentration of Gaseous Sulfur Compounds. *Environ. Sci. and Tech.* 1:479.
9. PESCAR, R.E., and C.H. HARTMAN. 1971. Automated Gas Chromatographic Analyses of Sulfur Compounds, preprint, 17th Annual Analysis Instrumentation Symposium, Instrument Society of America, Houston, Texas, April.
10. STEVENS, R.K., J.D. MULIK, A.E. O'KEEFFE, and K.J. KROST. 1971. Gas Chromatography of Reactive Sulfur Gases in Air at the Parts Per Billion Level. *Anal. Chem.* 43:827.
11. NATIONAL BUREAU OF STANDARDS. 1971. New

Sulfur Dioxide Permeation Tubes. NBS Technical News Bulletin, p. 106, April.
12. AMERICAN PUBLIC HEALTH ASSOCIATION, INC. 1977. Methods of Air Sampling and Analysis. 2nd ed. p. 722. Washington, D.C.

Subcommittee 1
D. F. ADAMS, *Chairman*
J. O. FROHLIGER
D. FALGOUT
A. M. HARTLEY
J. B. PATE
A. L. PLUMLEY
F. P. SCARINGELLI
P. URONE

711.

Tentative Method of Analysis for Sulfur Trioxide and Sulfur Dioxide Emissions from Stack Gases (Titrimetric Procedure)

12402-01-74T

1. Principle of the Method

1.1 Flue gas is withdrawn by means of a glass or glass lined probe equipped with a filter to remove particulate matter and a heating system to prevent condensation within sampling apparatus.

1.2 Sulfur trioxide is condensed as sulfuric acid by controlled cooling of the flue gas. The resulting sulfuric acid aerosol is collected on an acid washed or treated filter maintained above the water dewpoint. This procedure reduces the problem of oxidation of dissolved sulfur dioxide, and provides improved precision and accuracy in SO_3 collection. Sulfur dioxide is collected in impingers (in series with the SO_3 collector) and oxidized to sulfuric acid by aqueous 3% hydrogen peroxide absorbing solution. The sulfate concentration of each solution is determined separately by titration with alkali using bromphenyl blue indicator or titration with barium perchlorate using thorin indicator. The emission is calculated from the measured concentration of the sulfur oxides and the volumetric flow which is determined by a pitot tube traverse (**1**).

1.3 This method is intended for high precision determination of sulfur oxides emissions in flue gases.

2. Range and Sensitivity

2.1 This method is applicable to the determination of SO_2 (26 to 15,600 mg/m^3 or 10 to 6000 ppm) and SO_3 (33 to 19,800 mg/m^3 or 10 to 6000 ppm). Below 26 mg/m^3 SO_2 or 33 mg/m^3 SO_3 the color change at the endpoint cannot be visually detected.

2.2 Based on replication of stack tests, the method has a sensitivity of ± 3% at SO_2 concentrations of 1000 ppm and ± 10% at 10 ppm.

3. Interference (2,3)

3.1 Ammonia, ammonia compounds and fluorides are known interferences. High concentrations of HCl must be corrected for when alkali determination is used.

3.2 Particulate matter containing sulfur dioxide and/or metals lead to erroneous results. However, these compounds are usually removed by filtration through either a Kaolin or quartz wool plug in the sampling probe or a filter preceding the impingers.

4. Precision and Accuracy

4.1 SULFUR TRIOXIDE. This condenser method for SO_3 collection has been shown to have a collection efficiency of 98% (**4–6**), with quartz or Kaolin wool filter media and within a flow rate range of 1 to 20 l/min.

4.2 SULFUR DIOXIDE. The collection efficiency of SO_2 absorbed in 3% peroxide solution in impingers was determined (**3**). Over the temperature and concentration range examined, no change in the collection efficiency was observed. With 15 cc of 3% peroxide solution, 96% collection efficiency was obtained at sampling rates of 0.5 and 1 liter/min. Collection efficiency dropped to 90% at 3 liter/min.

4.2 The analytical precision is greater than ± 0.5%.

5. Apparatus

5.1 The following sections describe an integrated modular flue gas sampling apparatus for collection of SO_3 by the condenser method and collection of SO_2 in impingers.

5.1.1 *Probe and Probe Heating.* The probe is usually a 1- to 2-m length of 5 mm borosilicate or quartz* tubing with a 12/5 socket joint on the downstream end. The inlet of the probe is enlarged to 38-mm diameter, for a length of 4-cm and loosely packed with quartz or Kaolin wool for filtration of particles. The glass probe is inserted into a stainless steel shell which provides mechanical strength. This device will support probe lengths up to 4 meters. The glass probe is wrapped with asbestos-covered heating wire to heat the glass insert above the acid dewpoint. The minimum probe temperature is controlled by a variable transformer which is preset to 160 C.

5.1.2 *Sulfur Trioxide Condenser.* The condenser is constructed of a glass coil with a medium porosity sintered glass frit at the downstream end. The upstream end of the condenser assembly terminates in a ball joint which mates with the probe. The downstream ball joint mates with the socket joint in the SO_2 impinger. The condenser is maintained above the water dewpoint (usually 60 to 75 C) by immersion in an electrically heated, thermostatted water jacket. (Figure 711:1)

*In actual practice Vycor tubing is often used.

5.1.3 *Impingers.* Two impingers are modified by addition of 12/5 ball and socket connectors. (Plastic or rubber tubing must not be used because of absorption and desorption of gases.)

5.1.4 *Calibrated Dry or Wet Test Meter.* Accurate to ± 1%.

5.1.5 *Pump.* Leak-free vacuum type.

5.1.6 *Flow Meter.* Rotometer or equivalent to measure 0 to 5 l/min (0 to 10 scfh) range.

5.1.7 *Stop Watch.* Accurate to 0.1 min for measurement of sampling duration.

5.1.8 *Thermometer.* A dial thermometer or thermocouple, 100 to 260 C (200–500 F) for measuring the stack gas temperature.

5.1.9 *Plastic Bottles.* For storage of impinger and condenser samples.

5.2 Laboratory Equipment

5.2.1 *Shaker, wrist-action.*

5.2.2 *Centrifuge.* Small clinical type (capable of 2800 to 3000 rpm) with 15-ml tubes.

5.2.3 *Oven or muffle furnace capable of maintaining 250 C.*

6. Reagents

6.1 PURITY. Reagent grade chemicals shall be used in all tests. Unless otherwise indicated, it is intended that all reagents shall conform to the specifications of the Committee on Analytical Reagents of the American Chemical Society, where such specifications are available.† Other reagents may be used provided it can be demonstrated that they are of sufficiently high purity to permit their use without de-

†"Reagent Chemicals, American Chemical Society Specifications," American Chemical Society, Washington, D.C. For suggestions on testing reagents not listed by ACS, see "Reagent Chemicals and Standards," by Joseph Rosin, D. Van Nostrand Co., Inc., New York, N.Y.

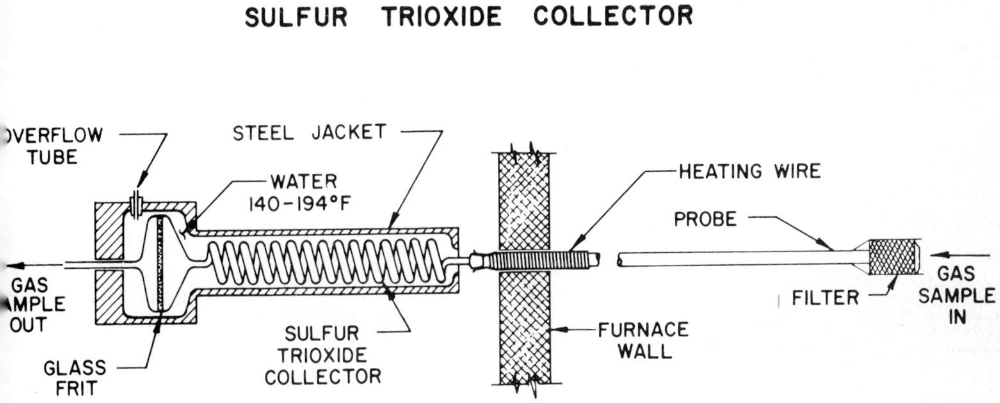

Figure 711:1.

creasing the accuracy of the determination.

6.2 PURITY OF WATER. Unless otherwise indicated, references to water shall be understood to mean reagent water conforming to ASTM specification D1193, Reagent Water.‡ Additionally, this method requires the use of sulfate-free water.

6.2.1 *Sulfate-Free Water.* Distilled water is poured through a column of mixed-bed ion exchange resin (Amberlite MB-3, etc.) contained in a large funnel having 150-mm diam and 100-mm stem. The stem is indented near the bottom to hold a plug of glass wool in place. The resin (no pretreatment) is packed to a depth of 4- to 5-cm in the stem. Another plug of glass wool is placed above the resin bed. The remainder of the funnel is used as a water reservoir. The distilled water used is usually quite low in sulfate; however, the mixed bed exchanger has a capacity of about 0.5 meq/ml, thus changing the resin bed after 25 to 30 l of water throughput is recommended. This volume may be adjusted after checking the effluent water for blank level. This deionizing equipment is also commercially available.

6.3 HYDROGEN PEROXIDE (3%) FOR SO_2 COLLECTION. Prepare by tenfold dilution of 30% hydrogen peroxide. This reagent should be prepared fresh daily and stored in polyethylene containers.

6.4 SULFURIC ACID, APPROX. 0.05N. Add 10 ml of conc, sulfuric acid (18 M H_2SO_4) to 600 to 700 ml of water in a 1000-ml volumetric flask and mix by swirling the flask. Dilute to the mark with water and mix well. Dilute 150 ml of this 0.18 M solution to 1000 ml and again mix well. Standardize against anhydrous sodium carbonate.

6.5 ISOPROPANOL.

6.6 THORIN INDICATOR. 1-(o-arsenophenylazo)-2-naphthol-3,6-disulfonic acid, disodium salt (or equivalent). Dissolve 0.20 g in 100 ml of distilled water.

6.7 BARIUM PERCHOLORATE (0.01N). Dissolve 1.95 g of barium perchlorate, $Ba(ClO_4)_2 \cdot 3H_2O$, in 200 ml distilled water and dilute to 1 l with isopropanol. Standardize with sulfuric acid.

6.8 SODIUM HYDROXIDE, APPROX. 1 N. Slowly add 40 g of NaOH pellets to 800 to 900 ml of water in a 2-l beaker with stirring until all pellets are dissolved. Dilute to 1000 ml with water and mix well. Store in a polyethylene or polypropylene container. Standardize against potassium acid phthalate.

6.9 BROMPHENYL BLUE INDICATOR. 0.1% in water.

6.10 PHENOLPHTHALEIN, 0.05%. Dissolve 0.05 g phenolphthalein in 50 ml ethanol and dilute to 100 ml with water.

‡Annual Book of ASTM Standards, Part 23, 1973.

6.11 ANHYDROUS SODIUM CARBONATE.

6.12 POTASSIUM ACID PHTHALATE. Primary standard grade dried to constant weight and stored in a desiccator. Weigh approximately 2 g to an accuracy of 0.0002 g and dissolve in 1 l of water.

6.13 MIXED INDICATOR. Methyl purple.

7. Procedure

7.1 SAMPLING RATES.

7.1.1 *Sulfur Trioxide*. For SO_3 collection by the controlled condensation method, the flow rate is not critical.

7.1.2 *Sulfur dioxide*. The recommended sampling rate for SO_2 is 1 l/min which is limiting for the sampling train.

7.2 SAMPLING TIME. Expected SO_2 conc should be calculated since sulfur oxide emissions depend primarily on the sulfur content of the fuel. For oil- and coal-fired units§ SO_2 concentration may be estimated (650 to 700 ppm/1% S) from the fuel analysis (C, H, S), the fuel feed rate and the amount of excess air. The SO_3 content is usually 1 to 3% of the SO_2 concentration. (see Figure 711:2).

7.3 SAMPLE COLLECTION. Fit the probe module to the stack flue. Connect power cords between the SO_x probe heater through the variable transformer to the power control panel. Heat the probe to the operating temperature of 160 C. Connect the SO_3 condenser electrically to the power control panel. After checking the water level in the SO_3 condenser jacket (adding water, if necessary), switch on the condenser heater and allow to come to operating temperature of 60 to 70 C. Charge SO_2 impingers with sufficient 3% peroxide solution to cover the bubbler frit. After the probe and SO_3 condenser have reached their respective operating temperatures, assemble the collector module as shown in Figure 711:3. Connect the pump to the second impinger with a vacuum hose and start from the switch on the control module. Record the initial reading of the dry gas meter and record the pressure and temperature values during sampling. Determine stack gas temperature and moisture content. At the end of the 20 to 30 min sampling period, switch off the pump and record the time. Determine the sample volume by difference from the dry gas meter final reading. Disassemble the sample module. Rinse the SO_3 collector (the water can be forced through the frit by applying a slight pressure from a squeeze bulb attached to a 12/5 ball joint) with several portions of sulfate-free water. Transfer the sample in a polyethylene bottle for transport to the laboratory. Rinse the SO_3 collector with isopropyl alcohol. Then draw clean air through the collector for a short period of time to dry and the system is ready to be used. Transfer the contents of the two midget impingers (which contain the SO_2 sample) into a polyethylene bottle. Rinse the impingers several times with sulfate-free water containing 5 to 10% isopropanol and add these washings to the contents of the polyethylene bottle. For immediate collection of additional samples, recharge the impingers.

§Sulfur compound emissions are normally zero for gas-fired units.

7.4 ANALYSIS.

7.4.1 *Analysis for Sulfur Trioxide*. Quantitatively transfer the contents of the polyethylene bottle into a 100-ml graduated beaker. Evaporate the solution to approximately 15 ml. Transfer to a 50-ml volumetric flask. Dilute to the 50-ml mark with sulfate-free distilled water. Pipet a 10 ml aliquot of this solution into a 125-ml Erlenmeyer flask. Add 40 ml of isopropanol and 2 to 4 drops of thorin indicator. Titrate to a pink endpoint using 0.01 N barium perchlorate or to a blue-gray endpoint using 0.01 N NaOH with bromphenyl blue. Run a blank with each series of samples. **(3, 7)**.

7.4.2 *Analysis for Sulfur Dioxide*. Quantitatively transfer the solution from the SO_2 impingers into a 50-ml volumetric flask and dilute to the mark with sulfate-free distilled water. Pipet a suitable sized aliquot into a 50-ml volumetric flask. Pipet a 10 ml aliquot of this solution into a 125 ml Erlenmeyer flask. Add 40 ml of isopropanol and 2 to 4 drops of thorin indicator. Titrate to a pink endpoint using

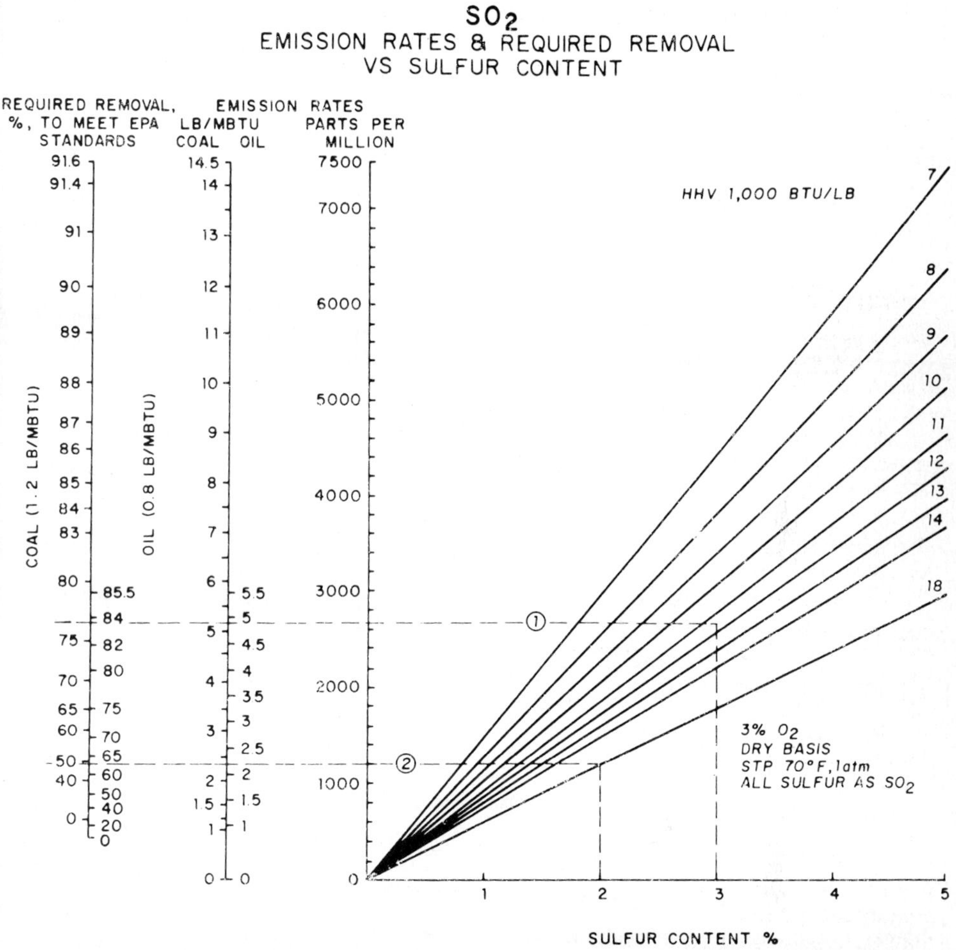

Figure 711:2—Sulfur Dioxide Nomograph—The sulfur dioxide nomograph (Fig. 3) yields the emission rate in pounds per million Btu when firing fuel of the given sulfur content and higher heating value. Further, the nomograph indicates the present removal required to meet the EPA standards of 1.2 lb SO_2/MBtu for coal and 0.8 lb SO_2/MBtu for oil. Note that both the lb/MBtu line and the required removal line have separate scales for coal and oil. Example 1: Coal with 3% sulfur, 11,500 Btu/lb. Construct a vertical line from 3% on the sulfur content scale and intersect the 11,500 Btu/lb HHV line. From this intersection, extend the line horizontally to the left. Read the emission rate as 5.2 lb/MBtu. Example 2: Oil with 2% sulfur, 18,000 Btu/lb. Emission rate: 2.2 lb/MBtu.

0.01 N barium perchlorate or to a blue-gray endpoint using 0.01 N NaOH with bromphenyl blue. Run a blank with each series of samples.

8. Calibration and Standards

8.1 Standardization of the 0.05 N H_2SO_4 is accomplished with anhydrous sodium carbonate. Heat 2.39 g of anhydrous sodium carbonate in a crucible for 4 hr at 250 C to remove water and decompose any residual bicarbonte. Cool in a dessicator. Accurately weigh 0.115 ± 0.005 g of the dried sodium carbonate into each of the three 250-ml Erlenmeyer flasks and dissolve the sample in 50 ml of water. A

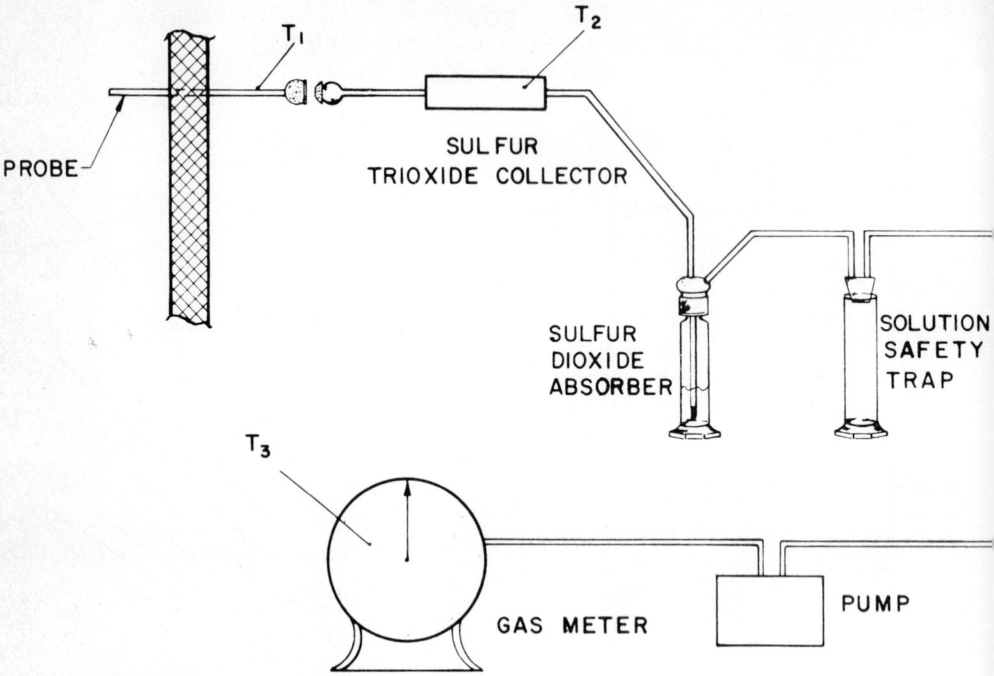

Figure 711:3—Sulfur Oxides Gas Analysis Apparatus.

blank containing no added sodium carbonate should be determined with each set of samples. Add 2 drops methyl purple solution and titrate with the 0.025 M H_2SO_4 in a 50-ml buret to a gray neutral shade. A color reference of 50 ml of the potassium acid phthalate solution containing 2 drops of indicator should be used to identify the endpoint. The normality of the H_2SO_4, N, is computed as follows (where V = ml H_2SO_4 used in titration):

$$N = g\ Na_2CO_3/0.053\ V$$

8.2 Standardize the barium perchlorate with standard sulfuric acid containing 100 ml of isopropanol and the sodium hydroxide against potassium acid phthalate.

8.3 The minimum detection limit for SO_3 can be increased by collecting a larger volume of flue gas or by concentrating the condenser washings.

9. Calculations

9.1 DRY GAS VOLUME. Correct the sample volume measured by the dry gas meter to standard conditions (70 F and 29.92 inches Hg) by using the following equation:

$$V_s = V_0 \frac{T_s}{T_0} \frac{P_0}{P_s} = 17.71 \frac{°R}{\text{in. Hg}} \frac{V_m P_0}{T_0}$$

where:

V_s = Volume of gas sample through the dry gas meter (standard conditions), cu. ft.
V_0 = Volume of gas sample through the dry gas meter (meter conditions), cu. ft.
T_s = Absolute temperature at standard conditions, 530°R.
T_0 = Average dry gas meter temperature, °R.
P_0 = Barometric pressure at the orifice meter, inches Hg.
P_s = Absolute pressure at standard conditions, 29.92 inches Hg.

9.2 SULFUR DIOXIDE CONCENTRATION.

$$C_{SO_2} = \frac{7.05 \times 10^{-5} \frac{lb.^{-1}}{g.-ml} (V_t - V_{tb}) \frac{NV_{soln}}{V_a}}{V_s}$$

where:

C_{SO_2} = Concentration of sulfur dioxide at standard conditions, dry basis, lb./cu. ft.
7.05×10^{-5} = Conversion factor, including the number of grams per gram equivalent of sulfur dioxide (32 g./g.–eq), 453.6 g./lb., and 1,000 ml./l., lb.-1/g-ml.
V_t = Volume of barium perchlorate or sodium hydroxide titrant used for the sample, ml.
V_{tb} = Volume of barium perchlorate or sodium hydroxide titrant used for the blank, ml.
N = Normality of barium perchlorate or sodium hydroxide titrant, g.-eq/l.
V_{soln} = Total solution volume of sulfur dioxide, 50 ml.
V_a = Volume of sample aliquot titrated, ml.
V_s = Volume of gas sample through the dry gas meter (standard conditions), cu. ft.

Table 711:I. Average F Factors

Fuel	F Factor scfd/10^4 Btu
Coal—anthracite	101.4
Coal—bituminous, lignite	98.2
Oil—crude, residual, distillate, fuel oil	92.2
Gas—natural, butane, propane	87.4

9.3 Similarly for sulfur trioxide concentration:

$$C_{SO_3} = \frac{8.82 \times 10^{-5} \frac{lb.^{-1}}{g.-ml} (V_t - V_{tb}) \frac{NV_{soln}}{V_a}}{V_s}$$

9.4 CALCULATION OF SO_x EMISSIONS. For each run, emissions expressed in lb per 10^6 B.T.U. may be determined by the following equation:

$$E = CF\left(\frac{2090}{20.9 - \%\ O_2}\right)$$

where

E = SO_x emission, lb/10^6 B.T.U.
C = pollutant concentration, lb/scfd.
F = factor from Table 711:I
% O_2 = % oxygen

10. Effects of Storage

10.1 Samples may be stored up to 2 weeks without affecting the final results.

11. References

1. See ASTM "Tentative Method of Test for Sampling Stacks—Particulate."
2. CORBETT, P.F. 1951. The determination of SO_2 and SO_3 in Flue Gases, *Journal of the Institute of Fuel*, 24:247–251.
3. DRISCOLL, J.N., and A.W. BERGER. 1971. "Improved Chemical Methods for Sampling and Analysis of Gaseous Pollutants from the Combustion of Fossil Fuels," Final Report, Vol. I—Sulfur Oxides, Contract No. CPA 22-69-95. June.

4. LISLE, E.S., and J.D. SENSENBAUGH. 1965. *Combustion* 36:12.

5. GOKSOYR, H., and K. ROSS. 1962. *J. Inst. Fuel (London)* 35:177.

6. HISSINK, M. 1963. *J. Inst. Fuel (London)* 36:372.

7. MATTY, R.E., and E.K. DIEHL. 1957. Measuring Flue-Gas SO_2 and SO_3, *Power*, 101:94–97.

Subcommittee 1
D. F. ADAMS, *Chairman*
J. O. FROHLIGER
D. FALGOUT
A. M. HARTLEY
J. B. PATE
A. L. PLUMLEY
F. P. SCARINGELLI
P. URONE

712.

Tentative Method of Analysis for Sulfur Trioxide and Sulfur Dioxide Emissions from Stack Gases (Colorimetric Procedure)

12402-02-74T

1. Principle of the Method

1.1 Flue gas is withdrawn by means of a glass or glass lined probe equipped with a filter to remove particulate matter and a heating system to prevent condensation within sampling apparatus.

1.2 Sulfur trioxide is condensed as sulfuric acid by controlled cooling of the flue gas. The resulting sulfuric acid aerosol is collected on an acid washed or treated filter maintained above the water dewpoint. This procedure reduces the problem of oxidation of dissolved sulfur dioxide, and provides improved precision and accuracy in SO_3 collection. Sulfur dioxide is collected in impingers (in series with the SO_3 collector) and oxidized to sulfuric acid by aqueous 3% hydrogen peroxide absorbing solution. The sulfate concentration of each solution is determined separately by reaction with barium chloranilate in a pH-controlled, 50% alcohol solution to yield the highly colored acid chloranilate anion (**1**). The concentration of the colored product is determined spectrophotometrically at 530 nm. The emission is calculated from the measured conc of the sulfur oxides and the volumetric flow which is determined by a pitot tube traverse (**2**).

1.3 This method is intended for high precision determination of sulfur oxides emissions in flue gases.

2. Range and Sensitivity

2.1 This method is applicable to the determination of SO_2 (26-7800 mg/m^3 or 10-3000 ppm) and SO_3 (33-9900 mg/m^3 or 10-3000 ppm). A 25-liter flue gas sample containing 10 ppm by volume of sulfur oxides will yield an absorbance at 530 nm of twice the blank value.

2.2 Higher concentrations of either SO_2 or SO_3 may be determined by suitable aliquoting provided that the $SO_4^=$ concentration in the final solution is within the region where Beer's law applies (up to 350 μg $SO_4^=$/ml).

3. Interferences

3.1 CATIONIC INTERFERENCES. Cations such as Al^{+++}, Ca^{++}, Fe^{+++}, Pb^{++}, Cu^{++}, and Zn^{++} cause interference by precipitation of the acid chloranilate ion. Particles containing K^+, Mg^{++}, Na^+, and NH_4^+ reportedly cause 1% or less interference (**1**). These cationic interferences are usually removed by filtration through a quartz or Kaolin wool plug in the sampling probe preceding the impingers.

3.2 ANIONIC INTERFERENCE. The following substances have been found to give the per cent interference listed (**3**) when added to a solution containing 250 μg/ml of sulfate. Of the anionic interferences, oxalate and phosphate are the most significant.

Interfering Substance	Concentration in Final Solution		Per cent Interference
Oxalate	0.01 M	880 μg/ml	86*
Phosphate	0.01 M	950	46*
Fluoride	0.02 M	380	2
Bicarbonate	0.01 M	610	2
Chloride	0.01 M	350	Nil
Chloride	0.02 M	700	2
Nitrate	0.02 M	1240	Nil
Formaldehyde	0.02 M	600	Nil
Hydrogen peroxide	0.18%	—	Nil

*Reported not to interfere at 100 ppm conc level (**1**).

If a treated Kaolin or quartz wool filter is used to remove particulate matter containing sulfates, there appear to be no major anionic interferences since phosphate or oxalate anions are not expected to be encountered in fossil fuel effluents.

4. Precision and Accuracy

4.1 PRECISION OF SAMPLING. Sulfur trioxide: This condenser method for SO_3 collection has been shown to have a collection efficiency of 98% (**4–6**), with quartz or Kaolin wool filter media and within a flow rate range of 1 to 20 liter/min.

Sulfur Dioxide: The collection efficiency of SO_2 absorbed in 3% peroxide solution in impingers was determined (**3**). Over the temperature and concentration range examined, no change in the collection efficiency was observed. With 15 cc of 3% peroxide solution, 96% collection efficiency was obtained at sampling rates of 0.5 and 1 liter/min. Collection efficiency dropped to 90% at 3 liter per min.

4.2 PRECISION OF ANALYTICAL PROCEDURE. The average absorbance of ten K_2SO_4 samples containing 250 μg of sulfate per ml, was determined to be 0.474, with a relative standard deviation of the mean of 0.4% (**3**).

4.3 REPEATABILITY OF OVERALL METHOD. The repeatability of SO_2 concentration determination in a flue gas has been found to be ± 2.6% at 1500 ppm (**7**). The precision of SO_3 determination is probably ± 5% at 10 ppm.

5. Apparatus

5.1 SAMPLING COMPONENTS. The following sections describe an integrated modular flue gas sampling apparatus for collection of SO_3 by the condenser method and collection of SO_2 in impingers.

5.1.1 *Probe and Probe Heating*. The probe is usually a 1- to 2-m length of 5 mm borosilicate or quartz† tubing with a 12/5 socket joint on the downstream end. The inlet of the probe is enlarged to 38 mm diam for a length of 4 cm and, loosely packed with quartz or Kaolin wool for filtration of particles. The glass probe is inserted into a stainless steel shell which provides mechanical strength. This device will support probe lengths up to 4 m. The glass probe is wrapped with asbestos-covered heating wire to heat the glass insert above the acid dewpoint. The minimum probe temperature is controlled by a variable transformer which is preset to 320 F.

5.1.2 *Sulfur Trioxide*. The condenser is constructed of a glass coil with a medium porosity sintered glass frit at the downstream end. The upstream end of the condenser assembly terminates in a ball joint which mates with the probe. The downstream ball joint mates with the socket joint in the SO_2 impinger. The condenser is maintained above the water dewpoint (usually 60 to 75 C) by immersion in an electrically heated, thermostatted water jacket (Figure 712:1).

5.1.3 *Impingers*. Two impingers are modified by addition of 12/5 ball and socket connectors. (Plastic or rubber tubing must not be used because of absorption and desorption of gases).

†In actual practice Vycor tubing is often used.

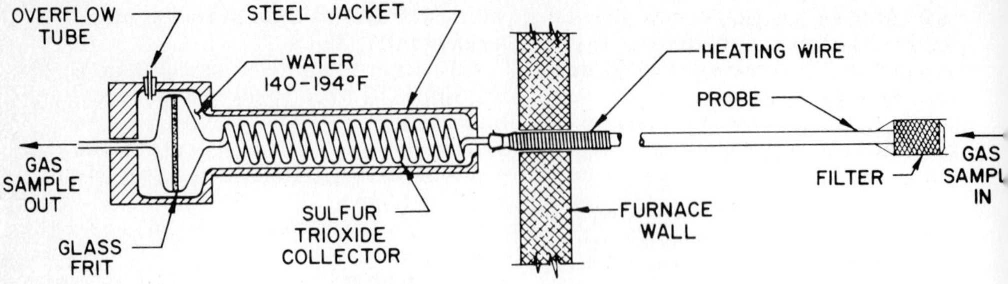

Figure 712:1—Sulfur Trioxide Collector.

5.1.4 *Calibrated Dry or Wet Test Meter.* Accurate to ± 1%.

5.1.5 *Pump.* Leak-free vacuum type.

5.1.6 *Flow Meter.* Rotometer or equivalent to measure 0 to 5 l/min (0 to 10 scfh) range.

5.1.7 *Stop Watch.* Accurate to 0.1 min for measurement of sampling duration.

5.1.8 *Thermometer.* A dial thermometer or thermocouple, 100 to 260 C (200 to 500 F) for measuring the stack gas temperature.

5.1.9 *Plastic Bottles.* For storage of impinger and condenser samples.

5.2 Laboratory Equipment.

5.2.1 *Shaker, wrist-action.*

5.2.2 *Centrifuge.* Small clinical type (capable of 2800 to 3000 rpm) with 15 ml tubes.

5.2.3 *Spectrophotometer.* Suitable for measuring light in the visible region (at 530 nm).

5.2.4 *Oven or muffle furnace.* Capable of maintaining 250 C.

6. Reagents

6.1 Purity. Reagent grade chemicals shall be used in all tests. Unless otherwise indicated, it is intended that all reagents shall conform to the specifications of the Committee on Analytical Reagents of the American Chemical Society, where such specifications are available.‡ Other reagents may be used provided it can be demonstrated that they are of sufficiently high purity to permit their use without decreasing the accuracy of the determination.

6.2 Purity of Water. Unless otherwise indicated, references to water shall be understood to mean reagent water conforming to ASTM specification D1193, Reagent Water.§ Additionally, this method requires the use of sulfate-free water.

6.2.1 *Sulfate-free Water.* Distilled water is poured through a column of mixed-bed ion exchange resin (Amberlite MB-3 etc.) contained in a large funnel having 150-mm diam and 100-mm stem. The stem is indented near the bottom to hold a plug of glass wool in place. The resin (no pretreatment) is packed to a depth of 4 to 5 cm in the stem. Another plug of glass wool is placed above the resin bed. The remainder of the funnel is used as a water reservoir. The distilled water used is usually quite low in sulfate; however, the mixed bed exchanger has a capacity of about 0.5

‡"Reagent Chemicals, American Chemical Society Specifications", American Chemical Society, Washington, D.C. For suggestions on testing reagents not listed by ACS, see "Reagent Chemicals and Standards", by Joseph Rosin, D. Van Nostrand Co., Inc., New York, NY.

§Annual Book of ASTM Standards, part 23, 1973.

meq/ml, thus changing the resin bed after 25 to 30 l of water throughput is recommended. This volume may be adjusted after checking the effluent water for blank level. This deionizing equipment is also commercially available.

6.3 HYDROGEN PEROXIDE (3%) FOR SO_2 COLLECTION. Prepare by tenfold dilution of 30% hydrogen peroxide. This reagent should be prepared fresh daily and stored in polyethylene containers.

6.4 SULFURIC ACID, APPROXIMATELY 0.05 N. Add 10 ml of conc sulfuric acid (18 M H_2SO_4) to 600 to 700 ml of water in a 1000-ml volumetric flask and mix by swirling the flask. Dilute to the mark with water and mix well. Dilute 150 ml of this 0.18 M solution to 1000 ml and again mix well. Standardize against anhydrous sodium carbonate.

6.5 BARIUM CHLORANILATE.

6.6 ISOPROPYL ALCOHOL (IPA).

6.7 BUFFER, pH 5.6. Add 50 ml of 0.2 M acetic acid (11.4 ml glacial acetic acid in 1000 ml of distilled water) to 500 ml of 0.2 M sodium acetate (27.2 g $NaC_2H_3O_2 \cdot 3 H_2O$ in 1000 ml of water).

6.8 SODIUM HYDROXIDE, APPROXIMATELY 1 N. Slowly add 40 g of NaOH pellets to 800 to 900 ml of water in a 2-l beaker with stirring until all pellets are dissolved. Dilute to 1000 ml with water and mix well. Store in a polyethylene or polypropylene container.

6.9 HYDROCHLORIC ACID, APPROXIMATELY 1 N. Add 90 ml of conc HCl to 800 to 900 ml of water in a 2-l beaker with stirring. Dilute to 1000 ml with water and mix well. This solution can be stored in glass.

6.10 PHENOLPHTHALEIN, 0.05%. Dissolve 0.05 g phenolphthalein in 50 ml ethanol and dilute to 100 ml with water.

6.11 ANHYDROUS SODIUM CARBONATE.

6.12 POTASSIUM ACID PHTHALATE. Primary standard grade, dried to constant weight and kept in a desiccator. Weigh approximately 2 g to an accuracy of 0.002 g and dissolve in 1 l of water.

6.13 MIXED INDICATOR. Methyl purple, pH 3.5 to 5.

7. Sampling Procedure

7.1 SAMPLING RATE. For SO_3 collection by the controlled condensation method.

7.1.1 *Sulfur Trioxide.* The flow rate is not critical.

7.1.2 *Sulfur Dioxide.* The recommended sampling rate for SO_2 is 1/l/min and is the controlling rate.

7.2 SAMPLING TIME. Since sulfur oxide emissions depend primarily on the sulfur content of the fuel, expected SO_2 concentrations can be calculated. For oil- and coal-fired units|| SO_2 concentration estimate (650 to 700 ppm/1%S) from the fuel analysis (C, H, S), the fuel feed rate and the amount of excess air. The SO_3 content is usually 1 to 3% of the SO_2 concentration. (See Figure 712:3).

7.3 SAMPLE COLLECTION. Fit the probe module to the stack flue. Connect power cords between the SO_x probe heater through the variable transformer to the power control panel. Heat the probe to the operating temperature of 160 C. Connect the SO_3 condenser electrically to the power control panel. After checking the water level in the SO_3 condenser jacket (add water to fill if necessary), switch on the condenser heater and allow to come to operating temperature of 60 to 70 C. Charge the SO_2 impingers with sufficient 3% peroxide solution to cover the bubbler frit. After the probe and SO_3 condenser have reached their respective operating temperatures, assemble the collector module as shown in Figure 712:2. Connect the pump to the second impinger with a vacuum hose and start from the switch on the control module.

Record reading on the dry gas meter and the pressure and temperature values during sampling. Determine the stack gas temperature and moisture content. At the end

||Sulfur compound emissions are normally zero for gas-fired units.

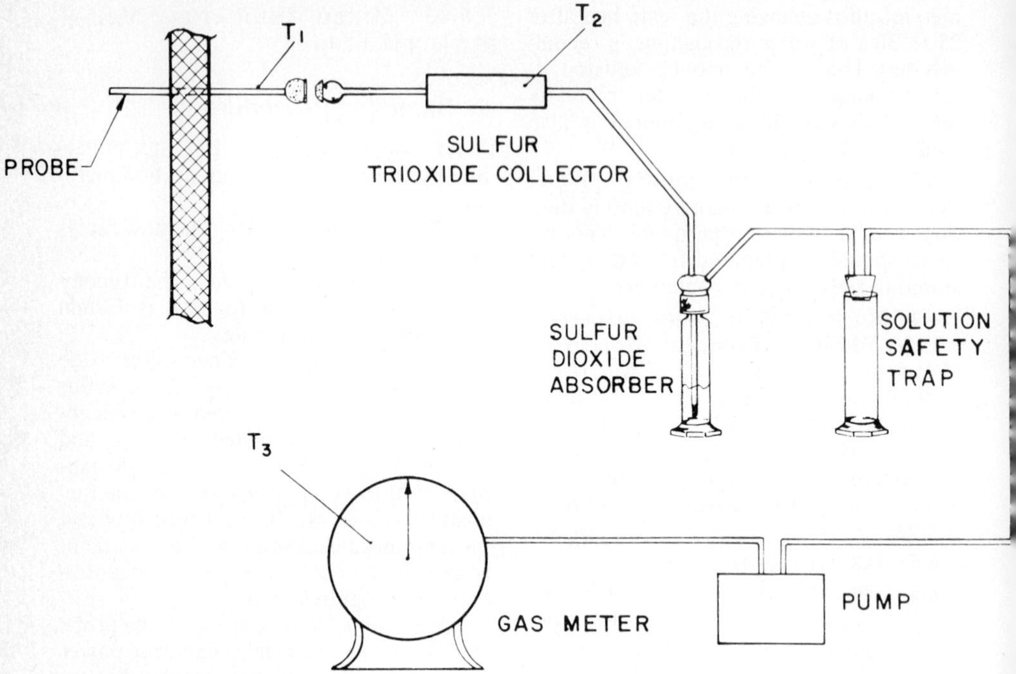

Figure 712:2—Sulfur Oxides Gas Analysis Apparatus.

of the 20 to 30 min sampling period switch off the pump and record the time. Determine the sample volume by difference from the dry gas meter reading. Disassemble the sample module. Rinse the SO_3 collector. The water can be forced through the frit by applying a slight pressure from a squeeze bulb attached to a 12/5 ball joint with several portions of sulfate-free water. Transfer the sample in a polyethylene bottle for transport to the laboratory. Rinse the SO_3 collector with isopropyl alcohol. Then draw clean air through the collector for a short period of time to dry and the system is ready to be used. Transfer the contents of the two midget impingers (which contain the SO_2 sample) into a polyethylene bottle. Rinse the impingers several times with sulfate-free water containing 5 to 10% isopropanol and add these washings to the contents of the polyethylene bottle. For immediate collection of additional samples recharge the impingers.

7.4 ANALYSES.

7.4.1 *Analysis for Sulfur Trioxide.* Quantitatively transfer the contents of the polyethylene bottle into a 100-ml graduated beaker. Evaporate the solution to approximately 15 ml. Pipet 5 ml of pH 5.6 sodium acetate buffer, then add 50 ml of isopropyl and mix well. Pour contents into a 100-ml volumetric flask containing 0.2 to 0.3 g barium chloranilate. Stopper the flask and shake for 20 min on a wrist-action shaker. Dilute to mark with sulfate-free water and mix. Decant into 15-ml centrifuge tubes and centrifuge at 3000 rpm

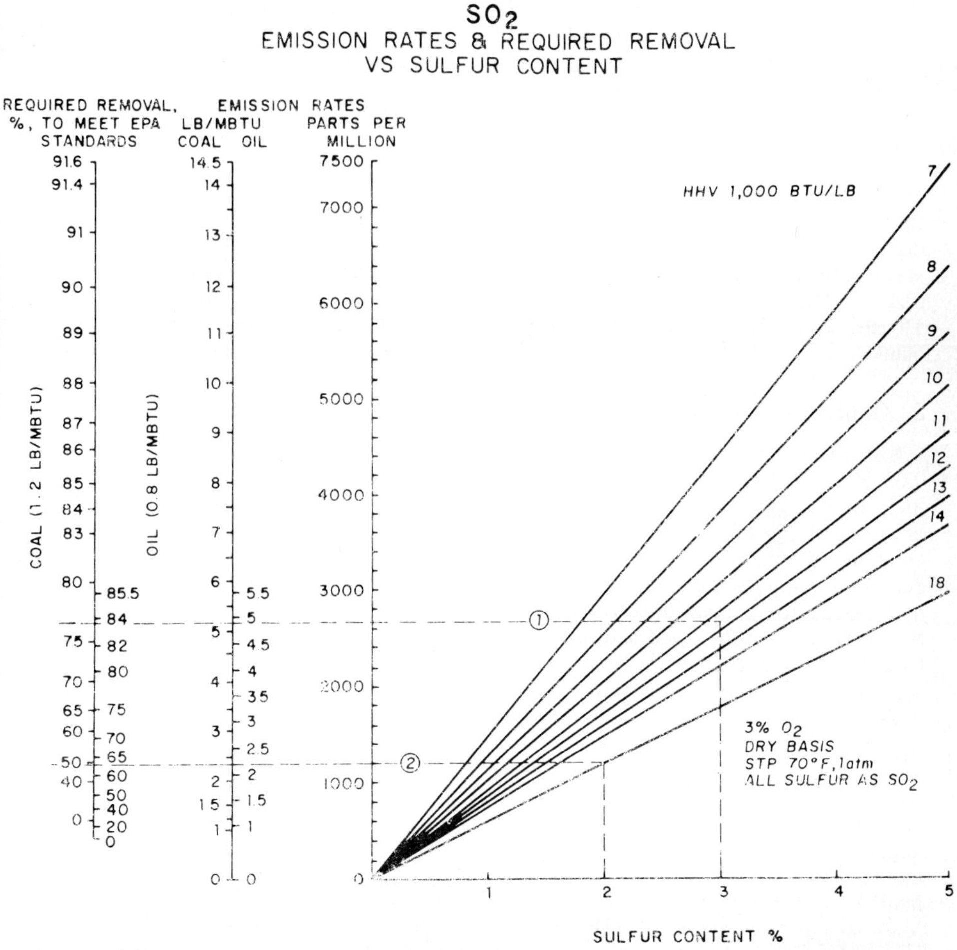

Figure 712:3—Sulfur Dioxide Nomograph—The sulfur dioxide nomograph (Fig. 3) yields the emission rate in pounds per million Btu when firing fuel of the given sulfur content and higher heating value. Further, the nomograph indicates the percent removal required to meet the EPA standards of 1.2 lb SO_2/MBtu for coal and 0.8 lb SO_2/MBtu for oil. Note that both the lb/MBtu line and the required removal line have separate scales for coal and oil. Example 1: Coal with 3% sulfur, 11,500 Btu/lb. Construct a vertical line from 3% on the sulfur content scale and intersect the 11,500 Btu/lb HHV line. From this intersection, extend the line horizontally to the left. Read the emission rate as 5.2 lb/MBtu. Example 2: Oil with 2% sulfur, 18,000 Btu/lb. Emission rate: 2.2 lb/MBtu.

for 5 min, decant the clear supernatant liquid into a spectrophotometer cell and read the absorbance versus a water blank at 530 nm. In addition, determine a reagent blank.

7.4.2 *Analysis for Sulfur Dioxide.* Quantitatively transfer the solution from the SO_2 impingers into a 250-ml volumetric flask and dilute to the mark with sulfate-free distilled water. Pipet a suitable sized aliquot into a 100-ml beaker. Add 1 drop of phenolphthalein solution to the flask, then add 1 N NaOH dropwise until the solution turns pink. Add 1 drop of 1 N

HCl to return the solution to colorless. Pipet in 5 ml of pH 5.6 buffer, then add 50 ml of isopropyl alcohol and mix well. Pour contents into a 100-ml volumetric flask containing 0.2 to 0.3 g of barium chloranilate. Stopper the flask and shake for 20 minutes on a wrist-action shaker. Dilute to mark with sulfate-free water and mix. Decant into 15-ml centrifuge tubes and centrifuge for 5 min at 3000 rpm, decant into a spectrophotometer cell, and read the solution absorbance versus a water blank at 530 nm. In addition, determine a reagent blank.

8. Calibration and Standards

8.1 SULFATE. Standardization of the 0.05 N H_2SO_4 is accomplished with anhydrous sodium carbonate. Heat 2 to 3 g of anhydrous sodium carbonate in a crucible for 4 hr at 250 C to remove water and decompose any residual bicarbonate. Cool in a desiccator. Accurately weigh 1 g to the nearest mg of the dried sodium carbonate into 250-ml volumetric flask and dissolve the sample and dilute to the mark in 250 ml of water. Add 25 ml aliquots to each of 3 125-ml Erlenmeyer flasks. Add 2 drops methyl purple solution and titrate with the 0.05 N H_2SO_4 in a 50-ml buret to a gray neutral shade. A blank containing no added sodium carbonate should be determined with each set of samples. The normality of the H_2SO_4, N, is computed as follows: (where V = ml H_2SO_4 used in titration and 1.886 = aliquot fraction/eq. wt.)

$$N = g\ Na_2CO_3 \times 1.886/V$$

8.2 A standard curve is prepared by pipetting 0.5, 1, 2, 5, and 7 ml of 0.05 N H_2SO_4 into 100-ml beakers. Add water to the first four to bring all volumes up to about 10 ml. Add 1 drop of phenolphthalein solution, then add 1 N NaOH dropwise to the appearance of a pink color. Add 1 N HCl dropwise to the disappearance of a pink color. (This will usually require just one drop). Pipet 5 ml of pH 5.6 buffer into each beaker. Add 50 ml of IPA to each flask. Mix well. Pour the contents of each beaker into a corresponding 100-ml volumetric flask containing 0.2 to 0.3 g of barium chloranilate. Shake for 20 min on a wrist-action shaker and dilute to mark with sulfate-free water and mix. Decant into 15-ml centrifuge and centrifuge for 5 min at maximum rpm. Decant the centrifugate into 1-cm cells and read the absorbance versus water at 530 nm. The blank (no sulfate) versus water should read no more than 0.01 to 0.03 absorbance units. Plot the absorbance units. Plot the absorbance versus sulfate concentrations in mg/ml of final solution. mg = 96/2 × N × ml; mg/ml = 48 N

8.3 SENSITIVITY. The lower limit of detection for SO_3 can be increased by collecting a larger volume of flue gas, by concentrating the condenser washings, by the use of longer path length cells (5-cm), or by using a different pH (1.8 with phosphate buffer) (**8**). It is simpler to increase the sampling time or use a longer path cell.

8.4 TEMPERATURE. The solution temperature of all standards and sample solutions must be between 20 and 30 C.

8.5 MIXING TIME. A 20-min mixing time is required for maximum color development. Contact time with barium chloranilate to 40 min will not affect results.

8.6 STABILITY. Once the colored solutions have been centrifuged, stoppered solutions are stable for one hr.

9. Calculations

9.1 DRY GAS VOLUME. Correct the sample volume measured by the dry gas meter to standard conditions (70 F and 29.92 inches Hg) by using the following equation:

$$Vm\left(\frac{T_S}{T_o}\right)\left(\frac{P_o}{P_s}\right) = \frac{17.71\ ^\circ R}{in.\ Hg.}\left(\frac{V_oP_o}{T_o}\right)$$

where:

V_s = Volume of gas sample through the dry gas meter. (standard conditions), cu. ft.

V_o = Volume of gas sample through the dry gas meter (meter conditions), cu. ft.

T_s = Absolute temperature at standard conditions, 530°R.

T_o = Average dry gas meter temperature, °R.
P_o = Barometric pressure at the orifice meter, inches Hg.
P_s = Absolute pressure at standard conditions, 29.92 inches Hg.

9.2 CALCULATE THE CONCENTRATION OF SO_x IN THE SAMPLE AS FOLLOWS:

9.2.1 ppm SO_3 (dry basis) = corrected absorbance × (slope)-1 (from calibration curve) × 50 × (24.4/96) × (1/Vg)

9.2.2 ppm SO_2 (dry basis) = corrected absorbance × (slope)-1 (from calibration curve) × 100 × (250/Va) × (24.4/96 × (1/Vg)

where

Va = aliquot volume.
Vg = sample volume of gas liters [sampling rate of orifice (corrected to S.T.P.) × time]. Standard conditions 70 F 29.9 in Hg.

9.2.3 SO_x (mg/m³) = 41.4 × MW × ppm SO_x.

9.2.4 SO_x (lb/ft³) = 41.4 × MW × ppm SO_x) × 62.4 × 10^{-9}.

9.3 CALCULATION OF SO_x EMISSIONS. For each run, emissions expressed in lb/10^6 B.T.U. may be determined by the following equation:#

$$E = CF\left(\frac{2090}{20.9 - \%O_2}\right)$$

Where

E = SO_x emission, lb/10^6 B.T.U.
C = Pollutant concentration, lb/scfd.

#U.S.-EPA Emission Standards and Engineering Division "Proposed Method of Calculating Power Plant Emission Rates" 1974.

TABLE 712:I. Average F Factors

Fuel	F Factor scfd/10^4 Btu
Coal—anthracite	101.4
Coal—bituminous, lignite	98.2
Oil—crude, residual, distillate, fuel oil	92.2
Gas—natural, butane, propane	87.4

F = factor from Table 712:I
$\%O_2$ = % oxygen.

10. Effects of Storage

10.1 Field samples are stable indefinitely.

11. References

1. BERTOLACINI, R.J. and J.E. BARNEY. 1957 *Anal. Chem.* 29:281.
2. See ASTM "Tentative Method of Test for Sampling Stacks—Particulate."
3. DRISCOLL, J.N. and A.W. BERGER. 1971. Improved Chemical Methods for Sampling and Analysis of Gaseous Pollutants from the Combustion of Fossil Fuels," Final Report, Vol. I—Sulfur Oxides, Contract No. CPA 22-69-95. June.
4. LISLE, E.S. and J.D. SENSENBAUGH. 1965. Combustion 36:12.
5. GOKSOYR, H. and K. ROSS. 1962. *J. Inst. Fuel* (London) 35:177.
6. HISSINK, M. 1963. *J. Inst. Fuel* (London) 36:372.
7. BERGER, A.W., J.N. DRISCOLL and P. MORGENSTERN. 1972. AIHA 33:397.
8. CARLSON, R.M. *et al.* 1967 *Anal. Chem.* 39:689.

Subcommittee 1

D. F. ADAMS, *Chairman*
J. O. FROHLIGER
D. FALGOUT
A. M. HARTLEY
J. B. PATE
A. L. PLUMLEY
F. P. SCARINGELLI
P. URONE

713.

Tentative Method of Analysis for Sulfate Aerosols Pyrolyzable below 400 C by Effluent Gas Analysis

12403-01-74T

1. Principle of the Method

1.1 The collection of aerosols containing sulfates on specially treated glass fiber filters or by impaction on copper discs separates particulate sulfur compounds from the higher concentration of gaseous sulfur compounds encountered in the atmosphere.

1.2 A portion of the filter or the disc is heated to 400 C under a stream of nitrogen. The liberated sulfur trioxide is swept through hot copper, which reduces sulfur trioxide to sulfur dioxide (**1–3**). The resulting sulfur dioxide is determined by spectrophotometry (**4, 5**), coulometry or flame photometry.

1.3 Pyrolyzable sulfate, as defined herein, are those sulfates which include for the most part sulfuric acid and ammonium sulfate that thermally decompose at or below 400 C. Figure 713:1 depicts the decomposition temperatures of some common sulfates as determined by Scaringelli et al. (**6**). These data were obtained with the equipment described herein, with the exception that the temperature of the furnace was programmed at 20 C per min rather than maintained at a constant temperature. Figure 713:1 shows that $(NH_4)_2 SO_4$ and H_2SO_4, decomposed quantitatively with only partial decomposition of Hg_2SO_4 and $Fe_2(SO_4)_3 \cdot (NH_4)_2SO_4$. Alkaline earth and alkali sulfates require extremely high temperatures for the liberation of sulfur trioxide. Hence, the method is applicable to the determination of H_2SO_4 and $(NH_4)_2SO_4$ in ambient air without substantial interferences.

2. Range and Sensitivity

2.1 The range of interest is usually 0.003 to 10 $\mu g/m^3$, expressed as sulfuric acid. The volume of air sample should be selected so that the quantity of pyrolyzable sulfates to be analyzed falls within the range of 0.1 to 40 μg.

2.2 The method has sufficient sensitivity to measure pyrolyzable sulfate content of aerosols separated by multi-stage impactors and by using only a small portion of the samples collected for 24 hr on high volume filters. The lower limits of detectability (LLD) and range for optimum precision (ROP) are as follows:

	LLD, $\mu g\ H_2SO_4$	ROP, $\mu g\ H_2SO_4$
Spectrophotometry	0.21	0.8 to 40
Coulometry	0.03	0.1 to 20
Flame Photometry	0.003	0.6 to 15

3. Interferences

3.1 If only H_2SO_4 and $(NH_4)_2 SO_4$ are of primary interest, then the only reported interference would be mercurous sulfate, which is not expected to be encountered in ambient air. Glass fiber filters contain residual alkali that creates a sulfuric acid demand (**1, 7, 8, 9**) and which must be removed by washing with acetic acid or other means. Untreated filters result in a loss of 0.4 to 7.8 μg H_2SO_4 per cm^2. Thorough washing with acetic acid reduces residual alkali to the equivalent of 0.4 to 0.8 μg H_2SO_4. Barton and McAdie (**8**) reported that treatment with acetic acid did not remove all of the interference from residual alkali.

3.2 Analysis. The spectrophotometric procedure is considered the most specific detection system for sulfur dioxide. The coulometric detector is fairly specific for sulfur dioxide, but will respond to compounds that are oxidized by halogens. The

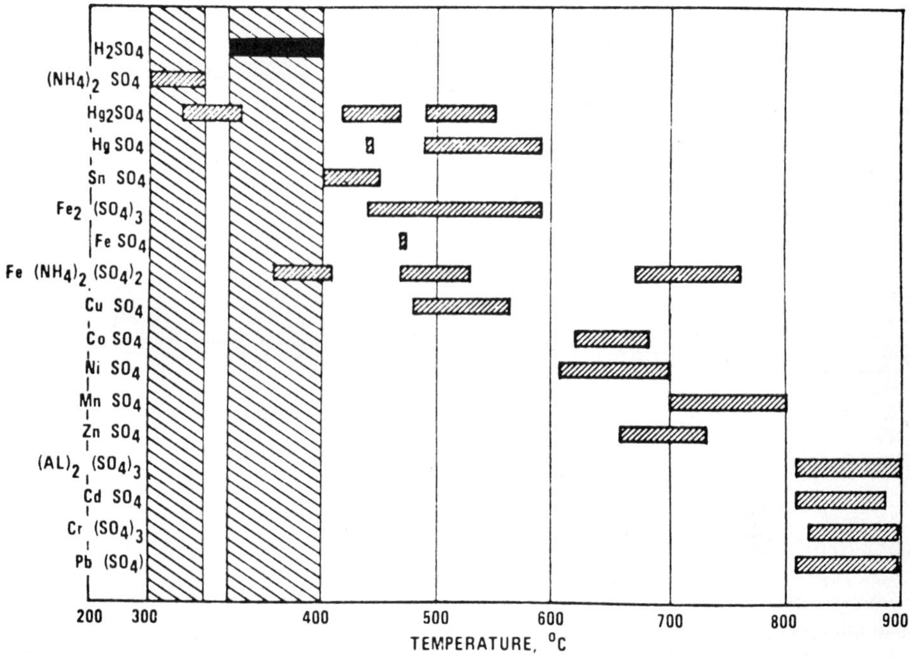

Figure 713:1—Decomposition ranges of metal sulfates. Lower values, initial decomposition temperature. Upper range temperature of energetic decomposition.

flame photometric detector is specific for sulfur compounds.

4. Precision and Accuracy

4.1 SAMPLING.

4.1.1 Glass fiber filters have a collection efficiency greater than 99% for aerosols of 0.3μ, or larger when measured by the dioctyl phthalate test. Size distribution of sulfate aerosols has not been definitively established but appears to be submicronic.

4.1.2 The efficiency of collection on copper disc and glass fiber filters was comparable when tested against H_2SO_4 aerosols generated in the laboratory.

4.2 ANALYSIS.

4.2.1 *Precision*. The repeatabilities, within laboratory agreement, expressed as relative standard deviation about the mean for the colorimetric, coulometric, and flame photometric procedure are 4.8%, 4.5% and 2.1%, respectively **(1)**.

5. Apparatus

5.1 SAMPLING APPARATUS. One of the following sampling systems may be used to collect the aerosol:

5.1.1 *Filtration Option A*. Stainless steel filter holder of 1″, calibrated rotameter or calibrated critical orifice and vacuum pump.

5.1.2 *Filtration Option B*. High Volume samplers can be used.

5.1.3 *Impaction Option*. Midget impinger with copper disc, calibrated rotameter, vacuum pump. Diam of the jet for the midget impinger is 1 mm yielding a maximum flow rate of 2.8 l/min.

5.1.4 For the determination of size distribution of the aerosol, one of the cascade impactors **(10)** is required using copper foils as the impaction surface.

5.2 ANALYSIS.

5.2.1 *Pyrolysis Apparatus*. Assemble the gas train as shown in Figure 713:2, consisting of the following: a cylinder of pure

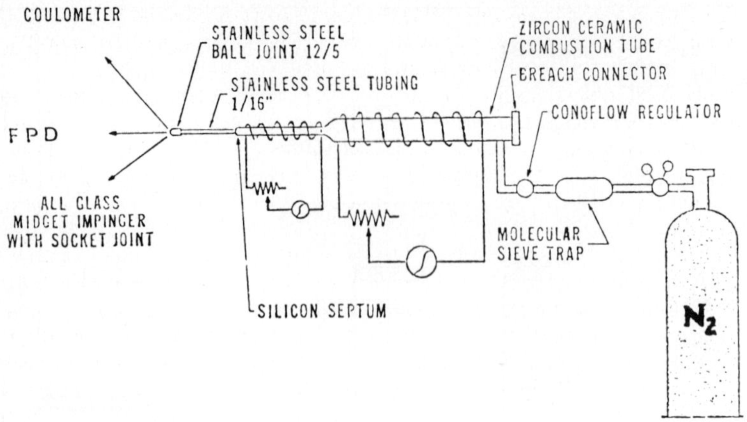

Figure 713:2—Gas train and pyrolysis apparatus.

nitrogen; a two-stage regulator; a purification trap containing molecular sieve 5A; a flow-regulating device; a breech connector with adapter for the rapid insertion of sample; a zirconium oxide combustion tube, which is attached to the breech, sealed and made gas-tight by filling the annular space between the tube and the breech with High-pyseal, a silicone septum, open-barrel type; and a stainless-steel tube, 203-mm long and 1.5-mm OD, with a 12/5 end of the combustion tube with degreased (washed with acetone) copper turnings to a depth of 3″.

5.3 DETECTION SYSTEM. One of the following systems may be used to determine sulfur dioxide in the gaseous effluent.

5.3.1 *Spectrophotometric Option A.* Midget impingers, spectrophotometer (suitable for measuring absorbancy at 548 or 575 nm).

5.3.2 *Coulometer Detector Option B.* Dohrman Microcoulometer with sulfur cell, T-300-P, or equivalent, Recorder 1 mV, preferably with an integrator.

5.3.3 *Flame Photometer Option C.* Melpar SO_2 analyzer, Model SR 1100-1, or equivalent.

6. Reagents

6.1 SAMPLING.

6.1.1 Glass fiber filters specially treated with acetic acid.

6.1.2 Copper, electrolytic grade, 2 mil thick, disc cut to fit the impactor used.

6.2 ANALYSIS.

6.2.1 Gas Cylinder of Reagent Grade Nitrogen.

6.2.2 Standardized sulfuric acid solution, 0.2 N H_2SO_4.

6.2.3 Hydrogen 5% in an inert gas.

6.2.4 Prepare working standard of approximately 1.5 H_2SO_4 mg/ml by diluting exactly 15 ml of the 0.2 N H_2SO_4 with water to the 100-ml mark in a volumetric flask. Calculate the exact concentration to 0.01 mg/ml of H_2SO_4.

6.2.5 *Spectrophotometry*. Details of this procedure are described in reference **(4, 5)**.

6.2.6 COULOMETRY. Electrolyte solution prepared according to manufacturer's instructions.

6.2.7 *Flame Photometry*. Cylinders of Reagent Grade Oxygen and Hydrogen. Electrolytic generation of hydrogen is preferable.

7. Procedure

7.1 COLLECTION OF SAMPLE.

7.1.1 *Filter*. To determine the conc of pyrolyzable sulfate in the atmosphere, collect the aerosol on a 1″ (25 mm) circular filter of glass fiber, which has been acid washed. Wash filters thoroughly in batch quantities, in turn with 1% acetic acid, distilled water, ethanol, and acetone. Place the filters on clean glass plates, dry with clean air and store in a petri dish or desiccator. Before use, check filters for uniformity and pin holes by holding them to the light. Set up sampling train consisting of, in order, a stainless-steel filter holder, flowmeter, flow control device, temperature and vacuum gauges, and vacuum pump. Calibrated hypodermic needles, 13 or 15 gauge, can be used as critical orifices to control the rate of flow at about 2 l/min. Insert a filter into the holder; start vacuum pump and timer. It is desirable to collect 0.8 to 50 μg of pyrolyzable sulfate for the colorimetric analysis and 1 to 20 μg for the instrumental procedures. For example, at a pyrolyzable conc of 5 $\mu g/m^3$ and flow rate of 10 l/min, sample the air sample for 100 min.

7.1.2 *High Volume Sampler*. Alternatively, a portion of a Hi-Vol, 8″ × 10″ filter can be used for the analyses if the filter has been specially pretreated as described above. Fold the filter and punch or cut out, with a cork borer or hole puncher, the required amount of sulfate. Measure the internal diam of the cutting instrument with a vernier caliper to calculate the fraction of the sample taken.

7.1.3 *Single Stage-Single Jet Sampler*. The standard midget is suitable as the impaction device. Place the copper disc that has been conditioned at 600 C in the impinger. Oxide coating can be removed by heating the copper with a flame until it glows, then carefully dropping the hot copper into methanol. Attach a vacuum pump that will maintain 0.6 atmospheres across the orifice of the jet. Typically, the orifice of 1 mm ID yields a maximum flow of 2.8 l/min. The collection efficiency for submicronic aerosols depends on the size of the orifice and the impaction distance **(11)**. The impaction distance should be between one and three times the ID of the orifice or 1 to 3 mm.

7.1.4 *Disc-Multi Stage, Multi Jet Sampler*. The Anderson sampler **(12)** can be used to measure the size distribution of the aerosol. Cut the copper disc to fit the dimension of the impaction plates and condition as described above. If replicate analyses are required the impacted spots can be cut from the copper disc and combined or analyzed individually.

7.2 Analysis. Insert the combustion tube into a furnace that can maintain a temperature of 400 C. Adjust the flow of nitrogen through the system to 200 ml/min, and check for leaks. Bring the combustion tube to 400 C and the reaction tube containing copper to 500 C. Condition the tube with sulfuric acid as follows. By means of a microliter syringe, place 25 μl of the standard sulfuric acid solution on a copper disc. Evaporate the excess water by placing the disc in a beaker, which is immersed in a bath of 70% phosphoric acid at 110 C. Place the disc in a copper boat. Quickly open the breech cap, insert the boat into the hot zone of the combustion tube and immediately close the breech. Repeat the conditioning operation at least six times or until the detection response becomes constant. Normally conversion of 80 to 85% should be expected. The system is now ready for calibration or for actual samples. Reconditioning of the system is required for each start-up or for large temperature changes. Sufficient 25 μl aliquots of the standard solution should be analyzed to restore constant response. Normally, no regeneration of the copper catalyst is required for periods of 3 months or longer. Regeneration is necessary whenever the conversion factor drops sharply. Regenerate the copper in place by passing a stream of 5% hydrogen in nitrogen at a rate of 50 ml/min for 1 hr at a temperature of 300 C.

7.2.1 *Spectrophotometric Analyses*. For spectrophotometric detection, connect a midget impinger (with 12/5 ball joint) comtaining 10 ml of tetrachloromercurate solution. Insert sample and collect the sulfur dioxide from the gaseous effluent for 10 min. Similarly, run one or more blanks and one or more control samples for each series of determinations. Analyze the collected sulfur dioxide samples by the Intersociety Method **(5)**.

7.2.2 *Coulometric Analyses*. Connect the ball joint from the combustion tube to the inlet of the detector cell. Follow the procedure outlined by the manufacturer's manual but use high gain, low sensitivity (6 or 10 ohms), and the sulfur cell. Allow the titration to go to completion; less than 10 min is normally required. Response of the instrument is rapid, but the titration is not completed rapidly because the capacity of the system is usually exceeded. A trapezoidal curve, modified by portions of a normal curve at the beginning and end, appears on the chart, see Figure 713:3. This phenomenon appears to be caused by polarization and depolarization of the electrodes. Compute the area under the curve after subtracting the area of the blank produced by the volatility of the iodine in the cell. A ball and disc integrator is convenient and satisfactory for measurement of these areas.

7.2.3 *Flame Photometric Analyses.* With a ball joint, connect the sample inlet of the flame housing to the outlet of the combustion tube. Follow the procedure detailed in the manufacturer's manual with the following modification. Reduce the amount of light supplied to the photomultiplier tube to maintain linearity and to guard against saturation of the detector. Apply black tape with sharp edges to the face of the interference filter to produce a slit approximately 3 mm by 2 mm. Adjust the electrometer to provide a sensitivity of 1.6 μa full scale. Set the flow rate of oxygen to 40 ml/min. First, press ignitor and then turn on the valve that controls the hydrogen to ignite the flame. Adjust the flow rate of hydrogen to 200 ml/min. **Caution!** If the flame goes out, turn off the flow of hydrogen gas and allow the gases to sweep out all traces of hydrogen before pressing the ignitor button again. Maintain temperatures of 100 C or above for the flame housing and for the exit port to avoid condensation or adsorption problems. If condensation occurs in the flame housing,

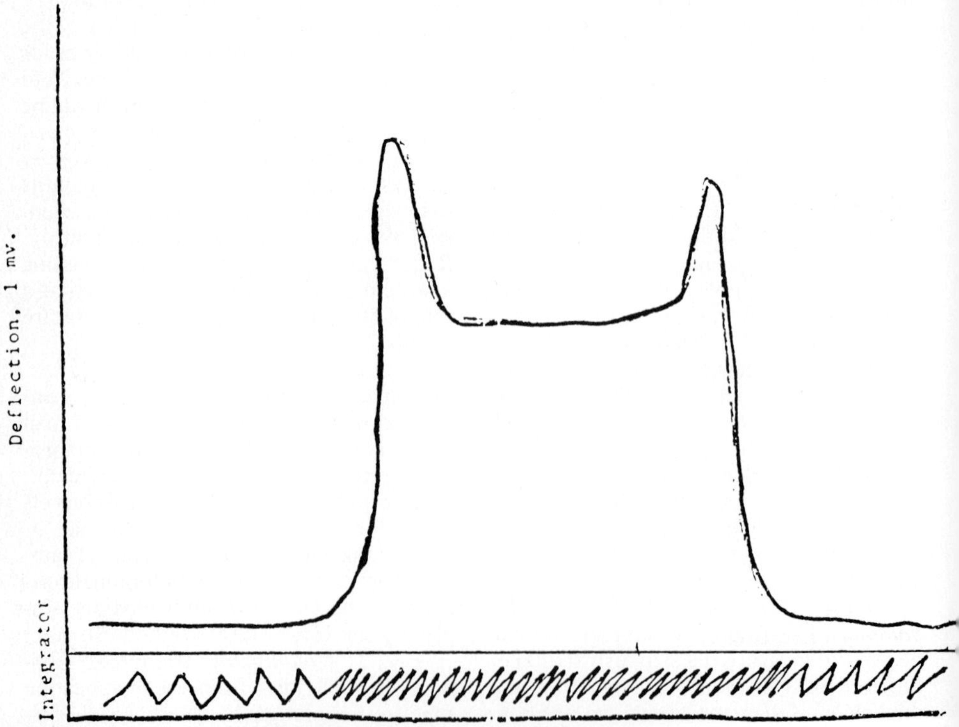

Figure 713:3—Typical Coulometric response curve.

losses of sulfur dioxide will occur. If water from the exhaust gases condenses in the exit port, momentary interruptions in the gas stream will result in spurious spikes on the recorder.

The photomultiplier tube responds rapidly to changes in ultraviolet light produced by sulfur compounds in the hydrogen flame. The electrometer measures the changes in current through the photomultiplier tube and produces a peak on the recorder. The area under this peak is a linear function of the amount of sulfuric acid in the sample.

Insert the sample as previously described and calculate the amount of sulfuric acid in each sample by multiplying the area by the appropriate factor obtained from the calibration curve.

8. Calibration and Standards

8.1 CALIBRATION OF FLOW MEASURING DEVICES. Calibrate the rotameter or the critical orifice, glass capillary or hypodermic needle, against a wet test meter or a calibrated rotameter.

8.2 CALIBRATION OF ANALYTICAL SYSTEMS. For coulometry and for flame photometry, accurately transfer with a microsyringe onto preconditioned copper discs graduated amounts (2, 4, 6, 8, 10 μl) of the standard sulfuric acid solution. For colorimetry, larger graduated amounts are required (5, 10, 15, 20, 25 μl). Remove the excess water in each sample at 100 C. Avoid prolonged heating. Determine the concentration of sulfuric acid as described in the analytical procedure. At each start-up, calibrate the analytical system.

8.3 CONTROL SAMPLES. Every tenth sample, use one of the working standards as a control sample to validate the accuracy of the analyses. Calibration with working standard and the control samples is needed for accuracy of the analysis.

9. Calculation

9.1 SAMPLING.

9.1.1 *Calculate the air samples as follows:*

Where:

V_a = Rt
V_a = Volume of the air sample, liters.
R = Flow rate in liters per minute.
t = Sampling time in minutes.

9.1.2 *Correct all volumes to standard conditions.*

$$V_c = \frac{V_a P_o T_s}{P_s (t_o + 273.1)}$$

Where:

V_c = Corrected volume.
V_a = Observed volume, computed above.
P_o = Observed pressure, Torr.
P_s = Standard conditions, 760 Torr.
t_o = Observed temperature, °C.
T_s = Absolute temperature 298.1° K Ambient. standard conditions.

9.1.2 *High Volume Samples.* Compute corrected volume for the entire filter, then compute the volume of air corresponding to the size of the sample taken as follows:

$$V_s = \frac{A_s V_c}{A_t}$$

V_c = Correct volume of entire sample, m^3.
V_s = Volume of air sample analyzed, corrected to standard conditions m^3.
A_s = Area of filter analyzed, cm^2.
A_t = Total area of filter, 17.78 × 22.86 = 406.45 cm^2.

9.2 ANALYTICAL CALCULATIONS.

9.2.1 *Spectrophotometric.*

$$\text{Sulfuric acid, } \mu g/m^3 = \frac{A\ B}{V_c L}$$

A = Net absorbance units.
B = Factor, reciprocal of the slope of the calibration curve, μg H_2SO_4 per absorbance unit.
L = Area of filter used divided by total area of filter.
V_c = Air volume corrected to standard conditions, m^3

For Example (Using the entire 25 mm filter);

A = 0.300 absorbance units.
B = 58.66 μg H_2SO_4 per absorbance unit.
V_c = 7.10 m^3
L = 1

$$\mu g\ H_2SO_4/m^3 = \frac{(0.300)\ (58.66)}{(7.10)\ (1)}$$

$$= 2.48\ \mu g/m^3$$

9.2.2 *Coulometric.*

$$\text{Sulfuric acid, } \mu g/m^3 = \frac{(A_1 - A_2) \times B_1}{V_c\, L}$$

A_1 = Area under curve in cm^2 or counts.
A_2 = Area of standing current, counts.
B_1 = Factor, reciprocal of the slope of the calibration curve, μg H_2SO_4 per unit area or μg H_2SO_4 per count.
L = Area of filter used divided by total area.
V_c = Air volume in m^3.

For Example:

A_1 = 1000 counts
A_2 = 0
B_1 = 0.0103 μg H_2SO_4/count
V_c = 2.00 m^3

$$\mu g\ H_2SO_4/m^3 = \frac{(1000)\ (0.0103)}{(2.00)}$$

$$= 5.15\ \mu g/m^3$$

L = 1

9.2.3 *Flame Photometric.*

$$\text{Sulfuric acid, } \mu g/m^3 = \frac{(A_1 - A_2) \times B_1}{V_c\, L}$$

A_2 = Area under curve in cm^2, counts from the integrator or microcoulombs.
A_1 = Background area, in counts or coulombs.
B_2 = Factor, reciprocal of the slope of the calibration curve, μg H_2SO_4 per cm^2, μg H_2SO_4 per count, or μg H_2SO_4 per microcoulomb.
L = Area of filter analyzed divided by total area of filter.
V_c = Air volume in m^3.

For example when using a portion of the high volume filter:

A_1 = 31.4
A_2 = 107 counts
B_1 = 0.02194 μg H_2SO_4/count
V_c = 2.00 m^3
L = 0.001

$$\mu g\ H_2SO_4/m^3 = \frac{[107 - (-31.4)](0.02194)}{(2.00)}$$

$$= 1.52\ \mu g/m^3$$

10. Effects of Storage

10.1 Samples should be analyzed within 24 hr after collection. The average rate of decay of filter sample for a period of several weeks is approximately 2% per day (**1**). Decay rates for shorter intervals of time are probably higher.

11. References

1. Scaringelli, F.P. and K.A. Rehme. 1969. Determination of Sulfuric Acid Aerosol by Spectrophotometry, Coulometry and Flame Photometry. *Anal. Chem.* 41:707.
2. Kiyoura, R., M. Hiroshi, I. Kunio, K. Harvo, and V. Yoshisuke. Review of Direct Measurement Method of Sulfuric Acid Mist in Atmosphere. Presented at Japan Chemical Society, 24th Annual Meeting, Tokyo, March, 1971, paper 3406.
3. Omichi, S. and V. Hiromi. 1971. The Determination of Sulfuric Acid Aerosol in Suspended Particulates. *Air Pollution Research* 6:62.
4. Federal Register, National Primary and Secondary Ambient Air Quality Standards, 36 No. 84, 8191 (1971).
5. American Public Health Association, Inc. 1977. Methods of Air Analysis and Sampling. 2nd ed. p. 696. Washington, D.C.
6. Scaringelli, F.P., E. Rosenberg, and K.A. Rehme. Separation of Sulfate Aerosols by Gas Effluent Analysis. Unpublished Data, National Environmental Research Center, Research Triangle Park, North Carolina 27711.
7. Dubois, L., A. Zdrojecski, T. Teichman, and J.L. Monkman. 1972. The Effect of Sampling and

Conditions on Sulfuric Acid Concentration Findings in Urban Air., *Int. J. Environ. Anal. Chem.* 1:259 (French).
8. Barton, S.C. and H.G. McAdie. 1970. Preparation of Glass Fiber Filters for Sulfuric Acid Aerosol Collection. *Environ. Sci. Technol.* 4:769.
9. May, K.R. The Cascade Impactor: An instrument for Sampling Coarse Aerosols. *J. Sci. Instr.* 22:187.
10. Ranz, W.E. and J.B. Wong. 1952. Jet Impactors for Determining particle-size Distributions of Aerosols. *Arch. Ind. Hyg. Occupational Med.* 5:464.
11. Anderson, A.A. 1958. New Sampler for the Collection, Sizing, and Enumeration for Airborne Particles. *J. Bacteriol.* 76:471.

Subcommittee 1
D. F. Adams, *Chairman*
J. O. Frohliger
D. Falgout
A. M. Hartley
J. B. Pate
A. L. Plumley
F. P. Scaringelli
P. Urone

PART III
METHODS FOR CHEMICALS IN AIR OF THE WORKSHOP AND IN BIOLOGICAL SAMPLES

801.

Analytical Method for Ammonia in Air

Analyte:	NH_3	Method No.:	P & CAM 205
Matrix:	Air	Range:	20-135 ppm
Procedure:	Midget Impinger/ Colorimetric	Precision:	Unknown
Date Issued:	1/1/75		
Date Revised:		Classification:	E (Proposed)

1. Principle of Method

1.1 Ammonia is collected in a dilute sulfuric acid solution in a midget impinger to form ammonium sulfate.

1.2 The solution is reated with Nessler reagent to produce a yellow-brown complex.

1.3 The ammonia concentration is determined by reading the absorbance of the yellow-brown solution at 440 nm and comparing it with a standard curve. Absorption peak may shift with concentration.

2. Range and Sensitivity

2.1 The Nessler reagent is said to be sensitive to as little as 0.002 mg of ammonia.

2.2 The range of application of this method is 0.10 mg to 0.80 mg of ammonia in a 10-l air sample corresponding to about one-third to twice the present limit of 35 mg per m^3 of air.

3. Interferences

3.1 Ammonium salts will react with Nessler reagent to give a false high reading. These can be removed by filtration of the air before its passage into the absorption tube.

3.2 Other interferences have not been reported.

4. Precision and Accuracy

4.1 The precision and accuracy of this method have not been reported.

5. Advantages and Disadvantages

5.1 The method is sensitive, but ammonium salts can interfere.

5.2 The method does not distinguish between free and combined ammonia.

6. Apparatus

6.1 BATTERY OPERATED PERSONAL AIR SAMPLING PUMP. Capable of drawing 1 l of air per min through 10 ml of absorbing solution.

6.2 IMPINGERS, STANDARD MIDGET. All glass, calibrated.

6.3 BECKMAN MODEL B SPECTROPHOTOMETER, OR EQUIVALENT.

6.4 SPECTROPHOTOMETER CELLS, 1-CM.

6.5 VOLUMETRIC FLASKS. 50-ml, 100-ml, 200-ml, 1000-ml, glass stoppered.

6.6 PIPET. 0.5-ml, 1.0-ml, 5-ml, 10-ml, 20-ml.

6.7 GRADUATED CYLINDERS. 25-ml, 50-ml, 100-ml.

7. Reagents

7.1 PURITY. All reagents should be analytical reagent grade.

7.2 Double distilled water free of ammonia.

7.3 Potassium iodide.

7.4 Mercuric chloride.

7.5 Potassium hydroxide.

7.6 Ammonium sulfate.

7.7 Nessler Reagent. Dissolve 35 g of mercuric chloride in 500 ml of hot water. Filter and allow to cool. Dissolve 62.5 g of potassium iodide in 260 ml of cold water. Gradually add the mercuric chloride solution to 250 ml of the iodide solution until a slight permanent red precipitate is formed. Dissolve the precipitate with the remaining iodide solution and again add mercuric chloride slowly until a red precipitate remains.

Dissolve 150 g of potassium hydroxide in 250 ml of distilled water. Add this solution to the potassium iodide-mercuric chloride solution and make up to 1 liter with distilled water. Stir thoroughly and allow to stand a day or so, and decant the clear liquid.

7.8 Warning. The Nessler reagent should be handled with caution because of its toxicity and corrosive properties.

7.9 Standard Ammonium Sulfate Solution. Dissolve 77.6 mg of ammonium sulfate in distilled water and make up to 1 l. 1 ml of this solution contains 20 μg of ammonia. Discard solution after a week.

7.10 Absorbing Solution. Dilute 2.8 ml of concd sulfuric acid (18M) to 1 l with distilled water to form 0.1 N sulfuric acid.

8. Procedure

8.1 Sampling. Add 10 ml of absorbing solution to each midget impinger including one for the blank. Attach the bubbler to a personal air sampling pump and draw air through the bubbler at a rate of 1 lpm for 10 to 15 min. Record the volume of air sampled.

8.2 Color Development. Dilute the sample to 50 ml with distilled water in a volumetric flask and shake well. Take 1 ml of this solution and make up to 50 ml with distilled water in another volumetric flask. Shake. Add 2 ml of Nessler reagent to the latter flask and determine absorbance after 10 min at 440 nm in a spectrophotometer using a 1-cm cell. Treat the blank in the same manner as the sample.

9. Calibration and Standardization

9.1 Dilute samples of 5 ml, 10 ml, 20 ml, 30 ml, and 40 ml of the standard ammonium sulfate solution and process in the same manner as described in section **8.2**. Determine light absorbance in a spectrophotometer at 440 nm. Prepare a standard curve of absorbance versus μg of ammonia.

10. Calculations

10.1 μg NH_3 per m^3 = (W/V) × 1000, where W = μg of ammonia from the standard curve minus blank value and V = Volume of air in liters, corrected to 25 C and 760 Torr

11. References

1. Goldman, F.H. and M.B. Jacobs. 1953. Chemical Methods of Industrial Hygiene. Interscience Publishers Inc., 250 Fifth Ave., New York, N.Y.
2. Elkings, H.B. 1959. The Chemistry of Industrial Toxicology, 2nd Ed. John Wiley and Sons, New York, N.Y.

Subcommittee 3
B. Dimitriades, *Chairman*
J. Cuddeback
E. L. Kothny
P. W. McDaniel
L. A. Ripperton
A. Sabadell

802.

Analytical Method for Antimony in Air and Urine

Analyte:	Antimony	Method No.:	P & CAM 107
Matrix:	Air, Urine	Range:	Air-10-50 μg/m³ Urine-1-10 μg/sample
Procedure:	Filter Collection of Air Sample/Colorimetric-Rhodamine B	Precision:	CV < 5% (analytical)
Date Issued:	4/13/73	Classification:	A (Recommended)
Date Revised:	6/4/75		

1. Principle of the Method

1.1 Pentavalent antimony in the presence of a large excess of chloride ion **(1)** reacts with Rhodamine B to form a colored complex, which may be extracted with organic solvents such as toluene xylene, or isopropyl ether **(2, 3, 4, 5, 6, 7)**. Trivalent antimony will not react with Rhodamine B and must be oxidized to the pentavalent state, (*i.e.*, Sb_2O_5) by the addition of perchloric acid after digestion in sulfuric and nitric acids. Phosphoric acid is used to minimize interference by iron. The absorbance of the pink-colored extract is measured in a spectrophotometer at a wavelength of 565 nm and the amount of antimony present determined by reference to a calibration curve prepared from known amounts of antimony.

2. Range and Sensitivity

2.1 The method covers a range of 1 to 10 μg of antimony in 10 ml of solvent when using a 1.0-cm light path cell. The range may be extended by dilution of more concentrated solutions with solvent.

2.2 One μg antimony is detectable. This corresponds to 0.1 μg of antimony per ml of solvent.

2.3 Air. In a 20-l sample of air, 50 μg of antimony per m³ can be determined.

2.4 Urine. If a 50-ml sample of urine is taken for analysis, the 1 μg detectable amount of antimony corresponds to 20 μg of antimony per liter of urine.

3. Interferences

3.1 Air. Of the commonly encountered ions, only iron is likely to interfere. Maren **(6)** suggests iron interference can be eliminated by extracting with isopropyl ether, however, this solvent may not provide complete separation due to emulsion formation. Results obtained by the ACGIH show interference from iron to be insignificant for 750 μg of iron/ml **(1)**. The presence of iron could be significant for air samples taken in areas of high airborne concentrations. The analysis will fail if any nitric acid remains in the digest. Other substances can interfere but are not expected to interfere at normally occurring concentrations **(8)**.

3.2 Urine. The normal iron content in urine of 7 μg/Kg body weight per day is well below the 750 μg of iron/ml level at which interference was found to be insignificant. As in the case of air samples, the analysis will fail if any nitric acid remains in the digest.

4. Precision and Accuracy

4.1 Three samples consisting of aqueous suspensions of antimony were analyzed in triplicate by 10 collaborating laboratories in addition to the referee, and the mean coefficient of variation calculated (**1**).

4.2 Seven tests were made to determine the recovery of antimony added in solution by pipet to 9-cm Whatman #1 filters which were then oven dried (**9**). The results showed that recovery of the antimony ranged from 95 to 103% of the 2 to 10 μg added, which is within the expected relative standard deviation of the method.

5. Advantages and Disadvantages

5.1 This method requires no special equipment except a good spectrophotometer.

5.2 Unless all antimony is present in the pentavalent state, recoveries cannot be complete. No less than 10 drops of perchloric acid are required to oxidize the antimony to Sb (V).

5.3 Operations must be performed at low temperatures throughout the analytical determinations. Samples, reagents and apparatus should all be kept cold in a refrigerator prior to use.

6. Apparatus

6.1 AIR.

6.1.1 *Spectrophotometer*. Capable of measuring absorbance at 565 nm.

6.1.2 *Erlenmeyer flasks*. 125-ml with ground glass stoppers.

6.1.3 *Squibb separatory funnels*. 125-ml with Teflon stopcocks.

6.1.4 *Boiling aids*. Perforated pyrex beads to promote smooth boiling after ascertaining that no measurable contribution to the blank results from their use.

6.1.5 *Cellulose acetate membrane filters*. Appropriately sized for sample volume and rate, with a pore size of 0.8 μ are satisfactory for air sampling.

6.1.6 *Battery operated sampling pump*. Sufficient capacity to operate continuously for an 8 hr workday in the range of 1 to 3 lpm (*e.g.*, Monitaire Sampler-Mine Safety Appliances Company, Pittsburgh, Pennsylvania).

6.2 URINE. In addition to the above the following apparatus is required.

6.2.1 *Bottles, urine specimen. 8-oz.* Polyethylene wide mouth type with Teflon lined caps for sample collection.

6.2.2 *Specific gravity meter, or urinometer*. Having a range of 1.000 to 1.060 units with a precision of ± 0.002 units.

7. Reagents

7.1 PURITY. Reagent grade chemicals should be used in all tests. Unless otherwise indicated, all reagents shall conform to the specifications of the Committee on Analytical Reagents of the American Chemical Society where such specifications are available (**10**).

7.2 DISTILLED WATER. Where the term distilled water is used, this should be taken to mean double distilled.

7.3 HYDROCHLORIC ACID—(36 to 38%), sp. gr. 1.19.

7.4 SULFURIC ACID—(95.0 to 98.0%), sp. gr. 1.84.

7.5 NITRIC ACID—(69.0 to 71.0%), sp. gr. 1.42.

7.6 PERCHLORIC ACID—(60 to 62%), sp. gr. 1.19.

7.7 PHOSPHORIC ACID—(85%), sp. gr. 1.70.

7.8 TOLUENE, XYLENE OR ISOPROPYL ETHER.

7.9 HYDROCHLORIC ACID, 6N. Prepare from conc hydrochloric acid by mixing with an equal volume of distilled water.

7.10 ORTHOPHOSPHORIC ACID, 3N. Dilute 70 ml conc phosphoric acid to 1 liter with distilled water.

7.11 RHODAMINE B, 0.02%. Prepare solution 0.02% w/v in distilled water by dissolving 20 mg of Rhodamine B in water and diluting to 100 ml.

7.12 URINE ANALYSIS. In addition to the above, the following reagents are needed.

7.12.1 *Capryl alcohol.*

7.12.2 *Thymol.*

8. Procedure

8.1 CLEANING OF EQUIPMENT. All glassware should be cleaned thoroughly.

8.1.1 Wash with a solution of detergent and tap water, followed by tap water and distilled water rinses.

8.1.2 Soak in chromic acid for 30 min and follow with tap and distilled water rinses.

8.1.3 Soak in 1:1 or conc nitric acid for 30 min and then follow with tap, distilled and double distilled water rinses. Any glassware that was cleaned initially by this entire procedure, and is re-used, need only be cleaned as in this Section. Polyethylene equipment can be cleaned by subjecting it to treatment as in Sections **8.1.1** and **8.1.3** only.

8.2 COLLECTION OF SAMPLES: AIR

8.2.1 Membrane filters of 0.8 μ pore size are suitable for collection of air samples. The sampling flow rate, times, and/or volume must be measured as accurately as possible. The samples should be taken at a flow rate of 21/m for a minimum of 1 hr. Larger sample volumes are recommended, provided the filters do not become overloaded. Blank filters should be checked for antimony content before use. Samples collected on filters may be stored indefinitely.

8.3 COLLECTION OF SAMPLES: URINE.

8.3.1 Urine samples should be collected in clean polyethylene bottles. A minimum of 100 ml should be collected and the samples must be preserved by the addition of 5 mg of thymol per 100 ml of urine. The urine should be kept cool during shipment and storage.

8.3.2 Specific gravities of urine samples are determined after the samples have been allowed to warm to room temperature. This may be done using a specific gravity meter or a urinometer. The specific gravity determination and analysis of samples should be performed within one week.

8.4 ANALYSIS OF SAMPLES.

8.4.1 *Antimony*, collected from the atmosphere must be oxidized to the pentavalent state before analysis with Rhodamine B is possible. The filter is placed in a 125-ml Erlenmeyer flask, 5 ml conc sulfuric acid is added, and followed by 5 ml conc nitric acid. After the addition of a few glass beads, the flask is heated on a hot plate or over a gas flame until brownish-red fumes of NO_2 are driven off, the solution blackens and dense white fumes of sulfur trioxide are driven off. The solution is allowed to cool slightly (2 to 5 min.) and 10 drops of nitric acid are added. If the solution is not clear and colorless, the HNO_3 acid addition must be repeated until a clear and colorless solution is obtained. The solution is then heated until white fumes of SO_3 are driven off. To insure complete removal of NO_2 fumes, the samples and standards should be purged with a stream of clean filtered compressed air for at least 1 min. The solution is allowed to cool, 10 drops of perchloric acid are added and the solution reheated to the appearance of white fumes.

8.4.2 The resulting solution is then cooled and placed in an ice bath. It is important to cool separatory funnels, hydrochloric acid, phosphoric acid, toluene, and Rhodamine B solution in a refrigerator before use. The analyst should try to maintain a temperature of 5 to 10 C during the analysis. Since the presence of phosphoric acid will lower the results if the samples are allowed to stand, the extraction should be completed within 5 min after the phosphoric acid is added.

8.4.3 After temperature equilibrium is established (at least 30 min) 5 ml of precooled 6 N hydrochloric acid is added slowly (taking 1 to 2 min.) by pipet to minimize a temperature increase in the solution. The solution is cooled in the ice bath for at least 15 min, after which 8 ml of 3 N phosphoric acid is added (precooled to the ice bath temperature) followed by the addition of 5 ml of precooled 0.02% Rhodamine B solution. Without delay, the flask is stoppered, shaken vigorously, and the contents transferred to a separatory funnel. Prior to use the separatory funnels should be cooled to 5 C in a refrigerator.

8.4.4 Ten ml of precooled toluene is added to the separatory funnel, the contents shaken vigorously for 1 min and after allowing the contents to separate, the aqueous layer is discarded and the toluene layer allowed to warm to room temperature. The toluene phase (colored pinkish-red if antimony is present) is collected in a centrifuge tube and allowed either to stand for a few minutes to permit water to settle out, or it may be centrifuged. A 1.0-cm light path cuvette is filled with several milliliters of the toluene extract. The absorbance of the sample is then determined at 565 nm against a blank which has been taken through the entire procedure.

8.3.5 If the color developed is too intense to measure, the extract may be diluted with toluene, and the antimony present found by multiplying by the appropriate dilution factor.

8.3.6 Antimony collected in the urine samples must be oxidized to the pentavalent state before analysis with Rhodamine B is possible. A 50-ml aliquot of urine is placed in a 125-ml Erlenmeyer flask and 5 ml conc sulfuric acid is added, followed by 5 ml conc nitric acid. A few glass beads and 1 drop of capryl alcohol are added. The flask is heated on a hot plate or over a gas flame until brownish-red fumes of NO_2 are driven off, the solution blackens (for organic materials) and dense white fumes of sulfur trioxide are driven off. To insure complete removal of NO_2 fumes, the samples and standards should be purged with a stream of clean filtered compressed air for at least 1 min. If the solution is not clear and colorless, the HNO_3 acid addition must be repeated until a clear and colorless solution is obtained. The solution is again heated until white fumes of SO_3 are driven off. After the resulting solution is allowed to cool, 10 drops of perchloric acid are added and the solution reheated to the appearance of white fumes.

8.3.7 Proceed as in Sections **8.3.2**, **8.3.3** and **8.3.4**.

9. Calibration and Standards

9.1 ANTIMONY STANDARD SOLUTION. Dissolve 1.329 g of reagent grade Sb_2O_5 in 25 ml of conc hydrochloric acid and dilute to 1 l with distilled water. This solution contains 1000 μg of antimony/ml.

9.2 DILUTE ANTIMONY STANDARD. Dilute 10 ml of stock antimony standard solution to 100 ml with distilled water. This solution contains 100 μg Sb/ml. Prepare fresh weekly.

9.3 WORKING ANTIMONY STANDARD. Dilute 10 ml of dilute antimony standard solution to 1 l with distilled water. This solution contains 1 μg of Sb/ml. Prepare fresh daily.

9.4 Standards of 1 to 10 μg of Sb are analyzed by carrying aliquots of the appropriate reference solutions through the digestion and analysis procedure. Since NO_2 fumes may not be driven off visibly when no organic material is present, the solutions should be heated until white fumes of SO_3 are driven off and the solution has stopped bubbling. Standards should be purged with a stream of clean filtered compressed air for at least 1 min for complete removal of NO_2. Obtain the calibration curve by calculating a least squares best fit line. The blank is subtracted from the value for each standard prior to the calculation.

10. Calculations

10.1 AIR. If an air sample of V cubic meters (at standard conditions) is found to contain X μg of antimony, then the concentration of antimony in the air is calculated as:

$$\mu g\ Sb/m^3 = \frac{X}{V};$$

(if an aliquot A is taken from the total sample T, then the result must be multiplied by $\frac{T}{A}$.)

10.2 URINE. If the specific gravity has been determined for a urine sample of V milliliters, found to contain X μg of antimony, then the concentration of antimony in the urine is:

$$\mu g\ Sb/ml\ urine = \frac{X}{V}\ \frac{(1.024 - 1.000)}{(\text{Specific gravity} - 1.000)}$$

The specific gravity adjustment factor adjusts all values to a mean specific gravity of 1.024 (**11**).

11. References

1. American Conference of Governmental Industrial Hygienists. 1963. Manual of Recommended Analytical Methods. Determination of Antimony in Air and Biological Materials.
2. Edwards, F.C., and A.F. Voigt. 1949. Separation of Antimonic Chloride from Antimonious by Extraction into Isopropyl Ether. Anal. Chem. **21**:1204–1205.
3. Fredman, L.D. 1947. Rhodamine Method for Microdetermination of Antimony. Anal. Chem., **19**:502.
4. Luke, C.L. 1953. Photometric Determination of Antimony in Lead Using the Rhodamine B Method. Anal. Chem., **25**:674–675.
5. Maren, T.H. 1947. Colorimetric Microdetermination of Antimony with Rhodamine B. Anal. Chem., **19**:487–491.
6. Ramette, R.W., and E.B. Sandell. 1953. Rationalization of the Rhodamine B Method for Antimoney. Anal. Chem., ACta. **13**:453–458.
7. Ward, F.N., and H.W. Lakin. 1954. Determination of Traces of Antimony in Soils and Rocks, Anal Chem., **26**:1168–1173.
8. Sandell, E.B. 1965. Colorimetric Determination of Traces of Metals, Third Edition, Interscience Publishers, Inc., New York.
9. American Public Health Association, Inc. 1977. Methods of Air Sampling and Analysis. 2nd ed. p. 431. Washington, D.C.
10. American Chemical Society Specifications. 1961. Reagent Chemicals, American Chemical Society, Washington, D.C.
11. Elkins, H.B., L.D. Pagnotto, and H.L. Smith. 1974. Concentration Adjustments in Urinalysis, Amer. Ind. Hyg. Assoc. J. **35**:559–565.

Subcommittee 6
T. J. Kneip, *Chairman*
R. S. Ajemian
J. N. Driscoll
F. I. Grunder
L. Kornreich
J. W. Loveland
J. L. Moyers
R. J. Thompson

803.

Analytical Method for Arsenic in Urine and Air

Analyte:	Arsenic	Method No.:	P & CAM 140
Matrix:	Urine and Air	Range:	Air—0.4 to 800 $\mu g/m^3$ Urine—0.01 to > 1 mg/liter
Procedure:	Filter Collection for Air Samples/Arsine Generation-Colorimetric	Precision:	4 to 9% Average deviation
Date Issued:	3/24/72	Classification:	A (Recommended)
Date Revised:	6/4/75		

1. Principle of the Method (1, 2)

1.1 Air samples collected on filters or urine samples are wet ashed with a mixture of nitric and sulfuric acids to destroy the organic matrix. The arsenic in the sample is converted to the trivalent form with potassium iodide and stannous chloride. The trivalent arsenic is further reduced to arsine, AsH_3, by zinc in acidic solution in an arsine generator. The evolved arsine is then passed through an H_2S scrubber which consists of glass wool impregnated with lead acetate and then into an absorber containing silver diethyldithiocarbamate dissolved in pyridine. In this solution, the arsine reacts with silver diethyldithiocarbamate forming a soluble red complex which is suitable for photometric measurement with a maximum absorbance at a wavelength of 560 nm (**3**).

2. Range and Sensitivity (4)

2.1 General. Quantities of up to 200 μg may be determined by this method.

Higher concentrations may be determined by dilution with pyridine.

2.2 AIR. The method is capable of determining conc as low as 0.4 μg of arsenic per m³ (μg/m³) of air for a 750-l air sample.

2.3 URINE. Arsenic conc as low as 0.01 mg/l of urine can be determined. Concentrations greater than 1 mg/l may be determined by taking smaller aliquots for analysis.

3. Interferences

3.1 Antimony present in the sample will form stibine, SbH_3, which may interfere slightly by forming a colored silver diethyldithiocarbamate complex which has a maximum absorbance at 510 nm. A complete scan of the spectral region from 400 to 600 nm should be made and recorded to determine the presence of interferences. The amount of antimony found in air or urine is usually small enough that it will not interfere. High conc of nickel, copper, chromium, and cobalt interfere in the generation of arsine (**5**). In addition, certain combinations of elements may produce a similar interference (**5**). The presence of interferences can be determined by adding known amounts of arsenic to the samples and determining recoveries. Hydrogen sulfide would interfere by producing a dark color, but it is removed by the lead acetate scrubber.

3.2 The filters used to collect atmospheric particulates may contain measurable quantities of arsenic. Unused filters from the same batch should be analyzed as part of the reagent blank procedure and corrections made if necessary.

4. Precision and Accuracy

4.1 PRECISION.

4.1.1 Spiked air samples containing 0.1, 1, 5, and 10 μg arsenic were determined with a precision of ± 0.04 μg based upon replicate determinations at each concentration (**4**).

4.2 ACCURACY.

4.2.1 Four air samples containing known amounts of arsenic were analyzed by 8 laboratories with the following results (**4**).

TABLE 803:1. Analysis of Samples Containing Known Amount of Arsenic

Sample	μg As/ml	Relative Average Deviation, %
1	0.05	9.1
2	0.50	3.5
3*	1.00	6.1
4	1.50	4.0

*Contained 0.5 μg Sb/ml to test for interference.

5. Advantages and Disadvantages

5.1 This is a reliable and well established method for arsenic analysis. Very little specialized equipment is required, except for the arsine generator.

5.2 Some minor disadvantages are that the method is tedious and the use of pyridine as a solvent involves a somewhat objectionable odor.

5.3 The air sampling method will not collect arsine gas if present.

6. Apparatus

6.1 AIR.

6.1.1 *Arsine Generator, Scrubber, and Absorber.* Fisher arsine generator, cat. no. 1-405 or equivalent. See Figure 803:1.

6.1.2 *Kjeldahl flasks*, 300-ml.

6.1.3 *Cellulose acetate membrane filters.* Appropriately sized for sample volume and rate, with a pore size of 0.8 μ are satisfactory for air sampling.

6.1.4 *A spectrophotometer.* Capable of measuring absorbance over the 400 to 600 nm spectral region.

6.1.5 *Battery operated sampling pump.* Of sufficient capacity to operate continuously for an 8 hr workday in the range of 1 to 3 l/m (*e.g.* Monitaire Sampler—Mine Safety Appliances Company, Pittsburgh, Pennsylvania, or equivalent type). The pump should be accurate to within ± 5%.

6.2 URINE. In addition to the laboratory apparatus required above, the following equipment will be required in the urine procedure.

6.2.1 *Specific gravity meter, or hydrometer.* Certified against NBS or other standard source having a range of 1.000 to

1.060 units with a precision of ± 0.002 units.

6.2.2 *Bottles*. Urine Specimen 250 ml. Polyethylene wide mouth type with cap is satisfactory.

7. Reagents

7.1 PURITY. Reagent grade chemicals should be used in all tests. Unless otherwise indicated, it is intended that all reagents shall conform to the specifications of the Committee on Analytical Reagents of the American Chemical Society where such specifications are available **(6)**.

7.2 DISTILLED WATER. References to water shall be understood to mean double distilled water.

7.3 SILVER DIETHYLDITHIOCARBAMATE (Ag $SCSN(C_2H_5)_2$ REAGENT. Dissolve 4.0 g of silver diethyldithiocarbamate in 800 ml of pyridine. Some pyridine may contain colored materials which can be removed by passing the pyridine through a 2.5-cm diam by 15-cm depth alumina column at 150 to 200 ml/hr. Other pyridine is sufficiently pure so that repurification is not necessary.* The useful life of this reagent can be extended to at least 2 months by storing in a dark brown bottle and refrigerating when not in use.

7.4 STANNOUS CHLORIDE REAGENT. Dissolve 10.0 g of $SnCl_2 \cdot 2H_2O$ in 25 ml of 12 N (sp gr 1.19) hydrochloric acid. Store in glass after adding 3 or 4 pieces of tin metal to minimize oxidation. This solution is stable for at least 2 weeks.

7.5 LEAD ACETATE SOLUTION. Dissolve 10 g of $Pb(C_2H_3O_2)_2 \cdot 3H_2O$ crystals in 100 ml of water. The solution may be slightly turbid as a small amount of the basic salt is formed, but this will not affect its usefulness.

7.6 POTASSIUM IODIDE SOLUTION. Dissolve 15 g of KI in 100 ml of water. The solution should be stored in a brown glass bottle.

7.7 SODIUM HYDROXIDE PELLETS.

7.8 SODIUM HYDROXIDE 10 N. Dissolve 400 g of NaOH pellets in 500 ml of freshly boiled and cooled distilled water and dilute to 1 liter. This solution should be stored in tightly stoppered, precleaned polyethylene bottles.

7.9 HYDROCHLORIC ACID. sp. gr. 1.19. (36.5 to 38.0%)

7.10 NITRIC ACID. sp. gr. 1.42 (69.0 to 71.0%)

7.11 SULFURIC ACID. sp. gr. 1.84 (95.0 to 98.0%)

7.12 HYDROCHLORIC ACID 6 N. Add 500 ml of conc hydrochloric acid, (sp. gr. 1.19) to 400 ml of water and dilute to one liter.

7.13 AMMONIUM OXALATE. Prepare a saturated solution in distilled water (about 6 grams per 100 ml).

7.14 ZINC. Granular, 20 mesh, low arsenic content.

7.15 DISODIUM ETHYLENEDIAMINETETRA-ACETATE (EDTA). (Used for preserving urine specimens.)

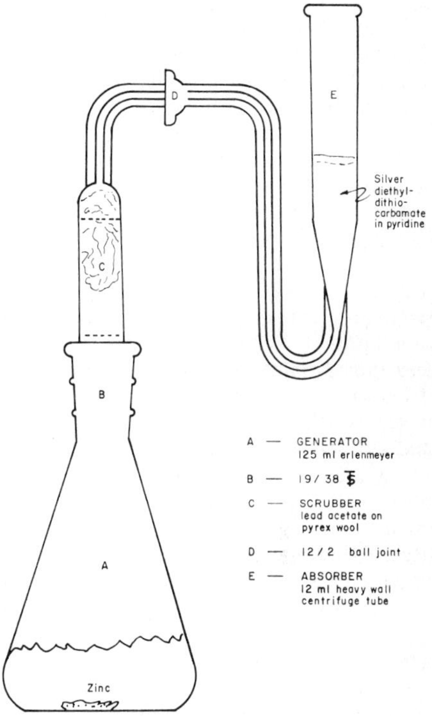

Figure 803:1—Arsine generator

8. Procedure

8.1 CLEANING OF EQUIPMENT. All glassware must be cleaned with a deter-

*__Caution__—Pyridine vapors are toxic. The analysis for arsenic should be carried out in the presence of effective local exhaust ventilation.

gent solution, followed by tap water and distilled water rinses. Then the glassware is treated with hot conc nitric acid and thoroughly rinsed with tap water followed by distilled water.

8.1.1 Arsine generators are rinsed with conc HCl after the nitric acid wash.

8.2 COLLECTION AND SHIPPING OF SAMPLES.

8.2.1 Urine Samples should be collected in polyethylene bottles that have been precleaned with 1:1 nitric acid. A minimum of 100 ml should be collected and the samples must be preserved with 0.1 g of EDTA per 100 ml of urine. The urine should be kept cool during shipment and storage.

8.2.2 Specific gravities of urine samples are determined after the samples have been allowed to warm to room temperature. This may be done using a specific gravity meter or certified hydrometer. Specific gravity measurements and analysis of samples should be done within one week after sample collection.

8.2.3 AIR SAMPLES should be collected on a membrane filter, pore size 0.8 μ, as follows;

a. Assemble the filter unit (37-mm Millipore filter, or equivalent) by mounting the filter disc in the filter cassette.

6. Connect the exit end of the filter unit to the pump with a short piece of flexible tubing.

c. Sample the air at a measured flow rate of 1 to 2/pm. Record the time and/or volume accurately and also the temperature and atmospheric pressure. A minimum sample of 60 liters should be collected. Avoid plugging the filter with an excessive load of dust.

d. The sample cassettes should be shipped to the laboratory in a suitable container. Avoid disturbance or loss of sample during shipment and storage. Samples may be stored indefinitely at room temperature without loss of arsenic.

e. With each batch of samples, one filter labeled as a blank, must be submitted. No air should be drawn through this blank filter.

8.3 ANALYSIS OF SAMPLES

8.3.1 *Urine Samples*. Place a 50 ml aliquot of the urine sample into a 300-ml Kjeldahl flask and add 25 ml of nitric acid and 5 ml of sulfuric acid. The flask should be kept at a 45° angle to help prevent losses of volatile arsenic compounds. Heat until dense fumes of sulfur trioxide are produced.† When the digestion is complete, cool, add 5 ml of a saturated solution of ammonium oxalate and again heat until fumes are given off. Cool and transfer to the flask of the arsine generator with 25 ml of distilled water.

8.3.2 Add 10 ml of 6 N hydrochloric acid, 5 ml of 15% potassium iodide solution and allow to stand for 5 min. Add dropwise 0.5 ml of the stannous chloride solution, mix, and allow to stand for 20 min for complete conversion of arsenic to the trivalent sate.‡ Cool the solution to 4 C in an ice bath.

8.3.3 Impregnate the glass wool in the scrubber with lead acetate solution. About 1.5 ml of lead acetate solution is sufficient to saturate a 4-cm length of glass wool packing. Drain off any excess of solution. Pipet 3.0 ml of the silver diethyldithiocarbamate solution into the absorber tube. Add 3 g of zinc to the flasks containing the previously cooled solutions and immediately connect the scrubber-absorber assembly, making certain that all connections are tight. Allow 90 min for complete evolution of the arsine. Pour the solution directly from the absorber into the spectrophotometer cell and determine the absorbance of the solution using a reagent blank as the reference.

8.3.4 *Air Samples*. The membrane filter samples are placed in a Kjeldahl flask and ashed using 10 ml of nitric and 5 ml of sulfuric acids. Heating is continued, with additional increments of nitric acid if necessary to completely digest the filters (see note Section **8.3.1**). When the digestion is completed, the samples are cooled and 5 ml of a saturated solution of ammonium oxalate is added. The samples are heated again to dense white fumes. Cool and

†**Caution** must be exercised to avoid signs of darkening. If the sample solution starts to darken, a few more drops of nitric acid must be added immediately. Only sufficient acid should be added to overcome the darkening.

‡A colorless solution should be obtained at this point.

transfer the samples to the arsine generator and proceed as indicated in the urine procedure (Section **8.3.2** and Section **8.3.3**).

9. Calibration and Standards

9.1 STOCK ARSENIC STANDARD SOLUTION. Dissolve 1.320 g of As_2O_3 in 50 ml of 10 N NaOH and dilute with distilled water to 1 l. This solution contains 1000 μg As/ml. (Toxic solution—care should be taken to avoid skin contact and ingestion.)

9.2 DILUTE ARSENIC STANDARD. Dilute 10 ml of stock arsenic standard solution to 100 ml with distilled water. This solution contains 100 μg As/ml. Prepare fresh weekly.

9.3 WORKING ARSENIC STANDARD. Dilute 10 ml of dilute arsenic standard solution to 1 liter with distilled water. This solution contains 1 μg As/ml. Prepare fresh daily.

9.4 Prepare a set of air or urine standard solutions depending on the type of sample being analyzed by pipetting 1, 3, 5, 7, 10, and 15 ml of the working standard arsenic solution (1 μg/ml) into a series of Kjeldahl flasks. Each standard used for preparing the curve for air should contain a filter from the same batch as used for the samples. The urine standard should be prepared using an arsenic free urine source. Include a reagent blank.

9.5 Follow the procedure as outlined under Sections **8.3.1**, **8.3.2** and **8.3.3** for the urine standard and Section **8.3.4** for air standard.

9.6 Prepare a calibration curve by calculating a least squares best fit line. The blank is subtracted from each standard prior to the calculation. (After a standard curve has been established, a few selected standards should be carried through the entire procedure with each set of samples. A serious decrease in standard recoveries may indicate degradation of the silver diethyldithiocarbamate reagent.)

10. Calculations

10.1 If an air sample of V cubic meters (at standard conditions) is found to contain X μg of arsenic, the concentration of arsenic in air is calculated as:

$$\text{mg As/m}^3 \text{ air} = \frac{X}{V\ 1000}$$

(if an aliquot, A is taken from the total sample T, then the results must be multiplied by $\frac{T}{A}$).

10.2 If a urine sample of U ml, having a specific gravity, Sp. G., is found to contain X μg of arsenic, then the conc arsenic in the urine is

$$\mu\text{g As/ml urine} = \frac{X}{U}\ \frac{(1.024 - 1.000)}{(\text{Sp. G.} - 1.000)}$$

The specific gravity adjustment factor adjusts all values to a mean specific gravity of 1.024 (**7**).

11. References

1. VASKA, V.; and V. SEDIVEC. 1952. Colorimetric Determination of Arsenic. *Chem. Listy*. 46:341–344.
2. JACOBS, MORRIS B. 1967. The Analytical Toxicology of Industrial Inorganic Poisons, Interscience Publishers, New York.
3. AMERICAN PUBLIC HEALTH ASSOCIATION. Standard Methods for the Examination of Water and Wastewater. 14th Edition, 1976.
4. AMERICAN CONFERENCE OF GOVERNMENT INDUSTRIAL HYGIENISTS MANUAL. Determination of Arsenic in Air 1955.
5. SANDELL, E.B. 1959. Colorimetric Determination of Traces of Metals. 3rd ed. pp. 278–280. Interscience, New York, New York.
6. AMERICAN CHEMICAL SOCIETY SPECIFICATIONS. 1961. Reagent Chemicals. American Chemical Society, Washington, D.C.
7. ELKINS, H.B., L.D. PAGNOTTO and H.L. SMITH. 1970. Concentration Adjustments in Urinalysis. American Industrial Hygiene Association J. 35:559.

Subcommittee 6

T. J. KNEIP, *Chairman*
R. S. AJEMIAN
J. N. DRISCOLL
F. I. GRUNDER
L. KORNREICH
J. W. LOVELAND
J. L. MOYERS
R. J. THOMPSON

804.

Analytical Method for Arsenic, Selenium and Antimony in Urine and Air by Hydride Generation and Atomic Absorption Spectroscopy

Analyte:	As, Sb and Se	Method No:	
Matrix:	Urine, Air	Range:	1–50 μg/l for As and Se in urine 10–500 μg/l for Sb in urine
Procedure:	Hydride generation and atomic absorption		
Date Issued:		Range:	0.4–50 μg/m³ for As and Se in air 4–500 μg/m³ for Sb in air
Date Revised:			
		Precision:	Unknown for urine and air samples 1–2% RSD for hydride generation procedure
		Classification:	D (Operational)

1. Principle of the Method

1.1 Urine and air particulate samples are wet ashed with a mixture of nitric, perchloric and sulfuric acid to destroy the organic matrix. The analysis is subsequently made by generating the hydride in an acidic solution with sodium borohydride and measuring the evolved gas by background corrected atomic absorption spectroscopy.

2. Range and Sensitivity

2.1 For a 25-ml urine sample and under optimum conditions, the optimum range extends from 1 μg/l to 50 μg/l for arsenic and selenium, and from 10 μg/l to 500 μg/l for antimony.

2.2 For a 60-l air sample and under optimum conditions, the optimum range extends from 0.4 μg/m³ to 50 μg/m³ for arsenic and selenium and 4 μg/m³ to 500 μg/m³ for antimony.

2.3 The sensitivity of the hydride generation and measurement system is such that 5 ng of As and Se and 50 ng of Sb may be determined with a relative standard deviation of 20% or better if interferences are not present.

3. Interferences

3.1 The presence of high concentrations of several metals may significantly reduce the efficiency of the hydride generation for As, Sb and Se (**1, 2**). Such interferences may be checked and corrected for by using the method of standard additions described in Section **9.1.6**. The sensitivity of the method will be reduced if metal interferences are present.

3.2 Background correction techniques (**3, 4**) must be used with this method since the changes in flame composition resulting from the H_2 generated by the addition of sodium borohydride to the sample will change the amount of radiation absorbed by the flame.

4. Precision and Accuracy

4.1 The precision of the method for As, Se and Sb has not been reported using urine and air samples. The precision of the hydride generation procedure has been reported to be as good as 1 to 2% RSD for 0.1 μg of Sb, Se and As **(5, 6)**.

4.2 No accuracy data are available at this time for As, Se and Sb in air and urine samples.

5. Advantages and Disadvantages of the Method

The hydride generation and measurement system is highly sensitive in the absence of interfering metals. Interferences may be checked and corrected by the method of standard additions. Untreated air filter samples may be stored indefinitely if sealed in air-tight containers. Urine samples treated with Na_2H_2-EDTA and refrigerated are stable biologically for 1 week. The specific gravity of the urine sample should be measured within 1 week of sample collection.

The sensitivity of the method is reduced by the presence of interfering metals. The addition of sodium borohydrate to the sample and resultant hydrogen generation can change the amount of radiation absorbed by the flame. Consequently, background correction techniques must be used with this method.

6. Apparatus

6.1 Air Sampling Equipment.

6.1.1 *Cellulose acetate membrane filters,* 37 mm, *e.g.* Millipore Type AA (0.8 μm pore diameter) or equivalent.

6.1.2 *Filter unit,* consisting of filter media (**6.1.1**) and appropriate cassette filter holder, either a 2- or 3-piece filter cassette.

6.1.3 *A vacuum pump such as a personal sampling pump.* The pump must be sufficient to maintain an air face velocity of approximately 7 cm/sec (flow rate of 2–5 lpm through a 37 mm diam filter) **(7)**.

6.1.4 Flow measurement devices such as a calibrated rotometer or a critical flow orifice are required to monitor or control the sampling rate.

6.2 Glassware, borosilicate. Before use, all glassware must be cleaned in hot 1 : 1 nitric acid and rinsed several times with distilled water.

6.2.1 *125-ml Phillips or Griffin beakers* with watch glass covers.

6.2.2 *25-ml volumetric flasks.*

6.2.3 *500-ml volumetric flasks.*

6.2.4 *1-l volumetric flasks.*

6.2.5 *125-ml polyethylene bottles.*

6.2.6 *Hydride generator* (see Figure 804:1, or equivalent).

6.2.7 *Additional glassware,* such as pipets and different sized volumetric glassware may be required.

6.3 Equipment.

6.3.1 *Atomic absorption spectrophotometer,* with burner head for argon-hydrogen flame, and equipped for simultaneous background correction. Electrodeless Discharge Lamps (EDL) are recommended (Hollow cathode lamps may be used with reduced sensitivity).

6.3.2 *EDL (or Hollow cathode) lamps* for each element and a continuum lamp or source for background correction.

6.3.3 *Hotplate* (suitable for operation at 140 C to 250 C).

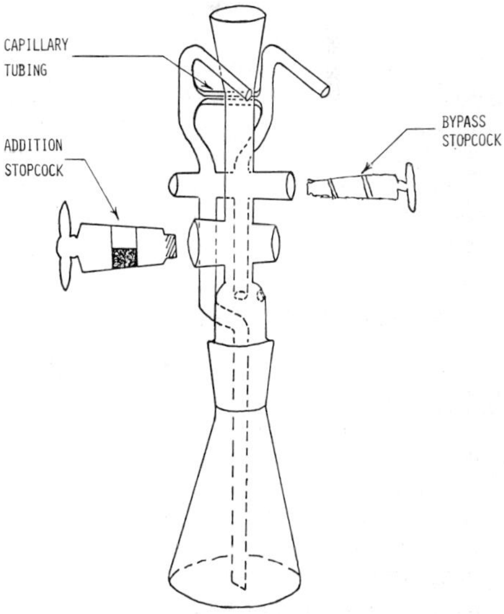

Figure 804:1—Hydride Generator

6.3.4 *Two stage regulators for argon and hydrogen.*

6.3.5 *Rotometer* for hydride generator (0 to 5 lpm).

6.4 SUPPLIES.

6.4.1 *Hydrogen gas (cyclinder),* electrolytic grade.

6.4.2 *Argon gas (cylinder),* high purity.

7. Reagents

7.1 PURITY. ACS reagent chemical or equivalent shall be used in all tests. References to water shall be understood to mean double distilled water or equivalent. Care in selection of reagents and in following the listed precautions is essential if low blank values are to be obtained.

7.2 CONC NH_4OH (28%), reagent grade, sp gr = 1.090.

7.3 CONC NITRIC ACID (68–71%), reagent grade, redistilled, sp gr = 1.42.

7.4 CONC SULFURIC ACID (90%), reagent grade, sp gr = 1.84.

7.5 PERCHLORIC ACID (72%), reagent grade, sp gr = 1.67.

7.6 CONC HYDROCHLORIC ACID (36–38%), reagent grade, sp gr = 1.18.

7.7 AMMONIUM OXALATE, saturated solution in distilled water.

7.8 SODIUM BOROHYDRIDE, 8 to 9 mm pellets (approximately 200 mg).

7.9 STANDARD STOCK SOLUTIONS (1000 μg/ml) for each metal element. Commercially prepared or prepared from the following after desiccator drying:

7.9.1 *Sodium arsenate* ($Na_2HAs_3O_4$), reagent grade.

7.9.2 *Antimony metal (Sb),* reagent grade.

7.9.3 *Selenium dioxide* (SeO_2), reagent grade.

8. Procedure

8.1 COLLECTION OF SAMPLES.

8.1.1 Urine samples are collected in polyethylene bottles which were precleaned in nitric acid. About 0.1 g EDTA is added as a preservative. At least 75 ml should be collected.

8.1.2 Air samples are collected on 37-mm membrane filters using a plastic filter holder in a closed-faced configuration. The flow rate, times, and/or volume must be measured as accurately as possible. Record atmospheric pressure and temperature at beginning and end of the sampling period. The sample should be taken at a flow rate of 1.5 to 3.0 lpm. A minimum of 60 l should be collected. Larger sample volumes are strongly encouraged provided the filters do not become loaded with dust to the point that either loose material might fall off or the filter becomes plugged.

8.2 STORAGE.

8.2.1 Urine samples are stored or shipped in the polyethylene containers. Care should be taken to maintain the containers in an upright position to avoid leakage.

8.2.2 Air filters are stored and shipped in the sample cassettes. The sample cassettes should be shipped in a suitable container designed to minimize contamination and to prevent damage in transit. Care must be taken during storage and shipping that no part of the sample is dislodged from the filter, nor that the sample surface be disturbed in any way. Loss of sample from extremely heavy deposits on the filter may be prevented by mounting a clean filter in the cassette on top of the sample filter (care must be taken to avoid contamination when this is done). When samples are analyzed the protective filter is included as part of the sample. Blank filters (minimum of 1 filter blank for every 10 samples) should be mounted in a cassette and stored or shipped in the same manner as the samples. **Note:** No air is drawn through the blank filter and blanks are treated exactly as samples in the storage, shipping, and analyses procedures.

8.3 PREPARATION OF SAMPLES.

8.3.1 Determine the specific gravity of the urine sample at room temperature. This may be done with the use of a calibrated specific gravity meter or urinometer.

8.3.2 Transfer 25 ml of the urine sample or the filter sample into a 125-ml Phillips or Griffin beaker. If Se is to be determined in urine samples 0.1 g of solid HgO should be added to the sample to prevent volatilization of Se during the ashing of sample.

8.3.3 Wet-ash the sample by adding first 4 ml of a mixture of 3 parts HNO_3 and 1 part H_2SO_4. Carefully add 1 ml $HC10_4$, swirl and heat on a hot plate at 130 to 150 C. Add dropwise small amounts of redistilled HNO_3, until the digest is colorless. If the sample approaches dryness and ashing is still incomplete, additional ashing acid mixture (3 : 1 : 1 = HNO_3, H_2SO_4, $HCLO_4$) can be added dropwise. (**Caution:** Standard safety precautions for perchloric acid digestions must be followed at all times.)

8.3.4 Continue heating to fumes of SO_3. After cooling the colorless or nearly colorless liquid, add 10 ml of distilled water and 5 ml of a saturated solution of ammonium oxalate. The mixture is again heated until dense white fumes of SO_3 appear, to free the ashed solution of all traces of nitric acid. The resulting solution should not be heated for more than a few minutes after SO_3 evolves to insure that none of the elements being analyzed are lost.

8.3.5 Allow the mixture to cool, transfer to a 25-ml volumetric flask, and make up to volume with distilled water.

8.4 ANALYSIS.

8.4.1 Set the instrument operating conditions as recommended by the manufacturer. The instrument should be set at the radiation intensity maximum for the wavelength listed below for the element being determined.

As—193.7 nm
Sb—217.6 nm
Se—196.0 nm

8.4.2 The position of maximum atom population (As, Se, or Sb) in the flame (thus position of maximum radiation absorption) when using the introduction system described in **8.4.3** will not necessarily be identical to that obtained when sample solutions are aspirated in the conventional manner. For this reason, it is necessary to adjust the radiation from the EDL (or hollow cathode) position in the flame using the sample introduction procedure described in **8.4.3**. This may be accomplished by adjusting the horizontal burner position using a high concentration solution and aspirating the sample by the conventional pneumatic nebulizer. This step assures centering of the radiation source output in the center of the flame. The vertical adjustment of radiation passing through the flame is made by introducing a series of identical standards (using procedures in **8.4.3** or **8.4.4**) at increasingly lower burner positions (these positions are noted). The vertical burner position is then adjusted to that position giving maximum absorbance for the analysis of samples and standards as described in the following section.

8.4.3 Be sure the stopcocks of the hydride generation equipment are turned such that the argon (Ar) flow bypasses the reaction vessel. (**Note:** The bypass tubing in Figure 804:1 must present a pressure drop in excess of that resulting from the liquid head present in the generation flask. If this is not the case, the argon used to purge the system will not be swept through the reservoir in a reproducible manner. This bypass pressure drop can be created by using capillary glass tubing as the connector between the inlet and outlet tubes). Add 50 ml of 6 N HCl and 5 ml of a standard solution (from section **9.1.3.a**) to the reaction flask. Connect the sample flask to the generating system and mix well with a magnetic stirrer. Turn the bypass stopcock such that the Ar passes through the reaction vessel. Add to the sample solution, via the addition stopcock, a single sodium borohydride pellet (8-mm diam, 200 mg). The absorption signal is recorded on a rapid response strip chart recorder (0.5 sec full scale response) or on an automatic peak integration facility. Standard solutions should match the sample matrix as clearly as possible and should be run in duplicate. Prepare a calibration graph as described in section **9.1.4** (**Note:** All combustion products from the AA flame must be removed by direct exhaustion through use of a good separate flame ventilation system.).

8.4.4 Recently automated reagent introduction systems based upon the use of aqueous $NaBH_4$ have become available (e.g., see reference **6**). Such systems can be used in place of that described in section **8.4.3**. Follow manufacturers procedures for the use of such automated systems. Care must be taken, however, that the sample is acidified with HCl prior to the addition of the reducing agent. Interferences are more pronounced if the reducing agent is added to the sample prior to the addition of HCl (**2**).

8.4.5 Analyze 5 ml of the sample from section **8.3.5** in the same manner as is used in **8.4.3** or **8.4.4** and record the absorbance for comparison with standards. Should the absorbance be outside the calibration range, analyze an appropriately smaller or larger aliquot. A mid-range standard must be analyzed with sufficient frequency (i.e., once every 10 samples) to assure the precision of the sample determinations. To the extent possible, all determinations are to be based on replicate analysis.

9. Standards and Calibrations

9.1 STANDARD SOLUTIONS.

9.1.1 *Standard stock solutions of each element*. Standard stock solutions are made to a concentration of 1.000 mg of element per ml. These standard stock solutions are stable for at least 3 months when stored in polyethylene bottles.

a. Arsenic Standard Stock Solution, 1000 ppm. Dissolve 0.4165 of $Na_2HAsO_4 \cdot 7H_2O$ in a mixture of water, 5 ml of conc H_2SO_4 and 20 ml of conc HCl and bring volume to 100 ml with water.

b. Selenium Standard Stock Solution, 1000 ppm. Dissolve 1.405 g of dry selenium dioxide in water containing 5 ml of 5 N NaOH. Dilute to volume in a 1 liter flask with water.

c. Antimony Standard Stock Solutions, 1000 ppm. Dissolve 1.000 g of antimony in a minimum volume of conc sulfuric acid. Dilute to volume in a 1-l flask with water.

9.1.2 *Dilute standard solutions of each element*. Dilute standards are made to a conc of 1 μg/ml for As and Se and 10 μg/ml for Sb. Dilute standards should be prepared weekly.

a. Dilute Arsenic Standard. Pipet 1 ml of the arsenic master standard into a 1000-ml volumetric flask and dilute to volume with distilled water. This solution contains 1 μg/ml As.

b. Dilute Selenium Standard. Pipet 1 ml of the selenium master standard into a 1000-ml volumetric flask and dilute to volume with distilled water. This solution contains 1 μg/ml Se.

c. Dilute Antimony Standard. Pipet 10 ml of the antimony master standard into a 1000-ml flask and dilute to volume with distilled water. This solution contains 10 μg/ml Sb.

9.1.3 *Working Standards*.

a. Working standards of 0.025, 0.05, .1, .25, .5 μg of As and Se in 25 ml of solution are prepared. Working standards of 5.0, 10.0, 30, 50, and 100 μg of Sb are prepared in 25-ml of distilled water. The working standards are prepared daily.

b. Add 0 ml, 0.5 ml, 1 ml, 2 ml, 5 ml, and 10 ml of the dilute standard for As, Se and Sb to separate 500-ml volumetric flasks. Add to these flasks 300 ml distilled water (approximately), 100 ml of conc HCl and 25 ml conc H_2SO_4. Dilute to volume with distilled water. These solutions will contain As, Se, and Sb in the concentrations specified in **9.1.3.a**. For urine samples in which Se is being determined, the working standards must have HgO added such that the concentration of Hg^{++} matches that added to the samples. The flask to which no dilute standards are added is used as the reagent blank.

9.1.4 The standard solutions are analyzed as described in **8.4.2** and the absorbance (or concentration) area (or peak height) recorded. If the instrument used displays transmittance, these values must be converted to absorbance. After the subtraction of the reagent blank signal from all standard signals, a calibration curve is prepared by plotting absorbance (concentration) versus metal concentration. The best-fit curve (calculated by linear least square regression analysis) is fitted to the data points. This line or the equation describing the line is used to obtain the metal concentraiton in the samples being analyzed.

9.1.5 To ensure that the preparation procedure is being properly followed, clean blank membrane filters are spiked with known amounts of the elements being determined by adding appropriate amounts of the previously described standards and carried through the entire procedure. The amount of metal is determined and the percent recovery calculated. These tests will provide recovery and precision data for the procedure as it is carried out in the laboratory for the soluble compounds of the elements being determined.

9.1.6 *Analysis by the method of standard additions.* In order to check for interferences (which may not have been fully investigated for use in this method), samples are initially and periodically analyzed by the method of standard additions and the results compared to those obtained by the conventional analytical determination. For this method the sample is divided into three 5-ml aliquots. One 5-ml aliquot is added to the reaction flask and analyzed as described in section **8.4.5**. The second 5-ml aliquot is added to the reaction flask along with an amount of element (As, Se or Sb) approximately equal to that in the sample (typically 0.5 to 5 ml of thc highest concentration working standard described in **9.1.3.1**) and analyzed. The third 5-ml aliquot along with an amount of element approximately equal to twice the sample concentration is added to the reaction vessel and analyzed. A 5-ml portion of the sample blank is also analyzed. The blank reading is subtracted from the three aliquot readings and these corrected values are plotted against metal added to the original sample. The line obtained from such a plot is extrapolated to 0 absorbance and the intercept on the concentration axis is taken as thc amount of metal in the original sample (**3**). If the result of this determination does not agree to within 10% of the values obtained with the procedure described in section **9.1.4**, an interference is indicated and standard addition techniques should be utilized for sample analysis.

9.2 BLANK. Blank filters must be carried through the entire procedure each time samples are analyzed. Reagent blanks are used for making blank corrections in the case of urine samples.

10. Calculations

10.1 The concentration of As, Se, and Sb in the urine sample should be expressed as mg element per liter of urine.

$$\text{mg element/liter} = \frac{\mu\text{g element}}{\text{ml of urine}}$$

where:

μg element = (μg of element in total sample used in **8.3.2**: normally 25 ml)—(reagent blank value)

ml of urine = volume of urine sample used in **8.3.2**: normally 25 ml.

The use of a specific gravity correction factor to normalize values of 1.024, the average specific gravity of urine, is recommended.

corrected mg element/liter =

$$\text{mg element/l} \times \frac{(1.024 - 1.000)}{(\text{Sp G} - 1.000)}$$

10.2 The concentration of arsenic in air can be expressed as mg As per cubic meter of air, which is numerically equal to micrograms As per liter of air.

$$\text{mg element/m}^3 = \mu\text{g element}/V_s$$

where:

μg element = (micrograms element in total sample)—(blank filter value)

V_s = volume of air sampled in l at 25 C and 760 Torr.

11. References

1. Smith, E.E., 1975. Analyst, 100:300.
2. Pierce, F.D., and H.R. Brown, 1976. Anal. Chem., 48:693.
3. Slavin, W., 1968 Atomic Absorption Spectroscopy, Interscience Publisher, N.Y., N.Y.
4. Ramirez-Munoz, J., 1968 Atomic Absorption Spectroscopy, Elsevier Publishing Co., N.Y., N.Y.
5. Vijan, P.N., and G.R. Wood, 1974 Atomic Absorption Newsletter, 13:33.
6. Fiorno, J.A., J.W. Jones, and S.G. Capan, 1976 Anal. Chem., 48:120.
7. Air Sampling Instruments for Evaluation of Atmospheric Contaminants, American Conference of Governmental Industrial Hygienists 1971.
8. Atomic Absorption Analytical Method No. As-3, Jarrell-Ash Co.
9. Atomic Absorption Application Study No. .468, The Perkin-Elmer Corp., June 1971.
10. Fernandez, Frank J. 1973 Atomic Absorption Newsletter, 12:93–97.
11. Schmidt, F.J. and J.L. Royer. 1973 Analytical Letters, 6:17.

Subcommittee 6

T. J. Kneip, *Chairman*
R. S. Ajemian
J. N. Driscoll
F. I. Grunder
L. Kornreich
J. W. Loveland
J. L. Moyers
R. J. Thompson

805.

Analytical Method for Chloride in Air

Analyte:	Chloride	Method No.:	P & CAM 115
Matrix:	Air	Range:	Lower Limit Air-18 $\mu g/M^3$
Procedure:	Midget impinger/ion Selective Electrode	Precision:	Unknown
Date Issued:	9/15/72	Classification:	D (Operational)
Date Revised:	1/1/75		

1. Principle of the Method

1.1 Atmospheric samples are taken using midget impingers containing 10 ml of 0.5 M sodium acetate

1.2 Samples are analyzed using the chloride ion selective electrode.

2. Range and Sensitivity

2.1 The range and sensitivity of this method have not been established. The electrode recommended range is 0.35 to 3,500 μg/ml.

3. Interferences

3.1 Sulfide ion must be absent because it poisons the chloride ion electrode. Touch a drop of the sample to a piece of lead acetate paper to check for sulfide ion.

3.2 If sulfide is present, it is removed by addition of a small amount of powdered cadmium carbonate to the sample. Swirl to disperse the solid and recheck a drop of the sample with lead acetate paper. Avoid a large excess of cadmium carbonate and long contact time with the solution. Filter the sample through a small plug of glass wool and proceed with the analysis.

3.3 For interference-free operation, the chloride ion level must be at least 300 times the bromide ion level, 100 times the thiosulfate level and 8 times the ammonia level.

4. Precision and Accuracy

4.1 The precision and accuracy of this method have not been established. No collaborative tests have been performed on this method.

5. Advantages and Disadvantages of the Method

5.1 Advantages are the simplicity, specificity, speed and accuracy of the method. Disadvantages are the interferences from sulfide ion and possibly ammonia.

6. Apparatus

6.1 ORION 94-17 CHLORIDE ION ELECTRODE, OR EQUIVALENT.

6.2 ORION 90-02 DOUBLE JUNCTION REFERENCE ELECTRODE, OR EQUIVALENT.

6.3 EXPANDED SCALE MILLIVOLT—PH METER.

6.4 MIDGET IMPINGERS.

6.5 SOURCE OF VACCUM, ELECTRIC OR BATTERY OPERATED PUMPS.

6.6 STOPWATCH.

6.7 ASSOCIATED LABORATORY GLASSWARE.

7. Reagents

7.1 PURITY. The reagents decribed must be made up using ACS reagent grade or better grade of chemical.

7.2 DOUBLE DISTILLED WATER.

7.3 SODIUM ACETATE 0.5 M. Dissolve 41 g sodium acetate in double distilled water and dilute to 1 l. Adjust pH to 5 with acetic acid.

7.4 SODIUM CHLORIDE STANDARDS.

7.4.1 Dissolve 5.84 g NaCl in double distilled water and dilute to 1 l for 10^{-1} M (Cl^-) (3500 μg/ml). This solution is stable for about 2 months. The following more dilute standards should be prepared fresh weekly.

7.4.2 Dilute 10 ml 10^{-1} M (Cl^-) to 100 ml with 0.5 M sodium acetate for 10^{-2} M (Cl^-) (350 μg/ml).

7.4.3 Dilute 10 ml 10^{-2} M (Cl^-) to 100 ml with 0.5 M sodium acetate for 10^{-3} M (Cl^-) (35 μg/ml.)

7.4.4 Dilute 10 ml 10^{-3} M (Cl^-) to 100 ml with 0.5 M sodium acetate for 10^{-4} M (Cl^-) (3.5 μg/ml).

7.4.5 Dilute 10 ml 10^{-4} M (Cl^-) to 100 ml with 0.5 M sodium acetate for 10^{-5} M (Cl^-) (0.35 μg/ml).

8. Procedure

8.1 CLEANING EQUIPMENT. All glassware is washed in detergent solution, rinsed in tap water followed by a rinse with double distilled water.

8.2 COLLECTION OF SAMPLES.

8.2.1 Atmospheric samples are collected in midget impingers containing 10 ml of 0.5 M sodium acetate solution.

8.2.2 A sampling rate of about 2.5 l/min is used.

8.2.3 A total air volume of about 200 liters drawn through the impingers.

8.2.4 After sampling, the openings of the midget impingers are sealed with masking tape and the impingers returned to the laboratory for analysis.

8.2.5 A "blank" impinger should be treated in the same way as other samples, except that no air is passed through this impinger.

8.3 ANALYSIS OF SAMPLES.

8.3.1 The volume of the sampling solution in the impinger is measured and recorded.

8.3.2 The solution is transferred to a 50-ml beaker and the pH adjusted to 5.0 to 5.2, if necessary, with 0.5 M sodium acetate or acetic acid. Dilute to a volume of 25 ml with double distilled water.

8.3.3 The chloride ion electrode and the double junction reference electrode are placed in the stirred solution and the resulting millivolt reading recorded. The reading should be taken to the nearest 0.5 mV after the meter has stabilized.

9. Calibration and Standards

9.1 Prepare a series of chloride standards solutions in 50-ml beakers by diluting 10 ml of each of the chloride standards, prepared as in **7.4.2** to **7.4.5**, to a volume of 25 ml with double distilled water, starting with the most dilute standard. Place the chloride ion electrode and the double reference electrode in the stirred solution. Record the resultant millivolt readings to the nearest 0.5 mV.

9.2 Plot the mV readings *vs* the chloride ion conc of the standards on semi-log paper. The chloride ion concentration in μg/ml is plotted on the logarithmic axis.

10. Calculations

10.1 The mV reading obtained from the analysis of the sample is translated to μg Cl per ml of solution using the calibration curve.

10.2 The μg content of the sample is multiplied by the sample volume to obtain the total μgCl in the sample. The blank analysis result, if any, is deducted from the total μgCl in the sample.

10.3 The total μgCl is divided by the volume of air sampled in liters to obtain μgCl per l or mgCl per M^3. (The volume of air sampled is converted to standard conditions of 25 C and 760 Torr.)

11. References

1. ORION RESEARCH INC.,Chloride Ion Specifications, Cambridge, Mass.
2. FRANT, M.S.1974 Detecting Pollutants with Chemical Sensing Electrodes. Environmental Science and Technology, 8:224.

Revised by Subcommittee 2
C. R. THOMPSON, *Chairman*
G. H. FARRAH
L. V. HAFF
A. W. HOOK
J. S. JACOBSON
E. J. SCHNEIDER
L. H. WEINSTEIN

806.

Analytical Method for Free Chlorine in Air

Analyte:	Chlorine	Method No.:	209
Matrix:	Air	Range:	0.05–1.0 ppm
Procedure:	Fritted Bubbler for sampling/Colorimetric Methyl Orange	Precision:	±5%
Date issued:		Classification:	E (Proposed)
Date Revised:	1/1/75		

1. Principle of the Method

1.1 Sampling is performed by passing a measured volume of air through a fritted bubbler containing 100 ml of dilute methyl orange.

1.2 Near a pH of 3.0 the color of a methyl orange solution ceases to vary with acidity. The dye is quantitatively bleached by free chlorine, and the extent of bleaching can be determined colorimetrically (**1, 6**). The optimum conc range is 0.05 to 1.0 ppm in ambient air (145 μg to 2900 μg/M^3 at 25 C and 760 Torr).

2. Sensitivity and Range

2.1 The procedure given is designed to cover the range of 5 to 100 μg of free chlorine/100 ml of sampling solution. For a 30-l air sample, this corresponds to approximately 0.05 to 1.0 ppm in air, which is the optimum range.

2.2 Increasing the volume of air sampled will extend the range at the lower end, but only within limits, since 50 liters of chlorine-free air produce the same effect as about 0.01 ppm of chlorine.

2.3 By using a sampling solution more

dilute in methyl orange, a concentration of 1 μg/100 ml of solution may be measured, but beyond this, problems are encountered because of the absorption of ammonia and other gases from the air and by the presence of minute amounts of chlorine-consuming materials even in distilled water.

3. Interferences

3.1 Free bromine, which gives the same reaction, interferes in a positive direction **(4)**. Manganese (III, IV) in concentrations of 0.1 ppm or above also interferes positively **(3)**. In the gaseous state, interference from SO_2 is minimal, but in solution, negative interference from SO_2 is significant. Nitrites impart an off-color orange to the methyl orange reagent. NO_2 interferes positively, reacting as 20% chlorine. SO_2 interferes negatively, decreasing the chlorine by an amount equal to one-third the SO_2 conc. Ozone may also interfere positively **(2)**.

4. Precision and Accuracy

4.1 The data available **(5)** indicate that 26 chlorine concentrations produced by two different methods (flowmeter calibrated by KI absorption, and gas-tight syringe) were measured by this procedure with an average error of less than ±5% of the amount present.

5. Advantages and Disadvantages of the Method

5.1 The method is relatively simple with direct bleaching of the methyl orange reagent.

5.2 Interfering substances in the air sample may affect the accuracy of the method. The sampled solutions must be protected from direct sunlight if the color is to be preserved for 24 hr.

6. Apparatus

6.1 SPECTROPHOTOMETER. Suitable for measurement at 505 nm, preferably accommodating 5-cm cells.

6.2 FRITTED BUBBLER. Coarse porosity, of 250 to 350 ml capacity. A small bubbler of 50 to 60 ml capacity may be more convenient for industrial hygiene sampling; volumes of reagents are then reduced proportionally.

7. Reagents

7.1 PURITY OF REAGENTS. Reagents must be ACS analytical grade quality. Distilled water should conform to ASTM Standard for Referee Reagent Water.

7.2 CHLORINE-DEMAND-FREE WATER. Add sufficient chlorine to distilled water to destroy the ammonia. The amount of chlorine required will be about ten times the amount of ammoniacal nitrogen present. In no case should the initial residual be less than 1.0 mg/l free chlorine. Allow the chlorinated distilled water to stand overnight or longer, then expose to direct sunlight for one day or until all residual chorine is discharged.

7.3 METHYL ORANGE STOCK SOLUTION, 0.05%. Dissolve 0.500 g reagent grade methyl orange in distilled water and dilute to 1 liter. This solution is stable indefinitely if freshly boiled and cooled distilled water is used.

7.4 METHYL ORANGE REAGENT, 0.005%. Dilute 100 ml of stock solution to 1 liter with distilled water. Prepare fresh for use.

7.5 SAMPLING SOLUTION. 6 ml of 0.005% methyl orange reagent is diluted to 100 ml with distilled water and 3 drops (0.15 to 0.20 ml) of 5.0 N HCl added. One drop of butanol may be added to induce foaming and increase collection efficien cy.

7.6 ACIDIFIED WATER. To 100 ml of distilled water, add 3 drops (0.15 to 0.20 ml) of 5 N HCl.

7.7 POTASSIUM DICHROMATE SOLUTION, 0.1000 N. Dissolve 4.904 g anhydrous $K_2Cr_2O_7$, primary standard grade, in distilled water and dilute to 1 liter.

7.8 STARCH INDICATOR SOLUTION. Prepare a thin paste of 1 g of soluble starch in a few ml of distilled water. Bring 200 ml of distilled water to a boil, remove from heat, and stir in the starch paste. Prepare fresh before use.

7.9 POTASSIUM IODIDE.

7.10 SODIUM THIOSULFATE SOLUTION, 0.1 N. Dissolve 25 g of $Na_2S_2O_3 \cdot 5H_2O$ in freshly boiled and cooled distilled water and dilute to 1 liter. Add 5 ml chloroform as preservative and allow to age for 2 weeks before standardizing as follows: To 80 ml of distilled water add with constant stirring, 1 ml conc. H_2SO_4, 10.00 ml 0.1000 N $K_2Cr_2O_7$, and approximately 1 g of KI. Allow to stand in the dark for 6 min. Titrate with 0.1 N thiosulfate solution. Upon approaching the end-point (brown color changing to yellowish green), add 1 ml of starch indicator solution and continue titrating to the end-point (blue to light green).

$$\text{Normality } Na_2S_2O_3 = \frac{1.000}{\text{mls or } Na_2S_2O_3 \text{ used}}$$

7.11 SODIUM THIOSULFATE SOLUTION, 0.01 N. Dilute 100 ml of the aged and standardized 0.1 N $Na_2S_2O_3$ solution to 1 liter with freshly boiled and cooled distilled water. Add 5 ml chloroform as preservative and store in a glass-stoppered bottle. Standardize frequently with 0.0100 N $K_2Cr_2O_7$.

7.12 CHLORINE SOLUTION, APPROXIMATELY 10 PPM. Prepare by serial dilution of household bleach (approx. 50,000 ppm), or by dilution of strong chlorine water made by bubbling chlorine gas through cold distilled water. The diluted solution should contain approximately 10 ppm of free (available) chlorine. Prepare 1 liter.

8. Procedure

8.1 Place 100 ml of sampling solution in the fritted bubbler. A measured volume of air is drawn through at a rate of 1 to 2 l/min for a period of time appropriate to the estimated chlorine concentration. Transfer the solution to a 100-ml volumetric flask and make to volume, if necessary, with acidified water. Measure absorbance at 505 nm in 5-cm cells against distilled water as reference.

8.2 The volume of sampling solution, the concentration of methyl orange in the sampling solution, the amount of air sampled, the size of the absorbing vessel and the length of the photometer cell can be varied to suit the needs of the situation as long as proper attention is paid to the corresponding changes necessary in the calibration procedure.

9. Calibration and Standards

9.1 Prepare a series of six 100-ml volumetric flasks containing 6 ml of 0.005% methyl orange reagent, 75 ml distilled water, and 3 drops (0.15 to 0.20 ml) of 5.0 N HCl. Carefully and accurately pipet 0, 0.5, 1.0, 5.0 and 9.0 ml of chlorine solution (approximately 10 ppm) into the respective flasks, holding the pipet tip beneath the surface. Quickly mix and make to volume with distilled water.

9.2 Immediately standardize the 10 ppm chlorine solution as follows: to a flask containing 1 gm KI and 5 ml glacial acetic acid, add 400 ml of chlorine solution, swirling to mix. Titrate with 0.01 N $Na_2S_2O_3$ until the iodine color becomes a faint yellow. Add 1 ml of starch indicator solution and continue the titration to the end-point (blue to colorless). One ml of 0.0100 N $Na_2S_2O_3$ = 0.3546 mg of free chlorine. Compute the amounts of free chlorine added to each flask in **9.1**.

9.3 Transfer the standards prepared in **9.1** to absorption cells and measure absorbance *vs.* μg of chlorine to draw the standard curve.

10. Calculations

10.1

$$\text{ppm } Cl_2 = \frac{\text{mg } Cl_2 \text{ found}}{\text{liters of air sampled}} \times \frac{24{,}450}{71}$$

For different temperatures and atmospheric pressures, proper correction for air volume should be made to standard conditions of 25 C and 760 Torr.

11. References

1. TARAS, M. 1917. Colorimetric Determination of Free Chlorine with Methyl Orange. *Anal. Chem.* 19:3–12.
2. BOLTZ, D.F. 1958. Colorimetric Determination of Nonmetals, p. 163, Interscience Publishers, New York.
3. AMERICAN PUBLIC HEALTH ASSOCIATION, INC. 1976. Standard Methods for the Examination of Water and Waste Water. 14th ed.
4. TRAYLOR, P.A. and S.A. SHRADER. Determination of Small Amounts of Free Bromine in Air. Dow Chemical Co., Main Laboratory Reference MR4N, Midland, Michigan.
5. THOMAS, M.D. and R. AMTOWER. Unpublished work.
6. AMERICAN PUBLIC HEALTH ASSOCIATION, INC. 1977. Methods of Air Sampling and Analysis. 2nd ed. pp. 381. Washington, D.C.

Revised by Subcommittee 2

C. R. THOMPSON, *Chairman*

G. H. FARRAH

L. V. HAFF

A. W. HOOK

J. S. JACOBSON

E. J. SCHNEIDER

L. H. WEINSTEIN

807.

Analytical Method for Chromic Acid Mist in Air

Analyte:	Chromium (Cr VI)	Method No:	P & CAM 152
Matrix:	Air	Range:	0.01 to 0.25 mg/m^3
Procedure:	Filter Collection, Atomic Absorption	Precision:	Unknown
		Classification:	E (Proposed)
Date Issued:	11/20/72		
Date Revised:	6/4/75		

1. Principle of Method

1.1 A known amount of air is drawn through a polyvinyl chloride **(2)** filter. The filter is washed with 0.005 M hydrochloric acid and the solution brought to a pH of about 2.8 after addition of ammonium pyrrolidine dithiocarbamate (APDC).

1.2 The Cr(VI) in the solution is separated from Cr(III) by extraction with a methylisobutyl ketone (MIBK)—hexane solvent which selectively extracts the Cr(VI) ion **(3)**.

1.3 The MIBK-hexane phase is aspirated into an atomic absorption spectro photometer and the absorbance at 357.9 nm is read and compared to known standards of Cr(VI).

2. Range and Sensitivity

2.1 This method is applicable to Cr(VI) over the range of 0.01 mgim^3 to 0.25 mg/m^3 (calculated as CrO_3); higher concentrations can be determined by dilution **(4)**.

2.2 The limit of detection for Cr(VI) (as CrO_3) is about 0.02 μg/ml of extract which corresponds to 0.01 mg/m^3 on the basis of a 20-liter sample of air **(1)**.

3. Interferences

3.1 Extraction with MIBK and ammonium pyrrolidine dithiocarbamate (APDC) complexing reagent avoids Cr(III) interference by transferring Cr(VI) into the organic layer. Under the conditions specified, the same reagent will extract zinc, cadmium, iron, manganese, copper, nickel, silver and lead. The interference from manganese can be minimized by extraction at a pH between 2.5 and 3.0 where manganese is not extracted to any great extent. The other metals do not affect the chromium determination unless concentrations are a few orders of magnitude greater than chromium in which case iron and nickel may suppress the chromium absorption (**5**). Where interferences are suspected, use the method of standard additions to correct for the inferences (**5, 6**).

3.2 Reducing agents will lower the result for Cr(VI). The most noteable interferences are hydrogen, olefins and acetylenes and hydrocarbons of C_3 or higher (**7**). Reduced valence states of most metals will also reduce the Cr(VI) to Cr(III) on the filter.

4. Precision and Accuracy

4.1 Precision of the method has not been reported. However, in a method similar to this, a relative standard deviation of 11% at a conc of 5 μg/l in water was reported (**8**).

4.2 Accuracy of the method has not been established.

5. Advantages and Disadvantages

5.1 Polyvinyl chloride filters are an advantage in collection of samples.

5.2 Cr(VI) is separated from Cr(III) by an extraction procedure.

5.3 No ashing procedure is required.

5.4 Reducing agents, such as unsaturated hydrocarbons, and reduced valence states of heavy metals will yield lower results for Cr(VI).

6. Apparatus

6.1 SAMPLING APPARATUS.

6.1.1 *Polyvinyl chloride filter*. Diameter to fit the filter cassette (37 mm suggested). (Gelman type VM-1, 5 μm pore size, or equivalent)

6.1.2 *Battery operated personal sampling pump*. To pull up to 21/m of air. (Monitaire Sampler, MSA, Pittsburgh, Pa. or equivalent)

6.1.3 *2-piece filter cassette*. (Polystrene, open face preferred). (Millipore M-000-037-00 or equivalent with cellulose backing pad) (AP-10-037-00 or equivalent)

6.2 ATOMIC ABSORPTION SPECTROPHOTOMETER. With an acetylene burner head.

6.3 VOLUMETRIC GLASSWARE.

6.4 PH METER.

6.5 FORCEPS. Teflon tipped

7. Reagents

7.1 PURITY. ACS reagent grade chemicals or equivalent shall be used in all tests. References to water shall be understood to mean double distilled water or equivalent. Care in selection of reagents is essential if low blank values are to be obtained.

7.2 CONC NITRIC ACID (69.0–71.0%). sp gr 1.42.

7.3 CONC HYDROCHLORIC ACID (36.5%–38.0%). sp gr 1.18.

7.4 STOCK DILUTE HYDROCHLORIC ACID 0.5 N. Dilute 43 ml of conc acid to 1 l with water

7.5 WORKING DILUTE ACID 0.005 N. Dilute 10 ml of 0.5 N HCl to 1 l with water.

7.6 AMMONIUM PYRROLIDINE DITHIOCARBAMATE (APDC) SOLUTION. Dissolve 1.0 g APDC in water and dilute to 100 ml. Prepare fresh solution daily. Filter if necessary.

7.7 3:1 METHYLISOBUTYL KETONE (MIBK)—HEXANE SOLUTION. Mix 1 part hexane with 3 parts of MIBK and shake to insure homogeneity of solvent. This solution should be saturated with 0.005 N HCl prior to use.*

7.8 ETHANOL.

7.9 BROMPHENOL BLUE INDICATOR SOLUTION. Dissolve 0.10 g bromphenol blue in 100 ml of 50-50 ethanol-water.

*Pure MIBK is slightly soluble in acidic aqueous solutions. Use of the 3:1 mixture decreases the solubility in the aqueous phase without any significant change in the extraction efficiency (**3**).

7.10 Sodium hydroxide solution, 0.1 N. Dissolve 4.0 g NaOH in 1 l of water.

7.11 $K_2Cr_2O_7$.

7.12 Cleaning Solution. Add 150 ml of conc hydrochloric acid to 200 ml of water, then add 50 ml of conc nitric acid with stirring. Use to clean all glassware and store in a borosilicate, glass stoppered bottle. (Never use chromic acid cleaning solution).

8. Procedure

8.1 Collect the chromic acid mist on the filter at a flow rate of 2 l/min for a minimum period of one hr. For longer periods of collection, flow rates less than 2 l/m can be used as dictated by the expected Cr(VI) concentrations. Some minimum sampling volumes for various conc levels based on the limit of detection are as follows:

Concentration to be Measured ($mg\ CrO_3/m^3$)	Minimum Required Sample Size (liters)
0.01	20
0.05	4
0.1 (TLV)	2
0.2	1

8.2 Prepare two filter blanks in the same manner as for sampling and transport to sampling area, but keep tightly capped at all times.

8.3 Remove the filter with forceps and place on the fritted glass of a funnel held in suction flask, apply suction and wash sides of holder and the filter with 3 separate 20-ml washings of 0.005 N hydrochloric acid solution. Be careful not to combine the backing pad with the filter.

8.4 Pour washings into a 200-ml volumetric flask. Rinse the funnel and suction flask with 2 separate 20-ml washings of 0.005 N hydrochloric acid and add washings to volumetric flask.

8.5 Add 2 drops of bromphenol blue indicator solution and add sufficient 0.1 N sodium hydroxide or 0.5 N HCl solution with stirring to bring the pH to between 2.2 and 2.4. Check the pH with a pH meter on the first few samples to insure proper indicator color change with pH value.

8.6 Add 5.0 ml of APDC solution and mix. This should raise the pH to about 2.8 ± 0.1. Check with a pH meter occasionally to insure the pH is within these limits.

8.7 Add exactly 10.0 ml of the MIBK-Hexane solvent saturated with 0.005 N HCl and shake vigorously for 3 min.

8.8 Allow the two layers to separate and add 0.005 N HCl until the MIBK-Hexane layer is completely in the neck of the flask.

8.9 Aspirate the MIBK-Hexane layer directly into the atomic absorption apparatus. Measure the absorbance or other suitable scale reading at 357.9 nm after adjusting the instrument to the operating conditions specified by the manufacturer.

8.10 Adjust the fuel to air ratio prior to running samples and standards with MIBK-Hexane solution until a steady blue flame is obtained. Do not reduce the fuel flow to the extent that the flame will blow out when no solvent is being aspirated.

8.11 Compare the reading to that of a calibration curve previously obtained as described in **8.2** after subtracting the average of the blank samples taken to the sampling site.

8.12 Should the reading be off scale, dilute the MIBK-Hexane with an appropriate amount of solvent which has been saturated with 0.005 N HCl and run again.

9. Calibration and Standards

9.1 Chromic acid standards.

9.1.1 *Master Solution A:* 1 ml = 1.000 mg chromium = 1.92 mg CrO_3. Dissolve exactly 2.829 g of dry potassium dichromate in 0.005 N HCl and dilute to 1 l with 0.005 N HCl. Keep for 1 year.

9.1.2 *Dilute Standard Solution B:* 1 ml = 0.010 mg chromium = 0.0192 mg CrO_3. Dilute 10 ml of solution A to 1 l with 0.005 N HCl. Prepare monthly.

9.1.3 *Working Standards Solution C: 1 ml = 0.1 μg chromium = 0.192 μg CrO_3*. Dilute 10 ml of Solution B to 1 l with 0.005 N HCl. Prepare weekly.

9.2 CALIBRATION PROCEDURE.

9.2.1 Pipet onto separate polyvinyl chloride filters in holders inserted into separate suction flasks the following amounts of working standards from Solution C: 0, 2, 5, 10 and 25 ml.

9.2.2 Wash each filter with 3 successive 20-ml portions of 0.005 N HCl as done in the procedure in step **8.3**.

9.2.3 Perform the analysis from step **8.3** through **8.10** of the procedure. Plot the calibration curve of μg of Cr(VI) versus absorbance or appropriate scale readings linear with absorbance using the best curve as determined by the least squares method. Subtract the calibration blank value so that 0 ml equals 0 μg $CrCO_3$.†

10. Calculations

10.1 Convert all readings to μg of chromium trioxide (CrO_3) per 10.0 ml of MIBK-Hexane solution by the following factor: $\mu g\ CrO_3 = 1.92 \times \mu g\ Cr(VI)$.

10.2 Conc of CrO_3 in air is calculated as follows:

$$mg\ CrO_3/m^3 = \frac{(\mu g\ CrO_3 \text{ found per 10 ml of solvent} - mg\ CrO_3 \text{ in sample blank})}{\text{Liters of air sampled}}$$

†The hydrochloric acid solutions used should be the same for samples and calibration and blanks. Use of a different bottle of concentrated hydrochloric acid or a new batch of filters requires recalibration of the procedure to insure that any variation in metal impurities does not affect the results.

11. References

1. U.S. DEPARTMENT OF HEALTH, EDUCATION AND WELFARE, Public Health Service. 1973. Occupational Exposure to Chromic Acid. NIOSH.
2. ABELL, M.T., and J.R. CARLBERG, 1974. *Am. Ind. Hyg. Assoc. J.*, 35:229–233
3. ENVIRONMENTAL PROTECTION AGENCY, NATIONAL ENVIRONMENTAL RESEARCH CENTER. 1971. Methods for Chemical Analysis of Water and Wastes, Analytical Quality Control Lab, Cincinnati, Ohio 45268.
4. MIDGETT, M.R., and M.J. FISHMAN, 1967. *At. Abs. Newsletter*, 6:128.
5. SLAVIN, W., 1968. Atomic Absorption Spectroscopy, Interscience Publishers, N.Y., N.Y.
6. DEAN, J.A. and T.C. RAINS, 1969. Flame Emission and Atomic Absorption Spectrometry, Vol. 1, p. 377, Marcel Dekker, New York.
7. PRIVATE COMMUNICATION, John Kralgic, Allied Chemical, Syracuse, N.Y.
8. BROWN, E., M.W. SKOUGSTAD, and M.J. FISHMAN. 1970. Techniques of Water Resources Investigations of the U.S. Geological Survey, Chapter A1, Methods for Collection & Analysis of Water Samples for Dissolved Minerals and Gases. Dept. of Interior, Laboratory Analysis, Book 5, pp. 76–80.

Subcommittee 6
T. J. KNEIP, *Chairman*
R. S. AJEMIAN
J. N. DRISCOLL
F. I. GRUNDER
L. KORNREICH
J. W. LOVELAND
J. L. MOYERS
R. J. THOMPSON

808.

Analytical Method for Cyanide in Air

Analyte:	Cyanide	Method No.	P & CAM 116
Matrix:	Air	Range:	0.013 to 13 mg/m^3
Procedure:	Collection via Impinger Ion Selective Electrode	Precision:	Unknown
		Classification:	D (Operational)
Date Issued:	9/8/72		
Date Revised:	1/1/75		

1. Principle of the Method

1.1 Atmospheric samples are taken using midget impingers that contain 10 ml of 0.1 M NaOH.

1.2 Samples are analyzed using the cyanide ion selective electrode.

2. Range and Sensitivity

2.1 The range and sensitivity of the method have not been established. The recommended range of the method is 0.013 to 13 mg/m^3 in air.

3. Interferences

3.1 Sulfide ion irreversibly poisons the cyanide ion selective electrode and must be removed if found to be present in the sample. Check for the presence of sulfide ion by touching a drop of sample to a piece of lead acetate paper. The presence of sulfide is indicated by discoloration of the paper.

3.2 Sulfide is removed by the addition of a small amount (spatula tip) of powdered cadmium carbonate to the pH 11 to 13 sample. Swirl to disperse the solid, and recheck the liquid by again touching a drop to a piece of lead acetate paper. If sulfide ion has not been removed completely, add more cadmium carbonate and long contact time with the solution.

3.3 When a drop of liquid no longer discolors a strip of lead acetate paper, remove the solid by filtering the sample through a small plug of glass wool contained in an eye dropper and proceed with the analysis.

4. Precision and Accuracy

4.1 The precision and accuracy of this method have not been determined. No collaborative tests have been performed on this method.

5. Advantages and Disadvantages of the Method

5.1 Advantages are the simplicity and speed of the method. Specificity depends upon the removal of sulfide prior to analysis.

6. Apparatus

6.1 SAMPLING EQUIPMENT. The sampling unit for the impinger collection method consists of the following components:

6.1.1 *A prefilter unit (if needed).* Consists of the filter media and cassette filter holder.

6.1.2 *A midget impinger.* Containing the absorbing solution or reagent.

6.1.3 *A pump suitable for delivering desired flow rates.* The sampling pump is protected from splashover or water condensation by an adsorption tube loosely packed with a plug of glass wool and inserted between the exit arm of the impinger and the pump.

6.1.4 *An integrating volume meter such as a dry gas or wet test meter.*

6.1.5 *Thermometer.*

6.1.6 *Manometer.*

6.1.7 *Stopwatch.*

6.2 ORION 94-06 CYANIDE ION SELECTIVE ELECTRODE, OR EQUIVALENT.

6.3 ORION 90-01 SINGLE JUNCTION REFERENCE ELECTRODE, OR EQUIVALENT.

6.4 EXPANDED SCALE MILLIVOLT—PH METER.

6.5 ASSOCIATED LABORATORY GLASSWARE.

6.6 PLASTIC BOTTLES.

6.7 MAGNETIC STIRRER AND STIRRING BARS.

7. Reagents

7.1 PURITY. The reagents described must be made up using ACS reagent grade or better grade of chemical.

7.2 DOUBLE DISTILLED WATER.

7.3 SODIUM HYDROXIDE 0.1 M. Dissolve 2.0 g NaOH in double distilled water dilute to 500 ml.

7.4 POTASSIUM CYANIDE STANDARDS.

7.4.1 Dissolve 0.65 g KCN in 0.1 M NaOH and dilute to 100 ml with additional 0.1 M NaOH for 10^{-1} M $[CN^-]$ (2600 μg/ml). This solution should be checked on a weekly basis. All other dilutions should be made up fresh daily.

7.4.2 Dilute 10 ml of 10^{-1} M $[CN^-]$ to 100 ml with 0.1 M NaOH for 10^{-2} M $[CN^-]$ (260 μg/ml).

7.4.3 Dilute 10 ml of 10^{-2} M $[CN^-]$ to 100 ml with 0.1 M NaOH for 10^{-3} M $[CN^-]$ (26 μg/ml).

7.4.4 Dilute 10 ml of 10^{-3} M $[CN^-]$ to 100 ml with 0.1 M NaOH for 10^{-4} M $[CN^-]$ (2.6 μg/ml).

7.4.5 Dilute 10 ml of 10^{-4} M $[CN^-]$ to 100 ml with 0.1 M NaOH for 10^{-5} $[CN^-]$ (0.26 μg/ml).

7.5 LEAD ACETATE PAPER.

7.6 CADMIUM CARBONATE.

8. Procedure

8.1 CLEANING OF EQUIPMENT. All glassware is washed in detergent solution, rinsed in tap water and then rinsed with double distilled water.

8.2 COLLECTION AND SHIPPING OF SAMPLES.

8.2.1 Pour 10 ml of the absorbing solution (0.1 M NaOH) into the midget impinger, using a graduated cylinder to measure the volume.

8.2.2 Connect the impinger (via the adsorption tube) to the vacuum pump and the prefilter assembly (if needed) with a short piece of flexible tubing. The minimum amount of tubing necessary to make the joint between the prefilter and impinger should be used. The air being sampled should not be passed through any other tubing or other equipment before entering the impinger.

8.2.3 Turn on pump to begin sample collection. Care should be taken to measure the flow rate, time and/or volume as accurately as possible. The sample should be taken at a flow rate of 2.5 lpm. A sample size of not more than 200 l and no less than 10 l should be collected. The minimum volume of air sampled will allow the measurement at least 1/10 times the TLV, 0.5 mg/m^3 (760 Torr, 25 C).

8.2.4 After sampling, the impinger stem can be removed and cleaned. Tap the stem gently against the inside wall of the impinger bottle to recover as much of the sampling solution as possible. Wash the stem with a small amount (1 to 2 ml) of unused absorbing solution and add the wash to the impinger. Seal the impinger with a hard, non-reactive stopper (preferably Teflon). Do not seal with rubber. The stoppers on the impingers should be tightly sealed to prevent leakage during shipping. If it is preferred to ship the impingers with the stems in, the outlets of the stem should be sealed with Parafilm or other non-rubber covers, and the ground glass joints should be sealed (*i.e.*, taped) to secure the top tightly.

8.2.5 Care should be taken to minimize spillage or loss by evaporation at all times. Refrigerate samples if analysis cannot be done within a day.

8.2.6 Whenver possible, hand delivery of the samples is recommended. Otherwise, special impinger shipping cases designed by NIOSH should be used to ship the samples.

8.2.7 A "blank" impinger should be handled as the other samples (fill, seal and transport) except that no air is sampled through the impinger.

8.2.8 Where a prefilter has been used, the filter cassettes are capped and placed in an appropriate cassette shipping container. One filter disc should be handled like the other samples (seal and transport) except that no air is sampled through, and this is labeled as a blank.

8.3 ANALYSIS OF SAMPLES.

8.3.1 The solution is quantitatively transferred from the impinger to a graduated cylinder and the volume recorded. The solution is made up to 25 ml with double distilled water and transferred to a 50-ml beaker.

8.3.2 The cyanide ion electrode and the single junction reference electrode are placed in the solution and the resulting mV reading recorded. The reading should be taken after the meter has stabilized. Both the samples and standards should be stirred while the readings are being taken.

9. Calibration and Standards

9.1 Pour 10 ml of each potassium cyanide standard, Sections **7.4.2** to **7.4.5**, into a graduated cylinder and make up to 25 ml with double distilled water. Transfer the standard solution in each case to 50-ml beaker and place the cyanide ion electrode and the single junction reference electrode in the solution. Obtain the mV readings from each of the cyanide standards, commencing with the weakest standard.

9.2 Plot the mV readings vs. the cyanide ion conc of the standards on semi-log paper. The cyanide ion conc in μg/ml is plotted on the log axis.

10. Calculations

10.1 The mV readings from the analysis of the sample are converted to μg CN/ml of solution using the calibration curve.

10.2 The μg content of the sample is multiplied by the sample volume to obtain the total μg CN in the sample.

10.3 Convert the volume of air sampled to standard conditions of 25 C and 760 Torr:

$$V_s = V \times \frac{P}{760} \times \frac{298}{T + 273}$$

where:

V_s = volume of air in liters at 25 C and 760 Torr.
V = volume of air in liters as measured
P = Barometric Pressure in Torr.
T = Temperature of air in degree centigrade.

10.4 The conc of CN in the air sampled can be expressed in μg CN per liter or mg CN per cubic meter.

$$mg/m^3 = \mu g/liter$$

$$mg/m^3 = \frac{\text{total } \mu g \text{ CN}}{V_s} \quad \begin{matrix}\text{(Section \textbf{10.2})}\\ \text{(Section \textbf{10.3})}\end{matrix}$$

10.5 The concentration of CN can also be expressed in ppm, defined as μl of component per liter of air.

$$ppm = \mu l\ CN/V_s = R/MW \cdot \mu g\ CN/V_s = 0.94 \times \mu g\ CN/V_s$$

where:

R = 24.45 at 25 C, 760 Torr.
MW = 26

11. References

1. CYANIDE ION SPECIFICATIONS. Orion Research Inc., Cambridge, Mass.
2. FRANT, M.S. 1974. Detecting Pollutants with Chemical Sensing Electrodes. *Environmental Science & Technology* 8:224.

Revised by Subcommittee 3
B. DIMITRIADES, *Chairman*
J. CUDDEBACK
E. L. KOTHNY
P. W. MCDANIEL
L. A. RIPPERTON
A. SABADELL

809.

Analytical Method for Fluorides and Hydrogen Fluoride in Air

Analyte:	Total Fluoride	Method No.:	P & CAM 117
Matrix:	Air	Range:	Lower Limit Air 0.009 mg/m^3
Procedure:	Collection via Impinger/Ion Selective Electrode	Precision:	Unknown
		Classification:	D (Operational)
Date Issued:	9/7/72		
Date Revised:	1/1/75		

1. Principle of the Method

1.1 Atmospheric samples are taken using midget impingers containing 10 ml of 0.1 M NaOH.

1.2 Samples are diluted 1:1 with Total Ionic Strength Activity Buffer (TISAB).

1.3 The diluted samples are analyzed using the fluoride ion selective electrode.

2. Range and Sensitivity

2.1 The range and sensitivity have not been established. The recommended range of the method is 0.009 to 95 mg/m^3 air.

3. Interferences

3.1 Hydroxide ion is the only significant electrode interference; however, addition of the TISAB eliminates this problem. Very large amounts of complexing metals such as aluminum may result in low readings even in the presence of TISAB.

4. Precision and Accuracy

4.1 The accuracy and precision of this method have not been determined. No collaborative tests have been performed on this method.

5. Advantages and Disadvantages of the Method

5.1 Advantages over previous methods include simplicity, accuracy, speed, specificity and elimination of distillation, diffusion and ashing of the samples.

5.2 No significant disadvantages are known at present.

6. Apparatus

6.1 SAMPLING EQUIPMENT. The sampling unit for the impinger collection method consists of the following components.

6.1.1 *A prefilter unit, (if needed).* Consists of the filter media and cassette filter holder.

6.1.2 *A midget impinger.* Containing the absorbing solution or reagent.

6.1.3 *A pump, suitable for delivering desired flow rates.* The sampling pump is protected from splashover or water condensation by an adsorption tube loosely packed with a plug of glass wool and inserted between the exit arm of the impinger and the pump.

6.1.4 *An integrating volume meter.* Such as a dry gas or wet test meter.

6.1.5 *Thermometer.*

6.1.6 *Manometer.*

6.1.7 *Stopwatch.*

6.2 Orion Model 94-09 Fluoride Selective Ion Electrode, or equivalent.

6.3 Reference Electrode, Orion 90-01 single junction, or equivalent calomel or silver/silver chloride electrode.

6.4 Expanded Scale Millivolt-pH Meter. Capable of measuring to within 0.5 mV.

6.5 Polyethylene Beakers 50-ml capacity.

6.6 Laboratory Glassware.

6.7 Magnetic Stirrer and Stirring bars, for 50-ml Beakers.

7. Reagents

Purity. All chemicals must be ACS reagent grade or equivalent. Polyethylene beakers and bottles should be used for holding and storing all fluoride-containing solutions.

7.1 Double Distilled Water.

7.2 Glacial Acetic Acid.

7.3 Absorbing Solution. 0.1 M Sodium Hydroxide Solution. Dissolve 4 g sodium hydroxide pellets in 1 l distilled water.

7.4 Sodium Hydroxide, 5 M Solution. Dissolve 20 g sodium hydroxide pellets in sufficient distilled water to give 100 ml of solution.

7.5 Sodium Chloride.

7.6 Sodium Citrate.

7.7 Total Ionic Strength Activity Buffer (TISAB).Place 500 ml of double distilled water in a 1-liter beaker. Add 57 ml of glacial acetic acid, 58 g of sodium chloride and 0.30 g of sodium citrate. Stir to dissolve. Place beaker in a water bath (for cooling) and slowly add 5 M sodium hydroxide until the pH is between 5.0 and 5.5. Cool to room temperature and pour into a 1-liter volumetric flask and add double distilled water to the mark.

7.8 Sodium Fluoride.For preparation of standards.

7.9 Standard Fluoride Solution.

7.9.1 Dissolve 4.2 g of sodium fluoride in double distilled water and dilute to 1 liter. This solution contains 10^{-1} M[F] (1900 μgF^-/ml). The 0.1 M fluoride solution may also be purchased from Orion Research, Inc., Cambridge, Mass. Store in plastic bottle. Reagent is stable for 2 months. Other dilutions should be prepared weekly.

7.9.2 Prepare 10^{-2} M [F] by diluting 10 ml of 10^{-1} M [F] to 100 ml with double distilled water (190 μgF^-/ml).

7.9.3 Prepare 10^{-3} M [F] by diluting 10 ml of 10^{-2} M [F] to 100 ml with double distilled water (19 μg F^-/ml).

7.9.4 Prepare 10^{-4} M [F] by diluting 10 ml of 10^{-3} M [F] to 100 ml with double distilled water (1.9 μg F^-/ml).

7.9.5 Prepare 10^{-5} M [F] by diluting 10 ml of 10^{-4} M [F] to 100 ml with double distilled water (0.19) μg F^-/ml).

8. Procedure

8.1 Cleaning of Equipment. All glassware and plastic ware are washed in detergent solution, rinsed in tap water, and then rinsed with double distilled water.

8.2 Collection and Shipping of Samples.

8.2.1 Pour 10 ml of the absorbing solution (section 7) into the midget impinger, using a graduated cylinder to measure the volume.

8.2.2 Connect the impinger (via the adsorption tube) to the vacuum pump and the prefilter assembly (if needed) with a short piece of flexible tubing. The minimum amount of tubing necessary to make the joint between the prefilter and impinger should be used. The air being sampled should not be passed through any other tubing or other equipment before entering the impinger.

8.2.3 Turn on pump to begin sample collection. Care should be taken to measure the flow rate, time and/or volume as accurately as possible. The sample should be taken at a flow rate of 2.5 lpm. A sample size of not more than 200 l and no less than 10 l should be collected. The minimum volume of air sampled will allow the measurement at least 1/10 times the TLV, 0.2 mg/m^3 (760 Torr, 25 C).

8.2.4 After sampling, the impinger stem can be removed and cleaned. Tap the stem gently against the inside wall of the impinger bottle to recover as much of the sampling solution as possible. Wash the stem with a small amount (1 to 2 ml) of unused absorbing solution and add the wash to the impinger. Seal the impinger with a hard, non-reactive stopper (preferably Teflon). Do not seal with rubber. The stoppers on the impingers should be tightly sealed to prevent leakage during shipping. If it is preferred to ship the impingers with the stems in, the outlets of the stem should be sealed with Parafilm or other non-rubber covers, and the ground glass joints should be sealed (*i.e.*, taped) to secure the top tightly.

8.2.5 Care should be taken to minimize spillage or loss by evaporation at all times. Refrigerate samples if analysis cannot be done within a day.

8.2.6 Whenever possible, hand delivery of the samples is recommended. Otherwise, special impinger shipping cases designed by NIOSH should be used to ship the samples.

8.2.7 A "blank" impinger should be handled as the other samples (fill, seal and transport) except that no air is sampled through this impinger.

8.2.8 Where a prefilter has been used, the filter cassettes are capped and placed in an appropriate cassette shipping container. One filter disc should be handled as the other samples (seal and transport) except that no air is sampled through, and this labeled as a blank.

8.3 ANALYSIS OF SAMPLES.

8.3.1 Transfer the sample from the impinger to a 50-ml plastic beaker; add 10 ml of TISAB and dilute to 25 ml with double distilled water. Stir the solution.

8.3.2 The fluoride ion electrode and the reference electrode are lowered into the stirred solution and the resulting mV reading recorded (to the nearest 0.5 milivolt) after it has stabilized (drift less than 0.5 mV per min).

9. Calibration and Standards

9.1 Prepare a series of fluoride standard solutions by diluting 5 ml of each fluoride standard (**7.9.2** to **7.9.5**) and 5 ml TISAB in a clear polyethylene beaker to 25 ml with double distilled water. Insert the fluoride ion electrode and the reference electrode into each of the stirred calibration solutions starting with the most dilute solution and record the resulting mV reading to the nearest 0.5 mV. Plot the mV readings vs the fluoride ion conc of the standard on semi-log paper. The fluoride ion conc in μg/ml is plotted on the log axis. The calibration points should be repeated twice daily.

10. Calculations

10.1 The conc (μ/ml) of fluoride in the sample solution is obtained from the calibration curve.

10.2 Total μg F^- in the sample = sample conc (μg/ml) × sample solution volume (ml).

10.3 The total μg F^- is divided by the volume in liters, of air sampled to obtain conc in μg F^-/liter or mg F^-/m^3.

$$\text{Mg } F^-/m^3 = \mu g\ F^-/\text{liter}$$

$$\text{mg } F^-/m^3 = \frac{\text{total } \mu g\ F^-}{V_s} \quad \text{(Section 10.4)}$$

10.4 Convert the volume of air sampled to standard conditions of 25 C and 760 Torr.

$$V_s = V \times \frac{P}{760} \times \frac{298}{T + 273}$$

where:

V_s = volume of air in liters at 25 C and 760 Torr

V = volume of air in liters as measured

P = barometric pressure in Torr

T = temperature of air in degree centigrade

10.5 The conc can also be expressed in ppm, defined as μl of component per liter of air.

$$\text{ppm } F^- = \mu l\ F^-/V_s = \frac{24.45}{MW} \times \mu g\ F^-/V_s$$

$$= 1.29\ \mu l\ F^-/V_s$$

Where:

24.45 = molar volume at 25 C and 760 Torr

MW = 19, weight of fluoride ion, (i.e., 19 μg F^- = 24.45 μl at 25 C, 760 Torr)

10.6 To calculate the conc of hydrogen fluoride as mg HF/m^3 or ppm HF, simply multiply the corresponding concentration of F^- (from **10.3** or **10.4**) by 1.05.

11. References

1. ELFERS, L.A., and C.E. DECKER. 1968. Determination of Fluoride in Air and Stack Gas Samples by Use of an Ion Specific Electrode. *Anal. Chem.* 40 (11), p. 1658.

Revised by Subcommittee 2
C. R. THOMPSON, *Chairman*
G. H. FARRAH
L. V. HAFF
A. W. HOOK
J. S. JACOBSON
E. J. SCHNEIDER
L. H. WEINSTEIN

810.

Analytical Method for Gaseous and Particulate Fluorides in Air

Analyte:	Gaseous and Particulate Fluorides	Method No.:	P & CAM 212
Matrix:	Air	Range:	Lower Limit 0.005 mg/m^3 air
Procedure:	Collection on membrane filter and alkaline filter selective electrode	Precision:	Unknown
		Classification:	E (Proposed)
Date Issued:	8/6/74		
Date Revised:	1/1/75		

1. Principle of the Method

1.1 Particulate material from a measured volume of air is collected by means of a membrane filter.

1.2 Gaseous fluoride, from the same sample of air, is absorbed by an alkali-impregnated cellulose pad placed immediately behind the membrane filter.

1.3 The membrane filter and collected solids are made alkaline, ashed, the residue fused with additional alkali, and fluoride determined in a solution of the melt by use of a fluoride ion selective electrode.

1.4 Gaseous fluoride is determined in an aqueous extract of the cellulose pad, also by means of the fluoride ion selective electrode.

2. Range and Sensitivity

2.1 The method is capable of collection and analysis of both particulate and gaseous fluorides over a wide range (**1, 2, 3, 4**). The recommended range of the method is 0.005 to 5 mg F^-/m^3 air. The sensitivity has not been established.

3. Interferences

3.1 Because a selective ion electrode responds to ionic activity, insoluble and complexed forms of fluoride must be released by appropriate combinations of fusion, adjustment of pH, and addition of complexing agent.

3.2 Acidity (pH) and ionic strengths of fluoride standard solutions must be matched to those of samples.

3.3 Temperature of sample and standard solutions must be controlled within ± 2 C.

4. Precision and Accuracy

4.1 Precision and accuracy of this method have not been determined.

5. Advantages and Disadvantages of the Method

5.1 Advantages include simplicity, elimination of distillation or diffusion, speed and specificity.

5.2 No significant disadvantages are known at present.

6. Apparatus

6.1 PERSONAL SAMPLING PUMP. Mine Safety Appliances Company Model G, or equivalent.

6.2 FILTER HOLDER. Plastic holders of the pre-loaded personal monitor type, which accept filters of 37 mm diam, are preferred. The holder is to be numbered for identification. (Millipore Corporation Field Monitor, Cat. No. MAWP 037 AO, or equivalent. This is supplied with a membrane filter and cellulose back-up pad).

6.3 MEMBRANE FITLER OF MIXED CELLULOSE ESTERS. 0.8 μm pore size, and of diam to fit the filter holder (**6.2**).

6.4 CELLULOSE PAD. Of size to fit the filter holder (**6.2**).

6.5 CRUCIBLES. 20 ML, OF NICKEL OR INCONEL.

6.6 FLUORIDE ION SELECTIVE ELECTRODE.

6.7 REFERENCE ELECTRODE, CALOMEL TYPE. Preferably combined with the fluoride ion selective electrode. (Orion Research Inc. combination fluoride electrode, No. 96-09, or equivalent).

6.8 AN ELECTROMETER OR EXPANDED SCALE PH METER. With a mV scale for measurement of potentials.

6.9 MAGNETIC STIRRER AND TEFLON COATED STIRRING BARS. For 50-ml plastic beakers.

7. Reagents

7.1 PURITY. Reagent grade chemicals, ACS specifications or equivalent, must be used in all tests. Polyethylene beakers and bottles should be used for holding and storing all fluoride-containing solutions.

7.2 PURITY OF WATER. References to water shall be understood to mean double distilled water.

7.3 ALKALINE FIXATIVE SOLUTION. Dissolve 25 g sodium carbonate (Na_2CO_3) in water, add 20 ml glycerol, and dilute to 1 l.

7.4 TOTAL IONIC STRENGTH ACTIVITY BUFFER (TISAB). Dissolve 37 g potassium chloride (KCl), 68 g sodium acetate ($NaC_2H_3O_2 \cdot 3\,H_2O$), and 294 g sodium citrate ($Na_3C_6\text{-}\ H_5O_7 \cdot 2\,H_2O$) in water, adjust to pH 8 ± 0.1 with addition of a small amount of 1:1 hydrochloric acid, and dilute to 1 l.

7.5 STANDARD FLUORIDE SOLUTION, 100 MG/LITER. Dissolve 0.2211 g sodium fluoride (NaF, dried at 105 C for 2 hr in water and dilute to volume in a 1-liter volumetric flask.

7.6 ETHANOL, DENATURED. Formula 30 denatured alcohol is satisfactory.

7.7 BORATE-CARBONATE FUSION MIXTURE. Intimately mix a 1:2 (w/w) combination of sodium tetraborate ($Na_2B_4O_7$) and sodium carbonate (Na_2CO_3).

8. Procedure

8.1 PREPARATION OF MONITORS. Disassemble the personal monitor, **6.2**, removing the membrane filter and cellulose pads. Moisten the pad with a measured volume of alkaline fixative solution, **7.3**; 0.8 ml is required for a pad of 37 mm diam. Dry the pad at 105 C for 30 to 45 min. Preparation of alkali-impregnated pads must be carried out in fluoride-free environment with minimum exposure.

8.2 Reassemble the filter monitor, inserting an impregnated pad and membrane filter, and closing with the filter retaining ring and front cover. The assembly is sealed against air leakage by a wrap of masking tape covering the crevice between retaining ring and back cover. Inlet and outlet openings of the monitor are closed with plastic plugs.

8.3 CALIBRATION OF PERSONAL MONITORING PUMP.

8.3.1 Select several of the prepared monitors at random, for calibration of air flow rate with the personal sampling pump **6.1**. Connect the monitor exit to the sampling pump by means of a 30″ length of hose. Connect the flowmeter (preferably a bubble flowmeter or wet test meter) to the inlet port of the monitor. Start the pump and adjust its rate, noting the position of the rotameter ball when a sampling rate of 2.5 l/min is indicated by use of the calibration flowmeter and a timer.

8.4 SAMPLING.

8.4.1 A worker whose exposure is to be evaluated is equipped with a personal monitor connected by a 30″ length of hose to a belt-supported sampling pump. The monitor is attached to the worker's collar and the cover removed. Air is drawn through the filter at the calibrated rate of 2.5 l/min and maintained at that rate by occasional checking and adjustment. On termination of sampling the duration of sampling is noted, the monitor is resealed and returned to the laboratory. A minimum air sample of 250 l should be filtered. This will allow the measurement of at least 1/10 times the TLV, 0.2 mg/m³ (760 Torr, 25 C).

8.4.2 Total particulate loading may be determined, if required, by pre- and post weighing of the membrane filter.

8.5 ANALYSIS OF SAMPLES.

8.5.1 *Particulate Fluoride*. Carefully remove the membrane filter from the filter holder and place it in a nickel or Inconel crucible containing about 0.5 g borate-carbontate fusion mixture. Transfer any dust from inside the filter cover and retaining ring to the crucible. Drench the filter with ethanol and ignite with a small gas flame. Heat the residue to fusion temperature for 1 to 2 min, then cool and dissolve in a few milliliters of water. Transfer the sample solution into a plastic beaker by means of 25 ml TISAB followed by a rinse of the crucible with a few drops of 1:1 hydrochloric acid. Adjust to pH 8 ± 0.1 by further addition of 1:1 hydrochloric acid. Dilute to 50 ml in a volumetric flask, mix, and bring to standard temperature (25 C). Pour about 20 ml of the solution into a plastic beaker and read the potential while the electrodes are immersed in the gently stirred solution. Convert potential (mV) to fluoride concentration (μg/ml) by means of the calibration chart determined from the standard fluoride series containing borate-carbonate flux **9.3**.

8.5.2 *Gaseous Fluoride*. Transfer the impregnated cellulose pad to a 100-ml plastic beaker containing 25 ml water and 25 ml TISAB. Allow the pad to soak for about 30 min with sufficient stirring to reduce it to a pulp. Bring the solution to standard temperature (25 C), insert the electrodes, and read potential of the gently stirred mixture after two minutes. Convert potential (mV) to fluoride concentration (μg/ml) by means of the calibration chart determined from the standard fluoride series **9.5**.

9. Calibration and Standards

9.1 FLUORIDE STANDARDS, PARTICULATES. Add 1.0 g borate-carbonate fusion mixture to each of four 250-ml beakers containing 10 ml water and 50 ml TISAB. Add various size aliquots (0.1, 1, 5, 10, and 25 ml) of 100 mg/l standard fluoride solution to produce a series of working standards (0.1, 1, 5, 10 and 25 μg F^-/ml). Adjust each solution to pH 8 ± 0.1 with 1:1 hydrochloric acid, transfer to a 100-ml volumetric flask, and dilute to volume with water. These standards may be stored for several months in tightly capped polyethylene bottles, under refrigeration.

9.2 For calibration of the electrode, pour about 20 ml of the working standard solution into a plastic beaker containing a Teflon-coated stirring bar. Adjust solution to within ± 2 C of the 25 C standard temperature. Insert the fluoride and reference electrodes into the constantly stirred solution and record the potential after 2 min. Repeat for each of the working standards.

9.3 Prepare a calibration curve, on three-cycle semi-log paper, relating potential, in millivolts (linear scale), to conc of fluoride, in μg/ml (log scale). Reproducibility of each point should be ± 1 mV. A linear calibration curve is obtained in the 0.1 to 25 μg/ml range, with a slope of between 57 and 59 mV per tenfold change in fluoride concentration. If solutions containing less than 0.1 μg/ml are measured, additional standards must be prepared since the calibration curve is not linear at low fluoride concentrations.

9.4 FLUORIDE STANDARDS, GASEOUS. Into each of four 250-ml beakers, place 10 ml water and 50 ml TISAB. Add various size aliquots (0.1, 1, 5, 10 and 25 ml) of 100 mg/standard fluoride solution to produce a series of working standards (0.1, 1, 5, 10 and 25 μg F^-/ml). Adjust each solution to pH 8.0 ± 0.1 with 1:1 hydrochloric acid, transfer to a 100-ml volumetric flask, and dilute to volume with water.

9.5 A separate calibration curve for gaseous fluoride is prepared from potential measurements of these standards **9.4**, by the same procedure as used for particulate fluorides, **9.2** and **9.3**.

10. Calculations

10.1 The concentration of particulate fluorides in air, integrated over the sampling period, is calculated as follows:

$$C_p = 0.05 \times \frac{C_1}{V}$$

where:

C_p = concentration of particulate fluoride, in milligrams per cubic meter

C_1 = concentration of fluoride in particulate sample solution, in micrograms per milliliter

V = volume of air sample, in cubic meters, corrected to 25 C and 760 Torr

10.2 The concentration of gaseous fluoride in air, integrated over the sampling period, is calculated as follows:

$$C_G = 0.05 \times \frac{C_2}{V}$$

where:

C_G = concentration of gaseous fluoride, in milligrams per cubic meter

C_2 = concentration of fluoride in gaseous sample solution, in micrograms per milliliter

V = volume of air sample, in cubic meters, corrected to 25 C and 760 Torr

10.3 If desired, gaseous fluoride concentration in air, in mg per m^3, may be converted to equivalent concentration expressed as parts per million: C_G mg/m^3 × 1.29 = C_G, ppm

11. References

1. ELFERS, L.A. and C.E. DECKER. 1968. Determination of Fluoride in Air and Stack Gas Samples by use of an Ion Specific Electrode. *Anal. Chem.* 40:1658.
2. ASTM METHOD D1606-60 (REAPPROVED 1967). Standard Method of Test for Inorganic Fluoride in the Atmosphere.
3. WEINSTEIN, L.H. and R.H. MANDL. 1971. The Separation and Collection of Gaseous and Particulate Fluorides. *VDI Berichte Nr.* 164, 53–63.
4. AMERICAN PUBLIC HEALTH ASSOCIATION, INC. 1977. Methods for Air Sampling and Analysis. 2nd ed. p. 426. Washington, D.C.

Revised by Subcommittee 2
C. R. THOMPSON, *Chairman*
G. H. FARRAH
L. V. HAFF
A. W. HOOK
J. S. JACOBSON
E. J. SCHNEIDER
L. H. WEINSTEIN

811.

Analytical Method for Fluoride in Urine

Analyte:	Fluoride	Method No.:	P & CAM 114
Matrix:	Urine	Range:	Lower Limit Urine 0.19 mg/liter
Procedure:	Ion Selective Electrode	Precision:	Unknown
Date Issued:	8/28/72	Classification:	D (Operational)
Date Revised:	1/1/75		

1. Principle of the Method

1.1 Urine samples are diluted with an equal volume of a pH-activity coefficient buffer solution.

1.2 A fluoride sensitive selective ion electrode and a reference electrode are inserted into the buffered sample. The observed potential difference between the electrodes can then be related to the fluoride concentraton in the sample via calibration curves prepared with the same buffer and electrodes.

2. Range and Sensitivity

2.1 The range and sensitivity of this method have not been established at this time. The electrode recommended range is 0.19 to 1900 mg/l urine.

3. Interferences

3.1 Hydroxide, the only positive interference in the measurement, is eliminated by use of the pH buffer. Negative interferences, which may result from complexion of fluoride by cations such as calcium, are minimized by the buffer and EDTA added as a stabilizer. The buffer also yields a solution of high ionic strength in which the activity coefficient of fluoride is relatively constant.

4. Precision and Accuracy

4.1 The accuracy and precision of this method have not been completely determined at this time. No collaborative tests have been performed on this method.

5. Advantages and Disadvantages of the Method

5.1 Advantages over previous methods include simplicity, accuracy, speed, specificity, and elimination of ashing, distillation or diffusion steps. Interfering ions are complexed.

5.2 No significant disadvantages are known.

6. Apparatus

6.1 FLUORIDE SENSITIVE SELECTIVE ION ELECTRODE. Model 94-09, Orion Research Corporation (Cambridge, Massachusetts) or equivalent.

6.2 REFERENCE ELECTRODE. Calomel or silver/silver chloride, sleeve junction preferable but not necessary.

6.3 Combination electrode can be used instead of **6.1** and **6.2**.

6.4 pH/MILLIVOLT METER. Capable of measuring to within 0.5 mV.

6.5 MAGNETIC STIRRER. Equipped with small stirring bars which can spin freely in the sample beakers.

6.6 PLASTIC BEAKERS (50-ML) AND SAMPLE COLLECTION BOTTLES (100-ML CAPACITY). Polyethylene or polypropylene.

6.7 LABORATORY GLASSWARE.

6.8 URINOMETER.

7. Reagents

7.1 PURITY. All chemicals used should be ACS Reagent grade or equivalent. After being prepared in clean glassware, all fluoride-containing solutions should be stored in plastic beakers and bottles.

7.1 DOUBLE DISTILLED WATER.

7.2 SODIUM CITRATE.

7.3 ETHYLENEDINITRILOTETRACETIC ACID (EDTA), DISODIUM SALT.

7.4 GLACIAL ACETIC ACID.

7.5 SODIUM CHLORIDE.

7.6 SODIUM HYDROXIDE, 5 M SOLUTION. Dissolve 20 g sodium hydroxide pellets in sufficient distilled water to give 100 ml of solution.

7.7 TOTAL IONIC STRENGTH ACTIVITY BUFFER (TISAB). Place 500 ml of double distilled water in a 1-liter beaker. Add 57 ml of glacial acetic acid, 85 g of sodium chloride, and 0.30 g of sodium citrate. Stir to dissolve. Place beaker in a water bath (for cooling). Slowly add 5 M sodium hydroxide until the pH is between 5.0 and 5.5. Cool to room temperature and pour into a 1-liter volumetric flask and add double distilled water to the mark.

7.8 SODIUM FLUORIDE. For preparation of standards.

7.9 STANDARD FLUORIDE SOLUTIONS. All solutions should be stored in plastic bottles. The 0.1 M fluoride standard is reported to be stable at least two months. All other dilutions should be prepared fresh weekly.

7.9.1 *0.1 M Fluoride (1900 μg F/ml).* Dissolve 4.20 g of sodium fluoride in water and dilute to 1000 ml. This solution can also be purchased from Orion Research, Inc., Cambridge, Massachusetts.

7.9.2 *10^{-2} M Fluoride (190 μg F/ml).* Dilute 10 ml of **7.9.1** to 100 ml with water.

7.9.3 *10^{-3} M Fluoride (19 μg F/ml).* Dilute 10 ml of **7.9.2** to 100 ml with water.

7.9.4 *10^{-4} M Fluoride (1.9 μg F/ml).* Dilute 10 ml of **7.9.3** to 100 ml with water.

7.9.5 *10^{-5} M Fluoride (0.19 μg F/ml).* Dilute 10 ml of **7.9.4** to 100 ml with water.

8. Procedure

8.1 CLEANING OF EQUIPMENT. All plastic and glassware must be rendered free of any fluoride before use.

8.2 COLLECTION AND SHIPPING OF SAMPLES.

8.2.1 Urine samples representing pre-exposure (early morning) and after exposure (end of working day) conditions should be taken. To minimize bacterial contamination, one should collect the mid-stream urine sample of a single voiding.

8.2.2 The urine samples are collected in chemically clean bottles containing 0.2 g of EDTA, disodium salt.

8.2.3 Upon receipt at the laboratory, the volume of the urine sample should be determined. If the volume is greater than 100 ml, an amount of EDTA equivalent to 0.2 g per 100 ml in excess of the first 100 ml should be added.

8.2.4 Samples that cannot be analyzed within 48 hr should be refrigerated.

8.3 ANALYSIS OF URINE SAMPLES.

8.3.1 Transfer 10 ml of well mixed urine sample and 10 ml of TISAB buffer solution into a 50-ml plastic beaker, and add a magnetic stirring bar.

8.3.2 Immerse the electrodes in the solution. Adjust stirring speed and stirrer position to avoid splashing and hitting the electrodes.

8.3.3 After the EMF (voltage) reading stabilizes (that is, EMF drift less than 0.5 mV per min) record the EMF to the nearest mV.

For most samples, the EMF will stabilize in a minute or two if the fluoride conc is 2 μg/ml or more. Stabilization may take slightly longer at lower concentrations.

8.3.4 Rinse the electrodes with double distilled water and blot with clean tissue before proceeding with next step.

8.3.5 Temperature of sample should be controlled within ± 2 C of standards.

9. Calibration Standards

9.1 Use the standard fluoride solutions prepared in **7.9** and treat them in the same way as the samples, Section **8.2** start with the most dilute solution **(7.9.5)** and proceed to **7.9.2** (10^{-2} M fluoride).

9.2 Prepare a calibration curve by plotting the concentration EMF data on semi-log paper with the fluoride concentration on the log axis.

9.3 Calibrations should be repeated at least twice daily or more frequently, if needed.

10. Calculations

10.1 The conc of fluoride is obtained by converting the mV readings into conc using the calibration curve.

10.2 The conc is reported as mgF per liter (numerically equivalent to μg per ml) of urine. Specific gravity correction for urine sample corrected

$$\text{mgF/l} = \text{mgF/l} \cdot \frac{(1.024 - 1000)}{(\text{specific gravity} - 1000)}$$

11. References

1. INSTRUCTION MANUAL FOR FLUORIDE ELECTRODE MODEL 94-09. Orion Research Incorporated, Cambridge, Massachusetts.
2. SUN, MU-WAN. 1969. Fluoride Ion Activity Electrode for Determination of Urinary Fluoride. *Amer. Ind. Hyg. Assoc. J.* 30:133.

Revised by Subcommittee 2
C. R. THOMPSON, *Chairman*
G. H. FARRAH
L. V. HAFF
A. W. HOOK
J. S. JACOBSON
E. J. SCHNEIDER
L. H. WEINSTEIN

812.

Analytical Method for Hydrogen Sulfide in Air

Analyte:	Hydrogen Sulfide	Method No.:	P & CAM 126
Matrix:	Air	Range:	0.008 to 50 ppm
Procedure:	Absorption—Methylene Blue—Spectrophotometric	Precision:	$CV_T = 0.121$
		Classification:	C (Tentative)
Date Issued:	6/9/72		
Date Revised:	1/1/75		

1. Principle of the Method

1.1 Hydrogen sulfide is collected by aspirating a measured volume of air through an alkaline suspension of cadmium hydroxide **(1)**. The sulfide is precipitated as cadmium sulfide to prevent air oxidation of the sulfide which occurs rapidly in an aqueous alkaline solution. STRactan 10 is added to the cadmium hydroxide slurry to minimize photo-decomposition of the precipitated cadmium sulfide **(2)**. The collected sulfide is subsequently determined by spectrophotometric measurement of the methylene blue produced by the reaction of the sulfide with a strongly acid solution of N,N-dimethyl-p-phenylenediamine and ferric chloride **(3, 4, 5)**. The analysis should be completed within 24 to 26 hr following collection of the sample.

1.2 Hydrogen sulfide may be present in the open atmosphere at conc of a few ppb or less. The reported odor detection threshold is in the 0.7 to 3.4 $\mu g/m^3$ (0.5 to 2.4 ppb) range **(6, 7)**. An exposure limit of 15 mg/m^3 (10 ppm) has been established **(8)** for work place areas.

1.3 Collection efficiency is variable below 10 $\mu g/m^3$ and is affected by the type of scrubber, the size of the gas bubbles and the contact time with the absorbing solution, and the concentration of H_2S **(9, 10, 11)**.

2. Range and Sensitivity

2.1 This method is intended to provide a measure of hydrogen sulfide in the range of 1.1 to 100 $\mu g/m^3$. For conc above 70 $\mu g/m^3$ the sampling period can be reduced or the liquid volume increased either before or after aspirating. (This method is also useful for the mg/m^3 range of source emissions. For example, 100 ml cadmium sulfide-STRactan 10 media in Greenberg-Smith impingers and 5 min sampling periods have been used successfully for source sampling.) The minimum detectable amount of sulfide is 0.008 $\mu g/ml$, which is equivalent to 0.20 $\mu g/m^3$ in an air sample of 1 m^3 and using a final liquid volume of 25 ml. When sampling air at the maximum recommended rate of 1.5 lpm for 2 hr, the minimum detectable sulfide conc is 1.1 $\mu g/m^3$ at 760 Torr and 25 C.

2.2 Excellent results have been obtained by using this method for air samples having a hydrogen sulfide content in the range 7 to 70 mg/m^3 (5 to 50 ppm).

3. Interferences

3.1 The methylene blue reaction is highly specific for sulfide at the low concentrations usually encountered in ambient air. Strong reducing agents (*e.g.* SO_2) inhibit color development. Even sulfide solutions containing several $\mu g\ S^{=}/ml$ show this effect and must be diluted to eliminate color inhibition. If sulfur dioxide is absorbed to give a sulfite conc in excess of 10 $\mu g/ml$, color formation is retarded. Up to 40 $\mu g/ml$ of this interference, however, can be overcome by adding 2 to 6 drops (0.5 ml/drop) of ferric chloride instead of a single drop for color development, and extending the reaction time to 50 min.

3.2 Nitrogen dioxide gives a pale yellow color with the sulfide reagents at 0.5 $\mu g/ml$ or more. No interference is encountered when 0.3 ppm NO_2 is aspirated through a midget impinger containing a slurry of cadmium hydroxide-cadmium sulfide-STRactan 10. If H_2S and NO_2 are simultaneously aspirated through cadmium hydroxide-STRactan 10 slurry, lower H_2S results are obtained, probably because of gas phase oxidation of the H_2S prior to precipitation as CdS (**11**).

3.3 Ozone at 57 ppb reduced the recovery of sulfide previously precipitated as CdS by 15% (**11**).

3.4 Sulfides in solution are oxidized by oxygen from the atmosphere unless inhibitors such as cadmium and STRactan 10 are present.

3.5 Substitution of other cation precipitants for the cadmium in the absorbent (*i.e.*, zinc, mercury, etc.) will shift or eliminate the absorbance maximum of the solution upon addition of the acid-amine reagent.

3.6 Cadmium sulfide decomposes significantly when exposed to light unless protected by the addition of 1% STRactan to the absorbing solution prior to sampling (**2**).

3.7 The choice of impinger used to trap H_2S with the $Cd(OH)_2$ slurry is very important when measuring concentration in the range 7 to 70 mg/m^3 (5 to 50 ppm). Impingers or bubblers having fritted-end gas delivery tubes are a problem source if the sulfide in solution is oxidized by oxygen from the atmosphere to free sulfur. The sulfur collects on the fritted-glass membrane and may significantly change the flow rate of the air sample through the system. One way of avoiding this problem is to use a midget impinger with standard, glass-tapered tips.

4. Precision and Accuracy

4.1 A relative standard deviation of 3.5% and a recovery of 80% have been established with hydrogen sulfide permeation tubes (**2**).

4.2 The overall sampling and analytical precision is 12.1% relative standard deviation (**18**).

5. Advantages and Disadvantages of the Method

5.1 Effect of Light and Storage.

5.1.1 Hydrogen sulfide is readily volatilized from aqueous solution when the pH is below 7.0. Alkaline aqueous sulfide solutions are very unstable because sulfide ion is rapidly oxidized by exposure to the air.

5.1.2 Cadmium sulfide is not appreciably oxidized even when aspirated with pure oxygen in the dark. However, exposure of an impinger containing cadmium sulfide to laboratory or to more intense light sources produces an immediate and variable photo-decomposition. Losses of 50 to 90% of added sulfide have been routinely reported by a number of laboratories. Even though the addition of STRactan 10 to the absorbing solution controls the photo-decomposition, it is necessary to protect the impinger from light at all times. This is achieved by the use of low actinic glass impingers, paint on the exterior of the impingers, or an aluminum foil wrapping.

6. Apparatus

6.1 ABSORBER—MIDGET IMPINGER. Should have a capacity of 25 ml and a standard glass gas delivery tube with a tapered nozzle.

6.2 AIR PUMP. With flowmeter and/or gasmeter having a minimum capacity of 2 lpm through a midget impinger.

6.3 COLORIMETER. With red filter or spectrophotometer at 670 nm.

6.4 AIR VOLUME MEASUREMENT. The air meter must be capable of measuring the air flow with ± 2%. A wet or dry gas meter, with contacts on the 1-ft^3 or 10-l dial to record air volume, or a specially calibrated rotameter, is satisfactory. Instead of these, calibrated hypodermic needles may be used as critical orifices if the pump is capable of maintaining greater than 0.7 atmosphere pressure differential across the needle **(12)**.

7. Reagents

7.1 PURITY. Reagents must be ACS analytical reagent quality. Distilled water should conform to the ASTM Standards for Reagent Water.

7.2 Reagents should be refrigerated when not in use.

7.3 AMINE-SULFURIC ACID STOCK SOLUTION. Add 50 ml conc sulfuric acid to 30 ml water and cool. Dissolve 12 g of N,N-dimethyl-p-phenylenediamine dihydrochloride (para-aminodimethylaniline) (redistilled if necessary) in the acid. Do not dilute. The stock solution may be stored indefinitely under refrigeration.

7.4 AMINE TEST SOLUTION. Dilute 25 ml of the Stock Solution to 1 l with 1:1 sulfuric acid.

7.5 FERRIC CHLORIDE SOLUTION. Dissolve 100 g of ferric chloride, $FeCl_3 \cdot 6H_2O$, in water and dilute to 100 ml.

7.6 AMMONIUM PHOSPHATE SOLUTION. Dissolve 400 g of diammonium phosphate, $(NH_4)_2HPO_4$, in water and dilute to 1 l.

7.7 STRACTAN 10, (ARABINOGALACTAN).

7.8 ABSORBING SOLUTION. Dissolve 4.3 g of $3CdSO_4 \cdot 8H_2O$, and 0.3 g sodium hydroxide in separate portions of water, mix, add 10 g STRactan 10 and dilute to 1 l. Shake the resultant suspension vigorously before removing each aliquot. The STRactan-cadmium hydroxide mixture should be freshly prepared. The solution is only stable for 3 to 5 days.

7.9 H_2S PERMEATION TUBE. Prepare or purchase* a triple-walled or thick-walled Teflon permeation tube **(11, 13, 14, 15, 16)** which delivers hydrogen sulfide at a maximum rate of approximately 0.1 μg per min at 25 C. This loss rate will produce a standard atmosphere containing 50 μg per m^3 (36 ppb H_2S) when the tube is swept with a 2 lpm air flow. Tubes having H_2S permeation rates in the range of 0.004 to 0.33 μg/min will produce standard air concentration in the realistic range of 2.7 to 220 μg/m^3 H_2S with an air flow of 1.5 lpm.

7.9.1 *Concentrated Standard Sulfide Solution*. Transfer freshly boiled and cooled 0.1 M NaOH to a liter volumetric flask. Flush with nitrogen to remove oxygen and adjust to volume. (Commercially available, compressed nitrogen contains trace quantities of oxgyen in sufficient concentration to oxidize the small concentrations of sulfide contained in the standard and dilute standard sulfide standards. Trace quantities of oxygen should be re-

*Available from Metronics, Inc., 3201 Porter Drive, Palo Alto, California 94304, or Polyscience Corp., 909 Pitner Avenue, Evanston, Illinois 60202.

moved by passing the stream of tank nitrogen through a pyrex or quartz tube containing copper turnings heated to 400 to 450 C.) Immediately stopper the flask with a serum cap. Inject 300 ml of H_2S gas through the septum. Shake the flask. Withdraw measured volumes of standard solution with a 10-ml hypodermic syringe and fill the resulting void with an equal volume of nitrogen. Standardize with standard iodine and thiosulfate solution in an iodine flask under a nitrogen atmosphere to minimize air oxidation. The approximate concentration of the sulfide solution will be 440 μg $S^{=}$/ml of solution. The exact concentration must be determined by iodine-thiosulfate standardization immediately prior to dilution.

For the most accurate results in the iodometric determination of sulfide in aqueous solution, the following general procedure is recommended:

a. Replacement of oxygen from the flask with an inert gas such as carbon dioxide or nitrogen.

b. Addition of an excess of standard iodine, acidification and back titration with standard thiosulfate and starch indicator **(17)**.

7.9.2 *Diluted Standard Sulfide Solution.* Dilute 10 ml of the conc sulfide solution to 1 liter with freshly boiled, distilled water. Protect the boiled water under a nitrogen atmosphere while cooling. Transfer the deoxygenated water to a flask previously purged with nitrogen and immediately stopper the flask. This sulfide solution is unstable. Therefore, prepare this solution immediately prior to use. This test solution will contain approximately 4 μg $S^{=}$/ml.

8. Procedure

8.1 Air Sampling.

8.1.1 Aspirate 10 ml of the absorbing solution in a midget impinger at 1.5 lpm for a selected period up to 2 hr. The addition of 5 ml of 95% ethanol to the absorbing solution just prior to aspiration controls foaming induced by the presence of STRactan 10. In addition, 1 or 2 Teflon demister discs may be slipped up over the impinger air-inlet tube to a height approximately 2.5 to 5 cm from the top of the tube.

8.2 Effect of Light and Storage.

8.2.1 Hydrogen sulfide is readily volatilized from aqueous solution when the pH is below 7.0. Alkaline, aqueous sulfide solutions are very unstable because $S^{=}$ ion is rapidly oxidized by exposure to the air.

8.2.2 Cadmium sulfide is not appreciably oxidized even when aspirated with pure oxygen in the dark. However, exposure of an impinger containing cadmium sulfide to laboratory or to more intense light sources produces an immediate and variable photo-decomposition. Losses of 50 to 90% of added sulfide have been routinely reported by a number of laboratories. Even though the addition of STRactan 10 to the absorbing solution retards photo-decomposition **(2)**, it is necessary to protect the impinger from light at all times through the use of low actinic glass impingers, paint on the exterior impingers or an aluminum foil wrapping.

8.3 Analysis. Add 1.5 ml of the amine-test solution to the midget impinger through the air inlet tube and mix. Add 1 drop of ferric chloride solution and mix. (Note: See section **3.1** if SO_2 exceeds 10 μg/ml in the absorbing media.) Transfer the solution to a 25-ml volumetric flask. Discharge the color due to the ferric ion by adding 1 drop ammonium phosphate solution. If the yellow color is not destroyed by 1 drop ammonium phosphate solution, continue dropwise addition until solution is decolorized. Make up to volume with distilled water and allow to stand for 30 min. Prepare a zero reference solution in the same manner using a 10-ml volume of unaspirated absorbing solution. Measure the absorbance of the color at 670 nm in a spectrophotometer or colorimeter set at 100% transmission against the zero reference.

9. Calibration and Standards

9.1 Aqueous Sulfide. Place 10 ml of the absorbing solution in each of a series of 25-ml volumetric flasks and add the diluted standard sulfide solution, equivalent to 1, 2, 3, 4, and 5 μg of hydrogen sulfide, to the different flasks. Add 1.5 ml of amine-acid test solution to each flask, mix, and add 1 drop of ferric chloride solution to each flask. Mix, make up to volume and allow to stand for 30 min. Determine the absorbance in a spectrophotometer at 670 nm against the sulfide-free reference solution. Prepare a standard curve of absorbance vs. μg H_2S/ml.

9.2 GASEOUS SULFIDE. Commercially available permeation tubes containing liquefied hydrogen sulfide may be used to prepare calibration curves for use at the upper range of atmospheric concentration. Triple-walled tubes, drilled rod and micro bottles which deliver hydrogen sulfide within a minimum range of 0.1 to 1.2 μg/min at 25 C have been prepared by Thomas **(10)**, O'Keeffe **(13, 14)**, and Scaringelli **(15, 16)**. Preferably the tubes should deliver hydrogen sulfide within a loss rate range of 0.14 to 2.9 μg/min to provide realistic concentrations of H_2S—70 to 1450 μg/m^3 (0.05 to 1.0) ppm)—within the range required for determining the collection efficiency of midget impingers. Analyses of these known conc give calibration curves which simulate all of the operational conditions performed during the sampling and chemical procedure. This calibration curve includes the important correction for collection efficiency at various concentrations of hydrogen sulfide.

9.2.1 Prepare or obtain a Teflon permeation tube that emits hydrogen sulfide at a rate of 0.14 to 2.9 μg/ml (at standard conditions of 25 C and 1 atmosphere). A permeation tube with an effective length of 3 to 10 cm and a maximum wall thickness of 0.318 cm will yield the desired permeation rate if held at a constant temperature of 25 C ± 0.1 C. Permeation tubes containing hydrogen sulfide are calibrated under a stream of dry nitrogen to prevent the precipitation of sulfur in the walls of the tube.

9.2.2 To prepare standard concentrations of hydrogen sulfide, assemble the apparatus consisting of a water-cooled condenser, constant-temperature bath maintained at 25 C ± 0.1 C, cylinders containing pure dry nitrogen and pure dry air with appropriate pressure regulators, needle valves and flow meters for the nitrogen and dry air diluent streams. The diluent gases are brought to temperature by passage through a 2-meter-long copper coil immersed in the water bath. Insert a calibrated permeation tube into the central tube of the condenser maintained at the selected constant temperature by circulating water from the constant-temperature bath and pass a stream of nitrogen over the tube at a fixed rate of approximately 50 ml/min. Dilute this gas stream to obtain the desired concentration by varying the flow rate of the clean, dry air. This flow rate can normally be varied from 2 to 15 lpm. The flow rate of the sampling system determines the lower limit for the flow rate of the diluent gases. The flow rate of the nitrogen and the diluent air must be measured to an accuracy of 1 to 2%. With tubes permeating hydrogen sulfide at rates of 0.14 to 2.9 μg/min., the range of conc of hydrogen sulfide will be between 70 to 1450 μg/m^3 (0.05 to 1.0 ppm), under these conditions a generally satisfactory range for workspace atmospheres. When higher concentrations are desired, calibrate and use longer permeation tubes.

9.2.3 *Procedure for Preparing Simulated Calibration Curves.* Obviously one can prepare a multitude of curves by selecting different combinations of sampling rate and sampling time. The following description represents a typical procedure for ambient air sampling of short duration, with a brief mention of a modification for 8 hr sampling.

The system is designed to provide an accurate measure of hydrogen sulfide in the 1.4 to 84 μg/m^3 (1 to 60 ppb) range without dilution. An aliquot of the absorbing solution can be taken easily to meet special needs.

The dynamic range of the colorimetric procedure fixes the total volume of the air sample taken at a maximum air concentration of 84 μg/m^3 to 186 l. To obtain linearity between the absorbance of the solution and the concentration of hydrogen sulfide, select a constant sampling time. Fixing of the sampling time is desirable also from a practical standpoint; in this case, select a sampling time of 120 min. Then, to obtain a 186 liter sample of air requires a flow rate of 1.55 liters/min.

A plot of the conc of hydrogen sulfide in ppm (x axis) against absorbance of the final solution (y axis) against absorbance of the final solution (y axis) will yield a straight line, the reciprocal of the slope of which is the factor for conversion of absorbance to ppm. This factor includes the correction for collection efficiency. Any deviation from the linearity at the lower concentration range indicates a change in collection efficiency of the sampling system. If the range of interest is above the dynamic range of the method the total volume of air collected should be decreased or an aliquot taken for colorimetric analysis to obtain a color within the dynamic range of the colorimetric procedure. Also, once the calibration factor has been established under simulated conditions, the conditions can be modified so that the concentration of H_2S is a simple multiple of the absorbance of the colored solution.

For longer-term sampling of 8 hr duration, the conditions can be fixed to collect 400 liters of sample in a larger volume of STRactan 10 cadmium hydroxide. For example, for 8 hr at 1.5 lpm approximately 720 liters of air is scrubbed. An aliquot is taken for the analysis containing a maximum of 15 μg H_2S.

The remainder of the analytical procedure is the same as described in the previous paragraph.

9.2.4 The permeation tubes must be stored in a wide-mouth glass bottle containing silica gel and solid sodium hydroxide to remove moisture and hydrogen sulfide. The storage bottle is immersed to two-thirds its depth in a constant-temperature water bath in which the water is controlled at 25 C ± 0.1 C.

Periodically (every 2 weeks or less), the permeation tubes are removed and rapidly weighed on a semimicro balance (sensitivity ± 0.01 mg) and then returned to the storage bottle. The weight loss is recorded. The tubes are ready for use when the rate of weight loss becomes constant (within ± 2%).

10. Calculations

10.1 STANDARD H_2S IN AIR. The conc of standard H_2S in air is computed as follows:

$$C = \frac{Pr}{F + f} \times 1000$$

where

C = Concentration of H_2S in $\mu g/m^3$.
Pr = Permeation rate in μg/min.
F = Flow rate of diluent air, lpm.
f = Flow rate of diluent nitrogen, lpm.

10.2 CORRECTION OF AIR SAMPLE VOLUME TO STANDARD CONDITIONS. Determine the sample volume (V) in liters from the gas meter or flow meter readings and time of sampling. Adjust volume to 760 Torr and 25 C (V_s) using the correction formula:

$$V_s = V \times \frac{P}{760} \times \frac{298}{(T + 273)}$$

where

V_s = Volume of air in liters at standard conditions.
V = Volume of air sampled in liters.
P = Barometric pressure in mm Hg
T = Temperature of sample air in °C.

10.3 Using the Beers-Law standard curve of absorbance vs. μg $S^=$ ion, determine μg $S^=$ ion in the sampling impinger corresponding to its absorbance reading at 670 nm.

10.4 Calculate the concentration of H_2S in the aspirated volume of air using the formula:

$$H_2S, (\mu g/m^3) = \frac{\mu g\ S^= \text{ ion} \times 10^3 \times V}{V_s A}$$

where

V_s = Volume of air in liters at standard conditions

V = Volume of absorbing solution, ml

A = Volume of aliquot, ml

11. References

1. JACOBS, M.B., M.M. BRAVERMAN and S. HOCHHEISER. 1957. Ultramicro-determination of sulfides in air. *Anal. Chem.* 29:1349.
2. RAMESBERGER, W.L. and D.F. ADAMS. 1969. Improvements in the collection of hydrogen sulfides in cadmium hydroxide suspension. *Environ. Sci. & Tech.* 3:258.
3. MACKLENBURG, W. and R. ROZENDRANZER. 1914. Colorimetric determination of hydrogen sulfide. *A. Anorg. Chem.* 86:143.
4. ALMY, L.H. 1925. Estimation of hydrogen sulfide in proteinaceous food products. *J. Am. Chem. Soc.* 47:1381.
5. SHEPPARD, S.E. and J.H. HUDSON. 1930. Determination of labile sulfide in gelatin and proteins. *Ind. Eng. Chem. Anal. Ed.* 2:73.
6. ADAMS, D.F., F.A. YOUNG and R.A. LUHR. 1968. Evaluation of an odor perception threshold test facility. *Tappi.* 51:62A.
7. LEONARDOS, G., D. KENDALL and N. BARNARD. 1969. Odor threshold determinations of 53 odorant chemicals. *J. Air Pollut. Contr. Assoc.* 19:91.
8. DANGEROUS PROPERTIES OF INDUSTRIAL MATERIALS. 1968. N.I. Sax, Editor, Third Edition, pg. 825, Van Nostrand Reinhold Co., New York.
9. BOSTRÖM, C.E. 1965. The absorption of sulfur dioxide at low concentrations (ppm) studies by an isotopic tracer method. *Air & Water Pollut. Int. J.* 9:333.
10. BOSTRÖM, C.E. 1966. The absorption of low concentrations (ppm) of hydrogen sulfide in a $Cd(OH)_2$ suspension as studies by an isotopic tracer method. *Air & Water Pollut. Int. J.* 10:435.
11. THOMAS, B.L. and D.F. ADAMS. Unpublished information.
12. LODGE, J.P., J.B. PATE, B.E. AMMONS, and G.A. SWANSON. 1966. The use of hypodermic needles as critical orifices. *J. Air. Poll. Control Assoc.* 16:197.
13. O'KEEFFE, A.E., and G.C. ORTMAN. 1966. Primary standards for trace gas analysis. *Anal. Chem.* 38:700.
14. O'KEEFFE, A.E., and G.C. ORTMAN. 1967. Precision picogram dispenser for volatile substances. *Anal. Chem.* 39:1047.
15. SCARINGELLI, F.P., S.A. FREY, and B.E. SALTZMAN. 1967. Evaluation of teflon permeation tubes for use with sulfur dioxide. *Am. Ind. Hyg. Assoc. J.* 28:260.
16. SCARINGELLI, F.P., E. ROSENBERG, and K. REHME. 1969. Stoichiometric comparison between permeation tubes and nitrite ion as primary standards for the colorimetric determination of nitrogen dioxide. Presented before the Division of Water, Air and Waste Chemistry of the American Chemical Society, 157th National Meeting, Minneapolis, Minn., April.
17. KOLTHOFF, I.M., and P.J. ELVING. Eds. 1961. Treatise on Analytical Chemistry, Part II, Analytical Chemistry of the Elements, V. 7. Interscience Publishers, New York.
18. Documentation of NIOSH Validation Tests. Contract No. CDC-99-74-45.

Revised by Subcommittee 1

D. F. ADAMS, *Chairman*
J. O. FROHLIGER
D. FALGOUT
A. M. HARTLEY
J. B. PATE
A. L. PLUMLEY
F. P. SCARINGELLI
P. URONE

813.

Analytical Method for Lead in Air

Analyte:	Lead	Method No.:	P & CAM 155
Matrix:	Airborne Dust and Fume	Range:	0.01 to 1.0 mg/M^3
		Precision:	5% RSD
Procedure:	Collection via Cellulose Filters/ Atomic Absorption	Classification:	D (operational)
Date Issued:	6/6/72		
Date Revised:	6/30/75		

1. Principle of the Method

1.1 Airborne dust and fume samples are collected on cellulose membrane filters.

1.2 The filter samples are ashed using nitric acid to destroy the organic matrix and solubilize the lead.

1.3 The lead content of the ashed material is determined by atomic absorption spectroscopy **(1, 2, 3, 4)**.

2. Range and Sensitivity

2.1 By atomic absorption the detection limit for lead in aqueous solution is about 0.1 μg/ml. Assuming the ashed sample is diluted to 10 ml this would correspond to 1 μg of lead/filter. For a 100 l air sample this would represent a lead conc of 0.01 mg/m^3.

2.2 The optimum working range for Pb extends up to about 10 μg/ml. If necessary the sample solution can be diluted to extend the upper limit of the analysis **(5)**.

3. Interferences

3.1 Calcium and high concentrations of $SO_4^=$ can interfere. Samples should be periodically analyzed by the method of additions to check for such interferences.

4. Precision and Accuracy

4.1 The precision of this method as determined by analysis of spiked, unexposed filters is 5% relative standard deviation. When precision was tested using duplicate fractions cut from filters, the average percent deviation for 11 pairs was ± 2% **(6)**.

5. Advantages and Disadvantages

5.1 The method is rapid and simple. Ashing of samples can be performed rapidly and the analysis by means of AA requires only a very short time per sample. Untreated filter samples may be stored indefinitely.

5.2 Atomic absorption equipment may not be available to some small laboratories.

6. Apparatus

6.1 Sampling Equipment

6.1.1 *Filter unit.* Consisting of the filter media and appropriate cassette filter holder, either a 2 or 3-piece filter cassette.

6.1.2 *Personal sampling pump.* This pump must be properly calibrated so the volume of air sampled can be measured as accurately as possible. The pump must be calibrated with a representative filter unit in the line. Flow rate, time, and/or volume must be known.

6.1.3 *Thermometer.*

6.1.4 *Manometer.*

6.1.5 *Stopwatch.*

6.1.6 *Cellulose membrane filters.* 0.8 μ pore diam.

6.2 GLASSWARE, BOROSILICATE. Before use all glassware must be cleaned in 1:1 diluted nitric acid and rinsed several times with distilled water.

6.2.1 *125*-ml *Phillips or Griffin beakers, with watch glass covers.*

6.2.2 *10*-ml *volumetric flasks.*

6.3 ATOMIC ABSORPTION SPECTROPHOTOMETER. With air-acetylene flame system and recorder or concentration readout system.

6.4 PB HOLLOW CATHODE, OR ELECTRODELESS DISCHARGE LAMP.

7. Reagents

7.1 PURITY. ACS reagent grade chemicals or equivalent shall be used in all tests. References to water shall be understood to mean double distilled water or equivalent.

7.2 Care in selection of reagents and in following listed precautions is essential if low blank values are to be obtained.

7.3 Conc nitric acid (68 to 71%) redistilled specific gravity 1.42.

8. Procedure

8.1 COLLECTION OF SAMPLES.

8.1.1 Assemble the filter unit by mounting the filter disc in the filter cassette.

8.1.2 Connect the exit end of the filter unit to the pump with a short piece of flexible tubing.

8.1.3 Turn on pump to begin sample collection. The flow rate, times, and/or volume must be measured as accurately as possible. The sample should be taken at a flow rate of 1 to 2 l/m. A minimum sample of 100 liters should be collected. Larger sample volumes are encouraged, provided the filters do not become loaded with dust to the point that loose material might fall off or the filter become plugged.

8.1.4 With each batch of samples, one filter, labelled as a blank, must be submitted. No air should be drawn through this filter.

8.2 SHIPMENT OF SAMPLES.

8.2.1 The sample cassettes should be shipped in a suitable container to minimize contamination and to prevent damage in transit. Care must be taken during storage and shipping that no part of the sample is dislodged from the filter, nor that the sample surface be disturbed in any way. Loss of sample from heavy deposits on the filter may be prevented by mounting a clean filter in the cassette on top of the sample filter.

8.3 PREPARATION OF SAMPLES.

8.3.1 Samples and blank filters are placed in clean 125-ml Phillips beakers, then 2 to 3 ml conc HNO_3 is added to each.

8.3.2 Each beaker is heated on a hot plate in a fume hood until a clear or slightly yellow solution remains and the volume has been reduced to less than 1 m*l*. (Some HNO_3 should be left in the flask to maintain low pH in resulting solution.) More than one addition of HNO_3 may have to be made to achieve the clear solution. At this point all organic matter has been adequately destroyed and the solution is allowed to cool. It is then quantitatively transferred (by rinsing with double distilled water) to a 10-ml volumetric flask and brought to the mark. Aliquots of this solution may be diluted should analysis indicate that the lead concentration is above the working range.

8.4 ANALYSIS OF SAMPLES

8.4.1 Set the instrument operating conditions as recommended by the manufacturer. The instrument should be set at the radiation intensity maximum (283.3 nm).

8.4.2 Standard solutions should match the sample matrix as closely as possible and should be run to duplicate. Working standard solutions prepared fresh daily, are aspirated into the flame and the absorbance recorded. Prepare a calibration graph as described in section **9.1.4**. (**Note:** All combustion products from the AA flame must be removed by direct exhaustion through use of a good separate flame ventilation system.)

8.4.3 Aspirate the appropriately diluted samples directly into the instrument and record the absorbance for comparison with standards. Should the absorbance be above the calibration range, dilute an appropriate aliquot to 10 ml. Aspirate water between each sample. A mid-range standard must be aspirated with sufficient frequency (*i.e.*, once every 10 samples) to assure the accuracy of the sample determinations. To the extent possible, all determinations are to be based on replicate analysis.

9. Calibrations and Standards

9.1 STANDARD METAL SOLUTIONS.

9.1.1 *Master lead standard.* Dissolve 1.598 g lead nitrate [$Pb(NO_3)_2$] in 2% (v/v) HNO_3. Dilute to volume in a 1-l volumetric flask with 2% (v/v) HNO_3. This solution is stable for about 6 months.

9.1.2 *Dilute lead standard (100 μg/ml).* Pipet 10 ml of the master lead standard into a 100-ml volumetric flask, add 10 ml of 2% HNO_3 and dilute to volume with water. Prepare fresh daily.

9.1.3 *Working lead standards. (0 to 10 μg/ml).* Prepare standards by pipetting 2 ml of the dilute standard into a 100-ml volumetric flask and diluting to volume with 2% (v/v) HNO_3. Repeat with 4, 6, 8 and 10 ml of dilute standard to prepare standards containing 2, 4, 6, 8 and 10 μg/ml repectively.

9.1.4 The standard solutions are aspirated into the flame and the absorbance (or concentration) recorded. If the instrument used displays transmittance, these values must be converted to absorbance. A calibration curve is prepared by plotting absorbance (concentrations) versus metal concentration. The best fit curve (calculated by linear least square regression analysis) is fitted to the data points. This line or the equation describing the line is used to obtain the metal concentration in the samples being analyzed.

9.1.5 To ensure that the preparation procedure is being properly followed, clean membrane filters are spiked with known amounts of the element being determined by adding appropriate amounts of the previously described standards and carried through the entire procedure. The amount of metal is determined and the per cent recovery calculated. These tests will provide recovery and precision data for the procedure as it is carried out in the laboratory.

9.1.6 *Analysis by the method of standard additions.* In order to check for interferences, samples are initially and periodically analyzed by the method of standard additions and the results compared to those obtained by the conventional analytical determination. For this method the sample is divided into three two ml aliquots. To one of the aliquots an amount of metal approximately equal to that in the sample is added. To another aliquot twice this amount of metal is added. (**Note:** Additions should be made by micropipetting techniques such that the volume does not exceed 1% of the original aliquot volume—i.e., 10-μl and 20-μl additions to a 2-ml aliquot.) The solutions are then analyzed and the absorbance readings are plotted against metal added to the original sample. The line obtained from such a plot is extrapolated to 0 absorbance and the intercept on the concentration axis is taken as the amount of metal in the original sample (**2**). If the result of this determination does not agree to within 10% of the values obtained with the procedure described in section **9.1.4**, an interference is indicated and standard addition techniques should be utilized for sample analysis.

9.2 BLANK. Blank filters must be carried through the entire procedure each time samples are analyzed.

10. Calculations

10.1 The uncorrected volume collected by the filter is calculated by averaging the beginning and ending sample flow rates, converting to cubic meters and multiplying by the sample collection time. The formula for this calculation is given in **10.1.1**.

10.1.1 $$V = \frac{F_B + F_E}{2 \times 1000} t$$

where:

V = uncorrected sample volume (m^3)
F_B = sample flow rate at beginning of sample collection (lpm)

F_E = sample flow rate at end of sample collection (lpm)
t = sample collection time (minutes)

10.2 The volume is corrected to 25 C and 760 Torr by using the formula given in **10.2.1**.

10.2.1 $V_{corr} = \frac{(298)(P)(V)}{(760)(T)}$

where:

V_{corr} = corrected sample Volume (m^3)
P = average barometric pressure druing sample collection period (mm Hg).
T = average temperature during the sample collection period (°K). (Note: °K = °C + 273.)
V = uncorrected volume calculated from **10.1.1** (m^3).

10.3 After any necessary correction for the blank has been made, metal concentrations are calculated by multiplying the μg of metal per ml in the sample aliquot by the aliquot volume and dividing by the fraction which the aliquot represents of the total sample and the volume of air collected by the filter. **10.3.1** indicates the formula for this calculation.

10.3.1 $\mu g\ metal/m^3 = \frac{(C \times V) - B}{V_{corr} \times F}$

where:

C = concentration (μg metal/ml) in the aliquot
V = volume of aliquot (ml)
B = total μg of metal in the blank
F = fraction of total sample in the aliquot used for measurement (dimensionless)
V_{corr} = corrected volume of air samples and calculated from **10.2.1**

11. References

1. SLAVIN, W. 1968. Atomic Absorption Spectroscopy. Interscience Publishers, N.Y., N.Y.
2. RAMIREZ-MUNOZ, J. 1968. Atomic Absorption Spectroscopy, Elsevier Publishing Co., N.Y., N.Y.
3. DEAN, J.A. and T.C. RAINS. ed. 1969. Flame Emission and Atomic Absorption Spectrometry—Volume I, Theory, Marcel Dekker, N.Y., N.Y.
4. WINEFORDNER, J.D. ED. 1971. Spectrochemical Methods of Analysis. John Wiley & Sons, Inc., N.Y.
5. BELCHER, C.B., R.M. DAGNALL and T.S. WEST. 1964. An Examination of the Atomic Absorption Spectroscopy of Silver, Talanta 11:1257.
6. EISENBUD, M. and T.J. KNEIP, eds. 1975. Third Annual Report: Trace Metals in Urban Aerosols, submitted to Electric Power Research Institute and American Petroleum Institute by New York University Medical Center, Institute of Environmental Medicine, in press.

Subcommittee 6
T. J. KNEIP, *Chairman*
R. S. AJEMIAN
J. N. DRISCOLL
F. I. GRUNDER
L. KORNREICH
J. W. LOVELAND
J. L. MOYERS
R. J. THOMPSON

814.

Analytical Method for Lead in Blood and Urine

Analyte:	Lead	Method No:	P & CAM 208
Matrix:	Blood, Urine	Range:	0.05 – >1.5 μg per gram of blood or per ml of urine
Procedure:	Chelation, Extraction, AA		
Date Issued:	1/1/75	Precision:	Relative standard deviation ± 5%, (analytical)
Date Revised:			
		Classification:	E (Proposed)

1. Principle of the Method

1.1 Whole blood is collected in a lead-free sampling tube containing heparin anticoagulant; urine is collected in a polyethylene bottle containing nitric acid preservative. Blood is hemolyzed, the lead complexed with ammonium pyrrolidene dithiocarbamate and the complex extracted into methyl isobutyl ketone. Urine is similarly extracted, with hemolysis not required. The lead content of the organic phase is determined by atomic absorption spectrometry at 283.3 nm using an air-acetylene flame (**1**).

2. Range and Sensitivity

2.1 This method is applicable to the determination of lead in blood and urine in an optimum range of 0.1 to 1.5 μg lead per g blood or per ml urine.

2.2 The sensitivity of the method is 0.05 μg lead per ml methyl isobutyl ketone (MIBK) solution for a 1% change in transmittance (0.0044 absorbance units).

2.3 The detection limit of the method is 0.05 μg lead per g blood or per ml urine.

3. Interferences

3.1 Phosphate buffers, EDTA and other complexing agents used as anticoagulants or preservatives may cause reduced recoveries. Addition of citrate will eliminate phosphate interference (**2**) and excess calcium will prevent interference by EDTA (**3**).

3.2 Blood specimens stored at 18 C and above for 5 days or more may show poor lead recoveries, typically 20 to 80% (**2**).

3.3 Ammonium pyrrolidene dithiocarbamate (APDC) is an unstable reagent, extraction efficiency should be checked before use.

3.4 The Pb-APDC complex in MIBK is unstable and must be analyzed within 2 hr of extraction.

4. Precision and Accuracy

4.1 Relative standard deviations of about 5% at the 0.45 μg lead per g blood are obtained (**2**).

4.2 Analytical data obtained by this procedure agree closely with that obtained by several other analytical procedures, including the carbon rod atomizer (**4**), Delves cup (**5**) and anodic stripping voltammetry (**6**).

5. Advantages and Disadvantages of the Method

5.1 The method is rapid and does not require a high level of technical skill. It is not necessary to ash the sample with a mixture of nitric and perchloric acid to destroy the organic matter. Over 99% extraction of lead by MIBK solvent from blood and urine samples has been achieved without wet ashing (3). Addition of a non-ionic surfactant, Triton X-100, to the APDC solution allows aspiration of the MIBK solvent layer into the flame of the atomic absorption spectrometer. The only known disadvantage of this method is the need for AA spectrometer equipment.

6. Apparatus

6.1 SAMPLE COLLECTION.

6.1.1 *Heparinized sample collection tubes, 10-ml.* Certified to contain less than 0.1 μg lead per tube (Vacutainer®, Becton-Dickinson, Rutherford, New Jersey,) or equivalent.

6.1.2 *Polyethylene or polystyrene syringes 5-ml.*

6.1.3 *Polyethylene bottles, 25-ml and 2 l.*

6.1.4 *25- × 100-mm swab tubes, or equivalent fitted with polyethylene caps.*

6.2 GLASSWARE. Borosilicate glassware should be used throughout the analysis. The glassware must be washed with 1:1 HNO_3 and rinsed with distilled water. Beakers, pipets and auxiliary glassware will be required.

6.3 EQUIPMENT.

6.3.1 *Centrifuge 2000 rpm.*

6.3.2 *Vortex mixer* (Vortex-Genie, Scientific Industries, Inc., New York or equivalent).

6.3.3 *Atomic absorption spectrometer.*

6.3.4 *Weighing syringe, polypropylene syringe with Teflon coated plunger.*

6.3.5 *Lead hollow cathode lamp.*

6.3.6 *Air supply.* Filtered to remove oil and water, and with a minimum pressure of 40 psi.

6.3.7 *Acetylene gas.* (Tank should be changed when pressure < 100 psi). The grade recommended by the manufacturer of the atomic absorption spectrometer should be used.

6.3.8 *Hydrometer.*

7. Reagents

7.1 PURITY. All reagents should be ACS reagent grade or the equivalent. All references to water are double-distilled water or the equivalent.

7.2 NON-IONIC SURFACTANT. Triton X-100 octyl phenoxy polyethoxyethanol (Sigma Chemical, St. Louis, Mo.), or equivalent.

7.3 AMMONIUM PYRROLIDINE DITHIOCARBAMATE (APDC). APDC is unstable and each new batch should be checked for extraction efficiency. The reagent is stable for about 6 months.

7.4 METHYLISOBUTYL KETONE (MIBK).

7.5 LEAD NITRATE $Pb(NO_3)_2$. Dry at 140 C for 1 hr prior to use.

7.6 NITRIC ACID, CONC. SP GR = 1.421.

7.7 MIBK—WATER SATURATED. Add 100 ml water to 900 ml MIBK, and saturate by shaking and allowing to stand 1 hr.

7.8 TRITON X-100/APDC SOLUTION (TX-APDC) 2.5% TRITON X-100, 2% APDC. Add 4 g APDC to about 40 ml water, then add 5 ml Triton X-100 and mix until dissolved. Make up to 200 ml. Prior to use check the reagent by analyzing standards using both the new and old batches of reagent.

7.9 CALCIUM CHLORIDE 1.5 M. Dissolve 16.648 g $CaCl_2$ in a few ml water, make up to 100 ml.

8. Procedure

8.1 SAMPLE COLLECTION.

8.1.1 At least 4 ml blood is collected by venipuncture in sample collection tubes. Immediately after collection, the tube is shaken vigorously to ensure thorough mixing with the anticoagulant. Samples are stored at 4 C prior to shipping to the laboratory and in the laboratory prior to analysis. During shipping, samples should not be exposed to high ambient temperatures (ideally maintained at temperatures below 25 C). Samples should be promptly analyzed, preferably within 3 days of collection.

8.1.2 Urinc samples (preferably 24 hr) are collected in a 2-1 polyethylene bottle containing *ca.* 0.5 ml conc nitric acid. After collection, the sample is thoroughly mixed and an aliquot of about 20 ml is transferred to a 25-ml bottle for shipping to the laboratory.

8.2 SAMPLE PREPARATION.

8.2.1 *Blood.* Shake the sample collection tube, draw the entire sample into a weighing syringe and weigh. Disperse a 2-g aliquot into a swab tube and reweigh. Disperse a second aliquot.

8.2.2 *Urine.* Filter the urine and measure two 2 ml fractions into a swab tube.

8.2.3 Add 0.8 ml TX-APDC solution to each swab tube, cap the tube and mix on a vortex mixer for 10 sec. Remove the cap, add 2.00 ml MIBK, cap the tube and shake for 2 min, inverting the tube at regular intervals. Centrifuge at 2000 rpm for 10 min.

NOTES: **(a)** Poor mixing will lead to low recovery.

(b) Do not allow aqueous phase from the MIBK to be included in the aliquot.

(c) If the patient is receiving EDTA therapy add 50 μl 1.5 M $CaCl_2$ immediately before adding the MIBK.

8.3 ANALYSIS.

8.3.1 Allow the atomic absorption spectrometer and hollow cathode lamp to warm up for 30 min and operate an air-acetylene flame for 15 min (aspirating water) to warm up the burner. Set the wavelength at 283.3 nm and set the instrument parameters according to manufacturers recommendations.

8.3.2 Aspirate MIBK. This will increase the fuel content of the flame so it will be necessary to reduce acetylene flow to produce a stoichiometric flame. After a further 30 sec to allow the nebulizer-burner to stabilize, take a background reading.

8.3.3 Aspirate the standards and record the absorbance. Prepare a calibration curve as described in Section **9.1**.

8.3.4 Aspirate the samples plus blanks and record absorbances for comparison with the standards. Should the absorbance be greater than 0.35, dilute an appropriate aliquot with MIBK to 10 ml. Be certain to aspirate MIBK between samples. Standards must be aspirated with sufficient frequency to assure the accuracy of the sample determinations. Aspirate a 0.80 μg lead/ml standard every tenth sample.

9. Standards and Calibration

9.1 STANDARD LEAD SOLUTION.

9.1.1 *Master Standard (1000 μg lead/ml).* Heat lead nitrate for 4 hr at 120 C, cool in a desiccator, dissolve 1.598 g lead nitrate in one liter of 2% nitric acid. This solution is stable for 1 year when stored in a polyethylene bottle.

9.1.2 *Working lead standards.* Each day, dilute 10.00 ml master standard to 1 liter using a volumetric flask. Aliquot exactly 2, 4, 6, 8, 10 and 15 ml respectively of this diluted solution into 100-ml volumetric flasks and dilute to volume. 2.00 ml standards are analyzed as described in Sections **8.2.3** and **8.3**.

9.2 BLANK. Blank water samples (2 g) must be carried through the entire procedure.

10. Calculations

10.1 The net signal for each sample, standard and blank is computed from:

$$\text{Net Signal} = \frac{S - (B_1 + B_2)}{2}$$

Where

S = Sample signal
B_1 = Background signal, read immediately before the sample signal.
B_2 = Background signal, read immediately after the sample signal.

10.2 A least squares curve of best fit is constructed from the duplicate measurements of standards and blank over the range 0 to 1.5 μg lead per ml. Blanks should contain less than 0.05 μg lead/ml.

10.3 If a 24 hr urine sample was not collected, the lead level can be corrected to a standard urine specific gravity of 1.024 by multiplying by the factor $\frac{(1.024 - 1000)}{(S.G. - 1.000)}$ where S.G. is the specific gravity of the urine sample.

11. References

1. HESSEL, D.W. 1968. *Atomic Absorption Newsletter* 7:55.
2. MITCHELL, D.G., F.J. RYAN, and K.M. ALDOUS. 1972. *Atomic Absorption Newsletter* 11:120.
3. ZINTERHOFER, L.J.M., P.I. JATLOW, and A. FAPPIANO. 1971. *J. Lab. Clin. Med.* 78:664.
4. ROSEN, J.F. 1972. *J. Lab. Clin. Med.* 80:567.
5. MITCHELL, D.G., K.M. ALDOUS, and F.J. RYAN. 1972. International Symposium, Environmental Health Aspects of Lead. Amsterdam, Holland.
6. SEARLE, B., W. CHAN, and B. DAVIDOW. 1973. *Clin. Chem.* 19:76.

Subcommittee 6
T. J. KNEIP, *Chairman*
R. S. AJEMIAN
J. N. DRISCOLL
F. I. GRUNDER
L. KORNREICH
J. W. LOVELAND
J. L. MOYERS
R. J. THOMPSON

815.

Analytical Method for Mercury in Urine

Analyte:	Mercury	Method No.:	P & CAM 165
Matrix:	Urine	Range, Urine:	0.003 > 0.3 mg/l
Procedure:	Flameless Atomic Absorption	Precision:	± 3% (analytical)
Date Issued:	2/13/73	Classification:	B (Accepted)
Date Revised:	1/1/75		

1. Principle of the Method

1.1 After initial decomposition of the urine sample with nitric acid, the mercury is reduced to its elemental state with stannous chloride.

1.2 Mercury vapor is bubbled through an absorption cell of an atomic absorption unit and the response at 253.7 nm is measured either as peak height or peak area.

1.3 The mercury content of the sample is calculated from calibration data for the sparger photometer system obtained by carrying known amounts of ionic mercury standards through the analysis procedure.

2. Range and Sensitivity

2.1 For a 1.0 ml urine sample the sensitivity is 0.003 mg/l or below. This corresponds to an absolute sensitivity of 3 ng of mercury.

2.2 The range extends up to 0.3 mg/l for a 1.0 ml aliquot. The range can be extended beyond 0.3 mg/l by taking an aliquot of urine that is less than 1.0 ml.

3. Interferences

3.1 Metals such as gold, platinum and copper will interfere since they form an alloy with the reduced Hg.

3.2 Certain organic solvents, such as benzene, that absorb 253.7 nm radiation would interfere if present in significant amounts as would any substance with broad absorption such as dust, water droplets, etc.

4. Precision and Accuracy

4.1 Comparisons of bubbler recoveries from water standards and urine samples spiked with identical amounts of radio or stable mercury indicate that the urine samples give a 10% lower signal (Figure 815:1). Although the use of water standards may reduce the accuracy of the method, results may be acceptable for routine analyses. Where more accurate values are needed a multiple addition technique using the same urine spiked with Hg standard is recommended.

4.2 Absolute deviations from the means between sets of identical samples varied from 0.000 to 0.005 mg/l. The average relative standard deviation (1σ) was ± 3%.

4.3 The temperature of the water used to dilute the sample should be controlled to within ± 1 C.

4.4 At least 5 ml of nitric acid/l ml sample should be used to insure the release of metabolized mercury.

4.5 Only aliquots of urine less than one ml should be used if water standards are used. *Larger* aliquots of urine would increase the difference in transfer efficiency between water standards and urine (Figure 815:1).

5. Advantages and Disadvantages of the Method

5.1 A trained technician can do 20 to 40 samples a day.

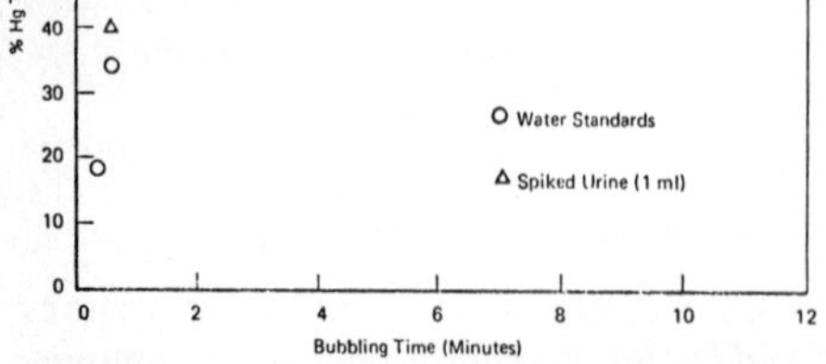

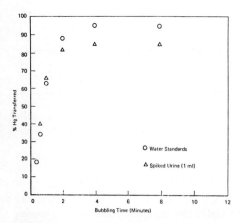

Figure 815:1—Transfer Efficiency in the Bubbler System

6. Apparatus

6.1 MERCURY EVOLUTION TRAIN.

6.1.1 *Bubbler Flask.* 300-ml BOD sample bottle with coarse fritted bubbler extending to within 0.3 cm of bottom of bottle.

6.1.2 *Connecting Tubing.* Minimum lengths of either glass or Tygon.

6.1.3 *Rotamater.* 0.5 to 3 lpm range.

6.1.4 *Drying Tube.* Approximately 8-cm length (optional).

6.1.5 *Midget Impinger.* To act as safety trap.

6.2 ANALYTICAL EQUIPMENT.

6.2.1 The photometric detector system consits of a mercury discharge lamp capable of isolating the 253.7 nm mercury emission line, the photometer cell and a photo-detector-amplifier system. Many commercial and laboratory constructed versions have been described (**1**). Typical units are manufactured by Perkin-Elmer Corporation, Coleman Instrument Corporation and Laboratory Data Contol, Inc.

6.2.2 *Pipets.*

7. Reagents

7.1 PURITY OF REAGENTS. All reagents used must be analytical reagent grade.

7.2 DOUBLE DISTILLED WATER.

7.3 NITRIC ACID, CONC.

7.4 STANNOUS CHLORIDE SOLUTION, 20% (w/v) in 6 N HCl. Freshly prepared.

7.5 STANDARD STOCK MERCURY SOLUTION. Add 0.1000 g metallic mercury or 0.1713 g mercuric nitrate. $Hg(NO_3)_2 \cdot H_2O$, into a clean, dry 100-ml volumetric flask. Add 10 ml conc HNO_3 dissolve the mercury and then dilute to the mark with double distilled water. Pipet exactly 10.0 ml of Hg into a one-liter volumetric flask. Add 50 ml conc HNO_3 and dilute to the mark with water. The final concentration is 10 μg/ml. This solution is stable for 6 months.

7.6 WORKING STANDARD MERCURY SOLUTION. Transfer 10.0 ml of the standard stock solution to a 1-l volumetric flask. Add 50 ml conc HNO_3 and dilute to the mark with double distilled water. The final conc is 0.1 μg/ml. This solution should be prepared fresh daily.

7.7 Potassium Persulfate. Low nitrogen.

7.8 Magnesium Perchlorate, anhydrous (for drying tube). The drying tube must be repacked with fresh magnesium perchlorate after 20 analyses.

7.9 Antifoam Solution (**1**). Suspend 5 g of General Electric "antifoam 60" in 95 ml of double distilled water. Antifoam 60 is available from General Electric Company, Silicone Products Division, Waterford, New York.

8. Procedure

8.1 Cleaning of Equipment. Acid-clean all glassware before use. This can be done by using the following procedure.

8.1.1 Wash in 1% Na_3PO_4 solution and follow with tap water.

8.1.2 Soak in conc HNO_3 for 30 min and follow with tap, distilled and double-distilled water rinses.

8.2 Collection and Shipping of Samples.

8.2.1 Urine samples must be collected in acid cleaned borosilicate bottles. At least 25 ml should be collected.

8.2.2 The samples must be preserved by the addition of 0.1 g potassium persulfate per 100 ml of urine. Urine treated with potassium persulfate is stable at room temperature for 2 weeks.

8.3 Analysis of Samples.

8.3.1 Transfer a 1.0 ml urine sample to the BOD bottle. If a sediment is present in the collection bottle, shake well immediately before removing the aliquot. Add 5 ml of conc HNO_3 to the BOD bottle and allow the sample to stand at room temperature for 3 min.

8.3.2 Add 95 ml of distilled water (at the same temperature) blowing out any mist that may form above the liquid with gentle puffs of air or nitrogen. Add 2 drops antifoam solution. Add 5.0 ml of $SnCl_2$ solution and then immediately connect the flask to the bubbler tube in the generating train, shown in Figure 815:2. The samples and standards should be at the same temperature (± 1 C) since variations in water temperatures produce variations in peak height of nearly 1% per degree Centigrade (**1**).

8.3.3 Turn on pump and blow air through the flask at a constant rate. Recommended air flow is between 0.5 to 3 lpm to obtain maximum peaks. Rate of flow must be determined experimentally and thereafter held constant.

8.3.4 Observe or record the maximum absorbance by measuring peak heights or areas. If the reading goes beyond calibration range, repeat the analysis using a smaller sample volume or if too low take larger quantity up to 5 ml of urine.

8.3.5 Remove the BOD bottle and allow the signal to return to zero. Analyze each sample in duplicate using two different volumes of urine.

9. Calibration and Standards

9.1 Standards for instrument calibration are analyzed exactly as samples except that known amounts of mercury are added to the sample.

9.2 The calibration curve for the multiple addition method is constructed by subtracting the reading of the unspiked urine from that of the spiked aliquots.

9.3 Prepare a calibration curve by plotting absorbance versus μg mercury added. Standards of 0.0, 0.05, 0.1, 0.2, 0.3 and 0.4 μg should be plotted.

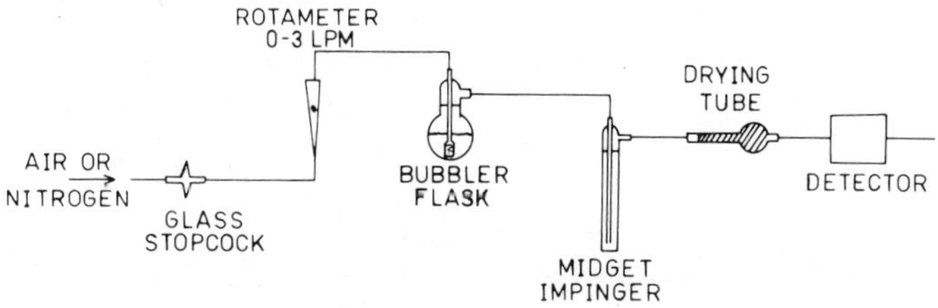

Figure 815:2—Schematic Diagram of Test Apparatus

9.4 Run a calibration before analysis of samples, then insert a standard between every 5th, or 6th sample to maintain a check on method integrity.

10. Calculations

10.1 Subtract the background signal, if any, from the sample signal.

10.2 Micrograms of mercury are determined directly from the calibration curve using the corrected signal peak height obtained for each sample.

10.3 The mercury concentration is calculated by dividing the μg of mercury by the sample volume analyzed to give μg Hg per liter.

10.4 The mercury level may be corrected to a standard urine specific gravity (S.G.) of 1.024 by use of the factor

$$\frac{(1.024 - 1.000)}{(S.G. - 1.000)}$$

11. References

1. RATHJE, ARNOLD O. 1969. A Rapid Ultraviolet Absorption Method for the Determination of Mercury in Urine. *Am. Ind. Hyg. Assoc. J.* 30:126–132.
2. TRUJILLO, P. *et al.* 1974. The Preservation and Storage of Urine Samples for the Determination of Mercury. *Am. Ind. Hyg. Assoc. J.* 35:257–261.

Revised by Subcommittee 6

T. J. KNEIP, *Chairman*
R. S. AJEMIAN
J. N. DRISCOLL
F. I. GRUNDER
L. KORNREICH
J. W. LOVELAND
J. L. MOYERS
R. J. THOMPSON

816.

Analytical Method for Nitric Oxide in Air

Analyte:	NO	Method No.:	P & CAM 218
Matrix:	Air	Range:	2 to 100 ppm
Procedure:	Collection via solid absorber/Colorimetric Manual	Precision:	Unknown
		Classification:	C (Tentative)
Date Issued:	5/15/75		
Date Revised:			

1. Principle of the Method

1.1 After removing nitrogen dioxide normally present in the atmosphere by a solid absorber containing triethanolamine, the nitric oxide is oxidized quantitatively to nitrogen dioxide.

1.2 Quantitative oxidation of nitric oxide in the gas phase can be accomplished using chromic oxide on an inert inorganic substrate (**1, 2**).

1.3 Resultant nitrogen dioxide is determined by method No. P & CAM 210 for NO_2.

2. Range and Sensitivity

2.1 This method is intended for the manual determination of nitric oxide in the atmosphere in the range of 2 to ~ 100 parts per million (ppm) by volume or 2 to ~ 123 mg/m³.

2.2 The sensitivity is 0.01 μg NO/10 ml of diluting reagent (**3**).

3. Interferences

3.1 Oxidizing vapors that are known to interfere with the NO_2-measurement part of this method cannot coexist with NO at significant conc, because of their rapid gas phase reaction with NO (**4**). Therefore, no significant interferences are expected from such vapors.

3.2 Nitrogen dioxide interferes by being partially converted (3 to 4%) into NO in the NO_2 absorbers (**5**).

3.3 Sulfur dioxide is removed by the oxidizer (**6**) and ordinarily produces no interference. However, if present at very high conc, it depletes the oxidizer rapidly and requires more frequent oxidizer changes. The oxidizer will indicate depletion by a change from orange to a brownish color.

3.4 Relative humidity higher than 70% (at room temperature) may reduce the efficiency of the oxidizer (**7**).

4. Precision and Accuracy

4.1 The precision of the method depends on the conversion efficiency of the oxidizer and other variables such as volume measurement of the sample, sampling efficiency of the solid triethanolamine (TEA) absorber and absorbance measurement of the color.

4.2 Under normal conditions, the conversion efficiency of the oxidizer varies within 98 to 100% (**1, 6**).

4.3 At present, accuracy data are not available.

5. Advantages and Disadvantages

5.1 The sampling method is simple and permits storage of the absorbed samples for periods of several weeks.

5.2 The analytical method has the disadvantages associated with wet chemical methods.

5.3 Exposure to light affects the developed color.

6. Apparatus

6.1 An approved and calibrated personal sampling pump for personal and area samples whose flow can be determined accurately at 50 ml per min (**8**).

6.2 NITROGEN DIOXIDE ABSORBER. A 20-mm ID × 50-mm long polyethylene tube with connecting caps at both ends is filled with pellets of nitrogen dioxide absorbent. The pellets are held in place with glass wool plugs.

6.3 HUMIDITY REGULATOR. In order to provide steady 40 to 70% relative humidity control for the efficient working range of the chromic oxide, a 20-mm ID × 50-mm long polyethylene tube is filled with constant humidity buffer mixture in the same manner as in Section **6.2** (**9**).

6.4 OXIDIZER (**1, 2, 7**). A 15-mm ID glass tube with connecting ends is filled with 10 cm in length of oxidizer pellets between two glass wool plugs. (**CAUTION:** Protect eyes and skin when handling this reagent.)

6.4.1 The combined train of **6.1, 6.a, 6.3, 6.4** can be used for at least $\frac{800}{\text{ppm NO}_2}$ hrs and should be changed whenever it becomes visibly wet or discolored.

7. Reagents

7.1 PURITY. All chemicals should be analytical reagent grade (**10**).

7.2 NITROGEN DIOXIDE ABSORBENT (**1, 5, 7**). 1.5-mm (1/16′) pellets of 10 to 20 mesh porous inert material, such as firebrick, alumina, zeolites, etc., are soaked for 30 min in 20% aqueous triethanolamine, drained, extended on a wide petri dish, and dried for 30 to 60 min at 95 C. The pellets should be free flowing. (**CAUTION:** Soda lime should not be used as an absorbent (**2**).)

7.3 CONSTANT HUMIDITY GRAINS. (Consisting of a 50 + 50% anhydrous and hydrated sodium acetate mixture). Stir slowly and add dropwise 13 ml of water into a beaker containing 40 g of anhydrous sodium acetate in order to obtain coarse grained crystal pellets.

7.4 OXIDIZER (**1, 2, 7**). Soak glass, firebrick, or alumina, mesh size 15 to 40 in a solution containing 17 g of chromium trioxide in 100 ml of water for 30 min. Then drain, dry in an oven at 105 to 115 C, and condition to 70% relative humidity. Conditioning can be done by exposing overnight a thin layer of pellets contained in a petri dish to a saturated solution of sodium acetate in a desiccator; the reddish color changes to orange when conditioning is complete.

7.5 DILUTING REAGENT. Add 15 g of triethanolamine to approximately 500 ml of distilled water; then add 3 ml of n-butanol. Mix well and dilute to 1 l with distilled water. The n-butanol acts as a surfactant. If excessive foaming occurs during sampling, the amount added of n-butanol should be decreased. The reagent is stable for 2 months if kept in a brown bottle, preferably in the refrigerator.

7.6 SULFANILAMIDE SOLUTION. Dissolve 10 g of sulfanilamide in 400 ml of distilled water, then add 25 ml of conc (85%) phosphoric acid. Make up to 500 ml. This solution is stable for several months if stored in a brown bottle and up to one year in the refrigerator.

7.7 N-(1-NAPHTHYL)-ETHYLENEDIAMINE DIHYDROCHLORIDE (NEDA). Dissolve 0.1 g of NEDA in 100 ml of water. The solution is stable for a month, if kept in a brown bottle in the refrigerator.

7.8 STOCK NITRITE SOLUTION (1.77 g/l). Before preparing this solution it is desirable to assay the solid reagent, especially if it is old. The stock solution is stable for 90 days at room temperature and for a year in a brown bottle under refrigeration.

8. Procedure

8.1 SAMPLING. Assemble a sampling train comprised, in order, of nitrogen dioxide absorber, humidity regulator, oxidizer, nitrogen dioxide absorber, and pump. Draw an air sample at a rate of 0.05 1/min. Record total air volume sampled, as well as temperature and pressure.

8.2 COLOR DEVELOPMENT. Transfer the nitrogen dioxide absorbent pellets and glass wool plugs from the second nitrogen dioxide absorber in the train to a 50-ml test tube. Wash the tube with about 10 ml of water and add washings to the test tube. Ignore volume occupied by the pellets and glass wool. Dilute to 50.0 ml with diluting reagent. Cap and shake vigorously for about one min. Set aside and shake again after 10 min. Allow the solids to settle and transfer 10 ml to a 25-ml graduated cylinder. Similarly treat an unexposed nitrogen dioxide absorber as blank. To each cylinder add 10 ml of the sulfanilamide solution and 1.0 ml of the NEDA solution, mix well, and dilute to 25 ml with distilled water. Allow the color to develop for 15 min and read the absorbance at 500 nm.

9. Calibration and Standards

9.1 The method may be calibrated with nitrite solutions **(11)** or with known gas mixtures. In the latter case, stoichiometric and efficiency factors need not be used in the calculations.

9.1.1 *Calibration with nitrite solutions* **(3)**. Assuming a quantitative conversion of NO to NO_2, the method can be calibrated as described in the Tentative Method of Analysis for Nitrogen Dioxide Content of the Atmosphere **(12)**.

Calibration with nitrite solution is based upon the empirical observation that 0.63 mole of sodium nitrite produces the same color as one mole of nitrogen dioxide **(13)** and this gas is obtained in a quantitative yield during oxidation of NO **(1, 2)**. Nitrite solution equivalent to 10 μl NO per ml (10^{-5} liter NO/ml) contains:

$$\text{g NaNO}_2 \text{ per liter} = \frac{10^{-5}\ 1\ \text{NO} \times 0.63 \times 69 \times 1000\ \text{ml}}{24.47} = 1.77 \times 10^{-2} \text{ or } 0.0177\ \text{g/liter}$$

where:

69 = molecular weight of $NaNO_2$
0.63 = efficiency factor NO/NO_2^-
24.47 = molar volume at 760 Torr and 25 C
1000 = conversion from ml to liter

This solution is prepared fresh just before use from a stronger stock solution containing 1.77 g $NaNO_2$ per liter **(7, 8)**.

Construct the calibration curve by adding graduated amounts of dilute nitrite solution, equivalent to concentrations expected to be sampled, to a series of 25-ml volumetric flasks. Plot the absorbances against μl of NO added to the 25 ml final solution. If preferred, transmittance in percent may be plotted on the logarithmic scale versus nitric oxide conc on the linear coordinate on a semilog graph paper. The plot follows Beer's Law up to absorbance one (or 10% transmittance).

9.1.2 *Dynamic calibration method.* Known quantities of nitric oxide in dilution air are made by the method outlined below. The known gas mixture is sampled using the sampling train described in **8.1** and analyzed by the procedure in **8.2**. From the known concentration of nitric oxide in the dilution system a calibration curve is constructed of absorbance vs. concentration of NO in dilution air. Several known conc of NO in air should be made to cover the expected range of the air samples. The dynamic calibration method enables the entire sampling system to be calibrated and eliminates the need to know the equivalence factor of nitrite to the initial NO present in the air sample.

9.1.3 *Dynamic calibration method.* Nitric oxide and argon or nitrogen is obtained from a specialty gas supplier and is certified as to NO concentration either by the supplier or by independent analysis. The high pressure cylinder should contain at least 300 ppm NO for calibration of the sampling system over the full range of 100 ppm to 2 ppm. Concentration of NO in the dilution system can be set up as follows:

$$\text{ppm NO} = \frac{CF_1}{F_1 + F_2}$$

where

C = concentration of NO in the cylinder

F_1 = flow rate of cylinder gas

F_2 = flow rate of dilution air

The dilution air flow, F_2, should be at least twice as large as the cylinder gas, F_1, in order to approximate ambient air. The greater the ratio of F_2 to F_1 the more the composition of the dynamic system corresponds to the major constituents of air. The dilution air should be humidified before mixing the NO cylinder gas into the stream; the relative humidity should approximate sampling conditions.

10. Calculations

10.1 For convenience, standard conditions are taken as 760 Torr and 25 C, where the molar volume is 24.47. Ordinarily the correction of the sample volume to these standard conditions is slight and may be omitted. However, if conditions deviate significantly, corrections might be made by means of the perfect gas equation.

10.2 From the plot obtained with nitrite solutions **9.1** read the amounts of μl NO at the intercept of the calibration curve with absorbance one or transmittance 10%. This amount can be used as factor K for calculating the concentration in ppm, rather than reading concentrations directly from the graph, according to the following equation:

$$\text{NO ppm} = \frac{A \times K \times 10}{V \times 25}$$

where:

A = absorbance of sampling solution made up to 10 ml after sampling

K = factor as described above

V = volume of air sampled in liters

10 = volume of sampling solution

25 = volume of calibrating solution

This technique gives at best an approximation of the true value.

10.3 When using the calibration with gaseous mixtures, the sampling must be made under similar conditions of time and flow rate as the calibration because the stoichiometric factor varies with incoming concentrations of NO and thence of NO_2 **(13)**. The concentrations of NO collected are read directly from the calibration curve.

11. References

1. Levaggi, D.A., E.L. Kothny, T. Belsky, E.R. de Vera, and P.K. Mueller. 1971. A precise method for analyzing accurately the content of nitrogen oxides in the atmosphere. Presented 17th Annual Meeting Institute of Environmental Sciences, Los Angeles, April.
2. Levaggi, D.A., E.L. Kothny, T. Belsky, E.R. de Vera, and P.K. Mueller. 1974. Quantitative analysis of nitric oxide in presence of nitrogen dioxide at atmospheric concentrations. *Env. Sci. and Tech.* 8:348–350.
3. Saltzman, B.E. 1954. Colorimetric microdetermination of nitrogen dioxide in the atmosphere. *Anal. Chem.* 12:1919.
4. Johnston, H.S. and H.J. Crosby. 1951; 1954. Kinetics of the fast gas phase reaction between ozone and nitric oxide. *J. Chem. Phys.* 19:799, 22:689.
5. Levaggi, D.A., W. Siu, M. Feldstein, and E. Kothny. 1972. Quantitative separation of nitric oxide from nitrogen dioxide at atmospheric concentration ranges. *Env. Sci. and Tech.* 6:250–252.
6. Saltzman, B.E. and A.F. Wartburg. 1965. Absorption tube for removal of interfering sulfur dioxide in analysis of atmospheric oxidant. *Anal. Chem.* 37:779.
7. Belsky, T. 1970. Experimental evaluation of triethanolamine and chromium trioxide in the continuous analysis of NO in the air. AIHL Report No. 85, California State Department of Public Health, June.
8. American Public Health Association, Inc. 1977. Methods of Air Sampling and Analysis. 2nd ed. Washington, D.C.
9. Huygen, I.C. 1970. Reaction of nitrogen dioxide with Griess type reagents. *Anal. Chem.* 42:407.
10. American Chemical Society. 1966. Reagent chemicals. American Chemical Society Specifications, Washington, D.C.
11. Scaringelli, F.P., E. Rosenburg, and K.A. Rehme. 1970. Comparison of permeation tubes and nitrite ion as standards for the colorimetric determination of nitrogen dioxide. *Env. Sci. and Tech.* 4:924.
12. American Public Health Association, Inc. 1977. Methods of Air Sampling and Analysis. 2nd ed. p. 527. Washington, D.C.
13. Blacker, J.H. 1973. Triethanolamine for Collecting Nitrogen Dioxide in the TLV range, Am. Ind. Hyg. Association Journal, p. 390, Sept.

Subcommittee 3

B. Dimitriades, *Chairman*
J. Cuddeback
E. L. Kothny
P. W. McDaniel
L. A. Ripperton
A. Sabadell

817.

Analytical Method for Nitrogen Dioxide in Air

Analyte:	Nitrogen Dioxide	Method No.:	P & CAM 108
Matrix:	Air	Range:	.01 to 10 μg/liter
Procedure:	Fritted bubbler for sampling/Colorimetric, azo dye	Precision:	< 10% RSD
Date Issued:	12/11/72		
Date Revised:	1/1/75	Classification	C (Tentative)

1. Principle of the Method

1.1 Air is drawn through a bubbler containing a frit of 60 μ maximum pore diameter.

1.2 The nitrogen dioxide is absorbed in an azo dye forming reagent (**1**). A stable pink color is produced within 15 min which may be read in a colorimeter or spectrophotometer at 550 nm.

2. Range and Sensitivity

2.1 This method is intended for the manual determination of nitrogen dioxide in the atmosphere in the range of 0.005 to about 5 parts per million (ppm) by volume or 0.01 to 10 μg/l, when sampling is conducted in fritted bubblers. The method is preferred when high sensitivity is needed.

2.2 Concentrations of 5 to 100 ppm in

industrial atmospheres and in gas burner stacks also may be sampled by employing evacuated bottles or glass syringes. For higher concentrations, for automotive exhaust, and/or for samples relatively high in sulfur dioxide content, other methods should be applied.

3. Interferences

3.1 A 10-fold ratio of sulfur dioxide to nitrogen dioxide produces no effect. A 30-fold ratio slowly bleaches the color to a slight extent. The addition of 1% acetone to the reagent before use retards the fading by forming another temporary product with sulfur dioxide. This permits reading within 4 to 5 hr (instead of the 45 min required when acetone is not added) without appreciable interferences. Interference from sulfur dioxide may be a problem in some stack gas samples (Section **2.2**).

3.2 A 5-fold ratio of ozone to nitrogen dioxide will cause a small interference, the maximal effect occurring in 3 hr. The reagent assumes a slightly orange tint.

3.3 Peroxyacylnitrate (PAN) can give a response of approximately 15 to 35% of an equivalent molar concentration of nitrogen dioxide (**2**). In ordinary ambient air the concentrations of PAN are too low to cause any significant error.

3.4 The interferences from other nitrogen oxides and other gases which might be found in polluted air are negligible. However, if the evacuated bottle or syringe method is used to sample concentrations above 5 ppm, interference from NO (due to oxidation to NO_2) is possible (Section **8.2.3**).

3.5 If strong oxidizing or reducing agents are present, the colors should be determined within 1 hr, if possible, to minimize any loss. Colors should be shielded from exposure to ambient light.

4. Precision and Accuracy

4.1 A precision of 1% of the mean has been reported by a single laboratory (**3**). However, multi-laboratory tests under ASTM Project Threshold (**14**) showed the following reproducibility relation between standard deviation and mean value of NO_2 in the range of 16 to 400 $\mu g/m^3$: $S = 0.517 + 1.27\sqrt{m}$, where m = NO_2 concentration $\mu g/m^3$.

4.2 ACCURACY. Based on simultaneous analyses of spiked and unspiked samples at three different locations, the average uncertainty of the mean at the 95% confidence level was ± 10% (**14**).

5. Advantages and Disadvantages of the Method

5.1 This is a simple method with direct coloration of absorbing reagent, which can be put directly into cuvettes and read, or diluted with absorbing reagent and read. Differences in accuracy may be caused by other interfering substances in the air sample.

6. Apparatus

6.1 ABSORBER. The sample is absorbed in an all-glass bubbler with a 60 μ maximum pore diameter frit.

6.1.1 The porosity of the fritted bubbler, as well as the sampling flow rate, affect absorption efficiency. An efficiency of over 95% may be expected with a flow rate of 0.4 l/min or less and a maximum pore diameter of 60 μ. Frits having a maximum pore diameter less than 60 μ will have a higher efficiency but will require an inconvenient pressure drop for sampling (see formula in Section **6.1.2**). Considerably lower efficiencies are obtained with coarser frits, but these may be utilized if the flow rate is reduced.

6.1.2 Since the quality control by some manufacturers is rather poor, it is desirable to measure periodically the porosity of an absorber as follows: Carefully clean the apparatus with dichromate-conc sulfuric acid solution and then rinse it thoroughly with distilled water. Assemble the bubbler, add sufficient distilled water to barely cover the fritted portion, and measure the vacuum required to draw the first perceptible stream of air bubbles through the frit. Then calculate the maximum pore diameter as follows:

$$\text{Maximum pore diameter } (\mu) = \frac{30\ s}{P}$$

where:

s = surface tension of water at the test temperature in dynes/cm (73 at 18 C, 72 at 25 C, and 71 at 31 C),

P = measured vacuum, mm Hg.

6.1.3 Rinse the bubbler thoroughly with water and allow to dry before using. A rinsed and reproducibly drained bubbler may be used if the volume (r) of retained water is added to that of the absorbing reagent for the calculation of results. This correction may be determined as follows: Pipet into a drained bubbler exactly 10 ml of a colored solution (such as previously exposed absorbing reagent) of absorbance A_1. Assemble the bubbler and rotate to rinse the inside with the solution. Rinse the fritted portion by pumping gently with a rubber bulb. Read the new absorbance, A_2, of the solution. Then:

$$10A_1 = (10 + r)A_2$$

or:

$$r = 10\frac{A_1}{A_2} - 1$$

6.2 Air Metering Device. A glass rotameter capable of accurately measuring a flow of 0.4 l/min is suitable. A wet test meter is convenient to check the calibration.

6.3 Sampling Probe. A glass or stainless steel tube 6 to 10 mm in diam provided with a downward-facing intake (funnel or tip) is suitable. A small loosely fitting plug of glass wool may be inserted, when desirable, in the probe to exclude water droplets and particulate matter. The dead volume of the system should be kept minimal to permit rapid flushing during sampling to avoid losses of nitrogen dioxide on the surfaces.

6.4 Grab-Sample Bottles. Ordinary glass-stoppered borosilicate glass bottles of 30 to 250 ml sizes are suitable if provided with a mating ground joint attached to a stopcock for evacuation. Calibrate the volume by weighing with connecting piece, first empty, then filled to the stopcock with distilled water.

6.5 Glass Syringes. 50 or 100-ml syringes are convenient (although less accurate than bottles) for sampling.

6.6 Air Pump. A vacuum pump capable of drawing the required sample flow for intervals of up to 30 min is suitable. A tee connection at the intake is desirable. The inlet connected to the sampling train should have an appropriate trap and needle valve, preferably of stainless steel. The second inlet should have a valve for bleeding in a large excess flow of clean air to prevent condensation of acetic acid vapors from the absorbing reagent, with consequent corrosion of the pump. Alternatively, soda lime may be used in the trap. A filter and critical orifice may be substituted for the needle valve (**4**).

6.7 Spectrophotometer or Colorimeter. A laboratory instrument suitable for measuring the pink color at 550 nm, with stoppered tubes or cuvettes. The wavelength band width is not critical for this determination.

6.8 Assorted laboratory glassware.

7. Reagents

7.1 Purity. All chemicals used should be analytical reagent grade (**5**).

7.2 Nitrite-Free Water. All solutions are made in nitrite-free water. If available distilled or deionized water contains nitrite impurities (produces a pink color when added to absorbing reagent), redistill it in an all-glass still after adding a crystal each of potassium permanganate and of barium hydroxide.

7.3 N-(1-Naphthyl)-Ethylenediamine Dihydrochloride, Stock Solution (0.1%). Dissolve 0.1 g of the reagent in 100 ml of water. Solution is stable for several months if kept well stoppered in a brown bottle in the refrigerator. (Alternatively, weighed small amounts of the solid reagent may be stored.)

7.4 Absorbing Reagent. Dissolve 5 g of anhydrous sulfanilic acid (or 5.5 g of $NH_2C_6H_4SO_3H \cdot H_2O$) in almost a liter of water containing 140 ml of glacial acetic acid. Gentle heating is permissible to speed up the process. To the cooled mixture, add 20 ml of the 0.1% stock solution of N-(1-naphthyl)-ethylenediamine dihydrochloride, and dilute to 1 liter. Avoid lengthy contact with air during both preparation and use, since discoloration of reagent will result because of absorption of nitrogen dioxide. The solution is stable for several months if kept well-stoppered in a brown bottle in the refrigerator. The absorbing reagent should be allowed to warm to room temperature before use.

7.5 STANDARD SODIUM NITRITE SOLUTION (0.0203 G/LITER). One ml of this working solution of sodium nitrite ($NaNO_2$) produces a color equivalent to that of 10 μl of nitrogen dioxide (10 ppm in 1 l of air at 760 Torr and 25 C (see Section **10.2.1**). Prepare fresh just before use by dilution from a stronger stock solution containing 2.03 g of the reagent grade granular solid (calculated as 100%) per liter. It is desirable to assay (**5**), the solid reagent, especially if it is old. The stock solution is stable for 90 days at room temperature, and for a year in a brown bottle under refrigeration.

8. Procedure

8.1 CLEANING OF GLASSWARE. All washed glassware should be allowed to stand awhile in chromic acid solution, and then thoroughly rinsed with single and then double distilled water.

8.2 COLLECTION AND STORAGE OF SAMPLES. Three methods are described below. Concentrations below 5 ppm are sampled by the bubbler method. Higher conc may be sampled by the evacuated bottle method, or more conveniently (but less accurately) by the glass syringe method. The latter method is more useful when appreciable conc (*e.g.*, 20 ppm) of nitric oxide are expected.

8.2.1 *Bubbler Method.*

a. Assemble, **in order**, a sampling probe (optional), a glass rotameter, fritted absorber, and pump. Use ground-glass connections upstream from the absorber. Butt-to-butt glass connections with slightly greased vinyl or pure gum rubber tubing also may be used for connections without losses if lengths are kept minimal. The sampling rotameter may be used upstream from the bubbler provided occasional checks are made to show that no nitrogen dioxide is lost. The rotameter must be kept free from spray or dust.

b. Pipet 10.0 ml of absorbing reagent into a dry fritted bubbler (Section **6.1.3**).

c. Draw an air sample through it at the rate of 0.4 l/min (or less) long enough to develop sufficient final color (about 10 to 30 min). Note the total volume of air sampled.

d. Measure and record the sample air temperature and pressure.

8.2.2 *Evacuated Bottle Method.*

a. Sample in bottles of appropriate size containing 10.0 ml (or other convenient volume) of absorbing reagent. For 1-cm spectrophotometer cells, a 5:1 ratio of air sample volume to reagent volume will cover a concentration range up to 100 ppm; a 25:1 ratio suffices to measure down to 2 ppm.

b. Wrap a wire screen or glass-fiber-reinforced tape around the bottle for safety purposes.

c. Grease the joint lightly with silicone or fluorocarbon grease.

d. If a source of vacuum is available at the place of sampling, it is best to evacuate just before sampling to eliminate any uncertainty about loss of vacuum. A three-way Y stopcock connection is convenient. Connect one leg to the sample source, one to the vacuum pump, and the third to a tee attached to the bottle and to a mercury manometer or accurate gauge. In the first position of the Y stopcock, the bottle is evacuated to the vapor pressure of the absorbing reagent. In the second position of the Y stopcock the vacuum pump draws air through the sampling line to flush it thoroughly. The actual vacuum in the sample bottle is read on the manometer. In the third position of the Y stopcock the sampling line is connected to the evacuated bottle and the sample is collected.

e. The stopcock on the bottle is then closed. Allow 15 min with occasional shaking for complete absorption and color development.

f. For calculation of the standard volume of the sample, record the temperature and the pressure. The pressure is the difference between the filled and evacuated conditions. The uncorrected volume is that of the bottle plus that of the connection up to the stopcock minus the volume of absorbing reagent.

8.2.3 *Glass Syringe Method.*

a. Ten ml of absorbing reagent is kept in a capped 50- (or 100-) ml glass syringe, and 40 (or 90) ml of air is drawn in at the time of sampling.

b. The absorption of nitrogen dioxide is completed by capping and shaking vigorously for 1 min. after which the air is expelled. (When appreciable concentrations—*e.g.*, 20 ppm—of nitric oxide are suspected, interference caused by the oxidation of nitric oxide to nitrogen dioxide is minimized by expelling the air sample immediately after the absorption period.)

c. Additional air may be drawn in and the process repeated several times, if necessary, to develop sufficient final color.

8.2.4 *Effects of Storage.* Colors may be preserved, if well stoppered, with only 3 to 4% loss in absorbance per day; however, if strong oxidizing or reducing gases are present in the sample in concentrations considerably exceeding that of the nitrogen dioxide, the colors should be determined as soon as possible to minimize any loss. (See Section **3** for effects of interfering gases.)

8.3 ANAYLSIS OF SAMPLES.

8.3.1 After collection or absorption of the sample, a red-violet color appears. Color development is complete within 15 min at room tempearture.

8.3.2 Compare with standards visually or transfer to stoppered cuvettes and read in a spectrophotometer at 550 nm using unexposed reagent as a reference. Alternatively, distilled water may be used as a reference and the absorbance of the reagent blank deducted from that of the sample.

8.3.3 Colors too dark to read may be quantitatively diluted with unexposed absorbing reagent. The measured absorbance is then multiplied by the dilution factor.

9. Calibration and Standardization

Either of two methods of calibration may be employed. The most convenient is standardizing with nitrite solution. Greater accuracy is achieved by standardizing with accurately known gas samples in a precision flow dilution system (**3, 6, 7**). The permeation tube technique (**8**) may also be used. If the gaseous method is used, the stoichiometric factor is eliminated from the calculations. Concentrations of the standards should cover the expected range of sample concentrations.

9.1 NITRITE SOLUTION METHOD.

9.1.1 Add graduated amounts of the $NaNO_2$ solution up to 1 ml (measured accurately in a graduated pipet or small buret) to a series of 25-ml volumetric flasks, and dilute to the mark with absorbing reagent.

9.1.2 Mix, allow 15 min for complete color development and read the colors (Section **8.3**)

9.1.3 Good results can be obtained with these small volumes of standard solution if they are carefully measured. Making the calibration solutions up to 25 ml total volume, rather than the 10 ml volume used for samples, facilitates accuracy. If preferred, even larger volumes may be used with correspondingly larger volumetric flasks.

9.1.4 Plot the absorbances of the standard colors against the μl of nitrogen dioxide per ml of absorbing reagent. The latter values are equal to the corresponding ml of standard nitrite solution times 0.4 (Section **10.2.2**). If preferred, transmittance may be plotted instead of absorbance, using semilogarithmic graph paper. The plot follows Beer's law. Draw the straight line through the origin giving the best fit, and determine the slope, K, which is μl of NO_2 intercepted at absorbance of exactly 1.0 or at 10% transmittance. For 1 cm cells, the value of K is about 0.73.

9.2 GASEOUS STANDARD METHODS. Two techniques are outlined below. Consult the original references for complete details, and Section **10.3** for calculations.

9.2.1 *Method 1* (**3**)*for gaseous standardization.*

a. About 5 ml of pure liquid nitrogen dioxide is placed in a small glass bubbler (10 mm in diam and 100 mm in length) provided with ground glass stopcocks and spherical joints on both intake and outlet tubes.

b. The bubbler is immersed in a thermos bottle ice bath and connected to an air line. A small pump with flowmeter delivers a steady stream of a few ml/min of air to the bubbler, thence through two flowmeters which permit the discarding of up to 90% of the NO_2 and finally to a large stream of carbon filtered air (1000 to 1500 liters/min) from a small blower. All this air passes through a 10-cm Biram anemometer mounted on the end of a pipe 10-cm in diam. It has been found empirically with this arrangement that the anemometer reading in ft/min times 1.64 is equivalent to liters/min.

c. The bubbler is weighed to 0.1 mg at the beginning and end of an accurately measured time period. The stopcocks are closed each time before the bubbler is removed from the ice bath and wiped dry for weighing.

9.2.2 *Method 2.* (**6, 7**) *For preparing known dilutions of nitrogen dioxide.* Consists in making a preliminary dilution (about 0.4%) of nitrogen dioxide in air in a stainless steel tank at 1000 lbs pressure. Subsequent dilution, by air in a flow system at atmospheric pressure, of the analyzed tank mixture controlled by an asbestos plug and manometer yields concentrations of 0.1 to 10 ppm NO_2.

9.2.3 *Method 3 (Reference* **15**, *Part I, Section* **3**) *for preparing gaseous nitrogen dioxide standards.* This consists of passing a metered flow of dry nitrogen over a permeation tube maintained at a constant temperature. This stream is then diluted with air to desired level. This is a very convenient method but requires protection of the permeation tube from humidity to prevent permanent changes in its calibration.

9.2.4 Sample the gas mixtures by the bubbler method (Section **8.2.1**), and read the colors (Section **8.3**). Select concentrations and sample volumes to produce colors covering the accurate absorbance range of the spectrophotometer.

9.2.5 Standardization by gaseous samples can be based either upon a weight-volume relationship if the source of nitrogen dioxide is weighed, or a volume-volume relationship if the source is an analyzed tank mixture.

9.2.6 Calculate the concentration of the sample air stream, C, in ppm by volume (μl NO_2/l of air).

$$C = \frac{10^6(W_1 - W_2)}{t} \times \frac{0.532}{F_1} \times \frac{F_2}{F_3}$$

or:

$$C = C_t \times \frac{F_2}{F_3}$$

where:

C = concentration of sample air stream in ppm.

$(W_1 - W_2)$ = difference between initial and final weights, in grams, of nitrogen dioxide bubbler (or of permeation tube); weight loss is usually 0.01 to 0.05 g.

t = time interval in minutes, between the weighings.

0.532 = (24.47/46.0), ideal volume in l at 25 C and 760 Torr of 1.0 g of nitrogen dioxide.

F_1 = flow rate, in lpm, of air passed through nitrogen dioxide bubbler (or over permeation tube), corrected to 25 C and 760 Torr.

F_2 = flow rate, in lpm, of concentrated gas mixture injected into sample air stream.

F_3 = total flow rate, in lpm, of sample air stream.

C_t = analyzed concentration of the tank mixture, in ppm by volume on an ideal gas basis.

9.2.7 For each standard color, calculate the μl of nitrogen dioxide per ml of absorbing reagent:

$$\mu l\ NO_2/ml = CV$$

where:

C = ppm concentration (Section **9.2.5**)

V = volume of air sample at 25 C and 760 Torr, in l/ml of absorbing reagent.

9.2.8 Plot the absorbances of the colors against the μl of gaseous nitrogen dioxide/ml of absorbing reagent. Draw the straight line through the origin giving the best fit, and determine the slope, K (the value of μl/ml intercepted at absorbance of exactly 1.0).

10. Calculations

10.1 For convenience, standard conditions are taken as 760 Torr and 25 C, at which the molar gas volume is 24.47 l. (This is identical with standard conditions for Threshold Limit Values of the American Conference of Governmental Industrial Hygienists; it is very close to the standard conditions used **(9)** for air-handling equipment, of 29.92 in Hg, 70 F, and 50% relative humidity, at which the molar gas volume is 24.76 l or 1.2% greater.)

Ordinarily the correction of the sample volume to these standard conditions is slight and may be omitted; however, for greater accuracy, it may be made by means of the perfect gas equation.

10.2 Standardization by nitrite solution is based upon the empirical observation **(1, 6)** that 0.72 mole of sodium nitrite produces the same color as 1 mole of nitrogen dioxide.

10.2.1 This factor is applied to calculate the equivalence of the nitrite solution to the volume of NO_2 absorbed as follows: One ml of the working standard solution contains 2.03×10^{-5} g $NaNO_2$. Since the molecular weight of NaNO is 69.00, this is equivalent to:

$$\frac{2.03 \times 10^{-5}}{69.00} \times \frac{24.47}{0.72} = 1 \times 10^{-5} \text{ l of } NO_2,$$

or

$$= 10\ \mu l \text{ of } NO_2$$

10.2.2 In Section **9.1**, the calibration standard containing 1 ml of nitrite solution (10 μl NO_2) per 25 ml total volume is equivalent to 10/25 or 0.4 μl of NO_2 per ml.

10.3 Compute the concentration of nitrogen dioxide in the sample as follows:

Nitrogen dioxide, ppm = AK/V

where:

A = measured absorbance

K = standardization factor from Section **9.1.4** or **9.2.7**

V = volume of air sample, at 25 C and 760 Torr, in liters/ml of absorbing reagent.

10.4 If preferred, the graph from Section **9.1.4** or **9.2.7** may be used instead as follows:

Nitrogen dioxide, ppm = μl NO_2 per ml/V

10.4.1 If V is a simple multiple of K, calculations are simplified. Thus, for the K value of 0.73 previously cited, if exactly 7.3 l of air are sampled through a bubbler containing 10 ml of absorbing reagent, K/V = 1, and the absorbance is also ppm directly.

10.4.2 For exact work, an allowance may be made in the calculations for sampling efficiency and for fading of the color using the following equation:

Nitrogen dioxide, ppm = A_cK/VE

where:

A_c = corrected absorbance; the absorbance is corrected for fading of the color as indicated in Section **8.2.4** when there is a prolonged interval between sampling and measurement of the absorbance.

E = sampling efficiency; for a bubbler, E is estimated from prior tests using two absorbers in series (Reference **7** and Section **6.1.1**); for a bottle or syringe, E = 1.0.

11. References

1. SALTZMAN, B.E. 1954. Colorimetric Microdetermination of Nitrogen Dioxide in the Atmosphere. *Analytical Chemistry*. 26:1449–1955.
2. MUELLER, P.K. F.P. TERRAGLIO and Y. TOKIWA. 1965. Chemical Interferences in Continuous Air Analyzers. Presented 7th Conference on Methods in Air Pollution Studies, Los Angeles, California. January.
3. THOMAS, M.D. and R.E. AMTOWER. 1966. Gas Dilution Apparatus for Preparing Reproducible Dynamic Gas Mixtures in Any Desired Concentration and Complexity. *J. Air Pollut. Contr. Assoc.* 16:618–623.
4. LODGE, J.P., JR., J.B. PATE, B.E. AMMONS and G.A. SWANSON. 1966. The Use of Hypodermic

Needles as Critical Orifices in Air Sampling. *J. Air. Pollut. Contr. Assoc.* 16:197–200.
5. AMERICAN CHEMICAL SOCIETY. 1966. Reagent Chemicals, American Chemical Society Specifications, Washington, D.C.
6. SALTZMAN, B.E. and A.F. WARTBURG, JR. 1965. Precision of Nitrogen Dioxide. *Anal. Chem.* 37:1261–1262.
7. SALTZMAN, B.E. 1961. Preparation and Analysis of Calibrated Low Concentrations of Sixteen Toxic Gases. *Anal. Chem.* 33:1103–1104.
8. O'KEEFE, A.E. and G.C. ORTMAN. 1966. Primary Standards for Trace Gas Analysis. *Anal. Chem.* 38:760–763.
9. ASTM COMMITTEE D-22. 1962. Terms Relating to Atmospheric Sampling and Analysis, D 1356–60, ASTM Standards on Methods of Atmospheric Sampling and Analysis, 2nd ed. Philadelphia, Pa.
10. STRATMANN, H and M. BUCK. 1966. Messung von Stickstoffdioxid in der Atmosphere. *Air and Water Pollut. Int. J.* 10:313–326.
11. STRATMANN, H. 1966. Personal Communication, September.
12. SHAW, J.T. The Measurement of Nitrogen Dioxide in the Air. *At. Envir.* 1:81–85.
13. AMERICAN PUBLIC HEALTH ASSOCIATION, INC. 1977. Methods of Air Sampling and Analysis. 2nd ed. p. 527. Washington, D.C.
14. FOSTER, J.F. and BEATTY, G.H. 1973. Final Report on Interlaboratory Cooperation Study of the Precision and Accuracy of the Measurement of Nitrogen Dioxide Content in the Atmosphere Using ASTM Method D-1607. Battelle, Columbus Laboratories, Columbus.
15. AMERICAN PUBLIC HEALTH ASSOCIATION, INC. 1977. Methods of Air Sampling and Analysis. 2nd ed. p. 527. Washington, D.C.

Revised by Subcommittee 3
B. DIMITRIADES, *Chairman*
J. CUDDEBACK
E. L. KOTHNY
P. W. MCDANIEL
L. A. RIPPERTON
A. SABADELL

818.

Analytical Method for Nitrogen Dioxide in Air

Analyte:	Nitrogen Dioxide	Method No.:	P & CAM 210
Matrix:	Air	Range:	0.01 – 10 ppm
Procedure:	Solid Sorbent Colorimetric	Precision:	Unknown
		Classification:	E (Proposed)
Date Issued:	1/1/75		
Date Revised:			

1. Principle of the Method

1.1 Nitrogen dioxide is absorbed from the air by a solid absorber containing triethanolamine (TEA). Subsequent analysis is performed using an azo-dye forming reagent **(1, 2, 3, 11)**. The color produced by the reagent is measured in a spectrophotometer at 540 nm.

2. Range and Sensitivity

2.1 The atmospheric concentration range for which this method may be used with confidence is 10 to 1000 $\mu g/m^3$ (0.005 to 0.50 ppm). Method performance at NO_2 levels between 1000 $\mu g/m^3$ (0.5 ppm) and 6200 $\mu g/m^3$ (3.1 ppm) has not yet been established **(2, 11)**.

2.2 The sensitivity of the method is dependent on that of the Griess-Saltzman reagent (**3**). For a one cm sample cell path and 0.1 absorbance units this is equivalent to 0.14 μg/ml of NO_2 in the absorbing solution.

3. Interferences

3.1 Sulfur dioxide in concentrations up to 250 $\mu g/m^3$ (0.13 ppm) does not interfere and negative interference caused by concentrations of up to 3000 $\mu g/m^3$ (1.6 ppm) is about 20% of the collected NO_2, if hydrogen peroxide is added after sampling and before color development (as described in Section **8.2**). Ozone causes no interference at atmospheric concentration ranges (up to 1000 $\mu g/m^3$). Nitric oxide at concentrations up to 800 $\mu g/m^3$ (0.6 ppm) as an 8 hr average causes no interference (**2**). The interference of TLV levels of SO_2 and nitric oxide have not been established.

3.2 Organic nitrites and peroxyacyl nitrate (PAN) which might be present in the air would produce a positive interference. There is no data on the magnitude of the interference; however, in view of the average concentrations of organic nitrites and PAN that have been reported in the literature, it would appear that the interference would be negligible (**2**).

4. Precision and Accuracy

4.1 The precision of the method in the field, expressed as standard deviation, was ± 4 $\mu g/m^3$ (± 0.002 ppm) on a 4-day sampling for a mean of 95 $\mu g/m^3$ (0.050 ppm) as compared with a continuous colorimetric analyzer (**2**). The precision at higher levels has not been established. The TLV is 5.0 ppm.

4.2 At present, accuracy data are not available. However, the method is comparable to that of continuous air monitoring instrumentation using the modified Griess-Saltzman reagent (**2, 4**). The NO_2 (gas) to NO_2 (ion) factor was found to average 0.85 in 25 tests, evaluated against the Saltzman factor of 0.72 (**2**), for ambient air sampling.

4.3 Absorption efficiency is over 95%, using either the liquid or solid absorbers (**2, 11**), in the concentration range up to 0.5 ppm.

5. Advantages and Disadvantages of the Method

5.1 The method is simple and samples collected in the field can be stored without loss for up to 4 weeks.

5.2 The samples will not spill.

5.3 The disadvantage of the method is the extracting step in the central laboratory.

6. Apparatus

6.1 SPECTROPHOTOMETER, B & L SPEC. 20, UTILIZING 10-MM CELLS. Any comparable instrument may be used.

6.2 OBSERVATION TUBE. Becton-Dickinson No. 420 LST, 55-cm × 5-mm glass, with a standard Luer lock fitting permanently attached to one end.

6.3 AIR METERING DEVICE. A rotameter capable of measuring flow rates up to 0.4 lpm., calibrated against a wet test meter, is suitable. Calibrated orifices, such as hypodermic needles can also be used for controlling the flow. In such case, a minimum pressure differential of 500 Torr across the needle is required. A membrane filter should be placed in front of the hypodermic needle to prevent water droplets from adhering to the needle, causing alteration of the gas flow.

6.4 SAMPLING LINES. Sampling lines should be constructed of Teflon, glass or stainless steel.

6.5 AIR PUMP. A vacuum pump capable of sampling air through the absorber at the required rate of 0.4 lpm for 8 hr, is suitable. The pump should be equipped with a needle valve for accurate control of sampling rate.

6.5.1 If a critical orifice is used as a flow control device, the pump should be also capable of maintaining a minimum vacuum of 500 Torr across the orifice at the required rate of 0.4 lpm.

7. Reagents

7.1 PURITY. Analytical reagent grade chemicals should be used (**6**). The water should be free of nitrite and conform to ISC standards **(7)**.

7.2 LIQUID ABSORBER. Add 15 g of triethanolamine to approximately 500 ml of distilled water; than add 3 ml of n-butanol. Mix well and dilute to 1 liter with distilled water. The n-butanol acts as a surfactant. The reagent is stable for 2 months if kept in a brown bottle, preferably in the refrigerator.

7.3 SOLID ABSORBER. Add 25 g of triethanolamine to a 250-ml beaker, add 4.0 g of glycerol, 50 ml of acetone, and sufficient distilled water to dissolve. Dilute to 100-ml total volume with distilled water. To the mixture add about 50 ml of 12 to 30 mesh molecular sieve 13×. Stir and let stand in the covered beaker for about 30 min. Decant the excess liquid and transfer the molecular sieve to a porcelain pan. Place under a heating lamp until the bulk of moisture has evaporated, and then in an oven at 110 C for 1 hr until dry. Store in a closed glass container.

7.4 HYDROGEN PEROXIDE. Dilute 0.2 ml of 30% hydrogen peroxide to 250 ml with distilled water. **Caution:** Hydrogen peroxide is a strong oxidant and will damage skin and clothing.

7.5 SULFANILAMIDE SOLUTION. Dissolve 10 g of sulfanilamide in 400 ml of distilled water, then add 25 ml of conc (85%) phosphoric acid. Make up to 500 ml. This solution is stable for several months if stored in a brown bottle and up to 1 year in the refrigerator.

7.6 N-(1-NAPHTHYL)-ETHYLENEDIAMINE DIHYDROCHLORIDE (NEDA). Dissolve 0.1 g of NEDA in 100 ml of water. The solution is stable for a month, if kept in a brown bottle in the refrigerator.

7.7 NITRITE STOCK SOLUTION. Dissolve 0.135 g of sodium nitrite in distilled water and make up to 1 liter. The sodium nitrite should be of known purity or analyzed before use. The solution contains 90 μg of nitrite per ml which corresponds to 100 μg NO_2 (gas) per ml (sec **9.2**). The solution should be made up once a month.

8. Procedure

8.1 SAMPLING. Fill an observation tube with the solid absorber, using small glass wool plugs for retainment of the solid. The sampling train consists of (in tandem) a sampling probe observation tube, a critical orifice (27 gauge hypodermic needle) and a pump having a capacity of at least 500 Torr vacuum. Turn pump on and rapidly check flow rate entering system and record. This flow rate should be between 150 and 200 cc/min. Sample for 8 hr. At the end of the sampling period recheck flow rate and record.

8.2 COLOR DEVELOPMENT. Transfer the molecular sieve and glass wool plugs to a 50-ml test tube. Wash the observation tube with approximately 10 ml of water and add washings to the molecular sieve. The volume occupied by the molecular sieve and glass wool is so small (~ 1%) it may be ignored. Dilute to 50 ml with absorbing solution. Cap and shake vigorously for about 1 min. Let stand and shake again after 10 min. Allow the solids to settle and transfer 10 ml to a 25-ml graduated cylinder.

Similarly, treat an unexposed molecular sieve as a blank as directed above. To each cylinder add 1.0 ml of dilute hydrogen peroxide solution and mix well. Then add 10 ml of the sulfanilamide solution and 1.0 ml of the NEDA solution to sample and blank, mix well and allow color to develop for 10 min. Measure the absorbance of the sample solution against the blank at 540 nm. The NO_2 is then determined from the calibration curve.

9. Calibration and Standardization

9.1 A calibration curve may be obtained by one of two different methods. The simplest and most convenient method is standardization with nitrite solutions. However, accurately known gas mixtures would provide more realistic standards for calibration of the entire analytical system. The known gas mixtures may be prepared by use of a flow dilution system **(3, 8, 9)**, or permeation tubes **(10)**.

9.2 STANDARDIZATION WITH NITRITE SOLUTION. Transfer 1.0 ml of the nitrite stock solution into a 50-ml volumetric flask and dilute to the mark with absorbing solution. The working solution so obtained contains 1.8 μg NO^{-}_2/ml (corresponding to 1 μg NO_2 (gas) per ml). To a series of 25-ml graduated cylinders introduce 0, 1, 3, 5, and 7 ml of the nitrite working solution. Dilute to 10 ml with absorbing solution. From this point treat the standards as described in **8.2**. A graph of absorbance versus NO_2 concentration is constructed. Weekly checks should be made to determine if the response of the reagents has changed.

10. Calculations

10.1 The air sampling volume is corrected to 25 C and 760 Torr. The normal deviations from these conditions add only small corrections.

10.2 Standardization by the nitrite solution requires use of an empirical factor that relates μg of NO_2 (gas) to μg of nitrite ion. The conversion factor in this method is 0.90, *i.e.* 90% of the NO_2 in the air sample is eventually converted to nitrite ion which reacts with the color producing azo-dye reagent (**2, 4**). This factor has been incorporated in the nitrite stock solution (Section **7.7**).

10.3 Computation of the average concentration of NO_2 in air for the 8 hr sample period may be made as follows:

Read the μg NO_2 directly from the graph.

$$c = \frac{\mu g\ NO_2 \times 10^3}{r \times t \times k}\ \mu g\ NO_2/m^3$$

$$p = \frac{\mu g\ NO_2 \times 0.532}{r \times t \times k} = \frac{c \times 0.532}{10^3}\ ppm$$

where:

c = concentration in μg NO_2/m^3
p = concentration in ppm NO_2 (μl/l)
r = sampling rate in liters/min
t = sampling time in min
k = dilution factor (i.e. 0.2 if one fifth of the sample is analyzed)
0.532 = μl NO_2/μg NO_2 at 25 C and 760 Torr
10^3 = conversion factor liters/m^3

11. References

1. SALTZMAN, B.E. 1954. Colorimetric microdetermination of nitrogen dioxide in the atmosphere. *Anal. Chem.* 26:1949.
2. LEVAGGI, D.A., W. SIU, and M. FELDSTEIN. 1973. A new method for measuring average 24-hour nitrogen dioxide concentrations in the atmosphere. *J. Air Poll. Cont. Assoc.* 23:30.
3. AMERICAN PUBLIC HEALTH ASSOCIATION, INC. 1977. Methods of Air Sampling and Analysis. 2nd ed. p. 527. Washington, D.C.
4. SCARINGELLI, F.P., E. ROSENBURG, and K.A. REHME. 1970. Comparison of permeation tubes and nitrite ions as standards for the colorimetric determination of nitrogen dioxide. *Env. Sci. Tech.* 4:924.
5. FEDERAL REGISTER. 1971. 36 (84), part II, 8201.
6. AMERICAN CHEMICAL SOCIETY. 1966. Washington, D.C. Reagent Chemicals, American Chemical Society specifications.
7. AMERICAN PUBLIC HEALTH ASSOCIATION, INC. 1977. Methods of Air Sampling and Analysis. 2nd ed. p. 53. Washington, D.C.
8. SALTZMAN, B.E. 1961. Preparation and analysis of calibrated low concentrations of 16 toxic gases. *Anal. Chem.* 33:1100.
9. AMERICAN PUBLIC HEALTH ASSOCIATION, INC. 1977. Methods of Air Sampling and Analysis. 2nd ed. p. 541. Washington, D.C.
10. SALTZMAN, B.E., W.R. BURG, and G. RAMASWAMY. 1971. Performance of permeation tubes as standard gas sources. *Env. Sci. Tech.* 5:1121.
11. BLACKER, J.H. 1973. Triethanolamine for collecting NO_2 in the TLV range. *Amer. Ind. Hyg. Assoc. J.* 34:390.

Revised by Subcommittee 3
B. DIMITRIADES, *Chairman*
J. CUDDEBACK
E. L. KOTHNY
P. W. MCDANIEL
L. A. RIPPERTON
A. SABADELL

819.

Analytical Method for Ozone in Air

Analyte:	Ozone	Method No.:	P & CAM 153
Matrix:	Air	Range:	0.01 to 10 ppm
Procedure:	Sampling with Midget Impinger/Neutral KI Absorption, Colorimetric	Precision:	± 5% Deviation from the mean **(1)**
		Classification:	B (Accepted)
Date Issued:	9/8/72		
Date Revised:	1/1/75		

1. Principle of the Method

1.1 Air containing ozone is drawn through a midget impinger containing 10 ml of 1% potassium iodide in a neutral (pH 6.8) buffer composed of 0.1 M disodium hydrogen phosphate and 0.1 M potassium dihydrogen phosphate.

1.2 The iodine liberated in the absorbing reagent is determined spectrophotometrically be measuring the absorption of the tri-iodide ion at 352 nm.

1.3 The stoichiometry **(1)** is approximated by the following reaction:

$$O^3 + 3\ KI + H_2O \rightarrow KI_3 + 2\ KOH + O_2$$

1.4 The analysis must be completed within 30 min to 1 hr after sampling.

2. Range and Sensitivity

2.1 The range extends from 0.01 ppm to about 10 ppm.

2.2 The sensitivity of the method is dependent on the volume of air sampled. When air sampling is conducted with a midget impinger containing 10 ml of absorbing solution, between 1 and 10 μl of ozone may be collected.

3. Interferences

3.1 The negative interference of reducing gases such as sulfur dioxide and hydrogen sulfide are very serious (probably on a mole-to-mole equivalency).

3.2 Interference from high concentrations of sulfur dioxide can be eliminated by using a prefilter consisting of a U-tube filled with strips of glass fiber paper impregnated with chromium trioxide.

3.3 The procedure is very sensitive to reducing dusts which may be present in the air or on the glassware. Losses of iodine occur even on clear glass surfaces, and thus the manipulations should minimize the exposure.

3.4 Halogens, hydrogen peroxide, organic peroxides, organic nitrites and various other oxidants will liberate iodine or cause positive interference by this method.

3.5 Peroxyacetyl nitrate gives a response approximately equivalent to 50% of that of an equimolar concentration of ozone **(1)**. Concentrations in the atmosphere may range up to 0.1 ppm.

4. Precision and Accuracy

4.1 The precision of the method within the recommended range is about ± 5% deviation from the mean. Major errors are from reducing impurities in the reagent, apparatus and air (see Section **3.3**) and from losses of iodine during longer sampling periods. The latter can be reduced by using a second impinger in series.

4.2 The accuracy of this method has not been established. Calibration is based on the assumed stoichiometry of the reaction with the absorbing solution.

5. Advantages and Disadvantages

5.1 The method is simple, precise and believed to be more accurate than the alkaline iodide method (2).

5.2 The analysis must be completed during the period of 30 to 60 min after sampling due to the instability of the colors.

6. Apparatus

6.1 SAMPLING EQUIPMENT. The sampling unit for the impinger collection method consists of the following components.

6.1.1 *A graduated midget impinger containing the absorbing solution or reagent.*

6.1.2 *A pump suitable for delivering desired flow rates.* The sampling pump is protected from splashover or water condensation by an absorption tube loosely packed with a plug of glass wool and inserted between the exit arm of the impinger and the pump.

6.1.3 *An integrating volume meter such as a dry gas or wet test meter.*

6.1.4 *Thermometer.*

6.1.5 *Manometer.*

6.1.6 *Stopwatch.*

6.2 SPECTROPHOTOMETER. Capable of measuring the yellow color at 352 nm.

6.3 MATCHED CELLS OR CUVETTES OF 2 CM PATH LENGTH.

6.4 ASSOCIATED LABORATORY GLASSWARE.

7. Reagents

7.1 PURITY. The reagents described must be made up using ACS reagent grade or better grade of chemical.

7.2 DOUBLE DISTILLED WATER. Double distilled water should be used for all reagents. The double distilled water can be prepared in an all-glass still by adding potassium permanganate to produce a faint pink color and barium hydroxide to alkalize the distilled water before the second distillation.

7.3 ABSORBING REAGENT. (1% KI IN 0.1 M PHOSPHATE BUFFER). Dissolve 13.61 g of potassium dihydrogen phosphate, 14.20 g of anhydrous disodium hydrogen phosphate (or 35.82 g of dodecahydrate salt) and 10.00 g of potassium iodide successively and dilute the mixture to exactly 1 liter with double distilled water. Age at room temperature for at least 1 day before using. This solution may be stored for several weeks in a glass stoppered brown bottle in the refrigerator or for shorter periods of time at room temperature. Do *not* expose to sunlight.

7.4 STANDARD IODINE SOLUTION, 0.025 MI_2 (0.05N). Dissolve successively 16.0 g of potassium iodide and 3.1730 g of iodine with double distilled water and make up to a volume of exactly 500 ml. Age for at least 1 day before using. Standardize shortly before use against 0.025 M $Na_2S_2O_3$. The sodium thiosulfate is standardized against primary standard bi-iodate, $KH(IO_3)_2$, or potassium dichromate, $K_2Cr_2O_7$.

8. Procedure

8.1 CLEANING OF EQUIPMENT. All glassware should be cleaned with *dichromate* cleaning solution followed by thorough rinsing with tap water and distilled water.

8.2 COLLECTION AND SHIPPING OF SAMPLES.

8.2.1 Pipet 10 ml of the absorbing solution (Section 7) into the midget impinger.

8.2.2 Connect the impinger (via the absorption tube) to the vacuum pump and the prefilter assembly (if needed) with a short piece of flexible tubing (not rubber). The minimum amount of tubing necessary to make the joint between the prefilter and impinger should be used. The air being sampled should not be passed through any other tubing or other equipment before entering the impinger.

8.2.3 Turn on pump to begin sample collection. Care should be taken to measure the flow rate, time and/or volume as accurately as possible. The sample should be taken at a flow rate of 0.2 to 1 lpm. If the sample air temperature and pressure deviate greatly from 25 C and 760 Torr, measure and record these values.

8.2.4 Sufficient air should be sampled so that the equivalent of 0.5 to 10 μl of ozone is absorbed. Sampling periods of longer than 30 min should be avoided. For a flow rate of 2 liters/min, a 30 min sample should yield a sensitivity of 0.01 ppm. Do *not* expose the absorbing reagent to direct sunlight.

8.2.5 After sampling, the impinger stem can be removed and cleaned. Tap the stem gently against the inside wall of the impinger bottle to recover as much of the sampling solution as possible. Wash the stem with a small amount (< 1 ml) of unused absorbing solution and add the wash to the impinger. Seal the impinger with a hard, non-reactive stopper (preferably Teflon). Do not seal with rubber.

8.2.6 The sample solutions must be analyzed within 60 min after sampling.

8.2.7 A "blank" impinger should be handled as the other samples (fill and seal) except that no air is sampled through this impinger.

8.3 ANALYSIS.

8.3.1 Record the total volume of sample solution present in the graduated impinger. This is usually 10 ml.

8.3.2 Transfer the exposed absorbing reagent (avoiding dilution with rinse water) to a clean cell or cuvette.

8.3.3 Determine the absorbance at 352 nm using a matched cell or cuvette freshly filled with distilled water as the reference. This measurement must be done within 30 to 60 min after sampling.

8.3.4 Determine daily the blank correction (to be deducted from sample absorbance) by reading the absorbance of the "blank" sample using distilled water as the reference.

8.3.5 Samples having a color too dark to read may be quantitatively diluted with additional absorbing reagent and the absorbance of the diluted solution read. The dilution factor must then be introduced into the calculations.

9. Calibration and Standards

9.1 Freshly prepare 0.00125 M (0.0025 N) iodine standard by pipetting exactly 5 ml of the 0.025 M (0.05 N) standard stock solution into a 100-ml volumetric flask and diluting to the mark with the absorbing reagent.

9.2 Pipet 0.2, 0.4, 0.6, and 0.9 ml portions of the diluted standard iodine in separate 25-ml volumetric flasks and dilute to the mark with absorbing reagent. Mix thoroughly. The concentrations of these solutions (must be based on exact molar concentration as determined from the molarity of the standard iodine solution) are respectively—0.25, 0.50, 0.75, and 1.125 μmoles I_2 per 25 ml; or more conveniently expressed as 0.10, 0.20, 0.30, and 0.45 μmoles of I_2 per 10 ml of absorbing reagent.

9.3 Immediately after the preparation of this known series, read the absorbance of each at 352 nm in the usual manner.

9.4 Plot the corrected absorbances (absorbance minus absorbance of reagent blank) of the known standards against the exact calculated molar concentrations (in μmoles/10 ml). Draw the straight line that gives the best fit. Beer's Law is followed.

10. Calculations

10.1 The absorbance of the "blank" sample (Section **8.3.4**) must first be subtracted from the sample absorbances.

10.2 As indicated by the stoichiometric equation in Section **1.3**, it has been empirically determined that 1 mole of ozone (O_3) liberates 1 mole of iodine (I_2) by this procedure. Thus, one can read directly from the calibration curve (Section **9.4**) the equivalent μmoles of O_3 per 10 ml of absorbing reagent. If the total volume of the sample solution is not 10 ml, or if the sample solution has been diluted prior to analysis, the dilution factor must be included in the calculation.

10.3 The concentration of ozone in the air sampled can be expressed as ppm of ozone, defined as μl ozone per liter of air.

$$\text{ppm, } O_3 = \frac{\mu\text{moles } O_3 \times 24.45}{V_s}$$

where:

μmoles O_3 = micromoles ozone determined from calibration curve (Section **9.4**) including the dilution factor, if any.

24.45 = molar volume of ozone (μl/μmole) at 25 C and 760 Torr.

V_S = volume of air sampled in 1 at 25 C and 760 Torr.

11. References

1. AMERICAN PUBLIC HEALTH ASSOCIATION, INC. 1977. Methods of Air Sampling and Analysis. 2nd ed. p. 556. Washington, D.C.
2. USDHEW PUBLIC HEALTH SERVICE. "Selected Methods for the Measurement of Air Pollutants," Publication Number 999-AP-11.

Revised by Subcommittee 3

B. DIMITRIADES, *Chairman*

J. CUDDEBACK

E. L. KOTHNY

P. W. MCDANIEL

L. A. RIPPERTON

A. SABADELL

820.

Analytical Method for Ozone in Air

Analyte:	Ozone	Method No.:	P & CAM 154
Matrix:	Air	Range:	0.1 to 20 ppm
Procedure:	Midget Impinger/ Alkaline 1% KI, Colorimetric	Precision:	Unknown
		Classification:	D (Operational)
Date Issued:	9/1/72		
Date Revised:	1/1/75		

1. Principle of the Method

1.1 Air containing ozone is drawn through a midget impinger containing 10 ml of 1% potassium iodide in 1.0 N sodium hydroxide. A stable product is formed that can be stored with little loss for several days.

1.2 The analysis is completed in the laboratory by the addition of phosphoric-sulfamic acid reagent, which liberates the iodine.

1.3 The yellow iodine color is read in a spectrophotometer at 352 nm.

2. Range and Sensitivity

2.1 The range extends from 0.1 ppm to about 20 ppm.

2.2 The sensitivity of the method is dependent on the volume of air sampled.

3. Interferences

3.1 Chlorine, hydrogen peroxide, organic peroxides, and various other oxidants will liberate iodine by this method.

3.2 The response to nitrogen dioxide is limited to 10% by the use of sulfamic acid in the procedure to destroy nitrite, thus minimizing any error due to the collection of NO_2.

3.3 The negative interferences from reducing gases such as sulfur dioxide and hydrogen sulfide are very serious (probably on a mole-to-mole equivalency).

3.4 The procedure is very sensitive to reducing dusts and droplets that may be present in the air or on the glassware. Losses of iodine also occur even on clean glass surfaces and thus the manipulations should minimize this exposure.

4. Precision and Accuracy

4.1 The precision and accuracy of this method have not been completely determined at this time. No collaborative tests have been performed on this method.

5. Advantages and Disadvantages

5.1 A delay of several days is permissible between sampling and completion of analysis.

5.2 The method is simple, accurate and precise.

5.3 Those compounds listed in **3.1** will liberate iodine by this method.

6. Apparatus

6.1 MIDGET IMPINGER WITH GRADUATION MARKS. (Fritted bubblers tend to give relatively low results).

6.2 AIR SUCTION PUMP, capable of drawing the required sample flow for intervals of up to 30 min.

6.3 AIR METERING DEVICE, capable of measuring a flow of 1 to 2 liters per min.

6.4 SPECTROPHOTOMETER, capable of measuring the yellow color at 352 nm with stoppered tubes or cuvettes.

6.5 ASSOCIATED LABORATORY GLASSWARE.

7. Reagents

7.1 PURITY. The reagents described must be made up using ACS reagent grade or better grade of chemical.

7.2 DOUBLE DISTILLED WATER. Double distilled water should be used for all reagents. The double distilled water can be prepared in an all-glass still by adding potassium permanganate to produce a faint pink color and barium hydroxide to alkalize the distilled water before the second distillation.

7.3 ABSORBING REAGENT. Dissolve 40.0 g of sodium hydroxide in almost a liter of water, then dissolve 10.0 g of potassium iodide and make the mixture to 1 liter. Store in a clean glass bottle with a screw cap (with inert liner) or rubber stopper (previously boiled for 30 min in alkali and washed). Age for at least 1 days before using.

7.4 ACIDIFYING REAGENT. Dissolve 5.0 g of sulfamic acid in 100 ml of water, then add 84 ml of 85% phosphoric acid and dilute to 200 ml.

7.5 STANDARD POTASSIUM IODATE SOLUTION. Dissolve 0.758 g of potassium iodate in water and dilute to 1 l. One ml of this stock solution is equivalent to 400 μl of ozone.

7.6 DILUTE STANDARD SOLUTION. Prepare, just before it is required, a dilute standard solution by pipetting exactly 5 ml of stock solution into a 50-ml volumetric flask and making to mark with distilled water. One ml of this solution is equivalent to 40 μl of ozone.

8. Procedure

8.1 CLEANING OF EQUIPMENT. All glassware should be cleaned with dichromate cleaning solution followed by 3 tap and 3 distilled water rinses.

8.2 COLLECTIONS OF SAMPLES.

8.2.1 Assemble a train composed of a midget impinger, rotameter, and pump. Use ground glass connections upstream from the impinger.

8.2.2 Pipet exactly 10 ml of the absorbing solution into the midget impinger and sample at a flow rate of 1 to 2 l/min. Note the volume of air sampled. If the sample air temperatures and pressures deviate greatly from 25 C and 760 Torr, measure and record these values.

8.2.3 Sufficient air should be sampled so that the equivalent of 1 to 15 μl of ozone is absorbed. If appreciable evaporation has occurred, add distilled water to restore the volume to the 10 ml graduation mark.

8.2.4 If the analysis is to be completed later, cap the impinger with the supplied stopper for shipment to the laboratory.

8.3 ANALYSIS.

8.3.1 Measure the volume of exposed absorbing reagent in a 25-ml glass stoppered, graduated cylinder (Do not use rinse water in any transfer).

8.3.2 Using a rapid (serological type), graduated pipet, add the acidifying reagent in the amount of exactly one-fifth of the volume of absorbing solution measured in **8.2.2** to the graduated cylinder containing the absorbing solution. Mix thoroughly.

8.3.3 Place the stoppered cylinder in a water bath at room temperature for 5 to 10 min to dissipate the heat of neutralization.

8.3.4 Transfer a portion of the sample to a cuvette and determine the absorbance at 352 nm. A cuvette containing distilled water is used as the reference. Do not delay reading since reducing impurities sometimes causes rapid fading of the color.

8.3.5 Prepare a reagent blank by adding 2 ml of the acidifying reagent to 10 ml of unexposed absorbing reagent. Cool and determine the blank absorbance. The blank absorbance should be determined before each series of measurement and should be subtracted from the absorbance of the samples.

8.3.6 Samples may be aliquoted before or after acidification if very large concentrations of oxidant are expected. In the former case dilute the aliquot to 10 ml with unexposed absorbing reagent and proceed in the usual manner. In the latter case dilute the aliquot to 10 ml with reagent blank mixture. Aliquoting after acidification is not as reliable as before acidification and should be used only to save a sample when unexpectedly large concentrations of oxidant are encountered. The calculations should include the aliquoting factor.

9. Calibration and Standards

9.1 Add the freshly prepared, dilute standard iodate solution in graduated amounts of 0.10 to 0.45 ml (measured accurately in a graduated pipet or small buret) to a series of 25-ml glass stoppered graduated cylinders. This corresponds to 4 to 18 μl O_3.

9.2 Add sufficient alkaline potassium iodide solution to make the total volume of each exactly 10 ml.

9.3 Add 2 ml of acidifying reagent and determine the absorbance of each standard as with the samples.

10. Calculations

10.1 For convenience, standard conditions are taken as 760 Torr and 25 C; thus only slight correction by means of the well known perfect gas equation is required to get V, the standard volume in liters of air sampled. Ordinarily this correction may be omitted. Quantities customarily expressed in terms of ozone may be expressed as μl, defined as V times ppm ozone. It has been determined empirically that 1 mole of ozone liberates 0.65 mole of iodine (I_2) by this procedure.

10.2 Plot the absorbance of the standard (corrected for the blank) against ml of dilute standard iodate solution. Beer's Law is followed. Draw the straight line giving the best fit and determine the value in milliters of the diluted potassium iodate solution intercepted at an absorbance of exactly 1. This value multiplied by 40 gives the standardization factor M, defined as the number of μl of ozone required by 10 ml of absorbing reagent to give a final absorbance of 1. For 2-cm cells this value is 9.13.

ppm oxidant (expressed as ozone) =

$$\text{corrected absorbance} \times \frac{M}{V}$$

If the volume of the air sample V is a simple multiple of M, calculations are simplified. Thus, for the M value of 9.13 previously cited, if exactly 9.13 l of air are sampled through the impinger, the corrected absorbance is also ppm directly. If other volumes of absorbing reagent are used, V is taken as the volume of air sample per 10 ml of absorbing reagent.

11. References

1. USDHEW PUBLIC HEALTH SERVICE. 1965. Selected Methods for the Measurement of Air Pollutants. Publication Number 99-AP-11.

Revised by Subcommittee 3
B. DIMITRIADES, *Chairman*
J. CUDDEBACK
E. L. KOTHNY
P. W. MCDANIEL
L. A. RIPPERTON
A. SABADELL

821.

Analytical Method for Phosgene in Air

Analyte:	Phosgene ($COCl_2$)	Method No.:	P & CAM 129
Matrix:	Air	Range:	0.005 to 5 ppm (.02 to 20 mg/m^3)
Procedure:	Midget Impinger/ Nitrobenzylpyridine Colorimetric	Precision:	Unknown
		Classification:	E (Proposed)
Date Issued:	6/1/75		
Date Revised:			

1. Principle of the Method

1.1 4,4′-nitrobenzyl pyridine in diethyl phthalate reacts with traces of phosgene to produce a brilliant red color. The addition of an acid acceptor such as N-phenylbenzylamine stabilizes the color and increases the sensitivity. The absorbance is determined at 475 nm (**1**).

2. Range and Sensitivity

2.1 Sampling efficiency is 99% or better. Five μg of phosgene can be detected; the minimum sample size is 25 liters. Therefore, 0.02 mg/m^3 phosgene should be detectable with a 250-liter sample. 20 mg/m^3 upper limit should be within the range of most laboratory photometers for a 25-liter air sample.

3. Interferences

3.1 Other acid chlorides, alkyl and aryl derivatives which are substituted by active halogen atoms and sulfate esters are known to produce color with this reagent. However, most of these interferences can be removed in a prescrubber containing an inert solvent such as "Freon-113" cooled by an ice bath.

3.2 This method is not subject to interference from likely concentrations of chloride, hydrogen chloride, chlorine dioxide or simple chlorinated hydrocarbons such as carbon tetrachloride, chloroform, and tetrachlorethylene. A slight depression of color density has been noted under high humidity conditions.

4. Precision and Accuracy

4.1 The accuracy and precision of this method has not been determined. Calibration is a major problem because of the reactivity of phosgene. Flow dilution systems can be used but require anhydrous air for dilution. Permeation-type calibration standards may be employed, if available.

5. Advantages and Disadvantages of the Method

5.1 Advantages include high sensitivity, standard laboratory equipment, and relative simplicity.

5.2 Disadvantages include potential interferences, relative changes in color formation with various lots of reagents and the need for frequent calibration checks.

6. Apparatus

6.1 MIDGET IMPINGERS.

6.2 VACUUM SOURCE, Aspirator or pump

6.3 FLOW METER (Wet test meter)

6.4 POLYETHYLENE BOTTLES AND BEAKERS

6.5 BAUSCH AND LOMB SPEC. 20, OR EQUIVALENT PHOTOMETER.

6.6 ASSOCIATED LABORATORY GLASSWARE AND TUBING. No plasticized tubing (such as Tygon) should be used.

7. Reagents

7.1 4,4′-NITROBENZYL PYRIDINE. (Aromil Chemical Co., Baltimore, Md., or Aldrich Chemical Co., Milwaukee, Wis.)

7.2 N-PHENYLBENZYLAMINE. (ACS reagent grade or better).

7.3 DIETHYL-PHTHALATE. (Special selection may be required since some lots of the phthalate produce unstable color).

7.4 COLOR REAGENT. Reagent is made up of a solution of 2.5 g 4,4′-nitrobenzyl pyridine, 5 g N-phenylbenzylamine and 992.5 g diethylphthalate. The concentration of color forming reagents in diethylphthalate is critical as less color is developed with either a decreased or increased conc.

8. Procedure

8.1 CLEANING OF EQUIPMENT. All glassware and plastic ware are to be doubly rinsed with isopropanol (IPA) and dried. If IPA rinse(s) does not thoroughly clean apparatus, use standard glassware cleaning procedures before IPA rinses.

8.2 COLLECTION OF SAMPLES.

8.2.1 Atmospheric samples are collected in midget impingers loaded with 10 ml of reagent prepared as described in Section **7.4** above.

8.2.2 A calibrated sample rate of 1 liter per min is used.

8.2.3 A known volume of about 50 liters is drawn through the impingers for a range of 0.04 to 1 ppm phosgene. (For ranges above 1 ppm, 25 liters should be drawn.)

8.2.4 The final solution volume is recorded and the sample transferred to a small polyethylene or glass bottle or directly to the cuvette.

8.2.5 The red color formed is stable for at least 4 hr, but should be measured within 9 hr as a 10% to 15% loss in color density after 8 hr has been reported.

8.3 ANALYSIS OF SAMPLE

8.3.1 The ample is transferred to a cuvette for the photometric measurement at 475 nm.

8.3.2 The photometer should be "zeroed" (at 100% T or 0 absorbance) with a matched cuvette containing unreacted reagent at the 475 nm wavelength.

8.3.3 The absorbance (or transmittance) of the sample is measured with the photometer at the 475 nm wavelength.

9. Calibration and Standards

9.1 Phosgene is highly reactive, and standards are difficult to prepare. However, dynamic standards may be prepared from a flow dilution system using a standard cylinder of 1000 ppm phosgene in nitrogen. (Lower concentrations of phosgene in cylinders are generally unstable. In any case, the "standard" cylinder concentration should be checked periodically by mass spectrometric or other methods.) A double dilution system with $COCl_2$-free air to dilute the phosgene about 1000/l should be used. Permeation-type calibration technique can be employed if phosgene standard permeation tubes are available.

10. Calculations

10.1 A calibration curve should be prepared of absorbance at 475 nm versus the phosgene concentration in air with ***50 liters*** of gas sample.

10.2 For a gas sample of volume X, the $COCl_2$ concentration as determined by the calibration curve should be multiplied by $\frac{50}{X}$ for the corrected value.

11. References

1. AMERICAN INDUSTRIAL HYGIENE ASSOC. 1969. Akron, Ohio. Analytical Guides.
2. LINCH, A.L., S.S. FORD, K.A. KUBTIZ, AND M.R. DEBRUNNER. 1965. Phosgene in Air-Development of Improved Detector Procedures, *Amer. Ind. Hyg. Assoc. J.* 26:465.
3. NOWEIR, M.H., AND E.A. PFITZER. 1964. The Determination of Phosgene in Air. Unpublished paper presented at the Amer. Ind. Hyg. Conference in Philadelphia, Pennsylvania April 28.

Subcommittee 2
C. R. THOMPSON, *Chairman*
G. H. FARRAH
L. V. HAFF
A. W. HOOK
J. S. JACOBSON
E. J. SCHNEIDER
L. H. WEINSTEIN

822.

General Atomic Absorption Procedure for Trace Metals in Airborne Material Collected on Membrane Filters

Analyte:	Trace Metals (see Tables 1 and 2)	Method No.:	P & CAM 173
Matrix:	Air	Range:	See Table 2
Procedure:	Membrane Filters for Sampling/Atomic Absorption	Precision:	± 3% RSD (Analytical)
Date Issued:		Classification:	D (operational)
Date Revised:	1/1/75		

1. Principle of Method

1.1 This procedure describes a general method for the collection, dissolution and determination of trace metals in industrial and ambient airborne material. The samples are collected on membrane filters and treated with nitric acid to ash the organic matrix and to dissolve the metals present in the sample. The analysis is subsequently made by atomic absorption spectrophotometry (AAS).

1.2 Samples and standards are aspirated into the appropriate AAS flame. A hollow cathode lamp for the metal being measured provides a source of characteristic radiation energy for that particular metal. The absorption of this characteristic energy by the atoms of interest in the flame is related to the concentration of the metal in the aspirated sample. The flames and operating conditions for each element are listed in Table 822:I.

2. Range and Sensitivity

2.1 The sensitivity, detection limit and optimum working range for each metal are given in Table 822:II. The sensitivity is defined as that concentration of a given element which will absorb 1% of the incident radiation (0.0044 absorbance units) when aspirated into the flame. The detection limit is defined as that concentration of a given element which produces a signal equivalent to three times the standard deviation of the blank signal (**Note:** The blank signal is defined as that signal which results from all added reagents and a clean membrane filter which has been ashed exactly as the samples.) The working range for an analytical precision better than 3% is generally defined as those sample concentrations which will absorb 10% to 70% of the incident radiation (0.05 to 0.52 absorbance units). The values for the sensitivity and detection limits are instrument dependent and may vary from instrument to instrument.

3. Interferences

3.1 In atomic absorption spectrophotometry the occurrence of interferences is less common than in many other analytical determination methods. Interferences can occur, however, and when encountered are corrected for as indicated in the following sections. The known interferences and correction methods for each metal are indicated in Table 822:I. The methods of standard additions and background monitoring and correction (**1-4**) are used to identify the presence of an interference problem. Insofar as possible the matrix of the samples and standards are matched to minimize the possible interference problems.

3.1.1. Background or non-specific absorption can occur from particles produced in the flame which can scatter the incident radiation causing an apparent absorption signal. Light scattering problems may be encountered when solutions of high salt content are being analyzed. Light scattering problems are most severe when measurements are made at the lower wavelengths (*i.e.*, below about 250 nm). Background absorption may also occur as the result of the formation of various molecular species which can absorb light. The background absorption can be accounted for by the use of background correction techniques (**1**).

3.1.2 Spectral interferences are those interferences which occur as the result of an atom different from that being measured absorbing a portion of the incident radiation. Such interferences are extremely rare in atomic absorption. In some cases multi-element hollow cathode lamps may cause a spectral interference by having closely adjacent radiation lines from two different elements. In such instances multi-element hollow cathode lamps should not be used.

3.1.3 Ionization interferences can occur when easily ionized atoms are being measured. The degree to which such atoms are ionized is dependent upon the atomic concentration and presence of other easily ionized atoms in the sample. Ionization interferences can be controlled by the addition to the sample of a high concentration of another easily ionized element which will buffer the electron concentration in the flame.

3.1.4 Chemical interferences occur in atomic absorption spectrophotometry when species present in the sample cause variations in the degree to which atoms are formed in the flame. Such interferences may be corrected for by controlling the sample matrix or by using the method of standard additions (**2**).

3.1.5 Physical interferences may result if the physical properties of the sam-

Table 822:I. Flame and Operating Condition for Metals.

Element	Type of Flame	Analytical Wavelength	Interferences†	Remedy†	References
		nm			
Ag	Air–C_2H_2 (oxidizing)	328.1	IO_3^-, WO_4^{-2}, MnO_4^{-2}	‡	**(5)**
Al*	N_2O–C_2H_2 (reducing)	309.3	Ionization, SO_4^{-2}, V	‡, §, ‖	**(4)**
Ba	N_2O–C_2H_2 (reducing)	553.6	Ionization, large conc. of Ca	§, #	**(1, 4)**
Be*	N_2O–C_2H_2 (reducing)	234.9	Al, Si, Mn	‡	**(4)**
Bi	Air–C_2H_2 (oxidizing)	223.1	None known		
Ca	Air–C_2H_2 (reducing) N_2O–C_2H_2	422.7	Ionization & chemical	§, ‖	**(1, 4)**
Cd	Air–C_2H_2 (oxidizing)	228.8	None known		
Co*	Air–C_2H_2 (oxidizing)	240.7	None known		
Cr*	Air–C_2H_2 (oxidizing)	357.9	Fe, Ni	‡	**(4)**
Cu	Air–C_2H_2 (oxidizing)	324.8	None known		
Fe	Air–C_2H_2 (oxidizing)	248.3	High Ni conc., Si	‡	**(1, 4)**
In	Air–C_2H_2 (oxidizing)	303.9	Al, Mg, Cu, Zn, $H_xPO^x{}_4^{-3}$	‡	**(10)**
K	Air–C_2H_2 (oxidizing)	766.5	Ionization	§	**(1, 4)**
Li	Air–C_2H_2 (oxidizing)	670.8	Ionization	§	**(11)**
Mg	Air–C_2H_2 (oxidizing) N_2O–C_2H_2 (oxidizing)	285.2	Ionization & chemical	§, ‖	**(1, 4)**
Mn	Air–C_2H_2 (oxidizing)	279.5	None known		
Na	Air–C_2H_2 (oxidizing)	589.6	Ionization	‖	**(1, 4)**
Ni	Air–C_2H_2 (oxidizing)	232.0	None known		
Pb	Air–C_2H_2 (oxidizing)	217.0 283.3	Ca, high conc. $SO^{4=}$	‡	**(7)**
Rb	Air–C_2H_2 (oxidizing)	780.0	Ionization	§	**(1, 8)**
Sr	Air–C_2H_2 (reducing) N_2O–C_2H_2 (reducing)	460.7	Ionization & chemical	§, ‖	**(1, 8)**
Tl	Air–C_2H_2 (oxidizing)	276.8	None known		
V*	N_2O–C_2H_2 (reducing)	318.4	None known in N_2O–C_2H_2 flame		
Zn	Air–C_2H_2 (oxidizing)	213.9	None known		

*Some compounds of these elements will not be dissolved by the procedure described here. When determining these elements one should verify that the types of compounds suspected in the sample will dissolve using this procedure. (See Section **3.2**)

†High concentrations of Si in the sample can cause an interference for many of the elements in this table and may cause aspiration problems. No matter what elements are being measured, if large amounts of silica are extracted from the samples the samples should be allowed to stand for several hours and centrifuged or filtered to remove the silica.

‡Samples are periodically analyzed by the method of additions to check for chemical interferences. If interferences are encountered, determinations must be made by the standard additions method or, if the interferent is identified, it may be added to the standards.

§Ionization interferences are controlled by bringing all solutions to 1000 ppm Cs (samples and standards).

‖1000 ppm solution of La as a releasing agent is added to all samples and standards.

#In the presence of very large Ca concentrations (greater than 0.1%) a molecular absorption from $Ca(OH)_2$ may be observed. This interference may be overcome by using background correction when analyzing for Ba.

ples vary significantly. Changes in viscosity and surface tension can affect the sample aspiration rate and thus cause erroneous results. Sample dilution andior the method of standard additions are used to correct such interferences. High concentrations of Si in the sample can cause an interference for many of the elements and may cause aspiration problems. No matter what elements are being measured if large amounts of Si are extracted from the samples, the samples should be al-

Table 822:II. Sensitivity detection limit, and optimum working range of metals.

Element	Sensitivity* μg/ml	Range μg/ml	Detection Limits† μg/ml	Detection Limits† μg/m³	Minimum‡ TLV μg/m³
Ag	0.036	0.5–5.0	.003	0.1	10 (metal and soluble compounds)
Al	0.76	5–50	.04	2	NL
Ba	0.20	1–10	.01	0.4	500 (soluble compounds)
Be	0.017	0.1–1.0	.002	0.08	2
Bi	0.22	1–10	.06	3	NL§
Ca	0.021	0.1–1.0	.0005	0.02	5,000 (CaO)
Cd	0.011	0.1–1.0	.0006	0.03	200 (metal dust & soluble salts) 100 (cadmium oxide fume)
Co	0.066	0.5–5.0	.007	0.3	100 (metal fume and dust)
Cr	0.055	0.5–5.0	.005	0.2	100 (chromic acid & chromates, as CrO_3) 500 (soluble chromic, chromous salts) 1,000 (metal and insoluble salts)
Cu	0.040	0.5–5.0	.003	0.1	100 (fume) 1,000 (dusts and mists)
Fe	0.062	0.5–5.0	.005	0.2	10,000 (iron oxide fume, as iron oxide) 1,000 (soluble compounds)
In	0.38	5–50	.05	2	100 (metal and compounds)
K	0.010	0.1–1.0	.003	0.1	NL§
Li	0.017	0.1–1.0	.002	0.08	25 (as lithium hydride)
Mg	0.003	.05–.50	.0003	0.01	10,000 (as magnesium oxide fume)
Mn	0.026	0.5–5.0	.003	0.1	5,000 (metal and compounds)
Na	0.003	.05–.50	.0003	0.01	2,000 (as sodium hydroxide)
Ni	0.066	0.5–5.0	.008	0.3	1,000 (metal and soluble compounds)
Pb	0.11	1–10	.02	0.8	150 (inorganic compounds, fumes & dusts)
Rb	0.042	0.5–5.0	.003	0.1	NL§
Sr	0.044	0.5–5.0	.004	0.2	NL§
Tl	0.28	5–50	.02	0.8	100 (soluble compounds)
V	0.88	10–100	0.1	4	500 (V_2O_5 dust) 100 (V_2O_5 fume)
Zn	0.009	0.1–1.0	.002	0.08	1,000 ($ZnCl_2$ fume) 5,000 (ZnO fume)

*Sensitivity data are taken from "Analytical Data for Elements Determined by Atomic Absorption Spectrometry", Varian Techtron, Walnut Creek, California, 1971.

†Solution detection limits are taken from "Detection Limits for Model AA-5 Atomic Absorption Spectrophotometer", Varian Techtron, Walnut Creek, California. The atmospheric concentrations were calculated assuming a collection volume of 0.24 m_3 (2 lpm for 2 hours) and an analyte volume of 10 ml for the entire sample. (NOTE: These detection limits represent the ultimate values since the blanks resulting from the reagents and the filter material have not been taken into account.)

‡Threshold limit values of airborne contaminants and physical agents with intended changes adopted by ACGIH for 1971. All values listed are expressed as elemental concentrations except as noted.

§NL signifies No Limit expressed for this element or its compounds.

lowed to stand for several hours and centrifuged or filtered to remove the Si.

3.2 This procedure describes a generalized method for sample preparation which is applicable to the majority of samples of interest. There are, however, some relatively rare chemical forms of a few of the elements listed in Table 822:I which will not be dissolved by this procedure. If such chemical forms are suspected, results obtained using this procedure should be compared with those obtained using an appropriately altered dissolution procedure. Alternatively, the results may be compared with values obtained utilizing a non-destructive technique which does not require sample dissolution (*e.g.*, x-ray fluorescence, activation analysis).

4. Precision and Accuracy

4.1 The relative standard deviation of the analytical measurement is approximately 3% when measurements are made in the ranges listed in Table 822:II. The overall relative standard deviation will be somewhat larger than this value due to errors associated with the sample collection and preparation steps.

4.2 No data are presently available on the accuracy of this method for actual air samples.

5. Advantages and Disadvantages of the Method

5.1 The sensitivity is adequate for all metals in air samples but only for certain metals in biological matrices. The sensitivity of this direct aspiration method is not adequate for Be, Cd, Ca, Cr, Mn, Mo, Ni, and Sn in biological samples.

5.2 A disadvantage of the method is that at least 1 to 2 ml of solution is necessary for each metal determination. For small samples, the necessary dilution would decrease sensitivity.

6. Apparatus

6.1 SAMPLING EQUIPMENT.

6.1.1 Membrane filters with a pore size of 0.8 μm appropriately sized for the sampling holder are satisfactory for air sampling. The filter holder contains a 2-piece filter cassette.

6.1.2 Sampling pump must be of sufficient capacity to maintain a face velocity of 7 cm/sec (**5**). A personal sampling pump must be calibrated with a representative filter unit in the line.

6.1.3 Flow measurement device such as a calibrated rotameter is required to monitor or a critical flow orifice is used to control the sampling rate.

6.2 GLASSWARE, BOROSILICATE. Before use all glassware must be cleaned in 1:1 diluted nitric acid and rinsed several times with distilled water.

6.2.1 *125-ml Phillips or Griffin beakers with watch glass covers.*

6.2.2 *15-ml graduated centrifuge tubes.*

6.2.3 *10-ml volumetric flasks.*

6.2.4 *100-ml volumetric flasks.*

6.2.5 *1-liter volumetric flasks.*

6.2.6 *125-ml polyethylene bottles.*

6.2.7 Additional auxiliary glassware such as pipettes and different size volumetric glassware will be required depending on the elements being determined and the dilutions required to have sample concentrations above the detection limit and in the linear response range (*i.e.*, see Table 822:I). All pipettes and volumetric flasks required in this procedure should be calibrated class A volumetric glassware.

6.3 EQUIPMENT.

6.3.1 *Atomic absorption spectrophotometer,* with burner heads for air-acetylene and nitrous oxide-acetylene flames.

6.3.2 HOLLOW CATHODE LAMPS, for each metal and a continuum lamp.

6.3.3 *Hotplate,* suitable for operation at 140 C and 250 C).

6.3.4 *Two stage regulators,* for air, acetylene and nitrous oxide.

6.3.5 *Heating tape and rheostat,* for nitrous oxide regulator (second regulator stage may need to be heated to approximately 60 C to prevent freeze-up).

6.4 SUPPLIES.

6.4.1 *Acetylene gas (cylinder).* A grade specified by the manufacturer of the instrument. (Replace cylinder when pressure decreases below 100 psi.)

6.4.2 *Nitrous oxide gas (cylinder).*

6.4.3 *Air supply,* with a minimum pressure of 40 psi filtered to remove oil and water.

7. Reagents

7.1 PURITY. ACS reagent grade chemicals or equivalent shall be used in all tests. References to water shall be understood to mean double distilled water or equivalent. Care in selection of reagents and in following the listed precautions is essential if low blank values are to be obtained.

7.2 CONC NITRIC ACID (68 TO 71%). Redistilled sp g 1.42.

7.3 STANDARD STOCK SOLUTIONS (1000 μG/ML), FOR EACH METAL IN TABLE 822:I. Commercially prepared or prepared from the following:

7.3.1 *Silver nitrate* ($AgNO_3$).
7.3.2 *Aluminum wire.*
7.3.3 *Barium chloride* ($BaCl_2 \cdot 2H_2O$).*
7.3.4 *Beryllium metal.*
7.3.5 *Bismuth metal.*
7.3.6 *Calcium carbonate* ($CaCO_3$).*
7.3.7 *Cadmium metal.*
7.3.8 *Cobalt metal.*
7.3.9 *Copper metal.*
7.3.10 *Potassium chromate.* (K_2CrO_4).
7.3.11 *Iron wire.*
7.3.12 *Indium metal.*
7.2.13 *Potassium chloride* (KCl).*
7.3.14 *Lithium carbonate* (Li_2CO_3).*
7.3.15 *Magnesium ribbon.*
7.3.16 *Manganese metal.*
7.3.17 *Sodium chloride* ($NaCl$).*
7.3.18 *Nickel metal.*
7.3.19 *Lead nitrate* [$Pb(NO_3)_2$].*
7.3.20 *Rubidium chloride* ($RbCl$).*
7.3.21 *Strontium nitrate* [$Sr(NO_3)_2$].*
7.3.22 *Thallium nitrate* ($TINO_3$).
7.3.23 *Vanadium metal.*
7.3.24 *Zinc metal.*

7.4 LANTHANUM NITRATE [$La(NO_3)_3 \cdot 6H_2O$].

7.5 CESIUM NITRATE ($CsNo_3$).

8. Procedure

8.1 CLEANING OF EQUIPMENT.

8.1.1 Before initial use, glassware is cleaned with a saturated solution of sodium dichromate in conc sulfuric acid (**Note:** Do not use for chromium analysis) and then rinsed thoroughly with warm tap water, conc nitric acid, tap water and deionized water, in that order, and then dried.

*These salts must be stored in a dessicator to avoid pick-up of water from the atmosphere.

8.1.2 All glassware is soaked in a mild detergent solution immediately after use to remove any residual grease or chemicals.

8.2 COLLECTION AND SHIPPING OF SAMPLES.

8.2.1 Ambient atmospheric particulate matter and industrial dusts and fumes are sampled with cellulose membrane filters of 0.8 μm average pore size (Millipore Type AA or equivalent). The pump used with any membrane filter must be sufficient to maintain a face velocity of at least 7 cm/sec throughout the sampling period. Sample flow rate is monitored with a calibrated rotameter (**5**) or the equivalent. The flow rate, ambient temperature and barometric pressure are recorded at the beginning and the end of the sample collection period.

8.2.2 For personal sampling 37-mm diam filters in holders (Millipore Filter Type AA or equivalent) are used. The personal sampler pumps for this application are operated at 1.5 lpm. In general, a 2 hr sample at 1.5 lpm will provide enough sample to detect the elements sought at air concentrations of 0.2 × TLV.

8.2.3 After sample collection is complete the exposed filter surface should be covered with a clean filter. Losses of sample due to overloading of the filter must be avoided.

8.3 SHIPMENT OF SAMPLES.

8.3.1 Filter samples (with clean filter covers) should be sealed in individual plastic filter holders for storage and shipment.

8.4 PREPARATION OF SAMPLES.

8.4.1 The samples including the clean filter covers and blanks (minimum of 1 filter and cover blank for every 10 filter samples) are transferred to clean 125-ml Phillips or Griffin beakers and sufficient conc HNO_3 is added to cover the sample. Each beaker is covered with a watch glass and heated on a hot plate (140 C) in a fume hood until the sample dissolves and a slightly yellow solution is produced. Approximately 30 min of heating will be sufficient for most air samples. However, subsequent additions of HNO_3 may be needed to completely ash and destroy high concentrations of organic material, and under these conditions longer ashing times will be needed. Once the ashing is complete as indicated by a clear solution in the beaker, the watch glass is removed; and the sample is allowed to evaporate to dryness.

8.4.2 Place the sample beaker on a hot plate controlled at 250 C for several minutes. If evidence of charring is observed, removed from the hotplate, cool, and repeat the procedure in **8.4.1**. If the residue in the beaker is a light whitish material the beaker is cooled and 1 ml of HNO_3 and 2 to 3 ml of distilled H_2O are added. The beaker is replaced on the hot plate and swirled occasionally until the residue is dissolved (as indicated by a light clear solution). The beaker is then removed and the solution is quantitatively transferred with distilled water to a 10-ml volumetric flask. If any elements are being determined which require the ionization buffer, 0.2 ml of 50 mg/ml Cs is added to the volumetric flask (see Table 822:I, footnote §). If any elements requiring the releasing agent are being determined, 0.2 ml of 50 mg/ml La is added to each volumetric flask (see Table 822:I, footnote ||). The samples are then diluted to volume with water.

8.4.3 The 10 ml solution may be analyzed directly for any element of very low concentration in the sample. Aliquots of this solution may then be diluted to an appropriate volume for the other elements of interest present at higher concentrations (**Note:** Approximately 2 ml of solution are required for each element being analyzed.) The dilution factor will depend upon the concentration of elements in the sample and the number of elements being determined by this procedure.

8.5 ANALYSIS.

8.5.1 Set the instrument operating conditions as recommended by the manufacturer. The instrument should be set at the radiation intensity maximum for the wavelength listed in Table 822:I for the element being determined.

8.5.2 Standard solutions should match the sample matrix as closely as possible and should be run in duplicate. Working standard solutions prepared fresh daily, are aspirated into the flame and the absorbance recorded. Prepare a calibration graph as described in section **9.2.4** (**Note:** All combustion products from the AA flame must be removed by direct exhaustion through use of a good separate flame ventilation system.)

8.5.3 Aspirate the appropriately diluted samples directly into the instrument and record the absorbance for comparison with standards. Should the absorbance be above the calibration range, dilute an appropriate aliquot to 10 ml. Aspirate water between each sample. A mid-range standard must be aspirated with sufficient frequency (*i.e.*, once every 10 samples) to assure the accuracy of the sample determinations. To the extent possible, all determinations are to be based on replicate analysis.

9. Calibration and Standards

9.1 IONIZATION AND CHEMICAL INTERFERENCE SUPPRESSANTS.

9.1.1 *Lanthanum solution (50 mg/ml).* Dissolve 156.32 g of lanthanum nitrate [$La(NO_3)_3\ 6\ H_2O$] in 2% (v/v) HNO_3. Dilute to volume in a 1-l volumetric flask with 2% (v/v) HNO_3. When stored in a polyethelene bottle this solution is stable for at least one year.

9.1.2 *Cesium solution (50 mg/ml).* Dissolve 73.40 g of cesium nitrate ($CsNO_3$) in distilled water. Dilute to volume in a 1-l volumetric flask with distilled water. When stored in a polyethylene bottle this solution is stable for at least one year.

9.2 STANDARD METAL SOLUTIONS.

9.2.1 *Standard stock solutions.* All standard stock solutions are made to a concentration of 1.0 mg of metal per ml. Except as noted these standard stock solutions are stable for at least one year when stored in polyethylene bottles.

a. Master silver standard. Dissolve 1.574 g silver nitrate (Ag NO_3) in 100 ml distilled water. Dilute to volume in a 1-l volumetric flask with 2% (v/v) HNO_3. The silver nitrate solution will decompose in light and must be stored in an amber bottle away from direct light. New master silver standards should be prepared every few months.

b. Master aluminum standard. Dissolve 1.000 g of Al wire in a minimum volume

1 : 1 HCl. Dilute to volume in a 1-liter flask with 10% (v/v) HNO_3.

c. Master barium standard. Dissolve 1.779 g of barium chloride ($BaCl_2 \cdot 2H_2O$) in water. Dilute to volume in a 1-l volumetric flask with distilled water.

d. Master beryllium standard. Dissolve 1.000 g of Be metal in a minimum volume of 1 : 1 HCl. Dilute to volume in a 1-l flask with 2% (v/v) HNO_3.

e. Master bismuth standard. Dissolve 1.000 g of bismuth metal in a minimum volume of 6 N HNO_3. Dilute to volume in a 1-l volumetric flask with 2% (v/v) HNO_3.

f. Master calcium standard. To 2.498 g of primary standard calcium carbonate ($CaCO_3$) add 50 ml of deionized water. Add dropwise a minimum volume of HCl (approximately 10 ml) to dissolve the $CaCO_3$. Dilute to volume in a 1-l volumetric flask with distilled water.

g. Master cadmium standard. Dissolve 1.000 g of cadmium metal in a minimum volume of 6 N HCl. Dilute to volume in a 1-l volumetric flask with 2% (v/v) HNO_3.

h. Master cobalt standard. Dissolve 1.000 g of Co metal in a minimum volume of 1 : 1 HCl. Dilute to volume in a 1-l flask with 2% (v/v) HNO_3.

i. Master copper standard. Dissolve 1.000 g of copper metal in a minimum volume of 6 H NHO_3. Dilute to volume in a 1-l volumetric flask with 2% (v/v) HNO_3.

j. Master chromium standard. Dissolve 3.735 g of potassium chromate (K_2CrO_4) in distilled water. Dilute to volume in a 1-l flask with distilled water.

k. Master iron standard. Dissolve 1.000 g of iron wire in 50 ml of 6 N HNO_3. Dilute to volume in a 1-l volumetric flask with distilled water.

l. Master indium standard. Dissolve 1.000 g of indium metal in a minimum volume of 6 N HCl. Addition of a few drops of HNO_3 and mild heating will aid in dissolving the metal. Dilute to volume in a 1-l volumetric flask with 10% (v/v) HNO_3.

m. Master potassium standard. Dissolve 1.907 g of potassium chloride (KCl) in distilled water. Dilute to volume in a 1-l volumetric flask with distilled water.

n. Master lithium standard. Dissolve 5.324 g of Li_2CO_3 in a minimum volume of 6 N HCl. Dilute to volume in a 1-l volumetric flask with distilled water.

o. Master magnesium standard. Dissolve 1.000 g of magnesium ribbon in a minimum volume of 6 N HCl. Dilute to volume in a 1-l volumetric flask with 2% (v/v) HNO_3.

p. Master manganese standard. Dissolve 1.000 g of manganese metal in a minimum volume of 6 N HNO_3. Dilute to volume in a 1-l volumetric flask with 2% (v/v) HNO_3.

q. Master sodium standard. Dissolve 2.542 g of sodium chloride (NaCl) in distilled water. Dilute to volume in a 1-l volumetric flask with distilled water.

r. Master nickel standard. Dissolve 1.000 g of nickel metal in a minimum volume of 6 N HNO_3. Dilute to volume in a 1-l volumetric flask with 2% (v/v) HNO_3.

s. Master lead standard. Dissolve 1.598 g of lead nitrate [$Pb(NO_3)_2$] in 2% (v/v) HNO_3. Dilute to volume in a 1-l volumetric flask with 2% (v/v) HNO_3.

t. Master rubidium standard. Dissolve 1.415 g of rubidium chloride (RbCl) in distilled water. Dilute to volume in a 1-l volumetric flask with distilled water.

u. Master strontium standard. Dissolve 2.415 g of strontium nitrate [$Sr(NO_3)_2$] in distilled water. Dilute to volume in a 1-l volumetric flask with distilled water.

v. Master thallium standard. Dissolve 1.303 g of thallium nitrate ($TlNO_3$) in 10% (v/v) HNO_3. Dilute to volume in a 1-l volumetric flask with 10% (v/v) HNO_3.

w. Master vanadium standard. Dissolve 1.000 g of vanadium metal in minimum volume of 6 N HNO_3. Dilute to volume in a 1-l volumetric flask with 10% (v/v) HNO_3.

x. Master zinc standard. Dissolve 1.000 g of zinc metal in a minimum volume of 6 N HNO_3. Dilute to volume in a 1-l volumetric flask with 2% (v/v) HNO_3.

9.2.2 *Dilute standards.* Diluted standard mixtures of the elements listed in **9.2.1** are prepared according to the directions in the following sections **a** to **c**. The mixed dilute standards are prepared such that the accuracy of the working standard preparation (section **9.2.3**) is maximized. Only those elements being determined in the samples need be prepared as dilute and working standards.

a. Mixed calcium, cadmium, potassium, lithium, magnesium, sodium and zinc standard (0.010 mg/ml for each metal). Pipet 10 ml of the master standard for calcium, cadmium, potassium, lithium, magnesium, sodium and zinc into a 1-l flask, add 100 ml of conc HNO_3, and dilute to volume with distilled water. Prepare fresh monthly.

b. Mixed barium, bismuth, cobalt, chromium, copper, iron, manganese, nickel, lead, rubidium and strontium standard (0.100 mg/ml for each metal). Pipet 10 ml of the master standards for barium, bismuth, cobalt, chromium, copper, iron, manganese, nickel, lead, rubidium and strontium into a 100-ml volumetric flask, add 10 ml of conc HNO_3 and dilute to volume with distilled water. Prepare fresh monthly (**Note:** Due to volume considerations, if more than 8 elements are to be prepared one must prepare 2 dilute standards.)

c. Dilute silver standard (0.100 mg of silver per ml). Pipet 10 ml of the master silver standard and 10 ml conc HNO_3 into a 100-ml volumetric flask, and dilute to volume with distilled water. Store in an amber bottle away from direct light. Prepare fresh weekly.

9.2.3 *Working standards.*

a. Mixed working standards. Working standards are prepared by pipetting appropriate amounts of the dilute standards from **9.2.2a**, **9.2.2.b** and the master standards for Al, In, Tl and V. Pipet 1 ml of the dilute standard from section **9.2.2a** and 1 ml from the dilute standard from **9.2.2b** into a 100-ml volumetric flask. Pipet into this same volumetric flask, 1 ml from each of the master standards for Al, In, Tl and V. Add 2 ml of 50 mg/ml Cs solution, 2 ml of 50 mg/ml La solution and 10 ml of HNO_3 to the volumetric flask and dilute to volume with distilled water.* This solution contains the following metals at the indicated concentrations: Ca, Cd, K, Li, Mg, Na and Zn—0.1 ppm Rb, Ba, Bi, Co, Cr, Cu, Fe, Mn, Ni, Pb, & Sr—1.0 ppm Al, In, Tl, and V—10.0 ppm (**Note:** Dilute and working standards need to be prepared only for the elements being determined in the sample. If Cr and K are being determined in the samples, separate working solutions must be prepared since the chromium standard contains K. Using the above described preparation procedures a standard containing 1.0 ppm Cr will also contain 1.5 ppm K). This procedure is repeated using 2, 3, 4, and 5 ml of the same standard metal solutions indicated above. These standards must be prepared fresh daily.

b. Working silver standard. Pipet 1 ml of the dilute silver standard (from section **9.2.2c**) into a 100-ml volumetric flask and dilute to volume with distilled water. This solution contains 1.0 ppm silver ion. Repeat this procedure using 2, 3, 4, and 5 ml of the dilute silver standard. The working silver standards must be prepared fresh daily.

9.2.4 The standard solutions are aspirated into the flame and the absorbance (or concentration) recorded. If the instrument used displays transmittance, these values must be converted to absorbance. A calibration curve is prepared by plotting absorbance (concentration) versus metal concentration. The best fit curve (calculated by linear least square regression analysis) is fitted to the data points. This line or the equation describing the line is used to obtain the metal concentration in the samples being analyzed.

9.2.5 To ensure that the preparation procedure is being properly followed, clean membrane filters are spiked with known amounts of the elements being determined by adding appropriate amounts of the previously described standards and carried through the entire procedure. The amount of metal is determined and the per cent recovery calculated. These tests will provide recovery and precision data for the procedure as it is carried out in the laboratory for the soluble compounds of the elements being determined.

9.2.6 *Analysis by the method of standard additions.* In order to check for interferences, samples are initially and pe-

*The procedure as described has been designed to match the nitric acid concentration of samples and standards (*i.e.*, 10% v/v HNO_3). If the sample solution from section **8.4.3** is to be diluted prior to analysis the amount of acid added to the standards must be reduced by an amount equal to the sample dilution factor.

riodically analyzed by the method of standard additions and the results compared to those obtained by the conventional analytical determination. For this method the sample is divided into three 2-ml aliquots. To one of the aliquots an amount of metal approximately equal to that in the sample is added. To another aliquot twice this amount of metal is added. (**Note:** Additions should be made by micropipetting techniques such that the volume does not exceed 1% of the original aliquot volume—*i.e.*, 10 μl and 20 μl additions to a 2-ml aliquot.) The solutions are then analyzed and the absorbance readings are plotted against metal added to the original sample. The line obtained from such a plot is extrapolated to 0 absorbance and the intercept on the concentration axis is taken as the amount of metal in the original sample (**2**). If the result of this determination does not agree to within 20% of the values obtained with the procedure described in section **9.2.4**, an interference is indicated and standard addition techniques should be utilized for sample analysis.

9.3 BLANK. Blank filters must be carried through the entire procedure each time samples are analyzed.

10. Calculations

10.1 The uncorrected volume collected by the filter is calculated by averaging the beginning and ending sample flow rates, converting to cubic meters and multiplying by the sample collection time. The formula for this calculation is given in **10.1.1**.

10.1.1 $$V = \frac{F_B + F_E}{2 \times 1000} t$$

where:

V = uncorrected sample volume (m^3)
F_B = sample flow rate at beginning of sample collection (lpm)
F_E = sample flow rate at end of sample collection (lpm)
t = sample collection time (minutes)

10.2 The volume is corrected to 25 C and 760 Torr by using the formula given in **10.2.1**.

10.2.1 $$V_{corr} = \frac{(298)\,(P)\,(V)}{(760)\,(T)}$$

where:

V_{corr} = corrected sample volume (m^3)
P = average barometric pressure during sample collection period (mm Hg)
T = average temperature during the sample collection period (°K). (Note: °K = °C + 273.)
V = uncorrected volume calculated from **10.1.1** (m^3).

10.3 After any necessary correction for the blank has been made, metal concentrations are calculated by multiplying the μg of metal per ml in the sample aliquot by the aliquot volume and dividing by the fraction which the aliquot represents of the total sample and the volume of air collected by the filter. **10.3.1** indicates the formula for this calculation.

10.3.1 $$\mu\text{g metal/m}^3 = \frac{(C \times V) - B}{V_{corr} \times F}$$

where:

C = concentration (μg metal/ml) in the aliquot
V = volume of aliquot (ml)
B = total μg of metal in the blank
F = fraction of total sample in the aliquot used for measurement (dimensionless)
V_{corr} = corrected volume of air sample and calculated from **10.2.1**

10.4 Untreated filter samples may be stored indefinitely.

11. References

1. SLAVIN, W. 1968. Atomic Absorption Spectroscopy. Interscience Publishers, N.Y., N.Y.
2. RAMIREZ-MUNOZ, J. 1968. Atomic Absorption Spectroscopy, Elsevier Publishing Co., N.Y., N.Y.

3. DEAN, J.A. and T.C. RAINS, ED. 1969. Flame Emission and Atomic Absorption Spectrometry—Volume I, Theory. Marcel Dekker, N.Y., N.Y.
4. WINEFORDNER, J.D. ED. 1971. Spectrochemical Methods of Analysis, John Wiley & Sons, Inc.
5. AMERICAN CONFERENCE OF GOVERNMENTAL INDUSTRIAL HYGIENISTS. 1971. Air Sampling Instruments for Evaluation of Atmospheric Contaminants.
6. Analytical Methods for Atomic Absorption Spectrophotometry. 1971. THE PERKIN ELMER CORPORATION, NORWALK, CONN.
7. BELCHER, C.B., R.M. DAGNALL and T.S. WEST. 1964. Examination of the Atomic Absorption Spectroscopy of Silver. *Talanta*. 11:1257.
8. MULFORD, C.E. 1966. Gallium and Indium Determinations by Atomic Absorption. *Atomic Absorption Newsletter*. 5:28.
9. ROBINSON, J.W. 1966. Atomic Spectroscopy. Marcel Dekker, Inc., N.Y., N.Y.
10. VARIAN TECHTRON. 1971. Analytical Data for Elements Determined by Atomic Absorption Spectroscopy. Walnut Creek, California.
11. VARIAN TECHTRON. 1971. Detection Limits for Model AA-5 Atomic Absorption Spectrophotometer. Walnut Creek, California.

Subcommittee 6
T. J. KNEIP, *Chairman*
R. S. AJEMIAN
J. N. DRISCOLL
F. I. GRUNDER
L. KORNREICH
J. W. LOVELAND
J. L. MOYERS
R. J. THOMPSON

823.

Analytical Method for Sulfur Dioxide in Air

Analyte:	Sulfur Dioxide	Method No:	P & CAM 163
Matrix:	Air	Range:	0.11–11 ppm
Procedure:	Impinger Collection/ Peroxide Absorption, Titration with Barium Perchlorate, Modified Thorin Indicator	Precision:	4% Relative Standard Deviation at 2.5 ppm
		Classification:	D (Operational)
Date Issued:	1/26/73		
Date Revised:	1/1/75		

1. Principle of the Method

1.1 Sulfur dioxide is absorbed and oxidized to sulfate by aspirating a measured air sample through a 3% hydrogen peroxide solution.

1.2 The aqueous sample is passed through a cation exchange column to remove interfering cation. Isopropyl alcohol is added to bring the alcohol concentration to 80% by volume. The resulting solution is titrated with 0.005 M barium perchlorate using the modified thorin indicator. The color change is from green to clear (**1**).

2. Range and Sensitivity

2.1 The procedure can detect a difference of 0.05 mg sulfur dioxide in the samples. The lower limit of detection is 0.03 mg sulfur dioxide. The upper limit of detection is limited by the amount of sulfur dioxide that can be oxidized by the peroxide solution. Aliquots which contain more than 3.2 mg will require more than 10 ml of titrant.

2.2 A 1 hr sample collected at 1.8 lpm containing 0.3 mg/m^3 (.11 ppm) to 30 mg/m^3 (11 ppm) can be analyzed directly by this procedure.

3. Interferences

3.1 Soluble sulfates present in the air samples will give erroneously high readings. Sulfate aerosols can be removed by filtration through a 0.8 μ cellulose ester filter in a Tenite plastic filter holder. This filter only reduced the SO_2 concentration 0.02 ppm over the range of 0.06 to 0.5 ppm **(2)**.

3.2 The presence of cations other than hydrogen tend to obscure the endpoint of the indicator. These interfering cations must be removed by passing the sample through a cation exchange column.

3.3 Interferences can also be due to the presence of anions, in particular, phosphate ions. Concentrations of phosphate ions greater than the sulfate ion concentration cause considerable interference **(3)**. The use of a prefilter should remove the phosphate ions.

4. Precision and Accuracy

4.1 At 2.5 ppm, the accuracy is 5% with a precision of 4% relative standard deviation. Accuracy and relative standard deviation can be improved at higher concentrations. The absolute precision of the method is ± 0.008 ml, which corresponds to 0.02 mg SO_2.

5. Advantages and Disadvantages

5.1 The sulfuric acid formed during sample collection is stable and nonvolatile. The 30% hydrogen peroxide reagent should be stored in a refrigerator for not more than 6 months.

5.2 The barium perchlorate solution is stable indefinitely, but should be restandardized monthly due to potential loss of isopropyl alcohol by evaporation. The standard sulfate solution is also stable indefinitely.

5.3 The analysis is relatively simple and rapid.

6. Apparatus

6.1 SAMPLING APPARATUS.

6.1.1 *Tenite plastic filter holder.* 37 mm (Millipore Corp.) or equivalent.

6.1.2 *Cellulose ester membrane filter.* (0.8 μ pore size) or equivalent.

6.1.3 *Glass midget impinger or fritted bubbler.*

6.1.4 *Personal sampling pump.* Capable of sampling at a rate of 1.8 lpm through the membrane filter.

6.1.5 *Calibrated flow meter.*

6.2 ANALYTICAL APPARATUS.

6.2.1 *Glass column for ion exchange.* 11 mm ID × 500 mm (a 50-ml buret serves well).

6.2.2 *Ten ml buret.* Graduated in 0.05 ml subdivisions.

6.2.3 *Erlenmeyer flask.* 250-ml.

7. Reagents

7.1 PURITY. Reagents must be A.C.S. analytical grade quality. Distilled water should conform to ASTM standards for Referee Reagent Water.

7.2 ISOPROPYL ALCOHOL.

7.3 ABSORBING SOLUTION—HYDROGEN PEROXIDE. Dilute 100 ml of 30% H_2O_2 to one liter with distilled water. Prepare fresh daily.

7.4 ION EXCHANGE RESIN. Amberlite IRC-50, cation exchange resin or equivalent.

7.4.1 *Preparation of Ion Exchange Column.* Place a small glass wool plug in the bottom of the 50-ml buret or chromatography column. Mix the ion exchange resin with distilled water and pour some of the mixture into the column. Drain the liquid level down to near the top of the resin. Do not let the liquid level fall below the top of the resin or air bubbles will be entrapped and the efficiency of the column will be impaired. Continue to add resin in the slurry until a column height of 15 cm is obtained. Back flush the column. Place a small piece of glass wool on top of the column and wash several times with distilled water.

7.4.2 To convert the resin to the hydrogen form, pass 20 ml of 4 M hydrochloric acid through the column. The column effluent will be acidic to pH indicator test paper. Drain the liquid level down to just above the top of the column. Wash with 5 or 6 five-ml portions of distilled water, draining the water wash down to the top of the column between each washing. The final wash solution should not be acidic to pH test paper. The column will have an exchange capacity of approximately 20 milliequivalents and should be regenerated after one liter of titrant has been used on the standards and samples passed through the column.

7.5 BARIUM PERCHLORATE, 0.005 M. Dissolve approximately 2 g barium perchlorate trihydrate in 200 ml of distilled water and dilute to 1 l with isopropyl alcohol. Adjust apparent pH to about 3.5 with perchloric acid, using a pH meter. Standardize against standard sulfate solution.

7.6 STANDARD SULFATE SOLUTION. Dissolve 0.7393 g anhydrous Na_2SO_4 in distilled water and dilute to 1 l. This solution contains 0.500 mg/ml of SO_4.

7.7 INDICATOR. Dissolve 0.2 g of Thorin indicator [1—(0-arsonophenylazo)—2—naphthol—3.6—disulfonic acid disodium salt] and 0.3 g of Xylene Cyanole FF dye in 100 ml distilled water.

7.8 PERCHLORIC ACID, 1.8%. Dilute 25 ml of reagent grade perchloric acid (70 to 72%) to 1 l with distilled water.

7.9 HYDROCHLORIC ACID, 4M. Add 300 ml of conc hydrochloric acid to 600 ml of distilled water.

8. Procedure

8.1 COLLECTION OF SAMPLE. Assemble the filter, midget impinger and pump. The filter should be connected to the impinger with a short piece of Teflon tubing. Place 10 ml of absorbing reagent in the impinger and sample at 1.8 lpm. A minimum of 100 l of air should be sampled. Samples containing between 0.3 mg per m^3 (0.11 ppm) and 30 mg per m^3 (11 ppm) can be titrated directly. If SO_2 concentrations greater than 30 mg per m^3 are expected, an aliquot should be analyzed or a smaller volume of sample collected. If the concentration is less than 0.3 mg per m^3 a longer sampling time should be used. For every 10 samples collected at least one blank impinger containing absorbing reagent should be included for analysis.

8.2 ANALYSIS OF SAMPLE. The sample is transferred quantitively to the ion exchange column. The column is drained to the top of the resin bed and collected in a 250-ml Erlenmeyer flask. The column is washed with five 5-ml portions of distilled water and the wash collected in the flask. Add 100 ml of isopropyl alcohol, 2 drops of the thorin-dye indicator and titrate with barium perchlorate until the solution is clear. If a smaller aliquot of the sample is taken for analysis, the column will still require five 5-ml washes with distilled water.

9. Calibration and Standards

9.1 STANDARDIZATION OF BARIUM PERCHLORATE. Pipet 5 ml of the standard sulfate solution onto the ion exchange column. Drain to the top of the column and collect the effluent in a 250-ml Erlenmeyer flask. Wash with five 5-ml portions of water, draining the liquid level to the top of the resin bed between each addition. Add 100 ml of isopropyl alcohol and 2 drops of the modified thorin indicator. Titrate with barium perchlorate solution to a change in the indicator from green to clear.

9.2 The air flow rate of the sample apparatus should be calibrated with a metric wet test meter before use. The apparatus is assembled and a clean filter, (if used), is placed in the filter holder. The apparatus is connected to a volume meter and the sampling pump is adjusted to a flow rate of 1.8 lpm and the flow meter marked.

9.3 STANDARDIZATION OF BARIUM PERCHLORATE. The barium perchlorate titer in mg SO_2/ml $Ba(ClO_4)_2$ is calculated from the equation

$$\text{Titer } Ba(ClO_4)_2 = (\text{mg } SO_2/\text{ml } Ba(ClO_4)_2) = \frac{W_s \times V_a \times F \times 1000}{V_s \times V_T}$$

where

W_s = weight of Na_2SO_4 in standard sulfate solution (grams)

V_s = volume of standard sulfate solution (ml)
V_a = volume of standard sulfate solution analyzed (ml)
V_T = volume of $Ba(ClO_4)_2$ titrant used (ml)
F = factor of conversion of Na_2SO_4 to SO_2 (0.4507)

10. Calculations

10.1 The volume of air samples is corrected to 25 C and 760 Torr as follows (Ideal gas law relationship)

$$V_o = \frac{P_l \times R_l \times t \times T_o}{P_o T_l}$$

where

V_o = corrected volume (liters)
P_l = observed pressure (Torr)
R_l = flow rate (liters/minute)
t = time of sample (minutes)
T_o = 298.1 K (temperature at 25 C)
P_o = 760 Torr (standard pressure)
T_l = observed temperature (°C + 273.1)

The equation simplified to:

$$Vo_o = \frac{P_l R_l t}{T_l} \times 0.392$$

10.2 The number of mg of SO_2 per cm^3 of air is calculated from the following:

$$\frac{mg\ SO_2}{m^3} = \frac{V_T \times T}{V_o} \times 1000$$

where

V_T = volume of titrant (ml)
T = $Ba(ClO_4)_2$ titer, (mg SO_2/ml $Ba(ClO_4)_2$
V_o = volume of air sampled corrected to 25 C and 760 Torr (liters)

10.3 The concentration of SO_2 in ppm, under standard conditions, can be calculated with the following conversion factor:

$$ppm\ SO_2 = \frac{mg\ SO_2}{m^3} \times 0.38228$$

11. References

1. GILLIES, D. 1974. Evaluation of Modifications of the Thorin Titration Method for Sulfur Dioxide. Thesis, University of Pittsburgh, Graduate School of Public Health, August.
2. DUCKWORTH, S. 1969. Particulate Interference in Sulfur Dioxide Manual Sampling by the West-Gaeke Method, 10th Conference on Methods in Air Pollution and Industrial Hygiene Studies, San Francisco, California, February.
3. FRITZ, J.S. AND S.S. YAMAMURA. 1955. *Anal. Chem.* 27:1461.

Subcommittee 1
D. F. ADAMS, *Chairman*
D. FALGOUT
J. O. FROHLIGER
A. M. HARTLEY
J. B. PATE
A. L. PLUMLEY
F. P. SCARINGELLI
P. URONE

824.

SULFURIC ACID AEROSOL IN AIR BY TITRATION—MIXED INDICATOR

Analyte:	H_2SO_4, Mist	Method No.	
Matrix:	Air	Range:	0.1 to 3 mg/m^3H_2SO_4
Procedure:	Collection on Filters $Ba(ClO_4)_2$ titration-mixed indicator (**2**).	Precision:	Unknown
		Classification:	D (Operational)
Date Issued:			

1. Principle of Method

1.1 SAMPLING. Sulfuric acid aerosol is collected by filtering at least one cubic meter of air through cellulose acetate or polycarbonate filter (**2**).

1.2 SAMPLING PREPARATION. The sulfuric acid is removed from the filter by dissolving it in a solution of 80% isopropanol in water.

1.3 ANALYSIS. The soluble sulfate is titrated with barium perchlorate to the mixed Xylene Cyanole FF-thorin indicator end-point. Barium ions react with sulfate forming an insoluble precipitate of barium sulfate. When a slight excess of barium ion is present in the solution, this barium reacts with the mixed thorin indicator forming a highly colored complex.

2. Range and Sensitivity

2.1 The range of interest of sulfuric acid mist for occupational safety is 0.1 to 3 mg/m^3 with maximum precision and accuracy at 1 mg/m^3 of H_2SO_4 (**3**). A 1000 l air sample of 1 mg/m^3 H_2SO_4 is sufficient for analysis by titration. With 0.01 N $Ba(ClO_4)_2$ this quantity, 1 mg, requires a titration volume of approximately 2 ml.

2.2 SENSITIVITY. A 5-ml microburet graduated in 0.01 ml will yield a sensitivity of ± 0.01 ml, which is equivalent to ± 4.9 μg H_2SO_4; therefore, the lower limit of detection for 1000-liter air sample is 0.01 mg/m^3.

3. Interferences

3.1 The method is not specific for sulfuric acid since it measures total soluble sulfate.

3.2 Generally, any substance that precipitates or complexes the barium cation is a potentially positive interferent and any substance that forms insoluble sulfates is a potentially negative interferent. Certain metallic cations interfere by forming colored complexes with the thorin indicator, particularly in alcoholic solutions.

3.3 Fluorides and nitrates cause minor interferences in concentrations greater than 2 mg/l. Phosphates interfere at concentrations in excess of 2 mg/l. Chlorides at concentrations greater than 1000 mg/l obscure the end-point, when the $SO_4^=$ concentration is less than 5 mg/l.

4. Precision, Accuracy and Stability

There are no available data for precision and accuracy for the procedure. The accuracy of the procedure depends on the size of the aerosol and the efficiency of the filter. Although this procedure has been tested collaboratively at higher concentrations, these results are not applicable to these low values.

5. Advantages and Disadvantages

5.1 Advantages.

5.1.1 The titrimetric procedure is very simple and is applicable to the collection of samples by personal dosimeters.

5.1.2 A mixed indicator is used to sharpen the end-point and to increase the accuracy of detecting the end-point.

5.2 Disadvantages.

5.2.1 This method is not specific for sulfuric acid.

5.2.2 A long sampling time is required to obtain an adequate sample size resulting in a long time weight average (TWA). Other more elaborate procedures may be required for a shorter TWA or for better selectivity.

5.2.3 Insoluble sulfates are not detected.

6. Apparatus

6.1 Sampling. The sample collection requires a sampling train consisting of the following:

6.1.1 *Filter cassette,* containing a cellulose acetate, or polycarbonate filter having a pore size of 0.2 μm and a diam of 37 mm.

6.1.2 *Sampling pump, personal type,* that is capable of maintaining a pressure drop of one inch of mercury and a flow of 1 to 2.8 l/min.

6.1.3 *Calibrated flowmeter,* 0.06 to 2.86 l/min.

6.1.4 *Timer,* to measure sampling time.

6.1.5 *Stop watch,* to calibrate flow meter.

6.1.6 *Wet-test meter or a calibrated flow meter.*

6.2 Analysis.

6.2.1 *Erlenmeyer flask,* of 250-ml capacity.

6.2.2 *Magnetic stirrer.*

6.2.3 *Microburet,* Class A with Teflon stopcock and a capacity of 5 ml with subdivision of 0.01 ml.

6.2.4 *Nalgene Mityvac hand pump.*

6.2.5 *Spectro-Electro Titrator,* or equivalent, with a 525 nm filter.

7. Reagents

7.1 Sampling. None.

7.2 Analysis.

7.2.1 *Isopropanol solution (80%) in water.*

7.2.2 *Water,* distilled or deionized.

7.2.3 *Mixed Indicator.* Dissolve 0.2 g of Thorin indicator-1-(0-arsonophenylazo)-2-naphthol-3,6-disulfonic acid, disodium salt and 0.03 g of Xylene Cyanole FF dye in 100 ml distilled water.

7.2.4 *Barium percholorate (0.01 N).* Dissolve 1.95 g of the trihydrate $[Ba(ClO_4)_2 \cdot 3H_2O]$ in 200 ml of distilled water and dilute to 1 l with isopropanol.

7.2.5 *Calibration standard, Na_2SO_4.* Dissolve 0.1440 g of anhydrous sodium sulfate (C.P.) in water and dilute to exactly one l. Each ml of this solution contains the equivalent of 0.1 mg H_2SO_4.

7.2.6. *Ion exchange resin.* Amberlite IRC-50, Mesh Size 20–50 or Dowex 50W-X8 or equivalent cation exchange resin.

a. Preparation of Ion Exchange Column. Place a small glass wool plug in the bottom of the 50-ml buret or chromatography column. Mix the ion exchange resin with distilled water and pour some of the mixture into the column. Drain the liquid level down to near the top of the resin. Do not let the liquid level fall below the top of the resin or air bubbles will be entrapped and the efficiency of the column will be impaired. Continue to add resin in the slurry until a column of 15 cm is obtained. Back flush the column using a Nalgene Mityvac hand pump. Place a small piece of glass wool on top of the column and wash several times with distilled water.

b. To convert the resin to the hydrogen form, pass 20 ml of 4 M hydrochloric acid through the column. The column effluent will be acidic to pH indicator test paper. Drain the liquid level down to just above the top of the column. Wash with 5 or 6 small portions of distilled water, draining the water wash down to the top of the column between each washing. The final wash solution should not be acidic to pH test paper. The column will have an exchange capacity of approximately 20 milliequivalents and should be regenerated after one liter of titrant has been used to titrate the standards and samples passed through the column.

8. Procedure

8.1 SAMPLING. Assemble the sampling train consisting of cassette, sampling pump, and flowmeter. Record starting time and flow rate. Collect sample 7 hr or until 1000 l of air have been sampled. Record final flow rate and ending time, attach data to the cassette and ship to laboratory in a suitable container.

8.2 ANALYSIS. By means of forceps transfer the cellulose acetate, or polycarbonate filter from the cassette to a 250-ml Erlenmeyer flask. Add 100 ml of 80% isopropanol, add stirring bar, and allow to stir for 5 min before commencing the titration.

8.2.1 In the absence of substantial quantities of cations, add 2 to 4 drops of indicator to the flask. By means of the microburet, titrate the sample with the barium perchlorate solution drop by drop until the indicator changes from green to clear. Record the initial and final volume of titrant. With each set of 7 determinations, analyze in a similar manner, unexposed filter to obtain a filter and reagent blank. If substantial quantities of interference are expected, they may be removed by ion exchange or by precipitation (**4**).

8.2.2 In the presence of substantial quantities of cations, transfer the sample quantitatively to the ion exchange column. Drain the column to the top of the resin bed while collecting elutriate and washings in a 250-ml flask. Wash the column with small portions (5 to 15 ml total) of distilled water. Titrate to a clear color as described in 8.2.1.

Accurately pipet 20 ml of the 0.1 mg/ml $SO_4^=$ to the ion exchange column. Wash with 5 ml of distilled water, collecting the elutriate in a 250-ml flask. Repeat this operation two additional times. To each of these 3 flasks add 100 ml isopropanol. Prepare a blank in each of the other 3 flasks by adding 20 ml of water and 80 ml of isopropanol. Add 2 drops of the mixed indicator and titrate with barium perchlorate to a clear solution. Record initial and final titer volumes for each flask.

8.2.3 If a titration accessory for a spectrophotometer is available, it may be convenient to titrate to a photometric end point at 525 nm.

9. Calibration and Standards

9.1 The standard solution of anhydrous sodium sulphate prepared as in Section **7.2.5** contains the equivalent of 0.1 mg H_2SO_4/ml. For calibration purposes, titrate a series of standards consisting 1, 5, 10, 20 and 30 ml, equivalent to 0.1, 0.5, 1, 2 and 3 mg H_2SO_4, as described in Section **8.2.1**, with the 0.01 N barium perchlorate solution.

9.2 If substantial quantities of interferences are present in the samples and require removal by means of the ion exchange column, the calibration standards should be passed through this column also, as described in Section **8.2.2**.

10. Calculations

10.1 SAMPLING. Compute the volume of air sampled as follows (corrected to 25C and 760 Torr)

$$V_a = Q_n t$$

Where:

V_a = volume of air sampled in liters
Q_n = Average flowrate, liters per minute = $(Q_2 - Q_1) \div 2$
t = Elapsed time in minutes

10.2 ANALYSIS. Compute the quantity of sulfuric acid as follows:

$$S = V_t \times N \times E$$

Where:

S = Quantity of H_2SO_4, mg.
V_t = Net volume of titrant used, ml.
N = Normality of $Ba(C10_4)_2$
E = Milliequivalent wt of sulfuric acid, 49 mg.

10.3 CALCULATE THE CONCENTRATIONS AS FOLLOWS:

$$C = \frac{S \times 10^3}{V_a}$$

Where:

C = concentration, mg/m^3
S = Wt. of H_2SO_4, mg
V_a = Volume of air sampled in liters.
10^3 = conversion factor from liters to cubic meters.

11. References

1. FRITZ, J.F., S.S. YAMAMURA, and M.J. RICHARDS. 1957. Titration of Sulfate Following Separation with Alumina, Anal. Chem., 29:158.
2. Private Communication, DR. J.F. FROHLIGER, Graduate School of Public Health, University of Pittsburgh, Pittsburgh, Pa., 15213.
3. BARTON, S.C., and H.G. MCADIE. 1970. A Specific Method for the Automatic Determination of Ambient Sulfuric Acid Aerosol, Environ. Science and Technol., 4:769.
4. HAMIL, H.F., D.E. CAMANN, and R.E. THOMAS. 1974. Collaborative Study of Method for the Determination of Sulfuric Acid Mist and Sulfur Dioxide Emissions from Stationary Sources, Environmental Monitoring Series, Pub. No. EPA-650/4-75-003, November.
5. American Public Health Association, 1977. Methods of Air Sampling and Analysis 2nd ed. p. 851. Washington, D.C. 20036.

Subcommittee 1
D. F. ADAMS, *Chairman*
J. O. FROHLIGER
D. FALGOUT
A. M. HARTLEY
J. B. PATE
A. L. PLUMLEY
F. P. SCARINGELLI
P. URONE

825.

Analytical Method for Acrolein in Air

Analyte:	Acrolein	Method No.:	P & CAM 118
Matrix:	Air	Range:	1–30 μg/10 ml
Procedure:	Collection via Midget Impinger/Complexation, Colorimetric	Precision:	Unknown for air samples, ± 5% for standards
Date Issued:	4/6/72	Classification:	C (Tentative)
Date Revised:	1/1/75		

1. Principle of Method

1.1 The air sample is drawn through a train of two midget impingers, with fritted glass inlets, containing mixed absorbing reagent.

1.2 Acrolein reacts with 4-hexylresorcinol in an ethyl alcohol-trichloroacetic acid solvent medium in the presence of mercuric chloride and forms a blue colored product. The strong absorption maximum at 605 nm is used as a quantitative measure of acrolein (**1, 2, 3**).

2. Range and Sensitivity

2.1 The linear range of the absorbance at 605 nm is at least 1 to 30 μg of acrolein in the 10 ml portions of mixed reagent (**4**).

2.2 A concentration of 0.01 ppm of acrolein can be determined in a 50-liter air sample based on a difference of 0.05 absorbance unit from the blank using a 1 cm cell. Greater sensitivity could be obtained by use of a longer path length cell.

3. Interferences

3.1 There is no interference from ordinary quantities of sulfur dioxide, nitrogen dioxide, ozone and most organic air pollutants. A slight interference occurs from dienes: 1.5% for 1,3-butadiene and 2% for 1,3-pentadiene. The red color produced by some other aldehydes and undetermined materials does not interfere in the spectrophotometric measurement (**4**).

4. Precision and Accuracy

4.1 Known standards can be determined to within ± 5% of the true value (**5**). No data are available on precision and accuracy of air samples.

5. Advantages and Disadvantages of the Method

5.1 The major advantage of this method is the sensitivity. A concentration of 0.01 ppm of acrolein can be determined in a 50-l air sample.

5.2 Both the solid trichloroacetic acid and the solution are corrosive to the skin. Mercuric chloride is highly toxic. This reagent and the former should be handled carefully. This method is not attractive as a field technique since the absorbing solution tends to evaporate upon collection of large volumes of air and the colored complex begins to fade after 2 hr upon completion of sampling.

6. Apparatus

6.1 SAMPLING EQUIPMENT. The sampling unit for the impinger collection method consists of the following components:

6.1.1 *A train of two glass standard midget impingers.* With fritted glass inlets containing the absorbing solution or reagent (Section **7.5**). The fritted end should have a porosity approximately equal to that of Corning EC (170 to 220 μ maximum pore diam).

6.1.2 *A pump.* Suitable for delivering desired flow rates; at least 2 l/min for 60 min. The sampling pump is protected from splashover or water condensation by an adsorption tube loosely packed with a plug of glass wool and inserted between the exit arm of the impinger and the pump.

6.1.3 *An integrating volume meter such as a dry gas or wet test meter.* A calibrated limiting orifice meter may also be used.

6.1.4 *Thermometer.*

6.1.5 *Manometer.*

6.1.6 *Stopwatch.*

6.2 WATER BATH. Any bath capable of maintaining a temperature of 58 to 60 C is acceptable.

6.3 SPECTROPHOTOMETER. This instrument should be capable of measuring the developed color at 605 nm. The absorption band is rather narrow and thus a lower absorptivity may be expected in a broad-band instrument.

6.4 MATCHED GLASS CELLS OR CUVETTES. 1-cm path length.

6.5 ASSORTED LABORATORY GLASSWARE. Pipets, volumetric flasks, graduated cylinders of appropriate capacities.

7. Reagents

7.1 PURITY. All reagents must be analytical reagent grade. An analytical grade of distilled water must be used.

7.2 ETHANOL (95%).

7.3 TRICHLOROACETIC ACID SOLUTION, SATURATED. Dissolve 100 g of the acid in 10 ml of water by heating on a water bath. The resulting solution has a volume of approximately 70 ml. Even reagent grade trichloroacetic acid has an impurity which affects product intensities. Every new batch of solution should be standardized with acrolein. It is convenient to prepare a large quantity of solution from a single batch of trichloroacetic acid to maintain a uniformity of response.

7.4 MERCURIC CHLORIDE SOLUTION (3%). Dissolve 3 g of mercuric chloride in 100 ml of ethanol.

7.5 4-HEXYLRESORCINOL SOLUTION. Dissolve 5 g of 4-hexylresorcinol (MP 68 to 70 C) in 5.5 ml of ethanol. This makes about 10 ml of solution.

7.6 MIXED ABSORBING REAGENT. Mix, in the specified order, the reagents in the following proportions: 5 ml ethanol, 0.1 ml 4-hexylresorcinol solution, 0.2 ml mercuric chloride solution, and 5 ml saturated trichloroacetic acid solution. The mixed reagent may be stored for a day at room temperature. Prepare the needed quantity by selecting an appropriate multiple of these amounts. Protect from direct sunlight.

7.7 ACROLEIN, PURIFIED. Freshly prepare a small quantity (less than 1 ml is sufficient) by distilling 10 ml of the purest grade of acrolein commercially available. Reject the first 2 ml of distillate. (The acrolein should be stored in a refrigerator to retard polymerization.) The distillation should be done in a hood because the vapors are irritating to the eyes.

7.8 ACROLEIN, STANDARD SOLUTION "A" (1 MG/ML). Weight 0.1000 g (approximately 0.12 ml) of freshly prepared, purified acrolein into a 100-ml volumetric flask and dilute to volume with ethanol. This solution may be kept for as long as a month if properly refrigerated.

7.9 ACROLEIN, STANDARD SOLUTION "B" (10 μG/ML). Dilute 1 ml of standard solution "A" to 100 ml with ethanol. This solution may be kept for as long as a month if properly refrigerated.

8. Procedure

8.1 CLEANING OF EQUIPMENT. No specialized cleaning of glassware is required, however, since known interferences occur with dienes, cleaning techniques should insure the absence of all organic materials.

8.2 COLLECTION AND SHIPPING OF SAMPLES

8.2.1 Draw through the assembled sampling unit measured volumes of the vapor-laden air at a rate of either 1 l/min for no more than 60 min or 2 l/min for no more than 30 min through 2 bubblers in series, each containing 10 ml of mixed absorbing reagent. An extra bubbler containing water may be added as a trap to protect the pump. A maximum of 60 l of air can be sampled before possible reagent decomposition may occur. Care should be taken to measure flow rate, time and/or volume as accurately as possible. Note also the atmospheric pressure and temperature.

8.2.2 This sampling system collects 70 to 80% of the acrolein in the first bubbler and 95% of the acrolein in the first 2 bubblers, using absorbers with EC fritted glass inlets. The absorption efficiency might be increased by use of a C porosity frit (60μ maximum pore diam).

8.2.3 Because of the 2 hr time limit for color development, it is probably best to analyze the samples soon after completion of sampling. For this reason shipping of the sample involving time periods of as much as 2 hr should not be practiced with this method.

8.3 ANALYSIS OF SAMPLE.

8.3.1 If evaporation has occurred during sampling the absorbing solution is diluted to its original 10-ml volume with ethanol.

8.3.2 Transfer the samples from each bubbler and backup bubbler to separate glass stoppered test tubes.

8.3.3 Immerse the tubes in a 60 C water bath for 15 min to develop the colors. A test tube containing only 10 ml of mixed absorbing reagent must be run similarly and simultaneously. This serves as the reagent blank.

8.3.4 Cool the test tubes in running water immediately upon removal from the water bath.

8.3.5 After 15 min read the absorbances at 605 nm in a suitable spectrophotometer using 1-cm cells. There is no appreciable loss in accuracy if the samples are allowed to stand up to 2 hr before reading the absorbances. For very low acrolein concentrations it may be convenient to use a longer path length cell.

9. Calibration and Standards

9.1 PREPARATION OF STANDARD CURVE.

9.1.1 Pipet 0, 0.5, 1.0, 2.0, and 3.0 ml of standard solution "B" into glass stoppered test tubes.

9.1.2 Dilute each standard to exactly 5 ml with ethanol.

9.1.3 Add in order, to each tube, exactly 0.1 ml of 4-hexylresorcinol solution, 0.2 ml of mercuric chloride solution, and 5 ml of trichloroacetic acid solution.

9.1.4 Mix, develop and read the colors as described in the analytical procedure (Sections **8.3.3** to **8.3.5**).

9.2 Construct a calibration curve by plotting absorbance against μg of acrolein in the color developed solution.

10. Calculations

10.1 Subtract blank values, if any, from each sample.

10.2 Determine from the calibration curve the concentration of the acrolein present in each of the two bubblers, and add the values to get the total μg of acrolein in the air sampled.

$$\mu g \text{ acrolein} = \mu g_1 + \mu g_2$$

where:

μg_1 = microgram concentration of acrolein in the first bubbler.

μg_2 = microgram concentration of acrolein in the back up bubbler.

10.3 The concentration of acrolein in the sampled atmosphere can be calculated in ppm, defined as μl acrolein per liter of air.

$$\text{ppm} = \frac{\mu g \text{ acrolein}}{V_S} \times \frac{24.45}{MW}$$

where:

μg acrolein = total μg concentration as determined in **10.1**. (Probably 95% of the actual concentration as stipulated in Section **8.2.2**. A correction for efficiency may be used accordingly if deemed necessary.)

V_S = volume of air sampled in l at 25 C and 760 Torr.

24.45 = molar volume of an ideal gas at 25 C and 760 Torr.

MW = molecular weight of acrolein, 56.06.

11. References

1. COHEN, I.R. and ALTSHULLER, A.P. 1961. A New Spectrophotometric Method for the Determination of Acrolein in Combustion Gases and the Atmosphere. *Anal. Chem.* 33:726.
2. ALTSHULLER, A.P. and MCPHERSON, S.P. 1963. Spectrophotometric Analysis of Aldehydes in the Los Angeles Atmosphere. *J. Air Poll. Control Assoc.* 13:109.
3. COHEN, ISRAEL, R. and BERNARD E. SALTZMAN. 1965. Determination of Acrolein: 4-Hexylresorcinol Method, *Selected Methods for the Measurement of Air Pollutants*, Public Health Service Publication No. 999, AP-11, p. G-1.
4. AMERICAN PUBLIC HEALTH ASSOCIATION, INC. 1977. Methods of Air Sampling and Analysis. 2nd ed. p. 297. Washington, D.C.
5. LEVAGGI, D.A. and FELDSTEIN, M. 1970. The Determination of Formaldehyde, Acrolein and Low Molecular Weight Aldehydes in Industrial Emissions on a Single Collected Sample. *J. Air Poll. Control Assoc.* 20:312.

Revised by Subcommittee 5
E. SAWICKI, *Chairman*
T. BELSKY
R. A. FRIEDEL
D. L. HYDE
J. L. MONKMAN
R. A. RASMUSSEN
L. A. RIPPERTON
L. D. WHITE

826.

Analtyical Method for Acrolein in Air

Analyte:	Acrolein	Method No.:	211
Matrix:	Air	Range:	1 to 30 μg/10 ml reagent
Procedure:	Collection in Midget Impinger with 1% aqueous $NaHSO_3$; reaction with 4-hexylresorcinol; colorimetry	Precision:	± 5% for standards, unknown for air samples
Date Issued:		Classification:	C (Tentative)
Date Revised:	1/1/75		

1. Principle of the Method

1.1 Air is drawn through two midget impingers containing 1% $NaHSO_3$.

1.2 The reaction of acrolein with 4-hexylresorcinol in an alcoholic trichloroacetic acid solvent medium in the presence of mercuric chloride results in a blue colored product with strong absorption maximum at 605 nm (**1–3**).

2. Range and Sensitivity

2.1 The absorances at 605 nm are linear for at least 1 to 30 μg of acrolein in the 10 ml portions of mixed reagent (**4**).

2.2 A concentration of 10 ppb of acrolein can be determined in a 50-l air sample based on a difference of 0.05 absorbance unit from the blank using a 1-cm cell. Greater sensitivity could be obtained by use of a longer path length cell.

3. Interferences

3.1 There is no interference from ordinary quantities of sulfur dioxide, nitrogen dioxide, ozone and most organic air pollutants. A slight interference occurs from dienes: 1.5% for 1,3-butadiene and 2% for 1,3-pentadiene. The red color produced by some other aldehydes and undetermined materials does not interfere in spectrophotometric measurement (**4**).

4. Precision and Accuracy

4.1 Known standards can be determined to within ± 5% of the true value (**5, 6**). No data are available on precision and accuracy of air samples.

5. Advantages and Disadvantages of the Method

5.1 The method is sensitive to the extent that 0.01 ppm of acrolein can be determined in a 50-l air sample.

5.2 Both the solid trichloroacetic acid and the solution are corrosive to the skin. Mercuric chloride is highly toxic. This reagent and the former should be handled carefully.

6. Apparatus

6.1 Absorbers. All glass standard midget impingers with fritted glass inlets are acceptable. The fritted end should have a porosity approximately equal to that of Corning EC (170 to 220 μ maximum pore diameter). A train of 2 bubblers is needed.

6.2 AIR PUMP. A pump capable of drawing at least 2 l of air/min for 20 min through the sampling train is required. A trap at the inlet to protect the pump from the corrosive reagent is recommended.

6.3 AIR METERING DEVICE. Either a limiting orifice of approximately 1 or 2 l/min capacity or a glass meter can be used. If a limiting orifice is used, regular and frequent calibration is required.

6.4 SPECTROPHOTOMETER. This instrument should be capable of measuring the developed color at 605 nm. The absorption band is rather narrow and thus a lower absorptivity may be expected in a broad-band instrument.

7. Reagents

7.1 PURITY. All reagents must be analytical reagent grade. An analytical grade of distilled water must be used.

7.2 $HgCl_2$-4-HEXYLRESORCINOL. Dissolve 0.30 g $HgCl_2$ and 25 g 4-hexylresorcinol in 50 ml of 95% ethanol (stable at least 3 weeks in refrigerator).

7.3 TCAA. To a 1 lb. bottle of trichloracetic acid add 23 ml of distilled water and 25 ml of 95% ethanol. Mix until all the TCAA has dissolved.

7.4 COLLECTION MEDIUM. 1% sodium bisulfite in water.

7.5 ACROLEIN, PURIFIED. Freshly prepare a small quantity (less than 1 ml is sufficient) by distilling 10 ml of the purest grade of acrolein commercially available. Reject the first 2 ml of distillate. (The acrolein should be stored in a refrigerator to retard polymerization.) The distillation should be done in a hood because the vapors are irritating to the eyes.

7.6 ACROLEIN, STANDARD SOLUTION "A" (1 MG/ML). Weigh 0.1 g (approximately 0.12 ml) of freshly prepared, purified acrolein into a 100-ml volumetric flask and dilute to volume with 1% aqueous sodium bisulfite. This solution may be kept for as long as a month if properly refrigerated.

7.7 ACROLEIN, STANDARD SOLUTION "B" (10 μg/ML). Dilute 1 ml of standard solution "A" to 100 ml with 1% sodium bisulfite. This solution may be kept for as long as a month if properly refrigerated.

8. Procedure (5, 6)

8.1 CLEANING OF EQUIPMENT. No specialized cleaning of glassware is required. However, since known interferences occur with dienes, cleaning techniques should insure the absence of all organic materials.

8.2 COLLECTION OF SAMPLES. Two midget impingers, each containing 10 ml of 1% $NaHSO_3$ are connected in series with Tygon tubing. These are followed by and connected to an empty impinger (for meter protection) and a dry test meter and a source of suction. During sampling the impingers are immersed in an ice bath. Sampling rate of 2 l/min should be maintained. Sampling duration will depend on the concentration of aldehydes in the air. One hour sampling time at 2 liters/min is adequate for ambient concentrations.

8.3 STORAGE OF SAMPLES. After sampling is complete, the impingers are disconnected from the train, the inlet and outlet tubes are capped, and the impingers stored in an ice bath or at 0 C in a refrigerator until analyses are performed. Cold storage is necessary only if the acrolein determination cannot be performed within 4 hr of sampling.

8.4 SHIPPING OF SAMPLES. Because of the 2 hr limit for initiation of color development, it is best to analyze the samples soon after completion of sampling. However, where a refrigerated sample can be shipped and analyzed within 48 hr after collection, then samples can be shipped or carried to a central lab at some distance from the sampling area.

8.5 ANALYSIS OF SAMPLES. (Each impinger is analyzed separately). To a 25-ml graduated tube add an aliquot of the collected sample in bisulfite containing no more than 30 μg acrolein. Add 1% sodium bisulfite (if necessary) to a volume of 4.0 ml. Add 1.0 ml of the $HgCl_2$-4-hexylresorcinol reagent and mix. Add 5.0 ml of TCAA reagent and mix again. Insert in a boiling water bath for 5 to 6 min, remove, and set aside until tubes reach room temperature. Centrifuge samples at 1500 rpm for 5 min to clear slight turbidity. One hr after heating, read in a spectrophotometer at 605 nm against a bisulfite blank prepared in the same fashion as the samples. Determine the acrolein content of the sampling solution from a curve previously prepared from the standard acrolein solutions.

9. Calibration and Standards

9.1 PREPARATION OF STANDARD CURVE.

9.1.1 Pipet 0, 0.5, 1.0, 2.0, and 3.0 ml of standard solution "B" into glass stoppered test tubes.

9.1.2 Dilute each standard to exactly 4 ml with 1% aqueous sodium bisulfite.

9.1.3 Develop color as described above, in section **8.5**.

9.1.4 Plot data on semi-log paper, of absorbance values against μg of acrolein in the color developed solution. The color after forming is stable for about 2 hr.

10. Calculations

10.1 The concentration of acrolein in the sampled atmosphere may be calculated by using the following equations.

$$\mu g/m^3 = \frac{\text{total } \mu g \text{ of acrolein in test sample} \times 1000}{\text{air sample volume in liters}}$$

$$\text{ppm} = \frac{\text{total } \mu g \text{ of acrolein in sample}}{2.3 \times \text{sample volume in liters}}$$

Correct the air sample volume to 25 C and 760 Torr.

11. References

1. ROSENTHALER, L. and G. VEGEZZI. 1954. Determination of Acrolein. *Z. Lebensm. Untersuch. Forsch.* 99:352.
2. COHEN, I.R. and A.P. ALTSHULLER. 1961. A New Spectrophotometric Method for the Determination of Acrolein in Combustion Gases and the Atmosphere. *Anal. Chem.* 33:726.
3. ALTSHULLER, A.P. and S.P. MCPHERSON. 1963. Spectrophotometric Analysis of Aldehydes in the Los Angeles Atmosphere. *J. Air Poll. Control Assoc.* 13:109.
4. AMERICAN PUBLIC HEALTH ASSOCIATION, INC. 1977. Methods of Air Sampling and Analysis. 2nd ed. p. 297. Washington, D.C.
5. LEVAGGI, D.A. and M. FELDSTEIN. 1970. The Determination of Formaldehyde, Acrolein and Low Molecular Weight Aldehydes in Industrial Emissions on a Single Collected Sample. *JAPCA* 20:312.
6. AMERICAN PUBLIC HEALTH ASSOCIATION, INC. 1977. Methods of Air Sampling and Analysis. 2nd ed. p. 300. Washington, D.C.

Revised by Subcommittee 5
E. SAWICKI, *Chairman*
T. BELSKY
R. A. FRIEDEL
D. L. HYDE
J. L. MONKMAN
R. A. RASMUSSEN
L. A. RIPPERTON
L. D. WHITE

827.

Analytical Method for Aromatic Amines in Air

Analyte:	Aromatic Amines (see Table 827:I)	Method No.:	P & CAM 168
Matrix:	Air	Range:	0.01–12 mg/sample at 0–100% humidity
Procedure:	Absorption on silica gel; elution by ethanol; GC analysis	Precision:	$2\sigma = \pm 9\%$
Date Issued:	11/1/73		
Date Revised:	1/1/75	Classification:	D (Operational)

1. Principle of the Method

1.1 A known volume of air is drawn through a tube containing silica gel to trap the aromatic amines present.

1.2 The silica gel in the tube is transferred to a glass-stoppered tube and desorbed with ethanol.

1.3 An aliquot of the resulting solution of desorbed aromatic amines in ethanol is injected into a gas chromatograph.

1.4 Peak areas are determined and compared with calibration curves obtained from the injection of standards.

2. Range and Sensitivity

2.1 The lower limit of this method using a flame ionization detector is 0.01 mg/sample of any one compound when the analyte is desorbed with 5 ml ethanol and 10 μl of the resulting solution is injected into the gas chromatograph. Sensitivity for p-nitroaniline is 0.002 mg/sample.

2.2 For air samples of 100% relative humidity and containing 140 mg aniline/m^3, at least 13 mg aniline/sample is retained without breakthrough by the primary absorbing section of sampling Tube A after 8 hr of sampling at 200 ml/min. For Tube B, the primary absorbing section will retain 0.9 mg aniline/sample without breakthrough for air samples of 100% relative humidity and containing 90 mg aniline/m^3 after 10 min of sampling at 1 l/min.

3. Interferences

3.1 The most common sampling interference is water vapor. The sampling Tube A has been designed so that 96 l of 100% humidity air can be sampled over an 8-hour period at 200 ml/min without displacement of collected aromatic amines from the initial sorbent section by water vapor at room temperature.

3.2 Any compound which has nearly the same retention time as one of these aromatic amines at the gas chromatograph analytical conditions described in this method is an interference. This type of interference can often be overcome by changing the operating conditions of the gas chromatograph or selecting another column. Retention time data on a single column, or even on a number of columns, cannot be considered as conclusive proof of chemical identity in all cases. For this reason, it is important that whenever practical a sample of the bulk compound or mixture be submitted at the same time as the sample tube (but shipped separately) so that chemical identification(s) can be made by other means.

4. Precision and Accuracy

4.1 The accuracy of the method depends appreciably upon collection efficiencies and desorption efficiences. If a negligible amount of aromatic amine is detected on the backup section, the collection efficiency of the tube must be essentially 100%. Desorption efficiencies for the range of 1 to 8 mg have been found to be 100% within experimental error of ± 5%.

4.2 The precision of the chromatographic analysis is quite dependent upon the precision and sensitivity of the technique used to quantitate gas chromatographic peaks of samples and standards. More accurate peak integration techniques (normal or electronic) can maximize precision, particularly at lower concentrations.

4.3 Precision of standard preparations can be ± 1% (calibration standards).

4.4 The precision of the overall method, 2 σ = ± 9%, has been determined from eight consecutive identical samples taken with a personal sampling pump.

4.5 Analytical precision can be improved by reducing or eliminating the error associated with syringe injection into the gas chromatograph. This is best accomplished by the addition of an internal standard to the ethanol used to both prepare standards and elute samples from the silica gel. Therefore, it is recommended that about 0.1% solution of n-heptanol in ethanol be used. For isothermal analyses at temperatures above 120 C, n-octanol may be preferable.

5. Advantages and Disadvantages

5.1 The sampling method uses a small, portable device involving no liquids. It is not affected by the humidity of the air. A sample of up to 8 hr can be taken for an average work day concentration or a 15 min sample can be taken to test for excursion concentrations. Desorption of the collected sample is simple and is accomplished with a solvent with low toxicity. The analysis is accomplished by a quick instrumental method. Most analytical interferences which occur can be eliminated by altering gas chromatographic conditions. Since several aromatic amines can be collected and analyzed simultaneously, this method is useful where the composition of the aromatic amine vapors may not be known.

5.2 A major disadvantage of the method is the limitation on its precision due to the use of personal sampling pumps currently available. After initial adjustment of flow any change in the pumping rate will affect the volume of air actually sampled. Furthermore, if the pump used is calibrated for one tube only, as is often the case, the precision of the volume of air sampled will be limited by the reproducibility of the pressure drop across the tubes.

6. Apparatus

6.1 SAMPLING EQUIPMENT. The sampling unit for the silica adsorption method has the following components:

6.1.1 *The glass—silica gel sampling tubes.* (Refer to Section **6.2**).

6.1.2 *One or more personal sampling pumps,* whose flow can be set at and maintained at 1 l/min for 15 min or 200 ml/min for 8 hr. A bypass attachment or a syringe needle may be needed, depending on the pressure drop of the pumps, to allow the lower flow rate.

6.1.3 *Thermometer.*

6.1.4 *Manometer.*

6.1.5 *Stopwatch.*

6.2 Two types of *pyrex glass sampling tubes* of the dimensions shown in Figure 825:1, packed with two sections of 45/60 mesh silica gel. The silica gel can be purchased from commercial sources presieved or can be ground from larger particles and sieved before use. No other preparation appears necessary. Glass wool or woven glass cloth should be used for the plug in front of the primary sorbent section. More porous urethane foam plugs should be used between the sorbent sections and after the backup section. For reproducible pressure drops, the consistency of plug size and packing is critical. A 12-mm long piece of 5-mm OD pyrex tubing between the primary and backup sections greatly reduces migration of the sample to the backup section prior to analysis. After packing, the ends of each tube should be flame sealed to prevent contamination before sampling.

6.2.1 *Tube A*. This sampling tube contains a primary adsorbing section of approximately 1 g of 45/60 mesh silica gel packed firmly into a nearly spherical volume of 1.4 cm^3. The backup section contains 100 mg of silica gel in 8-mm OD tubing. The pressure drop of such tubes should not exceed 85 mm of water at air flows of 200 ml/min. (Figure 827:1)

6.2.2 *Tube B*. This sampling tube contains a primary adsorbing section of 150 mg and a backup section of 100 mg of 45/60 mesh silica gel in a straight 8-mm OD pyrex tube. The pressure drop of such tubes should not exceed 85 mm of water at air flows of 200 ml/min or 25 mm Hg at 1 l air/min. (Figure 827:1)

6.3 GAS CHROMATOGRAPH, equipped with a flame ionization detector. Linear temperature programming capability is desirable but not essential.

6.4 COLUMN (4 FT × ⅛ IN OD S.S.). Packed with Silicone OV-25 liquid phase, 10% on 80/100 mesh Supelcoport (or equivalent support). A column (2 ft × ⅛ in OD) packed with Chromosorb 103 (80/100 mesh) will also work except for p-nitroaniline.

6.5 RECORDER, and some method for determining peak height or area.

6.6 GLASS-STOPPERED TUBES OR FLASKS, 1- AND 10-ML.

6.7 SYRINGES, 10-μl.

6.8 PIPETTES AND VOLUMETRIC FLASKS. For preparation of standard solutions.

7. Reagents

7.1 PURITY. All reagents used must be ACS reagent grade or better.

7.2 ETHANOL, 95%.

7.3 N-HEPTANOL OR N-OCTANOL.

7.4 DESORBING SOLUTION. Prepare a 0.1% solution of n-heptanol or n-octanol in ethanol.

7.5 AROMATIC AMINE STANDARDS.

7.6 BUREAU OF MINES GRADE A HELIUM.

7.7 PREPURIFIED HYDROGEN, "ELECTROLYTIC GRADE."

7.8 FILTERED COMPRESSED AIR.

8. Procedure

8.1 CLEANING OF EQUIPMENT. All glassware used for the laboratory analysis should be detergent washed followed by tap and distilled water rinses.

8.2 CALIBRATION OF PERSONAL PUMPS. Each pump must be calibrated with a representative tube in the line to minimize errors associated with uncertainties in the sample volume collected.

8.3 COLLECTION AND SHIPPING OF SAMPLES.

8.3.1 Immediately before sampling break the ends of the tube so as to provide an opening at least one half the internal diameter of the tube.

8.3.2 Attach the sampling tube (backup section nearer the pump) to the sampling pump using a short piece of flexible tubing. Air being sampled should not be passed through any hose or tubing before entering the sampling tube.

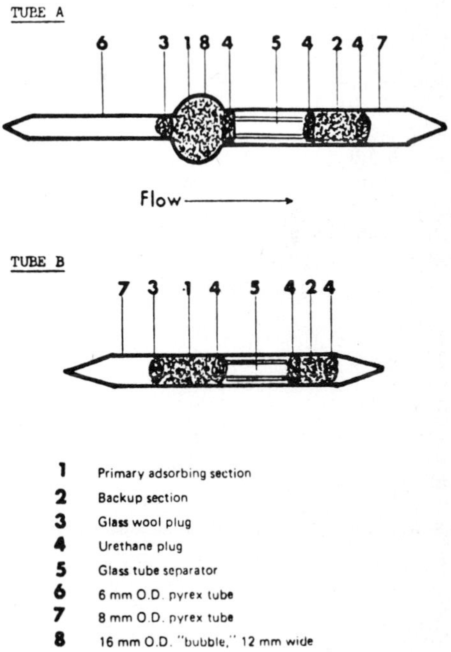

Figure 827:1—Silica Gel Sampling Tubes for Aromatic Amines.

Table 827:I. Aromatic Amines for Which the Method Has Been Tested

Compound	OSHA Standard*		15-min Tube	Sampling Rate (1/min)†	8-Hour Tube	Sampling Rate (1/min)†
	(ppm)	(mg/m^3)				
Aniline	5	19	B	0.2	A	0.2
N,N-Dimethyl-aniline	5	25	B	0.2	A	0.2
o-Toluidine	5	22	B	0.2	A	0.2
o-Chloroaniline	Not established		B	0.2	A	0.2
Xylidines (mixed)	5	25	B	0.2	A	0.2
o-Anisidine	0.1	0.5	B	1	B	0.2
p-Anisidine	0.1	0.5	B	1	B	0.2
p-Nitroaniline	1	6	B	0.2	A	0.2

**Federal Register, 37*, #202, 22139 (18 October 1972).
†See Section **8.3.3**.

8.3.3 Sample the atmosphere at the desired flow rate for the desired period of time. Suggested sampling rates and times are given in Table 827:I. The flow rate and sampling time or the volume of air sampled must be measured as accurately as possible.

8.3.4 Measure and record the temperature and pressure of the atmosphere being sampled.

8.3.5 Seal the sampling tubes with plastic caps immediately after sampling. Under no circumstances should rubber caps be used.

8.3.6 One tube should be handled in the same manner as the sample tube (break, seal, and transport), except that no air is sampled through this tube. This tube should be labeled as a blank.

8.3.7 Capped tubes should be packed tightly before shipping to minimize tube breakage during shipping.

8.3.8 Samples of the bulk liquids or solids from which the aromatic amine vapors arise should be submitted to the laboratory also, but *not* in the same container as the air samples or blank tubes.

8.3.9 *Storage.* Tubes after sampling should be tightly capped and not subjected to extremes of high temperature or low pressure, if avoidable. If the analysis is to be delayed beyond one week after sampling, each tube should be filled with an inert gas (helium, nitrogen, etc.) to retard the loss of sample by oxidation. Refrigeration is also recommended.

8.4 Analysis of Samples.

8.4.1 Carefully transfer the silica gel of the primary adsorbing section and the glass wool that precedes it to a glass-stoppered tube or flask (10 ml for Tube A or 2 ml for Tube B).

8.4.2 Remove and discard the separating sections of foam and glass tubing. Exercise due care to avoid any loss of silica gel particles.

8.4.3 Then transfer the silica-gel in the back-up section to a 2-ml glass-stoppered tube or flask.

8.4.4 These two silica gel sections are desorbed and analyzed separately according to the succeeding sections.

8.4.5 Pipet an aliquot of the desorbing solution—5 ml for the primary adsorbing section of Tube A and 1 ml for all other sections—into each tube or flask. Allow the samples to stand for 30 min. Tests indicate that desorption is complete in 30 min if the sample is stirred or shaken occasionally.

8.4.6 Inject an aliquot of the sample into the gas chromatograph.* With an internal standard in the eluent, direct injection of up to 10 μl with a microliter syringe is acceptably precise. At least dupli-

*Gas chromatograph conditions: Typical operating conditions for the gas chromatograph are: carrier flow (25 ml He/min); injection port (150 C); flame ionization detector (250 C, 50 ml/min H_2, 470 ml/min air); oven temperature program (100 C for 4 min, then increase at 8 C/min to 225 C).

cate injections of the same sample or standard are recommended.

8.4.7 Measure the areas of the sample peak and the internal standard peak by an electronic integrator or some other suitable method of area measurement. The ratio of these areas is calculated and used to determine sample concentration in the eluent by using a standard curve prepared as discussed in Section **9**.

9. Calibration and Standards

9.1 For good accuracy in the preparation of standards, it is recommended that one standard be prepared in a relatively large volume and at a high concentration. Aliquots of this standard can then be diluted to prepare other standards. The solvent and diluent used must be the same desorbing solution used for the elution of the samples. For example, to prepare 100 ml of standard corresponding to a 96-l sample of air containing 95 mg aniline/ml desorbed with 5 ml of desorbing solution, 178 μl[(1)] of aniline (density = 1.022 g/ml) is added to a 100-ml volumetric flask and diluted to the mark. The resulting concentration is 1.82 mg/ml[(2)]. If 2 ml of this solution is diluted to the mark with the desorbing solution in a 10-ml volumetric flask, the resulting concentration is 0.365 mg/ml[(3)].

$$\frac{95\ \text{mg/m}^3 \times 0.096\ \text{m}^3 \times 100\ \text{ml}}{5\ \text{ml} \times 1.022\ \text{mg}/\mu\text{l}} = 178\ \mu\text{l} \quad (1)$$

$$\frac{178\ \mu\text{l} \times 1.022\ \text{mg}/\mu\text{l}}{100\ \text{ml}} = 1.82\ \text{mg/ml} \quad (2)$$

$$1.82\ \text{mg/ml} \times \frac{2\ \text{ml}}{10\ \text{ml}} = 0.365\ \text{mg/ml} \quad (3)$$

When microliter pipettes are used instead of microliter syringes, it is better to prepare standards using a round number of microliters (*e.g.*, 200 μl aniline instead of 178 μl). For solids the amount of compound used for the first standard should be weighed on an analytical balance.

9.2 Prepare a series of standards of varying concentration over the range of interest.

9.3 The standards prepared as above should be analyzed under the same GC conditions and during the same day as the unknown samples.

9.4 Prepare a standard curve for each compound, plotting ratios of peak areas of the compound, to the internal standard against concentration of the compound. From the resulting curve the concentration of an eluted sample is determined. This concentration (in mg/ml) is then converted to total sample weight by multiplying with the volume of desorbing solution used for that silica gel section (1 or 5 ml).

10. Calculations

10.1 Corrections for the blank must be made for each sample.

$$\text{Corrected mg} = \text{mg}_s - \text{mg}_b$$

where:

mg_s = mg found in primary adsorbing section of sample tube

mg_b = mg found in primary adsorbing section of blank tube

A similar procedure is followed for the backup sections.

10.2 Add the corrected amounts present in the primary and backup sections of the same sample tube to determine the total measured amount (w) in the sample.

10.3 Convert the volume of air sampled to standard conditions of 25 C and 760 Torr.

$$V_s = V \times \frac{P}{760} \times \frac{298}{T + 273}$$

where:

V_s = volume of air in liters at 25 C and 760 Torr.

V = volume of air in liters as calculated (sampling time times correct flow rate)

P = Barometric pressure in Torr.
T = Temperature of air in degree C.

10.4 The concentration of the organic solvent in the air sampled can be expressed in mg per m³, which is numerically equal to μg per liter of air

$$\text{mg/m}^3 = \mu\text{g/l} = \frac{\text{w(mg)} \times 1000\ (\mu\text{g/mg})}{V_s}$$

10.5 Another method of expressing conc is ppm, defined as μl of compound vapor per liter of air

$$\text{ppm} = \mu\text{l of vapor}/V_s$$

$$\text{ppm} = \frac{\mu\text{l of vapor}}{V_s} \times \frac{24.45}{\text{MW}}$$

where:

24.45 = molar volume at 25 C and 760 Torr in μl per μmole.
MW = molecular weight of the compound, in μg per μmole.

11. References

1. E.E. Campbell, G.O. Wood, and R.G. Anderson. 1972, 1973, 1974. Los Alamos Scientific Laboratory Progress Reports LA-5104-PR, LA-5164-PR, LA-5308-PR, LA-5389-PR, LA-5484-PR, and LA-5634-PR, Los Alamos, N.M., Nov., Jan., June, Aug., Dec., and June.

Revised by Subcommittee 5
E. Sawicki, *Chairman*
T. Belsky
R. A. Friedel
D. L. Hyde
J. L. Monkman
R. A. Rasmussen
L. A. Ripperton
L. D. White

828.

Analytical Method for Bis (Chloromethyl) Ether (Bis-CME) in Air

Analyte:	Bis-CME	Method No.:	P & CAM 213
Matrix:	Air	Range:	0.5 to 10 ppb
Procedure:	Adsorption on chromosorb 101, desorption with helium through GC column, measurement by mass spectroscopy	Precision:	± 10% RSD
		Classification:	E (Proposed)
Date Issued:	1/1/75		
Date Revised:			

1. Principle of the Method

1.1 Bis-CME is concentrated from air by adsorption on Chromosorb 101 in a short tube. It is desorbed by heating and purging with helium through a gas chromatography column where it is separated from most interferences. The concentration of bis-CME is measured from the spectroscopic signal at m/e 79 and 81.

2. Range and Sensitivity

2.1 The range of the mass spectroscopic signal for the conditions listed corresponds to 0.5 to 10 ppb.

2.2 A concentration of 0.5 ppb of bis-CME can be determined in a 5-l air sample. Greater sensitivity can be obtained by using a larger sample.

3. Interferences

3.1 Interferences resulting from materials having retention times similar to bis-CME or simply background such as ions C_6H_7, $C_5C_1{}^{13}H_6$, C_4H_3Si and Br giving rise to m/e 79 can be encountered. However, the Du-Pont 21-491 mass spectrometer has sufficient resolution to resolve at least three of these ions at m/e 79, particularly the $C_5C_1{}^{13}H_6$ and C_6H_7 ions, that are normally encountered in the plant samples. These are completely resolved from the bis-CME $C_2H_4Cl^{35}O$ ion.

4. Precision and Accuracy

4.1 The precision of this method has been determined to be ± 10% of relative standard deviation when different sampling tubes were spiked with 133 ng (corresponding to 1.4 ppb in 20 l of air) at intervals of approximately 1 hr. These data were obtained using 4″ × ¼″ stainless steel packed with 60/80 mesh Chromosorb 101.

4.2 The accuracy of the analysis is approximately ± 10% of the amount reporte as determined from repeated analysis of several standards.

5. Advantages and Disadvantages of the Method

5.1 The gas chromatography-mass spectrometry technique interfaced with a Watson-Biemann helium separator is extremely sensitive and specific for the analysis of bis-CME. The gas chromatographic separation yields a retention time that is characteristic for bis-CME, but it is not highly specific for positive assignment of the signal as bis-CME. The mass spectrometer in combination with the gas chromatograph provides the high degree of specificity. The most intense ion in the mass spectrum of bis-CME is formed at m/e 79 with its chlorine-37 isotope at m/e 81. The chlorine isotopic abundances require that the intensity ratio of peaks at m/e 79:81 be approximately 3:1. The mass spectrometer is set to continuously monitor these ions as the sample elutes through the gas chromatograph. It is absolutely necessary that the retention time, observed masses at m/e 79 and 81 and the intensity ratio be correct in order to assign the signal to bis-CME.

5.2 Bis-CME, an impurity in chloromethyl methyl ether, has been reported to be a pulmonary carcinogen. Extreme safety precautions should be exercised in the preparation and disposal of liquid and gas standards and the analysis of air samples. Bis-CME may be destroyed in a methanol-caustic solution.

5.3 Sampling must be carried out away from the mass spectrometer.

5.4 GC-MS is not a good approach for continuous monitoring.

6. Apparatus

6.1 SAMPLING TUBES.

6.1.1 The sampling tubes are prepared by packing a 2¼″ × ¼″ stainless steel tubing with 1¾″ of 60/80 mesh Chromosorb 101 with glass wool in the ends. These are conditioned overnight at 200 C with helium or nitrogen flow set at 10 to 30 ml/min. The conditioned sampling tubes are then cooled under flow and capped immediately upon removal. An identification number is scribed on one of the nuts. The stainless steel tubing should be rinsed internally with water, acetone and methylene chloride before packing with Chromosorb 101.

6.1.2 Longer sampling tubes may be prepared with a proportional amount of Chromosorb 101.

6.2 GAS CHROMATOGRAPHY COLUMN.

6.2.1 A 4′ × ⅛″ stainless steel separator column is rinsed internally with water, acetone and methylene chloride and air dried. The cleaned tubing is then packed with Chromosorb 101. The column is conditioned overnight at 200 C and 10 to 30 ml/min helium or nitrogen flow.

6.3 A WATSON-BIEMANN HELIUM SEPARATOR.

6.4 GAS CHROMATOGRAPH.

6.4.1 A Hewlett-Packard 5750 gas chromatograph or equivalent. A gas chromatograph employing a single column oven and a temperature programmer is adequate.

6.5 MASS SPECTROMETER.

6.5.1 A mass spectrometer with a resolution of 1000 to 2000 and equipped with a repetitive scan attachment can be used in conjunction with a gas chromatograph. A CEC (now DuPont Analytical Instruments Div.) Model 21-491 modified by adding a follower amplifier with a long time constant (0.03 sec) to the output of the electron multiplier to increase the gain and signal to noise ratio has been found satisfactory for this purpose.

6.6 SYRINGES.

6.6.1 *Syringe,* 1-ml. Gas tight (The Hamilton Co., Inc.).

6.6.2 *Syringe,* 10-μl. (The Hamilton Co., Inc.).

7. Reagents and Materials

7.1 PURITY. All reagents must be analytical reagent grade.

7.2 BIS-CHLOROMETHYL ETHER (EASTMAN KODAK).

7.3 ACETONE.

7.4 DICHLOROMETHANE.

7.5 N-PENTANE.

7.6 METHANOL.

7.7 SODIUM HYDROXIDE.

7.8 PLASTIC FILM GAS BAG, 13 L. Saran brand (Trademark of The Dow Chemical Company abroad) plastic film gas bag has proven to be satisfactory for this purpose (The Anspec Co., Inc.). Polyethylene film bags are not suitable.

7.9 CHROMOSORB 101 (JOHNS-MANVILLE).

8. Procedure

8.1 CLEANING OF EQUIPMENT. NONE REQUIRED.

8.2 COLLECTION OF SAMPLES.

8.2.1 Spot samples may be obtained by attaching the sampling tube to a manual syringe pump. The sample is drawn into the end of the sampling tube with the marked nut.

8.2.2 Continuous sampling of air may be accomplished by attaching a vacuum pump to a throttling valve and manometer. Each sample tube is calibrated so that the pressure drop is determined. Then the flow is established by adjusting the pressure drop with the throttling valve.

8.2.3 Continuous or cumulative sampling at fixed points may be accomplished by attaching a vacuum pump to a throttling valve and calibrated rotameter. The flow is then adjusted with the throttling valve through the calibrated rotameter. This sampling is much preferred to **8.2.2**.

8.2.4 For larger sample size it is important to realize that larger flows of air over long period of time may cause elution of bis-CME through the sampling tube. It has been shown that a total of 25 l of air can be pumped through a 4″ × ¼″ packing of Chromosorb 101 over a period of 24 hr without any elution of the bis-CME. This corresponds to a flow of 17 cc/min maximum. A maximum air volume of 5 l at a rate of 80 cc/min has been found to be safe for the 2½″ × ¼″ sampling tube.

8.2.5 Bis-CME has been found to be stable in the sampling tubes for at least 10 days when kept tightly closed and at room temperature.

8.3 ANALYSIS OF SAMPLE.

8.3.1 *Instrument Conditions and Setup.* The mass spectrometer should be set to sweep peaks at m/e 79 to 81 with the visicorder speed at 0.8 in/sec. for recording the resulting signal.

8.3.2 Hold the temperature of the Watson-Biemann helium separator at 160 C. Set the helium flow at 30 cc/min. Balance the helium flow so that there is no vacuum applied to the outlet of the gas chromatographic column.

8.3.3 Attach the sampling tube to the gas chromatograph via an adapter to the septum nut. The inscribed sample tube nut should be closest to the inlet of the analytical column.

8.3.4 A ¼″ Swagelok Tee fitting is attached on the other end of the sampling tube to accommodate the carrier gas and a silicone rubber septum so that standards may be injected.

8.3.5 After thoroughly checking for leaks, wrap the sampling tube with heat tape and heat to 150 C for 5 min to assure complete desorption of the organic components onto the analytical column which is at ambient temperature.

8.3.6 Temperature program the analytical column to 130 C at 30 C/min and hold to yield a retention time for bis-CME of 8 to 9 min. Turn on recorder 30 sec before elution of bis-CME.

8.3.7 The analytical column should be cooled to ambient temperature before changing sampling tubes.

9. Calibration and Standards

9.1 PREPARATION OF GAS STANDARD.

9.1.1 Rinse out a 13-l Saran bag with prepurified nitrogen.

9.1.2 Introduce 5 l of prepurified nitrogen through a dry test meter into the Saran bag and cap it with a cork.

9.1.3 Place plastic tape on the Saran bag and inject 1 μl of bis-CME through the tape into the bag. The tape should prevent any cracking of the bag at the injection point.

9.1.4 Tape the injection point at once upon removing the syringe needle from the bag.

9.1.5 Clean the syringe immediately with methanol-caustic solution and then water and acetone.

9.1.6 Knead the bag for approximately 2 min.

9.1.7 Carefully connect the bag with the meter to introduce an additional 5 l of prepurified nitrogen (the end of the Saran tubing inside the bag should be kept tight with a finger from the outside in order to avoid any loss of bis-CME while connecting the bag with the meter).

9.1.8 Knead the bag for additional 2 min to assure complete mixing.

9.1.9 With the 10-l bag this procedure can prepare a ppm level standard (*e.g.*, 1 μl of liquid bis-CME yields a 28.3 ppm by volume standard). This standard should not be kept for more than three days.

9.2 PREPARATION OF LIQUID STANDARD.

9.2.1 Bis-CME Standard (0.132 μg bis-CME/μl). Dissolve 1 μl of bis-CME liquid in 10 ml n-pentane. This standard solution may be kept for as long as desired if avoiding evaporation.

9.3 CALIBRATION.

9.3.1 Inject a 1-ml portion of the 28.3 ppm standard bis-CME (0.132 μg) through a standard septum onto a sampling tube attached to the inlet of the analytical column. This practice may be done routinely after two or three samples to ensure accuracy. Also one μl of the liquid standard (0.132 μg/μl) could be used for the calibration.

9.3.2 Repeat steps as described in paragraphs **8.3.3–8.3.7**.

10. Calculations

10.1 The total ng of bis-CME in the air sampled is determined from the ratio of the peak height of m/e 79 of the sample and the standard multiplied by 132.

10.2 Concentration in ppb is calculated as follows:

$$\text{ppb} = \frac{\text{ng bis-CME}}{V} \times \frac{24.45}{MW}$$

where:

ng bis-CME = total ng concentration as determined in **10.1**

V = volume of air in liters sampled at 25 C and 760 Torr

24.45 = molar volume of an ideal gas at 25 C and 760 Torr

MW = molecular weight of bis-CME, 115

11. References

1. VAN DUNREN, B.L., A. SINAK, B.M. GOLDSCHMIDT, C. KATZ and S. MELCHIONNE. 1969. *J. Nat. Cancer Inst.* 43:481.
2. LASKIN, S., M. KUSCHNER, R.T. DREW, V.P. CAUPPIELLO and N. NELSON. 1971. Technical Re-

port, New York University Medical Center, Department of Environmental Medicine, August.
3. LEONG, K.J., H.N. MACFARLAND and W.H. REESE. 1971. *Arch. Environmental Health* 22:663.
4. WATSON, J.T. and K. BIEMANN. 1964. *Anal. Chem.* 36:1135.
5. DUPONT DE NEMOUR, E.I. Analytical Instrument Division, 1500 S. Shamrock Ave., Monrovia, Calif.
6. SHADOFF, L.A., G.J. KALLOS and J.S. WOODS. 1973. Analysis for bis-Chloromethyl Ether in Air. *Anal. Chem.* 45:2341.
7. FEDERAL REGISTER. 1974. Vol. 39, No. 20, Tuesday, January 29. pp. 3773–3776.

Subcommittee 5
E. SAWICKI, *Chairman*
T. BELSKY
R. A. FRIEDEL
D. L. HYDE
J. L. MONKMAN
R. A. RASMUSSEN
L. A. RIPPERTON
L. D. WHITE

829.

Analytical Method for Chloromethyl Methyl Ether (CMME) and Bis-Chloromethyl Ether (BCME) in Air

Analyte:	CMME and BCME	Method No.:	P & CAM 220
Matrix:	Air	Range:	0.5–7.5 ppb (CCME or BCME)
Procedure:	Collection via Impinger/GC Electron Capture Detector	Precision:	10% RSD
		Classification:	E (proposed)
Date Issued:	6/27/75		
Date Revised:			

1. Principle of the Method

1.1 A known volume of air is drawn through glass impingers containing a methanolic solution of the sodium salt of 2,4,6-trichlorophenol.

1.2 CMME and BCME react with the derivatizing reagent to produce stable derivatives.

1.3 Sample is heated on a steam bath for 5 min, cooled, diluted with an equal volume of distilled water and 2 ml of hexane for extraction.

1.4 An aliquot of the hexane is analyzed by electron capture gas chromatography.

1.5 The peak heights of the CMME and BCME derivatives are measured and the concentration determined from standard curves.

2. Range and Sensitivity

2.1 A relative standard deviation of 10% can be expected within the concentration range of 0.02 to 0.3 ng/μl of both CMME and BCME.

2.2 The sensitivity of the method is 0.5 ppb (v/v) when a 10-l air sample is used.

3. Interferences

3.1 The known components used in chloromethylation processes do not interfere with the determination of CMME or BCME.

3.2 Interferences can be expected from highly halogenated organic compounds or compounds that may produce same derivatives.

3.3 The quality of 2,4,6-trichlorophenol is important since impurities can be extracted with hexane and seriously interfere with the chromatographic analysis.

4. Precision and Accuracy

4.1 The precision of the sampling technique with an accurate, calibrated air sampling pump and analysis of the derivative samples is 10% relative standard deviation.

4.2 The accuracy of the method is typically affected by the efficiency in sampling, extraction, calibration and data handling.

5. Advantages and Disadvantages

5.1 The major advantage of this method is the sensitivity and the simultaneous analysis of both CMME and BCME.

5.2 The derivative stabilizes both CMME and BCME while significantly increasing the sensitivity.

5.3 Preparation of the derivatives eliminates further hazardous handling of both CMME and BCME during the analysis.

5.4 Disadvantages of the method are the handling of liquids, extractions and the dilution of the samples by the hexane extraction.

6. Apparatus

6.1 SAMPLING EQUIPMENT. The sampling unit for the impinger collection method consists of the following components:

6.1.1 *Two standard air impinger assemblies.* With fritted glass inlets.

6.1.2 *Calibrated battery-powered pump.* Capable of drawing an accurate and reproducible volume of air through the impingers at a flow rate of 0.5 l/min is required.

6.1.3 *Rotameter,* calibrated for air. Parts should include only the sapphire ball and small Teflon tubing, inside at both ends.

6.1.4 *Thermometer.*

6.1.5 *Stopwatch.*

6.2 STEAM BATH. Any bath capable of maintaining a temperature of 65 to 90 C is adequate.

6.3 GAS CHROMATOGRAPH. Equipped with a ^{63}Ni electron capture detector.

6.4 GAS CHROMATOGRAPH COLUMN. A 6-foot long (1.83 m) × ¼″ (6.35 mm) glass column is packed with 100/120 mesh textured glass beads (GLC-100) coated with a two component stationary phase consisting of 0.1% by weight of QF-1 and 0.1% by weight OV-17. The column is equipped for on-column injection. The packed column is preconditioned at 160 C overnight with nitrogen carrier at a flow rate of 30 to 50 cm^3/min.

6.5 STRIP CHART RECORDER, 1.0 MV FULL SCALE RANGE.

6.6 HAMILTON MICROSYRINGES.

6.7 ASSORTED LABORATORY GLASSWARE, pipettes, graduated cylinders, etc.

7. Reagents

PURITY. All reagents are analytical reagent grade. An analytical grade of distilled water must be used.

7.1 DERIVATIZING REAGENT, prepared in the laboratory as described in the procedure.

7.2 SODIUM METHOXIDE, AR GRADE.

7.3 2,4,6-TRICHLOROPHENOL, M.P. 67 TO 68 C.

7.4 METHANOL AND HEXANE, DISTILLED IN GLASS.

7.5 CHLOROMETHYL METHYL ETHER, B.P. 55 TO 58 C.

7.6 BIS-CHLOROMETHYL ETHER, B.P. 100 TO 102 C.

7.7 SODIUM HYDROXIDE.

7.8 PREPARATION OF DERIVATIZING REAGENT. Twenty-five g of sodium methoxide are weighed into a beaker and dissolved in 1 liter of methanol. Five g of 2,4,6-trichlorophenol are weighed into the beaker and allowed to dissolve in the methanol-sodium methoxide solution. The reagent is stable for 3 or 4 weeks when stored in a dark brown bottle. The sodium methoxide should be added slowly to the methanol since the reaction is very exothermic.

7.9 The sensitivity of BCME can be increased 6 to 8 fold by using stoichiometric quantities of sodium methoxide and 2,4,6-trichlorophenol (16.0 g of 2,4,6-trichlorophenol and 4.4 g of sodium methoxide in 1 liter of methanol). When using this formulation, 2.0 N sodium hydroxide is used in place of distilled water prior to the hexane extraction of the derivative (see Section **8.3**).

8. Procedure

8.1 Cleaning of Equipment. All glassware used for the analysis must be thoroughly washed, rinsed with distilled water and dried. The impinger assemblies can be rinsed with reagent grade methanol for repeated use.

8.2 Collection of samples.

8.2.1 CMME and BCME in air are sampled at a rate of 0.5 l/min, (up to 2 hr if necessary) through the two impinger assemblies each containing 10 ml of the derivatizing solution.

8.2.2 Teflon connections should be used for the attachment of the two impinger assemblies in series and the rotameter to the first impinger. Rubber tubing may be used for the connection of the second impinger to the intake of the pump.

8.2.3 A bypass flow of air can be established in case the pump does not operate at low flow rates. An appropriate syringe needle injected into the rubber tubing between the impinger and the intake of the pump should provide satisfactory flow rates. Also a needle valve connected through a T should produce adjustable flow rates.

8.3 Analysis of sample.

8.3.1 After sampling, the solutions are transferred to a vial that is capped loosely and placed on a steam bath for five minutes (any bath capable of maintaining a temperature of 65 to 90 C is suitable). The samples are allowed to cool and an equal amount of distilled water and 2 ml of hexane are pipetted into the vial. Then the sample is shaken for 5 min. The mixture is allowed to stand for a few minutes to allow the phases to separate.

8.3.2 *GC conditions.* The following are the recommended starting instrumental conditions. A gas chromatograph with a ^{63}Ni electron capture detector is equipped with a 6-foot long (1.83 m) by ¼″ (6.35 mm) glass column packed with 100 to 120 mesh textured glass beads (GLC-100) coated with a two component stationary phase consisting of 0.1% by weight QF-1 and 0.1% by weight OV-17. The column is equipped for on-column injection. The flow rate of the prepurified nitrogen carrier gas is set at 30 cm^3/min. The temperature of the sample injection zone is adjusted at 175 C and that of the detector at 250 C. The column oven is operated isothermally at 140 C. An oxygen filter is required on the carrier gas.

8.3.3 *Injection.* A 2-μl aliquot of the hexane extract is injected into the GC. A complete chromatogram should be obtained in about 6 min. Duplicate injections of each sample and a standard should be made.

9. Calibration and Standards

9.1 Preparation of Standard Curve.

9.1.1 Two microliters of CMME and BCME are added to 50 ml of hexane. The weights of the components are obtained by using their respective specific gravities, 1.03 g/ml for CMME and 1.33 g/ml for BCME. This concentrated standard is then used for preparing a standard curve. Both of these compounds should be handled in a good hood only!!

9.1.2 Ten ml of the derivatizing reagent (Section **8.3**) is pipetted into five 20-dram Kimble screw cap vials. Ten, five, two, one and zero microliters of the concentrated standard are added. These volumes are equivalent to 0.50, 0.25, 0.10, and 0.05 μg of bis-CME and 0.40, 0.20, 0.08 and 0.04 μg of CMME.

9.1.3 The vials are capped loosely and placed on a steam bath for 5 min. The standard is cooled and 10 ml of distilled water and 2 ml of hexane are pipetted into the vials. Then the standards are shaken for 5 min.

9.2 Standard curves are established by plotting the peak height of recorder response versus concentration in nanograms.

10. Calculations

10.1 Determine from the calibration curve the concentration of the desired component in nanograms.

$$\text{Component (ng)} = \frac{A \times B}{C}$$

where:

A = response for the component of interest in the sample

B = weight of CMME or BCME in the standard expressed in nanograms

C = response for the component of interest in the standard

10.2 The concentration of CMME and BCME in the sampled atmosphere can be calculated in ppb (v/v).

$$\text{ppb (v/v)} = \frac{D \times 24.45}{V_s \times MW}$$

where:

D = total ng concentration as determined in Section **10.1**

24.45 = molar volume of an ideal gas at 25 C and 760 Torr

V_s = volume of air sampled in liters at 25 C and 760 Torr

MW = molecular weight of CMME and BCME, 80.5 and 115, respectively.

11. Reference

1. SOLOMON R.A. and G.J. KALLOS. 1975. *Anal. Chem.*, 47:955.

Subcommittee 5
E. SAWICKI, *Chairman*
T. BELSKY
R. A. FRIEDEL
D. L. HYDE
J. L. MONKMAN
R. A. RASMUSSEN
L. A. RIPPERTON
L. D. WHITE

830.

Analytical Method for 3,3′-Dichloro-4,4′-Diaminodiphenylmethane (MOCA)* in Air

Analyte:	3,3′-Dichloro-4,4′-Diaminodiphenylmethane (MOCA)	Method No.:	P & CAM 207
Matrix:	Air	Range:	2 to 150 $\mu g/m^3$
Procedure:	Adsorption of Gas-chrom S, desorption with acetone/gas chromatography	Precision:	5% relative standard deviation (analytical)
Date Issued:	1/1/75	Classification:	E (Proposed)
Date Revised:			

1. Principle of the Method

1.1 A known volume of air is drawn through a Gas-chrom S tube with a personal air sampling pump.

1.2 Sections of the Gas-chrom S tube are transferred individually to stoppered, 1-ml volumetric flasks and extracted with 0.5 ml acetone.

1.3 Sample is permitted to stand for 25 min with occasional agitation, and an aliquot is analyzed by flame ionization gas chromatography.

1.4 The peak area of MOCA is determined and its concentration found from a calibration curve.

2. Range and Sensitivity

2.1 Within the concentration range 0.002 to 0.15 $\mu g/\mu l$ of MOCA, a relative standard deviation of 5% can be expected.

2.2 A detection limit with 1-μl aliquot of sample is 0.002 μg or about 2 $\mu g/m^3$ for a 500-l air sample.

3. Interferences

3.1 Fortuitously, no impurity yet encountered in field trials has the same retention time as MOCA to cause interference in this concentration range.

3.2 Interference can be expected from organic compounds with similar functional groups.

3.3 Isomers of chloraniline commonly associated with MOCA do not interfere as they are completely resolved by the gas chromatograph.

3.4 The solvent effect is pronounced since MOCA appears on the tailing edge of the acetone peak. It is therefore important to maintain an injection volume of less than 2 μl.

4. Precision and Accuracy

4.1 The precision of the sampling method with a personal air sampling pump and analysis of synthetic MOCA samples varied from 1.3 to 5% relative standard deviation.

5. Advantages and Disadvantages of the Method

5.1 The sampling device is small, portable, involves no liquids, and is relatively free of interference from commonly associated molecular species.

*MOCA—Trade name E.I. DuPont Co.

5.2 Both short and long term (8-hr) samples may be taken.

5.3 The collection efficiency of Gas-chrom S in the tubular configuration approaches 100% for MOCA in air, while the desorption efficiency is near 92%.

5.4 Disadvantages of the method are the exacting experimental conditions of hydrogen flow rate and the large solvent effect when the injection volume is greater than 2 μl.

5.5 MOCA is a carcinogenic substance and precautions must be taken to avoid skin or other contact.

6. Apparatus

6.1 Gas-chrom S (40/60 mesh) is sieved to 45 mesh and packed into 63 mm × 6 mm OD tapered pyrex tube as two separate sections 10 and 50 mg., Figure 830:1. These sections are held in place with a thin layer of quartz wool.

6.2 AIR-SAMPLING PUMP. An air pump capable of drawing a constant flow of air through a limiting orifice (1 lpm) is suitable for area samples. For personal samplers, a calibrated battery-operated portable pump capable of aspirating an accurate and reproducible volume of air through the S-tube is required.

6.3 GC EQUIPPED WITH FLAME IONIZATION DETECTOR.

6.4 GAS CHROMATOGRAPH COLUMN. A 30-cm long by 3-mm OD (2.34-mm ID) stainless steel tube is packed with 10% by weight of Dexsil 300 GC coated on 80/90 mesh ABS Anakrom. The packed column is pre-conditioned at 300 C for 24 hr with helium at a flow rate of 40 to 50 cm³/min.

6.5 Silylation of the packed column is accomplished by installing the column and a 3-mm OD pyrex injector liner on the GC. A column temperature of 150 C and a helium flow rate of 30 cm³/min is maintained while introducing three 10-μl samples of Silyl-8 with a Hamilton microsyringe. The detector is removed from the GC to avoid contamination.

6.6 A STRIP CHART RECORDER. 1 mV sensitivity is acceptable.

6.7 MICROSYRINGES (10- and 25-μl) and 1- and 10-ml stoppered volumetric flasks are required to handle samples.

7. Reagents

7.1 AR GRADE ACETONE AND SILYL-8.

7.2 HELIUM CARRIER GAS. HIGHEST PURITY.

7.3 HYDROGEN. Ultrapure hydrogen.

7.4 AIR. Prepared by mixing oxygen and nitrogen to 21% O_2.

7.5 MOCA, purified chromatographically to 99.5%.

8. Procedure

8.1 CLEANING OF GLASSWARE. All glassware used for the laboratory analysis should be thoroughly cleaned, washed and rinsed with distilled water. Volumetric ware should be dried in vacuo at 50 C, and allowed to cool.

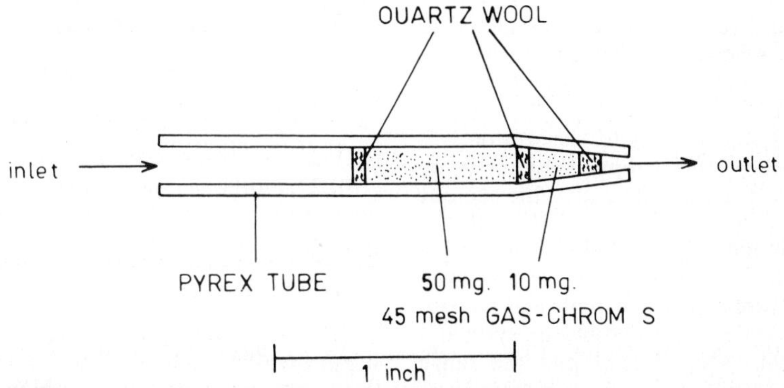

Figure 830:1—MOCA sampling tube.

8.2 CALIBRATION OF PERSONAL PUMP. Each personal pump must be calibrated with a representative Gas-chrom S tube.

8.3 COLLECTION.

8.3.1 MOCA in air is sampled at a rate of 1 lpm (up to 8 hr if necessary) through a Gas-chrom S tube with a personal air sampling pump.

8.3.2 The tapered end of the S-tube is attached to the pump so that the initial air sampling occurs in the 50-mg section.

8.4 ANALYSIS OF SAMPLE.

8.4.1 After sampling, each section of the Gas-chrom S tube with its plug is transferred into a 1-ml volumetric flask and 0.5 ml of acetone is added to initiate extraction of the collected MOCA.

The sample is allowed to stand 25 min with occasional agitation prior to analysis by gas chromatography. Acetone is a very volatile solvent, hence volume changes can occur during the desorption process depending on the ambient temperature and the seal of the volumetric flask. In all cases proper safety precautions should be taken while working with hazardous chemicals.

8.4.2 *GC conditions*. Operating conditions for the GC may vary from instrument to instrument and must be optimized. The following are the recommended starting instrumental conditions. A gas chromatograph with a flame ionization detector (FID) is equipped with a silylated glass liner and a 30-cm 10% Dexil 300 GC column operated at column and inlet temperatures of 200 and 250 C, respectively. The flow rate of the helium carrier gas is adjusted to 35 cm^3/min, and the flow ratio of helium:hydrogen:air is maintained at 1:.71:10. After the recorder is turned on, the electrometer is balanced and the FID is ignited.

8.4.3 *Injection*. After desorption, a 2-μl aliquot of the sample solution is injected into the GC. The 2 μl of the acetone extract is injected into the GC with the Hamilton microsyringe held in the sample inlet until the acetone peak, as displayed on the recorder, goes off-scale and then returns on-scale. At this point, the syringe is withdrawn. After 2 to 3 min the range switch is reset to 0.1 and a complete chromatogram is obtained in about 13 min. At low MOCA concentrations an increase in the sensitivity is achieved with a 1-μl sample, provided the calibration curve is prepared with the same volume, 1 μl. Triplicate injection of each sample and a standard should be made.

9. Calibration and Standards

9.1 MEASUREMENT OF AREA. The peak area of MOCA is measured by the peak height multiplied by peak width at half height. The unknown concentration is determined from a calibration curve prepared with identical injection volumes.

9.2 Since no internal standard is used in this method, standard solutions must be analyzed at the same time and with the same injection volume as the test sample. This will minimize the effect of known day-to-day variations of the FID response. Standard curves are established by plotting concentrations in μg versus peak area or integration output.

10. Calculations

10.1 It is convenient to express concentration of MOCA in terms of milligrams per cubic meter of air, mg MOCA per m^3. By use of the following equation the concentrations of MOCA in the air sampled can be expressed in mg per m^3, which is numerically equal to μg/l of air.

$$\frac{(\mu g \text{ MOCA})\,(A)\,(1.087)}{[\text{Sampling rate (lpm)}]\,[\text{Sampling time (min)}]} = \text{mg MOCA/m}^3$$

where

A = 250 for 2-μl, or 500 for 1-μl aliquot sample

10.2 Another method of expressing concentration is ppm, defined as μl of compound per liter of air.

$$\text{ppm} = \mu l \text{ of compound}/V_s$$

$$\text{ppm} = \frac{\mu g \text{ of compound}}{V_s} \times \frac{24.45}{MW}$$

Where:

24.45 = molar volume at 25 C and 760 Torr

MW = molecular weight of the compound

V_s = volume of air in liters at 25 C and 760 Torr

and where:

$$V_s = V \times \frac{P}{760} \times \frac{298}{T + 273}$$

V = Volume of air in liters as measured

P = Barometric pressure in Torr

T = Temperature of air in degrees C.

11. References

1. FEDERAL REGISTER 38. 1973. No. 85, 10929–10930 May 3.
2. A.L. LINCH, G.B. O'CONNOR, J.B. BARNES, A.S. KILLIAN, JR. and W.E. NEELD, JR. 1971. *Amer. Ind. Hyg. Assoc. J.* 802–819. Dec.

Subcommittee 5

E. SAWICKI, *Chairman*

T. BELSKY

R. A. FRIEDEL

D. L. HYDE

J. L. MONKMAN

R. A. RASMUSSEN

L. A. RIPPERTON

L. D. WHITE

831.

Analytical Method for P,P-Diphenylmethane Diisocyanate (MDI) in Air

Analyte:	MDI	Method No.:	P & CAM 142
Matrix:	Air	Range:	0.007–0.073 ppm
Procedure:	Collection via Midget Impinger, Colorimetric, Azo dye	Precision:	Unknown
		Classification:	D (Operational)
Date Issued:	9/13/72		
Date Revised:	1/1/75		

1. Principle of the Method

1.1 MDI is hydrolyzed by a solution of hydrochloric and acetic acids to methylene dianiline. The method is a modification of the Marcali method (**2**).

1.2 The methylene dianiline is diazotized by sodium nitrite-sodium bromide solution.

1.3 The diazo compound is coupled with N-(1-Naphthyl)-ethylenediamine to form the colored complex.

1.4 The amount of colored complex formed is in direct proportion to the amount of MDI present. The amount of colored complex is determined by reading the absorbance of the solution at 555 nm.

2. Range and Sensitivity

2.1 The range of the standards used is 1.5 to 15 μg MDI. In a 20-l air sample, this range is equal to 0.007 ppm to 0.073 ppm.

2.2 For samples of high concentration whereby absorbance is greater than the limits of the standard curve, dilution of the sample with absorber solution and rereading the absorbance can extend the upper limit of the range.

2.3 The amount of MDI that would saturate 15 ml of the absorber solution has not been determined. Therefore, it is possible that, in extremely high concentrations, some of the MDI would not be absorbed by the absorber solution. In such cases two impingers should be used in series, and the appropriate corrections made for efficiency.

3. Interferences

3.1 Any free aromatic amine may be diazotized and coupled forming a positive interference.

3.2 Toluenediisocyanate (TDI) also interferes.

4. Precision and Accuracy

4.1 The precision and accuracy are unknown.

4.2 There has been no collaborative testing.

5. Advantages and Disadvantages of the Method

5.1 The concentration of acidity for the coupling reaction has been changed in this method so as to reduce the final reaction time to 15 min compared to 2 hr for the Marcali method.

5.2 Since any free aromatic amine may interfere and TDI definitely interferes, the method is not specific for MDI.

6. Apparatus

6.1 SAMPLING EQUIPMENT. The sampling unit for personal samples by the impinger collection method consists of the following components:

6.1.1 *A midget impinger.* Containing the absorbing solution or reagent.

6.1.2 *Battery operated personal sampling pump—MSA Model G, or equivalent.* The sampling pump is protected from splashover or water condensation by an adsorption tube loosely packed with a plug of glass wool and inserted between the exit arm of the impinger and the pump.

6.1.3 *An integrating volume meter such as a dry gas or wet test meter.*

6.1.4 *Thermometer.*

6.1.5 *Manometer.*

6.1.6 *Stopwatch.*

6.1.7 *Various clips, tubing, spring connectors, and belt,* for connecting sampling apparatus to worker being sampled.

6.2 BECKMAN MODEL B SPECTROPHOTOMETER, OR EQUIVALENT.

6.3 CELLS, 5-CM MATCHED QUARTZ CELLS.

6.4 VOLUMETRIC FLASKS. Several of each: 100-ml and 1-l.

6.5 BALANCE, capable of weighing to at least three places, preferably four places.

6.6 PIPETS. Delivery: 0.5-, 1-, 2-, 5-, 10-, 15-ml; graduated: 2 ml.

6.7 GRADUATED CYLINDERS; 50-, 100-ML

7. Reagents

7.1 PURITY. All reagents must be made using ACS reagent grade or a better grade.

7.2 DOUBLE DISTILLED WATER. (See Preparation of Reagent Water, **3**)

7.3 SODIUM NITRITE.

7.4 SODIUM BROMIDE.

7.5 SULFAMIC ACID.

7.6 CONCENTRATED HYDROCHLORIC ACID, 11.7 N.

7.7 GLACIAL ACETIC ACID, 17.6 N.

7.8 N-(1-NAPHTHYL)-ETHYLENEDIAMINE DIHYDROCHLORIDE.

7.9 SODIUM CARBONATE.

7.10 METHYLENE DIANILINE (MDA).

7.11 SODIUM NITRITE–SODIUM BROMIDE SOLUTION. Dissolve 3.0 g sodium nitrite and 5.0 g sodium bromide in double distilled water and dilute to 100 ml. The solution may be stored in the refrigerator for one week.

7.12 SULFAMIC ACID SOLUTION. Dissolve 10.0 g sulfamic acid in 90 ml double distilled water.

7.13 ABSORBING SOLUTION. Add 35 ml conc hydrochloric acid and 22 ml glacial acetic acid to about 600 ml double distilled water and dilute to 1 l with double distilled water.

7.14 Coupling Solution. Dissolve 1.0 g N-(1-Naphthyl)-ethylenediamine dihydrochloride in 50 ml water, add 2 ml conc hydrochloric acid and dilute to 100 ml with double distilled water. This solution is stable for about 10 days.

7.15 Sodium Carbonate. Dissolve 16.0 g sodium carbonate in double distilled water and dilute to 100 ml.

7.16 Solution A. Dissolve 0.3000 g MDA in 700 ml glacial acetic acid. Dilute to 1 l with double distilled water.

7.17 Solution B. Immediately after making solution A, pipet 10 ml solution A into a 1-l volumetric flask. Add 35 ml conc hydrochloric acid, 15 ml glacial acetic acid, dilute to volume with double distilled water. This solution is equivalent to 3 μg MDI/ml.

8. Procedure

8.1 Cleaning of Equipment.

8.1.1 Wash all glassware in hot detergent solution, such as Alconox, to remove any oil.

8.1.2 Rinse thoroughly with hot tap water.

8.1.3 Rinse thoroughly with double distilled water. Repeat this rinse several times.

8.2 Collection and Shipping of Samples.

8.2.1 Pipet 15 ml of the absorbing solution (Section 7) into the midget impinger.

8.2.2 Connect the impinger (via the absorption tube) to the personal sampling pump with a short piece of flexible tubing. The minimum amount of tubing should be used between the sampling zone and impinger. The air being sampled should not be passed through any other tubing or other equipment before entering the impinger.

8.2.3 Turn on pump to begin sample collection. Care should be taken to measure the flow rate, time and/or volume as accurately as possible. Record atmospheric pressure and temperature. The sample should be taken at a flow rate of 1 lpm. Sample for 20 min making the final volume 20.1.

8.2.4 After sampling, the impinger stem can be removed and cleaned. Tap the stem gently against the inside wall of the impinger bottle to recover as much of the sampling solution as possible. Wash the stem with a small amount (1 to 2 ml) of unused absorbing solution and add the wash to the impinger. Seal the impinger with a hard, non-reactive stopper (preferably Teflon or glass). Do not seal with rubber. The stoppers on the impingers should be tightly sealed to prevent leakage during shipping. If it is preferred to ship the impingers with the stems in, the outlets of the stem should be sealed with Parafilm or other non-rubber covers, and the ground glass joints should be sealed (*i.e.*, taped) to secure the top tightly.

8.2.5 Care should be taken to minimize spillage or loss by evaporation at all times. Refrigerate samples if analysis cannot be done within a day.

8.2.6 Whenever possible, hand delivery of the samples is recommended. Otherwise, special impinger shipping cases designed by NIOSH should be used to ship the samples.

8.2.7 A "blank" impinger should be handled as the other samples (fill, seal and transport) except that no air is sampled through this impinger.

8.3 Analysis of Samples.

8.3.1 Remove the bubbler tube, if it is still attached, taking care not to remove any absorbing solution.

8.3.2 Start reagent blank at this point by adding 15 ml fresh absorbing solution to a clean bubbler tube.

8.3.3 To each tube, including the blanks, add 0.5 ml sodium nitrite solution, stir well, and allow to stand for 2 min.

8.3.4 Add 1 ml 10% sulfamic acid solution, to each tube, stir for 30 sec, and allow to stand for 2 min.

8.3.5 Add 1.5 ml sodium carbonate solution to each tube and stir.

8.3.6 Add 1 ml coupling solution to each tube, make up to volume of 20.0 ml with double distilled water and stir. Allow color to develop for 15 to 30 min.

8.3.7 Transfer each solution to a 5-cm quartz cell.

8.3.8 Using the blank, adjust the spectrophotometer to 0 absorbance at 555 nm.

8.3.9 Determine the absorbance of each sample at 555 nm.

9. Calibration and Standards

9.1 To a series of 5 impinger tubes, add the following amounts of absorbing solution: 15.0, 14.5, 14.0, 13.0, 10.0 ml respectively.

9.2 To each tube add standard solution B in the same order as the absorbing solution was added: 0.0, 0.5, 1.0, 2.0, 5.0 ml, so that the final volume is 15 ml (*i.e.*, 0.0 ml of standard is added to the 15 ml absorbing solution; 0.5 ml of standard is added to the 14.5 ml absorber solution, etc.). The cylinders now contain 0.0, 1.5, 3.0, 6.0, 15 μg MDI, respectively. The standard containing 0.0 ml standard solution is a blank.

9.3 To each tube, add 0.5 ml sodium nitrite solution, stir well and allow to stand for 2 min.

9.4 Add 1 ml of 10% sulfamic acid solution, stir for 30 sec, and allow to stand for 2 min.

9.5 Add 1.5 ml sodium carbonate solution and stir.

9.6 Add 1 ml coupling solution, make up to volume of 20.0 ml with double distilled water and stir. Allow color to develop for 15 to 30 min.

9.7 Transfer each solution to a 5-cm quartz cell.

9.8 Using the blank, adjust the spectrophotometer to 0 absorbance at 555 nm.

9.9 Determine the absorbance of each standard at 555 nm.

9.10 A standard curve is constructed by plotting the absorbance against micrograms MDI.

10. Calculations

10.1 Blank values (Section **8.2.7**), if any, should first be subtracted from each sample.

10.2 From the calibration curve (Section **9.10**), read the μg MDI corresponding to the absorbance of the sample.

10.3 Calculate the concentration of MDI in the air sampled in ppm, defined as μl MDI per liter of air.

$$\text{ppm} = \frac{\mu g}{V_s} \times \frac{24.45}{MW}$$

where:

ppm = parts per million MDI
μg = micrograms MDI (Section **10.2**)
V_s = corrected volume of air (Section **10.4**)
24.45 = molar volume of an ideal gas at 25 C and 760 Torr.
MW = molecular weight of MDI, 250.27.

10.4 Correct the volume of air sampled to standard conditions of 25 C and 760 Torr.

$$V_s = V \times \frac{P}{760} \times \frac{298}{T + 273}$$

where:

V_s = volume of air in liters at 25 C and 760 Torr.
V = volume of air in liters as measured.
P = Barometric Pressure in Torr.
T = Temperature of air in degree C.

11. References

1. Grim, K.E., and A.L. Linch. 1964. Recent Isocyanate-in-Air Analysis Studies, *American Industrial Hygiene Association Journal*, 25:285.
2. Marcali, K. 1957. Microdetermination of Toluene diisocyanates in Atmosphere, *Anal. Chem.* 4:552.
3. American Public Health Association, Inc. 1977. Methods of Air Sampling and Analysis. 2nd ed. p. 53. Washington, D.C.

Revised by Subcommittee 3
B. Dimitriades, *Chairman*
J. Cuddeback
E. L. Kothny
P. W. McDaniel
L. A. Ripperton
A. Sabadell

832.

Analytical Method for Nitroglycerin and Ethylene Glycol Dinitrate (Nitroglycol) in Air

Analyte:	Nitroglycerin (NG) and Ethylene Glycol Dinitrate (EGDN)	Method No.:	P & CAM 203
Matrix:	Air	Range:	NG, 1.0 μg to 1 mg/sample EGDN, 0.1 μg to 1 mg/sample
Procedure:	Sorption on Tenax-GC desorption with ethanol, gas chromatographic determination	Precision:	Analytical method, $2\sigma = \pm 6\%$; Overall sampling and analytical method, estimated, $2\sigma = \pm 20\%$
Date Issued:	9/6/74	Classification:	E (Proposed)
Date Revised:	1/1/75		

1. Principle of the Method

1.1 A known volume of air is drawn through a glass tube packed with Tenax-GC (a porous organic polymer) to collect the vapors of the nitrate esters.

1.2 The sorbent is transferred to a small test tube and extracted with ethanol.

1.3 An aliquot of the ethanol solution is injected into a gas chromatograph with an electron-capture detector.

1.4 The areas of the resulting peaks are measured and the amount of each nitrate ester is determined from calibration curves. The concentration of the nitrate ester in the air sample is calculated from the amount found by GC, the size of the aliquot, and the volume of air sampled.

2. Range and Sensitivity

2.1 The minimum amount of EGDN detectable by the overall procedure as described is substantially smaller than 0.1 μg. This sensitivity easily permits the detection of EGDN in a 10-l air sample containing 0.01 mg/m^3. The minimum amount of NG detectable by the overall procedure is smaller than 1 μg. This sensitivity permits the detection of NG in a 100-liter air sample containing 0.01 mg/m^3.

2.2 The sorbent tubes will retain at least 1 mg of either EGDN or NG.

3. Interferences

3.1 Ethylene glycol mononitrate, if present in a high concentration relative to that of EGDN, produces a GC peak that tails into the EGDN peak. This interference, although troublesome, does not prohibit a reasonably satisfactory analysis for EGDN. The combined selectivity of Tenax-GC and electron-capture detection practically eliminates other interference. Water vapor does not interfere.

4. Precision and Accuracy

4.1 The precision (2σ) of the analytical method is estimated to be about $\pm$ 6%.

4.2 The precision of the overall sampling and analytical method has not been adequately determined. However, since both EGDN and NG are sorbed on Tenax-GC at ambient temperature with 100% efficiency, it is believed that the overall precision will be determined by variations in the measurement of sample volume.

4.3 In laboratory measurements not involving a personal sampling pump, the average recovery of EGDN from a vapor sample was 104% and that of NG was 102%.

5. Advantages and Disadvantages of the Method

5.1 The general technique of sampling and analysis is similar to that already established for organic solvents in air in which a charcoal tube and gas chromatography are used (P & CAM 127). The advantages of that method also apply to the present method.

5.2 The principal difficulties are those associated with the use of electron-capture detection. They include the limited linear range of electron-capture detectors and operating variables such as the effects of column bleed, moisture, and oxygen. The effects of these variables may be minimized by analyzing standards at the same time as samples. The electron-capture detector is about 500 times as sensitive to the nitrate esters as a hydrogen-flame ionization detector, and it provides a substantial advantage in specificity.

5.3 The principal limitation on precision and accuracy in this method is that associated with the use of small personal sampling pumps for the measurement of sample volume.

5.4 The range of sample size is very wide. However, repeated dilutions of the ethanol extract may be necessary to bring the amount of sample injected into the proper range for the electron-capture detector.

6. Apparatus

6.1 An approved and properly calibrated personal sampling pump is required.

6.2 The sorbent tubes are 70-mm long and 5-mm ID. They contain two sections of 35/60 mesh Tenax-GC separated and held in place by glass-wool plugs. The front section contains 100 mg of sorbent and the backup section contains 50 mg. Since the pressure drop must be limited to 1 in. of Hg at 1.0 l/min, it is necessary to avoid overpacking with glass wool. Limited laboratory evaluation indicates that 40/60-mesh Porapak Q may be substituted for the Tenax-GC. (Tenax-GC is distributed by Applied Science Laboratories, Inc., State College, Pa., and Porapak Q by Waters Associates, Inc., Milford, Mass.)

6.3 A gas chromatograph equipped with an electron-capture detector, a 2.5-ft by 0.25-in. glass column packed with 10% of OV-17 on 60/80-mesh Gas Chrom Q, and an integrator are required.

6.4 A volumetric flask and pipets, a 10-μl syringe, and other normally available laboratory supplies and equipment are needed.

7. Reagents

7.1 The only reagent required is absolute ethanol for extraction of the sorbent. The gas chromatograph requires a supply of helium and an argon-methane mixture, or other carrier or purge gases as required for the particular instrument used.

8. Procedure

8.1 CLEANING OF EQUIPMENT. All glassware for the laboratory analysis should be washed with detergent and thoroughly rinsed with tap water and distilled water. Particular attention should be paid to the cleaning of the microliter syringe with ethanol.

8.2 CALIBRATION OF PERSONAL SAMPLING PUMP. Each pump should be calibrated with a Tenax-GC sorbent tube in the line. This will minimize errors associated with uncertainties in the volume of sample collected.

8.3 COLLECTION AND SHIPPING OF SAMPLES.

8.3.1 Immediately before sampling, break the ends of the tube to provide an opening of at least 2 mm.

8.3.2 The smaller section of sorbent

is used as a backup and should be positioned nearest the sampling pump.

8.3.3 The sorbent tube should be placed in a vertical direction during sampling.

8.3.4 Air being sampled should not be passed through any hose or tube before entering the charcoal tube. This is particularly important with the nitrate esters since they are strongly absorbed on most surfaces, including glass.

8.3.5 The flow, time, and/or volume must be measured as accurately as possible. The sample should be taken at a flow rate of 1.0 lpm or less to attain the total sample volume required. The minimum volume that must be collected to permit detection of NG at the TLV concentration (or EGDN at a still lower concentration) is only 0.1 liter. At the TLV concentration, sample volumes as large as 1000 liters can be collected without causing trouble in the analysis.

8.3.6 The temperature and pressure of the atmosphere being sampled should be measured and recorded.

8.3.7 The Tenax-GC tubes should be capped with the supplied plastic caps immediately after sampling. Under no circumstances should rubber caps be used.

8.3.8 One tube should be handled in the same manner as the sample tube (break, seal, and transport), except that no air is sampled through this tube. This tube should be labeled as a blank.

8.3.9 Capped tubes should be packed tightly before they are shipped to minimize tube breakage during shipping.

8.4 ANALYSIS OF SAMPLES.

8.4.1 The sorbent tube is scored with a file near the front end and is broken open. The first glass-wool plug and the front (100-mg) section of sorbent are transferred to a small vial or test tube. The remaining glass-wool plug and the backup section of sorbent are transferred to another small vial. Two ml of ethanol are added to each vial, the vials are stoppered, and the contents are shaken for about 1 min. Aliquots of the ethanol solutions are injected into the gas chromatograph. It is convenient to take a 5-μl aliquot, but it may be necessary to dilute the ethanol solution and take another aliquot if the concentrations of the nitrate esters are high.

8.4.2 Typical operating conditions for the gas chromatographic analysis are:

Column: 10% OV-17 on 60/80-mesh Gas Chrom Q, 2.5-ft by 0.25-in. glass column
Carrier gas: Helium, 100 ml/min
Purge gas: Argon/methane, 125 ml/min
Column Temperature: 130 C
Injection port temperature: 160 C
Detector temperature: 280 C

9. Calibration and standards

9.1 The use of pure NG and pure EGDN as reference standards is considered to be impractical for general application because of the safety hazard and the lack of availability of the compounds in pure form. In the course of the development of this method, several reference standards were obtained from the quality control laboratories of manufacturers of nitrate ester products. These materials included the following:

(1) A material called "Nitroglycerin Lactose Trituration" was provided by Eli Lilly and Company, Indianapolis, Indiana. This material is used for quality control of sublingual nitroglycerin tablets that are manufactured for medicinal use and has a stated nitroglycerin content. It consists of nitroglycerin dispersed in an inert, ethanol-soluble powder. To prepare a GC calibration curve, a weighed amount of the material is dissolved in a known volume of ethanol and appropriate aliquots are injected into the gas chromatograph.

(2) Solutions of known concentrations of NG and EGDN were provided by Hercules, Inc. These solutions were quantitatively prepared from analyzed mixtures of the compounds and are used for plant quality control in the manufacture of explosive products and in industrial hygiene applications. Aliquots of these solutions, appropriately diluted, may be injected directly into the gas chromatograph. Department of Transportation regulations may limit the shipment of such solutions by public transportation.

(3) A sample of analyzed dynamite containing EGDN but no NG was provided by Atlas Chemical Industries, ICI America, Inc. A weighed quantity of the dynamite is dispersed in ethanol, filtered, washed with ethanol, and diluted to a known volume. Aliquots of this solution are injected. Since dynamite is a heterogenous material, care must be taken to obtain a representative sample.

(4) The pharmaceutical-grade sublingual NG tablets may be obtained from wholesale drug dealers. These tablets are manufactured to meet United States Pharmacopeia specifications, which require that the actual NG content be between 80 and 112% of the stated volume. The actual NG content of a specific lot of tablets may be obtained from the manufacturer. For use, the tablets are weighed, crushed, and extracted with ethanol, the solution is filtered and diluted to a fixed volume, and aliquots are injected.

An experimental comparison of the NG and EGDN contents of these standard materials gave a relative standard deviation of 3.6% for the NG contents and 1.6% for the EGDN contents.

It is necessary that standards be analyzed concurrently with samples to minimize the effects of variations in detector response and other factors.

10. Calculations

10.1 From the measured peak areas for NG and EGDN and the calibration curve, determine the amounts of each compound in each section of the sample sorbent tube (usually, no NG or EGDN will be found in the backup section). Subtract from these values the amounts found in the corresponding sections of the blank solvent tube.

10.2 Add the corrected amounts found in the two sections of the sample tube to determine the total amount of each nitrate ester.

10.3 Calculate the concentration of each compound in the air sample by the following equation:

$$\text{Concentration in mg/m}^3 \text{ (or } \mu\text{g/l)} = \frac{\text{Total amount of compound in micrograms}}{\text{volume of air sampled in liters } (V_s)}$$

The volume of air sampled is corrected to standard conditions (defined as 25 C and 760 Torr) by the equation:

$$V_s = V \times \frac{P}{760} \times \frac{298}{T + 273}$$

where:

Vs = corrected volume of air sampled
V = volume of air in liters, as measured
P = pressure in mm Hg
T = temperature in °C

10.4 Concentrations in mg/m³ may be converted to parts per million (ppm) by volume by the following equations:

For NG, $\text{ppm} = \text{mg/m}^3 \times 0.108$
For EGDN, $\text{ppm} = \text{mg/m}^3 \times 0.161$

11. Reference

1. SOUTHERN RESEARCH INSTITUTE. 1974. Contract No. HSM-99-73-63, Final Report, September.

Subcommittee 3
B. DIMITRIADES, *Chairman*
J. CUDDEBACK
E. L. KOTHNY
P. W. McDANIEL
L. A. RIPPERTON
A. SABADELL

833.

Analytical Method for N-Nitrosodimethylamine in Ambient Air

Analyte:	DMN	Method No:	
Matrix:	Air	Range:	0.5 ppt–10 ppb
Procedure:	Adsorption on Tenax GC, thermal desorption with He purge, measurement by capillary gas-liquid chromatography/mass spectrometry	Precision:	±10%
		Classification:	E (Proposed)
Date Issued:			
Date Revised:			

1. Principle of the Method

N-nitrosodimethylamine (DMN) is concentrated from ambient air on Tenax GC in a short glass tube (**1, 2**). It is desorbed by heating and purging with helium into a liquid nitrogen cooled nickel capillary trap and then introduced onto a high resolution gas chromatographic column where it is separated from interferences. The concentration of DMN is measured from the mass spectrometric signal at m/e 74 (**3**).

2. Range and Sensitivity

2.1 The range of the mass spectrometric signal for the conditions listed corresponds to 0.5 ppt to 10 ppb.

2.2 A concentration of 0.5 ppt of DMN can be determined in a 150-l air sample.

3. Interferences

Interferences may result from materials having background ions of *m/e* 74 ($C_2H_8N_3$, $C_2H_4NO_2$, $C_2H_6N_2O$, C_3H_3Cl, C_3H_6S, $C_3H_6O_2$, or $C_3H_{10}N_2$) if at the same retention time of DMN.

4. Precision and Accuracy

4.1 The precision of this method has been determined to be ±10% of relative standard deviation when replicate sampling cartridges were spiked with 50 ng (corresponding to 10 ppb in 150 ℓ of air). These data were obtained using 10.0-cm long glass tubes (1.5-cm ID) packed with 35/60 mesh of Tenax GC (bed dimensions: 1.5-cm × 6-cm in depth).

4.2 The accuracy of the analysis is approximately ±10% of the amount reported as determined from repeated analysis of several standards.

5. Advantages and Disadvantages of the Method

5.1 The gas chromatography-mass spectrometry technique interfaced with a Finnigan glass jet separator (Model 01512-42158 Finnigan Corp., Sunnyvale, CA) is extremely sensitive and specific for the analysis of DMN. The high resolution gas chromatographic separation yields a retention time that is characteristic for DMN, and relatively specific for positive assignment of the signal as DMN. The mass spectrometer in combination with high resolution gas chromatography yields a very high degree of specificity. The base peak of DMN is at *m/e* 74 which is also the parent ion. In order to assign the signal at *m/e* 74 to DMN it is absolutely necessary that the retention time matches with the signal.

5.2 Collected samples can be stored up to 1 month with less than 10% losses.

5.3 Because DMN is a suspected carcinogen in man it is extremely important to exercise safety precautions in the preparation and disposal of liquid and gas standards, cleaning of used glassware, etc., and the analysis of air samples.

5.4 Since the mass spectrometer can not be conveniently mobilized, sampling must be carried out away from the instrument.

5.5 High resolution gas chromatography/low resolution mass spectrometry is not a convenient technique for handling a large number of samples (> 100/wk).

5.6 Efficiency of air sampling increases as the ambient air temperature decreases (*i.e.* sensitivity increases).

5.7 Ambient air sampling is limited to cases where the NO_x levels are less than 3 ppm when dimethylamine is also present.

6. Apparatus

6.1 SAMPLING TUBES.

6.1.1 The sampling tubes are prepared by packing a 10-cm long × 1.5-cm ID glass tube with 6.0 cm of 35/60 mesh Tenax GC with glass wool in the ends. Cartridge samplers are conditioned at 270 C with helium flow at 30 ml/min for 20 min. The conditioned cartridges are transferred to Kimax (2.5-cm × 150-cm) culture tubes, immediately sealed using teflon-lined caps, and cooled.

6.1.2 Cartridge samplers with longer beds of sorbent may be prepared using a proportional amount of Tenax GC.

6.2 GAS CHROMATOGRAPHIC COLUMN.

6.2.1 A 0.35-mm ID × 50-m glass SCOT capillary coated with DEGS stationary phase and 0.1% benzyl triphenylphosphonium chloride is used. The capillary column is conditioned (detector end disconnected) for 48 hr at 210 C @ 1.5 to 2.0 ml/min helium flow.

6.3 A Finnigan type glass jet separator on a magnetic or quadropole instrument is used at 200 C.

6.4 INLET-MANIFOLD.

6.4.1 An inlet-manifold is fabricated and employed (Figure 833:1, **1, 2, 4**).

6.5 GAS CHROMATOGRAPH.

6.5.1 A Varian 1700 gas chromatograph or equivalent. *A gas chromatograph employing a single column oven and a temperature programmer is adequate.*

6.6 MASS SPECTROMETER.

6.6.1 A mass spectrometer with a resolution of 500 to 2000 equipped with single ion monitoring capabilities must be used in conjunction with a gas chromatograph. A Varian-MAT CH-7 has been found to be satisfactory for this purpose (**2, 3**).

6.7 SYRINGES.

6.7.1 *Syringes,* I-ml gas tight (Precision Sampling, Inc.) and 10 μl (The Hamilton Co., Inc.).

7. Reagents and Materials

All reagents must be analytical reagent grade.

7.1 N-NITROSODIMETHYLAMINE

7.2 ACETONE

7.3 ISOCLEAN

7.4 TENAX GC (35/60 mesh, Applied Science)

7.5 TWO 2-L ROUND BOTTOM FLASKS, fitted with injection ports.

7.6 SOXHLET APPARATUS

8. Procedure

8.1 CLEANING OF GLASSWARE. All glassware, glass sampling tubes, cartridge holders, etc., should be washed in Isoclean/water, rinsed with doubly distilled water and acetone and air dried. Glassware is heated to 450 F for 2 hr.

8.2 PREPARATION OF TENAX GC.

8.2.1 Virgin Tenax GC is extracted in a Soxhlet apparatus overnight with acetone prior to its use.

8.3 COLLECTION OF DMN IN AMBIENT AIR.

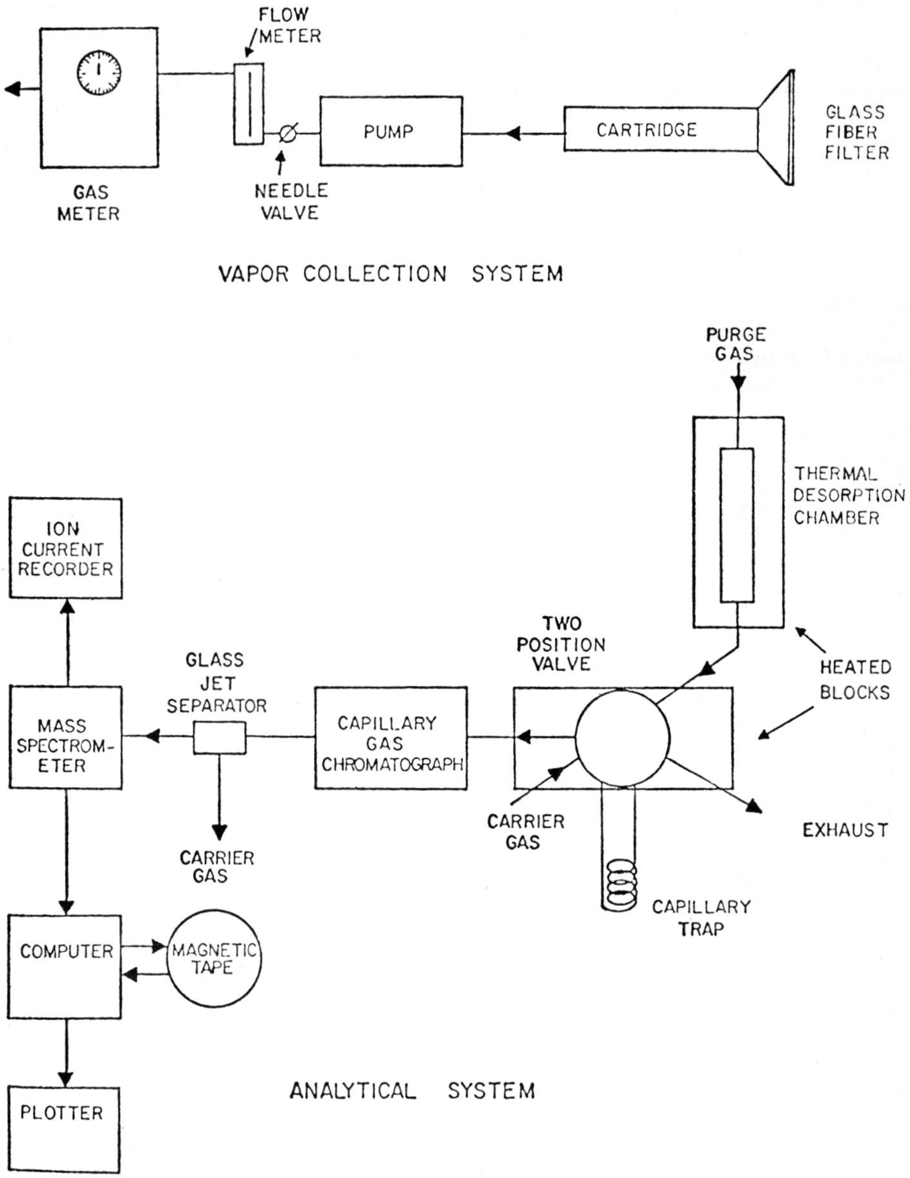

Figure 833:1. Vapor collection and analytical systems for analysis of hazardous vapors in ambient air.

8.3.1 Continuous sampling of ambient air may be accomplished using a Nutech Model 221-A portable sampler (Nutec Corp., Durham, NC) or its equivalent (**2**). Flow rates are adjusted with a metering valve through a calibrated rotameter. Total flow is registered by a dry gas meter.

8.3.2 For larger sample sizes it is important to realize that a larger total volume of air may cause elution of DMN through the sampling tube. It has been demonstrated that exceeding a total of 385, 332, 280, 242, 224, 204, 163, 156, 148, 127, 107, 93, or 79 l of air at ambient temperatures of 50, 55, 60, 65, 70, 75, 80, 85, 90, 95, 100, 105, or 110 F, respectively will result in elution of DMN from the cartridge sampler. A flow of 10 cc/min to 30 l/min may be used with the sampler described in **6.1**.

8.3.3 DMN has been found to be stable and quantitatively recoverable from cartridge samplers after 4 weeks when tightly closed in cartridge holders, protected from light and stored at 0 C.

8.4 Analysis of Sample.

8.4.1 *Instrument Conditions and Setup*. The thermal desorption chamber and six-port valve are set to 200 C. The glass jet separator is maintained at 200 C. The mass spectrometer is set to monitor *m/e* 74 (Figure 833:2).

8.4.2 Adjust the He purge gas through the desorption chamber to 50 ml/min. Cool the Ni capillary trap at the inlet manifold with liquid nitrogen.

8.4.3 Place the cartridge sampler in the desorption chamber and desorb for 5 min.

8.4.4 Rotate the six-port valve on the inlet-manifold to position "B", heat the Ni capillary trap to 180 C with a wax bath.

8.4.5 Temperature program the glass capillary column from 75 to 205 C at 4 C/min and hold at upper limit for 10 min. The retention time of DMN is approximately 26 min (Figure 833:3).

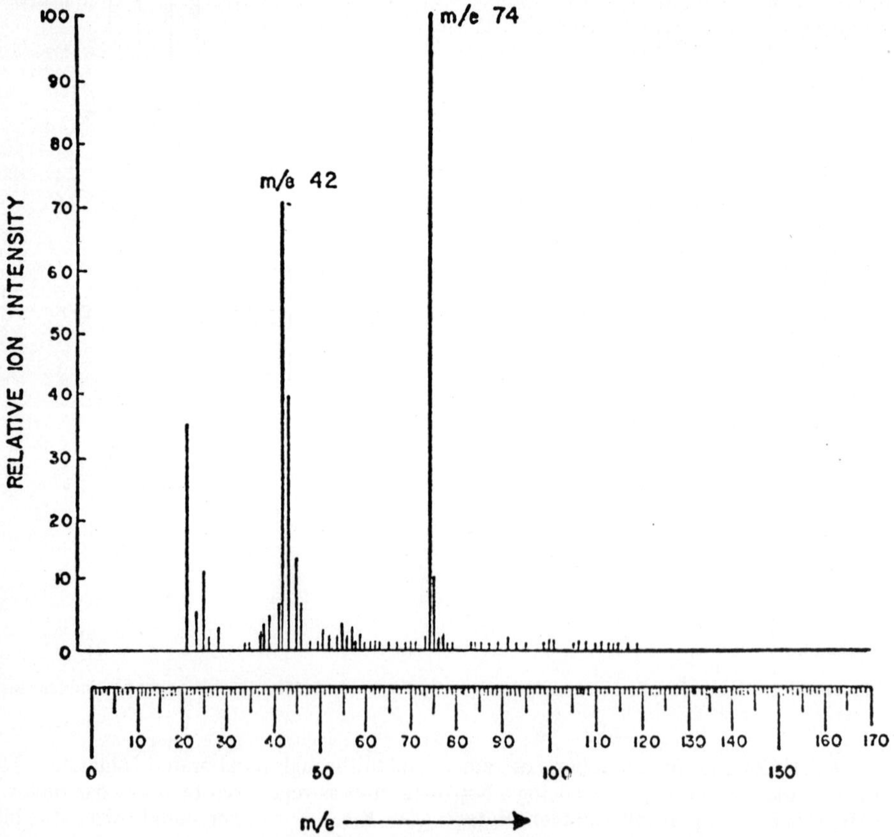

Figure 833:2. Mass spectrum of N-nitrosodimethylamine.

8.4.6 The analytical column is cooled to ambient temperature and the next sample is processed.

9. Calibration and Standards

9.1 PREPARATION OF GAS STANDARD.

9.1.1 Purge two 2-l round bottom flasks with helium, warm flasks to 50 C with heating mantels and use magnetic bar to stir vapors.

9.1.2 Inject 0.1 to 1 μl of DMN into flask and let stir for 30 min. Make further dilutions into second flask by transferring milliliter gas volumes as needed.

9.1.3 Purge air/vapor mixtures from second flask onto cartridge samplers.

9.2 CALIBRATION.

9.2.1 Prepare standard curve (with ten concentration points) by thermally desorbing cartridge samplers loaded with 3 ng to 30 μl of DMN. Plot *m/e* 74 response *vs* ng of DMN. A linear response is observed.

10. Calculations

10.1 The total quantity of DMN in ambient air is determined by comparing *m/e* 74 response for samples of DMN with standard curve.

$$\text{ppb} = \frac{\text{ng DMN}}{\text{V}} \times \frac{24.45}{74}$$

where:

ng DMN = total ng concentration as determined in **9.2.1**

V ~ volume of air in liters sampled at 25 C and 760 Torr

24.45 = molar volume of an ideal gas at 25 C and 760 Torr

MW ~ molecular weight of DMN, 74.

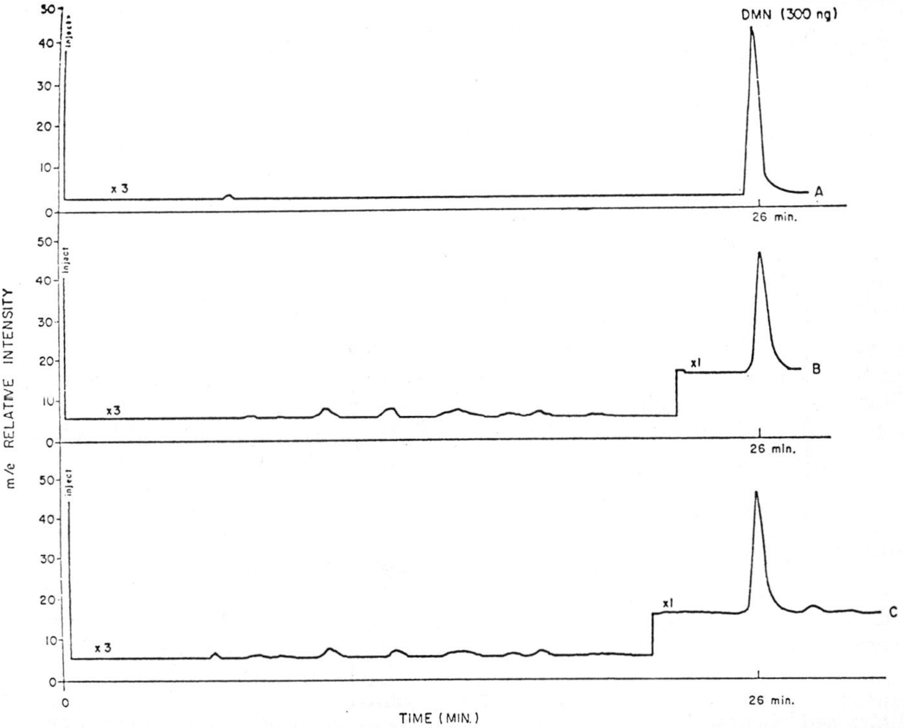

Figure 833:3—Mass (*m/e* 74) Chromatograms. A = standard DMN; B,C = replicate air samples.

11. References

1. PELLIZZARI, E. 1974. Development of Method for Carcinogenic Vapor Analysis in Ambient Atmosphere, Pub. No. EPA-650/2-74-121, 148pp.
2. PELLIZZARI, E. 1975. Development of Analytical Techniques for Measuring Ambient Atmospheric Carcinogenic Vapors. Pub. No. EPA-600/2-75-076, 187pp.
3. PELLIZZARI, E., J. E. BUNCH, R. E. BERKLEY and J. MCCRAE. Biomedical Mass Spec, submitted.
4. PELLIZZARI, E. D., B. H. CARPENTER, J. E. BUNCH, and E. SAWICKI. 1975. Environ. Sci. Tech., 9:556.

Subcommittee 5
E. Sawicki, *Chairman*
T. BELSKY
R. A. FRIEDEL
D. L. HYDE
J. L. MONKMAN
R. A. RASMUSSEN
L. A. RIPPERTON
L. D. WHITE

834.

Analytical Method for Organic Solvent Vapors in Air

Analyte:	Organic Solvents (See Tables 834:I and 834:II)
Matrix:	Air
Procedure:	Adsorption on charcoal, desorption with carbon disulfide G.C.
Date Issued:	9/15/72
Date Revised:	1/1/75
Method No.:	P & CAM 127
Range:	Refer to Tables 834:I and 834:II
Precision:	± 6% to ± 17% relative standard deviation
Classification:	See Table 834:I

1. Principle of the Method

1.1 A known volume of air is drawn through a charcoal adsorption tube to trap the organic solvent vapors present (**4, 5**).

1.2 The charcoal is transferred to a small stoppered tube and the organic compounds are desorbed with carbon disulfide (**5, 9**).

1.3 An aliquot of the desorbed sample is injected into a gas chromatograph (G.C.) equipped with a flame ionization detector (FID).

1.4 The area of the resulting peak(s) is (are) measured and compared with areas obtained from the injection of standards.

2. Range and Sensitivity

2.1 The approximate breakthrough capacity in milligrams (mg) of selected compounds per 100 mg of activated charcoal is given in Table 834:I. This value is the approximate number of milligrams of the compound which the front section will hold before a significant amount of compound is found on the backup section (The charcoal tube consists of two sections of activated charcoal separated by a section of urethane foam. (See Section **6.2.**). If a particular atmosphere is suspected of containing a large amount of contaminant, a smaller sampling volume should be taken.

Table 834:I. Selected Parameters

Organic Solvent	Method Classification	Approximate Breakthrough Capacity* † (mg/100 mg charcoal)	Suggested Sample† Volume (liters)	Federal Standard‖ mg/m³
Benzene	A	6	10	32
2-Butanone	B	9	10	590
1,2-Dichloroethane	A	—	10§	200
Dichloromethane	B	4.5‡	4‡	890
1,4-Dioxane	A	—	10§	360
Styrene	D	15	10	420
Tetrachloroethane	B	21	10	670
Tetrachloromethane	A	—	10§	65
Toluene	B	28‡	10‡	750
1,1,1-Trichloroethane	B	11	5	1900
1,1,2-Trichloroethane	B	10	10	45
Trichloromethane	A	—	10§	240
Trichloroethane	A	13	10§	535
Xylene	A	—	10§	435
(1,2,-dimethylbenzene, 1,3-dimethylbenzene, 1,4-dimethylbenzene) mixture				

*See Section 2.1.

†Based on breakthrough experiments with activated petroleum charcoal, unless otherwise stated.

‡Based on breakthrough experiments with activated coconut charcoal.

§Estimated.

‖Reference 3.

2.2 The lowest detectable limit in mg/sample for specific compounds using this method has not been determined. Values as low as 0.001 mg/sample (about 25 ppb in an integrated 10-l air sample) were reported during collaborative testing of this method **(1)**. It was also reported that when the weight of the compound on the charcoal section was less than 0.1 mg, analytical errors were substantially higher (greater than ± 20%). Therefore, if a contaminant is present at relatively low concentrations in the industrial atmosphere, a large volume of air should be sampled.

3. Interferences

3.1 When the amount of water vapor or mist in the air is so great that condensation occurs in the sampling tube, organic vapors will not be trapped. Preliminary experiments indicate that high temperatures, high humidity, and high sampling flow rates cause a decrease in the adsorption capacity of activated carbon for organic solvent vapors.

3.2 When two or more solvents are known or suspected to be present in the air, such information, including their suspected identities, should be transmitted with the sample, since, with differences in polarity, one may displace another from the charcoal.

3.3 It must be emphasized that any compound which has the same retention time as the specific compound under study at the operating conditions described in this method is an interference. Hence, retention time data on a single column, or even on a number of columns, cannot be considered as proof of chemical identity. For this reason it is important that a sample of the bulk solvent(s) be submitted

Table 834:II. Chemicals Which Have Greater than 80% Desorption Efficiency but Have not Been Thoroughly Tested by NIOSH

Class E	
Acetic Acid, ethyl ester	Isophorone
Allyl glycidyl ether	Isopropyl glycidyl ether
Butanoic acid, ethyl ester	
Butanol	3-Methylbutanoic acid, ethyl ester
2-Butoxyethanol	4-Methyl-2-pentanone
Butyl glycidyl ether	2-Methylpropanoic acid, ethyl ester
Chlorobenzene	3-Methyl-1-propanol
2-Chloro-1,3-epoxypropane	2-Methylpropanoic acid, methyl ester
Cyclohexane	α-Methyl styrene
Cyclohexanone	4-Methyl styrene
1,2-Dichlorobenzene	Octane
1,4-Dichlorobenzene	3-Octanone
Diethyl ether	Pentane
N,N-Dimethylaniline	2-Pentanone
2,6-Dimethylpyridine	Pentanoic acid, ethyl ester
2-Ethoxyacetic acid, ethyl ester	α-pinene
Ethylbenzene	Propanoic acid, ethyl ester
Formic acid, ethyl ester	2-Propanone
Furfural	Propenenitrile
Heptane	Propenoic acid, methyl ester
3-Heptanone	1,1,2,2-tetrachloroethane
Hexane	Tetrahydrofuran
2-Hexanone	1,1,2-Trichloro-1,2,2-trifluoroethane

at the same time so that identity(ties) can be established by other means (**2**).

3.4 If the possibility of interference exists, G.C. separation conditions (column packing, temperature, etc.) must be changed to circumvent the problem.

3.5 Activated charcoal cannot be stored, transported or handled in a contaminated atmosphere. The adsorbed compounds will contribute to the background and may prove to be an interference. The charcoal tubes should be flame sealed or otherwise sealed from the atmosphere.

4. Precision and Accuracy

4.1 Seven selected organic solvents (benzene, carbon tetrachloride, chloroform, p-dioxane, ethylene dichloride, trichloroethylene, and xylene) have been collaboratively tested (**1**). The precision of the sampling method using an approved personal sampling pump plus the analytical method varied from ± 6 to ± 17% relative standard deviation, depending upon the compound, mixture and concentration.

4.2 The accuracy of the sampling and analytical method is about ± 10% when the personal sampling pump is calibrated with a charcoal tube in the line. Accuracy is affected primarily by sampling error and the extent of desorption of each compound from the activated charcoal. Sampling precision can be improved by using a more powerful vacuum pump with associated gas-volume integrating equipment (*e.g.*, limiting orifice).

5. Advantages and Disadvantages of the Method

5.1 The sampling device is small, portable, and involves no liquids. The samples are analyzed by means of a relatively fast, instrumental method. The method is applicable for the sampling and analysis of a number of organic vapors. Compounds which have been determined by National Institute for Occupational Safety and Health (NIOSH) Laboratories using this method are listed in Tables 834:I and 834:II (**9**, unpublished data). A special analytical procedure has been developed for carbon disulfide (**6**), since it is used as the desorbant in this procedure.

5.2 The occupational exposure standards (**3**) established by authority of the Occupational Safety and Health Act (OSHA) of 1970 are classified as acceptable ceiling concentrations and as 8 hr time weighted averages. An acceptable grab sampling technique may be used to determine peak concentrations of chemicals for which a ceiling value has been specified. However, for those chemicals which have standards based on 8 hr time weighted average concentrations, a sampling procedure capable of collecting an integrated sample over a longer time span is required. The activated charcoal sampling tube is particularly suited for taking integrated solvent vapor samples. The Sipin pump (Section **6.2**) can be used to take an integrated charcoal tube sample over a time period of more than 3 hr (10-l sample).

5.3 Interferences are minimal, and most of those which do occur can be eliminated by altering G.C. operating parameters. The method can also be used for the simultaneous analysis of two or more solvents suspected to be present in the same sample by simply changing G.C. conditions from isothermal to a temperature-programmed mode of operation (**9**).

5.4 The greatest selectivity of activated charcoal is toward nonpolar organic solvent vapors. Organic compounds which are gaseous at room temperature, reactive, polar, or oxygenated (aldehydes, alcohols and some ketones) are either not adsorbed (relatively early breakthrough) or desorbed efficiently. Such compounds may require individual testing to determine if the charcoal tube may be satisfactorily employed for sampling.

5.5 A disadvantage of the method is that the amount of sample which can be collected is limited by the number of milligrams that the charcoal will adsorb before overloading. When the sample value obtained for the backup section of the charcoal trap exceeds 25% of that found on the front section, the possibility of sample loss exists.

5.6 The precision of the method is limited by the reproducibility of the pressure drop across the tubes. Variable pressure drops will affect the flow rate and cause the sampled volume to be imprecise, because the personal sampling pump is usually calibrated for one tube only.

5.7 The desorption efficiency of a particular compound can vary from one batch of charcoal to another. Whenever a new batch of charcoal is used, it is therefore necessary to determine at least once the percentage of each specific compound removed in the desorption process.

6. Apparatus

6.1 CHARCOAL TUBES. A glass tube with both ends flame sealed, 7-cm long with a 6-mm OD and a 4-mm ID containing 2 sections of 20/40 mesh activated charcoal separated by a 2-mm portion of urethane foam (**4**). The activated charcoal* is fired at 600 C with nitrogen purge for 1 hr prior to packing. The adsorbing section contains 100 mg of charcoal, the backup section 50 mg. A 3-mm portion of urethane foam is placed between the outlet end of the tube and the backup section. A plug of silylated glass wool is placed in front of the adsorbing section. The pressure drop across the tube must be less than one inch of mercury at a flow rate of 1 lpm. Such tubes are commercially available from SKC, Inc. and from Mine Safety Appliances Co., both of Pittsburgh, Pennsylvania.

6.2 AIR SAMPLING PUMP. An air or vacuum pump capable of drawing a constant air flow through a limiting orifice (1 lpm) is suitable for area samples. For personal samples, a calibrated battery-operated portable pump capable of aspirating an accurate and reproducible volume of air through a charcoal tube is required. A pocket-sized pump capable of accurately drawing from 50 to 200 cc of air per min through the charcoal tube is commercially available from Anatole J. Sipin Co., 385 Park Avenue South, New York City. OSHA approved coal-mine-dust personal sampling pumps are also recommended for collection of breathing zone samples.

*Activated charcoal is prepared from coconut shells.

6.3 GAS CHROMATOGRAPH (G.C.). Equipped with a flame ionization detector (FID) (**7, 9**).

6.4 GAS CHROMATOGRAPH COLUMN. Any column capable of separating the organic vapor mixture under study is acceptable. Stainless steel tubing is recommended, especially if reactive chemicals or high temperatures are involved. A stainless steel column with 10% FFAP stationary phase on 80/100 mesh AW-DMCS chromosorb W solid support has been used to separate fourteen selected organic solvents (**9**).

6.5 A strip chart recorder with a one mV and one second response is acceptable. A mechanical or electronic integrator is also recommended.

6.6 A 10-μl syringe is recommended for sample injection into the G.C., and convenient (10 to 100 μl) sizes for preparation of standards.

6.7 GLASS OR POLYETHYLENE STOPPERED TUBES. 2.5-ml graduated microcentrifuge tubes are recommended.

6.8 PIPETS. 0.5-ml delivery pipets or 1.0 ml type graduated in 0.1 ml increments.

6.9 VOLUMETRIC FLASKS. 10-ml or convenient sizes for making standard solutions.

6.10 A calibrated flowmeter is required to accurately measure the flow rate of the air drawn through the charcoal tube during calibration of the sampling pumps.

7. Reagents

7.1 PURITY. All organic compounds utilized to make G.C. standards should be of ACS analytical grade or spectrograde quality.

7.2 SPECTROQUALITY CARBON DISULFIDE.

7.3 CARRIER GAS. Compressed nitrogen, argon, or helium should be prepared from water pumped sources or be of guaranteed high purity.

7.4 HYDROGEN. High quality electrolytic generators which produce ultrapure hydrogen under pressure are recommended. They provide some degree of safety when compared to the explosive hazard of bottled hydrogen. All connections should be made with thoroughly cleaned stainless steel tubing and properly tested for leaks. If hydrogen in compressed gas cylinders is used, the hydrogen should be prepared from water pumped sources.

7.5 AIR. Filtered compressed air, low in hydrocarbon content.

8. Procedure

8.1 CLEANING OF GLASSWARE. All glassware used for the laboratory analysis should be detergent washed and thoroughly rinsed with tap water and distilled water.

8.2 CALIBRATION OF PERSONAL PUMPS. Each personal pump must be calibrated with a representative charcoal tube in the line. This will minimize errors associated with uncertainties in the sample volume collected.

8.3 COLLECTION AND SHIPPING OF SAMPLES.

8.3.1 Immediately before sampling, break the ends of the tube to provide an opening at least one-half the ID of the tube (2-mm).

8.3.2 The smaller section of charcoal is used as a back-up and should be positioned nearest the sampling pump.

8.3.3 The charcoal tube should be placed in a vertical direction during sampling.

8.3.4 Air being sampled should not be passed through any hose or tubing before entering the charcoal tube.

8.3.5 The flow, time, and/or volume must be measured as accurately as possible. The sample should be taken at a flow rate of 1 lpm or less. The suggested sample volume that should be collected for each solvent at its OSHA standard (**3**) or threshold limit value (**8**) is shown in Table 834:I.

8.3.6 An identification number is marked on the tube. The temperature and pressure of the atmosphere being sampled should be measured and recorded along with other appropriate field data (location of worker, distance from operation, type of operation, wind direction, etc.).

8.3.7 The charcoal tubes should be

capped with masking tape or plastic caps immediately after sampling. Under no circumstances should rubber caps be used. If the sample tube must be stored for more than a week, refrigeration is recommended (or the ends may be flame sealed).

8.3.8 One tube should be handled in the same manner as the sample tube (break, seal, and transport), except that no air is sampled through this tube. This tube should be labeled as a blank.

8.3.9 Capped tubes should be packed tightly before they are shipped to minimize tube breakage during shipping.

8.3.10 Samples of the suspected solvent(s) should be submitted to the laboratory (take appropriate safety precautions). These liquid bulk samples should **not** be transported in the same container as the samples or blank tube. If possible a bulk *air* sample (*i.e.*, a charcoal tube used to sample a *large* volume of air) should be shipped for qualitative identification purposes.

8.3.11 Due to the high resistance of the charcoal tube, this sampling method places a heavy load on the sampling pump. Therefore, no more than ten charcoal tube samples should be taken without fully recharging the battery.

8.4 Analysis of Samples.

8.4.1 *Preparation of Samples.* In preparation for analysis, each charcoal tube is scored with a file in front of the first section of charcoal and broken open. The glass wool is removed and discarded. The charcoal in the first (larger) section is transferred to a small stoppered tube. The separating section of foam is removed and discarded; the second section is transferred to another test tube. These two sections are analyzed separately.

8.4.2 *Desorption of Samples.* Prior to analysis, 0.5 ml of carbon disulfide is pipetted into each tube. All work with carbon disulfide *must* be performed in a hood because of its high toxicity. Tests indicate that desorption is complete in 30 min if the sample is agitated occasionally during this period. An ultrasonic probe suitable for agitation of the sample is available from Heat Systems Ultrasonics, Inc., New York. The use of graduated, stoppered microcentrifuge tubes is recommended so that the analyst can observe any apparent change in volume during the desorption process. Carbon disulfide is a very volatile solvent, so volume changes can occur during the desorption process depending on the surrounding temperature and how the tube is stoppered. The time allowed for desorption should be as consistent as possible and should not exceed 3 hr **(9)** to minimize volume and concentration changes.†

8.4.3 *G.C. Conditions.* Operating conditions for the G.C. may vary depending upon the type of instrument used and requirements of the sample being analyzed. It is recommended that the injector and detector temperatures be maintained at a level which will completely vaporize the sample and prevent undesirable condensation of combustion products (about 200 C). A complete description of the operation of the FID has been published **(7)** and may be consulted. Simple mixtures and individual compounds often may be determined with a column at constant temperature (isothermal). However, complex mixtures will require the testing of different column materials and temperature programs to resolve the components **(9)**. The analyst should use the bulk liquid sample or bulk air sample to establish optimum G.C. operating conditions and the column which will completely separate the components of the sampled mixture *prior* to the analysis of the charcoal sample tubes.‡

8.4.4 *Injection.* After desorption, an aliquot of the sample solution is injected into the G.C. To eliminate difficulties arising from blowback or distillation within the syringe needle, one should employ the solvent flush injection technique. The 10-μl syringe is first flushed with carbon disulfide several times to wet the barrel and

†The initial volume occupied by the charcoal plus the 0.5 ml CS_2 should be noted and corresponding volume adjustments made whenever necessary just before GC analysis.

‡Typical operating conditions for GC are:
(1) helium carrier gas flow, 85 cc/min (70 psig)
(2) hydrogen gas flow to detector, 65 cc/min (24 psig)
(3) air flow to detector, 500 cc/min (50 psig)
(4) injector temperature 200 C.

plunger. Three μl of solvent are drawn into the syringe to increase the accuracy and reproducibility of the injected sample volume. The needle is removed from the carbon disulfide, and the plunger is pulled back about 0.2 μl to separate the solvent flush from the sample with a pocket of air to be used as a marker. The needle is then immersed in the sample, and a 5-μl aliquot is withdrawn, taking into consideration the volume of the needle, since the sample in the needle will be completely injected. After the needle is removed from the sample and prior to injection, the plunger is pulled back a short distance to minimize evaporation of the sample from the tip of the needle. Triplicate injections of each sample and standard should be made. No more than a 3% difference in area is to be expected.

8.4.5 *Measurement of area.* The area of the sample peak is measured by an electronic integrator or some other suitable form of area measurement, and preliminary results are read from a standard curve prepared as discussed below.

8.5 DETERMINATION OF DESORPTION EFFICIENCY.

8.5.1 *Importance of determination.* The desorption efficiency of a particular compound can vary from one laboratory to another, from one batch of activated charcoal to another, and can also vary with the amount of compound adsorbed on the charcoal. Thus, it is necessary to determine at least once the percentage of each specific compound that is removed in the desorption process each time a different batch of charcoal is used. The Physical and Chemical Analysis Branch of NIOSH has found that the average desorption efficiencies for the compounds in Tables 834:I and 834:II are between 81% and 100% and vary with each batch of charcoal.

8.5.2 *Procedure for determining desorption efficiency.* Activated charcoal equivalent to the amount in the first section of the sampling tube (100 mg) is measured into a 2-in, 4-mm ID glass tube, flame-sealed at one end. This charcoal must be from same batch as that used in obtaining the samples and can be obtained from unused charcoal tubes. The open end is capped with Parafilm. A known amount of the compound is injected directly into the activated charcoal with a μl syringe, and the tube is capped with more Parafilm. The amount injected is usually equivalent to that present in a 10-l sample at a concentration equal to the OSHA standard **(3)**. (For the conversion formula, see Section **10.7**)

a. At least 5 tubes are prepared in this manner and allowed to stand for at least overnight to assure complete adsorption of the specific compound onto the charcoal. These 5 tubes are referred to as the samples. A parallel blank tube should be treated in the same manner except that no sample is added to it. The sample and blank tubes are desorbed with carbon disulfide (CS_2) and analyzed in exactly the same manner as the sampling tube described in Section **8.4**.

b. Two or three standards are prepared by injecting the same volume of compound into a 0.5 ml of CS_2 with the same syringe used in preparation of the sample. These are analyzed with the samples.

c. The desorption efficiency equals the difference between the average peak area of the samples and the peak area of the blank divided by the average peak area of the standards, or

$$\text{Desorption efficiency} = \frac{\text{Area sample} - \text{Area blank}}{\text{Area standard}}$$

9. Calibration and Standards

9.1 It is convenient to express concentration of standards in terms of mg/0.5 ml CS_2 because samples are desorbed in this amount of CS_2. To minimize error due to the volatility of carbon disulfide, one can inject 20 times this weight into 10 ml of CS_2. For example, to prepare a 0.3 mg/0.5 ml standard, one would inject 6.0 mg into exactly 10 ml of CS_2 in a stoppered flask. The density of the specific compound is used to convert 6.0 mg into μl for easy measurement with a μl syringe. A series of standards, varying in concentration over the range of interest, is prepared and analyzed under the same G.C. conditions and during the same time period as the unknown samples. Standard curves are established by plotting concentration in mg/0.5 ml versus peak area or integrator output.

Note: Since no internal standard is used in the method, standard solutions must be analyzed at the same time that the sample analysis is done. This will minimize the effect of known day-to-day variations and variations during the same day of the FID response. Standard solutions should be made fresh each day they are required for G.C. analysis. However, this does not preclude the use of an internal standard in this method.

10. Calculations

10.1 Read the weight, in mg, corresponding to each peak area from the standard curve for the particular compound. No volume corrections are needed, because the standard curve is based on mg/0.5 ml CS_2 and the volume of sample injected is identical to the volume of the standards injected into the G.C.

10.2 Corrections for the blank must be made for each sample.

$$\text{Corrected mg} = mg_s - mg_b$$

where:

mg_s = mg found in front section of sample tube

mg_b = mg found in front section of blank tube

A similar procedure is followed for the backup sections.

10.3 Add the corrected amounts present in the front and backup sections of the same sample tube to determine the total measured amount in the sample.

10.4 Divide this total weight by the determined desorption efficiency (See Section **8.5.2**) to obtain the total mg per sample.

10.5 Convert the volume of air sampled to standard conditions of 25 C and 760 Torr.

$$V_s = V \times \frac{P}{760} \times \frac{298}{T + 273}$$

where:

V_s = volume of air in liters at 25 C and 760 Torr

V = volume of air in liters as measured

P = barometric pressure in mm Hg

T = temperature of air in degrees C.

10.6 The concentration of the organic solvent in the air sampled can be expressed in mg per m³, which is numerically equal to μg per liter of air.

$$mg/m^3 = \mu g/l = \frac{\text{total mg (Section 10.4)} \times 1000\ (\mu g/mg)}{V_s}$$

10.7 Another method of expressing concentration is ppm, defined as μl of compound per liter of air

$$ppm = \mu l \text{ of compound}/V_s$$

$$ppm = \frac{\mu g \text{ of compound}}{V_s} \times \frac{24.45}{MW}$$

where:

24.45 = molar volume at 25 C and 760 Torr

MW = molecular weight of the compound

Note: Activated charcoal adsorbs nearly all organic vapors, thereby requiring precautions be taken to prevent it from becoming contaminated after the sampling process. Masking tape or plastic caps may be used to cap the ends for short term storage. If the tubes are to be stored for a week or longer, refrigeration is recommended to prevent loss of the more volatile adsorbed vapors or the ends be flame sealed. Many organic vapors are chemically stable on activated charcoal at room temperature and can be stored for months. Highly volatile, oxygenated, and reactive compounds should be analyzed as soon as possible; preliminary experiments indicate that recovery efficiencies of such compounds decrease with time.

11. References

1. SCOTT RESEARCH LABORATORIES, INC. 1973. Contract No. HSM 99-72-98. Collaborative Testing of Activated Charcoal Sampling Tubes for Seven Organic Solvents. SRL 1316 10 0973.
2. COOPER, C.V., L.D. WHITE and R.E. KUPEL. 1971. Qualitative Detection Limits for Specific Compounds Utilizing Gas Chromatographic Fractions, Activated Charcoal and a Mass Spectrometer. *Amer. Ind. Hyg. Assoc. J.* 32:383.
3. FEDERAL REGISTER. 1972. Vol. 37, No. 202, Part 1910, 93, Wednesday, October 18. Washington, D.C. Occupational Safety and Health Standards as required by the Occupational Safety and Health Act of 1970, Public Law 91-596.
4. KUPEL, R.E. and L.D. WHITE. 1971. Report on a Modified Charcoal Tube. *Amer. Ind. Hyg. Assoc. J.* 32:456.
5. OTTERSON, E.J. and C.U. GUY. 1964. A Method of Atmospheric Solvent Vapor Sampling on Activated Charcoal in Connection with Gas Chromatography. Transactions of the Twenty-Sixth Annual Meeting of the American Conference of Governmental Industrial Hygienists, Philadelphia, Pa., p. 37, American Conference of Governmental Industrial Hygienists, Cincinnati, Ohio.
6. QUINN, P.M., C.S. MCCAMMON and R.E. KUPEL. 1973. A Charcoal Sampling Method and a Gas Chromatographic Analytical Procedure for CS_2. Paper presented at the American Industrial Hygiene Conference in Boston, Masschusetts, May. *Amer. Ind. Hyg. Assoc. J.* to be published.
7. AMERICAN PUBLIC HEALTH ASSOCIATION, INC. 1977. Methods of Air Sampling and Analysis. 2nd ed. p. 257. Washington, D.C.
8. AMERICAN CONFERENCE OF GOVERNMENTAL INDUSTRIAL HYGIENISTS. 1973. Threshold Limit Values for Chemical Substances and Physical Agents in the Workroom Environment with Intended Changes for 1973. Cincinnati, Ohio.
9. WHITE, L.D., D.G. TAYLOR, P.A. MAUER, R.E. KUPEL. 1970. A Convenient Optimized Method for the Analysis of Selected Solvent Vapors in the Industrial Atmosphere. *Amer. Ind. Hyg. Assoc. J.* 31:225.

Revised by Subcommittee 5
E. SAWICKI, *Chairman*
T. BELSKY
R. A. FRIEDEL
D. L. HYDE
J. L. MONKMAN
R. A. RASMUSSEN
L. A. RIPPERTON
L. D. WHITE

835.

Analytical Method for Parathion in Air

Analyte:	Parathion	Method No:	P & CAM 158
Matrix:	Air	Range:	5 $\mu g/m^3$–250 $\mu g/m^3$
Procedure:	Collection via impinger/ethylene glycol, extraction into hexane, GLC analysis	Precision:	Unknown
		Classification:	C (Tentative)
Date Issued:	11/3/72		
Date Revised:	1/20/75		

1. Principle of the Method

1.1 SAMPLING. Parathion, 0-0 diethyl-0-p-nitrophenyl phosphorothioate, is effectively removed and retained from an air sample by passing a measured volume through an impinger containing ethylene glycol.

1.2 SAMPLE PREPARATION. The ethylene glycol is diluted with water and the parathion extracted into hexane. The hexane solution is concentrated to a small volume.

1.3 ANALYSIS. An aliquot of the concentrated hexane extract is gas chromatographed and the parathion determined as if elutes from the column by a flame photometric detector designed to measure the emissivity of phosphorus at 526 nm. When phosphorus compounds are burned in a hydrogen rich flame, the phosphorus atoms are excited to a higher energy state. Upon return of the atom to the ground state, a characteristic spectrum of light is produced. The quantity of light emitted is directly proportional to the amount of phosphorus present.

1.4 APPLICABILITY. The procedure as described herein is applicable for the measurement of parathion at 0.1 to 2.5 times the proposed OSHA Federal Standard of 100 $\mu g/m^3$.

2. Range and Sensitivity

2.1 The range of interest is 5 to 250 $\mu g/m^3$ of parathion in air with maximum precision and accuracy at 10 $\mu g/m^3$. A 50-liter air sample containing 50 $\mu g/m^3$ of parathion is concentrated in 1 ml of hexane. Take a 2 μl aliquot to yield 5 ng of parathion to the detector. This quantity is within the linear dynamic range of the detector of 0.5 ng to 400 ng.

2.2 SENSITIVITY. Maximum sensitivity and linearity of responses depend on auxiliary equipment. The sensitivity of the detector is 1×10^{-12} g/sec of phosphorus. The lower limit of detection in the described procedure, is 0.5 $\mu g/m^3$ of parathion. Although both the range and sensitivity may be raised or lowered by changing the volume of air sampled, the final volume of hexane or the size of the aliquot taken for analysis, this would deviate from the range of interest and the recommended method. Such changes could negate the precision and accuracy data when established.

3. Interferences

3.1 Phosphorus compounds having retention volumes close to that of parathion will interfere with the analysis. If interferences are anticipated a lengthy pretreatment of the sample is required. All equipment and reagents must be scrupulously free of any traces of phosphate detergents.

4. Precision and Accuracy

4.1 There are no data available for precision and accuracy for this procedure. Limited data, available from a comparable procedure that includes lengthy pretreatment of the sample, indicated a relative standard deviation of 25%.

5. Advantages and Disadvantages of the Method

5.1 ADVANTAGES.

5.1.1 The method is very sensitive (1×10^{-12} g/sec of P).

5.1.2 The detector is highly selective for phosphorus compounds and the G. C. column increases selectivity.

5.1.3 Separation and quantification are made in a reasonable length of time.

5.2 DISADVANTAGES.

5.2.1 Cost of equipment and supplies may tax the budget of some laboratories.

5.2.2 The sensitivity of the equipment depends on careful adjustments of operating parameter.

5.2.3 The procedure requires a skilled chemist or technician.

5.2.4 Equipment and reagents are easily contaminated.

5.2.5 If interfering compounds are anticipated a lengthy clean-up procedure is necessary **(3)**.

6. Apparatus

6.1 SAMPLING. Collection of the sample requires a sampling train consisting of the following:

6.1.1 *Filter cassette,* with 8μ glass-fiber filter with a diameter of 37 mm.

6.1.2 *Standard glass midget impinger.*

6.1.3 *Adsorption tube.*

6.1.4 *Glass wool.*

6.1.5 *Sampling pump, personal type.* Capable of maintaining a pressure drop of one inch of mercury and a flow rate of 1 to 2.8 l/m.

6.1.6 *Calibrated flowmeter,* 0.06 to 2.86 l/m.

6.1.7 *Stop watch.*

6.2 SAMPLE PREPARATION. The equipment needed for sample preparation is as follows:

6.2.1 *Kuderna-Danish evaporator-concentrator.* Consisting of 125-ml Erlenmeyer flask, 3-ball Snyder Column and a 10-ml receiver graduated in ml.

6.2.2 *Separatory funnels, 60 and 125-ml with teflon stopcocks.*

6.2.3 *Beakers, 100-ml.*

6.2.4 *Glass stirring rods.*

6.2.5 *Forceps.*

6.2.6 *Funnels, 65 or 74-mm diam at the top.*

6.2.7 *Glass wool.*

6.2.8 *Hot water bath.*

6.2.9 *Glass beads, 3-mm diam.*

6.3 ANALYSIS. The following apparatus is required for analysis:

6.3.1 *Gas chromatograph,* Tracor MT220 or equivalent with ancillary equipment including a phosphorus flame photometric detector.

6.3.2 *Syringes, 5, 10, and 100-μl capacities.*

6.3.3 *Gas chromatographic column, constructed of borosilicate glass.*

7. Reagents

7.1 SAMPLING.

7.1.1 *Ethylene glycol, chromatoquality.* Pretest this reagent by analyzing a portion of this stock reagent to insure the absence of extraneous material that has retention times close to parathion and are detected by the flame detector.

7.2 SAMPLE PREPARATION. The following reagents are required for sample preparation:

7.2.1 *Hexane, pesticide quality.*

7.2.2 *Distilled water, interference free.*

7.2.3 *Saturated solution of sodium chloride.*

7.2.4 *Sodium sulfate, anhydrous.*

7.3 ANALYSIS. The following reagents are required:

7.3.1 *Parathion of known purity.* A set of standards is available that contains known quantities of trithion, parathion, paraoxon, malathion and diazinon.

a. Prepare a standard stock solution, 250 μg/ml, by accurately weighing 25 mg of parathion to the nearest 0.1 mg. In a hood, transfer the parathion quantitatively to a 100-ml volumetric flask with hexane and bring to mark. **CAUTION:** Avoid contact with parathion and its solutions by using rubber or plastic gloves. From the standard stock solution, prepare a working standard by serial dilution of the stock solution with hexane to provide equally spaced working standards between 0.15 to 12.5 μg/ml.

b. Column material. 4% SE 30/6% OV-210 on 80 to 100 mesh Chromosorb W, H. P. Condition column for 3 days at 245 C under a nitrogen flow of 60 ml/min. Subsequently, a column 50-mm long and 4 mm ID containing 10% Carbowax 20 M on 80 to 100 mesh silanized support is inserted before the column and the entire assembly heated to 230 to 235 C for 17 hr under a nitrogen flow of 20 ml/min. Remove the Carbowax column before use.

7.3.2 *Cylinder of oxygen.*

7.3.3 *Cylinder of air.*

7.3.4 *Cylinder of hydrogen, or hydrogen generator.*

8. Procedure

8.1 COLLECTION OF SAMPLE. Assemble a sampling train consisting of a filter cassette, a midget impinger, adsorption tube and the sampling pump. Pack the adsorption tube loosely with glass wool to protect the pump from splash over or water condensate. Prior to use, charge the midget impinger with 15 ml of ethylene glycol. For simplified calculations, collect exactly 50 l of sample at a flow rate of 1 to 2.8 l. Turn on pump and start stopwatch. During the first minute of operation, check flow rate with the calibrated flow meter and compute the sampling time required to collect 50 l of sample as follows: Time = 50 l/flow rate. Recheck flow rate during the last minute of sample collection. After 50 l of air are collected, turn off pump. Remove stem from the impinger and wash the stem with 2 to 5 ml ethylene glycol. Record total volume (sample plus washing) in the impinger and seal the impinger with a plastic stopper covered with teflon. Total liquid volume is reported solely for the purpose of checking for sample loss during shipping. Remove filter cassette and stopper the openings. Place identifying number on both cassette and impinger. Tape the cassette to the impinger with masking tape and pack for transport. With each series of samples, provide one unexposed filter and 15 ml of unexposed ethylene glycol to serve as sample blanks.

8.2 SAMPLE PREPARATION.

8.2.1 Transfer quantitatively the sample of ethylene glycol of less than 20 ml to a 125-ml separatory funnel (#1). Wash the impinger with exactly 5 ml of distilled water and add additional water so that the total quantity of water added to the glycol in the separatory funnel is 70 ml.

8.2.2 Transfer the corresponding filter to a 100-ml beaker, with a clean forcep. Add 12 ml of hexane and agitate with a clean glass rod. Transfer the hexane quantitatively to separatory (#1) and extract by shaking vigorously for 2 min. If an emulsion forms, add 0.5 ml of saturated NaCl to the funnel. Shake again and allow layers to separate: transfer aqueous layer to a 125-ml separatory funnel #2, and the hexane layer to a 60-ml separatory funnel #3, re-extract the filter with a fresh 12 ml of hexane. Transfer the hexane extract to separatory #2 containing the water and shake for 2 minutes. Transfer the aqueous layer back to separatory #1 and the hexane layer to separatory #3. Again extract the filter and repeat the operations. This time discard the aqueous layer and transfer the hexane to separatory #3 as usual. To the hexane composite in separatory #3, add 10 ml of water and extract. After separation, discard aqueous layer. Repeat the extraction with 10 ml of water.

8.2.3 Pack the stem of a funnel with a plug of glass wool and 2.6 g of anhydrous Na_2SO_4. Dry the hexane by passing it through the funnel and collect the eluate in a 125-ml Kuderna-Danish flask fitted with a 10-ml receiver containing a glass bead of 3-mm diameter. Wash separatory #3 and the funnel with Na_2SO_4 quantitatively with 3 successive 2-ml portions of hexane. Finally, rinse the funnel and Na_2SO_4 with two additional 2-ml portions of hexane.

8.2.4 Place the Kuderna-Danish assembly in a boiling water bath and evaporate the solution to approximately 5 ml. Remove the assembly from the bath and allow to cool. Disconnect the receiving tube from the flask rinsing the joints with a little hexane. Place the receiving tube under a stream of nitrogen and concentrate to approximately 0.5 ml at room temperature. Rinse the walls of the tube with hexane using a 100-μl syringe. Dilute the extract with hexane to exactly 1.0 ml and stir with a clear glass rod.

8.3 ANALYSIS.

8.3.1 Prior to sample preparation, the gas chromatograph should be placed in operating condition. The critical parameters are as follows:

Column temperature: 200 C

Injection port temperature: 225 C

Detector temperature: 200 C

Transfer line and switching valves: 235 C

Carrier gas (N_2) flow rate: 60 ml/min.

Detector air flow rate: between 100 and 150 ml/min.

Flame gases: air + oxygen + nitrogen should have a composition of 80% N_2 + 20% O_2. For example, if carrier gas (N_2) flow is 60 ml/min, the O_2 flow is 15 ml/min.

Hydrogen flow rate: between 140 and 180 ml/min. After column has been conditioned at elevated temperatures as described previously, set column temperature at 200 C and nitrogen carrier flow at 80 ml/min. Set oxygen flow at 50 ml/min. for initial ignition and air flow at 30 ml/min. Open H_2 shutoff valve, set H_2 flow at 100 ml/min. Close hydrogen shutoff valve and allow H_2 gas to be swept out. Depress ignition button and slowly open hydrogen shutoff valve. A "pop" may be heard as the hydrogen ignites. Recorder will register and maintain an upscale reading when the flame is on. Also, check the deposit of moisture on a cold mirror, flat glass or metal surface when held above the exhaust tube to insure the flame is on. If the tests are negative, the flame is out. During sample injections, if the response to the recorder pen shifts radically below the base line and remains there, the flame has been extinguished. Repeat ignition procedure starting with *close hydrogen shutoff valve*". Note the flame photometric detector has a high background current, additional bucking voltage is required to zero the recorder when the flame is on.

8.3.2 Once the system is operating satisfactorily, the column requires additional conditioning or priming to parathion. Using one of the working standards of parathion in hexane, repeatedly inject this standard until a constant response is obtained on the recorder. **CAUTION:** the solvent should be vented so that the flame will not be extinguished. Continue priming the column until replicate responses agree within 5%. Then proceed to section **9** and calibrate the instrument to produce a reference curve.

8.3.3 Proceed to analyze samples by making at least duplicate injections of the samples into the chromatograph. A working standard that produces a deflection of 80% of full scale on the recorder should be interspersed among the regular samples 7% of the time to monitor detector response and validate the reference curve. If the results obtained with this standard do not agree within 5% of the reference curve, then corrections in the results should be made using the average values of the interspersed standards rather than the reference curve.

9. Calibration and Standards

9.1 CALIBRATION. Using the equally spaced working standards, inject 2 μl of each concentration in a random order at least in duplicate. Calculate area under the curve by multiplying the peak times the width of the peak at ½ the peak height. Then plot the quantity of parathion in nanograms against peak areas. The line of best fit drawn through the points should go through the origin. If this condition is not satisfied, then either the linear range of the detector has been exceeded or there is a system malfunction. If the latter, refer to operating manual for correction. Alternatively, analog to digital converter or signal processors are available which monitor the detector and perform all necessary calculations. These automated systems are particularly useful when a large number of determinations are to be made.

9.2 STANDARDS. Parathion samples of known purity are available to prepare standard solutions as described in Sec. **7.3.1.a**.

9.2.1 *Quality Control Standards.* Weigh 25 mg of parathion of known purity and transfer quantitatively to a 100-ml volumetric flask with ethylene glycol. Dilute to mark. Prepare a quality control standard by serially diluting to a final concentration of 0.2 μg/ml of parathion in ethylene glycol. This control sample should be treated as a regular sample to establish the accuracy of the analytical process exclusive of the sampling. With multiple analyses at least 7% of the samples should be quality control samples. Quality control charts may be prepared to establish the precision and accuracy of the method exclusive of the sampling.

10. Calculations

10.1 SAMPLING. Compute volume of air sampled using following formula:

$$V_a = Qt$$

Where:

V_a = Volume of air sampled, liters at 25 C and 760 Torr.
Q = Flow rate of liters/min.
t = time, min.

10.2 ANALYSIS. Compute micrograms of parathion in portion analyzed using formula **10.2.1** or **10.2.2**.

10.2.1

$$P_f = R \times S$$

P_f = Parathion in 2 μl, nanograms
R = Response signal, area
S = Slope of the line of best fit, nanograms/area

10.2.2

$$P_f = P_s \frac{R_f}{R_s}$$

Where:

P_f = Parathion of unknown, nanograms
P_s = Concentration of interspersed standard, nanograms
R_f = Response of the unknown, area
R_s = Response of the standard, area

10.2.3 Compute and report concentration of parathion in $\mu g/M^3$ as follows:

$$P = \frac{P_f \times F}{V_a} = \frac{500\ P_f}{V_a}$$

Where:

P = Parathion, $\mu g/m^3$
P_f = Parathion found (**10.2.1** or **10.2.2**), nanograms
F = Dilution factor,

$$500 = \frac{\text{Volume of hexane, } 1000\ \mu l}{\text{Volume of solution injected } 2\ \mu l}$$

V_a = Volume of air sampled (**10.1**) liters ≅ 50.

NOTE: ng/liter equivalent to $\mu g/M^3$.

11. References

1. ENOS, H.F., J.F. THOMPSON, J.B. MANN, and R.F. MOSEMAN. 1972. Determination of Pesticide Residues in Air, presented at the 163rd American Chemical Society National Meeting, Boston, Massachusetts, April.
2. MILES, J.W., L.E. FETZER, and G.W. PEARCE. 1970. *Environmental Science and Technology*, 4:420.
3. ANALYSIS OF PESTICIDE RESIDUES IN HUMAN AND ENVIRONMENTAL SAMPLES. ed. J. Thompson. 1972, Perrine Primate Research Laboratories, EPA, Perrine, Florida.
4. WATTS, R.R. and R.W. STOHRER. 1969. *Journal of the Association of Official Analytical Chemists*. 52:513.

Revised by Subcommittee 1
D. F. ADAMS, *Chairman*
J. O. FROHLIGER
D. FALGOUT
A. M. HARTLEY
J. B. PATE
A. L. PLUMLEY
F. P. SCARINGELLI
P. URONE

836.

Total Particulate Aromatic Hydrocarbons (TpAH) in Air Ultrasonic Extraction Method

Analyte:	TpAH	Method No.:	P & CAM 206
Matrix:	Air	Range:	Lower limit, 3 nanograms benzo(a)-pyrene
Procedure:	Sampling with glass fiber filter; extract ultrasonically; enrich and measure with HPLC	Precision:	± 1.33% RSD
		Classification:	E (Proposed)
Date Issued:	1/1/75		
Date Revised:			

1. Principle of the Method

1.1 Airborne particles collected from polluted atmospheres on glass fiber filters are extracted ultrasonically in the presence of silica powder (**1, 2, 3**). The TpAH in the filtered extract are separated by high speed liquid chromatography on a column of Corasil II with a non-polar solvent, and the absorbance is measured by a UV detector at 254 nm. Compounds responding to the detector are shown in Tables 836:I, II, III, IV. The extract is suitable also for the analysis of the aliphatic hydrocarbons (**4**).

2. Range and Sensitivity

2.1 Minimum reproducible level of standard benzo(a)pyrene at 254 nm is approximately 3 ng.

2.2 The minimum detectable TpAH (in terms of benzo(a)pyrene) for particulates collected on one glass fiber filter of approximately 452 cm^2 is approximately 5 μg, or 3.3 ng/m^3 air if 1500 m^3 of air are sampled in the ambient atmosphere.

2.3 The upper range of TpAH conc can be increased by dilution of the extract and/or analyzing smaller samples. Sensitivity for low concentrations can be increased by injecting larger samples into the chromatograph. Thus, very high levels of TpAH can be measured.

3. Interferences

3.1 Any compound which is not retained on the silica column and absorbs light at 254 nm is measured in this procedure.

3.2 Fluorene and some of its analogues and derivatives listed in Table 836:II, and polychloro derivatives of some di- and tricyclic hydrocarbons in Table 836:III are examples of such compounds.

3.3 Amino, carbonyl, hydroxy and nitro compounds elute after the PAH, so do not interfere. See Table 836:II.

3.4 Carbazoles and aldehydes are either retained or have retention times larger than the PAH, except N-alkyl substituted derivatives, which elute with the PAH. See Table 836:IV.

3.5 Oxygenated compounds, some phenols and aza and imino-heterocyclics (except some members of the indole series) are retained. Examples are benzoquinone, o-ethylphenol, acridine and quinoline.

3.6 Most interfering compounds have quite low peak area per μg values, which decreases their significance, as shown in Tables 836:II, III, IV.

4. Precision and Accuracy

4.1 Homogeneous glass fiber samples containing air particulates were analyzed by Soxhlet and ultrasonic extraction. See

Table 836:V. The relative standard deviation for 6 ultrasonic extracts was ± 1.33% and for 4 Soxhlet extracts ± 26.1%. The ratio of ultrasonic to Soxhlet recovery was 1.14.

4.2 Recovery of PAH added to glass fiber filter blanks and extracted ultrasonically was 95% for anthracene; 97.5% for phenanthrene; and 98.2% for benzo(a)pyrene, Table 836:VI.

5. Advantages and Disadvantages of the Method

5.1 The extraction is done at room temperature. Complete extraction of the TpAH is assured by the fine shredding of the glass fibers and the breaking up of clumps of particulates.

5.2 Only a relatively small sample of air particulates is required. Complete analysis time is well under an hour, most of which is waiting time.

5.3 Most of the polar constituents are removed by adsorption in the homogenizing vessel. The remainder are removed by the fast simple chromatographic analysis.

5.4 The method can accommodate a wide range of hydrocarbon pollution concentration, since sample extract volumes ranging from 0.1 to 2 ml can be chromatographed.

5.5 Time and work are saved by not weighing the particulates or soluble organics.

5.6 A disadvantage is that a blank correction must be made for the fiber glass filter. Also care must be taken to avoid evaporation of the extract to dryness.

5.7 A further disadvantage is that the ultrasonic extraction must be done in a sonabox to reduce the unacceptably high noise level.

Table 836:I. Elution of PAH*

Compound	% Eluted Through Column	PA‡/μg × 10^{-3}
Mono-, dicyclics		
Benzene	99	0.4
N-Hexylbenzene	100	0.5
N-Heptylbenzene	100	0.7
Naphthalene	101	0.7
Azulene	93	3.0
Tricyclics		
Anthracene	100	36.0
9-Methylanthracene	99	15.0
Xanthene	102	1.3
Phenoxethiin	92	0.2
Phenanthrene	100	10.0
Tetracyclics		
Naphthacene	95	4.7
Chrysene	105	4.5
Pyrene	96	3.6
4-Methylpyrene	100	1.7
1,3-Dimethylpyrene	96	0.9
Triphenylene	100	9.0
Benz(a)anthracene	96	4.3
7,12-Dimethylbenz-(a)anthracene	102	3.3
Pentacyclics		
Dibenz(a,h)anthracene	96	0.6
Benzo(a)pyrene	100	5.3
Benzo(e)pyrene	92	2.2
Picene	99	5.0
Perylene	96	5.8
Hexacyclics		
Benzo(ghi)perylene	99	1.8
Anthanthrene	93	2.6
Dibenzo(fg,op) naphthacene	93	0.6
Coronene	91	0.5
Dibenzo(g,p)chrysene	96	1.0
Naphtho(2,1,8-qra) naphthacene†	100	0.7

*Retention time is approximately 2 minutes.
†Or naphtho(2,3-a)pyrene.
‡PA = Peak Area.

6. Apparatus

6.1 SONIFIER CELL DISRUPTOR. 20 KHz power ultrasonic generator capable of dialing 70 W accurately, with a 1.27 cm (½″) horn disruptor and Sonabox.

6.2 LIQUID CHROMATOGRAPH. With stainless steel column 2.6 × 300 mm, UV Detector with 254 nm filter and loop injector with a capacity ranging from 0.1 to 2 ml.

6.3 STRIP CHART RECORDER WITH DISC INTEGRATOR.

6.4 An approved and calibrated personal sampling pump for collection of particulate matter. Any vacuum pump whose flow can be determined accurately to within 1 lpm or less.

6.5 COLUMN BYPASS.

6.6 FISHER FILTRATOR AND MEDIUM SINTERED GLASS FILTER.

Table 836:II. Elution of Fluorene, Analogues and Derivatives

Compound	t_R Min.	% Eluted Through Column	PA/μg × 10^{-3}
Fluorene	2.0	100	2.9
Dibenzothiophene	2.0	96	1.8
Dibenzofuran	2.0	98	0.3
Fluoranthene	2.0	95	2.5
Benzo(k)fluoranthene	1.8	97	1.4
Benzo(b)fluoranthene	1.0	99	1.6
2-Ethylfluorene	1.0	95	1.6
11 H-Benzo(b)fluorene	1.0	110	5.6
2-Nitrofluorene	4.8	104	0.2
2,5-Dinitrofluorene	7.0	71	0.3
9-Fluorenol	8.5	14	0.2
3,6-Dinitrodibenzoselenophene	18.2	38	0.2
3-Aminofluorene	18.2	68	0.4
4-Fluorenecarboxylic acid		Retained on column	
2-Hydroxyfluorene		Retained on column	
2-Nitro-7-hydroxyfluorene		Retained on column	
Fluorenone		Retained on column	

6.7 U.S. STANDARD SIEVE SERIES NO. 120, WITH 125 μ OPENINGS.

7. Reagents

7.1 CYCLOHEXANE, A.C.S. Spectroanalyzed, distilled once from glass.

7.2 POLYNUCLEAR AROMATIC HYDROCARBONS.

7.3 GLASS POWDER. Spherical, nonwettable 38 to 53 μ in diam.

7.4 CORASIL II.

8. Procedure

8.1 EXTRACTION.

8.1.1 The 1.27-cm horn of the sonifier cell disruptor is supported in a sonabox to reduce noise. The sonifying vessel is a beaker 3.8 cm ID × 10 cm tall. The end of the horn is set about 0.6 cm above the bottom of the beaker to insure adequate "stirring" of the mixture and equal exposure to areas of intense cavitation. Approximately 16 square cm of the exposed fiber glass filter and blank are cut in-

Table 836:III. Elution of Polychloro Derivatives of Di- and Tricyclic Hydrocarbons*

Compound	% Eluted Through Column	PA/μg × 10^{-3}
1,1-Dichloro-2,2-bis(p-chlorophenyl)ethane (p,p′DDD)	94	0.02
1,1-Dichloro-2,2-bis(p-chloropenyl)ethylene(DDE)	97	0.50
1,1,1-Trichloro-2,2-bis(p-chlorophenyl)ethane (p,p′DDT)	85	0.02
Aroclor 1260 (chlorinated biphenyls, 60% chlorine)	100	0.13
Aroclor 5432 (chlorinated triphenyls, 32% chlorine)	104	0.61
Halowax 1099 (mixture of tri- and tetrachloronaphthalenes, 52% chlorine)	101	0.25
1,2,3,4,5,6,7,8-Octachloronaphthalene	97	0.64
2,3,4,5,6,2′,3′,4′,5′,6′-Decachlorobiphenyl	95	0.19
1,2,3,4,5,6,7,8-Octachlorodibenzofuran	93	0.33
1,2,3,4,6,7,8,9-Octachlorodibenzo-p-dioxin	98	0.85
Tetradecachloro-p-terphenyl	95	0.22

*Retention times from 1 to 2 min.

Table 836:IV. Elution of Some Indoles, Carbazoles and Aromatic Aldehydes

Compound	t_R Min.	% Eluted Through Column	PA/μg × 10^{-3}
Indole	5.3	82	1.1
Carbazole	11.8	67	0.7
4-H-Benzo(def)-carbazole	8.0	98	2.0
11-H-Benzo(a)-carbazole	14.5	55	3.0
7-H-Dibenzo(c,g)-carbazole	18.0	92	2.1
N-Phenyl-carbazole	2.3	74	1.8
N-Ethyl-carbazole	2.5	98	0.5
5-Methyl-5,10-dihydroindeno-(1,2-b)indole	2.8	103	1.9
2,3-Dimethyl-indole	5.3	90	5.5
2-Methyl-carbazole	6.8	100	0.8
2-Hydroxy-carbazole		Retained on column	
N-Ethyl-3-amino-carbazole		Retained on column	
Benzaldehyde	12.8	56	1.1
2-Naphthalde-hyde	8.2	78	0.3

to roughly 1.3 cm squares to facilitate shredding. The sonifying vessel is surrounded by an ice water bath up to the level of the solvent mixture.

8.1.2 Homogeneous replicate samples of approximately 16 cm^2 of exposed and blank glass fiber filters are prepared and adjusted to exactly 100 mg. This weight necessarily includes both the particulates and the glass fiber. These samples were used to maximize parameters and for comparison of ultrasonic and Soxhlet extractions, shown in Table 836:V.

8.1.3 Samples for routine analysis are not weighed. Only the areas of the sample (16 cm^2) and the whole filter, the volume of air sampled and the volume of extract injected need to be determined. Sample at rate of at least 2 lpm for 1 hr or more.

8.1.4 *Extraction Procedure.* The sample, 60 ml cyclohexane and 5 ml silica powder are placed in the sonifying vessel, and sonified for 8 min at 70 W. The supernatant is decanted into the sintered glass filter supported on a Fisher Filtrator. Cyclohexane is added to the sonifying vessel to the level of the original mixture (usually about 50 ml). Sonification is carried on for an additional 4 min. The contents are filtered and combined with the first fraction, and rinsed with 50 ml cyclohexane. The filtrates and rinsings are collected in an Erlenmeyer flask and evaporated to about 5 ml, transferred quantitatively to a 10-ml volumetric flask and made to the mark.

8.1.5 Sample and blank filters, **8.1.2** are extracted by Soxhlet with 80 ml cyclohexane for 6 to 8 hr for comparison with the ultrasonic extraction. See Table 836:V after filtering, the extracts are evaporated in the same manner as the ultrasonic extracts.

8.1.6 The glass fiber filters used for air sampling should be as free as possible of soluble compounds which absorb at 254 nm. It may be necessary to flash fire or extract them and care should be taken to avoid contaminating them.

8.2 CHROMATOGRAPHIC SEPARATION.

8.2.1 A schematic of the chromatographic system is shown in Figure 836:1. The stainless steel column is 2.6 × 300

Table 836:V. Comparison of Ultrasonic and Soxhlet Extractions

	Ultrasonic		Soxhlet	
Sample No.	PA/μg	% Eluted*	PA/μg	% Eluted*
1.	0.575	51	0.449	28
2.	0.562	53	0.509	—
3.	0.567	50	0.500	—
4.	0.579	48	0.545	31
5.	0.560	44	—	—
6.	0.573	44	—	—
Average	0.569	49	0.509	30
Rel. Std. Dev.	± 1.33%		± 26.1%	
Ultrasonic/Soxhlet Recovery = 1.14				

*Refers to % of TpAH in the extracted material.

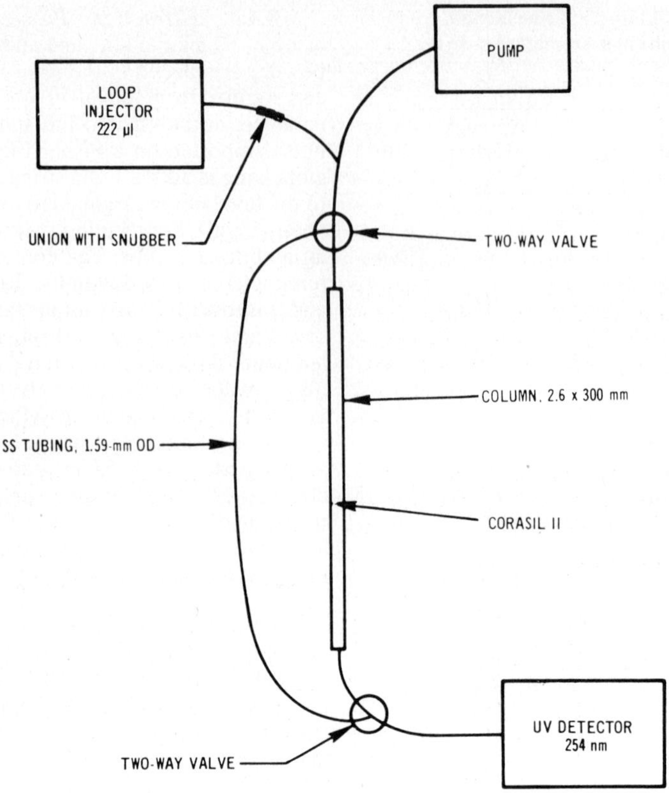

Figure 836:1—Schematic of Chromatographic System.

mm; the packing is Corasil II; the eluent is cyclohexane. Two two-way valves are installed in the chromatographic line, one before the column, the other after it. They are connected with stainless steel tubing. This enables the sample to be pumped through either the column or the tubing (column bypass) into the UV detector. A union with snubber may be placed in the line to prevent clogging.

8.2.2 To test the performance of the column, the percent of PAH which elutes is calculated from the peak areas through the column and the column bypass. Typical chromatograms from column and tubing are shown in Figure 836:2. Recovery of benzo(ghi)perylene was 99%. The percent of other hydrocarbons which eluted through the column ranged from 91 to 105, Table 836:I.

8.3 Analysis Procedure.

8.3.1 An appropriate volume of extract is injected through the loop injector. A flow rate of 1.6 ml/min gives a pressure drop of less than 200 PSI. The peak area is measured with a disk integrator, driven by 0 to 10 servo strip chart recorder with a 0.5″/min chart speed. The PAH elute in 3 to 5 min. Benzo(a)pyrene is used as the standard. Polar compounds are retained on the column. Samples can be chromatographed every 5 to 10 min.

8.3.2 The column bypass is also used to determine the per cent of PAH in the organic material of the extract. Chromatograms of sample extracts made on the col-

Table 836:VI. Recovery of Added PAH

Compound	Sample, μg	Peak Area: Sonified Filter + Std.	Peak Area: Standard Solution	% Recovery
Anthracene	0.035	1005	1055	95.0
Phenanthrene	0.147	1155	1185	97.5
Benzo(a)pyrene	0.355	1846	1880	98.2

umn and column bypass are shown in Figure 836:3. On the basis of absorbance measurements at 254 nm, approximately 50% of the organic material in the unchromatographed extract is PAH. This procedure is not necessary for routine analyses, but is helpful in elucidating the analytical situation.

8.4 EFFECTS OF STORAGE.

8.4.1 Urban particulates on glass fiber filters stored in the dark in an envelope for one year lost 32% of its benzo(a)pyrene. Losses of some other PAH ranged from 1 to 88% (**5**).

8.4.2 Benzene-soluble extracts evaporated to dryness and stored in closed bottles in a refrigerator were stable (in terms of benzo(a)pyrene concentrations) for 4 years (**6**).

8.4.3 The ultrasonic extract is stable in the dark at room temperature for several days, longer in the refrigerator. However losses usually occur after about 2 weeks.

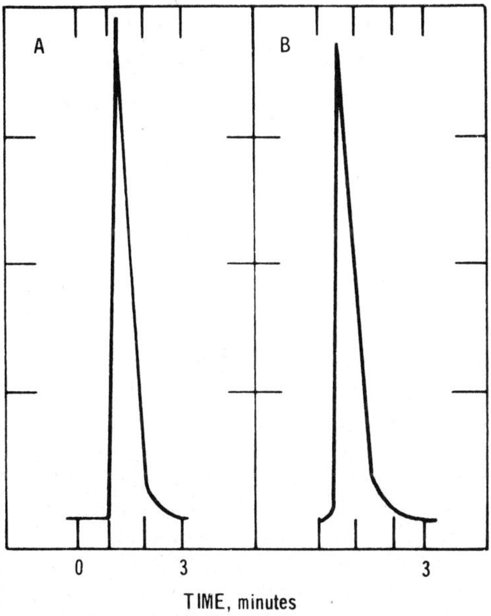

Figure 836:2—Chromatograms of benzo(ghi)perylene, 0.629 μg in cyclohexane, through the column (A) and through the column bypass (B). Stationary phase, Corasil II; eluent, cyclohexane; flow rate, 1.6 ml/minute. Peak area on column, 1165; on column bypass, 1180; recovery on column, 99%; peak area/μg, 1850.

Table 836:VII. Analysis of Particulate Samples

Description	Corrected Peak Area	M^3 Air Sampled	Pa/m^3 Air*	TpAH† (μg)m^3 air
Urban I	1200	1500	1120	.211
Urban II	620	1500	580	.109
Urban III	545	1500	509	.096
Mt. Storm	0	1673	0	.000

*See *Calculations*—Section **10.1**.
†See *Calculations*—Section **10.2**.

9. Calibration and Standards

9.1 The benzo(a)pyrene (BaP) standard is made in cyclohexane and is chromatographed when the samples are run, and repeated whenever a parameter such as solvent lot is changed. Both standard and samples are run at concentrations which do not overload the detector and give reproducible results when diluted. For example 0.4 μg BaP gave a peak area of about 2000 and fulfilled the above criteria.

The standard is expressed in terms of peak area per microgram (PA/μg). The unit of measurement for the samples is corrected peak area per cubic meter of air (PA/m^3). The BaP equivalent of the TpAH

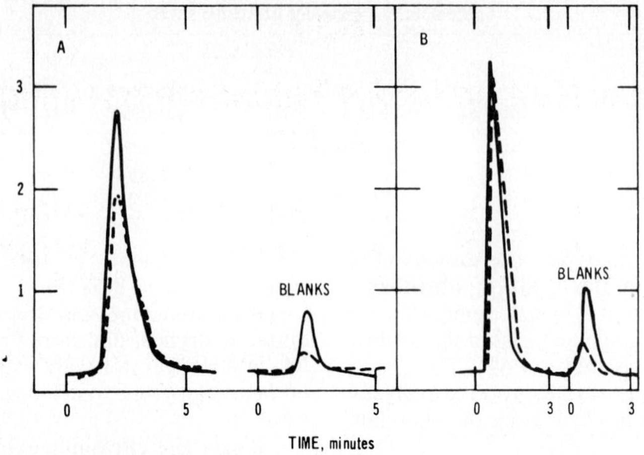

Figure 836:3—Chromatograms of ultrasonic and Soxhlet extracts of composited sample No. 1, Table 836:VI and blanks, through the column (A) and through the column bypass (B). Stationary phase, Corasil II; eluent, cyclohexane; flow rate, 1.6 ml/min. Solid lines are ultrasonic extracts; broken lines are Soxhlet extracts. Extracts were diluted × 3.3 for column bypass.

is calculated from these data (**10.2**). The standard is kept in the dark and is stable for more than 30 days when refrigerated nights and weekends.

10. Calculations

10.1 The peak area of the TpAH in a cubic meter of air is given by the equation.

$$PA/m^3 = \frac{PA \times A \times B}{V \times a \times b}$$

Where:

PA = Peak area, corrected for the blank.
V = Volume of air sampled in m^3, corrected to 25 C and 760 Torr.
A = Area of glass fiber filter in cm^2.
B = Volume of extract in ml.
a = Area of glass fiber filter sample in cm^2.
b = Volume of extract injected in ml.

10.2 The concentration of the TpAH may be expressed in terms of their equivalent in benzo(a)pyrene.

$$TpAH(\mu g)/m^3 \text{ air} = \frac{PA/m^3 \text{ air (See) Table 836:VII)}}{PA/\mu g \text{ benzo(a)pyrene (See Table 836:I)}}$$

11. References

1. BROWN, B. and J.E. GOODMAN. 1965. High Intensity Ultrasonics, Industrial Applications, Chapter 2, p. 30–35. Van Nostrand Co., Inc., Princeton, N.J.
2. CHATOT, G., M. CASTEGNARO, J.L. ROCHE, and R. FONTANGES. 1971. *Anal. Chim. Acta* 53:259.
3. CHATOT, G., R. DANGY-CAYE, and R. FONTANGES. 1972. *J. Chromatog.* 72:202.
4. WITTGENSTEIN, E. and E. SAWICKI. 1972. *Intern. J. Environ. Anal. Chem.* 2:11.
5. COMMINS, B.T. 1962. In Analysis of Carcinogenic Air Pollutants, E. Sawicki and K. Cassel, Jr., Eds. Natl. Cancer Inst. Monograph, No. 9, p. 225.
6. AMERICAN PUBLIC HEALTH ASSOCIATION, INC. 1977. Methods of Air Sampling and Analysis. 2nd ed. p. 216. Washington, D.C.

Subcommittee 5
E. SAWICKI, *Chairman*
T. BELSKY
R. A. FRIEDEL
D. L. HYDE
J. L. MONKMAN
R. A. RASMUSSEN
L. A. RIPPERTON
L. D. WHITE

837.

Analytical Method for 2,4 Toluenediisocyanate (TDI) in Air

Analyte:	TDI	Method No.:	P & CAM 141
Matrix:	Air	Range:	0.007 ppm to 0.140 ppm
Procedure:	Midget Impinger Colorimetric	Precision:	Unknown
Date Issued:	11/17/72	Classification:	D (Operational)
Date Revised:	1/1/75		

1. Principle of the Method

1.1 TDI is hydrolyzed by the HCl-acetic acid solution to the corresponding toluene-diamine derivative. This method is a modification of the Marcali method (**1**).

1.2 The diamine is diazotized by the sodium nitrite-sodium bromide solution.

1.3 The diazo compound is coupled with N-(Naphthyl)-ethylenediamine to form a colored complex.

1.4 The amount of colored complex formed is in direct proportion to the amount of TDI present and is determined by reading the absorbance of the solution at 550 nm.

1.5 Toluenediamine is formed via hydrolysis of TDI on a mole-to-mole basis. This amine is used in place of the TDI for standards. It has the advantages of being less toxic than the TDI and can be weighed because it is solid at room temperature. Both compounds have been tested by this method and the results compare favorably.

1.6 TDI kits based on the Marcali method are commercially available but have not been thoroughly tested to date.

2. Range and Sensitivity

2.1 The range of the standards used is equivalent to 1.0 to 20.0 μg TDI. In a 20-l air sample, this range is equivalent to 0.007 to 0.140 ppm.

2.2 For samples of high concentration whereby absorbance is greater than the limits of the standard curve (1.0 to 20.0 μg TDI), dilution of the sample with absorbing solution and rereading of the absorbance can extend the upper limit of the range.

2.3 The amount of TDI that would saturate 15 ml of absorbing solution has not been determined. It is possible that, in extremely high concentrations, some of the TDI would not be absorbed by the absorbing solution. Therefore, if a sample is diluted and reread, it could give an erroneously low value.

2.4 A single bubbler absorbs 95% of the diisocyanate if the air concentration is below 2 ppm. Above 2 ppm, about 90% of the diisocyanate is recovered. At high levels, it is suggested that two impingers in series be used.

3. Interferences

3.1 Any free aromatic amine may give a coupling color and thus may be a positive interference.

3.2 Methylene-di-(4-phenylisocyanate) (MDI) will form a colored complex in this reaction. However, its color development time is about 1 to 2 hr compared with 5 min for TDI. Therefore, MDI will generally not interfere.

4. Precision and Accuracy

4.1 The precision and accuracy of this method are unknown.

4.2 There has been no collaborative testing.

5. Advantages and Disadvantages of the Method

5.1 Few methods are available.

5.2 Any free aromatic amine will interfere. The method is not specific for TDI.

6. Apparatus

6.1 SAMPLING EQUIPMENT. The sampling unit for personal samples by the impinger collection method consists of the following components.

6.1.1 *An all glass, calibrated midget-impinger containing the absorbing solution or reagent.*

6.1.2 *Battery operated personal sampling pump—MSA Model G, or equivalent.* The sampling pump is protected from splashover or water condensation by an adsorption tube loosely packed with a plug of glass wool and inserted between the exit arm of the impinger and the pump.

6.1.3 *An integrating volume meter such as a dry gas or wet test meter.*

6.1.4 *Thermometer.*

6.1.5 *Manometer.*

6.1.6 *Stopwatch.*

6.1.7 *Various clips, tubing, spring connectors, and belt,* for connecting sampling apparatus to worker being sampled.

6.2 BECKMAN MODEL B SPECTROPHOTOMETER, OR EQUIVALENT.

6.3 CELLS, 1-CM AND 5-CM MATCHED QUARTZ CELLS.

6.4 VOLUMETRIC FLASKS, (several of each), glass stoppered: 50-, 100-, 1000-ml.

6.5 ANALYTICAL BALANCE, capable of weighing 200 g with a sensitivity of 1 mg or less.

6.6 PIPETS; 0.5-, 1-, 15-ML.

6.7 GRADUATED CYLINDERS, 25-, 50-ml.

7. Reagents

7.1 PURITY. All reagents must be made using ACS reagent grade or a better grade.

7.2 DOUBLE DISTILLED WATER (3).

7.3 2,4-TOLUENEDIAMINE.

7.4 HYDROCHLORIC ACID, CONC 11.7 N.

7.5 GLACIAL ACETIC ACID, CONC 17.6 N.

7.6 SODIUM NITRITE.

7.7 SODIUM BROMIDE.

7.8 SODIUM NITRITE SOLUTION. Dissolve 3.0 g sodium nitrite and 5.0 g sodium bromide in about 80 ml double distilled water. Adjust volume to 100 ml with double distilled water. This solution is stable for one week if refrigerated.

7.9 SULFAMIC ACID.

7.10 SULFAMIC ACID SOLUTION, 10% w/v. Dissolve 10 g sulfamic acid in 100 ml double distilled water.

7.11 N-(1)NAPHTHYL)-ETHYLENEDIAMINE DIHYDROCHLORIDE.

7.12 N-(1-NAPHTHYL)-ETHYLENEDIAMINE SOLUTION. Dissolve 50 mg in about 25 ml double distilled water. Add 1 ml conc hydrochloric acid and dilute to 50 ml with double distilled water. Solution should be clear and colorless; coloring is due to contamination by free amines, and solution should not be used. The solution is stable for 4 days.

7.13 ABSORBING SOLUTION. Add 35 ml conc hydrochloric acid and 22 ml glacial acetic acid to approximately 600 ml double distilled water. Dilute the solution to 1 l with double distilled water. 15 ml is used in each impinger.

7.14 STANDARD SOLUTION A. Weigh out 140 mg of 2,4-toluenediamine (equivalent to 200 mg of 2,4-toluenediisocyanate). Dissolve in 660 ml of glacial acetic acid, transfer to a 1-l glass stoppered volumetric flask, and make up to volume with distilled water.

7.15 STANDARD SOLUTION B. Transfer 10 ml of standard solution A to a glass-stoppered 1-l volumetric flask. Add 27.8 ml of glacial acetic acid so that when solution B is diluted to 1 l with distilled water, it will be 0.6 N with respect to acetic acid. This solution contains an equivalent of 2 μg TDI/ml.

8. Procedure

8.1 CLEANING OF EQUIPMENT.

8.1.1 Wash all glassware in a hot detergent solution, such as Alconox to remove any oil.

8.1.2 Rinse well with hot tap water.

8.1.3 Rinse well with double distilled water. Repeat this rinse several times.

8.2 COLLECTION AND SHIPPING OF SAMPLES.

8.2.1 Pipet 15 ml of the absorbing solution (Section **7**) into the midget impinger.

8.2.2 Connect the impinger (via the absorption tube) to the vacuum pump and the prefilter assembly (if needed) with a short piece of flexible tubing. The minimum amount of tubing necessary should be used between the breathing zone and impinger. The air being sampled should not be passed through any other tubing or other equipment before entering the impinger.

8.2.3 Turn on pump to begin sample collection. Care should be taken to measure the flow rate, time and/or volume as accurately as possible. Record atmospheric pressure and temperature. The sample should be taken at a flow rate of 1 lpm. Sample for 20 min, making the final air volume 20 l.

8.2.4 After sampling, the impinger stem can be removed and cleaned. Tap the stem gently against the inside wall of the impinger bottle to recover as much of the sampling solution as possible. Wash the stem with a small amount (1 to 2 ml) of unused absorbing solution and add the wash to the impinger. Seal the impinger with a hard, non-reactive stopper (preferably Teflon or glass). Do not seal with rubber. The stoppers on the impingers should be tightly sealed to prevent leakage during shipping. If it is preferred to ship the impingers with the stems in, the outlets of the stem should be sealed with Parafilm or other non-rubber covers, and the ground glass joints should be sealed (*i.e.*, taped) to secure the top tightly.

8.2.5 Care should be taken to minimize spillage or loss by evaporation at all times. Refrigerate samples if analysis cannot be done within a day.

8.2.6 Whenever possible, hand delivery of the samples is recommended. Otherwise, special impinger shipping cases designed by NIOSH should be used to ship the samples.

8.2.7 A "blank" impinger should be handled as the other samples (fill, seal and transport) except that no air is sampled through this impinger.

8.3 ANALYSIS OF SAMPLES.

8.3.1 Remove bubbler tube, if it is still attached, taking care not to remove any absorber solution.

8.3.2 Start reagent blank at this point by adding 15 ml fresh absorbing solution to a clean bubbler tube.

8.3.3 To each bubbler, including the blank, add 0.5 ml of 3% sodium nitrite solution, gently agitate, and allow solution to stand for 2 min.

8.3.4 Add 1 ml of 10% sulfamic acid solution to each tube, agitate for 30 sec and allow solution to stand 2 min to destroy all the excess nitrous acid present.

8.3.5 Add 1 ml of 0.1% N-(1-Naphthyl)-ethylenediamine solution to each tube. Agitate and allow color to develop. Color will be developed in 5 min. A reddish blue or pink color indicates the presence of TDI.

8.3.6 Add double distilled water to adjust the final volume to 20 ml in bubbler tube. Mix.

8.3.7 Transfer each solution to 1-cm or 5-cm quartz cell.

8.3.8 Using the blank, adjust the spectrophotometer to 0 absorbance at 550 nm.

8.3.9 Determine the absorbance of each sample at 550 nm.

9. Calibration and Standards

9.1 To each of a series of eight graduated cylinders add 5 ml of 1.2 N hydrochloric acid.

9.2 To these cylinders add the following amounts of 0.6 N acetic acid: 10.0, 9.5, 9.0, 8.0, 7.0, 6.0, 5.0 and 0.0 ml, respectively.

9.3 To the cylinder add standard solution B in the same order as the acetic acid was added: 0.0, 0.5, 1.0, 2.0, 3.0, 4.0, 5.0 and 10 ml, so that the final volume is 15 ml (*i.e.*, 0.0 ml of the standard is added to the 10 ml acetic acid; 0.5 ml of the standard is added to the 9.5 ml acid; etc. The cylinders now contain the equivalent of 0.0, 1.0, 2.0, 4.0, 6.0, 8.0, 10.0 and 20.0 μg TDI, respectively. The standard containing 0.0 ml standard solution is a blank.

9.4. Add 0.5 ml of the 3.0% sodium nitrite reagent to each cylinder. Mix. Allow to stand 2 min.

9.5 Add 1 ml of the 10% sulfamic acid solution. Mix. Allow to stand for 2 min.

9.6 Add 1 ml of the N-(1-Naphthyl)-ethylenediamine solution. Mix. Let stand for 5 min.

9.7 Make up to 20 ml with double distilled water.

9.8 Transfer each solution to a 1-cm or 5-cm quartz cell.

9.9 Using the blank, adjust the spectrophotometer to 0 absorbance at 550 nm.

9.10 Determine the absorbance of each standard at 550 nm.

9.11 A standard curve is constructed by plotting the absorbance against micrograms TDI.

10. Calculations

10.1 Subtract the blank absorbance, (Section **8.2.7**), if any, form the sample absorbance.

10.2 From the calibration curve (Section **9.11**), read the micrograms TDI corresponding to the absorbance of the sample.

10.3 Calculate the concentration of TDI in air in ppm, defined as μl TDI per liter of air.

$$\text{ppm} = \frac{\mu g}{V_s} \times \frac{24.45}{MW}$$

where:

ppm = parts per million TDI.
μg = micrograms TDI (Section **10.2**).
V_s = corrected volume of air in l (Section **10.4**).
24.45 = molar volume of an ideal gas at 25 C and 760 Torr.
MW = molecular weight of TDI, 174.15.

10.4 Convert the volume of air sampled to standard conditions of 25 C and 760 Torr:

$$V_s = V \times \frac{P}{760} \times \frac{298}{T + 273}$$

where:

V_s = volume of air in liters at 25 C and 760 Torr
V = volume of air in liters as measured
P = Barometric Pressure in Torr
T = Temperature of air in degree C.

11. References

1. MARCALI, K. 1957. Microdetermination of Toluenediisocyanates in Atmosphere. *Anal. Chem.* 4, 552.
2. LARKIN, R.L. and R.E. KUPEL. 1969. Microdetermination of Toluenediisocyanate Using Toluenediamine as the Primary Standard. *Am. Indus. Hyg. Assoc. J.* 30:640.
3. AMERICAN PUBLIC HEALTH ASSOCIATION, INC. 1977. Methods of Air Sampling and Analysis. 2nd ed. p. 53. Washington, D.C.

Revised by Subcommittee 3
B. DIMITRIADES, *Chairman*
J. CUDDEBACK
E. L. KOTHNY
P. W. MCDANIEL
L. A. RIPPERTON
A. SABADELL

PART IV
STATE-OF-ART REVIEWS

1. State-of-the-Art Review

Source Sampling for Fluoride Emissions from Aluminum, Steel and Phosphate Production Plants

1. Introduction

The purpose of this position paper is to present the state-of-the-art of source sampling for fluorides in both gaseous and particulate forms. The sources to be concerned will be primary aluminum plants, iron and steel-making and phosphate rock processing plants.

2. Methods for Primary Aluminum

2.1 PRODUCTION. The Hall-Heroult process is used almost exclusively to produce aluminum. In this operation alumina is dissolved in a bath of melted cryolite and the oxide is reduced by a heavy direct current to aluminum metal and oxygen. During the reduction, oxygen formed at the anode reacts with carbon in the anode to produce CO and CO_2. The overall reaction may be expressed as $Al_2O_3 + 2C \rightarrow 2Al + CO_2 + CO$. In addition, gaseous SiF_4 and HF plus particulate metal salts of fluoride such as CaF_2, AlF_3, and $Na_5Al_3F_{14}$ are evolved.

The top of the bath is covered with a frozen crust of alumina and bath mixture which dissolves into the cryolite bath as aluminum production proceeds (1). The carbon electrodes are renewed as they are consumed. There are two types of electrodes in use. One is the prebaked electrode which is periodically lowered into the bath until the stub end is reached. Another is the so-called Soderburg electrode which is charged to the operation as a paste and is baked into a hard electrode by the process heat as it is lowered into the pot. Make-up cryolite and fluorspar are also spread on top of the crust as required along with the alumina charge. The molten aluminum sinks to the bottom of the bath and is withdrawn.

An aluminum reduction potline is a very large-scale operation. The individual pots are about 16′ long, 8′ wide and 6′ deep. They are arranged side by side in the pot building which is of open high bay construction. It must be very well ventilated to dissipate the substantial heat and gases.

The pots are covered with hoods, but the CO_2 evolution, crust breaking, adding alumina, and stirring allow significant volumes of gas to escape to the building. Wet scrubbers are used to collect fluorides from the pot hoods, but to clean the large effluent from the entire building requires extensive installations. A fluid bed reactor devised by Alcoa in which fluorides are absorbed on the incoming alumina feed is coming into use.

2.2 PRIMARY ALUMINUM SAMPLING. Stack sampling to collect coarse and fine particulate fluorides plus gaseous compounds can be done with the apparatus illustrated in Figure 1:1. In this procedure a heated sampling probe is inserted in a stack and sampling is done isokinetically. A temperature sensor is inserted alongside the probe to measure temperature and the probe temperature, especially in wet scrubbers, is set high enough to prevent condensation and consequent blinding of the filters. The standard stack traverse sampling technique is used. Normal sampling time is about three hours.

As illustrated in Figure 1:1, the sample is aspirated through an alundum thimble with pore sizes of 5μ or a cellulose extraction filter, which serves to collect particulates. Cellulose filters are preferred below temperature of 125 C because they can be ashed to measure total fluorides and they have less tendency to blind. Following the filter, two Greenburg-Smith Impingers containing 0.1 N NaOH collect the gaseous fluoride. The dry impinger serves to trap spray and some fine particulate matter. The fourth impinger containing silica gel dries the airstream to protect the pump and gas meter.

There may be some retention of gaseous fluoride by the prefilter. A limited amount of testing under field conditions showed that the alundum thimble retained 10% to 12% of the total gaseous fluoride when the gas was about 35 ppm v/v fluoride (2). It is less likely to occur with cellulose.

Fluoride from particulate filter is ashed, fused and distilled from sulfuric acid at 160 C and measured by a fluoride sensitive membrane electrode. The gaseous fluoride sample is washed from the impingers, hose and probe and analyzed.

3. Methods for Iron and Steel Manufacture

3.1 PRODUCTION. An integrated steel mill is comprised of several combined facilities, including a sintering plant, blast furnaces, coke and by-products plant and several steelmaking shops which utilize open hearth, electric and basic oxygen (BOF) furnaces. Support services are also required, including boiler plants, furnace and annealing rooms, scarfing areas, water treatment, etc. Each of these facilities contribute to pollution emissions to some extent. The major fluoride emissions result from iron ore pelletizing and sintering, and

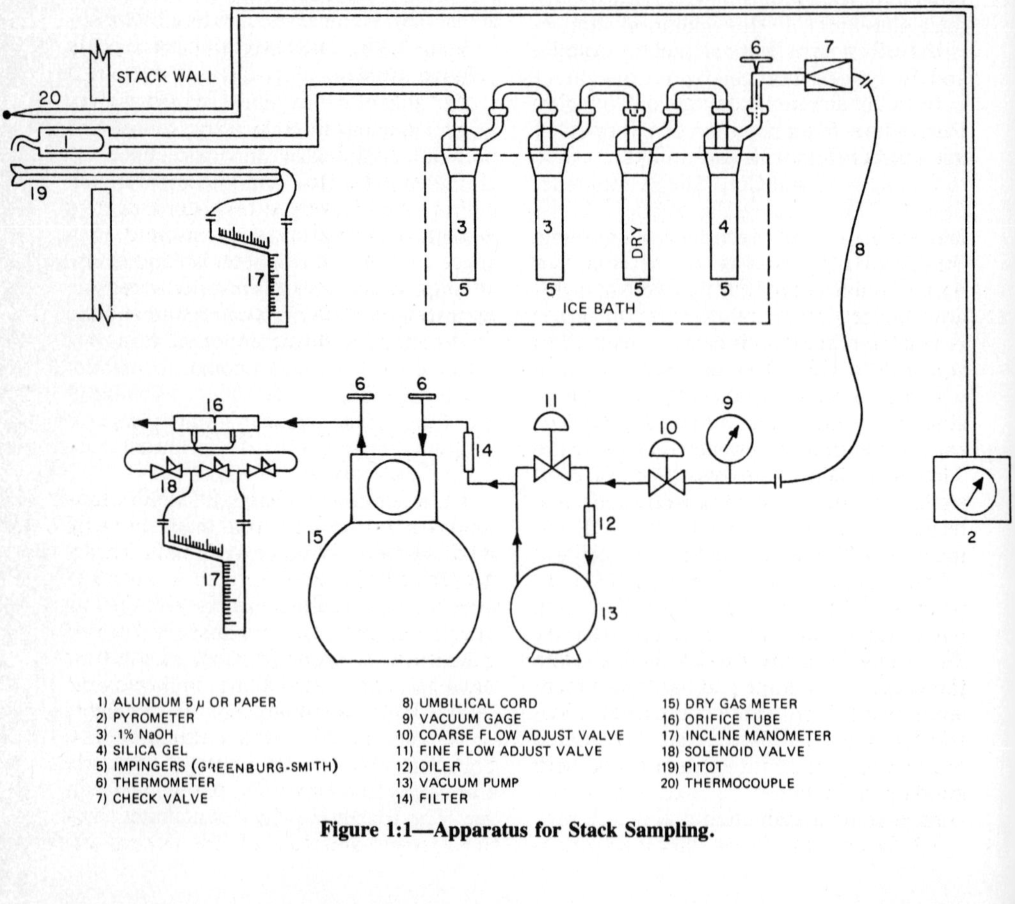

Figure 1:1—Apparatus for Stack Sampling.

open hearth and BOF steelmaking processes.

A typical steel mill emits 0.2 to 1 pound of fluorine (as fluorides) to the atmosphere per ton of steel produced. About 5% to 10% of the emitted fluoride is solid particulate and the remainder is gaseous.

The principal source of fluoride is fluorspar, CaF_2, which is employed in quantities of 2 to 15 pounds per ton of steel produced. The fluorspar is added as a slag conditioner to improve the collection of oxide impurities; the resulting increase in slag fluidity promotes desulfurization and increases the rate of heat transfer through the slag, improving process efficiency. Western ores, particularly those from southern Utah, include another source of fluoride. These ores contain 2,000 to 4,000 ppm fluorine present as fluorapatite, fluorspar and fluorosilicates. The remaining quantities of input fluorine are introduced as impurities in the other raw materials.

In steel manufacture by open hearth furnace, about 1% of the CaF_2 reacts with the silica according to the following reaction:

$$2\,CaF_2 + SiO_2 \rightarrow SiF_4 + 2\,CaO$$

and while this reaction occurs only to a small extent, it does create an emission problem due to the fact that the SiF_4 volatilizes. The presence of moisture in the combustion gases will tend to hydrolyze any SiF_4 to HF at the slag surface.

Basic oxygen steelmaking is quite recent in development, accounting for less than 1 per cent of the steel produced in 1960. The majority of new steelmaking installations, however, have been BOF, and basic oxygen steelmaking surpassed the open hearth in annual tonnage in 1970.

BOF steelmaking employs 12 to 15 pounds of CaF_2 per ton of steel, four to five times the amount used in open hearth. The majority of the fluoride is tied up in the slag and emissions to the atmosphere are only a few per cent of that used. Electric furnace steelmaking uses 3 to 15 lbs/ton of CaF_2, but with moderate gas input hoods and vents can be used with scrubbers to control emissions.

3.2 SAMPLING. Stack sampling equipment and procedures for fluoride in iron and steelmaking are very similar to that used in aluminum reduction. An additional impinger may be included to the train with water or dilute alkali to collect trace amounts of gaseous fluoride **(5)**. A small electrostatic precipitator **(9)** can be used before the impingers in stacks carrying a heavy load of particulate. Some operators use polyethylene impingers to avoid reaction of HF with glass.

Stainless steel (Type 316) or polytetrafluoroethylene, PTFE, sampling probes have been used over long periods of time (several years) with no evidence of degradation due to exposure to fluorides. This lack of reactivity with stainless steel may have resulted from the low water content in the gas stream. Glass probes have not been used because they are too fragile and the reaction of HF with glass at elevated temperatures can form gaseous SiF_4.

Sampling can be carried out at a single point of average velocity at a rate of about 1 cfm or preferably on a traverse pattern designed to obtain an average gas sample **(7)**. The sampling rate is adjusted for cyclic operations to maintain gas velocity within 10% of average velocity. Sampling is usually carried out for one hour. The probe and particulate collector (precipitator or filter) is maintained at a temperature greater than the water dewpoint and the impinger solutions are held in an ice bath.

4. Methods for Phosphate Rock Processing

4.1 PROCESSING. Some 25 million tons of phosphate rock are processed annually in the United States. The rock is mostly fluorapatite [$3Ca_3\text{-}(PO_4)_2 \cdot CaF_2$] and hydroxyapatite [$2Ca_3\text{-}(PO_4)_2 \cdot Ca(OH)_2$]; the fluoride content is generally 3% to 4% on a weight basis. Approximately one third of the rock is processed into phosphoric acid (by the wet process method) and a quarter is used for electric furnace production of elemental phosphorus and phosphoric acid. The remaining 40% is used mainly in the production of fertilizers, including triple superphosphate (TSP), normal superphosphate (NSP), diammonium phosphate (DAP), and dicalcium phosphate, and defluorinated rock which is employed in feed preparations for cattle and chickens.

Most of the phosphoric acid produced by the wet process is subsequently used in other processes, including the preparation of TSP and DAP. The major fluoride emissions from all of these processing operations are HF and SiF_4; the latter readily hydrolyzes to H_2SiF_6.

4.1.1 *Wet-Process Phosphoric Acid.* In the wet process, reaction of phosphate rock with an acid produces free phosphoric acid and the salt of the acid. In most commercial processes, sulfuric acid is used; the by-product salt is calcium sulfate which precipitates as gypsum and is removed by filtration. The reaction is described by the following equation:

$$3Ca_{10}(PO_4)_6F_2 + 30H_2SO_4 + SiO_2 + 58H_2O \rightarrow 30CaSO_4 \cdot 2H_2O + 18H_3PO_4 + H_2SiF_6.$$

There are a number of impurities in the phosphate rock which consume acid and affect the reaction. One of the principal impurities is fluoride. Generally, it is present in either of two forms, as a part of the fluorapatite mineral or as additional amounts of free calcium fluoride. When the rock is acidified, fluoride is converted to hydrofluoric acid. In most rocks, there is enough silica present to react with all of this hydrofluoric acid to form fluosilicic acid. At the temperatures and acidity conditions under which the digestion reaction must be carried out, some of the fluosilicic acid is vaporized as silicon tetrafluoride gas with traces of gaseous hydrofluoric acid.

In most plants, the substantial heat of reaction is removed by sparging the digester with air or by flash cooling of the hot liquor. In either case, there is substantial emission of air, steam, the fluorine containing gases, and P_2O_5 mist to the stack. The effluent gas is processed in a wet scrubber in which the fluorine compounds and the H_3PO_4 are absorbed in water to make dilute solutions. In some plants fluorine containing by-products are recovered from the scrubber liquor. In most cases, however, the liquor is discharged as an effluent to the gypsum disposal pond and acid water cooling ponds.

4.1.2 *Elemental Phosphorus.* In the production of elemental phosphorus, phosphate rock is reacted in an electric furnace with silica and coke, according to the formula $2Ca_3(PO_4)_2 + 6SiO_2 + 10C \rightarrow 6CaSiO_3 + P_4 + 10CO_2$. The volatized phosphorus is carried along by the gas stream, which is mostly CO and CO_2 and is condensed. The furnace effluent also contains nitrogen, hydrogen, methane, and fluorides.

4.1.3 *Triple Superphosphate (TSP).* A principal end product of the phosphate fertilizer industry is triple superphosphate, which is a solid product formed by acidulating phosphate rock with phosphoric acid. For run-of-pile triple, 50% acid as P_2O_5 and ground rock are mixed and the product, which is a pasty mass, is conveyed to storage, where reaction or "curing" continues along with drying. This simple process is difficult to control because of the process conditions and the nature of materials. The principal contaminants are gaseous fluorides and phosphate rock dust. All of the equipment must be closed or hooded and vented to a stack through suitable fume control equipment, which is commonly some type of one- or two-stage wet scrubber.

A special problem exists in the storage building where substantial quantities of gaseous fluorides are evolved during the curing of the triple superphosphate. To control this fume, the ventilating air must be collected and passed through a scrubber.

Almost all of the processes that utilize phosphate rock produce a very wet exhaust gas stream; most of the process streams carry entrained water droplets or contain large quantities of steam. As a result, particulate filters have not been used in stack sampling for even though they are heated, the amount of water entrainment is sufficient to cause plugging. Therefore, all of the particulate, as well as the soluble gaseous fluorides, are collected in the impingers, and analyses are limited to providing the total fluoride content.

Most phosphate rock processing companies employ the procedure adopted by the State of Florida for the sampling and analysis of water soluble fluorides in stack emissions (**6**). A typical sampling train for use with this method consists of a corrosion resistant probe such as PTFE, two impingers in series for the collection of fluorides, a dry impinger, a dry gas meter, and finally a vacuum pump. In some cases, a filter is included between the second impinger and the mist trap to collect any particulate or aerosol that passes through the impingers.

5. Summary of Methods Employed by Industry

The primary aluminum and steel industries often have "dry" emission streams and can therefore carry out separation of gaseous and particulate fluorides by straightforward methods. Several types of particulate collectors have been employed, apparently with good reliability, including electrostatic precipitators and high efficiency filters. The former tends to minimize particulate surface area which would be expected to reduce the potential for gaseous interactions with the collected particulate. Pretreated filters have been used efficiently for some applications where only light particulate loadings are encountered. In either case, some of the particulate is generally collected along the walls of the sampling probe. Attempts have been made to reduce the amount of particulate in the probe by employing an in-stack filter; although there is sufficient data reported to demonstrate the effectiveness of this approach, further consideration appears to be warranted. All sampling trains employ impingers to collect gaseous fluorides. There does not appear to be any difference in the collection efficiency by water or dilute caustic.

Phosphate rock processing operations are characterized by "wet" streams which generally preclude the use of a particulate collector before the impingers. Consequently, there is no provision for carrying out a separation of gaseous and water soluble particulate fluorides for this case. The apparatus generally used for sampling stack emission is therefore the same as for the "dry" stream industries except for elimination of the particulate collector.

The tendency of HF to react with particulate matter on filters can be minimized in some cases by converting the HF to silicon tetrafluoride with a heated glass probe (**3, 4**). The SiF_4 is then collected in the impingers, hydrolyzed with dilute alkali and is analyzed by an appropriate method (**8**).

Sampling practice within all industries is reasonably uniform, involving single point isokinetic sampling for a sufficient length of time (30 min to 2 hr) to average the effect of process variations. In cases where there is good evidence for large variations in stream velocity from point to point, or for stratification or "cycloning", sampling is carried out along a point traverse. Collected samples are generally brought directly to the analytical laboratory or, when necessary, are stored and sealed in plastic containers.

6. References

1. PETERS, E.T., OBENHOLTZER, J.E. and VALENTINE, J.R. 1972. Development of Methods for the sampling and analysis of particulate and gaseous fluorides from stationary sources. EPA Contract Report 68-02-0099. Arthur D. Little, Inc., Cambridge, Mass.
2. Private communication FLETCHER G. SMITH, Reynolds Metals Co., via H.F. Schiff, Gilbert Associates, Reading, Pa.
3. DORSEY, J.A., and KEMNITZ, D.A. 1968. A source sampling technique for particulate and gaseous fluorides. *J. Air Pol. Control Assoc.* **18**:12–13
4. ELFERS, L.A., and DECKER, C.A. 1968. Determination of fluoride in air and stack gas samples by use of an ion specific electrode. *Anal. Chem.* **40**:1658–61.
5. HAALAND, H.H. 1968. Methods for determination of Velocity, Volume, Dust and Mist content of gases. Bulletin WP-50, 7th Ed. Western Precipitation Division of Jay Mfg. Co., Los Angeles, CA.
6. FLORIDA DEPT. OF AIR AND WATER POLLUTION CONTROL. 1971. Standard sampling techniques and methods of analysis for the determination of air pollutants from point sources.
7. FEDERAL REGISTER. 1971. **36**:(#159), 15708. Part II, Tues. Aug. 17. Sample and Velocity Traverse for Stationary Sources. THOMPSON, C.R., FARRAH, G. H., HAFF, L.V., HILLMAN, H.S., HOOK, A.N., SCHNEIDER, E.J., STRAUTHER, J.D. and WEINSTEIN, L.H. 1972. Tentative Method of Anal-

ysis for Fluoride Content of the Atmosphere and Plant Tissues. Health Lab Science, **9**:304.
8. AMERICAN PUBLIC HEALTH ASSOCIATION, INC. 1977. Methods of Air Sampling and Analysis. 2nd ed. p. 417. Washington, D.C.
9. ROUNDS, G.L., and MATOI, H.J. 1955. Electrostatic Sampler for Dust-Laden Gases. *Anal. Chem.* **27**:829.

SubCommittee 2

C. R. THOMPSON, *Chairman*
G. H. FARRAH
L. V. HAFF
W. S. HILLMAN
A. W. HOOK
R. L. SALTZMAN
E. J. SCHNEIDER
J.D. STRAUTHER
L.H. WEINSTEIN

2. State-of-the-Art Review

Sampling and Analytical Techniques for Mercury in Stationary Sources*

1. Introduction

The purpose of this report is to review sampling and analytical techniques for mercury with respect to use in chloralkali and ore processing operations.

Mercury is a health hazard due to absorption of nearly 80% of the inhaled vapor at concentrations between 50 and 350 $\mu g/m^3$. Elemental and organic mercury compounds pose additional problems by being absorbed through the skin. The toxicity of mercury is due to the strong bonds formed by it with sulfur atoms in body components which results in an interference in the various body functions including synthesis and function of both enzymes and proteins. Mercury is a cumulative poison and concentrates in the brain, liver, and other organs (**1**).

2. Sources of Mercury

A summary of mercury emissions by source is given in Table 2:I. The largest sources of mercury released to the environment in 1971 were from paint, electrolytic chlorine, incineration, coal-burning power generation, and ore processing. By 1974, mercury will no longer be used as a fungicide in paints, therefore, this source will be eliminated (**2**). Any new chloralkali plants are being built with diaphragm rather than mercury cells in order to meet the EPA emission regulations of five pounds of mercury per day (**3**). However, the 27 tons of mercury per year from chloralkali plants in 1974 will still represent nearly 10% of the mercury emitted. Ore processing will remain a problem in 1974 although some reduction of mercury emissions in smelters will be effected by the SO_2 control system (sulfuric acid plant). Two major sources of mercury from coal combustion and incineration will be very difficult to control because of the inefficient scrubbing of low mercury concentrations, *ca.* 10 $\mu g/m^3$ in stack effluents.

A summary of EPA mercury concentration data (**4**) from various sources is given in Table 2:II. Note that the highest concentrations were found for mercury smelters and chloralkali plants. Chloralkali Plants 1, 3, and 4 are uncontrolled and would not meet EPA standards of five pounds per day, while the emissions for Plant 2 are considerably lower (**4**) and very close to the emission standards of five pounds per day. For the mercury smelters, Plants 1 and 2 have very high mercury concentrations while those of Plant 3 are much lower. The latter plant has a condenser on the stack to reduce the mercury concentration. The SO_2 concentration in these mer-

*Prepared for the Intersociety Committee by John N. Driscoll, Orion Research Corporation, 11 Blackstone Street, Cambridge, Massachusetts 02139

TABLE 2:I Mercury emissions by source

Category	Group	Emissions (tons) 1971	1974*
Mining	Mercury and non-ferrous	2.5	2.8
Processing	Primary mercury	33.5	5.5
	Secondary mercury		
	Non-ferrous:		
	copper	0.5	0.6
	zinc	11.0	11.4
	lead	5.0	5.2
Processing	Paint	0.8	0
	Electrical	2.6	2.9
Consumptive	Paint	229.0	0
	Agricultural	8.1	0
	Pharmaceuticals	4.2	4.7
	Electrolytic chlorine	150.0	26.9
	Instruments	2.6	2.9
	Dental preparations	0.9	1.0
	Other	6.9	7.7
	Coal-power plants	59.3	66.4
	Other	31.8	35.6
	Oil	10.2	10.4
Incineration and other disposal	Municipal	10.8	12.1
	Sewage sludge	4.4	4.9
	Other	77.5	86.8
		686.6	324.2

*Based on institution of EPA controls on fungicides and air emission, growth rate 4%/year. (Patrick, D. 1971. E.P.A., Research Triangle Park, N.C., 27711

cury smelters is approximately several thousand ppm.

3. Sampling Methods

The mercury composition, relative concentrations and potential analytical interferences for five different sources are given in Table 2:III. This table can be used to evaluate the adequacy of various sampling techniques for mercury collection.

Techniques for collection of mercury can be divided into three processes, *e.g.*, absorption, amalgamation, and adsorption. These are discussed in the following sections.

3.1 ABSORPTION.

3.1.1 *Aqueous iodine monochloride.* Elemental mercury is bubbled through an aqueous solution of ICl and is absorbed due to the following reaction (**3, 5**):

$$2ICl + Hg \rightarrow Hg^{++} + 2Cl^- + I_2$$

The attractive features are the high collection efficiency for elemental mercury (99%), collection of particulate mercury by impingement, and high efficiency by impingement for other solid mercury compounds. This method has been used successfully for sampling chloralkali plants by EPA (**4**).

The disadvantages are: (a) high and nonreproducible blanks, (b) failure to differentiate between gaseous and particulate mercury compounds, (c) instability of mercury compounds collected, and (d) mercury is not quantitatively recovered in the presence of high concentrations of SO_2.

Table 2:II. Summary of EPA mercury concentration data for various sources

Type of Plant		Site	Range of Total Hg Conc.
Chloralkali	1	H_2 stack	40–1600 mg/m^3
Chloralkali	2	H_2 stack	2.5–10 mg/m^3
		Fume system	4–12 mg/m^3
		Cell room	3–5 $\mu g/m^3$
Chloralkali	3	H_2 stack	1–3 mg/m^3
		End box vent	1–2.5 g/m^3
Chloralkali	4	Vent system (end room)	2.7 g/m^3
		H_2 stack	0.2–0.4 g/m^3
Coal-Fired Power Plant			~10 $\mu g/m^3$
Mercury Smelter	1	Stack	0.1–0.6 g/m^3
Mercury Smelter	2	Stack	0.9–1.1 g/m^3
Mercury Smelter	3	Stack	1.3 mg/m^3
Non-Ferrous Smelter		Stack	~50–250 $\mu g/m^3$

TABLE 2:III. Mercury composition and potential interferences for five stationary sources

	Form of Mercury		Mercury Conc.		Potential Interferences		
Source	Particulate	Gas	High	Low	SO_2	Cl_2	Hydrocarbons
Chloralkali							
End box vent	X	X	X			X	
H_2 stack		X	X				
Hg Smelter	X	X	X		X		
Non-Ferrous Smelter		X		X	X(VH)		X
Coal-Fired Power Plant	X	X*		X	X		
Incinerator		X		X			X

VH = Very High.
*90% is gaseous mercury.

This latter feature may be due to the formation of the $Hg(SO_3)_2^{=}$ complex which is not reduced to elemental mercury by hydroxylamine. This complex may be broken up by the addition of an excess of iodide solution which forms the more stable $Hg(I_4)^{=}$ complex. This could result in quantitative recovery of mercury in the presence of SO_2.

The sampling method could be modified by placing impingers containing aqueous sodium carbonate in front of the ICl impingers. This will remove SO_2 and certain mercury salts, but elemental mercury will be quantitatively passed (**6**). Analysis of the ICl and Na_2CO_3 impingers should give total mercury.

3.1.2 *Aqueous potassium permanganate*. Mercury is collected by bubbling through acidic solutions of potassium permanganate (**7–9, 16**). The collection efficiency of elemental or inorganic mercury is reported to be high (> 90%) (**8**), but organic mercury compounds are not efficiently collected (**5**). This latter feature is not a serious problem for chloralkali or ore processing effluents. This procedure has been used for collection of mercury in chloralkali plants (**9**) and coal-fired power plants (**7**). The advantages of the method include: (a) a low blank, (b) can be used to collect Hg in the presence of high concentrations (> 500 ppm) of SO_2, and (c) particulate mercury compounds are also absorbed. The disadvantages include: (a) the sensitivity of $KMnO_4$ to light, (b) auto-oxidation reactions can occur, (c) organic mercury compounds are not efficiently absorbed, and (d) possible losses of mercury by adsorption in MnO_2 or on walls of container.

This method should be further evaluated since it has been utilized for chloralkali plants, ore processing **(16)**, power plants, etc.

3.1.3 *Other liquids.* Nitric acid has been used in industrial hygiene studies for the collection of elemental and inorganic mercury compounds **(10)**. The efficiency is high for mercury vapor and no problems are encountered with decomposition or instability of the absorbing solution. Carr and Wilkness **(11)** have shown that in nitric acid (pH < 1), mercury losses in vessels are minimal with storage. The method has not, however, been used for ore processing or chloralkali plants and the efficiency of collection of organic mercury compounds may be low. No problems should be encountered with SO_2 since the nitric acid would oxidize sulfite ion, preventing the complexation with mercuric ion.

3.2 AMALGAMATION.

3.2.1 *Silver amalgamation.* Mercury forms amalgams with various noble metals such as silver and gold. Williston **(12)**, Corte, *et al.* **(13)**, and Long, *et al.* **(14)** have used silver gauze or silver wool for the collection of elemental Hg in ambient air. The Geomet instrument **(15)** utilizes a silver grid† for collection of mercury. Inorganic and organic mercury compounds are decomposed to elemental Hg by a high temperature catalytic converter, then collected on the silver grid. The elemental mercury is removed from the grid thermally and then analyzed by UV absorption.

This silver amalgamation procedure *cannot* be used in either chloralkali or ore process operations because of the reactions given below:

$$Cl_2 + 2Ag \rightarrow 2AgCl$$
$$SO_2 + 6Ag \rightarrow Ag_2S + Ag_2O$$

†A gold grid is also available.

These reactions poison the surface of the silver and reduce the collection efficiency. Thompson **(27)** has shown that a tube packed with ascarite will effectively remove SO_2 and pass mercury vapor. An ascarite prescrubber, used with silver wool, may then be a useful method for sources containing Cl_2 or SO_2.

3.2.2 *Gold amalgamation.* Tradet **(17)** has developed a gold amalgamation technique for the collection of Hg at nonferrous smelters. The major advantage of gold over silver is that it does not react with SO_2 or Cl_2. In the Tradet studies **(17)**, it was found that a prescrubber was necessary to (a) remove SO_3 (or H_2SO_4 mist) from the stream, and (b) lower the temperature. Apparently, the SO_3 attacks the surface of the gold and traps some mercury reducing the recovery. The Tradet system consisted of four impingers in series: first, containing acidic stannous chloride, second, empty for condensation, third and fourth, gold chips.

The advantages of this system are that one could determine vaporous and particulate mercury, however, the Tradet system will collect only total Hg. This technique has been used in smelters where high levels of SO_2 (up to 5%) are present, the elemental mercury is from silver operation, the gold can be regenerated by firing at 700 C for 3 hr.

The disadvantages include: (a) the need for a prescrubber to remove SO_3 and lower the temperature of the effluent; (b) the efficiency is a strong function of surface contamination; and (c) 30 to 35g of gold are required for each sample, making this procedure very expensive.

3.3 ADSORPTION. Solids such as charcoal have been used to collect mercury for industrial hygiene studies **(18)**. Iodine impregnated charcoal was employed by Stock **(19)** for increasing the collection efficiency for elemental mercury. Inorganic and particulate mercury compounds have been determined by Sergeant, *et al.* **(18)** using a glass wool filter and activated charcoal. The major difficulty with charcoal has been a failure to recover the mercury collected. This approach would not be very useful in effluents containing high concentrations of water vapor, SO_2 or Cl_2.

Palladium chloride has been used as an efficient adsorbent scrubber for mercury (**20**), but no studies on the recovery of mercury have been reported.

4. Analytical Methods

Mercury analyses have been performed by a variety of techniques including dithizone colorimetry (**8**), polarography (**21**), nephelometry (**21**), neutron activation (**17**), anodic stripping voltametry (**7**), and optical methods such as atomic absorption (**3, 5, 9, 11–14, 16–18, 22**) or atomic emission (**7, 23**). Additional procedures are discussed in References **1, 8, 10,** and **21**.

The majority of the optical methods are based on modifications of the Hatch and Ott flameless AA procedure (**23**). Here, an aliquot of the liquid sample (**3, 5, 7, 9**) is treated with a reducing agent such as hydroxylamine or stannous chloride, and the elemental mercury formed is swept with dry N_2 into the optical cell of a spectrophotometer. The mercury content is measured by attenuation of the 253.7 nm line from a low pressure mercury lamp. The advantages of this technique are: (a) high sensitivity (ppb levels of Hg), and (b) ease of use. The disadvantages are due to nonspecificity and subsequent interferences from SO_2, NO_2, hydrocarbons, or other species which absorb at 253.7 nm. Some of the problems can be resolved by the use of a spectrophotometer with background correction capability which will allow compensation for absorbing species. If a selective scrubber for mercury *only* is placed before the gas passes into the reference cell, a single beam system can be used. This background correction has been difficult to achieve. However, Vaughan (**24**) describes a system whereby mercury vapor is trapped on gold or silver leaf. Any interferences such as SO_2, hydrocarbons, etc., would be flushed through the system. The amalgamated silver or gold is heated via an rf field and the mercury, free of interferences, is passed into an absorption cell for determination by flameless AA.

In the procedures which collect mercury by amalgamation (**12–17**), the first step, *e.g.*, removal of interferences, is accomplished during the sampling period. The grid, gauze, or wool is then returned to the lab and heated in an induction furnace to release the mercury vapor collected into a spectrophotometer cell. Tradet Corporation (**17**) found that a single trapping on gold foil was not sufficient to ensure quantitative recovery of mercury from non-ferrous smelters. Their procedure required a double amalgamation technique to ensure complete and reliable recovery of mercury. Schlesinger and Schultz (**25**) evaluated a number of techniques for analysis of mercury in coals. They also found that a double gold amalgamation procedure followed by AA provided more reliable results than a single silver amalgamation followed by AA.

Hinkle and Learned (**26**) determine mercury in aqueous solutions by amalgamation on a silver screen. The mercury vapor is released by heating the dried screen in an rf coil and analyzed by AA.

A combination of the above technique with an aqueous absorption method such as permanganate might possibly eliminate many of the sampling and analytical problems associated with the aqueous absorption methods. If the mercury were collected in acidic permanganate, the excess permanganate and manganese dioxide would be decomposed by hydroxylamine. An excess of this reducing agent could release elemental mercury which could be collected on the silver grid in the field. This should eliminate problems with stability of the sample, as well as sampling and analytical problems imposed by the presence of high concentrations of sulfur dioxide.

5. Acknowledgment

This document was reviewed by the Mercury Task Group of Intersociety Committee Subcommittee 6 which consists of the following members:

J. N. Driscoll
L. Kornreich
J. W. Loveland
R. J. Thompson

6. References

1. Lawrence Radiation Laboratory. 1973. Instrumentation for Environmental Monitoring—Air-Hg.
2. Patrick, D. 1971. Private communication, EPA, Research Triangle Park, N.C. 27711.
3. EPA. 1973. National Emission Standards for Hazardous Air Pollutants, 38:8820.
4. McInnity, J. 1971. Private communication, EPA, 411 West Chapel Hill Street, Durham, N.C.
5. Linch, A.L., R.F. Stalzer, and D.T. Lefferts. 1968. *Amer. Ind. Hyg. J.*, 29:79.
6. Ehrenfeld, J.R., *et al.* 1973. Evaluation of Instrumentation for Monitoring Total Mercury Emissions from Stationary Sources. Final Report to EPA on Contract No. 68-02-0590.
7. Billings, C.E., *et al.* 1973. Air Pollution Control Assoc. J., 23:773.
8. Sandell, E.B. 1959. Colorimetric Determination of Trace Metals. Interscience Publishers, N.Y.
9. Atmospheric Emissions from Chloralkali Manufacture. 1971. Air Pollution Office Publ. #AP-80.
10. Jacobs, M.B. 1967. The Analytical Toxicology of Industrial Inorganic Poisons. Interscience Publishers, N.Y.
11. Carr, R.A. and P.E. Wilkness. 1973. *Env. Sci. & Tech.*, 7:62.
12. Williston, S.H. 1968. *J. Geophys. Res.* 73:7051.
13. Corte, G., L. Dubois, R.S. Thomas, and J.L. Monkman. 1972. Paper presented at 164th National ACS Meeting, N.Y.
14. Long, S.J., D.R. Scott, and R.J. Thompson. 1973. Paper presented at 165th National ACS Meeting in Dallas, Texas.
15. Geomet Corp., Rockville, Maryland.
16. Statnick, R. 1973. Sampling and Analysis of Mercury Vapor in Industrial Streams Containing Sulfur Dioxide, presented at the ACS Meeting, Chicago, Illinois.
17. Tradet Corp. Development of the Gold Amalgamation Technique for Mercury in Stack Gases, APTD 1171, PB 210–817.
18. Sergeant, G.A., B.E. Dixon, and R.G. Lidzey. 1957. *Analyst*, 82:27.
19. Stock, A., 1934. *Z. Angew. Chem.*, 47:64.
20. Corte, G. and L. Dubois. 1973. Paper #73-297 presented at 66th Annual APCA Meeting in Chicago, Ill.
21. U.S. Dept. of the Interior. 1970. Mercury in the Environment, Geological Survey Professional Paper #713.
22. Hatch, W.R. and W.L. Ott. 1968. *Anal. Chem.*, 40:2085.
23. April, R.W. and D.N. Hume. 1970. *Science*, 170:849.
24. Vaughan, V.W. and J.H. McCarthy. 1965. U.S. Geol. Survey Prof. Paper #501-D, p. D123–D127.
25. Schlesinger, M.D. and H. Schultz. 1971. An Evaluation of Methods for Detecting Mercury in U.S. Coals, R.I. 7609, USBM.
26. Hinkle, M.E. and R.E. Learned. 1968. U.S. Geol. Survey Prof. Paper #650-D, p. D251–D254.
27. Thompson, R. 1973. Private communication, EPA, Research Triangle Park, N.C. 27711.

3. State-of-the-Art Review

Methods for Measurement of Nitrogen Oxides in Automobile Exhaust

1. Introduction

The nitrogen oxides present in exhaust and of consequence to air quality are the nitric oxide (NO) and nitrogen dioxide (NO_2). In primary engine discharges the mixture of these two nitrogen oxides (NO_x) consists almost entirely of NO; NO_2 is a secondary product resulting from spontaneous reaction of NO with O_2. While the NO_x concentration in exhaust depends on engine operating conditions, the composition of the NO_x mixture in the exhaust sample depends on the sampling method and the time taken to collect and prepare the exhaust sample for analysis.

The problems in the measurement of exhaust NO_x differ for the two nitrogen oxides. Thus, direct measurement of NO is difficult owing to the lack of distinct properties associated with NO. For this reason most, but not all, earlier analytical methods for NO are based on conversion of NO into the more easily measured NO_2. Unlike NO, NO_2 does have distinct chemical and physical properties that permit its measurement; however, such measurement is complicated by the extreme instability of NO_2 in the exhaust environment. Because the analytical problems differ for the two nitrogen oxides, and because the relative levels of the two oxides vary depending on sampling procedure, the analytical methods reviewed here will be judged separately for the applications requiring different sampling techniques.

2. Methods for Measurement of NO_x in Auto Exhaust

2.1 PHENOLDISULFONIC ACID (PDS) METHOD **(2, 3)**. The method is based on contact of the gaseous batch sample with an oxidizing reagent (H_2O_2 + H_2SO_4) to convert the NO_2 into NO_3^-. The NO is also oxidized first into NO_2 (in reaction with O_2) and then into NO_3^-. Resultant NO_3^- is subsequently reacted with the PDS reagent to a colored product that is measured spectrophotometrically. The method has been used as a standard method for measurement of NO_x in combustion atmospheres for which NO_x concentrations exceed 20 ppm. The method's precision decreases rapidly below 200 ppm NO_x **(4)**. Large amounts of reducing compounds (e.g., SO_2) may interfere by consuming the H_2O_2 in the absorbing solution. Halides also tend to cause low results; however, this is not a problem in ordinary exhaust **(5)**. Main drawbacks of the method are the long analysis time (20 hr, approximately) and skill requirements. Some modifications have been proposed that cut the analysis time down to 1 to 2 hr **(6, 7)**. Also, the method does not discriminate NO from NO_2, which is another disadvantage for research studies. The method has been compared with other methods, and differences in results—less than 10%—were observed **(5)**; these differences, have not been explained adequately. Systematic comparison with other, more recent methods has not been reported.

2.2 SALTZMAN METHOD **(8)**. This method was the one most often used in the past, and involves contact of the gaseous batch sample with the Saltzman reagent and subsequent spectrophotometric measurement of resultant colored product. The method is specific for NO_2; however, procedure variations of NO_x measurement have been devised and used, that include oxidation of NO into NO_2 (with oxygen or air) prior to the NO_2 measurement **(5, 9)**. The method is applicable in the range 0.005 ppm to 5 ppm. For higher conc, such as in ordinary exhaust batch samples, sampling in evacuated containers or in syringes and subsequent dilution with nitrogen or oxygen or air may be necessary. The method, although simpler than the PDS method, is still tedious and time consuming relative to the instrumental methods. Further, it is reasoned that the method for NO_x should give somewhat low results because of NO_2 losses occurring during the NO oxidation step. Such losses are caused by reactions of NO_2 with other exhaust components **(5)**, and resultant products are not likely to be "seen" by the analytical method. No significant interference problems are believed to exist. The method is normally calibrated using standard NO_2^- solutions and assuming that 0.72 moles of NO_2^- produces the same color intensity as one mole of NO_2. However, the 0.72-value has been questioned; it is, therefore, preferable to calibrate the method using standard mixtures of NO or NO_2 or NO_x.

2.3 NON-DISPERSIVE INFRA-RED ABSORPTION METHOD (NDIR) **(5)**. The method is based on an infra-red absorption spectrophotometry principle. The spectrophotometer's detector is a chamber filled with NO, and is therefore capable of absorbing all radiation within the infra-red absorption spectrum of NO. A typical NDIR instrument uses a double-beam arrangement such that the IR radiation absorbed by NO in the sample-cell can be measured by the detector and be converted into ppm-concentration of NO. NDIR instruments are usually subject to interferences by exhaust components such as H_2O, CO, CO_2, and certain hydrocarbons.

*A considerable part of the information presented in this chapter was taken from a Lawrence Berkeley Laboratory report to be released in the near future **(1)**.

Such interferences can be minimized either by using special "interference filters" that remove those wavelengths that can be absorbed by NO and the interfering components or by scrubbing the interfering components from the sample. The latter technique has been used to remove the water vapor interference. However, some of the drying materials used were found to cause complications; namely, they inadvertently remove varying amounts of NO_2 and NO, or they convert NO_2 into NO. Such problems do not exist when the NDIR measurement is made upon samples of fresh exhaust where the NO_2-to-NO_x ratio is negligibly small.

NDIR instruments are typically operative in the 10^1 to 10^6 ppm—NO range; however, some manufacturers claim sensitivities in ranges extending down to 2 ppm. Response time varies depending on sample cell size, flow rate, and speed-of-response characteristics of the instrument; typically, it is 1 to 10 seconds.

Response varies non-linearly with NO concentration complicating the interpretation of response data. However, this complication can be removed by use of a signal-linearization device.

Manufacturers presently known to offer NDIR instruments are: Beckman Instruments, Inc., Bendix Corporation, Horriba Instruments, Inc., Intertech Corporation, and Mine Safety Appliances Company. Performance characteristics are specified by the manufacturers. Some evaluation studies of older models **(5)** and newer instruments **(10)** have been made; however, the reader is advised to consult the evaluation reports for details and evaluation results.

2.4 NON-DISPERSIVE ULTRA-VIOLET ABSORPTION METHOD (NDUV) **(5)**. The NDUV method is used for measurement of exhaust NO_2 only. The NDUV instruments operate based on an ultra-violet absorption spectrophotometry principle. Typically, measurement is based on absorption of radiation of 436 nm wavelength. Such instruments are somewhat more sensitive than the NDIR analyzers and do not suffer from the same interferences (*e.g.*, H_2O, CO_2) as NDIR analyzers do. However they are affected by presence (in the sample) of SO_2, particulate matter, and by NO_2 deposition on walls of the sample cell. However, electronic filtering, removal of particulates, and heating of sample cell do eliminate these problems.

Manufacturers presently known to offer commercial instruments are Beckman Instruments, Inc., and Peerless Instruments Company, Inc.

2.5 DISPERSIVE IR AND UV ABSORPTION METHOD. The Chrysler Corporation has introduced recently (1973) a vehicular emission instrument that measures, among others, NO via dispersive UV absorption **(11)**. By using a reactor converting NO_2 into NO, it is possible to measure NO_x also. Performance characteristics are specified by the manufacturer—Chrysler Corporation, Huntsville, Alabama. The instrument operates in scales from 0 to 100 ppm to 0 to 4000 ppm. With the instrument set on 0 to 3000 ppm full-scale NO, it is claimed that no detectable interference is caused by other exhaust components. However, interferences to the NO_2 measurement are unknown.

2.6 CHEMILUMINESCENCE METHOD **(12)**. The method is based on measurement of the chemiluminescence (600 to 3000 nm with maximum at 1200 nm) generated in reaction between NO and O_3. Therefore, the method is applicable to NO only. For NO_2 measurement, the NO_2 must be converted into NO, and NO_2 is recorded as the difference between the NO and NO_x measurements. For high sensitivity—needed *e.g.*, in ambient air measurement—the contact of sample with the O_3 reagent must be made under extremely low pressure conditions to minimize quenching of chemiluminescence. For exhaust measurements however, the system can be operated at higher sensor pressures obviating thus the need for a high vacuum pump. Even under such moderately reduced pressure conditions (~200 Torr), chemiluminescent instruments can detect NO at concentrations as low as a few parts per billion. Interferences can be of two types: (a) interference caused by sample compounds reacting with O_3 and emitting chemiluminescence similar to that emitted by the NO-O_3 system; (b) interference caused by sample components that quench the NO-O_3 chemiluminescence. No interferences were found of the first type. Quenching interferences have been suggested to be caused by CO_2 and H_2O (**13**); however, the problem was not verified by others (**12**).

In general, application of the method for NO measurement presents no problems. For NO_x measurement, however, problems may be caused with the O_2 concentration in the sample is low—e.g., in raw exhaust samples—or when there is NH_3 in the exhaust sample. Such problems are discussed in reference **12**.

Chemiluminescence instruments, in general, more than meet the sensitivity requirements of exhaust analysis. Their response time is in the order of 1 to 2 seconds, and response varies linearly with NO over a large concentration range. Their main disadvantages are the relatively high cost, and the need for generation and disposal of O_3 reagent.

Manufacturers presently known to offer commercial instruments for chemiluminescent measurement of NO and NO_x in exhaust are: Aerochem Research Laboratories, Inc., Beckman Instruments, Inc., Bendix Corporation, Environmental Tectonics Corp., Intertech Corp., LECO Corp., McMillan Electronics, Inc., Meloy Laboratories, Inc., Monitor Laboratories, Inc., REM Scientific, Inc., Scott Research Laboratories, Inc., and Thermo Electron Corp.

2.7 ELECTROCHEMICAL CELL METHOD (**14, 15**). Electrochemical cell instruments use a sealed module contacting the sample stream through a semi-permeable membrane. NO_x diffuses through the membrane into the module where it reacts with the electrolyte film on the surface of the sensing electrode. The electrochemical oxidation of NO and NO_2 into nitrate releases electrons that cause a current flow from the sensing electrode to the counter electrode. The current is proportional to the sample NO_x concentration. Response specificity is determined by the semi-permeable membrane, the electrolyte, the electrode materials, and the retarding potential. "Retarding" potential is used to retard oxidation of those sample components that are less readily oxidized than NO_x.

The measurement is affected by a positive SO_2 interference; therefore, SO_2 be scrubbed off the sample. Some evaluation data have been reported (**10, 16**); however, instrument design has since undergone modifications. The instruments tested thus far did not perform as well as the infra-red monitors. However, they are being continuously developed and they have a considerable compactness advantage over the IR analyzers. They are also simple to maintain and operate, and have a short response time (5 to 6 secs).

One manufacturer presently known to offer a commercial instrument for exhaust analysis in EnviroMetrics, Inc. The EnviroMetrics instrument can measure NO or NO_2 or NO_x in the 0.1 to 10,000 ppm range.

3. Applications of Methods for Measurement of Exhaust NO_x

The main application of NO_x measurement methods is in emission testing of motor vehicles by the Federal Testing procedures. The Federal procedures for testing automobiles call for collection of exhaust mixed with air in a plastic film bag (**17**). Resultant sample contains NO_x in the form of NO and NO_2 and must be analyzed immediately; upon standing, the sample loses its integrity as a result of conversion of NO into NO_2 and of loss of NO_2 on the walls and in reactions with other exhaust components (**5**).

Measurement of NO_x in such a sample can be made by one of the following ways: (a) by separate measurements of NO and of NO_2; (b) by a single NO measurement, made following complete conversion of the sample's NO_2 into NO; and (c) by a single measurement of NO_2, following complete conversion of the sample's NO into NO_2. Critiques of these methods are as follows.

Separate measurements of NO and of NO_2 can best, most simply, and most reliably be made using the chemiluminescence method for NO and the NDUV method for NO_2. The disadvantage of this NO_x measurement technique is that it requires use of two analytical methods. Conversion of NO_2 obviates the need for the NO_2 measurement; however, inclusion of this conversion step is not entirely without penalty. Uncertainties regarding the degree of NO_2 conversion and the response to other nitrogeneous components of the sample, either penalize the method's accuracy or they dictate inclusion of additional procedural steps that increase the method's complexity. Following NO_2 conversion, resultant total NO can be measured by chemiluminescence or NDIR or electrochemically. Of these, the NDIR has the disadvantage—a minor one—that it requires drying of the sample, which in turn requires that the measurement results be adjusted for the sample concentration caused by the water removal step.

Conversion of the sample's NO into NO_2 also simplifies the NO_x measurement in that it requires measurement of only one compound—NO_2. Resultant NO_2 can be measured reliably by NDUV. The disadvantage of this NO_x measurement technique is that the NO oxidation step is unavoidably accompanied by some NO_2 loss on the bag walls and/or in reactions with other sample components (**5**).

Besides the application upon samples collected by the Federal testing procedures, NO_x measurement is also called for in research studies of automotive emissions where NO_x must often be measured instantaneously. Such measurement can be made either upon the raw exhaust stream or upon the exhaust-air mixture stream delivered by a "constant volume" sampling device (**17**). Since in either case the proportion of NO_x present as NO_2 is negligibly small, the analysis is reduced to a relatively simple NO measurement. All instrumental NO-methods discussed in preceding sections can be used. However, the chemiluminescence and electrochemical cell methods have the relative advantage, although not an important one, of not requiring drying of the sample stream. The NDIR analyzer, operated under fast sample flow conditions to minimize response time, and equipped with a response-linearizer, can perform as well as the other instruments. However, the sample must be dried and this requires that a minor adjustment be made to the response data to compensate for the water removal effect.

One other application of the NO_x measurement methods can be made in analysis of *dried* exhaust-air mixture samples collected in bags. Such a sample collection technique is not normally applied in current exhaust emission studies. Nevertheless, it is mentioned here because it has certain advantages over the closely similar Federal testing procedures technique. Thus, by passing the "constant volume" exhaust-air sample through a drier prior to collection in the bag, it simplifies the NO_x measurement problems in at least two respects. First, under such dry conditions NO_2 losses in the bag are small. Second, the bag contents can be analyzed directly, *i.e.*, without drying, for NO and NO_2, a simplification that eliminates the dryer problems discussed in the preceding paragraphs. Note that these problems do not exist when drying is applied in the exhaust-air stream delivered by the "constant volume" sampler; this is because in such "fresh" exhaust the proportion of NO_x present as NO_2 is negligibly small.

4. References

1. INSTRUMENTATION FOR ENVIRONMENTAL MONITORING. 1974. Vol. 1: Air. Lawrence Berkeley Laboratory Report LBL-1, University of California, Berkeley, California.
2. BEATTY, R.L., L.B. BERGER, and H.H. SCHRENK. 1943. Determination of the Oxides of Nitrogen by the Phenol Disulfonic Acid Method. U.S. Bureau of Mines, Report of Investigations RI 3637 (February).
3. AMERICAN PUBLIC HEALTH ASSOCIATION, INC. 1977. Methods of Air Sampling and Analysis. 2nd. ed. pp. 578. Washington, D.C.
4. DRISCOLL, J.N., and A.W. BERGER. 1971. Improved Methods for Sampling and Analysis of Gaseous Pollutants from the Combustion of Fossil Fuels, Vol. II-Nitrogen Oxides. Report of Walden Research Corp., Cambridge, Mass., prepared for the EPA, EPA document APTD-1108; APTIC, EPA, Research Triangle Park, N.C. 27711.

5. DIMITRIADES, B. 1968. Methods for Determining Nitrogen Oxides in Automotive Exhaust. U.S. Bureau of Mines Report of Investigations RI 7133.
6. COULEHAN, B.A., and H.W. LANG. 1971. Rapid Determination of Nitrogen Oxides with Use of Phenoldisulfonic Acid. *Envir. Sci. Technol.*, 5:1963.
7. DAVIS, R.F., and W.E. O'NEILL. 1966. Determination of Oxides of Nitrogen in Diesel Exhaust Gas by a Modified Saltzman Method. U.S. Bureau of Mines Report of Investigations RI 6790.
8. SALTZMAN, B.E. 1954. Colorimetric Microdetermination of Nitrogen Dioxide in the Atmosphere. *Anal. Chem.* 26:1949.
9. SELECTED METHODS FOR THE MEASUREMENT OF AIR POLLUTANTS. 1965. Public Health Service document 999-AP-11, p. C-6, U.S. Dept. of Health, Education and Welfare.
10. SNYDER, A.D., E.C. EINUTIS, M.G. KONICAK, L.P. PARTS, and P.L. SHERMAN. 1972. Instrumentation for the Determination of Nitrogen Oxides Content of Stationary Source Emission. Vol. 2, Report of the Monsanto Research Corp., Dayton Lab., Dayton, Ohio, 45407.
11. LORD, H.C., D.W. EGAN, F.L. JOHNSON, and L.D. MCINTOSH. 1974. Measurement of Exhaust Emissions in Piston and Diesel Engines by Dispersive Spectroscopy. *JAPCA*, 24:136.
12. SIGSBY, J.E., JR., F.M. BLACK, T.A. BELLAR, and D.L. KLOSTERMAN. 1973. Chemiluminescent Method for Analysis of Nitrogen Compounds in Mobile Source Emissions (NO, NO_2 and NH_3). *Env. Sci. Technol.*, 7:51.
13. NIKI, H., A. WARNICK, and R.R. LORD. 1971. An Analysis in Piston and Turbine Engines. Paper presented at the SAE Automotive Engineering Congress, Detroit, Mich.
14. PLUG-IN SENSORS AND MEMBRANES PUT FINGER ON AIR POLLUTANTS. 1970. Product Engineering. 41:40.
15. SHAW, M. ENVIROMETRICS INC., Cartridge Sensors for Air Pollution Monitoring. Unpublished paper available from Manny Shaw, EnviroMetrics Inc., 13311 Beach Ave., Marina del Rey, Calif. 90291.
16. CHAND, R., and R.V. MARCOTE. 1971. Dynasciences Corp., "Evaluation of Portable Electrochemical Monitors and Associated Stack Sampling for Stationary Source Monitoring"; paper presented at the 68th Nat'l Mtg. of Am Institute of Chem. Engineers, Feb 28–March 4. Houston, Texas.
17. ENVIRONMENTAL PROTECTION AGENCY. 1972. Light Duty Vehicle Regulations for 1975 and 1976 Model Year Vehicles. *Fed. Register*, 37:(221), 24250. Nov. 15.

Subcommittee 3

B. DIMITRIADES, *Chairman*
W. A. COOK
J. E. CUDDEBACK
E. F. FERRAND
R. G. KLING
E. L. KOTHNY
P. W. MCDANIEL
G. D. NIFONG
F. T. WEISS

4. State-of-the-Art Review

Source Sampling for Mass Emissions of Particulate Matter

1. Introduction

The intent of this status report is to evaluate the present state-of-the-art of source sampling for particulate matter. The considerations presented herein are applicable to measurements of all categories of sources rather than to one or more specific industries.

The concepts and conclusions should be considered in standards writing, test method specification and emission data interpretation. It is particularly important to recognize that emission limitation regulations are based on source emission data which may or may not have been collected by source measurement techniques similar to those specified in the regulation. Changing the test method by which compliance is determined may result in a different numerical emission value. This factor must be considered when defining or interpreting regulatory standards for emissions.

*Prepared for the Intersociety Committee by Robert A. Herrick, Director, Environmental Engineering & Technology, Owens-Corning Fiberglas Corporation, Toledo, Ohio 43659

2. Historical

Most particulate control devices operate on only that material which is in the particulate phase at stack conditions. For many years source tests were designed to measure only these particulates. With the exception of data from a few local agencies, most pre-1970 emission test data are based upon the amount of particulate matter at stack temperature and pressure, as determined by an in-stack filter. The alundum thimble option from ASME's PTC-21 and 27 (**1, 2**) was the most common technique. This met the needs of the time by enabling the determination of the performance efficiency of such equipment as cyclones and precipitators.

With the passage of the Air Quality Act Amendments of 1970 (**3, 4**) a new issue came into focus. The ambient air quality standards promulgated pursuant to this Act defined standard conditions for ambient suspended particulate matter as 70 F. and 700 Torr; this raises a potentially serious ambiguity. Simple arithmetic can be applied to express source emission concentrations at these conditions. Depending on the chemical species of the emissions and the conditions at the point of measurement, more or less of the emissions may be in the gaseous phase at the point of measurement but in the liquid or solid particulate phase at 70F. and 700 Torr. The fraction of the emissions exhibiting this behavior has been called condensible particulate matter.

Several state and local air pollution agencies recognized this situation many years ago and developed source testing sampling trains which attempted to measure these condensibles, generally based on cooling the sampled gases. The condensation of moisture from the sampled gas required a condenser or impinger in the sampling train.

The predecessor agencies of EPA first dealt comprehensively with this problem when evaluating the emissions from incinerators. Incineration processes generate substantial amounts of condensible particulate matter, commonly greater than the amount of in-stack particulate matter. The sampling train selected for these studies was basically that used by the Los Angeles Air Pollution Control District (**5**). It embodies a heated probe, cyclone, filter, impinger train, desiccant pump and meter.

This same basic sampling train was used by EPA to gather the background particulate emission data used as the basis for the first set of Proposed National Standards of Performance for New Stationary Sources (**6**)—Fossil Fuel Stream Generators, Portland Cement Plants, and Incinerators. Controversy developed over the inclusion of the impinger catch as "particulate matter." The final standards (**7**) adjusted the numerical standards downward and specified a modified sampling train, reporting only the weight gain of a glass fiber filter maintained at a nominal temperature of 250 F. and the washings from the train upstream of the filter as the particulate weight. This is commonly called "front half" particulate matter.

The net result of this modification is that these standards are set on the basis of testing at compromised physical conditions which do not specifically meet the data requirements of the control equipment engineer *or* the scientist who is attempting to relate point source emissions to ambient air quality. They have the utility of a standardized test procedure, however, and the results of tests conducted by these methods can be used to evaluate the performance of new or existing installations.

3. Methods

3.1 DETERMINATION OF PARTICULATE MASS CONCENTRATION. Manual test methods are used for nearly all determinations of mass concentration at the present time. These methods consist of sampling the gas stream of interest typically for a period of one to three hours, then transporting the sample(s) to the laboratory for analysis.

The analysis of the samples collected during field test programs is as important as their collection. Laboratory gravimetric procedures concerning pre- and post-conditioning of samples, in particular, must be defined in specific terms. Attention must be given to the tare weight of the filter, beaker, or other container in terms of its statistical reliability in comparison to the sample weight.

Filters and impaction devices are the most important particulate collection elements used in current sampling trains. The ASTM procedure **(8)** includes recommended filter types to be used for ranges of dust concentration, temperature and moisture conditions. Staged impactors **(9, 10)** have been used to collect stack particulate samples in sized fractions. The Greenburg-Smith impinger is used as a collection element in several sampling trains.

Although the efficiency of some filters has been studied in depth, there is still some uncertainty in that some solid volatile or liquid particles may pass even the glass fiber filter used both in the ambient high volume sampler and many stack sampling trains. The selection of a filter depends in part on the purpose of the test. The alundum thimble, for example, is ideally suited to sampling upstream of a dust collector because of its capacity to collect over five grams of sample. The alundum thimble is unsuited to the measurement of the very low concentrations of particulate matter passing state-of-the-art control equipment because of its low initial efficiency and its high tare weight.

Impaction devices have the inherent characteristic of selective capture of particles above a certain range of aerodynamic diameters. For the Greenburg-Smith impinger, a wetted impaction device designed for the evaluation of respirable dust exposures in work places, performance parameters have been well defined. At an inlet volume of 1.0 cfm (a pressure drop of 2.3 in. Hg) the device collects essentially 100% of 0.8 micrometer silica particles **(11)**. The collection efficiency for other materials can vary widely **(12)**. Keenan and Fairhall **(13)** found when operating the Greenburg-Smith impinger at sonic velocity (about 1.6 cfm inlet volume) that 98.5% collection was attained for freshly formed lead fume, while only 60% collection of this fume was attained when operating the impinger at 1.0 cfm (0.0283 m^3/min).

The impaction devices require pre-and post-weighing of plates, or complete removal of the collected particulate matter for direct weighing. The Greenburg-Smith impinger's liquid medium, commonly water, wets the impaction plate and retains the impacted dust so it does not blow through the device. While the use of the wet impinger greatly increases the capacity of this particulate collection device, it leads to some other complications. Many gases, *e.g.*, SO_2, are at least partially soluble in water. These gases are absorbed in water and an ionic solution is formed. Soluble particulates also generate ions, and when the sample is evaporated the ions may recombine as compounds other than those which originally existed in the source emission being sampled **(14)**. This complicates the interpretation of the data.

There are several techniques which have been developed to the prototype stage which show promise in the evaluation of particulate emissions. The weight gain of a filter can be determined *in situ* by beta absorption. This approach has been used in continuous source monitors where a sample of stack gas is drawn through a membrane tape on which the particulate mass is continuously or intermittently measured with a beta gauge.

A technique potentially suited to continuous monitoring is the piezoelectric crystal method, where particles are deposited electrostatically or by impaction on a quartz crystal. The change in crystal resonance frequency is a direct measure of the increase in mass if certain specific criteria are met. Several developments in both sampling and analysis have been recently reviewed **(15)**.

3.2 MEASUREMENT OF IN-STACK PARTICULATE CONCENTRATION The data requirements of the control equipment engineer have been met for many years by procedures designed to evaluate the concentration, composition and chemical-physical properties of in-stack particulate matter, *i.e.*, those compounds which exist as solid or liquid particles at the temperature and pressure conditions at the point of measurement. Procedures for the selection of a sampling location, number of traverse points, and testing at isokinetic conditions are well described **(1, 2, 7, 8, 16)**.

The location of the sampling station is important in that velocity and particulate concentration profiles can be different. The selection of the sampling location, conduct of the velocity measurement and the maintenance of isokinetic sampling rates are frequently difficult. All of these procedures call for particulate collection on some type of filtration medium.

Other important considerations for in-stack sampling are the integrity of the filter, its collection efficiency, and the reliability of the weighing. Any particulate matter deposited in the nozzle and front portion of the filter holder must be included in the sample.

An extension of this concept is the transport of the sampled gases through a probe to a collection element located outside the stack. Filters, electrostatic precipitators and impaction devices have been used. This approach has both advantages and disadvantages.

The withdrawal of a sample of stack gases to an external measurement element, whether it be a pre-weighted filter or one of the new continuous mass meters, presents certain technical problems which must be addressed. There is the strong probability that some material will be deposited in the probe. Particular attention must be paid to the prevention of condensation at random points in the sampling train. Limits of temperature variation should be defined and maintained throughout the system in order to standardize the procedure.

Manual sampling techniques usually are adequate to meet these criteria. The use of continuous, automatic out-stack particulate mass sensors will require innovative mechanical design to overcome these several potential problems.

3.3 MEASUREMENT OF CONDENSIBLES. There is presently no standardized technique for measuring condensibles, *i.e.*, compounds which are in the gaseous state at the point of measurement in the stack but are solid or liquid particles at 70 F. and 700 Torr.

The moisture content of many stack gases is such that 70 F. is below the water dew point. Once condensate forms, most filters blind. The impinger overcomes this problem, but raises the question of the absorption of gases which become reported as particulate matter.

A total particulate continuous monitoring device which dealt with these problems was developed at Harvard University under an American Iron and Steel Institute grant **(17)**. This device uses clean, dry air dilution to reduce the dew point of stack gases without forming liquid droplets. The particle-laden sample gas stream is analyzed for mass in a tape sampler outfitted with a beta gauge. While not widely used, this system has been commercially available. EPA is currently funding contract research for similar approaches, for comparison with the standard EPA sampling train.

Adequate definition of condensibles is the weak link in source measurements for particulate matter at the present time.

3.4 COMPARISONS OF METHODS. Considerable time, energy and expense have been expended on demonstrating that there is little comparability between reported concentrations determined by in-stack, EPA "front half," and EPA full train methods **(18, 19)**. There is no substantive logic which would lead to the conclusion that the reported particulate concentrations should be the same, since the temperature and pressures are not standardized and some of the trains use impingers while others do not.

4. Conclusions

There are three purposes for source testing for particulate matter. The ability to accomplish these three purposes is at various levels of technology.

4.1 TO GAIN ENGINEERING DATA FOR THE DESIGN AND EVALUATION OF IN-STACK PARTICULATE AIR POLLUTION CONTROL DEVICES. The in-stack sampling trains typified by ASME and ASTM will, with appropriate filters and proper test execution, adequately provide mass concentration data for particulate matter at stack conditions.

4.2 TO TEST EMISSIONS FOR COMPLIANCE OR NON-COMPLIANCE WITH REG-

ULATORY STANDARDS. Any technique which is reproducible can accomplish this purpose; however, the standards must be consistent with the results obtained by the specified technique. Substantial test data should be available to assure that the method measures only the parameter of interest, without influence from other variables.

4.3 TO DEVELOP EMISSION DATA WHICH CAN BE CORRELATED WITH AMBIENT AIR QUALITY. There is no accepted method to accomplish this purpose at the present time. The best approach is the determination of particulate matter which exists after reduction to a standard temperature and pressure. This still does not take into account the myriad of atmospheric reactions which will ultimately result in ambient suspended particulate matter formation.

5. Summary

Significant confusion exists in the air pollution field regarding methods for the measurement of particulate emissions from point sources. The heart of this confusion lies in the lack of specificity of the temperature and pressure of the sampled stack gases at the point of measurement; *i.e.*, there is no good definition of particulate matter.

Several established methods are available for measuring the particulate content of stack gases at stack conditions.

At least one method, the EPA "front half" as adopted with the New Source Standards of Performance, is a specific method using arbitrary pressures and temperatures. While it correlates directly with neither control equipment engineering data nor ambient air quality measurement, it is a repeatable method which can be used to check emission levels against a numerical standard.

There are currently no fully adequate methods for measuring the condensible fraction of point source particulate emissions. The use of the Greenburg-Smith impinger is unsatisfactory because of its low particulate collection efficiency in the submicron range and the possible absorption of gases.

Until particulate matter is adequately defined there can be no fully satisfactory source measurement methodology. Until this time, an arbitrary method such as the EPA "front half" represents a satisfactory definition of source particulate emissions in terms of a specific measurement method.

6. References

1. DUST SEPARATING APPARATUS. 1941. Power Test Code 21, American Society of Mechanical Engineers, New York.
2. DETERMINATING DUST CONCENTRATION IN A GAS STREAM. 1957. Power Test Code 27, American Society of Mechanical Engineers, New York.
3. CLEAN AIR AMENDMENTS OF 1970. PL 91-604 (December 31).
4. NATIONAL PRIMARY AND SECONDARY AMBIENT AIR QUALITY STANDARDS. 1971. Federal Register, **36**:N84 (April 30). Washington, D.C.
5. DEVORKIN, H., *et al.* 1963. Air Pollution Source Testing Manual. Los Angeles County Air Pollution Control District.
6. STANDARDS OF PERFORMANCE FOR NEW STATIONARY SOURCES. NOTICE OF PROPOSED RULE MAKING. 1971. Federal Register, **36**:N159 (August 17). Washington, D.C.
7. STANDARDS OF PERFORMANCE FOR NEW STATIONARY SOURCES. 1971. Federal Register, **36**:N247 (December 23). Washington, D.C.
8. STANDARD METHOD FOR SAMPLING STACKS FOR PARTICULATE MATTER. Standard No. D2928-71, American Society for Testing and Materials, Philadelphia.
9. BRINK, J.A., JR. 1958. Cascade Impactor for Adiabatic Measurements, Ind. and Eng. Chem., **50**:N4.
10. ANDERSON, A.A. 1966. A Sampler for Respiratory Health Hazard Assessment. JAIHA **27**:2.
11. DRINKER, P., and HATCH, T. 1954. *Industrial Dust.* McGraw-Hill, New York.
12. SCHADT, C. and CADLE, R.D. 1957. Anal. Chem. **29**:864.
13. KEENAN, R.G., and L.T. FAIRHALL. 1944. The Absolute Efficiency of the Impinger and of the Electrostatic Precipitator in the Sampling of Air Containing Metallic Lead Fume, J. of Ind. Hyg. and Toxicol. **26**:7.
14. BARNES, E.T., and SANDERSON, J.G. 1971. Sampling and Analysis of Gas Streams for Particulate Matter. A State-of-the-Art Review. Canadian Pulp and Paper Association, Montreal.
15. WAGMAN, J. 1972. Recent Developments in Techniques for Monitoring Airborne Particulate Emissions from Sources, presented at the 72nd Annual Meeting of AICHE, St. Louis, Missouri.

16. "BULLETIN WP-50" Joy Manufacturing Company, Western Precipitation Division, Los Angeles.
17. BULBA, E., and L. SILVERMAN. 1965. A Mass Recording Stack Monitoring System for Particulates, presented at the 58th Annual Meeting of the Air Pollution Control Association, Toronto, Canada.
18. HEMEON, W.C. L. 1972. Stack Dust Sampling: In-Stack Filter or EPA Sampling Train, JAPCA **22**:7.
19. SELLE, S.J., and G.H. GRONHOVD. 1972. Some Comparisons of Simultaneous Gas Particulate Determinations Using the ASME and EPA Methods. ASME preprint 72-WA/APC-4.

5. State-of-the-Art Review

Source and Ambient Air Analysis for Sulfur Trioxide and Sulfuric Acid*

1. Introduction

This report presents a review of the current status of methods for sampling and analysis of emissions containing sulfur trioxide and sulfuric acid mist from various stationary sources and from ambient air. Two distinct operations are involved, (a) collection of the sample by a variety of techniques such as absorption, condensation or filtration; (b) subsequent analysis of the sample.

Presently, the greatest weakness in existing methods for emission sources lies in the sampling procedure. More research is required to improve sampling techniques and to determine the efficiency of sample collection under field conditions. The various analytical steps to analysis of samples seem to be satisfactory in the light of the limitations of sampling procedures.

2. Emissions from Sulfuric Acid Manufacturing Plants

The sampling and analysis of sulfur trioxide and sulfuric acid mist in emission sources has been the subject of a number of recent studies (**1–7**). Various forms of standardized and proposed alternate methods are being studied for their reproducibility, accuracy and performance under field sampling conditions (**1, 2, 7–9**). The results of a recent study by Walden Research Corporation for EPA† on the validation of improved chemical methods for sulfur oxides measurements has been published (**1**). Samples from a number of types of sources, including sulfuric acid manufacturing plants were taken simultaneously by up to three different trains and analyzed by several analytical techniques. EPA Method 8 for the analysis of H_2SO_4 mist has been compared with the Brink-Monsanto method (**7**). Figures 5:1 to 5 show the types of sampling trains used. A cursory examination shows major differences in the types of apparatus under study.

The most serious problem comes from sulfur dioxide which generally is present simultaneously in much higher relative concentrations. It tends to oxidize in solution leading to high sulfate results (**1, 10, 11**). It is soluble in aqueous solutions at pH values greater than 3. Depending on the method used, other interferences may come from particulates, hydrochloric acid, nitrogen oxides (or nitric acid), various sulfate compounds, and heavy metal ions which catalyze the oxidation of sulfur dioxide in solution by dissolved atmospheric oxygen (**1, 2, 10, 11**).

The more commonly used methods generally include some variation of two distinct and somewhat independent opera-

*Prepared for the Intersociety Committee by Paul Urone, Environmental Engineering Department, University of Florida, Gainesville, Florida 32611

†Environmental Protection Agency.

tions. One involves the sampling and the other involves the analytical, or measurement technique.

2.1 SAMPLING. The sample of the emission source must be taken so that it is representative of the gas stream, at isokinetic velocities and for a prescribed time to obtain a known total volume. Methods for selecting the number and positions of the transverse sampling points, and the techniques for isokinetic sampling have been officially specified and generally accepted **(12)**. The treatment of the sample gas as it is being withdrawn depends upon the analytical method to be used. As a rule, the probe and connecting tubing to the sampling train are heated to prevent unnecessary condensation.

Selectively removing the SO_3 and H_2SO_4 from the gas sample involves some variation or combination of two basic techniques: absorption or condensation followed by filtration.

2.1.1 *Absorption.*

a. Isopropyl Alcohol Solutions. The EPA acid mist analytical method **(12)** uses a heated sampling probe fitted with a stainless steel nozzle to obtain the sample (Figure 5:1). The sample gas is then passed through a standard Greenburg-Smith impinger containing 100 ml of 80% isopropyl alcohol (IPA) cooled in an ice bath. Presumably, the IPA absorbs the SO_3 and H_2SO_4 and prevents the oxidation of SO_2 **(13, 14)**. Any SO_3 which may tend to escape is assumed to have become hydro-

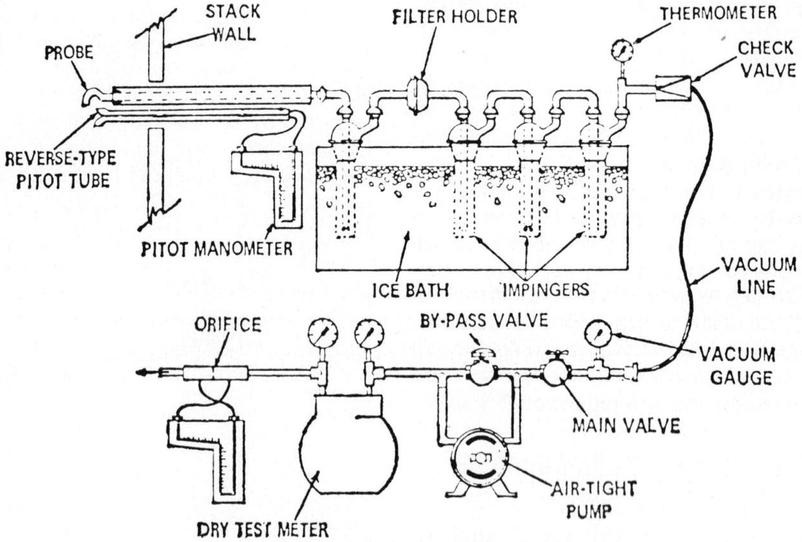

Figure 5:1—EPA Sulfuric Acid Mist Sampling Train.

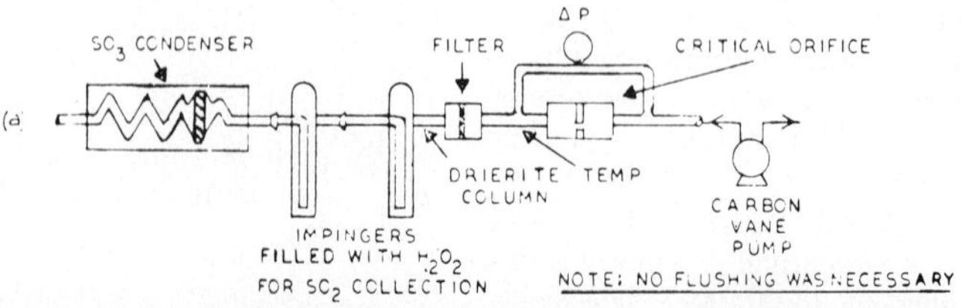

Figure 5:2—Train "A". Controlled Condensation. Walden Research.

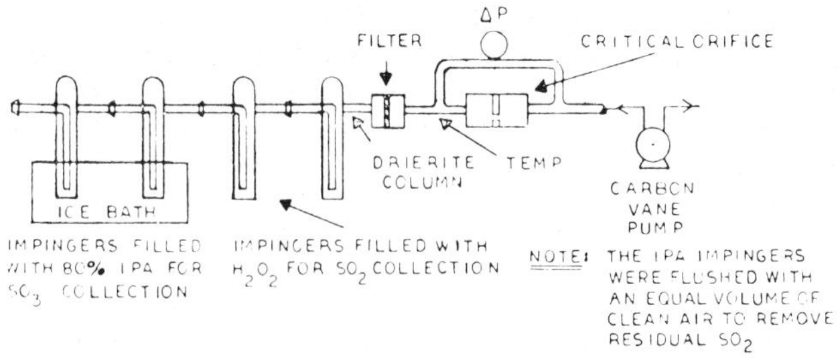

Figure 5:3—Train "B". IPA Collection. Walden Research.

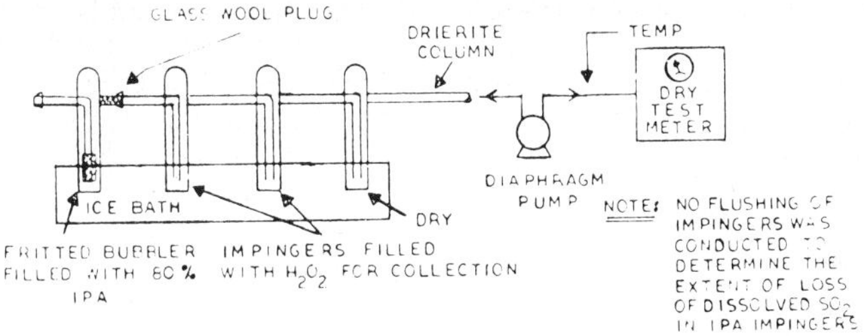

Figure 5:4—Train "C". Modified IPA Collection. Walden Research.

lyzed to H_2SO_4 mist, which is then collected with any mist not captured by the impinger by a glass fiber filter (Figure 5:1). Sulfur dioxide is not highly soluble in the IPA solution and is absorbed in the following two impingers which contain 100 ml of 3% H_2O_2. The fourth impinger contains 200 g of silica gel which strips off water and any residual polar gases or vapors.

Any sulfur dioxide retained by the IPA solution in the first impinger is purged by flushing with clean air at ambient temperature for 15 min or for a time equal to the sampling time.

Serious difficulties have been encountered with the IPA absorbing solution. Under field conditions it is difficult if not impossible to keep the solution from partially evaporating when sampling dry stack gases. Tests by Walden Research Corporation **(1)** showed that significant amounts of SO_2 remain in the IPA solution even after 15 min of purging. When air containing only SO_2 was sampled, 1 to 3% of the SO_2 was found as sulfate in the IPA solution. Impingers containing IPA were placed after the fiber glass filter in the EPA train (Figure 5:1), and as much as 50% additional sulfate was found in the second and third impingers as was found in the first **(7)**. It is difficult to ascertain at this point whether the IPA solution is not capturing the SO_3 or H_2SO_4 quantitatively or that SO_2 is becoming oxidized in the IPA solution with and without sufficient purging times or that some combination of both is happening.

b. Alkaline Solutions. Sodium hydroxide solutions have been used to absorb total SO_x compounds. There have been some efforts to analyze for SO_2 in separate or the same solution and to determine SO_3 and H_2SO_4 by difference **(15–17)**. However, SO_2 oxidizes somewhat readily in alkaline solutions, and inhibitors such as benzyl alcohol, mannitol, formaldehyde, and glycerol have not been found to prevent the oxidation **(15, 18, 19)**. Measurements of SO_3 and H_2SO_4 generally are high and erratic when sampled with alkaline solutions **(1)**.

2.1.2 *Condensation.* A considerable number of sampling methods are based on the condensation of sulfur trioxide as sulfuric acid mist followed by filtration, impingement, etc. **(20–26)**. Taylor **(20)** and Schneider **(21)** studied the condensation of SO_3 as a function of temperature. Seith **(22)** obtained a patent on a condensing water method of SO_3 analysis. Minzl **(26)** used a multiple stage system composed of impaction, filtration, and condensation with added water vapor. Since SO_2 is very soluble in water, care is taken not to condense water along with the H_2SO_4 mist. However, this precaution may not be entirely necessary as SO_2 is not very soluble in acidic aqueous solutions **(11)**, and the aerosol condensate is highly acidic over a wide range of water vapor pressure **(20)**. Lisle and Sensenbaugh **(23)** obtained quantitative recoveries from synthetic flue gas streams. Milligram amounts of H_2SO_4 were evaporated into the stream and recovered with a sampling condenser, similar to that shown in Figure 5:2, kept at 60 to 90 C. No additional sulfate was obtained when as much as 2000 ppm SO_2 was also present in the gas stream. Changes in water vapor content from 6.9 to 9.4% did not affect the results.

a. Controlled Condensation. The controlled condensation method used by Walden Research (Figure 5:2) was very similar to that of Lisle and Sensenbauch **(23)** and Wang **(25)**. The sample air stream was passed through the coiled condenser kept at 60 to 90 before passing through a fritted glass disc built into the condenser. A considerable amount of difficulty was experienced with the method and erratic results were obtained for various fossil fuel combustion sources (Ref. **1**, pp. 1–58 to 1–60). In some cases controlled condensation gave results considerably higher than simultaneous samples run by the IPA method **(12)**. In others, the opposite was true. Correlation coefficients were very poor, running from zero to 0.4.

For a contact sulfuric acid manufacturing plant, Walden Research found that the controlled condensation method gave results less than 50% (Ref. **1**, pp. 2–83 to 2–96). The following relation was found:

$$\text{ppm } SO_{3\ (\text{controlled condensation})} = 0.14 + 0.45 \text{ ppm } SO_{3\ (\text{IPA method})}$$

the correlation coefficient was 0.80.

Such results are in sharp contrast to those found by Lisle and Sensenbaugh **(23)** in their controlled experiments. They indicate serious losses caused either by failure of the SO_3 to condense and grow to a sufficiently large particle size to be collected by the fiber glass plug, or by oxidation of SO_2 in the IPA solution or some combination of both.

Stack gases from contact sulfuric acid manufacturing plants can be very dry, and satisfactory condensation may be more difficult to obtain under such conditions. Sulfuric acid aerosols can be quite small, with particle sizes ranging from 0.2 to 0.5 micrometers **(27)**.

b. Uncontrolled Condensation. The Brink-Monsanto method **(8)** uses a cyclone followed by a glass filter tube filled with pyrex glass wool to catch the sulfuric acid mist and whatever SO_3 gas will condense or adsorb on the glass wool and walls of the sampling train up to the filter (Figure 5:5). The box containing the cyclone and filter is heated to stack temperature plus 10 F which probably prevents further condensation and particle size growth.

Some methods **(14)** pass the flue gases directly into impingers containing IPA solution where condensation as well as absorption occur. The efficiency of IPA solutions for capturing all SO_3 and H_2SO_4 and preventing the oxidation of SO_2, as mentioned above, is in doubt. Changes in the amount and concentraion of the IPA solution during sampling is difficult to control. Evaporation occurs with dry flue gases and with sampling pressure drops. When the flue gases are wet, considerable amounts of water are collected, and the IPA solution becomes excessively dilute.

c. Filtration. It has long been recognized that fiber glass and other types of filters can adsorb and oxidize sulfur dioxide to give high sulfate and low SO_2 results **(28, 29)**. Barton and McAdie described a technique for treating glass fiber filters to prevent SO_2 oxidation, and recently proposed the use of Nucleopore filters for the analysis of sulfuric acid mist in ambient air **(4)**. Few if any of the methods used for source sampling of sulfuric acid mist prescribe an interference free filter or a method for treating the filter to prevent interference.

The stability of sulfuric acid mist on a filtering medium has not been well documented. Lisle and Sensenbaugh **(23)** obtained good recoveries with controlled laboratory experiments. On the other hand, field tests involving filtration of condensed sulfuric acid mist have given poor results **(1)**. It is not clear whether Lisle and Sensenbaugh's high recoveries were due to the nature of the experiments: *i.e.*, volatilization of milligram amounts of sulfuric acid into the air stream at perhaps high "plug" type concentrations; or high moisture content (6 to 9%) of the air stream; or cessation of sampling after injection preventing gradual evaporation; etc. **(23)**.

Richards **(5)** working with 0.08 to 16 μg of sulfuric acid reported successfully volatilizing most of the acid from a Teflon filter with warm dry air at 50 C. The purpose was to volatilize the sulfuric acid for flame photofluorescent measurement without decomposition of other sulfate compounds including ammonium sulfate. Scaringelli and Rehme **(30)** earlier introduced a somewhat similar method but used a temperature of 400 C for volatilization of the sulfuric acid. Dubois, *et al.* **(31)**, used a temperature of 195 to 200 C. In the latter two cases ammonium sulfate interferes and a temperature of 130 C has been suggested by Richards and Dubois, *et al.* **(5, 31)**. The point taken here is that sulfuric acid does volatilize at temperatures encountered under field sampling conditions and proper techniques should be developed to prevent it.

2.2 ANALYTICAL. A rather large number of methods have been used to measure the amount of SO_3–H_2SO_4 present in a sample after it has been collected. The methods range from simple acid-base titrations to more complicated spectrophotometric measurements. Selection depends upon the sampling method used, the amount of sample present, and the personal preference of the investigator with respect to convenience, dependability, and his experience. EPA adopted a barium ion titration technique in its Method 8 for sulfuric acid manufacturing plants **(12)**. Figure 5:6 shows the analytical procedures used by Walden Research in its validation studies **(1)**. Included are the more commonly used techniques for the measurement of sulfuric acid or sulfur dioxide emissions from stationary sources.

2.2.1 *Sodium Hydroxide Titration.*

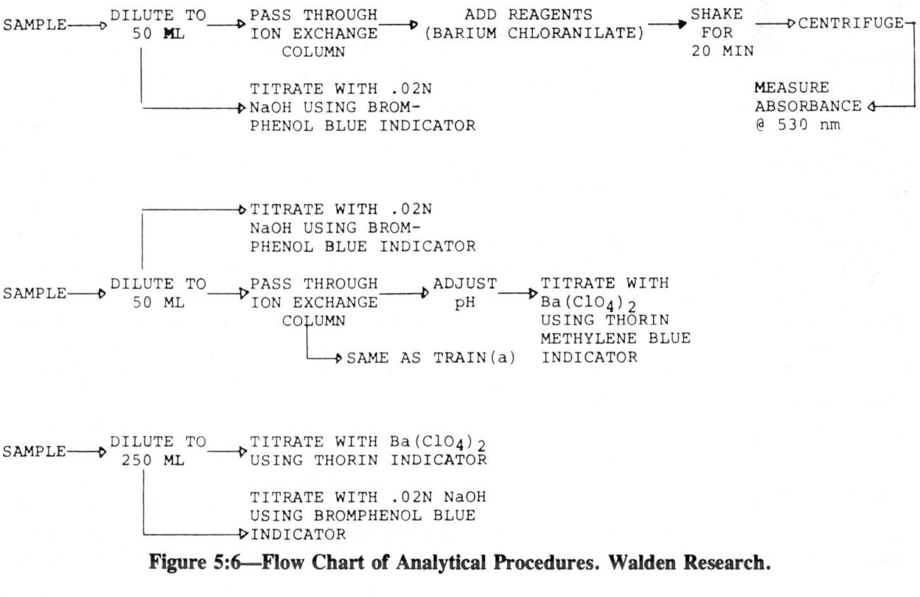

Figure 5:6—Flow Chart of Analytical Procedures. Walden Research.

Titration with sodium hydroxide continues to be popular with many laboratories (**1**, **2**, **8**, **23**, **25**, **32**). The method is simple, direct, and easy to use. However, it is not specific and depends upon the assumption that the only acid present is H_2SO_4 and that no basic substances are present to interfere. The Walden Research studies (**1**) compared titration with 0.02 N NaOH (bromphenol blue indicator) with barium ion titration and the chloranilate colorimetric method (see below). They found the sodium hydroxide method in itself reliable and accurate under the conditions used.

The Brink-Monsanto method also uses sodium hydroxide titration (phenolphthalein indicator) with good precision (**2**, **8**). Lombardo used 0.1 N NaOH and methyl red indicator (**32**).

2.2.2 *Barium Ion Titration.* EPA's Method 8 (**12**) specifies barium ion titration with thorin as an indicator. The reaction is reasonably specific for sulfate ion (**13**, **14**, **31**). Generally, thorin is used as the indicator; although "nitrochromazo" has been proposed as an alternate indicator (**33**). Any soluble sulfate will react. For those flue gases which may contain interfering particles (including sulfate salts and heavy metal ion catalysts) a glass wool plug is inserted in the probe. The effects of the glass wool plug have not been evaluated. The method may be used with IPA, alkaline, and aqueous sampling solutions. Barium perchlorate ($Ba(ClO_4)_2$) has displaced barium chloride ($BaCl_2$) (**13**) as the precipitating reagent. The danger of heating perchlorate compounds in the presence of organic substances has not been emphasized.

Field evaluation tests indicate that as an analytical method, barium ion titration in itself is reliable when properly used (**1**).

2.2.3 *Chloranilate Method.* This method depends on the release of the colored chloranilate ion from the insoluble barium chloranilate salt (**1**, **4**, **37**, **38**). Sulphate reacts with the barium chloranilate to form insoluble barium sulfate and release the chloranilate ion to color the solution in proportion to the amount of sulfate originally present. The method is sensitive enough to be used for ambient air sulfuric acid measurements (**4**), and it proved reliable as an analytical method for source emissions (**1**).

Heavy metal ions interfere, and they are removed by passing the adjusted sampling solution through an ion exchange column (**38**, Figure 5:6). Although good results have been reported on the use of the ion exchange column (**31**, **38**), Walden Research obtained very low results; consequently they did not use it for the sulfuric acid plant study (**1**). The procedure for the ion exchange step should be reviewed and written in more specific detail.

2.2.4 *Turbidimetric Methods.* Several investigators have used the turbidimetric method for measuring sulfate (**34**, **35**, **39**). Barium ion is used to precipitate the sulfate as a uniformly distributed suspension. The turbidity of the solution is measured spectrophotometrically. However, the precipitation process is not easily controlled and results can be highly variable (**2**, **14**). The method was not included among those studied for validation (**1**) and is not generally recommended at this time.

2.2.5 *Flame Photoluminescent Detector.* Sulfur compounds of all types fluoresce in hydrogen rich flames at a wavelength that is relatively specific for sulfur (**5**, **30**, **40**). The fluorescent light is passed through a narrow band optical filter and is measured with high sensitivity by means of a photomultiplier tube. The detection system is used for direct ambient air measurement (**41**) and can form a useful specific detection system for source sampling.

2.2.6 *Other Methods.* Other analytical or measurement techniques that have been used include an amperometric titration method using lead ion (**19**); a method using the reaction of excess sodium tetraborate with sulfuric acid and titration of the excess using bromthymol blue indicator (**42**); and a method using benzidine as a reagent (**2**, **16**).

3. Other Stationary Emission Sources

A brief word needs to be said about sampling and analysis for SO_3 and H_2SO_4 in the emissions from stationary sources in general. The methodology covers the same principles as discussed above. However, greater care has to be taken to prevent interference from the greater number and variety of particulate matter and gases that can be present. The emissions from cement plants, for example, contain limestone derived particles which interfere seriously by reacting to form insoluble calcium and magnesium sulfate during the air sampling step. Ammonia will interfere seriously with alkaline titrimetric methods.

In most instances, a glass wool plug, or an alundum thimble, is used to prevent the particulate matter from passing through the sampling probe. As mentioned above, the effect of such filters has not been properly evaluated.

4. Ambient Air Analysis

The analysis of ambient air for sulfuric acid mist includes the same general principles necessary for source analysis. The amount of mist present is much lower in concentration. The number of interfering substances can be the same, in lower concentration but greater in number and variety. Sampling of the ambient air, however, is much simpler. There is no need for a complex stack sampling train. The more recent proposed methods depend on filtering rather large quantities of air through glass fiber **(30)**, Teflon **(5)**, or Nucleopore **(4)** filters. Scaringelli **(30)** recommends evaporation of the H_2SO_4 from the glass fiber filters at 400 C in a combustion tube and converting to SO_2 which is analyzed by coulometry, flame photoluminescence or colorimetry.

Monkman recommends a micro-diffusion method from glass filters in an oven at 200 C followed by acid-base titration. Richards **(5)** recommends evaporation at 130 C and analysis by flame photoluminescence. Barton **(4)** rinses a Nucleopore filter with isopropyl alcohol (IPA) and analyzes sulfate by the chloranilate method.

Ammonia is a serious interferent in am-

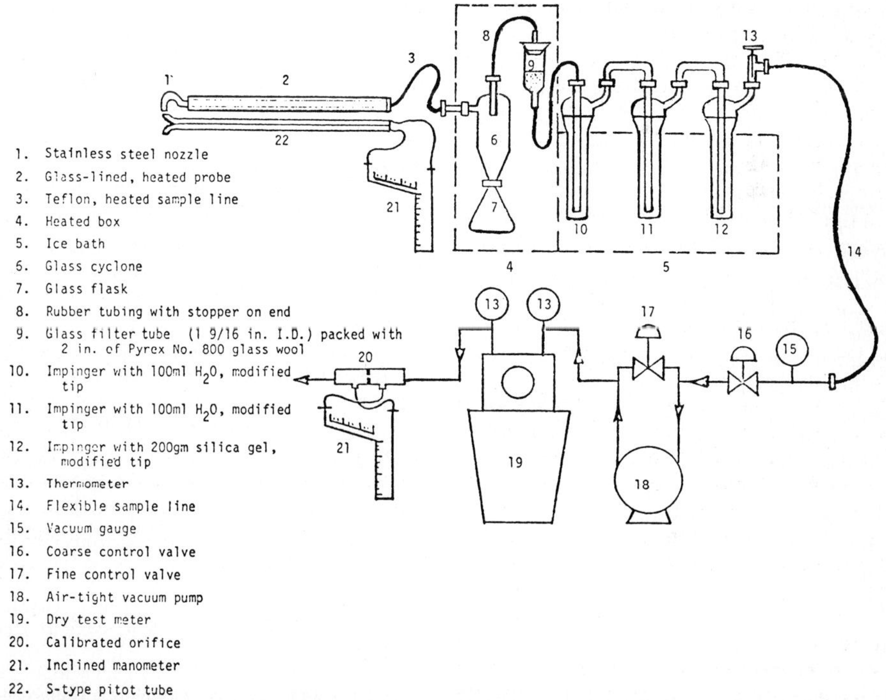

Figure 5:5—Brink-Monsanto Mist Sampling Train by Environmental Engineering, Inc.

bient air. It forms ammonium sulfate before, during and after the sampling period. At the higher temperatures, ammonium sulfate decomposes and volatilizes as sulfuric acid. The ammonium sulfate presumably results from what originally was sulfuric acid and may be assumed to properly represent the acid's earlier existence in air (**30**). Monkman suggests a lower microdiffusion temperature of 130 C if this were a problem (**31**). In Barton's method (**4**) sulfates including ammonium sulfate are assumed to be insoluble in nonaqueous isopropyl alcohol.

5. Summary

Probably the greatest weakness of the presently known and used methods for measuring sulfur trioxide and sulfuric acid in emission sources and in sulfuric acid manufacturing plants in particular is the sampling method. There is a need for more research and more field work to establish more exactly the conditions necessary for recovering a high percentage of the acid and its anhydride in the presence of relatively large amounts of sulfur dioxide and other interfering substances.

EPA's absorption method using isopropyl alcohol (IPA) is giving trouble in field trials. Maintenance of the concentration and total volume of the sampling solution is difficult. There also is evidence of oxidation of sulfur dioxide in the IPA solution. At this point it is not clear whether this does or does not depend on how well the SO_2 is purged from the sampling solution.

The condensation and filtering (or impingement) technique must be investigated more thoroughly. There are questions about the completeness of the condensation and the efficiency of collecting the condensate under field conditions which must be answered.

The analytical, or measurement, steps of the various methods in themselves seem to be satisfactory when allowances are made for the limitations of the sampling procedure used.

6. Acknowledgment

The assistance of Frank P. Scaringelli in helping obtain an APTIC literature review is gratefully acknowledged. A considerable portion of the ambient air review was performed by Roy L. Thompson.

7. References

1. J.N. Driscoll. 1972. Validation of Improved Chemical Methods for Sulfur Oxides Measurements from Stationary Source. Program Manager, Walden Research Corp., Nov. EPA-R2-72-105, National Tech. Information Service, Springfield, Va.
2. Berger, A.W., J.N. Driscoll, and P. Morgenstern. June 1970. Review and Statistical Analysis of Stack Sampling Procedures for the Sulfur and Nitrogen Oxides in Fossil Fuel Combustion. Presented before the Air Pollution Control Association, St. Louis, Missouri.
3. Air Pollution Aspects of Emission Sources: Sulfuric Acid Manufacturing. May, 1971. Office of Air Programs, EPA, Research Triangle Park, N.C.
4. Barton, S.C. and H.G. McAdie. Aug. 1972. An Automated Instrument for the Specific Determination of Ambient H_2SO_4 Aerosol, Presented before the Division of Water, Air and Waste Chemistry, ACS, New York.
5. Richards, L.W. April, 1973. A NEW Technique to Measure Sulfuric Acid in the Air, Ibid., Dallas, Texas.
6. Anfield, B.D. and C.G. Warner. 1968. A study of Industrial Mists Containing Sulfuric Acid, Ann. Occupational Hygiene, 11(3):185–194.
7. Environmental Engineering, Inc. Work in progress.
8. Patton, W.F. and J.A. Brink, Jr. 1963. New Equipment and Techniques for Sampling Chemical Process Gases. *J. of the Air Poll. Control Assn.* 13:162–167.
9. Morrow, N.L., R.S. Brief and R.R. Bertrand. 1972. Sampling and Analyzing Air Pollution Samples. Chem. Eng. 84–98.
10. Johnstone, H.F., and D.R. Coughanowr. 1958. Absorption of Sulfur Dioxide from Air: Oxidation in Drops Containing Dissolved Catalyst. *Ind. Eng. Chem.*, 50:1169.
11. Junge, C.E., and T.G. Ryan. 1958. Study of the SO_2 Oxidation in Solution and its Role in Atmospheric Chemistry. *Quart. J. Roy. Meterol. Soc.*, 84:46.
12. EPA, "Standards of Performance for New Stationary Sources," Federal Register (Dec. 23, 1971).
13. Seidman, E.B. 1958. Determination of Sulfur Oxides in Stack Gases. *Anal. Chem.*, 30:1680.
14. Fielder, R.S. and C.H. Morgan. 1960. An improved Titrimetric Method for Determining Sulfur Trioxide in Flue Gas. *Anal. Chim. Acta*, 23:538.
15. Berk, A.A. and L.R. Berdick. 1950. Bureau of Mines Report #RI 4618, Reference 2.
16. Smith, J.F., J.A. Hultz, and A.A. Orning. April, 1968, Bureau of Mines Report #RI 7108, Reference 2.
17. Steinke, Irmhild. 1969. Contribution to the Determination of SO_2 and SO_3 in Waste Gases (German), *Z. Anal. Chem.*, 244(4):253–254.

18. ALYEA, H.N. and H.L. BACKSTROM. 1929. Inhibitive Action of Alcohols on the Oxidation of Sodium Sulfite. *J. Amer. Chem. Soc.*, 51:90.
19. AXFORD, D.W. and T.M. SUGDEN. 1946. The Estimation of Sulfur Trioxide in its Mixtures with Sulfur Dioxide by the Method of Amperometric Titration. *J. Chem. Soc.*, 901.
20. TAYLOR, H.D. 1951. The Condensation of Sulfuric Acid on Cooled Surfaces Exposed to Hot Gases Containing Sulfur Trioxide (Trans.), *Faraday Soc.*, 47:1114–1120.
21. SCHNEIDER, WERNER. 1971. The Automatic Determination of the SO_3 Content and Sulfuric Acid Dew Point of Flue Gas (German), *Energie Tech.*, 21(11):505–509.
22. SIETH, JOACHIM and HANS-GUNTER HEITMANN. Feb. 6, 1968. Apparatus for Continuously Measuring the Concentration of a Gas-Mixture Component. U.S. Pat. 3,367,747, 5 p.
23. LISLE, E.S. and J.D. SENSENBAUGH. 1965. The Determination of Sulfur Trioxide and Acid Dew Point in Flue Gases. *Combustion*, 36:12–16.
24. GOKSOYR, H. and K. ROSS. 1962. *J. Inst. Fuel*, 35:177, Ref. 2.
25. WANG, G.K. 1967. An Instrument for Determining Sulfur Oxides in Flue Gases. *Combustion*, 38:46.
26. MINZL, E. 1972. Determination of H_2SO_4-Spray, H_2SO_4-Mist and Unabsorbed SO_3 in Waste Gases from H_2SO_4 Contact Plants (German), *Chem.-Ing.-Tech.*, 44(13):858–860.
27. GERHARD, EARL R. and H.F. JOHNSTONE. 1955. Photochemical Oxidation of Sulfur Dioxide in Air. *Ind. Eng. Chem.*, 47(5):972–976.
28. LEE, R.E. and J. WAGMAN. 1966. A Sampling Anomaly in the Determination of Atmospheric Sulfate Concentration, *Amer. Ind. Hyg. Ass. J.*, 27(3):266–271.
29. BARTON, S.C. and H.G. MCADIE. 1970. Preparation of Fiber Filters for Sulfuric Acid Aerosol Collection. *Env. Sci. & Tech.*, 4:769.
30. SCARINGELLI, F.P. and K.A. REHME. 1969. Determination of Atmospheric Concentrations of Sulfuric Acid Aerosol by Spectrophotometry, Coulometry, and Flame Photometry. *Anal. Chem.*, 41(6):707–713.
31. DUBOIS, L., C.J. BAKER, T. TEICHMAN, A. ZDROJEWSKI and J.L. MONKMAN. 1969. The Determination of Sulfuric Acid in Air: A Specific Method. *Mickrochimica Acta*, 269.
32. LOMBARDO, J.B. 1958. Analysis of Sulfuric Acid Contact Exit Gas. *Anal. Chem.*, 25:154.
33. BASARGIN, N.N. and N.K. OLEINIKOVA. 1966. Determination of Sulfuric Acid Mist in the Gases From Sulfuric Acid Contact Plants. *Ind. Lab.*, U.S.S.R. (trans.), 32:1118–1119.
34. CORBETT, P.F. 1953. *J. Inst. Fuel*, 26:92. Reference 2.
35. FIELDER, R.S. and D. JACKSON. 1960. ibid., 33:229. Reference 2.
36. CRUMLEY, P.H., H. HOWE and D.S. WILSON. 1958. ibid., 31:378. Reference 2.
37. BERTOLACINI, R.J. and J.E. BARNEY. 1957. Colorimetric Determination of Sulfate with Barium Chloranilate. *Anal. Chem.*, 29(2):281.
38. SCHAFER, H.N.S. 1967. An Improved Spectrophotometric Method for the Determination of Sulfate with Barium Chloranilate as Applied to Coal Ash and Related Materials. *Anal. Chem.*, 39(14):1719.
39. ROESLER, J.F., H.J.R. STEVENSON and J.S. NADER. 1965. Size Distribution of Sulfate Aerosols in the Ambient Air. *J. Air Pollution Control Assoc.*, 15:576.
40. CRIDER, W.L. 1965. Hydrogen Flame Emission Spectrophotometry In Monitoring Air For SO_2 and Sulfuric Acid Aerosol. *Anal. Chem.*, 37:1770–1773.
41. STEVENS, R.K. and O'KEEFFE. 1970. Modern Aspects of Air Pollution Monitoring. *Anal. Chem.*, 42(2):143A.
42. COMMINS, B.T. 1963. Determination of Particulate Acid in Town Air. *Analyst*, 88:364.

PART V
RECOMMENDED FACTORS AND CONVERSION UNITS FOR ANALYSIS OF AIR POLLUTANTS

PART V

Recommended Factors and Conversion Units for Analysis of Air Pollutants

1. Introduction

The results of methods of sampling and analysis of air pollutants are expressed in a variety of units that are derived from a number of different scientific disciplines. There is little evidence of uniformity or standardization in the units presently employed to report observations and analytical data for the assessment of air quality or for control purposes. The present compilation of factors and conversion units is intended to assist those interested in this field in efforts to standardize the reporting of data and to enable observations presented in differing units to be compared more easily.

The factors presented here are derived mainly from the Recommended Practice of the American Society for Testing and Materials issued under the designation D1914-68. All units and factors are based on information from the 10th General Conference on Weights and Measures held in 1954. Use of these factors facilitates the change from measurement units in any one system to related units in other systems or to larger or smaller units in the same system. To convert values of units in the left hand column of the following Table to values of units in the right hand column, multiply by the conversion factors given in the center column.

In Intersociety Committee methods for analysis of air pollutants, the concentrations of gases, vapors and particulates are reported at standard conditions of 25 C and 760 Torr, in metric units, such as micrograms per cubic meter of air sample.

2. Factors and Conversion Units

Multiply Unit	By Factor	To Obtain Conversion Unit
	Temperature	
Degrees Fahrenheit (F) + 459.72	1	Degrees Fahrenheit Absolute or Rankine (R)
Degrees Fahrenheit (F) − 32	5/9	Degrees Celsius (C)
Degrees Celsius (C) + 273.16	1	Degrees Celsius Absolute or Kelvin (K)
Degrees Celsius (C) + 17.78	1.8	Degrees Fahrenheit (F)
Degrees Rankine (R) − 459.72	1	Degrees Fahrenheit (F)
Degrees Kelvin (K) − 273.16	1	Degrees Celsius (C)
	Pressure	
Dynes per square centimeter	1.4504×10^{-5}	Pounds per square inch
	10.197×10^{-4}	Grams per square centimeter
	1×10^{-6}	Bars
Pounds per square inch absolute (psia)	70.307	Grams per square centimeter absolute
	51.715	Millimeters of mercury absolute
	144	Pounds per square foot absolute
	1	Pounds per square inch gauge + 14.696

Multiply Unit	By Factor	To Obtain Conversion Unit
	Pressure (Cont.)	
Pounds per square inch gauge (psig)	70.307	Grams per square centimeter
	51.715	Millimeters of mercury at 0 C (Torr)
	27.673	Inches of water at 4 C
	1	Pounds per square inch absolute − 14.696
Inches of water (at 4 C)	0.03614	Pounds per square inch
	0.07355	Inches of mercury
	0.57818	Ounces per square inch
	25.399	Kilograms per square meter
	2490.8	Dynes per square centimeter
Inches of mercury (at 32 F)	0.49116	Pounds per square inch
	13.595	Inches of water at 4 C
	345.31	Kilograms per square meter
	3.3864×10^4	Dynes per square centimeter
Millimeters of mercury (at 0 C)	0.01934	Pounds per square inch
	1.3595	Grams per square centimeter
	1333.2	Dynes per square centimeter
Centimeters of mercury (at 0 C)	1.3332×10^4	Dynes per square centimeter
	135.95	Kilograms per square meter
	27.845	Pounds per square foot
Atmosphere (normal)	760	Millimeters of mercury at 0 C (Torr)
	1.0133	Bars
	14.696	Pounds per square inch
	29.921	Inches of mercury at 32 F
	1033.2	Grams per square centimeter
	1.0133×10^6	Dynes per square centimeter
Bars	14.504	Pounds per square inch
	1.0197×10^4	Kilograms per square meter
	1.000×10^6	Dynes per square centimeter
	750.06	Millimeters of mercury (0 C) (Torr)
	0.98692	Atmosphere
	Density	
Grams per cubic centimeter	1	Grams per milliliter
	0.03613	Pounds per cubic inch
	8.3452	Pounds per gallon (U.S.)
	62.428	Pounds per cubic foot
Pounds per cubic foot	0.01602	Grams per cubic centimeter
	5.7870×10^{-4}	Pounds per cubic inch
	Concentration (See also Section I at end of Table)	
Gases in Gas		
Parts per million by volume (ppm)	1	Micromoles of gas per mole of gas
	1×10^{-4}	Per cent by volume
	$\frac{\text{Molecular weight}}{24.45}$	Milligrams of substance per m³ of air at 25 C and 760 Torr
	1×10^{-6}	$\frac{\text{Partial pressure of one constituent}}{\text{Total pressure of mixture}}$
Parts per billion by volume (PPB)	1×10^{-3}	Parts per million
One per cent by volume	10,000	Parts per million
Milligrams per liter	1000	Milligrams per cubic meter
	1×10^6	Micrograms per cubic meter
Milligrams per cubic meter	1×10^{-3}	Milligrams per liter
Micrograms per cubic meter	1×10^{-6}	Milligrams per liter

Multiply Unit	By Factor	To Obtain Conversion Unit
	Concentration (Cont.)	
Liquid and Solid Particles in Gas		
Milligrams per liter	1×10^{3}	Milligrams per cubic meter
	1×10^{6}	Micrograms per cubic meter
Milligrams per cubic meter	1×10^{-3}	Milligrams per liter
Micrograms per cubic meter	1×10^{-6}	Milligrams per liter
Ounces per thousand cubic feet	1.0021	Grams per cubic meter
Grains per cubic foot	2.2883	Grams per cubic meter
Particles per cubic centimeter	2.8317×10^{4}	Particles per cubic foot
	1×10^{6}	Particles per cubic meter
Particles per cubic meter	1×10^{-6}	Particles per cubic centimeter
	0.02832	Particles per cubic foot
Millions of particles per cubic foot	35.314	Millions of particles per cubic meter
Gases, Liquids, and Solids in Liquids		
Gram molecular weight per liter	1	Moles per liter
Parts per million by weight	1	Milligrams per liter (where specific gravity of dispersion medium is 1.00)
Radioactivity		
Millicuries (mCi)	1×10^{-3}	Curies (Ci)
Microcuries (μCi)	1×10^{-6}	Curies
Nanocuries (nCi)	1×10^{-9}	Curies
Picocuries (pCi)	1×10^{-12}	Curies
Femtocuries (fCi)	1×10^{-15}	Curies
	Length	
Angstrom units	1×10^{-10}	Meters
	3.9370×10^{-9}	Inches
	1×10^{-4}	Microns (Micrometers, μm)
	1×10^{-8}	Centimeters
	0.1	Millimicrons
Millimicrons	1×10^{-9}	Meters
	1×10^{-7}	Centimeters
	10	Angstrom Units
Microns (μm)	3.9370×10^{-5}	Inches
	1×10^{-6}	Meters
	1×10^{-4}	Centimeters
	1×10^{4}	Angstrom Units
Millimeters	0.03937	Inches (U.S.)
	1000	Microns
Centimeters	0.39370	Inches (U.S.)
	1×10^{4}	Microns (μm)
	1×10^{7}	Millimicrons
	1×10^{8}	Angstrom Units
Meters	6.2137×10^{-4}	Miles (statute)
	1.0936	Yards (U.S.)
	39.370	Inches (U.S.)
	1×10^{9}	Millimicrons
	1×10^{10}	Angstrom units
Kilometers	0.53961	Miles (nautical)
	0.62137	Miles (statute)
	1093.6	Yards
	2280.8	Feet
Inches (U.S.)	0.02778	Yards
	2.5400	Centimeters
	2.5400×10^{8}	Angstrom units

Multiply Unit	By Factor	To Obtain Conversion Unit
	Length (Cont.)	
Feet (U.S.)	0.30480	Meters
	30.480	Centimeters
Yards (U.S.)	5.6818×10^{-4}	Miles
	0.91440	Meters
	91.440	Centimeters
Miles (nautical)	1.1516	Statute miles
	2026.8	Yards
	1.8533	Kilometers
Miles (U.S. statute)	320	Rods
	0.86836	Nautical miles
	1.6094	Kilometers
	1609.4	Meters
	Area	
Square millimeters	0.00155	Square inches
	1×10^{-6}	Square meters
	0.01	Square centimeters
	1.2732	Circular millimeters
Square centimeters	1.1960×10^{-4}	Square yards
	0.00108	Square feet
	0.15500	Square inches
	1×10^{-4}	Square meters
	100	Square millimeters
Square kilometers	0.38610	Square miles (U.S.)
	1.1960×10^{6}	Square yards
	1.0764×10^{7}	Square feet
	1×10^{6}	Square meters
	247.10	Acres (U.S.)
Square inches (U.S.)	0.00694	Square feet
	0.00077	Square yards
	6.4516×10^{-4}	Square meters
	6.4516	Square centimeters
	645.16	Square millimeters
Square feet (U.S.)	3.5870×10^{-8}	Square miles
	0.11111	Square yards
	144	Square inches
	0.09290	Square meters
	929.03	Square centimeters
	2.2957×10^{-5}	Acres
Square miles	640	Acres
	3.0976×10^{6}	Square yards
	2.7878×10^{7}	Square feet
	2.5900	Square kilometers
	2.5900×10^{6}	Square meters
	Volume	
Cubic millimeters	6.1023×10^{-5}	Cubic inches
	1×10^{-9}	Cubic meters
	0.001	Cubic centimeters
Milliliters	0.03381	Ounces (fluid, U.S.)
	0.06102	Cubic inches
	0.001	Liters
	1	Cubic centimeters

Multiply Unit	By Factor	To Obtain Conversion Unit
	Volume (Cont.)	
Liters	0.00131	Cubic yards
	0.26418	Gallons (U.S.)
	0.03532	Cubic feet
	33.815	Ounces (fluid, U.S.)
	61.025	Cubic inches
	0.00100	Cubic meters
	1000	Cubic centimeters
Cubic meters	1.3079	Cubic yards (U.S.)
	35.314	Cubic feet (U.S.)
	264.17	Gallons (U.S.)
	6.1023×10^{4}	Cubic inches
	1000	Liters
	1×10^{6}	Cubic centimeters
	1×10^{9}	Cubic Millimeters
Gallons (U.S.)	0.00495	Cubic yards
	0.13368	Cubic feet
	128	Ounces (fluid)
	231	Cubic inches
	0.00378	Cubic meters
	3.7854	Liters
	3785.4	Milliliters
	3785.4	Cubic centimeters
Cubic inches (U.S.)	2.1433×10^{-5}	Cubic yards
	5.7870×10^{-4}	Cubic feet
	0.00433	Gallons (U.S.)
	0.55410	Ounces (fluid)
	1.6387×10^{-5}	Cubic meters
	0.01639	Liters
	16.387	Milliliters
	16.387	Cubic centimeters
	1.6387×10^{4}	Cubic Millimeters
Cubic feet (U.S.)	0.03704	Cubic yards
	7.4810	Gallons (U.S.)
	1728.0	Cubic inches
	0.02832	Cubic meters
	28.316	Liters
	2.8316×10^{4}	Cubic centimeters
Cubic Centimeters	1	Milliliters
	1.3079×10^{-6}	Cubic yards
	3.5314×10^{-5}	Cubic feet (U.S.)
	2.6417×10^{-4}	Gallons (U.S.)
	0.03381	Ounces (fluid, U.S.)
	0.06102	Cubic inches
	1×10^{-6}	Cubic meters
	1000	Cubic millimeters
	Time	
Seconds (mean solar)	1.1574×10^{-5}	Days
	2.7778×10^{-4}	Hours
	0.01667	Minutes
Minutes (mean solar)	9.9206×10^{-5}	Weeks
	6.9445×10^{-4}	Days
	0.01667	Hours

Multiply Unit	By Factor	To Obtain Conversion Unit
	Time (Cont.)	
Hours (mean solar)	0.00595	Weeks
	0.04167	Days
	3600	Seconds
Days (mean solar)	1440	Minutes
	8.6400×10^4	Seconds
Weeks (mean solar)	168	Hours
	1.0080×10^4	Minutes
	6.0480×10^5	Seconds
Months (mean calendar)	30.42	Days
	730	Hours
	4.3800×10^4	Minutes
	2.628×10^5	Seconds
Years (calendar)	8760	Hours
	5.256×10^5	Minutes
	Velocity	
Feet per second	0.01136	Miles per minute
	0.68182	Miles per hour
	1.0973	Kilometers per hour
	18.288	Meters per minute
	30.480	Centimeters per second
Feet per minute	0.00508	Meters per second
	0.01136	Miles per hour
	0.01667	feet per second
	0.01829	Kilometers per hour
	0.30480	Meters per minute
	0.50800	Centimeters per second
Centimeters per second	3.7282×10^{-4}	Miles per minute
	0.02237	Miles per hour
	0.03281	Feet per second
	0.3600	Kilometers per hour
	0.60000	Meters per minute
	1.9685	Feet per minute
Miles per hour	0.01667	Miles per minute
	1.4667	Feet per second
	1.6093	Kilometers per hour
	26.822	Meters per minute
	44.704	Centimeters per second
	88	Feet per minute
Meters per minute	0.03728	Miles per hour
	0.05468	Feet per second
	0.06	Kilometers per hour
	1.6667	Centimeters per second
	3.2808	Feet per minute
Kilometers per hour	0.27778	Meters per second
	0.62137	Miles per hour
	0.91134	Feet per second
	16.667	Meters per minute
	27.778	Centimeters per second
	54.681	Feet per minute

Multiply Unit	By Factor	To Obtain Conversion Unit
	Mass (or Weight)	
Milligrams	2.2046×10^{-6}	Pounds (avoirdupois)
	3.5274×10^{-5}	Ounces (avoirdupois)
	0.01543	Grains
	1×10^{-6}	Kilograms
Micrograms	1×10^{-6}	Grams
Grams	0.00220	Pounds (avoirdupois)
	0.03527	Ounces (avoirdupois)
	15.432	Grains
	1×10^{6}	Micrograms
Kilograms	0.00110	Tons (short)
	2.2046	Pounds (avoirdupois)
	35.274	Ounces (avoirdupois)
	1.5432×10^{4}	Grains
Grains	1.4286×10^{-4}	Pounds (avoirdupois)
	0.00229	Ounces (avoirdupois)
	0.06480	Grams
	64.799	Milligrams
Ounces (avoirdupois)	3.1250×10^{-5}	Tons (short)
	0.06250	Pounds (avoirdupois)
	437.50	Grains
	28.350	Grams
Pounds (avoirdupois)	5×10^{-4}	Tons (short)
	16	Ounces (avoirdupois)
	7000	Grains
	0.45359	Kilograms
	453.59	Grams
Tons (short, U.S.)	2000	Pounds (avoirdupois)
	3.200×10^{4}	Ounces (avoirdupois)
	907.19	Kilograms

3. Conversion for Gases and Vapors

To convert concentrations of gases and vapors from parts per million by volume to milligrams per cubic meter and vice versa at any condition of temperature and pressure, the following expressions will be found useful:

Concentration, ppm =

$$\frac{C_1 \times 24.450 \times T \times 760}{\text{mol wt} \times 298 \times P}$$

Concentration, mg per cubic meter =

$$\frac{C_2 \times \text{mol wt} \times 298 \times P}{24{,}450 \times T \times 760}$$

where:

C_1 = concentration in milligrams of substance per cubic meter,

C_2 = concentration in parts per million by volume (ppm),

T = absolute temperature in degrees Kelvin, and

P = absolute pressure in Torr of the sampled air stream

Note—It is usually necessary to convert the measured air flow to the temperature and pressure of the sampled air stream.

4. Sieve Number versus Particle Size

The following are excerpts from the Specifications for Wire Cloth Sieves for

Testing Purposes (ASTM Designation: E 11).* For complete information on this subject, refer to ASTM Specifications E 11.

Sieve Number	Sieve Opening, μm	Sieve Number	Sieve Opening, μm
80	177	200	74
100	149	230	63
120	125	270	53
140	105	325	44
170	88	400	37

*Annual Book of ASTM Standards, Part 30.

INDEX

A

B

C

D

E

G

H

K

L

M

N

Q

R

S

T